NATURAL SCIENCES IN AMERICA

NATURAL SCIENCES IN AMERICA

EVOLUTION EMERGING

WILLIAM KING GREGORY

VOLUME I

ARNO PRESS
A New York Times Company
New York, N. Y. • 1974

Reprint Edition 1974 by Arno Press Inc.

Reprinted from a copy in the Newark
Public Library

NATURAL SCIENCES IN AMERICA
ISBN for complete set: 0-405-05700-8
See last pages of this volume for titles.

Manufactured in the United States of America

Publisher's Note: The illustrations in this book have been reduced by 15%.

Library of Congress Cataloging in Publication Data

Gregory, William King, 1876-
Evolution emerging.

(Natural sciences in America)
A collaborative work of the American Museum of Natural History and Columbia University.
Reprint of the ed. 1951 published by Macmillan, New York.
Bibliography: v. 1, p.
1. Phylogeny. 2. Evolution. 3. Paleontology.
I. American Museum of Natural History, New York.
II. Columbia University, III. Title. IV. Series.
[QH367.5.G73 1974] 575 73-17822
ISBN 0-405-05738-5

EVOLUTION EMERGING

In Two Volumes

VOLUME I: Text

VOLUME II: Illustrations

THE MACMILLAN COMPANY
NEW YORK • BOSTON • CHICAGO
DALLAS • ATLANTA • SAN FRANCISCO

MACMILLAN AND CO., LIMITED
LONDON • BOMBAY • CALCUTTA
MADRAS • MELBOURNE

THE MACMILLAN COMPANY
OF CANADA, LIMITED
TORONTO

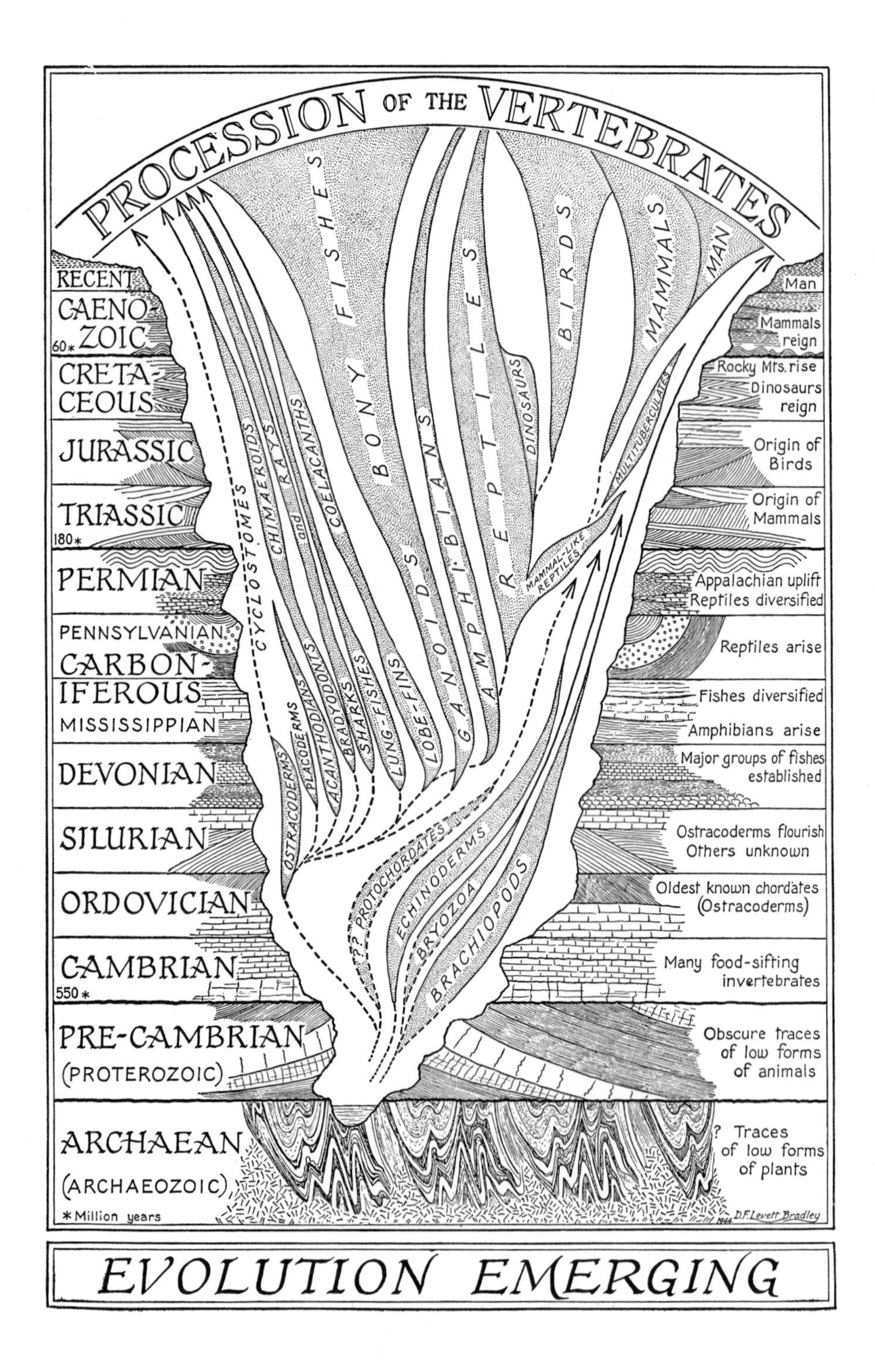

EVOLUTION EMERGING

EVOLUTION EMERGING

A Survey of Changing Patterns from Primeval Life to Man

WILLIAM KING GREGORY

A Collaborative Work of

THE AMERICAN MUSEUM OF NATURAL HISTORY

and

COLUMBIA UNIVERSITY

VOLUME I

THE MACMILLAN COMPANY

New York

PRINTED IN THE UNITED STATES OF AMERICA

First Printing

IN MEMORIAM

Bashford Dean

Edmund B. Wilson

Henry Fairfield Osborn

Frederick J. E. Woodbridge

James Hyslop

Professors of Zoology, Palaeontology, Philosophy

COLUMBIA UNIVERSITY

PREFACE

THE BEGINNINGS of the present work date back to the decade of 1890–1900, when at Columbia University Professors Bashford Dean and Henry Fairfield Osborn gave their lecture courses on "Fishes, living and fossil," "Mammals, living and fossil," and "The Evolution of the Vertebrates." Under the latter title Professor Osborn projected a general book to follow Dean's "Fishes, living and fossil" (1895) in the Columbia University Biological Series. He and Doctor McGregor did indeed prepare notes and illustrations for the chapters on amphibians and reptiles but they soon became so much occupied with their special studies on fossil vertebrates at the American Museum of Natural History that the proposed general text gradually receded into the background.

Although the project of a single general textbook on the evolution of the vertebrates, with special reference to the skeleton, was put aside for several decades, the resulting efforts to clarify the main outlines led to Professor Osborn's well known works, "The Reptilian Subclasses Diapsida and Synapsida" (1903), "The Evolution of the Mammalian Molar Teeth" (1907), "The Age of Mammals" (1910), "Men of the Old Stone Age" (1915), "The Origin and Evolution of Life" (1917), as well as to many of my own researches and publications and those of my graduate students.

Beginning in 1896, Professor Dean inspired me with an abiding curiosity as to the origin and evolution of the main groups of fishes; a little later, Professor Osborn opened to me the evolving orders of fossil and recent reptiles and mammals. Sitting reverently in the class of Professor E. B. Wilson, I beheld his amazing vision of the emergence of one great invertebrate phylum after another, while Professors James Hyslop and Frederick J. E. Woodbridge showed me that a naturalist may profitably endeavor to look below the surface phenomena of evolution and try to generalize the ways in which reality behaves, or seems to behave, through the ages.

In 1931, Professor Osborn suggested that I should revive the projected general work on the evolution of the vertebrates and soon afterward I began to prepare preliminary outlines and manuscripts. But, as in the earlier effort, as the work progressed so did its scope. For example, it came to seem futile to introduce suddenly a review of the various theories of the origin of the vertebrates until earlier chapters had sketched, in a comparative way at least, a few characteristic samples of the vast world of invertebrates; for some invertebrates, whether known or unknown, supplied the source of the vertebrates, while others have served either as food, as competitors, or as enemies. Similar difficulties had to be met all along the branching lines.

In attempting to survey within practicable limits the age-long transformations of the vertebrates I have centered attention mainly upon the emergence of new skeletal patterns. The study of the skeleton as a whole, including not only the endoskeleton but the integumentary exoskeleton and its diverse derivatives, makes it possible to combine and integrate the knowledge of both fossil and still existing animals. Upon the skeleton is imprinted a legible record of the medium in which the animal lived, of the mode of locomotion, of the ways in which food was sought for, seized, divided, ingested; while the endocranial cast often affords a record of the form and pattern of the main parts of the brain, of the location of the cranial nerves and in some cases of the main blood vessels of the head.

The geneticists deal largely with the crossing of individuals as representatives of strains, populations, varieties, subspecies and species, but in the present work individuals are considered as representatives of genera. Here we must, so far as possible, avoid descriptive detail as an end in itself and hold to the aim of a broad survey, which may indicate within each larger group the general range of adaptations in body form, locomotor system, and in the organs for seizing, subdividing and ingesting the food, as well as the most likely source of that group and the probable stages that led to its oustanding characteristics.

For a half century or more, Columbia University and the American Museum and their leaders have been building up the resources, assembling the opportunities and training the individuals that have made possible the present book and hosts of others. I have already spoken of the inspiration and training I received from my professors and predecessors at Columbia. With equal pleasure and gratitude I recall my close association at the Museum with Professor Osborn, Doctor Matthew and other members of his staff, as well as with my colleagues, Henry C. Raven, Dr. Milo Hellman, and many others. Nor do I forget the long line of my graduate students who contributed so much by their researches and publications to the material used in this book. An inspection of the bibliography will suggest to how great an extent this work is indebted to the authors listed, but I cannot forbear to mention especially Dr. Erik Stensiö, Professor D. M. S. Watson, Dr. Robert Broom, Dr. F. von Huene, Professor A. S. Romer and Professor George Gaylord Simpson.

To President Butler I am indebted for the opportunity to study living gorillas in Africa, in company with Messrs. Raven, McGregor, and Engel. I desire to thank Professor Dudley S. Morton both for his effective promotion of that expedition and for his illuminating studies on the evolution of the human foot.

To the late President Henry Fairfield Osborn and to his successors, President Trubee Davison and Acting President Perry Osborn, the Trustees of the American Museum, and Director Albert Eide Parr, I owe continued access to all the facilities and treasures of the Museum, opportunities for field studies in North America, Australia, the Sargasso Sea, Africa, New Zealand, the West Indies; as well as unfailing and generous support during more than five decades of education, research and publication.

In conclusion, I thank all my present colleagues, friends and assistants at the Museum for innumerable courtesies and much helpful discussion of problems connected with this work. For many years past, Mrs. Helen Ziska has made the excellent line drawings that bear her name or initials, and for the past eight years Miss D. F. Levett Bradley has skillfully prepared the comparative outlines and charts. During all the years in which this work has gone through its successive revisions and enlargements, Mrs. Clara Platt Meadowcroft has put into it an incalculable measure of constructive effort. Its development and completion indeed owe very much to her vision, her determination and her always generous spending of her time and effort.

W. K. G.

The American Museum of Natural History,
New York

ACKNOWLEDGMENTS

THE ILLUSTRATIONS (Volume II) are arranged along comparative-systematic lines, to indicate the range and evolution of form in a given group, with the more primitive or generalized forms near the stem and the more specialized set out on diverging branches. Credit to the author or other source of the illustration is given in the legends and on the copyright page.

CONTENTS

VOLUME I

PART THREE. Air-breathing Fishes (Crossopterygii, Dipnoi) —Amphibians—Reptiles —Birds

CHAPTER X

CHAPTER XI

CHAPTER XV

PART FOUR. The Rise and Branching of the Mammals

CHAPTER XVI

CHAPTER XX

CHAPTER XXI

* For subject index and contents of bibliography see pp. xxiii–xxvi.

CONTENTS

SUBJECT INDEX and BIBLIOGRAPHY

PART THREE

AIR-BREATHING FISHES, AMPHIBIANS, REPTILES, BIRDS

(Cf. Vol. I, Chapters X–XV)

PART FOUR

MAMMAL-LIKE REPTILES, ORIGIN, RISE AND BRANCHING OF MAMMALIAN ORDERS

(Cf. Vol. I, Chapters XVI–XXII)

PART ONE

(*Introduction, Chapters I–III*)

TIME AND EVOLUTION · THE OUTER SKIN OR ARMOR · SHELL FORM AND COLOR PATTERNS · EXTERNALLY JOINTED SKELETONS

INTRODUCTION

THE COSMIC CINEMA

Contents

INTRODUCTION

THE COSMIC CINEMA

THE PHILOSOPHER Hobbes tried to convey an image of the multitudinous powers and attributes of the "leviathan" state by figuring a gigantic crowned king, bearing a scepter and a sword; the king's appurtenances and his sovereignty were made up of innumerable people, many villages, churches, castles, and towns. Some such method of illustration may be useful in order to suggest the general intent and scope of the present work, which also deals with a complex pattern made up of innumerable pieces, the whole, nevertheless, being greater than the sum of its parts.

Let us assume that there has been and will be a sort of cosmic cinema, or slowly turning kaleidoscope. This history of all things has been turning continuously for untold ages. Let us concede that considerable bits and stretches of this drama have been discerned and described by many faithful and generally reliable historians and that their testimony is recorded in thousands of special reports and books and written in many languages. Moreover, just as there are sermons in stones and books in running brooks, so we ourselves have been free to draw from the original sources such knowledge as we could.

Both the cinema itself and the written records of it abound in repetitive appearances. How are we to interpret or gain useful information from these rhythmic or cyclic phenomena in time and space? First then we may note that some pairs or sequences of phenomena are so regularly associated that if the presence of one is established, the existence or approach of the other may with varying degrees of probability be taken for granted. For example, except in the polar regions, if one sees the sun go down in the west, one knows that night is coming on. Less regular, however, is the association of an east wind with a storm; again, the local weather forecasts are usually right, but unless we always carry an umbrella we may get caught in an unpredicted shower. Thus, at least in the cinema of ordinary terrestrial life, we learn by experience to estimate degrees of probability in coming events; in other words, we strive to make profitable responses even though we can be sure of only a few terms of any complex association formula; indeed we often have to guess the rest correctly or pay the consequences.

Nature has fortunately provided us with almost fool-proof neuro-muscular mechanisms which automatically provide the right response in the presence of many physical dangers. But as human beings our response to any given complex situation will be conditioned by our own past. Thus when we are confronted by the cinema of evolution, our egoistic habit may block our ability to see it as it is and we may cherish the illusion that we are the center of all things, including the drama of creation. Thus our answers regarding evolution may be dictated by prejudice rather than by knowledge. However indifferent we may be as to the evolution of the horse, we may not like to admit any such unpleasant creatures as monkeys to the list of "begats" in our own genealogy of distinguished ancestors; hence passionate preachers, like medicine men trying to scare away devils, continue to cry out against the "Darwinian hypothesis."

Obviously, if we have always been interested only in our own ego and in the things of the moment, or if we have been too heavily conditioned by ignorance and prejudice, we shall scarcely be aware of the supreme epic of evolution. If, on the other hand, life has taught us to keep on comparing one thing with another and to try consistently for the long view, then we shall seek some way or method to peer beyond the range of our passion-colored spectacles, in order that we may ascend step by step from subjective to objective levels; consciously we shall ever strive to advance from information about particular things to knowledge of principles and uniformities, from learning to wisdom and from wisdom to science.

During the long and perilous ascent from science to philosophy many innocents fall into deep pits set for the unwary by medicine men disguised as mathematicians or by clerics posing as philosophers and scientists. These adept rhetoricians often succeed in slipping into their philosophies a virtual negation of the foundation of historical science by blandly assuming that "Darwinism" has been deprived of its sting by modern zoologists.

Although there are many reliable uniformities, repetitions, recurrences of similar situations in the cosmic stream, there are also innumerable deceptive appearances; hence we must constantly search for trustworthy methods of grading truths, half-truths, ambiguities, illusions, self-deceptions, wilful distortions and complete fictions.

Even after we are conscious that there is something in the epic more enduring than the business of saving our own skins or feeding our own ego, our efforts to visualize the larger story will be along several paths which have themselves had long and complicated histories. If we approach along the path of tradition, as did, for example, the Maoris of New Zealand, we shall be taught an age-old story of man-like gods, animated by long hatred and jealousy, improvising the human race in order to spite another god, commanding the earth and the heavenly bodies into existence as a stage for the all-engrossing drama of the sins, frustrations and final triumphs of the tribal gods and ancestors.

Coming to more rational levels, we may approach certain parts of the cosmic story with various measuring rods, ranging from a human forearm and hand (cubit) to the latest astronomic apparatus for measuring the red shift in the spectra of receding stars. To many it gives such great satisfaction to determine exactly how many times the unit is contained in the dimension measured that, by the common error of confusing the part with the whole, Science, in which the interpretative function is the end and the measurement the means, is often thought of as limited to quantitative determinations.

It was not by measurements, however, that the riddle of the Rosetta stone was deciphered, but by the method of matching like with like, of starting with the known and gradually decoding the unknown. This was the method used by Darwin in his great delineation of the outlines of evolution and it has continued to yield abundant results in the hands of a goodly fellowship of geologists, palaeontologists, zoologists, botanists, physiologists, ecologists, students of heredity and others.

EVOLUTION THE MASTER KEY

The word evolution has been defined in many different ways. Its most literal meaning is of course simply an unrolling or unfolding, as of a bud into a flower. Herbert Spencer (1867–1872) defined the term in an elaborate and somewhat obscure way so as to limit evolution to cases of progressive or upward change, that is, in the direction of increasing complexity. For the opposite process of degeneration or degradation from the complex to the simple, he used the term "dissolution." This use of evolution to mean only progressive advance has sometimes been the source of confusion to antagonistic critics, who have seldom realized that progressive specialization and complexity in certain directions can take place only if there are compensatory reduction, degeneration or loss of parts and apparent simplification in others.

Evolution to many laymen is synonymous with "Darwin's theory of evolution." Sometimes it is taken to imply the theory that *all* forms of life have descended from a single or few ancestral forms and since this can never be proved mathematically, it presents a favorite angle of attack by anti-evolutionists.

To practical botanists or zoologists, who are concerned with the classification of plants or animals, the term carries the connotation that the members of any systematic group, such as the orchids among plants, have been derived by "descent with modification" from an ancient central stock which may in itself have included many interbreeding strains or varieties.

The use of the term evolution in a broad sense was discouraged by T. H. Morgan (1932), who implied that when the single word "evolution" is stretched to cover such widely different phenomena as the historical development of armor in mediaeval Europe and the evolution of the horse from its remote ancestors, the door is opened for loose thinking and reasoning by analogy.

No doubt persons that are frequently deceived by the classification of different logical species under one generic term may be misled by a broad use of the term evolution into the almost universal fallacy of the "undistributed middle term" of the syllogism. For example, they might wrongly infer that in the later stages of its evolution the armor of the Devonian armored fishes assumed a decorative value and a social significance, as did the later stages of mediaeval armor in Europe. But the term "defensive armor" can surely be applied legitimately to the protective covering of both Devonian arthrodiran fish and mediaeval knight and the facts are well established that in both cases this protective armor developed (or evolved) along somewhat similar lines. This very fact leads to the interesting question, why human tradition expressed in a manufactured human product so closely paralleled the part of heredity.

Thus, notwithstanding the tendency of specialists to use the word evolution only in some restricted sense such as "organic evolution," I maintain that the modern developments of atomic physics, chemistry, astronomy, and other sciences, justify a broadening rather than a restriction of its meaning. As the boundaries of the fields of chemistry, physics and biology are no longer fixed but overlap in many places, evolution may be usefully defined in a broad sense as the history of energy systems and of all their derivatives. So far as we know, the physical universe as a whole has had no real beginning and will have no cessation; its spatial and temporal boundaries, if any, seem to recede before us as our powers of measuring space and time increase. Although its fundamental properties and laws appear to remain unchanged, the patterns of its parts do change and so far as such patterns or systems can be distinguished they have an origin, rise, decline and dissolution in time and space. "Evolution" as here understood involves the study of both the relatively fixed background and the changing pattern.

That the evolutionary concept is by no means to be discarded is indicated by the fact that the physicist, chemist, astronomer, geologist and palaeontologist are gradually recognizing the steps by which simple units are built up into complex organizations, and how the latter in turn break down into their elements. To begin with, the physicists have come to the conclusion that energy and mass are convertible; indeed, Sir William Bragg (1933) in his "Universe of Light" shows that we may regard light as the primal form of energy. In another direction the students of atomic physics headed by Lord Rutherford (1937) have traced the building up of the heavier kinds of atoms out of hydrogen units, so that Eddington has called hydrogen the "raw material of the universe." Meanwhile the study of astrophysics, according to Henry Norris Russell, indicates that hydrogen is the "fuel of the stars," which emit light as a result of the energy generated by the building up of heavier atoms out of hydrogen under the force of gravitation. Nor does the Old Master, Einstein, relinquish his efforts to reach a "unitary field formula" by which the laws of light, electromagnetism and gravitation can be harmoniously and consistently described.

The earth itself, from the convergent evidences of geology, seismology, and astrophysics, is made of the same materials as the sun, at least in its core, and as we shall later see, there is good reason to regard all the planets as the children either of the sun itself or of a parent mass. Even the great gap which was formerly supposed to separate the organic from the inorganic has been, to say the least, lessened as the steps in the building up of proteins out of amino-acids and of the latter out of nitrates are being discovered.

Thus in spite of the opposition of dualists or polygenists the essential unity of the astronomical cosmos in basic composition and behavior is becoming ever clearer to students of the natural sciences. Moreover there is increasing evidence that whatever the ultimate nature of time, space, energy, mass, may be, the events and patterns of nature in so far as they can be perceived and investigated by organic beings have cumulative, extended and interweaving origins, even though they may appear to be separate, catastrophic and spontaneous.

In the present work we shall not be concerned primarily with the classic problems whether the course of organic evolution has been guided by natural selection or by some other natural system. Even the hereditary and statistical mechanisms through which evolution works will be taken for granted here because they are being dealt with (as in the growing volumes on evolution) by many who are competent to do so.

Still less are we interested in *proving* evolution. In this respect we shall claim the privilege of modern mathematicians, who insist on the right to choose their own postulates. Our postulates (which are chosen because they appear from experience to be truthful statements of fact) are as follows:

1) Evolution, the natural history of the universe and its parts, deals with the processes and results of the transformation of energy and matter, viewed against the background of time and space.

2) Our sense organs and measuring instruments give only partial information of the complexities of the subject matter; our mental pictures, written records, symbols and measurements of evolution, having been made by fallible human beings, are often anthropomorphic or at best approximate. Nevertheless, since scientists are always subjecting their colleagues' results to the test of experience, there is in the long run a rapidly growing residue of increasing approximation to the facts of evolution.

Although evolution is concerned with the results of the transformations of matter and energy, we may be pardoned for excluding from the objectives of the present review the problem of the ultimate fate of the universe, or even the question whether it will continue indefinitely to "explode," as it is now said to be doing, or eventually simmer down through entropy or heat dissipation into a gray uniformity. Nor are we competent to guess whether organic evolution is limited to this tiny terrestrial ball, or whether among so many myriads of suns there may not be other planets which could support living things. In fact the "why-may-not-there-once-have-been" type of reasoning is to be avoided if possible, as coming too close to wishful thinking.

THE ORIGIN OF THE EARTH

Terrestrial organic evolution then will be our main theme and even of this limited field we can deal with only a few samples. First, to consider the general background of terrestrial evolution we may refer briefly to the problem of the origin of the earth. In 1755 Immanuel Kant of Königsberg presented his theory (Schuchert, 1931, p. 69) that the universe had been developed out of highly attenuated material, which in time, due to the Newtonian law of gravitation, segregated into spinning masses of rarefied gas, or nebulae. These in turn gave rise to hot stars, of which the sun is one. He further imagined the solar nebula, or ancestral sun, to have given rise to the planets. Laplace, a French astronomer, in 1796 and again in 1824 proposed what has since become known as the "nebular hypothesis" of earth origin. According to this view the ancestral sun was a rotating nebula of gas that was slowly contracting through loss of heat by radiation. In the course of time it left behind nine rings of gas liberated from the equatorial region of its mass; these gaseous rings condensed into the planets and the many much smaller asteroids. According to Jeans, this theory was developed by Laplace with great mathematical detail. Nevertheless, after much further study it has been definitely rejected by modern astronomers and geologists, who have until recently adopted one or another of the several "tidal" theories.

The planetesimal, or cold-earth theory, of Chamberlin and Moulton is one form of tidal theory (Chamberlin and Salisbury, 1904, vol. I). Its most essential feature is that during the approach of a "stranger" star the sun threw out two immensely long arms of hot gas; this cooled quickly into dust-like materials called planetesimals, which were the raw material of the planets and satellites. According to this theory, the planets originated as relatively small, internally hot masses called knots, and these attracted to themselves the remaining planetesimals. The planetoids represent some of the smaller masses which so far have escaped falling into the planets. The falls of meteorites, according to this view, illustrate the later stage in the growth of the earth, which was far smaller at the time of the first disruption.

The more recent theory of Jeans (1929) also holds that the approach of a "stranger" star set up huge tides of hot gas on the sun. By centrifugal action the ends of these tides were thrown off into space and eventually became the planets. Meanwhile the disturber had moved off. Later variants of this theory are: that the "parent" star disrupted one of the companions into the planets, leaving the other companion as the sun, or that the sun, passing through an enormous cosmic dustcloud, attracted the raw material of the planets, which eventually revolved around the sun.

Every modern theory of the origin of the solar system has required years of patient mathematical analysis, during which the necessary implications of the hypothesis regarding the changes in the paths of the planets and many other factors in this complex problem have been considered by astronomers. When apparently fatal objections to certain parts of the hypotheses are uncovered, these hypotheses, having served their purpose, are cast aside or modified. Whatever may be the final form of the theory of the origin of the earth, it seems safe to assume for our present purpose that the earth is either a fragment of the sun, or possibly of its companion or of some other common cosmic mother.

It would be almost impossible to overemphasize the importance of the sun itself in a broad review of terrestrial evolution. The sun, besides being the parent, or at least the uncle, of the planets, is their great central generator of electronic energy (including light) and by means of his mass he keeps the planets revolving around him where they can benefit by his lavish expenditure of almost inexhaustible energy, while the clouds of his making protect the earth against the deadly cold of interstellar space.

The sun is also the prime mover of all the earth-building forces, such as the drawing up of moisture from the ocean and its subsequent deposition on the land, the erosion of the surface of the earth, the laying down of the stratified rocks, and the like.

The vast numbers of chloryphyll-bearing plants drink in, as it were, the sun's energy, and use it to make their own food and the food of animals. In short, the Egyptian king, Aken-aten, was far beyond his time when he selected the image of the sun's disc as the symbol of his primal deity.

According to Dr. Leason H. Adams (1937), Director of the Geophysical Laboratory of the Carnegie Institution of Washington, the evidence obtained through study of earthquake waves, supplemented with data gathered by geologists, supports the view that a cross-section of the earth would show three principal regions, or zones; a core, or central region, about half the diameter of the earth; a crust, or superficial layer, from twenty-five to thirty miles thick; and an intermediate zone of two thousand miles. (See also Macelwane, 1946.) The central core is believed from various lines of evidence to be composed of metallic iron under enormous pressure. Now iron is also the dominant constituent of meteorites and of the Sun itself. Silica, which is by far the most abundant element in the crust of the earth, is believed by some to have been derived eventually from iron in combination with oxygen. Although the crust of the earth, estimated at twenty-five to thirty miles, is extremely thin in proportion to the diameter of the earth, it is only the outer few miles of this crust that are at all accessible to direct observation, except in so far as material has been brought up by pressure from deeper levels, during the great mountain-forming movements, and subsequently exposed by erosion.

GEOLOGIC TIME

The fossil record of organic evolution practically begins with the Palaeozoic era and therefore covers only the last quarter of the entire geologic record. Nevertheless this quarter is estimated to have commenced some five hundred million years ago and, as we shall see later, that time has been sufficient for a great deal of change within many large groups, such as the molluscs, the insects, and the vertebrates.

The great eras of geologic history, with revised estimates by Schuchert and Dunbar (1941, pp. 70, 71), are as follows:

	DURATION OF EACH ERA
Caenozoic era	60* millions of years
Mesozoic era	130 millions of years
Palaeozoic era	300 millions of years
Proterozoic era Archaeozoic era Cosmic era	1300 millions of years

These estimates are for the most part based on the so-called radioactive method of measuring the age of radium-bearing rocks, including pitchblende. Uranium, the parent substance of radium, is found in uraninites and related minerals. These are always disintegrating or giving off rays of different types called alpha, beta, gamma, rays. As the rays are given off, the uranium and thorium break down and the residue is lead. Holmes, Boltwood and later workers (cited in "The Age of the Earth," Bull. Nat. Res. Council, No. 80, Washington, 1931), have shown that this disintegration proceeds at measurable rates and that the age of the rock (that is, since its crystallization) can be determined by a formula that involves among other constants the known rate of disintegration and of the amounts of uranium and of lead now present. In practice the matter is very complex. Nevertheless out of the large number of age determinations based on uranium rocks from different geologic periods there has emerged a fairly consistent body of data, which, in combination with data based on the maximum thickness of the rocks of the different eras, is the basis for the estimates by Schuchert given above.

The oldest rocks thus studied, giving an estimated age of 1,800 millions of years, have been injected as dikes into still older rocks also of the Archaeozoic era. Hence, writes Schuchert (*op. cit.*, p. 8), "it must be concluded that the age of the earth is in round numbers at least 2,000 million years." (Cf. Holmes, 1947.) Whatever may be the judgment of future scientists concerning the details of the radioactive method, it seems to be established that the radioactive disintegration goes on at a constant rate which resists the utmost efforts of the physicists to accelerate it by excessively high temperatures or to slow it up by a combination of extreme cold and huge pressures.

Thus we should never for a moment lose sight of the stupendous stretches of time which have been available for organic evolution.

"A Million Years."—We shall not be wasting time if we pause now to struggle with the modern unit of geologic time, which is one million years. The idea of one million years implies several other ideas that are important for students of evolution. Every student may be supposed to know these things, but for the present purpose it may be useful to state them, because we learn to count at such an immature age that numbers may continue to be among the many things which are ordinarily taken for granted with little or no reflection upon their real meaning.

In the present case "one million years" implies first that time can be divided into successive pieces called years, which appear to be equal to each other (polyisomeres); second, that

* Add 17 for Paleocene epoch (cf. Simpson, G. G., 1949, p. 12).

we have a recognized sequence of verbal sounds and written symbols to designate the place in relation to its neighbors of every particular year in a series. If we are counting cigarettes in a pack, we can see and count them all at once, but if we are dealing with years we must *represent* them by numbers beginning at an assumed datum or date. Again, the word *year* is a symbol for a class of variable objects which are made to conform to an "average" only by abstraction, definition and convention. We must further assume that counting is in itself a form of measurement; that one million years is the *present sum* of one million present or past units; also that there is a way of measuring one million years of geologic time, etc., etc. But after all, what is a year and what is a million years?

We use the word *year* a great deal. We say: "A year ago we were in South Africa"; "It will cost you six years of strenuous work to go through college and medical school." But nearly always we are taking the meaning of a year for granted and trying to measure, evaluate, recall or describe something else in terms of the year, rather than *vice versa*. And when we come to think of it, the *sense of time* is a peculiarly elusive and difficult thing to describe or standardize.

In one sense time is a subjective or mental feeling. Everyone knows how time lags if we are bored and how it slips away when we are interested. And our internal time-keepers, heartbeat and pulse which say, *now, now, now,* are variable and not consciously correlated with each other. Then there are recurrent physiologic phenomena (e.g., the feeling for mealtime in infants and goldfish) which would serve for dividing the day into periods before, during and after meals. Such recurrences are occasionally said to be "as regular as clockwork," but the phrase itself suggests that in most people they are not as good time-keepers as clocks are. Perhaps we may discover something about the nature of time by considering one or two of the ways in which we ourselves build up a generally recognized abstraction called time.

A young baby evidently feels discomfort and pain or their infantile equivalents and these sensations soon arouse his breathing and struggling equipment, including his very potent voice. Everyone knows that, at least on our parents' cultural level, a crying baby usually stimulated relief measures by mother or aunt. As the baby's memories developed the association of crying followed by relief; this, it is now said, encouraged more vigorous crying. In any case a baby seems to feel increasing discomfort as the moments pass before relief. (Here perhaps is another basis for the recognition of a sense of *before* and *after*.) Similarly the desire for food may be supposed to increase if meals are delayed. Thus a physiological time sense slowly becomes part of the psychic background.

Du Noüy (1937) has shown that the rate of healing in wounds and other automatic physiologic mechanisms are normally more rapid in childhood and slow down with age. And from the previous examples it is evident that the conception of time is largely associated with memory and with recurrence of similar events or situations. Later we learn to "tell the time" on a clock or watch, upon which the sequential seconds, minutes and hours are indicated in ever recurrent days.

Thus the child gradually accumulates experience of *events* in time, which start, go on, get better or worse, stop and pass away. But to such a degree is the background of time taken for granted that even when in school or college we are seldom taught that in general, verbs have tenses *because they are time words* expressing relations both to the subject and the object. All this implies, on the part of the users of language, the *acts* in time of (a) perception, (b) comparison, (c) the passing of judgments according to learned or conventional scales of values. But once again time itself eludes us, except as measured by its effects.

In short, there are two more or less contradictory or mixed impressions of time: first, its variable physiological sense and, secondly, its supposedly invariable objective or universal sense.

That animals know when feeding time is near is ancient common knowledge and is illustrated in the story of domestic carp coming up to the sound of a dinner bell and in similar modern laboratory experiments. Underneath the whole field of the conditioned reflex and of the nature of learning lies the problem of the origin and workings of the time sense. The latter has been shown to vary with differences in age and health. It is accelerated by pleasure and retarded by pain. No doubt also there are several levels, from the most elementary physiological time-space sense of the amoeba and other beings that merely wrap themselves around their food, up to creatures with a brain crowded with association networks of ideas and motor responses. On the level of dreaming in man the time-sense seems to be latent in the recognition of events or changing patterns, especially in the sense of approaching danger, based on confused memories of past dangers, pains and pleasures.

It will be repeatedly shown in this book that simple repetition (polyisomerism) and repetition with varying emphases (anisomerism) are basic concepts of indefinitely wide scope and that there has been no evolution without them. Still wider and more elementary concepts are those of permanence *versus* variability. These opposite aspects of the outer world are communicated to sentient beings by organs that respond to vibrations of different frequencies and patterns. The recognition of time and space is latent in the ability to feel a present sensation and to compare it with the memory of past ones of the same general kind; in other words, to distinguish identity *versus* variations.

In many low forms of life there are not two different sets of sense organs, one for measuring space, the other for measuring time, but one organic complex that takes in both. In man, however, the eyes and the sense of touch have become the dominant receptors for space, while the ears (which have evolved from tactile organs of the lateral line system), together with the sense of hunger and other visceral drives, respond more directly to time. But through the networks of associated memories and because of the repetitive nature of phenomena, successive units of time can be recorded by ticks of a pendulum on an extended tape, while spatial displacements (as of earthquakes) can be measured and their times determined by the amount of deflection of the needle from the straight line of continuous time.

Time is commonly measured by its relation to space in the concepts of rate, speed, acceleration, and the like. We can *define* a year as the time it takes for the earth to make one complete trip around its orbit but nobody but the astronomically minded would think of this in the examples given above. We can measure the difference that a year makes in the growth of a tree, after the year has passed, but the year itself is only manifested in its traces. How much less then can we *feel* a century, a thousand years, a million years! Nevertheless the years leave their marks in many ways, so we must try to apprehend them indirectly by symbolizing their value in terms of something else.

Let us turn next to examples of "one million." Suppose that a passenger plane could make a trip of three thousand miles every day. Then at the end of 333⅓ days it will have travelled one million miles. At this rate it could cover a distance equal to that from the earth to the sun (about ninety-three million miles) in about eighty-five years. But this sort of calculation barely if at all helps us to get the feeling of *one million years,* because it attempts to visualize time in terms of space, while time itself is still the invisible partner of space.

Another illustration: Since there are three thousand six hundred seconds in every hour, we live through one million seconds every 11.57 days, and in 365 days we have used up 31,536,000 seconds of our total life span.

Another way to represent one million years is in terms of human generations. Let us suppose that there are four human generations in a century. Then in ten thousand centuries, or one million years, we would have a procession of forty thousand successive generations! But that would take us back only to our already human or nearly human ancestors just before or during the oncoming of the Glacial epoch; this in turn would only be less than one one-hundredth of the time back to the days of our early mammalian relatives and ancestors of the Age of Reptiles.

THE BUILDING-BLOCKS OF MATTER

Sir Isaac Newton conceived atoms to be excessively minute indivisible pellets, the true building-blocks of matter. The science of crystallography also suggests that crystals are built up of piles of little pellets arranged according to definite ground plans and elevations, much as the great pyramids of Egypt were made up of superposed layers of blocks, each layer above the bottom being one block shorter along each of the four sides than the preceding layer. When X-rays are projected through crystals upon sensitive plates they record upon an enormously magnified scale the regular spacing of the atoms. This and much other evidence suggests that the atoms, although confined to definite limits, are really not inert blocks but are more like whirlpools of excessively thin smoke. Nor is the atom, according to the findings of contemporary physicists, in itself the ultimate unit of matter. Indeed atoms are composed of still smaller elements, especially the following: [1]

- the *electron,* of excessively small mass ($1/2000$ unit), bearing a charge of one unit of negative electricity;
- the *proton,* of far greater mass (1 unit), bearing one unit of positive electricity;
- the *neutron,* a combination of 1 proton and 1 electron, with a unit mass and zero charge.

An atom of hydrogen, which is regarded as the basic unit of atoms of greater mass, consists of a proton, forming the nucleus, and an electron, whirling around the nucleus but at a relatively very great distance from it (Slack, 1937). And yet the atoms as a whole are in themselves so small that, according to Karlson (1936, page 37), there are 606,000 billion billion hydrogen molecules in a small teaspoonful (2 grams). "But it's impossible," he continues, "to form any notion of such inflated figures. Suppose the whole earth were made into little billiard balls; the number would be about the same." *

An atom of helium consists of two atoms of hydrogen plus two protons, all revolving in a complex system. The steps by which Lord Rutherford "sat in his laboratory and built up a world" out of the hydrogen atoms have been summarized both by Karlson (*op. cit.,* pp. 21–32) and by Rutherford himself (1937). His general conclusions seem to be widely accepted by his colleagues. However, the complexity of atomic structures and movements becomes compounded when we are dealing with molecular structure, until among the higher hydrocarbons, with long chemical formulae, the intricacy of the swirling systems far surpasses the present powers of human imagination. Even in such a relatively simple substance as graphite the general shape of the molecule is still being discussed by chemists.

POLYISOMERES AND ANISOMERES: THE ROLE OF REPETITION AND EMPHASIS

The slats in a Venetian blind, a set of encyclopaedias, a string of synthetic pearls, a list of addresses, the spokes of a wheel, the notes in an octave and millions of other similar parts of wholes, may be called *polyisomeres,* and so may the inches in a yard stick or any other units of measurement. These are artificial or man-made units or polyisomeres, but in the following chapters we shall see that evolution emerging has involved an infinite number and variety of natural polyisomeres in both space and time (Gregory, 1934, 1935).

There are indeed so many kinds of polyisomeres in so many different worlds of human interest that we usually fail to recognize them as such and think only of the special word for each kind. Even in the few preceding pages of this introduction we have spoken of such sets of polyisomeres as people, villages, churches, towns, historians, languages, weather forecasts,

[1] From "Nuclear Physics Chart," prepared by Dr. Charles M. Slack of the Westinghouse Laboratories, published in the New York Times, Sunday, May 16, 1937.

* "The World Around Us: A Modern Guide to Physics," New York, 1936, p. 23.

measurements, systematic groups, building blocks, atoms, nebulae, planetesimals, waves, days, years and the like as if they had no common source in our experience. We now realize that it is the repetitive character of polyisomeres which permit beings with memories to develop mental images of each kind of polyisomere and to refer any given example to its right class.

As units of structure, polyisomeres are called *primary polyisomeres* if they represent an older or more primitive state, and *secondary polyisomeres* when they have been derived by the subdivision or budding of primary polyisomeres. For example, the primitive perch-like, spiny-finned fishes were short-bodied and probably had only twenty-four vertebrae, which in this sense would be considered to be primary polyisomeres, but in some of their elongate eel-like descendants the number of vertebrae rises above ninety (*Mastacembelus*); these would be secondary or derived polyisomeres. Similarly, the main cheek plate of the garpike (*Lepidosteus*) is derived from one of the primary polyisomeres of the covering plates of the skull, but it has become subdivided into a mosaic of small plates which in this sense are secondary polyisomeres. In a somewhat different sense secondary polyisomeres may result from the convergent evolution between different parts of the same animal. For example, the lumbar and sacral regions of the vertebral column were fully differentiated in the ancestral mammals, but in whales, which have no doubt been derived from mammals, they have become de-differentiated, or secondarily similar to each other, thus giving rise to convergent or secondary polyisomerism between lumbar and sacral vertebrae.

Most polyisomeres exhibit variability. If the diameters of certain polyisomeres become conspicuously different from those of their fellows, we call such altered polyisomeres *anisomeres* (not equal parts); the evolutionary process of altering such measurements is called anisomerism. For example, the teeth on the outer upper jaw of the more primitive teleost fishes are fairly uniform in size, but in some forms (e.g., *Portheus molossus*, p. 152) one or more pairs of the front teeth become enlarged into tusks (positive anisomerism), while other specialized forms (e.g., carps) reduce or eliminate the teeth (negative anisomerism).

Patterns, both man-made and natural, are obviously made up of polyisomeres, or repetitive units, which are arranged on or with repetitive intervals in space or time. Simple patterns result from using only a single kind of polyisomeres and uniform repetitive spaces, as, for example, the general arrangement of the scales on the sides of a tarpon or the checker-like color pattern of a sole that is resting on a dappled background.

Mixed, or poly-anisomerous, patterns arise when the units used are either different sets of polyisomeres (as in the general pattern of the side view of the teleost skull) or when there is a combination of polyisomeres and poly-anisomeres, as in a chain of *Salpa,* of which each unit is a poly-anisomere.

In short it is but natural that polyisomeres of a given set should be so much like each other: (1) because they are derived from the same mother stock; (2) because rhythmic spasms of budding exert similar influences on each successive polyisomere; (3) because the environmental influences (both external and internal) which affect the first polyisomere of the series very often persist when the second, third, etc., appear.

The reason why most human products exhibit both poly- and anisomerism is primarily not because man has consciously imitated nature but because man himself and all his mental behavior are a part of nature and therefore must follow and parallel the methods of nature. And since man has evolved in a space-time universe of polyisomeres and their derivatives, he has gradually developed systems of representation, including such polyisomeres and their derivatives as are called signals, symbols, numbers, words. Tacit recognition of polyisomerism is implied in the use of plurals to designate more than one thing (or concept) of the same kind, while anisomerism is implied in all words that suggest more or less, how many, none. Words can then be arranged in shifting patterns which to a greater or less degree represent code records of nature.

CONVERGENT EVOLUTION, HABITUS AND HERITAGE

People who believe firmly that "bats will get in your hair" will probably never learn that bats are not birds, and in the fish markets neither sellers nor buyers hesitate to lump lobsters with clams and oysters under the word "shellfish." Whales are still called fishes by the majority of citizens and the discovery that whales are only fish-like in external form but are really aquatic mammals has not yet reached the newspapers, although clearly recognized by scientists since the time of Linnaeus (1758).

In the following chapters we shall deal with many such cases of convergent evolution, in which two animals that may look alike belong to widely different zoological groups. The kangaroo, for example, was at first classified with the jerboas among the rodents because of its general form and its habit of making great leaps upon its long hind legs. Later it was discovered that many Australian mammals, including the kangaroo, carried their young in a pouch and were alike in many features of their reproductive systems although quite unlike in external form; also that the various Australian mammals that are popularly called "tigers," "cats," "squirrels," "beavers," etc., all belong to a large group, the Marsupials, and that their resemblances to the familiar animals of the northern world whose names they bear are all due to convergent evolution.

Botanists long ago used the word *habitus* or habit to denote the general form and appearance of a plant in its relation to the environment, and I have used it in zoology (1914, 1936) in a somewhat similar sense for animals, as in the eel-like habitus of the lamprey, the fish-like habitus of ichthyosaurs and whales, the bird-like habitus of the flying reptiles. I have also used the word *heritage* to mean the older or deeper seated characters which reveal the animal's true place among the classes and orders of the animal kingdom. In whales, for example, the mammalian heritage is retained in many parts of the internal anatomy and especially in their way of feeding the young with milk from the maternal breast. Thus the heritage of whales is mammalian but their habitus is fish-like.

ARGUMENT: EVOLUTION EMERGING

I

The green leaves, bathing in the solar rays,
Transmute their food-stuff into chlorophyll.
For this, the timid doe plucks up her fare;
For this, the wolves pursue her and devour;
And from them all, man draws his daily food.
Hence I proclaim a basic law of life:
Apart from certain forms of low degree,
Dwelling in darkness or within the soil,
All creatures owe their being to the sun.
From him the flame of life glows in us all.
He is the source and fountain of our power.

II

Deep in the womb the living globule waits
The bliss of union with her spermal mate.
Dormant awhile, the procreative sphere
Divides itself and grows by subdividing
To daughter cells, their progeny in turn
Rival the desert sands in multitude.
So the mass grows, taking its proper shape.

III

As the cells multiply, the total form
Changes; reduplication makes new parts,
Many from one, arrayed in lines diverse,
Curving or straight or spiral, global or flat.
Sometimes the whole divides itself in parts
Equal or nearly equal as may be,
As when a cloud mass into raindrops falls.
In English, such are "polyisomeres."

IV

The dolphin's equal teeth have been derived
From very unequal teeth of carnivores
Unequals changing into equal parts
Are "secondary polyisomeres."

V

But equal parts to parts unequal changed
Are, on the contrary, called "anisomeres."
As when the rows of teeth almost alike
Gave rise in crocodilians to festoons.

VI

When equals with unequals are conjoined,
We have to do with "poly-anisomeres."
As when a salp buds off her progeny,
Each bud's composed of several different parts.

VII

When parts are lost and simpler forms appear,
"Degeneration" or "reduction" note.
Thus long-tongued ant-bears all their teeth have lost;
Their jaws look simple, but were once complex.

VIII

Better in some, but worse in other ways,
When one is substituted for another,
Or Peter robbed in order to pay Paul,
Call it the "law of compensation."

IX

The stronger or more useful part prevails
Over the weaker or less useful one;
The most adaptive and enduring live;
The others die by "Natural Selection."

X

The whole contains the parts, the greater rules
Over the less, in varying degree;
But when some parts outstrip or lag behind
Their team-mates, they are called anisomeres.

XI

From many, one is made: this is the source
Of federal power, as when native sons
With fellows from abroad together serve
The commonwealth. 'Tis then the whole appears
Greater than all its parts: the word is "holism."

XII

Of earlier forms the habitus, or mask,
That fits them for a special way of life,
To all their seed becomes prerequisite,
The basic portion of their heritage—
"Preadaptation" but not predestination.

XIII

And yet the later "habitus" conceals
Part of the "total heritage," as when bats,
Flying like birds, are proven to be mammals.
Thus heritage and habitus intertwine.

XIV

The living multiply at compound rates.
Productive lands, alas, have limits fixed.
Malthus saw there the threat of mass starvation
Worsened ten-fold by war and pestilence.

XV

Innumerable species have succumbed,
Their dwindling numbers crowded to the wall
By hardier or more fortunate contenders.
But some will live and, living, branch again,
Replenishing the earth and passing on
To ages yet unsung the light of life.

XVI

Statistics "Indeterminism" proves;
All her equations bear a hidden X,
The Lady Luck of gamesters. Chance into Law,
And Fate's kaleidoscopes new patterns throw.
Ifs become *whens* and seeming wonders happen.

XVII

Thus may the "Irreversible" be turned
Into new paths; "Evolution Emerging"
Looses a few from overspecialization,
Opens the door to save them from extinction.

XVIII

The lower limit of a given species
May be defined as where it forked away
From the meandering, vagrant stream of life
Flowing from Time's immeasurable past.

XIX

But who can tell the span or know the limits
In Time's equation of the Past and Future?
What! will the players quit, with coin in hand,
Tipping the cosmic table in their anger?
'Tis idle to imagine; they'll stay in,
"Law" against "Chance," so spin the wheel again!

CHAPTER I

THE FIRST LIVING THINGS

Contents

CHAPTER I

THE FIRST LIVING THINGS

THE SUN, CHIEF SOURCE OF LIFE ENERGY

MAN, LIKE other animals, draws his supply of energy from the sun.

The sun is the source of the energy stored up in plants. It is the leading factor in the forces which have made the earth fit for human habitation. Without the sun man could never have appeared; without it he could not survive a moment.

Man, however, cannot absorb the sun's energy directly, as the plants do; he must take it in his food, thus appropriating it from other animals and from plants.

The green coloring matter (chlorophyll) of plants has the power of absorbing some of the red, blue and violet rays of the sun, using them to transform raw materials into food for the plant's growth.

These raw materials are carbon dioxide, which the plant takes from the atmosphere, water, dissolved nitrates and other salts, which it draws up from the soil.

When the chlorophyll, with the aid of the sun's rays, breaks up these substances, the carbon is pulled out of the carbon dioxide and the hydrogen out of the water.

After a complex series of chemical reactions the finished products appear as sugar, starch and other carbohydrates, fats and proteins. These are the foodstuffs of both plants and animals.

Thus the energy of the sunlight, stored up in the substance of plants and animals, becomes a hidden treasure of great worth, to obtain which all animal life labors and struggles unceasingly.

THE EMERGENCE OF LIVING MATTER

With regard to the classic problem of the ultimate origin of living matter, it may be assumed that further studies in the same general direction will throw more light. Chemists understand, at least in a general way, the steps by which plants build up water and carbon dioxide into sugars and starches, as well as the elaboration of nitrates, carbonic acid and water to form amino-acids, and the working up of the latter into proteins (Sherman, 1925). Chemists also know a great deal about the composition and action of chlorophyll, which acts like a photochemical dye in catching and storing up the sun's energy (Encycl. Brit., "Photochemistry," p. 787). The students of the millions of hydrocarbon compounds have also been able to build up many organic products which are indistinguishable chemically from those produced by living organisms.

One reason why no one has yet been able to put together a combination which may fairly be called a living thing is that the protoplasm of even the simplest of living creatures today contains qualities which have been gradually built into it during millions of years of hereditary history. "Chemically," writes Calkins (1933, p. 43), "protoplasm is not a substance but a harmoniously working aggregate of different interacting substances which have been identified in general as nucleins, nucleo-albumins, nucleo-proteins, lipoproteins, fats, carbohydrates, salts and the almost endless varieties from these and from their combinations." Also, "The almost infinite variety of form and structure represented by the Protozoa must be traced back to the chemical nature of the proteins and to their relations and interactions with other substances in protoplasm." Moreover the various families of Protozoa have probably existed for enormous stretches of time, during which new or emergent features have been accumulated and specific and generic branching has taken place on a vast scale.

NITRIFYING BACTERIA

Whether living matter was first elaborated on the soil of the Archaeozoan lands or along its streams and sea coasts or in some other way is not yet clear. Osborn, in his book "The Origin and Evolution of Life" (1917, pp. 81–84), summarized the story of the simplest known organisms, which are the "primitive feeders" (*Nitrosomonas,* etc.). These nitrifying bacteria live in the soils and are able to feed on ammonia compounds, including ammonium sulphate, from which they form nitrites, while their closely related partners (*Nitrobacter*) oxidize these nitrites into nitrates. Osborn, following I. J. Kligler, thought that such bacterial organisms may have flourished on the lifeless earth and chemically prepared both the earth and the waters for the lowly forms of plant life.

Nevertheless these organisms, although exceedingly simple in appearance, are open to the suspicion of being adapted to live on the products of decomposition including nitrogenous material from pre-existing plants and animals. And not even the possibility that some pre-Cambrian iron ores indicate the presence of certain iron-oxidizing bacteria annuls the claim of the blue-green algae (Cyanophyceae) to be the oldest still existing forms of true plants.

Even the "filterable viruses," which were at first regarded as ultramicroscopic substances that are possibly the simplest forms of living matter, may also prove to be a secondary product of reduction. The hypothesis that life originated on some other world and that the germs of life arrived here in the cold interior of a stony meteorite seemed to gain some support through the reputed finding of bacteria in the center of meteorites. But it is difficult to be certain that the bacteria had not seeped in from the outside.

PROTOPHYTA, PRIMITIVE SOLAR ENGINES

The Protophyta (Encycl. Brit., vol. 18, p. 615) are for the most part simple one-celled organisms that obtain their nourishment after the manner of a plant; that is, by photosynthesis. In this process the energy of the sunlight, which is absorbed by the pigments in the chlorophyll, is used to build up food substances from carbon dioxide and water. The Protophyta may thus be regarded as "living fossils" which have come down to us almost unchanged through the geologic ages.

The most primitive of all true plants, which are to be found among the algae, are formed from rows or clusters of one-celled units which are practically indistinguishable from Protophyta. In the more primitive of these, each individual is equipped with one or two delicate protoplasmic threads called cilia or flagella; by the vibration of these threads the little cell moves through the water. This appears to be the most ancient and primitive of all forms of locomotor apparatus. The famous *Volvox* arises as a colony of single-celled Protophyta, which, growing at equal rates in all directions, assumes the form of a hollow sphere.

The kingdom of plants, with the exception of the bacteria and some others, emphasized the method of obtaining food by photosynthesis in the manner noted above. On the other hand, the ancestors of the animal kingdom were cells which found it easier to absorb the products of their neighbors' efforts. Between the plant and animal kingdoms are many forms like *Euglena* (fig. 1.1A), which were formerly classed among the animals because they are otherwise like certain Protozoa but which retain enough chlorophyll to entitle them to be classed as plants.

MICROSCOPIC HOSTS OF PROTOZOA

Let us begin by asking the question: Which is the most primitive body-form among the

Protozoa? Probably most former students of biology would immediately name the familiar *Amoeba proteus* of the classroom as the most primitive of all known animals; but for many reasons this claim has been disputed, apparently with good reason, by specialists on Protozoa (e.g. Calkins, 1933, pp. 140, 141). Many laymen seem to have the idea that because an organism has a relatively simple construction it is therefore equally primitive. But in literally hundreds of cases there is good evidence that animals whose remote ancestors have become parasitic have at the same time become progressively simplified in structure because they derive their food and shelter ready-made from their hosts. While *Amoeba proteus* is not in itself parasitic, many of its near relatives are so and at present it seems not improbable that all the naked amoebae have been derived eventually from what have been called "amoebae with shells." The most primitive, or at least the most central forms, among the Rhizopoda, the class to which *Amoeba* belongs, may perhaps be the diphasic rhizopods, including *Dimorpha mutans* (fig. 1.1C). These apparently motile cells have the two flagella characteristic of the primitive Protophyta combined with the radiating "axopodia" or protoplasmic rays of Heliozoa.

Heliozoa or "sun animalcules" are for the most part spherical, floating forms with numerous "axopodia" radiating from the center (fig. 1.1D, E, F). A few (e.g. *Clathrulina elegans*) are attached by means of a stalk which may represent the concrescence of ancestral axial filaments (Calkins, 1933, pp. 148, 149). Probably in this case the stalked or attached stage came later than the free-swimming stage.

While the Heliozoa are typically fresh-water forms, their relatives the Radiolaria are mostly marine. A central capsule, presumably of chitin or pseudochitin, separates the inner and outer spheres of protoplasm. In it is the nucleus, the center of growth. It is elaborately perforated and forms the basis of a complex system of spicules forming the skeleton. Pseudopodia radiating from the sphere are essentially threads of protoplasm used to entangle the food. These pseudopodia are usually supported on needle-like rods.

This spherical-radiate pattern with its principal features is no doubt conditioned in large part by hereditary forces, transmitted through the nucleus; and doubtless these hereditary forces, conditions and processes are of great complexity; moreover their operations are extended in space and time and are constantly reacting with the conditions and forces of the environment; so that we can scarcely hope to visualize the process more clearly without extended and many-sided research. Nevertheless we may reason by analogy with purely physical processes as follows:

The nucleus of a heliozoan or radiolarian may owe its spherical or oval shape primarily to the internal molecular force of cohesion plus the even pressure of a relatively homogeneous medium. The presence of pseudopodial needles radiating from the nucleus may imply a quasi-explosion of material pent up in the nucleus and perhaps suddenly released as a result of accumulated solar energy. The presence of skeletal material along the axis of the needle-like axopodium involves the process of osmosis and may be due to alternating periods of growth and deposition. To a similar alternation but in a larger cycle may be due the deposition of inner and outer spheres. The spicules, as D'Arcy Thompson has shown (1917, 1942), are deposited at the intersections of protoplasmic bubbles, while bubbles, spicules and radiating pseudopodia increase the flotative power and offset the tendency to sink resulting from the deposition of the skeleton.

While a globular or spherical symmetry is characteristic of the skeleton of more typical Radiolaria, the emphasis of growth of one part or another produces a great variety of forms, some of which are bilaterally symmetrical. Many of the Radiolaria produce spores that are provided with flagella and are thus suggestive of both the primitive Protophyta and the Flagellata (Mastigophora).

The class Rhizopoda (including the *Amoeba* group, the Heliozoa, Radiolaria, Foraminifera) has possibly been derived from primitive Heliozoa by the loss of the internal skeletal structures, and in the more degenerate forms the radiating axopodia are replaced by thick, irregular pseudopodia. In accordance with semi-parasitic habits, the body often becomes more or less irregular and movement is by flowing rather than by the lashing of flagella. A protective test of attached grains of quartz or silica is often developed. As noted above, the Amoebae may be regarded as degenerate derivatives of this type.

The Mycetozoa or "slime molds" are terrestrial parasites or semi-parasites, which seem to be related on the one hand to the Amoebae and on the other to the Proteomyxa, which in turn have ray-like pseudopodia recalling the appearance of Heliozoa.

The Foraminifera (fig. 1.1G) are geologically ancient Protozoa, in which, for the most part, the living gelatinous protoplasm is provided with a shell or test usually composed of calcium carbonate. The pseudopodia tend to anastomoze, as they do in the Rhizopoda, and the protoplasm is relatively simple. By different methods of budding more and more complex shells are built up, some of which present a curious convergence in general appearance to the shell of the nautiloid molluscs.

The great classes of Protozoa noted above may have all been derived eventually from simple motile cells provided with two flagella (cf. Calkins, 1933, p. 141), as in the primitive Protophyta. The organs of locomotion become greatly emphasized in the Zoömastigophora (or animal Flagellata), which range in form from simple, irregular cells (protomonads) with pseudopodia in addition to flagella, up to the highly differentiated Hypermastigida with segmented structure and many flagella, culminating in the annelid-like *Cyclonympha mirabilis.* At first sight it is difficult to realize that this form is really a protozoan, composed of a single cell, so much does it resemble superficially many segmented animals of the Metazoa.

The Infusoria, or infusion animalcules, include two great classes, the Ciliata and the Suctoria. The Ciliata (fig. 1.1J, K, L) are provided with simple cilia throughout life and are for the most part actively motile forms, but the Suctoria (fig. 1.1M) have cilia only in the developmental stages and in the adults are highly specialized, usually with suctorial tentacles. They may be either attached, with or without a stalk, or free. Even the parasitic ciliates may develop elaborate organoids for locomotion and feeding.

The Sporozoa represent an extreme form of adaptation to internal parasitic life, and they prey upon other Protozoa and higher forms, including man. They vary in form from simple amoeboid cells to star-shaped, dagger-shaped, or branched forms. They multiply by spore formation and often have complicated life histories, passing from one host to another and producing pernicious diseases of the blood and intestinal tract.

This rapid survey of the Protozoa may be sufficient to indicate the wide range of adaptations in regard to different methods of locomotion. Apparently the central stock were free-swimming, single cells, varying from globular to oval, spindle-like to irregular in form, propelled by the vibratile action of one or more flagella. In many Protozoa of different classes the spores revert to this simple type, which is also not unlike the spermatozoa of higher animals. In the typical Ciliata, such as the slipper animalcule, *Paramecium,* synchronous lashing of rows of cilia often propels the animal in an elongate spiral. Some of the Ciliata carry the locomotor organoids to a high degree of specialization, developing small flaps or membranelles which are used in jumping through the water. In this group the organoids of aggressive pursuit and attack by predators seem to reach their highest development. At the other extreme in the parasitic Sporozoa the cells float in a nutritive fluid and the organs of attack and pursuit are reduced or absent, although the spores may retain one or more flagella. Attached or sessile forms which draw in a food-bearing current by the lashing of cilia are very numerous and diversified.

Since the Protozoa are definitely animals and therefore do not manufacture their own food as do the plants, they must have something which corresponds to the organs of prehension, digestion and excretion among the higher animals. As we should expect, these organoids are highly developed in those predatory Ciliata (fig. 1.1L) which attack, ingest and digest living prey. Even some of the parasitic Sporozoa develop sucker-like processes by which they tap the food supplies in the cells of the host.

As to the geological antiquity of the Protozoa, it has been claimed that fossil remains of Radiolaria have been found in Pre-Cambrian deposits but, as we shall presently see, a critical examination of the materials has failed to support these claims. Typical Foraminifera date only from the late Palaeozoic (early Carboniferous). The known Protozoa (Radiolaria and Foraminifera) of the Cambrian fauna are estimated by Raymond (1935, p. 384) to make up only two-fifths of one per cent of the total fauna but, as he points out, this is no indication that the animals themselves were not abundant at that time. "The animals may have been naked; at best their minute skeletons are ill adapted for preservation or recovery."

Thus in spite of the lack of good fossil records below the Lower Carboniferous (Mississippian), when the Foraminifera began to pile up vast deposits of their skeletons, it is not improbable that the earliest Protozoa became separated from the single-celled plants long before the opening of the Cambrian record and that since then the Protozoa have enjoyed an evolution that in some respects is parallel to that of the Metazoa. Those numerous Protozoa which live as parasites in or upon the Metazoa have, with the pathogenic bacteria, doubtless played a great and devastating part among the death-dealing legions.

The Protozoa show us several grades in the process of reproduction. In the simplest case as in *Amoeba,* the single cell after growing to a certain point pulls itself apart, dividing in the middle into two equal cells, each containing half the original nucleus. Protozoa may thus reproduce for many generations by simple fission or by budding, but it has been shown by many investigators that eventually, at least in *Paramecium,* the stock deteriorates unless conjugation, or the union of the nuclei of two individuals, takes place. Here then are the remote beginnings of sexual reproduction and of all the complexities that result therefrom.

In the Foraminifera the subdivisions of a single cell remain together. This complex may be regarded as a single individual cell, which is divided into successive similar parts. On the other hand, in *Volvox globator* some of the subdivisions are much smaller than the others and have specially assigned to them the function of asexual reproduction. In this form there is a close approach to the formation of eggs (larger globular bodies with yolk) and spermatozoa, small motile cells which fuse with the ova (Parker and Haswell, vol. I, pp. 47–77).

In short, the Protozoa, although exclusively single-celled animals, are literally a world in themselves, in which many of the problems of locomotion and food-getting have been solved in more or less the same ways as in the Metazoa, or many-celled animals. One great difference, however, between the microscopic Protozoa and ordinary or macroscopic animals is that the former, on account of their minute size, present a high relation of surface tension to mass. Therefore they tend both to lose and to absorb solar energy more quickly than do larger many-celled animals and are less free or independent of the variations of the medium in which they live.

THE SPONGE-STATE AND ITS CITIZENS

Among the Protozoa certain types are named Choanoflagellata because each vase-shaped individual cell bears a large lash or flagellum, which is surrounded by a funnel or delicate protoplasmic collar (fig. 1.1I). This collar is contractile and by its slow but rhythmic movements it draws in minute food particles which become embedded in the absorptive protoplasm at its base (Encycl. Brit., article on Sponges).

In *Proterospongia* (earlier sponge) the simple collar cells have become divided into many individuals that stick together in an amorphous jelly-like mass (fig. 1.1I). In the true sponges, however, the same general sort of collar cells are combined as citizens in a corporate body or state.

Apparently the very remote ancestral sponges must have consisted of globular colonies of collar cells with their collars and flagella pointing inward toward an empty space in the middle of the sphere. By the lashing of the flagella water was drawn in through pores between the cells (whence the class name Porifera). Where the cells were in contact with each other they secreted spicules, originally of silica, which served to hold the colony together. In a later stage of evolution there are many subspherical groups connected by tubes and all opening into a central chamber which, in turn, has a circular chimney or exit at the top. The base of the sponge is now fastened on the sea floor and although the sponge becomes irregular in shape it is definitely oriented in reference to the top-and-bottom axis. Its inside surface is that toward which the collar cells are directed but it lacks a many-layered integument or shell and has no integrated nervous system, each individual collar cell being more or less an isolated living unit embedded in a huge community dwelling house constructed chiefly of spicules. In other words, the primitive sponge-state stands on a low stage of evolution, with relatively slight differentiation of its individual workers. As such it offers an extremely sharp contrast with the typical ant- or bee-state.

In one grand division of the sponges (the Silicispongiae or Glass sponges) the spicules are composed of silica; in the other division (Calcispongiae) of lime (calcium carbonate). Throughout the five hundred-odd million years of recorded fossil history since the Lower Cambrian the sponges have, as it were, played innumerable variants upon the simple structural theme noted above, so that several thousands of fossil and recent species of sponges have been recorded.

With regard to embryonic development, after the early cleavage stages there is a ciliated, free-floating embryo (fig. 1.6A) which later settles down and becomes attached. To this extent the sponges agree with many other groups of marine invertebrates in which the free-floating ciliated larvae insure a wide distribution of the species, while the sessile adults occupy the best available spots in the littoral zone where floating food bits are abundant.

PRE-CAMBRIAN LIFE AND THE ORIGIN OF SKELETAL ARMOR

Thousands of square miles of Pre-Cambrian rocks have been exposed by the erosion of later, overlying formations in the great Canadian Shield, in the Adirondacks and in not a few other regions, especially in Northern Europe, China, Australia and New Zealand, but it is in only a few localities that the Pre-Cambrian rocks have escaped thorough reorganization or metamorphosis due to the enormous pressures and heat to which they have usually been subjected. In all but a very few thousands of feet of Pre-Cambrian strata the fossils, if there ever were any in them, have been hopelessly squeezed, melted and twisted along with the original bedding planes. However, indirect evidence of the presence of living organisms during Pre-Cambrian times is afforded by the presence of carbonaceous material, of bedded iron ores that were once carbonates, of cherts, and probably of limestones (Chamberlin and Salisbury, 1904).

Many kinds of organisms have been reported as represented by fossil remains from Pre-Cambrian rocks but few of these have withstood the test of critical examination. The most famous was the *"Eozoön canadense"* of Sir William Dawson from Pre-Cambrian rocks about eighty miles west of Montreal. According to Professor Percy E. Raymond (1935) of Harvard University, the best specimens have been carefully investigated by C. A. Osann (quoted by Raymond), whose chief conclusion he endorses, namely, that the *Eozoön* "is the product of two periods of alteration of the original sediment and can by no possibility represent an original structure. There seems not the slightest chance that it can be organic." The author of the article on Marble (Sir John Smith Flett) in the Encyclopaedia Britannica (vol. 14, p. 861) states that "The much discussed *Eozoön,* at one time supposed to be the earliest known fossil and found in Archaean limestones in Canada, is now known to be inorganic and to belong to the ophicalcites" (marble containing much serpentine, which has been formed by the decomposition of forsterite, or diopside).

From the Newland limestones of the Pre-Cambrian Beltian series in Montana the late Dr. Charles D. Walcott (1899), a very eminent palaeontologist, collected a great series of supposed fossils which he described as new genera and species of blue-green algae. Professor Raymond (1935), who has made a most careful study of Pre-Cambrian fossils, points out that while some of Walcott's specimens retain a general similarity to the "lake-balls" or "water-biscuits" which at the present day are formed in lakes by the blue-green algae, yet they are not very convincing as evidences of genuine fossil algae for the reason that when lime-secreting algae are found in true fossils in the Ordovician (Palaeozoic) and more recent deposits, it is always possible to identify the plant at least generically by means of its internal structure, which does not show in the case of the Pre-Cambrian "lake-balls."

Dr. Raymond has subjected many other supposed Pre-Cambrian fossils to a critical and close examination. He gives good reasons for suspecting that the alleged Pre-Cambrian Radiolaria and Foraminifera are at least open to serious doubt. The famous Pre-Cambrian "spongoid" named *Atikokania* by Walcott is very similar, he states, to forms found in the Permian dolomites at Sunderland, England, which have been shown by G. Abbot and by Olaf Holtedahl to be concretionary structures resulting from the replacement of limestone by dolomite. Another lot of "Atikokania" from Walcott's Canadian locality was closely studied by Raymond, who states that thin sections show that it is composed of aggregates of quartz crystals embedded in a matrix of limestone and is therefore purely of inorganic origin. Nevertheless, the striking resemblance between such inorganic forms as "Atikokania" and the algal "lake-balls" may serve to show that even in living things the mechanical processes that produce limiting membranes or zones also set up differential or osmotic pressures on either side of such a membrane and encourage the deposition of skeletal or inert material in zones or rings.

In Australia a well known geologist, Sir Edgeworth David, collaborated with Dr. R. J. Tillyard, a botanist, in writing a small book on fossils from the Adelaide Series of South Australia (1936). In this work certain supposed fossils from a Pre-Cambrian formation near Adelaide were interpreted as evidence of a peculiar form of arthropod that had some features in common with the eurypterids. Both authors died soon after the publication of their work but not before one of them (Dr. Tillyard) upon further study of the material rejected part of it as unsatisfactory evidence for his former thesis. I have seen some of the type specimens in the Australian Museum at Sydney but they did not look convincing, being largely composed of ovoid and irregular dark blotches. The authors gave no evidence of the internal microscopic structure of the supposed fossils and some Australian scientists have now definitely denied the validity of the evidence submitted by Messrs. David and Tillyard. On the other hand, Sir Edgeworth David expressly stated (1936) that he was constantly on his guard against being deceived by either mineral or algal pseudomorphs, and several of his colleagues, including a specialist in algae, after careful and critical examination of the evidence, fully endorsed his conclusion that these fossils do not represent algae but an arthropod (in the widest sense) which has basic resemblances to the *Nauplius* larvae of the lowest existing crustaceans, together with other features suggesting relationship with the common ancestors of the trilobites, limuloids and eurypterids.

In spite of the failure of any one individual fossil of Pre-Cambrian age to yield final proof that it was really a living organism and not a mere "imitation" of mineral origin, like "moss agate," Professor Raymond (1935) concludes that there is strong, although indirect, evidence for the conclusion that long before the opening of the Cambrian record the following classes of animals were already in existence: (1) naked Protozoa, (2) siliceous sponges, (3) primitive coelenterates, (4) segmented worms, (5) inarticulate brachiopods and (6) trilobites. These are the older groups, whose evolution was already well advanced at the opening of the Cambrian record.

On the other hand, he notes, there was a series of groups of later origin which enjoyed the greater part of their deployment or evolution after the opening of the Cambrian. Among this second lot are the articulate brachiopods, the typical molluscs, the echinoderms, the higher arthropods, including the insects, and the vertebrates.

Raymond (1935), after examining the six main theories that have been advanced to account for the relatively sudden appearance of fossil remains in the Lower Cambrian formations, sets forth the view that the Pre-Cambrian animals had no skeletons and therefore were excessively rare as fossils, and that skeletons appeared as a physiological result of the precipitation of more calcareous material than could be dissolved by the body fluids; this condition in turn was both the cause and the result of the adoption of a sessile or sluggish mode of existence.

Raymond says:

"That a definite relationship between activity and skeletal armor exists is obvious. The common symbols of sluggishness are the snail and the tortoise; such animals as the corals and the bryozoans, fixed always in one spot, show a maximum of calcareous shell and a minimum of flesh. It is commonly said that armored animals are sluggish because well protected, and that the unarmored are active because they must seek safety in flight. As a matter of fact, cause and effect are reversed in this oft-repeated remark, for animals are probably armored because of their sluggishness or entire lack of movement. Calcium is present in greater or less quantities in all water and in many kinds of food. While in solution, it enters the alimentary tracts of animals, but within them it is converted into solid calcium carbonate, either by the action of ammonia produced by putrefactive bacteria, or by the effect of the nascent methane formed during the digestion of the cellulose of plants. This is harmless, but small amounts, still in solution, get into the body-fluids and reach the protoplasmic cells. Some of these cells apparently pass on the dissolved calcium to the excretory system. Others cause its precipitation *in situ* and so build a skeleton. All animals and many plants are confronted with the necessity of getting rid of surplus calcium carbonate. Man is no exception to this rule, and as with other animals, those most active seem best to solve the problem. Men leading an active existence get rid of the excess lime, but those engaged in sedentary occupations find sooner or later that calcium carbonate has hardened the walls of the veins and arteries, so that an unwanted skeleton has been built to shut out life. It should be mentioned that this is not the common form of 'hardening of the arteries' or 'arteriosclerosis,' but may be an end stage in such a history, for it commonly follows fatty or fibroid changes in the walls of arteries or veins. The production of a calcareous skeleton is, therefore, an involuntary chemical function which will take place in animals in any environment. In one sense, it may be thought of as a pathologic condition, brought on by inactivity. . . ."

His final conclusion is as follows:

"That the pre-Cambrian animals were all motile, seems, therefore, to explain their lack of hard parts, and hence to solve the question as to why so few pre-Cambrian fossils are found. That more than are now known will eventually be discovered, is, however, very probable, but it may confidently be predicted that they will prove to have either no skeletons, or else thin ones, composed of silica or chitin."

Although it has been widely assumed that skeleton-bearing forms were derived from older free-swimming creatures without skeletons, I have very gradually come to the conclusion that in many cases the opposite may be true.

1) In many-celled animals the very multiplication of cells forces some into immediate contact with the environment and favors the development of a cuticle or limiting membrane. The resulting differences in osmotic pressures lead to the deposition of heavier materials, at least in the basal layers of the skin.

2) Without relatively tough or stiff masses or surfaces of chitin, silica or lime, it would be

hard for an animal to find an effective fulcrum on which to base the contractile force of his muscular tissue. This principle applies both to external skeletal parts (such as the horny jaws of annelids, the jaw-like pincers of crustaceans, etc.) and to all internal skeletal parts, as the dental masses in the gizzard of a lobster, and especially to locomotor appendages.

3) Many naked forms, as will be frequently shown in this work, are not primitive but highly specialized members of their groups.

FROM STINGING CUPS TO WORMS THAT BITE

In the coelenterates (hollow intestine), including the highly varied hydroids, jellyfishes (medusae), corals, seafans, etc., the body is essentially cup-like, derived from the outer and inner layers of the gastrula stage of the embryo. This cup-like condition in the gastrula stage is due to the faster growth of the smaller ectoderm cells, which grow around and enclose the larger, nutriment-bearing endoderm cells. Thus the adult coelenterates may be regarded as forms which have never gone far beyond the gastrula stage. The coelenterates have no true mesoderm (middle layer) but there is a layer of contractile fibres, collectively called mesogloea, connecting the sensitive epidermis and the nutritive cells of the primitive gut.

In the fresh-water *Hydra,* for example, the opening of the cup or sac forms a functional mouth (fig. 1.2) and is surrounded by a ring of tentacles. These tentacles arise from folds of the body wall and are both sensitive and contractile. Each tentacle bears thousands of microscopic dart-threads, or nettle-threads (fig. 1.2), which are a unique feature of the general coelenterate plan. The nettle-thread, although incredibly fine, is really a tube coiled up in a ball and enclosed in its own cell in the epithelium of the tentacle. A cnidocil or sensitive hair on the cell projects above the ball. According to Storer (1943, p. 306), the cnidocil "was formerly thought to act like a trigger but direct mechanical stimulus, as by touching with a glass rod or by the protozoans that live and move about on the surface of hydra, is not ordinarily effective. Substances diffusing in the water from the small crustaceans, worms and larvae on which the hydra feeds will usually provoke discharge, as will acetic acid added to water. Eversion of the thread evidently is caused by increased osmotic pressure within the capsule." Thus if any luckless small snail or floating insect larva happens to come near, the dart is ejected with explosive violence and hundreds of other dart guns near by explode also. Even a human being, touching the trailers of a floating jellyfish, may receive a poisonous shock which may be almost fatal. Strange to say, the point or barbed end of the dart is at the bottom of the thread-ball and the tube-thread is turned inside out like the finger of a glove. The principle of the blowgun or of the harpoon gun is apparently being applied here, and before the explosion the effective internal pressure must accumulate through a lowering of the external osmotic pressure due to the presence of the intruder.

Presumably the velocity of ejection is greatly increased by the narrowing of the neck or core of the coil toward the aperture. Moreover the lower osmotic pressure in the direction of the intruder may actually help to draw the thread toward him. In some coelenterates such thread cells are also borne on convoluted cords on the inner septa surrounding the gut, where they doubtless serve to kill recently ingested prey.

What were the stages leading to this perfected dart-cell mechanism? After formulating and rejecting several provisional hypotheses, I note that a satisfactory hypothesis must take into account all the following facts:

(1) That the coiled thread is laid down within a single cell which has its own nucleus; (2) that the containing cell is filled with liquid and surrounded by a tough membrane; (3) that one end of the cell is invaginated, forming a hollow central pillar, in the middle of which is the dart; (4) that the rest of the thread is coiled around the central axis and is ejected first, before the dart; (5) that coiled thread-cells are found also on the convoluted coils of the intestine; (6) that, according to Storer (1943, p.

306), a cnidoblast, containing a nematocyst, does not originate on the tentacle but in the body epidermis, "whence it migrates into and through the fluid in the enteron and again passes through the body or tentacle wall to its final location in the epidermis."

How did this coiling come into existence?

1) It may be suggested that the coiled threads are at least partly analogous with the trichocysts of ciliated Protozoa, which alternate between the bases of the cilia and may be discharged as long threads to serve perhaps in attachment or defense (Storer, 1943, p. 277). This fact gives a clue that trichocysts and cilia may be regarded as homologous opposite polyisomeres, the former inverted, the latter everted. In the hydroids the flagella of the cells of the digestive tract suggest the everted tubes of the discharged nematocyst, while the coiled dart threads of the nematocysts are inverted. This inversion may perhaps be due to transverse expansion or swelling of the growing wall of the cnidoblast, which may exert suction on the thread tube. After inversion is completed more liquid may seep into the interior of the cnidoblast by osmosis, thus increasing the pressure and preparing for the discharge.

2) According to Storer (*op. cit.*, p. 306), there are four kinds of nematocysts in *Hydra,* of which only the *penetrant* kind bears at its base three long spines and three rows of small thorns. The pear-shaped *volent* contains a short thick thread in a single loop, which upon discharge coils tightly around bristles or hairs on the prey. In the third and fourth types, or *glutinants,* the threads produce a sticky secretion possibly used in locomotion as well as in feeding.

Nematocysts are derived from small, round, undifferentiated interstitial cells found between the bases of the epidermal cells. These interstitial cells also give rise to buds and sex cells (Storer, *op. cit.*, p. 305), so that the nematocysts are, so to speak, the siblings of the sex cells. Some of the ciliated protozoans have nematocysts (fig. 1.1L) but no certain homologues of the nematocysts are found in the phyla above the coelenterates.

Some of the stalked hydroids, instead of reproducing by buds which are like the parent, bud off little floating jellyfishes which progress by rhythmic contractions of their parachute-like cups. The common jellyfish (*Aurelia*) eventually throws off eggs and sperm from the genital ridges of the entoderm, and from the union of these elements are developed small ciliated sacs which later become pear-shaped, attach themselves to rocks, etc., and develop into a hydra-like form. Later this stage subdivides itself into a series of cup-shaped discs, or strobilae, which in turn grow and are transformed into the sexual generation called *Aurelia.* We may then raise another question: Which is the phylogenetically older stage, the stalked hydroid or the free-swimming *Aurelia*?

One of the most noteworthy features of the jellyfish is that, as seen from above, it forms a nearly perfect circle which has the appearance of being divided into quadrants, each of which in turn is again divided into two equal parts. This beautiful symmetry recalls that of the first four segmentation cells of the early embryo of the sponge and from other evidence it would seem that the ancestral stock of all coelenterates were deeply stamped with such a combination of bilateral and radial symmetry as is most perfectly developed in the free-swimming, slowly pulsating medusae. Nevertheless this entire symmetry is essentially goblet-like since its center is homologous with the stalk of the fixed hydroid. Moreover, the attached hydroid with its tentacle-trap and sac-like stomach remains in the crowded shallow-water zone where food is abundant, while its offspring, the free-floating medusa, must take its chances of being washed up on shore by storms.

Among the higher coelenterates are the Siphonophores, including the Portuguese man-of-war (*Physalia*) and related forms; here we find a relatively high differentiation of function on the part of groups of cells; some serving as flotative organs, others as stinging and killing organs, others for digestion, reproduction, etc.

The graptolites (fig. 1.3) are very numerous and important time-markers in the Cambrian, Ordovician and Silurian ages. Some of these fossils suggest little fan-like bushes (fig. 1.3A). Each branch bears along one side a great many little cups (thecae); these were the seats of unknown zooids, which may have been more or less like tiny medusae. The bush was made of chitinous stuff and may have floated with the nema, or parent thread at the top. Some forms were attached to the bottom or to floating objects by a stalk. From the central fan-like type were derived five main lines, with a great number of genera and species; the evolution of these lines has been traced by many palaeontologists. Some have thought that the graptolites were essentially bryozoans, but later studies by Bulman (1936, p. 62) and others tend to confirm the older views that the graptolites were a very distinct class of coelenterates, allied remotely to the Hydrozoa and Scyphozoa. Some of the later species were equipped with a vesicle or float, beneath which the main branches hung down somewhat as in the modern "man-of-war" (*Physalia*).

In the more primitive sea-anemones and corals the polyps are seated like jellyfishes turned upside down, with the bell forming a cylinder and the mouth and tentacles directed straight upward. The ciliated free-swimming embryo (Planula) has "an elongated ovoidal body with an outer layer of ciliated ectoderm and an inner layer of large endoderm cells surrounding a closed enteric cavity, usually filled with a mass of yolk which serves as a store of nutriment. In this condition the embryo escapes from the parent through the mouth, swims about for a time and then settles down, becoming attached by its broader or anterior end." (Parker and Haswell, vol. I, p. 179.) After the establishment of a digestive tract the interior of the late larva becomes subdivided by more or less elaborately folded septa radiating around the digestive tube. The walls and bottom secrete a limestone base which bears the imprint of some of the body folds. From this simple plan has evolved an enormous number of patterns, many of which have endured through the ages, while the myriad little creatures that follow these ways of life have built up tremendous ramparts of coral reefs.

The typical comb-jelly (ctenophore) is essentially spherical (fig. 5.11C1), with the mouth at the upper pole. The gastro-intestinal canals branch symmetrically from the axial tube into a right and a left set, each with four branches. Eight vertical bands of ciliated patches (or combs) extend along equally spaced meridians. Two long trailing tentacles issue from pits on opposite surfaces of the sphere. A paired statocyst, or elementary sense organ, bestrides the aboral pole. The thread-cells of the hydroids (cf. pp. 26, 27) may perhaps be represented by the lasso or glue-cells (collo-blasts) on the tentacles. These "secrete an adhesive material to entangle small animals, which are then conveyed to the mouth" (Storer, 1943, p. 323). This is evidently a more complex highly organized type, with only distant relationships with the true coelenterates. The ctenophores are indeed structurally annectent between the coelenterates and the flatworms (cf. Allee, 1926–1927, pp. 263, 269–270). One family of ctenophores, instead of being content to die like most jellyfish when cast on shore, have become flattened and crawl on the ground like flatworms.

As compared with a simple sponge, the coelenterates have made far greater progress in the evolution of an integrated individual composed of well differentiated living units. The ectodermal layer of cells produces sensitive tentacles which have the power of contractility and which discharge thousands of dart-cells at living victims. Eye spots are present between the bases of certain tentacles. The rudiments of a contractile neuromuscular system are present. The endoderm is elaborately folded and the incipient vascular system produces the materials for the base of the corals.

In the flatworms (fig. 1.4) appears a new and important structural element (mesoderm) that was budded off near the blastopore between the ectoderm and the entoderm. The contractile neuromuscular cells of coelenterates became differentiated into muscle cells (fig. 1.4C), each with its own nerve fibre, the muscle coming from the mesoderm, the nerve from the ectoderm. The new layer soon assumed a great variety of functions, entering into coöperation with the ecto- and entoderm in the development of efficient systems of locomotion, feeding, alimentation, distribution, excretion, sensation, response, growth, reproduction.

The primitive three-layered (triploblast) animal was presumably a creeping, unsegmented form of Pre-Cambrian date, essentially like a free-living flatworm in habit. Equipped with the highly potential three layers (fig. 1.4C), this lowly but prolific race proceeded to give rise to the myriad special types of triploblastic animals which for several hundred millions of years have swarmed in the seas, on land and in the air.

The free-living turbellarian flatworms are leaf-like to ribbon-like in body form (fig. 1.4A) and are able to glide on the ground by means of the cilia on the ventral surface, which lash against a trackway of sticky mucus laid down by epidermal glands (Storer, 1943, p. 329). In aquatic forms the margins of the body are often extended like fins and they swim like skates. They have a many-layered skin and a muscular body wall with circular and longitudinal muscles which can give rise to creeping movements.

The free-living types of flatworms (planarians) recall the radiate jelly-fishes in the central position of the mouth on the ventral surface (fig. 1.4B); but the flatworm ground-plan is bilaterally symmetrical, elongate, with a pair of longitudinal nerve cords and a pair of eye-like spots crowded with light-sensitive cells. The opposite eyes are connected by transverse nerve threads with each other and by longitudinal threads with the nerve cords and other nerves to various surface areas.

The two ventral nerve cords and their ganglia are ventral to the gut. The eyes vary from simple retinal cells to complex eyes with a lens; they rest on two fairly large cerebral ganglia on the ventral surface of the head. Dorsal to the nervous system are the digestive tract and its numerous side branches. The mouth, opening on the ventral surface, is surrounded by a muscular pharynx which can usually be everted and retracted.

All this represents an early stage in the evolution of a head and central control system which reached a far higher stage of complexity in scorpions, flies and other animals with a jointed cuticular skeleton. The parasitic flatworms are far more specialized than the turbellarians.

Among the hosts of creatures called worms, which belong to several widely different phyla and classes, many have become either external parasites (such as the leeches) or internal parasites (such as the liver-flukes, roundworms and tapeworms).

Some of the parasites (such as the liver-flukes) during their development from fertilized egg to adult go through an amazingly complex series of transformations, living during successive stages in such different hosts as snails, mites, and sheep.

The various classes of flatworms, roundworms, hook-headed worms, have feeble locomotor powers and little or no skeleton, inner or outer, apart from cuticular spines around the mouth or on a proboscis.

One very curious side branch (Chaetognatha) of the roundworm phylum are the arrow-worms (*Sagitta, Spadella*). These are free-swimming marine forms which suggest the fast-darting lancelet *Amphioxus* among the chordates. They are elongate, bilaterally symmetrical, cylindrical, with two or three pairs of "lateral fins" formed by keels from the epithelium. These are placed horizontally and suggest that the wide tail moved up and down rather than from side to side; this is also indicated by the arrangement of the muscles in four longitudinal bands. The distinct head is provided with paired eyes, a pair of "olfactory" nerves and organs, a circlet of chitinoid hooks around the anteriorly placed mouth, a comparatively large brain and a large ventral ganglion with nerves branching out on each side over the muscular body wall. The superficial resemblance of these arrow-worms to *Amphioxus* seems to be entirely convergent, for in their method of development and anatomic plan they seem nearer to the roundworms (Parker and Haswell, vol. I, pp. 293–296).

In the marine annelids we find fairly advanced types of aquatic locomotion effected by means of serially arranged projections from the body wall named parapodia, which combine locomotor with respiratory functions.

In *Nereis,* which is a central type of the annelid (or annulate) worms, the small head (fig. 1.5B) is followed by a large number of similar segments or metameres, each containing a chamber or compartment of the body cavity and a section of the alimentary canal and other organs (*op. cit.,* p. 403). The metameres are separated by transverse creases and partitions; the latter are composed of folds of the peritoneal or coelomic epithelium containing muscular fibres. The partitions are pierced by the gut-tube, to which they are attached. In the body-wall two dorsal and two ventral longitudinal muscles flex the body up and down; two oblique muscle bands passing from the mid-ventral line to the dorsal surface just above the parapodia would tend to constrict or warp the upper surface, while the superficial circular muscles, running transversely in each segment, would tend to squeeze the body and lengthen the segments. If used alternately or in combinations, all these muscles could impart varied thrusts to the projecting lateral appendages or parapodia.

From each metamere (fig. 1.5B) projects laterally a pair of movable flaps called parapodia, which are subdivided into two main lobes supported by stiff cirri or bristles. One thin part of the parapodium may have a respiratory function. In writhing movements of the body the parapodia are pushed against the water, sand, mud, or seaweed, and thus they foreshadow somewhat the movable appendages of crustaceans. In both cases such appendages consist essentially of protuberances enclosing muscles and covered with chitinoid cuticle, the latter being the source of the exoskeleton of the externally jointed crustaceans, insects, scorpions and spiders.

The earthworms (*Lumbricus*) seem to be derived from marine annelids which have invaded the soil, perhaps originating along the shores and gradually becoming highly specialized for burrowing. In so doing they have acquired a glandular slime-secreting epithelium and lost all trace of the parapodia except for a few short bristles on certain segments. Meanwhile the circular and longitudinal muscles have become powerful and the transverse septa and body-wall are capable of distension and contraction. The worm takes in the earth at the mouth, extracts the food particles in the long digestive tube and leaves the residue behind it. Its power to push itself through the subsoil suggests that one part of the fluid-filled body is inflated and pushed forward while the rest of the body is being used as a fulcrum.

The metameres of annelids are "polyisomeres," not in the chemical but in the morphological sense, since they consist of many units all budded off from a parent mass; in the case of marine annelids the parent mass is a free-swimming, more or less sac-like larva. "Anisomeres," which are specially emphasized or differentiated polyisomeres, are well illustrated in the parchment worm (*Chaetopterus*). This creature lives along the seashore in a U-shaped tube which it has dug for itself in the muddy sand and which it lines with a somewhat parchmentlike excretion from its own skin. It has a series of movable ventral appendages which are greatly enlarged on three segments of the middle region of the body. Here the whole segments form flap-like valves and by their rhythmic movements the water, bearing organic food particles, is pumped in at one end of the tube and pushed out at the other. This gives a good example of the frequent association of anisomerism or differentiation with a change of function. In the case of *Chaetopterus* the change is from the swimming-aerating function of the small parapodia of free-swimming annelids to the pumping function of the enlarged mid-segments and their valve-appendages.

The annelids exhibit an early stage in the evolution of a head, which eventually becomes the center of control for all parts of the body. The head is built up around the mouth, which is the port of entry for all food; the sense organs, located in or connecting with the head (fig. 1.5A, B), serve to direct the locomotor apparatus toward the food, while the motor nerves direct the mouth, jaws and throat in the seizure and ingestion of the food. The highly protrusile jaws of typical marine annelids are located in a muscular sac or proboscis (fig. 1.5B). They are formed by a transverse pair of horny or chitinous hooks on either side of the mouth and are operated by special muscles in the wall of the buccal cavity. When retracted they are located in the pharynx behind the mouth.

The annelid body is segmented, the successive segments being nearly alike, and many of the segments include reproductive cells. The sensory and motor nerves of the segments converge into a paired ventral nerve cord (fig. 5.12A), which continues forward beneath the digestive tract to the throat (oesophagus); there it divides into a right and a left half which together form a nerve ring connecting the ventral nerve cord with the dorsal ganglion; the latter in turn is the central station for nerves running to the eyes and other parts of the head.

This general type of nervous system is found with appropriate modifications in many other groups of invertebrates, especially trilobites, crustaceans, eurypterids, centipedes, scorpions, spiders, insects, *Peripatus,* and allied groups. Traces of it may be detected in the molluscs but, as we shall see later, it differs so radically from the type of nervous system that is found in the vertebrates that it seems improbable that the vertebrates have arisen from any of the annelid-arthropod stock.

The Archiannelida are a division of presumably primitive annelids in which the front end of the body is developed from a ciliated sac-like structure which represents the trochophore, or larval stage (fig. 1.7D). Apart from its possessing the beginnings of the mesoderm, this trochophore larva is not far above the sac-like, two-layered coelenterate which may have given rise to all the triploblastic, or three-layered groups of animals. A trochophore larva is also seen among the true annelids (fig. 1.7E), while more or less similar larvae, but specialized in various ways, are found among such widely diverse phyla as the crustaceans, rotifers (fig. 1.7C), brachiopods, molluscs, echinoderms and others. If ontogeny would always repeat phylogeny, as indeed it undoubtedly does in some cases, and if nature did not so often produce similar forms by convergence from widely dissimilar ancestors, then the possession of a trochophore larval stage by all these phyla would constitute striking evidence of their common origin; but even with these reservations it appears that other evidence (such as the formation of the mesoblast from pole cells) links the annelids, crustaceans, arachnoids and insects, and other arthropods into a group (superphylum Trochozoa) related by derivation from some unknown common Pre-Cambrian stock (see Allee, 1926, 1927, p. 293).

Peripatus, a worm-like creature (fig. 1.7) for which no common name is known, looks somewhat like a naked caterpillar but its stumpy "feet" terminate in small claw-like hooks (whence the name Onychophora, meaning claw-bearing). Although living in rotten wood, the animals are carnivorous. Their mouth is armed with hook-like, transversely placed jaws and there are jointed tentacles above and in front of it.

From the Burgess Shale of Middle Cambrian age Walcott found among other beautifully preserved invertebrates a form which he named *Aysheaia,* that in some respects is strikingly like *Peripatus,* which survives today on the high plateau of Africa, in South America, Australia and New Zealand, and may be among the oldest existing land animals of those regions.

The amazingly generalized character of this lowly form is summarized thus by Parker and Haswell (vol. I, p. 566):

"*Relationships.—Peripatus* is the most primitive of existing Arthropods, and presents some striking points of resemblance to the Chaetopoda. The development is in the main arthropodan, especially as regards the mode of segmentation (at least in the forms with much food-yolk, which are probably the more primitive), the mode of closure of the blastopore, and the mode of development of the mesodermal strands. Arthropodan also are the relatively large size of the brain and the presence of tracheae, the character of the heart with its pairs of ostia, together with the clawed appendages, and the jaws in the form of modified limbs. The nephridia on the other hand, and their modification in certain segments to form the gonoducts, which are ciliated internally, are annulate in character, and in all probability the slime glands and coxal glands correspond to the setigerous glands of the Chaetopoda. The nervous system is peculiar, and is most nearly paralleled among the Platyhelminthes and the Mollusca. Also peculiar, and serving to distinguish *Peripatus* from the rest of the Arthropoda, are the large number of stigmata with their irregular arrangement, the presence of only a single pair of jaws, and the nature of the cuticle." *

According to Snodgrass (1938), and to Stormer (1944, p. 143), the Onychophora are in general intermediate between generalized annelids and the common stem of the crustaceans, arachnids and insects. Thus they belong with the arthropod, or trochozoan, series of phyla.

* T. J. Parker and W. A. Haswell, *A Textbook of Zoology,* First Ed., Vol. I. Copyright 1897 by Macmillan and Co., Ltd., London. Reprinted by permission of the publishers.

CHAPTER II

THE WORLD OF MOLLUSCS

Contents

CHAPTER II

THE WORLD OF MOLLUSCS

THE SEA-SNAILS AND THEIR ALLIES

The Limpet and Other Primitives.—A living limpet (*Patella,* fig. 2.1) is of high interest to the student of evolution because the earliest known fossil shells tentatively referred to the limpet order have been found in rocks of the Lower Cambrian period, the age of which, as estimated by the radium emanation method (p. 9), would be about 500 million years. But the limpet himself, as he clings tightly to his rock, exhibits only a strong negative interest in the activities of an amateur conchologist and unless he can be suddenly caught off guard, it is almost useless to try to pull or push him off his firm base. The best way to dislodge him is to push the edge of a sheath knife suddenly under his shell and break the suction exerted by his powerful oval base or foot. This so-called foot (fig. 2.1C) is a thick muscular plaque or fold lying beneath the head and digestive tract, whence the name Gastropoda (belly-foot) applied to that class of molluscs which includes the limpets, sea-snails, land-snails and many others. The foot is well supplied with a branching system of nerves; these in turn are connected with the main "cerebral ganglia"; these central stations doubtless correlate the nervous messages from the sense organs and effect appropriate responses of the foot and other movable parts.

The limpet's shell seems to be well adapted for the protection of the animal. A certain long-dried limpet shell was wrapped in a cloth, placed flat on the floor and trod upon; it successfully bore the experimenter's full weight of one hundred and sixty-five pounds. Another limpet shell of a different species gave way under similar conditions, crumbling into thin pearly layers like isinglass; these in turn were composed of exceedingly thin translucent flakes. These thin flakes were probably squeezed out of the shell glands in a minutely fibrous and plastic condition. The shell-glands are especially abundant at the lower outer edge of the shell-secreting mantle.

The mantle of the limpet is like a tent (fig. 2.4G) which grows from the top outward and downward as the animal itself increases in size. Each very small flake of which the shell is composed is not wholly flat but has little creases and humps along its outer lower edge. The steady accumulation of these creases and humps where those of successive layers overlap each other builds up into a system of larger and smaller ribs (figs. 2.1A; 2.4B,F), which run up the surface of the tent toward the top of the shell and add greatly to its strength. The dome of the shell is thick and strong; it is made up of a thin, dense, polished outer layer. In many other kinds of shells there is a horny layer (periostracum) covering the surface of the shell, but in most limpets, even in the living state, the horny layer is thin on the sides of the shell and often worn off, perhaps by weathering, from around the tip.

The waves at high tide pass lightly over the limpet's convex shell. It is at this time that the animal must raise its shell a little to admit the fresh water to its gill; after which it is said to move about slowly until it reaches a patch of algae on the rock; upon these it feeds, rasping the weed off the rock with the ribbon-like band-saw (radula) which it carries in its throat. Nothing exactly like this band-saw or radula is found outside the molluscan phylum.

The ovoid contours of most limpet shells (fig. 2.4) are not wholly symmetrical and some species show possible traces of a former spirality (fig. 2.4E) in the arrangement of the spots and swellings on the outer surface. The color patterns are obviously dependent primarily on the ovoid contour and radial ribs.

Some of the fossil shells (e.g., *Scenella*) of Cambrian age that are referred to the limpet family (Patellidae) can scarcely be distinguished from those of the recent *Acmaea* (Zittel, 1924, p. 460). Nevertheless there is a possibility that the simple conical or tent-like shape of the limpets may have been derived from a still earlier, as yet undiscovered ancestor in which the shell was secondarily symmetrical (fig. 2.14E), like those of the Capulidae (Zittel, *op. cit.,* p. 463).

Very probably the conical or low tent-like limpet form has been produced independently in several other families of gastropods (Fissurellidae, fig. 2.5; Capulidae, fig. 2.14E; Siphonariidae) by the gradual loss of an earlier spiral tip through acceleration of the lower whorls. In the "keyhole limpets," or chink shells (family Fissurellidae), the finally symmetrical oval tent-like shell of the adult is developed from a young state (fig. 2.5A) in the form of a shallow spiral or helicoid closely resembling that of the ear-shell (fig. 2.5F) or abalone (family Haliotidae). But in view of the relatively late (Upper Cretaceous) appearance of the latter family, we may suspect that in the remote adult ancestors of both the Fissurellidae and Haliotidae the shell was essentially like that in *Pleurotomaria,* whose ancestors (fig. 2.2B) are recorded as far back as the Cambrian, that there was a very rapid radial increase in the width of the foot, and that the haliotids (fig. 2.5F) may have emphasized the helicoid stage which was preserved in the Triassic *Fissurella emarginula münsteri* (fig. 2.5E).

MAIN GENETIC FACTORS OF SHELL FORM

Polyisomeres, Anisomeres.—In almost every kind of shell, after the minute bead-like infant stage or protoconch has been left behind, the successive ridges, knobs, spikes, etc., tend to resemble each other, so that even a broken fragment of a shell is often sufficient to identify its genus and species (cf. "polyisomerism," p. 12). On the other hand, by acceleration or retardation of growth, certain units of a series of similar parts may be selected for special emphasis (cf. "anisomerism," p. 13).

If we look at a snail preserved in the extended position after the removal of the shell (fig. 2.7aA1), we see that the head and foot are fairly symmetrical on either side of the midplane but that the visceral mass projects as a coil on the right side above the foot (A2). This coil is enclosed in a thin membranous bag, the upper part of the mantle, which is spirally wrapped around it.

When the young animal was hardly larger than the head of a pin, the distal border of the embryonic mantle began to secrete the shell around the visceral mass in a clockwise direction (as seen from the apex), always growing faster on the outer or distal side than on the inner or proximal side. No doubt the mass of the visceral coil and of the foot must be one of the factors that determine the size of the shell, at least whenever the snail can be withdrawn into the shell, yet a review of the numerous figures given in Bronn's Tierreich (III Bd., II Abt., Taf. I–XIII) shows that there are very wide diversities in the form and proportions of both the snails and the shells. Thus one gets the impression that in many cases the evolution of the shell has been but loosely correlated with the general form. In other cases, however, it seems evident that the form of the shell has been greatly influenced by the shape of the snail. For example, in the limpets (fig. 2.1) the oval tent-like shell is entirely filled by the animal, and, as noted above, the extreme width of the foot in the haliotids must have been an important factor in the wide lateral spread of the mantle and shell (fig. 2.6).

While the raw material for typical shells is carbonate of lime, small quantities of other salts (calcium phosphate and magnesium carbonate) are also found. The outer layer of the typical shell is composed of calcite (calcium carbonate), the inner of aragonite—a slightly heavier and harder form of the same chemical formula. In the Indian apple snail (*Pila*), according to Prashad (1932), the calcareous granules of the shell glands consist of a double organic salt of calcium. The organic matrix in which the shell granules are laid down is formed of conchyolin, a substance allied to chitin. The strength of the shell is largely due to the criss-cross arrangement of the minute columns in the different layers (fig. 2.7B1).

Relative Growth Rates: Height *versus* Spread. —The data assembled in figure 2.10A, B, C, D, show that the height of the spire above the aperture increases directly with both the number of whorls and the downward pitch or inclination of the whorls, inversely with the relative width of the whorls as compared with their height. Whenever helicoid growth occurs, it is a sure sign that one side or part of a rod, cylinder or tube has grown faster than the other. This has long been known to mathematically inclined students of growth (D'Arcy Thompson, 1917, 1942). It gives us an important key to the causes of many of the conspicuous differences between different shell forms. If the coil is very regular and open below (fig. 2.9A, B, C), it is because the growth of the mantle on the lower outer side of the tube has been but little faster than that of the lower inner side. Such is the case in the existing sundial shell, *Solarium* (*Architectonium,* fig. 2.9A, D) and its fossil predecessors. On the other hand, in tightly wound shells (fig. 2.10A-D) the outer side has evidently grown much faster than the inner. This tends to crowd the inner side of any given whorl toward the center and is doubtless partly responsible for the appearance of a columella or corkscrew (fig. 2.10C, D), at the core of many kinds of shells. In fact, the columella is essentially the inrolled, twisted left side of the tunnel.

If a shell has a tall slender spire, as in *Turritella* (fig. 2.10B), the pitch or inclination of the growing shell is pronounced, and this means that the roof of the tube has grown forward and downward a little faster than the floor and outer side. On the other hand, if the shell forms a nearly flat coil, as in certain pulmonate snails (fig. 2.10E), it is a sign that the floor of the tube has grown almost as fast as the roof.

If the transverse diameter of the shell tube (as seen from above, fig. 2.9D) increases but gradually with the successive turns or whorls, a correspondingly regular slow increase in growth rates is evident. If, on the contrary, there is a rapid increase in the transverse diameters as we follow the whorls outward (figs. 2.9E; 2.6E) a correspondingly rapid acceleration in growth rates is implied.

Almost at the dawn of fossil history in the Cambrian period we find low tent-like forms (e.g., *Scenella*, see p. 35), symmetrical pillbug-like forms, which are spirally rolled up in a fore-and-aft direction (Bellerophontidae, fig. 2.2A, C), and true helicoid forms (Pleurotomariidae, figs. 2.2B; 2.11A) with slightly unequal growth of opposite sides of the shell tube.

The helicoid spiral form followed many divergent paths. In the sundials (fig. 2.11I), tops or trochids (fig. 2.11B-H), turbans (fig. 2.12) and others, the shell spread radially and the spire remained moderate; at the other extreme in the wentletraps or epitoniids (fig. 2.10A), towers (turritellids, fig. 2.10B), cerithiids (fig. 2.16A, B) and others, the spire heightened as the downward pitch of the screw increased. In the worm-shells (vermetids, fig. 2.15B) the tall spire loses its regularity and gives rise to a meandering tube (fig. 2.15C), simulating those of the annelid *Serpula*.

Reduction of Spire and Development of Globose, Biconical and Obconical Forms.—A fairly well developed spire seems in many cases to be a relatively primitive character, and the spire was often reduced as the lower whorls increased in width (figs. 2.16; 2.25A-F). In the neritids (fig. 2.13B-E) the spire was much shortened and the lower whorl became globose; in the ear shells (haliotids) the shell, growing rapidly in a radial direction, becomes flattened into a most beautiful example of the logarithmic spiral (fig. 2.6D). By reducing the spire, emphasizing the "shoulder" of the whorls and tapering the lower or anterior end, the biconical (fig. 2.16B-C), and eventually the obconical forms (figs. 2.28D; 2.36; 2.37) were developed.

Flattening Due to Pressure of Foot on Columella.—When the shell-secreting mantle is reflected over either or both lips of the shell, as in certain helmets (fig. 2.17C) and cowries (fig. 2.20), it lays down a smooth shiny surface wherever it overlaps the shell. The substance deposited by overlapping flaps of the mantle may be dense and cement-like, in which case it is called a *callus,* or it may be thin and glassy, as in the cowries.

The broad callus on the columella of the neritid shells has been progressively emphasized (fig. 2.13), while the inner edge of the columella, squeezed against the dorsal side of the foot, has produced a convergent imitation of the deck shells, or crepidulids (fig. 2.14).

The latter have probably arisen independently of the Neritidae (fig. 2.13B-D) and in a different way (fig. 2.14). From the facts compiled by Zittel (1924, pp. 463–464) we may infer that the earlier Palaeozoic ancestors of the capulids may have been essentially like the Lower Cambrian *Stenotheca* (*Heliconella*) *rugosa* Hall, which was a low, oval conical form with a backwardly curved apex. From this stage is but a short structural step to the Hungarian cap shells, *Capulus,* and thence to *Hipponyx* (fig. 2.14E). Another line led perhaps from some form not unlike the Devonian *Platyostoma,* a normally helicoid shell, thence to the Trias to Recent Calyptraeidae (fig. 2.14A-D). The modern deck or slipper shell (*Crepidula,* fig. 2.14C) may represent a Tertiary branch from the calyptraeid stem not far from *Galerus* (*Calyptraea*) *trochiformis.* The "deck" in all these cases represents an extension of the flattened side of the columella, to which certain specialized slips of the muscular foot are attached. The inner cup of the crucible shells (fig. 2.14D) seems to represent a further specialization of the "deck"; this has become separated from the outer rim of the shell except on one side and folded around a central cavity to which the foot is attached.

Thus the slipper or deck shells (Crepidulidae) may be added to the list of gastropods, including patellids, fissurellids, neritids (*Pileolus*), capulids, calyptraeids, siphonariids, limnaeids (*Ancylus*), which, although belonging to different families or even orders, have assumed a low and more or less conical form and cling like limpets to the rocks or to other shells.

Continuous *versus* "Stop-and-Go" Growth.—In most shells the forward spread of the mantle is recorded by lines that are parallel to the outer lip of the shell (e.g., fig. 2.6D), each line representing what was at one time the outer edge of the mantle. The lines mark the rhythmic and regular addition of new deposits of shelly material. At varying angles to these peripheral lines are long spiral or revolving lines (fig. 2.6A), "cords" (fig. 2.11D) and revolving ridges (fig. 2.25G) running up to the tip of the spire; these record the spiral growth of the mantle. In many cases the two sets intersect, forming small, more or less rectangular cells (cancelli) or lattice-work, as in the Cancellariidae (fig. 2.26B, C). The emphasis of the grooves may produce a bead-like surface, as in *Nassarius gemmulatus* (fig. 2.26D).

Whenever the forward growth of the mantle is arrested for some time, the marginal ridges are often built up into axial folds, or varices (figs. 2.22A; 2.23A). In the larger helmet shells (fig. 2.17C) the outer lip of the shell during each retarded period becomes thickened and everted, while the inner lip of the mantle slowly overspreads the preceding body-whorl and the columella. Then there seems to be a period in which the animal grows forward out of its cave, the roof of the cave being the thickened, everted outer lip of the shell. Meanwhile the outer flap of the mantle has been developing a new crease or rim in the middle of the everted outer lip near the lower or anterior end. This new flap then extends upward and backward just beneath the everted lip. The new flap as a whole keeps on growing forward, following the inner lip but at a great distance. After a long while there is another retardation and the once new crease gives rise to a new everted lip. We have called this the "stop-and-go" principle of growth but it is of course more correctly designated as retardation and acceleration of growth. This method is recorded whenever humps, ridges, spikes, etc., are developed (figs. 2.19C; 2.10A).

While the big helmet shells carry the "stop-and-go" method to an extreme degree, some of their relatives the tuns (Doliidae or Tonnidae) have emphasized continuity and have largely overcome discontinuity of their spiral growth (fig. 2.17E-F). In other tuns (e.g., *Malea pomum, Malea ringens*) there is a sharp eversion of the outer lip (fig. 2.17D), somewhat as in the helmets; in others (*Tonna*) the steady forward growth of the mantle (E) is resumed after a brief check, which causes a crease behind the rim but does not last long enough to form an everted lip. The paper fig-shells (fig. 2.17F) are highly specialized tuns partly convergent toward *Fasciolaria* (fig. 2.10D), in which there has been a marked reduction of discontinuity, so that even the spiral folds are diminished. In view of the relatively late age (?Cretaceous to Recent) of this type there is no reason to regard it as in any way a primitive shell, so that its apparent simplicity is very probably a sign of reduction or secondary loss of differentiation.

Elongation of Siphon.—In the Archaeogastropoda, or geologically older families of gastropods, the shell rode above the middle of the animal and the aperture or mouth of the shell was more or less oval as it is in the modern apple snail (fig. 2.7). In many later families (e.g., fig. 2.22B) the shoulder or widest part of the shell was shifted toward the rear end (the "upper" end of the shell), while the lower or anterior rim of the shell was prolonged forward into a canal covering the siphon. The latter is a fold of the mantle that leads backward to the mantle cavity containing the gill or the lungs. The siphonal canal reaches an extreme length in many spindle shells (Fusidae) and related families (fig. 2.25A-C). The lengthening of the siphonal canal is often associated with a lengthening both of the lower or body whorl and of the aperture of the shell. Still later (figs. 2.25F-I), the siphonal canal often became secondarily shortened and twisted into a curved notch and spout.

Specializations of the Outer and Inner Margins of the Aperture.—The outer lip of the shell in the older families is usually not excessively different from the inner lip but in the strombs (fig. 2.19) and their fossil relatives (fig. 2.18) the outer lip gives rise to a prominent lateral process (fig. 2.19C), which by successive spreading and subdivision culminates in the remarkable form of the "spider shell" (*Pteroceras,* D). In certain stromboids (E-G) the prominent lateral processes have apparently been lost and the tendency to form a tall spire, seen in the earlier aporrhaids (A) and indicated in young strombids, is emphasized in the adult.

In the muricids (fig. 2.23) the outer lip of the mantle finally attains such rapidly increasing powers of growth and deposition as to give rise to a wide display of simple to compound folds (fig. 2.23A-C) or to fin-like crests (fig. 2.23E, F) or to sharp spikes (fig. 2.24A-C).

PHYLOGENY AND CLASSIFICATION OF SEA-SNAILS

From the foregoing viewpoint almost every shell will yield to comparative analysis a good deal of the story of its own individual and racial history. The shell is merely the outer covering of the animal to which it belongs and, as we have seen in the case of limpet-like shells, similar-looking shells sometimes cover very different kinds of animals. Hence the "families" and still more the "orders" and "subclasses" of the shell world are based not on the shell characters but on different anatomical characters of the animals themselves, such as the character and arrangement of the gills, the heart, the nerve loop, or on the number and arrangement of the teeth in each cross row on the radula. Yet even a comparative study of the shells alone, in the light of what is easily accessible knowledge of the other characters, will in the majority of cases be sufficient for the purpose of identifying the shell as to its order, family, genus and species.

The present inquiry into the origin, evolution and individual development of the forms and patterns of sea shells (chiefly of the prosobranchiate order) is based primarily upon close and prolonged personal studies of the author's collection; but such a collection, although including some thousands of shells, representing all the major divisions and more common families, is but a meager sample of the vast hosts of living and extinct species described by conchologists and palaeontologists. Therefore the fossil record of molluscan life, especially as set forth and summarized by Zittel and more recently by Wenz, was constantly studied, as were also its surviving representatives. At the same time the major classifications of the encyclopaedists, based upon characters of the radula, upon the number, position and interrelations of parts of the heart, brain, nerves, gills, upon the characters of the eyes, tentacles, snout, proboscis, siphon, etc., were all carefully considered, especially in the compilation of the Family Tree of Common Sea-Shells (fig. 2.38).

From this general background of evidence it appears that all the diverse hosts of recent prosobranchiate gastropods and their shells may be traced back to the symmetrically coiled slit shells represented by the Bellerophontacea (fig. 2.2) of the older Palaeozoic ages and by the simple conical helicoids of surviving Pleurotomacea (fig. 2.11A). From this general source hundreds of lines streamed onward and outward through the ages, many becoming extinct but others branching again and again; all the while the several branches and twigs produced innumerable divergences, parallels and not a few convergences, as to shell forms and color patterns.

The symmetrically coiled Bellerophontacea (fig. 2.38) gave rise among many others to the asymmetric conical and trochoid Pleurotomacea; the latter in turn, by radial acceleration of the body whorl, gave rise to the fissurellids and haliotids (fig. 2.5) and by globose expansion to the Neritacea (fig. 2.13B-D). In another direction they gave off the conical Trochacea (fig. 2.11) and their derivatives, the star shells (fig. 2.9B), tops (fig. 2.11), turbans and pheasants (fig. 2.12); while a side branch led to the periwinkles (littorinids), viviparids, apple snails (fig. 2.7), and eventually to the lung-bearing land-snails. Meanwhile the central branch (fig. 2.38) produced the Cerithiacea, essentially high-spired helicoids (fig. 2.16A, B), which stand near to or on the stem of all the higher types of sea-snails. An early branch from the cerithioid stem are the moon shells (Naticacea), which are prevailingly globose, with a few depressed forms (fig. 2.13F-H).

All the higher types of sea-snails give evidence of derivation from snails that were provided with an elongate siphon or breathing tube (fig. 2.7A2), which was eventually protected by a gutter (figs. 2.10D; 2.25) at the anterior (lower) end of the shell. The beginnings of such a gutter are shown in some of the high-spired cerithioids (fig. 2.16A), but at least traces of it occur in nearly all the swarming tribes of Strombacea (fig. 2.18), Tonnacea (figs. 2.17; 2.22), Cypraeacea (fig. 2.20), Muricacea (figs. 2.23; 2.24), Buccinacea (figs. 2.25; 2.26), Volutacea (figs. 2.27B; 2.31), and Conacea (figs. 2.35–2.37).

Frequently associated with the siphonal groove are various localized expansions of the "outer lip" of the shell, eventually producing the fantastic contours of the strombs and spider shells (fig. 2.19), and rock shells (figs. 2.23; 2.24). In the cowries (fig. 2.20) and their allies (Cypraeacea) both the outer and the inner lobes of the mantle are greatly expanded over the body whorl and these two flaps often meet on top of the dome of the shell.

Hitherto, growth of the shell, while variously accelerated, had been nearly continuous; but some of the Tonnacea (tuns, frog shells or ranellids, tritons, helmets) exhibit the "stop-and-go" method, often with extreme results (figs. 2.17; 2.22).

The deck shells and crucibles (Calyptraeacea, fig. 2.14) are judged by leading authorities to be a highly specialized offshoot from the common stock of the Strombacea and Tonnacea, in which a more or less depressed trochoid form had greatly expanded the foot, eliminated the siphonal canal and flattened the inner lip into a "deck."

In spite of enormous diversity of shell forms the Muricacea, Buccinacea, Volutacea and Conacea are all rather closely related to each other and the common stock was probably a primitive branch of the Tonnacea (fig. 2.38). The typical rock shells (Muricacea) are noted for the varying folds of the outer lip at successive stages of growth, culminating in the many-branched *Murex palmaerosae* (fig. 2.23C).

Starting from or near the spindle shells (fusids) with long proboscis canal, the Buccinacea (figs. 2.38; 2.25) vary in many directions, including the thaisids, fasciolariids, neptuneids, nassariids and others. The probable derivations of all these shell forms are suggested in the illustrations.

COLOR PATTERNS

The colors of shells and parts of shells range from the black of certain nerites (fig. 2.13D) to the white interior of many species, from the lustrous mother-of-pearl colors of *Haliotis* and certain trochids to the dark stains on the inner and outer lips in the big helmets. Golden yellow is displayed in the interior of some turbans, while red, green and blue dashes occur on the revolving cords on the outer side. In the buccinid *Eburna* (*Babylonia*) *japonica* (fig. 2.25 K) the interior of the shell and the callus area are bluish white, while the pigmented outer surface is marked with "oil painted" brown spots and lakes.

All the diverse factors of shell structure have their influences on the color patterns but owing to the fact that the shell-depositing cells may stop while the pigment cells work on, and *vice versa,* there are varying degrees of independence between parts of the color patterns and the forms upon which they are developed.

The color of the earliest Palaeozoic shells, even if known in sporadic cases, would hardly be a safe guide to the intervening evolution of the color patterns of their existing direct descendants. We may nevertheless venture to generalize the observations recorded in the present work in stating the following principles:

1) If the pigment cells work on the stop-and-go method, they may follow the revolving cords (fig. 2.12C) and give rise to dots, dashes, etc.

2) The pigment-releasing stimulus may pass longitudinally (axially) along the secreting mantle margin even while the latter is growing spirally forward; the record of such impulses will appear as axial bands (fig. 2.28D, F-H). After the bead-like protoconch or baby stage (which is usually retained in adults at the tip of the spire) some young shells become conical, with their whorls slightly inclined to the long axis (figs. 2.10C; 2.11; 2.15). But when the visceral coil (fig. 2.7A) increases rapidly in diameter, so do the successive revolving cords or spiral bands (fig. 2.11) and the aperture of the shell bulges outward (fig. 2.11F, H). Thus the pigment spots of adjacent rows may touch their nearest neighbors (figs. 2.12B, D; 2.11B, H, F, E). This may be one of the factors in forming oscillating bands (figs. 2.11E; 2.12B), V's (figs. 2.28G; 2.30A; 2.34), zigzags (figs. 2.34C; 2.36B; 2.12B), "zebra" stripes (figs. 2.29E; 2.36F), irregular streaks (fig. 2.31A), or oblique color bands (fig. 2.11H).

3) Unless there is an abundance of pigment, any product of the stop-and-go method of building up the shell itself is likely to affect the deposition of pigment, so that axial panels (figs. 2.28A, D; 2.37F, I) are often marked by pigment deposited along the axial sutural lines or by the rhythmic cessation of pigment deposition.

4) If there is a repeated scarcity of pigment, what remains may form bars, spots, crescents or knees (figs. 2.12E, G; 2.13G), "partridge feathers" (figs. 2.22C; 2.32B), "ocelli" (fig. 2.28H), etc.

5) In some cases where the patterns are exceedingly variable (as in certain *Phasianella* shells, fig. 2K), the successive "electric discharges" and oscillations (fig. 2.37bG) suggest that the sensitive mantle has reacted to some oscillating stimuli.

6) The fortuitous juxtaposition of oscillating streaks and/or unpigmented areas in adjacent shell zones sometimes gives rise to fantastic

"animals" (figs. 2.12G, H; 2.36E; 2.37bH), "letters" (fig. 2.12F), "musical staffs with notes" (fig. 2.28E) and other cases of "unnatural history resemblances."

In summary, the color patterns, although always dependent upon the directions and extent of growth of the shell-secreting parts of the mantle, often show considerable self-assertiveness and very complex and little understood reactions to rhythmic growth periods. In the more simple cases, as in *Stomatella imbricata* (fig. 2.6A), or in *Calliostoma* (fig. 2.11B, C), the pigment cells work on a simple stop-and-go schedule, producing a revolving dot-and-dash record. The juxtaposition of corresponding spots or nodes in adjacent revolving cords (fig. 2.12A) leads to revolving color bands (fig. 2.30B), axial lines (fig. 2.29C, E), oblique stripes (fig. 2.11H), panels (fig. 2.28A, D), zigzags (fig. 2.11E). The breaking up of color bands leads to irregular and confused patterns (figs. 2.29B; 2.30C), "corybantic dancers" (fig. 2.37J) and monsters (fig. 2.37A, H). Crescents arise from more or less rhythmic interruptions of spiral color bands (fig. 2.12B, G), as do also "partridge feathers" (figs. 2.22C; 2.32B), while horizontal or oblique V's (figs. 2.13G; 2.34D) as well as reticulate patterns (figs. 2.37B, C; 2.34A) suggest the "geodesic" winding and intersection of two sets of color-forming bands. Similar reticulate patterns in lines of fish scales are known to follow the "geodesic" growth forces (Breder, 1947).

Among the cowries (fig. 2.20) and some other shells the dorsally extended flaps of the mantle give rise to folds, swellings, projections and tubes. These print later marks on top of earlier ones, as in a palimpsest, and seem to be able to absorb and redeposit pigment, thus giving rise to round spots (fig. 2.20C), rings (fig. 2.20E) and the complex "Arabic writing" in *Cypraea arabica.*

In general, the simpler forms of shells and color patterns may be compared with a kaleidoscope in which the number of bits of glass is very small, so that only a few combinations and permutations are possible. On the other hand, elaborate and, especially, confused patterns may be compared to a kaleidoscope with many units, so that the particular pattern appears to arise, as it were, by chance. On the whole, all these results seem consistent with the view that in any individual shell both structure and color pattern are determined primarily by interaction of the genetic patterns of the parents and to a lesser degree by physiologic and environmental conditions, such as the ripening of the gonads, the seasonal changes in tides, etc.

Finally, the shell may be regarded as having probably originated not as a protective shield but as the result of the adherence of the cuticular secretions to each other to form a crust. Growing crustaceans and insects escape from their crusts by bursting them open, but the typical snail merely grows spirally away from the older parts of its shell, which is carried on its back and has come to serve as a protective turret.

NERVE LOOPS

The tendency toward spirality which is so clearly expressed in both the construction and color patterns of typical gastropods is also manifested during embryonic development in the twisting of the visceral nerve loop (fig. 2.8) into a figure 8 (whence the name, Streptoneura, twisted nerve). As a part of this twisting the "pallial complex" (including the gill of the morphologically right side) moves up the right side over the intestine and across to the left side; somewhat as in the developing flatfish, the originally left eye moves up over the forehead and comes to rest on the right side. Meanwhile the embryonic gastropod shell twists around so that its spire falls over toward the right side of the foot. Most ordinary gastropods go through this streptoneural stage. The opposite tendency, involving progressive incompleteness of the shift from right to left and eventual loss of the streptoneural condition, culminates in the "detorted" seahares (Aplysiidae) and other families, which are grouped in the subclass Euthyneura. Thus evolution is not always "irreversible" and many later specializations obliterate earlier ones.

GASTROPOD AND VERTEBRATE

Although the gastropods are perhaps as far removed from the vertebrates as any animals could be, they nevertheless exhibit many ways of transformation which are familiar to students of the evolution of the vertebrates. Since their more fundamental needs and reactions are the same as in the vertebrates, the gastropods have developed various organs or organ systems to which we can readily apply some of the names used in vertebrate anatomy without implying special homologies between the two. Thus both groups have organ systems that provide for feeding, locomotion, control, protection, secretion, excretion, reproduction. Moreover, a study of the evolution of the gastropods permits constant application of evolutionary concepts that have been developed in the study of vertebrates: here belong such concepts as habitus and heritage, polyisomeres and anisomeres, parallelism, convergence, divergence and others.

The history of the gastropods offers some interesting contrasts to that of the vertebrates, of which Man (*Homo sapiens*) regards himself as the most noble example. The typical gastropods were primarily clinging and creeping animals, the dominant organ being the muscular base or foot; their locomotor apparatus was not constructed on the metameric or many-unit plan; so that few gastropods became free swimmers and none ever equalled the vertebrates, which very early adopted a metameric system and became active swimmers and, we must admit, highly successful robbers. The gastropods, being clinging animals, early developed the radula or flexible band saw as an efficient tool for rasping a hole in their neighbor's armor. The vertebrates, on the other hand, disdaining such patient abrasive tactics, developed devouring jaws and swallowing gill arches. When in danger from foes the typical gastropod shuts himself into his strong tower and awaits safer times; most vertebrates, on the contrary, either flee quickly away or make sudden and aggressive counter-attacks.

In keeping with their slow movements the more typical gastropods have not developed an advanced type of brain, involving elaborate integration and subordination of lower to higher control systems; in the very nervous, not to say jumpy vertebrates, on the contrary, the brain becomes a vastly complex control center for the elaborate activities of feeding, locomotion and breeding.

However, in the deeper levels of being, limpet and man are fundamentally much alike. Indeed a cynic might name some points in which the limpet would have the advantage.

It would be easy to extend such comparisons and contrasts, for the Tree of Life among the gastropods has branched and branched again through the ages, much as it has among the vertebrates but through a far longer period.

NAUTILOIDS, CUTTLES, SQUIDS AND OCTOPODS

The cephalopod division of the molluscs contrasts with the gastropods in that with few exceptions their shells tend to coil up symmetrically, so that they form a true spiral rather than a helicoid. They also have better developed heads, large eyes, relatively large brains and a circle of tentacles with suckers around the mouth.

In contrast with the gastropods, the foot of cephalopods is hardly differentiated from the head (whence the name, cephalopod—head-foot). On the other hand, the cephalopods inherit a radula, a mantle and mantle cavity, ctenidia and other gastropod features. Their nervous system is basically the same as in the gastropods but with far more complex brains, possibly correlated with the advanced stage of both the eyes and the tentacles. The suckers on these tentacles are to some extent analogous but not at all homologous with the tube feet of starfishes. Thus they are practically unique organs and apparently nothing is known as to their stages of evolution. Presumably the suckers must have arisen from contractile loops or muscular flounces which first gave rise to figures of 8 and then to rows of circles along the borders of the tentacles.

The group had an enormous range and diversity through nearly all the fossil record of marine life from the Lower Cambrian upward. In the oldest fossil cephalopods we note a sort of metamerism or serial repetition of the shell as it increases in size, giving rise both to the straight-tubed *Orthoceras* and to the spirally coiled *Cyrtoceras* and later nautiloids (fig. 2.39B, D). Thus the cephalopod shell may have arisen not from a gastropod crawling type, but by a sort of strobilization of a proto-molluscan type, discussed below (p. 51). The shell-secreting mantle of the Palaeozoic and Mesozoic cephalopods must have been an extremely active organ in order to provide material for the numerous internal partitions, external ridges and knobs and enormously infolded sutures, the latter with primary, secondary and tertiary lobes and lobules (fig. 2.39D).

Of the living cephalopods the pearly *Nautilus* is the least specialized, since it retains a complete, well developed shell. In *Spirula* the shell is reduced to a delicate and small chambered spiral, which is mostly embedded in the mantle. In the cuttlefish the shell has degenerated into the porous cuttle "bone"; in the squids the shell is a delicate horny "pen" and in the typical *Octopus* it has disappeared entirely, leaving a naked and molluscous body. The modern cuttles and squids have become free-swimming and predatory, propelling themselves by jets of water expelled through a nozzle which protrudes from the mantle cavity. Their locomotor pulsations also bring fresh water to the gills. All this represents a rather belated attempt, so to speak, to evolve a free-swimming, naked predator out of a castle-bound ogre. The growth energy that formerly was expended in protective and strengthening armor is now spent in swift pursuit and strangulating attack.

THE MAIL-CLAD CHITONS (AMPHINEURA) AND THEIR WORM-LIKE RELATIVES (SOLENOGASTRES)

When seen from above (fig. 2.40), the oval contour of a chiton, as well as its clinging motionless to the rocks, suggest a limpet, while its series of broad, closely articulating valves vaguely recall the broad abdominal segments of lobsters and crayfishes or the strong median plates on the back of certain fossil amphibians (*Cacops, Otocoelus, Dissorophus*). Laterally the mantle is prolonged into a tough marginal border, sometimes bearing "ossicles" and short rods. Internally in the arrangement of the reproductive tubes the chiton somewhat resembles the primitive flatworms. On the other hand, the ground plan of the nervous system is not dissimilar to the embryo gastropod type prior to the torsion of the visceral loop and pallial complex; some of the chitons and related forms possess a radula, while larvae of *Chiton* pass through a trochosphere stage—free-floating, with ciliated band, apical tuft and retort-shaped gut (Kowalewsky)—all fundamentally similar to those of both annelids and gastropods.

A small group of shell-less wormlike creatures, the Aplacophora (Solenogastres), on the one hand retain certain marks of relationship with the chitons and on the other hand possess internal anatomical features of the reproductive tubes which again recall the conditions in primitive flatworms. For these and other reasons both Thiele and Odhner regard the Solenogastres as being more closely related in certain respects to certain groups of flatworms than to the Mollusca. Other malacologists, however (cf. Encyclopaedia Britannica, 14th Ed., Art. Mollusca), have not accepted Thiele's hypothesis. We may also suggest that the Solenogastres may well be regarded as degenerate worm-like derivatives of the Amphineura, analogous in part with the "shipworms" (*Teredo*) among the bivalve molluscs.

The typical Amphineura (chitons) also may be the highly specialized descendants of early limpet-like gastropods. Such clinging forms may have experienced a secondary subdivision of the growth centers in the shell-producing mantle and of the muscles for bending the joints and producing a tight suction against the dry rocks. Such a subdivision of the adductor muscles is clearly indicated in the Cambrian *Palaeacmaea* (cf. Wenz, 1938, pp. 89–91) and others (fig. 2.40B1). Together with this there might well supervene a limited reorganization of the internal anatomy along pseudo-metameric lines. Nor do we find anywhere else in the entire range of the molluscan class any other striking indications of close relationship with the flatworms. Moreover the absence of known ancestors of the chitons below the Ordovician period, together with other palaeontologic data, is not inconsistent with a very different hypothesis of the origin of the Mollusca (p. 51).

THE BIVALVES—BRAINLESS BUT SUCCESSFUL

While the clam is often taken as a type for introductory study of the molluscs it is here reserved for almost the end of the chaper because its extreme specializations may be better understood by viewing them against the background already set forth.

The shell-producing mantle in bivalves forms a folded hood around the animal and bears on its inner side the respiratory lamellae or gills, while secreting the growing shell along its outer border (fig. 2.41). In the most primitive existing clams the mantle is not completely divided into right and left halves, its undivided rear end being prolonged into a siphon. The latter in turn is subdivided into two long tubes set one above the other, the lower for the incurrent, the upper for the excurrent streams that are set in motion by the cilia on the gills. The circulation of water that is set up by the ciliated tracts in the siphon may perhaps supplement the normal circulation through the heart and blood vessels, which appear to be relatively weak.

Thus the clam's siphon is at the opposite end of the body from the sea-snail's siphon, both organs being independently developed from a common ancestral source, which is the highly versatile and adaptable molluscan mantle. The clam's siphon is at the rear end because the clam dives headfirst into the mud, either to escape predatory enemies or to avoid being dragged away by the undertow from the best feeding ground. To put the case according to the theory of preadaptation, in the ancestral clam, as a result of genetic mutation, the mantle, arising ultimately from right and left buds, retained its bilaterality and held its potential spirality in abeyance, at least until the right and left shell glands were laid down in the embryo.

Foot and Shell Form.—Meanwhile by hybridization and selection of other genetic mutations, the front and rear ends of the animal were becoming differentiated in such a way that the foot, originally a flat crawling organ as it still is in *Nucula* and other Protobranchia, was becoming compressed into a flexible ploughshare pointing forward and downward (fig. 2.41). This muscular organ, in allusion to which the class name Pelecypoda (hatchet-foot) was given, must be able to produce a partial vacuum below and in front of the shell, the semi-fluid sand or mud into which the shell sinks, flowing around the streamlined shell valves.

As a result of a third but simultaneous series of mutations, hybridizations and selections, there was a cumulative improvement in the creative qualities of the shell-producing parts of the mantle, which was constantly experimenting with detailed structural changes that enabled the shells better to withstand the diverse tensions, pressures and shearing stresses of a dangerous life on the shifting floor of the tidal zone. For example, there were contrasting lines of selection, at one extreme toward the very ponderous, massive, slow-moving, strongly-braced shells, as in the quahog (*Venus mercenaria,* fig. 2.48E), and other species (fig. 2.42C) able to withstand the pounding and sucking of a heavy surf. At the other extreme stands the delicate thin-shelled, narrow and relatively fast-moving razor clam (fig. 2.42A). By suddenly withdrawing its foot, this clam ejects a stream of water through the siphon and thus jerks forward (Parker and Haswell, I, p. 66). The smooth surface of the razor shell may favor easy progress through the mud or sand, enabling the animal to make a rapid shift according to the changes in the environment, at high and low tide, in rough and quiet waters, and the like. On the other hand, a rough, folded, flounced or angulated shell is variously useful to sedentary types (figs. 2.52B, 42D, 51B). For example, in the "spiny oysters" (*Spondylus*) the mantle gives rise to many sharp-edged, outwardly turned flounces (fig. 2.52A), which form a sort of *chevaux-de-frise*. These look as if they would cut into the arms of an enveloping starfish as soon as the latter began to tighten up against the efforts of the mollusc to keep its valves shut.

The shell itself, as in the sea-snails, consists chiefly of calcium carbonate laid down in two layers, the outer more prismatic, the inner, wavy and nacreous (pearly). Especially among the fresh-water bivalves, some (e.g., *Quadrula quadrula*) develop very massive shells with a thin horny layer or periostracum, while others (e.g., *Anodonta cataracta*) are thin and fragile, indicating corresponding differences in the calcium metabolic cycle.

The latent tendency toward spirality which seems inherent in the molluscan organization reasserts itself especially in the varying size and curvature of the umbo, a boss or growth center at the apex of each half, or valve, of the bivalve shell (figs. 2.42, 44, 47C). These opposite growth centers are fastened together across the midline of the back of the shell by the external ligament. The line of contact between the opposite umbones is the hinge line between the right and left valves (figs. 2.42; 2.48–52). The free, or lateral, borders of the growing mantle expand radially away from the umbones below and laterally to the hinge line. The form of the adult shell is determined by the differential rates of growth of parts of the right and left mantle edges moving away from the umbones in the three planes of space (length and breadth as in figs. 2.42, 2.43, 2.44; height, fig. 2.45C). If the dorsal hinge line be taken as the main axis, in a horizontal position, the contours of each half of the shell could be projected on the vertical, anteroposterior plane passing through the mid-dorsal line and the line of the aperture or gape. In *Isocardia* and *Hemicardium* (fig. 2.43) the umbones form equal and very beautiful but short spirals.

In an extreme case both umbones may grow into a spiral "ram's horn," as in certain fossil Chamidae. A very curious convergence is that of the Cretaceous ram's horn shell, *Requenia ammoni,* also of the family Chamidae, in which the right umbo has given rise to a coiled ram's horn, the left to a flat spiral lid, the whole looking amazingly like a sea-snail shell. In many other bivalves also the right and left valves are unequal; especially in sedentary types, including the oyster, one valve may form a sort of basin in which the animal lies on one side, while the opposite valve forms a lid. In the Cretaceous Hippuritidae (Rudistidae) the right valve forms a high coral-like column tapering at the base, while the left valve forms a tight-fitting lid articulating with the lower by strong vertically directed "teeth" and sockets and movable only in the vertical direction. Among the brachiopods the Permian family Richthofeniidae consists of forms that superficially resemble the family of lamellibranchs mentioned above, a clear case of convergent evolution.

The adductor muscles (fig. 2.41), which hold the opposite valves of the clam together, are perhaps not homologous with those which tend to pull the sea-snail into his shell and close its lid, but in both cases the muscles are an important part of the mantle-shell-foot complex. Even the oldest known Palaeozoic bivalves were already subdivided into (1) the Heteromyaria, in which the anterior adductor is small or wanting, including the wing-shells (fig. 2.45A and pearl-oysters (B), or Aviculidae, the ear shells or Pinnidae (E), the pectens or scallops, Pectinidae (fig. 2.46C), the spiny oysters or Spondylidae (fig. 2.50A, C), the mussels or Mytilidae (fig. 2.45D), etc., and (2) the Homomyaria, or those with two equal or nearly equal adductors, one anterior, the other posterior, including all sorts of ark-shells (fig. 2.47), clams (figs. 2.48; 2.49), fresh-water clams, cockles or Cardiidae (fig. 2.43), etc. Since these muscles coöperate with the movements of the digging foot, perhaps by adducting the valves as the foot is retracted, the adductor muscles are dorsal to the foot (fig. 2.41) and below the hinge line between the opposite valves of the shell (figs. 2.47; 2.48).

There is often a correlation between the relative sizes of the anterior and posterior adductor muscles and the form of the shell. If they are subequal and nearly equally distant from the umbo, as in *Glycimeris* (fig. 2.47D), then the shell tends to be symmetrical and equilateral; but if the posterior adductor muscle is very large, as in *Meleagrina* (fig. 2.46B1), the posterior part of the shell expands accordingly, while in the oyster (fig. 2.51D) the umbo moves forward and the anterior part of both valves narrows.

Hinge-joints and Evolution.—The joint between the right and left valves of the shell is of great mechanical importance, since its characters determine the limits of movement between the two valves. One of the most primitive forms of hinge-joint is that of the Lower Silurian *Ctenodonta* (family Nuculidae). Each half consists of an arc or band, bearing two parallel rows of low comb-like "teeth" beneath the umbones, those of the right interlocking with those of the left. Different types of hinge have been evolved in different groups and the classification of fossil lamellibrachs is partly based upon characters of the hinge in relation to the ligaments, external and internal, which tie the two valves together along the mid-dorsal line. The following tentative phylogenetic-taxonomic treatment of the hinge-types is based chiefly on the data in Zittel's Grundzüge (Broili, 1924), Henry Woods' Palaeontology (7th ed., 1937) and Twenhofel and Shrock's Invertebrate Paleontology (1937).

The early Palaeozoic to recent Anisomyaria are forms in which the posterior adductor muscle is large and the anterior adductor small or wanting. The Aviculidae or wing-shells (fig. 2.45A) and the pearl-oysters (fig. 2.46B, B1) have a strong longitudinal ligament located in a shallow groove running horizontally along the mid-line, above the small anterior adductor and the much larger posterior adductor muscle. In some of the Palaeozoic members of this family there was a prominent fold on the inner side of the left shell, this fold being subdivided into several small "teeth" lying above the small anterior adductor muscle and one long narrow ridge running obliquely downward and backward to the scar caused by the large posterior adductor muscle. On the inner side of the flatter right shell there were also elevations and depressions which fit into those of the left shell. This type of hinge has been classified under the Prionodesmacea, or saw-band division. The modern *Pteria* (fig. 2.46A1) retains most of the essential features of this *Aviculoid* or prionodesmodont type but the "teeth" are reduced to slight swellings. In the fossil *Aviculopecten* the hinge-line was shorter and the external ligament concentrated into a triangular pit, one on each valve just below the umbo, the ligament radiating outward from this point. This triangular pit is present also in the Limidae (fig. 2.46G), and Pectinidae (fig. 2.46C1), but in the latter family two or four prominent ridges form an inverted V above the triangular pit. The spiny oysters, or Spondylidae (figs. 2.46E, 2.51A, C) are specialized derivatives of the pectens, in which the lower ends of the inverted V become swollen into two subequal condyles (called "teeth"), which are received into corresponding pits on the opposite valve. This arrangement is called *isodont*.

The type of hinge in the true oysters (fig. 2.51D), in which there is a prominent apical pit or groove for the large ligament, but no teeth, seems to be a derivative of the prionodesmodont type, which has lost the teeth. In the mussels (Mytilidae) and allied forms the long and thin hinge margin is either toothless (fig. 2.45D) or there may be small weak folds. This is the *dysodont* type.

Among the Homomyaria (with two subequal adductors), the apparently most primitive hinges are seen in the ark shells, or Arcidae (figs. 2.44A; 2.47). Here the external ligament on each side radiates toward the midline from a point beneath the umbo (fig. 2.44A). The hinge line just beneath this is usually straight but may be elevated or curved toward the umbo (fig. 2.47C). In some Arcidae there are many very small "teeth," increasing slightly in size both ways beneath the umbo (fig. 2.47D). This is the *taxodont* type, which is also found in the most primitive clams (Nuculidae).

From this or some similar variety of the taxodont type were probably derived the highly differentiated *heterodont* or *teleodont* hinge-joint (fig. 2.48) of the higher clams (Veneridae). The hinge plates, crests ("teeth") and slots (fossae), as well as the ligaments, grow radiately from the opposite mantle folds beneath the umbo. The three main crests beneath the umbo are called "cardinals." These have a central pit for the internal ligament or resilium, a pair of cardinal crests or teeth, one in front and one behind this pit and several other oblique ridges and accessory teeth. In one extreme variety of the heterodont type seen in *Spisula* (fig. 2.49), the usual groove for the resilium is expanded into a large triangular basin or chondrifer and the cardinal teeth are quite small. A distorted form of the heterodont type with extremely thick, swollen "teeth" and ridges is called the *pachyodont* type (e.g., in the fossil Megalodontidae).

In another direction the entire hinge line may become very narrow, with small, sharply divergent, narrow teeth (and grooves) in front of and behind the pit for the resilium and reduced or feeble diverging lateral ridges. Such is the *asthenodont* type in the delicate shells of the Tellinidae. In the Trigoniidae, the anterior and posterior moieties mentioned above diverge sharply below the umbo and each moiety may bear a deep longitudinal sulcus dividing the little teeth on its margins into two parallel rows, whence the name *schizodont*. Other types (*desmodont, cyrtodont, diagenodont*) have been named but enough has been said to suggest the great variety of hinges, which in a complete description would require almost as many special names as there are families.

A strong hinge-joint (figs. 2.50A; 2.51A) is found in a robust shell which is subject to the buffeting of the waves or to the action of strong currents or to the possibly disruptive efforts of such enemies as starfishes, which try to pull the valves apart, and mollusc-eating fishes and other vertebrates with nut-cracker jaws and crushing dentitions.

Diverse Devices for Living.—Except for *Nucula* which has a flat, crawling foot, the most active lamellibranchs are digging forms. Apparently from this central type one branch, culminating in *Lima* and *Pecten,* became free-swimming merely by snapping their shells open and shut. Certain descendants of the free-swimming *Pecten* settled down on the right side and gave rise to the spiny oysters, or Spondylidae (fig. 2.50A, C). A certain branch of the Homomyaria, by way of the cockles (fig. 2.43E) and *Hippopus* (fig. 2.52A), finally rested on their backs with the aperture of the shell turned upward, as in the so-called giant clam *Tridacna* (fig. 2.52B). Here a part of the foot, protruding in front and reflected toward the umbo, serves as a holdfast and is analogous with the byssus or adhesive organ of the mussels.

Tridacna, like *Pecten,* has a row of "eyes" or light-sensitive organs all along the outer edge of the mantle. On a calm day in the Great Barrier Reef, Australia, I had an opportunity to observe the living *tridacnas,* which would close the shell as soon as one's hand was held out in the air, well above the submerged shell but within sight of the mantle eyes. Thus it is evident that these light-sensitive organs are connected by nerves with the powerful adductor muscles.

The piddocks, or boring shells (fig. 2.53), have small projections or serrations on the edges of the shells, which are used for cutting grooves and boring tunnels in sedimentary rocks. We may perhaps infer that the rasping is effected by the animal's holding fast by one side and twisting or rocking the opposite valve by means of the adductor muscles. In some of the piddocks the resourceful mantle produces an extra pair of valves. In the shipworms (*Teredo*) the mantle itself becomes the dominant organ, while the shell valves are reduced to a gnawing beak. The very remarkable angel-wing shells (fig. 2.44F, F1) are large piddocks (pholads) in which the "myophore," a part of the hinge line on each side, has been reflected above the umbo, while the articulation of the two valves has become limited to one point.

Perhaps the most surprising specialization is that of the "watering-pot," *Aspergillum,* in which the siphon builds a long straight tube with a many-holed strainer at the distal end, while the shell itself is represented by two very small, widely open valves that are embedded on the roof of the tube.

Another remarkable organ system derived from the mantle is that of the gills, which range from simple ridges in *Nucula* to the elaborately subdivided lamellae of the higher bivalves, to which the class name Lamellibranchiae refers. There is probably some correlation of the degree of elaboration of the branchial lamellae with the activity of the ciliated tracts, the amount of energy spent by the animal in moving about, the oxygen content and temperature of the medium, the duration of exposure to the air at low tide, etc. We can readily suppose that these environmental and metabolic factors in turn condition the rates of growth, the texture and form of the shell, the relative thickness of the horny covering or periostracum and other conspicuous features.

Equipped with these and many other remarkable devices for living, the diversified bivalves flourish in many sorts of environments, from freshwater ponds, lakes and rivers, to estuaries, muddy flats, sandy ocean beaches, down the long slope of the continental shelf to almost abyssal depths, and ranging from the polar seas to the tropics.

Contrasts with Other Molluscs.—In contrast with other molluscs, there is, with certain exceptions, a singular uniformity both in the methods of feeding and in the construction of the digestive system throughout the lamellibranch class, in which true jaws, teeth and radula are conspicuously absent, and even the mouth lies deep within the shell. The wood-boring *Teredo,* as above noted, has succeeded in making a gnawing beak out of the shell valves and it is said (Encycl. Brit., 14th ed., vol. 21, p. 946, art. *Teredo*) that these wormlike bivalves even succeed in digesting woody fibre as well as minute floating organisms drawn in with the respiratory current. But almost all other bivalves depend upon microscopic food particles which are drawn in through the gill-sieve by the lashing of cilia.

Another point of contrast between the bivalves and most other molluscs is the absence of a head and the very poor development of a central nervous system. Nevertheless there is enough nervous correlation of different parts to insure efficient adjustments, especially to the cyclical changes in the environment induced by diurnal and seasonal changes in tides; certain clams, equipped with an otocyst for detecting slow vibrations, respond to a disturbance of their ground by digging in rapidly toward the quiet substratum.

The lack of a head in bivalves may be correlated in part with the dominance of the sensitive and responsive mantle, which envelops the entire digestive tract and has largely taken over the functions usually performed by lips, jaws, teeth, branchial, and even locomotor apparatus; in other kinds of animals all these are more immediately connected with the digestive tract and their nerves form important components of the brain and central nervous system. This protean mantle is very sensitive to contact with the environment, taking the place of the antennae of snails, which are borne by the head; in the pearl oyster it precipitates iridescent tissue around an irritating point (fig. 2.46, B1); in *Pecten, Lima,* and *Tridacna* it bears elaborate eyes.

But in spite of the dominance and adaptability of the mantle, the pallial, visceral and other nerves and ganglia in the shell-protected bivalves are never brought together into a controlling compound brain as similar nerves are in the gastropods and cephalopods. There is a still wider contrast between the headless, brainless, non-metameric bivalves and the well cephalized metameric arachnids, in which the entire organization is more directly motivated by the hungry digestive tract.

The color patterns of bivalves are doubtless just as much a result of spatial arrangements of growth forces as are the forms of the shells. Pigment-producing cells are located in the edge of the mantle just beneath the under surface of the horny outer covering. The widening color rays of the sunrise shell (*Tellina* sp.) must have been produced in much the same manner as the radiating ridges and points in the angel-wing (fig. 2.44F).

SOME QUEER SHELLFISH

(Heteropods, Pteropods, Airbreathing Land-Snails. Tusk-Shells and Others)

The heteropods are a branch of the snails which took to a floating life in the sea, frequently reducing or thinning the shell, swelling the body with sea water, developing a ventral sucker on the foot, especially in males, and compressing the foot into a paddle, sweep or float, but retaining the deep-seated heritage of the Streptoneura, or forms in which the visceral commissures are twisted into a figure of 8. Among the living heteropods *Atlanta* and *Oxygyrus* have a symmetrical spiral shell which so closely resembles those of certain palaeozoic Bellerophontidae and Porcellidae that the derivation of the modern heteropods from this source seems probable (Zittel, 1924).

In the tectibranchs, including the sea-hares, the neck often becomes long and the body slug-like, the foot develops a pair of lateral lobes which act as fins. The shell may be reduced or wanting and the visceral commissures are "detorted" (untwisted from a figure 8), these forms therefore being referred to the subclass Euthyneura.

The marine nudibranchs are euthyneuran gastropods which in the adult stage have lost the shell, together with the primitive respiratory organs (ctenidia) and breathe through secondary branchial outgrowths from the mantle.

The pteropods (wing-foot) are for the most part minute marine creatures which go in schools of countless millions, their dead shells forming the pteropod ooze. The foot is produced laterally into a pair of "wings" (fig. 2.39F). The shell varies greatly: some snail-like to globose, others tube-like, some conical to calyx-like, often with two distinct shells, a dorsal and a ventral, with lateral keels and posterior projections. The growth stages are often marked by numerous lines or ribs. The head is not clearly differentiated and the eyes are vestigial. The visceral commissure is symmetrical (orthoneural). They may be specialized Hyolithids (p. 50).

The Pulmonata, or true land-snails and slugs, are Euthyneura, or "detorted" gastropods in which the walls of the mantle cavity form a pouch with a narrow contractile aperture and a vascular network leading through the short pulmonary vein into the heart. This sac functions as a lung and with it the true snails have been able to spread from their ancestral home on the seashore to the fresh waters and high-lands. Convergent development of the mantle cavity into a functional lung is found in certain marine and fresh-water Streptoneura (e.g. in *Littorna, Cerithium.* Encycl. Brit., vol. 20, p. 63). A splendid account of the amphibious habits and respiration of the Indian apple-snail (*Pila*) is given by Prashad (The Indian Zoological Memoirs, IV, 1932). This snail, although one of the Streptoneura, has a well developed pulmonary sac much like that of a land-snail.

In the Scaphopoda (boat-foot), or elephant's-tusk shells (fig. 2.39E), the mantle folds are almost completely united to form a cylindrical tube, which secretes the long tusk-like shell on its outer surface. The shell is open at both ends, the head, like the heart and viscera, is simplified (or perhaps, poorly differentiated) and gills are wanting; but the molluscan buccal cavity is bordered with tentacles and a radula is present; there is a narrow trilobate foot which can be extruded through the oral opening of the shell and is used for burrowing in the sand. The principal ganglia and connectives of the molluscan nervous system are also present and the larvae resemble the veliger stage of other molluscs. Fossil shells referred to the Scaphopoda are recorded from the Silurian and later ages. In *Cadulus* of this group the tube is greatly shortened and swollen in the middle. Certain authors have emphasized some points in common with the gastropods (e.g. the radula), others have stressed those in common with the lamellibranchs, but the prevailing custom is to treat the scaphopods as a distinct class from either.

ENIGMATICAL CONULARIDS

The Conularida (fig. 2.39A) are an important group of fossil molluscs(?) of somewhat doubtful relationships, which range from the Cambrian to the Cretaceous. One family (Styliolinidae) are very small needle-like conical or curved shells which closely resemble the shell of the pteropod *Styliola.* In another family (Hylolithidae) the shells may be elongate, conic, pyramidal, or sharply curved, in cross-section triangular, elliptical or lense-shaped, frequently flattened on one side, curved on the other. The growth lines may be clearly seen, the surface smooth or ribbed. The aperture of the shell bears a well fitting operculum. The derivation of the modern pteropods from Cambrian forerunners of the family Hyolithidae was supported by Walcott's discovery of his *Hyolithes carinatus,* with "wings" or lateral extensions of the foot, like those of pteropods (Boucek, 1939, p. 8). In the Tentaculitidae the shell is elongate conical, with round cross-section, the surface often ringed, the hinder part of the shell often filled in with a chalky deposit or closed off with concave cross-partitions. The true Conularidae have straight shells with rhombic to quadrate cross-section. Long furrows subdivide the side plates into halves. The aperture of the shell is restricted by the overgrowth of the side plates. The tapering end in young stages bears an adhesive disc (Haftscheiben), but later stages are free-swimming (Zittel, 1915, p. 477; 1924, p. 506). Both Kiderlen (1937) and Knight (1937) cited by Boucek (1939, p. 127) refer the conularids to the coelenterates, near the Scyphozoa, with which they agree in the four-sided, radial symmetry of their cross-sections.

PHYLOGENY OF THE MOLLUSCA

Bateson in 1913 (1914) specifically banned the study of the phylogeny of the Mollusca in the following words:

"Naturally . . . we turn aside from generalities. It is no time to discuss the origin of the Mollusca or of Dicotyledons while we are not even sure how it came to pass that *Primula obconica* has in twenty-five years produced its abundant new forms almost under our eyes."

Bateson's objective and method were judged to be both restrictive and obstructive by Gregory (1917, p. 623).

Of course the problem of the origin of the Mollusca deals with unknown events of the geologically distant past, for which direct evidence from fossils may never be available. Nevertheless there is cogent evidence for a remote common ancestry of all the diversely modified major divisions, whose subsequent evolution is fully documented. Only the trunk or central stem of the tree is lacking and it is highly likely that among the divergent derivatives some have retained more of the primitive stem characters than others. In short, was the protomolluscan more like either a nautiloid or a clam, or a sea-snail or a *Dentalium?* Even a tentative solution of this problem may lead to further comparison and correction.

The actively swimming squid propels itself by a stream of water ejected from the mantle cavity through a nozzle-like fold of the mantle; it has large paired eyes and well developed brain and contrasts most widely with the oyster, which in the adult stage is completely stationary and has no eyes and no brain. In between these extremes would stand the typical sea-snail, such as the whelk *Buccinum,* which creeps slowly on its muscular foot and has small eyes and a fairly developed complex of ganglia and nerves in its head.

In general the molluscs are far inferior to the jointed animals (Arthropoda) in regard to speed of movement, complexity of locomotor apparatus and development of eyes and brain.

SUMMARY

In typical molluscs as compared with other invertebrates several important features are conspicuously lacking or poorly developed.

Although different kinds of molluscs have achieved their own methods of swimming, crawling, climbing, digging, etc., no mollusc possesses the paired jointed levers called legs which are characteristic of trilobites, crustaceans, arachnids, insects and related groups of arthropods. Secondly, in molluscs the body itself is not divided into metameres or neuromuscular segments and although the shell of the chitons is subdivided into a longitudinal series of segments, the nervous system is not so divided and the jointed shell may have been derived ultimately from a single undivided shell. Thirdly, the head of molluscs, when present, differs from that of the higher arthropods in lacking paired jointed "foot-jaws." The remote origin of the molluscan foot is obscure. Nevertheless, since its muscular layer in the gastropod is continued behind and underneath the digestive tube on to the inner wall of the shell, we may infer that part of the "foot" originally served to pull the head back into the shell and that it later became differentiated into an elongate functional foot and a retractor muscle. Thus all the diversified methods of locomotion in the Mollusca may have started from sedentary adults.

Since the known classes of the Mollusca had already become separated from each other at the virtual beginning of the fossil record in the Cambrian period, we are forced to make tentative phylogenetic hypotheses based on embryological and morphological data. From embryological evidence it has long been known that the molluscs belong to the teloblastic series of invertebrates, in which the mesoderm originates from a single center or cell, much as it does in the annelids, crustaceans and other arthropods. It has also been known that the free-floating veliger and trochosphere larval stages of various molluscs are fundamentally similar to the trochophore larvae of annelids, so that a distant connection with the annelid stem has been widely admitted. Thiele, however, from the characters of the reproductive tubes of certain worm-like relatives (Solenogastres) of the chitons, assumes a close connection at least between these molluscs and the polyclad flatworms. But on the whole, the anatomy of the chitons seems to tie this class firmly in with the molluscan stem, and their bilaterality may be no more primitive than it is in *Patella,* while the subdivision of their shell into a series of articulated valves may be due to a secondary budding of the shell-producing mantle. On the other hand, the mantle of the chitons may have retained some of its primitive ability to give rise to segments or polyisomeres and this power is still manifested whenever the mantle produces serially similar characters in the many shells of cephalopods, gastropods and others.

According to the hypothesis set forth above, the dominant organ in the Mollusca is the highly variable and adaptable mantle, which produces the myriad forms of shells and the diversified ctenidia, lamellae, respiratory sacs and other organs. We may infer that while the larval Protomollusca were ciliated floating forms, the adults were inshore, bottom-living animals which were already using the products of their own calcium metabolism as a protective covering. Among living molluscs the tube-dwelling *Dentalium* (fig. 2.39E), which digs into the sandy bottom by means of a small fleshy foot, may be nearer in general habitus to the Protomollusca than are the rock-clinging limpets. The "foot" and its muscle may originally have been somewhat like a muscular lower lip, serving the triple purpose of testing and drawing desirable particles into the mouth, closing the mouth and retracting the incipient "head." Possibly the tapering orthoceratids of the Cambrian may really be among the Protomollusca or near to them and they have merely seemed to be specialized to zoologists who had chosen the rock-clinging limpet as the most central and primitive type.

We may further suppose that the diverging possibilities of the future molluscan phyla began to be realized when regular strobilization became accelerated so that the animal frequently outgrew its shell and moved forward to build a larger chamber. At such times the animal would be more exposed and in the effort to dig in, the incipient head would be turned downward. If the mantle followed this cue, as it were, a symmetrical coil as in the most primitive cephalopods (fig. 2.39B-D) would be developed.

In the cephalopods symmetrical spirality became the ruling habit and asymmetrical spirals or helicoids were rare. Eventually the cephalopods became floating marine forms in which the primitive "foot" muscle gave rise to a ring of strong tentacles around the mouth. In the earlier gastropods, as we have seen, the helicoid or laterally developed spiral early appeared and the foot muscle became specialized for creeping along the shore. From these early amphibious gastropods some branches developed lungs and ventured inland, giving rise to the pulmonates (snails and slugs); others, losing their shells, became strictly marine, as the sea-hares; still others clung to the rocks and gave rise at different times to the sedentary *Patellas,* fissurellids, calyptraeids.

The headless bivalves may be interpreted as excessively specialized, primarily digging derivatives of a very early tube-living type in which the animal in the terminal chamber finally lost his stalk, settled on the bottom in shallow water and began to dig in. The mantle at this time divided into right and left halves. Thereafter the latent and opposing forces of symmetry and asymmetry, spirality and what may be called helicoidality, were, so to speak, let loose to contend with each other and to produce the bewildering variations of the shell in the thousands of species of known fossil and recent lamellibranchs. That the lamellibranchs might be related to tubed or stalked forms was long ago implied to some extent by Lacaze Duthiers (cited by Zittel, 1924, p. 434), who noted the important characters that tend to connect them with the *Dentalium* group (Scaphopoda).

According to the present hypothesis, the pteropods, *Orthoceras, Dentalium,* retain traces of the tube-dwelling stage in the adult. The many-chambered stalk has been wholly eliminated in the lamellibranchs. The larval stages were, it may be supposed, always free-swimming forms even in the stalked or tube-living forms, as they are in the long-shelled *Dentalium.*

The foregoing hypothesis seems consistent with the following facts:

1) the great development and adaptability of the shell-producing mantle, which in one way or another retained the power of producing polyisomeres, or similar segments, either in the shell itself or in parts derived from the mantle, such as the gills;

2) the early appearance of such forms as the hyolithids, pteropods, *Orthoceras* and scaphopods, which retain more of the tube pattern than do any of the gastropods or lamellibranchs or amphineurans;

3) the relatively low development of the head in lamellibranchs, which may be due to direct inheritance from a tube-living ancestor rather than to degeneration of a better developed head;

4) in the typical snails the sensitive organs of the head often serve to start the machinery for retreating into the shell and closing the lid; this would be a deeply hereditary habit in such tube-living animals as the scaphopods;

5) the presence among the Crustacea of such forms as the Cladocera, with paired valves formed from lateral wings of the shell mantle; these forms are merely convergent toward the bivalve molluscs;

6) the free-swimming veliger and trochophore larvae are common to all the diverse teloblastic groups regardless of the adult habitus.

Perhaps one reason why the head is very poorly developed in the lamellibranchs is that they took to a digging habit too soon after they had been set free from the tube-living habit; another, that the dominance of the protective mantle and shell encouraged digging in and holding still rather than sustained flight. Among the gastropods the "foot" had by hypothesis formerly served chiefly as part of the musculature for drawing the animal back into the shell and the habit of shutting up, as in tube-living animals, was and is still very strongly entrenched in the majority of the shelled gastropods. It was also to be expected perhaps that a sedentary ancestry would predispose a slow creeping form to overcome its prey by prolonged and deliberate rasping rather than by rapid pursuit and devouring jaws. Jet propulsion in squids, etc., came later.

The foregoing hypothesis even if lacking direct palaeontologic evidence has the advantage of bringing forward several fundamental morphologic and phylogenetic considerations. First, in the animals described above we see the complex called a head in a low stage of integration of its separate components (mantle, sense organs, mouth, retractile lid, respiratory tufts); while at the other extreme, in the large-eyed, swift-swimming and predaceous decapods and octopods, the head-foot is well advanced and supplied with a large brain, controlling elaborate motor apparatus, and able to make quick and effective responses to a fleeing, struggling prey or to an attacking enemy. Obviously the complex called "head" is merely analogous, not homologous with the head of the advanced arthropods and still less with the head of vertebrates. The same is true of the locomotor apparatus.

CHAPTER III

EXTERNALLY ARTICULATED ANIMALS (ARTHROPODS)

Contents

CHAPTER III

EXTERNALLY ARTICULATED ANIMALS (ARTHROPODS)

THE MANY-LIMBED TRILOBITES AND CRUSTACEANS

THE TRILOBITES were actively moving jointed animals, capable of exploring their environment in search of animal food, and they contrast widely with the brachiopods, which were stalked or sessile food-sifters. Although Pre-Cambrian trilobites are not definitely known, the group was the dominant form of animal life in the Lower Cambrian and flourished exceedingly during the Cambrian, Ordovician and Silurian periods; they declined in the Devonian and "a few last survivors are found in the Carboniferous and Permian" (Goldring). Possibly they could not bear the pressure of competition with more advanced types, especially the eurypterids and true crustaceans.

The trilobites were formerly regarded as a special subclass of the Crustacea but they are on the whole far more primitive than the true Crustacea and appear to have been the descendants of a Pre-Cambrian central stock which was ancestral to all the arthropods (including, besides the Crustacea, the eurypterids, king crabs, scorpions, spiders and insects)

The oldest known trilobites (fig. 3.1) already exhibited a sharp dorso-ventral asymmetry or contrast between the upper and lower surfaces; that is, their oval bodies were convex above and concave below. On the ventral or concave side of the body there was a series of paired and jointed appendages, extending from beneath the head backward (fig. 3.1A). All this indicates that the typical trilobite had a strong sense of attachment for the rocky or sandy bottom on which he lived and that he could both crawl and swim.

The name trilobite refers to the triple longitudinal division of the body as seen from above into a long and relatively narrow median lobe, with a broad lateral lobe on either side. There were also three lobes transversely, since the body in dorsal view was divided into a head-shield, a thorax and a pygidium (fig. 3.1C). The head-shield, although bearing traces of former segmentation, in most cases functioned as a single piece; the thorax was made up of a varying number (e. g., 8, 13, 20) of successive articulated segments; the pygidium likewise consisted of a variable number of segments but these were crowded together or partly coalesced. The whole body was covered with a tough chitinous-calcareous crust, essentially like that of crustaceans. This was thicker on the back and very thin on the ventral surface, which was protected by the appendages. As in crustaceans, this crust was split (probably along the mid-ventral line) and sloughed off at intervals, whenever the growth pressure increased to a critical point—another contrast with sedentary or slow-moving forms with cumulatively increasing armor. The more median and less movable parts of the body were doubtless the fulcra upon which the more distal and lever-like parts moved, and the fact that some trilobites are found in a rolled-up condition indicates the possession of strong flexor muscles beneath the jointed median axis, probably similar in principle to the muscles that flex the abdominal segments in the lobster. In this connection, it is noteworthy that in the trilobites and other articulated animals the muscular system was completely covered, being attached to the inner wall of the exoskeleton.

The paired appendages, one pair to each thoracic segment, are best known in *Triarthrus becki* (fig. 3.2), as described by Beecher, later by Leif Stormer. In this genus they were exceptionally elongated, projecting far beyond the plates of the lateral lobes. Each appendage was divided near the base into two nearly equal branches, the ventral one of which consisted of about seven articulated segments; the extreme outer tip of this branch bore one or two minute spikes. This was evidently the locomotor branch of the appendage. The dorsal or respiratory branch in the very primitive *Neolenus* consisted of two subequal sections, medial and distal, each bearing a comb-like gill (Stormer, 1944, fig. 5, *19*). Four pairs of similar but smaller gill-bearing appendages were located beneath the head-shield but they showed only minor modifications into jaw parts and in this respect again the trilobites were more primitive than typical crustaceans. As in existing crustaceans, each hollow segment of the paired appendages may have contained obliquely arranged muscles that stretched across the joints to effect the flexure thereof in swimming and crawling movements. On the other hand, it is quite possible that in trilobites the muscles of the appendages were as yet mainly slips of the ventral body wall which were attached chiefly to the finger-like processes at the proximal ends of the appendages.

For the efficient operation of this many-pieced locomotor system the trilobites doubtless had a correspondingly subdivided and integrated neuro-muscular system, which must have been capable of sending successive waves of flag-like movements along the series of lateral appendages, such as may be seen in the movements of existing millipeds. On the ventral surface there was a long median furrow or depression lying between the opposite rows of appendages. The very small mouth lay at the anterior end of this furrow, the anus at its posterior end. Hence we may suppose that as the animal moved forward over the mud, by wave-like movements of the appendages, small food particles were directed into the mouth, while rejected or waste material was worked backward through the relatively straight digestive tube and along the median ventral groove.

The trilobite head was covered with a dorsal shield (fig. 3.1), which was subdivided into a large median convexity called the "glabella," and this in turn was flanked by the "fixed cheeks," while to the latter were articulated the "free cheeks," which covered the cephalic appendages.

In the general direction and coördination of these activities the large eyes of the typical trilobites doubtless played a conspicuous part. The median convexity of the head-shield often bore a large field crowded with minute eyes, not unlike the compound eyes of insects. Obliquely behind and laterad to the glabellar eye-field were the usually conspicuous paired eyes (fig. 3.4A), each of which consisted of a kidney-shaped or crescentic field containing a multitude of small eyes that, at least in some forms, seem to have been in the process of integration into large eyes. These paired eyes lay in the territory of the "fixed cheeks" or near their outer border and were dorsal to the "free cheeks." Here we have a certain analogy to the situation of the eyes in early vertebrates, in which the paired eyes lie beneath the supraorbital and above the lateral cheek-plates, but a true homology of parts is contraindicated by the fact that in the trilobites the heart tube was doubtless just beneath the functionally dorsal surface of the animal, as in annelids and crustaceans, while in the vertebrates this surface covered the nerve tube.

In early larval trilobites (Stormer, 1942, pp. 52, 57) the cephalon or prosoma includes five subdivisions of the median lobe, perhaps corresponding to originally separate segments. The later larvae already foreshadow the adult forms to which they gave rise and show but little resemblance to the larvae of either annelids or modern crustaceans.

Although the trilobites are restricted to Palaeozoic times, they produced an array of "over two thousand species and nearly two hundred genera" (Goldring, 1929, pp. 187, 188). Nevertheless the fundamental plan of the group is relatively stable and the differentiation of species and genera involves chiefly the emphasis, or anisomerism, of certain polyisomeres (genal or caudal spines, cheek-plates, median and paired eyes, number of body segments and appendages and the like). But if the range of major structural patterns within the group seems at first rather meagre in comparison with the more fundamental variations in such groups as the echinoderms and molluscs, we should note first, that trilobites were formerly given only the rank of a subclass (of the class Crustacea), and that what are probably the more profoundly modified descendants of the primitive trilobite stock are referred to other subclasses, especially the king crabs, eurypterids, scorpions and spiders.

Stormer (1939, 1944, p. 119), in opposition to other authorities, maintains that, while the pectoral appendages of primitive trilobites and primitive crustaceans show a certain fundamental arthropod resemblance, they differ especially in the location and serial homology of the gill-bearing branch of the appendage. For this and other reasons he believes that the primitive trilobitic and crustacean appendages have been derived independently from those of as yet undiscovered Pre-Cambrian primitive arthropods, which in turn were derived from primitive annelids.

Even in early Palaeozoic times a marked structural gap existed between the annelids as a class and the trilobites, which were the central group of the arthropods or animals with articulated chitinous exoskeletons. Worm tubes probably made by annelids are found in Pre-Cambrian strata but the thin cuticular coverings of the worms themselves was not preserved. Nor were the possible links between annelids and trilobites preserved, although the connection of the two groups is strongly evidenced by the fundamental identity of their respective ground-plans. The earlier larval stages of trilobites consisted mostly of the enlarged head, containing three small segments. This stage has been compared with the free-swimming "Nauplius" larva of crustaceans, but the connection lacks definite fossil evidence (Twenhofel and Shrock, pp. 424–425).

The trilobites were already well established in Lower Cambrian times and had advanced beyond the annelid plan (p. 30) in the following features: (1) their cuticular skeleton was locally thickened and hardened with calcium carbonate; (2) their back was covered by a median and two lateral rows of fairly thick plates; (3) their head was much larger and more complex than in annelids and was covered dorsally by a large median plate (glabella), paired fixed and free cheek-plates; (4) usually large compound paired eyes were on the head-shield between the fixed and the free cheeks; (5) on the lower side of the head there was a fairly large median plate (the hypostoma) in front of the mouth, which was continuous above with the head-shield; (6) their paired appendages, one pair in each of the usually very numerous segments, consisted of a basal piece and an axially jointed rod with six segments; plus a dorsal branch plate bearing bristles or setae on its rear edge; (7) the first five pairs of appendages including the antennules were attached beneath the head-shield; (8) on the under side of trilobites there was a long median depression or axial groove flanked on either side by the inwardly projecting basal pieces of the paired appendages. If ripples of up-and-down movements were to pass along the paired appendages from behind forward, a current of water bearing small food particles might be driven toward the downward and backwardly directed mouth; (9) the proximal or basal pieces of the paired appendages beneath the head were modified for cutting food and are designated "gnathobases" (Twenhofel and Shrock, 1935, p. 423), but there is no evidence that the trilobites ever became swift and aggressive predators against large prey and they were indeed unable to endure the competition and/or attacks of the crustaceans, eurypterids and fishes of the later Palaeozoic, at the end of which period they became extinct.

Fundamentally similar to the trilobites but already established in Lower Cambrian times are the phyllopods (branchiopods) or brine-shrimps (fig. 3.3), which are the most primitive of existing crustaceans. They differ from the trilobites especially in having a transverse pair of furcae or feelers on the end of the tail. They live in fresh and brackish-to-salt lakes and brine pools. Some (*Apus*) have a domed chitinous carapace which covers a large pair of shell glands. The latter may secrete a pair of lateral valves, meeting in the mid-dorsal line and simulating the shells of some bivalve molluscs (e.g., Middle Cambrian *Walcottella,* Triassic and Recent *Estheria*). Another has no carapace (Middle Cambrian *Yohoia*) and somewhat suggests a marine annelid but has a more distinct head and stalked eyes. In the modern *Apus* (fig. 3.3A) and its relatives the paired appendages on the ventral surface range from simple antennae to hook-like jaws and to "feet," the latter with five short toe-like branches and two flat leaf-like expansions which may have a respiratory function.

Another Palaeozoic (Ordovician, Silurian) but still surviving group of the crustaceans are the usually minute ostracods, which have bivalve shells, sometimes simulating those of certain bivalve molluscs but often perforated by the paired eye-spots. Internally the ostracods are much like phyllopods except that they have only seven pairs of appendages and frequently raised lobes, grooves or nodes or processes on the surface of each valve.

The barnacles (Cirripedia) are attached head downward to rocks or logs or ships, and are enclosed in a bilateral shell composed of five or six pieces. Their slender thoracic appendages include not more than six pairs, ending in curved antenna-like feelers. These keep up a waving, grasping movement and sweep the food toward the jaws and mouth.

The free-swimming copepods or water-fleas are shrimp-like oceanic crustaceans. They float near the surface in countless millions and form one of the main food items of the whalebone whales. The parasitic copepods, which are attached to the gills and bodies of fishes, have reduced their chitinous exoskeleton, while their paired appendages, if any, have lost locomotor functions and the body assumes a wide range of bizarre forms, which retain little resemblance to that of the ancestral shrimp-like, free-swimming copepods.

The acme of parasitism is found in the tumor-like *Sacculina,* which is attached to the abdomen of crabs. Little is left but a digestive sac with root-like processes which grow into all parts of the host's body and absorb nutriment. But this parasite is shown to be a crustacean by the fact that it has developed from a shrimp-like, free-swimming young, which attaches itself to a crab and is gradually transformed into a *Sacculina.*

The Malacostraca, another subclass of crustaceans, includes a great diversity, from essentially crayfish-like forms to the most specialized lobsters and crabs. Among living malacostracans, the pill-bugs and wood-louse (*Oniscus*), representing the order Isopoda, are completely terrestrial. They are wide transversely and somewhat suggest a trilobite, but they lack the three-lobed division of the body and have other characters that connect them remotely with the crayfish group. One primitive forerunner (*Hymenocaris*) of the subclass Malacostraca, dating from the Mid-Cambrian, bore a bivalved shell somewhat like those of the branchiopods, and this feature persists in the modern *Nebalia;* but in the more advanced forms the carapace is undivided and extended forward over the fused head and thorax.

The enlarged chelae, or pincers (fig. 3.2B) of lobsters and crabs are especially remarkable on account of their well known power. The two arms of this pincer seem to represent respectively the fixed distal end of the appendage and a movable branch; the movable arm is attached to the base by hinge-like joints which limit motion nearly to one plane. Adduction of the movable arm is effected by powerful muscles which arise from a median crest-like plate from the fixed arm and are inserted on either side of the fulcrum of the movable arm. Tubercles of the crust along the opposing edges of both arms increase the hold of the pincers on the unlucky enemy or prey, while sharp points on the ends give this formidable weapon considerable penetrating power.

In the crabs there has been a functional integration of the head and thorax (fig. 3.2B), which have fused into a strong cephalo-thorax sometimes very wide in proportion to its length. At the same time there has been a folding under of the abdominal segments. Locomotion is thus centered on the thoracic appendages, which are often hook-like for climbing on rocks. In well-swimming crabs the hinder pair bear flap-like swimmerets, which are moved obliquely in a fan-like manner. Perhaps these and other improvements have contributed to the almost world-wide success of the crabs, some of which have even become adapted to living on land by enclosing their gills in a water-containing chamber.

Some of the fiddler crabs, living in mud-holes between tide marks, come forth in great hosts and appear to move like an army of ants. Probably this apparently social behavior is at best only in the incipient stage. In a certain fiddler crab (*Uca*) rather elaborate courtship antics, involving specifically characteristic motions of the enlarged chela, are described by Jocelyn Crane (1941).

"LAW *VERSUS* CHANCE" IN THE EVOLUTION OF LOBSTERS

The male lobster, with his huge fighting claws, his very large gills, well articulated abdominal scutes and expansible tail fins, is much more advanced than the relatively primitive brine-shrimps. "The ventral and lateral regions of the thoracic exoskeleton are produced into the interior of the body in the form of a segmental series of calcified plates, so arranged as to form a row of lateral chambers in which the muscles of the limbs lie, and a median tunnel-like passage or *sternal canal,* containing the thoracic portion of the nervous system." (Parker and Haswell, vol. I, p. 499.) Thus "the entire endophragmal system, as it is called, constitutes a kind of internal skeleton," which arises merely from an extension or curling inward of the exoskeleton. Similarly the embryonic stomodaeum, a cup-like insinking of the ectoderm forming the mouth, in the lobster extends further inward and develops into an elaborate "gastric mill." This consists of three chitinous teeth, one median and two lateral, suspended in a complex articulated framework and drawn together by strong muscles arising from the carapace.

This arrangement somewhat recalls the "Aristotle's lantern" of the sea-urchin and the pharyngeal teeth of carps. It gives another example of several well established principles of evolution, namely, that:

1) the three primary germ layers (ecto-, meso-, and endoderm) have similar potentialities in different groups of animals (e.g., to secrete chitinous, muscular and nervous tissues);

2) two or more animals belonging to different phyla but having similar needs have not infrequently evolved more or less similar mechanisms to meet those similar needs;

3) historical geology and palaeontology show that during the known fossil record of the crustaceans there has been an immeasurably long succession of mountain-building, erosion and subsidence, upheaval of certain land-bridges (e.g., Isthmus of Panama), incursions and retreats of the sea, climatic changes from tropical to glacial, together with corresponding changes in vegetation from swamp lands to forests; again, to the grassy open plains, or to cactus-bearing deserts. Meanwhile the tree of animal life was spreading, branching, invading new lands and becoming exterminated in others. Due to hereditary inertia, each strain yielded, but slowly, to the chance-determined vagaries induced by mixture of strains, isolation, "crossing over" of genes, and similar genetic events.

4) When comparisons are made between convergent resemblances of widely unrelated animals, as between the somewhat similar dental apparatus of sea-urchin, lobster and carp, such convergent resemblances are but limited and imperfect. The similarities between them are due to the repetition in different times and places of similar general forces or conditions (e.g., the similarity of chitinous tissue wherever and whenever produced, the similar need for the energy stored in food, etc.); whereas at least some of the structural differences between

convergent characteristics or organs are associated with differences in hereditary potentialities, with differences in opportunities and needs experienced in different environments, in different ecologic backgrounds and times.

5) When a new hereditary structural feature becomes widely distributed in a large population, its further progress or decline may be due to generally uniform or recurrent preëxistent forces or conditions; in other words, it may follow certain precedents, general patterns, rules or laws observable in many similar cases. But the same new characteristic or organ, in so far as it is new, arises not from any one of the various forces or conditions that were prerequisite for its existence but from the new and repeated intersections of all these prerequisite series of forces and conditions at particular moments of time and locations in space. It is the concurrence of such unpredictable intersections of various forces and conditions at given moments that give rise to what is called chance or luck, which has operated in the world of animals with chitinous articulated skeletons much as it has in the world of articulately speaking men.

To return to our male lobster: his great fighting claws or pincers are serially homologous with, but much larger and more powerful than, his other paired appendages. Their great size and strength are somehow associated with his maleness and aggressiveness. The same association is observed in crabs, and it seems probable that it is caused by forces that emanate from the chromosomes and genes that predetermine the development of primary and secondary sexual differences in males and females. This association of maleness, strength, aggressiveness, and the possession of an effective weapon of offense, apparently happen to be of selective value in lobsters, crabs and some other crustaceans, as well as in the scorpions among the arachnids.

The mechanism of the fighting claws, while merely emphasizing that of the smaller claws, affords a clear example of one of the many ways in which a cuticular exoskeleton has been sufficiently plastic to yield such diverse appendages as the following: (1) long and sensitive antennae; (2) stalked compound eyes; (3) several maxillipeds which help to pull the food into the mouth; (4) the dental mill described above; (5) the straining sieve behind the dental mill; (6) the large, well folded gills; (7) the elaborately jointed and freely movable abdominal flaps; (8) the expansible tail flaps; (9) the infolded lateral chambers of the thorax, etc. In general, the appendages are moved by opposing sets of muscles which arise from the inside of the thorax, or within one of the chambers or segments, and extend over the external cuticular joints to be inserted into the next segment. The joints, while permitting movement in certain directions, strictly limit and confine movement in others and thus they are severally adapted to concentrate and apply the forces involved in ways that in the long run prove to be statistically contributory to the protection of the species.

ARACHNIDS

(King Crabs, Eurypterids, Scorpions, Spiders)

The humble horseshoe- or king-crab, *Limulus* (fig. 3.4E), which was long common in certain bays near New York, has irrefutable claims to a high rank in the ancient and honorable order of living fossils; for though its own genus *Limulus* dates back only to the Triassic, its Devonian ancestors (*Protolimulus*) carry the line well down into the middle Palaeozoic; while several Silurian members, including *Bunodes* (fig. 3.4E) and *Pseudoniscus,* belonging to the same subclass (Merostomata), afford a practically perfect transition to the Cambrian *Molaria* of Walcott, which was an extremely generalized form related to the trilobite stock. Moreover, every *Limulus* today goes through a "trilobite larval stage," in which the trilobite relationships are at once obvious and deep-seated.

The horseshoe crab, as its name implies, has a streamlined horseshoe-shaped contour, broadly curved in front, narrowing behind, strongly convex above and concave to flat below. Superficially it is similar to some of the round-disked skates and is likewise mostly a bottom-living form. The paired appendages of *Limulus* are hidden from above beneath the large overhanging head-shield. The body ends behind in a long, straight, tapering median spine, the telson.

When we compare the various limuloids with the trilobites (figs. 3.1; 3.4) we find many resemblances significant of relationships, and differences that indicate the more advanced specializations of the former. The trilobite headpiece comprises five larval segments and bears five pairs of appendages (fig. 3.2A), counting the antennae as the first pair; the limuloid headpiece (fig. 3.4F) is called a cephalo-thorax and it bears six pairs of appendages, of which the first pair sometimes end in pincer-like chelicerae; the sixth pair in *Limulus* (fig. 3.4F) "is surrounded by whorls of plates which aid in pushing the animal over or through the mud" (Goldring, p. 209); in the eurypterids (fig. 3.4D) the sixth pair normally form the large flattened paddles, but in *Stylonurus* the paddles are replaced by very long, many-jointed rods. By comparison with the trilobite *Triarthrus becki* (fig. 3.2A), it seems as if this sixth pair in the eurypterids, and therefore also in *Limulus,* has been derived from the long first thoracic pair in trilobites. Moreover, in certain trilobites (fig. 3.4A-C) the last segment of the head-shield looks as if it were really a first thoracic which was being appropriated by the head-shield.

Thus these trilobites may suggest a certain stage in the union of originally free thoracic segments with the head, while the merostomes (limuloids plus eurypterids) have certainly in other respects also carried forward the principle of progressive cephalization.

The very numerous and delicate thoracic appendages of the trilobite *Triarthrus becki* (fig. 3.2) may have been difficult to protect against the nibbling attacks of annelids and small crustaceans. Thus the limulids were selected, so to speak, for the developing of a series of horny flaps or opercula (fig. 3.4F), representing parts of the locomotor appendages but now serving for the protection of the gills. Meantime the gill plates by multiplying provided better respiration, until more than a hundred pairs of gill leaves may be present (Goldring). At the same time, the jointed median and dorso-lateral plates or pleurae of the trilobite thorax fused into the compact abdominal portion of the shield of *Limulus,* while a strong transverse joint, permitting marked dorso-ventral flexion, was developed between the head-shield and the abdominal shield. Thus the "thorax" of the trilobite (or part of it) gave rise to the "abdominal" segments of *Limulus*. Another strong joint, this one of the ball-and-socket type, developed between the "abdominal" box and the telson or tail rod (fig. 3.4F). The possession of these two joints must enable the animal to extend both the head-shield and tail-piece and thus to turn over like a turtle when placed on its back.

The mouth, both in the limuloids and the eurypterids, forms a small median opening on the under side of the head and it is surrounded by the proximal joints of the paired appendages (fig. 3.4F). These features have doubtless been inherited from the trilobite ground plan but the merostomes are more advanced in that the basal joints of the cephalic appendages bear toothlike processes and are beginning to function as masticatory organs.

The paired eyes in *Limulus* are small but the central nervous system represents a highly "cephalized" derivative of the basal arthropod (and presumably trilobite) type. There is a thick nerve ring around the oesophagus, which gives off branches to the paired appendages; the ring is connected dorsally with very large coördinating ganglia; ventrally it is produced posteriorly into the ventral nerve chain.

Strikingly different in appearance from the limuloids are the typical eurypterids (fig. 3.4D), which have a relatively small head-shield, a much longer, sometimes almost tubular-looking, "abdominal" region and six pairs of appendages protruding from beneath the head-shield, the sixth pair usually forming huge flat jointed paddles. The telson or tail-piece may be spatulate, slightly sickle-shaped or very long, straight and pointed. The eurypterids appear in the Cambrian (the reported Pre-Cambrian eurypterids are of very doubtful validity); they are represented in the Ordovician, but become very abundant in the Silurian; some are found in the uper Devonian; the last survivors occur in the Permian.

The eurypterids followed different paths of adaptive evolution from those of the typical limuloids and most of them appear to have been more active swimmers, using their jointed cephalic paddles rather than the flapping of thoracic appendages.

The eurypterid head (often called a "prosoma") is comparatively small, its shield being somewhat quadrate but with the front corners rounded. The paired eyes are derived doubtless from the compound eyes of trilobites; they are sometimes very large and directed laterally; in other cases they are reduced or absent. A pair of very small median dorsal eyes seems to represent part of the glabellar eye field of trilobites.

The eurypterid "abdomen" corresponds to that of the limuloids but consists of twelve segments instead of six. It is noteworthy that twelve is the number of "thoracic" segments in some of the Cambrian trilobites (*Olenellus*), which also bear a caudal spine (fig. 3.1A). The eurypterid gill-leaves on the ventral surface are protected by plate-like appendages. A median abdominal plate behind the enlarged basal plates of the paddles is believed to cover the genital opening, as in *Limulus*. "The larval forms [of eurypterids] have bigger heads, immense eyes situated near the margin, smaller number of segments and broader bodies, as compared with the full grown animals, and the swimming legs are usually larger" (Goldring, 1931, I, pp. 212, 213). Such advanced larval forms often become the starting-point for new lines of evolution (see below).

The scorpions are recognized as being the land-living cousins of the eurypterids and have progressed beyond them in the development of book-gills for respiration on land. They date from the Silurian period, when the eurypterids were enjoying their widest expansion. The telson is modified into a fang, grooved for injection from a pair of poison glands. As modern scorpions are able to paralyze animals that are many times larger than themselves, the Silurian scorpions may have attacked some of their larger contemporaries such as the eurypterids. Their minute mouth and tubular oesophagus seem to be adapted for blood-sucking. Apparently the second cephalic appendages are both enlarged and modified into powerful claws like those of lobsters. The sixth are not enlarged as paddles but serve with three other pairs as walking limbs.

The scorpion's brain (fig. 5.12C) is even more highly developed than that of *Limulus* (fig. 5.12B), as both its mode of locomotion and its feeding habits are more complex. The more precise derivation of the scorpions from the eurypterids is not clear. On the whole, it seems probable that they have arisen from some narrow-headed, small-eyed relative of *Eusarcus scorpionis,* developing the curved tail-piece into a poison sting, greatly enlarging the second pair of cephalic appendages into chelicerae, and elongating the swimming paddles into walking feet (cf. *Stylonurus*).

The spiders do not come into the fossil history until late in the Palaeozoic period, by which time they had attained the essentials of their present very advanced type of structure. They are recognized by all authorities as being related to the scorpions but their exact origin is unknown. The existence of such groups as the Pedipalpi, or scorpion-spiders, and the sun-spiders (Solpugida) strongly favor the derivation of the spiders from the scorpion stock. This may have come about by a process of paedogenesis, followed by intensive specialization of the web-making organs. The larval eurypterids differed remarkably from the adults in their large eyes, smaller number of segments and broader bodies. The same sort of changes applied to larval pro-scorpions might carry them far toward the basic spider type. It has been suggested that the web-making of spiders may have originated through earlier habits of wrapping the eggs in silky threads and lining their retreats with silk; "later they built platforms outside of their retreats and from these have developed the snares which have been the wonder and admiration of all ages, humanly speaking" (F. E. Lutz).

Most of the spiders lead strenuous and exemplary lives, according to arachnid standards; doubtless animated by a Spartan ideology, they sometimes even kill their own diminutive husbands in the interests of the family. Some of the descendants of this superior race, however, have fallen so far below its proud standards as to prefer a safe niche and a full belly to glorious deeds. Consequently they fasten themselves preferably to some weak or inaccessible spot in the integument of larger animals and by happily eating their way inward soon find themselves in a land of plenty. After this, why should they care if their locomotor apparatus degenerates and they themselves look like newly filled wineskins?

Spiders differ sharply from insects in possessing four pairs of legs instead of three pairs. In general, their anatomy is widely different from that of insects and the latter have probably been derived from a very different ancestral source. Thus the ability to secrete a viscous fluid which hardens into silk has been independently acquired by spiders and insects.

PERIPATUS AND THE INSECTS

One of the most remarkable of all "living fossils" is *Peripatus* (fig. 1.7), a caterpillar-like animal found in South Africa, Australia, New Zealand, South America, West Indies. Its direct ancestors, of essentially identical structural plan, were discovered in the Middle Cambrian Burgess shale of British Columbia. The Cambrian form (*Aysheaia*), although marine in habit, apparently differed little from the surviving *Peripatus,* which lives in rotting logs.

In the modern *Peripatus* the tubular antennae and the legs are covered with transverse rings of conical papillae, bearing little chitinous tips. Each leg ends in a little foot, bearing a pair of horny claws (whence the class name, Onychophora, or claw-bearers). The cylindrical body is not divided transversely into segments, but the integument is thrown into a number of fine transverse wrinkles. Collectively these permit considerable bending of the body, so that oblique thrusts could be exerted by the muscles of the body-wall against the medium. The small hooks on the feet imply that the appendages are used for pulling rather than pushing. They may be useful in crawling inside of rotten logs.

Peripatus seems to be remotely related, on the one hand, to the annelid worms (fig. 1.5), from which it may have inherited its basic ground-plan, and, on the other, it anticipates in many features the centipedes, millipedes, and insects. It agrees especially with the insects in breathing by means of tracheal tubes lined with a thin chitinous layer by which air is conveyed to all parts of the body. Since the tracheae are functionally correlated with the tubular heart, and since they have small pores (stigmata) opening on the surface, it has been suspected that the tracheal system arose as a series of inpocketings not unlike the pulmonary sacs of the scorpion, and that the branching of the tracheal system came later. Its cylindrical body and numerous paired appendages suggest those of the millipedes.

In living millipedes locomotion is effected by successive waves of movement passing along the line of appendates, as if successive scales were being played down the keyboard of a piano. Presumably this is the method of locomotion in *Peripatus* but perhaps the movement of the waves is much faster in the millipede. Doubtless in both cases there are transverse nerve threads connecting the right and left nerve cords, so that the locomotor waves on opposite sides may be regulated or correlated efficiently.

The insects, according to good authorities, probably arose quite independently of the scorpions and spiders by way of relatives of *Peripatus* and of the centipedes and millipedes, rather than by way of the trilobites and crustaceans, king crabs, scorpions and spiders. In contrast with *Peripatus* and the millipedes, the insects have evolved a new and more efficient locomotor system by reducing the number of locomotor segments to three and combining these three in a new entity, the thorax. All the abdominal appendages, which are well developed both in *Peripatus* and the crustaceans, are sacrificed in the insects. This gives a good example of negative anisomerism, the reduction or loss of certain members of a series of polyisomeres, which often accompanies the positive anisomerism of other members, in this case, those of the three thoracic segments. The thorax of insects, located as it is between the large head and the still heavier abdomen, becomes the complex framework to which are delivered the thrusts from the three pairs of legs below and from the wings above.

The legs of typical insects, as compared with those of *Peripatus* are excessively long and slender. Moreover, they are subdivided into five main segments, of which the flexible distal unit, called the foot, includes five very short segments ending in a pair of claws. The various segments of each leg are movably connected with each other by joints in the chitinous cuticle which permit movement in some directions but limit it in others. From the proximal segments of the legs, muscles pass inward and are fastened to the stiff plates of the thorax. Thus each anterior leg can be jerked forward like a fishing-rod-and-line and then pulled backward, thus pulling the body forward. The third legs have the appearance of pushing but it may be suspected that what they really do is to lift the thorax or to snap it upward by extending the femur upon the tibia, the latter being momentarily fixed by the foot on the substratum. In walking, the opposite sets of three legs each move in such a way that at any moment two on one side and one on the other form a supporting tripod. Thus the legs of typical insects are compound levers and springs which are often widely spread apart at the feet, the body being suspended between the main knee-like joints. Since the entire mass of the legs is usually small in proportion to the combined mass of the head, thorax and abdomen, the controlled pendulum effect, which is conspicuous in the swing of the human leg, must be very small in a typical insect.

The well-braced thorax and effective unique method of walking have together made possible a wide diversity of adaptive types, from the stout-limbed, digging mole-cricket to the long-and-slender-legged mosquitoes, wasps, ichneumons and water-striders. Enlargement of the rear pair of legs enables the grasshoppers and similar forms to make relatively huge leaps in proportion to their size.

On the top of the thorax in typical insects are two pairs of wings. The wing is moved by two sets of muscles of three each, one set above, the other below, the wings. The six muscle strands converge toward the base of the wings and are severally inserted around the oval section of the base, just beyond its socket. C. H. Curran (1948) has given a splendid description of "How flies fly," in which the methods and mechanism of flying used by various insects and humming birds are clearly contrasted.

A third pair of wings, inserted on the first segment of the thorax (cf. fig. 3.5B), is recorded in one genus of the Palaeodictyoptera of Palaeozoic times, the other two pairs being borne by the second and third segments as in later insects. It was formerly supposed that the tiny "halteres" or balancers, which are located behind and below the main wings in flies and mosquitoes, represent vestigial remnants of the second pair of wings of modern insects. But Curran (*ibid.,* p. 63) identifies the second pair of wings with the "squamae" or folds which are found immediately behind the wings in flies. He regards the "halteres" as organs which developed de novo in the common ancestors of the flies, mosquitoes and other Diptera. These minute organs consist of a pair of tiny stems, capped by knobs, and inserted on the lower half of the thorax. During flight, as shown in high speed pictures, the "halteres" vibrate up and down with extreme rapidity and though they do not spin like a rotary gyroscope, they have a gyroscopic or balancing effect, both in keeping the fast moving fly on a straight course and in helping it bank on turns. The extremely fast moving pictures taken for the Sperry Gyroscope Company have also shown that the wings in flight do not describe a figure of eight but that the plane of the wing is changed during its up or down movement, thus greatly increasing its propulsive efficiency (Curran).

Insects are the only arthropods that can fly, and the wings of insects are unique in form and structure. Nothing suggestive of wings is known in centipedes, scorpions or eurypterids. To some extent wings seem to be analogous with the enlarged pairs of swimming appendages that occur in the Cambrian entomostracan *Waptia* and in the recent *Daphnia;* but these organs are attached ventrally and are jointed muscular legs, while insect wings are exclusively tegumentary and are unjointed except at the base. Their muscles may be serially homologous in part with some of the thoracic muscles connected with breathing, in part with the muscles of the legs.

If insects had been derived ultimately from some sort of terrestrially adapted trilobites, as was formerly supposed, the three pairs of wings might well represent either the lateral plates of the trilobite dorsal shield or the dorsal respiratory division of the paired appendages. But if the trilobites died without issue, as now held by leading authorities, there are no known separate plates in *Peripatus* and the millipedes which have been surely equated with the wings. In either case, insect wings afford an excellent example of emergent or "creative" evolution, since whenever they first appeared or were sufficiently organized to be called wings, they were then "new" products of the outer skeleton. In the young cockroach the wings, which are at first absent, subsequently grow out from the terga (dorsal tegumentary plates) of the second and third segments of the thorax (Parker and Haswell, vol. I, p. 583). This fact is of course not conclusive, because of the uncertainty in many cases whether the relative time of development does or does not follow the order of evolution; but the fact suggests either that the wings have sprung from the dorsal tegumentary plates of those segments of the thorax upon which they are based, or that the wings have been, so to speak, blown out from the chitinous skin, perhaps as aerating appendages connected with the well developed breathing tubes on the roof of the thorax. This possibly new idea may be worth further consideration.

The highly effective locomotor apparatus of primitive insects has been readapted by the various orders and families of insects in very numerous ways. In the mole-cricket, for example, the legs have become very short and strong for digging. In the water-striders they have become excessively long and slender. In one genus of water-beetles (*Halobates*) the legs have been modified for swimming, but these are the only known oceanic insects out of all the half million or more species of an otherwise inland or continental class. The anterior wings are gossamer-like in the May-flies, thick and leathery in the beetles, covered with minute feathery scales in the butterflies and moths, or lost entirely in parasitic bugs. The second pair of wings, which are large in butterflies and moths, may vibrate very rapidly. Flight may be amazingly fast in bulldog-flies or very slow and clumsy in the larger beetles.

Fortunately for the vertebrates, the body, even in the largest known insect, is hardly as large as that of a rat; the vast majority are small and there are some whose length must be measured in tenths of a millimeter. Thus there seem to be some limiting factors, perhaps in the strength of the chitinous exoskeletons, in the locomotor and aerating systems, which have prevented the insects from evolving successful competitors with the many gigantic vertebrates, such as dinosaurs, moas, mammoths and whales; these could all support their huge bulk on an endoskeleton made of bone.

INSECTS AND MAN

People who live in modern city apartments and work in tall office buildings may usually ignore the existence of the whole world of insects; on rare occasions they may slap at a stray mosquito or buy a can of something to prevent moths in a clothes closet. Such people would not dwell long on the newspaper report that government estimates for the past year state that insects inflicted damage to crops and other property to the extent of so-and-so many hundreds of millions of dollars.

Not thus can it be with the modern farmer and fruit grower or with those who raise the anmals that yield beef, milk, wool; but many, even of these out-of-doors people, value chiefly concise information on what to do about this or that nuisance; they feel that the most desirable quality in such obnoxious forms of life, next to a ready yielding to inexpensive treatment, would be complete and permanent disappearance from the neighborhood.

The health officers and their staffs, who must try to protect an unappreciative public against insect-borne yellow fever, bubonic plague, African sleeping-sickness and other terrible diseases, display a more scientific interest in insects; they know that effective ways to keep down the insects that carry the diseases can be devised only in the light of knowledge of the foe's breeding habits, food and enemies.

More directly concerned with insects *per se,* in all their complex biotic relations, are the government bureaus of entomology, the numerous field stations for experimental studies on insects and many museum departments of entomology with collections running into millions of specimens, vast reference libraries, staffs of specialists, laboratories and exhibits.

From all the sources named above we begin to appreciate the major importance of the insect world to man. On the negative side we may well believe that if throughout the United States all human checks against insects were allowed to lapse over a long term of years, the damage would be cumulative and incalculable. Crops would be limited to such an extent that if under these conditions trade with the rest of the world were cut off, famine would greatly reduce the population, yellow fever and bubonic plague would again scourge the cities, flies would spread disease from house to house, and the people would be weakened in many ways against the attacks of bacteria, protozoa and other hordes of death. When we consider that insects have crowded the lands for at least one hundred times longer than has the present human species, that they outnumber mankind by thousands to one, and that there is only one species of man against perhaps a million species of insects, it is not surprising that some pessimists have predicted the final fall of man and the dawn of a new Age of Insects.

Nevertheless there is still something to be said in favor of the insects (cf. Lutz, 1941, pp. 93–97). If they did not carry the pollen from flower to flower, our gardens would soon be empty, our forests vanish and there would be no crops to feed man or beast. Thus the warfare against insects should not be indiscriminate, lest we kill friends as well as foes. Many an unexpected misfortune has come from man's upsetting the balance of nature and letting loose some new pest, which under more natural conditions would ordinarily soon be checked by its own insect foes. All this and much more constitutes the need for modern entomology, nor are we likely to forget the contributions which have been made to the science of heredity through breeding experiments on the humble fruitfly.

THE SCIENCE OF INSECTS

The foundation for this modern superstructure is the study of systematic entomology, the correct identification and classification of the individual insect among the estimated half-million to million species already listed. And to those who are interested in the scientific problems involved either in the breeding or in the killing of particular species, it is of great practical importance, if only in the interest of clearness, to have none other than the correct technical name label attached to their sample specimens.

Systematic entomology has, however, a broader aspect. After all the binomial or trinomial labels are checked and rechecked, what then? Then it is obviously desirable to make general surveys of larger units and so in the course of time the modern classifications of families and orders of insects have gradually been built up. Doubtless the insects of the world constitute such an appallingly vast field that entomologists inevitably specialize in one group or another, but it is fortunate that there are still a few entomologists whose vision is broad enough to present a general view of the main divisions of the insect Tree of Life.

It is with the ancient roots of this tree that this chapter is particularly concerned but in order to avoid the defect of vagueness and to have some basis for theoretical considerations, let us look first at a few representative insects belonging in widely different orders, and showing the coädaptation of form and habit.

INSECT SPECIALISTS

On a bright day in August I am sitting under an apple tree with a table and my manuscript before me. Immediately in front of me I notice a bright green, worm-like caterpillar climbing up one of the scaly main stems of the tree. I lift him off and place him on the table where I can examine him with a hand glass and then put him back on the tree. He progresses by the inchworm method and seems to be a larva of one of the geometrid moths. Grasping the support with his front feet, he arches up his body so as to bring the rear props up close to the front legs; taking a fast hold with this stout hind pair of props, he suddenly rears up, raising the fore part of the body, straightens out the curve and again clasps the support with his front feet. This must be an efficient mode of locomotion because he not only hangs on easily to the under side of a branch but raises himself to a point about six feet above his starting-point in less than an equal number of minutes. Evidently the advantages of the peculiar locomotor method which he has adopted spring in part from the fact that the distance along the arc of a curve is much longer than the chord, especially when the ends are brought together. Each time he inches upward he raises himself by an amount which may be roughly equal to one-half the total body length minus the combined lengths of the fore and hind props. By squeezing himself with his transverse muscles as he reaches forward, he probably increases his length and when the hind supports let go, the tension in front brings the rear end quickly forward.

This caterpillar is endowed with a certain power of discrimination of a simple *yes* or *no* type. When he comes to the dead end of one of the scale-like projections on the bark, he reaches forward while holding firmly with his hind claspers but, finding nothing to grasp in front, he twists his flexible body sidewise so that the front feet are placed almost parallel to the rear props rather than in front of them. Thus grasping with his forefeet, he lets go with the hind and climbs down the projection. When he reaches the main track his evident preference for going upward reasserts itself and he turns around and climbs up and up. Soon he meets a rough and aged pine crosspiece which I have placed in his way. Climbing out on this, he hesitates, moves uneasily and evidently does not like either the feel or the smell of it. He climbs down, gets on the main trunk, again climbs upward and soon reaches a small green branch which leads to a fresh leaf; after going around the under surface of this leaf, which exactly matches his own color, he settles down and is so quiet that half an hour later when I search for him he is hard to find. Brought down again to the level of the table and placed on the bark, he climbs quickly upward but, missing the twig on which he formerly rested, he goes on and finds one a couple of feet higher up, where he stops. I think he has earned a rest and let him stay there.

These movements, together with the circular and oblique movements which he must perform later in weaving his cocoon, are effected by a system of longitudinal, transverse and oblique muscles of the body wall, which are stated to number more than 50,000. Each of these is doubtless controlled by a tiny nerve thread and these nerve threads build up into larger strands along the ventral nerve chain, which eventually leads to the brain. No wonder that when climbing the caterpillar can discriminate small differences in the quality of the surface and in the stresses and strains set up in his own body. And, when he reached the green leaf, its surface, smell and color had an apparently soothing influence on his nerves which soon calmed him down, stopped his movements and prevented him from moving away from the spot for which, as a sensitive, responsive automaton, he was best fitted.

This incident perhaps illustrates the high degree of competence of these insect specialists, each in his own peculiar way. It is the more wonderful when we reflect that after the caterpillar has absorbed enough solar energy from the apple tree leaves he will let himself down from the tree by a silken thread, burrow in the ground and start spinning a tough cocoon; here he will remain through the winter, during which time he will undergo a complete transformation, emerging the next spring either as a wingless female moth or a winged male moth (Lutz, 1935, pp. 192–197). The wings of the latter bear an elaborate pattern including prominent shaded crossbars, the pigment extending across hundreds of the minute scale- or feather-like hairs to which the name of the entire order (Lepidoptera) refers. Most of the heavy eating is done by the caterpillars, which have strong transversely working grasping jaws. In the adult Lepidoptera the mouth parts are sometimes fused into a long tube, often for sucking the nectar of flowers.

When we ask, whence came all the lepidopteran hosts, we learn (Lutz, 1918, p. 57) that they may have been derived from relatives of the caddice flies (order Trichoptera), which in the adult stage have four pairs of wings with a primitive type of venation (the "veins," needless to say, are scaly strengthening ridges upon which the delicate web of the wing is stretched). The caddice flies in turn seem to lead back toward the salmon- or stone-flies (Plecoptera, in part), which are among the more primitive of existing flying insects (*op. cit.*, pp. 50, 51).

THE COLOR PATTERNS OF MOTHS AND BUTTERFLIES

The immense diversity of coloration and color patterns in the wings of butterflies and moths yielded to Eimer (1889–1895, 1897) the materials for his monographs on orthogenesis. Süffert (1929) gave a relatively brief review of some of the probable stages by which the more specialized patterns have been derived from earlier stages. Among the more primitive patterns are those which are conditioned directly and simply by other structures, such as streaks along the "veins," limited patches between veins, and those which are conditioned by the cross-veins of the pupal wing. Autonomic wing patterns, on the other hand, are not simply and directly limited by veins or other structures. The entire wing, for example, may be crossed by a great number of similar undulating bands. Then these bands, becoming irregular, may form closed lakes from the union of adjacent waves. Different stages of symmetry may be developed: in primary symmetry the pattern as a whole is arranged with reference to the transverse and longitudinal axes of the wing; secondary, tertiary and double symmetries may then be developed. Zigzags resulting at first in "lakes" may eventually give rise to hollow bars in rows, symmetrical rings, like contour lines, and finally to discoidal ocelli or giant ocelli, or to rows of "eyes." Such variations are to a certain extent analagous to the zigzags, bars, spots, etc., in the patterns of some snail shells and show still closer analogies with color patterns of fishes and birds, which likewise run across individual scales or feathers and form zigzags, bars, lakes, ocelli, etc.

A far more extensive literature has grown up around the subject of "mimicry" in butterflies (for a brief historical review, see Goldschmitt, 1943). Here the problem has been not by what steps a given color pattern was reached, either by the original or the mimic, but how natural selection or other evolutionary forces have effected selective breeding in the mimic to make its pattern look like that of the original.

SOCIAL INSECTS

But let us return to our field studies. Another day I am sitting at work in a summerhouse, the four columns of which are made of rough sapling trunks of white birch. Immediately in front of my eyes are several circular holes, into which at intervals some kind of wood-boring, wasp-like insects come and go. I stop to watch one of them. When he comes out of the hole he circles around just outside the summerhouse and then darts off and is gone for some time. After a while he returns, flies around a yard away from the birch bark and then suddenly alights not far from the hole. He wanders around as if searching until he comes near to the hole, then suddenly enters it. All this he does again and again on the same afternoon, so there is no doubt that this particular insect can leave his hole, go on fairly long excursions, find his way back to the summerhouse, locate his birch stem and then find his own hole, at least by the method of elimination.

On other days I watch the nest-making activities of a wasp (*Polistes* sp.) that is all too common for the comfort of certain members of the family. I have never been stung by them but perhaps that is because I take care to stand or sit perfectly still or move very gently even if they happen to alight on me. This species builds a sort of communal nest, consisting of about thirty tubes made of chewed-up wood pulp. The tubes are clustered together, so that the inner ones are pressed into irregular hexagons. The tubes open downward and all converge above into one short stem which is glued to the under side of a beam. The adult wasps, rarely more than a few at a time, keep coming and going at irregular intervals. According to Lutz (1918, p. 429), each colony of these wasps *Vespes* and *Polistes*

> ". . . is a single family, in which unmarriageable daughters help to build the house, keep it clean, and feed the younger children. The food consists of chewed-up animal matter, such as caterpillars, but some species use honey and pollen also. The larvae are fed from day to day, or oftener, no food being stored for them."

Here we have a suggestion of the beginning of social habits, which in the bees culminates in the development of large-scale honey factories. In the ants, which may be thought of as specialized ground-living wasps, social organization finally attains a degree of regimentation which far surpasses that of any human totalitarian state.

But an ant does not have to be painfully molded by years of propaganda and intensive training to do any of the things which are the predestined duties of his, her or its particular caste. Among the soldier caste, for example, the tread of an invading human foot on the ground will bring out the warriors, who circle about, snapping their formidable jaws, raising their heads and trying to locate the invader. Meanwhile thousands of the nurses may be streaming through a long narrow trench on their way to a new ant city which is being prepared by other thousands for the reception of the heavy white pupae; these the nurses carry in their jaws but tucked between their straddling six legs, making excellent progress all the while. As for the queens, it is well known that by special feeding the queen grows into a huge egg-factory and does all the reproduction for the whole city.

INSECT ORIGINS

Enough has been said perhaps to lend further interest to the question, whence came all this vast insect world? Although the fossil record for insects does not compare in completeness with that of animals which have a more massive and enduring skeleton, yet quite enough has been found to indicate that insects were conspicuously absent as such during all the long ages of the Cambrian, Ordovician and Silurian; in the Devonian there were as yet no fossil insects found among the remains of primitive trees and other land plants. It has been suggested by the late Dr. F. E. Lutz that insects must have existed in Silurian times because scorpions did exist then and scorpions feed on insects. But it seems more in accord with the negative palaeontologic evidence, which is fairly impressive, to infer that the scorpions of the Carboniferous age shifted their attacks from eurypterids or other arthropods to insects as the former became scare and the latter appeared and multiplied.

The long reaches of the Lower Carboniferous or Mississippian are mostly known from marine limestones, but in the true Carboniferous or Pennsylvanian we find insects which, by comparison with later forms, were exceedingly primitive. The Palaeodictyoptera were in fact not unlike the very primitive existing insects called salmon-flies (Plecoptera), except that in front of their two pairs of wings they had a pair of projecting discs, like small wing-covers. The patterns of the "veins" or strengthening ridges on the wings were very primitive as compared with those of other insects (Handlirsch, 1908) but it is remarkable that many of the latter retain much of the primitive basic pattern, although it is modified by the loss of certain veins and the fusion of others.

The earliest known insects resembled the Silurian scorpions and the eurypterids in that they had eliminated free appendages from the abdominal segments, but they differed conspicuously in having three pairs of thoracic appendages instead of four.

The name *insect* is based on the usual division into sharply divided head, thorax and abdomen; but, especially in the larvae of the Plecoptera and Megaloptera, the thorax is subdivided into three segments named pro-, meso- and metathorax, each bearing one pair of the walking appendages. In a typical adult insect when standing, the more or less centrally placed thoracic segments bear the weight of both the head and the abdomen. These segments also take the thrusts of the wings in flight and when there are long piercing ovipositors at the end of the abdominal segment as in certain ichneumonoid wasps, or when the abdomen bears either a pair of "saws" or a wood-boring blade as in the sawflies and horntails.

Grabau has suggested that the insects arose directly from the trilobites (fig. 3.5) and certainly they exhibit a derived but far more advanced basal plan of the head, thorax, paired appendages and posterior thoracic (or abdominal) segments. Their paired eyes are of the compound type but more highly organized than those of trilobites; their jaw parts, in the more primitive insects, retain clear traces of derivation from the paired appendages of the head of trilobites. The relations of the primitively straight gut, of the main parts of the nervous system, and of the primitive heart are all likewise of the fundamental arthropod type. After extensive and judicial studies of the evidence from embryology, comparative anatomy and taxonomy, Snodgrass (1938) places the "Prototrilobita" as the oldest main branch of the "Protoarthropoda." This branch leads to the trilobites, limuloids, eurypterids and arachnids, and related groups. The second main branch gives rise to the crustaceans, the third, by way of the "Protomyriapoda" to the Hexapoda, or insects.

The highly segmented Chilopoda (centipedes and millipedes) are the land-living analogues of the marine annelid worms. According to Snodgrass, their relationship is closer to the stem of the insects than to the arachnids. The poisonous centipedes live among rotting wood and leaves; their body is flattened and consists of many segments. In the millipedes the body is cylindrical, the segments annular. The body is propelled by undulation of the line of paired feet.

Lief Stormer in a highly important monograph, "On the Relationships and Phylogeny of Fossil and Recent Arachnomorpha," likewise treats the insects as derivatives of the stem of the Chilopoda or centipede-myriapod group, that is, as a main branch of the undiscovered stock which also gave rise to the trilobites and crustaceans. These Pre-Cambrian forerunners of the entire arthropod super-phylum are regarded as very probably derived from remote ancestors of the existing annelid worms. Thus the worm-like segmentation which persists in centipede and many insect larvae may be in part the heritage received from remote annelid-like ancestors.

The insects share also with the trilobites, crustaceans and primitive arachnids the habit of molting. An emphasis of the molting habit in the insects has given the opportunity for the amazing metamorphoses from larva to pupa and from pupa to adult in such advanced orders as the Lepidoptera and the Hymenoptera. But in the most primitive Plecoptera and allied groups the adult differs from the larva chiefly in the possession of wings.

There remains the question, By what steps did the wings attain their status as such? The insects at first may have been an inland-water or swamp-living offshoot of some late and advanced arthropod stock. The future wings were evolved out of certain of the dorso-lateral plates or pleura. Already vascular, they served at first for aeration, in order to take the place of the former branchial branches of the paired appendages. Meanwhile respiratory tufts were developed along the abdominal segments, as in the nymph of *Ephemera varia* (cf. Lutz, 1935, Pl. VIII).

As the animals took to climbing up the primitive trees, whose roots came down into the marsh, these abdominal branchial tufts (cf. *Corydalis cornuta,* Lutz, Pl. XIII) began to proliferate inside the body, and by further extension (somewhat after the manner of the proliferating air-sacs of certain birds) presently gave rise to tracheae. This left the future wings with reduced respiratory function. Meanwhile these future wings had acquired their own set of muscles from the dorsal body wall; perhaps they shivered or were rattled. As the respiratory tubes withdrew from the wings, the venation was retained for strengthening purposes. The art of flying was initiated by dropping off from the trees and falling into the water.

Against this hypothesis it might be said that the existing "silver-fish" insects (Thysanura), which even today crawl up out of bathtub pipes, have no wings; but this could be countered by the suggestion that the Thysanura rather closely approach the ephemerids except for the lack of wings and that they may be neotenic larvae.

VARIED BY-PRODUCTS OF THE EXOSKELETON IN ARTHROPODS

Up to the present page our story has dealt mainly with the origin, rise and branching evolution of the outer skeleton among the invertebrate animals. This outer skeleton, like the bark and outer woody layer of trees, marks the limiting membrane of the organism and thus helps to contain, restrain, and protect the growing forces within and to regulate the flow of heat between the organism and its environment. In the invertebrate animals, on account of its tough resistent nature, the exoskeleton acts as a frame upon which the weight is supported, while the appendages act as levers to propel the body.

But what happens when the growth forces within increase the bulk of the organism until they tend to disrupt the cuticular envelope? Under such growth pressure the outer bark of the tree is cracked and shed in small pieces, while new layers are forming beneath it. The sea-snail secretes its envelope only on part of its surface and, even while building a new rim on the shell, the animal grows away from the shell without losing its own hold upon it. At intervals the chitinous envelope of the externally jointed lobster cracks open along the back and the animal works itself free from its old shell. The same method is used by the grasshopper, cicada and many other insects. But during this period of skin-shedding the animal is encumbered with its old skin, while its newly forming shell is too thin to afford much protection and too weak for swift flight. In many insects these dangers are reduced by the creature's ability to secrete viscous threads, with which it covers itself by turning around and around. Thus it spins a protective cocoon in which it can lie torpid during the winter and later be transformed by its own growth force from some sort of larval stage into an adult insect.

An even more amazing use of a silk-like product of certain glands of the exoskeleton is seen in the spiders, which apparently at first wove a lair for themselves and for their young ones; then they continued the web in front of the lair so as to form a cobweb to entangle passing insects. Later the web was expanded and refined into the wonderful spiral network of the orbicular spiders. Perhaps most remarkable of all are the lassoo-throwing habits of the Bolas spider (Gertsch, 1947).

Among many other surprising by-products of the exoskeleton of arthropods may be mentioned: the pollen-carrying bags on the legs of honey bees; the sucking-tube, lancets, pump and poison glands of the mosquitoes; the anal stings, probes and egg-tubes of ichneumon flies; the swiftly vibrating halteres of flies; the spirally wound proboscis of moths; the feather-like wing-scales and the "eyes" and other brilliant features in the color patterns of moths and butterflies. All these and thousands more known to entomologists are fine examples of "emergent" or "creative" "designs" which have been evolved through the ages within their own taxonomic groups by the cumulative interaction of hereditary variation, mutation, inbreeding, isolation and natural selection.

PART TWO

(*Chapters IV–IX*)

EMERGENCE OF THE VERTEBRATES · BRANCHING EVOLUTION OF GILL-BREATHING FISHES

CHAPTER IV

COELOMATE FOOD-SIFTERS OF DIVERSE TYPES

(*BRACHIOPODS, POLYZOA, ECHINODERMS*)

Contents

CHAPTER IV

COELOMATE FOOD-SIFTERS OF DIVERSE TYPES

BRACHIOPODS, THE SHELLS WITH INNER FOOD-COILS

THE EVIDENCE cited by W. K. Brooks (1890, 1899) and extended by Percy E. Raymond (1935, 1939) suggested that the Pre-Cambrian animals were for the most part motile, naked forms. Although this idea has been widely accepted, it has been suggested above (p. 25) that the different osmotic pressures on opposite sides of enclosed sacs of membranes may have encouraged the differentiation of materials of different density and tensile strength on opposite sides of the membranes. Thus the excretions of calcium carbonate and phosphate of lime, originating as a result of primary physiologic response, may have acquired a secondary value as materials for a firm base or for support and protection. As soon as the organisms took in more calcium than they could eliminate, they began to form a skeleton, mostly of calcium carbonate. With increasing weight, many sank to the bottom. Some, becoming stationary, specialized in drawing the food-bearing current into their mouths by the lashing of rows of cilia. If their skeletal secreting activity continued, they eventually found themselves covered with a shell and attached in some way, either to a stony base (as in the corals) or to a stalk (as in the lamp-shells and sea-lilies).

Among the oldest known of the stalked animals are the brachiopods, or lamp-shells. These may have acquired their general or class characters not very far back in Pre-Cambrian times; but by the beginning of the Cambrian they were represented by many families. Survivors of six superfamilies of brachiopods, including more than one hundred species, persist to the present time; of these superfamilies one dates back to the beginning of the Devonian, two to the base of the Ordovician and two to the upper Cambrian. Thus the brachiopods furnish extreme examples of persistence with but few basic changes through long geological ages.

An adult brachiopod is enclosed in a pair of shells (fig. 4.1A, C, D) which have a superficial, convergent resemblance to those of clams cockles (B) and other bivalve molluscs and are likewise secreted by a mantle-fold; but brachiopod shells differ profoundly from those of bivalve molluscs in that they are placed respectively over the "back" and "belly" of the animal instead of on the right and left sides. The brachial or dorsal valve (C) of a brachiopod shell is usually smaller than the peduncular or ventral valve. The opening or mouth of the shell is in the transverse plane (F). Originally this opening seems to have been directed upward (as in *Rustella*), the animal being fastened to the bottom by a stalk coming from its lower end. In most brachiopods, however, the shell is bent upon the stalk so that the "mouth" of the shell is turned forward and the peduncular or ventral valve is uppermost (C).

Brachiopods may be regarded as creatures without any real head but with enormously developed "lips" (fig. 4.1F), produced above and on either side in front of the small mouth-opening and bearing rows of ciliated tentacles. These "lips," called "arms," are curled up in several ways, often in a spiral (helicoid) on either side of the mouth (E, F). The loops or spirals bear many tentacles with ciliated grooves and they draw the food-bearing water toward the mouth. They were usually supported by skeletal loops which are often preserved in the fossils (G).

The cavity of the shell (E) is divided by a transverse mantle-septum into an anterior chamber containing the loops or lophophores, and a posterior chamber containing the digestive tract, the other viscera and the five or six pairs of muscles that move the valves of the shell. The true or internal mouth is a small slit located in the transverse septum behind the lophophores (E, F). The muscles are arranged in opposing pairs and are able not only to open and close the shell but to move the upper and lower shells in several directions and to turn the shell as a whole on the stalk. In the still existing *Lingula* (D) the stalk or pedicel is modified into a burrowing organ, which is doubtless wriggled by some of the muscles inside the shell. In *Crania* (A), on the other hand, the stalk is absent and the shell becomes cemented to some foreign object.

The brachiopods are among the oldest of the sedentary animals that have built for themselves protective shells—usually of carbonate of lime. They parallel many other such fixed, defensively armored forms in being limited to microscopic food drawn in by the action of ciliated grooves or tentacles, which in this case are borne by greatly elongated, coiled brachia or "arms."

The larvae of brachiopods (fig. 4.2) are free-swimming organisms provided with cilia and presenting a remote general resemblance to the free-swimming larvae called trochospheres, which are found among annelid worms, rotifers, molluscs, echinoderms and other groups. In certain important features of their embryonic development the brachiopods present significant points of resemblance with the echinoderms and the chordates, but the relationship, if any, can only be remote and must date back to Pre-Cambrian times.

Within the class of brachiopods the course of evolution from Lower Cambrian to Recent times was marked by many changes, which have been studied in thousands of fossils from many successive horizons, for example, by Fenton (1931). The latter confirms the conclusion of Grabau that during ontogeny or individual development of several members of the *Spirifer* group the shells increase in relative width up to a certain point, after which they increase more rapidly in length, and that similar changes take place in the adult forms as we follow this and allied races from earlier to later horizons (fig. 4.3). In individual development the ventral or pedicel valve protrudes much beyond the dorsal one, forming the umbo or hump. The same thing happens in phylogeny. In a front view of the *Spirifer* shell there is a prominent fold in the middle. This fold becomes accented both in ontogeny and in phylogeny. The detailed history of the surface folds and ridges of the shell also shows a close parallelism between ontogeny and phylogeny. Fenton therefore concludes that, at least for such relatively short periods of time, the law of recapitulation does hold good here, as in many other cases among invertebrates; but he notes that in vertebrates the correspondence between the embryonic development of the individual and the evolutionary history of the race is masked and distorted by numerous secondary adaptations in the embryonic stages.

Assuming then that we may use the principle of recapitulation as a tentative guide to the prehistory of the brachiopods before they developed shells, let us consider briefly the anatomy of the larval stage of the existing brachiopod *Cistella,* as summarized and figured by Parker and Haswell (vol. I, pp. 341–343). At a certain stage (fig. 4.2A) the embryo includes chiefly a mesenteron, or sac-like primitive gut, surrounded by coelomic or mesoderm pouches derived from the gut. The surface of the embryo is divided by transverse grooves into three parts: a head region, a body region and a peduncular region.

In the larval stages the head region is surmounted by four eye-spots and is reminiscent of the prostomial plate of annelids (fig. 1.5B). Its margin is produced in front into an umbrella-like expansion overhanging the mouth and bearing cilia on its periphery (fig. 4.2A1). Meantime the mantle fold grows out from the sides of the body and then grows backward or downward like a skirt, which ends below in four tufts of long setae, or stiff hairs (A1). In this condition the larva swims freely like the trochophore larvae of annelids, molluscs and other groups. After a time it comes to rest and fixes itself by its peduncular segment (A2). The dorsal and ventral lobes of the mantle-fold then become reflexed, like an umbrella that has turned inside out. In this way they cover the primitive head and true lip. The dorsal lobe then secretes the dorsal valve on its "outer" surface (which was formerly on the inner side) and the ventral lobe similarly gives rise to the ventral or pedicel valve. Meantime the lophophore has been growing out from the inner surface of the dorsal mantle lobe at first above the true lip but later extending in median dorsal and lateral loops entirely around the mouth. The mantle fold itself contains a prolongation of the primitive coelomic pouches (A), as do also its derivatives including the lophophore. The muscles that move the shell are derived from parts of the primitive coelomic pouches.

From the data of embryology, if the doctrine of recapitulation applies in this case as it seems to do, we may conclude that the remote Pre-Cambrian ancestors of the brachiopods were free-swimming, sac-like organisms, possessing a dorso-ventral asymmetry and a bilateral symmetry, having at least a rudimentary head provided with eye-spots and swimming by means of folds from the body wall ending in tufts of setae. The most characteristic parts of adult brachiopods are the dorsal and ventral valves, the lophophore and the peduncle; presumably these have been developed during the transformation from a motile to a fixed mode of life.

The vast numbers, the great diversity and immensely long history of the brachiopods, especially in Palaeozoic times, indicate a correspondingly wide economic opportunity for such an attached, stalked or sessile food-trap, containing an exquisitely regulated collecting and filtering system for minute floating food particles. Perhaps their near competitors were the attached pelecypods and stalked sea-lilies. In the submarine gardens these innumerable "animal flowering plants" tended to absorb the organic food energy that was diffused in the water and they, in turn, built up incredible stores of food for such roving and predatory organisms as the trilobites, the annelid worms, the crustaceans, the eurypterids, the sea-snails, the starfishes and eventually the vertebrates, while the disintegration of the brachiopod skeleton must have afforded raw material for the skeletons of many other types of animals both fixed and motile.

POLYZOA, COLONIAL MOSS-ANIMALS

There are no colonial forms known among brachiopods but it seems not improbable that the Polyzoa or Bryozoa (moss animals), which are placed in the same phylum, Molluscoidea, with the Brachiopoda by most authorities, represent a derivative of the brachiopod stock which has become excessively specialized for colonial life. The Bryozoa are not known until the Ordovician, by which time they had attained their definitive status. This circumstance combined with others suggests that the Bryozoa are not the little-modified survivors of some unknown Pre-Cambrian group but the greatly disguised derivatives of some older known group such as the brachiopods. The fossil Polyzoa from the Ordovician and later ages show only the skeletal bases, not the animals themselves.

In *Bugula* (fig. 4.4A), which is chosen by Parker and Haswell (vol. I, pp. 314–319) as a relatively central example of the group, each zooid or unit of the colony may be described as a minute cylindrical sac containing a U-tube, the digestive tract. The mouth of the tube is surrounded by a crown of many tentacles. The tentacles and the U-tube can be withdrawn into the sac by strap-like muscles fastened to the inner wall and base of the sac. From the side of the sac projects a most strange apparatus called an avicularium (fig. 4.4A *avic*) from its resemblance to a bird's head and beak. This structure, although minute in *Bugula,* is provided with relatively powerful muscles by which its "jaws" are snapped vigorously. It is suggested by Parker and Haswell (*op. cit.,* p. 324) that the avicularia represent modified zooids.

According to Harmer (Article on Polyzoa in Encyclopaedia Britannica, 14th ed., vol. 18, p.

196), the avicularia in some Polyzoa are in series with the zooids, which they may surpass in size; or they may appear as appendages of ordinary zooids. They are either "sessile," closely attached to the zooid, or stalked, and they show a wide range of form in different species.

In spite of certain embryological difficulties to be noted below, it may be worth while to suggest that the foregoing facts may be tentatively interpreted to mean that the ordinary zooid of *Bugula* may represent a greatly modified brachiopod, while an avicularium represents a zooid that has lost its lophophore, digestive and reproductive organs while retaining the shell and some of its muscles. The vibracula may perhaps be regarded as degenerate avicularia in which the movable part has become elongated into a lash. Moreover in those cases in which the avicularium bears a "vestigial polypide," the two together may represent an entire although excessively distorted brachiopod. In those Bryozoa in which the polypides can be completely withdrawn into a tubular sheath (fig. 4.4B) the latter may perhaps be considered in part as the equivalent of the body wall and stalk of the pedicel of the embryo of such a brachiopod as *Lingula* (fig. 4.2D). In this case perhaps the operculum of the polypide would likewise correspond with the dorsal shell of the brachiopod, while the ventral shell and body wall of the embryo brachiopod would be represented by the ventral body wall of the polypide.

The embryology of *Phoronis* (fig. 4.4D1-D3), a marine worm-like and tube-living animal with an elaborate double-spiral lophophore (Parker and Haswell, vol. I, p. 329), suggests that this form is practically a brachiopod which has eliminated the shell entirely but which retains some evidence of relationship with the *Plumatella* division of the Bryozoa. The nonchalant way in which brachiopods and *Phoronis* turn their mantle lobes inside out during individual development seems to be consistent with the intro- and extraversion of the zooid in the adult bryozoan.

But whether the Polyzoa be derived from the stem of the Brachiopoda, as supposed above, or whether they are related rather to the stem of the annelid worms (on account of the basic resemblance of the polyzoan trochophore larva to those of certain annelids), we may be confident that both their peculiar adult structure and their highly metamorphosed mode of development are related to, or have in turn made possible, their prevailingly colonial mode of life.

Recapitulation is obscured among them by the profound metamorphosis that supervenes after the larva becomes attached. One outstanding feature is the incomplete separation of the zooids from each other and the appearance of "castes" or differentiated zooids with a division of labor, including the ordinary zooids and the avicularia.

Within the group there were many lines of divergent evolution according to the different ways in which the single ancestral zooid branched and gave rise to subsequent individuals, but it is not necessary at present to enter into this detailed history through the long ages from the Ordovician to the present time.

The typical "moss-animals" (Bryozoa or Polyzoa) stand in contrast to the vertebrates, since in the adult stages they are very small fixed and colonial organisms with no locomotor apparatus and no head. On the other hand, the embryonic polyzoan approximates to that ciliated floating sac with a cup-shaped digestive sac which Haeckel called the "gastraea" stage, lying between the single-celled animals and all the higher groups, including the worms, arthropods, molluscs, echinoderms and vertebrates.

Whatever may be the final solution of the origin of the moss-animals, their very name suggests their economic importance in the incredibly extensive business of catching and storing up floating food particles. Upon these living pastures grazed the countless tribes of limpets and other gastropods with fine-toothed radulae, while they in turn were attacked by still other robbers of high and low degree. It was in such a world of give and take that the vertebrates eventually exploited the opportunity to become the master bandits.

ECHINODERMS, FIVE-ARMED SKELETONS IN THE STARFISHES, SEA-LILIES AND SEA-URCHINS

In the echinoderm phylum (including starfishes, sea-lilies, sea-cucumbers, sea-urchins, cake-urchins and their numerous fossil relatives) the whole organism attains a high degree of complexity. In spite of their diversity in adult body form (figs. 4.5–4.7), all are built upon a radiate ground-plan, with five arms or sides or other main subdivisions. Moreover the echinoderms have advanced much beyond the coelenterates in having a highly differentiated system of coelomic sacs, or outgrowths from the primitive gut or archenteron. These organizing sacs, lined with endoderm cells, give rise to many important parts of the echinoderm life-machine, especially to the elaborate hydraulic or water-system, including the tube-feet and the cavities of the arms.

The larvae of modern echinoderms (fig. 4.8A, B) are bilaterally symmetrical, the primitive gut being in the midplane and flanked by paired pre- and post-oral folded outgrowths of the body-wall bearing ciliated bands. Certain of these free-floating larvae rather closely resemble the free-floating "Tornaria" larvae (fig. 4.8C) of the so-called "acorn worm," which is believed to be related remotely to the lancelet (*Amphioxus*) and the vertebrates.

The known echinoderms were widely different from the vertebrates at least as early as Ordovician times (400± million years ago). Nevertheless some facts suggest that the two groups may be divergent offshoots of a primitive group of three-layered animals with a U-shaped gut in the midplane and probably with lateral coelomic sacs.

Certain of the least specialized echinoderms (fig. 4.5A, A1) of the Cambrian and Ordovician ages were covered with a test of thin polygonal plates of carbonate of lime. Many of the plates were perforated with small pores leading to little tubes in the plate and possibly serving to draw sea-water into the body. On the rounded end of the pear was a gash-like opening interpreted as a mouth, while a smaller near-by opening is regarded as the anus. The gut tube was probably U-shaped as it is in many floating or stalked invertebrates and larval echinoderms. In a related genus (fig. 4.5B, B1) there is a five-rayed branching system of supposed food-grooves radiating out from the mouth and running along the sides of the body. These grooves, since they converge toward the mouth, are suggestive of the ciliated grooves of the arms of a crinoid or sea-lily.

According to Swinnerton (1923), the five-rayed plan was probably preceded by one in which there was a median anterior groove, the hind part of which divided into two; these in turn both divided again, making five in all.

The pear-shaped or carpoid echinoderms of Cambrian and Ordovician ages are related to the cystoids, blastoids and crinoids or sea-lilies, the majority of which are attached to the bottom by a jointed stem (fig. 4.5C). All these have a very complex jointed exoskeleton divided into: (1) a branching, root-like system for holding fast to the substratum; (2) a jointed stem or column consisting of polygonal or ring-like pieces more or less analogous with the ring-like vertebral centra of vertebrates; (3) a calyx or body which is cup-like, its walls consisting of three concentric rows of rectangular or polygonal plates; the calyx is covered by a curved vault or tegmen composed of many small pieces; (4) the five arms or brachia, which often divide into many branches.

The ordinary starfish may be regarded as a sort of crinoid which has lost its stem and turned over so that the mouth faces downward rather than upward (fig. 4.7A2 *mth*). The calcareous plates have been reduced to ossicles, which are more or less buried in a leathery, flexible integument. On the under side of each arm (fig. 4.7A) is a long "ambulacral groove" running out from the central mouth cavity to the tip of the arm. On either side of each groove are the long rows of little suckers, or "tube-feet." By means of these the starfish can slowly pull itself along.

Each of these tube-feet is continuous above with a bulbous syringe-like ampulla (fig. 4.7A) provided with muscles for squeezing it. Radial tubes connect all the tube-feet on one arm, while looping vessels pass from arm to arm and are connected with five very large ampullae called Polian vesicles (A1). This water system ends above in a short canal and a small perforated top called the madreporite (A1 *madr*). Although the sucking strength of each individual tube-foot is small, yet collectively they exert force enough to pull open the valves of an oyster after a long tug of war against the oyster's thick adductor muscle.

The entire water system is controlled by radial and loop-like nerves which somehow help the coördinated action of the five arms of the innumerable tube-feet, of the muscles of the digestive tract and other parts. And yet there is no central brain of crowded ganglia and interconnected centers, merely rings of nerves and tracts of sensitive epithelium.

In the brittle stars (ophiuroids) each of the five arms is highly flexible and serpent-like and in some forms the arms divide and subdivide into a writhing mass. The arms are supported by a segmental series of exoskeletal ossicles; each of these originates from the fusion of four adjacent plates and it articulates with the ones next to it by means of dome-like elevations on each side (Twenhofel and Shrock, 1935, p. 196). These are called "vertebral ossicles" because they suggest the vertebral centra of fishes and thus afford a good example of the phenomenon of convergent evolution.

Still more complicated in construction are the sea-urchins (fig. 4.7B), sea biscuits and heart-urchins or spatangoids (C), collectively belonging in the class Echinoidea (Sea-urchins). In the Indian sea-urchin (*Salmacis*), as described by Gopala Aiyar (1938), the body is more or less globular in shape, flattened at one pole and conical at the other. The test represents the five arms of a starfish or sea-lily which have been curved downward to enclose a sphere, the mouth being at the lower end and the anus at the top. The test is composed of very many calcareous dermal plates fitting closely and rigidly against one another by means of sutures.

The surface of this strong spherical skeleton supports a vast forest of outwardly pointing, sharp-tipped, needle-like spines, each of which is actively movable upon a raised tubercle and a cup-and-ball joint. These spines are moved by two concentric muscular sheaths, each provided with a nerve ring. The forest of spines has an undergrowth of still more wonderful microscopic organs called pedicellariae, which are often provided with a ciliated receptor organ and relatively large pointed jaws operated by adductor and flexor muscles. Some of the pedicellariae also bear poison glands and are able to paralyze small organisms. The ophicephalus or bulldog pedicellariae grip larger prey such as small worms; they hold fast the prey till the nearby tube feet take it up and convey it to the mouth. Other remarkable organs amid the waving forest of spines and tube feet around the mouth are the sphaeridia, minute glassy club-shaped structures covered with cilia and possessing a rich supply of nerves at their bases. "In a living urchin," writes Aiyar (*ibid.*, p. 16), "the sphaeridia are observed to be in a state of constant movement and they are almost certainly balancing organs."

Even more amazing is the complex apparatus known as Aristotle's lantern. The essential feature is the set of five long incisor-like teeth which converge to a point and can be protruded from the mouth (fig. 4.7B). Each tooth is fixed in a highly pyramidal frame or alveolus, and the five frames are connected by small pieces called respectively epiphyses, rotulae and compasses, so that there are no less than twenty ossicles and five teeth. The lantern is worked by four groups of muscles called adductor, divaricator, radial and masticatory muscles; some of these are attached to the "perignathic girdle," consisting of five arches upon which the lantern is supported. And yet the entire complex machinery of the body is controlled only by ring-like and radial nerves and sensitive areas, with no concentration of nervous tissue that can be called a brain.

But although all the echinoderms are headless and brainless, each kind has its own "wisdom of the body," meaning chiefly an appropriate set of interacting reflex responses, so that it will retreat from a noxious environment, or right itself when turned upside down, or start the killing and feeding mechanisms when properly stimulated.

Out of all the hundreds of known forms of fossil and recent echinoderms, only certain early Palaeozoic cystoids (e.g., *Placocystites, Mitrocystites*) even remotely approach any of the known vertebrates in body form. The echinoderms have used their coelomic pouches as parts of an elaborate water tube system; while the vertebrates have used somewhat similar sacs as a point of departure for the evolution of an axially segmented locomotor system, which has been accompanied by the elaboration of a highly complex brain, both being conspicuously lacking in the echinoderms.

The embryology of representatives of the principal divisions of echinoderms has been eagerly studied for clues to the more remote and as yet undiscovered ancestors of the group. In the earlier stages of development a gastrula, or pushed-in ball, is formed as in many other groups. A single pouch or body cavity, corresponding to the coelome or mesoderm pouches of other groups, is given off from the primitive gut, which is bent up into a U. This pouch becomes divided into three pairs of body cavities, of which the middle pair eventually develops into the complex water system of tube feet with their reservoirs (ampullae) and canals.

The free-swimming larvae of different classes of echinoderms (fig. 4.8) differ considerably in the arrangement and extent of the bands of cilia and folds from the surface, but at least in early larval stages all have the more or less U-shaped gut with the primitive mouth opening beneath the apical plate, or frontal field. It is supposed that in the more immediate ancestors of the echinoderms such free-swimming ciliated sacs settled down and became attached by the apical plate, or praeoral lobe. At this stage the larva is called a Dipleurula (in allusion to its paired coelomic pouches) and it is supposed to represent in its general ground plan the remote ancestral adult stage. From such a form it is supposed by Bather that the pear-shaped primitive cystoid (fig. 4.5A, A1) has been derived by the migration of the mouth away from the praeoral lobe up along the side of the test until it rested at the top, not far from the openings of the front pair of body cavities. This migration of the mouth would cause the primitive gut to be looped around the left hinder body cavity, while the left water cavity would be looped around it. Partly from the twisted relations of the gut and the body cavities, not only in the fixed crinoids but in the freely moving starfishes, sea-urchins and holothurians, it is assumed that the ancestors of all classes of echinoderms at one time were more or less fixed and radially symmetrical animals.

In short, the echinoderms no less than other groups illustrate the extraordinary transformations from a relatively primitive central type, which has supplied the fauna of all ages with the astounding diversity of classes, orders, families, genera and species within each larger group. It is only when we look too exclusively at the relatively smaller divisions such as the orders, that by following through the recorded fossil history for comparatively short periods the course of evolution appears to be largely orthogenetic or predetermined.

The echinoderms began in a small way as sac-like animals feeding on minute organic matter ingested by the lashing of cilia. Some of the oldest of them are still in this primitive stage but at the extreme end of one branch we find the aggressive and predacious starfishes and along other lines we encounter the mud-sucking, worm-like holothurians or the hard-biting echinoids.

The possible relations of the echinoderms to the vertebrates will be considered below.

CHAPTER V

THE PROCHORDATES AND THE PROBLEM OF THE ORIGIN OF THE VERTEBRATES

Contents

CHAPTER V

THE PROCHORDATES AND THE PROBLEM OF THE ORIGIN OF THE VERTEBRATES

THE AMBIGUOUS LANCELET (*AMPHIOXUS*)

THE FISH-LIKE lancelet, *Amphioxus lanceolatus* (fig. 5.1), found in the North Sea, the English Channel and the Mediterranean, is truly a "Basal Chordate" in so far as it exhibits: (1) a longitudinal dorsal or spinal nerve cord, supported by (2) a long axial rod or notochord; the latter lies above (3) a straight gut-tube—all in diagrammatic simplicity. Moreover, the lancelet has what long seemed to be only the beginning of a heart and of a brain, and its locomotor system consists of simple repetitive muscle-segments, which are all much alike and extend from snout to tail.

If the lancelet had been designed by a zoologist for the classroom, instead of having been evolved by the long and often devious ways of evolution, it would have almost nothing but the truly generalized features noted above (except, of course, for the inevitable gonads and their ducts). But there are indications that some of its present simple features, including the small size of the brain, may be due to retrogression and that others may be the result of quite secondary specialization.

The notochord in the lancelet is relatively large and is prolonged into a spear-like tip at the front end of the animal (fig. 5.1A1). The frontispiece of Willey's book on Amphioxus (1894), a picture of living lancelets in the Naples Aquarium, shows some of them darting into the sand head first and others with their heads sticking up out of the sand and their circlet of cirri or feelers extended around the mouth. The lancelet is at least momentarily protected by the sand in which it is buried and if disturbed it darts away by means of its relatively high-powered locomotor system.

In Willey's time zoologists expected really primitive forms to have a thin pliable epidermis, but later evidence suggests that this is usually a later condition. The quick-darting habit would make a tough armored integument both unnecessary and hampering. Thus the epidermis of the lancelet is very thin and the dermis is formed mainly of soft connective tissue (Parker and Haswell, vol. II, p. 42). But we have seen above that among the invertebrates naked forms have often been derived from well armored ancestors (e.g., slugs, sea-hares, earthworms, ship-worms). Among vertebrates, lack of armor is sometimes associated with specializations for high speed (e.g., swift-darting bonitos, tunnies, flying fishes). Again, among the oldest known fossil fishes (ostracoderms) the more slowly moving forms (fig. 6.12A, B) had large tough head-shields, the more slender active forms had small heads and narrow flank-scales (fig. 6.12C-F).

The idea that *Amphioxus* in its present form really gives us a fair picture of an early vertebrate is due in part to the lack of known or recognized direct fossil ancestors of the lancelet, in part to the wide influence of Haeckel, who made his "Urwirbelthier" a shortened, simplified and partly larval *Amphioxus,* and in part to the influence of embryologists, who expected embryonic development to "recapitulate" adult evolution. But *Amphioxus* may well be both a vertebrate reduced to its simplest terms and in some ways a highly specialized derivative of the earliest known vertebrates, which were armored fishes.

The embryonic development of a given organ or group of organs may or may not give reliable indications of the remote history of the animal; yet it may also give fairly indirect evidence as to the relationships of the animal to other systematic groups having similar embryonic stages or processes.

The notochord of the adult lancelet (fig. 5.10A1) is a long cylindrical tube filled with large thin, "vacuolated" cells and surrounded by a strong notochordal sheath of connective tissue. "The notochord, like the parenchyma of plants, owes its resistant character to the vacuoles of its component cells being tensely filled with fluid, a condition of turgescence being thus produced" (Parker and Haswell, vol. II, p. 43). Apparently the embryonic notochordal cells have the power to absorb fluid and expand radially, while the connective tissue sheath has restrained the expansion and produced the internal pressure of this axial rod, which is truly the centerpiece of the internal skeleton of vertebrates.

It has been shown by Conklin (1932, pp. 100–102 and figures) that: (1) in the gastrula stages in the embryonic development of *Amphioxus* the cells which later give rise to the notochord develop in the roof of the gastrocoele (= inner cavity of gastrula) and in front of the dorsal lip of the blastopore; (2) that the young chorda cells (pl. 7, figs. 82, 84) closely resemble endoderm cells of the gut, but that when a longitudinal notochordal groove appears the cells of the right and left halves of the notochord begin to interdigitate across the midline (fig. 5.2B2, 3, 4). The notochord is pushed forward by the growth and division of cells at its posterior end, while the cells swell and are later transformed into the "vacuolated" turgid supporting condition. The neural plate of ectoderm cells later rolls up to form the spinal nerve cord; it rests on the notochord, while the mesodermal somites or coelomic pouches abut against the notochord on either side. These notochordal cells, by folding up into a tube, lose their connection with the gut but they retain their functional association with the lateral mesodermal masses which give rise to the muscle-segments, and with the dorsal nerve tube, which is supported by the notochord. Unlike the lateral muscle masses, the notochord arises in the embryo as an unsegmented structure, but in most later vertebrates it is replaced by the segmental and articulated spinal column.

The exceptionally large size of the notochord in the lancelet is functionally associated with the relatively great mass of the dorsal portions of the myomeres or V-shaped muscle-segments (fig. 5.6B *somites*). These arise from the outer walls of the coelomic sacs in the late embryos and early larvae (Conklin, 1932, pls. 14, 15). They lie on either side of the notochord. The muscle-segments above the very large pharynx or gill chamber do not extend downward to the mid-ventral line (fig. 5.14C), but behind that chamber they nearly or quite meet below the body cavity, as their derivatives do in more advanced vertebrates.

Possibly their failure to meet in the midline below the gill chamber may be associated with the large size not only of the gonads but also of the ventral atrial chamber (figs. 5.1A1; 5.14C) or atrium which extends beneath the body cavity, the gonads and the gill chamber. This atrium arises on the ventral aspect of the body, behind the incipient gill openings, from longitudinal right and left folds of the epidermis. These "metapleural folds" (fig. 5.1A2) turn outward a little to form longitudinal keels or stabilizers. On their inner sides they send inward small secondary folds which meet and fuse in the midline, and the cavity thus enclosed becomes the atrium (fig. 5.14C).

The lancelet's mouth and gill openings begin to appear in the young larval stages on the under side of the body (fig. 5.1A3) as holes surrounded by hoops of thickened epidermis. The mouth opening and its hoop appear on the left side and rapidly grow very large. The hoops around the gill openings increase in number posteriorly and, as further growth and multiplication proceed, the primary gill folds become subdivided by secondary down-growths or so-called tongue bars (A4). The bars between successive gill-slits bear cilia and by the lashing of the cilia the water is drawn into the gill chamber, carrying both food and oxygen. The food particles are drawn into a ventral ciliated groove or endostyle (A1 *endst*) leading to the very simple straight gut. After going through the gill-sieve, the water escapes into the atrium and is passed out at the rear through the atrial pores (A2).

Perhaps the atrium acts also as a storage tank. The strong dorsal muscle-segments, by bending the body from one side to the other, may exert a pumping action by squeezing water out of the atrium through the atriopore (A2) and drawing water into the mouth and gill-sieve, essentially as in Anaspid ostracoderms (fig. 6.12C-F).

The branchial or gill systems of *Amphioxus* (fig. 5.1A1) is to some extent analogous to the food-and-oxygen filtering gills of oysters and clams, in which the current is likewise drawn in at the front by ciliary action and passed out through a rear tube. This food-and-oxygen supply system seems more primitive in principle than the complex mouth-and-gill mechanism of modern fish (fig. 9.9B); in the latter the water is drawn in by an elaborate pumping apparatus of articulated skeletal pieces operated by special gill muscles, while the food is seized by articulated jaws and by parts of the gill arches. Thus *Amphioxus* at present stands far below the morphological level of the recent fishes.

The general result of Conklin's study (*ibid.*, pp. 116, 117) of the normal embryology of *Amphioxus* was "to show that the localization pattern in the egg and embryo is essentially the same in *Amphioxus* as in ascidians." The relationship of *Amphioxus* to the ascidians has indeed long been recognized, but the series may conceivably have run either this way:

basic chordates → *Amphioxus* → primitive fishes (ostracoderms)
↓
Ascidians

or more probably this way:

basic chordates → primitive fishes → *Amphioxus*
(ostracoderms) ↓
Ascidians.

THE BAG-LIKE SEA-SQUIRTS AND THEIR ALLIES (ASCIDIANS)

The ascidians are of interest in the present connection, first, because they combine basic chordate characters in the larva with invertebrate-like features in the adult; second, because they supply a dramatic instance of a complete transformation during individual history, from the active, free-tailed tadpole-like larva (fig. 5.3A1-A4) to the fixed, bag-like adult ascidian (A), in which the larval notochord has been entirely broken down and resorbed. As noted above, they share with *Amphioxus* the basic plan of the earlier embryonic development, as well as the larval notochord, the rapidly growing mesoderm cells, and other features. The adult ascidian (A) has a relatively enormous pharynx or branchial chamber, with a large number of vertical slit-like openings arranged in horizontal zones and lying between a grid-like system of horizontal and vertical vessels. The adult fixed ascidians (fig. 5.3A), as well as the secondarily free *Appendicularia* (fig. 5.4B, B1), contrast with *Amphioxus* in having a loop-like gut, instead of a straight one, the intestinal part behind the stomach having been drawn forward into the atrial chamber which surrounds the great pharynx.

The entire food-sifting and aerating complex of ascidians, while sharing basic features with that of *Amphioxus,* also has functional parallels among invertebrates, such as sponges and oysters. Other invertebrate-like features of ascidians are: (1) larval eyes (fig. 5.31A1) and otoliths, (2) the colonial adaptations of the simple and composite fixed ascidians (fig. 5.4C, C1), (3) the habit of budding off chains (fig. 5.5A1) or groups (fig. 5.5) of zooids or new individuals, and (4) the filmy pulsating cylinders of the secondarily free-swimming salps (fig. 5.5A). But none of these convergent resemblances to invertebrates appear to be inconsistent with derivation of ascidians by transformation from an *Amphioxus*-like ancestor through the enormous enlargement of the pharynx and loss of the notochord and its muscle segments. Therefore the ascidians do not seem to help us much in determining which great group of invertebrates gave rise to the vertebrates.

THE MISNAMED "BARNACLE-WORMS" (*BALANOGLOSSUS*)

The shore-living acorn or barnacle-worms, *Balanoglossus* (fig. 5.6A) are more like invertebrates than the ascidians are. The name *Balanoglossus* (meaning barnacle-tongue) refers to the protruding "proboscis" with muscular walls at the front end of the cylindrical worm-like body. Around the stem of the proboscis is a circular slit called the mouth. The mouth is surrounded by a wide circular band or "collar." The stem acorn is supported by a small fold from the wall of the oesophagus (fig. 5.6A *ntch*), which is doubtfully supposed to represent the notochord of *Amphioxus*. There is a large pharynx with a double row of small gill slits. The animal forces its way through the sand, engulfing it through the permanently open mouth, the nutrient matter being extracted in the long, nearly straight gut and the refuse left behind, like worm-castings. Thus the habits of the barnacle-worm as well as its general form suggest the earthworm, but there is no sign of relationship.

The *Balanoglossus* family includes two different genera, each with its own way of development. In the first mode of development, after the impregnated egg has divided into ecto-, meso-, and endoderm cells, an externally thimble-shaped, ciliated gastrula is formed (fig. 5.6A1). The top of the thimble bears a median tuft of cilia; the middle of it has an equatorial groove, and the bottom, a circle of cilia. Later a second groove arises, parallel to the first. The dome-like part above the first groove later gives rise to the "proboscis" (A), the second groove defines the "collar"; below the second groove the first gill-slits appear, while the lower end later becomes greatly elongated to form the tubular body. This is called the direct mode of development because there is a gradual progress in one direction from the dividing egg to the adult stage. In the larval stage noted above the body is already divided into three segments which give rise directly to the proboscis, collar and cylindrical body (A2). This stage is commonly equated with the alleged "three-part" larva of *Amphioxus,* but Conklin (1932) found no enlarged collar cavity there.

In the second mode of development among the balanoglossids the free-swimming "Tornaria" larva (fig. 5.6C) consists chiefly of a curved gut enclosed in a sac which is somewhat constricted in the middle; the upper half of the sac is folded into ciliated loops; near the bottom of the sac there is a ciliated girdle; the anus is in the center of the lower half, the mouth opens on the left side of the upper half; the rudiment of the proboscis cavity is in the mid-frontal plane, an eye-spot is at the top of the dome-like upper part of the sac. This "Tornaria" larva of the acorn-worm was long regarded as the larva of certain echinoderms which it resembles (fig. 4.8A, B). It supplies the chief point of resemblance of the balanoglossids to the echinoderms. Moreover the balanoglossids, like the echinoderms, arrow-worms, brachiopods, *Amphioxus* and the vertebrates, all form the tissues of their coelome or body-cavity and mesoderm by outgrowth from the gasterocoele or endoderm-lined gut of the gastrula stage (fig. 5.2B2); whereas in the annelid worm, representing a primitive stage preceding the crustaceans, insects, etc., E. B. Wilson and others found that the mesoderm organs develop from a certain cell called the second somatoblast (fig. 5.2A); this cell (*som* 2) lies below the small upper cells (which give rise to the ectoderm) and above the larger lower cells (which give rise to the endoderm).

Cephalodiscus (fig. 5.7A), a deep-sea relative of the balanoglossids, lives in colonies which secrete a common case or investment (A1). From the collar region eight fleshy "arms" protrude, each with numerous fine plume-like branches (A). The "proboscis" is a thick fold, somewhat like an operculum, overlapping the slit-like mouth (A, A2). It may perhaps be no more than a coincidence that the larva of *Cephalodiscus* is said to bear a striking resemblance to that of certain Polyzoa or moss-animals (Parker and Haswell, vol. II, 1940, p. 10). But these in turn seem to be remotely connected through *Phoronis* (figs. 4.4D-D4) with the brachiopods (figs. 4.1, 2), and other coelomate or "enterocoelic" invertebrates. So far as it goes, the embryological evidence suggests that the vertebrates are distantly connected not at all with the externally-jointed annelid-arthropod series but rather with the assumed Pre-Cambrian roots of the "enterocoelic" major division of many-celled animals.

Thus the balanoglossids in their adult forms seem to lead far away from the direction of the vertebrates and so do the ascidians. All their invertebrate-like adaptations may, however, have been assumed perhaps at a relatively late period, when they began to be shut off from further direct advance toward the vertebrate grade and were compelled by competition from the fish-like vertebrates to adapt themselves to the relatively safer life in the muddy sand.

In the case of *Amphioxus* this competition from the more progressive or fish-like branches of the early chordates (fig. 5.9A) may have been one of the factors (through the coöperation of genetic variation and natural selection) in the great over-development of the notochord (fig. 5.10B) in the larval stage, when this dominant organ pushed forward to the tip of the snout and apparently retarded the development of the neural plate into a normal brain, the brain of *Amphioxus* being remarkably small and undifferentiated. Meanwhile the great increase in the mesodermal muscle-plates supplied the power to dart the notochordal head-spear and the body into the sand (fig. 5.20E1). The same processes encouraged the increasingly reflex action of the spinal nerves, while the loss of encumbering armor and thinning of the integument may have opened the way for the dissolution of the paired eyes and the distribution of their light-sensitive cells along the spinal cord.

EMERGENCE OF THE NOTOCHORD

The notochord, which is the leading one of three insignia of chordate status, seems to have been a wholly new or emergent organ in the remote ancestors of *Amphioxus,* not inherited as such from invertebrate predecessors, but evolved within the lower limits of the chordates.

It is apparently a somewhat new idea to regard the ascidians and balanoglossids as relicts of some very early group or groups of enterocoelates, which were severally driven by the pressure of the arising vertebrates to retreat into various modes of life that were and are characteristic of invertebrates. Perhaps it is no longer necessary to seek further among invertebrates for a direct homologue of the notochord, as Gaskell did when he attempted to derive the notochord of vertebrates from the ventral groove of primitive arthropods (fig. 5.18A, B).

The ectoderm and endoderm of vertebrates may be more or less homologous respectively with those of invertebrates, but as noted below (p. 99) the mesoderm and its derived coelomic tissues have arisen in different ways in the annelid-arthropod series on the one hand and in the brachiopod-echinoderm-vertebrate constellation on the other. The multiplication of metameric segments in *Amphioxus* and in the earliest vertebrates seems indeed to have taken place quite independently from the multiplication of somites in the annelid-arthropod series. In both cases the very process of increasing the length of the body by the multiplication of segments opened up the possibility of articulating the segments with each other in a bilateral, anteroposterior series, thus increasing flexibility in locomotion and making it possible for each segment to act interchangeably either as a lever to move the adjacent segments or as a fulcrum upon which they could be moved. But the detailed ways of building up a locomotor system in the vertebrates were quite radically different from the ways followed in the annelid-arthropod series and, as will presently be shown, all attempts to derive the former from the latter have had to make unverifiable assumptions. Possibly the notochord may have arisen as a supporting core of the neural plate (fig. 5.2B2) which was later seized by the muscles of the body wall, at first as a fulcrum for movements of the primitive digestive, respiratory and urinogenital sacs.

Whatever its remote origin may have been, the notochord came to be surrounded by the segmentally deposited skeletal tissues of the locomotor system. It proved to be of great importance as a sort of prerequisite temporary scaffolding for the vertebral column, and it is remarkable that this essentially early Palaeozoic organ persists in the human embryo no less than in the embryos of man's humbler relatives.

THEORIES OF THE ORIGIN OF THE VERTEBRATES

The problem of the origin of the vertebrates was bound to emerge sooner or later after the idea of the "stairway of beings" began to be applied in detail to the study of comparative anatomy. Etienne Geoffroy St. Hilare (1818) is generally credited with the idea that a vertebrate represents an insect lying on its back (fig. 5.10), although, as he conceived the situation, the insect represents a reversed vertebrate. His colleague Cuvier showed that there were grave objections to this theory. But the fundamental idea of reversal was applied with various details by Dohrn (1875), Semper (1875–76), Minot (1897), Patten (1891, 1912), Owen (1883), Gaskell (1896, 1898–1906), who all held, either explicitly or implicitly, that one or another type of jointed invertebrate with ventral nerve-chain and a dorsal heart was transformed into a vertebrate in which the heart is on the functionally ventral side and the nerve tube on the back.

The method of proof followed by each of these writers was in short to assume that there was a true homology between similarly named parts possessed in common by the vertebrate and the supposed invertebrate "ancestor," and then to trace the possible steps by which the various invertebrate organs might have been transformed into their supposed vertebrate equivalents.

The opposing method would try to show, for example, that the legs of insects are not homologous with those of vertebrates and that the two kinds of "legs" are so profoundly different from each other that we have no right to assume that one is genetically identical with the other.

After more than a century of debate the methods on both sides remain fundamentally the same but the amount of detail has become enormous.

Meanwhile the triumphs of experimental science turned men's minds away from what seemed to be merely academic discussions to the more immediate problem of heredity. Such broader morphological problems as the one under consideration were indeed officially laid on the shelf by the President of the British Association for the Advancement of Science, who at the meeting in 1914 proclaimed: "This is no time to discuss the origin of the Mollusca or of Dicotyledons while we are not even sure how it came to pass that *Primula obconica* has in twenty-five years produced its abundant new forms almost under our eyes." Another preëminent zoologist, E. B. Wilson, put an end to his superb graduate course on invertebrate morphology and phylogeny, because he seems to have felt that the problems with which it dealt were not only unsolved but insoluble.

But time has to some extent already brought its compensations because both Bateson's studies on the embryology of *Balanoglossus* and E. B. Wilson's masterly but unpublished lectures on the morphologic interrelationships of the various groups of invertebrates still call for further consideration of the grand problem of the origin of the vertebrates.

In the present comparative survey of evolution we are maintaining: (1) that though much remains to be learned in detail, we can at least narrow the problem by the elimination of most of the great phyla of typical invertebrates from the list of possible ancestors of the vertebrates; and (2) that further analysis of the data already in hand puts both the problem and the older proposed solutions in a somewhat new light and renders one or two new and long unsuspected solutions (see p. 98) at least worthy of further study. Let us then first review briefly the diverse theories or hypotheses of the origin of the vertebrates, and record various criticisms or comments. For convenience we may name each theory from the type of animal which it assumed as representing, at least to some extent, the invertebrate ancestor of the vertebrates.

Coelenterate Theory. — Vertebrates like all other animals begin their development from a single fertilized cell or zygote, which represents the union of two parental cells. There is general agreement that all many-celled animals were in all probability derived from colonies of single cells, and as we have seen in Chapter I, it is further probable that some of the colonial flagellate Protozoa represent somewhat closely the remote ancestors of all the Metazoa or many-celled phyla. As noted above (p. 26), the most central and primitive of the many-celled phyla are the coelenterates, which are two-layered or diploblastic animals essentially cup-like with an inner or nutritive layer and an outer protective and sensitive cell layer. The vertebrates, on the other hand, are three-layered (or triploblastic) animals and they are also provided with coelomic cavities developed in the mesoderm. Between the radial free-floating coelenterates and the most primitive three-layered forms, which are the flatworms, stand the ctenophores, or comb-jellies (fig. 5.11C, C1). In view, however, of the relative complexity of the vertebrate groundplan, it seems to be an over-simplification if we assume with Masterman (1897) that the bilaterality and triploid subdivision of an early embryonic stage of *Amphioxus* has been derived *directly* from the secondary bilaterality of a circular coelenterate (fig. 5.11C), resulting from contact with the substratum and from subsequent efforts to creep forward (fig. 5.11D).

Nemertean Worm Theory.—Hubrecht (1887) took the long proboscis and its sheath (fig. 5.11E) of these animals as the primal source for the vertebrate notochord and shifted the horizontal lateral blood vessels (E1) to form the main dorsal and ventral vessels. No transitional stages were cited nor even imagined. In view of the vast intricacy of the anatomy of even the earliest known vertebrates, Hubrecht's theory appears to take far too much for granted as to the unknown intermediate stages.

Annelid and Arthropod Theories.—The marine annelid worms (figs. 1.5; 5.14A) are coelomate, triploblastic Metazoa, as are the vertebrates, and they broadly resemble the latter in possessing a metameric locomotor apparatus controlled by a segmental nervous system. They also show an early stage in the elaboration of a head and brain, and their paired appendages fulfill the functions of locomotion and aeration. Hence Semper (1875–76), Dohrn (1875), Minot (1897) and others, who upheld the annelid theory, sought to derive the vertebrates directly from some primitive annelid stock. The annelids, like the crustaceans, arachnids, insects, have a nerve ring around the gullet or oesophagus (fig. 5.12A) which connects the dorsal ganglia for the eyes with the main ganglia on the under side of the gullet; these ventral ganglia serve, so to speak, as the special brains of the paired ventral nerve-chain. This in turn gives off segmental nerves to each body segment. Thus in the annelid the primitive gut passes through a "circumoesophageal" nerve-ring. In vertebrate embryos it was found that at an early stage of development a pocket or sac, the infundibulum (fig. 5.13A, B *inf*) extends downward from the floor of the brain and meets a small finger-like hollow process (the hypophysis, *hyp*) that extends upward from the roof of the stomodeum or embryonic mouth pouch. From the union of the infundibulum and the hypophysis arises the pituitary body. Richard Owen (1883) in his paper on the "Conario-hypophyseal tract" pointed out that at this stage the cavity of the gut was continued upward by a narrow tube, the hypophysis, leading into the brain cavity. This arrangement then offers a certain degree of resemblance to the condition in the annelids and throughout the Arthropoda, in which the tube from the mouth pierces the circumoesophageal nerve-ring. Thus if the comparison be admitted as valid, we have in the vertebrate embryo the anterior end of the digestive tract connected with the brain cavity.

But where was the "old mouth"? This was

usually placed on the dorsal side, in the region of the medulla oblongata (fig. 5.13B *"mth"*) at the site of the "fossa rhomboidea."

Patten's Arachnid Theory.—The supposed closure of this "old mouth" is ascribed by Patten (1912, p. 26) in part to the pressure of the rapidly enlarging brain and to the backward growth of the rostral border of the carapace (cf. fig. 5.15). The "new mouth," he supposed, broke through at the site of the "cephalic navel," lying on the haemal or originally ventral side between the bases of the radially diverging paired appendages (*op. cit.,* p. 40). The latter were supposed to lose their locomotor function and to emphasize their respiratory function, and at the same time to form the framework of the future branchial skeleton of vertebrates.

This bold and highly ingenious comparison captivates many minds, especially those that appreciate whatever force it may have; but neither this nor most of the other numerous supposed equivalences between annelid and arthropod on the one hand and vertebrate on the other would have any compelling force if one could show that the two sides of the supposed equation are so profoundly different in other respects that such resemblances with each other as they do show are more likely to be expressions of convergent evolution.

Patten assumed that the higher Arthropoda on the one hand and the Vertebrata on the other are brought together by the common possession of a complex head formed by the coalescence of enlarged nerve segments or neuromeres (fig. 5.12B, C). He therefore (*op. cit.,* p. 471) brackets the vertebrates and the entire arthropod series under the term Syncephalata (complexly headed), to which he awards the rank of a superphylum. He supports this classification by a mass of evidence which it took him several decades to collect; but fortunately for the reader of his truly great work, "The Origin of the Vertebrates" (1912), he has summed up his theory in a series of, at first sight, amazingly convincing comparative diagrams (especially those on pages 8, 17, 67, 69, 466–467). Nevertheless, after repeated and prolonged consideration of this evidence, it may be asserted that his method consisted in supporting his main hypothesis by the assumption of a host of implied hypotheses and by the citation of analogies in assumedly similar cases.

For example, the assumption that the invertebrate lateral paired eyes in the course of their supposed transformation into vertebrate eyes were shifted around on to the opposite side of the head (fig. 5.15) might be admitted if direct evidence for it were available; even the further idea (*op. cit.,* pp. 156–158, figs. 107, 108) that the retinal and pigmented layers were turned inside out as the result of the folding over of the neural plate into a brain tube is conceivable, provided that an eye in the transitional stage could still be functional. But paired eyes have surely been developed independently in cephalopod molluscs and vertebrates; moreover, eye-like organs sensitive to light have been developed along the edge of the mantle in certain members of the clam class (e.g., *Pecten, Lima, Tridacna*), while eye-like glow lamps have been developed quite independently in deep-sea cephalopods and deep-sea fishes. Accordingly the first question is: Are the paired eyes of Patten's *Limulus* truly homologous with those of vertebrates? If we assume that they are, we are denying the possibility that paired eyes may have been developed independently in *Limulus* and the vertebrates, which in view of the cases just cited is evidently an unsafe assumption.

Patten further assumed (*op. cit.,* pp. 198, 199, 208) that the *Limulus* "heart" is homologous with the vertebrate ventricle and the pericardial chamber with the atrium, but, as he well recognized, comparison with such very primitive existing arthropods as the phyllopods (fig. 3.3B, *ht*) shows that the arthropod heart (figs. 5.14B, B1) is essentially a muscular tube pierced by ostia, or intake valves; through these blood enters from other parts of the body, including the paired appendages, which in turn carry respiratory combs or gills. Rhythmic contractions of a muscular coating of the dorsal pump drive this mixed blood forward to the head and backward to the body; the blood returns eventually to the heart through the pericardial sinus and the intake valves above noted. Thus in arthropods the heart does not deliver the blood directly to the gills and the connection between them is only by a circuitous route (5.14B). In the vertebrates, on the other hand, the ventricle (D) is derived from a localized swelling of the basal aorta immediately associated with the gills, to which it delivers its blood directly. Nor do the conditions in *Amphioxus* (fig. 5.1A1), the cyclostomes (fig. 6.19A1), or the sharks (fig. 5.14D) support the assumption that the main pump of vertebrates is either homologous with or derived from the heart of arthropods.

Patten made ingenious use of the available palaeontologic evidence in favor of his view that the existing *Limulus* is the nearest living survivor of the invertebrate ancestors of the vertebrate. The primitive ancestors and relatives of *Limulus* (figs. 3.4D; 5.15A) have a broad head-shield which is distinctly suggestive of the head-shield of certain cephalaspid ostracoderms; he assumed then that the eurypterid stock, which also gave rise to the scorpions, was directly ancestral to the vertebrates. He also drew a most skilfully conceived diagram (fig. 5.15C-F), showing how the cephalic shield of a eurypterid might have been transposed into that of an ostracoderm. Since by hypothesis the eurypterid would be swimming on its back, the originally anteroventral rim of its cephalic shield would extend backward in such a way as to encroach gradually on the originally ventral (now dorsal) surface; thus it would eventually close over the "old mouth." Meanwhile the originally dorsal side of the cephalic shield retreated toward the front rim, carrying the paired eyes with it; these were thus gradually carried around (fig. 5.15C, D) from the old functionally dorsal part of the shield to the new dorsal part formed from the originally ventral rim.

The jointed paired appendages, being deprived of their original function, meanwhile were supposed to have grown toward the future ventrolateral side of the head; the respiratory branches gave rise to external gills (5.15F) or "balancers" (such as are present in embryo *Polypterus,* dipnoans and amphibians), while the locomotor branches turned into the branchial arches.*

The general answer to Patten's theory is that although he most ingeniously and consistently conceived how a primitive eurypterid *might* be transformed into a vertebrate there has been a permanent lack of real transitional stages showing that it did so happen. That the pretrilobite stock gave rise to the limuloid, eurypterid and arachnoid branches, as well as to the crustaceans, rests on the cumulative concurrence of morphologic and embryologic evidence. But the supposed transformations of eurypterid organs and systems into those of chordates are severally open to such grave objections as have already been cited: on the palaeontologic side there is an absence of real intermediates between eurypterids and ostracoderms, while the morphologic data yield direct evidence that the resemblances between eurypterids and ostracoderms form an ideal example of convergent evolution.

In his comparison of the minute structure of the carapace of *Limulus* with that of the cephalic shield of certain ostracoderms, Patten brought out a striking similarity (*op. cit.,* p. 295) in the presence of both groups of a stellate surface pattern with inner and outer tabulae (fig. 5.16A, A1) supported by vertical trabeculae (*trab*). A comparison of the sections of certain ostracoderm head-shields (B) later published by Kiaer and Heintz (1935) with Patten's figures of the integument of *Limulus* abundantly confirms Patten's observations, but does not necessarily prove his conclusions. For in spite of the general resemblance in these figures there is no convincing evidence that the vertically striate "dentine" layer (*dent*) of *Limulus* has given rise to the true cosmine denticles of the ostracoderm (B), or that the basal lamellae of *Limulus* have given rise to the basal lamellae of the ostracoderm. The independent development of basal lamellae and dentine-like surface, of vertical trabeculae alternating with large vascular

* Grave objections to these supposed homologies are considered in my recent article (1950) on parallel and diverging evolution in arthropods and vertebrates.

channels, in limuloid arthropods and certain ostracoderms may rather have been in part a response of the growing skeletal tissue to pressures exerted by the circulating fluids against the limiting surfaces offered by the epidermis on the outside and the visceral mass on the inside.

The suggestion of Dr. Homer Smith (1939) that the primitive vertebrates, coming up from the sea into the fresh-waters of the land, acquired their exoskeleton as a response of the excretory system to the need for a waterproof covering, is equally applicable to such invertebrates as the eurypterids, which were derived from the marine trilobite stock but were invading at least the brackish waters of the rivers (Dr. Marjorie O'Connell). Indeed the large cavities between the outer and inner tabulae of the exoskeleton in both *Limulus* and the ostracoderms may well serve as buffers to regulate the varying salinity due to the alternation of incoming ocean tide and outgoing fresh-water.

Since the publication of Patten's work a wider and more detailed knowledge of the anatomy of the earliest known chordates, the Palaeozoic ostracoderms, has accumulated. This great body of facts amply supports the conclusion that the resemblance between the Palaeozoic limuloid *Bunodes* (fig. 3.4E) and the cephalaspid ostracoderm type (fig. 5.15F) was due to convergent evolution.

Since *Bunodes* and other Palaeozoic limuloids were admittedly close to *Limulus,* they probably grew by periodically casting off the entire exoskeleton in the manner generally characteristic of arthropods. In the numerous cephalaspids, on the contrary, there was no evidence of such molts, and both peripheral and internal growth kept even pace with each other.

Bunodes and *Cephalaspis,* belonging in widely different superphyla (pp. 61, 106) of coelomate animals, were independently rising from the status of food-sifters to that of predators and were therefore each synthesizing a complex directional center or head from originally distinct sets of poly-anisomeres. But Patten's attempts to equate corresponding poly-anisomeres on the two sides of the column have resulted only in forcing analogies to assume the status of homologies. In a word, the habitus of *Bunodes* and *Cephalaspis* was in certain respects similar but their phyletic heritage was widely different.

Another group of Palaeozoic vertebrates which Patten used for the support of his "arachnid theory" was that called Antiarchi by Cope. (See p. 112.) This group included among others: (1) *Pterichthys* (fig. 7.2A) from the Devonian of Europe, which was made celebrated by Hugh Miller; (2) its American relative *Bothriolepis* (fig. 5.17A); and (3) some north European and Greenland forms named *Asterolepis* (fig. 7.2B, C), *Phyllolepis* (fig. 7.6B), etc. The morphology of *Bothriolepis* was well described and figured by Patten (*op. cit.,* pp. 367–380) and the group has since been the subject of important monographs by Stensiö (1931) and Gross (1931). In broadest terms, a pterichthyid consists of four principal parts (fig. 5.17A, A1): (1) a rather small convex head-shield, bearing near its center a pair of large circular orbits; (2) a box-like cuirass, carapace, or buckler, and plastron, corresponding to those of a tortoise; (3) from a notch on either side just behind the head-shield and at the base of the carapace springs a pair of long appendages tapering distally and articulating with a complex joint at the "shoulder" and with a slightly movable joint suggesting an elbow; these appendages are covered with a surface layer of many plates; (4) a long flexible tail containing a well developed notochord and bearing on its under side distally a caudal web. The head-shield, carapace, plastron and the shell of the paired appendages together comprise a bony exoskeleton composed of stratified layers of bony tissue containing bone cells.

In earlier days before the true relationships of the Antiarchi were worked out, *Pterichthys* was at various times thought to be related to: (1) the chelonians, on account of its bony carapace and plastron; (2) the arthropods, on account of its articulated paired appendages; (3) to the ostracoderms, because its head-shield somewhat resembled that of *Cephalaspis.*

Patten, in the work under consideration, assumed (p. 10, fig. 4) that the paired appendages of *Bothriolepis* were evidence of eurypterid relationship; and that the small curved upper and lower jaw plates (fig. 7.16B) were homologous with the mouth parts of arthropods and that they moved "after the manner that prevails among the arthropods" (p. 374). He also regarded the head-shield of *Bothriolepis* as closely comparable with that of the cephalaspid ostracoderm *Tremataspis* (*op. cit.*, p. 359).

On the same slab with a *Tremataspis* head-shield he found certain small fragments, which he interpreted as parts of paired pectoral appendages. Hence he restored *Tremataspis* (p. 359) with a large pair of paired jointed appendages, and from that it was an easy step to infer that the little round openings (fig. 5.17B1) forming a curved series on each side of the plastron of *Tremataspis* ". . . may have served for the attachment, or for the exit, of other organs of a similar nature, as for example, external gills" (*op. cit.*, p. 362). This interpretation was consistent with his view (*op. cit.*, pp. 265–270) that the "balancers" and external gills of amphibian larvae were "one set of serially homologous structures, that are in turn comparable with the segmental appendages of arthropods" (p. 270).

Against this interpretation of the relationship of the pterichthyids as intermediate between eurypterids on the one hand and ostracoderms on the other may be advanced the following considerations:

1) Through the researches of Stensiö (1931, pp. 158–160), the pterichthyids have been shown to be not ostracoderms but an early side branch of the higher or jaw-bearing vertebrates.

2) Their paired pectoral appendages, while very probably homologous with those of acanthaspids, arthrodires, acanthodians and higher fishes, contrast widely with those of eurypterids in detailed construction.

3) Professor George M. Robertson, who carefully examined Patten's cited specimen of *Tremataspis*, identified the supposed pieces of a paired appendage as parts of median fins, and Patten himself (1903, p. 4) had corrected this error.

4) The little round openings in *Tremataspis* (figs. 5.17B1; 6.14) were homologous with those of other ostracoderms, which in turn lead into branchial chambers (6.14A) and are assuredly not sockets for paired appendages.

5) The entire exoskeleton of the pterichthyids (figs. 7.2; 5.17A, A1), being composed solely of articulated bony plates and not of a chitinous-calcareous continuous tegument, is merely analogous, not homologous, with the cephalo-thorax and abdominal exoskeleton of eurypterids.

6) The caudal region of *Bothriolepis* (fig. 5.17A) is completely chordate in character and contrasts widely with the ringed abdominal segments and spine-like telson of eurypterids (fig. 3.4D).

We may indeed affirm that the more we learn of the detailed construction of the ostracoderms, antiarchs, and other earlier chordate groups, the less we can rely upon the hypothetical intermediate stages (fig. 5.15D, E) which Patten invented to bridge the great gap between eurypterid and limuloid arthropods on the one hand and ostracoderms and antiarchs on the other.

Whether or not *Amphioxus* be regarded as a completely primitive chordate, its general morphology and embryology constitute a serious stumbling block to Patten's theory. It must be admitted that neither in the embryonic (fig. 5.2B) nor the adult stage (fig. 5.1) does *Amphioxus* afford any encouragement to the idea that its structures were derived from an arthropod source. What could be less like arthropod paired appendages than the bent-in cups in the embryonic pharynx (fig. 5.1A3, A4) which develop into the multitudinous branchial basket and gill-slits of the adult? And where is the evidence that the myomeres (somites) of *Amphioxus* (fig. 5.6B2), arising from the so-called mesoderm groove (fig. 5.2B2–4), are at all homologous with the body segments of eurypterids and scorpions? What has become of the annelid-arthropod circumoesophageal ring (fig. 5.12A)? That it may not be safely homologized with the annular nerves of the buccal cavity or of the velum of *Amphioxus* (cf., Delage and Hérouard, pl. 16, figs. 1, 2, 3) is indicated by the difficulty in identifying either the dorsal or the ventral ganglia of arthropods, which indeed appear to be absent in *Amphioxus*. Why is the "heart" complex (of ventral aortae) in *Amphioxus* (fig. 5.1A1) so directly connected with the respiratory basket? Where are the traces of its derivation from an arthropod haemal sinus (cf. figs. 5.1A, A2; 5.14B)? Why is it that the early embryo of *Amphioxus* (fig. 5.6B1) presents some evidence of relationship with that of the acorn-worms but contrasts very widely with the embryos of modern scorpions (*op. cit.,* figs. 15, p. 20; 16, p. 22; 17, p. 24; 46, p. 56), which are almost the true lineal descendants of the eurypterid stock?

The way that Patten (*op. cit.,* pp. 397, 471) disposed of such obvious objections was to treat *Amphioxus* as a highly specialized "acraniate" and in one of his diagrams (p. 271) he seems to suggest its derivation even from such a relatively complex type as *Bothriolepis*. Although we may admit that the adult *Amphioxus* is retrogressively specialized in many respects (see p. 98), it still seems to have an irrefutable claim (p. 84) to be an offshoot of the basal chordate stock (p. 99). Thus if Patten's main thesis were true, we should expect the embryology of *Amphioxus* (fig. 5.2) to retain some traces of derivation from remote arthropod ancestors. But here the contrast is so strong that he seldom refers to it; indeed he looks rather to such advanced vertebrates as amphibians for the main data for his comparative embryologic diagrams (*op. cit.,* figs. 30, 34; 159, 173; 307, 308).

***Ammocoetes-Limulus* Theory.**—Gaskell (1895–1910) started from the same general base as did Patten, namely, a comparison between *Limulus,* the scorpions and other arthropods on the one hand and of various vertebrates on the other; but the paths that he followed were on the whole widely different, as were also his assumptions and results. One of his most disputed assumptions was that if we take corresponding cross-sections of a "trilobite-like animal" and a larval lamprey (*"Ammocoetes"*), *without turning the arthropod upside down,* then we may equate the organ systems or structures as in his diagrams (cf. fig. 5.18). For example, the longitudinal groove between the paired appendages in trilobites and in *Limulus* is assumed to correspond with the notochord; the inner branches of the paired arthopod appendages, growing downward, meeting and fusing ventrally, would give rise to a new tube corresponding to the gut of the vertebrate; the trilobite pleura covering these appendages would then correspond to the somatic layer of the mesoderm and are in position to give rise to the muscular body wall of the vertebrate.

Another assumption was that, presumably in accordance with greatly increased activity of the entire animal, the right and left halves of the ventral nerve cords grew laterally, like the neural plate of the vertebrate embryos, and then curved dorsally around the arthropod gut tube. When they met and fused at the top, the latter would be inside the nerve tube (fig. 5.18C, D). Thus the gut tube, losing its primary function, would become the cavity of the brain and the lumen of the spinal cord, but at the hypophysis it would be connected below with the new gut which was being formed in the manner noted above. Among other astonishing results of this theory was the supposed transformation of some of the paired dorsoventral muscles of the carapace of the scorpion into the six pairs of eye muscles of typical vertebrates (Gaskell, 1900, p. 506).

Gaskell's theory grew primarily out of his amazingly detailed comparisons of the cranial nerves in the larval lamprey (*"Ammocoetes"*) with those of scorpions, of *Limulus* and of other arthropods. He also was impressed with the resemblances between Palaeozoic eurypterids and ostracoderms.

Between 1895 and 1910 Gaskell's theory was bitterly attacked, especially in scientific meetings, by nearly all his co-workers but was defended by him with immense learning and skill (see the "Discussion of the Origin of Vertebrates," Proc. Linn. Soc., 1909–1910). The theory has sometimes been favorably regarded by physiologists but since the partial eclipse of phylogenetic morphology by experimental biology, it has been rarely even mentioned.

It is, of course, not a decisive objection to Gaskell's theory to point out that at many points it contradicts other theories that start from annelid or arthropod bases; because conceivably it might be the only theory that is right. All theories of the origin of the vertebrates from any other known phylum must rest in part upon the closely correlated principles, evolution by internal reorganization and evolution by change of function. It is therefore not necessarily a final answer to Gaskell's theory to allege that its demands upon these generally sound principles are pushed to fantastic limits and to a degree unparalleled elsewhere, because the origin and the evolution of the vertebrate type involved a unique and astonishing transformation even if we go back to the single cell for the common ancestor of all phyla.

Perhaps the best reasons at the present time for denying either Patten's or Gaskell's theory seem to be: (1) that in spite of the resemblance between *Bunodes* and cephalaspids, the cumulative evidence indicates that arthropods and vertebrates arose respectively from the two already noted primary divisions of the triploblastic Metazoa and (2) that either general or detailed resemblances between any two individual members of the two series seem to be convergent features, developed independently among arthropods and ostracoderms and related to the synthesis of originally distinct metameres into a complex cephalothorax; (3) that the various "homologies" of the several parts of the brains, cranial nerves, eye muscles, etc., between arthropods and vertebrates as differently worked out by Gaskell and Patten require the forcing of the same names to cover things that are so widely different in the two phyla that specialists in these fields deny the supposed homologies, while the general zoologists demand more unequivocally intermediate stages, which are lacking.

Echinoderm-*Tornaria* Theory.—In previous pages (82, 87, 88) we have already noted: (a) that in spite of the extreme dissimilarity in appearance between typical sea-lilies and other echinoderms on the one hand, and acorn-worms, *Amphioxus* and the vertebrates on the other, all belong to the same enterocoelic superphylum; (b) that the free-swimming larvae of echinoderms resemble the free-swimming *Tornaria* larva of the acorn-worms; (c) that what seem to be the most primitive known echinoderms, the Cambrian and Ordovician Carpoidea (fig. 5.20B), might be conceived as Tornarian-like forms which had begun to settle down to a sedentary life, acquiring an exoskeleton of many plates, together with a water-pore system developed out of ciliated grooves.

Torsten Gislen (1930) pointed out that in the Lower Silurian *Cothurnocystites Elizae* Bather (cf. fig. 5.19A, A1), one of the carpoid echinoderms, the calyx was markedly asymmetrical and was perforated with a series of sixteen small openings which suggested the asymmetry and arrangement of the larval gill openings of *Amphioxus*. In this view the stalk of the curious lopsided sea-lily would form the tail or locomotor portion of the body. Another misshapen echinoderm (fig. 5.19B) might also be conceived as the motile ancestor of the ascidians. In others of these widely diversified carpoids (fig. 5.19C, D) the armored "calyx" bears an astonishing but perhaps quite superficial resemblance to the cephalothorax of *Drepanaspis* (fig. 5.19E), one of the pteraspid ostracoderms. Whether or not the carpoids stand near to the ancestors of the prochordates, it is noteworthy that some of them show a functional dorsoventral asymmetry and bilateral symmetry that might be expected in a formerly radiate form that had taken to lying on that side which was henceforth to become the ventral surface.

It has been objected by Professor J. H. McGregor (personal communication) that partly because the larva of certain echinoderms (fig. 4.8D1) becomes attached by its head-end, the stalk of sea-lilies and other echinoderms cannot be homologous with the larval tail of ascidians and *Amphioxus,* and that for other reasons also the general resemblances between certain echinoderms and certain vertebrates (figs. 4.6; 5.19) are merely convergent, especially since Bather and others regarded the flattened surface of the carpoid test as homologous with the left side rather than the ventral surface of echinoderms. However, in the Middle Cambrian *Gyrocystis* (fig. 4.6C) the jointed stalk is at the opposite end from the mouth and tentacles, as it is also in other carpoid echinoderms (fig. 4.5C). The anus in *Gyrocystis* (fig. 4.6C) opens near the mouth and the gut as a whole is somewhat retort-shaped. In holothurians (fig. 4.7D) the stalk has been lost, the anus has been displaced far to the rear and a secondary bilateral symmetry has been attained. In the development of typical ascidians (fig. 5.3) the embryonic tail becomes far displaced sidewise with reference to the apical adhesive processes and to the essentially U-shaped gut, while the enormous pharynx is turned nearly at right angles to the oesophagus and stomach. In the *Salpa*-like ascidians (fig. 5.5) the atrial aperture (which in attached forms is near the oral aperture) becomes displaced to the rear end, thus establishing a secondary anteroposterior differentiation and bilateral symmetry. In vertebrates both mouth and anus remain ventral to the notochord as they are in *Amphioxus* (fig. 5.1) and the ostracoderms (fig. 6.12). Thus the long distance between the mouth and the anus and the lengthening of the gut may be due in part to the multiplication of the locomotor somites.

In view of such profound and diverse shiftings of organs in the development of the known groups, it seems that, whatever may be the explanation of the apparent reversal of larval "head-end" into adult stalk in certain crinoid echinoderms (fig. 4.8D2), the stalk of carpoid echinoderms (fig. 4.6A-C) corresponds *functionally*, but probably not homologically, with the notochord of *Appendicularia* (fig. 5.4A) among the ascidians.

In the echinoderms the water-tube system (fig. 4.7) is derived from mesoderm sacs, similar in principle to those which give rise to the myotomes or body segments of vertebrates. The echinoderms, however, had taken only the first step in the development of a head, namely, that of surrounding the mouth with tactile or prehensile "arms," and so far as we know the primitive carpoids lay so far below the morphologic level of *Amphioxus* that it is risky to assume a connection between them. To McGregor indeed the pores in *Cothurnocystites* are part of the water system and not homologous with the branchial pouches of *Amphioxus*.

Garstang's Neotenous Larva Theory (fig. 5.6C-D1).—De Beer writes (1930, p. 65; 1940, p. 52):

> "Garstang [1894] was the first to look for the trace of the ancestors of the vertebrates in early instead of adult stages of invertebrates; and he focused his attention on the larvae of Echinoderms (starfish, sea-urchins, sea-cucumbers, etc.). He showed that if the ciliated bands on the larva (auricularia) of a sea-cucumber were to become accentuated and rise up as ridges leaving a groove between them, and if these ridges were to fuse, converting the groove into a tube, a structure would be produced which has all the relations of the vertebrate nervous system, including such details as the neurenteric canal."

De Beer then points out that this theory of the origin of the vertebrate nervous system has several advantages, which he discusses (*ibid.*, pp. 52–54). He concludes that the adult structure of the starfish, sea-urchin or sea-cucumber is so special and peculiar that no one would dream of regarding the adult form of any echinoderm as ancestral to anything at all. He does not propose that the chordates were derived from the larvae of echinoderms as they now are, but only that there must have been animals whose adults were less special and peculiar than the adults of existing echinoderms, and whose larvae resembled those of existing echinoderms. He cites Garstang's theory that if the larval form of such animals persisted and they became sexually mature in this state, such neotenous forms would provide ". . . exactly the necessary material for the evolution of the chordates on the lines suggested" (fig. 5.6A). However, it must be noted that this theory still lacks further evidence.

CONCLUSIONS: AUTHOR'S ECLECTIC THEORY

(Predators from Food-Sifters; *Amphioxus*, an Archaic but Specialized Vertebrate)

Out of the shadows of Pre-Cambrian time many major groups of invertebrates emerged and were already clearly recognizable in the early Cambrian ages (fig. 6.1): here were ancestral sponges, coelenterates, annelids, molluscs; here were ancestors of the thrice-venerable *Peripatus,* connecting link between annelids and arthropods; here were trilobites in great numbers and diversity, survivors of the ancestral stock of the crustaceans, limuloids and later arthropods. The vertebrates did not appear in the record until the Middle Ordovician and are not abundantly represented until the Upper Silurian. From these fossil forms and from the anatomy and embryology of their modern descendants an enormous amount of evidence has been gathered, analyzed and compared, and this evidence seems to support the following observations and conclusions:

1) The loss of chlorophyll in the single-celled ancestors of the animal kingdom imposed on them essentially dependent modes of subsistence, their food ranging from microscopic floating particles to large, actively resisting prey.

2) Food-sifting, the extraction of very small organic particles from the water, became very highly developed among the sponges, brachiopods, bryozoans, pelecypods, echinoderms; while killing the prey in one way or another was practiced by many coelenterates (with their dart cells), by the primitive annelid worms (with chitinous jaws), by the sea-snails (with their rasps), and by the arthropods (by modifications of their paired appendages).

3) In most of the invertebrates the larvae are free-swimming, moving by means of ciliated bands. The adults may be either sedentary, stalked or secondarily free.

4) In food-sifting forms, such as clams, brachiopods, and some echinoderms, there is little or no need of a head, and the nervous system is often very simple.

5) The development of a "head" containing large ganglia and connectives for the eyes and other sensory organs, together with extensive motor nerve tracts, reaches its climax in predatory, freely moving animals with an elaborate locomotor system, such as squids, limuloids, scorpions, spiders, insects.

6) Among the prochordates food-sifting, usually without a real head, is practiced by the following groups: (a) the acorn-worms (fig. 5.20C1) and supposedly related hemichordal forms, which are either very slowly moving or sessile and semi-colonial; (b) the ascidians (fig. 5.20D1), which have developed an enormous food-sifter in the pharynx and are typically sessile, but only in the adult stage, the larvae (D) having a well developed locomotor system and a much better "head" than that of the adults; (c) *Amphioxus* (E), which has the quite exceptional combination of adult food-sifting and a very low grade of head, with extremely rapid locomotion, the latter made possible by a great number of myomeres or muscle segments.

7) *Amphioxus* may well be specialized in its nakedness, in the prolongation of the notochord beyond the brain, in the great multiplication of the gill bars and many other features; yet its food-sifting habit may still have been inherited in part from pre-vertebrate days, before such predatory structures as jaws made from gill-arches had been developed.

8) Among living forms the grade above *Amphioxus* is represented by the "*Ammocoetes*" (figs. 6.19A; 6.20B1), or larval stage of the lamprey; this has an extensive food-sifting-aerating apparatus in its pharynx which is gradually modified as the predatory habits and suctorial rasping jaw-parts of the adult (fig. 6.20A1) are developed.

9) According to Patten (1912), the "grand strategy of evolution" involved the transformation of predatory, large-brained arthropods into predatory vertebrates; but, according to the preceding analysis, the more remote ancestors of the vertebrates will more probably be found to be related to the food-sifting brachiopods and echinoderms.

10) In the previous chapter it has been sug-

ested that many of the resemblances between ertain of the advanced members of the arthro- od series and the vertebrates are associated vith the independent convergent transformation f stationary, headless food-sifters into actively 1oving, predatory animals with large brains nd relatively complex habits and intelligence.

11) The earliest known vertebrates, which vere the ostracoderms (figs. 6.2–6.18), by Aiddle Ordovician times had lost all indisput- ble external resemblances to echinoderms or rachiopods and had attained the general verte- rate plan.

In short, so far as the evidence is available, : tends to show that the vertebrates belong to n "enterocoelic" superphylum (fig. 5.2) which vas entirely different from the "teloblastic" eries named Trochozoa, which includes the nnelids and arthropods. Moreover, the curious ouble relationship of the prochordates on the ne hand to stalked or sessile food-gatherers uch as *Cephalodiscus* (fig. 5.7), equipped with iliated, food-collecting lophophores, and on he other to the highly motile, fish-like *Amphi- xus* (fig. 5.9A), seems quite consistent with he evidence that, whereas some of the trilobites f the teloblastic series took up predatory, free- 1oving habits at an extremely early (Pre-Cam- rian) date, the enterocoelic series either did ot do so at all (as in brachiopods) or became ctively predatory only at relatively late periods 1 the histories of their several phyla. This is nown to be the case in the asteroid and echi- oid echinoderms and, as will be shown in the ext chapter, is also indicated by good palaeon- ologic evidence in the vertebrates. But since a rge part of the transformation probably oc- urred during the stage of the free-swimming rvae, which were devoid of skeletons, it is not urprising that intermediate stages are lacking 1 the known fossil record.

Evidence bearing on the beginnings of the keleton among the vertebrates comes in gen- ral from three main sources: comparative anat- my, comparative embryology and palaeontol- gy. No one of these sources by itself has as yet ielded enough unequivocal evidence to win eneral acceptance among specialists. For example, the evidence of comparative embryology has been interpreted by Garstang, De Beer and others as indicating that the Tornaria larva of the balanoglossids and the Dipleurula and Pluteus larvae of echinoderms are living relicts of a very remote period antedating the divergence of echinoderms and chordates. But H. B. Fell (1948), at the conclusion of a most comprehensive review of the comparative embryology and classification of the echinoderms, shows that parallelism in development has often taken place among the different subclasses of the echinoderms and he concludes that the similarity of the free-swimming larvae of certain echinoderms and of *Balanoglossus* supplies no trustworthy evidence of common ancestry. On the side of Palaeontology, Torsten Gislén noted that certain asymmetrical Silurian carpoid echinoderms show some curious points of resemblance to *Amphioxus,* while the writer (1935, fig. 6) has called attention to the resemblance between the plated calyx of other fossil echinoderms (*Placocystites, Mitrocystella*) and the armor of a Devonian ostracoderm, *Drepanaspis*. Such separate items of resemblance have not yet escaped the suspicion of being both fortuitous and convergent.

We may even concede, at least provisionally, that *Balanoglossus* may not be a chordate at all and that its baglike, ciliated, free-swimming larva is merely a parallel adaptation among different groups for securing a wide distribution in the search for suitable localities for the sessile or slow-moving adult. Nevertheless the general drift of evidence still suggests: (1) that although *Amphioxus* may be secondarily simplified in some features and peculiarly specialized in others, yet it is, after all, a true archaic chordate; (2) that the oldest known subclass of fishes, the ostracoderms of Ordovician to Upper Devonian age, although diversified in form and habits, had not yet attained the structural grade of the swift-swimming and predaceous jaw-bearing fishes which succeeded them in later ages; (3) that they retained indications of having been derived from a jawless, bottom-living, slow-moving enterocoelic stock, of which *Amphioxus* is perhaps a much modified survivor;

(4) that in the ostracoderms we find at last many reliable data as to what the outer and inner skeletons of the primitive vertebrates were actually like and how they functioned se eral hundred millions of years before the ance tors of modern man walked upright.

CHAPTER VI

THE VERTEBRATES COME FORTH: SHELL-SKINNED ANCESTORS OF THE VERTEBRATES AND SOME OF THEIR BRANCHES

Contents

CHAPTER VI

THE VERTEBRATES COME FORTH

THE EARLIEST KNOWN VERTEBRATES (OSTRACODERMS) AND THEIR MANY-LAYERED ARMOR

THE OLDEST admitted traces of the ostracoderms consist of an imperfect dorsal head-shield (fig. 6.2) and hundreds of fragments containing tesserae or minute units of the head-shield and detached "scales," from the Upper (?) Ordovician of Canyon City, Colorado. But even these fragments yield striking evidence as to the great complexity of the outer shell that enveloped the head and thorax of *Astraspis desiderata* and *Eriptychius americanus,* which were both named by their discoverer, Charles D. Walcott, a famous geologist. Otto Jaekel referred them to the Crossopterygii or fringe-finned ganoid fishes, but microsections by Bryant (1936) proved that the plates and scales belonged to ostracoderms that were not fundamentally different from the better known pteraspids (figs. 6.5–7) of Silurian and Devonian ages.

Each minute tessera of *Astraspis* (fig. 6.3) somewhat resembled a tooth in so far as it was composed largely of a bone-like or ivory-like substance, which, however, lacked true bone cells. Evidently this once living organ grew by depositing concentric layers of ivory-like material. Presumably the foci of growth were the tips of the numerous little star-like "enamel caps" or tubercles on the outer surface (fig. 6.3B). Each living osteoblast-like cell first excreted the tip of the "enamel cap" and then spread out radially like a growing many-rayed patella-shell (fig. 2.4F), depositing successive layers which were spread laterally while the tip was pushed outward. The growing edges received their material from the membrane-lined pulp-cavities and nutrient canals. The straight or meandering tunnels (figs. 6.3, 4, 8) must have originated from the living film of cells, which, by squeezing outward the deposits already laid down, made room for the vital pulp keeping it open, while narrowing its walls into tubes and irregular caverns.

The basal layer of the growing exoskeletal membrane was bounded internally by the outer wall of the mesoderm system, which in *Amphioxus* and true fishes gives rise to the muscle plates and the coelomic cavity. The successive basal layers of the exoskeletal membrane did not penetrate the muscular layer of the body wall but grew in curved lamellar surfaces as if it were enclosed by a limiting membrane. Thus in cross-section of a single tessera of *Astraspis* (fig. 6.8) four zones may be distinguished: (1) the basal zone (isopedin) of curved lamellae; (2) the middle zone of meandering nutrient vessels called the trabecular (beam-like) layer; (3) the denticle-like "cosmine" layer (fig. 6.4D), essentially like a single denticle of a shark's skin; (4) the glass-like enamel layer (fig. 6.3A, A1, A2). To these may possibly have been added: (5) an outer somewhat horny epithelial layer covering the enamel cap and corresponding to the horny outer layer of later fishes and higher vertebrates.

Exactly the same layers may be identified in later pteraspid ostracoderms, except that in many of the pteraspids (fig. 6.8C) the meandering and irregular passages in the middle layer have been organized into fewer and larger cavities separated by more or less straight partitions.

To some extent the sequence of layers noted

above in the shield of *Astraspis* is also comparable with that in the shell of the king crab (*Limulus*) (Patten). But in *Limulus* (fig. 5.16A) the horny epithelial layer is predominant, as it is in all the arthropods or externally jointed animals.

In *Eriptychius,* the second Ordovician ostracoderm whose surface tubercles were studied by Bryant, the units (fig. 6.4A, B) look more like scales and partly overlap each other. In microsection (C) these scales lack the radiate enamel-like tips of *Astraspis* tesserae. Somewhat like those of some Silurian pteraspids, they are ornamented with short subparallel dentine ridges (D) which tend to break up into low-crowned tubercles. The laminated basal layer (isopedin) abounds in large vertical canals (E). Bryant found that *Eriptychius* is closely related to several later pteraspids, although as noted by Jaekel its microstructure is also basically that of the older crossopterygian fishes (cf. figs. 6.4C, D and 10.7A-E).

FROM THICK ARMOR TO THIN SCALES

These two samples (*Astraspis* and *Eriptychius*) of the exoskeleton of chordates from the latter part of the first half of the Palaeozoic era emphasize the high complexity of their exoskeletons in contrast with the greatly thinned, reduced and delicate skin of *Amphioxus* (fig. 5.14C). They also tend to support the modern theory of Stensiö that the primitive fishes were not nearly naked but very well armored forms. The heavy armor of the known earliest ostracoderms was formerly the chief reason for regarding them not as ancestral vertebrates but as highly specialized side lines. In later ages such heavily armored forms as armored dinosaurs and glyptodonts were in a sense over-specialized and are now extinct side branches of their respective classes, whose primitive members were at least not massively armored. But that extinction always followed the acquisition of heavy armor is proved to be an unwarranted "extension" fallacy by every oyster and clam. Homer Smith, assuming that the ancestors of the ostracoderms were naked marine forms, suggested that when they invaded fresh-water, the salt in their own blood stream would draw in so much fresh-water that they would die of oedema and that the secretion of a thick and nearly impervious surface would have eliminated this source of danger. At present, however, we cannot confidently assert that the chordates (including the ostracoderms) were derived from marine prochordates or invertebrates, because the earliest chordates may well have lived in brackish and/or fresh-water. Indeed it seems highly probable in the light of Stensiö's extensive evidence that those ostracoderms which had very thick complex exoskeletal units were more primitive than the relatively thin-scaled anaspid ostracoderms (fig. 6.12C).

LIME *VERSUS* HORN

The fact seems to be that some organisms, such as corals, oysters, typical clams and rock-shells (*Murex palmarosae,* fig. 2.23C), select and excrete lime in large quantities and thus rapidly build up a thick laminated exoskeleton, and that other organisms living in the same general environment somehow get rid of most of the lime and gradually thin out the solid parts of the shell, as in the "window-glass" shells (*Placuna,* fig. 2.48G). Again, some river shells (Unionidae) have massive, thick lime shells and others have very thin lime layers. In short, at any given time each type of organism seems to have its own range of lime secretion, but during the long reaches of geologic time these "constants" vary and shift so that in many families of marine snails, such as the rock-shells, olives and cones, the shell becomes thicker, while the reverse, from thick shells to thin shells, has also occurred repeatedly in different groups.

In invertebrates the outer layer of the integument usually consists of chiton, and when there is a shell beneath it usually consists of calcium carbonate. In the vertebrates the two corresponding layers consist of keratin and bony plates (mostly of calcium phosphate). The relative thickness of the outer to the inner layer varies in both invertebrates and vertebrates but usually within narrow limits when closely related forms are compared.

Thus among sea-snails the outer or chitinous layer is thinner and more delicate in the larger bailer-shells (*Cymbium*, fig. 2.29B), but it is very thick and resistant in the claw-like operculum of the aggressive conch (*Strombus pugilis*). Similarly among fishes, the primitive ganoids had thick bony scales with a thin horny epidermis, but among normal teleosts this condition gradually gave place to thin horny scales (as in *Tarpon*) with no bony base. In the puffer-fishes (fig. 9.106B) a new generation of horny prickles grew up in the otherwise soft and flexible skin; in the coffer-fishes (Ostraciontidae) the integument (fig. 9.105A) gives rise to a strong box-like bony armor.

The foregoing digression suggests at least that the thickness of the exoskeleton of most ostracoderms does not *per se* eliminate them from ancestry to the later fishes; and, as noted below, Stensiö has shown that in respect to the construction of their nervous, vascular and respiratory systems, the thick-shelled ostracoderms retain many characters which are believed to be primitive on other grounds.

BODY-FORM AND HABITS IN OSTRACODERMS

Before following further the problem of the evolution of the exoskeleton in the ostracoderms, it will be well to sketch very briefly the range of body-form and probable habits in the best known ostracoderm faunas.

The most simple-looking ostracoderm, from the Upper Silurian of Europe, is *Phlebolepis* Kiaer (fig. 6.6A, A1) a kind of tadpole-like fish with a small mouth, a rotund body and a rather short tail. There were small mediandorsal and caudal fin-folds not supported by individual fin-rays. The body was covered with small separate scales or denticles. If there be those who still think of evolution as always leading from the simple to the complex, such persons might well take this innocent-looking bag as an ideal starting point for all the ostracoderms. And such it may indeed eventually be shown to be. But since there are so many cases in which it appears that apparent simplicity is not a reliable indication of true primitiveness, it seems safe merely to record the following considerations:

1) The Ordovician *Astraspis* (fig. 6.2A) although much older than the form mentioned above (*Phlebolepis*) has a far more complex exoskeleton, in which the individual units or tessarae were closely appressed into a relatively large cephalo-thoracic shield.

2) Among the pteraspid subclass there are gradations, which are interpreted by Stensiö, Kiaer and Heintz as indicating the breakdown of a cephalo-thoracic shield into minute individual denticles or scales (figs. 6.5A, B; 6.7A, A1).

3) In the coelolepid ostracoderms the surface denticles of the Upper Silurian *Lanarkia* (fig. 6.6C) were reduced to delicate thorns which so closely resemble the shagreen denticles of the shark's skin that Traquair at first referred this ray-like fish to the shark class (Elasmobranchii).

4) Such denticles, as long ago interpreted by Williams and by Goodrich, probably rested upon a thick skin, which we may interpret as the uncalcified embryonic scaffolding of the many-layered exoskeleton of *Astraspis*.

5) What looks like the "body" of *Phlebolepis* (fig. 6.6A) may be in reality chiefly the outer cover of an expanded chamber essentially like the expanded "pharynx" (fig. 5.1A1) of *Amphioxus* and the "oralo-branchial chamber" (fig. 6.15A) of cephalaspids.

6) We may conceive all known ostracoderms as being essentially self-propelled food traps, with an expanded mouth-gill apparatus and with various mouth openings adapted for scooping in different kinds of food; also with varying body-forms implying different ways of approach and pursuit in waters of varying movements. In the classic *Pteraspis* (fig. 6.7) the body as a whole was fusiform and streamlined, adapted for relatively fast swimming. The mouth of small or moderate size was bordered by movable exoskeletal plates and, according to Kiaer, there were thorn-like den

ticles on the roof of the mouth and on the opposing lower piece of *Pteraspis vogti* (fig. 6.13A).

7) Although such dermal plates around the mouth of ostracoderms may have functioned as jaws, the entire group of ostracoderms is often brigaded as a superclass or subphylum "Agnatha" (jawless), because they contrast with the "Gnathostomata" (jaw-mouthed) fishes, including sharks, ganoids, teleosts, etc.), in which the jaws are of complex type with an endoskeletal core and usually with exoskeletal jaw plates.

8) In *Drepanaspis* (fig. 6.7A), representing an adaptive extreme of the pteraspid order, the mouth was very wide, bordered by toothless exoskeletal plates with rounded rims. The cephalo-thoracic or oralo-branchial chamber was very wide and depressed. Probably this rounded, somewhat ray-like form rested on the bottom, perhaps more or less concealed in the mud but able to dart forward by means of its stout tail.

9) The oralo-branchial chamber in *Pteraspis* (fig. 6.7B2), *Cyathaspis* and allied genera (fig. 6.9A) had a more or less tunnel-like central passage bordered on each side by a row of small gill-pouches. These opened laterally in a duct which was covered by a long horizontal surface plate (figs. 6.7 *lateral;* 6.10A-E), behind which was a single orifice or branchial opening (fig. 6.9A). Thus the water was apparently drawn in either by ciliary action or by a pumping movement of the floor of the oralo-branchial chamber and passed out behind through the post-branchial apertures.

10) In the anaspid ostracoderms (fig. 6.12C-F) the body was relatively long and undulatory, the cephalic part small; the expanded oralo-branchial chamber may have been continued posteriorly into metapleural folds (D, *lat fin fold*) as in *Amphioxus*. There were median plates at the lower front end of the mouth-scoop which may have served as nippers. These Anaspida might be the direct ancestors of *Amphioxus* (fig. 5.1A-A3) by further reduction of the head, forward extension of the notochord, expansion, multiplication and folding of the gill-pouches, transformation of the metapleural folds into lateral fin-folds, multiplication of the body segments, etc.

11) Postscript, 1950. In 1946 Dr. E. I. White of the British Museum (Natural History) described a very remarkable ostracoderm from the late Silurian shale of Lanarkshire, which he named *Jaymotius kerwoodi.* The generic name was given in honor of his friend and colleague, J. A. Moy-Thomas, who was killed in World War II. As later restored by White (1949), on the basis of further studies, *Jaymoytius* was somewhat similar in general outline to *Amphioxus,* and in other respects it resembled some of the anaspid ostracoderms (figs. 6.12D, F; 6.18B). It had long horizontal lateral fin-folds, running from behind the very small head to the base of the tapering tail, which, however, was not turned downward as it was in Anaspids. There was a long spineless dorsal fin and a shorter anal fin. The skin was apparently very thin and without armor. The internal skeleton seems to have been cartilaginous.

Taken altogether this very ancient type tends to connect *Amphioxus* with the anaspid ostracoderms. It is regarded by White as one of the most primitive known fossil vertebrates and it does indeed appear to lend some support to the older view that the ancestral vertebrates were naked or thin-skinned forms, rather than heavily armored as in the Ordovician *Astraspis* and *Eriptychius.* Quite conceivably, however, the boneless *Jaymoytius* may well have been on the way toward *Amphioxus,* while the bony *Cephalaspis* may have been on or near the line leading to the lampreys.

ANATOMY OF THE HEAD-SHIELD IN CEPHALASPIDS AND LAMPREYS

In the typical cephalaspid ostracoderms (fig. 6.12A, B) the very large cephalic shield has a more or less lunate or semicircular outline as seen from above (fig. 6.11D). The body of *Cephalaspis* was also large, somewhat flattened below but with high sides, well filled with very numerous muscle-segments, covered by thin plates (fig. 6.12B). In some species the body was much higher than the head-shield, which sloped downward and forward like a snow-plough. Presumably *Cephalaspis* moved forward on or near the bottom by powerful thrusts of its muscular segments. In contrast with the laterally placed eyes of pteraspids, the large paired eyes of cephalaspids (which presumably arose in the embryo from the neural plate) have been shifted up on to the top of the head-shield (fig. 6.11D), which on the whole suggests that of the king crab *Limulus;* but, for reasons elsewhere set forth (pp. 91–93), the resemblance is probably entirely due to convergence.

The oralo-branchial pouches of the Cephalaspids (fig. 6.14A) were arranged in a more or less circular row (as seen from below), decreasing in size posteriorly. They may well have been essentially similar to the gill-pouches of the living lampreys (fig. 6.19A). In front of the anterior gill-pouches of *Cephalaspis* is the median oral cavity (fig. 6.14A), which may possibly have resulted from the junction of a pair of gill-cavities, as long ago inferred by students of the embryology and comparative anatomy of sharks and other fishes.

In some of the beautifully preserved cephalaspid shields discovered by Danish and Swedish expeditions to Spitzbergen and Greenland, fine silt had seeped into the interior of the brain-case (fig. 6.15A, A1) and into the numerous blood vessels and nerve tubes (fig. 6.16) of the shield itself; also into the walls and cavities of the oralo-branchial chamber and into the partial partitions between the gill-pouches (fig. 6.14A). After long ages the silt had changed into or been replaced by mineral substance. Stensiö, after years of comparative study and skillful preparation of this material, was enabled to publish a superb series of plates (1927, 1932) illustrating these uniquely important fossils, and to identify and name every part of the braincast and every tube for the cephalic nerves, blood vessels, etc.

Since the structural details of the cephalaspid head differ in important features from those of the higher fishes, it would have been impossible for Stensiö to do this, at least so completely, if the comparative neurologists and students of vertebrate brains and cranial nerves had not already identified and equated the corresponding parts in the higher vertebrates, the fishes and especially the living lampreys (fig. 6.21B).

The lampreys (fig. 6.19A1) have become highly specialized for attacking living fishes, to which they attach themselves with bilaterally folding lip-jaws studded with sharp cuticular thorns. They have similar thorns on their highly protrusile muscular tongue, which is used as a rasp to cut through the body-wall of the victim. This carnivorous specialization has carried the adult lampreys far beyond any of the known ostracoderms; but each lamprey still goes through a larval stage (fig. 6.19A) which recalls *Amphioxus* (fig. 5.1A1), especially in lacking jaws, and in drawing in, by ciliary action, water containing food particles and oxygen. In spite of all their specializations the lampreys have retained certain primitive features (such as the separateness of the ophthalmic (fig. 6.21V1) from the rest of the trigeminal nerve (V2 + V3) which were likewise present in the ostracoderms (fig. 6.16V1) but have been lost in higher vertebrates.

BONE CELLS IN CEPHALASPIDS, AND THEIR POSSIBLE ORIGIN

The cephalaspid ostracoderms had taken a momentous step in that they or their ancestors had evolved true bone cells (fig. 6.17A1). Although the higher vertebrates were probably not derived directly from known cephalaspids, they shared at least in the inheritance of this once new tissue, which eventually, according to some authorities, became "twice as strong as oak," so that "a cubic inch of it would support 5,000 pounds."

Presumably the bony tissue of these excessively ancient pre-vertebrates was built up by living osteoblasts which in turn had been derived from connective tissue cells. These osteoblasts may have found their way into the interior of the many-layered integument of the embryo through the numerous tubes and irregular spaces, especially in the middle layer of the cephalaspid armor (fig. 6.17A1, m.l.). We may perhaps conceive the primitive osteoblast as related to blood corpuscles, but with a strong affinity for calcium salts dissolved in the blood stream. The bony tissue of cephalaspids probably represents an advance beyond the dense dentine-like tesserae of *Astraspis* (fig. 6.2) in that the bone cells were laid down concentrically around the minute blood vessels which carried them and that these vessels were perforated by a system of Haversian canals and radiating canaliculi. Possibly also some of the living osteoblasts may seep through these canals and form a new bone layer outside or around the first one, somewhat as the living cells of the Foraminifera do. Perhaps later on in vertebrate history the bone-secreting units, instead of all being formed *in situ,* may have been turned out as osteoblasts at some nearby center of osteoblast formation and carried through the lymph spaces or the blood vessels to the parts that attracted them, somewhat as white blood corpuscles are attracted to invading foreign cells. Moreover, it is not improbable that one great advantage of numerous bone-cells, with connecting Haversian canals and canaliculi, over mere layers of ivory-like tissue laid down in the primitive tesserae, was that the bony deposit, being penetrated by canaliculi, could be more easily dissolved by the osteoclasts, which continually remodel the bone during its growth and changes of curvature.

In the skeleton of adult cephalaspids, Stensiö has observed bone cells in all parts of the cephalo-thoracic shield, not only in or near the surface but in such deep structures as the partitions between the gill-pouches (fig. 6.14A) and the "endoskeletal shoulder girdle" (fig. 6.15A2).

THE ORALO-BRANCHIAL CHAMBER PROBABLY DERIVED FROM THE STOMODEUM

Probably the entire oralo-branchial cavity (fig. 6.15A, A1) and all its subdivisions may represent a huge stomodeum or inpocketing of the outer surface of the embryo at the end opposite to the blastopore. This invagination of the stomodeum would carry inward ectoderm on its inner surface, mesoderm beneath the ectoderm, and perhaps endoderm in its deepest parts. In *Cephalaspis* inturned mesoderm may have given rise to the partial septa between the gill-pouches, the ectoderm and mesoderm to the gill-pouches themselves, while invading skeletal cells may be the source of the endoskeletal frame (fig. 6.15A) around the brain and sense organs.

MAIN SKULL PARTS

Thus the cephalaspids foreshadowed the later vertebrates in possessing a complex skull, consisting of exoskeletal and endoskeletal regions and comprising: (1) capsules or cavities for the olfactory, optic and otic (inner ear) organs, arranged in anteroposterior series (fig. 6.16); (2) a median trough (fig. 6.20D) for the brain, perforated by foramina for the cranial nerves and chief blood vessels; (3) an occipital base pierced by a groove for the notochord (D) (that structure, however, did not extend forward to the front end of the head as it does in *Amphioxus* (fig. 5.1A1), but was posterior to the hypophysis, as it is in all higher vertebrates); (4) the roof of the oralo-branchial chamber was strengthened in part by the floor of the braincase, in part by the bony transverse partitions between the gill-sacs (fig. 6.13A), which extended downward and probably supported the gills; (5) forward thrusts from the strong locomotor muscles (fig. 6.12B) were transmitted to the large and well built skull ("head-shield") partly through the "endoskeletal shoulder girdle" (fig. 6.15A2), partly through the anterior thoracic plates, which were behind the head-shield (fig. 6.11) but often fused with it.

No traces of vertebral centra around the vertebrae or of neural arches have been recorded in ostracoderms, but there must have been strong "myocomata" or double septa between the adjacent muscle segments.

The rear border of the shield in typical cephalaspids is curved forward (fig. 6.11) on each side to make room for the pectoral fins. These were paired extensions of the muscles of the body wall forming mobile lobate organs (fig. 6.12A, B) which may have been useful both in steering the large head-shield and in digging in the mud. They were based upon the "endoskeletal shoulder girdle" (fig. 6.16A2) and were at least potentially homologous with the pectoral paddles and fins of later fishes.

THE CYCLOSTOMES, EEL-LIKE SURVIVORS OF THE OSTRACODERMS

As noted above, the modern lampreys may well represent direct derivatives of the cephalaspids, but it was not until Stensiö showed that the highly complex relations of the brains, cranial nerves and blood vessels of cephalaspids could be identified by reference to their evident homologues in the lamprey, that the true relationships of the latter were recognized. But the lampreys have progressed much further than the cephalaspids in their specializations for attacking modern fishes and for rasping holes in them. This specialization has involved a great increase in the size and complexity of the muscles of the thorn-bearing tongue and of its partners, the laterally moving "lip-jaws," together with their supporting plates and bars. This region is unknown in the cephalaspid ostracoderms and was in them probably far less complexly organized. It was long ago noted by Gaskell that the crescentic plate of mucocartilage which braces the sucking-disc of the lamprey (fig. 6.20B) bears a striking resemblance to the crescentic head-shield of *Eukeraspis* (B), one of the cephalaspids. In that case, however, it must be supposed that the bony interior of the head-shield was gradually replaced in embryonic stages by mucocartilage. If a small-headed cephalaspid were to try to attach itself to a relatively large fish, it is not unlikely that the strong rim of the head-shield would be used to press against the victim and to help the suction set up by the flexible floor of the mouth. Thus it is hardly surprising that the lampreys, which agree so closely with the ostracoderms in the elaborate pattern of their cranial nerves and blood vessels, should have adopted a method of obtaining food in which suction by the floor of the mouth was supplemented by improved devices for seizing and pulling in large pieces of food. Nor does it now seem strange that during their embryonic development the lampreys pass through a larval stage ("Ammocoetes") in which the entire anatomy of the head (fig. 6.19A) is fundamentally similar to that of cephalaspids, or that the food consists of minute particles drawn in by ciliary action somewhat as in *Amphioxus*.

In the lampreys (fig. 6.20) body-scales have been eliminated and the cephalo-thoracic shield reduced to the supporting plate for the oral sucker. Meanwhile the bone-cells also have disappeared and the entire oralo-branchial skeleton consists of mucocartilage; in other words, it has retained the larval condition. The "branchial basket" of the lampreys consists of a thin, trellised, elastic framework (fig. 6.20B1) between and around the gill-openings attached to the reduced skull above and to the floor of the mouth below. In appearance it is quite unlike the beautifully articulated bent-lever system (fig. 7.27C) of the sharks, but would be readily derived from the conditions in *Cephalaspis*.

Still more specialized are the hagfishes (or myxinoids), which agree with the lampreys in fundamental characters but show further reduction of the cranial skeleton to a thin tube around the brain and various bits of the gill skeleton. But the stiff cartilaginous median tongue-bar is still functional and the very large and cylindrical fleshy tongue can be worked forward and backward like a rasp. Each of the numerous gill-sacs has a lateral tubular exit, but these tubes run together into a common duct which opens at the rear, as in the pteraspid ostracoderms (figs. 6.7B, B1; 6.9A). The tongue-bar and small rods of cartilage between the gill-opening are connected by narrow strips of muco-cartilage, but except for the tongue-bar, these endoskeletal pieces are much reduced as compared with those of the lampreys. The hagfishes also differ from the lampreys in that the large nasal naso-hypophysial sac penetrates the roof of the mouth (figs. 6.19B; 6.22B2) and is not shifted on to the top of the head as it is in lampreys (figs. 6.20C *naso hyp;* 6.22A4).

The lampreys as noted above pass through a free-swimming larval stage which draws in food particles by ciliary action; but the hagfishes develop directly into miniature hagfishes, within the shelter of their large-yolked eggs, which have strong cylindrical horny shells. As described by Bashford Dean (1898), the embryo of the California hagfish, after a long period of dependence upon the abundant food-yolk, comes forth from the egg and is soon ready to begin the predaceous habits of its race. The exoskeleton in the hagfishes is represented only by a pliable skin, whereas in the oldest ostracoderms it consists of very hard bone-like tesserae. Along each side of the body of the hagfish runs a line of glands, the mucous secretions of which take up an enormous amount of water, enveloping the fish in a jelly-like mass and assisting it in pushing itself into the bodies of the prey through holes bored by the dental rasp. Thus the hagfishes are at least temporary parasites on fishes and have indeed taken on some of the worm-like, slimy habitus of certain intestinal worms.

On account of these and other differences, Stensiö separates the myxinoids very widely from the cephalaspids, deriving them not from the latter but from the pteraspids.

Although the lampreys, and still more the hagfishes, are highly specialized side branches, which in their adult habitus are far from representing the direct ancestors of higher vertebrates, yet they alone among modern forms retain some primitive ostracoderm features, such as the separateness of the ophthalmic nerve (fig. 6.21B, A, VI) from the rest of the trigeminal complex, the failure to evolve all three of the semicircular canals of the inner ear (the external canal being absent), the pouch-like character of the gill-sacs, the unjointed endoskeletal tissue between the gill-sacs, the failure to evolve the complex jaws with endoskeletal core and dermal covering bones of jaw-bearing vertebrate type.

Goodrich (1909, figs. 20, 22 and page 38) notes that since the large muscles of the rasping tongue of both lampreys and hags are supplied by branches of the fifth cranial nerve (fig. 6.21V2, *ram md.*), it may be concluded that they represent the so-called visceral muscles of the mandibular region and that the cartilages to which they are attached correspond to the lower part of the mandibular arch (Meckel's cartilage) and perhaps to the hyoid arch as well (Ayers and Jackson). Stockard (1906) also found that in the embryo of the California hagfish the cartilages of the "tongue," in their relations to surrounding structures, correspond to the mandibular arch of jaw-bearing vertebrates.

"JAWLESS" AND JAW-BEARING VERTEBRATES

In view of the foregoing considerations, the practice of putting the ostracoderms and marsipobranchs (lampreys and hags) in a superclass Agnatha ("jawless") over against the Gnathostomata (with typical vertebrate jaws), including all the higher vertebrates, may indeed eventually be proved to be inadvisable, especially in view of the following considerations: (1) From the evidence supplied in cephalaspids, especially by the topographic relations of the cranial nerves (fig. 6.21A) and blood vessels and their branches in relation to the skeletal parts, Stensiö (1927) identified the series of transverse ducts and endoskeletal septa that depend from the roof of the cephalaspid oralo-branchial cavities (fig. 6.14A) as the paired oral, hyoid and eight branchial ducts. This would account for the basic conditions in the modern lamprey (fig. 6.19A1), save for the changes in the latter that are associated with the great enlargement of the thorn-studded upper "lips" and rasp-like "tongue." (2) It is highly probable that the oralo-branchial chamber as a whole is homologous in the "agnathous" ostracoderms and in such "gnathostomes" as the placoderms (fig. 7.3C, B); that is, it seems likely that the differences between them were originally not qualitative but rather quantitative (anisomerous). (3) Within the class of placoderms there was evidently a wide range in food habits and feeding adaptations, from the ingestion of small organisms or water algae in *Bothriolepis* (fig. 7.3B) to aggressive predation in the shear-toothed *Dinichthys* (fig. 7.13). (4) Since there is good evidence (fig. 7.27A, B, C) that jaw muscles in fishes are essentially locally specialized gill-pouch muscles, it is not surprising that in the placoderms the originally supporting tissues of the oral pouches should have increased and become further differentiated along with the muscles that moved them, or that they early began to support the superficial "lip plates" around the mouth.

CHAPTER VII

EARLY EXPERIMENTS AMONG STREAMLINED PREDATORS (PLACODERMS)

Contents

CHAPTER VII

EARLY EXPERIMENTS AMONG STREAMLINED PREDATORS (PLACODERMS)

THE ANTIARCHS AND THE PLACODERMS

MORE THAN a century ago Hugh Miller gave wide publicity to a certain Devonian fish-like fossil named *Pterichthys* (fig. 7.2), which had a distinct rounded head enclosed in a bony helmet, a tortoise-like domed carapace and flat plastron, and a pair of fairly long jointed appendages deceptively suggestive of the appendages of both sea-turtles and crayfishes. After some decades this puzzling combination of fish, turtle and crayfish features, which was seen also in its North American relative *Bothriolepis* (fig. 7.4A), led Gaskell (1896) and Patten (1912) to regard these forms as relicts of an intermediate stage connecting the vertebrates and the arthropods. Cope (1885) inferred from certain specimens of *Bothriolepis* that the anus did not open on the dorsal side, as it does in ascidians (fig. 5.3A), but in the normal vertebrate position. Nevertheless he took the bold but quite unwarranted step of referring the pterichthyids to a new order, of the class Urochorda (ascidians), which he named Antiarchi, in reference to the supposed opposite position of the anus (*anti*—opposite, *archon*—anus). Although Cope's name Antiarchi has been retained for the order that includes *Pterichthys* and allied genera (fig. 7.2), their bony head-shield and thorax are basically like those of cephalaspids in construction; but they differ in having the head-shield composed of separate bones connected by sutures (fig. 7.7A), while the externally jointed pectoral appendages of antiarchs (fig. 7.3) are unique.

Thanks to early exploratory mistakes, to the clash of independent minds, to many new factual discoveries and to the highly constructive work of a few leaders, the fundamental plan and varied manifestations of the placoderm class are coming into clearer view.

Careful analysis of an extensive series of facts led Hussakof (1906, pp. 134, 135) to re-adopt the classification of M'Coy, who in 1848 bracketed *Pterichthys* with the Arthrodira under the name Placodermata.

Patten (1912) demonstrated that *Bothriolepis* had small but functional upper and lower jaws made of dermal bone plates. Stensiö (1934) confirmed Jaekel's conclusion that the functional jaw-plates of certain arthrodiran placoderms were supported by cartilages (fig. 7.17) that were homologous with the Meckel's cartilage of sharks and of higher vertebrates. Stensiö was also able to show in detail that the upper and lower jaw-plates of the Antiarchi were severally homologous with those of placoderms (figs. 7.16; 7.17). Therefore he removed the Antiarchi from the Agnatha (ostracoderms and related forms) and transferred them, along with the various orders of placoderms, to the superclass Gnathostomata (jaw-mouths). The recognition of the Placodermi by Stensiö (1934, 1936) as a group distinct from the ostracoderms on the one hand and from the true fishes on the other has definitely advanced our knowledge of the evolution and interrelationships of all the earlier vertebrates.

Behind the small head of *Pterichthys, Bothriolepis* and allied forms was a deep transverse crease (figs. 7.2–4) and immediately behind this crease was the anterior rim of the bony thorax, which was apparently homologous with the "endoskeletal shoulder girdle" of the ephalaspid ostracoderms (fig. 6.15A2). Denison (1941) found that the gill-pouches of *Bothriolepis* (fig. 7.3B) were small and that the whole oralo-branchial chamber lay beneath the small head-shield and in front of the pectoral appendages. The capacious thorax contained the gut and other viscera.

The principal locomotor organs of the Antiarchi were the pectoral appendages, which suggest the well-jointed armor covering the shoulder, arm and hand of a knight, except that the distal end was pointed rather than hand-like. There was a complex ball-and-socket-like joint at the shoulder (fig. 7.5); the fist-like pedicle of the socket projected from the bony shoulder-girdle and the hollow "ball" was at the proximal end of the arm. The double relation of the ball to the socket in *Bothriolepis* was only partly like the simple relation of the ball-like head of the human femur to the socket (acetabulum) of the hip, because the outer surface of the ball was in sliding contact with the inner surface of the socket, while the inner surface of the ball was in sliding contact with the outer surface of the fist-like pedicle. Evidently the pectoral limb of *Bothriolepis* was able to move through considerable arcs and to rotate in its axis so as to perform fairly complex sculling movements.

It is commonly assumed that *Bothriolepis* and its allies are "highly specialized," the implication being that they are also aberrantly specialized and very far from being ancestral to any other kinds of vertebrates. But there are not wanting indications that retrogression has frequently played an important part in the emergent evolution of new types. Heintz (1931), for example, has indicated the high probability that the later arthrodires, which have a large pectoral girdle (thorax) but only a vestigial nubbin to represent the pectoral fin, have been derived from the acanthaspids by progressive reduction of the latter's massive pectoral spikes (fig. 7.9). There is still a wide gap between the freely movable pectoral appendages of *Bothriolepis* and the fixed spines of the acanthaspids, but in the Lower Devonian *Lunaspis* (fig. 7.4B) the large pectoral spikes, although fastened to the plastron, are apparently homologous on the one hand with the lobate, unarmed pectoral flaps of *Stensioella* (fig. 7.23B) and on the other with the movable appendages of *Bothriolepis* (fig. 7.4A).

The deep narrow cleft between the head and thorax in *Bothriolepis* and *Lunaspis* is more highly evolved in the head-neck hinge-joint of arthrodires (figs. 7.9C, D; 7.15; 7.19A1). The plastron of the antiarchs (fig. 7.11A) has approximately the same elements as in the arthrodires (fig. 7.11B-D); but, with the assumption of predaceous habits, the skull-roof of the arthrodires (fig. 7.15A) could be swung upward on the peg-and-socket joints between the skull and thorax; whereas in the Antiarchi this joint, although present, was not concentrated but extended across the occipital region, affording a firm base for the undulations of the tail and the thrusts of the pectoral appendages.

The Joint-necked Fishes (Arthrodira).—This diversified Devonian group began as small ostracoderm-like forms (acanthaspids, fig. 7.9A) with small head-shields and very large pectoral spikes affixed to the strong flat plastron (fig. 7.11). During the evolution of the main group the general size became gigantic, the jaws highly predatory (fig. 7.9D) and then variously retrogressive, the pectoral spines almost disappeared (fig. 7.11H). The head-shield was connected by a pair of peg-and-socket hinges with the thoracic buckler and could be swung upward as the jaw was depressed, but the jaws, as already stated, were broadly homologous with those of true fishes and were operated by a basically similar muscular system. The transverse joint between head and shoulder probably permitted waving movements in the vertical plane—a rare method of locomotion in fishes. The thoracic cuirass (fig. 7.12C-F) became shortened anteroposteriorly in the later forms until it resembled the dermal shoulder-girdle of typical fishes (fig. 7.19B).

In the macropetalichthyids the cephalic shield (fig. 7.8B) was homologous with that of the arthrodirans but lacked the movable hinge between the skull and thorax. The braincase (figs. 7.20B; 7.21A), the braincast (Stensiö, 1925) and the branchial arches (fig. 7.23A, A1) were fundamentally like those of the elasmobranchs, as were also the tribasic pectoral and multiradial, metapterygial pelvic fins (fig. 7.23).

The Upper Devonian Rhenanida (fig. 7.24) had widely expanded pectoral fins, like those of a skate, but the general shape of the body was not unlike that of a Port Jackson (heterodont) shark. The brain-case, on the other hand, recalled the *Macropetalichthys* type. The large flattened mandible articulated above with what seems to be a potential palatoquadrate, perhaps not yet separated from the brain-case.

The Devonian acanthodians (figs. 7.25; 7.26A, B) were completely predatory, fusiform fishes with fish-trap jaws and minute scales resembling those of earlier ganoids. The hyoidean arch (fig. 7.26B) was still like the branchial arches and its lower half had not yet lost its gill (hence the name "Aphetohyoidea" with evident hyoids given by Watson to a class including all the placoderms plus the Acanthodii). The oral arch (palatoquadrate plus Meckel's cartilage) bore an opercular flap (fig. 7.26 "mand rays") like those of the hyoid and branchial arches and presumably retained the primitive half-gill. Meckel's cartilage and the palatoquadrate were subdivided like the branchial arches (fig. 7.26B). The shoulder-girdle (fig. 7.25) probably represented the remnants of a thoracic buckler plus an internal extension of the fin base. The fin-spines in the older genera were large and fixed, but became needle-like in the latest eel-like acanthodians of Permian age (figs. 7.27; 7.28). The pectoral fins in certain cases had minute tribasic supports (fig. 10.18A), foreshadowing those of sharks (fig. 10.18B-D). The exoskeleton consisted of a mosaic of small plates covering the head and an armor of minute quadrate, many-layered scales on the body. The tail was heterocercal (fig. 7.27). The more primitive acanthodians had posteriorly converging rows of five pairs of fins between the pectoral and pelvic fins (fig. 7.25). These secondary polyisomeres may have developed as a remnant of a once continuous fin-fold (fig. 7.27A) as the need for stabilizers increased with the swiftness of forward darting.

ASCENDING GRADES FROM JAWLESS MUD-GRUBBERS TO WELL-JAWED PREDATORS

The main stages and branches from the basal ostracoderms to the chief classes or subclasses of fishes may be summarized as follows:

Grade I. AGNATHA. Functional jaws, if present, composed only of exoskeletal plates.

Class 1. Ostracodermi, with oralo-branchial series appearing as primary polyisomeres (fig. 6.14A), not or but little differentiated *inter se*. Potential growth centers for exoskeletal pectoral girdle or thoracic buckler represented in part by: (a) transverse zone of exoskeleton covering first postbranchial myomeres in *Anglaspis* of the Heterostraci (fig. 6.9A); (b) the ventral disc-plate of pteraspids (fig. 6.7B2); (c) the oblique row of small supra-postbranchial plates in anaspids (fig. 6.18C1); (d) the "trunk division" of the cephalic shield (fig. 7.1C) of *Cephalaspis* (Stensiö, 1927, pp. 231–235; 1932, p. 47).

Potential growth centers for pectoral appendages represented by: (a) cornual processes and pectoral fins of cephalaspids, e.g., *Hemicyclaspis* (fig. 6.12B) (Stensiö, 1932, pp. 53, 64); (b) "pectoral lappets" of *Thelodus* (fig. 6.6), and cornual plate of pteraspids (figs. 6.7B1; 6.10); (c) the pair of slender spines at the anterior end of pectoral-pelvic fin-folds (fig. 6.12C, E, F) in anaspids (Kiaer, 1924, p. 103).

Scapulo-coracoid arch not yet differentiated into scapulo-coracoid bar but possibly represented (fig. 6.15A2) by "endoskeletal shoulder-girdle" of cephalaspids (Stensiö, 1927, p. 45, fig. 12, *p. sh.*).

Pelvic fins feebly or not at all developed as part of paired pectoral-pelvic fin-fold, cf. *Aceraspis, Hemicyclaspis* (fig. 6.12B), as shown by Kiaer (1924, p. 103), Stetson (1927, 1928), and Stensiö (1932, pp. 65, 66). "Pelvic" bones not yet differentiated.

Class 2. Cyclostomata (lampreys and hagfishes). Eel-like Agnatha with suctorial mouth (fig. 6.19A1), bearing horny teeth supported (in lampreys) by a system of sliding cup-like cartilages abutting on the under side of a preolfactory extension of the chondrocranium. Tongue protrusile, bearing horny, rasping teeth and supported by long, median lingual cartilage (fig. 6.19A). Branchial basket (in lampreys) surrounding the small, round gill openings and purse-like gill-sacs, and consisting of slender bars and irregularly curved connectives (fig. 6.20B1), the whole forming a resilient framework without articulations or division into segments, continuous posteriorly with a cup-like pericardial cartilage.

Larval lamprey (fig. 6.19A1) lacking the rasping tongue and provided with a ciliated peripharyngeal and median groove. Anterior end of pituitary-olfactory fold of larval lamprey displaced during development (fig. 6.22A) to the dorsal surface (? in association with the great increase in size of the dorsal cartilages of the sucking disc). Auditory capsule lacking the external semicircular canal. Hag-fishes (myxinoids) with secondarily reduced chondrocranium and branchial cartilages.

Grade II. PLACODERMI. Oral arch more or less differentiated (figs. 7.17; 7.27) to form inner jaws (palato-pterygoquadrate + Meckel's cartilages). Outer or exoskeletal jaws well developed (figs. 7.16 *asg, psg, ig;* 7.17; 7.27A). Hyoid gill-slit complete, not changed into a spiracle, upper segment of hyoid arch or hyomandibular (fig. 7.27B) not modified to support oral arch ("Aphetohyoidea").

Branch A. Antiarchi. Oral arch small (fig. 7.3B), branchial arches increasing posteriorly, dermal jawbones (fig. 7.16B) consisting of anterosuperognathals, posterosuperognathals, inferognathals, presumably supported by pterygoquadrate and Meckel's cartilage (Stensiö). Head sharply demarcated from thorax (fig. 7.4). Pectoral girdle represented by anterior portion of thoracic shield, including a postbranchial septum, essentially like the "endoskeletal shoulder-girdle" (fig. 6.15A2) of *Cephalaspis.* Pectoral outgrowths (figs. 7.2; 7.3) forming long branchial appendages, covered with a definite pattern of exoskeletal plates, provided with a pedunculate ball-and-socket joint (fig. 7.5) at the base and an elbow joint (fig. 7.3B) between the longer proximal, and the shorter distal, segments.

Branch B. Arthrodira (figs. 7.9–7.17). Anterior and basal portion of thoracic shield serving as a pectoral girdle (fig. 7.9). Pectoral outgrowths (if present) variable, primitively in the form of recurved conical spines with wide base (figs. 7.11, 12), immovable, fastened to thoracic shield. Head sharply demarcated from thorax. Jaw plates fundamentally as in Antiarchi but finally becoming very large (fig. 7.15A-A2).

Sub-branch 1. Acanthaspida (Dolichothoraci). Pectoral spines large (figs. 7.9A, B; 7.11B-F). Jaw plates small.

Sub-branch 2. Euarthrodira. Pectoral spines variable (fig. 7.12B-F), finally vestigial or absent. Jaw plates becoming large. Thoracic shield becoming similar to the exoskeletal girdle of typical fishes (fig. 7.12). Joint between head and thorax finally giving rise to paired peg-and-socket articulations, permitting the head to be raised and lowered upon the pectoral girdle (fig. 7.15).

Branch C. Acanthodii (figs. 7.25–27). Oralobranchial arches (fig. 7.27B) consisting of a system of articulated rods and copulae; as seen from below, arranged *en chevron,* the large oral arch overlapping the diminishing ones behind it. Dermocranium (fig. 7.27A) consisting of many small pieces. Endocranium partly cartilaginous. Pectoral girdle (fig. 7.26) including vertical bars attached below to a base of several paired plates and one median anterior plate, which extend forward below the floor of the mouth.

Pectoral appendages represented by long spines (fig. 7.25), originally with widely conical bases, sometimes with a very small tribasic endoskeleton (fig. 10.18A). Originally with paired pectoral-pelvic fin rows (fig. 7.25), with several intermediate pairs of spines between the pectoral and pelvic pairs.

Branch D. Macropetalichthyida (Stegoselachia). Dermocranium and endocranium both well ossified (fig. 7.20B); patterns of skull roof (figs. 7.8; 7.4) somewhat like that of Arthrodira (fig. 7.7). Pectoral fins without spines, tribasic (fig. 7.23A), shark-like. Pelvic fins also shark-like, with many metapterygial segments (fig. 7.23A1).

Branch E. Rhenanida. Pectoral fins much widened (fig. 7.24) with long fan-like radials; body-form somewhat intermediate between sharks and rays. Jaws and hyoid arches near to macropetalichthyid type.

Grade III. EUGNATHOSTOMATA. Oral arch more sharply differentiated from remaining visceral arches; upper segment of hyoid arch (hyomandibular) early specialized for support of oral arch; spiracular gill-cleft closed below.

Branch A. Elasmobranchii (Chapter VIII). Paired fin spines absent, median fin spines often retrogressive or absent; pectorals tribasic, with prominent radials; horny rays conspicuous; endoskeleton cartilaginous, more or less calcified; cartilaginous endocranium dominant, dermocranium reduced to shagreen-bearing skin. No lungs or air-sacs.

Branch B. Osteichthyes (Bony Fishes, *sens. lat.*, Chapter IX). Primary fin spines replaced by enlarged fulcral scales, pectorals primitively with tripartite inner girdle ("coracoid," "mesocoracoid," "scapula"). Short radials ("baseosts," "pterygials"); dermal girdle usually predominant; lepidotrichs early becoming predominant in fins. Endoskeleton progressively ossified. Dermocranium formed of many suturally connected derm bones. Derm-jaws predominant. Body covered with ganoid to horny scales.

Branch C. Choanata (Chapter X). Pectorals without trace of spines, early becoming biserial, with jointed mesomeral axis; scales ganoid to cycloid. Originally with paired lungs.

Sub-branch 1. Dipnoi. Pectorals biserial, becoming delicate to vestigial (fig. 10.10); radiating dental plates on roof of mouth and inner sides of mandible (figs. 10.11–13); lungs progressive. Skull roof (fig. 10.16) early breaking up into many small elements; later reduced in number.

Sub-branch 2. Crossopterygii. Pectorals lobate (fig. 10.10A-E), with preaxial radials only. Dentition predatory.

CHAPTER VIII

ADVANCED PREDATORS OF THE SHARK TRIBE

Contents

CHAPTER VIII

ADVANCED PREDATORS OF THE SHARK TRIBE

THE SILVER-SHARKS (CHIMAEROIDS) AND THEIR ANCESTORS (BRADYODONTS)

In the Hopkins Marine Laboratory at Pacific Grove, California, on the afternoon of a certain day in the late summer of the year 1899, Dr. Bashford Dean beheld with his own eyes the authentic vision of a living *Chimaera collei.* With reverent care he floated his living treasure out of a pailful of sea-water into the aquarium. And there this radiant marine creature (fig. 8.1) swiftly vibrated her wide pectoral wings, turning her great head slightly from side to side, swaying her long tapering caudal train in living curves, advancing with erect dorsal sail, glowing with metallic luster, silver and blue and dark brown. Thus she moved majestically and with her large eyes she gleamed like the Athenian goddess.

After such a vision Dr. Dean had felt well rewarded for his arduous labors in securing this fish, but how much more was he elated the next morning; for, though his captive Queen of the Silver Sharks was dead, she had left behind her a priceless relic, since she had expelled from her body in her last hours a curiously shaped egg (fig. 8.2B), horny in texture and with flowing contours. When the side of this egg was carefully cut through and lifted off, Dr. Dean found inside of it a still living embryo, nourished by the remains of a large yolk (fig. 8.2A). This embryo was immediately drawn by him in color and it is figured in Plate VIII, Figure 49, of Dean's monograph on "Chimaeroid Fishes and their Development." Thus it became an important part of the material which enabled him to prove from their embryology that the chimaeroids were to be classified essentially as sharks in the wider sense, or more technically as elasmobranchs.

The very large rounded head and tapering body (fig. 8.1) of the adult *Chimaera* in the top and side views conform to the teardrop or streamlined pattern which slips through the water with a minimum of turbulence or backward drag. But such a general shape has been evolved several times in widely different groups of vertebrates; for example, among the Devonian placoderms (*Rhamphodopsis,* fig. 7.19); among the teleost fishes something of the same sort appears in *Coilia,* a highly specialized offshoot of the herring family, as well as in the deep-sea grenadiers or macrurids, belonging in the same order with the codfishes. Therefore this general form of body is no authentic clue to relationship, and we must as usual compare different parts of the anatomy for clues to the origin of the group.

The occurrence of a large persistent notochord is apparently a primitive feature inherited from a very early shark stock before the evolution of elaborately calcified vertebrae. On the other hand, the presence for each myomere or muscle segment of several thin central rings (fig. 8.3A) around the notochord is probably an expression of early secondary polyisomerism, producing an unusually flexible but reasonably strong axis. Nevertheless this type of vertebral column is a unique character among fishes and together with others it indicates a long separation of the chimaeroid stem from that of other elasmobranchs.

The dentition of chimaeroids (fig. 8.4), although widely varying in detail, includes a pair of tritoral teeth at the front end of the upper jaws which are opposed by the procumbent front ends of the lower dental plates, somewhat like the nipping and gnawing teeth of rodents. It is perhaps because of this faint suggestion of a rodent that the beautiful *Chimaera collei* is known to some fishermen as the rat-fish. Beneath the cheeks in most chimaeroids there are molar plates; in *Harriotta* (fig. 8.4B) these tooth plates are coiled around the upper and lower jaws somewhat after the manner of the dental pavements of the Palaeozoic cochliodont sharks (fig. 8.5C). In *Chimaera* itself, however, the molar masses (fig. 8.4C) are bordered by sharp, almost serrated edges with low ridges or convexities on their inner surfaces—an indication that this genus does more cutting and less crushing than the remotely ancestral *Myriacanthus* (fig. 8.4A).

Smith Woodward (1920) and Moy-Thomas (1936, 1939) have shown that a group of Palaeozoic sharks called Bradyodonti (fig. 8.5) have good claims to be regarded as the source of the chimaeroids. In the typical Lower Carboniferous cochliodonts (fig. 8.5C) of this group the teeth form dental plates arranged in whorls around the upper and lower jaws and the upper jaws are completely fused with the braincase as in chimaeroids. In vertical sections through the jaws of *Janassa* (B), representing another family of bradyodonts, it is seen that the dental plates are formed by fusion of a vertical succession of teeth. It is also known that these teeth succeeded each other very slowly (whence the name Bradyodonti), while the characteristic microscopic structure was "without an outer enamel-like layer, the crown of the tooth being formed of numerous vertical parallel tubes of dentine [fig. 8.6B], which gives the worn tooth a peculiar pitted appearance" (Moy-Thomas, 1939, p. 69). The same author has also shown that the cochliodont tooth plates resemble those of the later chimaeroids and that in its body form, fins and dentition the Upper Carboniferous shark named *Helodus simplex* (fig. 8.11A) is structurally intermediate between the Palaeozoic cochliodonts and the Mesozoic chimaeroids.

The necessary strength and rigidity for the support of this powerful dental apparatus is secured in part by the massiveness of the braincase itself, by the thickening of the lower jaw and by the complete union of the upper jaw (palatoquadrate) with the braincase (figs. 8.3; 8.7). This mode of attachment of the upper jaw to the skull has been called holostylic (complete column) and holautostylic (complete self-column) and it is another feature that is highly characteristic of chimaeroids, although paralleled in the dipnoans.

Behind the upper and lower jaws comes a decreasing series of branchial arches (figs. 8.3; 8.7C), of which the hyoid arch, according to Dean, is entirely free from attachment to or support of the jaws and is just like the true branchial arches in having two pieces above the main bar or ceratohyal, identified by Dean (1906, p. 120) as the epi- and pharyngohyal (*ph. hy.*). According to him also, the epihyal corresponds to the hyomandibular of other sharks. In this respect the chimaeroids are more primitive than other sharks, in which the uppermost segment (pharyngohyal) has disappeared. Goodrich (1909, p. 171), on the other hand, suggests that "from the evidence of embryology and the course of the hyomandibular branch of the facial nerve [cf. fig. 8.7D], it would appear more probable that the hyomandibular is indistinguishably fused both with the auditory capsule and with the quadrate, the spiracle being suppressed." Watson (1937, p. 141) also adopts this view, which is, however, again opposed by De Beer (1937) and Moy-Thomas (1935). Whichever of these two interpretations may be adopted finally, either one of them is consistent with the derivation of chimaeroids from Palaeozoic sharks of some sort. The more immediate question, however, is, sharks of *what* sort?

The modern chimaeroids (including about four genera) were essentially hold-overs from Mesozoic times, when the group was in its prime. Dean was inclined to derive the older chimaeroids from the stem of the Devonian ptyctodonts (cf. fig. 7.19). Subsequent discoveries, however, have led Watson (1937) and others to conclude that the ptyctodonts were not elasmobranchs at all but specialized arthrodires with no near relations with the chimaeroids; but it now seems probable that the ptyctodonts may be related on one side to the placoderms and on the other to the chimaeroids.

Perhaps in connection with their omnivorous feeding habits (the food including crustaceans, shellfish, worms, echinoids, small fish), the chimaeroids have an unusually complex system (fig. 8.8) of oronasal folds and grooves supported by labial cartilages (cf. Dean, 1904, pp. 9–11; Allis, 1919, pl. 4). The main or posterior lip fold is supported on each side by a chain of three articulated cartilages (C′, A, B), two in the upper (C′, A), and one (B) in the lower lip fold. The inner lip fold is anterior to the outer one and is supported by one principal cartilage and by a minor fold connecting with the main outer one. On the whole these are homologous with the labial folds and cartilages of sharks (figs. 8.7B; 8.42). It was long ago suggested that these labial cartilages represent pre-oral gill arches but decisive supporting evidence is still lacking and they may be secondary polyisomeres forming the cores of the lip folds.

The chimaeroids differ widely from all other sharks in the enlargement of the opercular flap that is attached to the posterior border of the hyoid arch (fig. 8.3A *operc*) and in the backward extension of this flap so that in the living fish it covers all the gill slits on the outer side, leaving only a single slit at the posterior end of the branchial chamber, as in the higher fishes.

Another curious feature is the presence of a tenaculum or forwardly curved and movable frontal spine, ending in a burr-like knob, on the foreheads of male chimaeras (fig. 8.9E). As shown by Dean (1906, p. 141), this sharply curved spine, which is used in mating, appears to have been derived from the straight spike-like anterior fin spine of *Myriacanthus* (A), a Jurassic chimaeroid. Moy-Thomas (1936, pp. 784, 785) notes that in *Oracanthus armigerus* Traquair, a cochliodont from the Lower Carboniferous of Scotland, there is a pair of large backwardly directed spikes on the postero-external corners of the head and another pair on the rear of the mandible, and that in these and several other features of the body spines, as well as in the bradyodont structure of the teeth, *Oracanthus* rather closely approached *Myriacanthus* and *Menaspis,* which were early chimaeroids.

The erectile dorsal fin of *Chimaera* and its relatives is mounted on a cartilaginous basal piece (fig. 8.3), which in turn is fused with the anterior end of the vertebral column and forms a fulcrum for the braincase. These conditions are more advanced than any that are met with in other sharks; the nearest approach to them being found in the heterodont sharks (fig. 8.13C), where, however, the basal plate of the first dorsal fin does not form a fulcrum for the skull. On the other hand, the relations of the dorsal fin spine to the surrounding parts would appear to be derivable from the conditions in the ptyctodonts (fig. 7.19) as a result of the reduction of the upper median part of the pectoral girdle (fig. 7.18A) and the shifting of the support of the fin spine to the underlying neural arches (fig. 8.13C).

The paired fins of *Chimaera* (figs. 8.1; 8.3) and its allies (fig. 8.10D) are of interest in connection both with general theories of the origin of paired fins and with the origin of the group itself. The pectoral fin is apparently a modification of the tribasic type (A), in which the small propterygium (*pro*) has withdrawn from the glenoid articular surface, and the mesopterygium forming the main contact with the coracoscapula (fig. 8.3) has left the enlarged metapterygium (*met*) to support the great majority of the radials (fig. 8.10D). Meanwhile the wide wing-like form of the dermal rays (fig. 8.1), together with a fan-like spreading of the fin muscles has encouraged some of the radials to curve around on to the postaxial border of the fin (fig. 8.3). In the Upper Carboniferous shark *Helodus* (fig. 8.11A), which (as restored by Moy-Thomas, 1936; 1939) is in skull characters the connecting link between the chimaeroids and the Palaeozoic bradyodonts, the pectoral fins were still tribasic, with the propterygium well developed, the small mesopterygium at the apex and the radials disposed in such a way as would readily be crowded into the observed pattern in chimaeroids.

The fan-like pelvic fins of chimaeroids (figs. 8.2A; 8.10H) have the radials radiately arranged about a rounded basal piece; the males (H) have also a pair of jointed metapterygial appendages, corresponding with the claspers of typical sharks (G). There is little or no suggestion of the so-called archipterygial or biserial pattern (fig. 8.12A) of the dipnoans (fig. 10.9) in the pelvic fins of adult modern and Jurassic chimaeroids. Dean (1906, p. 104) has shown that the *Anlagen* of the pectoral and pelvic fins in the late embryos of *Chimaera* (fig. 8.2A1) and *Callorhynchus* appear as long based fin-folds and are wholly unlike the biserial or "archipterygial" fins of dipnoans (fig. 10.18I).

While the evidence from embryology strongly favors Dean's view (1906, p. 152) that the chimaeroids are derived from a primitive elasmobranch stock, later palaeontological and morphological studies tend to show that the chimaeroids were derived not from such central elasmobranchs as the Mesozoic hybodonts (fig. 8.13), as Dean supposed, but from that broad division of the elasmobranchs which of late years has been named Bradyodonti (figs. 8.5, 6; 8.11A). Moreover, it even seems possible that the chimaeroids may be related to such arthrodires as *Leiosteus,* as intimated by Jaekel (1907), Broili (1933), Stensiö (1934, p. 68) and further that some of the chimaeroid characters which Dean regarded as highly specialized may be far more primitive than he suspected: as, for example, the retention of strong spines on the dorsal fin, the union of the palatoquadrate and hyomandibular with the skull, the presence of vomerine tooth plates and of elaborate preoral labial cartilages.

The Lower Carboniferous *Orodus* (mountain-tooth) is known only from its teeth (fig. 8.6E), in which the crown is shaped like a long mountain with a central peak and sloping valleys on either side of the high divide. These mountain-like teeth are basically like the cheek teeth of *Edestus* (D). In this strange shark and its relatives there was a spiral row of teeth on the symphysis of the mandible; these teeth became extremely large and compressed, with great cutting ridges and elongated bifurcated roots which overlap each other from front to rear. In the evolution of these spirals it looks as if unchecked reduplication (polyisomerism) had been stopped only by the extinction of the group. However, it would not be safe to assert that such spirals were useless, as the protruding front teeth might have been used either for attack upon other edestids or in some way as food cutters. It was formerly supposed that the edestids were related to the late Palaeozoic hybodonts (fig. 8.13) but it has been shown by Nielsen (1932) that the microscopic structure of the teeth of some of the less specialized edestids was of the bradyodont type (fig. 8.6B). The edestids were therefore tentatively assigned by Moy-Thomas (1939, p. 74) to the bradyodont division of the elasmobranchs.

FINFOLD SHARKS (CLADODONTS) AND THE ORIGIN OF PAIRED FINS

In the Upper Devonian cladodonts (fig. 8.14B, C, D) the teeth as seen from above were long ovals, which were arranged in whorls around the jaws essentially as in the protodonts (8.14A), orodonts (8.14G), and heterodonts (8.14J, K, L). In the side view the crown was produced upward into a high central cusp and a variable number of lateral cusps, of which the outer were sometimes longer than the intervening ones. However, the expanded base of each tooth was "overlapped by its successor behind" (Zittel, vol. II, 1932, p. 57). One of the British "cladodont" teeth belonging to *Ctenacanthus costellatus* (fig. 8.14E) has been sectioned and figured (fig. 8.6A) by Moy-Thomas (1936, p. 763), who notes (p. 765) that "In microscopic structure . . . these teeth can be seen to consist of a central mass of osteodentine, surrounded by a layer of fine-tubed unvascular dentine (pseudo-dentine, Thomasset, 1930), outside which lies a layer of enamel. The structure is therefore essentially the same as that of the Hybodonts and modern sharks like *Lamna.*" Thus there is a wide contrast between the microscopic structure of the cladodont-shark tooth series (fig. 8.6A) and the bradyodont-chimaeroid series (figs. 8.5, 6B).

For a long time but little was known of the body form and other characters of the sharks that bore these teeth, but thanks to the field labors of many collectors and to the investigations of Claypole (1893, 1895), Newberry (1890), and especially Dean (1909), one of the cladodonts of the Cleveland shales in Ohio, named *Cladoselache fyleri* (fig. 8.13A), is now known from numerous fossils, which have yielded the scalation, general body form, vertebral column, dentition, jaws and gill arches, together with the outer form and skeleton of all the fins. Still more surprising was the preservation in some specimens of fossilized portions of the myomeres, which in microsections even revealed the remains of red muscle fibres (Dean, 1909, p. 235), while other microsections across the kidneys showed clear traces of the glomeruli. Possibly this remarkable replacement of organic material was brought about by the presence of an abundance of oil from the liver of the shark, together with carbon from the vegetation which has blackened the Cleveland shales.

Notwithstanding the fact that the general morphology and many structural details are more fully known in *Cladoselache* than in any other fossil shark, its exact relationship with sharks of other groups is still a subject of great importance in a general history of vertebrate evolution; for during the past half-century as the knowledge of other sharks has widened, both the apparent phylogenetic and the morphologic values of different features of *Cladoselache* have shifted accordingly.

Cladoselache is most widely known as the perfect example and proof of the Thacher-Balfour "finfold theory" of the origin of the paired fins (fig. 8.12B), in opposition to the "archipterygium theory" of Gegenbaur (fig. 8.12A). The latter (1898) supposed that the pectoral fins had been derived by change of function from the cartilaginous branchial rays, which in sharks form long fringes around the outer margins of the gill-arches (fig. 8.12A). Later, the middle one of these rays became larger, while the ones above and below it were drawn outward by its growth, on to its basal portion; if this process continued there would result a sort of leaf-like appendage having a central axis with a diminishing series of rods along its upper border and a similar series on its lower border. The bending of this leaf-like appendage by muscles would encourage the fragmentation or budding of the central axis and one would thus readily come to the "archiptergium" of the living Australian lungfish (figs. 10.10; 10.18I). Since the pelvic fins of sharks showed evidence of having arisen in much the same way as the pectoral fins, Anton Dohrn (1876) in a classic essay entitled "Der Ursprung der Wirbelthiere und das Prinzip des Functionswechsels" did not hesitate to assume that the pelvic no less than the pectoral appendages had grown from gill arches; these had formerly extended along the sides of the body much as the gill-bearing parapodia do in the annelid worms, which he conceived to represent the invertebrate ancestors of the vertebrates.

The annelid theory of the origin of vertebrates led to Patten's theory in which serious defects have been shown above (p. 91). Consequently the very foundation of the Gegenbaur-Dohrn theory was illusory and for several decades past it has had chiefly an historic interest. Thacher and Balfour, on the other hand, relying upon evidence from the embryology of modern sharks, maintained that the paired fins were not essentially different in origin from the median fins and that both had come from folds or projections of the body wall (fig. 8.12B).

Cladoselache then supplied Dean with the evidence that, in the then oldest known Palaeozoic shark, the paired fins (fig. 8.15A3–A8) were essentially like the median fins (fig. 8.15A–A2) and that all had arisen from "fin folds."

That *Cladoselache* (fig. 8.13A) was altogether more primitive than other sharks in the structure of both its paired and median fins has been widely assumed; but Smith Woodward (1898) long ago suggested that its elongate body form and curiously specialized tail indicate some peculiar habits and that after all its pectoral fins might have been derived by reduction from a "biserial" or archipterygial type. I have recently suggested (1936, p. 331; 1941, pp. 279–281) that while there can be no reasonable doubt that median and paired fins have alike arisen from humps, projections, or folds from the body wall, this does not prove either that the cartilaginous radials of the first sharks extended to the outer margins of the fins as they did in *Cladoselache* (fig. 8.15A3–A6), or that the usual absence of spines in the latter's fins is necessarily a primitive shark character.

As to the character of the endoskeleton of the paired fins in *Cladoselache,* it was formerly assumed that they must be primitive because their constituent parts were so little differentiated and were composed of so many similar pieces. But a long experience with "secondary polyisomeres" which look primitive, but are merely of generic or subfamily age led me to the suspicion that the numerous very long and delicate radials of the pectoral fins of *Cladoselache* are no more primitive than those of its admittedly specialized tail (fig. 8.15A2). In this connection we may well compare the short radials in the pectorals of relatively primitive modern sharks, such as *Chiloscyllium* (fig. 8.16C), *Catulus* (A), *Halaelurus* (B), with the exceedingly long radials in *Isurus* (D), *Carcharinus* (E), and *Galeus* (F), which all belong within a single order (Galea).

In defense of the traditional view that *Cladoselache* is quite primitive in the construction of its fins, Moy-Thomas (1939, pp. 4–6) writes:

"The cladoselachian type of paired fin has long been used as evidence of the finfold theory of the origin of paired fins, but recently this has been challenged by some writers (Gregory, 1936; E. G. White, 1936, 1937), who maintain that the view is untenable since the geologically older acanthodians and macropetalichthyids . . . have definitely concentrated fin skeletons. From a study of the elasmobranchs themselves there appears to be no doubt that the cladoselachian fin is primitive for the group, and its late appearance in this primitive condition may be explained on the assumption that these forms have evolved from an unknown ancestral group which retained the larval fin-fold. A further possible explanation is that the elasmobranchs are derived paedomorphically from bony ancestors retaining the larval characters of cartilage and fin-folds. Whichever of these views is correct it does not alter the fact that the fin-fold is certainly primitive in the elasmobranchs, and probably in all fishes."

Gregory and Raven (1941, pp. 278–282) in reply to Moy-Thomas, wrote as follows:

"Although heavy spines and body armor are excellent for defence, they may become a hindrance to mobility and swiftness. Certain it is that in not a few evolutional series either the fin spines or the body armor or both underwent a reduction, often to the point of elimination. In the later eel-like acanthodians, for example, the spines became very slender and needle-like; in the squaloid sharks the spines show various stages of reduction until in *Laemargus* they are eliminated, as they are in the modernized families of galeoids. The hybodonts and heterodonts, or Port Jackson sharks, are an exception in retaining spines on the dorsal fins. Moy-Thomas (1936), developing the work of Smith Woodward, has shown that the hybodonts may be derived from the spine-bearing ctenacanthids of the Lower Carboniferous. He believes that these in turn were derived from spineless cladodonts of the Upper Devonian. But at least one of the cladodonts (*Ctenacanthus clarki*) had a well developed dorsal spine (Dean, 1909: 251). This makes it quite possible that the absence of this spine in *Cladoselache* may be secondary.

"Between the dermal fin rays and their rod-like supports there is an intimate relation. This may be seen most readily in the skeleton of a skate but is also evident in the pleuracanths. Dean (1909: 245, 246) considered that the acanthodian fin spine had arisen through the clustering of radials on the anterior border of the fin and that as the dermal denticles had coalesced into a spine, the radials beneath them had retreated, after the analogy of the retreat of the Meckel's cartilage beneath the encrusting derm bones of the jaw in ganoids and teleosts. The anterior margins of median and paired fins are indeed usually strengthened by the coalescence of small elements, either by shagreen denticles coalesced into spines, as in acanthodians, or by scales uniting into lepidotrichs of varying stiffness, as in typical ganoids and teleosts. Dean

(1909: 245) also cited evidence in support of his inference that the numerous long radial rods in the fins of *Cladoselache*, which extend almost to the margins, were in a more primitive state than the few and very short radial rods that are barely indicated in certain of the well-spined, webbed fins of acanthodians. However, it has been shown by Heintz (1938: 23) that among the Arthrodira the reduction of the pectoral spines was accompanied by a development of the fin web and of its supporting rods.

"With regard to the origin of paired fins in the shark group, we infer that in response to physiological requirements mineral matter was deposited successively in the integument after the manner of scale growth; these deposits were not evenly distributed but were concentrated at nodal points in a dorsal median row extending along the back and tail, and in converging lateral rows extending from the widest points behind the head to the anal region. If there were great crowding and concentration toward the anterior end of each node, spines were produced, as in acanthodians and as in the dorsal fins of ctenacanths, hybodonts, cestracionts, and squaloids. Less crowding at the base encouraged the spreading out of the inner layers of the supporting tissue into basals and radials. At any time retrogression of the spines would be favorable to extension of the basals and radials. Consequently we would be inclined to reconstruct the ancestral shark neither as a full-fledged acanthodian nor as a 'primitive cladoselachian' as restored by Dean (1909: 244) but as much nearer to the *Ctenacanthus costellatus* type [fig. 8.19A] as restored by Moy-Thomas (1936: 764).

"Whatever the exact form of the primitive shark may have been, we have much evidence that at an early date the metameric muscles seized hold of the bases of the nodal ridges and began to use them and their accompanying fin folds as movable keels and rudders. As the fin musculature began to attain its own individuality apart from its parent metameric mass, it spread radiately outward from the body toward the margin of the fins, carrying with it a set of stiffening endoskeletal rods which it had evoked between the opposing muscle masses either of the upper and lower faces of the paired fins or on the right and left sides of the median fins. From this viewpoint the extension of the radial rods to the tips of all the fins would be no more primitive in *Cladoselache* than it is in the pectoral fins of skates and rays. The large number and similarity to each other of the radial rods in *Cladoselache* used to be taken as a sign of primitiveness but it may rather be an example of what one of us (Gregory, 1934) has called 'secondary polyisomerism.'

"Most authors have assumed that an anteroposteriorly wide base, with little or no axillary incisure (as in Dean's 'primitive cladoselachian'), is the primitive condition for the pectoral fins, but this is often not so in the Anaspida, Acanthaspida, Acanthodii, *Macropetalichthys;* the base is of variable width in the ctenacanthids and bradyodonts, and becomes very narrow in the pleuracanths and crossopterygians. And in *Cladoselache* itself the skeletal base of the pectoral fin narrows rapidly as we pass inward toward the pectoral girdle (Dean, 1909: 229, fig. 28). Indeed in *Cladoselache acanthopterygius* (op. cit., p. 228) the pectoral fin has a very narrow base."

As noted above, it has been assumed that *Cladoselache* was primitive because it has no spines in the dorsal fin, although a review of the swarming Palaeozoic genera of bradyodonts and sharks that are known chiefly from fin-spines may indicate that the remote ancestral stock for all these spiny forms ought to show at least the beginning of spines. Finally, it has already been noted that Dean described a large Upper Devonian shark whose pectoral fins were identical with those of *Cladoselache* but in which the first dorsal bore a large spine, of the type known as *Ctenacanthus clarkii* (fig. 8.15B). For these and other reasons it has been suggested that the older members of the spiny-finned Upper Palaeozoic sharks called ctenacanthids, which have tribasic pectorals with short radials and well developed spines in both dorsals, may have been more primitive than *Cladoselache* itself. Dr. J. E. Harris (1938) has indeed testified that in the Cleveland Museum of Natural History there is a well preserved *Cladoselache fyleri* bearing a peculiar wide spine on the first dorsal fin, a feature which tends to bring it in line with many other early sharks and bradyodonts.

But whether or not *Cladoselache* has been derived from an earlier shark with dorsal fin spines, it does make an almost ideal structural ancestor for the spineless frilled shark and the notidanids, just as the related *Ctenacanthus* does for the spiny-finned hybodonts and heterodonts.

SWAMP-LIVING PLEURACANTHS OF THE COAL AGES

The pleuracanths were fresh-water and swamp-living sharks, of late Palaeozoic age (Carboniferous and Permian), with a sinuous tapering body (fig. 8.21) and a predaceous type of head (fig. 8.20). A long, sharp, movable spike, with its rear border denticulated on each side (whence the name *Pleuracanthus*), was borne on a socket on top of the occiput (fig. 8.22). This spike was probably a relic of the first dorsal fin. The stout, spreading pectoral fins (fig. 8.27B) had a jointed main axis and feather-tipped radials, the preaxial ones larger and more numerous than the postaxial; the pelvic fins were approaching the modern shark type, with long segmented clasper appendages in males (fig. 8.11C). A single dorsal fin (homologous with the second dorsal of other sharks) with very long, low base was supported by very many short cartilage rods (fig. 8.21). The caudal fin also was very long, not heterocercal, but elongate leaf-shape (diphycercal). The anal fin was represented by two small feather-like appendages. The notochord was persistent but neural and haemal arches were present. The shagreen denticles were reduced and very fine, the skin nearly naked. The teeth had two large divergent blades, often with a minute cusp between them (fig. 8.14H); the base of the tooth was swollen and projected backward. The jaws (fig. 8.22) were amphistylic, the much enlarged palatoquadrate having a strong postorbital articulation, while the large hyomandibula assisted in the support of the mandible.

The name Ichthyotomi, often applied to this group, was given by Cope in allusion to the supposed subdivision of the well calcified braincase into segments. This appearance seems, however, to have been due to post-mortem fractures in the matrix.

The pleuracanths show a surprisingly wide distribution of resemblances to other orders of cartilage fishes in combination with their own unique characters. Their skull and jaws (figs. 8.22; 8.20) were thoroughly shark-like, with special resemblances to the notidanids (fig. 8.25) and frilled shark (fig. 8.20C). The males had shark-like pelvic claspers (fig. 8.12C), this implying the practice of internal fertilization. Their peculiar teeth were possibly derived from some cladodont type (*vide infra*); but in the "archipterygial" character of their pectorals (fig. 8.27B) they were surpassed only by the dipnoans (fig. 10.10F), with which they had no near relationship. They slightly resembled the chimaeroids in the diphycercal tail, long low dorsal fin and high nuchal spine (fig. 8.1), but differed widely in the relations of the upper jaw to the skull and in the characters of the dentition.

The pleuracanths lived during a period (the Permian) of intense orogenic disturbance, when, so far as the fossil record indicates (Hay, 1900), the cartilage fish as a class suffered a decimating extinction (fig. 8.23) and when the only other now known cartilage fishes were the bradyodont ancestors or relatives of the chimaeroids and the last of the edestids; but the ctenacanthid-hybodont line must have been in existence somewhere, as it is known from earlier and later ages. Thus the pleuracanths were intermediate in time between the earlier period of the prosperity of the Devonian-Carboniferous cladodonts and the Mesozoic era, when the ancestors of all the modern orders of sharks were recognizable. Hence it is not surprising that while inheriting some primitive features from their older Palaeozoic ancestors, they had also become specialized in such features as the excessive length of the body, with multiplication of segments, the development of accessory anal fins and of other secondary polyisomeres, such as the rows of denticles on the edges of the prominent dorsal spike and the feather-like dermal rays of the pectoral fins. Some of these features, especially the lengthening of the body and the fin-fold form of the median fins, were probably retained from a larval stage.

That the pleuracanths were derived from some far older stock, such as the cladodonts and ctenacanthids, is indicated not only by the fundamentally shark-like characters of their skull and jaws but even by their peculiar teeth (fig. 8.14H), since the two lateral blades, although greatly enlarged, seem to correspond with the main lateral cusps of the cladodonts, while the minute central cusp may have been derived by anisomeral reduction from the cladodont high central cusp. Although the pectoral fin of *Pleuracanthus* was widely different from the long based pectoral of the cladodont *Cladoselache,* there is less difficulty in deriving it from the tribasic pectoral of *Tristychius* (fig. 8.19B); again the long tapering caudal fin of *Pleuracanthus* is widely different from the sharply upturned caudal of *Cladoselache* (fig. 8.13A), but it is not nearly so different from the tapering and not upturned caudal of the bradyodont *Helodus* (fig. 8.12A). But pleuracanths are excluded from close relationship with the bradyodonts, chimaeroids and edestids by the characters of their dentition and skull. They are quite different from the Triassic hybodonts in their jaws and fins; and they converge toward the notidanoids and frilled shark (fig. 8.25) in having an amphistylic skull and a not upturned caudal fin.

In conclusion, there seems to be but little reason apart from tradition for the current practice of treating the pleuracanths as if they represented a widely distinct subclass of cartilage fishes. Surely the construction of their skull and jaws entitles them to be admitted to the true elasmobranchs and to form at most another order (Pleuracanthea) of E. G. White's system (1937, p. 36), coördinate with the Cladodontea (Pleuropterygii), Heterodontea, Hexanchea, Galea, etc.

ARCHAIC RELICTS OF THE FINFOLD STOCK (HEXANCHEA)

The Cow-sharks or Indian Grays (Notidanidae).—These sharks are so named from their large cow-like eyes and gray color (figs. 8.24A, B). They hunt in the semi-darkness at moderate depths below the levels of the surface sharks. Their fairly long sinuous bodies evidently do not require radially braced vertebrae. The vertebral centra accordingly tend to revert to the embryonic ring-like form and to divide into anterior and posterior rings, especially in the tail region, whence the ordinal name Diplospondyli.

The notidanids are a very distinct and relatively ancient (Middle Jurassic) branch of the shark tribe. In all their modern members there are at least six gill-clefts and in some there are seven; but as most sharks have but five, this relatively high number of gill-slits may possibly be due to incipient budding.

The notidanids and the frilled shark (fig. 8.24C) may also be distinguished from all other sharks by the presence of only one dorsal fin. As two dorsal fins are present in all the Palaeozoic sharks and in most of the recent families, we may assume that the presence of a single dorsal fin in notidanoids is due to genetic reduction and loss of either the first or the second. It occupies the position of the second dorsal (fig. 8.13) of Palaeozoic sharks and I can find no evidence that it is a posteriorly displaced first.

The teeth of notidanid sharks are extremely jagged and of peculiar type (fig. 8.25). As in sharks generally, the teeth are wrapped spirally around the long axes of the primary upper and lower jaws and are grouped into dental families, each family comprising many generations (fig. 8.46D). The symphyseal teeth form a symmetrical coil but remain small; those on either side have their bases elongated in the direction of the jaws (fig. 8.25E). The upper teeth (E) of *Notidanus* (*Heptranchias*) have one high recurved cusp with a notch and several small cusps on the posterior side of the crown; the lower teeth are lateroposteriorly elongated, sharply serrated, the high first cusp followed by four cusps gradually decreasing posteriorly. Thus the upper and lower teeth are unlike and the principal cusp of the upper teeth is sharply recurved toward the outer side. But the lower teeth of the notidanoids, unlike those of typical sharks or charcharinids, have longer based crowns than the uppers and the highest point is not at or near the middle of the base but at the inner or anterior end. This peculiar arrangement has apparently been brought about by a shift in the growth gradient in the lower teeth, so that the point of fastest growth was shifted from the middle to the anterior border (fig. 8.25A-D). In this way the notidanid teeth may have been derived from a primitive shark type with a high cusp in the middle of the crown (fig. 8.25A). With this dental equipment the notidanoids have become fiercely predatory.

The notidanoids, dating from the Middle Jurassic, were at that time already widely separated from any other sharks of the modernized and galeoid division. Where then shall we seek their ancestors? As we shall presently see, there is no evidence that they were derived from either of the other surviving orders (Heterodontea, Squalea), and after long consideration I endorse the conclusion of E. G. White (1937, pp. 48, 49, 66, 67, 88, 89, 92, 100, 131) that they are one of the primary offshoots of the Devonian cladodont stock.

The Frilled Shark.—Before discussing further the origin of the notidanoid group, we must face the classic problem of the frilled shark (*Chlamydoselachus anguineus* Garman). The discovery of this shark, its description by Garman (1885), the subsequent investigations of its embryology and anatomy by Bashford Dean (1909) and by his successors, E. W. Gudger (1940) and B. G. Smith (1937), the beautiful monograph by Allis (1923) on the anatomy of its head, the work of Hawkes (1906) and others on its brain and cranial nerves, the interest in this fish as a supposed survivor from the dawn of shark evolution—all have put the frilled shark in the foreground of attention; if only to that small band of zoologists and palaeontologists who still deal with the major problems of the evolution of the vertebrates.

The frilled-shark occurs in moderately deep water off the coast of Japan and like many other deep-water fishes it has a very long body in relation to its height, so that it is usually described as eel-like (fig. 8.24C). Its long heterocercal tail is not upturned and the rather long single dorsal is paired with the ventral fin. The essentially tribasic pectoral (fig. 8.27C) has a reduced propterygium and an elongate metapterygium much as in the notidanids (fig. 8.27D) and equally derivable from the generalized cladodont type (fig. 8.27A). The head is long, low and serpent-like, the rostrum barely overhanging the large mouth, the upper and lower jaws prolonged backward beneath the braincase (fig. 8.20C). The teeth (fig. 8.14I), quite different from those of any other living shark, are delicate tridents with very fine points, sometimes with one minute basal point on either side of the central cusp; the whole set at an angle on a short, medially directed base with bifid root. Cope suggested that these teeth were a sign of derivation from the Permian pleuracanth sharks, but in the latter (fig. 8.14H) the teeth had two large cusps, sometimes with a small basal cusp between the large ones, and though they have an inwardly extended base not wholly unlike that of the frilled shark teeth, they rarely (H) show any tendency to develop a trident form. In 1885 J. W. Davis (cited by Gudger and Smith, 1933, p. 309) made a careful comparative study of the teeth of *Pleuracanthus laevissimus* and of cladodonts with those of the frilled shark. He concluded that the supposed resemblance of the teeth of the latter to those of the former "is only a superficial one and rests simply on the accidental circumstance of each having three denticles." Moreover, in the rest of its skeleton the frilled shark is much nearer to the notidanoid sharks than it is to the pleuracanths. Finally, trident-like teeth with or without vestigial basal cusps are not found in the fossil record until the Oligocene and Pliocene and are therefore many millions of years younger than the Middle Jurassic notidanids.

The most probable hypothesis seems to be that the three high points in the teeth of the frilled shark have been selected genetically in late Mesozoic and Caenozoic times for anisomeral emphasis and that the early Mesozoic ancestors of the frilled sharks may have had five cusps more or less equal on each tooth

(E. G. White, 1937, p. 37 and Table VI). Such a five-cusped pattern could readily be derived eventually from the many-cusped Carboniferous and Upper Devonian cladodonts (fig. 8.14D-F), which there is good evidence for regarding as representatives of the common ancestral stock of all sharks.

In the loose mode of attachment of the jaws to the skull (fig. 8.20C) and in their relative length the frilled shark recalls *Cladoselache* (fig. 8.13A). Especially important is the long hyomandibular bar, from which the rear parts of the upper and lower jaws are slung; this is the extreme hyostylic (hyoid support) method, whereas in the notidanids (fig. 8.20B) the palatoquadrate is much deepened posteriorly, its upper border being firmly attached directly to the postorbital process of the skull, while the hyomandibular is reduced to a thin rod covered externally by the palatoquadrate but still connected below by ligaments to the back of the mandible. This is the so-called amphistylic method. Which of these two is the more primitive? Apparently the notidanoid or amphistylic condition is on the whole the more primitive, the pronounced hyostylic suspension in the frilled shark being associated with the great enlargement of the jaws which grew far backward beneath the braincase. The unusually large palato-basal or ascending process of the palatoquadrate (fig. 8.20C) gives a strong sliding pivot for the upper jaw, resting against the side of the brain trough. The elimination of the firm attachment between the postorbital process and the palatoquadrate probably allows the rear end of the jaws to be lowered and the mouth to gape very widely. Thus the deep-water frilled shark, in retreating from competition with surface lining sharks, has evolved its own specializations for seizing and swallowing its prey.

The gill-slits, six in number (figs. 8.24C; 8.30A), decrease rapidly from the first backward and are covered with long frilled gill flaps (whence the name, frilled shark). At first sight these greatly enlarged gill-clefts with their frilled gill covers appear to be a unique specialization, but conceivably it might be an inheritance from an early shark condition similar to that which Watson (1937) made known in the acanthodians; for in the latter (fig. 7.27A) each gill-slit was covered with a dermal flap that was strengthened by horizontally placed exoskeletal plaques or rods. In the frilled shark there are no exoskeletal rods but they are represented functionally by cartilaginous rays attached to the branchial arches. However, the large gill-slits of both the frilled shark and the notidanoids may well be a sign of wide ranging predatory habits, and greater need for oxygen, as they are in the mackerel sharks (fig. 8.31E, G).

The notochord of the frilled shark is remarkably large and its vertebral centra are either truly primitive or retrogressive toward the larval condition. The centra are represented by ring-like thickenings and calcifications of the chordal sheath (fig. 8.26A). In the cervical and dorsal regions there is one of these curving cylindrical or ring-like centra to each neuromere or myomere, but in the region of the dorsal and anal fins there is a transition from the "monospondylic" to the "diplospondylic" condition (fig. 8.26A1), the latter becoming more perfectly expressed toward the tip of the tail (B. G. Smith, 1937, pp. 364–369). The nearest resemblances occur in the notidanoids.

In the pectoral fins (fig. 8.27C) the propterygium is much reduced, as it is in the notidanoids (fig. 8.27D), but the metapterygium bears a distal segment—perhaps inherited from ancient cladodonts (fig. 8.27A).

The spiral valve in the intestine of sharks is one of the numerous polyanisomeres which have been reviewed by E. G. White (1937) and shown by her to have value in classifying and defining the family and superfamily groups. Briefly, this valve has literally evolved from a fold in the wall of the intestines, which by continuing to grow faster on one side than on the other, rolls itself into a spiral. There are three main types of spiral valves. Of these, the more primitive ones of the type called "spiral" have from two to four valves (fig. 8.28A), the intermediates vary from five to thirty, while further multiplication produces a large number of "rings" arranged across the long axis of the intestine (B); these lack a central rolled axis, except near the upper end. In the opposite direction, atrophy of the spiral flanges leaves a long "scroll" attached on one side to the wall of the intestine (C). The frilled shark agrees with the notidanoids in having an advanced "spiral" with a relatively high number of folds (11–30), but it has not reached the "ring" type of the isurid sharks.

Another group of secondary polyisomeres in sharks is afforded by the valves in the conus arteriosus of the heart, which are arranged in two to seven rows. Dr. White concludes that the most primitive condition is that in which there are two rows of valves, as in the spiny-finned dogfish (*Squalus*), the Port Jackson shark (*Heterodontus,* fig. 8.29A), the nurse sharks or orectoloboids (B), and the less advanced cat-sharks or Catuloidea (C). The frilled shark, like the notidanoids, has four to five rows (E) and is thus relatively advanced in this feature.

Internal fertilization is effected in all modern sharks by means of elaborate cartilaginous appendages of the pelvic fins of the males. In the frilled shark these structures exhibit certain details which are present in all more primitive sharks but which have been modified in the galeoid order (cf. E. G. White, 1937, pp. 93–95).

The frilled shark is noted for the enormous size of its eggs (fig. 8.30aA1), due to the relatively huge supply of yolk which is laid down by special yolk glands of the uterus. Encapsuled eggs ranging from 90 to 116 mm. long are recorded by Gudger (1940, p. 570). But relatively gigantic eggs are also characteristic of the isurids and orectolobids (Gudger, 1940, p. 575).

In brief, the available evidence, as analyzed by Gudger and Smith (1933, p. 374) and by Smith (1937, pp. 492–495), tends to show that the frilled shark is in some respects a specialized side branch of the notidanoid stem.

MODERNIZED SHARKS (GALEA)

The classification of sharks has gone through several centuries of development and is still a live topic. Among the more recent contributors to the subject we may mention Tate Regan (1906), Garman (1913), E. G. White (1937) and Leo S. Berg (1940). The clearest and in our judgment most useful review of this great field is by E. G. White, whose classification and phylogenetic tree of the cartilage fishes are here largely followed. According to White, the modern sharks may be divided into four orders: Hexanchea, Galea, Heterodontea, Squalea. The order Galea (from Gr. *gale,* shark) is divided by her into two suborders: (I) ISURIDA (fig. 8.31), containing the nurse-sharks (Orectolobidae, fig. 8.31A, B), the whale-shark (*Rhineodon,* H), the sand-sharks (carchariids, C, D), the lamnids or isurids, including (a) porbeagle, mackerel (E) and mako sharks, (b) man-eater (*Carcharodon,* G), (c) basking shark (*Cetorhinus,* H), (d) thresher (*Vulpecula,* F); (II) CARCHARINIDA (fig. 8.32), including "cat sharks" (Scylliidae or Catulidae, J), the true carcharinids (requiem sharks, M, tiger-sharks, N, blue shark), as well as the smooth dogfish (*Galeorhinus,* L, or *Mustelus*) and the hammerhead (*Sphyrna* or *Zygaena,* figs. 8.30; 8.32O).

Of all these the nurse-sharks (orectolobids, fig. 8.31A, B) and the cat-sharks (catuloids, fig. 8.32J) are the most ancient, as their forerunners date back to the Upper Jurassic. At that time they were represented by both the depressed, bottom-living nurse-sharks (*Crossorhinus*) and the elongate inshore and bottom-living forerunners of the cat-sharks (*Palaeoscyllium, Pristiurus*). Next in order of antiquity came the ancestors of the fully marine and oceanic sand-sharks (carchariids, fig. 8.31C), mackerel-sharks (isurids or lamnids, E, F), which were dominant in the Cretaceous period (fig. 8.35). Latest in time were the carcharinids, including requiem sharks (fig. 8.32M), tiger sharks (N), hammer-heads (O), with Eocene and later Tertiary representatives. Thus the order Galea includes the highest or most progressive of all sharks and it is now also the dominant and most diversified of the existing orders.

CAT-SHARKS AND DOG-SHARKS

The "smooth dogfish" (*Galeorhinus* ("*Mustelus*") *laevis,* fig. 8.32L) is related to the requiem sharks (carcharinids, M, N), but has small blunt teeth forming a pavement. The "spiny dogfish" (*Squalus acanthias,* fig. 8.50B) is the type of the Squalea, or spiny-finned order (see below). The "spotted dogfish" (*Catulus* (*Scyllium*) *canicula,* cf. fig. 8.32J) belongs in the cat-shark tribe (Catuloidea).

The cat-sharks (Catulidae or Scylliidae) are mostly small and slender sharks, typically living on the bottom. They are variously colored, often with spots or blotches. The body is long and slender (J), the tail long and but little turned up. E. G. White (1937) has reviewed the evidence that these small "grovelling" sharks and not the big sea rovers are nearer the stem of the modern sharks and she has traced the transitional stages from the lower catuloids (J) to the higher carcharinids (N): in the patterns of the vertebrae (fig. 8.33), teeth (figs. 8.34, 8.35), skulls and rostral cartilages (fig. 8.48B, C), dermal denticles (fig. 8.36), pectoral fins (figs. 8.37; 8.38), spiral valves (fig. 8.28), heart valves (fig. 8.29) and other structures.

REQUIEM SHARKS

The centerpiece of the shark exhibits in the American Museum of Natural History is a large mounted group (fig. 8.39) entitled "The Sea Rovers," in obvious allusion to their far-ranging habits and piratical character. A gang of them, directed into converging paths by a common attraction, are swooping down on a large sea-turtle whose remaining minutes are evidently numbered.

The leader of the gang is a tiger-shark (*Galeocerdo tigrinus*), representing the Carcharinida, a burly monster with a rounded head, a rotund body and a tapering tail. His gray sides are marked with dull spots but the name *tiger* may have been given in allusion to his ferocity. His teeth, however, have another tiger character in so far as they bear short blades with jagged edges. A close view of the upper teeth (fig. 8.34K) would show that the blade is sharply curved toward the outer and rear sides and that there is a deep notch on that side of the blade.

Behind and above the tiger-shark (fig. 8.39) are three smaller gray sharks, often called *requiem* sharks; this is regarded as the equivalent of the French word *requin,* but the suggested allusion to a requiem for the dead may be accidental. They are represented by the genus *Carcharinus* (fig. 8.32M) and belong with the tiger-shark (fig. 8.32N) in the family Carcharinidae. The upper teeth (fig. 8.34K) in this genus and its allies (H, I, J) often have the tip bent toward the outer or posterior side, which causes a notch on that side, but the blade is usually more triangular (H, L). The lower teeth (fig. 34G) in most members of this family have smaller and more slender points than the upper ones, although occasionally the lower teeth may be almost exactly like the uppers. Thus the teeth afford very convenient characteristics for expressing specific and generic differences in this family as in others.

The blue shark, *Galeus* (*Prionace*) *glaucus,* another well known member of the requiem shark family, is viviparous (or rather, ovoviviparous) and one of the shark exhibits in the same hall shows a large female blue shark closely followed by its numerous newborn young. The upper teeth are jagged triangles, the lower vary from simple points to triangles (fig. 8.34H).

In the hammer-head (*Sphyrna zygaena,* figs. 8.32O; 8.39; 8.30b) the front part of the head is greatly flattened and widened across the snout and eyes. This remarkable effect of anisomeral transverse growth is initiated in the bonnet shark (*Sphyrna tiburo,* fig. 8.30b) and carried to an extreme in *S. blochii.* The flattened head in these strange forms serves as a bow rudder or plane, which permits sudden turns in pursuit of fish. The triangular upper teeth of the hammerhead (fig. 8.34L) have the tips recurved, much as in other carcharinids, and so do the notched lower teeth, which have much smaller points and blades. All the carcharinids have a third eyelid, or nictitating membrane. As noted by E. G. White (1937), this character seems to be correlated genetically with variants of the "Maltese cross" type of vertebral centrum (fig. 8.33C-H).

The nurse-sharks (Orectolobidae) are relatively ancient and primitive members (fig. 8.31A, B) of the suborder Isurida. In general appearance they resemble overgrown catsharks but may be recognized by the presence of projecting rods or tactile flaps or cirri around the mouth. The teeth are usually small, ranging from multicuspid (fig. 8.34N, O) to tricuspid (M, P) and are crowded together into a pavement, usually with low central cusps. Some of the Australian species of "wobbegongs" or carpet-sharks (*Orectolobus*) are recorded as attacking bathers or fishermen (Whitley, 1940, pp. 78–83). E. G. White has shown that in many anatomical characters the orectolobids, whose ancestors date from the Jurassic, are more primitive than the lamnids or isurids (fig. 8.31B-H), which arose in the Cretaceous.

In the right lower corner of the Sea-Rovers group (fig. 8.39) the sand-shark (*Carcharias taurus*) looks at first sight (fig. 8.31C) as if it were closely related to the carcharinids noted above, and indeed the similarity of their names, *Carcharias* and *Carcharinus,* has been and still is the source of confusion to students and others. But Garman, author of the definitive systematic monograph on the sharks, maintained that these two genera represent two widely separated series, and his results have been extended and confirmed by E. G. White (1937). In brief, *Carcharias* and its allies of the isuroid series have no nictitating membrane over the eye and their vertebral centra are of the radiating asterospondylic type (fig. 8.33K), while the carcharinid group have the contrary characters. Externally the sand-shark (*Carcharias*) can be distinguished from most of the carcharinid group by the relatively large size of the second dorsal, which is nearlv as large as the first (fig. 8.31C).

The teeth also afford clear evidence of the great differences between the carchariids and the carcharinids, since in the former (fig. 8.34Q) they have long awl-like points with two sharp little thorns springing from either side of the base, the whole crown being supported below by two widely separated root-like processes, while the typical carcharinids have blade-like (G), triangular (H) or recurved (J) teeth on a wide short base without roots. Both are derivable ultimately from the cladodonts of the Devonian (fig. 8.14B).

A long-snouted deep-sea relative (fig. 8.31D) of the carchariids (variously named *Odontaspis, Scapanorhynchus, Mitsukurina*) has surpassed even *Carcharias* in the length and acuity of its teeth (fig. 8.35). Its pectoral fin pattern (fig. 8.37D) closely resembles that of *Carcharias* (fig. 8.37C), but the snout (fig. 8.31D) is long, as in the deep-sea chimaeroids *Harriota* and *Rhinochimaera* (fig. 8.8A), while the skeleton has become cartilaginous and the integument is very delicate.

The famous man-eating or white shark, *Carcharodon rondeleti* (fig. 8.31G), reaches a length of 36½ feet (Whitley, 1940, p. 126). It represents the present-day giant of the carchariid-isurid series and far surpasses the tiger-shark (fig. 8.32N), which is the giant of the carcharinids; but it was in turn surpassed by its extinct cousin *Carcharodon megalodon* (fig. 8.40), which, from the huge size of its broadly triangular teeth (ranging up to six inches high), may have reached a length of more than forty feet. The recent man-eater belongs to the family of mackerel sharks or porbeagles (Isuridae or Lamnidae), including the mako (*Isuropsis*) of Australia and New Zealand. These are the speediest of all sharks. They are readily identified (fig. 8.31B) by the large vertical lunate tail, delicate caudal peduncle and lateral horizontal keels on either side of the tail. The second dorsal and anal fins are greatly reduced, while the first dorsal is very large and located on the middle of the back. The gill-slits are remarkably large. The teeth, as noted above, vary from the very narrow triangles of the mako shark to the wide triangles of the man-eater (fig. 8.35).

Attacks by sharks on bathers and fishermen are very rare in northern waters. However, in 1916 a single maneater (*Carcharodon*), ranging along the coast of New Jersey, killed three men and bit off a boy's foot (Murphy and Nichols, 1916, "The Shark Situation," Brooklyn Museum Quarterly, pp. 145–160). Along the coasts of Australia, some sharks, according to Whitley (1940), are justly feared by bathers, among the most savage and dangerous being the mako or blue pointer (*Isuropsis*), the mackerel shark or porbeagle (*Lamna*), the common "whaler" (*Galeolamna macrurus*), the "gray nurse" (*Carcharias avenarius*). Several species of wobbegongs or carpet sharks (especially *Orectolobus maculatus*) are known to have inflicted severe bites on wading fishermen.

The basking shark, *Cetorhinus* (*Selache*) *maximus,* is a specialized branch of the mackerel sharks (fig. 8.31H), with enormous gill-slits and very small teeth (fig. 8.35) derived from the *Corax* type. This large sluggish shark follows the schools of herring; its gill-rakers form long flexible plates (fig. 8.44B), which suggest the baleen of whales and likewise serve as a strainer to let the water out and keep the herring in.

The thresher shark (*Vulpecula,* fig. 8.31F) has an unusually small head, a very robust body and an exceedingly long tail, which is used as a lash in rounding up a school of small fish. It is a specialized offshoot from the mackerel shark stem.

THE WHALE-SHARK

The largest and in some respects most aberrantly specialized of all living sharks is the whale-shark (*Rhineodon,* fig. 8.31I), which may grow to more than forty, and possibly even sixty, feet in length (Smith, Hugh M., 1925).

Described in 1829 and 1849 by Andrew Smith, surgeon to the British troops at Capetown, South Africa, this great shark long remained but little known to naturalists; but, thanks largely to the consistent and long-continued researches and writings of Dr. E. W. Gudger, many specimens of the whale shark have been photographed and the anatomy has been partly studied (See the bibliography in the paper by E. Grace White, 1930, pp. 159, 160).

Gudger (1935) has collated numerous reports which prove that the whale shark wanders extensively in tropical and semitropical seas and is often seen in the Gulf of California.

This harmless and sluggish leviathan (fig. 8.31I) has an exceedingly wide terminal mouth but the teeth are reduced to minute denticles on a narrow band. It has large gill-slits and an excessively specialized sponge-like apparatus for sifting the water and at the same time retaining the food, which may consist of jellyfish and other plankton (fig. 8.44A, A1).

From the observations of Andrew Smith (1849), E. G. White (1937, p. 61) and E. W. Gudger (1941, pp. 90–96) on the sifting apparatus of *Rhineodon* (checked also by personal study of the material), we may draw the following notes:

1) The sponge-sieve, composed chiefly of modified "dermal denticles" (stomodeal denticles), fills the proximal parts of the gill pockets and lies on the pharyngeal side, covering the main gill arches. It is attached to a system of flexible integumental folds, the "transverse bands" of Gudger, which run transversely along the main gill arches and connect with a median longitudinal raphe on the floor of the pharynx (fig. 8.44A).

2) On the floor of the pharynx the sieve appears as a system of numerous closely packed, somewhat wavy rows of "fine-meshed denticles" (fig. 8.44A).

3) White (*op. cit.,* p. 61) notes the wide contrast between the closely knit gill-rakers of *Rhineodon,* forming the sieve of denticles, and the elongate, stiff and hair-like, parallel gill rakers of *Cetorhinus* (fig. 8.44B), which project into the pharynx and resemble whalebone in both appearance and feeling. Nevertheless she rightly implies that the two systems are homologous with each other.

4) On the ventral side of the sieve (fig.

8.44A1), the "gill bars" (gill arches) support oblique curved integumental blades ("ladder rung partitions" of Gudger) which spring downward and outward from the lower surface of each main gill-arch and form a dozen or more little arches or "curved partitions"; these in turn are crossed by still smaller partitions, forming a "secondary grid" (Gudger). Thus the water taken into the huge mouth cavity is forced through the sieve and sprayed down the sides of the gills.

5) As observed by Andrew Smith (1849), White (1937) and Gudger (1941), the very large gills lay beyond or distally to the complex sieve, which covered the proximal parts of the gill arches at the bottom of each gill pocket.

6) As a whole the apparatus involves polyisomeres of several orders: first, the main arches and gill pockets; second, the "ladder rung partitions," a dozen or more on each main gill arch; third, the small "secondary units"; fourth, the "denticles" or sieve units.

7) On its ventral side the apparatus at first sight also suggests such aërating structures as the lungs of *Ceratodus* and the epibranchial organs of the labyrinth fishes; but since its much branched and folded surface has been derived from the layer of horny stomodeal denticles, and since small sections of it do not indicate any extensive branching of blood vessels, a respiratory value is doubtful.

8) We may suppose that the "ladder rung partitions" (see above) of *Rhineodon* arose from tubercles or folds of integument which grew from the pharyngeal surface of each gill-bar across the gill pocket to the next gill-bar, the whole system of tubercles or flaps finally coalescing into a continuous sieve.

9) Each "rung" is possibly homologous with a single gill-raker of *Cetorhinus;* it is also analogous with the shaft of a feather, while the secondary grid is comparable with the barbs and barbules.

10) A very early stage in the evolution of these gill-raker folds may be represented by the large papillae on the surface of the oesophagus, which are continuous with the folds and furrows on the wall of the stomach (cf. White, 1937, p. 62). From some of the larger folds there are side branches and accessory buds, the whole covered with minute denticles but less complexly than on the "sieve."

E. G. White (1930, 1937) has shown that the whale-shark (fig. 8.31I) is allied on the one hand to the nurse sharks or Orectolobidae (A, B) and on the other to the stem of the mackerel sharks or Isuridae (E). Its huge lunate tail and enormous gill-slits suggest that it belongs with the mackerel sharks and the same is true of the construction of its complex claspers. On the other hand, its skull (figs. 8.42, 43), jaws, color pattern, and some other features, are more readily derivable from the nurse-shark type. Thus it represents a distinct family but tends to unite the orectolobids, carchariids and isurids into a single suborder (Isurida).

AUSTRALIAN BULL-HEADS AND THEIR FOREBEARS (HETERODONTS)

The living Port Jackson shark (named from the port of Sydney, Australia) is sometimes called the bull-head, in allusion to his massive short head, large eyes and heavy jowls (fig. 8.46aA). His mouth is very small and the sides of his short strong jaws (B) are covered with a breaking and crushing type of dentition. This consists of parallel whorls of large and small squarish dental plates curled transversely around the jaws. In the vertical plane of the symphysis, or junction between the right and left halves of the mandible, is a spiral of small pointed teeth, while on either side of the symphysis the whorls afford transitional stages from low-cusped to rounded pebble-like teeth. Obviously this is a "durophagous" dentition, contrasting widely with the sharp-toothed predaceous teeth of the typical Hexanchea (fig. 8.25) and Galea (fig. 8.34) and well adapted to the crunching of crustaceans and of the smaller shell-covered molluscs. In very young stages (fig. 8.46A1) the many-cusped teeth recall those of the cladodonts (fig. 8.14) and hybodonts (fig. 8.46b).

In order to support this crushing dentition the massive palatoquadrate (fig. 8.46C) articulates by a broad palato-basal facet on the under side of the stout braincase; the hyomandibular, which is unusually stout, also does its part in supporting the heavy mandible and is tied by ligament to the skull. In the axis of the stout body there is a strong vertebral column, in which the centra are strengthened by irregular folds radiating from the outer border of a small inner circle, the notochordal sheath (fig. 8.47C); this pattern is often incorrectly called the tectospondylic type.

The two large dorsal fins, as well as the pectorals and pelvics, are also strongly built (fig. 8.46aA), both dorsals having along their front margins stout sharp spines which extend down to the vertebral column and are backed by large and thick triangular basal plates.

The earlier members of the Port Jackson family (in the restricted sense) date back only to the Cretaceous (fig. 8.46bD). Their predecessors in turn were the hybodonts (fig. 8.13C), a long lived family which ranged from the Upper Devonian through the late Palaeozoic, up through the Triassic and Jurassic into the Lower Cretaceous. In the more primitive of these forms (*Hybodus*) the cheek teeth (fig. 8.46bA) were not yet expanded into rounded crushers but each tooth bore a stout central cusp with several smaller cusps on either side. As seen from above (B), these teeth formed long narrow ovals arranged in transverse whorls in the cheek region. In the backbone of *Hybodus hauffianus* (fig. 8.13C) the notochord retained its embryonic size and functional importance but there were stout neural arches and even ribs, an unusual feature in sharks. The dorsal fins were strengthened with massive spines and supported by triangular bases, much as in the modern Port Jackson shark, but the spines retained surface ridges and denticles. The pectoral fins of *Hybodus* (fig. 8.45B) were apparently more primitive than those of the modern *Heterodontus* (C) in retaining a backwardly prolonged and jointed metapterygial axis, essentially like that of the Devonian cladodont *Cladodus neilsoni* (A).

SHARKS WITH NO ANAL FINS (SQUALEA)

The members of this order may be at once distinguished from all other sharks by the absence of the anal fin. In the Jurassic *Palaeospinax* (fig. 8.50A) an anal fin was still present but as it has vanished in all known later squaloids (figs. 8.50; 8.51), as well as in all skates and rays, its permanent disappearance is apparently due to a genetic cause.

In the possession of spines on both dorsal fins the spiny dogfish (*Squalus acanthias,* fig. 8.50B) retains a primitive shark character which may well have been inherited through *Palaeospinax* (fig. 8.50A) from the stock that gave rise to the hybodonts (fig. 8.13C), ctenacanthids (8.13B) and cladodonts (18.13A). In some of the more specialized spiny-finned family, e.g., *Echinorhinus* (fig. 8.51C), *Somniosus* (D), the dorsal fin spines are much reduced or absent and they are also absent in the monk-fishes (Rhinidae, fig. 8.56C), as well as in the Japanese saw-fishes (*Pristiophorus*). The typical spiny dogfishes (*Squalus,* fig. 8.50B), apart from the presence of spines on the dorsals and the absence of anal fins, look so much like other "dogfishes" (e.g., the "smooth hound," *Galeorhinus mustelus,* fig. 8.32L), that it is no wonder they were long classified with sharks belonging to another order; but in addition to the absence of the anal fin the centra of their vertebrae are either cyclospondylic (fig. 8.47), tectospondylic (E), or pseudosterospondylic (D); the cartilaginous core of the rostrum is trough-like (fig. 8.48A), or reduced to a small knob (fig. 8.49), in contrast with the triradiate rostrum (fig. 8.48B) of the Galea. In some species the snout projects far in front of the mouth (fig. 8.49C). The jaw-suspension is hyostylic, that is, the hyomandibular forms the main connecting link between the jaws and the skull, the articulations of the palatoquadrate either with the cranial base or with the postorbital process of the braincase, being absent (E. G. White, 1937, p. 100). In the latter feature the Squalea parallel some of the Galea (fig. 8.46aD) and at the same time they fore shadow the skates and rays and contrast widely first with the typical Hexanchea (figs. 8.20B; 8.25), in which the postorbital contact predominates, and secondly, with the Heterodontea (fig. 8.46aC), in which the palato-basal contact is emphasized (E. G. White, 1937, p. 100).

The body form of the typical squalids varies from the normal dogfish type to the high-backed compressed *Oxynotus* (fig. 8.51B) and to the round-bodied Greenland shark (D).

In the Japanese saw-fish (*Pristiophorus japonicus*) the body is elongate and not unlike that of some of the grovelling catuloid sharks but the rostrum (fig. 8.49B) is produced into a long serrated weapon with enlarged tooth-like denticles set along its margins. One genus of this family (*Pliotrema*) has six gill slits. The family was well established in the Cretaceous period and may be a specialized offshoot from the grovelling or partly depressed stem-form of the Squalea.

Perhaps the least specialized dentition among the modern squaloids is found in *Centroscyllium fabricii* (fig. 8.53B), in which the upper and lower teeth both bear a prominent central cusp and a basal cusp on either side of it. Such a tooth crown could readily be derived from the cladodont type (fig. 8.14B). In *Etmopterus hilliani* the upper teeth (fig. 8.53B) have a high central cusp with two small ones on each side of it, but in the lower teeth the low central cusp is sharply deflected toward the outer side, somewhat as in certain Galea. In most of the squaloids the teeth are small and very numerous, collectively forming a pavement, usually with sharp points on each unit. These dental pavements stand in wide contrast to the great lacerating awls and knife blades of more predatory, free-swimming sharks and afford a transition to the crushing pavement teeth, without points, of the skates and rays. The large Greenland shark or sleeper (*Somniosus*), called the "gurry shark," likewise has a multitude of small close-set teeth (fig. 8.53D).

The dermal denticles (fig. 8.55A-F) usually bear keels but the central keel is never complete and never recurved or shell-like, in contrast with most Galea (fig. 8.36) (E. G. White, 1937, pp. 56–64).

The intestinal spiral valves (fig. 8.28) of most of the Squalea fall under the central spiral type (A) with 5–10 valves as in the Heterodontea and in wide contrast with either the numerous ring valves of the isuroids or the "scroll" type of the more advanced carcharinoids (E. G. White, pp. 82–86). The claspers of the males retain certain primitive features in common with Hexanchea and Heterodontea and in contrast with the Galea. The Squalea so far as known are viviparous, or rather ovo-viviparous, like most other sharks. They are highly variable in their dentition, surface denticles and other features and apparently their evolutionary possibilities are still unexhausted (E. G. White, p. 101).

Thus the place of the typical Squalea on the phylogenetic tree of the sharks lies next to the Heterodontea, both probably being offshoots of the cladodont-ctenacanthid stock.

The monk-fishes (Rhinidae) afford an almost perfect connecting link between the spiny dogfishes (Squalidae) and the skates and rays (Batea), but owing to the fact that the anterior borders of their pectoral fins are separated from the wide head on each side by a notch (fig. 8.56C) and have not grown forward along the snout, the gill slits continue to open rather laterally than ventrally, so that these fish have usually been classed with the Pleurotremata (with lateral openings) rather than with the Hypotremata (openings inferior), or skates and rays. The modern monk-fishes are the practically unchanged descendants of Upper Jurassic ancestors. While they inherit from the squaloid stem the fundamental construction of the skull, vertebrae and fins, they have progressed in the direction of the skates and rays in the widening of the body across the pectorals, in the shifting of the eyes on to the dorsal surface of the head, in the enlargement of the spiracle, flattening of the pelvic fins and displacement of both dorsal fins on to the caudal stem. But having reached this transitional stage in Upper Jurassic times, that part of the group which is represented by the existing monk-fishes became stabilized, at least in reference to the features named above, while some of its relatives (such as *Palaeospinax* (fig. 8.50A) gave rise to the free-swimming predatory dogfishes (squaloids); others, not unlike the Japanese sawfish in basic features, changed into the rhinobatids (fig. 8.61I) and thence into the skates and rays.

At this point it may be well to note that some members of several major groups of fishes tend to widen out and to cling to the bottom, while others lengthen and pass through first the grovelling or dogfish stage and then become streamlined, free-swimming predators. Even among the Agnatha, the depressed *Drepanaspis* (fig. 6.7A) contrasts with the fusiform *Pteraspis* (fig. 6.7B), while among the placoderms we have seen the depressed and skate-like *Gemündina* (fig. 8.57) and the swift-swimming fusiform acanthodians (fig. 7.28M). Obviously the adults of the bottom-living forms must have some sort of thigmotactic response which makes them feel comfortable when they are flattened down against a rocky or sandy bottom. Perhaps they have different sorts of sensory organs on their ventral and dorsal surfaces. On the other hand, the pelagic forms evidently feel more comfortable when they are in open water. There must of course be a complex network of genetic factors for these different responses, involving especially anisomeral emphasis, that is, acceleration or retardation of many growth gradients. But there must also be some degree of priority or dominance among these growth gradients and it is evident that in the origin of the skates and rays both the transverse and the longitudinal growth forces of the pectoral fins were the leaders.

WINGED SHARKS (SKATES AND RAYS)

The next stage in the evolution of the true skates is seen in the guitar-fishes (Rhinobatidae, fig. 8.60I), in which the already dominant pectorals extended across the former notch between the head and the pectorals and, growing forward above the gill openings, caused the latter to be visible only on the ventral surface. At the same time the rostrum, perhaps already somewhat elongate in a still earlier stage, grew far forward as the front border of the pectoral crept along its sides. Meanwhile the respiratory current of water passed out through the enlarging spiracle and the body behind the pelvic tended to become slender. In the saw-fishes (Pristidae, Pristiophoridae), the long rostrum of the rhinobatids (fig. 8.49C) was further elongated (B), while its marginal denticles became greatly enlarged into socketed teeth. The dominance of the pectoral fins culminates in the butterfly rays (Pteroplatea, fig. 8.60H), in which the width across the pectorals is much greater than the length from the tip of the snout to the proximal end of the almost vestigial tail. Thus the muscles of the pectoral appendage have become far larger than the primitive axial muscles. The depressed shape of the body and mode of locomotion in the monk-fishes, skates and rays makes it possible for them to seek their food along the rocky, muddy or sandy bottoms which abound in oysters, clams and other molluscs.

The dentition in *Rhina* (E. G. White, pl. xiv) consists of a pavement of little teeth with median points (fig. 8.53A) derived probably from the primitive squaloid type (B). In the saw-fishes (Pristidae) the numerous small teeth are without points, forming a pavement somewhat paralleling those of the smooth dogfishes (*Galeorhinus* or *Mustelus*) of the order Galea. In many of the true skates (Rajidae), although the little teeth form a pavement, they retain traces of sharp cusps, at least in the males of *Raja,* though the females have blunt cusps (E. G. White, p. 71). Among the sting rays (Dasybatidea) in *Rhinoptera polyodon* (fig. 8.54A) the teeth form a tesselated pavement of polygonal plates; the two middle rows are but little wider than the others. In the one of the eagle rays, *Myliobatis aquila* (B), the middle row becomes many times wider than the others, and in another (C), the middle row is extremely wide while the other rows have disappeared. By this time the dental apparatus has evolved into a relatively huge dental mill (fig. 8.58B) with broad upper and narrow lower plates, both curved vertically. The lower plat rubs against the upper and is moved by very strong jaw muscles; this mechanism is evidently well adapted for squeezing and smashing the shells of bivalves. Meanwhile the primary upper jaw (palatoquadrate) has further modified its loose connection with the skull by enlarging the hyomandibular sling; the lower segment (ceratohyal) of the hyoid arch has also become completely separate from both the mandible and the hyomandibular.

An apparent reversal of evolution seems to be indicated in the feeble dentition of the manta (*Mobula*), in which the numerous teeth (fig. 8.59) are very small and form a narrow band around the inner border of the upper and lower jaws, at least in very young specimens. This is in wide contrast to the huge dental mill of the eagle rays (Aëtobatidae). A possible origin for the cow-nosed rays (fig. 8.59A) and mantas would be directly from the sting-ray (dasybatoid) stem (fig. 8.60C) before the elaboration of large dental mills; however, in view of the close relationships of the cow-nosed rays with the eagle rays (Myliobatidae) as well as of the other cases where giant forms, feeding on plankton, have reduced the dentition to feeble bands of small teeth (as in the whale-sharks and basking sharks), it is probable that the minute teeth of the mantas are marginal denticles retained from a larval or young stage and that the median widened plates of the dental mill have been reduced and eliminated as the minute pointed denticles replaced them.

In the cow-nosed rays (Rhinopteridae, fig. 8.59A) the anterior ends of the opposite pectoral fins have become separated as "rostral" fins, which project well beyond the median end of the rostrum. In the adult manta (fig. 8.60F) these cephalic fins are twisted downward and forward and have developed a sort of elbow-like bend in front of the mouth, so that their spatulate ends form a pair of movable scoops or flaps, which are probably used in shovelling mud or in feeding on jellyfishes and other forms of plankton. These pectoral scoops are moved by modified fin muscles and afford a remarkable example of selective emphasis (anisomerism) which has modified a polyanisomeral complex to produce a new organ with a marked change of function: in this case the shift is from the original locomotor function to an accessory feeding function.

The sting-ray division (Dasybatoidea) are viviparous or ovoviviparous but the true rays (Rajidae) are oviparous, the eggs being protected by a tough horny shell which is laid down by shell glands and molded by special folds in the uterus. Since the practice of internal fertilization, as indicated by the presence of claspers in the male, probably dates back at least to the early Mesozoic stem-forms of all modern sharks and rays, and since all shark eggs have large yolks, it may well be that these early sharks were ovoviviparous but that the true rays (Rajidae) have tended to lay the eggs after a shorter intrauterine development and to strengthen the eggshell, while the sting-rays may have prolonged the internal development and thinned or eliminated the eggshell.

An early stage in the evolution of electric organs is preserved in the tail of the electric ray, *Raia batis* (fig. 8.61A). As described by Ewart (cited by Daniel, 1928, p. 107), the electric organ of this fish lies behind and is continuous with the lateral row of muscle cones. Daniel (1928, p. 107) notes that the organ itself is formed as a series of cones, which are essentially similar to those of the muscle except that the direction of the muscle fibers (*m.f.*) in the muscle is more oblique to the myosepta than are the discs (*ds*) of the electric organ. Ewart (1888, cited by Daniel, p. 108) found that the electric organ in *Raia batis* first appears when the embryo is about an inch long. It is confined to the tail and only those muscle fibres are affected which belong to the lateral bundles. These fibres undergo complete change of form and assume an entirely secondary functional rôle. Ewart figures three stages (fig. 8.61A1–A3) in the transformation of a striped muscle fibre into an electric disc. In brief, he finds that the nerve terminals of the muscle give rise to the outer electric layer, the muscle fibrillae of the distal end of the muscle give place to the thick striated layer of the "electric disc," while the main spindle of the muscle is changed into the inner or alveolar layer of granular tissue with long backward projections. At the base of the three layers is a thick cushion of gelatinous tissue which is contained in the connective tissue walls of the electric cone (Daniel, p. 109). The ability to accumulate a considerable difference of potential at the opposite poles of each minute electric cell must require the separation of the modified nervous elements of the outer layer from the modified muscle fibre of the inner layer by the presumably insulating, striate middle layer. It has been estimated that as many as 20,000 discs are present in an adult *Raia batis* (Daniel, p. 108).

The most remarkable of all rays is the electric torpedo (*Narcine* or *Torpedo*), in which part of the muscles and nerves of its pectoral fins have been converted into a voltaic pile capable of generating, storing and releasing a strong and painful shock. According to Daniel (1928, p. 110), "Removal of the skin from this [pectoral] area [cf. fig. 8.61B] shows the organ to be made up of multitudes of hexagonal columns resembling the cells in a honeycomb. Each column further consists of a series of discs piled one upon another, ten to twelve of these being present in each column. Each disc may be considered as having two surfaces, one ventral, the other dorsal. The ventral surface bears a negative charge of electricity, while the dorsal is positively charged." The little nerves that go to each disc are collected into larger nerves that supply each column and are in turn combined into the main electric nerves: these are from the hyomandibular branch of the seventh, glossopharyngeal and vagus (Daniel, *op. cit.*, p. 242).

SUMMARY

In the preceding chapter we have seen that the antiarchs or lower placoderms inherited many basic characters from the mud-grubbing cephalaspid ostracoderms, while the higher placoderms (apart from the arthrodires) approached the elasmobranchs: thus *Rhizodopsis* (fig. 7.19) in its general appearance strongly suggested the chimaeroids; *Macropetalichthys* (fig. 7.23) foreshadowed the sharks, especially in its brain and paired fins; *Gemündina* (fig. 7.24) was very like a ray, while the older acanthodians (fig. 7.25) combined an essentially pre-shark plan of the skull and jaws with an almost ganoidean type of scales. Indeed it seems possible that, while some of the foregoing may represent ancestors of sharks, rays, or chimaeroids, yet during the long Devonian ages the elasmobranch grade may have been attained more than once, and by different paths, from the placoderm stock.

Stensiö's evidence (1927, pp. 30–32, 333, 334, 374) against the traditional view that the skeleton of primitive vertebrates was cartilaginous in adults has cleared the way for an appreciation of the true place of the sharks and rays as specialized side branches of the earlier vertebrates. On the other hand, Watson's discovery of the many primitive and peculiar characters of the acanthodians, which tend to ally them with the placoderms, do not seem to be inconsistent with the view that the placoderms ("Aphetohyoidea") as a group are far more primitive than they have usually been supposed to be; moreover, the combination in "*Macropetalichthys*" of deep-seated coccosteoid with shark-like characters further indicates that in the primitive gnathostomes the head and thorax retained a bony armor which was lost in the sharks; while the conditions of the oral and hyoid arches in the acanthodians appear to have been more primitive than, and structurally precedent to, those found in the chimaeroids, sharks and rays.

Nor need we reconstruct the primitive shark as already a fast-swimming fusiform fish, rather than a partly depressed, grovelling placoderm with moderately wide pectoral fins, well developed fin spines and clusters of small teeth in coils around the jaws. Such types could readily pass along divergent lines to the bradyodonts and chimaeroids, edestids, squatinids, dasybatids, pleuracanths, ctenacanthids, heterodonts, cladodonts, notidanoids and galeoids. Finally the apparent relationships between squaloids, *Rhina*, rhinobatids, skates and rays (fig. 8.60) suggest that the latter two have been derived from squaloid sharks and that their resemblances to the placoderm *Gemündina* are convergent.

CHAPTER IX

THE BRANCHING TRIBES OF BONY FISHES

Contents

CHAPTER IX

THE BRANCHING TRIBES OF BONY FISHES

FIRST PRINCIPLES

It has been estimated that there are more than 30,000 named species and subspecies of recent fishes; and these are classified under several thousand genera, over 600 families, and, in one extreme system, 114 "orders." One main cause of the bewildering extent and complexity of this "family tree" is that it has been growing and producing more and more twigs and branches for at least 400 million years. But there are many indications that the classification is far harder to grasp than the outlines (fig. 9.7) of the tree itself.

The principal motives for attempting to classify the data of any branch of science are of course to divide the field into a descending series of units and to permit the student to gain a fair idea of the whole by the examination of samples of the different divisions; but in ichthyology the "splitter" school of specialists have made this task appallingly difficult through their custom of continually dividing up old genera, promoting species to be genera, and genera to families; so that some descriptions of collections omit the higher groups (orders, suborders) and catalogue the genera and species solely under families, listing them in a sequence that has become traditional. If this keeps on, we may eventually see families, genera and species severally arranged solely in alphabetic order. This would complete the ruin of a general classification, the upbuilding of which has resulted from several centuries of herculean labors by men who could still see the field as a whole.

This deplorable state has arisen partly because collections were and are often ranked according to the number of their type specimens; thus the effort to discover more and more new species and subspecies, although entirely necessary, has forced students to focus intently upon the smaller differences and to break down larger groups into smaller and smaller ones. And so in many papers on recent fishes the phylogenetic tree of earlier decades has been replaced by the analytical key, the chief end of which is to make it easy to determine the correct names of the subspecies and species. Meanwhile the demand for the identification of new collections emphasizes the need for more and more data relating to the proper names and exact geographic locality of the subspecies and species. Thus the men who have the most detailed knowledge are often prevented from attempting any constructive reviews of the recent and fossil members of any of the larger groups. In this way the very idea of evolution or interfamily relationships has been gradually squeezed out of many taxonomic papers on recent fishes. In short, the motto "analysis must precede synthesis" has become a reason for deferring synthesis.

While all this has been happening in the study of recent fish collections, the students of both fossil and recent fishes have gone right on through the decades, discovering more new forms, obtaining a more comprehensive knowledge of the structure of old forms, and constantly reappraising their interpretations of the phylogenetic relations of the older to the later groups and of the latter to each other. It is true that one consequence has been a shifting panorama of classifications, often of increasing complexity; but whenever new material has been studied, both intensively and broadly, improved

.nalyses and syntheses have brought into clearer ocus the relationships of the large and small ranches of the phylogenetic tree. This may be een in Moy-Thomas's "Palaeozoic Fishes," in he sixth edition of Parker and Haswell's "Text-ook of Zoology," Volume II, and in Romer's 'Vertebrate Paleontology" (1945).

"FIRST FAMILY" OF THE DEVONIAN AGE AND ITS DESCENDANTS (PALAEONISCOIDS)

The oldest known representative of the bony ishes (in the broad sense) is the genus *Cheiro-epis* of the Middle Devonian of Great Britain. ts well streamlined contour was shark-like, specially in its large predatory mouth, single lorsal fin and heterocercal tail (A. S. Wood-vard, 1906). It had inherited from a remote ast: (a) the basic chordate features which are een in their most primitive form in *Amphioxus* fig. 5.1A1) and in the ostracoderm stock (fig. .12A), plus (b) the complex jaws of bran-hial arch origin which were first clearly at-ained among the higher placoderms, especially he acanthodians (fig. 7.27A, B). Even the ninute quadrate scales which covered the sur-ace of the body of *Cheirolepis* were basically imilar to those of the acanthodians. But in the rrangement of the surface plates on the head fig. 9.6A) *Cheirolepis* differed in many im-ortant features from any of the acanthodians ind was essentially identical with the later gan-ids and teleost fishes.

The generic name *Cheirolepis* refers to the eculiar hand-like sculpture on the surface of he minute scales; but in the other members of he group, especially *Palaeoniscus* (fig. 3B), he scales were much larger, rhombic and cov-red with a shiny surface layer of ganoine. It ook many millions of years to transform this asic type into the thin, horny, rounded scales f modern teleost fishes, which have com-letely lost both the bony base and the ganoine urface. This process took place independently n numerous phyletic lines and many interme-liate conditions are known.

Unlike the acanthodians and earlier placo-derms, the early palaeoniscoids had no spikes on their fins (fig. 9.3B), which were supported on their front border by a row of thick, bent-up ridge scales called fulcra (fig. 9.4A, B). Pre-sumably the median fins could be raised and lowered only to a slight extent but could be warped a little to one side or the other. The fin-webs were supported on both sides by horny rays and by rows of narrow scales set end to end. In later forms (fig. 9.4D-G) these scales fused into bony rods called lepidotrichs, for the support of the fin, which could eventually be raised or lowered and folded up. But whereas the scales on the body finally lost their basal bony layer, those which were transformed into bony fin-rays retained it even in normal teleost fishes.

Beneath the single dorsal fin of the palaeonis-coids (fig. 9.2) there was a row of short pieces which were fewer in number than the scaly fin rays. These supporting pieces or pterygiophores are referred to the endoskeleton but this does not connote that they had any direct connection with the vertebral axis. To judge from condi-tions in recent sharks and ganoids, these pieces were laid down in or between the muscle fasciae in the space between the right and left myo-meres, immediately beneath the fin; they prob-ably served as bases not only for the fin itself but for the little buds of the myomeres which alternated with them and were beginning to move the fin. In later times there was a reduc-tion in number of these rods until there was only one for each bony ray; finally as the fins became able to execute more complex move-ments, the supporting rods (fig. 9.122A) were shaped by the muscles so as to permit the fin to be raised or lowered, turned to one side or the other and even sent into well controlled ripples.

In the early palaeoniscoids the notochordal axis (fig. 9.4A, B) diminished very gradually toward the tip of the tail, as did also the myo-meres (cf. fig. 9.5) that moved it. But in the catopterids (fig. 9.4D), and still more in the holosteans (E, F), the long dorsal lobe of the caudal fin became reduced to a blunt end and there was a pronounced swelling at the base of the tail; this indicates that the muscles were increasing in size and becoming subdivided into the complex arrangement seen in the existing garpikes (fig. 9.5B). Thus the tail was transformed from a nearly simple trailer into an accessory locomotor organ capable of elaborate movements in coöperation with the other fins.

The paired fins (figs. 9.2; 9.3) of the earlier palaeoniscids, like the dorsal, anal and caudal fins, were provided with surface rays made of conjoined scales set end to end; they also had fulcral scales on their convex preaxial borders and were doubtless moved a little by muscle slips from the underlying myomeres.

The pectoral girdle of palaeoniscids (figs. 9.2; 9.3; 9.6), like that of their descendants the later ganoids (fig. 9.7), and teleosts (fig. 9.31) was quite complex and had a double function. It consisted of an endoskeletal and an exoskeletal series. The outer or exoskeletal girdle (fig. 9.6A, B) was made of the same material as the scales, namely, a bony base and a ganoine surface. It was like a crescent turned forward and inward, curving around the posterior wall of the gill chamber and reaching forward to the rear outer corner of the skull. The lower horn of the crescent (*clv.*) nearly touched its fellow of the opposite side, beneath the throat. This outer girdle comprised five pieces on each side (from above downward): (1) the posttemporal (*ptm*) touched the tabular (*tab*) or dermal epiotic bone of the skull; the rear end of the posttemporal overlapped, (2) the lath-like supracleithrum, while the lower end of the latter crossed obliquely the top of (3) the cleithrum (*clt*); this in turn formed the main part of the crescent and supported the inner or endoskeletal girdle; the lower part of the cleithrum was tapering and curved inward as well as forward to its contact with (4) the infracleithrum or true clavicle (*clv*), a flat piece on the ventral wall extending to near the mid-line; beneath the supracleithrum and behind the cleithrum was (5) a thin vertical strip, commonly called the postclavicle (fig. 9.6B, *pcl*) but more correctly named postcleithrum XX.

The first and most important function of the exoskeletal or dermal girdle was to afford a fairly firm curving wall or base for the great muscle segments of the flanks. The cleithrum, forming the main part of the crescent, usually has a firm, inwardly directed flange along its front or concave border. The front borders of the supracleithrum and cleithrum served both as the rear wall of the gill chamber and as a rim or stop for the gill-cover or opercular flap (fig. 9.6A).

The corner or notch between the inner flange and the outer blade of the cleithrum serves above and below for the attachment of the flank muscles and in the middle for the endoskeletal piece or scapulo-coracoid with its attached pectoral fin (fig. 9.31). Not much is known about the scapulo-coracoid in *Cheirolepis,* but in one of the remote descendants of the palaeoniscoid stock, namely, the modern sturgeon, the scapulo-coracoid has become a large and complex base for the pectoral fin (figs. 9.7; 9.31A) and its muscles. In general the postcleithrum of ganoids and bony fishes (figs. 9.31; 9.32b) marks the dividing line between the superficial muscles of the throat and those of the flanks, and it has little to do with the pectoral fin although attached to the pectoral girdle. It originated as a row of enlarged scales parallel to and behind the cleithrum and onto its inner surface (fig. 9.31F) parts of the flank muscles were attached (fig. 9.75A).

The scapulocoracoid or inner girdle (fig. 9.31) seems to have been foreshadowed by the "endoskeletal shoulder-girdles" of the ostracoderms (fig. 6.15A2) just as the outer or dermal girdle was by the thoracic part of the head-shield (figs. 6.12B; 7.2C). In placoderms the thoracic or pectoral girdle as a whole (figs 7.4B; 7.9; 7.12; 7.14; 7.18; 7.23A, A1) begins to be shark-like; but in the sharks themselves (fig. 7.27C) the dermal part of the girdle is represented only by the skin and the scapulocoracoid has become more or less cartilaginous In the sturgeon (fig. 9.31A) the inner girdle (scapulocoracoid) is remarkably large, but in the teleosts (fig. 9.31E, F) the scapula becomes smaller than the coracoid.

The lower horn of the inner girdle, namely,

the coracoid process (figs. 9.7A1; 9.31 *cor*), extending forward toward the floor of the throat, gives attachment to the coraco-mandibularis and other muscles that are useful in the increasingly predatory character of the jaws and branchial arcade, while the top of the inner girdle has been seized, as it were, by the trapezius muscle (fig. 9.12A). This, being innervated by the vagoaccessorius nerve, doubtless tended to pull the top of the shoulder girdle forward and thus to coöperate with other muscles in swallowing movements. But such an innervation of this muscle lends no real support to the old theory that the entire coraco-scapular arch represents a former branchial arch (Miner, 1925, pp. 190–196).

The pelvic fins of *Cheirolepis,* according to Traquair, Woodward and Watson, were remarkably elongated anteroposteriorly. They were supported by numerous small rods (cf. fig. 9.2) which were fewer in number than the dermal rays (Woodward, 1906). In *Cheirolepis* there is no evidence of a basal piece or pelvic rod, beneath the small rods, such as there was in certain placoderms (fig. 8.10E), bradyodonts (F, G), or pleuracanths (fig. 8.21). In Palaeozoic sharks the pelvic pieces are restored as triangular (fig. 8.13). The pelvic endoskeleton arose in the core of a cone-in-cone outgrowth from the muscle plates or myomeres, which gave rise to the diverse pelvic bones of bony fishes (e.g., figs. 9.54; 9.75A), crossopterygians (fig. 11.12), and eventually to the pelvis of amphibians (figs. 11.1B; 11.12C) and higher vertebrates. In later embryonic growth bone cells are deposited in the connective tissues in and around the areas of stress from the muscles.

The skull of palaeoniscids (figs. 9.6, 8) like those of typical fishes and of tetrapods was a syncranium (fig. 9.9B) or complex structure with two closely integrated main divisions: (A) a neurocranium and (B) a branchiocranium, both comprising endo- and exoskeletal elements.

The neurocranium of *Palaeoniscus* (fig. 9.8A), like that of all typical vertebrates, was formed around the olfactory, optic and otic sense organs; there was also a median brain trough which in modern fish is developed from the embryonic trabeculae and parachordals. The neurocranium formed the fulcrum for the large notochord (fig. 9.8A1), which was inserted into the occipital (parachordal) base and extended forward to the border of the pituitary foramen. The structure of the neurocranium of several kinds of palaeoniscoids was most carefully explored by Watson (1925, 1928). In several of the endocrania described by him in 1925 the process of ossification was so complete that it was impossible to discover the sutural limits between the individual bones. He later (1928) described the braincase of a small palaeoniscoid (*Cosmoptychius*) in which ossification was incomplete and the location of a number of separate skull elements could be determined. In the Jurassic *Chondrosteus* (fig. 9.14B) the endocranium was becoming cartilaginous, a process which culminates in the modern sturgeons (fig. 9.14C), and at the same time it was breaking up into separate growth centers. This is in accord with the conclusions of Stensiö (1927, 1934), who has adduced much proof that the "cartilaginous" skeleton of recent sharks, sturgeons and other vertebrates is retrogressive, due to delay in the ossification of the endoskeleton and consequent retention of larval characters.

The rostrum of *Cheirolepis* (figs. 9.2; 9.8A) contained the small olfactory capsules and was very short, while the orbits were large. This contrasts with the expanded olfactory parts and relatively small eyes of typical sharks but agrees with the conditions in the acanthodians (fig. 9.11A) and other placoderms. In all more typical fish the posterolateral corner of the olfactory capsule is produced into a prominent horizontal buttress just in front of the eye, named parethmoid (fig. 9.8B). The parethmoids were evidently small in the palaeoniscoids. The optic division of the braincase (formed from the embryonic trabeculae) is called the sphenoid bone, its base being the basisphenoid and its ascending wall the orbitosphenoid (cf. fig. 9.8B, *osp*); the latter is pierced by the olfactory foramen (I).

The otic capsule in the palaeoniscoids, as in

all other vertebrates, contained the ventral vestibule (fig. 9.8A1) and the semicircular canals; it may have served for sensory and muscular coördination and balance rather than for hearing. In modern fish the otic capsule develops in the embryo from a simple pouch in the ectoderm and it becomes subdivided horizontally, giving rise to a lower chamber or sacculus (fig. 9.8B1) and an upper division, shaped like a hollow pyramid and containing the utriculus and the three semicircular canals. The sacculus was on the level of the cranial floor formed by the embryonic parachordals. The canals and utriculus were on the sides of the skull. The inner, or medial side of the otic capsule has a wide opening which surrounds the eighth or acoustic nerve and its several branches. The proötic bone center (fig. 9.8B, B1 *prot*) lies in the anterior half, the "opisthotic" (*opo*) plus exoccipital center is in the posterior half of the capsule. These two bones, extending upward, covered the lower part of the three canals, while the undivided bone covering the upper part formed the epiotic (fig. 9.8B1 *epiot*).

The primary braincase or chondrocranium was invested with an exoskeletal or dermocranium (fig. 9.6) divided into many named bones or regions, and these investing bones were basically like the scales, containing a bony base and a ganoine-covered surface. The lateral line grooves (fig. 9.6C) are continued forward from the body, passing over the supracleithrum, post-, supra-, and intertemporals and the postfrontals, continuing down the postorbital and extending forward beneath the orbit across the jugal to the lacrymal; thence the line may sometimes again continue forward, curving upward and backward across the antorbitals and frontals and ending on the parietals. Usually a side branch leaves the main line on the supratemporal, runs down the preopercular and is continued forward on the lower border of the mandible (fig. 7.22B). The general course of these lateral line canals is remarkably constant and assists in the correct identification of the bones that they cross and especially in the comparison of skull patterns of different classes of fishes (figs. 7.21, 7.22) with each other and with the earliest tetrapods. But, as Moy-Thomas has shown (1941), the idea that the surface bones owe their development to the lateral line organs is not confirmed by experimental evidence.

The surface bones (fig. 9.6) covering the neurocranium of primitive palaeoniscoids and their nearer derivatives may be readily classified into the following functional groups: (1) rostrals and nasals (antorbitals); (2) sclerotics, circumorbitals, postorbitals (*y, x*); (3) roofing bones (frontals, parietals, dermosphenotics (*dsph*), pterotics (*pto*); (4) the floor of the braincase is strengthened in front (fig. 9.8) by the vomer and behind by the parasphenoid; these median bones also form the mid-roof of the oral cavity. More in detail: (1) the rostral plates cover the ethmoid or olfactory region of the endocranium; they form the prow or entering angle of the entire fish as seen from above and are buttressed posteriorly by the ectethmoids (parethmoids), as noted above; (2) the supra- and circumorbital bones surround the sclerotics of the eyeball; they are typically five in number (lacrymal, jugal, postorbital, postfrontal, prefrontal) but are subject to subdivision or reduction; behind them come the two postorbitals (*y, x*), which cover the anterior part of the cheeks; (3) the roofing bones, where they overlapped the rostrum in front, were called "nasals" or antorbitals (possibly equivalent to the septomaxillae of tetrapods). Behind the antorbitals came the "frontals," which form the main part of the interorbital bridge. Posterolateral to the frontals are the intertemporals or dermosphenotics (*dsph*). Behind the frontals are the "parietals," forming the roof of the cranial vault between the otic capsules. Lateral to the parietals are the supratemporals or pterotics (*pto*). Behind the parietals is a transverse strip, often bearing a transverse commissure of the lateral line and consisting of a medial and lateral pair of extrascapulars. The median pair are sometimes called postparietals or dermosupraoccipitals; the lateral pair are called lateral extrascapulars or tabulars ("scale bones"). To the posterior margin of the extrascapular strip was attached the posttemporal (*ptm*), from which the pectoral girdle was suspended.

The branchiocranium is very poorly known in the palaeoniscoids but in a typical fish (fig. 9.9B) it includes the oralomandibular, hyoid and branchial arches and all their covering plates, which collectively surround the oralobranchial chamber and gill-clefts. The oralomandibular arch includes the palato-pterygoquadrate above and Meckel's cartilage below. Watson has shown that in the palaeoniscoids (fig. 9.10) the dorsal vertical flange of the palato-pterygo-quadrate is subdivided into several areas, or suprapterygoids. The palatal part of the arch bears dermal, often tooth-bearing plates named palatines, ento- (meso-) and ectopterygoids (pterygoids). This arch (fig. 9.32) in modern fish articulates in front with the parethmoid buttress and is supported in the rear by the enlarged hyomandibular (*hyom*). The latter in turn articulates above with the sphenotic and pterotic bones. The hyomandibular also supports the opercular series, as noted below. The hyomandibular is normally tied into a deep notch in the posterior part of the quadrate bone by a narrow strip, the symplectic (*sym*); on the medial surface, at or near the junction of the symplectic and the hyomandibular, is inserted a small bar, the interhyal (figs. 9.9B; 9.32B), which suspends from the hyomandibular the epi- and cerato-hyoids and with them the whole branchial basket. The deep jaw muscles (fig. 9.12A) were originally wrapped around the vertical or suprapterygoid (*s.pt.*) flange of the pterygo-quadrate (fig. 9.10) and inserted into the medial surface of the Meckel's cartilage.

An early stage in the evolution of the oralomandibular arch from one of the branchial arches is illustrated in the acanthodians (fig. 7.27), in which the oralomandibular arch and its muscles were evidently constructed more nearly on the plan of a branchial segment. In the bony fishes (including palaeoniscoids), however, the dermal plates covering the oralomandibular arch were already plainly homologous with those of bony fishes and early tetrapods.

In the typical palaeoniscoids the lateral surface of the oralomandibular arch (fig. 9.6) bears a series of jaw plates. Of these, the maxilla (*mx*) is much prolonged behind the small premaxilla (*pmx*), the suspensorium (hyomandibular, beneath *pop*) being directed sharply backward and the angle of the mouth far behind the eye. There are pointed teeth along the outer border of the premaxilla and maxilla, extending to the posterior limit of the maxilla where it overlaps the lower jaw. In the mandible the principal tooth-bearing plate is the dentary (*dn*), opposing the maxilla (*mx*). On the medial surface there were probably representatives of the articular ("Meckel's cartilage") splenial, coronoid and prearticular bones of other fishes.

The preopercular (*pop*), opercular (*op*), subopercular (*sop*) and branchiostegals (*brstg*) form a crescentic series of plates and slats originally on the surface of the opercular fold that overlaps the gill slits. The lower part of this series may correspond (fig. 9.11B) to the first opercular fold of acanthodians containing the "mandibular rays" (B), which grew backward from the oralomandibular arch, overlapping the hyoid arch. The uppermost member of the opercular series in all bony fishes was a bony plate, the opercular, convex externally and concave internally. This bone may correspond with the "hyoid operculum" of acanthodians (A). Near the upper corner of the inner surface of the opercular in typical fishes was a concave articular depression fitting on to a convex oval surface at the extremity of a process from the rear of the hyomandibular bone (fig. 9.32). Above the opercular were several muscles (fig. 9.12) which passed from the inner edge of the skull-roof to the hyomandibular and served to raise, lower, open and close the whole opercular flap. The opercular in the palaeoniscids (fig. 9.6) was smaller than the bone below it, which was the subopercular; the branchiostegal slats curved downward and forward and served to tie the opercular flap to the lower border of the mandible. The old postmandibular cleft of the acanthodians (fig. 7.27B) followed the course of the preopercular branch of the lateral line (fig. 9.11A), behind the preopercular and the posterior border of the maxilla. This cleft in acanthodians and sharks (fig. 7.27B, C) terminated above in the spiracle, or so-called hyoidean cleft; in palaeoniscoids it may be covered (fig. 9.6A, B) by the neighboring bones.

Thus the early palaeoniscoids had a nearly full complement of fish-skull elements and both the mode of growth and the basic pattern and arrangement of these elements were destined to be handed on in varying degree to the higher bony fishes.

A conspicuous difference between this basic fish skull and the human skull is that in the former the separate elements are several times more numerous than in the latter. Including all the elements of the branchial arches and their dermal covering plates as parts of the complex skull, there were about one hundred and forty-one separately named pieces in the skull of the palaeoniscoids (fig. 9.13) and but twenty-seven or twenty-eight (counting the hyoid bones) in man and other mammals. In earlier papers (1933, 1935) I have referred to the progressive reduction in the number of skull-elements as we pass from the lower to the higher vertebrates as "Williston's law," since the late Professor S. W. Williston supplied palaeontological evidence in proof of this principle. But Williston, as I later discovered, was long antedated in the publication of this principle by Professor Stromer von Reichenbach of Munich, who in 1912 (p. 282) very clearly recognized the principle and cited numerous examples of it. Apparent exceptions to it, as we pass from 141 bones estimated for the palaeoniscoids to 160 in the generalized amioids and 160 in the generalized isospondyls, are due in part to the unossified or cartilaginous condition of the neurocranium in the palaeoniscoids, in part to a secondary increase in the number of branchiostegals of the generalized isospondyl (Gregory and others, 1935, p. 133).

The enlargement of the eyes in the early palaeoniscoids (fig. 9.8A) has crowded the intervening brain-trough into a narrow interorbital bridge. In a typical modern fish (figs. 9.8B; 9.9B) this bridge widens out and passes into the cranial vault. The skull is braced below by the keel bone, or parasphenoid, which runs backward from the under side of the ethmo-vomer block to the cranial base. The latter is composed on either side of the expanded bases of the otic capsules and in the middle by the osseous parachordal tissue surrounding the notochord.

This complex neurocranium (fig. 9.8) has to sustain thrusts from various directions. It must in the first place equalize the fore-and-aft thrusts of the locomotor muscles against the pressure of the water; for this purpose the ethmo-vomer block, interorbital bridge and cranial vault and base are admirably built. But even more strenuous are the wrenching and shearing stresses and strains resulting from the sudden alternation of left and right movements of the fore part of the body in swimming. Finally we have the numerous thrusts and wrenching strains induced by the attachments of the jaws and branchial arches in forms which by this time had become capable of overcoming the death struggles of smaller fishes that were often armed with defensive spines.

Such was the efficiency of this type of skull (already established in its essentials among the palaeoniscoids) that it afforded a complete protection from jarring to the delicate brain and cranial nerves, which were peacefully suspended in the midst of these contending streams of balanced stresses and strains.

The early palaeoniscoids are truly primitive, structurally ancestral to the sturgeons (fig. 9.14C, D) and at least close to the ancestry of all the higher bony fishes. They are indeed so amazingly primitive that one is at first inclined to expect to find many of their primitive characters in still earlier and more remote ancestors. But, from very many lines of evidence noted in this book, it appears that great advances in evolution were marked by the budding of new sets of polyisomeres, that after these new polyisomeres were produced some of them became larger and others were reduced or eliminated; finally that still later crops of polyisomeres often appeared in the midst of the older crops. For example, among that line of the palaeoniscoids which produced the sturgeons, an older generation of polyisomeres, represented by the ganoid scales, gradually lost their ganoid character and became horny, while a later generation, taking the form of bony plaques, appeared among them and finally became dominant (fig. 9.14C). Many similar instances will be cited from time to time in this book. Hence, it would be unsafe to impute to the still unknown very remote Silurian ancestors of the palaeoniscoids the same sets of polyisomeres which characterize the palaeoniscoids themselves. As noted above, the highly complex skull of palaeoniscoids is basically so close to that of the acanthodians that the palaeoniscoids may have been derived from some pre-acanthodian stock. Such a primitive acanthodian as *Euthacanthus macnicoli* (fig. 7.26) of the Lower Old Red Sandstone of Scotland, although only a stage older than *Cheirolepis,* suggests that some of the older polyisomeres such as fin-spikes and accessory paired fins between the pectorals and pelvics were being reduced and partly eliminated. Meanwhile the palaeoniscoid series were developing their own characters as already summarized.

The palaeoniscoids share many basic skeletal features with their contemporaries the crossopterygians, or lobefins (fig. 10.1). This suggests that the two classes Actinopterygii and Crossopterygii were perhaps independently derived from different members of some unknown common ancestral class. The most important difference between them is in the construction of the paired fins (fig. 10.18H, E); but according to the evidence cited below, the pectorals in both groups were derived eventually from the tribasal type (with pro-, meso-, and metapterygia) which is best expressed in the sharks and is well developed in the placoderm *Pseudopetalichthys* (fig. 7.23A). We may suppose that in the line leading to the palaeoniscoids the emphasis was laid upon the production of the typical palaeoniscoid scales, strong lepidotrichs and marginal scales or fulcra, that the three basals of the pectoral fins gave rise by subsequent division to the four or more baseosts of the pectoral fin of sturgeons and teleosts, but that *Polypterus* (fig. 9.18E) retains the tribasal ground plan. The pelvic fins in most cases were less progressive than the pectorals and in the earliest palaeoniscoids (fig. 9.2) were essentially like the anals.

Accordingly the palaeoniscoids and the crossopterygians may have been derived from undiscovered primitive jaw-bearing fishes, which resembled the acanthodians in the general construction of their skull and oralo-branchial system but retained the tribasic stout pectoral lobefins of *Stensioella* (fig. 7.23B) and *Pseudopetalichthys* (A). In the skull-roof the median elements above the frontoparietal region (cf. figs. 10.20; 10.21) were dividing in the mid-line. The cranial lateral-line pattern was probably more or less intermediate between the radiate branching types of placoderms (figs. 7.21A, B; 10.21A-C) and the longitudinal and transverse arrangement of primitive fishes (fig. 7.21C, D).

The palaeoniscoids were essentially a middle and late Palaeozoic group ranging from the Lower Devonian into the Lower Carboniferous (Mississippian), Carboniferous (Pennsylvanian) and Permian, branching out in many directions and forming a prelude as it were to the increasingly wide branching of their descendants in the early and late Mesozoic and Caenozoic periods. Moy-Thomas (1939) has admirably summarized the principal divisions of the palaeoniscoids, the knowledge of which has been recently expanded by important discoveries in East Greenland described by Stensiö, and by an illuminating memoir by Westoll (1944).

In all these branchings the fossils and the few surviving representatives reveal many correlated changes in (1) the locomotor, (2) the feeding, (3) the respiratory and (4) the controlling systems. (1) The locomotor system may be taken to include body-form, integument, fins, the axial and appendicular skeletons and their related musculature; (2) the feeding and (3) respiratory system in so far as it directly affects the skeletal parts includes the teeth, jaws, hyoid and branchial arches and their muscles, together with the related parts of the skull. (4) The control system is indicated chiefly in the neurocranium, which is also the fulcrum of the locomotor system.

Among the palaeoniscoids in the narrower

sense the body was elongate and shark-like as in *Cheirolepis* and possibly almost eel-like in Stensiö's python-headed *Birgeria*. On the other hand, the body became extremely short, high and compressed in *Cheirodus* of the Platysomidae (fig. 9.6E). In such forms the body is rhombic or kite-shaped in side view, with a high apex on the back and a much depressed gasterion (the lowest point) on the belly. The dorsal and anal fins, both with very long bases, slope toward the tail and by waving simultaneously become important locomotor accessories in this quick-darting form of body, which has been evolved independently in many groups of bony fishes. The steepness of the body brings about detailed changes in the surface pattern of the skull bones, such as the deepening of the opercular flap, the crowding of the cheek bones and the sharp upward slope of the supraoccipital region; meanwhile the mouth has become very short and even beak-like in *Cheirodus,* while blunt rounded teeth on the splenials and pterygoids, on the roof of the mouth and inner parts of the mandible, coöperate in the habit of nibbling and crushing hard material, such as coral heads, which often harbor small crustaceans, etc. Intermediate conditions between the shark-like *Cheirolepis* (fig. 9.2) and the kite-shaped *Cheirodus* (fig. 9.6E) are represented by *Rhadinichthys, Palaeoniscus, Mesolepis* (fig. 9.19A), *Platysomus* (fig. 9.2D).

The Triassic Saurichthyidae were formerly a stumbling-block in the classification of the ganoids, since they were thought to be connected in some way with the garpikes; but the intensive monographic studies of Stensiö (1921–1925) show that the morphology of the skull (fig. 9.14A) is identical, except in detail, with that of the sturgeons and thus readily derivable from the palaeoniscoid stock. The rostrum is long and pointed like that of a swordfish, the mouth large, somewhat like that of a spoonbill sturgeon. The body scales, as in the sturgeons, are more or less completely reduced and there are several longitudinal series of scutes along the body. Some of these retain vestiges of the ganoine layers (Stensiö, 1925, p. 131). Stensiö regards the saurichthyids as one of the major divisions of the Chondrostei, but in view of the closeness of their skull structure to that of the sturgeons we doubt that they deserve more than family rank.

Still nearer to the sturgeons are the Jurassic Chondrosteidae (fig. 9.14B). Some of these attained gigantic size, the length of *Gyrosteus mirabilis* being estimated by A. S. Woodward (Cat. Foss. Fishes, Brit. Mus., Pt. III, 1895, p. 35) as not less than six or seven metres.

As noted above, the existing group named Chondrostei (sturgeons and spoon-bill sturgeons) are in many respects highly specialized descendants of the palaeoniscoid series. In the sturgeons (fig. 9.14C) the rostrum has become produced into a more or less elongate spade-shaped or pointed disc, often provided on the under side with sensory barbels, while the mouth parts are greatly reduced in size and have been changed into a dredging and sucking organ. In the spoon-bill sturgeons (D) the rostrum becomes long and spatulate, richly supplied with sensory organs, while the large toothless mouth, bag-like throat and long narrow gill-slits suggest a habit of netting schools of small creatures.

It is noted above that the endoskeleton of Chondrostei becomes cartilaginous and the investing bones porous and more or less horny. The cartilaginous condition of the skull formerly caused the sturgeons to be regarded as especially primitive. They do, however, retain typical palaeoniscoid characters in their fin-fulcra, heterocercal tail (fig. 9.4C) and in the construction of the endoskeleton of their median and paired fins.

Polypterus, a pillar of vertebrate morphology.—The bichir (*Polypterus,* fig. 9.15A) of Africa has been the subject of several classic memoirs: by Budgett (1902) on its embryology; by Allis (1922) on its cranial anatomy; by Klaatsch (1896) on the musculature of its paired fins, and others. It was formerly regarded as a survivor of the lobefins or Crossopterygii but Goodrich (1909, 1928) after a careful analysis of its skeletal characters and soft anatomy concluded that its supposed crossopterygian characters are due to convergence and in part to parallelism; that it has inherited its ganoid scales, paired gular plates, infraclavicles and the more primitive features of its skull from the palaeoniscoid stock. Its strong pectoral fins, which have a large rounded muscular lobe and a spreading web of stout lepidotrichs, externally suggest those of the Devonian crossopterygian *Osteolepis* (fig. 10.18G), but internally (fig. 9.15A1) they differ widely from the crossopterygian type (see below) and although specialized in certain features, distinctly suggest the tribasal type which, as we have seen, occurs in *Macropetalichthys, Acanthodes* and the sharks. The pelvic fins and their supporting rods also differ widely from the crossopterygian type and agree rather with the holostean type. In the microscopic construction of its scales *Polypterus* agrees closely with the true ganoids and differs widely from the crossopterygians and the same is true of many features in the soft anatomy. Hence *Polypterus* and its still more specialized and eel-like relative *Calamoichthys* are assigned to a distinct order, Polypterini (Cladistia, Brachyopterygii). A probable ancestor of *Polypterus* is the Jurassic *Cornuboniichthys* (fig. 9.16) described by E. I. White (1939), which tends to connect it with the palaeoniscoid stem. So much as to its remote heritage.

In its habitus *Polypterus* is a typical predator, with a cylindrical body (fig. 9.15) not unlike that of the garpike (fig. 9.17C) but with a short rostrum and a greatly elongate dorsal fin subdivided into many finlets (whence the name *Polypterus*). The rounded tail has a stout muscular base and a vestigial extension of the notochord. Its dermal rays are continuous with the dorsal fin around the end of the vertebral column; the anal fin is quite near to the caudal and if it joined with it the entire complex would come under the gephyrocercal (bridge tail) type of Dollo. Thus the most conspicuous newer polyisomeres (the dorsal finlets) of *Polypterus* tend to disguise its relationship with the old palaeoniscoid stock. The bilobed air bladder of *Polypterus* (cf. Goodrich, 1909, pp. 223, 224, fig. 197A) is supposedly more primitive than that of ordinary bony fishes in that it is ventral to the alimentary canal and may represent a pair of diverticula in series with the gill pouches. Its duct opens upward into the oesophagus, whereas in teleosts this duct opens downward.

RISE OF THE ARMORED HOLOSTEAN GANOIDS (HOLOSTEI)

As compared with modern bony fishes, *Cheirolepis* and other members of the earlier palaeoniscoids were, as we have seen, far more primitive.

In certain palaeoniscoids (fig. 9.19A) and derived forms (B) there is a gradual transition in the reduction of the dorsal or notochordal axis of the caudal fin base, which later is indicated by a single row of scales; the ventral lobe in the Catopteridae (fig. 9.4D) and others become swollen, indicating the progressive differentiation of the old hypaxial muscles into elaborate tail muscles (fig. 9.5). This half-way condition, which is called the hemiheterocercal stage, persists in the older Holostei and even in the modern garpike (figs. 9.4E, F; 9.5B). The order Holostei is an assemblage of many families, chiefly of Mesozoic age, in which the endoskeleton was more fully ossified than in the palaeoniscoids. The oldest known representative of the Holostei, *Acentrophorus* (fig. 9.17A) of the Permian of Europe, already foreshadows its descendants the Semionotidae, which have a high, well rounded back, small nibbling mouths often with rounded crushing molars (*Lepidotus,* B). The dorsal and anal fins are opposed, the wide caudal has a much shortened hemiheterocercal base; fin fulcra are often retained. The pectoral fins have reduced baseosts and the scapulo-coracoid is small. The infraclavicles (infracleithra or clavicles) are greatly reduced or eliminated. The pelvic fins are supported by a pair of basipterygial rods and by very small baseosts. The notochord is persistent but surrounded by bony half-rings, while large neural and haemal spines give the requisite strength for the support of the vertically extended muscle segments.

Evidently derived from the Holostei but distinguished by several sets of new polyisomeres is the garpike (*Lepidosteus,* figs. 9.17C; 9.18B) of Eocene to Recent age. Notwithstanding its long cylindrical body and long well-toothed predatory jaws, *Lepidosteus* retains many features that are reminiscent of the holostean Semionotidae and it may even be closely related to *Lepidotus* (fig. 9.18A), as clearly suggested by Goodrich (1909, pp. 335, 344). Among its relatively new polyisomeres are: (1) its numerous and completely ossified vertebrae, the centrum of each vertebra articulating by a convex anterior surface with the concave rear end of the preceding centrum; (2) the fragmentation of the opercular bone into a mosaic of small irregular polygons; (3) the forward extension of the anterior circumorbital row to form a series of plates (fig. 9.18B*mx*1–5) covering the maxilla and fused with it; (4) the multiplication of the teeth and infolding of their bases which form a pseudolabyrinthodont pattern. Derivation from the small-mouthed semionotids (or even from *Lepidotus,* fig. 9.18A) is indicated by the forward prolongation of the suspensorium so that the jaws, although secondarily elongated, open in front of the eyes as in semionotids, and not far behind them as in primitive palaeoniscoids (fig. 9.6A); the dorsal and anal fins, although short-based, are near the tail and opposed to each other as in semionotids.

The pycnodonts (fig. 9.19C) are usually regarded as a very specialized side branch of the Holostei. They are more or less kite-shaped or orbicular and some even exceed the platysomids in the great height and compression of the body. They are nibblers and crushers; their remarkably specialized dental mill consists of a long cylindrical or slightly conical mass set along the roof of the mouth; it is studded with rows of pebble-like teeth and works against two similar but concave surfaces set on the inner sides of the stout lower jaw. The framework of the body is stiffened by strong neural and haemal rods, the bases of which are more or less expanded around the notochord.

Resembling both the Platysomidae and the pycnodonts in general body form are the Triassic Dorypteridae (*Dorypterus, Bobasatrania,* fig. 9.19B). Their caudal base has the elongate single-scale row of the later palaeoniscoids plus the expanded basal swelling of the Holostei and in many parts of the skeleton they show a similar mingling of conservative and progressive characters. Thus they probably should be classed with the palaeoniscoids, although in certain respects they have advanced into a holostean stage of evolution.

THE DAY OF THE PROTOSPONDYLS (PROTOSPONDYLI)

The typical Jurassic Eugnathidae (fig. 9.20A), in contrast with the small-mouthed Semionotidae, have fairly large predatory jaws and swift-darting bodies with stout tail bases almost suggesting a brook trout. Brough (1939, pp. 89–97) has shown that the eugnathids differ widely from the semionotids in important characters, that they are structurally derivable from the palaeoniscoids and ancestral to the amioids. This family shows gradations in the transformation of stiff rhombic ganoid scales to thin scales with less bone and ganoine, rounded posteriorly like the cycloid scales of primitive teleosts. The group was named Protospondyli because the notochord is surrounded and gradually reduced by the simple checker-like centra which are pierced by it. These bony rings were evidently the dominant polyisomeres in this group. In the caudal region there is an alternation of rings, one set bearing the neural arches the other, the haemal arches. Transitional conditions in which the rings are represented by crescentic pieces are seen in several genera. During the Jurassic and Cretaceous periods the Protospondyli enjoyed considerable branching; some of the later forms multiplied the vertebrae and acquired a fast-swimming body with lunate caudal fin (e.g., *Hypsocormus*). The very large tail in the hypsocormids probably served as a pivot for turning quickly.

The only living representative of the Protospondyli is the bowfin, *Amia calva* (fig. 9.20C) a North American freshwater fish whose immediate ancestors date from the Eocene of North America and Europe. Its more remote ancestor *Megalurus* (fig. 9.20B) dates from the Jurassic of Europe. In its general characters *Amia* is on the whole nearer to the teleosts than it is to the palaeoniscoid stock from which it was eventually derived but it retains some important primitive characters, especially in its soft anatomy, and is usually classed as a ganoid. The bowfin is a more or less cylindrical, moderately elongate fish with lurking predatory habits. Its most distinctive polyisomeres are the much lengthened dorsal fin with its very numerous dermal rays; the anal fin is less elongate but coöperates with the dorsal in sending waves along the flexible fin web. The ring-like form of the short vertebrae contributes to the flexibility of the vertebral column. The muscular base of the tail is broad but the haemal rods in this region have not yet expanded into fan-shaped hypural bones as they have in typical teleosts. *Amia* has progressed in the direction of the teleosts in some characters, especially the loss of ganoine on the surface of the skull and scales and in the cycloid form of the scales.

The skull of *Amia* (figs. 9.21; 9.22B) is transversely rounded and contrasts with the narrow compressed skull of teleosts. The wide throat is protected below by a large median gular plate, as in typical protospondyls. In its broad rounded skull roof and round cheeks with expanded suborbital bones *Amia* recalls the Triassic *Perleidus* (fig. 9.22A), which is connected with the catopterid division of the palaeoniscoids. From this and other evidence it seems probable that the Protospondyli were derived not from any of the Holostei but from some more progressive branch of the palaeoniscoid stock, probably the Perleididae.

Under the term Subholostei, Dr. Brough (1936, 1939) has recently brought together a series of Triassic families of intermediate or transitional characters, which may have been derived from the catopterid division of the palaeoniscoids. The more primitive family Perleididae has been noticed above. Some other families of Subholostei had very deep scales on the sides of the body and were convergent toward the Pholidophoridae, which were almost teleosts. *Thoracopterus* of this family convergently resembled a teleost flying fish in the great size of its pectoral fins and strong downturned lower lobe of the tail.

DAWN OF THE MODERN TYPES (TELEOSTEI)

Among the direct or nearly direct ancestors of the lower teleosts were the Leptolepids of the Jurassic and Cretaceous (fig. 9.23B). As indicated by their name, meaning *delicate scale,* the scales had eliminated the bony base and greatly reduced the ganoine surface, the horny epidermal layer being dominant. The vertebral centra were still pierced by the notochord but, as in the teleosts, there was no alternation of centra and intercentra in the caudal region, each centrum carrying both a neural and a haemal arch. The haemal rods beneath the tip of the tail were long and narrow but in *Leptolepis dubius* (fig. 9.24A), as figured by Smith Woodward (1895, pl. XIV, fig. 7), two of them were widening out into fan-shaped hypural bones, as in teleosts. In this case as in many others a higher evolutionary grade was reached not by the production of a new crop of polyisomeres but by the selection and differentiation of certain polyisomeres of an older set, so that the favored ones acquired greater functional value. No doubt the expansion of these hypural bones is partly due to the differentiation of the deep subcaudal muscles (fig. 9.5C *ventr. flex*). The fan-like interradial muscles were composed of many little slips, each of which was inserted into the base of a single lepidotrich or dermal fin ray. Thus a waving motion could be imparted to the caudal fin web while the base of the fin as a whole was moved by the axial muscles (fig. 9.5C). The leptolepids (fig. 9.23B) were slightly elongate herring-like fishes with large mouths and small teeth. Presumably they were plankton feeders like the herrings, to which they were more or less closely related. The very primitive skeleton of *Luisichthys* (fig. 9.23B), a leptolepid from the Jurassic of Cuba (T. E. White, 1942), is a good structural intermediate between the eugnathid protospondyls (fig. 9.23A) and the isospondyl teleosts (figs. 9.26, 29, 33, etc.).

By Upper Cretaceous times teleosts of many orders had already become well differentiated and among their crowding lines parallelism and convergence were at work to confuse the palaeichthyologists of the then remote future. To what order, for example, should *Protosphyraena* (fig. 9.25) be referred? Is it merely an advanced protospondyl which had acquired certain teleost characters or an aberrantly specialized teleost which had paralleled certain contemporary protospondyls in some features? Its rostrum, like that of some of the protospondyls (e.g., *Belonostomus*) was long and pointed; one pair of the upper teeth grew downward into large curved tusks, while the pointed cheek teeth, like many small daggers, closely resembled those of the teleost *Sphyraena* (fig. 9.91). In this strange fish, which grew to have pectoral fins a yard long, the anterior edges of the fin produced a new set of polyisomeres in the form of a sinuous line of blades developed on the outer margin of alternate fin rays, the result being a long curved cutting edge like that of a huge bread knife. Although this family is referred to the Protospondyli, the construction of its skull is very close to that of true teleosts of the family Ichthyodectidae (fig. 9.27A), order Isospondyli. One of its isospondyl features was the presence of a massive mallet-like process on the antero-dorsal tip of the palatine bone (*pal*). This process overlapped and articulated with the slightly movable anterior end of the maxilla and prevented that bone from being dislodged when the mouth was widely opened.

The tyrant of the American Cretaceous Mediterranean sea was the twelve- to fifteen-foot *Portheus molossus* (fig. 9.26), called the bulldog fish in allusion to its short muzzle, upturned jaws and projecting teeth. Its smaller relative, *Ichthyodectes hamatus* Cope, is represented by a beautifully preserved specimen exhibited in the United States National Museum, which is less specialized than *Portheus,* especially in its jaws and teeth.

After extensive comparison with photographs of the *Portheus* skeletons in the British Museum (Natural History), the American Museum of Natural History, the Walker Museum (University of Chicago) and others, it was seen that there is no good evidence for the high dorsal and deep anal fins, curved back and almost scombroid contours of the restored skeleton in the American Museum of Natural History. The present restoration is the survivor of a series of tentative ones and is believed to be more accurate than any hitherto published. The contour as a whole is like that of *Ichthyodectes,* a primitive relative of *Portheus;* it also recalls those of the modern isospondyls *Chirocentrus* and *Tarpon.*

The only survivor or close relative of this family is the dorab (*Chirocentrus dorab*), a small fish of the Indian Ocean with similar jaws and body form (fig. 9.33B). The bulldog effect in the *Portheus* skull (fig. 9.27B) was conditioned by certain dimensional factors which have operated independently whenever predatory fish move up from below just before engulfing their prey. The first factor is the shortening of the distance from the anterior tip of the upper jaw to the line from the middle of the hyomandibular to the quadrate-articular joint. This we may take as the base of an inverted triangle (A, B, C) whose apex (C) is at the quadrate-articular joint. The second factor is the marked lowering of the latter joint so as to increase the dimension B C. The third factor is the degree of forward shifting of that joint, usually by the forward bending and lengthening of the lower end of the suspensorium (hyomandibular plus symplectic). A short mouth (A C) can be directed increasingly upward either by the retraction of the point A to A′ or by the forward shifting of the quadrate-articular joint (as from C to C′). Conversely an upwardly turned mouth (A C) can be turned forward either by (1) the forward growth of the upper jaw tip (A″), (2) the backward displacement of the quadrate-articular joint (C C″) or (3) by raising the latter (C to C″) or (4) by combinations of these factors. Obviously the lengthening or shortening of the dental border of the mandible (A C) is another growth factor which further complicates the diverse results, as do also the changes in the transverse diameters of the several regions. If, for example, the body as a whole becomes high and compressed, the line A B often shortens and the line B C increases both relatively and actually. The shape and position of the mouth are thus determined by the length, shape and angulation of its bounding bones, namely, the premaxillae, maxillae, dentaries, vomers, palatines, etc. (cf. fig. 9.28 A, C), each of which is subject to individual and phyletic anisomerism or variation.

The form of the body (fig. 9.29) as a whole in typical bony fishes is more or less rhombic in the side view, and since a rhomb may be conceived as including two opposed triangles with a common base (which is the length from the tip of the snout to the root of the tail), we can describe the differences in different regions of the body by comparing the proportional differences of the same part, either as between an ancestral form and its descendants or between diverse types of a given group (Gregory, 1928a, b).

Numerous examples of the foregoing principles may be observed in the herrings (Clupeidae, fig. 9.29) and allied families. The primitive clupeid of the North American Eocene (*Diplomystus*) was a compressed rhombic fish (fig. 9.29A) in which the gasterion (lowest point of the belly) was unusually depressed and armed with a sharp keel of serrated scales much as in the modern menhaden. These polyisomeres form a new and distinctive feature. The herrings are typically plankton feeders and have long hair-like sifters on the inner borders of the gill arches. Several of the highly varied mouth forms are shown in figures 9.28, 9.29.

THE TARPON: A STUDY IN SKELETAL ENGINEERING (ISOSPONDYLI)

The tarpons belong to a family (Elopidae) of isospondyls dating from the Lower Eocene of the London Clay. The recent tarpon, the giant of the tribe, is a powerful fish which can project itself high out of the water (fig. 9.30). Its skeleton is well fitted to resist the resultant stresses and to form a flexible but strong frame for the great muscles of the flanks. The vertebral centra are completely ossified and their sides are stiffened by longitudinal ridges. Above the ribs delicate intermuscular bones (epipleurals) are laid down in the fascia between the subdivisions of the myomeres. The muscles of the tail are well developed, likewise their supporting hypurals (fig. 9.24C); but the relatively primitive status of the tarpons among teleosts is made evident by the retention of a clear remnant of the old dorsal lobe of the tail. Other primitive features are: (1) the pelvic fins remain in the abdominal region and are never connected with the shoulder-girdle; (2) the lepidotrichs or fin-rays retain their jointed and distally branching character and are not fused distally into spikes or spiny rays as they are in many more advanced teleosts; (3) there is a narrow vestige of the median gular between the opposite halves of the lower jaw; (4) and the primitive open duct connecting the air-bladder with the throat is retained; (5) the presence of a mesocoracoid arch in the pectoral girdle (fig. 9.31). In this figure A belongs to a modified survivor of Devonian Palaeoniscoid stock, with enlarged mesocoracoid arch (from Goodrich, after Gegenbaur). B, A modified survivor of Triassic Protospondyl branch, showing the passage of the brachial nerve through the scapula and coracoid (after Goodrich, 1930). C, Another young fish of the same genus, showing the relations of the mesocoracoid arch to the scapular and supraglenoid foramina. The supra- and postcleithral plates (*po.cl. 1, 2, 3*) are shown, extending mesially to the pectoral fin. The girdle is viewed more obliquely and from the rear than it is in B. D, Relatively advanced modern survivor of the amioid Protospondyls, approaching the primitive teleost grade (after Goodrich). E, Relatively primitive isospondyl Teleost, retaining strong mesocoracoid arch and large supraglenoid foramen (with arrow). The anteroventral process of the coracoid is enlarged and extends to the lower end of the cleithrum. Slightly posterior view, showing the trochlea and glenoid facets, which are arranged so that the pectoral fin can be extended in a nearly horizontal plane. F, Advanced central percoid grade, showing loss of mesocoracoid and flattening of coracoid and scapular plates. The trochlea, glenoid facets and pterygial bones (*1–4*) are so arranged that the pectoral fin can be folded back in a nearly vertical position.

The skull of a large tarpon affords an ideal object for the study of the comparative anatomy and functional significance of its numerous parts. The neurocranium and its investing bones (figs. 9.32; 9.33A) form the entering wedge and fulcrum by means of which the powerful muscles of the flanks force the whole body through the water. In the tarpon the connection between the skull and the body has been strengthened by the forward growth of the dorsal muscles beneath the surface roof bones and above the inner roof of the braincase, to which they are fastened. Thus these muscles are lodged in a pair of arcades (fig. 9.32A1), which are conspicuous in the rear view of the skull. The outer or investing skull roof is supported posteriorly by a median plate, the supraoccipital (*soc*), separating the arcades and extending forward until it gains contact with the inner surface of the frontals. This median supraoccipital element was formerly thought to represent the neural spine of a vertebra which had been incorporated with the occiput, but a study of the tarpon skull suggests that it is, as it were, a new element, perhaps representing a separated crest of the neurocranium which has arisen between the fasciae of the opposite dorsal muscles plus a surface derm bone representing the single or paired median extrascapulars (*dsoc*) of the semionotids, the paired "parietals" of *Eoeugnathus* and the median "parietals" of *Heterolepidotus* (cf. Brough, 1939, text figs. 41, 42). In other families of teleosts, as in the Ichthyodectidae, the supraoccipital grows forward above the skullroof, more or less overlapping the parietals and gaining a stronger contact with the upper surface of the frontals.

The rear view of the cranium (fig. 9.32A1) of a large tarpon also shows how well the cranial wedge is braced transversely around the occipital segment, while the side view (A) reveals the function of the parasphenoid, which, although a derm bone forming in the roof of the mouth, ties the ethmovomer block to the cranial vault (cf. fig. 9.32b).

The large inner, or primary, upper jaw (fig. 9.32) constitutes an inverted truss based on the neurocranium as a whole; the front limb of the truss is made up of the palatine, the mesopterygoid (*mspt*) and pterygoid (*pter*) bones of the palatoquadrate arch; the apex is at the articular surface of the quadrate (*qu*) and receives the upward thrusts from the mandible; the rear limb of the V is composed of the posterior ridge of the quadrate plus the symplectic (*sym*) and hyomandibular (*hyom*); a thin bony web, the metapterygoid (derived from the posterior part of the palatoquadrate) connects the two limbs of the V. The front limb of the truss bears the prominent mallet-like process of the palatine bone; the under surface of the mallet articulates with and holds down the outer upper jawbone or maxilla (*mx*), while in the rear the mallet is overlapped by the parethmoid buttress of the olfactory capsule, to which it is also tied by ligament (fig. 9.32b *pl*).

The floor of the cranium, especially the parasphenoid bone, receives the thrusts from the front limbs of the opposite jaw-trusses, while the upper outer edges of the cranial vault on both sides take up the thrusts of the rear limbs of the trusses coming up through the hyomandibulars (fig. 9.32A, A1). The dorsal end of the hyomandibular (*hyom*) is extended anteroposteriorly and has a long convex articular facet, the front half being received into a horizontal groove on the sphenotic (intertemporal), the rear half into a corresponding socket borne by the pterotic (supratemporal). By means of this articulation the lower end of the hyomandibula can be swung outward so that the inverted V formed by the opposite lower jaws, as seen from below, can be spread laterally, thus increasing the transverse diameter of the throat during the act of swallowing. This transverse movement of the suspensorium is facilitated by the flexible joint between the lower end of the hyomandibula and the symplectic, while the ball-and-socket joint on the inner side of this junction permits a limited freedom of movement between the inner upper jaw and the branchial basket.

The core of the mandible of the tarpon, like

that of other teleosts, consists of the rear portion of Meckel's cartilage (fig. 9.9 *Meck. cart.*), ossified to form the articular bone (fig. 9.32B *art*). The front part of Meckel's cartilage dwindles into a thin sliver adhering to the inner side of the expanded dentary (*dn*) bones which together form much the greater part of the jaw in teleosts. In the ganoids (figs. 9.18B; 9.21 *sa*) other investing elements, especially the coronoid, surangular and splenial, were also present but in the teleosts these have been eliminated. This case furnishes another example of the principle that after a new set of polyisomeres (the originally numerous investing bones of the lower jaw) has been developed, genetic selection operates through anisomerism to emphasize and differentiate certain ones and eliminate the rest.

The outer upper jaw (fig. 9.32A) consists on each side of two investing bones, the premaxilla (*pmx*) and the maxilla (*mx*). The former rests on the upper front face of the ethmo-vomer block (fig. 9.9) by an incipient ascending process. Its short oral border is overlapped by a short process of the maxilla (fig. 9.33A). In contrast with more advanced teleosts the premaxilla does not send backward a long process beneath the maxilla (fig. 9.77A, B). In the earlier and more primitive ganoids the maxilla was much longer than the premaxilla and its tooth row extended to its posterior border beyond the angle of the mouth (figs. 9.6; 9.21, 22). This condition persists in many isospondyls (e.g., figs. 9.23; 9.33B, D; 9.34A-F), but in some others (e.g., fig. 9.33C, E) the maxillary teeth are feeble or wanting and the bone is beginning to lie behind and parallel to the premaxilla (fig. 9.33F).

Originally, that is, in the lower ganoids (figs. 9.6; 9.17A, B), both the premaxilla and maxilla formed the stiff and nearly immovable outer rim of the primary upper jaw. Perhaps the increasing lateral swing of the hyomandibular and associated parts, as described above, tended to push the rear end of the maxilla outward. In response perhaps to natural selection, the rear end of the maxilla became free from the cheekbones above it, or rather carried with it two of them, which thus gave rise to the two supramaxillary plates (*smx 1, smx 2*) of the tarpon (fig. 9.33A) and other primitive teleosts. In the primitive Triassic subholostean *Meridensia meridensis* of the family Perleididae, as described by Brough (1939, text fig. 1, p. 7), the supramaxilla is clearly in series with the small ventral postorbital plates, while the maxilla is in series with the preoperculum.

The tarpon is exceptional in its lack of large teeth but the dentigerous surfaces of the outer and inner jaws and of the tongue and branchial arches are beset with fine dental papillae; and *Elops,* a small, less specialized member of the family, retains small teeth on its outer jaws. In the tarpon the tongue bones (glossohyals) and the basibranchials bear patches of very fine teeth; in *Albula* (fig. 9.35A) some of the teeth on the basibranchials are like small pebbles and they oppose similar teeth on the mesopterygoids.

In *Hyodon* (fig. 9.35B) and in the Osteoglossidae (fig. 9.34A, B) similar new and distinctive polyisomeres take the form of a conspicuous median series of teeth borne on a bony plate covering the tongue bones, which can be closed against a similar series on the parasphenoid. The osteoglossids are of exceptional zoogeographic interest since their ancestors are represented in the Eocene of North America (*Dapedoglossus*) and England (*Brachyaetus*), while the few surviving members are found in South America, Africa, and the Indo-Pacific region. According to W. D. Matthew (1915), such a distribution may indicate the northern origin of the group, its subsequent spread to the southern continents and its extinction in the northern homeland.

ABYSSAL RELATIVES OF THE HERRINGS AND TARPONS (ISOSPONDYLI CONTINUED)

In the tarpons, as in many other isospondyls, the scales abound in argenteum, a silvery layer covering nearly the whole body. The mirror-like quality of this layer has possibly been one of the prerequisite conditions for the evolution of a new set of polyisomeres, namely, the photophores or glow-lamps on the surface of the head and body in various families of deep-sea fishes. The glow-lamps, which seem to attract the prey of these predators, are typically provided with a silvery reflecting surface, a lens, a black pigmented background containing glands which secrete a luminous slime or photogenic substance, luciferin, together with nutrient blood vessels and nerves. The latter are said to be connected with the lateral line system at least in some instances. It is known that the luciferin is somewhat like the visual purple of the eye in so far as it is a self-renewing substance. In brief, the luciferin seizes oxygen from the blood and becomes luciferase, a phosphorescent glow being emitted during the process of oxidation. The luciferase then loses its oxygen, again becomes luciferin and is ready for the next cycle (Harvey, 1918). It was suggested by Buchner (cf. H. M. Kyle, 1926, p. 156) that these glow-lamps originated in fishes of the upper layers of the warmer waters, where photogenic bacteria abound and that at first the fish merely reacted to their presence, but while this may have been the case in the ancestors of *Anomalops* and *Photoblepharon* (East Indian percoid fishes with brilliant subocular light organs), it may not apply at all to other light-bearing fishes in which photogenic bacteria are not the means of luminescence.

A comparison of the light-bearing organs in different parts of the body, even in the highly specialized stomiatids, shows that even in a single individual we may find a graded series indicating successive steps in the evolution of these typical polyisomeric systems, involving the progressive contrast between light and dark pigmented areas, the differentiation of the lens from a translucent spot in the parent scale, and the production of luciferin glands (T. W. Bridge, in Cambridge Natural History, Vol. VII, 1904, pp. 179–181).

The luminous organs doubtless represent the end of a long, cumulative evolution, of which various morphologic and physiologic stages are still preserved among different types of fishes. Unfortunately the subject has usually been treated only from the descriptive and physiologic viewpoints and the evolution of these organs has been considered, if at all, chiefly from random samples drawn from highly specialized fishes. Possibly the most primitive known stage is found in the order Heteromi, a deep-sea group of obscure affinities, but supposed to be more or less intermediate between the Haplomi and the spiny-rayed fishes. Boulenger states (Cambridge Natural History, Vol. VII, 1904, p. 624): "In *Halosaurus* [fig. 9.36A] the scales of the lateral line, which runs near the lower profile, are scarcely enlarged, and are destitute of luminous organs. . . . In *Halosauropsis* the scales of the lateral line are strongly enlarged and pouch-like and bear photophores." In their general grade among teleosts, as well as in the regular scalation of their body, these halosaurs are assuredly far less peculiarly specialized than the stomiatids cited by Boulenger. The association in them of photophores with a "lateral line" system tends to confirm the suggestion that the most essential features of the photophores are the luminous slime-producing glands, sunk into pits and covered by transparent scales. The presence of photophores in certain sharks, as well as in different orders of teleosts, indicates parallel or convergent evolution.

That light will attract toward it many sorts of creatures from moths to men needs no special documentation. It is, however, suggestive of the reality and omnipresence of natural selection through the survival of those best equipped for the struggle that in the abysmal darkness of the great depths of the ocean the "jack-light" method of stalking game has been worked out quite independently by many different families of teleost fishes.

Among the isospondyls there are two major types or suborders (Alepocephaloidea, Stomiatoidea) of deep-sea or abyssal families, of which the former is apparently derived from the clupeoid or herring group. *Alepocephalus* itself (fig. 9.34C) is a fairly normal looking fish, not unlike a trout, but with large eyes and moderate to small mouth, which is directed mostly forward since the quadrate-articular joint is not depressed. Each maxilla bears two supramaxillae (*smx 1, smx 2*) as in primitive Cretaceous clupeoids. The tail is well developed, with a wide muscular base. The dorsal and anal fins also are well developed, posterior, more or less opposed. Abyssal habits are suggested by the black pigment and large eyes but there are no luminous organs and the lateral line is normal. The cranium, as in the salmonids, is more or less cartilaginous, due to retention of larval characters.

This general alepocephalid stamp persists in many species of the family but anisomerism, or the tendency to produce differences in proportional emphases or diameters, runs riot in this as in other deep-sea families (fig. 9.37). Thus the mouth varies greatly in size, becoming very large in *Macromastax*. The head and eyes become abnormally large in *Anomalopterus* and very small in *Photostylus*. The two last named genera possess in common a unique structure called a "predorsal adipose fin," which is a translucent strip, possibly luminous. *Dolichopteryx,* which is referred to this group, is exceedingly specialized: the body is rod-like but the caudal, dorsal and anal fins retain the general family stamp. The pectorals (in one species) have enormously long rays, which may serve partly as tactile organs. The mouth is very small. The very large eyes are turned directly upward.

Also referred to the alepocephalid group are the highly specialized genera (fig. 9.37) *Aulastomatomorpha, Xenodermichthys, Leptoderma*. Of these, *Xenodermichthys* conforms in a general way to the alepocephalid type but has the anal and dorsal fins equally elongate. *Leptoderma* is an almost eel-like relative of *Xenodermichthys,* with excessively elongate dorsal and anal and minute pectoral, pelvic and true caudal fins. The eyes are very large and the mouth extremely small. *Aulastomatomorpha* has a fairly normal body, with long anal and normal caudal, but the dorsal fin is very short and the minute mouth is at the end of a tube in front of the large eyes.

Opisthoproctus (fig. 9.38A) is referred to a distinct family or superfamily of isospondyls. Its body is very short and several times as high as in *Dolichopteryx*. It also has an adipose fin behind the large dorsal fin, the latter being just behind the center of the body, whereas the alepocephalids have no true adipose fin and the dorsal is typically just in front of the peduncle of the tail.

The enormous orbits of *Opisthoproctus* nearly meet above in the midline because the eyes are like huge field glasses, directed upward. Their excessively black and silvery pigment suggests that the eyes themselves may be luminous. The absurdly minute and beak-like mouth is at the end of a long oropharyngeal tube passing beneath the huge orbits. In association with the extreme forward position of the quadrate-articular joint, the lower end of the suspensorium (symplectic, *sym*) is pulled out into a long tract beneath the orbit. The ascending ramus of the dentary bone is extremely large, as is also the space for the jaw muscles along the cheeks. The main otoliths are huge stony lenses set vertically on either side of the otic capsules. Probably they are important as range-finders and stabilizers and possibly they coöperate with the "telescope" eyes in directing the fish toward its prey. The large lower limbs of the shoulder-girdle (*clt*) are produced far forward beneath the skull and perhaps they make a sort of marsupium or pocket opening forward below and behind the mouth. There is a suggestion of this peculiar condition in the alepocephalid genus *Platytroctes* (fig. 9.37). Nevertheless the resemblance to *Dolichopteryx* noted above may be no more than mere convergence and *Opisthoproctus* may belong near the salmonoids (Trewavas, 1933).

The second main group of deep-sea isospondyls constitutes the suborder Stomiatoidei (figs. 9.36B, C; 9.38B; 9.39), including at least eight families. With the exception of a single family (Sternoptychidae), these range from pike-like to viper-like predators, typically with large to very large mouths, needle-like teeth, and jaws produced backward well behind the eyes. The eyes are usually far forward, typically small, rarely large. An adipose fin (a primitive salmonoid character) is retained in all but the more advanced stages of specialization. The main dorsal fin, originally near the middle of the back (in *Photichthys*), is often shifted caudally as the adipose fin is lost. Photophores are unusually well developed (in contrast with the alepocephaloids), often in a double row from the throat along the side of the belly to the tail (e.g., *Astronesthes,* fig. 9.39).

The stomiatoids, as well shown in the monographs by Brauer, Regan and Trewavas, Parr, Beebe and others, abound in special adaptations for their somewhat viper-like way of striking suddenly at the prey. They have become far more specialized than the gonostomids, for to the general equipment of a double row of photophores along the sides of the body they have added long barbels under the chin and in not a few cases a "cheek-light" (fig. 9.36C) of complex construction. In certain forms the anterior vertebrae are enlarged (fig. 9.34E) and in *Eustomias* there are one or two marked loops in the vertebral column immediately behind the skull, the vertebral centra being incompletely ossified. This arrangement probably acts as a spring or shock absorber when the fish strikes at the prey with his widely open jaws and long needle-like teeth. In *Chauliodus* (fig. 9.34F) the compact skull is braced by the greatly enlarged "first" vertebra.

In the stomiatid family (in contrast with the Gonostomids (figs. 9.36B, 34D) the tooth-bearing border of the premaxilla is prolonged backward beneath the maxilla (fig. 9.34E) but not to the same extent as in the Iniomi (fig. 9.62B), because the maxilla still retains a tooth-bearing strip at its posterior end, where it enters the angle of the mouth (fig. 9.39D, E, F). Another primitive isospondyl character persisting in the midst of extreme specializations is the retention of the adipose fin in *Astronesthes, Borostomias* and *Chirostomias.*

In the subdivision of the stomiatoids into superfamilies there is room for differences of judgment regarding the degree of relationship between the groups. It is conceivable, for example, that while the true stomiatids may have been derived from the upper Cretaceous Enchodontidae, as suggested by Tate Regan, the Gonostomidae may have been derived from the elopine branch of the clupeoids, as also held by that author. But a nearer relationship between the two branches is suggested by the annectant characters of *Chauliodus* and *Borostomias.*

Another view is that of Garstang (1932), who unites all the stomiatoids with the scopeloids (see below), calling this comprehensive group Lampadephori in allusion to the highly developed photophores which are developed in both groups.

Quite a different conception of relationships is implied in the classification proposed by Parr (1927, 1930), while Gregory and Conrad (1936) have attempted to reinterpret the evidence with the results shown in the chart (fig. 9.39). Space is lacking for an extended discussion of this problem, but as the case may be of considerable interest to students of evolution the following brief comments may be made.

In *Photichthys* (fig. 9.39) of the Gonostomatidae, according to Regan (1923, p. 613), there is a striking agreement with *Elops* (of the tarpon family) in the entire form of the skull and in the relations of its constituent bones. But these genera are far more specialized than *Elops.* In *Gonostoma* and still more in *Cyclothone* (fig. 9.38B) the jaws, hyoid and branchial arches have greatly lengthened into a very slender fish-trap set with needle-like teeth; the eyes and the neurocranium are small and the opercular elements reduced to a narrow strip. The scales in *Gonostoma* are large and cycloid, recalling those of *Elops.* In *Cyclothone* there is a row of large photophores on a flap of skin borne by the ceratohyal (fig. 9.38B *cerhy*). The relations of the premaxilla and maxilla are not unlike those in *Elops,* the premaxilla (*pmx*) having a short dentigerous portion which is overlapped by an anterior process of the maxilla (*mx*); the latter occupies most of the upper jaw and its teeth extend to the corner of the mouth. In specimens examined by the writer only the posterior supramaxilla (*smx*) was present, the anterior one of primitive isospondyls was not visible.

Apparently derived from the gonostomatid stem are the silver hatchet-fishes (Argyropelyidae, fig. 9.40). In these strange, very short, somewhat rhombic forms the belly and throat are excessively deepened, apparently to carry the greatly heightened photophores on the flanks. The small mouth is turned upward, the operculars (A, B *op*) much deepened but reduced to a vertical strip, the maxillae retain their place on the oral margin. The eyes are large. The narrow shoulder-girdle is much extended vertically and stiffens the entire skeleton. According to Garman, the very young *Argyropelecus* strongly resemble the adult of *Valenciennellus* or *Maurolicus* (C). The latter is a relatively quite normal-looking little fish except for its double ventral row of well developed photophores. It somewhat suggests the myctophids (fig. 9.63) of the order Iniomi but, especially in the arrangement of its luminous organs and the form of the mouth and eyes, it closely resembles the hatchet-fishes and differs widely from all the Iniomi.

One of the most striking examples of development by metamorphosis resulted from the long continued annual dredging of a particular deep-sea region in the Atlantic off Bermuda by William Beebe (1934), who was able to show that the minute fish named *Stylophthalmus,* in which the eyes are on long stalks (fig. 9.41A) growing out from the sides of the head, was only a larval form of a very different-looking stomiatoid fish named *Idiacanthus* (fig. 9.41C), because after the long-stalked stage, later stages show the eye-stalk in various degrees of shortening, until finally the eyes are withdrawn into their usual place in the orbits. Here too is almost the extreme of sexual dimorphism, since the males are only about one-seventh as long as the adult females (C).

THE VENERATED MORMYRIDS AND OTHERS (ISOSPONDYLI CONCLUDED)

Although many isospondyls have penetrated to great depths in the ocean, evidence is not wanting to suggest that the salmonoids and several other main divisions of the order originated in the lakes and rivers, up which they travel each year to their spawning grounds in the highlands. Among these fresh-water forms none is so weird as the mormyrid (fig. 9.42C), venerated by the Egyptians as having nibbled a bit of the dismembered Osiris. Perhaps the most surprising feature of the mormyrid is the great size of the brain, accompanied by the poor development of the eyes. Between the forks of the posttemporal bone there is a large cavity in the skull into which a branch of the air-bladder extends, covered by a thin scale-bone. The minute mouth in extreme mormyrids is at the end of a long decurved tube covered by the ethmoid and by the much prolonged roof of the mouth. The functional suspensorium (hyomandibular plus preopercular plus quadrate) is prolonged to a point well in front of the eyes but most of the lower half of the proboscis is supported by a greatly elongated and recurved articular bone.

Thus the order Isospondyli includes a great number of families which, in spite of their diversity, retain a few primitive characters from the stem teleosts of the later Mesozoic era. The duct between the air-bladder and the oesophagus remains open, although it may become small, the fin-rays (with few exceptions) retain their jointed condition, the first four vertebrae of the neck, although highly modified in the stomiatoids, do not assume the complexity attained in the derived order of Ostariophysi, the pelvis (pelvic pterygiophores) is never fastened by ligament to the lower part of the pectoral girdle (in contrast with the spiny-rayed fishes). The premaxilla rarely extends so far backward beneath the maxilla as to exclude the latter from the gape. The orbitosphenoid is retained and there are no marginal spikes or projections on the opercular and preopercular. The larval stages of isospondyls are often elongate, as in the herrings, tarpons and albulids, with very many segments, and in the later larvae the body often becomes at least relatively shorter. Apart from the deep-bellied herrings, even the adults are usually long-bodied, with rare exceptions including the high-backed *Notopterus* (fig. 9.35C) and the compressed and distorted hatchet-fishes (*Argyropelyx,* fig. 9.40A). Silvery and dark brown or black colors predominate and the scales are often large and cycloid, rarely ctenoid.

THE MASTER WRIGGLERS (APODES)

The eels (Apodes) are a highly specialized branch of some early teleost stock which have lost their pelvic fins, reduced their median fins and scales and greatly multiplied their vertebrae. In the morays these newer polyisomeres (or polyanisomeres) assume a large and complex form, which probably indicates a corresponding complexity of the axial muscles. There are few but sharp teeth on the vomer, maxillae, and mandible and the (?)maxilla is thinned to a narrow strip (fig. 9.43B). Their skulls also have been simplified, the premaxilla having been lost in the snake-like morays. The bones of the braincase fuse into a strong tube, which is also somewhat like that of a snake. In the morays (B) the very large hyomandibular has been produced far backward, thus lengthening the gape. The opercular bones as well as the gill-arches and the gills themselves are reduced but the opercular fold and throat have become very distensible and form an elastic chamber which pumps the water into the mouth and out through the small opercular spout. The naked-skinned morays living among the coral reefs display a wide range of bold color patterns of sharply contrasting patches (fig. 13), possibly of concealing value. Owing to the great length and flexibility of their bodies, eels progress by a literally sinuous motion, since two or more sine curves or waves in succession can be seen passing along the body at the same moment. Hence this wriggling method of locomotion has been named the *anguilliform* type, in contrast to the carangiform type of movement (Breder, 1926) in shorter and less flexible fish, where at any given moment there is room for only one sine curve, as seen in top view.

The ancestral eels (*Anguillavus*) of the Upper Cretaceous retained small pelvic fins, which may have previously impeded the development of the double wave movement. Although the morays look indolent in an aquarium and do spend much time in relaxing, yet they are able to spring into high speed when sufficiently stimulated. The locomotor method of the common eel must be pretty efficient, for by its means the young eels which are hatched on the floor of the Atlantic in a large area southeast of Bermuda are enabled to travel toward the distant land and to ascend the rivers into fresh water; later they make the return journey to the breeding grounds.

FISHES WITH AUTOMATIC PRESSURE RECORDERS (OSTARIOPHYSI)

The order Ostariophysi with its five thousand species of carps, characins and catfishes predominates in the fresh-waters of the present epoch much as the order Acanthopterygii does in the ocean. Notwithstanding the great diversity of its members, they are certified as descendants of a single ancestral stock by the possession of a concealed token called the "Weberian apparatus." This consists of a connected series of small bones immediately behind the skull, serving to connect a snared-off division of the air bladder (fig. 9.44) with the membranous wall of the otic capsule. Although there have been different interpretations of the functions of the Weberian apparatus and even of the air-bladder itself, it seems highly probable that the varying pressures of gas in the air-bladder (see p. 166) affect the tensions of this chain of ossicles upon the region of the inner ear. Such stimuli may set up reflex actions in the muscles that expand or contract the air-bladder, so that the tendency to rise or fall with change of density of the water may be compensated. The fact that some marine fishes have dispensed with an air-bladder does not prove that this organ is of no use to those in which it is well developed.

The basal polyisomeres of this remarkable apparatus (fig. 9.45) are of six general sorts: first, the neural arches and parapophyses of the four anterior vertebrae; second, the modified myomeric muscles; third, their ligaments; fourth, the exploratory pouches and tubes of the embryonic otic capsule to which the anterior member of the Weberian chain is applied; fifth, that part of the air bladder to which the posterior end of the chain is appressed; sixth, the appropriate contacts and connections between the several parts of the central and peripheral nervous system. Doubtless before this poly-anisomeric system could be started toward its final perfection, the "primary polyisomeres" (neural arches, parapophyses, muscles, ligaments, nerves, etc.) severally functioned in the ways that are normal for other isospondyl fishes. It was the gradual differentiation and integration of new functions (involving anisomeric emphasis of certain parts and reduction of others) that resulted in the transformation of ribs, etc., into parts of a complex stethoscope.

Among the diversified hosts of the bony fishes from the Devonian palaeoniscoids to the latest twigs of the teleosts this poly-anisomeric Weberian system is found only in the order to which it gives its name (Ostariophysi, "little bone-bladder"). A single prerequisite condition for it (namely, some sort of connection between the air-bladder and the otic region) is found in such widely different groups as the mormyrids and the physiculine cods, in both of which lateral pouches from the air bladder have gained contact with the otic capsules, but in these cases the Weberian apparatus itself is conspicuously absent. The group is sufficiently old to have divided itself into three suborders: the characins (Heterognathi, figs. 9.45–55), the carps (Eventognathi, figs. 9.44; 9.56–58), and the catfishes (Nematognathi, figs. 9.59–61). The sharp differences among them might well raise the question whether the group as a whole might not be polygenetic (that is, of different origins) and thus whether the resemblances between them are due to "convergence." The great monograph by Bridge and Haddon (1889, 1893) supplies good evidence that the common possession of the Weberian apparatus, with its severally identifiable ossicles and all their complex connections, is convincing evidence of a descent from a remote common ancestral stock; for here there is a basic agreement in plan with differences in details, whereas in true convergence there is agreement in many or few details with radical contrast in basic plans. Moreover the comprehensive studies of the otoliths of this order by Frost (1925–1930) indicate that two of the suborders named above (Eventognathi, Heterognathi) are still closely related to each other, while the third (Nematognathi) has become specialized in a different direction.

Nothing that has been written above seems inconsistent with the observations by Cockerell (1925) that while the four anterior vertebrae of *Lycoptera middendorffi* from the Jurassic of China were not modified into a Weberian apparatus, yet this family (Lycopteridae) approach the carp type in the microscopic structure of their scales. Moreover the skeleton of the short-skulled *Lycoptera sinensis,* as restored by A. S. Woodward (1901, Pt. IV, p. 3) would seem to be a favorable starting-point for the order. Finally, Woodward has noted that the Jurassic family Lycopteridae are closely related to the Leptolepidae, which on many grounds appear to be ancestral to the primitive order Isospondyli.

The available evidence therefore suggests that the Ostariophysi as a group are an offshoot of the early isospondyl stock, but that the Weberian apparatus itself was not yet developed in the ancestral Lycopteridae of the Jurassic but may have evolved in the long ages between the Jurassic and the Eocene. Berg (1943, pp. 417, 418) confirms the absence of Weberian apparatus in *Lycoptera* but also shows that it agrees with *Polypterus,* Amia, Characinidae, Cyprinidae, Gymnonotidae, in certain important characters of the otoliths and that in many other features it agrees with the Clupeiformes (Isospondyli).

By Lower Eocene time the catfishes (Nematognathi) were thoroughly differentiated from the other suborders and are represented by fossils from Wyoming, North America, England, Egypt and Nigeria. By late Tertiary times the catfishes had spread to India, Sumatra, Brazil.

In view of the probable antiquity of the order it is noteworthy that neither the carps nor the characins have been found as fossils in any pre-Tertiary formations. The carps, which are not found in South America, are represented by a doubtfully referred fragment from the Lower Eocene of England and in later Tertiary and Recent formations of Europe and North America by representatives of more than a dozen still existent genera. They are believed to have originated somewhere on the North American-Eurasiatic land mass at a relatively recent date, possibly from a survivor of the Lycopteridae. The characins are definitely known only from formations of Upper Tertiary age in Brazil and Peru, but several families of recent characins are found both in South America and tropical Africa. In view of the relatively close structural relationship between certain African and South American characins (*Alestes Brycon*) it seems probable that the order originated in the northern hemisphere in Cretaceous times and that parts of it spread either or both ways across a narrow, Eocene isthmus connecting northern South America and northern Africa.

The catfishes are almost cosmopolitan and in the eastern hemisphere have reached as far south as Tasmania. The catfish group, being older and essentially Mesozoic in origin, perhaps first spread around the world along the paths followed by other fresh-water fishes (e.g., dipnoans, osteoglossids). The pleurodiran chelonians, the crocodilians, pythons and sauropod dinosaurs offer analogous distributions.

Characins of South America and Africa.— Although all the groups of Ostariophysi are specialized in various ways, yet if we choose as a criterion the general construction of the skull, jaws and dentition, the characins as a whole appear to be distinctly less specialized than the carps, and far less specialized than the catfishes, and the same is true with regard to other skeletal features as well as to scalation and general body form. We may therefore consider the groups in that order in spite of the poverty of the fossil record.

Segemehl (1885) in his important paper on the cranium of the characins concluded that the characins in the majority of their structural relations must be closely connected with *Amia* and that among them the subfamily Erythrininae (cf. fig. 9.46) show the greatest correspondence with that form; further, that it is only with regard to a few points of their morphology that we must seek for ancestors that were even earlier and lower forms than *Amia,* and that in any case the stem form of the characins stands not far from *Amia.* No subsequent authors however, so far as we know, accepted Segemehl's conclusion and there is good evidence for the conclusions that *Amia* and the erythrinine characins are rightly classified under widely different orders of bony fishes and that such resemblances as there are between them are largely convergent. Eigenmann, who devoted many years to the description and classification of the South American characins, came to the conclusion that the least specialized of all known characins are certain genera of the subfamilies Cheirodontinae and Tetragonopterinae.

According to present evidence, the most primitive known characins are neither the amioid (cyclindrical) types like *Erythrinus* (fig. 9.46), nor the eel-like gymnotids (fig. 9.55), but the slightly compressed, almost rhomboid *Moenkhausia* (fig. 9.47) and the small unspecialized *Cheirodon* (figs. 9.47, 48).

Eigenmann, Tate Regan (1912) and others recognized numerous subfamilies and not a few families of characins, but Gregory and Conrad (1938), emphasizing the basic unity underlying great diversity, referred the entire group to one family comprising nine subfamilies (figs. 9.47, 48).

G. S. Myers (1943, pp. 60, 61) states that he "... is perfectly aware of the improvements in characin classification effected by Gregory and Conrad," but he asserts that "any attempts to fit numerous South American genera into their system, on an osteological or any other basis, results only in chaos." However, after renewed consideration of the problem we must recall that Eigenmann and others who grappled with the thousands of individual specimens and hundreds of characin species and genera were also those who gradually produced the broader classification into subfamilies; and that the typical genera of these subfamilies were considered by us as samples of the vast mass. We can only ffirm our confidence in the value of comparaive studies on the skeletons of representatives f the main subfamilies, and we await Dr. lyers' own treatment of the problem with inerest.

The tooth-forming forces among the characins (fig. 9.49) have produced a reckless proision of polyisomeres. Many of the more primive characins had small pleuricuspid teeth, specially in the premaxillae, grading either oward nibbling, crushing, or cutting extremes; i some (e.g., *Cynodon*) a few teeth became ery long and piercing. Eigenmann found that ie variations in the minor polyisomeres gave seful specific and generic characters. The charcins also show a wide range in the relations of ie premaxillae and maxillae. In the more primive forms the maxillae were of moderate ength, the teeth small or wanting. Extremes ad either very long, well-toothed maxillae or ery short, toothless ones.

In the large, strong-jawed, predaceous *Hyrocyon* (fig. 9.47) of the central African lakes nd rivers the sharp conical teeth (fig. 9.50) re attached to the jawbone by a basal zone of ttle folds, which are often indicated by the raiating septa left when a tooth has been broken ff. But this double convergence toward the onditions in the garpike on the one hand and the labyrinthodont amphibians on the other oes not prevent the teeth from being easily roken off, at least in the dried skull. The pseuolabyrinthodont mode of tooth attachment is ssociated with the lack of dental roots and in is case each tooth is fastened on top of a flatpped stump which projects from the jawbone. 'ithin the recesses of the dentary bone lie a ew set of teeth, large, sharp and directed back'ard. The teeth increase in size toward the free nd of the jaw and this must subject the joint etween the right and left halves of the jaw to evere stresses.

A remarkable display of polyisomeres has een revealed by close study of the complex inge (figs. 9.50, 51) which in *Hydrocyon* conects the front end of the right and left halves the lower jaw (Gregory and Conrad, 1937). he movement and general structure of this hinge may be illustrated in part as follows: Let the reader first stand with arms straight down at the sides; keeping upper arms still, bring hands up in front of body, with opposite palms opposed. The fore arms will then represent the right and left halves of the lower jaw of *Hydrocyon,* and the elbows its joints with the upper jaw. Next, oppose the tip of the thumb to that of the index finger of the same side, first of the right, then of the left hand, then bring the opposed tips of the thumb and index fingers of the right hand and shove them down into the ring formed by the thumb and index fingers of the left hand. Let the other fingers imitate this intertwining arrangement as well as they can. Now, keeping the hands clasped in this way, move them up and down and spread the elbows apart. This will illustrate the movements of the lower jaw of *Hydrocyon,* which can be spread apart at the rear without the two halves becoming separated in front. But in spite of the intertwining of the fingers the two hands can easily be pulled apart and so might the front ends of the jaw of *Hydrocyon* if the force of its jaw muscles, reacting against resistance, produced sufficiently strong outward thrusts. This condition is met in the following way: first there is a series of radiating bony spokes (fig. 9.51 VII) separated by corresponding slots growing from opposite sides of the front ends of the lower jaw; second, the curving spokes of the one side bend over or under those of the opposite side, invading the spaces between the latter and apparently reuniting beyond the midplane to form the so-called "hinge-stop" (fig. 9.51 *S, s*). Thus the hinge-stop (S) at the left of the mid-plane has been produced by joining the tips of the bony spokes from the right half of the jaw and the one on the right (s) from those of the left side.

By what steps did this unique arrangement arise? Other characins have already acquired a system of curved crests which project from the symphyseal surface of each half of the lower jaw, the crests of one side fitting into the spaces between the crests of the other (fig. 9.51B-F). But in some of the other genera examined (fig. 9.51B-D) the crests merely diverge backward and outward from the front edge of the symphysis and there are no interlocking radiating spokes, such as we have found in *Hydrocyon.* This is true even in *Alestes* (fig. 9.51C), which in most features is a close relative of *Hydrocyon.* In *Hoplias* (fig. 9.51E) and *Erythrinus* (fig. 9.51F), however, what may be called pivot-cones are developed and are seen in cross-section in the form of inverted V-like hillocks in the middle of the interlocking ledges. Unfortunately we cannot definitely assert either that the conical ledges of *Erythrinus* gave rise to the radiating spokes of *Hydrocyon,* or that merely by joining the tips of the cones of *Erythrinus* we could derive the hinge-stops (*S, s*) of *Hydrocyon.*

But fortunately we have a series of young *Hydrocyon lineatus* collected by Herbert Lang in the waters of the Belgian Congo; in the first two stages of this series (fig.9.51I-II) the hinge-stops have not yet been developed and the symphyseal parts of the jaw are essentially like those of *Alestes* and other characins in having merely projecting ledges and depressions, those of the left side interlocking with those of the right and *vice versa.* In the third stage (III) the middle projections of two of the ledges are closely appressed to each other. Although there is a long gap in the series at this point, representing the stages of growth between 5.5 cm. and 21 cm. in total length, yet in the latter stage (IV) the radiating ridges are not yet complete.

Further study and preparation of the larger *Hydrocyon* jaws (1943) confirm Gregory and Conrad's figure 2 (*op. cit.,* p. 977), which shows branch *D* looping over and beyond branch *d* and joining branch *E* below in a continuous link, which interlocks with a similar link formed from *d* and *c*. But in the lack of stages between our figures 9.51-III and IV we can only infer that the vertically continuous links as well as the hinge-stops (S, s) have been formed by the successive fusions of upper and lower spokes belonging to the same half of the mandible (cf. fig. 9.51IV, V).

In the upper jaw the opposite maxillae are also movably articulated with each other by means of a series of oblique flanges and grooves; these have remained far more prin tive than have those of the lower jaw.

Thus in the derivation of the highly compl hinge-joint of *Hydrocyon* from the much le complex one of *Alestes* we have a singular clear illustration of the principle that new pol isomeres are first developed in an older gen (e.g., *Alestes*) and are then variously differe tiated and emphasized in a derived genus (e. *Hydrocyon*). In the light of the preceding d cussion it would evidently be a mistake to pre icate the existence of an elaborate *Hydrocyo* hinge in the jaws of remote pre-characin ance tors, or to give a superfamily value to a chara ter that has so far been developed only with one genus.

The evidence analyzed by Gregory and Co rad shows that, while body-form, jaws and tee must obviously coöperate to form a viable co bination, yet any given combination attained a stem form may be very diversely modified the descendants and that parts of the "habitu of the ancestor are retained among the "he tage" features of its diversified descendan The premaxilla, for example, is the predon nant tooth-bearing member of the upper jaw most characins, perhaps because the ancest cheirodonts had small nibbling mouths. T premaxilla is relatively large even in such pik like derivatives as *Luciocharax* (fig. 9.52C); is only in extreme predators such as *Sarcodac* (fig. 9.52A) that some of the maxillary tee become large and prominent near the front e of the maxilla.

The jaws and teeth in the so-called herbiv ous subfamilies (figs. 9.48; 9.49; 9.53) a mostly of the small-mouthed nibbling type, wi the lower bar of the suspensorium produced fo ward below or in front of the orbit. In *Lepo nus* (fig. 9.53A) sharp front teeth become pr cumbent and almost nipper-like; in the Distic odontinae (C, D) the jaws vary in length b are peculiar in possessing a more or less mo able bend between the dentary and the articul The very numerous genera and species of th division retain clear traces of the characin bod form and fins.

The principle noted above is clearly seen

the variations of the body-form and fins. For example, the elongation of the anal fin, which is associated with a sub-rhombic form of body in such primitive cheirodonts as *Moenkhausia* (fig. 9.47), is progressively emphasized in the increasingly deep-bodied Serrasalmoninae; it is retained in varying degrees in the Characinae, Citharininae (fig. 9.48), and to some extent in other subfamilies. Occasionally it shortens its base and simulates a paddle (*Erythrinus, Curimatus*). The dorsal fin, originally near the apex of the back (fig. 9.55), persists there in many characins and never succeeds in becoming equal and opposed to the anal as it does in platysonids and other families. Forms with widely depressed heads, like the catfishes, seem to be without representatives among the characins.

Electric "Eels" (Gymnotids).—Eel-like forms appear only in one very highly specialized side branch, the gymnotids (fig. 9.54); this includes the electric "eels" and their allied genera, in which the electric organ is absent. The powerful electric organ of *Electrophorus* has been shown by Coates, Cox and Granath (1937, 1946) to be able to generate and discharge a current of one ampere at a pressure of 500+ volts. The larger eels in the New York Aquarium measured more than 100 inches long and were able to stun a scoop-net full of small carp at a distance of twenty feet. The basic polyisomeres in their electric battery are said to represent greatly hypertrophied muscle fibres and their nerve terminals, which with their intervening septa form a voltaic pile. As this structure is unknown in other characins, it probably originated within this genus.

Although now superficially eel-like, the gymnotids retain many evidences of derivation from primitive short-bodies characins: (1) the anal fin, elongated in the ancestral Cheirodontidae, has here become the dominant organ for slow locomotion; (2) the body cavity is short and in some gymnotids the vent opens under the chin; (3) in the mouth of *Electrophorus* the well-toothed premaxilla is dominant, the maxilla being reduced to a small appendix at the corner of the mouth; but the small-mouthed *Eigenmannia* is edentulous, the premaxilla is vestigial and the small maxilla supports the lip; (4) in all the gymnotids the suspensorium of the lower jaw is inclined sharply forward and the conditions of these parts as well as of the opercular region indicate derivation from a primitive characin type.

In view of the great extent of the adaptive branching of the characins, it seems strange that in their exploitation of many types of habitat, they should be wholly lacking in marine representatives. A possible exception is the presence in the Upper Cretaceous of North America, Europe and Egypt of certain teeth named *Onchosaurus,* which, as Eastman (1917) has shown, resemble those of the characins *Hydrocyon* and *Hoplias*. But even if all the formations in which *Onchosaurus* teeth have been found are of marine origin, these isolated teeth may have been swept in from the mouths of rivers; moreover it is doubtful that their owners were characins, especially since they have much better claims to be the "saw" teeth of the rostrum of sawfishes (A. S. Woodward). In another direction the fact that the maxillary teeth extend to the angle of the mouth in certain characins (e.g., *Sarcodaces*) much as they do in certain deep-sea isospondyls (e.g., *Gonostoma*), is pretty clearly a case of convergence; for no trace of the characin Weberian apparatus nor of any other definitely characin character is seen in any of the deep-sea families of isospondyls.

With regard to the degree of relationship between the South American and the African characins, Gregory and Conrad concluded (1938, p. 356):

"(1) that the African and South American characins are closely related (fig. 9.55);

"(2) that so far as known characins are wholly absent from ancient freshwater deposits of North America;

"(3) that there is much evidence analyzed by C. W. Andrews (1906), Schuchert (1932) and others for the reality of a narrow isthmian land or archipelago connecting Brazil with West Africa even in possibly late Tertiary times.

"Dr. Bequaert, however, in the light of his wide

knowledge of the faunae and florae of South America and Africa, tells us that neither the botanists nor the entomologists would favor the assumption of an extreme or prolonged contact between the two continents in Tertiary times on account of the large number of endemic families on either side of the Atlantic."

K. P. Schmidt (1943) also strongly supports Matthew's conclusions from many sides but does not deal especially with the characins. On the whole, the weight of authority is definitely against the hypothesis of even an isthmian bridge between South America and Africa in Tertiary times.

Carps, Suckers, Loaches.—The "peaceful carp," living by preference in quiet waters, roots for its food near the bottom among the weeds and sucks in tasty morsels with its toothless mouth (fig. 9.57). If attacked by a pike, its only defense is to dash about wildly and erect its dorsal and anal fins, each bearing a strong saw-edged spine in front. We wonder how the pike manages to prevent these formidable hooked spears from getting entangled in his own bristling fish-trap of a mouth and throat. Perhaps he only tries to cut open the body cavity and tear out the contents of that well-stocked storehouse, packed perhaps with a million or more eggs and carbohydrates, proteins, fats, oils and vitamines. But does the carp avoid its dangerous neighbor? And if so, does its Weberian apparatus pick up the faint but characteristic thrusts of the fins as the pike sneaks up among the weeds? According to the evidence summed up in Parker and Haswell, Sixth Edition, Volume II, pages 278–280, the Weberian apparatus "merely controls by reflex action the passage of gas to and from the [air] bladder, tending to maintain the volume of this organ constant at all depths. This ensures that the equilibrium will not change during a sufficiently slow migration to shallower or deeper water." Thus we may infer that this apparatus probably assists the carp in dozing comfortably near the bottom in spite of changes in the height of the column of water above it.

Nevertheless it has been proved by physicists that many kinds of fishes send out strong vibra tions under water which register on sound recording mechanisms. It has also been credibl reported that Malay and other native fisherme can locate schools of fish by listening unde water. It would be surprising if the fishes them selves, especially those provided with the Web erian apparatus, could not hear the sound made by their own kind and by other fishes.

With this suggestion of the carp as a livin organism in mind, let us glance briefly at som of the outstanding features of its locomotor an feeding systems. In body-form the typical car is a stout fish with a slightly arched back, a lon dorsal fin, short but prominent anal, wide cau dal, rounded pectorals and fan-shaped pelvi fins. This heavy form and strong body wit ample fins give the fish a firm stance when it i rooting in the mud and a margin of power t cope with swift-flowing streams or to leap awa under attack.

The protrusile mouth of the carp moves s quickly and easily that it might be supposed t be a very simple affair requiring little explana tion; but its apparent simplicity is the fruit o long ages of evolutionary history, which w know only in a general way. Broadly speaking the normal jawbones (premaxilla, maxilla, den tary) have been reduced to become a toothles frame for the thick lips (fig. 9.56A), whic may have a sensory function, as do the paire barbels around the mouth. When the mouth i closed the whole apparatus is folded up. Th action starts behind the eye and in front of th hyomandibular, when the protractor muscle that bone pulls it obliquely outward and for ward. This transmits a forward thrust to th rear parts of both the upper and the lower jaw At the same time certain throat muscles pu the lower jaw sharply downward. This causes right-angled projection (ascending ramus) o each side of the lower jaw to push forward th maxillary bones of the upper jaw, which ar tied by ligament to the ascending rami. Th thrust of the maxilla releases the attached pre maxillae, which before this were bent down ward to close the mouth. The premaxillae the spring forward, partly impelled by a media bony rod and ligament that are fastened above to the bony rostrum. Sudden reverse action dilates the throat, sucks the water into the mouth and closes the mouth.

These protrusile lips and mouth are thus very far from the primitive condition seen in the earliest known forerunners of the teleosts, namely, the palaeoniscoids, in which, it will be recalled, the premaxilla and maxilla were inert strips of tooth-bearing bone forming the fixed outer margins of the inner upper jaw. The rear end of the maxilla was first freed from the cheek-bones in some of the higher palaeoniscoids and their descendants; by the time of the earlier isospondyls the front end of the maxilla was beginning to become movable and to be held in place by an overlapping process of the palatine or anterior inner jawbone. The premaxilla was the last bone to become movable but in the carps, to prevent its dislocation, it is tied to the rostrum by a median ligament (fig. 9.56A) and by the rod-like rostral bone (B, B1) of Starks.

But after the vegetable or animal food is sucked into the mouth, what happens to it? The carp, being devoid of teeth on the premaxilla, maxilla and dentary, has produced a new set of polyisomeres, the tooth-like processes on the fifth ceratobranchials or the right and left lower pharyngeals. These "teeth" (fig. 9.44) are drawn upward by muscles that extend from the roof of the mouth on to the sides of the skull; the food is held in place against a small raised plaque (*dental plate*) from the basioccipital. In the different genera of carps and minnows the pharyngeal teeth assume a variety of shapes: "conical, hooked, spoon-shaped, molariform, etc." (Boulenger). In the suckers (Catostomidae) the pharyngeal teeth are in a simple comb-like row and are therefore supposed to be less specialized than the fewer larger teeth in the true carps. But as we shall see in many other cases, such numerous, very similar pieces are often "secondary," not primary, polyisomeres and lack of differentiation should no longer be taken in itself as evidence of primitiveness.

In this case, however, the thick-lipped "stone-sucking" catostomids may possibly be nearer the ancestral cyprinoid stock than the thin-lipped true carps with fewer pharyngeal teeth. This view is supported on extensive zoogeographic evidence by J. T. Nichols in his monograph on the Freshwater Fishes of China (1943). In North America, north of Mexico, there are now about twenty-one quite different types of catostomids and about seventy-five of cyprinids, whereas in Asia, as noted by Nichols, the catostomids are almost absent and the cyprinids remarkably diversified. According to Nichols, "the presumption is that suckers are ancestral to the carps and originated in Asia, where they have now been superseded by more specialized carps except for one or two arctic forms and a single peculiar species in China."

The fossil evidence indicates that the suckers (Catostomidae) and carps (Cyprinidae) of the present day belong to a relatively recent branching of the ostariophysan order from the second half of the Tertiary period, in contrast with the history of the catfishes, which were well developed in Eocene times. Nor were the carps able to colonize South America since the Lower Pliocene when the two continents became connected. The possible indication is that they did not arrive in North America until late Tertiary times. Notwithstanding the huge gaps in the fossil records of the carps, Cockerell's view that the group may be derived eventually from the Jurassic Lycopteridae and through the latter from the Leptolepidae still seems probable.

Although they are of relatively recent origin, the American carps (including both suckers, true carps and minnows) have enjoyed a wide adaptive branching. They range in body-form (fig. 9.57) from the massive arch-backed "mullet" (*Carpiodes cyprinus,* A) to the delicate, elongate "sucker-mouthed minnow" (*Phenacobius uranops,* D). The mouth varies between the very small and oblique mouth of the "pug-nosed shiner" (*Opsopoeodus,* E), the very large forwardly directed mouth of the pike-like squaw-fish (*Ptychocheilus,* G) and the enlarged down-turned sucker of *Catostomus griseus* (F). The head varies from very small in *Cycleptus elongatus* (C) to very large in *Ictiobus cyprinella* (B). In body size the range is from the pigmy minnow (*Iotichthys*), length one and one-half inches, to the Colorado River squawfish, *Ptychocheilus lucinus,* which reaches a length of five feet or more and a weight of eighty pounds. Even a wider range of anisomerism or differentiation is found in the Asiatic carps as described by Nichols (1944).

That genetic factors have in part determined the so-called adaptive branching of body-form and fins in carps is evident from the results of artificial selection in producing many strange and even monstrous varieties of the common goldfish.

We may safely ascribe to convergence the superficial resemblance to the carps of the small-mouthed edentulous characin *Curimatus* (fig. 9.48). On the other hand, certain carps (fig. 9.56B) and characins have a peculiar median fontanelle between the frontal bones, which is bridged by a special brace from the opposite frontals; some catfishes also have a similar structure, while Sagemehl (1891) held that the cyprinoid skull as a whole gives evidence of derivation from characinid ancestors. Evidence of the relationship of carps to characins is cited by Frost (1925, p. 561) on a basis of extensive comparisons of the otoliths.

The loaches (Cobitidae) of Europe and Asia are highly specialized and sometimes eel-like derivatives of the carp stock, with long compressed or cylindrical bodies and spineless median fins (fig. 9.58I). The pelvic fins are often small or wanting. In some of the loaches the scales are minute or absent and the skin abounds in mucus. They have down-turned sucking mouths and fleshy lips surrounded by barbels. The air-bladder is reduced and wholly or partly enclosed in a bony capsule. In stagnant waters they come to the surface and swallow air, the intestine serving as an organ of respiration (Encycl. Brit., 14th ed., vol. 14, p. 258). Probably by means of their Weberian apparatus they are very sensitive to changes in atmospheric pressure and have been called "weather fishes." They also respond quickly to sound waves and learn to come to signals in laboratory experiments. In some features the loaches are more or less intermediate between the catfishes and the carps, but this may b partly convergent.

Related to the loaches are the Asiatic fresh water Homalopterinae, in which the body i widened and the paired fins greatly expande laterally, serving as suckers (fig. 9.58).

In another very specialized subfamily o carps, the Rhodeinae, there is a long tubula outgrowth from the urinogenital opening of th female, by means of which the eggs are depos ited between the gills of a fresh-water mussel.

Catfishes (Nematognathi).—The name "catfish evidently refers to the "whiskers" or "feelers" (fig. 9.59A) of these otherwise very uncat-lik creatures. The scientific name Nematognath (thread-jaws) also directs attention to th thread-like "feelers" or barbels. The tactile sen sitivity of the barbels may perhaps be supple mented by the sense of taste, which is said to b evidenced by the reaction of the entire smoot skin of the fish to a nearby source of food.

Doubtless the older polyisomeres of the skin namely, the scales, have dwindled away an finally disappeared as the newer polyisomeres including the sensory and mucous cells of th epithelium, increased, so that probably sinc early Eocene time the skin has been scaleless The skin also abounds in black color organ (melanophores). Under the opposing action of different nerves and neurohumors (see p 185) the pigment is either concentrated i small spots, with resulting paleness of the skin or widely spread, producing a dark skin (G. H Parker, 1943). The deeper layers of the skin especially above the pectoral girdle and on th surface of the skull, have secreted still anothe set of polyisomeres, namely, bony plaque which bear on their outer surface the imprint o the lower polyisomeres of the epithelium. Thes bony plaques in some cases are new structures in others (fig. 9.59B, C) they are made fron the old roof bones of the skull; sometimes the form a sort of cephalo-thoracic shield, which in the days before convergent evolution wa recognized, conveyed a deceptive resemblanc to the cephalo-thoracic shield of the ostraco derms (fig. 5).

The skull of the typical catfish is sharply depressed and wide (fig. 9.59B), especially in front. The bony maxilla (*mx*) on each side is reduced to a short column serving as a movable base for the principal barbel; it is articulated on the inner side to a longer column formed from the palatine bone (A, *pl*) and this in turn is tied to the rest of the primary upper jaw, namely, the pterygo- quadrate arch. The mouth, although secondarily widened, may have been derived from that of some small-mouthed fish like a carp in which the maxilla was small and freely movable at the posterior end. Other evidence pointing to the same conclusion is found in the marked forward swing of the suspensor elements (hyomandibular + preopercular). With the flattening and widening of the skull-roof may be associated the reduction of the opercular, the loss of the subopercular and the reduction in the number of the branchiostegal rays (*brstg*). The mandible is composed of the standard bones (dentary, articular) and from its rear border (*an*) extends a bony tracker which communicates part of its movements to the interopercular (*iop*) and opercular bones. Teeth when present on the jaws are delicate and villiform and they are probably new polyisomeres derived from the skin, not homologous with the normal teeth of primitive teleosts.

The braincase forms a long narrow trough in which the orbitosphenoid (*orbsp*) bones are exceptionally large (fig. 9.59A). The occipital region (fig. 9.60), which evidently forms the fulcrum of the entire skeleton, is strengthened by the close union with a complex including the centra and expanded transverse processes of four or more anterior vertebrae.

The Weberian apparatus here reaches a highly specialized stage. The posterior division of the air bladder is lost and the anterior division (fig. 59A1) forms on each side a transversely widened funnel, the front end of which is enwrapped by the greatly expanded lateral flange from the transverse process of the fourth vertebra. In the dried skull of a large catfish the tripus of each side (fig. 9.60B), derived from the autogenous parapophysis of the third vertebra, forms a conspicuous movable horizontal lamina connected in front with the smaller movable units which transmit the changing pressures from the air bladder to the otic region of the skull (cf. fig. 9.45).

The under-side of the skull of certain catfishes displays a peculiar general appearance which bears, especially when seen at a distance, a certain superficial resemblance to a crucifix (fig. 9.60). The arms are suggested by the lower branches of the posttemporal bones which attach the shoulder-girdle to the skull. The opposite tripus, together with the obliquely directed projections from the fused vertebral complex, vaguely suggests a crown with radiating light. Even the rattle of dice—"goddes bones," as dice were called in Chaucer's time—is supplied by the enclosed otoliths.

"No better example perhaps could be found of a class of fortuitous resemblances between wholly unrelated objects, which Bashford Dean (1908) called 'unnatural history resemblances.' Is it any wonder then that still closer resemblances are often produced from patterns that *are* related?" (Gregory, "Fish Skulls . . . ," p. 196.)

The bony spines on the pectoral fins are prominent in typical catfishes, the mad tom (*Schilbeodes*) and the related stone cats (*Noturus*); these spines become strongly serrated and provided with a basal poison gland. "A prick from this spine," writes Schrenkeisen (1938, p. 165), "is like the sting of a bee but more powerful." Another novel invention is the lock-and-trigger device at the base of the great spine of the dorsal fin, a convergence toward a somewhat similar device in the trigger-fishes (figs. 9.102, 103).

The greatest height of specialization within the entire order Ostariophysi is attained among the armored catfishes (fig. 9.61) of Africa and South America (Loracariidae and related families). Somewhat as in the sturgeons, after the original scales have disappeared, a new generation of polyisomeres in the form of bony scutes has come up to perform the same function. These plates also form a covering mosaic over the skull. The bony funnel for the lateral branch of the air bladder is covered by the secondarily expanded posttemporal bone (A, A1, *ptm*) of the shoulder-girdle; the posttemporal plate also acts as a secondary cover for the opercular chamber, the true opercular bones being reduced. The very small, almost carp-like mouth (A) is provided with several rows of minute, curved, rod-like denticles on the premaxillae and dentaries. In view of the loss of true teeth in all the more typical catfishes, these denticles in *Plecostomus* are probably new polyisomeres like the bony scutes. In certain forms that cling to rocks the pelvic fins have been changed into suckers.

Although the catfishes are almost exclusively continental fishes, presumably originating in the northern land mass, the sea-cats *Arius, Felichthys* and related genera have succeeded in invading the ocean without undergoing much radical change in general structure. Both South America and Africa may have derived their catfishes, as well as the characins, cichlids, osteoglossids and dipnoans, from the late Mesozoic stocks, which ranged from North Africa to Europe, Asia and North America. In southeastern Asia most of the catfishes stop at Wallace's line. The few Australian catfishes may have come in from the ocean.

Notwithstanding the many deep-seated agreements between catfishes, loaches and carps, the more precise relationships are still veiled in the mists of Mesozoic time.

FISHES WITH "SHOULDERS" (INIOMI, HAPLOMI)

To the world of mankind hardly any representatives of the order Iniomi are widely known, so that really popular names for them, as opposed to the English names invented by scientists, are scarce. An exception is the lizard fish. The young of this species (*Saurus ophiodon*), taken in nets along the coasts of India, form a sauce to go with curry and rice. But presumably people would not care to eat a "lizard" fish, so it is further disguised under the name of "Bombay duck." In the dark world of the ocean depths the lantern fishes (Myctophidae) and other relatives of the lizard fish flash their glow lamps and devour their prey in successful competition with their rivals, the gonostomids, stomiatoids and other deep-sea isospondyls.

The more primitive scopeloids (such as *Synodus*) show a certain similarity to pikes but there are many significant differences in the skull (fig. 9.62). The premaxillae extend far backward and are well toothed even to the corner of the mouth (fig. 9.62A, B), whereas in the pike (fig. 9.62C) the premaxilla remains small and does not invade the oral margin of the maxilla. Both the scopeloid and the pike have long predatory jaws, but in the primitive scopeloids the greater part of the upper jaw is behind the anterior border of the orbit and the hyomandibular (*hyom*) is directed downward and backward, whereas in the pike the greater part of the upper jaw is in front of the anterior border of the orbit and the hyomandibular is directed downward and slightly forward. From this and many other comparisons emerges the suggestion that whereas the scopeloids (fig. 9.63) have been derived from very primitive Cretaceous offshoots of the herring group (such as *Sardinoides*) with large mouths, two supramaxillae and an adipose fin, the pikes have been derived from small-mouthed umbrids with forwardly-directed hyomandibular and no adipose fin. But eventually the Iniomi and Haplomi are to some extent parallel offshoots from different families of Cretaceous isospondyls.

The names Iniomi, meaning "nape-shoulder," and Haplomi, meaning "simple shoulder," refer to relatively slight differences in the form of the posttemporal bone, which connects the shoulder-girdle with the skull. In primitive isospondyls the upper end of this bone is forked (fig. 9.32b *ptm*), the outer fork being fastened to a backwardly-directed process of the pterotic (supratemporal), the inner dorsal form overlapping the epiotic and touching the parietal. A third or deep fork starts from the inner side of the basal plate of the posttemporal, runs inward and is connected with the strong exoccipital bone (*exoc*) below the epiotic. In the typical Iniomi both the upper and the lower outer forks of the posttemporal are present (fig. 9.62B *ptm*), but the deep fork, if present at all, must be partly ligamentous and does not show in the dried skull. In the Haplomi, as represented by the pike (fig. 62C), the deep fork to the exoccipital is completely wanting and to this extent the name Haplomi, interpreted as "simplified shoulder," seems appropriate.

In 1902 Smith Woodward showed that the existing scopeloid genera *Aulopus* (fig. 9.62A) and *Chlorophthalmus* (fig. 9.63) were but little modified survivors of the Cretaceous *Sardinoides,* which in turn are offshoots of the clupeoid isospondyl stock. The studies of Regan (1911) and Parr (1929) afforded the chief material for the accompanying provisional phylogeny of the Iniomi (fig. 9.63) by Gregory and Conrad (1936), in which *Aulopus* and *Chlorophthalmus* were taken as the respective stem forms for extensive branching. The end forms in the top row have obviously attained to extreme stages of divergent specialization. Anisomerism, or the differential emphasis of growth forces at different foci, has resulted in such bizarre structures as the sailfish-like dorsal fin of *Alepisaurus,* the excessively long dagger-teeth of *Omosudis* and *Evermanella,* the great rostral glow-lamp of *Diaphus,* the extreme elongation of the pectoral and pelvic barbels in the blind *Bathypterois,* the huge whale-like head and mouth of *Cetomimus* and the very long anal fin and partly cartilaginous skeleton of *Ateleopus.*

Still more fantastic forms are the gulpers (figs. 9.64; 9.65D), often referred to a separate order (Lyomeri) but possibly derived from a relative of *Cetomimus.* Here the scoop-net method of engulfing prey reaches a climax.

Tchernavin (1947), in a fine monograph on the anatomy of six specimens of Lyomeri in the British Museum, shows that these gulpers possess what may be interpreted as the ventral part of the hyoid gill slits and a "mandibular operculum"; there is one more than the usual number of gill slits, no hyoid opercular, and numerous other characters not, or but very rarely, found in other bony fishes (Actinopterygii). He concludes that the systematic position of the group is most uncertain and notes its contrasts, especially in the jaws, hyoid and branchial arches, with other specialized abyssal bony fishes with enlarged jaws.

It may be noted, however, that in spite of the extreme specializations of their jaws and associated parts, no satisfactory evidence has been advanced for homologizing the ventral hyoidean gill slit, the extra gill-cleft and the mandibular operculum with the so-named structures in acanthodians, or for removing the Lyomeri from the teleosts. Moreover, Dr. Tchernavin himself has shown that other large-mouthed, deep-sea swallowers have developed one or another adaptive feature that has been carried to greater length in the Lyomeri.

As an example of relatively new and distinctive polyisomeres we may cite the generically and specifically diagnostic numbers and arrangements of the photophores in the very numerous forms of myctophids.

The order Haplomi is close to the Isospondyli and its few known members appear to be descended from Cretaceous derivatives of the isospondyl stock. The Esocidae, which is the typical family of the Haplomi, includes only the fresh-water pikes and pickerels. The pike is famous as a lurking predator of the lakes and ponds, ready to dash out from under the water plants and seize the fat carp in his slender, sharp-toothed jaws. To some extent the pike parallels the ganoid garpike (*Lepidosteus*) but with far thinner armor and generally more delicate, swift build.

"If we consider the skull of the muskellonge, *Esox musquinoni* (fig. 9.62 C), with special reference to its 'adaptive' features, we shall note that the dominant physiological feature is the very large lower jaw with its row of high, well-spaced, great laniary teeth pointing upward and inward, in front of which is a series of small sharp teeth arranged along the upward curve of the undershot jaw. Obviously the big teeth pierce the prey, which is held fast by the *chevaux-de-frise* of smaller inwardly-directed sharp teeth arranged in long rows on the dermopalatines and vomer. The premaxiliary teeth are practically vestigial. The maxillary, although edentulous, serves to press the prey against the lower dagger-like teeth. The supramaxillary is a souvenir of much earlier times. As the dominant elements of the lower jaw are the great laniary teeth on the sides, so the most important elements of the upper jaw are the large dentigerous areas on the dermopalatines and entopterygoids, which run in nearly parallel antero-posterior tracts. These are supported and braced dorsally by the large true palatines and posteriorly by the ectopterygoid, quadrate and metapterygoid, which also receive the heavy thrusts from the lower jaw and transmit the whole load to the strongly-braced hyomandibular." (Gregory, 1933, p. 214.)

The chief distinctive and apparently new polyisomeres in the skulls of this order are the paired "proethmoid" bones (fig. 9.62C1, D), which coöperate with the anterior prolongations of the lacrymal bones in covering the dorsal side of the elongate palatines. These "proethmoids" seem to be new ossifications in the skeletal-producing layer lying on the side and roof of the elongate muzzle. They seem to represent new anterior extensions of the frontal tracts, which they overlap posteriorly.

The mud-minnows (Umbridae) are small and primitive relatives of the pikes, living in fresh-water, with small mouths, feeding on insects, small molluscs, crustaceans and vegetable matter.

LIVELY FISH SPECKS (MICROCYPRINI, PHALLOSTETHI)

The tooth-carps (fig. 9.67a), top-minnows or fresh-water killifishes (Poeciliidae) are basically similar to the mud-minnows (Umbridae) and may indeed be derived from them but they have progressed much further toward the spiny-rayed grade of evolution in the predominance of the premaxillae, the exclusion of the maxillae from the gape, the loss of the mesocoracoid arch, the form and arrangement of the branchiostegals. Some of the genera (e.g., *Gambusia, Molliensia,* etc.) are viviparous. The typical killifish (*Fundulus*) has a very small, slightly upturned mouth (fig. 9.66A) and is a voracious devourer of mosquito larvae. Its relatively stout body, prominent dorsal and anal fins and wide muscular base of its broad caudal fin indicate very quick and active movements. The Microcyprini are primarily small-mouthed nibblers but *Belonesox* (fig. 9.66B) is a pygmy predator with bristling teeth. The premaxillae of *Fundulus* and its allies have marked dorsal or ascending processes, as in the spiny-finned fishes, and the small maxillae are likewise pushed to the rear. The lower pharyngeal bones in different species of *Fundulus* show progressive evolution of a median dental mill, as in the synentognaths (p. 173).

In the four-eyed fish (*Anableps tetrophthalmus*) of this order the protruding eyes (fig. 9.66C) are divided horizontally into an upper segment, with the lens and accessory parts adapted for vision above the surface, and a lower segment, adapted for vision beneath the surface. No doubt the beginnings of such eyes must be sought in a stage in which the dorsal and ventral parts of the eye were becoming differentiated in reference to light coming either from above or below (see p. 185). The orbits of *Anableps* (fig. 9.66C) are placed near the skull roof and are provided with a large bony flange behind each eye but the skull is otherwise basically similar to the small-mouthed microcyprinid type.

The phallostethids are minute minnow-like fishes of the Malayan-Philippine region, the males of which possess extremely specialized generative organs; these contain complex skeletal structures for effecting internal fertilization of the females. These "new polyisomeres" have been shown to be derivatives of the anal and pelvic fins and associated ducts and other parts, which have all been displaced far forward, along with the cloaca, the whole complex being enclosed in a prominent swelling beneath the throat. The females show correspondingly modified generative structures. The phallostethids were formerly thought to belong with the viviparous killifishes (Microcyprini) but the studies of Myers (1928) indicate that they are more nearly related to the Percesoces (p. 193), to which he refers them as a suborder.

SKITTERING AND FLYING FISHES (SYNENTOGNATHI)

The common name "skipjack" is appropriate to the central type (*Scomberesox*) of the Synentognathi in allusion to its habit of leaping out of water and skipping or skittering like a bouncing stone. "Needle fish," often applied to this and related forms, refers to the long sharp-tipped beaks on the upper and lower jaws. Gar (= spear), obviously referring to the pike-like general appearance, is used for this and the related "half-beak" (*Hemirhamphus*). "Scizzors," as used for one of the half-beaks, suggests the lower jaw of the half-beaks. The scientific name, *Scomberesox,* literally mackerel-pike, perhaps hints at the difficulty of classifying these fishes, which are neither mackerels nor pikes. "Hound-fish," as applied to *Tylosurus* (fig. 9.70A), may refer to the way these fish go loping over the water in packs. In order to burst through the surface tension of their water ceiling and propel themselves into the air, they must be able to develop great speed and since they are long-bodied and have long-based dorsal and anal fins and strong caudal fins, they deliver a sustained sequence of thrusts from all their muscle segments arranged in series. The skeleton is light but strongly built. The braincase, especially in *Tylosurus,* is a narrow strong box (fig. 9.70A), forming a thrust-block between the well-braced spear-like jaws in front and the stout pectoral girdle and vertebral column in the rear. As in other pike-like forms, the jaws are produced far in front of the eyes and the suspensorium is inclined forward. The mouth opens forward both because of the length of the rostrum and because the quadrate-articular joint is but slightly depressed below the level of the eye. Numerous features of skull structure suggest derivation from the *Fundulus* group (fig. 9.66A) (Gregory, 1933, "Fish Skulls . . .," pp. 221–224). In that group can be traced the gradual development of triangular tooth-plates on the opposite lower pharyngeals and their close appression at the midline. In the present group they have fused to form a single median plate, whence the ordinal name Synentognathi. Similar plates have been developed independently in the wrasses, cichlids, demoiselles (pomacentrids).

The flying-fishes (Exocoetidae) are generally regarded as large-finned derivatives of the skipjacks, to which they are closely related (fig. 9.70). Contrary to popular impressions, flying fishes do not really fly. Although the "wings" appear to vibrate, it has been abundantly proved by Hubbs (1933–37) and Breder (1926, 1932, 1938), Nichols and Breder (1927) that after the initial leap the fish propels itself by vibrating the body and transmitting the thrusts to the water through the lower lobe of the tail, which is longer than the upper lobe. Dissection of a flying fish shows that the bulk of its pectoral muscles, which move the "wings," is only a small fraction of the bulk of the body muscles, which push the tail against the water. The pectoral muscles doubtless extend or fold up the wings but are merely accessory, not primary, organs of flight. The same is even more true of the pelvic fins, which in certain flying fishes are comparatively large.

Although the group is a relatively compact one, the exact taxonomic relations of its families, subfamilies, genera and species are quite complex. They are clearly set forth in the monographs by Hubbs and by Breder.

The fossil record of the group dates back only to the Mid-Tertiary.

AGGRESSIVE STICKLEBACKS AND PEACEFUL TUBE-MOUTHS (THORACOSTEI)

Fin spines have been but seldom developed in the lower teleosts (notable exceptions being the pectoral spines of catfishes and the dorsal spines of carps), but in the stickleback they appear as new and diagnostic polyisomeres, doubtless of genetic origin, the different number of spines being characteristic of different species. Apparently their tiny owners have been quick to make use of these gifts and many of the family have become celebrated for their pugnacity. The sticklebacks (fig. 9.69A) are small-mouthed nibblers and the lower part of the suspensorium has been produced forward so that the quadrate-articular joint is well in front of the eye. In *Gasterosteus spinachia* (fig. 9.69B) the entire preorbital region is elongate and in *Aulostomus* this tendency is further emphasized. It reaches a climax in *Fistularia* (D), which is weakly named "cornet fish." In this amazing creature the mouth has remained minute at the end of a long tube, stiffened and roofed above by the ethmoid and by the bones of the inner upper jaws (palatopterygoid); below it is braced by long narrow ridges from the preopercular and interopercular. Behind the head there is again an extreme emphasis upon fore-and-aft growth components in proportion to vertical and transverse ones. The shoulder-girdle is much drawn out posteriorly and to prevent dislocation of the skull upon the vertebral column the four anterior vertebrae and their transverse processes are fused into a rigid tube. Even the tail wisp enjoys a great elongation.

Another main line of specialization starting from the sticklebacks leads to *Centriscus* (fig. 9.68G), whose quite short body has acquired a huge serrated dorsal spine and an expanded shoulder plate. The long mouth tube now ends in almost microscopic upper and lower jaws and the food must be minute. *Amphisle* (H) of this group may be regarded as a wayward *Centriscus*, which has found safety in standing as it were on its head while hiding among the reeds. Especially in this creature the old exoskeleton of scales has been largely replaced by new polyisomeres in the form of bony plates covering the back and sides.

Much less specialized than either of the foregoing but classified in the same suborder (Hemibranchii) and superfamily is *Macrorhamphosus*. In a general way this form tends to bridge the gap between the stickleback-tube-mouth (*Aulostomus*) series (fig. 68A-C) and the pipefish-seahorse branch (Syngnathidae). *Solenostomus* (fig. 9.68E) again appears to lead back toward *Aulostomus* (fig. 9.69C) and *Fistularia* (D) both in skull and body-form. In the pipefishes (*Syngnathus*), seahorses (*Hippocampus*) et al. (F) the exoskeletal polyisomeres tend to take on a new and secondary squareness and regularity so that the creature is beginning to suggest some sort of invertebrate. The dorsal and pectoral fins finally become the main motors and their extremely rapid vibrations culminate in the graceful movements of the seahorses, which are really upright-swimming pipefishes with distended abdomens and prehensile tails. In an ingenious analysis of the movements of the fins in *Hippocampus*, Breder and Edgerton (1942) have shown that although the vibratory movements of the wing-like pectorals are very rapid, yet movement of the body through the water is quite slow. They suggest that the speed has been, as it were, geared down for the needs of these pygmy fishes that do not move far from the shelter of their home in the seaweed. In the seahorses the males receive the young into their abdominal pouch and after rearing them, set them free in an explosive spasm.

When the ordinary seahorse was reached one would think that evolution could go no further, but in the Australian *Phyllopteryx* (fig. 9.68F) of this family the dermal plates have produced still another set of polyisomeres in the form of long tabs and foliate growths which collectively simulate the water-weeds in which the creature hides.

The origin of the Thoracostei has been discussed by many earlier authors and more recently by Swinnerton (1902), Tate Regan (1906–1908), Jungerson (1908), Starks (1902, 1926) and Gregory (1933), but the problem is too extensive and confusing to be dealt with here at length. The opinion may, however, be expressed that while the sticklebacks indicate no definite relationship with the scorpaenoids or mail-cheeked fishes (fig. 9.139), the skull does show many points of agreement with that of *Belone* (*Tylosurus*, fig. 9.70A) of the Synentognathi, as pointed out by Swinnerton. Moreover, as noted above, *Macrorhamphosus* tends to connect the centriscid stock with the sticklebacks, while the aulostomids and fistularids tend to connect the pipefishes and seahorses with the same stickleback stem. Although the latter have evolved a new type of gills, referred to in the ordinal name Lophobranchii (crest-gills), this appears to be merely one more stage of specialization which is already prepared for in the reduced gills of the Hemibranchii (centriscids, fistularids).

LINKS BETWEEN LOWER AND HIGHER TELEOSTS (SALMOPERCAE, BERYCOIDEI)

In the Great Lakes and tributary streams lives the trout-perch (*Percopsis*, fig. 9.71A), a six-to-eight-inch-long fish, which, as its English name implies, combines the features of both trout and perch. Since fishermen with few exceptions are unaware of the existence of evolutionary links, we may suspect that the name trout-perch is a book name, a literal translation of the scientific name (Salmopercae) of the small group which includes this relic of an earlier fish fauna. Among its trout-like features are the presence of an adipose fin, the retention of a small open duct connecting the air-bladder with the oesophagus, the great predominance of soft rays in the dorsal and anal fins. For these and similar reasons Boulenger concluded that the trout-perch and its allies were in reality not primitive forerunners of the spiny-finned or percoid group but progressive members of the Haplomi (pikes) which had evolved in the direction of the spiny-finned forms. On the other hand, Hubbs (1919) noted that they have six branchiostegals arranged exactly as in the spiny-finned fishes.

In a young percopsid skull the characters of the jaws point in the same direction, since the premaxillae have long ascending processes and the smaller maxillae act as levers for the protrusion of the premaxillae, as in the typical percoids. The preopercular in this young skull is large and bears a spine, as in many percoids, and the condyles or articular surfaces at the back of the skull are tripartite, also as in percoids. The otoliths or earbones, according to Frost, show a strong resemblance to those of the cardinal fishes (*Apogon*) among the percoids.

The pirate or mud perch (*Aphredoderus*, fig. 9.71C) of the coastwise streams and bayous from Texas to New York and in the Mississippi basin north to Michigan is more definitely perch-like in appearance but is assigned to a separate family (Aphredoderidae). This mud perch is a descendant of one of the old settlers in North America, for fossil members of the same family have been found in the Green River shales of Eocene age in Wyoming.

In the seas of the English Chalk (Upper Cretaceous time) lived numerous berycoid fishes which had nearly attained the status of spiny-finned fishes but had already retreated to considerable depths, perhaps in competition with the more direct ancestors of the true percoids. The true berycoids (fig. 9.72A) had fairly short, compressed and rather deep bodies with twenty-four vertebrae as in primitive percoids; but in the related Polymixiidae the number of vertebrae rises to thirty-four (A. S. Woodward). Apparently the primitive berycoids were quick dodgers, moving by sudden lateral twists, pivoting easily on the dorsal and pelvic fins. There were a few spines in front of the soft rays in the dorsal and anal fins. Deep-sea survivors of the primitive Upper Cretaceous berycoid *Hoplopteryx* (A) are found in the seas off South Australia and New Zealand but many other existing berycoids are aberrently specialized, including *Monocentris*, which is enclosed in a stiff leathery skin. In at least two of the surviving berycoids (*Beryx* and *Holocentrus*) the primitive duct between the air-bladder and the digestive tract persists (Boulenger, 1910, p. 635).

The squirrel-fishes (*Holocentrus*) of the coral reefs of the West Indies and other semi-tropical and tropical seas parallel the true percoids in their strong development of spines in the fins and spikes on the preopercular and opercular (fig. 9.72B). But on the whole there is a conspicuous inconstancy in the number and degree of development of the spines in the anal fin, in contrast with the three anal spines which run through many families of percoids. For example, in *Holocentrus meeki* of Bermuda the anal spine is very short or barely indicated, whereas in *Holocentrus vexillarius* of the same island the second anal spine is very large (Beebe and Tee-Van, 1933). The jawbones of *Holocentrus* (B) are of primitive percoid type but in the related *Myripristis* (C) there is a sort of burr on the tip of the lower jaw which fits into a corresponding depression in the premaxillae.

As to the origin of the spiny-finned fishes (fig. 9.73), the broad structural stages of evolution that lead to them are recorded in different members of the orders Palaeoniscoidei, Protospondyli, Isospondyli, Salmopercae, Berycoidei, Percoidei. In this series (Table I and fig. 9.74) taken as a whole there was a gradual loss of primitive palaeoniscoid features, the duct from the air-bladder to the oesophagus becoming closed; meanwhile merely by changing the ganoid fulcral scales into spines (A. S. Woodward, 1942, p. 905; cf. fig. 9.73) the dorsal and anal fins acquire spines, the opercular develops a spike (fig. 9.72A), the pelvic bones (*pelv*) become tied to the shoulder-girdle (B), the premaxillae acquire long ascending processes (A) and the maxillae (B) become levers in sliding the premaxillae forward, the supraoccipital extends forward, shoving aside the parietals to gain firm contact with the frontals (fig. 9.76A), the branchiostegals (fig. 9.32b *brstg*) are reduced to six on each side, the scales pass from cycloid to ctenoid stage; in both pelvic fins the number of rays is finally reduced to one spiny ray and five soft rays.

As yet only one of the known genera of the foregoing orders can be considered to be in or near the direct line of ascent from primitive isospondyl to central acanthopt (spiny-finned). This, for us, fortunate exception is the extinct *Ctenothrissa*.

TABLE I—CONSPECTUS OF MAIN CHARACTERS FROM PALAEONISCOID TO PERCOID

Primitive or palaeoniscoid characters	PALAEONISCOID (*typical*)	PROTOSPONDYL	ISOSPONDYL	BERYCOID	PERCOID	*Advanced or percoid characters*
Earliest geological date	Mid-Devonian	Triassic	Jurassic	Cretaceous	Cretaceous	
Surviving relatives	Sturgeons Spoonbills	Garpike *Amia*	Clupeoids	*Beryx*	Basses	
Open duct air bladder to gut functional	+*	+	+	+	0	
Notochord in adult	+	±	0	0	0	
Spiral valve in intestine	+*	±	0	0	0	
	0	±	+	+	+	Bony vertebral centra
Tail shark-like in adult	+	±	0	0	0	
	0	±	+	+	+	Hypurals becoming fan-like
Fulcral or ridge scales in median fins	+	±	±	0	0	
	0	0	±	+	+	Fulcral ridges becoming spikes
Scales, ganoid, rhombic	+	±	0	0	0	
Scales with bony base	+	±	0	0	0	
	0	±	+	+	+	Scales horny
	0	0	0	+	+	Scales ctenoid
Fin-rays grading into scales	+	±	0	0	0	
All median fin-rays jointed	+	+	±	0	0	
	0	0	±	+	+	Some fin-rays spiny
Radial bases of dorsals fewer than their rays	+	0	0	0	0	

ABLE I—CONSPECTUS F MAIN CHARACTERS ROM PALAEONISCOID O PERCOID (continued)

rimitive or palaeoniscoid characters	PALAEONISCOID (typical)	PROTOSPONDYL	ISOSPONDYL	BERYCOID	PERCOID	Advanced or percoid characters
edian gular plate beneath jaw	+	±	±	0	0	
eparate coronoids, etc., in mandible	+	±	0	0	0	
	0	0	0	+	+	Spikes on operculum or preoperculum
ranchiostegals	9±	9±	(36± −3)	?6	6	
teroperculum not differentiated from branchiostegals	+	±	±	0	0	
	0	0	+	+	+	Interoperculum well differentiated
	0	0	±	+	+	Maxillae toothless, becoming levers for sliding premaxillae forward
upramaxillae part of cheek	+	±	0	0	0	
	0	0	+	+	+	Supramaxillae attached only to maxillae
	3	?	2	2	1–0	Supramaxillae, number
remaxillae small, not excluding maxilla from gape	+	+	±	0	0	
	0	0	0	+	+	Premaxillae with ascending rami
	0	0	0	+	+	One or more spikes on operculum
	0	0	±	+	+	Supraoccipitals keel-like
	0	0	±	+	+	Supraoccipital in contact with frontals, displacing parietals frontally
arietals large	+	+	+	0	0	
rbitosphenoid	+	+	+	+	0	
asisphenoid flat, not projecting downward	+	+	0	0	0	
	0	0	+	+	+	Basisphenoid V-shaped, projecting sharply downward
entral fins abdominal	+	+	+	0	0	
	0	0	0	+	+	Ventral fins subpectoral, attached to cleithrum
entral rays more than 5	+	+	+	0	0	
	0	0	0	+	+	Ventral rays I 5
nfraclavicles (clavicles)	+	0	0	0	0	
ostclavicles (postcleithra)	>2	>2	>2	2	2	
lesocoracoid arch	+*	+	+	0	0	
	0	0	0	3	3	Anal spines
ectoral pterygials	>4	>4	>4	?4	4	
UMMARY						
rimitive or palaeoniscoid characters	43	26	16	3	0	
Indetermined		1	0	2	0	
ransitional or variable	0	14	10	0	0	
dvanced or percoid	0	2	17	38	43	
otal number character pairs listed	43	43	43	43	43	

+ = present, ± more or less, 0 absent.
Inferred because of its presence in modern survivors of Palaeoniscoid stock.

Smith Woodward in his monograph on the Fishes of the English Chalk gave a careful reconstruction (fig. 9.73) of *Ctenothrissa microcephalus*. This fish is still classified as a member of the clupeoid (herring) division of the order Isospondyli but, as noted by Tate Regan, it is an ideal intermediate between the primitive isospondyl stock on the one hand and the berycoid stage of the spiny-finned fishes on the other. For like some other herrings it was already a fairly deep-bodied fish with cycloid scales and a single dorsal fin at the summit of the back and its jaws were also clupeoid in stamp with small premaxillae, the large curved maxillae forming most of the upper border of the mouth. As in other primitive isospondyls, each maxilla carried two supramaxillae and these two bones are still borne by the primitive berycoids and squirrel fishes. On the other hand, *Ctenothrissa* had advanced in the direction of the spiny-finned fishes in several important particulars: its large cycloid scales had acquired comb-like or ctenoid rear margins, the first three rays of its dorsal fin, the first rays of its pelvic and anal fins had pointed, not segmented, tips and its pelvic fins had moved forward to be immediately beneath the pectorals. Thus the quick dodging or "swivel-chair" arrangement which is characteristic of Upper Cretaceous berycoids was already initiated in the contemporary clupeoids.

SOME LOWER BRANCHES OF THE BASS-PERCH GROUP (PERCOIDEI)

In a crude way the skeleton of any bony fish may be likened to a many-pieced ship made of concrete that has been poured into a mold and allowed to set. In this case the concrete is represented by the bony precipitate (mostly calcium carbonate and calcium phosphate) which has oozed out from the gland-like osteoblasts into a jelly-like matrix (the hyaline cartilage of young stages). This matrix in turn has been confined within limiting walls, such as the periosteal membranes, the connective tissue septa between the muscle plates, the inner and outer membranes of the body. But the skeleton of a fish differs from a rigid cast of concrete in that it is not inert and static but a growing entity with many growth centers; also in that it is consumed or modified by special cells called osteoclasts. Moreover it responds differently to the influence of other parts: for example, in response to the stresses induced by the activities of the muscular system, the bone produces crests, ridges and strong bases; its trabeculae or strengthening strands dispose themselves in well organized systems so as to form the most effective resistance to normal pressures or tensions; the bony plates even develop thin areas or openings where stresses are light. Bony tissue also behaves as if it were sensitive to different kinds of structures that lie in its path. Thus it readily overlaps or adheres to other bony tissue but avoids and grows around the sense organs, nerves, blood vessels and viscera.

The axial skeleton of a bass (fig. 9.75) or that of any other bony fish, comprises two main divisions: (1) the first, or head-piece, is composed of many pieces that are united into a fixed whole called the neurocranium; this is the fulcrum or spearhead upon which the second, or movable part, is based; (2) the second or vertebral part in turn is composed of a series of jointed pieces called vertebrae, together with their processes (such as the neural and haemal arches, the transverse processes and ribs); all these parts are laid down between the septa covering adjacent or opposite myomeres or muscle plates. The vertebral column and its extensions constitute: (a) the framework upon which the motor organs are stretched; (b) the medium by which their thrusts are communicated to the spearhead or cranium in front and to the water all around; (c) the many arched tunnels for the spinal cord and blood vessels; (d) the many-ribbed body cavity for the viscera.

In the bass the vertebral centra in the adult are fully ossified and not perforated by the remnants of the notochord, which are found only in the intervertebral discs or buffers. In the true basses (Serranidae) there are but twenty-four vertebrae, including ten in the thorax and fourteen in the "caudal" portion behind the body cavity. This is believed to be the primitive number for all Percoidei but in many specialized derived forms there has been a tendency to increase the number until it rises from twenty-eight to forty in cichlids and thirty to forty-eight in the perches (Percidae). Primitive members of the percoid order with only twenty-four vertebrae, such as the true basses, tend to retain the quick-dodging method of locomotion of the stem berycoid forms, while those with a high number of vertebrae move in a more sinuous or eel-like way.

The bony bases of the caudal fin muscles (fig. 9.5C, C1), representing fan-shaped expansions of two of the haemal arches, are properly reckoned with the axial skeleton. Their origin in the bony fishes has already been considered (fig. 9.24). The bass, like other members of its order, has progressed beyond the primitive isospondyl stage by the further reduction of the upturned rear end of the vertebral column into a small tube called the urostyle (E), which lies concealed beneath the epineural rod (*epn*).

The appendicular skeleton of the bass includes the bony spines, bases and supports of the dorsal, anal, pectoral and pelvic fins, as well as the shoulder-girdle itself. The soft rays of the fins, which tend to become subdivided at the distal ends, thereby retain a slight trace of their ultimate derivation (see p. 141) from ganoid scales fused into rows. The spines in front of them, derived phylogenetically from the fulcral scales of early ganoids (fig. 9.73A-G), have at the same time greatly increased the number and density of their bone cells and fused their component particles into a stiff needle-like point. The physical causes and developmental steps in the formation of conic to needle-like bony points are not clear, but perhaps high surface tension of limiting membranes, together with rapid centrifugal growth and multiplication of bone cells may be assumed.

Some of the surface bones of the skull, especially the opercular and preopercular bones, likewise often produce similar spikes on their margins and among the scorpaenoid fishes such spikes also develop in fixed places also on the neurocranium (figs. 9.139, 140, 141).

The number of spikes or spiny rays on the dorsal fin varies from species to species not only among the perches but in most if not all of the numerous families of spiny-rayed fishes; the number of spines on the anal fin, on the other hand, is three in the typical families of percoids, labrids, scorpaenoids and other groups of spiny-rayed fishes but becomes variable among the sciaenids, centrarchids, scarids, and other specialized groups. On each pelvic fin in all the range of the percoids there is never more nor less than one spine and five soft rays. In the pectoral fin there is rarely any spine and the number of soft rays varies widely. Obviously genetic factors underlie these and similar polyisomeres of the skeleton and there is good comparative and palaeontologic evidence to prove that some of these factors, including those that determine the presence of three spines on the anal, one spine and five soft rays on the pelvic fins, one main spike on the opercular bone, are both older and more resistent to disturbing influences than the varying numbers of spines and soft rays on the dorsal fin, which are often no older than the species itself. The subject needs investigation by a combination of experimental, taxonomic and phylogenetic methods and data.

The web and horny parts of the fin rays are derived from the more superficial layers of the general integument, to which the ectoderm contributes, while the bony supports of the fin are regarded as parts of the endoskeleton and are exclusively mesodermal in origin. Nevertheless there is a most intimate relationship of these superficial and deep parts, primarily to the muscles that move them and secondarily to other parts of the skeleton, such as the skull or the dermal shoulder-girdle. For example, both the bony and the soft dorsal rays of the bass and other percoids are movably supported by endoskeletal pieces called pterygiophores, which usually have a one-to-one relation with the fin rays (fig. 9.110A). At the top of each pterygiophore is a transverse ridge bearing a rounded articular surface which permits the fin ray to move forward, backward or sidewise under the action of the several muscles.

In the base of the pectoral fins of the bass and in percoids generally there are four short to moderate rods, called baseosts, constricted in the middle and widening at the ends (fig. 9.31F). They regularly increase in size from above downward and articulate by their expanded proximal ends with a long, narrow, nearly vertical joint-surface borne mostly by the scapula and partly by the coracoid. The lower or fourth baseost articulates with the coracoid, the third at the junction between the coracoid and the scapula; the upper two baseosts articulate with the scapula. The outer ends of the baseosts (1, 2, 3, 4) and their common epiphyseal rim are clasped by the proximal ends of the dermal fin rays. The upper ray of the pectoral fin is much larger than those below it and it articulates directly with the scapula by a concavo-convex surface, here called the trochlea. The complex system of muscles that move the pectoral fin are based mostly on the scapulo-coracoid plate and are inserted into the outer ends of the baseosts and the proximal ends of the dermal rays. The foregoing arrangements of the pectoral baseosts are remarkably constant in the percoid series and clear traces of them may be seen in many derived groups.

The pectoral girdle (fig. 9.31), as in all other bony fishes, consists of a superficial or sheathing series, the so-called outer or dermal girdle, and of a deep or endoskeletal series, constituting the so-called primary shoulder-girdle. In describing the skeleton of the tarpon as a typical teleost fish (E), we have noted that the shoulder-girdle is fastened above and on each side to the skull by a three-forked posttemporal bone and that its main crescentic element, the cleithrum, bounds the branchial chamber in front and affords insertion for the inner or endoskeletal girdle just beneath its rear margin. In the bass (F) the principal advance of the shoulder-girdle beyond that of the ancestral isospondyl stage is the absence of the mesocorocoid arch (A-E), which had disappeared before the grade of the Iniomi and Haplomi was reached. The loss of the mesocoracoid arch in the percoids is associated with a nearly vertical insertion of the pectoral fin, whereas in the tarpon and other isospondyls which retained the mesocoracoid the articular facets for the pectoral fin were more nearly horizontal.

The two bones (scapula, coracoid) that remain in the bass form a thin bony plate which is attached vertically to the cleithrum, its lower end being prolonged into a ploughshare-like process, which is fastened by ligament to the cleithrum (fig. 9.76B1). Extending downward obliquely from above and overlapped by the upper rear corner to the cleithrum is a thin elongate bone, the upper postclavicle or postcleithrum, which is fastened below to the inner side of the nearly vertical bȧse of the pectoral fin, while a second sliver-like bone, the lower postcleithrum, continues the postcleithral curve and in the dried skeleton if it is not torn off it projects far below the inner surface and lower border of the pectoral fin (e.g., figs. 9.77B, D; 9.84A, C).

Perhaps one reason why the scapula, coracoid and postcleithra are all such thin and often delicate bones is that they are wedged in between thick and strong muscle masses which in turn are fastened mainly to the very strong cleithrum; thus perhaps most of the stresses generated by the pectoral muscles are transmitted to the cleithrum and among the stresses that remain for absorption by the plate-like and rod-like elements tension predominates over compression.

In the bass, as in most other members of the spiny-finned order, the pelvic fins are beneath the pectoral fins and the conjoined pelvic bones or basipterygia are produced forward into a long tapering apex; this passes between the converging forks of the coracoid bones and is inserted between and tied by ligament to the V-like limbs of the opposite cleithra (fig. 9.77B-D).

In forward locomotion the pelvic fins are held in a nearly horizontal position (fig. 9.77D) or appressed against the sides of the body, but a sudden stop is effected in part by pulling the pelvic fins sharply downward and spreading them. If there were no counterbalancing forces, this would tend to elevate the tail and depress the nose, but by bringing the pectoral fins sharply downward and forward at the same moment, the tendency to pitch downward is corrected and both pectoral and pelvic fins coöperate in checking forward motion. The main stresses delivered from the pelvis to the cleithra would tend to pull backward the lower ends of the latter, but they in turn are tied by the muscles of the throat to the lower part of the branchial arch system. In this connection the bass, like other spiny-finned fish, has a prominent vertical plate, the urohyal bone (fig. 9.9B) attached to the rear of the base of the tongue bones. From this as a base a strong pair of muscles pass backward to be inserted on the lower tips of the cleithra.

The cranium of the bass, as noted above, must be strongly built, since it forms the entering wedge (fig. 9.75) or spearhead of the forwardly-darting fish. As in other percoids, it bears a keeled supraoccipital bone (fig. 9.76B, A) which transmits some of the stresses from the dorsal extension of the myomeres to the well-braced roof, sides and base of the braincase. A second crest, borne (A) by the frontal (*fr*), parietal (*pa*) and epiotic (*epiot*), strengthens the braincase against stresses from the dorsal fork of the posttemporal (B1, *ptm*), while a third (A, A1), on the sphenotic (*sphot*) and pterotic (*pto*), is connected with the lower or lateral fork of the supratemporal (fig. 9.78C, E).

The orbitosphenoid bone (fig. 9.8B1 *orbsb*), which in more primitive orders surrounds the olfactory nerve where it emerges from the cranium, has disappeared. The basisphenoid (fig. 9.76B *bsp*) is reduced to a small narrow brace resembling a flying buttress, which transmits stresses from the parasphenoid (*pas*) to the cranial wall. This buttress also passes between the enlarged optic funnels. The latter, containing the recti muscles of the eyes, are continued downward and backward beneath the floor of the cranium but are in turn partly covered ventrally by the parasphenoid. The chamber thus enclosed is called the myodome or muscle-house and in the bass and other percoids it is even more conspicuously developed than it was in the palaeoniscoid (fig. 9.8A1) the tarpon (B1) and other isospondyls.

The occipital condyle of the percoids is sharply tripartite (fig. 9.76B); on each side the exoccipital forms a strong lateral facet for the first vertebra; this prevents lateral dislocation, while the upward slant of the condylar surface of the basioccipital prevents downward dislocation of the column.

The bass, like other spiny-finned fishes, has a protrusile mouth, in which the principal dorsal elements are the premoxillae (fig. 9.76B1). These have ascending processes which slide over the ethmoid and are pushed forward by the lever-like and toothless maxillae (fig. 9.77C). In *Scorpaena,* a derivative of the percoid type, Allis (1909) has described the complex system of cheek-ligaments (A, A1) whereby the premaxillae are tied to each other, to the maxillae and to the nasal and lacrymal bones. This system prevents the premaxillae from slipping too far forward, insures that their ascending processes will roll forward and downward parallel to the median vertical plane on either side of the median keel of the ethmovomer block, and in general prevents dislocations in any direction. Other ligaments tie the inner surfaces of the maxillae to the ascending rami of the dentary bones, so that when the lower jaw is depressed the maxillae are pulled downward, while the articular surfaces on the anterior inner forks of the maxillae permit the protruding premaxillae to roll downward and forward upon them. Primitive percoids, unlike the berycoids (fig. 9.72), have but one supramaxilla on each maxilla (fig. 9.78B, C). This represents the second or posterior supramaxillary bone of the primitive berycoids and clupeoids, the anterior one having disappeared.

Some of the surface bones, notably the preopercular and the lacrymal, bear more or less conspicuous depressions or pits for enlarged lateral line organs, which are disposed along the same general paths as in other body fishes. The two spikes on the border of the opercular (fig. 9.78E), when present, form the distal ends of strengthening ridges on the inner surface, which converge toward the center of growth of the opercular.

Among the near relatives of the bass are the darters, which are generally referred to the Percidae. These darters, including *Etheostoma, Percina* and many other genera and species, are small, long-bodied perches which have enjoyed an abundant speciation in the fresh-waters of North America. They somewhat resemble gobies but are far less specialized and have normal, not sucker-like, pelvic fins and normal perch-like mouths.

From the ancestral percoid stock came the fresh-water sunfishes (Centrarchidae), in which the number of vertebrae varies from twenty-eight to thirty-five (Jordan and Evermann, 1896, p. 985). Their skulls are essentially bass-like (cf. fig. 9.78B). The adaptive branching of this family has been well studied by Schlaikjer (1937, pp. 9–13).

Thus we observe several gradations between the primitive short, compressed, quick-dodging Centrarchidae and the somewhat more fusiform, large and small-mouthed black bass (*Micropterus,* fig. 9.78B), which with its fossil relatives points the way toward the marine basses or Serranidae (in the broad sense). In a splendid monograph on the latter group, Boulenger showed that the numerous genera all have many features in common, but later systematists have promoted Boulenger's subfamilies to family rank, so that this very natural group has been divided into several "families" (Moronidae, Epinephelidae, Serranidae, etc.)

Beebe and Tee-Van in their excellent "Field Book of the Shore Fishes of Bermuda" (p. 118) note the remarkable unity in general type underlying great diversity of details in this group and for want of an orthodox scientific name to include the "families" Epinephelidae and Serranidae they give the English name "Rockfish" to the entire group. They write:

"The fish included under this title are variously known as Hinds, Jewfish, Groupers, Soapfish and Sea-bass. . . . The name Rockfish is appropriate and the usual haunt of these fish is about the outer reefs, well down below the surface.

"In size they vary from the tiny, two-inch Gramma or Blue and Gold Basslet to the great Jewfish with a record of eight feet and seven hundred pounds. In shape they are remarkably alike, there being a sort of Grouper or Sea-bass type of outline which all the species approximate."

Thus these observant naturalists came to feel the need of an inclusive group beyond the confines of the now commonly used families of serranoid fishes.

A leading character that has been used by systematists to separate the Epinephelidae from the Serranidae is the presence of a single supramaxilla (on each maxilla) in the former and its absence in the latter. But the transitional condition of a reduced supramaxilla is present (fig. 9.76A *smx*) in *Roccus lineatus,* the striped bass of the Serranidae (in the wider sense). This bone is present (fig. 9.78B) also in the large-mouthed centrarchids (Jordan and Evermann, 1896, Part I, p. 985), but has disappeared entirely in the great majority of more advanced spiny-finned families (e.g., figs. 9.78D, E; 9.80–84).

Rather closely related to the serranids and epinephelids is a series of three families, the snappers (Lutianidae), the grunts (Haemulidae, fig. 9.80A), the porgies and breams (Sparidae, fig. 9.80B); these are readily distinguished from the rockfish by the fact that the upper rear corner of the maxilla, instead of being conspicuously outlined in the side view, slips under a prominent oblique fold which is continuous with the lower border of the expanded lacrymal or first suborbital bone. A possible early stage in the slipping if the maxilla under the suborbital fold is seen in one of the true serranids, *Paralabrax humeralis* (Jordan and Evermann, 1896, Pl. CXC, fig. 498). The snout of the lutianids and haemulids (fig. 9.80A) is longer than in the rockfishes and the mouth is consequently directed forward rather than upward (cf. fig. 9.78D). These families, like the true basses, have lost all trace of the primitive supramaxilla. The ascending process of the premaxilla is long and consequently the mouth is highly protrusile. The marginal teeth (on the premaxillae and dentaries) in the more primitive slender-jawed snappers and grunts (fig. 9.80A) are delicate and little differentiated, but in the breams (fig. 9.80B) they culminate in a remarkable dentition, with front teeth like human incisors and crushing cheek teeth, both on the premaxillae and dentaries; the crushing teeth have rounded oval or hemispherical crowns and massive short roots. To support this nipping and crushing dentition the jaws themselves and all the supporting bones of the suspensorium and inner jaws are very massive. The same is true of the cranium, which has to resist the stresses originating in the dental mill, while the very deep compressed body and large strong fins give a firm stance to the feeding fish.

As noted by Regan (1913), the lutianid stock, with concealed upper corners of the maxillae, seems to have given rise to various curiously specialized side branches: (1) The croakers or weak-fishes (Sciaenidae, fig. 9.78E), with large pits for muciferous glands of the lateral-line system on the preopercular, lacrymal and skull roof. The anal fin spines are weak and reduced from the typical three of primitive percoids; the body-form (fig. 9.79) varies widely from that of the massive but fairly normal drums (*Pogonias* and *Aplodinotus*) to the large-headed, tapering *Eques lanceolatus*. In the drums (*Pogonias* and others) the lower pharyngeal bones are united and bear pebble-like crushing teeth, as in the synentognaths, pomacentrids, wrasses (Fig. 9.84B) and other families. In the sciaenids the grinding of this dental mill gives forth croaking sounds that can be heard by fishermen. (2) The mullets or goatfishes (Mullidae) have barbels under their chins and feed on or near the bottom. (3) The mojarras, "shads" (Gerridae, fig. 9.81C), are extremely silvery compressed fishes with very long ascending processes of the premaxilla and very protractile mouths.

Less closely related to the Lutianidae are the African and South American cichlids, which resemble the wrasses in having strong pharyngeal teeth but their lower pharyngeal bones are merely attached by their inner edges or united by suture, not coalesced as in the wrasses (Tate Regan, 1913, p. 130). The presence of closely allied cichlids in South America and Africa has often been cited among other evidence for a former transatlantic land-bridge (see p. 162). Nor is the presence of a possible cichlid (*Priscacara*) in the Eocene of North America necessarily fatal to that evidence.

ANIMATED STONE-CRUSHERS (LABROIDEI)

The demoiselles or Pomacentridae are small-mouthed nibblers (fig. 9.83C) with close-set minute teeth in the jaws and tooth-bearing coalesced pharyngeal bones, but they differ from the wrasses in their vertebral column and rib attachments. The demoiselles are very active little dodgers around coral reefs in many parts of the world. The sergeant-major (*Abudefduf marginatus*) of this family can change its ground color from very dark to gold with dark vertical stripes.

Even in the less specialized of the wrasse series (Labridae) the opposite tooth-bearing lower pharyngeal plates have already fused into a single piece, as they have in several other groups mentioned above. The mouth is sharply protractile (fig. 9.82A), with very long ascending branches of the premaxillae, while the teeth on the premaxillae and dentaries remain simple. This line, represented by *Lachnolaimus* (fig. 9.82B) and *Bodianus,* culminates in *Epibulus* (fig. 9.82C), in which the entire inner upper jawbones (palatine, pterygoid, quadrate) have been reduced to long, loosely-connected bars which can be slung far forward to permit the protruded premaxillae and lower jaw to form a folding tube (C1, C2). The opposite line of specialization is initiated in the Coridae (fig. 9.82D), or slippery-dick wrasses (*Iridio*); here the mouth parts are shortened and vertically deepened; one pair of lower front teeth become like small curved canines, the other marginal teeth remaining small; the stout dentary tends to develop a movable joint with the articular.

In the parrot wrasses (Scaridae, fig. 9.84B) the dental strips of the premaxillae and dentaries have produced innumerable potential tooth buds which have fused into a very strong beak above and below. Meanwhile a movable joint has been developed between the dentary and the articular, while the ligament connecting the maxilla and the ascending process of the dentary has been replaced by a real joint. Thus both the upper and lower jaws of the scarids have been converted into very powerful double-pointed nutcrackers. Not less wonderful is the pharyngeal mill (fig. 9.84B) of these fishes; the lower pharyngeals bearing a median elongate rasp set with coarse transversely oval teeth; these oppose a similar set on the upper pharyngeals, which in turn slide on a special pedicle on the basioccipital. No better example could be found of the close coöperation of the forces that produce new polyisomeres and anisomeres (specially emphasized parts).

A fusion of teeth and bones to form a powerful nipper-like beak occurs also in *Hoplegnathus* (fig. 9.83A), the type of a quite different family of percoids.

DIGRESSION ON COLOR-PATTERNS (Physiology, Field Studies, Evolution)

The skin of groupers and rockfishes (family Serranidae) is abundantly supplied with cells called chromatophores, which are the main elements of the color patterns of fishes. These chromatophores are of several different kinds: melanophores (producing dark brown or black pigment), xanthophores (yellow), erythrophores (red), leucophores (white). Besides the chromatophores there are structures called iridocytes, which consist of clusters of semi-crystalline plates, the tints of which are of jewel-like brilliancy. According to G. H. Parker (1943b, p. 207), these structures [called by him, chromatosomes] "are usually surrounded more or less completely by melanophores, whose activities may expose or cover the glistening bodies. . . . They are concerned chiefly with the bluish and greenish tints of fishes, whereas the other types of color cells have to do with the red, yellow and gray tones."

Embryology and Heredity.—According to H. B. Goodrich (1935, p. 270), studies on *Fundulus* by Stockard (1915) have shown

". . . that chromatophores in fish arise from wandering mesenchyme cells that have their origin in the region of the closing blastopore and the mesenchyme cell mass lying within the embryo. In *Fundulus* unpigmented but recognizable cells migrate by amoeboid motion from these areas and later assume pigment. In *Oryzias* Mendelian phenotypes may be distinguished when this pigment first appears. The evidence from *Fundulus* indicates that the prospective fate of such cells is determined before the pigment is formed. We are therefore dealing with a case of 'Chemo-differentiation' (Huxley, 1924). It then seems reasonable to advance the hypothesis that an embryonic segregation of melanophore-producing cells has occurred and these cells migrate to various areas. In variegated fish [such as the variegated Medaka *Oryzias latipes*, fig. 9.85, 3] there has evidently been a further segregation of cells which have the power to produce the full amount of melanin from those capable of forming only a small amount of that pigment. These cells then migrate and by their aggregations mark the fish with the irregular, apparently chance patterns found in these types."

In the same comprehensive paper on hereditary color patterns in fish, Goodrich (p. 274) concludes that:

". . . this survey of conditions found in fish may well serve to illustrate the complexity and variety of formative processes that may operate in color pattern formation. In *Oryzias* [order Microcyprini] there may be embryonic segregation of cells having different capacities of melanin formation followed by cell migration (fig. 9.85, 3). In *Carassius* [order Ostariophysi, family Cyprinidae] there is the partial destruction of a uniform color [leaving more or less of a color pattern, fig. 9.85, 2]; in *Platypoecilus* [order Microcyprini] there occurs the formation of two different types of melanophores and their accumulation in definite areas in certain types [fig. 9.85, 4]; in *Lebistes* [order Microcyprini] there is a similar segregation of varied types of chromatophores in definite predetermined areas [fig. 9.85, 5] and in *Betta* [order Labyrinthici] there occurs the formation of crystals refracting different colors. In all the above-mentioned cases, except in *Brachydanio* [order Ostariophysi, family Cyprinidae], there is evidence that the color patterns are Mendelian characters."

Gordon (in a long series of papers, 1926 to date) has shown that at least in some of the Microcyprini certain sexual differences in color patterns are sex-linked.

In his paper on "Chromatophores in relation to genetic and specific distinction" Goodrich (1939) suggested that chromatophore cells "may show recognizable characteristics correlated with the taxonomic position of the species"; for example, among several species of wrasses of the genus *Sparisoma,* "all show the disc type of chromatophore; in the closely allied family of Scaridae *Scarus vetula* has a 'branching' type, while the related *Pseudoscarus* has very elongate processes."

From what precedes it is known that the color patterns of fishes are partly controlled by hereditary influences which operate especially during the developmental stages and that at least in some cases certain types of chromatophores predominate in related species of the same genus. In many fishes also the larvae and young stages are quite differently patterned from the adults, which attain their adult patterns partly under the influence of hormones which are associated with the ripening of the gonads.

Physiology.—The typical melanophore of the perch is a contractile cell surrounded by a cluster of nerve fibres; the latter include two opposing sets, one serving to dilate the chromatophore, the other to contract it. These fibres, like those for the heart muscle, are from the autonomic system (G. H. Parker, 1943b, p. 211). Contraction of the melanophore, with shrinking of the pigmented areas, thereby causes a blanching of the skin, while expansion of the melanophore spreads the pigment and causes darkening of the skin.

The activities of the melanophores of teleosts are also stimulated by a special group of hormones called chromatic neurohumors (Parker, *op. cit.,* p. 215): (a) *intermedine,* from the intermediate lobe of the pituitary gland, is carried in the blood to the melanophores, where it causes expansion with darkening of the skin; (b) *acetylcholine,* a neurohumor found in extremely small quantities in the skin of catfishes and others, also disperses melanophore pigment and thereby darkens the skin; (c) *adrenaline,* from the chromaffine tissue, scattered or segregated in the tissues of lower vertebrates; it induces the concentration of melanophore pigment with consequent blanching of the animal (G. H. Parker, *op. cit.,* pp. 216–218).

Fishes of many families and orders are able to change their colors so as to make themselves inconspicuous. Here the eyes of the fish play an important part. The complexity of the melanophore system of the eel is indicated in the diagram (fig. 9.86A) adapted from Parker (1943b, p. 220).

In the eel (*Anguilla*) the outer field (fig. 9.86A) is represented at the right, the central nervous connections in the center and the melanophores at the left. The reflex arcs are numbered and the special parts of the diagram are lettered. 1, Retino-pituitary arc; 2, Retino-cholinergic arc; 3, Retino-adrenergic arc; 4, Retino-tuberal arc. Abbreviations; A, adrenergic nerve-fibers; B, blood and lymph; (B) black background; C, cholinergic nerve-fibers; CM, melanophore with concentrated pigment; CNO, central nervous organs; D, beam of direct light; DM, melanophore with dispersed pigment; DR, dorsal retina; L, source of light; P, pituitary gland; S, beam of scattered light; VR, ventral retina; W, white background.

Here as in other fishes the dorsal retina is especially concerned with contraction of the melanophores and blanching of the skin; the ventral retina with the expansion of the melanophores and darkening.

In the catfish (*Ameiurus*) the arrangement of the diagram is the same. In addition to arcs 1, 2, and 3, all of which are excitatory, there are in *Ameiurus* arc 5, the dermo-intermediate excitatory, and arc 6, retino-intermediate inhibitory. The lettering is the same, with the addition of I, integument.

Parker notes (1943a, p. 207) that "the common catfish, possessing only melanophores, has a limited color range from a pale greenish yellow to a coal black. In the killifish [*Fundulus, op. cit.,* p. 205] in addition to melanophores there are xanthophores, leucophores and green-blue chromatosomes. These fishes change not only from dark to pale ". . . but they may assume yellow, green, blue or even red tints. The red coloration is due to an enlargement of the integumentary blood vessels, a true blush" (p. 205).

The color changes in tropical fishes, according to Parker (*idem*): "are like those in the killifish except that they are much more exaggerated. In some of these fishes even the general pattern seems open to change, almost to obliteration, as in the Nassau grouper."

Field Studies.—W. H. Longley, who spent a quarter of a century in diving at Tortugas and made careful notes of the color patterns and changes of fishes living in their natural environments, notes (1941, p. 95) that in the Nassau groupers (*Epinephelus striatus*), in addition to the ordinary changes in coloration response to change in color or shade of the environment, or to change from rest to active motion, one fish "was observed repeatedly displaying, when a red grouper came near it, a phase never seen in any other Nassau grouper. This distinctive phase appeared five or six times and passed in perhaps a minute at each showing. The stripe through the eye, instead of being one of the darkest, became one of the lightest on the body, and the color on the side above the level of the pectoral was abruptly replaced by white, on which were only a few scattered dark marks." This probably means, in the light of the data set forth by Parker, that the sight of the fish of another species received through part of the retina caused a sharp contractile spasm of the melanophores in the stripe through the eye and on the sides of the body. In other words, this display of sudden blanching was an indication of a particular emotional reaction of the fish. As a result of their observations, Beebe and Tee-Van (1933, p. 118) state that among the rockfish family "there is not only a great diversity of color but individual color change is very marked, spots, bands and flat ground colors appearing and disappearing with the changing emotions of the fish. . . ."

Evolution.—The literature relating to the coloration of fishes is very extensive and there are many important papers by experimentalists, field observers and those who are interested in the survival values of color patterns of animals. When experimental analysis is the sole aim, recent investigators seem to be concerned almost exclusively in what happens now and what causes it to happen now and they refrain from speculation about the possible stages in evolution which led to the observed facts in the species under consideration. Even those who seek to determine whether and in what way particularly striking patterns may be useful to their owners do not seem to be interested in what may be called the phylogeny of color patterns, while those systematists who might be interested in undertaking general reviews rarely have an opportunity to pause from the urgent business of fixing the limits of species and subspecies from particular localities. I have therefore been compelled to make a belated and perhaps futile attempt to sketch a preliminary outline of the broader stages of evolution of coloration in fishes, or at least to suggest some of the questions that may have to be answered along the road.

QUESTION 1: Since all chromatophores and chromatosomes are part of the skin, how are they related to the scales, and what was their remote origin?

According to the data summarized by Bridge (Cambridge Natural History, 1910, Section on Fishes, Chapter V, p. 183): "In the great majority of fishes the skin becomes the seat of calcareous deposit, and gives rise to such diverse exoskeletal structures as the varied forms of spines and scales. . . ." The structures are probably "the most ancient form of Vertebrate skeleton owing its existence to the presence of lime salts in the tissues of the body. . . ." A vertical section through the skin of an embryo shark (Bridge, *op. cit.*, p. 185) shows the shagreen denticle being formed as an outgrowth of the dermis or deep layer; it is covered by a layer of epidermis. Its projecting papilla or nutritive pulp cavity is the main agent in depositing the calcium carbonate of the denticle.

Bridge also states (pp. 166, 167): "The coloration of fishes is due to the presence in the dermic portion of the skin of (a) special pigment-containing cells (. . . chromatophores), and (b) a peculiar reflecting tissue composed of iridocytes [referred to above, under "chromatosomes," p. 184]. . . . Iridocytes consist of guanin, which, in its chemical reactions, closely resembles the guanin obtained from guano, and therefore is to be regarded as a further illustration of the utilisation of waste excretion products for the production of colour in animals." From what precedes it may be suggested that both scales and chromatophores are at least to some extent by-products of the excretory and circulatory systems and that their values as parts of a protective exoskeleton and as a means of disguise or warning were added to their primary functions in the process of calcium metabolism. A somewhat similar conclusion was suggested above (p. 41) with regard to the exoskeleton and color patterns of molluscs.

QUESTION 2: What kinds of coloration were displayed by the older groups of fishes, especially (a) ostracoderms and cyclostomes, (b) sharks, (c) ganoids?

We have noted above that beneath the scales the skin is supplied with blood vessels, by means of which the stimulating adrenalin is conveyed to the chromatophores.

(a) *Ostracoderms and cyclostomes:* Among the Palaeozoic ostracoderms, which are at least near to the basic vertebrates, the trunk-scales of the cephalaspid *Hemicyclaspis* (fig. 6.17), as described by Stensiö (1932, pp. 60, 61) were developed on the corium (dermis) and consist of a basal layer, a middle layer and a superficial layer. "The basal layer is always well developed and like that in the cephalic shield is perforated by ascending vascular canals. . . . The middle layer is likewise rich in canals and radiating canals are very well developed in it. . . . Finally, the superficial layer is present only on the exposed areas of the scales." The pigment in the scales of ostracoderms is not preserved, but presumably it was abundant in what Stensiö calls the "superficial layer" of the corium, which, as noted above, was developed only on the exposed parts of the scale, somewhat as it is in the scales of the modern salmon. In *Hemicyclaspis* there was also a mucous canal system in both the head shield and the trunk-scales.

In the modern cyclostomes, which are believed to be the highly specialized descendants of the ostracoderm class, the mucous-producing glands have become extremely abundant but they do not prevent the production of varied pigments. The marine lampreys are more or less silvery but when they come up the rivers to spawn, various species develop dark brown, black, bluish and silvery spots and colors in the skin.

(b) *Elasmobranchs:* According to Parker (1943a, pp. 214, 215), no evidence of chromatic nerves has been found in a considerable number of sharks and rays (except *Mustelus canis*), their color changes being probably effected chiefly by the action of certain neurohumors. In the smooth dogfish (*Mustelus canis*) the neurohumor selachine (derived from nerves of the skin) is believed to cause a contraction of the melanophores, while dispersal is probably effected by a pituitary neurohumor (*op. cit.*, p. 219).

In oceanic sharks the skin may be gray on the back, white below, but *Prionace glauca* wears the deep blue livery of the Gulf Stream. The carpet sharks are so named from their florid, dappled brown colors but *Stegostoma tigrinum* of this family displays dark transverse bands or spots on a brownish-yellow ground and is therefore called tiger- or zebra-shark (Boulenger, 1910, p. 447). Some of the catsharks (Catulidae), especially those living among coral reefs, bear large brown spots on a light field. A strange and unique pattern is that of the whale shark *Rhineodon,* with its transverse and longitudinal rows of sharp white spots and lines on a dark ground.

The skates and rays are dark above and white below and a few display light spots or circles on their backs. In the modern chimaeroids, which are a highly peculiar deep-water side branch of the sharks, the scaleless skin abounds in silver, blue, black and white pigments, sometimes with large brown blotches, but there are no very elaborate or sharply defined patterns.

Therefore, although the elasmobranchs are by no means directly ancestral to the bony fishes, their coloration may have remained in a relatively primitive stage.

(c) *Ganoids:* The coloration of the palaeoniscoids, which were basic teleostomes, is unknown, but their descendants, the modern sturgeons, are rather inconspicuously colored. The spoonbills (Polyodontidae) are gray or dark above, light below, and therefore not unlike sharks. In *Polypterus ornatipinnis* of the Congo basin, as figured from the living fish by Dr. James Chapin (Nichols and Griscom, 1917, pl. XXXIII, fig. 3), the body is dark greenish but the fins bear many dark bands on pale purple-red fin-webs. The fresh-water garpikes (*Lepidosteus*) are more or less greenish-silvery with brown blotches, especially in the breeding season. The bowfin *Amia,* which in its general anatomy is on the threshold of the teleosts, is "dark olive, the sides with greenish reticulations, the belly whitish; dark round spots on the lower jaw and gular plate. The male has a roundish black spot with an orange border at the base of the caudal fin" (Bean, 1903, p. 75).

Thus the existing ganoids, without developing any very bizarre designs, approach the teleosts in their general color patterns. Comparative experimental studies of their chromatic apparatus would be highly desirable, especially in order to test the assumptions implied by Parker (1943b, pp. 220–222) that probably the sharks represent a more primitive phase and that the chromatic apparatus of the eel is relatively more primitive than that of the catfish. But since these two fish are extremely specialized in quite different ways, it would be helpful if each of them could be studied against a wider taxonomic and experimental background.

QUESTION 3: How do the ranges of color patterns of the existing (a) isospondyls and (b) Ostariophysi compare with each other? Are they on the whole more advanced than those of the surviving ganoids?

(a) *Isospondyli:* The herrings (Clupeidae) and allied families, the lady-fishes (Albulidae) and the tarpons (Elopidae) are mostly oceanic and shore fishes, in which silvery, blue and green colors predominate; in some of the dwarf herrings (Dussumieridae) there is a silvery lateral stripe; the typical whitefish (Coregonidae) is also silvery. The silvery marine Salmonidae make long journeys from the sea up the rivers to their high breeding-grounds in the streams and lakes; they often display a profusion of brilliant colors, especially in the breeding season; but their brightest color spots are more or less scattered and are seldom grouped into well defined bands, stripes or sharply contrasting "cut out" areas. The graylings (Thymallidae) of the cold streams of the north have high spotted and striped dorsal fins.

Of the deep-sea to abyssal isospondyls, the more primitive ones are silvery; the stomiatids are often black, or at least very dark. On the whole the isospondyls, although doubtless possessing all standard kinds of melanophores, do not differ profoundly in their more usual types of coloration from the few surviving ganoids and, though their color patterns exhibit a wider range, that is probably because the isospondyls now far outnumber the ganoids in families, genera and species, as well as in the types of habitat from the high mountain streams to the abyssal depths of the ocean.

(b) *Ostariophysi:* The teeming hosts of the Ostariophysi dominate in the fresh-waters so that "oceanic" colors of silvery white and bluish-green give place in the carps and suckers to silvery yellow and olive; some of the minnows are reddish, with red-tipped fins; lateral stripes are conspicuous in *Brachydanio* (fig. 9.85.6). The catfishes are typically dark but the few marine ones are silvery. One of the characins, the *Dorado* (*Serrasalmo*) is a golden yellow; the characins also include some with longitudinal stripes (*Phago*), some with cross bands (*Distichodus*), both converging toward different percoids; some of the smaller characins (e.g., the "Neon tetra") are exceedingly brilliant.

QUESTION 4: Do the so-called intermediate groups supply a transition to the percoids?

Of the "intermediate" orders (including

Haplomi, Iniomi, Microcyprini) the only representative that seems to call for special notice here is the killifish (*Fundulus*), which, although well supplied with the usual types of chromatophores (Parker, 1943a, p. 205), doubtfully lacks a dispersing pituitary neurohumor (Parker, 1943b, p. 219).

It seems highly probable that the leptolepids (fig. 9.23B) and herrings (fig. 9.29B, C), including the Cretaceous *Ctenothrissa* (fig. 9.73), stand relatively near to the ancestry of all the orders of teleosts, and further that the predominately marine percomorphs were derived from the berycoid stem (fig. 9.72); a silvery bluish-green and white coloration, with darker sides and light belly, may therefore be tentatively assigned to the remote marine ancestors of the percoid group.

The fresh-water percopsids (fig. 9.71) and centrarchids (fig. 9.77B) may be regarded as more or less successive offshoots from primitive leptolepid isospondyls, which came in from the sea and ascended the rivers to breed, some of their descendants remaining in inland waters to compete with the dominant Ostariophysi. Meanwhile the "swivel chair" or quick dodging ancestors of the berycoids and primitive serranids either gave up the inland journey or never made it, but moved further inshore toward the rocks and coral reefs, which abound in invertebrates for food but are badly lashed in heavy storms.

QUESTION 5: Are there any observable correlations between systematic groups and types of coloration, especially among the percoids?

The plates and text of Dr. W. H. Longley's great work (1941), together with many other valuable sources and some personal experience, suggest that there could be several correct partial answers to the foregoing question. In such a high percoid type as the Nassau grouper (*Epinephelus striatus*) the color apparatus is so complex that it may be likened to a concert organ, with many stops. The performer in this case will play widely different color tunes, according to whether he is asleep or awake, standing still or in motion, resting over a light bottom or in the deep shade, dashing after food or suddenly confronted by a red grouper, which apparently he fears. Nevertheless the patterns which are ordinarily produced by this species are not too numerous to have been summarized by Dr. Longley.

In the red grouper (*Epinephelus morio*) the general type of color apparatus may be the same, but dark reddish brown is its most common phase, even though this is "one of the most changeable of fishes." In spite of the changeability of this group, Longley, in describing the coloration of *Epinephelus adscensionis,* states (p. 93): "The elements of a color pattern common among the species of its genus are visible in this one. An oblique line from snout through eye to posterior margin of the preopercle; . . ." etc. So that even in this genus a basic generic pattern may be dimly discerned amid wide specific and individual differences.

A very different-looking set of polyisomeres is displayed by the yellow-fin grouper (*Mycteroperca venenosa*) also of the same family (Serranidae). In one of its dark color phases (Longley, pl. 6, fig. 1) the sides abound in large spots; in a certain light phase (*ibid.*, pl. 7, fig. 2) these spots break up into leopard-like clusters. In *Mycteroperca tigris* of the same genus "the fish was changeable in shade but showed always the same pattern. Entire side of head covered by netted light lines, surrounding darker spots of the average size of the pupil . . . nine light bars, narrower than the brown interspaces, crossing dorsal surface . . ." etc. In the spotted jewfish (*Promicrops itaiara*) the somewhat blotched dark color seems to be less specialized than that of its relative the black jewfish (*Garrupa nigrita*). In wide contrast with these is the golden grouper (*Mycteroperca pardalis* Gilbert) of the Galapagos Islands.

While many serranids display transverse bands, the striped bass (*Roccus lineatus*) bears parallel longitudinal stripes; its relative the white perch (*Morone americana*) is silvery, with faint longitudinal stripes, especially in the young.

Eighteen genera, including thirty-six species, of Serranidae that occur at Tortugas were studied by Dr. Longley and it is evident that, while definitions of the color patterns could be framed for a few of the genera, it would be very difficult to make an effective definition for the coloration of the family as a whole, except to indicate its wide range and numerous adjustments to different shades of the environment and to different emotional states.

These are only a few of the wide range of patterns in the Serranidae, but it is probable that if a review of the entire family could be made, especially in connection with Boulenger (1895), and if furthermore his beautiful figures of the skulls could be compared at the same time, it would then be feasible to make at least a tentative outline of the evolution of adult serranid color patterns.

Color plates of six species of the lutianid genus *Neomaenis* are figured by B. W. Evermann in his work (1902, pls. 17–22) on the fishes of Porto Rico. In four of these (*N. griseus, N. jocu, N. apodus, N. analis*) the color patterns are evidently related and pass from a faint suggestion of vertical bands (*N. griseus*) to well marked bands (*N. apodus*). Here is almost a generic pattern. In the sixth species, *N. synagris,* however, only a trace of the bands remains and thin longitudinal strips are present. One of the latter is greatly emphasized in *Ocyurus chrysurus* (pl. 23), which, as noted by Evermann (*op. cit.,* p. 180), is allied to *Neomaenis* on structural evidence. The large horizontal stripe in *Ocyurus* may serve as a recognition mark in this schooling fish.

The grunts (Haemulidae) are closely related to the Lutianidae and even in their patterns (see the excellent brief descriptions and diagrams by Beebe and Tee-Van, 1933, pp. 151, 152) some of them distinctly suggest this relationship. Among the typical grunts curved oblique or reticulate arrangements of the scales on the sides are often evident, overlaid by more or less regular longitudinal stripes. In the porkfish (*Anisotremus*) the horizontal stripes on the sides run up against a most conspicuous vertical band running from the nape downward, while another dark oblique band runs from the forehead down through the eyes.

The wide variability of pattern and color in certain percoid fishes is shown in Beebe's (1943) description of seven pattern and color phases in a single individual fish, a Venezuelan cichlid, *Aequidens tetramerus,* all observed and recorded within twenty-four hours (fig. 9.87). The "typical diurnal pattern" includes a conspicuous lateral ocellus and five or six narrow longitudinal stripes of greenish-brown, alternating with equally narrow stripes of grass green. At night one extreme stage (pl. 1, fig. 5) presents seven transverse bands, no longitudinal stripe. Intermediate patterns combine longitudinal and several vertical bands, only the ocellus remaining conspicuous.

In short, the species of percoid fishes parallel and converge toward those of many other groups, so that a classification based on color patterns or range of color patterns alone would often include members of quite different orders. Within a given genus related species often show related color patterns and if many more cases were studied from a phylogenetic-taxonomic angle, the combined lines of evidence might well yield some of the probable stages in the evolution of particular color patterns, just as the embryologic and experimental data have already shown exactly how certain color patterns are developed and are variously conditioned by genetic factors, by neurohumors and by direct nervous response.

We shall return to this theme later, especially in dealing with the color patterns of angel- and butterfly-fishes. Meanwhile it may be noted that among percoids as well as in other groups, adult color patterns include, among others, the following categories:

1) Polyisomerous: more or less uniform coloration in which one of several kinds of polyisomeres (e.g., melanophores) predominate; e.g., dark or silvery, often grading into

2) Poly-anisomerous: with uneven or emphasized repetitive parts.

The latter in turn include:

a) *Banded:* more or less dark vertical bands, alternating with blanched areas;

b) *Striped:* longitudinal dark stripes alternating with blanched or contrasting areas or rims;

c) *Concentric:* concentric bands which seem to originate in given centers and to spread outward like successive waves;

d) *Spots:* often resulting from further blanching of (a) or (b), sometimes giving rise to ocelli;

e) *Aggregations:* due to the shifting and coalescence of formerly separate dark areas;

f) *Residuals,* or remains of an earlier phase;

g) *Mixed* or *confused,* containing remnants of larval, juvenile, or nuptual colors.

Doubtless the foregoing categories are not all mutually exclusive.

THE ARCHER FISH (TOXOTES)

One of the most curious of all percoids is the archer fish (*Toxotes*) of Siam, which shoots living insects with a squirt of water. All doubts and scepticism as to the achievements of this fish have been cleared up in an article by an eminent ichthyologist, Dr. Hugh M. Smith (1936). The archer fish (fig. 9.88) can and does shoot insects that may be hovering over the water or spiders hanging from a branch, and it can shoot very accurately up to three and a half to four feet. The "pellet" of water is strongly ejected from a long median groove on the palate (fig. 9.89), against which the tongue and jaws forcefully close as the fish sticks its jaws out of the water. The vision of the *Toxotes* is very keen and its eyes enable it to gauge accurately the distance, size, etc.

As to the steps which may have led to this highly specialized habit, it is probably significant that the related, less specialized, perch-like Nandidae, which are allied to *Toxotes* by many skeletal characters, likewise have a very protractile mouth and are stated (Stoye, 1935, p. 244) to be actively carnivorous. We may thus suppose that prior to the development of "sharp-shooting" the ancestral archer fish simply rushed at the insect prey and knocked it over with a spasmodically ejected splash of water. Among the necessary conditions for the foregoing stages were possibly the following: (1) such changes in the oral valves (which are flaps of skin dependent from the upper and lower jaw) as would favor the ejection of the pellet of water without hindrance or even with added force; (2) since the sides of the palatal tube (fig. 9.89) diverge slightly toward their rear ends, there may be an analogy with certain types of high velocity gun barrels in which the diameter of the tube diminishes slightly toward the muzzle, thus compressing the enclosed gas and greatly increasing the velocity of the projectile. Gradually the fish learned to stop suddenly (by extending its fins) and to throw the splash of water further ahead of it. Then would follow the gradual reduction of the quantity of water splashed, the narrowing of the median groove through which the water was ejected, the strengthening of the contracting muscles of the throat and opercular region, the improvement of the tip of the tongue, which acts as a trigger-valve for releasing the "pellet" of water when the pressure reaches the right level; with all this came the gradual lengthening of the effective range (H. M. Smith, *op. cit.*).

When the two eyes of the fish are turned directly toward the prey, the median groove on the roof of the mouth also points toward the same spot, so that correct aiming is secured merely by pointing straight toward the prey.

The skeleton of *Toxotes* (fig. 9.90) indicates great strength both of the lateral muscles which drive the body forward and of the fins which check or steer the flight of this living arrow. That the fish has become highly skilled in manoeuvering and in adjusting its aim is well attested.

Such perfection of fitness of means to end, in view of cumulative evidence of genetics, has been the result of millions of years of natural selection in which profitable genetic variations affecting the eyes, mouth, palate and nervous reactions of the fish have been gradually assembled and intensified under the stimulus of higher survival value for better and better marksmanship. Fortunately for the present inquiry, the existing percoid fishes of the family Nandidae supply good structural links between the archer fish and such primitive percoids as the Centrarchidae, or fresh-water basses.

THE DREADED BARRACUDA AND HIS PEACEFUL RELATIVES (PERCESOCES)

The pike is represented in fish lore as an exceptionally bold and ferocious fish and a glance at his yawning trap of a mouth would suggest that the following tribute to his villainy is not undeserved: "Look at any member of the Pike Family, and tell me whether it does not make you think of a pirate. Observe that yawning sepulchre of a mouth, that evil eye and low, flat forehead—all indicating a character replete with cunning and ferocity." (Hornaday's American Natural History, vol. IV, p. 214.)

But if the pike deserves this opprobrium, what shall we say of his marine counterpart, the barracuda, whose reputation for evil deeds far surpasses that of the pike? Things are not always as they seem, however, and Beebe and Tee-Van (1933) say this: "They [barracudas] share with sharks the reputation of being dangerous to bathers, but we have had no unpleasant experiences although they are not uncommon in many places where we swim and dive and often come quite close to a diver." Dr. Gudger (1918, 1928), on the other hand, quotes much irrefutable evidence that the barracuda has inflicted terrible wounds on people. Probably the great majority of bathers will continue to give the barracuda the benefit of the doubt.

From an engineering viewpoint the construction of the skull, jaws and teeth of the barracuda (fig. 9.91) indicate that this fish has a highly efficient equipment for swift pursuit and deadly attack. The strong and well-based jaws are armed with many sharp-pointed teeth, of which the larger ones have very sharp-cutting edges. A peculiar feature is that at the front end of the jaw the teeth are not alike on opposite sides; one of the largest lower teeth is received into a pit at the tip of an inter-premaxillary, more or less dermal element, which represents a fusion of the opposite ascending branches of the premaxillae.

Much that has been said above (pp. 143, 154) about the mechanism of the fish skull is evident also in the construction of the upper and lower jaws, hyomandibular, opercular region and neurocranium of the barracuda. The inverted V-truss that supports the lower jaw is especially evident in this skull; its apex being at the quadrate-articular joints, its front limb (the palato-pterygoid tract) based on the parethmoid buttress, its rear limb, consisting of the symplectic, metapterygoid and hyomandibula, being movably based in a longitudinal groove borne by the sphenotic and pterotic bones.

An outstanding character of the *Sphyraena* skull, which it shares with its relatives, the silversides (Atherinidae) and the mullets (Mugilidae), is the presence on the posterior corner of the epiotic bones of long, backwardly-developed processes (fig. 9.92) which are subdivided into branches or even into delicate bristles toward the rear ends, extending well behind and medial to the posttemporal bones of the shoulder. These peculiar polyisomeres, which afford a diagnostic character for the group, probably serve to tie the dorsal myomeres to the skull.

While the sphyraenids have a very long body and narrow skull with long pointed rostrum, the typical mullets (mugilids) have a stout body, a short wide skull and a very large wide rostrum (fig. 9.92C). The mouth is small and the jaws short but strong. The mullets grub in the mud for vegetable matter and have a tough gizzard-like organ in their throats. Their pectoral fins are set high up on the side of the body, but this and other points of resemblance to the flying-fishes and skipjacks are regarded by Tate Regan (1910) as being convergent characters of no value in indicating the ancestry of this group.

Highly placed pectoral fins are found also in the silversides or whitebait (Atherinidae), which are small elongate fishes with rather small upturned mouths (fig. 9.93A). In general body-form and skeletal construction the silversides seem to be intermediate between the very narrow-skulled barracudas and the wide-skulled mullets and no one has doubted that these three families belong to a natural subordinal or ordinal group. But what was the derivation of this group?

The ordinal or subordinal name Percesoces, meaning perch-pikes, suggests either that these forms are intermediate between perches and pikes or that they are pike-like perches or possibly the reverse. A suggestion of annectant characters is given by the position of the pelvic fins, which are subabdominal in most of the Percesoces, as in the Microcyprini and Synentognathi. In *Sphyraena,* however, they vary from almost the midpoint of the ventral contour in *Sphyraena borealis* to a more sub-pectoral position in *Sphyraena barracuda.* Dollo (1909) indeed suggested that the subabdominal position of the pelvic fins in *Sphyraena* is a convergence toward the isospondyl type, implying a secondary backward displacement of the pelvic fins. This inference is supported by the following facts: (a) that in *Sphyraena ideastes* I found the muscular cone of the pelvic fins continued forward into a long ligament which was inserted into the cleithral symphysis; (b) that in the relatively primitive family of the silversides the pelvic bones are connected by ligament with the cleithral symphysis as in typical spiny-finned fishes (Boulenger, 1910). The suggestion that the Percesoces have been derived from the percoid family of the cardinal fishes (Cheilodipteridae or Apogonidae) was carefully analysed from a comparison of the skeletal features by Starks (1899) but rejected by him on the ground of numerous differences between these families.

The otoliths of the Mugilidae and Sphyraenidae as described by Frost (1929, pp. 120–129) resemble in general the serranid and percid types and Tate Regan classifies the group (under the name Mugiloidea) as a division of the Percomorphi.

In seeking for their nearer relationships, it seems that in view of the retention of a conspicuous supramaxilla in *Sphyraena* (fig. 9.91A *smx*) and of the fact that the hinder border of the maxilla does not slip under an extension from the lacrymal, we can exclude from consideration the Lutianidae and all their diverse relatives (Haemulidae, Sparidae, etc.), while this maxillary region is a conspicuous point of resemblance to the basses (fig. 9.93C) and perches.

CLIMBING PERCH, SNAKEHEADS AND BUBBLE-MAKERS (LABYRINTHICI)

Although the climbing perch (fig. 9.94A) is said to be able to climb trees by using its spinose operculars, dorsal and anal fins as holdfasts, it presumably could not live very long out of water were it not for its remarkable accessory respiratory organ, whose labyrinthine folds doubtless give it a relatively great contact with the oxygen-containing moisture. This labyrinth is located in an accessory chamber dorsal to the gill chamber and medial to the wide hyomandibular and opercular regions and has been derived from the embryonic Anlage of one or more of the gill folds.

The climbing perch has a quite normal-looking percoid skull and the evidence of its otoliths (cited by Frost, 1928, p. 330) supports Tate Regan in referring it to the percomorph series. *Luciocephalus* of this group is more elongate and pike-like.

The snakeheads (*Ophiocephalus,* fig. 9.94B) of the fresh-waters of Africa, Southern Asia and the East Indies are appropriately named. They are regarded as an elongate snake-headed derivative of the climbing perch stock. An apparently related family includes the fighting bubble-makers, *Betta pugnax* (fig. 9.94C) and its allies, which also have a suprabranchial respiratory organ. These have a widely protrusile small mouth and skull of modified percoid type.

FISHES WITH THREE PAIRS OF APPENDAGES (POLYNEMOIDEI)

Doubtfully related to the labyrinthine fishes but undoubtedly of general percoid derivation are the marine threadfins (Polynemidae). Their skull characters (fig. 9.95D) seem consistent with derivation from the labyrinthine stem but the mouth and jaws have become secondarily enlarged. The unique feature to which the name *Polynemus* (many threads) refers is the fact that a few of the lower rays of the pectoral fins have separated from the main part of the fin and have moved down on to the lower limb of the coracoid, where they may function as feelers. Brazier Howell (1933) refers to them as if they represented another pair of appendages, otherwise not preserved in vertebrates, but in view of the general skull characters, which place the fish as a percoid teleost, there is good reason to infer that these structures are new polyisomeres produced within this small group of derivatives of the percoid stem. Somewhat analogous structures are the anterior rays of the pectoral fin of the gurnards (fig. 144B), which are used for crawling in a vaguely insect-like manner, but they should hardly be taken to indicate that the vertebrate stock once had insect-like appendages!

THE JOHN DORY (ST. PETER'S FISH) AND THE RHOMBOID BOARFISH (ZEOMORPHI)

The etymologist no less than the ichthyologist may well find in the John Dory (fig. 9.96A) a subject worthy of investigation. Are the dictionaries right in deriving the word from "John + Fr. dorée"? Is it, really, a "gilded Johnny"? And surely this fish bears no resemblance to a fisherman's flat-bottomed dory? But why did Linnaeus give him such a godlike name as *Zeus faber* (Zeus creator)? Again, was it an humble fisherman or a cloistered compiler of a mediaeval bestiary that selected this fish as the one from whose mouth St. Peter extracted the tribute money? Was it because the alleged mark of Peter's thumb and index finger can still be seen as a conspicuous spot on either side of the fish (fig. 9.96A)?

On the scientific side, the John Dory has had a somewhat checkered career even in recent times. Thus Smith Woodward referred it to the berycomorph division of the spiny-finned fishes. Boulenger (1910, p. 682) held that *Zeus* was related to the Eocene *Amphistium* and indirectly to the flatfishes (Heterosomata), with which he united it in the division Zeörhombi. Tate Regan (1929, vol. IX, p. 320) grouped the Zeidae with the Caproidae (fig. 9.96B), or boarfishes, in an order Zeomorphi, coordinate in rank with, and placed between, the Berycomorphi and the Percomorphi. Starks (1898) after a careful analysis of its skeletal features concluded that the Zeidae showed many important points in common with the butterfly fishes (Chaetodontidae). After renewed comparisons of the skull of *Zeus* with those of the others mentioned above, it seems to me that its relationship to the chaetodonts is but indirect and distant. The proportions of its jaws, opercular, preopercular and hyomandibula are notably different also from those of the boarfishes (fig. 9.96B, C), as well as the general contour of the face, the development of the supraoccipital crest and the number of the vertebrae (31–46 in the Zeidae, 22 in the Caproidae). Nevertheless *Zeus* agrees with the boarfishes in several significant points noted by Regan, including the rigid connection of the posttemporal bone with the skull. According to Regan, the group Zeomorphi has probably been derived from the Cretaceous berycoids, and has inherited from them many features of the skull and jaws, while becoming specialized in others.

The boarfishes (Caproidae, fig. 9.97B, C) somewhat resemble the chaetodonts (fig. 9.97D) in general appearance but differ widely from them in many details, including the shape of the maxilla and the greater emphasis of the anterior part of the spinous dorsal and of its bony supports. On the other hand, Starks (1902) cites several characters in which *Antigonia* (C) resembles the chaetodonts and the Ephippidae.

NIBBLING "ANGELS" AND PUGNACIOUS "BUTTERFLIES" (CHAETODONTOIDEI)

This series may have begun with the quite primitive *Scorpis* (fig. 9.98A), which retains such ancestral percoid characters as twenty-four vertebrae and pelvic fins with one spine and five soft rays. Such a type affords a favorable starting-point for the entire group. Let us review first the more specialized forms and work backward to the more primitive ones. An excessively deep-bodied branch culminates in *Psettus,* in which the depth is much greater than the length and the pelvic fins have disappeared. A related but less specialized form is *Platax;* here also the excessive deepening of the body is reflected in the extreme shortening of the skull. Less specialized than either of the foregoing are the spadefishes (C, C1) of the family Ephippidae. These, although more or less circular, have retained most of the features of a normal percoid fish; among their most conspicuous peculiarities are the great size and inflation of the supraoccipital crest, the lesser inflation of the supratemporal and first interhaemal bones and the appearance of many pits for muciferous glands on the thickened frontals.

The spadefishes and *Platax* have numerous close-set pencil-like teeth and small mouths. In the angel-fishes (D) the teeth become still more numerous and bristle-like; the preopercular is often margined with spikes, one of which becomes very prominent. In some genera the dorsal and anal fins acquire more or less elongate streamers.

The butterfly fishes (Chaetodontidae, fig. 9.97D) are classified with the angel-fishes (figs. 9.97E; 9.98D), which they resemble in their small mouths and bristle-like teeth and in the union of the posttemporal with the skull. They lack the spike on the preopercular but their color patterns are plainly related to those of the angel-fishes.

In the young forms of two species of angel-fishes (fig. 9.99A, D) and one species of butterfly fish (fig. 9.100A) figured by Weber and De Beaufort (1936, Vol. VII) the color patterns include a series of curving bands more or less concurrent with the contours of the fins and scale rows, with shadow bands between the main ones (fig. 9.99A, D). We may provisionally assume this to be a central type of color pattern for both the angel-fishes and the butterfly fishes, from which may have been derived the following principal variants: (1) by emphasis of the vertical components, those with vertical bars (figs. 9.99B, C; 9.100A1, G-I); (2) by emphasis of the axial components, those with anteroposterior stripes (figs. 9.99D-F; 9.100B, C); (3) by combination of vertical and horizontal components, the chevron-like pattern of certain forms (fig. 9.100D, E); (4) by segregation of pigment, those with large areas of sharply contrasting colors (figs. 9.99H; 9.100K-M); (5) by partial suppression of original bands and emphasis of oblique scale rows, the fine oblique to longitudinal stripes of some forms (figs. 9.100I1, G, H1, J; 9.98A).

A MOTLEY CREW: MOORISH IDOLS, DOCTORS (TEUTHIDOIDEI), TRIGGERS, COFFER-FISH, PUFFERS, PORCUPINES, HEADFISH (PLECTOGNATHI)

The moorish idol, *Zanclus cornutus* (fig. 9.101B), is usually classed with chaetodonts, perhaps because it recalls them in body form and in striking color pattern, but in the general construction and details of its skull, jaws and dentition it is much nearer to the doctors (acanthurids), with which it is grouped by Regan (Encycl. Brit., vol. 9, 1929, p. 320) in the suborder Teuthidoidea of the order Percomorphi. Thus the face of *Zanclus* below the orbit is drawn downward and forward into a tube that includes the ethmoid and lacrymal region, the suspensorium and the roof of the mouth. In this preorbital elongation it differs from the true butterfly fishes (Chaetodontinae), in which the ethmoid or preorbital region is very short (fig. 9.97D). In its bold vertical bands extending on to the dorsal and anal fins, *Zanclus* is somewhat like *Platax* (fig. 9.98B) of the family Ephippidae; in other ways it suggests *Heniochus* (fig. 9.100I) of the true chaetodonts.

The doctors or surgeon-fishes (Acanthuridae) must have been named by sailors in allusion to their sharp knife-like weapons borne on either side of the pedicle of the tail. These modified spiny crests have become erectile; they resemble the blades of a jackknife and are received into longitudinal grooves on either side of the tail. When erect they point forward and are certainly capable of inflicting wounds on incautious hands. Presumably they are used most effectively when the fish darts forward and suddenly swings the tail along its adversary's flanks.

Among the acanthurids the caudal spines are subject to some variation and in *Xesurus* (fig. 9.101C) they take the form of a longitudinal row of several small hooks on either side of the tail. Possibly these were once the new polyisomeres from which the "knife" of the typical surgeons has been selected for emphasis. This would be consistent with the fact that the skull and dentition of *Xesurus* are also a little more primitive in details than those of *Hepatus sandvicensis.*

The coloration of the doctor fishes is reminiscent of the *Platax-Zanclus* type but the vertical bars are variable.

The family Siganidae ("Teuthidae" *partim*) includes a considerable variety of marine fishes of the Indian and Pacific oceans. At first sight (fig. 9.101A) they are difficult to identify, as they do not exhibit any conspicuous specializations. The compressed body is symmetrical above and below the horizontal line, the face moderately long, the mouth small and of nibbling appearance; basic percoid features are the spiny rays of the dorsal, and position of the pelvic fins below the pectorals and the presence of one spine and five soft rays on each pelvic fin. The anal fin, however, has five or more spines in contrast with the three spines that are characteristic of primitive percoids. The color varies in different species; in *Siganus* ("*Teuthis*") *virgata* (fig. 9.101A) the general appearance suggests that of *Megaprotodon strigangulus* (fig. 9.100E), a chaetodont, but this resemblance is probably convergent. Examination of the skull of *Teuthis virgata* (fig. 9.101A) indicates that, except for the fact that its teeth are already fused into a beak in both the upper and the low jaw, it makes an ideal structural ancestor for the triggers (Balistidae, fig. 9.101D-F). This is especially evident in the forward curving of the suspensorium and the incipient elongation of the ethmoid region.

The genus *Triacanthus* (D) is far more specialized than *Teuthis* in many respects: (1) the general body-form, which is becoming more rhomboid; (2) the dorsal fin is now subdivided into an anterior division with one very high spine followed by smaller ones in a descending series, and a soft dorsal which is opposed to the anal; (3) the first spine of the pelvic fin is very long and strong; (4) in the skull the preorbital elongation, downward and forward, is much more pronounced than in *Teuthis* but distinctly less than in *Balistes.*

The trigger-fish (*Balistes,* fig. 9.101E; 9.102) is so named from the trigger-like relation of the second fin spine to the enlarged first, because when all the spines are erect the high first one can not be depressed as long as the base of the second short one is wedged in behind it (fig. 9.103C). The basal pterygiophores of the dorsal spines have enlarged lateral rims (b^3, d^3) which have fused together to form a rigid trough; this serves both as a strong base for the series and as a groove for the spines when depressed. The main spine is raised or drawn forward by a pair of muscles (fig. 9.102 *m.arrect.*) that run on either side of the low occipital crest and it is depressed by paired muscles on top of the frontals that are inserted on the rear surface of its pterygiophore (*m. depr.*). The arrangements for the other spines are similar.

The long tract on the face in front of the forwardly-swung preoperculum and below the eye is occupied by jaw muscles (*m. add. mand.*) The opercular bones are sunk beneath the thick leathery integument, which also covers the branchiostegal fold and reduces the respiratory opening to a small cleft. The body cavity is crowded forward by the great development of the locomotor muscles and the front end of the swimbladder is closely appressed to the base of the occiput and the rear side of the shoulder-girdle. There is also a patch of thick skin (*tym*) which overlaps the swimbladder, immediately above the respiratory cleft. The pelvic bones (fig. 9.101E) are united into a long median curved rod, bracing a ventral keel-like extension of the body. The pelvic fins are reduced to a vestigial nubbin.

In the filefishes (*Aleutera*) the excessive lengthening of the preorbital face (fig. 9.101F) is accompanied by a weakening of the erectile spine (fig. 9.103D) and by a secondary reduction of body-height as compared with length. This tendency is carried to still greater extreme in *Anacanthus* (fig. 9.104E), which is secondarily assuming the very shallow body-proportions of the tubemouth *Solenostomus* (fig. 9.68E). The spiny dorsal is represented only by a vestige of the first spine, which has retained its old position just above the eye, and the elongate soft dorsal remains paired with the anal. Thus there is good evidence that the apparent elongation of the body in *Anacanthus* is entirely secondary and that its remote ancestors passed through the following stages (fig. 9.104): (1) ancestral percomorph with moderately short high body and obliquely inclined face; (2) typical balistid, with emphasis of these features; (3) rapid reduction of body height with apparent lengthening of face and body, reduction of trigger, spine, and apparent lengthening of soft dorsal, anal, and caudal fins. A clear example of evolution by metamorphosis or transformation, which includes the subsidence of earlier specializations and the rise of later ones.

The construction of the coffer-fishes (Ostraciontidae, fig. 9.105A) is easily understood if we assume that their ancestors were primitive balistoids that had retained the compressed body, the high position of the orbits and a moderate elongation of the preorbital face, but in which the leathery skin had given rise to new hexagonal bony polyisomeres; these tended to immobilize the whole body except at the joint between the cephalo-thorax and the tail piece. Meanwhile we may reasonably suppose the trigger itself had been completely eliminated. Presumably the ancestral coffer-fish was a secondarily inshore form that frequently rested on the bottom. The vertical depth of the ventral keel rapidly diminished and the ventral edge rapidly became a wide base for the triangular cross-section of the body.

In spite of these radical changes in the surface features, the skull of the coffer-fishes has retained convincing evidence of close relationship with the balistid type. We need cite only the very peculiar relations of the supracleithrum (fig. 9.105A), which is nearly vertical and below the skull, not behind it. The posttemporal, which tends to become fused with the skull in the entire series from *Platax* (fig. 9.98B) throughout the spadefishes and chaetodonts, through *Zanclus* (fig. 9.101B) and the doctor-fishes to the balistids, finally becomes difficult to recognize in *Aleutera* (fig. 9.101F). In the ostracionts it has either disappeared entirely or may be fused with the pterotic (fig. 9.105A *pto*).

Since the coffer-fishes (fig. 9.105A) are enclosed in an almost fixed shell, the normal bending movements of the body are at least greatly diminished and the units of the vertebral column are closely united into a stiff but somewhat elastic axis. The myomeres on the sides of the body have accordingly changed into long cones wrapped one within the other, tips of successive cones uniting into thick longitudinal tendons; these pass backward and are inserted on either side of the flexible root of the tail, thus forming a powerful sculling organ. At least an early stage of this system is attained in the trigger-fish *Balistes*; but in the coffer-fishes the fixation of the body and the sculling action of the tail in fast movements give rise to a type of locomotion which Breder (1926, pp. 169–174) named "ostraciiform." In leisurely movements the flexible tail is used more as a rudder than as a sweep and the pectorals, vigorously moving up and down, supply the chief propulsive force.

In the coffer-fishes each half of the pectoral girdle at the same time supports and is supported by the rigid bony carapace, which takes the form of a triangular truss, the beams of which are supplied by the cleithrum and the blade-like coracoid.

The carapace of *Lactophrys tricornis* (fig. 9.105A) is composed of a mosaic of fairly large irregular polygons, mostly six-sided; on the outer side they are crowded with flat low papillae or gemmules arranged concentrically and/or radially. Each polygon, on both the outer and the inner surfaces, is well defined by straight sutures; on the inner surface of each there is a small shallow central basin surrounded by three relatively large concentric and parallel polygons of shallow grooves; beyond these the light-colored outer border, just inside the suture, is flat. These polyisomeres have evidently evolved from the deep layers of the corium within the limits of the family.

The puffers (fig. 9.106B, C), or swell-fishes (Tetraodontidae), appear to be a somewhat less specialized side branch of the balistid stock in so far as they have retained a flexible skin and unfused thoracic vertebrae, but in the following ways they are far more specialized than the coffer-fishes; although they still retain the primary marks of derivation from very small-mouthed nibbling balistids, their teeth have been fused into a large beak with two blades on both upper and lower jaws, and the opposite premaxillae (fig. 9.106B1) are connected above by an interdigitating, slightly movable symphysis. For the support of this powerful cutting apparatus the skull is widened and strengthened and those processes of the palatines which ordinarily overlap the maxillae have been enlarged and bear transversely extended facets permitting a vertical movement of the upper beak; meanwhile the lateral ethmoid buttresses have in turn become enlarged to support the palatines; the parasphenoid has also become enlarged and serves as a strong mid-ventral brace between the beak and the base of the endocranium.

The opercular region has been changed considerably, possibly in connection with the development of the puffing apparatus, the most essential part of which is a distensible bag, an outgrowth of the gullet. As described by Thilo (1899), the broad ventral and oblique muscles of the body wall were said to force water or air into the sac, the greatly elongated supracleithra and postclavicles serving to work the bellows. Parr (1927, p. 260), however, in spite of numerous observations on the swelling of live specimens of the puffer *Spheroides maculatus,* "was never able to observe any actions supporting the theory advanced by Thilo (1899) that the very large postclavicles of the Tetraodontidae through their spreading should serve to enlarge the capacity of the body cavity, this act being among the main factors of the pumping activity producing the inflation of these fishes. The postclavicular apparatus [of bones and muscles], on the contrary, seems quite passive during the swelling." He did find, however, that the postclavicular apparatus was spread laterally just before the fish descended to the bottom to rest; when it was oscillated rapidly on alternate sides, it helped the fish in shovelling movements and in squirming into the sandy bottom.

Although the act of puffing may not be due to the postclavicular apparatus, it would seem that at least in the puffer *Spheroides* the greatly enlarged ventral muscles, which are called "recti" muscles, must play an important part; for in pulling the clavicular arch backward the coracoanalis muscles would depress the floor of the throat and thus draw in water past the oral valves. Perhaps the bony interopercular tracker that extends from the posterior angle of the mandible to the inner side of the opercular (fig. 9.106B) may help to open the mouth quickly when the levator operculi muscle contracts. When inflated the tough skin is stretched and the spherical body becomes too big to be engulfed by most predators. We may suspect, however, that a large barracuda with his excessively keen erect lower front tooth could puncture this bubble.

In order to accommodate this swollen bag, the distance from the rear of the shoulder-girdle to the front of the hyomandibular has been greatly increased and the flexibility of its cover increased. Consequently the supracleithrum, the opercular, subopercular, and the lower border of the expanded preopercular have all extended backward, leaving flexible areas between them, and the branchiostegals have been reduced to narrow strips.

It will be noted that the puffers contrast with the coffer-fishes in the nearly horizontal, instead of oblique, position of the face. Morphological evidence indicates, however, that this horizontality is entirely secondary and that the remote ancestors of the puffers had a downwardly extended preorbital face and very small mouths, as in the coffer-fishes and balistids.

Some of the puffers (e.g., *Canthigaster rostratus*) retain a suggestion of the triangular cross section of the coffer-fish, since the upper part of the body is compressed and much narrower than the inflatable lower part. In others the body is more rounded and when somewhat elongate it has a deceptive resemblance in side view to a more normal fish body and but little resemblance to its compressed ancestors. In none of the puffers does the smooth skin retain normal scales but the ventral parts of the fish are covered with small spines. These secondary polyisomeres may well be derived from an embryonic stage of the hexagonal plates of the coffer-fishes before they become ossified.

In short we have suggested above that the puffers (Tetraodontidae) have not been derived altogether independently of the coffer-fishes from a trigger-like (balistid) stock, but that their ancestors, like the coffer-fishes, were small-mouthed nibblers with a triangular cross section of the body. The puffing-sac, we may infer, arose as a response to respiratory difficulties due to the reduction of the gills and the general tendency toward fixation of the thorax, accompanied by emphasis of paired dorsal and anal fins acting in conjunction with pectorals and caudals. As the puffing-sac gained genetic momentum the parts around it were greatly modified, the skin began to retain its embryonic flexibility and the incipient bony plaques which appeared after the disappearance of the scales were arrested in development and dwindled into small spines.

Meanwhile the great widening of the body that was associated with the evolution of the puffing apparatus spread to the head, the mouth cavity becoming enlarged, the small teeth fusing into upper and lower beaks. In order to support the powerful beak all the supporting parts were strengthened as described above, and the palatines developed obliquely transverse hinges for the up-and-down movement of the upper beak. Thus the short preorbital face of the puffers is a secondary derivative from the long preorbital face of the triggers and coffer-fishes.

Still greater heights of specialization are reached by the porcupine fishes (Diodontidae). For the separate right and left upper beaks of the puffers have here fused into a single massive nipper and crusher (fig. 9.108), which in addition to its cutting edge bears a median pair of crushing parts made up of many coalesced flat dental plates. Similar cutting and crushing parts are developed on the dentary bones of the mandible. These relatively new and dominant polyisomeres are supported by marked anisomeral emphasis of the width of the head, especially in the palatine and frontal regions. In the rear part of the skull-roof projecting wings of the pterotic accommodate the shoulder-girdle, which is widely spread around the expanded body. From small spines like those of the puffers new polyisomeres have developed in *Diodon* into spiny quills, but in *Chilomycterus* these spines are more or less immovable.

The climax of specialization in this order is attained by the so-called ocean sunfishes (Molidae), which are sometimes more descriptively named "headfishes." The "head" is propelled by huge dorsal and anal fins and by a flexible caudal flap (fig. 9.109). We might at first suppose that the headfishes have been derived directly from deep-bodied triggers (balistids) which had lost the spiny dorsal and greatly increased the height of the soft dorsal and anal, at the same time converting the tail into the high caudal flap of *Mola*. But intensive and extensive comparative surveys of the anatomy of the molids (Gregory, 1933, p. 294; Gregory and Raven, 1934; Raven, 1939) have brought out much evidence in support of Regan's conclusion that such resemblances as they bear to the balistids are partly convergent and that they have been derived from, or are very closely related to, the porcupine fishes (Diodontidae).

The first step toward the molid type was initiated by the obscure forces inherent in the hereditary mechanism, which unloosed a rapidly accelerating emphasis on the dorsal and anal fins and their constituent and auxiliary parts. For in the molids we find that a certain set of the muscles which in normal fishes move the individual fin rays has here become enormously developed in the dorsal and anal fins, the dorsal muscle mass spreading forward on the top of the skull and the ventral mass growing forward and upward, driving the body cavity in front of it and pushing the fore part of the kidneys on to the top of the occiput! Not content with this, these conquering polyisomeres have sent twigs into the region of the tail, driving out the normal tail muscles and sprouting a new tail flap. However, a transitory trace of the true tail appears in the larval young of *Mola* and in the median caudal tip of *Masturus* (fig. 9.110F).

Meanwhile the spiny skin of the porcupines has been replaced by a thick tough blubber-like integument, which serves as an effective armor. The larval *Mola,* however, looks like a porcupine fish and has very long spikes. In a fish in which the dorsal and anal fin muscles have usurped the function of the true caudal muscles, the latter have disappeared, while the virtual fixation of the body under the pachydermal integument deprives the old axial muscles of even the little usefulness that they retained in the porcupine fishes. Consequently the vertebral or axial muscles, which are the basal feature of all chordates and even older than the vertebrae, are lacking in the molids. The vertebrae themselves are secondarily reduced in number (dorsals 8, caudals 8) and the centra are relatively small. Moreover the whole skeleton is becoming largely cartilaginous and all the derm bones thin and streaky, so that the skeleton dries into curled-up and fragile remnants.

The skull of the molids (fig. 9.111), although largely cartilaginous, is more conservative than the vertebral column and retains the most definite evidence of derivation from diodont ancestors. The beak, although relatively feeble and probably able to deal only with soft squid and jellyfish, is essentially diodont, consisting of a single fused upper beak, moving vertically on the palatine hinges, and a single dentary beak. The opercular region is modified by the tough integument, the opercular and subopercular being greatly reduced and the branchiostegal slats covered by the integument. The climax of revolutionary advance in this group is attained by *Ranzania* (fig. 9.112), a small fish in which the body is becoming secondarily elongated and streamlined for faster swimming.

Thus the molids appear to have attained their most remarkable specializations not by a straight line of ascent but by a path that has followed an erratic course (figs. 9.107, 110).

This general conception of the origin and evolution of the plectognaths finds further confirmation in the excellent contribution of Breder and Clarke (1947) to the visceral anatomy, development and relationships of the Plectognathi; except that in their phylogenetic diagram (*op. cit.,* p. 315) the very wide divergence of the trunkfishes (Ostraciidae) from the other plectognaths, although correctly reflecting certain structural differences, does not seem to accord well either with their embryologic evidence or with the evidence from the skull, both of which indicate a much closer relationship to the puffers and diodonts than to the more ancient balistid forerunners.

SILVERY CARANGINS AND SWIFT-LINED MACKERELS (SCOMBROIDEI)

In the Upper Eocene of Europe have been found the well preserved body-forms and skeletons of *Acanthonemus* (fig. 9.113A), *Semiophorus* (B), *Amphistium* and other genera, with high compressed bodies; the dorsal fin is usually very high in front, the tail symmetrical, fan-like, the tail and caudal peduncle delicate. *Aipichthys* of this group dates from the Upper Cretaceous. The vertebrae are relatively few in number (ten in the abdominal, usually fourteen or fifteen in the caudal, region). These Cretaceous fossils are referred by Smith Woodward (Cat. Foss. Fishes, Brit. Mus., pt. IV, 1901, pp. 425–437) to the family Carangidae including the pompanos, or jacks, moonfishes, etc., which they indeed foreshadow in basic features. A possible trace of the stage (fig. 113A-C) when the front or spiny part of the dorsal fin was very high and far forward may be indicated in the forward extension upon the frontal region (fig. 9.114 *fr., soc*) of dorsal metameric muscles which, although now without fin-rays are in line with the true fin muscles. *Caranx* itself (fig. 9.113E) would be readily derived from such a Cretaceous fish as *Acanthonemus* (A), which in turn would be not far beyond the beryco morph- percoid stem (fig. 9.73C1). Other considerations also suggest that the carangid stem was distinctly short-bodied, compressed, with small silvery scales, small mouth and delicate skeleton and that the longer-bodied carangids such as *Seriola lalandi* (fig. 9.113F), the pilot fish (*Naucrates ductor,* G) and the dolphin (H) are more specialized than *Caranx*. The deep-bodied *Selene* (figs. 9.113I, 114D) goes to the opposite extreme. *Chloroscombrus* (J) with its upturned mouth is a plankton feeder (Breder, 1929, p. 138).

A relatively primitive short-bodied carangoid is *Brama* (fig. 9.114A), whose skull would make a good starting-point for that of the many-hued dolphin (*Coryphaena,* fig. 9.114B). As noted above, the connection of the mid-frontal crest with the muscles of a formerly high spiny dorsal fin is suggested in all the carangids.

The bluefish (*Pomatomus*) somewhat resembles the carangins in its general features but is far less specialized, more percoid, throughout its skull (fig. 9.115A) and skeleton. It may well be not their surviving ancestor but a true percoid which is converging toward them.

In the skull of the mackerels and tunas the ascending processes of the premaxillae (fig. 9.115C, D) and the protrusility of the mouth are much reduced and the teeth small. In the trichiurids (fig. 9.115E), however, the teeth become elongate and more or less barbed. In the wahoo (fig. 9.130A) the jaws are prolonged into a short beak. In contrast with the carangins, the relatively long forehead is never surmounted by a high median crest.

The skull as seen from above (fig. 9.116) ranges from excessively long and narrow (A) to short and very wide (F). The skull top in more primitive forms bears well marked pits, presumably for mucous lateral line organs. In the tuna (*Thunnus,* F) the paired holes in the skull which lead into the braincase appear to be due to secondary thinning of the bone and strengthening of surrounding crests.

The mackerel group is split into various families by systematists who refuse to use the subfamily grade, but it bears a well recognized stamp, suggesting the acme of fierce swiftness. Their skeleton is thin, the scales small, delicate, or wanting, the skin oily and abounding in blue, black and silver pigments; the crescentic symmetric tail is completely "homocercal," both internally and externally; it is usually braced laterally by more or less conspicuous horizontal keels, originally made of large ridged scales on either side of the delicate caudal peduncle. Highly characteristic is the presence of a row of detached finlets behind the well matched soft dorsal and anal fins. The body-form (fig. 9.117) ranges from the stout tunas (tunnies) to the elongate kingfish (Acanthocybiidae) and the eel-like cutlass fishes (Trichiuridae, fig. 9.115E).

The tunas (fig. 9.117III) oceanic bonitos and frigate mackerels (IV) develop to a high degree a unique specialization of the vascular system, which breaks up into a plexus developed in the lateral muscles, especially beneath the scaly "corselet," while a similar plexus appears within the liver or in the haemal canal; these are associated with a body temperature which may be 9° C. higher than that of the sea (cf. authors cited in Berg, 1940, p. 491). Chiefly on this account Kishinouye (1923) proposed to separate these fishes in a "subclass Plecostei," as opposed to all the rest of the bony fishes ("subclass Teleostei"). But these plexuses are evidently comparatively new polyisomeres developed within the suborder Scombroidei and the "Plecostei" are connected with the other scombroids by the striped bonito (figs. 9.115D, 116E, 117IV *Auxis*).

The problem of the relationship of the mackerels to the carangins has been differently answered by Tate Regan and Starks. The former puts the carangins among the order Percomorphi, the latter emphasizes the evidence for the retention of the group Scombroidei in the broad sense. Apparently transitional conditions between the two groups are supplied by *Scomberoides* (*Chorinemus*), which, however, appears to be a true carangin (fig. 9.115B). Both the Scombridae in the broad sense and the Trichiuridae date back to the Eocene epoch.

The louvar (*Luvarus imperialis*) is a rare, pelagic fish (fig. 9.118) chiefly from the Mediterranean but sparsely reported from the coasts of Australia, California and Long Island, New York. It has a very small mouth and weak toothless jaws and probably feeds on jellyfishes and other plankton. In a very puzzling way it combines the low vertebral number (23–24) and other features of the carangids with a tunny-like body-form and skull characters, besides possessing many curious specializations of its own (fig. 9.119). There is a muscular hump on top of the head, which, as in the dolphin fish (*Coryphaena*) has been developed out of the anterior part of the protractor dorsalis series that formerly moved the dorsal fin, even though the dorsal fin itself in *Luvarus* has become limited to the posterior half of the body. This dorsal hump is supported by a bony keel borne chiefly by the supraoccipital bone, which has grown forward with the muscle mentioned above until it surmounts the nasal region (fig. 9.120). The dorsal and ventral contours of the body are each supported by a longitudinal bony truss (fig. 9.119), the units of which represent parts of greatly enlarged pterygiophores, or bony fin supports (fig. 9.122). These trusses are fastened to the vertebral column (fig. 9.119) at the eighteenth or "anchor vertebra." Then comes the "pivot vertebra" (No. 19), followed by two "rudder-stock vertebrae" (Nos. 21, 22), which form a slightly movable base for the "hypural fan" (Nos. 23–24) and on to this in turn are fastened the strong bony rays of the very large lunate caudal fin (fig. 9.134E, E1). The pelvic fins are much reduced and fused to form a movable operculum or lid for the vent, which is far forward under the throat (fig. 9.119). The pelvis is fastened to the inner side of the coracoids and forms a long rod bearing channels for the muscles that open and close the anal operculum.

As a whole, the skeleton is but weakly ossified except the occipital base and adjacent parts of the cranium, which form the pivot for the vertebral column and are composed of dense ivory-like bone.

All parts of the skull (figs. 9.120, 121) are highly specialized, but may well have been derived either from a tunny-like (fig. 9.120B) or a caranx-like (fig. 9.114C) ancestral stage. The brain and spinal cord were minute, and sc were the parts of the inner ear.

The early larval stages of *Luvarus* (fig 9.123), as described by Roule (1924), differed so extremely in appearance from the adul that the process of transformation from larva to adult was called by Roule a "hypermetamorphosis."

Paralleling the carangids, or derived from related Upper Cretaceous berycoids, are the stromateoids, including: the European coverle fish *Stromateus,* the North Atlantic harvest fish (*Peprilus,* fig. 9.124E), the dollar- or butterfish (*Poronotus,* D), the man-of-war fishes (Nomeidae, B, C), the rudder fishes (Centrolophidae, F, G) and others. Many of these fish feed on medusae or swim about beneath the stinging tentacles of the Portuguese man-of-war. The stromateids have a pair of large sacs in the oesophagus bearing internal papillae o folds, often beset with setiform teeth. The pelvic bones are sometimes free from the pectora arch, sometimes more closely attached but only by ligament and movable (Boulenger, 1910, p 643). The skull of *Poronotus* (fig. 9.125B) i in general similar to the carangid and *Bram* types (figs. 9.114; 9.125). The large pelvic fins of the man-of-war fishes (fig. 9.124B, C) may enable them to stop or turn quickly so as to avoid contact with the stinging tentacles. Enlarged pelvic fins are also developed in *"Gasteroschisma,"* which, according to Rega (1902, cited by Berg, 1940, p. 484) is the young of *Lepidothynnus,* a true mackerel allie to the Cybiidae. The pelvic fins are vestigial i the deep-bodied *Poronotus* and *Peprilus* (fig 9.124D, E), which feed mostly on plankto and seek the protection of the man-of-war' tentacles only when young.

The Centrolophidae, or rudder fishes (fig 9.124G) indicate progressive elongation of th body, culminating in the excessively specialized fishes (Icosteidae, H, I, J). *Tetragonuru* (K) seems to be related to the Icosteidae but is much less specialized. All these families may have been derived from Upper Cretaceous berycoids (A).

The anatomy of *Acrotus* (fig. 9.126) ha

been investigated by H. C. Raven (1944). The skeleton is largely cartilaginous, its large empty spaces filled with collagenous tissue. So far as can be seen, the skull structure is consistent with derivation from a primitive stromateid type.

THE IMPETUOUS SWORDFISH AND THE LEAPING MARLINS (XIPHOIDEI)

In any fish or fish-like vertebrate the entering wedge of the living spear is formed either by the top of the nasal rostrum or by the nose and the conjoined ends of the upper and lower jaws. In the vast majority of fishes this entering wedge remains short or blunt but in not a few cases (fig. 9.127) it has been prolonged into a "spear" when rounded in section or a "sword" when flattened dorsoventrally. Far back among the Palaeozoic ostracoderms a long pointed spear was formed on the front end of the head shield in *Boreaspis rostrata* (A) but rostral spears are notably absent in the acanthodians, arthrodirans and related families. In some of the elasmobranchs a prolonged but flattened and sensory rostrum is developed, as in *Rhinobatus* (B), but in the sawfishes (Pristidae) and pristiophorids it becomes a toothed sword. Of the extinct chimaeroids the Jurassic *Squaloraja* (C) received from Agassiz the expressive name *Spinacorhinus* (spiketip nose). The crossopterygians and dipnoans avoided this line of specialization but, among the descendants of the palaeoniscoids, the Belonorhynchidae (including *Saurichthys,* D) were armed with very long pointed snout and jaws.

In the immediate ancestors of the sturgeons the rostrum, although pointed, emphasized its sensory functions; this line gave rise to the flattened snout of the sturgeons and the long spatulate feeler of the spoonbill (fig. 9.14D). The holostean ganoids started their career with a small nibbling mouth and very short rostrum; this initial obstacle was only partly overcome by the garpikes (*Lepidosteus,* fig. 9.17C), which developed elongate duck-like bills but no spear or sword. A close approach to the marlin type of head-spear appeared among the Jurassic-Cretaceous amioids or protospondyls in the ferocious *Protosphyraena,* (fig. 9.127F). In another family (Aspidorhynchidae), also of the Jurassic-Cretaceous, the marlin type was again foreshadowed, as well as the long-billed form with elongate upper and lower jaws. Among the lower teleosts long pointed snouts were developed independently in several families, notably in the Upper Cretaceous scopeloid *Rinellus* and in the modern snipe-eels (Apodes), also in the skittering gars and hound-fishes of the order Synentognathi (fig. 9.127G). We do not include in this list cases in which a prolonged tube is produced behind the small mouth, as in the mormyrids and seahorses.

In the spiny-finned order spear-fishes (fig. 9.133B) did not appear, so far as known, until Upper Cretaceous and Eocene times when, to judge from the considerable number of named genera and species, there must have been favorable conditions for them. Thus the true swordfish (*Xiphias,* figs. 9.127I, 128C), the sailfishes (*Istiophorus,* figs. 9.127H, 9.128A), and the marlins (*Makaira,* fig. 9.128B) of the present day are using an old and often independently made product of anisomeral evolution. In most cases spears or pointed bills go with elongate bodies; swords or long flattened bills may be carried either by flattened or stout bodies.

In general the true swordfish (fig. 9.128C) and the sailfish-marlin family (Istiophoridae) are superficially alike, not only in possessing a rostral weapon but also in leaping out of the water and "walking" on their tails when making desperate efforts to cast off the hook. The sword or bill in both marlins and true swordfishes is excessively tough and strong and the fish does not hesitate to use it either in straight thrusts as a ram or in vigorous side strokes. Dr. E. W. Gudger in his memoir "On the Alleged Pugnacity of the Swordfish and the Spearfishes as Shown by Their Attacks on Vessels" assembles and analyses evidence for the penetration of 5 inches to 18.5 and even 30 inches of copper sheathing, wood sheathing and solid oak timbers of various ships by charging swordfish and marlins. He estimates (pp. 286, 287) that in the case of a 600-pound swordfish moving at thirty miles per hour the striking force of the sword or ram would be 1650 pounds.

It has been assumed by ichthyologists that the two existing families (Xiphiidae, Istiophoridae) belong together in a single division (Xiphiiformes) of the mackerel-like fishes; but as sharp spears and swords have often been evolved independently, we may well inquire whether the relationship between the two families may be far less close than it has been supposed to be.

The two families differ in several deep-seated characters. Thus in the broadbills or true swordfishes the centra of the vertebrae (fig. 9.129B, B1) are short and stout, with rather inadequate-looking zygapophyses or interlocking processes. In the marlins, on the other hand, the centra (fig. 9.129A) are very long and narrow and have extremely long blade-like and closely overlapping zygapophyses, neural and haemal spines, which contribute greatly both to the stiffness and to the spring-like quality of the column as a whole. There are also differences in the relations of the nasals and premaxillae in the two families (figs. 9.130, 131). On the other hand, the skull in both families is adapted to resist stresses from the long rostrum in much the same ways: for example, the ethmovomer block is expanded anteriorly and bears special facets for the rostrum on either side. The rear end of the mesethmoid (dermethmoid) forms a great spongy buffer, filled with oil, while the tripartite character of the condyle of the primitive percoids (fig. 9.76) is emphasized to prevent dislocation (fig. 9.131A1). The tail bones in both families are essentially scombroid in type (fig. 9.134).

However, all such differences and resemblances obscure but do not completely obliterate a sort of basic mackerel heritage, derived, it was formerly thought, from some beaked mackerel with an elongate dorsal fin, as in *Acanthocybium* (figs. 9.132B, 9.130A). It is only when the recent and fossil xiphiid and histiophorid skeletons are compared with those of the Upper Cretaceous and Eocene families Blochiidae, Palaeorhynchidae, Trichiuridae, that we fully appreciate the suggestions of Smith Woodward and Tate Regan that the broadbills may have been derived from the upper Eocene *Blochius* (fig. 9.133A) or some nearly related genus, while the sailfishes (including the marlins) may have been derived from the Eocene-to-Miocene Paleorhynchidae (for a discussion of this problem see Gregory and Conrad, 1937). *Blochius*, unlike the modern broadbill (*Xiphias*) has a long, very slender body, while the vertebral centra are certainly far more slender than those of *Xiphias*. But in these features *Blochius* approaches its relative *Palaeorhynchus* (fig. 9.133B), and even the ancestral carangids one genus (*Acanthonemus*) had slender centra (fig. 9.113A), while a near relative, *Semiophorus* (B), had much shorter, vertically deeper vertebrae. In brief, it seems that the relative length and height of the centra, as well as their number, may be subject to fairly quick changes in related genera.

The cylindrical rostrum in *Blochius* (fig. 9.133A) is very probably more primitive than the flattened rostrum of *Xiphias* (fig. 9.131B). In *Cylindracanthus* or *Coelorhynchus* of the same family the cylindrical or somewhat depressed rostrum is strengthened, as seen in transverse section, by deep folds and ribs radiating from a central cavity (A. S. Woodward). This rostrum is said to consist "entirely of dermal tissue (cosmine)" so that it would really be a sort of dense dermal sheath on the end of the ethmoid bone. Such may well be the origin of the central or proximal portion of the sword in all the modern swordfishes but the sides of the sword are made from the enormously enlarged premaxillae and maxillae (fig. 9.131). The latter are squeezed against the ethmovomer block (*pareth. vo*) essentially as they are in the mackerels and cutlass fishes (Trichiuridae, fig. 9.115E, Gempylidae).

The pelvic fins of *Blochius* are absent (fig. 9.133), as they are also in the modern *Xiphias* (fig. 9.128C). In the marlins and sailfishes, on the other hand, the pelvic fins are retained, as they were in *Palaeorhynchus* (fig. 9.133B) although reduced to a few long slender rays. The dorsal fin of *Blochius*, although not high extends along the very long back from the occiput to near the base of the tail. The fin is supported by "flexible spines." In *Palaeorhynchus* the dorsal is equally long-based but also extremely high. Its rays are nearly fifty in number and each corresponds to a vertebra below it. Thus the very high and large dorsal fin of the sailfishes (*Histiophorus*) instead of being a relatively new feature may be a heritage from Eocene swordfishes. The anal fin is elongate both in *Blochius* and *Palaeorhynchus* but in all the modern swordfishes (fig. 9.128) the rear parts of the dorsal and anal fin have been reduced to a single fin comparable to the finlets of the mackerels (fig. 9.117).

The tail in all the Eocene forms is large and symmetrical and could readily give rise to the type seen in the moderns (fig. 9.134). The general construction of the bony bases of the tail is in fact almost identical in the true scombroids (fig. 9.134A), in the sailfish-marlin family (Histiophoridae, fig. 9.134B, C), in the true swordfishes (Xiphiidae, fig. 9.134D) and in the louvar (fig. 9.134E, E1) and it is hard to accept the implied suggestion of Roule (1924) that this is all merely an indication of convergence in widely unrelated families.

In short, the available evidence tends to confirm A. S. Woodward in placing the Xiphiidae (in the wide sense) immediately after the Palaeorhynchidae and Blochiidae; it suggests further that the true swordfishes or broadbills (*Xiphias*) may be more nearly related to *Blochius,* the marlins and sailfishes to the, in general, less specialized Palaeorhynchidae. Thus the stoutness of the body in the broadbills, as well as in the black marlin, is a late acquirement and both families are more nearly related to the slender *Ruvettus* and the cutlass-fishes than they are to the more normally shaped mackerel. The mackerels themselves, along with *Ruvettus,* the trichiurids and the swordfishes, may have been derived ultimately from such a compressed, deep-bodied carangid as *Acanthonemus* (fig. 9.113A); but in the ancestral scombroids there was a rapid multiplication in vertebrae and increase in body length; from this primitively elongate form the ancestors of the swordfishes first became very slender, as in *Palaeorhynchus;* much later, as they attained gigantic size, the body thickened and shortened.

THE "KING OF THE HERRING" AND OTHER "SEA-SERPENTS" (ALLOTRIOGNATHI)

The order Allotriognathi (strange jaws) was one of the last ones to be recognized and it was, so to speak, put together by Tate Regan (1907), who assembled within it such exceedingly different-looking fishes as the orbicular moonfish (*Lampris luna,* fig. 9.135b), the ribbon-like "King of the Herrings" (*Regalecus*) which is the so-called oarfish or sea-serpent (fig. 9.135a), the very short, compressed, high-finned *Velifer* (fig. 9.125C) and the elongate, excessively specialized deep-sea *Stylophorus.* The most outstanding feature which nearly all these forms hold in common is the peculiar construction of the maxillae, which have escaped from the normal overlapping processes of the palatines (which processes are lacking) and are drawn forward and downward with the premaxillae when the lower jaw is depressed (fig. 9.125C). At the same time they have developed a special groove and process upon which the ascending processes of the premaxillae slide; while at their upper rear ends they themselves slide over the vomer or on a preëthmoid cartilage (Regan).

Probably the least specialized of the group is *Velifer hypselopterus* (fig. 9.125C), which has a compressed, somewhat carangid-like body, with extremely high dorsal and anal fins, somewhat suggesting the very high dorsal and anal fins (fig. 9.113) of certain berycoids and carangins but directed more toward the rear. The skull is short and deep, surmounted with a huge crest overriding the frontals. The jaws combine a very small mouth with extreme protrusility, somewhat as in *Capros* (fig. 9.96) but more advanced in respect to the features of the maxilla already noted.

The moonfish, *Lampris luna* (fig. 9.135A), is a gigantic derivative of a *Velifer*-like stage in which the ventral half has been excessively deepened while the protrusility of the mouth and the size of the supraoccipital crest have been reduced. By a very peculiar change the lower tips of the cleithra are curved backward to receive the ligament of the basal bone of the large pelvic fins, while the coracoids, greatly expanded and deepened, suggest a rapid up-and-down vibration of the relatively narrow pectoral fins.

In *Trachypterus* the compressed body is deep in front but tapering toward the tail. The narrow, fan-like caudal fin is not in line with the long axis of the body as in normal fish but is set off at a sharp angle to the slightly downturned axis. Probably from the trachypterid stock issued the ribbon or oar-fish, sometimes called the sea-serpent (fig. 9.135A). The Japanese name for it is said to mean "Cock of the palace under the sea," presumably in allusion to the erectile crest of elongate fin rays on top of the head. In spite of its extreme length, the body is so much compressed that the name "ribbon-fish" is quite descriptive. The remains of the pelvic fins, in the form of two very delicate long filaments, are supported by a relatively large pelvic bone, which retains its primitive attachment to the cleithrum.

One of the few naturalists who have ever seen an oarfish alive is Professor Frederick Wood Jones, whose account of the fish is quoted by Weber and DeBeauffort (1929, pp. 92, 93). This oarfish came up beside a ship which was laying a cable in 1906, south of the Island of Sumbawa, and it was eventually caught and hauled on board. From a point just behind the vent the tail had been bitten off, but the remainder of the head and body was 11 feet 9 inches long and weighed 140 pounds. The greatest depth was 13 inches. While in the clear water the fish was a very beautiful sight. On top of the blue head was a very tall crest of vivid red, which was erected whenever the fish was touched by a hook or a rope. The body shone like bright silver and there was a pair of long scarlet streamers (the remnants of the pelvic fins) on the sides. Thus this fish was essentially similar to the one from the coast of New Zealand, described by T. J. Parker (1884).

In the deep-sea *Stylophorus,* as described by Starks (1908), the protrusility of the mouth reaches a new high level of specialization, for now the posterior half of the mandible has grown far backward and the hyomandibula has been secondarily lengthened and swung backward with it; meanwhile the pterygoids have been reduced to vestiges. Thus a great collapsible bag with a small mouth opening works on the principle of a bellows.

DISK-HEADED SUCKING-FISHES (DISCOCEPHALI)

A sand shark is weaving his way about in a large tank at the Aquarium, unmindful of the shark-sucker (*Echeneis*) that clings to him. At feeding time, however, the sucker may momentarily let go of his huge friend, dash for a fragment of food and then swiftly overtake his host. Another shark-sucker is attached to the glass and we can see that his sucker is an oval disc with numerous cross slats. But is it on the top of his head or on his throat? At first glance this is not altogether clear, as the rounded throat may be mistaken for the upper surface of the head and the general contour and colouring of the body is nearly the same both above and below a longitudinal black band which extends nearly the full length of the body and passes through the small eye. If, however, we have had some experience with fishes, we will notice that the sucker is on the opposite side from the paired ventral fins, that is, it is on the top of the head (fig. 9.136A). How does it work and by what steps did it come into being?

There can be little doubt that the sucking-disc represents part of the spiny dorsal fin. It was held by Storms (1888) that the right and left halves of the disc, which are divided by a median groove, represent the separated right and left halves of the spiny rays, which must arise in the embryo from paired Anlagen. But did these rays split at the top or at the bottom? Possibly the movable slats of the disc represent not the median dorsal part of the spines themselves but the paired bases of those spines of which the tips have disappeared. In either case the bony pieces upon which the transverse slats rest represent the basal processes of the normal pterygiophores. The rim surrounding the disc may represent the sides of a groove or slit on either side of the base of the fin. In any event, we must suppose that the dorsal fin of the ancestral sucking-fish was not unlike that of the sergeant-fish, or crab-eater (*Rachycentron*), which has extremely short dorsal fin spines (fig. 9.136B). There are of course numerous precedents for the forward migration of part of the dorsal fin on to the top of the head, and the separation of the dorsal fin into right and left halves in the sucking-fish is not unlike the separation of the caudal fin into right and left halves in certain breeds of goldfish. Moreover, an oval sucker with serially jointed parts adaptable to differently curved surfaces has been evolved elsewhere, as in the chitons (fig. 2) and in the patella-like Palaeozoic Tryblidiacea among the molluscs. The ancestral sucking-fish may have followed underneath sharks both for protection and for the opportunity of getting scraps of food dropped by the shark. The habit of pressing upward against the lower side of the shark would open up the opportunity for sticking the very short spines against the shagreen-covered surface of the shark. As the short spines of the sucking-fish tended to be invaginated into the back, the marginal parts of their bases spread apart and eventually took over the adhesive function, the resulting partial vacuum being secured by the action of the pterygiophore muscles in pulling downward the medial portion of the incipient sucking-disc.

But from what family or suborder of the vast assemblage of the spiny-finned fishes and their allies did the ancestral sucking-fishes branch off? Unfortunately no structurally linking forms exist and there are several widely different fishes which might claim to be related to the sucking-fish stem. The best-looking candidate so far proposed is the sergeant-fish (*Rachycentron*), a percoid of doubtful relationships in which the general form of the body and the color pattern (fig. 9.136A) would seem an ideal starting-point for the sucking-disc fishes (fig. 9.136A). Moreover, as pointed out by Gudger (1926), the young of the sergeant-fish resemble a small shark-sucker. On the other hand, as implied by Boulenger (1904, pp. 652, 691), a comparison of the skulls (fig. 9.137), vertebral columns, rib attachments, etc., of *Rachycentron* (B) and the shark-suckers (C) reveals many differences and we can not safely assume that these differences are due merely to the high specialization of the sucking-fish. Boulenger (1904, p. 651), in his diagram showing the probable interrelationships of the various groups of spiny-finned fishes, put the Discocephali as a branch from the stem of the Scleroparei, or scorpaenoids, but in his key definition of Discocephali (*op. cit.*, p. 652) he noted "a perforate scapula; three pterygials [baseosts] in contact with the coracoid," whereas in the Scleroparei the relation of the pterygials varies (cf. Boulenger, *op. cit.*, fig. 423, p. 693) from the relatively primitive condition in *Sebastodes*, with but two pterygials in contact with the coracoid, to the much more advanced stage of *Scorpaenichthys*, with three greatly shortened and widened pterygials in contact with the coracoid; but these are wholly unlike those of *Echeneis*.

In short no fish of the many representatives of different orders with which we have compared the suckers for relationship with them seems to have any better claims than does the sergeant-fish (Gregory, 1933, pp. 320, 321). Moreover the evidence from the otoliths cited by Frost (1930, p. 621), combined with the normal attachment of the pelvic bone to the shoulder-girdle, indicates that the sucking-fishes (*Echeneis, Remora*) are modified percoids of some sort. The case offers a good example of the principle that new polyisomeres, arising in a certain genus, may become a dominant structure around which other structures (e.g., the skull) tend to be modified, sometimes to such an extent as to conceal or wipe out the more ancient heritage characters.

FISHES WITH TWISTED SKULLS (HETEROSOMATA)

In the world of fishes truth is often stranger than fiction and man's feeble inventions and legends are far surpassed by the plain facts. It would not be difficult to imagine that a compressed, deep-bodied fish might turn over on one side when resting on the bottom, but to what piscatorial Münchausen would it have occurred to have the fish twist his head internally and shift his left eye over to the upper or right side? How would he have solved the problem of making room for both eyes on the same side? This, however, is exactly what has been done, both during the individual development and in the evolution of the race. Also, in response to the habit of lying on one side, the underside becomes pale and the upper side deeply pigmented. Moreover, the fish has a nice color sense and can match his environment closely.

In spite of the twisting of the eyes, the skull as a whole retains the normal basic pattern for teleosts, the only part that is seriously displaced being the interorbital bridge (fig. 9.138A fr^1), which is usually twisted on to the right side or the left, as the case may be, while a new postorbital process of the left frontal joins a similar bar from the left prefrontal to form a strengthening brace around the displaced left eye (A1, *fr.* 2). The other parts of the skull, especially the jaw parts, are readily identifiable as derivatives of the percoid type. The same derivation, according to Frost, is indicated by the characters of the otoliths, while Tate Regan (1929, pp. 214, 324) holds that *Psettodes,* the most primitive member of the group, except for its asymmetry and the long dorsal and anal fins, is a typical perch [in the wide sense].

The most conspicuous secondary polyisomeres in the flatfishes are the very numerous dorsal and anal rays, none of which are spiny. It will be noted that in this group the absence of spiny rays is not to be interpreted as a primitive heritage from the lower teleosts. The delicate myomeres enable the fish to progress along the bottom by undulatory movements of the body and fins. Within the group there is a considerable range in body-form from the almost percoid *Psettodes* to the rhombic halibut and thence to the secondarily elongate sole.

A STRANGE COLLECTION OF EVILDOERS (SCORPAENOIDEI): THE PERNICIOUS ROSEFISH AND OTHER TOUCH-ME-NOTS OF THE SCORPION-FISH KIND

"Nemo me impune laccessit" might well be the motto of the central stock of this evil brood, damned by any unlucky fisherman that gives them a chance to express their venom. Spiniest of the spiny fishes, they bristle with sharp spikes (fig. 9.139): on their fins, on their cheeks and gill covers, above their large eyes and even on top of their heads. Moreover, their skin is well supplied with pits exuding poisonous slime which may cause extreme pain and dangerous swelling in the hand or foot of the incautious fisherman. Nor do they lack sense organs to warn them when to get ready for trouble.

It may be noted that these spines and slime pits are the product of the integument, which in this group is the fertile source of new and dominant polyisomeres; these have stimulated outgrowths on the older generation of derm bones that cover the braincase and its appendages.

Disregarding for the moment these surface specializations, we may see that these fishes, both in their general configurations and in their deep-seated skeletal patterns, are thoroughly bass-like. This is shown also in the possession of two spikes on the opercular, one spine and five soft rays on the pelvic, and three spines on the anal, fin. The rear corner of the maxilla is fully exposed, that is, not covered by the lacrymal, a feature suggesting affinity with the true basses (serranids) rather than with the snappers and related families. Not basslike, however, is the marked vertical spread of the pectoral fins, the loss of the first pectoral baseost and the downward displacement of baseosts 3 and 4 on to the coracoid. These features are distinctly foreshadowed in the Cirrhitidae or Haplodactylidae (fig. 9.139D), which in turn were regarded by Boulenger (1910) as derivatives of the Serranidae, and by Jordan (1923) as forerunners of the trachinoid series.

The one outstanding feature which persists throughout the scorpaenoid assemblage is the presence of a horizontal "bony stay" extending across the cheek below the eye and usually braced against a prominent rim on the preopercular (fig. 9.139, *so 3*). This structure is referred to in the names Scleroparei (hard cheeks) and Pareiopliteae (cheek armed), which are often used instead of Scorpaenoidei. How did this diagnostic feature arise?

As a whole, this bony stay is a simple polyanisomere, for it is one of a series of polyisomeres, one of which has been selected for anisomeral emphasis. The polyisomeres in this case are the row of four closely connected suborbital bones and of these the third is enlarged, with its rear border attached to the rim of the preopercular. This extension of the third suborbital and contact with the preopercular is not found in any of the swarming families of true percoids but it is to a slight extent foreshadowed in *Cirrhitus* (fig. 9.139D), which also has other features that may have given rise to scorpaenoid characters. It is not a solution of the origin of the scorpaenoids to suggest, as Jungersen (1908) did, that the sticklebacks (fig. 9.69A) have also established a slight contact between the third suborbital bone and the preopercular, because the sticklebacks themselves appear to be in many features far more highly specialized than the basic scorpaenoids, secondly, because other evidence indicates a widely different origin for them. One possible solution of the origin of the scorpaenoid contact between the suborbital stay and the preopercular was that the scorpaenoids may have been derived not from normally bass-like fishes with moderately fusiform bodies, but from some earlier berycoid stage (fig. 9.72) in which the body was deep, the eye large and the head short. This would bring the third suborbital bone into contact with the rim of the preopercular and if the genetic fates permitted it, a subsequent elongation of the head and body would involve the pulling out of the third suborbital into an extended bony tract.

However, it is not necessary to push the scorpaenoid stem back to the berycoid stock, because the stages in the development of the rosefish (fig. 9.140), as figured by Bigelow and Welsh (1925), indicate that all we need to assume is that there was a great increase in the size of the eye and an accelerated individual development of it, so that at a relatively early stage it became the dominant organ, around which the preopercular and suborbital Anlagen formed a continuous tract. Possibly the emphasis of the suborbital and preopercular tubes of the lateral line system may have encouraged the close association of the suborbital and preopercular rim. The subsequent downward growth of the hyomandibular and preopercular, together with the forward growth of the jaws, plus the necessary growth force of the suborbital itself, all coöperated to produce the observed result.

The rosefishes (*Sebastes*) and scorpion-fishes (*Scorpaena,* fig. 9.139A), although already well specialized beyond the primitive percoid stage, seem to have retained more nearly than any others the ancestral characters of this highly diversified assemblage, only a few of which will be mentioned here. In *Pelor japonicum* (C) the mouth is turned sharply upward, the body is moderately elongate. The skull is that of a slightly modified scorpaenoid. In the Australian *Synanceja horrida* (fig. 9.141A) the mouth is turned upward because the preorbital region is extremely short and the quadrate-articular region markedly depressed. The top of the skull is like a "Turkish saddle" with "horns" above the eyes and a steeply upcurved occipital roof. But in these and other peculiarities this genus is merely an overstressed *Pelor.*

In another Australian genus *Pataecus* (C) the entire face is twisted around in such a way that the small mouth is below and behind the eye, while the dorsal fin rays increase in length from the tail forward, becoming very long above the forwardly inclined forehead, the effect of the whole being ludicrously suggestive of the head of a Sioux Indian chief with a feather headdress. *Aploactis* (B) seems to be a related and less distorted genus showing the initial stage in the evolution of the headdress.

The inadequately named zebra-fish (*Pterois*) is a peculiarly baleful scorpaenid (D) concealed in the midst of a supreme display of delicate streamers and alternating vertical bands of dark and light. Competent observers (Breder, 1932) have reported that in the presence of this infernal marine basilisk an intended victim may pause and hesitate; then the floating mouth-trap is sprung, the victim is sucked into the yawning gate. The skull of the zebra-fish is a simplified edition of the central scorpaenoid type, the forces that produce polyisomeres having run into tegumentary appendages rather than skeletal spikes.

From the central rosefish type the ever-interweaving forces that produce polyisomeres and anisomeres variously moulded the body-form, fins and integument into even more bizarre creatures, a few of which may now be mentioned. Along one branch of descent the creative genes or jinns, tiring as it were of bristling spikes and heavy armor, reduced or eliminated them, lengthened the body and shaped it more toward the "codfish" type. Thus arose the tribe of *Hexagrammos* (fig. 9.141E), which however retains the bar suborbital, the heraldic symbol of their order.

The guardian daemons of the "sons of *Cottus*" (Cottidae) chose to stress the spines on the lower rear corners of the opercular and subopercular and made thereof both a double-barbed weapon and a tribal mark. However, the eighteen-spined *Cottus* (F) received a fair heritage of other spines, as his name implies. His kinsman, the elongate serpent-tooth (*Ophiodon elongatum,* G), took on the swift lineaments of a free-swimming predator, while the broad flathead (*Platycephalus fuscus,* H) pressed himself down against the mud and made sudden forays upon the helpless.

The Obese Lump-sucker, the Crawling Sea-robin and Others.—The habit of flattening down and broadening out was further emphasized in the jaoks (*Myoxocephalus jaok,* fig. 9.142B), enfeebled offspring of aggressive forebears, whose very bones have become porous. The obese lump-sucker (*Cyclopterus lumpus,* fig. 9.142C) lacks the spikes and poisonous qualities of his remote scorpaenoid ancestors but his thick hide is covered with warts and bosses and his colors and habits are equally inconspicuous. He spends much of his time clinging to the rocks by means of a sucking-disc. The skull of the lump-sucker (C1) represents a much widened and shortened derivative of that of the jaok. The braincase has become largely cartilaginous and the bones are thin and delicate, somewhat as in the case of the ocean sunfish *Mola.* Although the spines have disappeared, the suborbital stay is unimpaired (C), while the premaxillae have retained not only their ascending rami but even the cheek ligaments, which are so well developed in the ancestral rosefish (figs. 9.139B1; 9.77A1).

The construction of the skull in the cottids *Cottus* (fig. 141F), *Hemitripterus* (fig. 143A-A2), and *Myoxocephalus* (fig. 143C) reveals very clearly the way in which each bone, whether it belongs to the overlying or to the deep series, starts from a separate center, producing concentric peripheral zones and radiating ridges and spines not unlike the shells of sea-snails. The occipital view of the skull of *Hemitripterus* (fig. 9.143A) brings out the even balance of the growth forces of opposite sides and the keystone function of the tripartite occipital condyles. In the side view of *Myoxicephalus* (C) we note the triradiate contact between the three equally diverging growth centers around the otic capsule and another triradiate contact between the otic capsule and the frontal-sphenotic segment. Here we begin to realize by what steps the undivided oval, invaginated otic capsule of the embryonic stage first divides into anterior and posterior growth centers (called proötic and opisthotic in the adult skull) and then grows vertically, giving rise to the epiotic; also how the sphenotic center grows dorsally away from the so-called alisphenoid in front and the proötic behind.

In *Aspicottus* (D), another curious side branch of the cottid stock, what we may call the growth-center organizers in certain bones (D1) have produced a new set of polyisomeres, consisting of innumerable papillae arranged in widening fan-stick rows with intervening grooves. Perhaps from *Aspicottus* or its kin stemmed the marine *Pegasus,* or "winged seahorse," which bears only a remote resemblance to the true seahorse (*Hippocampus*), except in so far as it has acquired a secondary, segmental exoskeleton.

In another direction, starting from some *Aspicottus*-like cottid with radiating bony papillae, the genetic "daemons" of the gurnards (Triglidae) took advantage, as it were, of the opportunity to encase at least the heads of their protégés in armor plates of extremely compact heavy bone (fig. 9.144). Luxuriating polyisomeres covered the greatly expanded areas of the suborbital-preopercular series and of the enlarged naso-ethmoid region (fig. 9.144B, C1). Meanwhile the opercular bone became quite small and its ancestral spike diminished. The marked elongation of the opposite lacrymal region equips the bottom-loving *Peristedion cataphractum* (C, C1) with a two-horned probe and scoop which far overhangs his small mouth and sturgeon-like barbels. Not less curious is the humble sea-robin (B), which can crawl on the bottom like a marine arthropod, by using the lower three rays of each of his pectoral fins. These lower three rays at first sight appear to be "preaxial" rays and have been inadvertently so called even by Broom (1930, p. 18) and the present author (Gregory and Raven, 1941, p. 323). But comparison of the pectoral fins of scorpaenoids with those of other less specialized teleosts shows that the pectoral fin has been turned upward and over, so that what was formerly, in primitive isospondyls, the postero-ventral border has become the antero-ventral border. These newly endowed polyisomeres, acting like fingers, twist and bend over each other in purposeful ways and in definite rhythms which must be controlled by new centers in the central nervous system. The experimentalist, desiring to cut the nerve fibres leading to a single one of these finger-like organs in order to study the effect on the controlling nerve centers, would naturally approach the subject by exploring first the actions and innervation of the less modified muscles that move the more normal part of the pectoral fin rays.

The Versatile Flying Gurnard.—Perhaps the most amazing fish of the entire order is the "flying gurnard" (*Dactylopterus volitans,* fig. 9.145). Beebe and Tee-Van, who have studied this fish in its native element, write thus (1933, pp. 184, 185):

> "These are probably the most versatile fish in the world for they can swim, glide, float, fly, and walk, using their fins in turn as hands, feet, oars, and wings. The flight is less strong than that of flying fish, the head being encased in almost solid bony armour. The moment the Gurnard touches the bottom the two pelvic fins are lowered and the fish steps daintily ahead or to one side, one fin after the other exactly like legs. It is not a common fish anywhere."

The creative forces of evolution have added to this fish a huge buckler covering the top of the shoulder-girdle and uniting with the expanded skull roof into a bony cephalo-thorax or carapace; this is a convergent resemblance to the petalichthyid placoderms. As a curious convergence with the ceratopsian dinosaurs, which likewise had a great bony hood over the nape, the first three vertebrae of the dactylopterids are elongate and have united into a strong tube or pivot for the support of the huge skull. Meanwhile the lower corner of the enlarged preopercular has been sharply produced backward to form a marginal framework upon which the heavy body may rest. The opercular and subopercular, overshadowed by the adjoining elements, have become very small. Comparison with the skull pattern of *Peristedion* (fig. 9.144C1) is made difficult by the great expansion of the shoulder plates, but the distribution of the lateral line tubes remains fundamentally the same and the ancestors of *Dactylopterus* must have passed through a stage that much resembled that of *Peristedion.* Starks, however, points out that it is "difficult to understand why this evidently aberrant trigloid form should possess the typical percoid shoulder-girdle of the main line of descent rather than that of its immediate relatives of the family Triglidae, which have the cottoid shoulder-girdle." In view of this and other important peculiarities that tend to separate the Dactylopteridae from the typical scorpaenoids, Regan (1929, vol. 9, p. 323) set them apart as a superfamily Dactylopteroidae, of the order Scleroparei; later, in the classification used in the British Museum (Natural History), they are segregated as a "Suborder Dactylopteroidea" of the order Percomorphi.

In brief, from the data reviewed in my monograph on "Fish Skulls" (1933, pp. 321–344) it appears that the cheek-armored fishes may be divided into five main series (fig. 9.146), including: (1) true scorpaenids, (2) triglids, (3) cottids, (4) hexagrammids and *Cyclopterus*; (5) *Dactylopterus,* which may have been derived quite independently of the triglids, possibly from near *Aspicottus*. *Scorpaena* is surely near or on the common stem.

THE VENTURESOME GOBIES AND CLING-FISHES (GOBIOIDEI, XENOPTERYGII)

By contrast with some of the huge and widely diversified groups already described, the gobies are a very small and compact tribe, all conforming to an easily recognizable type suggesting excessive alertness and great speed in darting about. Beebe and Tee-Van ("Bermuda Fishes," p. 215) state that "Gobies are adapted to withstand very unfavorable conditions of living, such as vitiated tidepools and thick mud, and can survive in a neglected aquarium after all other fish have died."

In side view the goby body (fig. 9.147) is moderately elongate with nearly parallel or but gently sloping dorsal and ventral lines; the posterior half with long-based, large and nearly equal dorsal and anal fins, the tail long, but little extended vertically and with convex or pointed tip. Spiny dorsal usually high and conspicuously distinct from soft dorsal. Pectorals spreading, inserted vertically. Pelvics usually appressed or united into a sucker by means of which the little fish clings to the rocks even when the waves dash over him.

In the mud-skipping goby (*Periophthalmus,* fig. 9.147B) the very large pectoral fins with strong muscular lobes are bent downward, serving as springs as the fish hops and wriggles along the muddy shores or banks. The huge eyes protrude above the head. There are no very evident substitutes for the gills but perhaps the fish does not stay out of water long enough for the gills to dry up.

Some of the smallest known fishes belong to the gobies, fully adult individuals of the Philippine *Mistichthys* measuring from 9–10 mm. in total length. The anatomy of these fishes as described by TeWinkel (1935) shows that they resemble larval stages of other fishes in the relatively swollen brain and braincase and in the slenderness of the opercular and suspensorial bones.

The skull of a typical goby (fig. 9.147C, C1) is of modified percoid type: the preorbital part is short, the premaxillae relatively large and percoid, the orbits large and displaced dorsally; the hyomandibular is remarkably wide (anteroposteriorly) and so is the preopercular; the symplectic and metapterygoid together form a strong buttress between the quadrate and the hyomandibular; the quadrate V has widely spread limbs and leaves an unossified space behind and below the symplectic; the opercular is V-shaped, without spine; scapula reduced and widely separated from the small coracoid by a cartilaginous tract; four large pectoral pterygials, posttemporal lacking pterotic fork.

The sleepers (*Eleotris,* A, A1) are large predatory gobies with small eyes and wide frontals. The pelvic fins are not united into a sucking-disc, a condition which may be secondary because their skulls are essentially goby-like in many features.

The clingfishes (Gobiesocidae, fig. 9.148) have a perfected pelvic-pectoral sucking-disc which has attained a relatively huge size and induced many new anisomeral changes in the neighboring parts of the skull. The same may be said of their almost rodent-like grasping front teeth. Nevertheless these fish are tentatively entered here with the gobies because it seems that their remarkable poly-anisomeres are derivable from a gobioid stem.

It is true that Starks's laborious and intensive osteologic comparisons with cottoid, blennoid and gobioid fishes left him in great doubt as to the derivation of these cling-fishes and that ichthyologists continue to give them the rank of a separate order (Xenopterygii), partly because of the apparent greatness of the differences between this family and all other percomorph fish. But we may here advance the hypothesis that though later poly-anisomeres have disguised the gobioid heritage, there are still indications that the cling-fishes have sprung not from typical gobies but from some genus allied to *Eleotris*.

In the first place, although the cling-fishes have lost their spiny dorsal fin, we may safely eliminate all isospondyls because the cling-fishes have a completely percomorph premaxillary-maxillary pattern and their pelvis and pelvic fins are attached to the pectoral girdle, which also lacks the mesocoracoid. The vertical position of the pectoral fins with their many rays suggests a search among some of the later derivatives of the percoid stem, such as the groups in which Starks (1905) sought in vain. But with the greatest respect for his immense labors, we may venture to suggest that his method of elimination may not have been wholly conclusive: first, because it was so strictly objective and secondly, because it relied upon a direct point-by-point checking of miscellaneously compared "single" characters observed without consideration of their functional significance. For example, Starks records the fact that in the cling-fishes three occipital condyles are present, one on the basioccipital, the other two on the exoccipitals, so that the three parts of the condyle are in a horizontal (transverse) line. This is a conspicuous objective difference from most other fishes but it would seem to be immediately derivable from the tripartite basi-exoccipital condyles of any percomorph merely by the downgrowth of the exoccipitals. In view of the enormous increase in size and power of the pelvic suckers (as compared with that of gobioids), it seems not unreasonable to infer that the three condyles in a transverse line enables the fish to move his head slightly up and down in managing his relatively massive nippers, while maintaining a secure stance with his pelvic sucker. As it happens, this particular feature of the three horizontal condyles would be very easily derivable from the conditions in the gobies, where the exoccipital condyles are lateral rather than dorsal in position.

Without delaying here for full analysis of Starks's list of differences between the cling-fishes and other families, we may return to the comparison of this skull with that of the gobioid *Eleotris* (fig. 147A). The nipping apparatus of *Gobiesox* is obviously far the more specialized of the two but both in the side and top views the structural differences are functionally related to the emphasis of the nipping or cutting front teeth of *Gobiesox* and to the retention of less specialized, more percomorph conditions in *Eleotris*. Again, the very short wide skull roof of *Gobiesox* (fig. 9.148A1) is merely an emphasis of the conditions in *Eleotris*. The small nipper-like mandible of *Gobiesox* is very firmly based on the relatively huge quadrate. This quadrate in turn is braced anterosuperiorly by the large symplectic, while behind the symplectic is an unossified space somewhat as in *Eleotris*. In the latter, however, the metapterygoid is exceptionally large but in *Gobiesox* it is reduced or wanting and the emphasis is shifted to the symplectic. One conspicuous difference is that in *Gobiesox* the nasals are exceptionally large, serving as a cover for the large ascending rami of the premaxillae, whereas in *Eleotris*, as in other gobies, the nasals seem to be absent or vestigial. It is hardly possible that these are not true nasals but new polyisomeres, similar to the proëthmoids of the Haplomi but their association with the greatly enlarged premaxillae suggests a secondary enlargement of the nasal Anlagen.

The backward prolongation of the preopercular and opercular in *Gobiesox* is plainly associated with the bracing of both the huge pelvic sucker and the skull itself.

In *Gobiesox* the cleithrum is inclined backward, the supracleithrum is a slender antero-posterior rod, the posttemporal a single short piece connected with the epiotic but with little trace of its originally three-branched condition. These conditions are foreshadowed in *Eleotris*, whose posttemporal, however, retains also its exoccipital branch. One of the apparent difficulties in trying to fix the derivation of *Gobiesox* from the gobies lies in the character of its pectoral pterygials (fig. 9.148), of which two articulate with the scapula, two with the coracoid, whereas in *Eleotris* (fig. 9.147A) three articulate with the scapula and one with the coracoid. Nevertheless it would seem that the huge size of the pelvic sucker in *Gobiesox* might well cause a dorsal displacement in the Anlagen of the pectoral pterygials, lying as they do directly above this predominant organ. In brief, although the hypothesis proposed above is not without its difficulties, it still seems that *Gobiesox* is a highly specialized derivative of the goby stock.

ANOTHER BAD LOT (TRACHINOIDEI): DOUR WEAVERS, LURKING STARGAZERS, OGLING DRAGONETS AND DARTING SAND-LANCES

Trachinus draco (fig. 9.149A), the weaver, is a mean-looking fish with a sharp spine, said to be poisonous, pointing straight backward from his opercular lid. His mouth is drawn down at the corners and his eyes are mostly directed upward, which suggests that he sneaks up on his victim from below. Otherwise his traits are unusual in small predators. He has a somewhat elongate tapering body, with very long-based, nearly equal, soft dorsal and anal fins. His spiny dorsal, above his opercular spine, can probably be suddenly erected when he is in a rage.

From such an ancestor may have been derived the family of flat-headed stargazers (Uranoscopidae); among their redeeming features are one or two novel points in the architecture of their skulls (fig. 9.149B, C, D, D1). In normal fishes the top of the braincase as seen from above forms a narrow triangle somewhat rounded at the apex or rostrum. In the stargazers the skull roof is widely expanded behind the orbits, contracted between the very large orbits and suddenly expanded across the ethmovomer block; moreover, the mesethmoid, usually conspicuous on top of the skull anterior to the frontals, is depressed into the bottom of a trough, while the parethmoid or prefrontal processes, usually lying behind it, are now pushed in front of it. Dr. Starks seemed inclined to place a high diagnostic value on this exceptional position of the mesethmoid, which obviously has resulted from a shift in relative growth rates whereby the mesethmoid lost its usual "drive" and allowed itself to be outgrown by its own children, which are the prefrontal or parethmoid processes. Moreover, after the mesethmoid also allowed itself to be overgrown by the large ascending processes of the premaxillae, it began to subside into the bottom trough that was being formed by the orbital rims of the frontals to enclose those processes.

Meanwhile in *Uranoscopus* the basal layers of the integument have begun to sprout a new crop of polyisomeres which are recorded on the top and sides of the skull by innumerable small pits and irregular papillae. Although this fish has lost the spike on his opercular (fig. 9.149B), no improvement of his disposition is indicated as he has developed a much larger one on the rear upper corner of his shoulder-girdle. The opercular is evidently a bone that has passed the climax of its career for it has not only lost its spike but it has allowed the preopercular to surpass it in size and to develop a vigorous array of ridges and spikes alternative with depressed areas that correspond to the sac of the preopercular branch of the lateral line in other fishes. This fish, growing old and rotund, has a barrel-like cross-section across the frontal region and a capacious rounded mouth with strong jaws. In his relative *Kathetostoma* (fig. 9.149C) the skull-roof is flattened and much wider than it is long—an example of intense anisomeral growth.

It is said that barefooted humans walking along the beaches south of Cape Hatteras sometimes receive a sudden and severe electric shock from an electric stargazer (fig. 9.149D) that is lying hidden in the sand. Doubtless this protective apparatus was evolved by the ancestors of the stargazer untold ages before human feet came to interrupt his astroscopic visions. Not against man was this keen trap directed but against prowling giant conchs or grubbing sting-rays or idly resting flatfishes. And if the intruder were temporarily knocked out by the shock, then the patient but hungry astronomer sprang into furious action.

But by what magic were two large and highly charged storage batteries inserted immediately behind the eyes? The answer is that the batteries were not inserted but grew up out of something very different in the places where we now find them. Electric batteries in fish of several families always represent greatly transformed muscles and their associated nerves, the minute subdivisions of the muscles and nerves, with alternating thin septa, supplying differences in electric potential, which has been built up in series and multiple units until charges of dangerous intensities have been accumulated. It is known that very minute electric discharges accompany all neuromuscular activities. In the case of the electric stargazer, natural selection, the guardian daemon of this race as of every other, working against and with the forces and conditions imposed by heredity, growth and environment, has stressed the electric side of the neuromuscular activities of the recti muscles of each eye, at the same time depriving these elements of muscular force but greatly increasing their size (E. Grace White, 1918).

In the case of other fishes we have seen that the "optic cones," consisting of the eyeball and its associated nerves, muscles and wrappings, have a way of impressing themselves upon the skull, the eyeballs fashioning, as it were, the orbits and the eye muscles penetrating the floor of the braincase in a cave or tunnel called the myodome (fig. 9.76B). In the electric stargazer the eyes remain small but the huge electric batteries, made out of transformed eye muscles and their nerves, have squeezed the interorbital bridge (fig. 9.149D1) into a very narrow strip and pushed back the normal side walls of the braincase behind the eyes into a nearly transverse position. Meanwhile the myodome, vacated by the eye muscles which have moved upward, has been erased.

Not far off to one side of the stem of the trachinoid series are the nototheniids of the antarctic and far southern coasts. Of these, *Cottoperca gobio* (fig. 9.149E), as its name implies, combines features of the cottoids and perches but is structurally nearer to the weavers (*Trachinus*). A cottid-like spike is sometimes borne on the opercular (figs. 9.149E; 150A, B, E), but the relationship to the cottids is not at all close.

As in all this suborder (Trachinoidei), the pelvic fins instead of being immediately beneath the pectorals (as in primitive spiny fins) have been pushed forward under the throat (figs. 9.149, 150), whence the name Jugulares, used to include this group, with others, such as the blennies. One of the more specialized of the nototheniids is *Parachaenichthys georgianus* (fig. 9.150B, B1) from the deep sea off the island of South Georgia, with its somewhat duck-like nonprotrusile snout. In this family anisomeral selection has produced an array of long heads (B1), broad heads (A1), and in-betweens (C), much as it has in dogs, pigs, and humans.

In *Pinguipes* (D) the skull tends to connect the nototheniids with the percoid stem, and the relations of the pectoral baseosts or pterygial bones testify to the same effect. In short, it seems probable that the nototheniids, and with them the entire trachinoid series, have been derived from such fundamentally perch-like fishes as *Percis nebulosa* (E).

The dragonets (Callionymidae, F, F1) may have sprung from some grim and dour weaver stock but they are now dazzling personalities among the stars of the world of fishlets. Their huge protruding eyes dominate the head and profoundly influence their responses, especially those of the female dragonets, to the courtship antics of their brilliant mates. And not in vain do these little marine peacocks strut and display the eyespots on their palpitating fins, for courtship in fishes is usually followed by parental care of the young. Thus the reproachful name of Jezebel should not be applied to these delightful sprites of the marine underworld and even "dragonet" seems inappropriate except as it suggests the quick-darting brilliance of the dragonfly.

But to return to our dry bones. The typical dragonets have everted lips and apparently small mouth openings but at least in *Callionymus lyra* (F, F1) the bony mouth-parts form a relatively huge but lightly built protrusile framework which is nearly as long as the braincase. After comparisons of this fish with many other types, it seems that however emphasized its anisomerism may be, with the mouth-parts hugely increased and the operculars greatly reduced, there is nothing in either the skull or body-form and fins of *Callionymus lyra* which could not easily be derived from the conditions in the primitive nototheniids *Cottoperca* and *Percis* (C, E). These fishes already have large mouth-parts, with the ascending processes of the premaxillae (C) received into a fossa lying between the projecting prefrontal (parethmoid) processes, the latter having extended forward in front of the reduced mesethmoids. The dorsally-placed eyes of *Cottoperca* have already begun to restrict the interorbital bridge and forced the braincase to spread laterally. The prominent pelvic fins (E) of *Percis* are in the right position to give rise to the sucking-disc of *Callionymus* (F). The scapula of *Notothenia* forms a bridge between the three expanded actinosts of the pectoral fin much as it does in *Callionymus* (Starks, 1930, pp. 222, 223). We therefore follow Schultz (1943, p. 267), who refers the Callionymidae to the order Jugulares and lists them next to the weavers (Trichodontidae). The probably convergent resemblances between the Callionymidae and *Platycephalus* (fig. 9.141H) of the scorpaenoid group include not only the general body form but the presence of spikes on a backwardly prolonged preopercular.

The sand lances (*Ammodytes lanceolatus,* fig. 9.150G, G1) seem to be appropriately named, since they have a sharply pointed head and a long body with an arrow-like caudal fin. The dorsal fin is very long-based but short vertically. The anal fin is about half as long as the dorsal fin. These external characters are not sufficient in themselves to permit us to refer the sand lance to any of the suborders of teleost fishes. Nor do the skull characters, which very often give such information, help decisively in this case except to indicate that the sand lances are derivatives of the percomorph stock.

Among the peculiarities and distinguishing features of the skull, the ascending branches of the premaxillae are very small but rest on narrow projecting extensions from the mesethmoid. The latter bone is also prolonged far backward between the frontals. Although the long and slender maxillae are excluded from the gape of the mouth in the usual percomorph way, they expand broadly at their upper front ends and cover over the premaxillae. The latter are toothless and very delicate. The dentary is also toothless; the symphysal regions of the opposite dentaries are expanded in a very unusual way. Although the suspensorium is produced forward, the mouth points almost forward, both because the vertical depression of the quadrate-articular joint is at a minimum and because the preorbital diameter is not reduced. The preopercular-opercular and subopercular are all "unarmed" and expanded anteroposteriorly. After considerable search for suitable structural ancestors, I would suggest for future investigation the possibility that the sand lance may be a specialized relative of the nototheniids (in the wide sense) derived remotely from a form like *Percis* (fig. 9.150E) but with individual features recalling those of *Eleginops* (triangular lacrymal, anteroposteriorly elongate hyomandibular and opercular complex) and *Parachaenichthys* (reduced ascending processes of premaxillae, fig. 9.150B1, delicate maxillae, elongate mesethmoid, divergent maxillary processes of palatines, very small supraoccipital). The body- and fin-form of the sand lance could readily be derived from the trachinoid stem.

THE SHARP-EYED BLENNIES AND THEIR SLITHERY RELATIVES (BLENNOIDEI)

The technical language of ichthyology is by no means lacking in convenient and accurate descriptive terms, especially for recording the minutiae of specific differences; but we can find in it no word or words to convey the impression of a living blenny (fig. 9.151A). This alert and determined speck challenges us to throw off the trammels of established scientific usage and find a new name for him and his tribe of round-eyed, thick-lipped fish-brownies of the shores and tidepools. Perhaps a better name was given long ago by some unknown fisherman, who called one of these sprites "Molly Miller" because of its impish resemblance to a certain saucy fish-wife.

The blennies (fig. 9.151) all are highly specialized and they belong to a superfamily or suborder of numerous and strange genera and species which seem to be now in the high tide of their history, but in the fish world of the future they may be expected to give rise to still more bizarre offshoots. For they are now experimenting with several new sets of polyisomeres, of which the functions are still unknown to us if not to their owners. For of what use are the various specifically differing small tentacles and cirri that are found on the nape and above the eyes and nostrils? Do these have a social value, like tribal marks, a sex value, or are they sensory organs connected with the lateral line system? And if so, what do they detect? Another novelty in polyisomeres is that blennies differ conspicuously from gobies and other fishes in having many more spiny rays in the dorsal fin, the spiny part of which is extending its territory backward at the expense of the soft rays. Thus even in the seven species of blennies of Bermuda, as described by Beebe and Tee-Van, the numbers range from XI (spiny rays) 16 (soft rays), XII 16, XVIII 12, XXII 14 or 15, XXIX 1. In Mark's blenny (*Emblemaria marki*), in which the dorsal fin is adorned with black blotches, the males have the 6th, 7th, and 8th spines very long, their height being about 3 inches (*op. cit.*, p. 229); thus in this instance at least there are sexual differences in the dorsal fin. In another blenny (*Auchenopterus fajardo*) the dorsal fin sometimes displays an ocellus or eyespot (*op. cit.*, p. 226) and these features are numerous on the sides of the body in *Clinus despicillatus* and *C. elegans* (fig. 9.151B).

The pelvic fin, which in many otherwise specialized offshoots of the percoid stock retains the primitive one spine and five soft rays, is here reduced to one spine and two or three, possibly tactile, rays. This constitutes another conspicuous difference from the gobies (fig. 9.147), in which the pelvic fins are more normally developed and may act as hold-fasts.

The skull and jaws of the less specialized or clinoid blennies (fig. 9.151A, B) present but few conspicuous departures from the basic percoid type, save perhaps that the mesethmoid forms a tube for the reception of the long ascending processes of the premaxillae, the strong metapterygoid reaches up to touch the side wall of the braincase and the interopercular connecting rod between the articular and the subopercular is unusually stout. At least in some blennies the mouth is very small and the teeth numerous and crowded, like those of chaetodonts. Beebe and Tee-Van (*op. cit.*, p. 227) state that in the soft-toothed tide-pool blenny (*Salariichthys textilis*) the teeth are very small, freely movable, implanted on the skin of the lips, not on the bone of the jaw.

Here then is another new departure in polyisomeres almost without precedent in the bony fishes, although the ceratioids *Rhynchoceratias, Haplophryne* and *Laevoceratias* (figs. 9.158B, 156E1) also appear to have little dermal denticles in front of the toothless premaxilla.

The blennies have enjoyed a wide adaptive branching, especially in northern and arctic waters, but we shall mention here only the cusk-eels (Zoarces, fig. 9.151C) and the sea-wolf (*Anarrhichas,* fig. 9.151D). The cusk-eels are long-bodied, almost eel-like blennies with greatly elongate dorsal and anal fins, the latter continuous with the small caudal fin; the pelvic fins are vestigial. The skull and dentition are essentially like those of the clinid blennies but a few of the more anterior teeth on both the premaxillae and mandible are somewhat stouter, with fairly short conical crowns. In the sea-wolf, on the other hand, the dentition is differentiated into a few strong procumbent, conical pointed teeth (D) in front and crushing dental plates on the vomer and inner sides of the mandible. While the exact relations between the blennies and the gobies is not clear, certain of the latter, such as *Eligniops* (fig.

9.151F) of the nototheniids are far more primitive than the blennies and tend to connect the trachinoid and blennioid groups with the percoid stock. *Malacanthus* (fig. 9.151E) of the latter is related to the serranids and its general resemblances to the blennies may be convergent.

THE OILY CODS AND GRENADIERS (ANACANTHINI)

Concerning the cods and their allies as the base of a huge fishing industry and as source of the cod-liver oil, there is a vast literature and much public information. As an indication of the historic and economic importance of the cod in New England, we have only to cite the words "sacred cod" and "codfish aristocracy." But with reference to the remote founders of the cod family there is and has long been a deadlock between two very different theories. According to Garman (1899), the families of the cusk-eels (fig. 9.151C), or Zoarcidae, often classed with the blennioids, the serpent fishes (Ophidiidae, fig. 9.152B) and the deep-sea brotulids (A) are related to the cods (C, D, E) and grenadiers (F), and therefore belong in the order Anacanthini (fish without spines). Cockerell (1916) finds that the scales of the brotulids resemble those of the serpent fishes and certain genera of the cods, while Trotter (1926) records several physiological resemblances between brotulids and grenadiers (macrurids). On the other hand, Regan (1910, p. 11) considers that "the absence of spinous fin-rays, the large number of rays in the pelvic fins and the indirect attachment of the pelvic bones to the clavicles are evidences that the Anacanthini are much more generalized than the ophidioids, near which they have been placed by some authors. They are perhaps derived from generalized scopeloids, such as the Aulopidae." Berg (1940, p. 455) cites the presence of supposedly primitive characters in this group (e.g., opisthotic distinct and very large, pierced by a foramen for the glossopharyngeal nerve, the position of olfactory bulbs close to the nasal capsules, the absence of spines from fins, the presence of cycloid scales) accompanying numerous admittedly high specializations. "As a whole," he writes, "I am inclined to regard the Gadiformes as a lowly organized order derived from forms allied to Pachycormidae, probably at the end of the Cretaceous."

The cods and their immediate allies, like many other fishes, have chosen, as it were, to multiply the rays of the dorsal fin and among these the dolphins (Coryphaenidae), the flatfishes (Heterosomata), for example, have greatly increased the number of functionally soft, or not spiny, rays. Some of the blennies, however, show a marked increase in the number of "spiny" rays, which in one case extend backward until only one "soft" ray is left. But it is a well established principle that when secondary polyisomeres evolve, their peculiar characters usually spread along a series from one to another, often lessening the initial differences between an older and a later set—as in teeth, vertebrae, dermal ornaments, etc. Also it sometimes happens (as in certain sciaenids) that between the spiny and the soft rays there are one or two rays of transitional character, as occasionally shown also in the anal fin. Hence the fact that there are no spiny rays on any of the fins of the cods does not in itself prove whether these soft rays are "primary" or "secondary" polyisomeres. That the dorsal fin of the cod (fig. 9.152D) is divided into three parts instead of the two in related genera, also suggests a high degree of secondary polyisomerism and this is not favorable to the implied claim that the cods are more primitive than typical percomorphs because they have only soft rays in their fins. The same objection applies to the argument that cods must be relatively primitive because in some of them the number of rays on the pelvic fins is much more than the six (one spine, five soft rays) of typical percoids. On the contrary, it would seem that the cods have reached such a stage of instability that the old and hitherto reliable numbers are varying.

The indirect attachment of the pelvic girdle to the shoulder may perhaps be rather consistent with the generally weak and variable character of the skeleton as a whole. And surely no one who had witnessed, so to speak, the great multiplication of the caudal vertebrae in other specialized fish, such as the eels and brotulids, would regard as a primitive character the absence of normal-looking teleost hypural bones in this group.

The cod skull is also evidently specialized in many ways but I can not see in it any significant resemblances either to *Aulopus* or other members of the Iniomi. On the other hand, the premaxillae of *Lota* (fig. 9.152E), although with more or less reduced ascending rami, recall those of *Zoarces, Clinus* and other blennioids. In a young *Melanonus* (fig. 9.152C) the opercular has the conspicuous central spike seen in many percoids, while the relatively small size and shape of the opercular as compared with the subopercular in *Gadus* and *Lota* can be nearly matched in *Zoarces* (fig. 9.151C). Moreover, the pedicle for the opercular, which comes off from the hyomandibular, is elongate and exposed to view in the gadoids much as in the blennies. The interopercular also is remarkably large, as in *Clinus* (fig. 9.151B) and *Zoarces* (fig. 9.151C).

In the deep-sea grenadiers (Macruridae), which are commonly referred to the Anacanthini, the skull (fig. 9.152F) is excessively specialized in its almost paper-like quality, as well as in being marked with great depressions for huge lateral line or mucous sacs. But it retains an almost typically percoid premaxilla with a long ascending process which is received into a tunnel in the rostrum much as it is in the blennies. The body-form of the grenadier *Coryphaenoides carapinus* (F) somewhat suggests that of a chimaeroid (fig. 8.1) with its large rounded head, prominent rostrum, high feather-like dorsal fin, greatly elongated anal fin and tapering tail wisp—a good example of both convergent evolution and deep-water habit.

The brotulids (fig. 9.152A) mentioned above tend to clarify the problem of the origin of the cods, for their body-form is somewhat like that of *Melanonus* (fig. 9.152C) among the cod group, while their skulls, including the opercular-branchiostegal series, are much more primitive-looking, with more normal percoid premaxillae and maxillae, the latter even retaining a small single supramaxilla. The relations between the opercular, with its prominent posterior and inferior spikes, and the subopercular of the brotulid *Dicrolene* also seem to present a favorable stage pointing toward the cods.

In *Fierasfer* (fig. 9.152B), the famous little fish that makes its home in a sea-cucumber, the long and wonderfully colored body tapers gently to a point, as do the greatly elongated dorsal and anal fins. The skull is evidently close to the less specialized brotulids. The connection between the brotulids and the blennies seems to be by way of *Zoarces* (fig. 9.151C) and its allies.

In short, the available evidence still seems to indicate that Garman (1899) was right in referring the brotulids to the same order with the cods (Anacanthini) and that the cods are secondarily soft-rayed derivatives of the spiny-finned stock and not far from the blennies except in adaptations to deep-water life.

THE UNSOCIAL TOADFISH (BATRACHOIDEI)

In a large aquarium tank with many strange and highly colored fishes moving about, the visitor may not notice an almost shapeless, dull colored and warty mass that is sulking in a corner and looking out with a spiteful eye on his less inhibited fellow citizens. Such will be the toadfish (fig. 9.153A). But if through sudden change in temperature or other causes the glass in the aquarium cracks, the water seeps out and the gay tropical fishes gasp and die violent deaths by suffocation, the toadfish will be there, sulking and gloating as usual and ready to bite the net that tries to lift him into better quarters.

The toadfish has a corpulent, tadpole-like body, long-based, flap-like dorsal and anal fins and miscellaneous tags and excrescences around his wide mouth. The color pattern is very irregular, with dark brown to olive streaks and blotches, more or less changeable to match the background.

The skull of the toadfish (fig. 9.153A, A1)

affords a brilliant example of a system of balanced, V-shaped trusses based on the braincase. The inner upper jaws of the vomero-palatine tract and their teeth are much better developed than the premaxillae and their teeth; they probably work together with the teeth of the mandible for grasping and holding the prey.

As to the origin of the group, the hypural bones have certain peculiarities which led Regan (1912, p. 279) to suggest the possibility that the batrachoids may have come off from a prepercomorph stock, such as the Salmopercae. But it hardly seems that this one point should outweigh the features in which they recall the trachinoid stargazers (fig. 9.149B, C, D) and other advanced derivatives of the percoid stem; for example, in the peculiar pattern of the top view of the skull, in which the broad cranial table suddenly contracts into a trough-like depression which receives the ascending rami of the premaxillae. Other suggestions of relationship are found in the general body-form of the nototheniids among the trachinoid group (fig. 9.150A-E). But I have been unable to find any one fish that has all the requisite features for connecting the toadfishes with some less specialized family.

GOBLINS WITH FIN-HANDS (PEDICULATI)

The world of fishes is doubtless made up of countless smaller worlds of more or less enduring association systems. Each of these has its limiting conditions which isolate it sufficiently for us to distinguish it and name it as a system. For instance, in many kinds of fishes the young live in different "worlds" and feed upon different foods from those of their parents, so that as they grow up they are graduated, so to speak, from one world into another. Again, some indefatigable fish-travellers actually propel themselves from their homes on the ocean floor up the long slopes and submarine canyons to the rivers, and thence against the current and even up over the falls to their breeding-grounds in the highlands of the interior.

The sargassum fish (*Histrio* or *Pterophryne* or *Antennarius,* figs. 9.154A; 156A, B), on the contrary, is content to be a passenger in a floating world, a clump of sargassum weed. This seaweed flourishes mightily along the shores of Bermuda and the West Indies and sends forth myriads of its clustering fronds. Supported by little round gas balls, they float outward into the far-flung ocean currents and some of them survive the long journey to the Sargasso Sea or other parts of the tropical Atlantic.

The sargassum weed is an alga of the brown class and as seen against the intensely blue water it is a mass of yellow, dark brown and paler tints combining into high lights, blotches and shadows. Its varied inhabitants likewise are clothed in similar colors and blotches, so that it must be difficult for our little sargassum fish, himself disguised in the same way, to detect the outlines of the naked sea-hare or of the shrimp or the tiny pipefish that are also amazingly well camouflaged in the weed. During the New York Zoological Society's expedition to the Sargasso Sea in 1925, under the leadership of Dr. William Beebe, many live sargassum fish were kept in the aquaria on board the *Arcturus* and from studies made by Mr. Dwight Franklin and myself at that time the following passages may be quoted (Gregory, 1928, pp. 408, 409):

"*Pterophryne* is a short and fairly deep, thickset, carnivorous little fish, with an upturned mouth and great hand-like pectoral fins with movable elbows; it has a prominent backwardly extended dorsal fin and downwardly projecting ventral fins that end below in large white 'feet.' Its golden-brown ground-color with irregular patches of dark brown, flecked with little white circular spots, forms a perfect camouflage as the fish lurks on the gold and brown weeds.

"When swimming slowly the principal thrusts were caused by the rhythmic jets of water from the small rounded gill openings, modified by the gentle undulations of the pectoral fin membranes.

"When crawling along the branches of the weed the *Pterophryne* sometimes moved as if it were stalking the alert little fishes and crustaceans upon which it feeds. One long pectoral flipper would be slowly swung forward while the opposite one was moving backward, the body being supported below by the large white feet, which turned outward and shuffled away in the well known manner of the cinema comedian.

"When resting in the weed the fish maintained his position with all four paired fins and with as many median fins as could reach parts of the weed. One huge pectoral 'arm' would be extended almost straight upward, the finger-like tips of the dermal rays clutching a branch of the weed that hung down above the fish; the opposite pectoral was thrust downward and reflected at the 'wrist,' the 'palm' turned outward and forward and the palmar side of the 'fingers' touching weed. One long 'foot,' following another branch of weed, was cocked forward and upward; the other, reaching still another branch, was directed backward and downward. The posteriorly elongate part of the dorsal was folded over and served as another prop, and at other times the caudal and anal fins also cooperated in keeping the fish securely placed in spite of the movements of wind or wave.

"But it would be a mistake to infer that *Pterophryne* was always a sluggish, slow-moving fish. When one of these fishes was placed in a large pan and attempts were made to catch it by hand, it made great flying leaps, such as it may have made in sudden dashes after its prey or in overtaking the weed after brief excursions."

The ability of the sargassum fish to use its pectoral fins almost as arms and hands doubtless depends primarily upon an unusual degree of differentiation of their muscles and nerves. Even the skeleton reflects this ability inasmuch as the joint between the basal piece and the elongate pterygials (fig. 9.156A, B) simulates a shoulder, while the next joint, between the largest pterygial rod and the fin rays, suggests an elbow (B). This "elbow bone" also has a curved expansion bearing the articular border for the fin rays and this border is set at such an angle to the shaft of the "ulna" that the fin rays are directed upward and backward, or, when the fin is turned over, forward and downward. A third and more slender pterygial rod (B) is parallel to the two main elements, which for convenience may here be called ulna and radius. These fan-like or foot-like pectorals are remarkably persistent throughout the order, to which they give the name Pediculati. Thus two out of the four pterygial polyisomeres of primitive percoids have been selected for emphasis in the pediculates.

Another instance of anisomeral selection from a series of polyisomeres has given rise to the most remarkable of all the curious features of the pediculate fishes. This is a complex apparatus of which the most conspicuous part is called an illicium, or lure. This represents the first ray of the dorsal fin (cf. figs. 9.154). At an earlier stage of evolution this fin had already extended forward on to the top of the snout. The third ray, above the eye, is much larger than the illicium but is not as freely movable. The upper end of the illicial ray often ends in a little tag-like or bulbous tip. The illicial ray rests below on a basal rod (A) which represents a modified pterygiophore. Three pairs of muscles arising from the sides of the illicial trough (fig. 9.158B1) are inserted on the sides of the illicial pterygiophore: the first pair elevate it, the second pull it to one side or the other, the third depresses it. By means of these muscles the little tabs or filaments on the illicial ray can be agitated, probably in a very delicate way, to focus the attention of the victim until the crafty sargassum fish has crept up from below near enough to strike. We may infer that the fish strikes by lunging forward and at the same instant jerking down his low jaw and expanding the cavity of his throat so that the victim, if not too large, is instantly sucked into the interior. It may be noted that the three sets of muscles which agitate the illicium appear to be severally homologous with the three sets (fig. 9.110A1) which in teleosts generally elevate, depress and wave each dorsal fin ray and which arise from the front and sides of its basal pterygiophore.

The skull (fig. 9.154) of the sargassum fish abounds in special adaptations for the support of the illicium, for the sudden capture of the prey and for the delivery of the latter to the globose abdominal sac. First, the illicium and its basal pterygiophore are supported in a large wide trough on the front half of the skull roof. The rear part of the trough is bordered by raised rims on the frontal bones; the middle of the bottom of the trough is contributed by the depressed mesethmoid (*eth*) and by a median projection of the frontals. The trough is bounded laterally by the parethmoids, or prefrontals.

Second, the strong jaws are supported on each side by a stout truss comprising the hyomandibular and the palato-pterygo-quadrate (A). The mouth is directed upward as a joint result of (a) an extreme depression of the quadrate-articular joint; (b) the forward swing of the lower part of the suspensorium and (c) the relative shortness of the distance from the tip of the premaxilla to the middle of the hyomandibular attachment to the cranium. The outer jaws are of modified percoid type, but the ascending branches of the premaxillae are rather short and when the mandible is depressed they doubtless roll down over the transversely-expanded vomer (A1, A2).

Third, the braincase is braced in the rear by having the large first vertebra united with it (A1). The supraoccipital is expanded laterally and extends forward to its strong union with the frontals, drawing the enlarged epiotics with it on to the cranial roof; meanwhile the expanded posttemporal (*ptm*) is closely appressed to the braincase in order to afford a firm anchor for the shoulder-girdle, which is greatly elongated downward (A). The frontals, parietals, sphenotics and pterotics together form the stiff base for the very large suspensoria.

Fourth, the very small opercular bone (*op*) is directed backward and downward and the forked subopercular is tightly appressed to it (fig. 9.154A). This peculiar arrangement is conditioned by the fact that the opercular-branchiostegal flap is greatly enlarged and produced backward into a tube. Thus most of the opercular slit, which was normally in front of the shoulder-girdle, has been obliterated and the respiratory current is carried out through a groove or tube that passes around under the subopercular and opens behind the cleithrum and even underneath the base of the pectoral fin. By the rhythmic contraction of the expanded throat the water is shot out from these openings at each "breath" and these streams seem to have considerable propulsive force (Breder).

The different species and genera of sargassum fishes show several stages leading toward the ceratioids or oceanic anglers. In these animated black bags (fig. 9.155) the illicium reaches almost unbelievable heights of specialization. Even in *Antennarius stellifer* of the sargassum-weed fishes, at the tip of the illicium there is a tiny sphere from which project numerous delicate filaments. The next step was to intensify the dark and light polarity of the pigments in this sphere and to improve the skin-glands that produce a substance (? luciferin) that glows in the abyssal darkness. Next, the little secondary polyisomeres near the tip began to acquire anisomeral emphasis and specific individuality; in some ceratioids (fig. 9.158E, F) they cooperate to give a general effect of a luminous shrimp with many paired appendages—probably a sure bait for the fish sought by their owners; in others (C, D) they give rise to a glowing bulb with one or more crests, finger-like processes or hooks or filaments. The height of apparent impossibility is attained in an abyssal fish described by Regan as *Lasiognathus saccostoma* (fig. 9.157B), in which the basal pterygiophore serves as the fishing-pole, the illicial ray as the line, which ends in three little hooks. The huge lip of this floating fish-trap is eversible and bordered with a row of curved bristle-like teeth which may be useful in raking in the catch. In another direction, *Linophryne arborifer* (fig. 9.155), not content with a stout light bulb, has developed a new set of luminous polyisomeres in a branching imitation of seaweed depending from his lower jaw. The ceratioids also seem to be in the midst of a period of genetic branching, in which new variations are being played in such polyisomeres as the number of rays (fig. 9.155) in the dorsal, anal and caudal fins, the number of branchiostegals (6 or 5), the number of pectoral pterygials (2, 3, or 4), the number of separate heads on the hyomandibular (2 or 1), the presence or absence of stalks, papillae, etc., on the lateral line organs, the presence of secondary "teeth" or denticles on the rostrum in front of the premaxillae in *Rhynchoceratias* (fig. 9.158B) and *Haplophryne* (fig. 9.156E).

Even more amazing than any of the traits described above is the severe reduction in size and almost parasitic habit of the males (figs. 9.156E, F; 9.158B) in certain genera of ceratioids. In fact the males, in comparison with their relatively huge mates, look like tiny larvae or fry of some totally different kind of fish and even when full grown they may at first sight look more like a dermal appendage than like a fish. In most cases the mouth of these little creatures has become secondarily reduced in size and provided with a nipping apparatus (fig. 9.156F) made out of thorny projections on the rostrum and tip of the lower jaw (fig. 9.158B). The more or less globose (fig. 9.158A) body-form is assuredly secondary and so is the great enlargement of the "olfactory" organs, the lack of an illicium, the hypertrophy of the testes and the reduction of the other viscera.

Small free-swimming males which have not yet assumed such extreme specializations are recorded in several families of ceratioids by Regan and Trewavas (1932). These authors close their account of sexual dimorphism in the ceratioids with the following interesting passages:

"Some males, except that they have no illicium, scarcely differ from the females, whereas others may differ from them in various ways. In some families the males become sexually mature as free-swimming fishes, in others they do not become mature until they have attached themselves to the females and have become parasitic on them.

"As an illicium, or line and lure, is characteristic of the Angler-fishes, it is evident that it is the female Ceratioids that most nearly retain the structure and habits of the original members of the group. But in the darkness of the ocean depths, if the fishes of both sexes floated about alone, attracting prey by means of a lure, it would be difficult for them to find one another. It is the males who have changed, have lost the lure, have adopted more active habits, and seek the females."

Let us now return almost to the base of the pediculate order and consider the fishing-frog, angler or goose-fish, *Lophius* (figs. 9.153B, B1), which as compared with any of the sargassum-fish or ceratioids attains gigantic size. These creatures sometimes come up into shallow waters and it is said that they have even engulfed a live duck. From the author's work on "Fish Skulls" (1933) we may quote the following passages (pp. 394–397) relating to *Lophius:*

"The skull of *Lophius* is much more specialized than that of *Antennarius* in many details connected with the marked benthonic habitus, but the heritage is evidently antennariid. I thought at first that *Lophius* stood nearer to the starting-point than did *Antennarius* but, as noted above, further study has convinced me that the opposite is the case.

"In the lophiids the fishing habits of the group attain their typical development. The successful fisherman is one who knows how to sit still and wait, while keeping his eye steadily on the bait, and for this congenial task the lophiids are eminently well adapted. In the first place, their enormously wide heads are flattened beneath so that they can rest comfortably on the sand, while the powerful pediculate pectorals and advantageously placed pelvic fins doubtless enable the fish to spring suddenly upward at the critical moment. As in the antennariids, the exhalent current, instead of escaping in front of the pectoral girdle in the ordinary way, is led around through a special tunnel in the skin, which in the lophiids opens in the lower axil, just behind the 'fore-arm' of the pectoral fin. Meanwhile the strong development of the pulsing opercular flap, together with the immense deepening and widening of the mouth and throat, has caused the extension and marked narrowing of the opercular apparatus into a tracker-like interopercular, a slender subopercular with a forked posterior end and a narrow opercular. The branchiostegals, sharing the excessive expansion of the throat, have become very long and slender, while the supracleithrum is pulled out into a narrow rod.

"The neurocranium of *Lophius* (fig. 9.153B) seems to me to be more specialized than that of *Antennarius* in the following features:

"(1) it is much widened and flattened, in connection with the increase in the transverse diameter of the mouth;

"(2) while the eyes retain their position immediately in front of the hyomandibular, the preorbital portion of the neurocranium has been lengthened and widened into a shallow trough, floored chiefly by the anterior wings of the frontals, which are secondarily widened;

"(3) perhaps in consequence of the flattening of the head the parasphenoid has established a broad contact with the frontals in front of the orbits, which braces the enlarged interorbital bridge;

"(4) thorn-like processes have been developed on many points on the surface of the skull.

"As *Lophius* dates back to the Upper Eocene, the time of the supposed origin of the lophiid from the antennariid stock must probably be not later than Basal Eocene or Upper Cretaceous."

The sea-bats (Ogcocephalidae) are excessively specialized for shuffling about on the sea bottom (fig. 9.159). Their very large pectoral fins and girdle are functionally and structurally combined with the head into a broad cephalothorax convergently resembling those of certain cephalaspids (fig. 6.11), armored catfishes and skates (fig. 8.60). They differ widely from *Lophius* in their small mouth parts, which are essentially like those of the sargassum-fishes (fig. 154A). Their illicium, if retained at all, is much reduced; it is essentially of sargassum-fish type but is often overarched by a secondary rostrum (fig. 9.159A, A1). This rostrum seems to have been derived from the enlarged dorsal fin ray which in the sargassum-fishes (fig. 9.154A) forms a great dermal horn above the eyes. We may suppose that in the ancestors of the batfishes this horn came to bend forward rather than backward and so began to overhang and overshadow the diminishing illicium; meanwhile this pseudo-rostrum attracted support from the lateral flanges of the frontal (fig. 9.159A-A2), which had formerly served as the sides of the illicial trough.

The genus *Chaunax* (fig. 9.159B) tends strongly to bridge the structural gap between the batfishes, the fishing-frogs (*Lophius*) and the sargassum-fishes (Antennariidae): its small illicium (fig. 9.159B, B1) is received into a crypt, representing the usual illicial trough, but the secondary rostrum is not developed; on the other hand, its pectoral fins are enlarged and directed toward the ventral side of the body as in *Lophius* (fig. 9.153D1) and the sea-bats (fig. 9.159A). So also the operculum (fig. 9.159B1) bifurcates distally and articulates with the enlarged suboperculum; in *Lophius* (fig. 9.153B1) and the sea-bats (fig. 9.159A, C) the subopercula (*sop*) become still further enlarged to form the lateral margins of the cephalo-thoracic shield. The climax of specialization in the sea-bats is attained in *Halieutichthys* (fig. 9.159C), in which the cephalo-thoracic disc has become subcircular, while the enlarged pectorals project like legs from the rear of the disc; meanwhile the pelvic fins on the under side of the disc (fig. 9.159C2) look like another pair of legs in front of the main pair, so that the batfishes like other pediculates (figs. 9.156A, B; 159B, B1, C2) is a pseudo-quadruped in which the hind feet are in front of the fore feet!

Gregory and Conrad (1936) have elsewhere analyzed more fully the evidence which tends to show that among the existing pediculates the sargassum-fishes stand at least near to the structural base of this order, which divided on one side into the sea-devils (ceratioids) and on the other side into the fishing-frogs (lophiids) and sea-bats (ogcocephalids). But whence came the ancestral sargassum-fish? One of the lowest of the sargassum-fish series seems to be the *Sympterichthys verrucosus* of Australia (fig. 9.154B), in which the second and third rays of the dorsal fin are still connected with a common web. But this fish already has the pedunculate elbow in his pectoral fin. The peculiar fish named *Tetrabrachium ocellatum* (fig. 9.154C), which seems to be related to the stem of the pediculates (B), suggests that the obliteration of the upper part of the postopercular cleft was accompanied by a great backward expansion of the flap that contained the branchiostegal rays (fig. 9.153B), which must have formed a pulsating throat-pouch not unlike that of eels and morays. A fold in the skin between the lower and upper sets of branchiostegals (fig. 9.154C) formed a vent for the pouch and became the branchial spiracle of pediculates, while the upper part of the branchiostegal flap overlapped the cleithral bar (fig. 9.156) and the skin on the flanks.

That the toadfishes (Batrachoidei, fig. 153A, A1) are related to the stem of the pediculates seems probable. The stock of the weavers, notothenoids and stargazers (Trachinoidei) seems to afford a favorable combination of characters: in general body-and-fin-form, in the distinctness and individuality of the spiny dorsal, in the presence of a trough above the reduced mesethmoid, etc., all of which might be expected in the remote common ancestors of both the toadfishes and the pediculates.

EEL-LIKE LEFT-OVERS (OPISTHOMI, SYMBRANCHII)

If Sir John Mandeville or some equally credulous early traveller had reported the presence in India of a fish which may be drowned in the water, subsequent commentators might well have classed this story as just another myth. But in Francis Day's monograph on the Fishes of India he states that the Indian fish *Rhynchobdella aculeata* "conceals itself in the mud and becomes drowned in water if unable to reach the surface, as it apparently requires to respire air directly." In this peculiar behavior this highly specialized and eel-like teleost resembles the eel-like African lungfish *Protopterus,* which also suffocates in water if prevented from breathing air (Homer Smith, 1932). This is a clear case of convergent evolution in widely different organisms.

While the general body-form of these Asiatic and African fishes is eel-like (fig. 9.160A, B, B1), the presence of spines in the dorsal fin justifies us in calling them eel-like spiny-fins rather than spiny-finned eels. They have a peculiar long nasal tube, which is covered by the elongate nasal bones. We may suspect that these tubes harbor accessory respiratory sacs, as the fish is supplied with branchial arches of percoid type. The neurocranium itself forms a tube, tapering gradually in front. The combination of elongate snout and slight depression of the quadrate-articular joint causes the moderately-sized mouth to open forward, while the overhang of the premaxillae slightly beyond the front of the lower jaw gives the mouth a slight but unusual downward tilt. The posttemporal bone is absent and the pectoral arch is unusually far behind the skull, whence the name Opisthomi (behind shoulder). The technical details of the skull bones merely indicate that *Mastacembelus* (fig. 160A) is a derivative of the percomorph group, but have not yet yielded decisive evidence as to which one of the scores of "families" of percomorphs may claim to be its next of kin.

Another deceptively eel-like type of fish is the cuchia (*Amphipnous*) of India and Burma, which according to Tate Regan (1929, p. 327) "spends the greater part of its life out of the water, wriggling along the banks in which it burrows during the dry season. It visits the water in search of food [including] worms, crustaceans and small molluscs." Boulenger (1910, p. 599) states that this amphibious fish, when in the water, constantly rises to the surface for the purpose of respiration, and it is often found lying in the grassy sides of ponds after the manner of snakes. Thus it is one of several fishes of different orders which have been able to come up out of the water or burrow in the mud by virtue of possessing a respiratory air-sac of some sort. Regan (1929, p. 327) notes that in the *cuchia,* the air-breathing sacs are a pair of diverticula of the pharynx which lie on each side of the back-bone above the gills. According to Boulenger (1910, p. 598), only second of the three branchial arches bears gill-filaments; the third supports, in their place, a thick and semi-transparent tissue. He also states that the principal organs of respiration are two small bladders, resembling the posterior portions of the lungs of snakes, which the animal has the power of filling with air immediately derived from the atmosphere.

The skull of *Amphipnous* is stated by Tate Regan (1912, p. 390) to differ from that of the Symbranchidae only in features which are connected with the presence of these respiratory sacs, which have pushed away the pectoral arch from the skulls, so that the posttemporal is reduced or absent. On the outside the sacs are covered by the operculum and suboperculum, which are enlarged and form thin, almost membranous laminae (Boulenger).

In the related family of Symbranchidae the generally eel-like form (fig. 9.160B, B1) is associated with a tubular skull of a quite different type from that of "eel-like spiny fins" (*Mastacembelus*). In a few ways it suggests that of the so-called "electric eel," which is a highly specialized offshoot of the characin stock (fig. 9.54). The resemblance, however, is surely an example of convergence. This skull (fig. 9.160B, B1) is remarkable for having lost more than six of the bones of the normal fish skull (Regan, 1912). Thus there is little evidence in it for derivation from the percomorph stock, as all the usual landmarks have disappeared. The supraoccipital (fig. 9.160B1) is minute and the large parietals meet in front of it and separate it widely from the frontals. This is usually the sign of a pre-percoid stage but in the peculiar Australian genus *Alabes* (fig. 9.160C), which has some claims to ordinal relationship with the symbrachoids, the parietals are widely separated by the supraoccipital, which is in contact with the frontal (Regan). Another percoid character of *Alabes* is that the ascending rami of the premaxillae (C1) are very large and the toothless maxillae are excluded from the gape.

The geographic range of the symbranchoids (Central and South America, West Africa, India, Southeastern Asia, New Guinea) suggests that they are relicts from Mesozoic times when the connections of continents and seas favored east-west distribution. The fact that one of them (*Macrotrema* from Penang and Singapore) is a marine form does not necessarily offset the other evidence that these air-breathing fishes, like the lung-fishes and characins, are essentially creatures of the inland waters and incapable of traversing long stretches of open seas. On the other hand, the extremely thorough way in which they have succeeded in eliminating definitive traces of their ancestral origin does not necessarily indicate that they are remnants of some single exceptionally ancient teleost stock, because the independent assumption of a more or less eel-like habitus has been accompanied by profound and relatively rapid modifications of the skull and branchial region in several other instances (e.g., the true eels and morays, the "electric eels" among the characins, the eel-like spiny fins or mastacembelids). In these cases a marked lessening of the vertical diameter of the head is accompanied by a relative increase in the anteroposterior spread of the hyomandibular-opercular region, with diverse adjustments of many bones of the neurocranium and the palatopterygoid arch.

So this long review of the bony fishes, albeit including only relatively few examples of the principal groups, comes to an end with a miscellaneous lot of indeterminate, false eels. But they, like many another tribe, still bear a stimulating challenge to students of evolution. For while we may record how a fish behaves and learn by experiment and comparison what are the nearer sources of such behavior, we can not be fairly said to understand the fish until we have settled his approximate place on the genealogical tree and at least the broader phases of his ancestral history.

RETROSPECT: EVOLUTION EMERGING—FROM PALEOZOIC GANOIDS TO RECENT TELEOSTS

The cave-man who scratched amazingly good pictures of salmon on an antler (Osborn, Men of the Old Stone Age, 1915, p. 406) unconsciously recorded the fact that the dorsal contour of a fish differs from its ventral contour. But this dorso-ventral differentiation is so nearly universal in animals and plants (to say nothing of man-made objects) that it usually passes unnoticed. We may therefore fail to ask ourselves why and how so many things come to have tops and bottoms.

The first *why* is explained by the fact that fish, like other earthly things, are adjusted to maintain normal spatial relations with the terrestrial ball on which they live. Dorso-ventrally flattened fishes (figs. 9.154D; 9.159A2, C) look indeed as if they had been widened in response to the habit of resting on or clinging to the bottom, although possibly they may so rest only because they have become widened. In either case it is the all-pervasive force of gravitation to which elaborate adjustments have been made.

Since Lamarckism, or immediate transmission of use-effects from parent to offspring, is ruled out by consensus of modern judgment, we may well ask how it is that free-swimming fishes that do not rest on the bottom also have a ventral surface which differs considerably from the dorsal surface. The first answer is that the embryo fish (fig. 9.161D) is developed on the outer surface of a yolk sphere and that the digestive tract develops next to the yolk, while the muscle-segments, notochord and nerve-tube grow dorsally, or away from the yolk. Lateral spreading between the yolk and the bounding outer egg membrane would, if continued, produce a widening of the body across the pectoral region, as in the earliest known ostracoderms and in the skates and rays. Marked vertical growth (fig. 9.161B), carrying the back far away from the belly, typically occurs after the yolk is reduced and the fish has become a free-swimming larva (fig. 9.140C-E).

Another fact that was recorded in the art of the cave men is that animals have a head end and a tail end; their fish drawings also reveal the streamlined contour of the salmon. But the cave men presumably did not ask *why* or *how,* and only those who have struggled with these problems can begin to realize how relatively simple are the general principles and how diverse and complicated the factual details. Streamlining (figs. 9.163, 164) is of course superposed upon both bilateral symmetry and dorso-ventral asymmetry. A streamlined body often approximates a falling raindrop in shape and its smooth contours are favorable to speed and economy of effort; that is, they ease the burden of pushing aside the water in the forward movement and reduce the whirlpools on the sides and in the rear, which by suction retard forward movement.

Streamlining in every case is a particular result of the differentiation and integration of growth factors operating between definite limits in the three planes of space. The primordial center of the system is the dorsal lip of the blastopore and the longitudinal and vertical growth gradients may be projected conveniently upon the long axis of the fish (fig. 9.164). Thus the lateral contours (fig. 164A, B) may usually be drawn around quadrilateral frames of reference (Gregory, 1928a, 1928b). As the vertical diameter increases in relation to the length, the simple rhomb (fig. 9.96C) may expand into a rounded pentagon (fig. 9.113I) or change into a circular disc (fig. 9.101B). On the other hand, as the length increases, the dorsal and ventral contours tend to become parallel or to taper slightly (fig. 9.165). In such a way an essentially rhomboid form (fig. 9.165A) may lengthen into an extremely elongate one (C). Overemphasis of the heights of the dorsal and anal fins has been a factor in B. Great increase in width is shown in C.

Growth pressure in any direction is usually dependent upon the absorption of the yolk or the transformation of food material taken in at the mouth. In some way the relatively simple mechanics of cell growth and division are influenced by the enzymes and other chemical substances that are produced by the chromosomes. These seem to accelerate, retard or distort simple cell division. To make a long story unduly simple, chemical and physical forces coöperate to induce the formation of a series of relatively simple polyisomeres such as the protovertebrae or body segments of larval fishes (fig. 9.140). Later they give rise to paired olfactory, optic and otic capsules along the margins of the neural crests; the oralo-branchial clefts arise, which predetermine the location of the primary jaws and visceral arches.

Evidently there are hereditary priorities and differential growth rates, varying no doubt in different types and resulting in wide differences in body form, size of the different fins, etc. In a typical percoid (fig. 9.140) segmentation and longitudinal growth of the body are accelerated within the egg, but in the larval stages the head grows very fast. In such a fish vertical growth is retarded until post-larval stages, when it becomes greatly accelerated, it often transforming an originally eel-like embryo into a high-backed adult.

In the preceding pages we have come across very many examples of polyisomeres in the adult, such as the successive branchial arches, the vertebral centra, the neural and haemal spines, the fin rays and their pterygiophores, or bony supports, the teeth, the scales, the melanophores and other units of the color patterns. The olfactory rosettes of teleosts are radial polyisomeres, supplied by polyisomeral nerve branches.

In embryonic and larval stages each of the repetitive units seems to arise from its own focus or center of organizing forces; these are subject to rhythmic accelerations, climaxes and interruptions, and thus recall cell division and the processes of regeneration and reproduction.

Secondary polyisomeres are those which result from the fragmentation or disintegration of large units, such as the ossicles covering the cheeks of *Lepidosteus* (fig. 9.18B) or the row of supramaxillary plates of the same fish. Very likely the mosaic of ossicles on the rostrum of the sturgeon *Scaphirhynchus* has likewise resulted from the fragmentation of a few pairs of rostral polyisomeres in the primitive palaeoniscids. Secondary or new polyisomeres often arise within the limits of a single species, either by budding or by subdivision of older polyisomeres. For the present purpose we may consider the five branching rays of each of the pelvic fins in the ancestral percoid as primary polyisomeres. By a somewhat arbitrary definition all those families which have more than five branched rays in the ventral fins are excluded from the Percoidei and are often placed in separate orders. But this hardly settles the question whether in a given case the higher number (e.g., nine ventral rays in Zeidae, fifteen to seventeen in Lampridae) is due to a secondary increase or to inheritance from prepercoid (e.g., berycoid) ancestry. The presence in certain flatfishes (Berg, 1940, p. 492) of thirteen ventral rays on the ocular side and only three to six on the blind side is one of several items that suggest selective response to different conditions and the unreliability of the numbers as major taxonomic characters.

Secondary polyisomerism frequently occu in the vertebral column; for example, the a cestral percoid had ten "abdominal" and fou teen "caudal" vertebrae but in certain scon broids the numbers rise to sixty-eight; th vertebrae of morays are very many and ve complex in form.

Anisomeres are usually parts of polyis meres, or groups of them, which have exper enced either a marked increase or decreas thus producing emphasis or sudden inequaliti in an otherwise graded series. Such are the fa shaped tail-bones (fig. 9.24), representir greatly enlarged hypural rods, or the various emphasized and distorted accessory parts of th vertebrae which give rise to the Weberian a paratus (fig. 9.45). Among countless observe positive anisomeres we may mention: a towe ing occipital crest (fig. 9.96C), an enormou mouth (fig. 9.64), very high dorsal and an fins (figs. 9.109; 9.113C), a very long rostr spear (fig. 9.133A), a huge pelvic sucker (fi 9.148). Negative anisomerism, *i.e.* reductio or loss, is perhaps equally frequent and ofte compensatory to positive anisomerism in som other part: e.g., reduction or loss of pelvic fi in eel-like forms, which also exhibit reductic or loss of caudal fins (figs. 9.36A; 9.43; 9.5 9.64), with multiplication of vertebrae. I short much may often be learned as to the prob able mode of locomotion, feeding habits an environment of any given fish by study and a alysis of the more conspicuous polyisomer and anisomeres exhibited in the general bod form, skull and locomotor skeleton.

In either a phylogenetic or a comparativ series the corresponding parts of successi stages or different individuals may be consi ered to be homologous polyisomeres and th divergence in proportions of given measure ments of such parts falls under the principle c anisomerism; as, for example, the wide varia tions in the size of the head to body length i various cyprinoids (fig. 9.57). In this sens anisomerism is synonymous with "heterog ony."

Since anisomeres are essentially distorted polyisomeres, there are no hard and fast lines between the two and since poly- and anisomeres often combine into larger repetitive units, such complexes may be called polyanisomeres. It is often difficult to distinguish the latter from organs. For example, the pectoral girdles and fins are obviously polyanisomeres which are serially analogous but not wholly homologous with the pelvic bones and fins. The optic organs include several series of polyisomeres belonging to distinct systems but organized into a single pair of polyanisomeres. The same may be said of the balancing and acoustic system; this has resulted from the emphasis and interaction between different parts of the lateral line systems, which in turn have been joined with motor nerves and muscles to produce more or less automatic adjustments of the locomotor apparatus. A somewhat similar combination of poly- and anisomeres gives rise to the remarkably complex chromatic systems (pp. 184–187) controlling the colors and shifting color patterns of fishes.

As a general rule neither polyisomeres, anisomeres, nor any of their combinations are immune either to retrogression, reduction and disappearance or to further emphasis and/or organization. Such typical polyisomeres as bone cells, centra, neural and haemal arches, ribs, hypurals and urostyles, scales, lateral line organs, melanophores, and other chromatophores, muscles, motor and sensory nerves, teeth and jaw parts, may be reduced and even disappear in part or all of the territory where they once ruled. The same is true of such elaborate polyisomeres as eyes, which may lose their function and become reduced and degenerate within the limits of a single genus (*i.e.,* the blind catfishes and characins). Indeed it seems that progressive evolution, implying emphasis in one direction, is usually accompanied by loss of characters in other directions. Thus it is that so many diagnostic or key characters of families and orders are negative ones. In the skull, for example, the orbitosphenoid is lacking in all the higher teleosts, and the same is true of the gular plates, spiral valve, conus arteriosus and other parts.

Nevertheless not a few polyanisomeres have persisted even from the Devonian to the present time. Among all the hosts of ganoids and teleosts the hyomandibular, the quadrate, and several of the branchial arches persist, while no fish is known to be devoid of semicircular canals.

PHYLOGENETIC REVIEW

Let us now outline in the broadest terms the general evolution of the ganoids and bony fishes in geologic time.

Cheirolepis (fig. 9.2), the Devonian patriarch of the swarming tribes of bony fishes, was already a true bony fish and had inherited from his remote undiscovered ancestors many structural and functional patterns which were destined to be transmitted with more or less modification to his modern descendants. Thus his body was already streamlined and spindle-shaped, the "entering angle" in the vertical midsection being greater than the "run," and the greatest cross-section was probably located near the beginning of the second third of the body length. These results were doubtless brought about by corresponding adjustments of metabolic gradients in the longitudinal, transverse and vertical axes.

Next, the bilateral symmetry of his body was doubtless due to the symmetrical budding of metameric segments on either side of the blastopore, with resulting pushing of the older buds toward the rear (fig. 9.161D). This also permitted the differentiation of the head-end from the tail. The longer the period of growth and multiplication of the proto-vertebrae, the greater was the number of vertebral segments and the approach toward an eel-like larval stage. Thus a simple prolongation of a certain stage of growth may in the course of time induce striking changes in body form.

The Devonian palaeoniscoids, of which *Cheirolepis* (fig. 9.2) was the oldest and most primitive, were distinguished from the contemporary crossopterygians or lobe-fins (fig. 10.1) especially by the limitation of the muscular part of the paired fins to the basal portion, and by the great extension of the scaly integument, in which rows of scales were being joined into lepidotrichs or fin rays (fig. 9.4A). By a rapid multiplication of the distal segments the tail had become heterocercal or shark-like; at this stage the superficial tail muscles were probably only beginning to wave the fin rays, and the marked emphasis of certain deep flexors and extensors, which culminates in the teleosts (fig. 9.5C), had not yet begun. In the dorsal and anal fins small slips of the myomeres were probably inserted into the base of the fin but the complex musculature and flexible interrelations of each fin ray to its own supporting rod were at best only incipient.

The entire skull (syncranium) of the palaeoniscoids was basically identical with that of the lobe-fins but there was no transverse line of articulation between the anterior and posterior moieties of the cranium as there is in the lobe-fins (fig. 11.23); there was also a closer integration of the ethmo-optic with the spheno-occipital sections. In all these and many other features the palaeoniscoids were much nearer to their descendants the varied ganoids and teleosts than they were to their distant relatives the lobe-fins.

During the long ages of the Devonian, Mississippian (Lower Carboniferous), Pennsylvanian (Carboniferous) and Permian the palaeoniscoid stock branched and rebranched (fig. 9.166) into many so-called families, representing a wide diversity of body form, jaws and dentition. During the Permian and Triassic they were gradually crowded out by some of their more progressive derivatives, the Holostei (figs. 9.17–19) and Amioidei (figs. 9.20–23) and finally by the teleosts. Meanwhile the more conservative families of the palaeoniscoids ran up into the saurichthyids (fig. 9.14A), the chondrosteids (B), the "modern" sturgeons (C) and the spoonbills (D). In so doing the endoskeleton became largely cartilaginous; they also lost most of their scales but retained the heterocercal tail and the small rhombic scales of its dorsal lobe (fig. 9.4C). Meanwhile the sturgeons were acquiring a new set of polyisomeres, which are the large and small bony scutes borne by the leathery skin.

Advance toward the modernized teleost types was initiated by the anisomerous expansion of the deep flexors of the tail (fig. 9.5) and by the shortening and upturning of the caudal lobe (fig. 9.4). While these improvements were made at different rates in different families, the vertebral centra became more firmly ossified and united with the neural and haemal arches.

In our review of the sequence from primitive palaeoniscoid to central percoid we were able to show (fig. 9.74) that the 43 known primitive or paleoniscoid characters were gradually reduced to 26, 16, 3, 0; while the advanced or percoid characters rose from 0 to 8, 17, 38, 43.

Each successive major group (Palaeoniscoidei, Protospondyli, Amioidei, Isospondyli, Haplomi, Percoidei, etc.) branched into many types of body-form: fusiform, eel-like, discoid, depressed, etc. Flexible, intermediate and fixed or armored forms were also again and again produced independently. Similarly the mouth and jaws varied in the different periods, from primitive predaceous to small-mouthed, sucking, nibbling, biting, crushing, shearing, toothless and other types. In some the mouth was shifted forward to the end of a long preorbital tube, in others the jaws extended backward, finally producing the super-pythons or gulpers; in still others the upper jaw and the rostrum grew forward into a spear or ram.

As a result of this riotous branching through the ages the modern student of fossil and recent fishes has to deal with perhaps thirty thousand named kinds (species and subspecies) distributed under six hundred and thirty-eight families of which the teleosts alone claim five hundred and eleven families. The families in turn are grouped by Berg (1940) under no less than one hundred and fourteen orders! In chapters VI to IX inclusive we have noticed some of the better known representatives of this vast array and we have seen that, notwithstanding their diversity, many of the larger branches may be followed from their inception in the Cretaceous period to the present time (fig. 9.166).

HAS EVOLUTION CEASED?

Some years ago Dr. Robert Broom of South Africa, M.D., F.R.S., the eminent historian of the South African fossil reptiles, published a startling pronouncement to the effect that at least in its broader results, progressive evolution has come to an end and that man is after all the main objective and climax of evolutionary advance. In a later chapter we shall return to consider this idea in its bearing on man's origin and destiny; but in the present section we shall content ourselves with the testimony that as we look over the vast fields of systematic ichthyology there is cumulative evidence that the fundamental processes of variation and selection (by isolation and otherwise) are still coöperating to produce new races, subspecies, species, genera and presumably, in the long run, new groups of higher rank. This is true in spite of the fact that in contemporary systematic ichthyology the emphasis is usually laid upon the differences or gaps that seem to warrant the naming of new subspecies and species.

In answer to our argument, Dr. Broom might well refer to the facts of palaeontology which prove that most of the existing orders and suborders of teleosts had produced recognizable representatives of their respective groups at least by Upper Eocene times and often before the close of the Cretaceous period. On the other hand, among all the hosts of fossil teleosts only one known family, the very primitive Leptolepidae, ranges from the Lower Cretaceous down into the Jurassic. Presumably therefore the main adaptive branching of the teleosts took place during the long ages of the Lower, Middle and Upper Cretaceous, while the isospondyl family of the tarpons (Elopidae) serves to connect the leptolepid-clupeoid division with the more advanced isospondyl divisions, which are found in the Upper Cretaceous and Eocene. Thus some of the oldest families of teleosts may date back almost one hundred million years; but this is less than one-fourth of the total time from the oldest known Ordovician pre-fishes (ostracoderms) to the present epoch. Unless the earth should be destroyed within another hundred million years (which period is probably less than one-twentieth of the age of the earth) it is difficult to imagine what can stop the age-old forces of isolation and variation from accumulating into new and strange results the differences which, as noted above, are now being initiated in all known parts of the great field.

To put the matter in a different way, we have seen in the preceding chapter the astonishing differences between ancestors and descendants that have been brought about by the constant interaction of the genetic and environmental forces that produce new polyisomeres (such as the high number of vertebrae in eels or the great number of bristle-like teeth in chaetodonts) and which then select certain polyisomeres for more or less extreme anisomeral emphasis. As these once new polyisomeres become old, such novelties emerge as the enormous enlargement of the dorsal and anal fins in the ocean sunfishes, or the great elongation of the skull in the tubefishes, or the transformation of a normal percoid fish into a seabat. A close study of the humanly unpredictable sequence of stages from bass-like fishes to triggerfishes and thence onward to the coffer-fishes, porcupine-fishes and ocean sunfishes, suggests that even such relatively advanced types as the triggers may still possess the potentiality for another series of transformations.

Even if we grant (unnecessarily perhaps) that *Ranzania,* the most specialized of the known ocean sunfishes, is nearing the end of its evolutionary career, and that the same is true of most of the peculiarly specialized fishes that we have described in the foregoing pages, we would still have left hundreds of genera with a greater or less reserve of primitive teleost characters; this fund would serve them as a stable physiological source for the support of new and unpredictable polyisomeres and anisomeres, affecting the integument, skeleton, muscles, brains, nerves, and complex behavioral patterns. Moreover the detailed studies of taxonomists upon particular populations, subspecies and species prove that variation in number and emphasis is still going on; while the foregoing review indicates that there are still abundant materials for the emergent evolution of new eel-like forms, of new types with tube-like mouths, of new wrass-like fish with crushing molars, of new flying-fishes, etc.

And while the evidence is all against the possibility that any fish stock will again give rise to a line leading to amphibians, reptiles, birds and mammals, with the past class characters of those groups, yet, as long as fish increase and multiply, vary and get separated into isolated castes or communities, new genetic mutations will supply the raw materials for selection and combination of new advantageous features.

PART THREE

(*Chapters X–XV*)

AIR-BREATHING FISHES (CROSSOPTERYGII, DIPNOI) · AMPHIBIANS · REPTILES · BIRDS

CHAPTER X

THE AIR-BREATHERS STRUGGLE FOR LIFE

Contents

CHAPTER X

THE AIR-BREATHERS STRUGGLE FOR LIFE

HOLD-FASTS AND PAIRED LIMBS

THE NUMEROUS evolving hosts of animals so far considered in this book have all been confronted with the alternative either of working out an effective locomotor system for pursuit and for flight or of holding fast to a fixed base and readapting their entire economy to a sedentary way of life. Sometimes the larval stages coped successfully with the problem of carrying the species far and wide in search of suitable dwelling places (fig. 5.20). Then the larvae underwent a transformation and the adults were immobilized, as in the ascidians and oysters. Among lepidopterus insects the actively travelling larvae, coming to a suitable spot, make fast and enclose themselves in an outer shell, inside of which they undergo a transformation into a winged adult, as in the moths. Of course the boundary between a sedentary and an active way of life is not always impenetrable, as when nestling birds grow into able fliers; neither are the obstacles in passing from one medium to another, as from water to air or the reverse, always insuperable, as shown by various creatures which have succeeded in such attempts.

Among fishes, those which cling to the rocks, as the gobies do with their sucker-like pelvic fins (fig. 9.147), might seem to be in danger of becoming sedentary, but the mud-skipping goby (*Periophthalmus*) likes to slither around on wet rocks or to climb up the rootlike branches of the mangroves by means of its remarkable limb-like pectorals. These have a bend that simulates an elbow, so that the pectoral fins can be used to raise the body-weight and propel the fish forward. But the mud-skipping goby has solved less than half the problem of transforming a fish into a quadrupedal land animal, for only its pectoral appendages serve as limbs and it has no lungs.

The Sargasso fishes (*Pterophryne*) have gone further in the way of the quadrupeds in so far as the pelvic fins also are somewhat footlike as the fish climbs through the branches of the seaweed (fig. 9.154).

In the sea-bats (Ogocephalidae) the pectoral fins are very large and armlike and the well developed pelvic fins rather resemble the large webbed feet of seals (fig. 9.159). As seen in an aquarium, a sea-bat can move somewhat ponderously about on these appendages but, being without lungs, it is improbable that they will ever come up on shore and make a place for themselves among the watchful crabs and alert gulls.

Not even the African lungfishes (*Protopterus,* fig. 10.1) with their good lungs have emancipated themselves from a fishlike way of life, except during their sedentary periods in the dry season, when they wrap around them a thick cocoon of mud in which they lie dormant, waiting for the next season of floods. Moreover the pectoral and pelvic appendages, although they can be moved in a rather limblike way, are reduced to long flexible threads and have no power to support the body on land. Much larger, well muscled fore-and-hind paddles (fig. 10.1) are found in the Australian lungfish. Nevertheless their tapering leaflike forms, although capable of being bent and twisted (fig. 10.9), are again too flimsy to make effective legs. Only the lobe-fins **(p. 241)** had paddles fit to become legs.

FAN-LIKE DENTAL PLATES, THEIR ORIGIN AND MECHANISM

All the lungfishes are peculiar in their dentition, which consists originally of fan-shaped clusters of small dental units coalesced into radiating rows (figs. 10.11–13). These dental complexes are located on the bony roof of the mouth and on the inner sides of the lower jaw, and even in Palaeozoic times the lungfishes had lost the teeth on the lateral margins of the jaws which gave rise to those of land-living vertebrates.

No adequate story of the evolution of the living frame among vertebrates could well omit: (1) some of the outstanding features in which both the exoskeleton and the endoskeleton of dipnoans are adapted to their peculiar ways of life, nor could it fail to refer to the evidence that (2) in spite of their specializations, the dipnoans were partly successful in evolving certain prerequisities or preadaptations for quadrupedal land-living.

In dipnoans the leading feature of the exoskeleton is the peculiar dentition noted above, which affords a clear example of emergent or creative evolution—the evolution of a once uniquely new organ system. How did this radiating fanlike dentition happen to arise? Embryologic evidence convinced Semon (1901) that in the Australian lungfish each of the little tubercles on the surface of the radiating ridges represents a single tooth, and fossil evidence from the Devonian *Dipterus* (fig. 10.7E) figured by Pander clearly supports Semon's conclusion. But again, what caused these teeth to be arranged in radiating rows, and how is it that when the upper and lower teeth are brought together the ridges or crests of the upper set fit so neatly into the depressions between the crests of the lowers (fig. 10.13), and *vice versa?* Unfortunately the pre-dipterine stages of dipnoan evolution are as yet undiscovered, so we are forced to search for analogous instances in other groups which may supply significant clues. Among the spiny-finned order (Acanthopterygii) many develop frontals with stiffening ridges radiating from the center of growth and ossification. In some of the cottids the radiating pattern (fig. 9.143) is conspicuous not only in the frontals but in many other bones both of exo- and endo-skeletal origin (Gregory, 1935, pp. 215, 216). Radiating patterns are found also on the skull roof of placoderms (fig. 10.21A-C).

These analogies, however, supply only a hint that each of the radiating tooth-plates of dipnoans have probably started from a single small patch. The paired dipnoan tooth-plates (fig. 10.11) do not start from the middle of an area but from the medial side of a somewhat lunate or crescentic area. The upper dental plates face outward and away from each other, their convex border being on the medial side and separated by the diamond-shaped parasphenoid (fig. 10.11A). They plainly represent the exoskeletal tissue covering the inferior surface of the palato-pterygoid arch to which they are attached, and they follow in general the transverse curvature of the palato-pterygoid arch. Moreover each ridge or radial row of little tubercles may be analogous with a single "dental family," which were formerly successional teeth but later came to stick together. These corrugated upper dental plates (fig. 10.12A) are somewhat analogous with the curving banks of teeth in the heterodont sharks (fig. 8.46). Rather close analogies seem to be afforded by the curving shells of certain bivalve molluscs, especially among the cockle-shells. If the opposite valves of a young *Tridacna* (fig. 2.53) be spread apart and viewed from above, the analogy with the radiating dipnoan tooth-plates is rather striking. The clue is that the ridges and valleys in both cases have resulted from the lateral compression and radiate folding of a rapidly and radiately growing flexible membrane, that is, the mantle in the clam and the dental tissue in the dipnoan. The lower jaw-plates of dipnoans must have arisen by a similar process.

There remains the problem of securing a perfect fit between the crests of the upper plates and the grooves of the lower ones. Here a close analogy to both the upper and lower radiating dental plates of dipnoans and the way in which their two sets fit into each other is again found not in any other vertebrates, but in the hinge "teeth" of certain Palaeozoic pelecypods of the family Lyradesmidae (cf. Zittel, 1924, Grundzüge der Palaontologie, p. 378, fig. 628). Here the five to nine radiating "teeth" of one side fit into grooves between the corresponding ridges of the opposite side. In the bivalve the radiating system started from the umbo or growth center of its own side, the radical difference being that in the clam the fitting is between right and left, whereas in the dipnoans, as in all vertebrates, it is between upper and lower sets.

The fact that the upper and lower tooth plates of dipnoans and the right and left hinge "teeth" of pelecypods, respectively, fit between their opposites, although a familiar feature in the dentitions of vertebrates (e.g., figs. 16.44; 17.13A2), requires further consideration. In the case of the pelecypods it seems to have resulted from the folding up of the shell-secreting mantle where the opposite halves come together in the midline, so that the convexities of one side fit into the concavities of the other. In the case of the dorso-ventral fitting of vertebrate teeth, the inverse relationships seem to be due to the sinking in of the stomodeum or embryonic mouth-pocket, which is lined with tooth-bearing material. This in-sinking would cause the surface layer of the upper tooth-bearing tissue to face downward, that of the lower to face upward (fig. 20.7A). Moreover, the originally anterior face of the uppers would be homologous with the inverted posterior face of the lowers. Thus appropriate folding of a once continuous plug of stomodeal material would easily cause the alternation and inversion of the parts of their upper and lower units in dipnoans.

In the earlier dipnoans these radiating dental plates apparently would have served for the crushing of shells, either of molluscs or of crustaceans, but in the modern African and South American dipnoans (fig. 10.12D) the radiating plates have been reduced in number and their ridges sharpened into knife-like edges that suggest a predatory flesh-shearing habit—thus affording an excellent example of correlated changes in structure and habit. It will also be noted that in this case the shell-crushing structures and habits of the older dipnoans were prerequisite, preadaptive stages for the transformed structures and habits of their surviving descendants.

At the front end of the palato-pterygoid arches of dipnoans there are a pair of compressed, obliquely placed cutting teeth (fig. 10.12C, D); these, however, have no counterparts in the lower jaw, which ends anteriorly in a wide channel apparently for a thick tongue (fig. 10.11A1). These anterior upper teeth do not represent the premaxillary teeth of normal fishes but may well be analogous with the vomerine teeth of crossopterygians (fig. 11.32) and with the "tritors" of chimaeroids (fig. 8.4), and possibly with the beaklike upper front teeth of arthrodires (fig. 7.16).

DIPNOAN SKULL FORMS

The massive dentitions of early dipnoans required correspondingly massive jaws and could be operated only by jaw muscles with thick cross-sections giving great power in proportion to length. Support for these massive jaws and jaw muscles might have been sought for in the exoskeletal skull roof, but in the dipnoans what may have once been a continuous cephalic shield, inherited from some far earlier placoderms, had already become fragmented into a mosaic of relatively small, more or less polygonal plates (fig. 10.16A) arranged in rows around the orbits, over the cheeks and on the skull roof. The dipnoan response to this situation was to enlarge and make denser the cartilaginous cranium (figs. 10.17D; 10.16C). In the embryo this endoskeletal tissue fills the space around the olfactory, optic and otic capsules; it also forms the parachordal tracts on either side of the notochord which develop into the base of the cranium. At the same time the large rhomboidal parasphenoid bone (fig. 10.11) on the roof of the mouth (derived from the in-sunken surface of the stomodeal pouch) strengthened the base of the cranium, while similarly derived pieces braced the inner sides of the lower jaws (fig. 10.13) and joined with the derm bones on the outer sides of the jaw to form a stiff sheathing for the massive Meckel's cartilage in the center. These changes adversely affected the hyomandibular bone, which in sharks, ganoids and teleosts forms the chief link suspending the upper and lower jaws from the skull and supporting the opercular covering the branchial chamber. Accordingly the hyomandibular, at least in modern Dipnoi, became reduced to a vestige (fig. 10.17a *hyom*). Especially in the later dipnoans many of the small plates on the roof of the skull have been eliminated, and in the African and South American genera the few that remain have been elongated and stiffened (fig. 10.15), while the underlying braincase has been molded into a strong tube, partly ossified, partly calcified.

Secondary Polyisomerism.—Since the dermal skull roofs of *Dipterus* and of related early dipnoans consist of many polygonal or small pieces, these seem to have been regarded as indicating more primitive conditions than those found in the crossopterygians. Some authors, indeed, assume that evolution *always* passes from the homogeneous to the heterogeneous and neglect the many obvious instances in which, long after earlier stages of polyisomerism and anisomerism have been passed through, some of the anisomeres give rise to secondary polyisomeres, which may then in turn spread and crowd out the older anisomeres. Of such examples we may cite here: (1) the breaking up of the dorsal shield in certain cyathaspid ostracoderms to form small secondary polyisomeres (fig. 6.5B); (2) the secondary breaking up of one of the cheek plates of the garpike *Lepidosteus* to form a mosaic of irregular secondary polyisomeres (fig. 9.18); (3) the regional de-differentiation and secondary multiplication of the vertebrae in eels; (4) the armor of secondary polyisomeres (fig. 9.70) in pipefishes and sea-horses (Syngnathidae).

Other chief reasons for regarding the skull pattern of dipnoans as largely secondary are: that it is associated with a highly specialized stage of the dentition, jaws, hyomandibular and endocranium and that the mosaic of derm plates on the skull of *Dipterus* (fig. 10.16A) are not individually homologous with those of acanthodians (fig. 7.27) even though they may be collectively homologous by regions, especially those located on the line of the lateral line system, as shown by Romer (1936). The evolution of paired bones (fig. 10.16B-D) on the skull roof of later Dipnoi (secondary anisomerism) may have been merely parallel to a similar but earlier tendency in the Crossopterygii (fig. 10.21E).

EVOLUTION OF BODY-FORM AND TAIL FIN

The rest of the exoskeleton in dipnoans shows several stages in reduction and elimination. In all probability the scales of the unknown Pre-Devonian ancestors of the dipnoans were not widely different from those of their remote relatives, the earlier Crossopterygii. That is, the scales were very numerous and small and were of complex construction (fig. 10.7D), with the same series of layers as in the Crossopterygii (B) and basal ostracoderms (figs. 6.2; 6.4). In the ceratodont line, leading to the Australian lungfishes, the scales became fewer and larger, their bony layers diminished and were replaced by horny layers and at least the anterior part of each scale sank into a pocket of skin and was overlapped by the horny posterior margins of its neighbors. In the line leading to the African and South American dipnoans the scales became very small and sank into the thick slippery skin.

Meanwhile, as shown from good fossil evidence, the dipnoans lost their originally heterocercal caudal fins (fig. 10.1), the anterior and posterior dorsals came together, were elongated and fused, the second dorsal losing its high web but meeting the similarly extended anal to form a pointed "gephyrocercal" or bridge-tail. This trend culminated in the eel-like bodies with low dorsal and anal fins of the African and South American genera.

The least specialized pectoral and pelvic appendages among the modern dipnoans are those of the Australian lungfish (fig. 10.10F) but, as already noted, although they can be freely moved under water in many directions (fig. 10.9), they could not support the weight of the body on land. Nevertheless they foreshadow the limbs of quadrupeds in several important features:

1) The internal skeleton of the paired appendages (fig. 10.10F, G) in dipnoans is well developed and the exoskeletal dermal rays are delicate, which is the reverse of the condition in bony fishes.

2) Each of the pectoral (fig. 10.10) and pelvic (fig. 10.9) paired appendages articulates with its girdle by means of a single proximal piece, which foreshadows the humerus or the femur respectively of land-living quadrupeds (fig. 11.1).

3) The myomeric muscles of the flanks are extended outward on to the paired fins in a cone-in-cone series, the dorsal set tending to extend and raise the paddles, the ventral set to adduct and lower them; this arrangement is somewhat suggestive of the musculature of the paired limbs of quadrupeds.

The pectoral girdle of dipnoans was, as in cephalaspid ostracoderms and palaeoniscoid ganoids, a complex of exo- and endo-skeletal elements. The former included a large crescentic cleithrum, which supported the small, chiefly cartilaginous scapulo-coracoid, the latter in turn receiving the thrusts from the pectoral fin. In the predominance of the cleithrum the dipnoans remained on the fish level, in contrast with the land-living quadrupeds or tetrapods, which early reduced the cleithrum and enlarged the scapulo-coracoid (fig. 11.5).

The ultimate source of the dipnoans is still unknown. In spite of their wide differences from the Crossopterygii, they are linked by Romer (1945) with that group under the term Choanichthyes, partly for the reason that in both groups there is a groove or tube leading from each nasal capsule down to the palate, which conducts air into the mouth cavity, whence it can be forced by the throat muscles into the lung or swim-bladder. But the marked differences in the skull and dentition between the dipnoans and crossopterygians may indicate a very remote common source and subsequent divergence. Since the dipnoans have specialized complex jaws, their origin may perhaps be sought somewhere among the well-jawed macropetalichthyid placoderms rather than in the pre-gnathostome ostracoderms. It is true that the skull (fig. 10.15B) and teeth (fig. 10.13) of the African lungfish *Protopterus* somewhat suggest the *Dinichthys* (fig. 7.16) type of arthrodire but the relationship could only be very remote. In another direction, however, such a ptyctodont placoderm as *Rhamphodopsis* (fig. 7.18A) may perhaps ultimately supply the clue to the origin of the dipnoans.

CONVERGENCE OF EEL-LIKE LUNGFISHES AND AMPHIBIANS

In spite of the overspecialized characters of adult dipnoans, the embryonic stages much resemble those of the tailed amphibians, especially in the circulatory system (Kellicott, 1905). However, the Australian lungfish, and even its remote Devonian ancestor *Dipterus*, have gone far out on a line of specialization which led away from the well-limbed, land-living tetrapods and culminated in the eel-like and almost limbless South American lungfish (*Lepidosiren*), which convergently resembles the so-called conger eel (*Amphiuma*) among Amphibians. Nor is there any suggestion that some of the earliest lungfishes abandoned their dipnoan specializations and reacquired the many features which their still more remote ancestors must have had in common with the crossopterygians or lobe-fins. All this may be affirmed with considerable confidence and in spite of certain mistaken attempts to show that the lungfishes gave rise to the urodeles or tailed amphibians.

THE CROSSOPTERYGIANS, OR LOBE-FINS, AND THEIR WELL-MUSCLED PAIRED PADDLES

Much nearer in most respects to the highly probable starting-point of the four-footed land animals are the crossopterygians, i.e., the fringe-finned, or lobe-finned ganoids (fig. 10.1). These range from the Devonian period to the present time, their last living representative being the fish called *Latimeria,* of which more anon. Of these, the earliest yet known is *Osteolepis* from the Middle Devonian of Europe. This was a small, somewhat shark-like predaceous fish with an internally heterocercal tail; but the body was covered with shiny ganoid scales, the fins were provided with dermal rays and the tail showed the beginning of the tripartite condition, which became more prominent in later members of the group. The scales (fig. 10.7A) retained a thin outer layer of ganoine and the cosmine, trabecular and isopedine layers were present as in all the older vertebrate stocks. The same complex, hard tissue covered the roof and sides of the cranium, the jaws and opercular series, and, inside the mouth, gave rise to the teeth. In other words, the exoskeleton was still dominant even in the cranium, of which the endoskeletal tissue was somewhat spongy. The notochord also was dominant, the vertebral centra being thin and more or less crescentic. The neural arches were slender, as were the haemal and interhaemal rods. The pectoral fins of *Osteolepis* had a short muscular lobe (fig. 10.18G) covered with ganoid scales and a rounded fin web supported by close-set dermal rays. The pelvic fins were smaller, with a lobate fleshy base and a rounded contour. The endoskeletal structure of the paired fins is little if at all known and was possibly not well ossified. However, the resemblance and relationship between *Osteolepis* and *Eusthenopteron* are so close that no fundamental difference in the pectoral endoskeleton may be assumed without direct evidence, especially since other genera of crossopterygians (e.g., *Diplopterax*) are more or less intermediate between *Osteolepis* and *Eusthenopteron.*

Eusthenopteron (fig. 10.4), chiefly from the Upper Devonian of Canada, had advanced beyond *Osteolepis* in its much greater size, the largest known specimen measuring *circa* 863 mm. (4 ft. 3¾ ins.) in overall length. The vertebral column was better ossified, especially in the anterior part, where the neural arches were large and crowded and the central crescents were well developed, although the notochord was still very large. The two dorsal fins were supported by stout, bony basal pieces and their webs were strengthened by large dermal rays. The caudal fin was very large, tripartite, and supported by long dermal rays. There were no internal bony fanlike supports of the very large tail, such as evolved in the teleosts, and the haemal and interhaemal rods remained narrow. The pectoral girdle agreed in basic features with those of the oldest dipnoans and palaeoniscoid ganoids and this fact in connection with many others, indicates that the dipnoans, the crossopterygians and the palaeoniscoid ganoids were partly parallel, partly divergent, offshoots of some hitherto unknown or unrecognized placoderm which had reduced its thoracic buckler to the status of a shoulder girdle and had retained the complex exoskeletal tissue of the Ordovician ostracoderms.

The scapulo-coracoid of *Eusthenopteron,* according to Bryant, was relatively quite small in proportion to the very large cleithrum, and somewhat like a horizontally placed Y (figs. 11.5A; 11.6A). The short thick stem of the Y received the thrusts from the pectoral fins and transmitted them to the large crescentic cleithrum. On the inner side of the cleithrum two fairly large triangular depressions, one above, one below, the scapula-coracoid, probably indicate the origin-areas of large muscle masses which ran obliquely downward and backward or upward and backward respectively, to be inserted along the dorso-posterior or ventro-anterior surfaces of the jointed endoskeleton of the pectoral appendages.

On the whole, the pectoral paddle of *Eusthenopteron* remained in the fish stage but it had evolved several critical features which were prerequisites or preadaptations for the evolution of land-living, quadrupedal appendages. The scapulo-coracoid has every appearance of being merely the innermost member of a series of six polyisomeres (fig. 10.19B, A-E), which decreased in size distally and formed the jointed axis of the appendage. In some ways these successive diminishing segments are analogous with the vertebrae which diminish in the tail region; the processes on their dorsal or postaxial side simulate primitive neural arches, while the "radial" rods on the ventral side recall the haemal arches of the proximal part of the tail. This analogy with vertebral structures has been hitherto overlooked, perhaps partly because the names and associations of the structures concerned have stemmed ultimately from human anatomy, where the analogies are far more remote; but at least in dipnoans this similitude is further heightened by the fact that the muscles of the pelvic fins exhibit a cone-in-cone arrangement not unlike that of the vertebral myotomes of vertebrate embryos. And in them this cone-in-cone structure is further reflected in the derivation of the nerves that supply limb muscles from several spinal nerves, other branches of which go to successive vertebral segments. In *Eusthenopteron* the axial segments, including the scapulo-coracoid, are limited to six, but in *Ceratodus* (fig. 10.18I) the dwindling secondary polyisomeres extend out to the end of the long tapering tip.

Much more elongate and narrow are the pectoral fins of the Upper Devonian *Holoptychius* (fig. 10.1). These bore a fringe of long dermal rays on their antero-ventral and postero-dorsal edges and thus gave rise to the name Crossopterygii (tassel fins), which was used by T. H. Huxley in that sense as an ordinal name. *Holoptychius* has probably converged toward one or another of the dipnoans in its tassel-like paired fins, shorter, rounder skull and large cycloid scales. But it was probably even further removed from the dipnoan stock than were its older relatives, *Glyptopomus* and *Osteolepis*.

The endoskeleton of the pelvic fins, best known in *Eusthenopteron* (figs. 11.11; 11.12), was much like that of the pectorals, with the probable exception that the morphologically preaxial border was turned inward toward the midline and that the fins as a whole during forward locomotion were probably held in a nearly horizontal position. The endoskeletal pieces (A, B, C; a, b, c) also were much like those of the pectorals. The paired pelvic bones were rodlike, with their anterior ends directed forward and inwardly. A low dorsal process near the posterior end of each pelvic rod (fig. 11.12B) possibly represents the beginning of the ilium, while the shaft of the rod suggests the combined pubo-ischial plate of tetrapods.

CROSSOPTERYGIAN SKULLS AND THEIR EVOLUTION

The exoskeletal shell of the skulls of *Osteolepis* (figs. 10.8; 10.21) and other *Eusthenopteron* opterygians agree in their main divisions with those of the palaeoniscoid ganoids (fig. 10.22B), which lie in or near the stem of the bony fishes. There are, however, many differences in detail, a few of which will presently be noted. But first it is necessary to point out that such well known names of the human skull-bones as nasals, frontals, parietals, supraoccipitals, were applied also to somewhat similarly placed bones in the skulls of lower vertebrates. Thus the system was gradually extended downward by Cuvier, Owen, W. K. Parker and their successors, from man down to the lower mammals, thence to crocodilians, lizards, tortoises, frogs, salamanders. The skulls of teleost fishes (figs. 9.9; 9.32; 9.76) are much more complex than that of man, which, as we shall see later (chapters 23, 24), has been greatly simplified in general appearance even while becoming more complex in detail. In fact the early palaeoniscoid skulls were composed of several times as many elements (fig. 9.13) as are found in the human skull, so that the names of human skull bones were not nearly numerous enough to go around when transferred to the fish skull. Thus many new names, such as quadrate, metapterygoid, epiotic, pterotic, etc., were invented for the extra bones of the fish skull and are indeed recognized by most authors. Much later it was realized (*vide* Stromer von Reichenbach, Williston, Gregory) that in passing from fishes and other earlier vertebrates to mammals many skull elements had dwindled away and disappeared, while some had fused with their neighbors. In a few cases new polyisomeres appeared by fragmentation of larger bones, as in the supraspiracular ossicles of *Po-*

lypterus (fig. 9.15), the circumorbitals and postorbitals of *Lepidotus* (fig. 9.17), the very numerous rostral ossicles in *Scaphirhynchus,* the cheek-plate mosaic of *Lepidosteus* (fig. 9.17C), etc. For the most part, however, the general arrangement of the bones of the teleost skull roof was quite uniform, conspicuous differences between skulls of different families being usually due rather to anisomerism, involving increase or decrease in size of certain units, than to the creation of new elements.

Through the labors of several successive generations of comparative osteologists, many errors, inconsistencies and discrepancies in earlier systems were corrected; but one outstanding problem remained unanswered until recently. How is it that among the reptiles and fossil amphibians the median or pineal eye always lies between the opposite parietals (figs. 11.1, 19, 20; 12.4, 25, 41), whereas in the crossopterygian fishes (fig. 11.19) it was located between the bones commonly called "frontals"? The fortunate discovery by Westoll (1938) of an Upper Devonian fish allied to *Eusthenopteron,* which he named *Elpistostege* (fig. 11.19C), because it gave him the desired skull room, supplied him with evidence that the fish bones called frontals really represent the parietals of amphibians (D) and higher vertebrates and should therefore be named parietals.

Although it may be inconvenient to remember that the fish bones called frontals in hundreds of special papers ought to have been called parietals, it seems much less inconvenient and confusing than the current procedure of several European authors (*e.g.,* Säve-Söderbergh), who have extended Allis's system of names (1898–1905, 1917–1935) for the elements of the fish skull, especially in the supratemporal and occipital regions, and have combined them with Ecker and Weidersheim's names (Gaupp, 1896–1904) for the "frontals" and "parietals" of the frog. The frog, however, has a much specialized skull (fig. 11.32) in comparison with the Permo-Carboniferous labyrinthodonts; the latter, indeed, come far nearer than the frog does to affording a reliable bridge (fig. 10.21F) from the skulls of higher vertebrates down to the fish skulls. For the present work it may suffice to retain the names of the older system, as set forth for example in Parker and Haswell's Textbook of Zoology (Vol. II, revised edition), except when there is convincing evidence for certain of the newer names proposed by Allis, Stensiö, and Säve-Söderbergh.

Another fish-skull bone that has been much contested is the "vomer." For many years Broom has maintained that the true vomer of mammals has been derived from the parasphenoid of lower vertebrates, including fishes, and that the name parasphenoid should therefore be replaced by vomer; also that the "vomer" of fish represents a fused pair of "prevomers." But von Huene and others, notably Parrington and Westoll (1940), have cited what appears to be cogent evidence that the vomer of mammals was derived from the vomer of reptiles, amphibians and fishes (Broom's "prevomer") and that the parasphenoid was reduced in the later mammal-like reptiles and eliminated in the mammals.

Still another "bone of contention" is the lacrimal. Gaupp (1898) found that in the lizard skull the larger bone at the anterior border of the orbit corresponded in position, on the one hand with the "prefrontal" of reptiles and birds and on the other with the true lacrimal of man and other mammals. Therefore he transferred the name lacrimal to the "prefrontal" from reptiles down to amphibians and eliminated the name prefrontal entirely. In this he was followed by Wiman and many other European authors. But Watson (1911, 1913, 1920) and Gregory (1920) showed that in the mammal-like reptiles the true or mammalian lacrimal is the lower of the two bones at the anterior end of the orbit, the upper one being the prefrontal. In the higher mammal-like reptiles the correspondence of the lower element to the mammalian lacrimal was so close as to leave no reasonable doubt as to their homology. Comparative evidence suggests also that the prefrontal of mammal-like reptiles, along with its partner, the postfrontal, were lost in the earliest mammals. After the loss of the prefrontal the mammalian lacrimal would agree somewhat in appearance with the prefrontal of reptiles and birds, but without being homologous with the prefrontal. In fishes the lacrimal appears to be represented by the first suborbital bone (fig. 10.21A). This bears on its outer surface the anterior end of the suborbital branch of the lateral line system, located in a groove which is regarded by Watson (1913) as the probable forerunner of the lacrimal duct of land animals.

LOSS OF PARTS IN PASSING TO AMPHIBIAN STAGE

As already noted, the skull of *Eusthenopteron* contains all the elements found in the early amphibians, plus those which were lost in passing from the fish to the higher grade. Notable among the lost bones are: (1) the entire opercular series (fig. 11.18), including a row of narrow lateral gulars or branchiostegals (fig. 11.8), which cover the lower part of the gill region; (b) one or more of the elements above the operculum, especially the lateral extrascapular (fig. 10.8), supraopercular or "scale bone" (fig. 9.32b); (2) the very short rostrum of *Osteolepis* and *Eusthenopteron* bore on its upper surface a mosaic of small pieces (fig. 10.21A), all of which have received names in Stensiö's system. In the earlier amphibians (figs. 10.21F; 12.3), on the other hand, these small elements had been replaced by large paired nasals.

The palatal side of the skull of *Eusthenopteron* (fig. 11.21) agreed in most essentials with that of *Osteolepis* and to a less degree with that of the palaeoniscoid ganoids, the forerunners of the teleosts. All these, however, differed greatly from those of later ganoids on the one hand, and of early amphibians on the other (fig. 11.21B, C), in that their parasphenoid did not extend backward to the basioccipatal but left between them a large gap which was very probably bridged by connective tissue or cartilage in the living skull. This curious feature on the ventral surface of the skulls of *Osteolepis* and *Eusthenopteron* was associated on the dorsal surface (fig. 10.21E) with a transverse groove behind the true parietals ("frontals"), which was continued backward and outward on either side above the cheek plate (squamosal). Apparently these fishes could elevate their snouts a little without raising the rear part of the cranium. On the implied but never proved principle that no specialization once evolved can ever be replaced, it has been assumed (e.g., Parker and Haswell, 1940, p. 286) that the foregoing "specialization" of these known crossopterygians would have debarred them from being ancestral to the amphibians, where the parasphenoid (fig. 11.23aA-C) extends to the base of the occiput and there is no transverse crease on the top of the skull. But by the same reasoning the palaeoniscoids would be debarred from being ancestral to the later ganoids, because their parasphenoid was also confined to the anterior part of the skull. It seems far more probable that, when the passage from crossopterygians to amphibians took place, the anterior section of the skull (fig. 11.23aA), comprising essentially the ethmo-vomer block and the anterior border of the orbits, was extended backward until it was telescoped on to the rear section or cranial vault, while the prevomers, palatines and pterygoids formed a new and strong palato-pterygoid brace; at the same time on the dorsal surface the orbits were moving backward and the nasal region expanding (fig. 11.19C, D). Thus a reorganization was effected by combining the olfactory and sphenethmoid parts and the cranial vault into one strong and more or less tubular endocranium strengthened with exoskeletal sheathing bones on top of the skull and on the roof of the mouth.

CHAPTER XI

THE AIR-BREATHERS COME UP ON LAND

Contents

CHAPTER XI

THE AIR-BREATHERS COME UP ON LAND

TRANSFORMATION OF MUSCULAR PADDLES INTO LIMBS

IN THE last chapter it was shown that the dipnoans and the crossopterygians (lobefins) both evolved toward the amphibious way of life, but independently, and in quite different ways. The dipnoans developed very good lungs and reduced the efficiency of their gills but their paddles remained too flexible to serve as legs and in the African and South American forms dwindled into thread-like appendages. The lobefins, on the other hand, presumably had good lungs and certainly some of them had strong muscular paddles, but none of the known members of the lobefin group had yet taken the decisive step of transforming their paddles into fore- and hind-legs suitable for walking on land.

It could only be through a profound reorganization in habits and structure that streamlined fishes (fig. 11.1A) with their weight supported by the water and with smoothly curved flexible fins and paddles ever gave rise to land-living tetrapods with sharply angulated limbs serving as compound levers and springs to push against the firm ground (fig. 11.1B). The earliest amphibians, however, solved the problem by retaining the fish-like way of life in their larvae and by adapting the adults for life in either or both media.

To judge from its shoulder girdle (fig. 11.5B), Watson's *Eogyrinus* of Carboniferous age ought to have had transitional characters in its pectoral limbs; but very unhappily they are completely unknown. In the embryonic stages of newts and frogs the pectoral appendages (fig. 11.10bD, E) are indeed somewhat paddle-like in outline, the web ending in a transitory pointed tip beyond the space between the second and third digits. This tip is in line with the jointed central axis of the paddle. Watson (1913) attempted to homologize the distal pieces of the jointed central axis of the lobefin with the fourth digit of the amphibian hand; but Gregory and Raven (1941) showed that if a *Eusthenopteron*-like pectoral paddle (figs. 11.9; 11.10b) were bent around to form the successive angles at the shoulder, at the elbow and at the upper, middle and lower wrist joints, there would apparently be no separate short, postaxial rods left to be changed into the metacarpals and phalanges of the third, fourth and fifth tetrapod digits. They therefore suggested that these parts had arisen in embryonic stages from extra buds which were given off along the curving outer border of the hand (fig. 11.10aB, C) from the embryonic continuous skeleton-forming tissue of the central axis. The origin of new polyisomeres by budding, in the embryonic or foetal stages, is the rule in many cases of increase in the number of vertebral segments, of teeth, fin-rods, etc., and has evidently occurred in the axial segments of the pectoral fins of dipnoans (fig. 10.10F). Such a transformation, although still inferential in the present case, was certainly no more profound than that which must have occurred in the reorganization of the skull, shoulder girdle, pelvis, vertebrae, etc., in passing from the air-breathing lobe-fin fishes to the amphibians. And there are much greater difficulties in the possible derivation of amphibians from the dipnoans or from any other known group except the crossopterygians.

The oldest known amphibian fossils, from the Upper Devonian of East Greenland, consist chiefly of a series of excellent skulls (fig. 11.20A) described by Säve-Söderbergh. These retained certain clear traces of derivation from earlier crossopterygians, but were already nearing the grade of organization that was reached by some of the Carboniferous labyrinthodonts. It is very unfortunate that no limb bones were found with these skulls; but not improbably the limbs also would have been more amphibian than fish-like. Nevertheless it is very unlikely that the varied labyrinthodonts (stegocephalians) of the Carboniferous and Permian arose "suddenly" or by gigantic mutations (saltation) from their fish-like ancestors.

Long before the transformation of paddles into limbs the air-breathing lobefins must sometimes have found themselves left by receding floods in muddy waters laden with carbonic acid gas from decaying vegetation. Many died and a few later became fossilized in black shales. But some of the luckier or more vigorous ones wriggled about on the muddy flats and escaped desiccation in the air by getting into larger pools or active streams, or else by wriggling into shady niches on the banks of the streams. Presumably after many millenia of Natural Selection their powers of staying out of water for longer periods increased, at least in the adults. The wriggling habit inherited from fish days eventually led either to better wriggling or to better walking powers, or not infrequently to combinations of these methods (fig. 11.27), which in such later amphibians as the frogs (fig. 11.31) and "conger eels" (fig. 11.37) are found in widely different animals.

EEL-LIKE AMPHIBIANS AND OTHERS

Some of the confirmed wrigglers among these Palaeozoic amphibians multiplied their vertebrae as their bodies were lengthened and reduced the limbs to vestiges. Thus they set a precedent which in later ages was followed again and again on the road toward serpentiform, limbless "racers." Here belong several of the numerous families of swamp-living amphibians (fig. 11.28) that swarmed in the "coal forests" near Linton, Ohio, as well as similar serpentiform creatures in the Lower Permian of Europe. But there were also more newt-like animals (Nectridea) with rather small but well formed pectoral and pelvic limbs. Some of the newt-like fossil forms retained scales on the sides of the body, but the serpent-like wrigglers evolved out of their scales large ventral "ribs" or gastralia (fig. 12.45B, C), which gave a better grip upon the ground.

The modern newts, salamanders, and mud puppies, etc., have no doubt been derived from some of the less specialized, small-limbed amphibians of the Coal ages. Eel-like bodies retaining minute limbs have been evolved in the sirens of Europe and the "conger eel" (*Amphiuma*) of North America (fig. 11.37). On the other hand, the flattened bodies and very wide low heads (fig. 11.36A) of the mud puppies or hellbenders (*Cryptobranchus*) parallel some of the flat-skulled labyrinthodonts of the European Triassic (fig. 11.28). But the skulls, vertebrae (fig. 11.37) and limbs (when present) of these modern urodeles show a basic unity of pattern which is more specialized than those of their Carboniferous and Permian relatives.

The larval stages of the typical modern amphibians have well developed gill-arches equipped with external gills, as in the tadpoles of frogs. As their lungs develop, they depend less upon their gills, but many amphibians later learned, as it were, to breathe chiefly through their skins.

SCALY TO NAKED AMPHIBIANS

The scaly exoskeleton of their fish ancestors long ago began to thin out and the deeper layers of the skin gave rise to various kinds of glands, including those that secrete pigments. The chemical history here is probably very complex but, in short, some of the soft-skinned amphibians began to take oxygen from the water through their skins. Thus tadpoles that live in swift mountain streams, which are well stocked with oxygen, can live happily with little or no use for either gills or lungs.

Naked amphibians, having lost their defensive armor, have usually remained small animals and cannot well afford to stray far from good hiding places, such as the underside of rotting logs, frequented by some newts, or quiet pools, the favorite haunts of many frogs. The newts under the logs find insect larvae, while frogs in and near the pools can flip out their sticky tongues to catch insects. When a snake comes, the newts presumably get into the smallest possible crevice and the frogs instantly dive to the bottom.

PRIMITIVE QUADRUPEDALISM OF SALAMANDERS AND ITS SKELETAL CORRELATES

The locomotor system of the newt-like or tailed amphibians (figs. 11.27A; 11.37) has lost various parts that were retained in the earlier amphibians, including the clavicles, cleithra and interclavicle of old ganoid fishes, and there has been great expansion of both coracoid plates, which serves for the origin of the wide pectoral and coracoid muscles, but remain cartilaginous. The loss of the clavicles has freed the shoulder girdle from its formerly nearly immobilized state and enables it to slip forward and backward, like the sliding seat in a rowing-shell, thus lengthening the stride of the relatively small limb.

A small newt in walking across a white kitchen table makes surprisingly good speed and gives a beautiful exhibition of the "Australian crawl," a swimming stroke applied to quadrupedal locomotion. Here also is an effective combination of quadrupedal walking and fish-like wriggling. The right elbow moves backward as the right knee is flipped forward, and vice versa. The wriggling movement not only lengthens the stride but augments the thrusts and pulls of the extremities by the much greater forces of the long muscular sides. In forms with very long bodies these lateral muscles alone suffice for both eel-like swimming and snake-like crawling.

In the newts the small ilia or ascending bars of the pelvic girdle are tied not too firmly to the flexible vertebral column. The small femora or thigh bones are tied to the tail by the tendons of the long and strong tail muscles named caudifemoralis or coccygeo-femoralis; these have played an important part in primitive quadrupedal locomotion. The sinuous movements of the long-tailed amphibians are facilitated by the elastic qualities of their cylindrical vertebral centra and of the intervening buffers formed from the remnants of the fish notochord.

LEAPING ON ALL FOURS

Whereas the limbs and extremities of newts are small and the axial muscles play a large part in locomotion, the frog's legs are relatively very large and the principal locomotor thrusts have been transferred from the reduced axial musculature and vestigial tail to the thick arms, thighs and lower legs (fig. 11.31). The vertebral column accordingly has been greatly shortened and nearly immobilized, the sacral attachment has moved far forward and the remnants of the caudal region have fused into a slender bony tube.

The frog's hind limbs, by lengthening of the entire hind foot and especially of certain bones of the ankle, have added an extra segment to this live jumping machine. At any second the crouching frog, using the massive muscles of his thighs and calves, can suddenly snap open this system of bent levers and with a powerful kick from both sides at once, leap for safety. However, this convulsive method of locomotion is not effective for long distances and the cold of winter is met not by migration but by hibernation.

The pectoral limbs of frogs are not as large as the pelvic limbs, yet they do have their share in raising the trajectory as the fore part of the body bounces upward, impelled by the thrusts of the well developed pectoral and coracoid muscles. In the toads and some other families of the frog order the enlarged opposite coracoid plates overlap along the ventral midline (fig. 11.34B), thus producing what is called the arciferous condition. In the typical frogs, however, the opposite halves of the coracoid have fused into a single median piece, with an anterior and a posterior annex. This median brace is strengthened by lateral surface bones which are somewhat doubtfully homologized with the interclavicle and clavicles of the earlier amphibians.

The strong thrusts from the hind limbs are delivered from the rodlike ilia to the vertebral column by way of two more or less fan-shaped lateral outgrowths of the single sacral vertebra (fig. 11.31B), which are tied by ligament to the ilia. The fan-like sacral attachment has gradually been pushed forward along the intervening vertebrae until only eight to five vertebrae are left between the skull and the sacral attachment. This is due in part to fusion of some of the presacral vertebrae. Thus the frogs and their allies have the shortest backs and the fewest presacral vertebrae of all known land-living vertebrates. In correlation with the forward displacement of the sacral attachment, the ilia have been pulled out into long rods (fig. 11.31B1).

The East Indian "flying frog" *Rhacophorous* uses his large webbed feet in making long sailing leaps among the trees. They serve partly to increase the area of the supporting column of air over which the animal is sliding, partly to check the speed before landing and to brake the impact. Evidently prerequisite for this once new method of locomotion and escape were the large feet and associated characters of the normal frog stage.

Toads are more terrestrial than frogs and less dependent upon water pools; some of them are even fairly good diggers. They have retained the overlapping or arciferous relations of the opposite coracoids and to that extent are less specialized than frogs.

CLIMBING TREE-TOADS

The frog-like tree toads (Hylidae) have used their relatively long limbs to climb up into the arboreal world, far above the ground and above the pools and swamps, but in spite of their arboreal life their locomotor skeleton, except in the flexible joint surfaces between the vertebrae, differs but little from the normal frog type. Their tadpole stages retain more or less of the frog heritage, which has been prerequisite to many amazingly new, diverse and complex breeding habits (cf. Noble, 1922) in the different families and subfamilies of the tree frogs and their allies. The so-called Surinam toad, *Pipa americana,* which is famous for developing and carrying the young in little cells on the back of the female, belongs to a very ancient branch of the frog order. It is thoroughly aquatic, its large and fan-like hands and feet serving as flexible paddles.

WORM-LIKE CAECILIANS

The most specialized of all living amphibians are the worm-like caecilians (Gymnophiona), which live underground in Mexico and South America, Africa, India and the East Indies. All traces of limbs are absent and the cylindrical body is externally zoned or segmented by deep transverse creases and overlapping shields. Each of these includes numerous small scales. This flexible construction probably assists the segmental muscles of the body in finding fulcra or points of leverage in pushing the animal forward. However, the lower jaw is very strongly built and it has an unusually long heel-like projection behind the quadrate-articular joint. The very strong depressors of the mandible may push the dirt backward and away from the head, and by twisting movements of the body the jaw could work effectively in enlarging the cavity around the head.

The skull of caecilians suggests derivation from the snake-like "Permian urodele" *Lysorophus,* but connecting stages are lacking (Romer, 1945, p. 159). The young of the Ceylon *Ichthyophis* have external gills and gill slits and live for a time in the water. This is one of the reasons for classifying the caecilians with the Amphibia rather than with the snake-like "blindworms" (Amphisbaenids) among the true lizards and other reptiles with which the caecilians were long confused.

Certain Palaeozoic amphibians have been regarded as in or near to the lines leading respectively to: (1) the Urodela or Caudata, comprising the salamanders and newts (commonly, but erroneously, called "lizards"), and (2) to the Anura (or Salientia), viz., the frogs, toads, tree frogs, etc.

GILL-BEARING BRANCHIOSAURS AND THEIR POSSIBLE RELATIONS WITH OTHER AMPHIBIANS

The branchiosaurs (fig. 11.28) of the Upper Carboniferous and Permian of Europe and North America are much more primitive than the newts and salamanders. The limbs of branchiosaurs had a cartilaginous core surrounded by a sheath of bone, with both ends remaining cartilaginous. The clavicles, cleithra and interclavicles all inherited from ancestral fishes, were reduced to splints but still present. In modern salamanders, and their derivatives, especially those which are thoroughly aquatic (fig. 11.37), the shafts of the limb bones also become cartilaginous and the ends (epiphyses) remain cartilaginous even in the adult, but the cleithra, clavicles and interclavicles have disappeared entirely. This means that the modern newts have indeed progressed further than did the Palaeozoic branchiosaurs along the path of adaptation to aquatic life, but not necessarily that the branchiosaurs were ancestral to them. The vertebrae of branchiosaurs were partly cartilaginous but there were also bony pieces or shells representing collectively dorsal and ventral half-centra, the neural arch and a large transverse process (parapophysis) on each side near the middle of the centrum. The latter received the thrusts from the rod-like ribs.

Partly on account of the characters of their vertebrae, the branchiosaurs used to be regarded as adult forms that represented a separate order (Phyllospondyli) of extinct Amphibia. But Romer (1947) has shown that in many cases the branchiosaurs were larval forms retaining gill arches and external gills and that they grew up to be labyrinthodonts.

The skull patterns of branchiosaurs were basically the same as those of both Urodeles and Anurans, but without the specializations of either. The same is true of the adult skulls of many different genera of Permian labyrinthodonts. It is certain that parallel changes in the skull, as for example, in certain lines the delay in ossification in the occipital segment, took place independently in different families and suborders of amphibians. This makes it difficult to find the direct ancestors of existing groups of amphibians among their swarming Palaeozoic predecessors.

PERMIAN *DIPLOCAULUS* COMPARED WITH MODERN *CRYPTOBRANCHUS*

The problem is further complicated by the existence of *Diplocaulus,* a Permian amphibian (fig. 11.28I) with the following interesting combination of characters.

1) In general body-form it was somewhat like the existing hellbender (*Cryptobranchus*), that is, with both skull and body depressed and with a long laterally compressed tail.

2) The very flat skull, as seen from above, was like an inverted V with widely divergent limbs and bluntly rounded tip; the "horns" of the skull projected backward along the sides of the neck, with which they may have been united.

3) The eyes were far forward, the mouth not very large, the lower jaw short and well curved in front.

4) Owing to the extreme size and backward direction of the "horns," the pattern of the skull roof, as compared with those of less specialized amphibians, shows minor or secondary readjustments.

5) The dorsal vertebrae (fig. 11.2F) in the adult agree with those of urodeles in that there is no suggestion of separate blocks and that there were large laterally projecting processes with dorso-ventrally bifid ends which received the thrusts of the short ribs.

This combination of characters raises the question whether *Diplocaulus* may not be directly ancestral to the hellbenders (*Cryptobranchus*). Those who adhere to the dogma that a specialization once gained can never be lost would point to the large occipital "horns" of *Diplocaulus* as at once excluding it from such an affiliation, but the European *Keraterpeton* and other related genera suggest: (1) that the occipital "horns" were part of the old connecting tract that tied the shoulder girdle to the occiput (see p. 142 above); and (2) that their widening in *Diplocaulus* was associated with the marked dorso-ventral flattening and widening of the head, neck and thorax; further that (3) there may easily have been a recession of these extreme features in the line leading to *Cryptobranchus*.

SKELETAL EVOLUTION OF THE NEWTS AND SALAMANDERS

Thus there would arise the further question, whether the *Cryptobranchus* line has been separated from the rest of the urodeles since Permian times. Fortunately at this point one may turn to Noble's excellent book on the Biology of the Amphibia (1931) and study there the evidence: (1) that the family of salamanders called Hynobiidae (*"Ranodon"*) are the most generalized of all living urodeles and (2) that they tend to connect all urodeles in one comprehensive order; (3) that the cryptobranchids are rather closely related to the hynobiids.

On the whole, the work of Noble and his numerous colleagues and predecessors also suggests: (4) that the modern urodele families have not arisen separately from different Palaeozoic families, but (5) that they are the diverse derivatives of some as yet unknown or poorly known Triassic salamanders. It also seems probable (6) that the adults of that ancestral Triassic family or families had retained the tubular braincase and flattened skull contour of branchiosaurs, together with large cartilaginous epiphyses of the limb bones, reduced clavicles and episterna, enlarged coracoids and a large anterior or procoracoid process of the scapula; but that the vertebrae had already attained the narrow constricted cylindrical form, perhaps as direct ossifications of the notochordal sheath, with large transverse processes for the rod-like ribs.

These somewhat technical but necessary details should not obscure the significance of the urodeles in our effort to present an over-all and necessarily somewhat simplified picture of the evolution of the outer and inner skeletons among vertebrates. The aquatic life of the larval stages has perhaps influenced the retention of an essentially fish-like pattern of the branchial arches, especially in *Hynobius*. On the other hand, the urodeles had lost the scaly exoskeleton of fishes and had acquired a soft glandular skin, which was important in producing concealing pigmentation as well as in breeding. In eliminating the cleithra and clavicles from the shoulder girdle, the urodeles made it impossible for them to be ancestral to the main line leading to the reptiles and higher vertebrates. They also became aberrantly specialized in many skull features (fig. 11.36), such as the flattening of the skull, the development of paired occipital condyles, the too early fenestration of the temporal region and the failure to evolve a complex pterygo-palatal arcade for the internal narial passage. In short, the skeleton of a modern urodele, such as that of *Cryptobranchus* is very far from being morphologically ancestral to those of primitive reptiles and higher vertebrates, as fully shown by Gaynor Evans (1944).

THE ANCESTRY OF THE FROGS AND THE DIVERSE EVOLUTION OF AMPHIBIAN VERTEBRAE

These complex questions are entwined with the problem as to the values, in the major classification of amphibians and reptiles, of the contrasting ways in which Gadow's "primary vertebral blocks" have been combined to produce the diverse patterns of vertebrae seen in adult fishes, amphibians (fig. 11.2) and higher vertebrates, as shown by Cope, Gadow (1896, 1933) and their successors (notably Watson and von Huene).

As noted above (p. 85), the notochord of vertebrates is a cylindrical rod enclosed in tough, impervious tissue. It is kept in a turgid state, apparently by a one-way system of absorption without leakage, and serves: (a) partly as a stiffening axis for the entire body and (b) as a fulcrum against which the myomeric muscles could react, (c) partly as a base for the support of the longitudinally extended spinal nerve, (d) partly as a non-rigid ridgepole for the roof of the body cavity. Bone cells invading this region settled in the following places (figs. 5.1; 10.4; 11.2; 11.29): (1) sometimes in the perichordal sheath, presumably as in the ancestors of the urodeles; (2) in the cartilaginous neural arches, which already served for the insertion of the tough, obliquely transverse septa covering the muscle segments and receiving from them the pulls that helped to warp the body into transverse undulations; (3) on the sides of the notochordal axis, especially at the points of intersection of the oblique septa; (4) on the ribs, which impinged against the sides of the notochordal axis and arched transversely over the body cavity and viscera. Obviously if the bone cells had been deposited evenly and continuously along the notochordal axis, the latter would have solidified into a rigid tube and locomotion by lateral undulation, as in fishes, would have ceased. But the presence of the transverse and oblique septa, running in from the lateral myomeres toward the axial rod, tended to segregate the centers of cartilage and bone into anchoring pieces that alternate with the myomeres (cf. Breder, 1947, fig. 33).

Given this new set of polyisomeres, or really poly-anisomeres, for they were already unequal with respect both to size and to location, further anisomerism or unequal emphasis among them opened the ways for the emergent evolution of Gadow's various types of vertebrae. In all such cases the diversifying mechanism of heredity produces new units, new accelerations and new retardations of growth rates, while the independently varying conditions of the environment and the degree to which a given change happens to be in the direction of improvement, all coöperate through long ages in producing statistically cumulative differences.

Already at the stage of *Eusthenopteron* (figs. 11.1; 11.2) among the crossopterygians, there were: (a) a pair of thin bony half-crescents ventral to the large notochord, and (b) another very small pair at the bases of the stout neural arch. The neural arch in Gadow's nomenclature would represent the fused pair of basidorsals, the basal crescents would be the basiventrals (called by Cope, intercentra). The pair of little ossicles behind the lower ends of the neural arches would be Gadow's interdorsals (Cope's pleurocentra), while the interventrals in this fossil fish were probably cartilaginous and hence are now represented by blank spaces.

The vertebrae of the East Greenland Devonian labyrinthodonts are unfortunately unknown. According to Watson (1926a), the most primitive, or "adelospondylous" type of amphibian vertebrae is found in certain fossil amphibians from Carboniferous horizons in Ireland and Hungary, for which he proposed the ordinal name Adelospondyli, presumably in allusion to the unknown (*adelos*) composition of the centrum (in terms of Gadow's blocks). C. Forster Cooper (in Parker and Haswell's Zoology, II, 1940, p. 337) figures the adelospondylous type as having a large neural arch, suturally united to a single centrum (presumably surrounding the notochord). On the side of the centrum near its lower border is a fairly large excavation. There is a small rib facet near the base of the neural arch, behind and below the anterior zygapophysis. This adelospondylous type is placed, with a query, next to the lepospondylous (cf. fig. 11.2E) which it approaches in general form and in the lack of distinct intercentra. Romer (1945, p. 591) refers the Adelospondyli to the Order Microsauria of the Subclass Lepospondyli (see p. 256 below) on the basis of his numerous monographic studies on fossil amphibia.

Among the modern frogs the neural arches in foetal stages grow downward around the notochord and take over the position and functions of the interdorsals (or pleurocentra), while the basiventrals are represented only by small nodules beneath the notochord. Such a type of vertebra is called notocentrous because it consists chiefly of the dorsal elements (Gadow). Watson, who with Cope, von Huene and others, set a very high taxonomic value on the vertebral characters, in his memoir on The

Origin of Frogs (1940) described two Carboniferous amphibians in which the construction of the vertebrae approached the frog type. They also agreed with the modern frogs in that the supratemporal and other bones covering the temporal region of the skull were absent, as in the frogs. The body form (known in one of these genera) was much more newt-like than frog-like and the tail was long, as in the modern urodeles, and the fossil Lepospondyli.

EVOLUTION OF AMPHIBIAN SKULLS

The Permo-Carboniferous Texas genus *Cacops* described by Williston suggests the frogs, especially in the shape of its large skull, the shortness of its body and the strengthening of its hind legs. Its vertebrae consisted of large basidorsals (neural arches), large basiventrals (intercentra) and small interdorsals (pleurocentra) and were thus of the type called rhachitomous (with cut-up axis) but they could readily be transformed into the adelospondylous or frog-type, chiefly by reducing the pleurocentra. Even the Permo-Carboniferous *Eryops megacephalus* (fig. 11.31A, A1), the best known of all rhachitomous labyrinthodonts, could readily be transformed into a frog-like type (B, B1), chiefly by shortening the tail and by modifying the larval vertebrae and ribs in the manner noted above. The ilia would have to be lengthened and the sacral attachment moved forward, as the hind limbs became larger and better adapted for leaping. The pubi-ischiadic plates on either side would have to be reduced to a small subcircular plate and the acetabular socket made concave to receive the spherical head of the femur (fig. 11.35D). The ribs would be shortened and as the flattening and widening of the body increased, the transversely directed parapophyses would functionally replace the ribs (fig. 11.31B, B1).

In order to transform the skull of *Eryops* into that of the frog (fig. 11.32B), it would be necessary to reduce the squamosal to a sickle-like bone lodging the head of the quadrate and to slenderize the bones of the upper and lower jaws. It would also be necessary to fenestrate the skull in the temporal region (B1), that is, to thin out and absorb the roofing bones which in more primitive forms (fig. 11.32A1) lie on either side of the narrow braincase.

The process of fenestration, that is, the creation of windows or apertures often surrounded by strong frames, is indeed one of the chief ways in which transformations of the skull as well as of vertebrae and girdles of vertebrates have occurred. It involves the thinning out and eventual perforation of flat or curved bones or bony processes and is often correlated with changes in the size or direction of stresses from the muscles of locomotion or of mastication.

Cacops and *Eryops* are very probably not in themselves immediately ancestral to the frogs; but they do indicate what the pre-frog skeleton was like before it became as highly specialized for leaping as it is in the frogs. Watson's genera, on the other hand, foreshadowed the frogs in their vertebral composition and in the reduction of the temporal roof but had not yet become frog-like in body-form and locomotor structure.

THE ROLE OF PAEDOGENESIS IN THE ANCESTRY OF MODERN URODELES

In the present chapter the recent surviving Amphibia have been considered first, partly because the correlations between body-form, skull-form and habit are all available for study. The fossil predecessors of the modern Amphibia have so far been mentioned here chiefly in their possible relationships to the moderns. The evidence suggests: (1) that none of the wriggling, eel-like Palaeozoic Amphibia of the orders Aistopoda, Microsauria were ancestral to the quadrupedal salamanders; (2) that the eel-like forms died out long before the modern Amphibia branched out in Jurassic and Cretaceous times; (3) that it is quite unlikely that the eel-like modern *Proteus* was derived, independently from the rest of the urodeles, from any eel-like Palaeozoic amphibian.

Viewing the modern urodeles as a whole, it seems probable that all are to a varying extent neotenic or paedogenetic forms which have retained more or less the larval adaptations for water-living, including the larval cartilaginous skeleton.

EMERGENCE OF NEW LOCOMOTOR TYPES

The famous axolotl (*Ambystoma*), which completes its metamorphosis and becomes a well developed terrestrial tetrapod, may be either retaining an otherwise unknown terrestrial ancestral stage, or pushing ahead into a new habitat after the manner of newly emergent types. The small, semi-terrestrial newts, hiding under damp stones or logs and searching for insect larvae, are very unlike the flat-bodied, flat-headed hellbender (*Cryptobranchus*) which lies on the bottom in streams, alert to spring up like a living trap at passing fishes. This flat-bodied type resembles the later flat-headed labyrinthodonts and were it not for the evidence cited by Noble (1931, pp. 465–469) that *Cryptobranchus,* along with more typical urodeles, has been derived from the stem of the hynobiid salamanders, one might raise the question of its possible relationships with the labyrinthodonts. From the latter, however, it differs radically in its vertebrae, which as noted above (p. 250) are much more like those of the Permian *Diplocaulus*. Serpent-like forms, as noted above, emphasize wriggling movements by multiplying their body segments and reducing their paired limbs. Worm-like digging forms may use the conical head or even the lower jaw to push the earth to one side.

CENTRAL POSITION OF THE LABYRINTHODONTS

The labyrinthodonts, which we have referred to only incidentally thus far, were the dominant group of Palaeozoic amphibians and included a considerable range in body-form and skull-form. The name labyrinthodont referred to the pattern of the teeth, as seen in cross-section near the base, which shows primary and secondary folds radiating from the pulp cavity. These labyrinthine folds were surrounded by a tooth cavity which must have confined the rapidly expanding tooth and favored the process of infolding. This prolific growth of the dental tissues is nearly matched among the Devonian crossopterygians (fig. 10.7C, C1), which from much other evidence are regarded as the nearest known piscine relatives of the amphibian class.

One of the most primitive of the labyrinthodonts was Watson's *Eogyrinus* from the Carboniferous of Great Britain. In the dorsal vertebrae (fig. 11.2D), both the intercentra (basiventrals) and the pleurocentra (basidorsals) were checker-like. This was the embolomerous (interjected part) type and the name Embolomeri is used to designate a suborder of Labyrinthodonts. Romer has noted that the skulls of some Embolomeri are essentially similar to those of certain Rhachitomi (in the latter, the intercentra are crescentic and the pleurocentra are small pieces) and he infers from much evidence (1947, fig. 11 and p. 69) that the protorhachitomous type with single crescentic intercentra and paired pleurocentra, was the more primitive of the two. The body in *Eogyrinus* and allied forms was long and somewhat newt-like, but they ranged from small (*Diplovertebran,* fig. 11.1B) to large size and in general habits were probably more like crocodilians, which lie in wait in the water and suddenly spring open their jaw-trap as they rush at the prey.

The skulls of the more primitive labyrinthodonts (figs. 11.32A; 12.2, 3) were of moderate length and height on the sides, not so much flattened dorso-ventrally as in later members of the group. The occipital base and roof were well ossified, not cartilaginous as in many later forms. The occipital condyles (figs. 11.32; 12.3, 4) ranged from a subcircular, almost single, medial depression to a wide, almost dicondylic stage on the exoccipitals, with reduced median contact on the basioccipital. The skulls of the later Triassic labyrinthodonts, especially of the suborder Stereospondyli (fig. 12.28D), were very flat on top and rounded to parabolic in shape, with the small eyes directed upward. The gigantic *Mastodonsaurus* and others of this suborder lurked in the streams of the Trias in Germany as *Buettneria* did in North America.

This apparently was an extinct branch. In the vertebrae the basiventrals (intercentra) were large and checker-like, the interdorsals (pleurocentra) being vestigial.

On the under side of the skull the process of fenestration (see p. 253), beginning in the older labyrinthodonts as a small median slit between the opposite pterygoids (fig. 12.4A), culminated in the Stereospondyli in a very large median gap between the opposite sides of the upper jaw, much as in the frogs (fig. 11.32B2).

The inner skull of labyrinthodonts (fig. 11.32A1) was very frog-like, including a tubular braincase covered by a flat skull roof. But the fenestrae above the temporal muscles had not yet begun to appear; while the large tympanic ring of the frogs (fig. 11.32B) was foreshadowed at first only by the otic notch (fig. 11.32A1), which in *Cyclotosaurus* was being converted into a circular opening. Nevertheless the labyrinthodonts and still more the modern frogs have preserved an early stage in the evolution of the drumhead type of mechanism for transmitting sound waves to the inner ear.

In some of the Triassic labyrinthodonts from East Greenland the front part of the skull was pulled out into a long snout (fig. 11.28F), as in many fish-catching reptiles.

THE OLDER LABYRINTHODONTS AND THE EMERGENCE OF THE REPTILES

Thus the class Amphibia, the oldest of the quadrupedal vertebrates, was once dominant in the swamps of late Palaeozoic and early Mesozoic times and is represented today by the frogs, toads, newts, hellbenders, tree frogs, caecilians, etc. Most of these modernized forms lost the chance of evolving into reptiles and higher vertebrates when they reduced their exoskeletal armor and began to specialize in skin glands. As already noted, these are important, especially in the production of concealing pigments and in breeding processes, but along with the retention of gills in the larvae and other conservative features, they encouraged many peculiar later specializations rather than an advance toward the reptilian grade. The eel-like forms and the worm-like burrowing caecilians also very slowly eliminated themselves from the lines leading to more advanced grades of locomotion.

Somewhat the same conditions debarred most of the Palaeozoic amphibians from attaining the higher grades. The limbless Aistopoda and many others retreated from what may be called the main line of advance, as did many of the reptiles in later ages. Even the majority of the labyrinthodonts, of which the earlier forms had strong limbs and well ossified skeletons (figs. 11.27; 11.31), succumbed, as it were, to the habits: (a) of retaining too long the cartilaginous endoskeleton of their own larval stages and (b) of remaining till maturity in or near the pools and swamps which were at once their breeding places and their hunting grounds. The saving minority, without which the long and branching roads toward the higher vertebrates would never have been developed, belonged to the labyrinthodont order and possibly to the suborder Embolomeri (cf. Romer, 1947, p. 306). *Eryops megacephalus* Cope from the Permo-Carboniferous of Texas and Pennsylvania was, so to speak, one of the uncles of this chosen tribe, but he was somewhat too far to the right, that is, on the conservative amphibian side, to be in the forefront of the proreptilian progressives. Still, unlike most of his relatives which he had left in the swamps and pools, he could amble about on land on his own sturdy, bony frame (fig. 11.31) and his vertebrae (fig. 11.29) indeed contained the pieces out of which reptilian and even mammalian vertebrae were later to be made.

As it now seems, Dr. Gadow's apparently rash proposal (1909, p. 285) to make *Eryops* the type of a new order, "Proreptilia," was indeed a revealing if not wholly correct suggestion. For, as hinted above, *Eryops* himself was still fundamentally and primitively amphibian in general structure, in both his exo- and endoskeletal patterns, but certain of his older embolomerous relatives were structurally nearer to the "stem reptiles" or seymouriamorphs and cotylosaurs, may best be dealt with in the next chapter.

THE MICROSAURS

Another group of Carboniferous age which has been regarded as possibly ancestral to the reptiles were the small lizard-like "microsaurs," some of whose remains were found in the interior of tree trunks in the coal mines of Ireland and Nova Scotia and in the fossil coal swamps of Bohemia. These microsaurs were long ago set apart in an order named Lepospondyli (husk vertebrae) because their vertebral centra took the form of constricted cylinders, which may have resulted from the ossification of the perichordal sheath. On a small scale they resemble the constricted cylindrical vertebrae of the smaller Pelycosaurian reptiles rather than those of the labyrinthodonts, which have the centra divided into pieces. Nevertheless their skull, so far as known, was not fundamentally dissimilar from those of other amphibians of the same age. Romer suggests (1945, p. 159) that their affinities lie rather with the modern salamanders and Apoda. However, it is noteworthy that in Permo-Carboniferous times these creatures had essentially amphibian skulls and reptile-like vertebral centra, while their contemporaries, the loxommid labyrinthodonts, combined very generalized skull patterns with "cut up" (temnospondyl) vertebrae of primitive amphibian type.

CHAPTER XII

FROM LOWLAND MARSH TO FOREST AND DESERT

Contents

CHAPTER XII

FROM LOWLAND MARSH TO FOREST AND DESERT

CONTRASTS BETWEEN TYPICAL AMPHIBIANS AND REPTILES

WHEREAS THE typical amphibians spawn in the water and pass through a water-living, tadpole larval stage with external gills, reptiles lay their eggs on land and the larval stages are passed through wholly within the relatively large yolked egg. While this contrast between amphibians and reptiles exists with partial exceptions today, it was probably already noticeable in Carboniferous and Permian times. Romer (1947) indeed has examined much evidence that the diversified adult labyrinthodont amphibians had aquatic larvae, which were formerly called branchiosaurs and referred to a distinct order, Phyllospondyli; he has also found in the Texas "Permian" a fairly large fossilized egg with a tough shell, which resembles the eggs of some recent reptiles and was probably laid by one of the pelycosaurian reptiles that lived in the same period with the amphibians. Thus the reptiles as a class early freed themselves from the water-bound life of the amphibians.

Among modern reptiles the true lizards (not the newts, which are often miscalled lizards) contrast with amphibians in retaining and developing a scaly exoskeleton on the head, body, tail and limbs, and their skin, while not wholly lacking in glands, is usually dry and never slimy as it is in many amphibians.

In typical adult reptiles even in Permo-Carboniferous times, the vertebral centra were like constricted cylinders, with no indication of the separate pieces which formed the centra of labyrinthodonts (fig. 12.14aB). Nevertheless there was a certain group of Permo-Carboniferous tetrapods, typified by *Seymouria* (fig. 12.2), in which the basiventral pieces or intercentra were quite distinct but the pleurocentra were enlarged and together formed the body of the centrum (fig. 12.6A). The neural arch had a short median spine and a pair of swollen shoulders suggesting those of *Diadectes* (fig. 12.14b). In *Seymouria* the ends of the limb bones (figs. 12.2; 12.5B, B1) long remained cartilaginous and the skull patterns (figs. 12.3, 4) were fundamentally the same as in the labyrinthodonts. Indeed all recent students of the group have assigned *Seymouria* and its Russian allies (fig. 12.7) to the Amphibia as a suborder (Seymouriamorpha) of Labyrinthodonts, next to the Loxommids.

But in another direction *Seymouria* approached its contemporary *Diadectes* (fig. 12.9), a large reptile which is the type of the order Cotylosauria, or "stem reptiles." *Diadectes* had a single median occipital condyle of subcircular form (fig. 12.11) and was therefore a reptile "by definition," but so did the labyrinthodont *Orthosaurus* (fig. 12.4) and some other labyrinthodonts. The primitive condyle of *Seymouria* (fig. 12.48) was borne on the basioccipital and was flanked by a pair of small condylar facets on the exoccipital. In the typical amphibians, as the occiput was widened, the exoccipital condyles grew larger and the median condyle was diminished (figs. 11.22A, B; 11.23bA, B). In the typical reptiles the reverse took place, the lateral facets diminishing and the median one becoming dominant and spherical. The pattern of the bones on the under side of the skull of *Seymouria* was essentially the same as in *Orthosaurus* (fig. 12.4) and likewise primitive as compared with the same region in typical amphibians (fig. 11.32) and typical reptiles (fig. 12.29). In brief, *Seymouria,* as shown in T. E. White's excellent monograph, forms an ideal structural link between labyrinthodont amphibians and primitive reptiles and is also definitely less specialized than *Diadectes* in its dentition (figs. 12.4B; 12.11) and vertebrae (figs. 12.6; 12.14).

DIADECTES, A SPECIALIZED SIDE BRANCH OF THE STEM REPTILES

Diadectes (fig. 12.9) was a large, clumsy and bent-legged animal in which the cheek teeth formed transversely placed, cross-ridged ovals (fig. 12.11). The construction of the jaws and temporal region (figs. 12.10–13) indicate very massive thick jaw muscles that were able with the assistance of the teeth to crack bones or perhaps mollusc shells. Thus the dentition was far more advanced than that of *Seymouria,* in which the conical teeth were essentially like those of labyrinthodonts and were adapted for a carnivorous diet (fig. 12.4). The skull roof and cheeks of *Diadectes* were covered with thick dermbones and the endocranium was provided with strong walls and braces (figs. 12.12–13).

The vertebrae of *Diadectes* (fig. 12.14a, b) had very large neural arches with great rounded shoulders above their anterior and posterior zygapophyses or articulating processes. They also had a second pair of articulating processes medial to the lateral pair. As a result of these closely interlocking relations, some anteroposterior and lateral sliding movements between adjacent vertebrae were possible but hardly any dorsoventral movements.

This construction of the vertebrae prevented dislocation and strangulation of the spinal cord and spinal nerves (fig. 12.14a). It also successfully met the great bending movements caused by the very thick muscles of the vertebral column and ribs and by the thrusts and pulls of the short but excessively thick muscles of the limbs. The shaft of the humerus (fig. 12.5a) was strongly twisted and the planes of the very wide proximal and distal ends were nearly at right angles to each other. The deltopectoral crest was very large; the head or articular facet formed a narrow curved trackway which enabled the sharply everted humerus to slide, roll and partly twist on its socket. The very wide distal end provided strong crests for the extensor, pronator and supinator muscles which twist and untwist the huge forearm. The ball-like capitellum or facet for the radius was located on the ventral surface of the humerus, as the elbow was sharply bent. The five-toed hands and feet were wide and strongly built (fig. 12.9), with thick digits ending in blunt claws. The shoulder girdle (fig. 12.9) formed a large, strong and wide U-shaped sling for the support of the wide body. The glenoid facet of the coracoscapula (fig. 12.20aA) which received the thrust from the humerus formed a horizontally elongated warped band, as in *Eryops* and the larger labyrinthodonts.

The ventral or pubi-ischiadic plates of the pelvis (fig. 12.24A) were long and thick but the ilia were rather short with no great expansion for the deep gluteal muscles. T. E. White, in describing similar conditions in *Seymouria,* suggests that in these wide-backed animals the exceptional strength of the vertebral column and of the axial muscles compensated for the weakness of the gluteal system, which in more advanced types of reptiles serves to hold up the hip region when the leg of that side is raised off the ground. The sacral attachment of the ilium, although limited to two vertebrae (fig. 12.9), was not loose as it was in *Eryops* (fig. 11.17), but fairly close and immobile.

Accordingly *Seymouria* and *Diadectes* may have waddled along like heavy tortoises but with a stronger transverse rocking movement and with somewhat of the "Australian crawl" method of swinging the lower arms and legs quickly forward and slapping the wide hands and feet down at the beginning of each stride.

Unlike the humerus, the femur was not twisted but formed a nearly straight shaft (fig. 12.24D) somewhat arched in section, with the facets for the lower leg so turned downward that the everted knee was sharply bent (fig. 12.9). Powerful adductor muscles converged from the pubi-ischiadic plate to the fossa on the under side of the femur (fig. 12.5aC1).

Apparently both *Seymouria* and *Diadectes* combined slow walking, swimming and digging habits.

Seymouria was one of the many natural experiments in quadrupedal locomotion. *Diadectes* was somewhat more tortoise-like and his order (Cotylosauria) may have indeed been on or near the road to the remote ancestry of the chelonians.

CAPTORHINUS AND ITS GENERALIZED REPTILIAN SKELETON

Not very far removed from *Diadectes* and related genera was the somewhat lizard-like *Captorhinus,* the skeleton of which (fig. 12.8), as figured by Williston, showed none of the marked specializations noted above in *Diadectes;* probably because of its small size its limbs were relatively much less massive, although still of the bent-limbed crawling type.

The skull of *Captorhinus* (fig. 12.29) was extremely generalized in its main features, that is, it was basic in pattern for the diverse experiments tried by the Permian pelycosaurs and mammal-like reptiles and by the ancestors of all the higher reptilian orders. Because of these primitive skull features, especially in the temporal region, the captorhinomorphs were formerly brought together with the diadectomorphs and pareiasaurs to form the Cotylosauria, or stem reptiles; but Olson (1947) removed the cotylosaurs and pareiasaurs to a new subclass which he named Parareptilia, referring the Captorhinomorpha to the subclass Eureptilia, comprising all the more typical reptiles. This concept seems to clarify the virtually ancestral position of *Captorhinus* to the typical reptiles (fig. 12.1), but it also obscures the underlying relationship between *Captorhinus* and the more specialized *Diadectes.*

In short, the seymouriamorphs and diadectomorphs, although each had evolved its own emergent novelties and specializations, show unmistakably that basic amphibian skeletal patterns, as exemplified in the older embolomerous labyrinthodonts, were gradually replaced b basic reptilian characters, as exemplified in th captorhinomorphs. And basic reptilian pat terns in their turn were the starting-point fo the immense diversity evolved in later reptiles birds and mammals. Hence not even a con densed history of the evolution of the inner an outer skeletons from the simplest forms to ma can well omit the outstanding features in th grand transformation from embolomeres t captorhinomorphs.

THE SO-CALLED "PRIMORDIAL CRANIUM" AND ITS PHYLOGENETIC SIGNIFICANCE

The results and conclusions of the presen work are based primarily on comparativ studies of adult skulls as preserved in fossil an recent stages. It may be noted also that out standing works on the embryology of typica vertebrate skulls (e.g., Howes and Swinnerto on *Sphenodon,* De Beer on all orders of verte brates) have by no means been neglected. O the other hand, some of the methods and con clusions of specialists in endocraniology, al though admirable in their descriptive data seem to have led, and to be leading still, int blind alleys. In brief, it may here be affirmed, a a result of experience, that the embryology c the chondrocranium alone may be no closer guide to past adult evolutionary stages than i the embryology of the entire organism.

Specialists in the morphology of the inne skull or chondrocranium, starting from th highly fenestrated chondrocranium of the lizar described by Gaupp, have extended his elabo rate terminology outward and upward to th chondrocrania of other reptiles, birds and mam mals, and backward to the skulls of existin amphibians, sharks and bony fishes. Thus ha accumulated an already formidable nomencla ture in the literature of what might be calle chondrocraniology. But this method led to th erroneous identification of the reptilian prefron tal with the mammalian lacrimal and of th tuberabasiphenoidea of reptiles with the alis phenoid of mammals. However, Reichert an Gaupp did correctly solve the problem of th origin of the mammalian auditory ossicles chiefly on the basis of embryological evidence.

Some contemporary chondrocraniologists, apparently following the methodological principle, which has been more or less unconsciously pursued also by many systematists, that "analysis must precede synthesis," have pushed analysis so far beyond synthesis that their specialty seems often to ignore the following considerations:

1) that the cartilaginous chondrocranium of larval and foetal stages is merely a prerequisite core or matrix upon which the permanent skull as a whole is going to be built later.

2) that each chondrocranial type is of necessity placed and molded in such ways that the infantile, young and adult stages can be built around it. This is because it is obviously the adult skull which carries the development to its climax in the reproductive stage, without which the chondrocranial pattern itself would immediately cease to exist.

3) Consequently the chondrocranium during its own development is bound to foreshadow increasingly the skull-form of the adult and to be influenced by the interacting forces that will finally finish the adult skull. For example, if the adult skull has a very long face, as in the horse and mole, even the earlier foetal stages will often foreshadow that condition. Or if the eyes be reduced or absent in the adult, the orbital fossae will be very small in the foetus, as in the Cape Golden moles. Again, if the adult cranium is to be greatly fenestrated, these fenestrae will be at least begun in early foetal stages.

4) Even such primitive structures as Meckel's cartilage and the remnants of the branchial apparatus, may be useful in the embryo as the pathway, cores, growth centers from which quite different adult structures may develop.

5) Hence features which have been inherited from very ancient vertebrate foetal stages may still retain their anticipatory or preparatory values in so far as they are prerequisites for later stages.

6) The implied assumption, in some papers on the chondrocranium, that it has a greater phylogenetic significance than the dermocranium, is not compatible with other evidence.

7) The foregoing considerations suggest that chondrocranium and dermocranium ought always to be considered together, with reference to their functional relations to the entire adult skull and in comparison with other appropriately chosen fossil and recent skull forms.

8) The adult skulls of vertebrates, recent and fossil, when studied on a wide comparative basis, indicate that the chief progressive line in or on either side of the "main line" from ostracoderm to man are recorded in the following series.

THE CRITICAL POSITION OF *CAPTORHINUS* IN THE "MAIN LINE" STAGES FROM EARLIEST FISH TO MAMMAL

GEOLOGIC AGE	GROUP	EMERGENT FEATURES
Ordovician	I. Ostracoderms	Basic chordate relations of spinal nerve-cord, notochord and gut. Streamlined form, armored head and thorax
Silurian	"	Bone cells in exo- and endo-skeleton. An oralo-branchial chamber
Silurian-Devonian	II. Placoderms (Aphetohyoids)	Exoskeletal plates; rise of complex (inner plus outer) jaws
Devonian-Carboniferous	III. Crossopterygians	Lobate paired fins; lungs, marginal choanae; predaceous jaws
? Upper Devonian, Carboniferous	IV. Labyrinthodonts	Tetrapod limbs, enlarged scapulo-coracoids, reduced cleithra, rise of sacro-iliac contact; loss of opercula; aquatic (tadpole) larvae; vertebrae temnospondylous
Upper Carboniferous	V. Captorhinomorphs	Terrestrial reptiles with imperforate temporal shell; reduction of supratemporal bones; four-way pterygo-palatine brace; vertebrae holospondylous, intercentra reduced or wanting

THE CRITICAL POSITION OF *CAPTORHINUS* IN THE "MAIN LINE" STAGES FROM EARLIEST FISH TO MAMMAL

GEOLOGIC AGE	GROUP	EMERGENT FEATURES
Upper Carboniferous	VI. Pelycosaurs	Predatory reptiles, with lower temporal openings
Permian	VII. Older Therapsid reptiles	Secondary bony palate forming; pterygoids losing dominance; running powers improving
Triassic	VIII. Advanced therapsids (cynodonts, bauriamorphs)	Ascending ramus of dentary enlarging and nearing contact with squamosal; occipital condyle becoming double; secondary palate progressive, pterygoids, quadrates and post-dentary bones of mandible diminishing; dorsolumbar region becoming differentiated
Jurassic	IX. Primitive mammals (triconodonts to pantotherians)	Mammalian status established with temporo-mandibular joint and transformation of quadrate into incus, malleus into articular. Presumably with high and relatively stable body temperature

In this list, comprising nine major stages, *Captorhinus* is about halfway up from the bottom and has already reached the status of a primitive land-living quadruped. We are therefore now in a position to consider the leading skull characters of *Captorhinus* in their proper geologic and morphologic perspectives.

1) Considering the series as a whole, polyisomerism, or the budding off of new morphologic units has seldom occurred; but

2) negative anisomerism, or reduction in number of skull elements, has been conspicuous, especially in passing from Stage III to IV, owing chiefly to the loss of the numerous elements of the opercular apparatus.

3) Anisomerism in general, involving the increase in dimensions of certain parts and the compensating reduction of others, has been predominant at all stages.

4) The foregoing methods of transformation, both in the inner skull (endocranium) and its outer sheathing (dermocranium), were correlated with corresponding and compensatory changes: (a) in the sensory, (b) the coordinating and (c) the motor systems. Incidentally these points further reinforce the argument set forth above (p. 261), which suggests that too intensive insistence upon minor topographic details of the chondrocranium may actually obscure the major evolutionary advances or divergences between the entire skulls of different animals.

WHAT *CAPTORHINUS* OWED TO HIS REMOTE ANCESTORS

The important position of *Captorhinus* may be realized more concretely by referring first in a rather summary way to what he owed to his successive ancestors of the major grades I to V.

1) Indeed *Captorhinus* owed to his earliest ostracoderm ancestors all the major basic vertebrate characters, such as bilaterality, anteroposterior locomotion by means of metameres, a notochord and a nervous system of vertebrate type, including sensory, motor, directive and correlating networks.

2) Moreover, the presence in *Captorhinus* of paired endocranial capsules for the nose, eyes, and semicircular canals arranged in series from front to rear, was already established in the Ostracoderms and formed the basic plan of all later vertebrate skulls.

3) *Captorhinus* had also inherited from the earliest ostracoderms the ground-plan of his dermocranium on the surface and of his chondrocranium underneath; for though the chondrocranium as such is not preserved in Heterostraci, which lacked bone cells, its general shape in some pteraspids (e.g., *Cyathaspis*) is preserved in natural molds bearing traces of the semicircular canals and of the gill pouches. Moreover, a firm inner skull was necessary as buffer for the notochord and myomeres as they propelled these very primitive fishes through the water.

4) *Captorhinus* owed to inheritance from the earliest ostracoderms the basic plan of his oralo-branchial system, although, as will be shown, that system had later been profoundly modified.

5) The oralo-branchial system from the earliest vertebrates onward is always morphologically ventral to, and partly posterior to, the neurocranium, from which it is separated by the base of the cranium and the roof of the mouth.

6) *Captorhinus* next inherited from some early ostracoderms of the grade of *Cephalaspis* the presence of bone cells both in the outer covering or mask and in the inner parts of the skull.

7) *Captorhinus* might well have thanked his remote aquatic forebears among the ostracoderm (I) and placoderm (II) grades for effecting the shift from food-filtering whereby microscopic food particles were drawn into the mouth by the lashing of cilia, to food capture by the active pursuit of smaller creatures, which were grabbed perhaps by the whole forepart of the oralo-branchial system and pushed into the stomach by the throat muscles. Outer or lip jaws of derm-bone were near their beginnings in the antiarch stage and were supported by partly evertible bars of the oralo-branchial series, which have been homologized by Stensiö respectively with the palato-pterygo-quadrate and Meckelian cartilages.

8) *Captorhinus* owed to his early amphibian (IV) ancestors the elimination of the ganoid surface layer from his bony facial and temporal mask and the subsequent evolution of a true leathery and perhaps partly horny surface layer covering the bony face and cheeks, but stopping suddenly at the posteriorly truncated occiput.

9) To his crossopterygian (III) and amphibian (IV) ancestors, our stem reptile was indebted for the in-sinking of the posterior part of the olfactory capsule into the antero-lateral parts of the palate and the opening up of internal choanae or nares at these points.

10) The union of the anterior and posterior sections of the endocranium was due to his earliest amphibian (IV) ancestors, in which the median gap between these sections had been bridged and the parasphenoid had spread backward over the occiput.

11) From his more immediate ancestors he inherited his single median occipital condyle, in contrast with the tripartite condyle of early amphibians.

12) His general equipment of vomers, palatines, pterygoids, epipterygoids, quadrates, and the full complement of derm bones covering the upper and lower jaws likewise had been handed down to him from the earlier amphibians (IV) and through them from the crossopterygian (II) fishes.

13) To the same sources also he owed his columella, or stapes. This had been derived from the upper segment of the gill-bearing hyoid arch of the earliest fishes but it had very early been modified to assist in the support of the jaws. In the amphibians it served to transmit sound waves to the capsule of the inner ear in the chondrocranium.

No doubt the lower segment of the hyoid arch, as in other reptiles, had been modified to serve as part of the larynx, or vocal organ, but these parts are not known in *Captorhinus.*

WHAT *CAPTORHINUS* BEQUEATHED TO HIS SUCCESSORS

But *Captorhinus* and his nearer reptilian ancestors had not failed to make numerous improvements upon the equipment received from more remote stages. One of the most remarkable of these was the paired five-branched pterygoid bones (fig. 12.29A2), which, together with connecting bars from the surrounding bones, formed a strong system of struts, strengthening the complex upper jaws against the pulls and thrusts of the powerful jaw muscles and the varying impacts of the strong lower jaws and numerous cheek teeth against the rather tough food, whatever that may have been. The first branch of the pterygoid ran forward on either side of the interpterygoid slit and was connected with the vomer and the palatine; the second ran outward and downward and was connected with the transverse or ectopterygoid; the third, or basipterygoid process, was a short stem and ran upward and inward to meet the basisphenoid bone; the fourth, or quadrate branch, ran backward and outward, overlapping the pterygoid branch of the quadrate; the fifth, or epipterygoid, ran from the dorsal surface, above the basipterygoid junction, forming an anteroposterior flange on either side of the tubular braincase. The system as a whole suggests an opposed pair of kingposts (fig. 12.29A2). It stiffens the entire skull and jaws, including the strong downwardly inclined upper incisor teeth, which may have been used as biting weapons, picks, or bill-hooks. As will be shown later, this five-branched pterygoid system was transmitted with numerous modifications to practically all later reptiles, as well as to birds and mammal-like reptiles. And even the mammals, including man, retain traces of it. In *Captorhinus* it probably served as a stiffening framework, which may have received heavy shocks from the bony facial mask, possibly as that was pushed forward into the burrows dug by the stout hands. The five-rayed pterygoid brace may also have been useful in fighting, when the whole skull would be wrenched by thrusts from the powerful muscles of the neck and back and by the counter-moves of the foe.

Another indication that the *Captorhinus* skull had to withstand strong thrusts from the rear is the truncation or transverse flattening of the upper and outer margins of the occiput, in which the paired dermosupraoccipitals (or postparietals) and tabulars were pressed flat against the cranial table. In connection with this flattening, *Captorhinus* and his relatives, *Labidosaurus* and others, had as it were squeezed the otic notch out of existence and had thus advanced beyond *Diadectes,* in which the old labyrinthodont otic notch, or one with the same function, still supported a tympanum or drum-membrane (fig. 12.30). Thus the question arises, was *Captorhinus* deaf to airborne sound waves? Or did he hear by bone conduction through the surrounding elements, or in some other way? In *Seymouria,* as observed by T. E. White (1939), the dorsal end of the slender hyoid bone (fig. 12.4B1) was bent sharply around the otic notch and was fastened to the side of the braincase. But *Seymouria* also had a stout stapes (fig. 12.29B2 *stp*) or columella, which was apparently directed externally toward the otic notch and its tympanum and connected internally with the capsule of the inner ear, ventral to the semicircular canals. Thanks to invaluable information obtained from Dr. Charles Bogert, curator of reptiles in The American Museum of Natural History, it can be said briefly that among recent reptiles the crocodilians, which have the stapes connected with the tympanum, are very alert to airborne sound waves. But turtles, which also have a tympanum and a stapes, often seem to be indifferent to sounds. Lizards, which have lost the otic notch and drum-membrane, but still have a stapes, show little or no response to sound waves—except possibly some tree-living geckos. Snakes, which have no tympanum and have the stapes attached to the long backwardly produced quadrate (fig. 12.51), may catch some sound-waves by vibration of the substratum transmitted possibly through the body and neck to the stapes and inner ear; but rattlesnakes are said to be indifferent to the piercing sound of their own or others' rattles. The larynx, which in the mammals becomes an efficient voice organ, emitting airborne sounds which are received by the tympanum of other mammals, does not seem to serve this way in the lizards, except possibly the geckos.

Thus it is conceivable that in losing its otic notch *Captorhinus* was evolving toward the lizards, in which airborne sound waves play but little part in the game of life. But several considerations suggest caution in prematurely excluding *Captorhinus* from relationship with the "main line of ascent" by appeal to the dogma of Irreversibility of Evolution. Gaupp indeed suggested with considerable reason that the tympanum of mammals was a new structure evolved within that class. *Captorhinus* is well recognized as a structural ancestor of the pelycosaurs and these likewise had no otic notch and apparently no tympanum; but they led more or less directly to the series of mammal-like reptiles, whose auditory region undoubt edly progressed from a somewhat lizard-like to a submammalian stage.

A basic consideration is that the internal ear, the essential organ of hearing in the higher vertebrates, was represented in primitive reptiles (fig. 16.49C) by a cartilaginous pouch containing the mass of the auditory nerve, whereas in the mammals (fig. 16.49E, F) the terminal twigs of the auditory nerve are wound into a spiral coil enclosed in the membranous and cartilaginous cochlea. Beginning at the top and running down the scale of the cochlea, vibrations of diminishing rate and increasing amplitude are believed to be received by successive twigs of the cochlear nerve. Therefore, since the reptiles possess only the beginnings of this system, it is unlikely that their discrimination of airborne sound-waves is either as wide or as refined as it is in mammals.

In brief, we may tentatively infer that in *Captorhinus,* which had a well developed stapes (fig. 12.29A2) but at most a very small tympanum, the sense of hearing was represented by vibrations received perhaps through the hands or other parts in contact with the substratum and transmitted to the inner ear by the hyoid and stapes.

DETOUR FROM THE "MAIN LINE"

But now we must leave the "main line" of advance (fig. 12.1) that leads "to the right" through the pelycosaurs and the mammal-like reptiles to the mammals, and follow along one after another of the branching highways "to the left" leading respectively to the turtles, the fish-like ichthyosaurs, plesiosaurs, etc., the lizards and snakes, the tuatara, the crocodiles and dinosaurs, the bat-winged reptiles and the birds.

THE BONE-AND-SHELL FORTRESS OF THE TORTOISE AND OTHER CHELONIANS

Outer Form, Locomotion and Skeleton of the Tortoise.—If a lumbering tortoise (fig. 12.16) never seems to be in a hurry, it may be partly because he lives in an armor-plated shell and has little to fear from dogs or cats, but it is also because his rather short legs are too far apart and too much crooked at the elbows and knees to make running possible. However, in order partly to offset this disadvantage, he swings his elbows far forward in front of the shell and parallel to the neck, where they look like knees (fig. 12.34). As the elbow on one side is swung backward, the body especially on that side is drawn obliquely forward. Meanwhile the femur, swinging almost horizontally backward, gives a vigorous push from the rear. At the end of the stride on one side, the shell lurches a little toward the opposite side, and so by alternate pulls and pushes the heavy body weaves and twists along its way. This mode of locomotion was basically like that of the clumsy, short-legged *Diadectes* (fig. 12.9), but the presence of the stiff shell necessitated many new adaptations, such as the extreme forward swing of the elbows noted above.

The shell itself (fig. 12.16C) is a wonderfully complex and unique creation within the order Testudinata (Chelonia). It is composed of an outer or horny layer and an inner or bony layer. Both layers are subdivided into well defined areas, each with its own growth center. The domed upper part or carapace is covered with polygonal or rectangular horny scutes, arranged in median-dorsal, dorso-lateral and marginal rows; the scutes of the flat base or plastron are arranged in six triangular or rectangular pairs. The growing margins of the horny scutes (fig. 12.16B) overlap the sutural boundaries of the bony plates beneath them, thus strengthening the resistance of the shell to crushing or distortion. The bony plates comprise: (a) polygonal plates on top of the neural arches, (b) oblong plates enclosing the ribs, (c) smaller marginal plates, and (d) plates of the plastron. The median part of the plastron is occupied by a polygonal to dagger-like plate, the entoplastron (pls. 12.I, II), which may correspond to the interclavicle of other reptiles, while the epiplastra in front of it may represent the clavicles.

In this rigid shell the dorsal vertebrae (pls. 12.I, II) are completely immobilized and serve chiefly as the keel of an anteroposteriorly arched ridge-pole (fig. 12.34). They are anteroposteriorly elongated and, in front view, sharply concave dorsally. This dorsal concavity might at first be mistaken for the ventral part of a notochordal tube, but the latter has disappeared entirely, its place being usurped by the downwardly pushed tube for the spinal nerve cord.

The neck of the tortoise can be withdrawn and folded vertically within the shell (fig. 12.34) and in extreme retraction the cervical vertebrae are bent into an S-curve. This curve is, however, an over-emphasis of a U-shaped depression behind the head when the head was drawn sharply backward. More in detail, this unique type of neck (which is characteristic of the whole chelonian suborder Cryptodira) has been modelled through the coincidence and interaction of the following and other phylogenetic-morphogenetic factors:

1) The lengthening of the entire neck, especially in the middle regions.

2) The upward arching of the dome of the carapace and the consequent downward inclination of the ninth vertebra (counting from the atlas backward) which serves as the first dorsal.

3) This ninth vertebra has become the fulcrum of the neck (fig. 12.37C) and has required strengthening and remodelling in the following ways: (a) by the enlargement of its neural arch; (b) by the bracing of its short ribs against the rib-plate of the vertebra behind it; (c) by the remodelling of its large forward-and-downwardly directed anterior zygapophysial processes which developed smoothly rounded articular convexities for the corresponding concavities on the under surfaces of the posterior zygapophyses of the preceding (eighth) vertebra.

4) Appropriate but complex changes and adjustments were necessary in the zygapophyses of all the cervical vertebrae so as to permit sharp dorsoventral flexure and moderate lateral movements.

5) The articular surfaces between the cervical centra were also variously modified (fig. 12.36): for spreading apart ventrally and crowding dorsally, and vice versa, to permit the limits of extension and flexure necessary for each vertebra, in the fully extended and fully retracted positions of the entire neck.

6) Moreover, a moderate degree of lateral movement of the head and neck had to be made possible. Thus the anterior and posterior articular surfaces of the cervical centra may become concave or convex or concave-convex as may be necessary at successive intercentral joints, to permit efficient total extension, retraction and abduction, adduction and torsion. Indeed E. Williams (1950) reports considerable intraspecific and interspecific variability in certain details, combined with fixity in others.

No less amazing is the fact that whereas in normal reptiles the scapulae are outside of the dorsal ribs, in chelonians the tops of the opposite scapulae are brought near to the midline (fig. 12.34), so that in front view the scapulae are underneath the ribs and entirely inside the carapace (fig. 12.37C). This result is achieved among recent turtles in late embryonic stages (fig. 12.18), in which the elevated anterior rim of the carapace grows laterally and anteriorly above the scapulae, which are at that stage very low (Ruckes, 1929b). With the widening of the carapace, the scapulae and coracoids are left relatively near the midline and the scapulae, growing upward as slender columns, are connected with the under side of the nuchal plate of the carapace. The rod-like scapula, the fan-like coracoid and the long horizontal proscapular (acromial) rod (fig. 12.20aD), together form the supporting frame of a muscular pyramid (fig. 12.19) which operates the humerus and the forearm. Between the pyramids of opposite sides is slung the long cylindrical neck, which can be fully retracted into the median fossa between the pyramids.

Posteriorly the carapace and some of its rib plates overlap the ilia or the dorsal blades of the pelvis (fig. 12.34), but the two pairs of rod-like sacral ribs have retained their attachment to the ilia (pl. 12.I). The tops of the ilia abut against the inner surface of the plastron, which is lifted by the upward thrusts from the femora. The fenestrated pubic and ischial plates are strongly braced for the support of powerful adductor and other muscles of the femora but in Cryptodira they are not fastened to the plastron.

The hind limbs as a whole seem to be somewhat larger and more massive than the fore-limbs and the carapace widens out a little toward the rear (pl. 12.I). This may be useful perhaps in digging, or in turning around, or in preventing the tortoise from pushing himself too far into a hole from which retreat would be impossible.

PROBABLE ORIGINS AND COMPARATIVE OSTEOLOGY OF THE CHELONIANS

What were the sources, origins of, and reason for, these and many other peculiar and wonderful adaptations of the living frame in the turtles? From the Permian of South Africa two somewhat turtle-like fossils were described by Seeley, who named them *Eunotosaurus,* or the well-backed reptile. Much later they were restudied to good effect by Watson (1914). Already the back was arched (fig. 12.17) and the ribs were very wide, suggesting the condition in the turtles, but the pectoral girdle was nearer to the normal reptilian type and the scapula was not yet overlapped by the ribs.

The Permian pareiasaurs (fig. 12.15) of South Africa and Russia appear to supply an ideal structural starting-point for the peculiar specializations which were in an early stage in *Eunotosaurus* and completed in the tortoises. For already at that remote period the backs of pareiasaurs were arched and dome-like, and in some pareiasaurs the thick leathery skin was studded with bony nodules arranged in median and transverse series. Moreover, as we shall see presently, the skull, vertebrae, pectoral and pelvic girdles, limb and foot bones of pareiasaurs all tend to connect the chelonians and pareiasaurs with the *Didectes*-like cotylosaurian stock.

True chelonians date back to the Triassic of Germany, in which they are represented by *Triassochelys dux* Jaekel, belonging to the Mesozoic suborder Amphichelydia. As restored by Jaekel, *Triassochelys* (fig. 12.16) had already succeeded in extending the overhanging rim of the carapace above the scapulae. The pattern of the carapace also conformed, except in minor features, to the basic pattern for turtles: namely, with five large median dorsal horny scutes, four laterals plus a supramarginal and a marginal row of smaller scutes. The bony plates of the carapace and plastron, so far as known, were essentially chelonian, and the plastron included two pairs of additional transverse plates, which have been retained in Pleurodira but not in the Cryptodira. The shell was deeply notched at both ends for the exit of the stout limbs.

The vertebrae of Triassochelys (figs. 12.35, 36), as described and figured by Jaekel, had not acquired the advanced specializations of the tortoise vertebrae; in the other direction, they were not radically different from the vertebrae of pareiasaurs, if due allowance be made for the much earlier grade of evolution of the latter.

The pectoral girdle of *Triassochelys* (fig. 12.20), as figured by Jaekel, was far less specialized than those of modern turtles and seems to represent a pareiasaurian girdle in which the coracoscapula had been narrowed anteroposteriorly and the glenoid face elongated and tilted downward posteriorly.

The humeri of *Diadectes* and *Pareiasaurus* (*s.l.*) taken together seem to supply two successive prerequisite stages for the peculiar humerus of the turtles (fig. 12.20), with its sigmoid curve, large transversely convex facet for the radius and ulna, and folding-up of the delto-pectoral crest and large internal tuberosity to form an incipient depression for the biceps.

The bones of the forearm and hands of chelonians appear at first sight to be inherited almost directly from a primitive pareiasaurian type and the same is true of the pelvis, femur, lower leg and foot. Allowance should be made, however, for the possibility of convergence in the extreme shortening of the digits and elephant-like form of the feet (fig. 12.22) in both pareiasaurs and the modern *Testudo.*

The skull of *Triassochelys* (figs. 12.25; 12.27, 28), according to Jaekel's description and figures, as seen from above was roundly triagonal and covered with a mask of derm bones. The temporal arcade (above the jaw muscles) was not perforated dorsally as it is in the modern tortoise but, as shown in the occipital view, the temporal muscles were beginning to lift up the entire temporal roof and depress the opisthotic bars, thus opening up the posttemporal fossae. This tendency has been carried much further in the sea-turtle. Thus the temporal roof of *Triassochelys* was essentially like that of pareiasaurs, in which the jaw muscles had made only small posttemporal openings. The occipital condyle of *Triassochelys* was essentially similar to that of *Pareiasaurus* but somewhat more convex. It was much less specialized than that of the sea turtle, in which the exoccipitals and basioccipital form a tripartite condyle.

On the palatal side the skull of *Triassochelys* likewise agreed in principle with the same forms. In short the construction of the skull of *Triassochelys* is exactly what one would expect in an early chelonian which, by hypothesis, had been derived from the cotylosaurian stock either directly or by way of some small and early pareiasaur.

According to Jaekel, vestiges of teeth were retained in *Triassochelys,* but they have been completely lost in all modern turtles, their function having been taken over by the massive horny beaks which cover the front parts of the upper and lower jaws in turtles. Especially in the sea-turtles the beaks have, as it were, molded all the supporting bones, especially the premaxillae and maxillae in the upper jaw and the dentary in the lower jaw. Together with the enlarging jaw muscles, they have required corresponding enlargement and strengthening in the palatopterygo-quadrate complex above and in the articular region below. A strong bridge over the internal narial tubes (fig. 12.28C) has been built by the palatines, braced by the maxillae and pterygoids. The pterygoids have coöperated (a) by widening or (b) by obliterating the interpterygoid vacuity, (c) by forming a strong longitudinal and transverse brace beneath the basisphenoid, (d) by reducing their epipterygoid column to a basal remnant (fig. 12.32C), (e) by sacrificing their ectopterygoid branch and (f) by bracing their diverging quadrate branches medially against a strong inverted V ridge (fig. 12.28C), which has been developed from the basipterygoid tuberosities of the basisphenoid. A prerequisite stage for this transformation is preserved in *Triassochelys* (fig. 12.28B), in which the palatines were developing small processes which were evidently curling around the paired narial duct but were still far from the midline. The palatal region of *Triassochelys* may have been derived either from the pareiasaur (fig. 12.28A) or the *Diadectes* (fig. 12.11) pattern.

The great strength of the jaws is reflected also in *Chelone* by the stiffening of the occiput and by the great development of the opisthotic (figs. 12.28, 32), which remains suturally separate from the exoccipital and forms a massive beam on each side, between the squamosal arcade and the braincase.

In the region of the drum-membrane or tympanum in *Triassochelys* (fig. 12.25C), the rim on the rear edge of the quadrate and squamosal was less incurved than is the corresponding ridge in the modern sea-turtle. In this feature *Triassochelys* was more like *Diadectes* (A) and quite unlike *Pareiasaurus* (B), in which the surface plates (*sq., qj.*) have grown backward over the curved rim of the quadrate and there is no indication of an otic notch. This situation may be in line with other evidence that a subcircular rim for the tympanum has been acquired independently in the sea-turtles and certain mosasaurs and that its absence in the captorhinomorphs (fig. 12.29) is, for reptiles, a primitive character.

A noteworthy feature of the skull of most turtles is the marked upward and backward growth of the supraoccipital crest, under the influence of the massive jaw muscles. Its middle area in large specimens of *Chelone* (fig. 12.32C) is so thin as to be translucent but its borders are enlarged and stiffened. Further emphasis of this condition would result in a window or fenestra analogous to those of the temporal region of many reptiles and likewise surrounded by strong rims. This is a fine example of the principle of fenestration (p. 253) which has operated in various parts of the skeleton in all classes of vertebrates.

BRANCHING EVOLUTION AMONG TORTOISES, POND-TURTLES, TERRAPINS, SNAPPERS

Since the true tortoises (Testudinidae) are known as fossils only from the Eocene, while the modern sea-turtles (Chelonidae) date from the Cretaceous, and since older marine turtles date from the Upper Jurassic, an attempt to derive the general plan of the skeleton of the sea-turtle from that of any recent tortoise or pond-turtle (both being merely single representatives of large assemblages) might well distort or invert the true picture of evolution. However, both groups were preceded in the Cretaceous, Jurassic and Triassic by an older and more primitive ordinal assemblage known as the Amphichelydia. This includes a wide range of body-forms, of which some (Baenidae) were quite tortoise-like. The oldest form of this order, *Triassochelys* from the Upper Trias, as we have seen, was definitely more primitive than the modern tortoise and its skull was in certain features more like that of the marine tortoise *Chelone*. Much older, in turn, was the Permian *Eunotosaurus,* the body-form of which, as noted above, was of a very generalized tortoise-like type and suited to give rise to any of the extreme later forms. Since the existing marsh-turtles (Dermatemydidae) and terrapins (Emydidae) show intergrading shapes, as in the wood tortoise (*Chelyopus*) between the high-backed tortoises and the depressed sea-turtles, it has been suggested both by Ditmars (1907) and by Romer (1945, p. 181) that the marsh-turtles and terrapins have retained more of the generalized build of the cryptodiran order than have the true tortoises. This may well be true even though these modern families may be now undergoing wide adaptive branching to aquatic and terrestrial life, in which they are to some extent paralleling older branchings.

SEA-TURTLES

The skeleton of the sea-turtles, including *Chelone* (pl. 12.II), is the result of another transformation but in a quite different direction from that which produced a tortoise-like form. This transformation has resulted in a complex series of adaptations for life in the ocean and the entire organism had to be made seaworthy. At the same time, due to retention of the habit of laying eggs in the sand, it was necessary also to retain the ability to crawl upon shore and scrape a hole in the sand above tide level. Especially in the wing-like form of its pectoral flippers, the sea-turtle *Chelone* has gone further in its adaptation to marine life than did its gigantic relative *Archelon* of the Upper Cretaceous. But even so, the origin of certain details of the skeleton of *Chelone* would be much more difficult to understand if it were not for the data already summarized bearing on the origin of the tortoise skeleton. The morphologic evidence indicates that the almost wing-like pectoral paddles of *Chelone* have been derived distantly from the moderately short-fingered feet of some form like the pond turtles, which have paw-like webbed and clawed feet. By greatly lengthening the second, third and fourth digits (particularly the third), the hand acquired a narrow tip, and by thickening the metacarpal of the thumb and its first joint, the paddle gained a firm anterior edge ending in a claw, which may be useful in digging or in shuffling about on shore. At the same time the fifth digit was slightly abducted from the others. Very thick leathery skin covered the entire paddle. The humerus (fig. 12.20bH) was even more profoundly changed by the great enlargement of the internal tuberosity (to which were attached very powerful muscles for pulling the paddle backward) and by the reduction of the medial branch of the deltopectoral crest (for the smaller muscles which pulled it forward and downward). The webbed and spreading hind-feet (pl. 12.II) were less modified from the pond tortoise type. The strong first digit bears a stout claw.

The bony plates of the carapace and plastron were lightened as much as possible, probably by thinning their texture and certainly by arresting the process of ossification of some of the plates around the margins or at the ends. As a result the ribs stick out from the shortened rib plates, the wide middle plates of the plastron have embayed borders, except on the medial side where they end in jagged points radiating from the growth center. These changes in the shell must result from the slowing-up of the process of ossification in very young stages of development. The anterior and mid-plates of the plastron were reduced to narrow thin pieces, which, perhaps convergently, resemble the interclavicles and clavicles from which they were remotely derived. The marginal plates of the carapace, although slenderized, were retained to stiffen the heart-shaped outline of the entire shell. All the plates are embedded in a thick leathery and horny exoskeleton which affords sufficient protection to the adults. The tender young, however, have to cross the no-man's land of the beach on their way to the water, assailed by gulls and terns by day and by crabs and prowling fishes by night. Here is where Chance on both sides competes with Law, but even in spite of the destructive activities of *Homo sapiens,* the sea-turtles in remote localities still meet the statistical minima for survival.

Further advance in marine specialization has been made by the great leather-backed sea-turtle (*Dermochelys*). Here the entire carapace has been broken up into a multitude of small secondary polyisomeres forming polygonal ossicles embedded in a massive leathery skin. In the days when secondary polyisomeres were frequently mistaken for primitive ones, the leather-back was made the type of a separate order, Athecae (without shells). But the skull of the leathery turtle is essentially similar to that of *Chelone* (except for the slenderness of the lower jaw) and the bones of the girdles and limbs give confirmatory evidence that this Eocene to Recent group is merely a separate branch of the marine Cryptodira.

THE RIVER-DWELLING SOFT-SHELLED TURTLES (TRIONYCHIDS)

A quite different line of specialization has culminated in the river-dwelling soft-shelled turtles (*Trionyx*) in which the body has greatly widened and flattened and the margin of the bony shell has been covered by a thick leathery flap. This group dates back to the Cretaceous, but its specialized skull, neck and other features ally it with the Cryptodira.

THE SIDE-NECKED TORTOISES OF AFRICA AND SOUTH AMERICA (PLEURODIRA)

These voracious chelonians differ widely from the Cryptodira in turning their long snake-like neck sidewise instead of retracting it vertically. Their neck vertebrae (figs. 12.35E, E1; 12.36I-I3) accordingly differ radically from the cryptodiran plan (C, D, H1-H3) and the emphasis upon lateral flexure has resulted in a peculiar outward twisting of the anterior zygapophyses of certain vertebrae (I 1–3) so that their articular surfaces facilitate this lateral movement. The fossil history of this group, which forms the suborder Pleurodira (side necks) is known from the Lower Cretaceous onward and they may be an independent derivative of the Amphichelydia.

The matamata (*Chelys fimbriata*) of South America is the most specialized of the Pleurodira. Its skull (fig. 12.33B) is greatly flattened dorsoventrally, perhaps in accord with the habit of lying still on the bottom of streams; its circular drum membranes are stretched upon a pair of large trumpet-like projections from the side of the skull; the small eyes are at the front end and look mainly upward; the temporal muscles rest chiefly on top of the wide pterygoid bones and occupy a deep embayment on the upper surface of the skull. The beginning of these specializations is seen in the Upper Cretaceous *Podocnemis* (A), a member of the same suborder Pleurodira.

The pelvis of the matamata and of other Pleurodira forms a stoutly built inverted V or truss, which is firmly based upon and attached to the narrow plastron floor (Ruckes, 1929a). But why do these creatures need this special bracing? Presumably when their long muscular necks quickly snap to one side or the other, there would be a tendency to wrench the shell loose at the sacral attachment; but the fixation of the pelvis to the shell may enable the widely spread, strong hind legs and wide paddles to resist these thrusts. And perhaps a firm stance is also needed for accurate snatching movements by the neck and jaws.

DARWIN'S METHOD AND THE EVOLUTION OF THE CHELONIA

Darwin said in effect that there are two ways of looking at a natural mechanism: one is to gaze at it in either astonishment or indifference, "as a savage looks at a full-rigged ship"; the other is to try to grasp it as a whole, to learn how its parts work, and to seek for probable evolutionary stages through which it may have passed in attaining its present status. It was by the consistent application of this inductive-deductive method of study that Darwin gained, for example, even from the meager evidence then available, his amazing insight into the probable origin and subsequent stages of evolution of the baleen plates of the whalebone whales. This method has here been followed in all humility in the search for even first approximations to correct answers to the questions: (1) How and by what steps has the tortoise-race built its shell-and-bone fortress out of the materials bequeathed to it by primitive reptiles? (2) How are its skull, vertebrae, girdles and limbs adapted for efficient living under the limitations imposed by the protective shell? (3) What special adaptations of the chelonian ground-plan have produced such extremely specialized forms as the soft-shelled turtles, the matamata and the sea turtles?

ADAPTIVE BRANCHING AMONG LIZARDS AND SNAKES

Lizards, Real and So-called.—Lizards of any kind mean little if anything in the lives of most people and in the United States the word "lizard" is more often applied to one of the newts or salamanders, which have naked, glandular skins and develop from gilled polywogs, as do frogs, toads and other members of the Class Amphibia; but the true lizards belong to the Class Reptilia and they have scaly skins and lay eggs with a parchment-like or hard shell; the young are never like polywogs and when hatched (or in a few cases "born alive") they closely resemble their parents. The very name lizard is an essentially foreign word in English and the dictionaries tell us that it was borrowed from the Portuguese *lagarto* and Old French *lesard* from Latin *lacertus*. Thus Gilbert White, the famous naturalist of Selborne, in the second half of the eighteenth century used the term "eft or common lizard" in describing newts, and referred to the frogs and toads as "reptiles" (pp. 74, 75). He was also familiar with "our green lizards" (*op. cit.*, p. 97), which was possibly the common English hedge-lizard, *Lacerta agilis* (Lydekker, p. 161) or even the true green lizard (*Lacerta viridis*) which he noted had been introduced from Germany.

True Lizards Not Amphibia.—Linnaeus (1759) included crocodiles, lizards, snakes and turtles in his group Amphibia. Brogniart (1800) was the first to distinguish the frogs and salamanders as a class (Batrachia) from the reptiles but it was not until 1825 that Latreille restricted the name Amphibia to the frogs, toads and salamanders, leaving the caecilians with the reptiles (Noble, 1931, p. 2). Thus it is hardly surprising that this confusion of the terms salamander, newt, lizard, Amphibia, Reptilia, should persist today, especially among people who are not students of zoology.

According to Darlington (1948, p. 23), the suborder Lacertilia (or Sauria) of the order Squamata includes nearly 3,000 recognized species, which are referred to some 300 genera; these are distributed among about 25 families; the families are built up into seven infraorders (Romer, 1945, p. 596) and these into two grand divisions, Ascalabota and Autarchoglossa (Camp, 1924, p. 296).

The aims of the present section are: first, to portray clearly the outstanding types and groups of lizards and snakes; second, to show how closely their outer features and skeletal characters are related to special modes of life; third, to indicate the main probable stages by which older structural patterns, each adapted to given habits and environmental stations.

have branched out into later, more or less modified structures and changed ways of life in new environments.

New World Iguanids and Old World Agamids.—The family of iguanid lizards (pl. 12.V) is characteristic of Middle and South America, but it has a few outliers in the north and extreme south. The more common body-form of iguanids is long, somewhat compressed on the sides, tapering to the tip of the long tail. The strongly angulated hind limbs are usually larger than the forelimbs and in the resting position often sprawl widely with the long five-toed hind feet directed outward and forward. This posture makes it easy for the lizard to spring forward suddenly, to turn quickly or to climb rapidly. The fifth or outer digit of the hind feet is often turned outward, "like a thumb on the wrong side of the hand," and is useful in scratching, digging and climbing. The hands are usually smaller and more spreading than the hind feet and the fourth digit is the longest.

The dorsal vertebrae of iguanids are procoelous or hollow in front (figs. 12.43, 44), with a convexity or condylar ball at the rear end, which gives sufficient flexibility to the vertebral column (Camp, 1923, p. 307).

In iguanids, as in reptiles generally (fig. 14.12), the tail is equipped with long muscles on its lower half; their tendons run forward and are inserted into the femur, with a branch running down the shank to the heel. These coccygeo-femoralis or caudi-femoralis muscles pull the femora backward or the tail to one side or the other, as in all primitive quadrupeds. The dorsal tail-muscles either raise the tail or pull the back up, especially in such lizards as *Cnemidophorus,* which can rear up and run on its hind legs. In some lizards the tail, notwithstanding its functional importance in wriggling movements of the body, can be violently snapped off by its owner in a moment of sudden fright. The bloody tail then wriggles about as if trying to distract the enemy's attention, giving the owner a chance to slink out of sight. This dramatic instance of escape through self-mutilation is paralleled also in certain crabs, which snap off their large claws if roughly handled. The breaking of the tail in lizards is made possible by the presence of a transverse plane of weakness in the middle of the vertebral centra near the root of the tail. Afterward a new tail grows out from the rod-like stump of the old one. Apparently the breaking joints must have some relation to the ring-like overlapping or alternating arrangement of the muscle segments and their septa with the slender centra of the vertebrae.

In *Iguana* and allied genera the front teeth are somewhat conical and the cheek teeth consist of an even series of small cylindrical teeth fastened to the sides of the jaws and with compressed, more or less serrated crowns. The smaller iguanids are active insect-hunters, the larger ones favor leaves, flowers and berries. The big *Conolophus* lizards of the Galapagos Islands, with jaws strong enough to chew through shoeleather, feed mostly on flowers and fruits, especially the fruit and spiny pads of the cactus (*Opuntia*), but with an occasional grasshopper (Beebe, 1924, pp. 246, 250, 251). Their famous relative, the Galapagos sea-lizard (*Amblyrhynchus*) feeds only on the tips and sprouts of the glutinous olive-green Sargassum seaweed (Beebe, *op. cit.,* pp. 116, 117).

The recent iguanids mostly inhabit tropical forests or grasslands in South and Middle America, including the West Indies; but with a few outliers that penetrate as far south as Tierra del Fuego and others that extend north to British Columbia. Two relict genera of the family Iguanidae are found in Madagascar and one on Fiji (Camp, 1923, p. 308; Darlington, 1948, p. 134). This remarkable case of discontinuous distribution may, as Lydekker notes (The New Natural History, no date, vol. V, sect. IX, p. 129), be due to the common derivation of North American, Madagascan and Polynesian members of the family from a common ancestral stock in the Eocene of Europe, where fossil jaws of *Proiguana* and *Agama* have been found (Camp, 1923, p. 309). Darlington (1948, p. 24), after a comprehensive study of the distribution of all recent lizard families, tentatively suggests that: the [family] Iguanidae was probably once cosmopolitan and the Agamidae has apparently been derived from it and replaced it on Old World continents.

Ancestral lizards or pro-lizards (fig. 12.41b) have been found in the Triassic of South Africa. In the Upper Jurassic of Europe they are represented by a few rare forms. By Upper Cretaceous times lizards had reached North America and the true iguanids are represented there and in the Tertiary of North America; but apparently they did not reach South America until the late Tertiary, perhaps accompanying the mammalian invaders from North America.

Thus most of the adaptive branching of the iguanids may have occurred in the western hemisphere since the Upper Cretaceous. As noted by Cope (1887), the American iguanids (pl. 12.IVa) branched into many body-forms and habits in which they parallel their Old World relatives, the agamids (pl. 12.V). None of the latter has ever been found in America. For example, the basilisk (*Basiliscus,* 1), an iguanid of Tropical America, parallels the sail-tailed lizard 1, an agamid of the East Indies, in the presence of a high crest on the tail. The wrongly named "horned toad" (*Phrynosoma,* 5), an iguanid of Western North America, is armed with sharp spikes on the head and body, somewhat like the Moloch lizard, an agamid of Australia (5). But, for reasons noted by Camp (p. 335), neither the Iguanidae nor the Agamidae ever developed limbless or snake-like forms such as were frequently evolved among other families of lizards (cf. p. 275 inf.).

Agamids differ from iguanids in that their teeth, instead of being pleurodont (i.e., fastened on the sides of the jaws, as in the latter) are acrodont (fastened to the tips or edges of the jaws). Possibly intermediate conditions were found in the North American Upper Cretaceous *Chamops* (Camp, op. cit., p. 309). Various minor differences between the Iguanidae and Agamidae are noted by Camp (*ibid*), but the two families are closely related.

Skull Structure of Lizards.—The upper jaws of iguanids (as of all lizards) are technically described as "streptostylic," that is literally, with a twisted column, and this word was invented to denote the fact that the quadrate bone, to which the lower jaw is attached, is free to move a little at the upper end in the socket borne for it by the squamosal bone (figs. 12.39C; 12.51A). This arrangement contrasts with that in the sea-turtle (fig. 12.25D), in which the quadrate is immovably fixed, or "monimostylic." This partial loosening of the quadrate in lizards is partly caused by the failure of the lower cheek bones (jugal, quadratojugal) to extend back to overlap the quadrate (fir. 12.39C). It will be recalled that in primitive amphibians and reptiles (fig. 12.39A) these surface bones completely covered the lower half of the temporal region, as the postfrontal, postorbital, supratemporal and squamosal covered its upper part; but in the lizards the jaw muscles have pierced the upper part of the temporal covering, above the postorbital and squamosal and have loosened the lower end of the quadrate by eliminating the quadratojugal and the posterior extension of the jugal. Thus in the lizard skull there is a deep embayment on either side behind the postorbito-jugal bar. Long ago Broom (1903b) noted that in an embryo lizard there was a thin remnant of the jugal-quadratojugal tract below the embayment and he therefore concluded that the lizards and snakes were specialized offshoots from the base of Osborn's subclass Diapsida, or reptiles with two temporal arches on each side (including the dinosaurs, crocodilians and others). Broom's conclusion has been backed with new evidence by Camp (1945), who described a beautifully preserved skull (fig. 12.41b), from the Triassic of South Africa, in which there was a clear remnant of the lower posterior bar of the jugal. This skull also resembled in different ways the skulls of several other orders of Triassic reptiles, including the Protorosauria (fig. 12.38), the Eosuchians (fig. 12.39B), and the Pseudosuchians (fig. 14.3A), the latter being structurally ancestral to the two-arched reptiles.

The pterygoid bones of the iguanids and other lizards (fig. 12:51A) are less tightly fastened to the skull-base than was the case in the Permian captorhinomorphs (fig. 12.29) and they also permit a little fore-and-aft movement of the quadrates at the lower end (fig. 12.51A). In the upper temporal arch the rather narrow contact of the postorbital (*po*) and squamosal (*sq*), together with a thick suture running straight across behind the frontals (*fr*), permits a slight raising of the bony face (fig. 12.41B). Thus the lizard skull has advanced beyond the primitive captorhinomorph archetype chiefly in the opening up of the upper and lower temporal fenestrae and in the subsequent loss of the lower temporal bar, with the consequent loosening of the lower end of the quadrate.

The hyobranchial apparatus in lizards is much more reduced than it is in the primitive salamander *Hynobius,* and Camp (1923, p. 308) states that in the arboreal Iguania, the ceratobranchials of the third arch are joined and produced, sometimes extending ventrally to dilatate the throat fan frequently present in tree-living forms of this group but unknown elsewhere. The habit of displaying the gorget, which is often brilliantly colored, is associated with a rapid emotional play of colors in the skin occurring in certain arboreal forms and in these alone. In the peculiar arboreal Rhiptoglossa [true chamaeleons] the changing but concealing skin-colors aid in stalking the prey.

Amazing Adaptations of Geckos and Chamaeleons.—The highest degrees of adaptation to arboreal life are displayed on the one hand by the geckos (figs. 12.52a, *a*; 49.A, B) and on, the other by the true chamaeleons (fig. 12.52*b*) which are both related, at the base, to the stem of the iguanids and agamids (Camp, 1923, pp. 303, 312, 332, 382–385).

The geckos are well known to travellers in tropical countries through their habit of running upside down on walls and ceilings. Although absolutely innocuous ". . . they have been credited from the earliest times with ejecting venom from their toes and of poisoning whatever they crawled over; while the teeth of one species have been asserted to be capable to leaving their impression on steel" (Lydekker, *op. cit.,* p. 110). Outstanding features are their short, spreading hands and feet, which are usually provided with sharp claws (fig. 12.49B) and adhesive discs (A1). The latter are composed of flexible folds or dermal ridges on the under surface of each toe. The lower ends of the ridges bear small bristles; these are many-jointed and by catching hold of slight irregularities they increase the friction of the foot against the substratum. The ridges are arranged on either side of the claw in parallel rows, or they may form one series which is later folded along the middle. When pressed against a rock, wall or tree and partly pulled back, the ridges create a partial vacuum sufficient to hold up the small animal against its own weight. Alternate adhesion and release is effected by the flexion and over-extension of the digits. These highly adaptive adhesive discs convergently suggest the large sucking-disc on top of the head of the *Remora,* or shark-sucker, but that is a greatly modified dorsal fin (figs. 9, 136, 137).

Many of the tree-frogs have terminal discs on their digits but not the enlarged lamallae or plates of geckos. In the fringe-toed lizards (*Acanthodactylus*) of the family Lacertidae the toes are fringed on the side and keeled below (Lydekker, New Natural History, *op. cit.,* p. 166) and such a condition as this might have been prerequisite for the more specialized stage attained in the geckos. In desert-living geckos, according to Lydekker (*op. cit.,* p. 111), the toes are of normal form, being often nearly cylindrical and keeled on their lower surfaces, but these forms may have lost the plates or ridges, although retaining the radiating digits of their arboreal ancestors. Probably the digital ridges in the geckos have arisen from an extreme overgrowth of the horny epidermis on the ball of the digit beneath the large vertically compressed claws. In this feature the geckos seem to be more specialized than any other lizards, although they retain primitive lizard features, especially in the branchial arches, vertebrae, skull, throat musculature, etc. (Camp, 1923).

The true chamaeleons, with headquarters in Africa, are of all lizards the most highly and wonderfully specialized for arboreal life. Very gradually with their clamp-like extremities and skinny, sharply crooked limbs, they pull themselves along the slender branches and twigs and push their huge helmeted head cautiously forward (fig. 12.52b, *S*). Meanwhile first one and then the other protruding globe-like eye independently turns this way or that or rolls upward toward the near-by branches. If a fly is sighted through the round "pin-hole" aperture on the thick granular eyelid, the chamaeleon stops or creeps quietly within range and looks intently at the target, as the curling tail clings to the branch and the strong foot-clamps steady the aim. In a flash the jaw drops (fig. 12.50) and the tongue is shot out straight at the target, which may be six inches away. The fly is hit by the sticky tip of the tongue and instantly snapped back into the closing gape. The tongue is shot forward by an elaborate inner mechanism of the muscles and skeleton of the jaws, tongue and hyobranchial complex which the chamaeleons have elaborated from the simpler conditions found in certain arboreal iguanids and agamids. (Camp, 1923, pp. 311, 312.)

The Lizard Hosts Divided into Ascalabota and Autarchoglossa.—The chamaeleons share with certain arboreal iguanids such significant characters as independent mobility of the eyes, a casque-shaped head, anterior position of the pineal foramen (Camp, 1923). The agamid *Calotes* "equals or exceeds the chamaeleons in its great power of color change, while agamids generally share with them the highly important characters, detailed elsewhere, of dentition, hemipenes, composition of lower jaw and pattern of throat musculature." In short, Camp's thorough comparative analysis of the soft anatomy and osteology of the geckos, iguanids, agamids and chamaeleons (*ibid.,* pp. 297, 313–312) led him to the important conclusion that these families were probably diversified main branches of an Upper Cretaceous or early Eocene common stock or grand division of the lizard tribe, for which he used the name Ascalabota (lizards) proposed by Merrem in 1820 to include the same families. This grand division (fig. 12.52a, b) includes a wide diversity of body-forms with skeletons and musculature adapted respectively for ground-living, semi-arboreal, arboreal, digging and swimming ways of life as well as for various intergrading conditions, but it has no limbless, worm-like or snake-like members. He connects this outstanding fact with the absence (except in a few agamids) of the rectus superficialis muscle and with the reduction or absence of certain other body muscles, which are of great importance in limbless forms (*op. cit.,* pp. 303, 377–385).

Skinks, True Lizards.—In contrast with the Ascalabota, its opposite division, the Autarchoglossa (self-leader tongue) of Wagler, 1830, retain all the muscles mentioned above, which are prerequisite for a slinking, worm-like burrowing, or a limbless, snake-like, grass-living habitus.

"Time and again in various parts of the world scincs, teiids, and anguids appear to have gone off on such a course, lost the limbs and girdles in varying degree and developed even more highly and in a number of different ways, sets of muscles already present in their more normal ancestral forms. The ascalabotids constitute half of the entire lizard population of the world. They have never developed limbless, burrowing genera and seemingly cannot do so on account of the arboreal specializations and reductions in their locomotory musculature.

"If for any reason whatever the limbs should degenerate in geckos and iguanians (including agamids) the lizards would find themselves helpless; if, however, such a thing happened in a scincoid or anguimorph (as indeed it seems to be happening in frequent instances), the lizard finds itself still capable of locomotion, and if grass-land or humus-soil environments happen to be on hand, such limb reduction may become favorable" (Camp, p. 385).

Thus among the Division Autarchoglossa limbless members are present in 10 out of the 17 families. The serpents, which are also an offshoot of this group, have carried this line of specializations in body musculature and locomotion to new lengths.

Reduction of limbs leading to snake-like forms has occurred most frequently among the skinks (Scincidae, fig. 12.52a, *b*) and related families, including among others the plated lizards (*Gerrhonotus*). Among those skinks which retain limbs, the stout, round-bodied, stump-tailed lizard (*Trachysaurus rugosus,* C) is covered with an armor of large imbricating scales somewhat resembling the scales of a pine cone. It is said to kill snakes with its powerful jaws.

The true lizards (Lacertidae, fig. 12.52a; *d*), although mostly living on the ground, never develop snake-like forms but some of their American relatives (*e*) of the family Teidae do so.

Intermediate between the skinks and the true lizards are the plated or keeled lizards (*Gerrhosauridae*), which like the Lacertids, have large dermal scales or plates covering the top and sides of the head and several rows of keeled scales along the back. The limbs are small but functional.

Among the many-ringed worm-lizards (Amphisbaenidae, *f*) the genus *Bipes* retains two short pectoral limbs immediately behind the head, while the Florida worm lizard (*Rhineura*) lacks all traces of external limbs. The pectoral girdle is entirely wanting in some amphisbaenids and the "rudimentary" (vestigial) pelvis resembles that of degenerate scincomorphs, to which this group, in spite of its high specializations especially in the skull, is apparently allied (Camp, 1923, pp. 316, 317).

The typical blind-worms, slow-worms, or glass-snakes (Anguidae, *k*) are extremely long and superficially snake-like, with no trace of limbs. They are also called "brittle snakes" because when attacked by a snake they can break off their tails.

Monitor and Alligator Skeletons Contrasted—an Exercise in Functional Osteology.—The monitor lizards (Varanidae) of Africa, India, Indonesia and Australia, living mostly in grasslands and deserts, represent a quite different division (Platynota) of the Autarchoglossa, with well developed limbs and fierce predaceous habits. At present the largest of them (fig. 12.52a, *h*) is the "Dragon Lizard" of Komodo Island (one of the Lesser Sunda Islands south of Celebes) which measures nine feet, six inches or more in length; but the greatest of all known monitors was the giant *Varanus* (*Megalania*) *priscus* of the Pleistocene of Australia, whose length as calculated by E. R. Dunn (1927) may have been about fifteen feet.

A splendid skeleton of the Komodo monitor in the American Museum of Natural History is mounted alongside of the skeleton of an alligator of nearly the same length. The basic agreements in general skeletal pattern are inherited from remote common ancestors of the lizards and snakes on the one hand and of the crocodilians and dinosaurs on the other. The contrasts between these two skeletons are mostly in the relative sizes of homologous parts (anisomerism). Thus as the voracious monitor slinks up through the grassland toward its prey, its small head and supple neck give it a wide range in striking; whereas the alligator swims up silently near to its victim, with only its eyes above water, ready to snap with its huge jaws; these are pivoted through the skull upon its very muscular thick neck and body. Many of the skeletal features become intelligible in the light of such differences in habits. The dorsal vertebrae of the monitor (fig. 12.44B, B1) are transversely widened in front by massive transverse processes which support the stout single-headed ribs below and the obliquely facing anterior zygapophysial facets above; the articular cup on the front end of the centrum is deeply concave and faces obliquely downward; it is supported by the upturned ball-like condyle that projects from the rear surface of the preceding vertebra. This interlocking arrangement permits a wide range of twisting and sliding movements of one vertebra upon the next. It strengthens the joints between the vertebrae because vertical dislocation is prevented by the upwardly facing balls and downwardly facing sockets, while oblique transverse dislocation is prevented by the upwardly facing anterior zygapophysial joint planes, sliding upon their downwardly facing partners on the posterior zygapophyses. Such an arrangement had its beginnings in the geckos (fig. 12.43B); it is developed in other lizards (Camp, 1923, figs.

1–26), and culminates in the snakes (fig. 12.14bC).

The dorsal vertebrae of the alligator are also technically procoelous but they differ sharply from those of the monitor in that the anterior concave facet on the centrum faces more forward than downward, and the posterior ball on the preceding centrum faces backward rather than upward. The conspicuous but thin and flat transverse processes of the dorsal vertebrae of the alligator are set relatively high up, above the centra, and project straight outward. To them are attached the heads or upper curved parts of the double-headed ribs, which arch above the wide lungs. Thus the alligator's body between the pectoral and pelvic girdles should be less flexible than that of the monitor and, to judge from photographs and observations, this is indeed the case. Moreover, both the head and the tail of the alligator are more massive than those of the monitor, so that the segment between them which serves as the fulcrum is more compactly built.

The pectoral girdles of monitor and alligator also contrast widely. In the monitor the widely curved, fan-like scapula is continued above by the much larger calcified suprascapula, which is arched inwardly over the back where it forms beautifully the dorsal curve of the shoulder muscles. The scapula-suprascapula in the monitor as in *Sphenodon* (fig. 12.19aA) slides beneath the sphincter colli muscle (*sphc*) in front and affords attachment on its dorsal border to the levator scapulae superficialis (*lsspfs*) and to the trapezius (*cu*); the scapula on its outer surface bears the dorsalis scapulae (*dsc*), on its inner surface the serratus anterior and subscapularis; while posteriorly it overlaps the latissimus dorsi (*ld*). Ventrally the coracoid plates are widely expanded and there are two large anterior fenestrae on the scapula-coracoid complex (cf. fig. 12.46C). The pectoral and coracoid muscles, like those of the scapula and supra-scapula, are correspondingly wide and fan-shaped (cf. fig. 12.19bA2) as they converge to their insertions on the humerus.

In the alligator, although the shoulder musculature (fig. 14.9ab) is basically the same as in *Sphenodon* and the lizards, the bones of the pectoral girdle (fig. 14.8C, F) are widely different. The stiff and strong scapular blade is somewhat narrowly V-shaped above and topped by a small suprascapula. The clavicles are absent and the coracoid of each side forms an anteroposteriorly narrow column with expanded cartilaginous lower end, which fits into a long oblique groove on the side of the sternum.

These differences in the pectoral girdle between the monitor and the alligator are no doubt connected in part with the much greater size of the head and neck, and the relatively smaller size of the pectoral limb in the alligator, together with differences in modes of walking and swimming. Among lizards there is a marked development of the clavo-deltoideus muscle (cf. fig. 12.19a *dcl*) as in forms with a bar-like (fig. 12.46B, C) clavicle (Camp, 1923, pp. 368, 369). The loss of the clavicles and the anteroposterior shortening of the coracoids in the alligator leaves a deep embayment (figs. 14.8C, F; 14.9b) on either side of the narrow interclavicle. This may also be due in part to the backward encroachment of the throat-pouch and its muscles (fig. 14.9b) which is necessary for a saurian that can swallow a heron or a man. In the monitor, on the other hand, the throat-pouch must be relatively longer and more extensible vertically, while the relative width inside the coraco-scapular arches is greater.

Both the monitor and the alligator have strong limbs upon which they can stride forward when necessary. The alligator can also straighten his hind limbs and really run on all fours (Colbert, 1947). The Komodo monitor can, it is true, rear up on his hind legs to look over the tops of the tall grass, but quadrupedal running with straight legs would seem not to be an efficient method of progression for the monitor, partly because the pelvis is but loosely connected with the vertebral column by one larger and one smaller pair of sacral ribs; also partly because the ilium is a narrow backwardly directed rod, as it is in the older crawling reptiles (cf. fig. 14.10A), whereas in the alligator (C) the more helmet-shaped ilium is more firmly attached at the sacrum and has a larger insertion area for the ilio-tibial and ilio-femoral muscles (fig. 14.12C1) which help to hold up the body on one side while the other leg is off the ground.

The hands of the monitor (fig. 14.11A) are of normal lizard type, adapted for ground-living and scratching, with strong flexors of the radiating, large-clawed digits and divergent fifth digit. The carpals also are of normal, primitive reptilian type. The hands of the alligator (B) are relatively narrow, slender and partly webbed. The chief features of its carpus are two stout rod-like bones, corresponding apparently to the radiale and intermedium or ulnare. No doubt they not only lengthen the stride but permit a more effective thrust both in running and swimming.

The hind-feet of the monitor also conform to the primitive lizard type and are thoroughly plantigrade. Those of the alligator, on the contrary, are relatively long, narrow and partly webbed, the first metatarsal thick and strong; and there are two proximal large tarsals (fig. 14.11I) bearing roller-facets, which permit the foot to be twisted obliquely outward, inward and backward in swimming.

Although the skulls of both the monitor (fig. 12.39C) and the alligator (fig. 14.3E, 5C, 6C) seem to have been derived from Triassic reptiles (Eosuchia) with two temporal openings on each side (fig. 12.39B), they have diverged in quite different directions. Thus the monitor, or rather, its remote *Prolacerta*-like ancestors (figs. 12.41b) reduced and finally eliminated the lower temporal bar and left in its place a great lateral embayment, the lower temporal opening (fig. 12.39C). The pterygoid bones (fig. 12.51) as noted above, have been simplified chiefly into narrow anteroposteriorly curved bars with rather loose connection with the expanded basi-pterygoid processes. Also a transverse joint behind the frontals (fig. 12.41C) permits the upper jaw to be raised a little and the lower one to reach forward. The jaws in the monitor were not much lengthened either in front or in the rear and the suspensor (quadrate) was directed forward as well as downward. In the alligator, on the other hand, the lower temporal bar (fig. 14.7B) was retained, but by the backward swinging of the quadrate (with corresponding lengthening of the gape) the lower temporal opening was squeezed vertically and the upper one was reduced to a small round hole in the top of the skull (fig. 14.3E). Meanwhile, with the great forward prolongation of the jaws, the pterygoid muscle increased in size and power and was prolonged in a bony tunnel in front of the eye (fig. 14.7B). A secondary palate (fig. 14.5C) was formed from plates of the premaxillae, maxillae, palatines and pterygoids and the latter were folded around this new internal narial tube in almost the same way that they were in the sea-turtle (fig. 12.28C); then, extending still further backward, the flaring pterygoids formed a transverse brace for the rear border of the palate (fig. 14.5C). All these and many other emergent features in the alligator were entirely unknown in lizards (fig. 12.51A1), which in this region, except chiefly for the fenestration of the palate and for the changes noted above in the pterygoid, retained the basic skull features of remote captorhinomorph ancestors (fig. 12.29).

The exoskeleton of reptiles generally includes the horny surface, often with its spikes or scales, and the subjacent bony plates or scutes ("osteoderms"), as well as the older derm bones such as the nasals, parietals, etc., on the roof and sides of the inner skull or chondrocranium as well as on the jaws. These were once just beneath the surface but are now overlaid by a later generation of horny or bony plates. The monitor's back is covered with innumerable horny scales which do not overlap; but it has no osteoderms. In alligators, as in other crocodilians, the back is covered with an armor of leathery skin, on the surface of which are thick quadrangular horny shields arranged in regular longitudinal and transverse rows. Beneath these horny shields, especially in old animals, are deeply pitted bony scutes.

The Parasternum in Lizards and Other Reptiles.—Similar but flatter horny plates and bony scutes extend, on the ventral surface of crocodilians, from the pubic bones forward, between the ventral prolongations of the ribs, to the sternum. Thus they appear to correspond collectively with the "abdominal ribs" (gastralia or parasternum) of *Sphenodon* (fig. 12.46B) and other primitive reptiles. On the other hand, the parasternum has been eliminated in the monitor and its allies among the Anguimorpha, including the glass-snakes, helodermids, etc. (Camp, *op. cit.,* p. 392). This is a good example of the evolutionary principle that such presence-or-absence characters in the adults are usually the resultant of opposing genetic factors continued over thousands of generations, but, as it were, "subject to change without notice." The parasternum was, it will be recalled (fig. 12.45B, C), a product of the ventral body wall of primitive amphibians, possibly inherited in part from the scales of fishes (A). In general the lizards of different families which became modified for a burrowing or worm-like way of life retained or strengthened the parasternal "ribs"; but in the usually well-limbed anguimorphs (including among others the monitors) a well developed parasternum is rarely present (Camp, *op. cit.,* pp. 300, 386).

To the median parasternal pieces in burrowing lizards are attached slips of the rectus superficialis muscle, while the rectus profundus lies dorsal to the median pieces of the parasternum. To the lateral extensions of the median pieces are attached slips of the superficial external oblique muscles (Camp, 1923, pp. 389–391). The crisscross arrangement of these abdominal muscles facilitates a creeping action of the abdominal wall and its scaly covering. In well-limbed forms these creeping muscles, and with them the entire parasternum, can be sacrificed without preventing improvement in other directions.

In short, genetic changes predetermine adult changes, which may facilitate or retard further adaptations in already assumed ways of life; or they may open up new ways of either utilizing the opportunities offered by the environment, or of shifting successfully from one environment to another. Thus new bits get into the over-all pattern of form and function.

Water-monitors and Sea-lizards (Mosasaurs).— The shore-living, partly aquatic sea-lizards, *Amblyrhynchus* (fig. 13.1A) of the Galapagos Islands, although well able to swim in saltwater, are but little modified either in their inner skeletons or dermal covering for an aquatic way of life, and their well-clawed limbs and feet are used as grappling hooks for clinging to the rocks and for walking under water rather than as paddles for swimming (Beebe, 1924, pp. 111–125). The tail, however, is somewhat compressed in its distal half and in swimming is undulated laterally, the feet being held at the sides (*ibid.,* p. 115).

The water-monitors, *Varanus salvator* (fig. 13.1B), which range from India through the Malayan region and China to Australia, likewise retain separate, not webbed, well-clawed digits and use the undulations of the compressed long tail for fast swimming (Lydekker, New Natural History, vol. V, Sect. IX, p. 152). The water-monitors usually live along the river banks but in the Nicobar Islands one of them, in order to escape capture, was seen to dive into the surf and disappear (*ibid.,* p. 153).

The monitors also have large cavities within the snout (fig. 13.4A) and when the apertures are closed these pouches serve as reservoirs of air (*ibid.,* p. 152). The monitors are the little changed survivors of the Lower Cretaceous stock which gave rise to the European aigialosaurs and through them to the mosasaurs or true sea-lizards (fig. 13.2C, D). In *Aigialosaurus* the skull (fig. 13.4B) was intermediate in form between those of *Varanus* (A) and the mosasaurs (C), but the hands and feet (fig. 13.5B, B1) had not yet become paddle-like. The mosasaurs were named in Cuvier's time from the river Meuse, which cut through the chalk beds containing the fossil skull and other parts of mosasaurs. Mosasaurs were later discovered in the Upper Cretaceous of Kansas, Europe, North Africa, East Indies, New Zealand, so that, like many other marine organisms, they probably spread along the coasts to many parts of the world.

The skeleton of ***Tylosaurus*** (fig. 13.2D), from the Upper Cretaceous of Kansas, mounted in the American Museum of Natural History, is about 26 feet long, the lower jaw alone being more than a yard long and armed with conical teeth. The pectoral limbs (fig. 13.5D, D1) were about as long as the jaw, the extremities large and supported by very long, thin, clawless digits, which were doubtless enclosed in a leathery paddle. All the limb bones had cartilaginous ends, as is usual in marine reptiles, and the numbers of the phalanges in the respective digits had increased beyond the formula of 2 3 4 5 3 for the first to fifth digits, which is the primitive series for lizards. The scapulae as compared with those of the monitor were small but the coracoids were expanded, as is often the case in marine reptiles. The hind-limbs (fig. 13.5D1) were similar to the fore-limbs but somewhat narrower.

The ilia were slender and rod-like, evidently connected only loosely with the sacral region, and in these undulating sea-reptiles the sacral vertebrae had become desacralized as their primitive role of transmitting thrusts from the hind limbs to the vertebral column was lost. Especially the mid-dorsal and caudal neural spines were large, indicating the presence of strong epaxial muscles for undulating the long tail. The cylindrical vertebral centra differed from those of the monitors (fig. 12.44B, C), in the lack of upturning of the ball-facets on the rear ends and of the associated downturning of the cup on the anterior ends; this arrangement had been replaced in the mosasaurs by a simplified, more anteroposteriorly directed joint. From the lower ends of the neural arches the normal anterior and posterior zygopophysial facets were supplemented by a zygosphene-zygantrum joint, as in monitors. The neck, consisting of seven cervical vertebrae, as in the monitors, was quite short. The skull of mosasaurs (fig. 13.4C) was essentially monitorlike in its general build but with such adaptive modifications as the marked forward elongation of the jaws and the development of a vertical joint in the middle of each half of the lower jaw, which permitted transverse expansion to assist in raking in large fish, marine birds or smaller marine reptiles.

Several authors have disputed the relationship of mosasaurs to monitors but after a most thorough and incisive comparative analysis, Camp (1923, pp. 322, 325) supports Williston (1904) that the aquatic mosasaurs of the Upper Cretaceous evolved from the semi-aquatic aigialosaurs of the Lower Cretaceous, which in turn had been derived from terrestrial varanoids (monitors) of the Lowermost Cretaceous or Upper Jurassic.

How Can "Primitives" Remain Unchanged While Related Forms Become Specialized?— The following questions might be asked: (1) How have the recent water-monitors retained, at least in all essential features, the ancestral monitor heritage of Cretaceous times? (2) Had their relatives the aigialosaurs started in the direction that led to the mosasaurs? And if so, why?

Answers: (1) The numerous existing species of monitors presumably represent those surviving populations of the supposed ancestral monitor stock which have, so to speak, escaped the genetically predetermined changes in the limbs and other parts that enabled others to shift from the original mixed environment to a more and more marine way of life. (2) That the aigialosaurs had already taken earlier steps away from the standard monitor skulls and limbs is evident from comparative study of these parts in monitors, aigialosaurs and mosasaurs. Presumably the aigialosaurs took those steps because: (a) some of their populations were in environments (such as oceanic islands) in which the opportunities, the pressure of competition and other factors of Natural Selection in general, gave higher survival values to those subspecies which ventured further offshore in pursuit of schools of fishes; (b) the more progressive populations and subspecies succeeded in adapting themselves more and more perfectly to a marine life, because they inherited a progressively adaptive genetic mechanism for predetermining such adaptation in the adult. That there always has been such a cumulative genetic mechanism which presents to the relatively fixed sieve of a given sort of environment, mixtures of varying degrees of adaptive and disadvantageous features, has been the growing conviction of naturalists from Darwin's time to the present day. Able geneticists such as Dobzhansky (1941) and Sewell Wright (1932) are gradually discovering that this mechanism operates according to statistical methods and that through thousands of generations it establishes hereditary associations between favorable characters in different parts of the organism (Simpson, 1947).

The Serpents (Ophidia, Serpentes) and Their Branching Evolution.—To most people any kind of snake is a fearful and repulsive object. Boys and men, with some honorable exceptions, show their fearlessness by killing eight-inch garter snakes, and girls and women make horrid grimaces and start writhing if they chance to see a live one in the garden. But naturalists who have gained even a little of Darwin's insight find new revelations of nature's laws and principles in the study of the comparative anatomy, the mechanisms, the evolution, and the ways of life of snakes.

"The Serpentes, or snakes," writes Darlington (1948, p. 24), "form only about eleven families but in genera (about 300) and species (about 2,600) they are almost as numerous as lizards. They occur in all habitable continents, most close-lying islands, and a few not too isolated oceanic islands, including the Mauritius group and the Galapagos, but not New Zealand. The majority are tropical but they are common enough in the temperate zones." In the Old World a few snakes penetrate to northern latitudes. In Africa they are numerous to its southern tip; in South America they extend into southern Argentina. Notwithstanding their great number and variety and almost world-wide range, their skeletons present few major differences among themselves.

Adaptations of the Boas and Pythons.—The Boidae, including the boas and pythons, are on the whole the least specialized of the living serpents (pl. 12.IV). They are famous among students of zoology for retaining the last remnants of pelvic limbs in the form of small movable, conical spurs on the outside near the vent; these are supported internally by bony nodules representing the femora and are operated by vestigial limb muscles (New Internal Encyclopaedia, vol. 3, p. 143). The pelvic spurs are said to be useful as holdfasts in climbing up large tree trunks (*ibid.*). But all vestiges of the pectoral limbs, arch and breastbone have disappeared. This sacrifice, as Lydekker remarks (New Natural History, vol. V, sect. IX, p. 179) has set free the ventral surface of the throat and the muscular wall of the stomach, so that they can be greatly stretched to engulf birds or mammals whose vertical and transverse diameters may be many times greater than those of the head and throat of the snake when in the undilated state. Other related adaptations to the swallowing of relatively large objects are as follows:

1) The loosening of the opposite bars of the lower jaw at the front ends, which are connected not by a firm symphysis but only by an elastic ligament, which permits them to separate widely as well as to move independently of each other.

2) Further mobility of the front or tooth-bearing part of each half of the lower jaw is added by the new joint between the dentary and surangular bones, which is shaped like a V turned to the rear (fig. 12.51B).

3) The upper jaws in the boas and pythons (fig. 12.51B, B1) are reduced on their outer sides to vertically low, maxillary bars, which are attached antero-internally to the rigid braincase beneath the olfactory capsules.

4) The paired palatine bones (fig. 12.51B1) which brace the maxillary bars, likewise abut anteriorly on the under side of the stiff braincase, but postero-medially they join the even longer bars of the pterygoids. These bend inward and are connected loosely with the braincase; but, continuing backward, their quadrate rami diverge laterally and are attached to the inner sides of the quadrates.

5) The latter are directed downward (B) to meet the articular facets of the lower jaw.

6) The quadrates are supported above by

a pair of narrow lath-like bones which, in lizards (figs. 12.41aB; 12.51B) and snakes, are identified by Camp (1923, pp. 302, 349, 350) as a modified occipital element (tabulare), the squamosals having been lost in snakes.

The braincase (fig. 12.51B, B1), in order to resist the bending and wrenching stresses resulting either from the struggles of the prey or the action of the snake itself, is modelled in its rear half into a firmly built tube which is stiffened above, behind and below by dense bony crests or ridges and firm side walls; anteriorly it widens out to support the maxillary or outer upper jaw bones, the orbits and the nasal capsules.

As a result of these and other adaptations of the bones and muscles of the pythons and boas, the gape of the mouth can be very widely opened and the opposite halves of the upper jaw can be advanced a little first on one side, then on the other, as the snake uses its upper teeth like grappling hooks and gradually pulls his head and jaws over and around the prey—a process which is facilitated by copious outpouring of mucous from the large salivary glands in the cheeks.

Especially among boas and pythons, which squeeze their prey to death within their mighty coils, there is need for extra strong bracing in the vertebral column to protect the spinal cord from strangulation. This need is met first by making the bony tissue of the vertebrae very dense, second, by multiplying the segments so as to distribute the loads over a wider number, third, by partly telescoping successive vertebrae without sacrificing the mobility of one upon the other.

These seemingly contradictory requirements were met by the following simple steps: (1) by the outgrowth of a pair of dense, V-shaped projections called the zygosphene on either side of the anterior slope of the neural arch (fig. 12.14bC, C1); (2) by the in-sinking of the rear wall of the neural arch of the preceding vertebra, so as to make practically a mold, called zygantrum (C2) of the intruding zygosphene; (3) by extending and infolding the lubricating membranes covering the bony surfaces that slide upon each other.

Since zygosphenes and zygantra are present at least in some of the fossil and recent varanoids (Camp, 1923, p. 321), their presence in the snakes may be part of the heritage from the Cretaceous monitor stock. Thus they would be "preadaptations" in the boas and pythons, whose ancestors may well have had these very strong vertebral columns before they became large enough to crush the prey in their coils. On the other hand, the high number of body segments, each including muscles, spinal nerves, ribs, vertebrae, etc.), which is no less than 412 in a certain python (pl. 12.IV), may well be a relatively recent progressive or emergent feature in these gigantic constrictor snakes, some of which are reliably stated to have exceeded 30 feet in length.

Another partly preadaptive feature in snakes is the loose attachment of the single-headed ribs, which permits wide lateral expansion of the gullet, stomach and intestine as the prey is pulled into the mouth and squeezed down the gullet into the stomach. The single-headed condition of the ribs was probably inherited from the ancestral monitor stock, which were already able to pull large objects into their widely gaping mouths.

In their locomotor methods the boas and pythons and indeed all land-living snakes combine lateral undulation of the body, as in other limbless vertebrates, with an important supplementary creeping system on the ventral surface. This surface is covered with transverse scales or shields, each one extending entirely across the body and moved by slips of the rectus superficialis, obliquus externus abdominis and intercostal muscles, the latter two being attached to the ribs. Thus the ribs can be pulled backward and forward and they act somewhat like the legs of a millipede as the rear edges of the ventral scale-plates catch hold of any irregularities of the substratum. As each plate is pulled backward by the crisscross system of muscles noted above, it tends to shove its own segment forward a little. At the end of the stroke it is lifted slightly off the ground and then set down again for a new stroke as the body moves forward. This supplementary system of small motors can be used either by itself or in combination with the main method of movement by lateral undulations of the main axial muscle segments, as when one part of the body is holding fast or pushing and the other part is sliding forward with its ventral scales relaxed. Thus, while adopting an eel-like mode of locomotion and sacrificing two pairs of limbs, the boas and pythons have developed a highly flexible supplementary locomotor system in which thrusts not essentially different from those which would be excited by many small limbs can be given to the substratum from any part of the long body. This system is adaptable to widely different kinds and shapes of substratum, including the trunks and branches of trees and the irregular surface of the jungle floor. In swimming, however, only the primary system of lateral undulations is effective. Thus in the sea-snakes (Hydrophiinae), which with one exception never leave the water, the ventral shields are either reduced or wanting, as the water affords them no effective holdfast, so that these snakes use only the primary axial system of lateral undulation and are quite helpless if cast up on land (Lydekker, *op. cit.,* p. 226).

Blind Ends of the Boa-Python Stock.—The cylinder snakes (Ilysiidae) of southeastern Asia and tropical America, like their relatives the boas and pythons, retain traces of the hind limbs; but unlike the boas and pythons, their "supratemporal" (tabular) bones form part of the wall of the braincase and do not stand out laterally from the skull as supports for the quadrate bone (Lydekker, *op. cit.,* p. 195). They are burrowing snakes, with small ventral scales. One of their members, the coral cylinder snake (*Ilysia scytale*), a harmless snake, in tropical South America is coral red with black rings and superficially resembles the poisonous coral snake (*Elaps corallinus*) of America and the Old World.

The burrowing shield-tails (Uropeltidae) have advanced beyond the boas and cylinder snakes in the complete loss of all traces of the limbs, by the loss of the "supratemporal" (tabular) and by the firm union of all the bones of the skull—as in several other burrowing reptiles. The chief novelty is the development, in some of the species, of a rough naked disc at the end of the tail; in one species it is covered with keeled scales. The cleft of the mouth is comparatively small and "the jaws have lost the great power of dilatation so characteristic of serpents in general. Their food seems to consist solely of earthworms." (Lydekker, *op. cit.,* p. 197.)

From Harmless Colubrids to Deadly Pit-vipers.—The last two families and several other relicts represent "blind ends" from the boa-python stock, but the swarming colubrine group (Colubridae) includes most of the 2,600-odd species of snakes and culminates in the vipers (Viperidae) and pit-vipers (Crotalinae). These exhibit three main stages in the evolution of the skull beyond the primitive boa-like ancestral pattern.

Colubrine snakes belonging to the first, most primitive stage, are grouped as Aglypha (without groove) in reference to the fact that all of the teeth have solid crowns, without any trace of a poison-groove. None of these snakes are poisonous. Well known examples are the ringed water-snakes (*Tropidonotus natrix*), the tesselated snakes, the common garter snake (*Eutaenia sirtalis*) and many kinds of climbing snakes, including the Aesculapean snake (*Coluber longissimus*). At this stage the outer upper jawbone or maxilla is still fixed and in a horizontal position, and the pterygoids reach backward to the quadrate bone or even behind it.

The second section, or Opisthoglypha, comprises all those colubrine snakes in which ". . . one or more pairs of the hinder maxillary upper teeth are longitudinally grooved and thus capable of acting as poison-fangs. Many of these snakes are indeed extremely venomous, their bite being capable of producing death in a few minutes" (Lydekker, *op cit.*, p. 215). The opisthoglyphs include among others the moon-snakes (*Scytale*) of South America, the nocturnal tree-snakes (*Dipsos*) of the Old World, the Indian whip snakes (*Dryophis*), the Oriental fresh-water snakes (*Homalopsis*).

The third section of colubrine snakes are the Proteroglypha (front grooves), in which the front pair of teeth of the main upper jaw bones or maxillae are grooved and the posterior ones simple and solid. Here belong the Elapidae, including the beautiful and conspicuously banded coral snakes (*Elaps corallinus*) and others, including the craits of India, China and Ceylon. The craits have from one to three small solid teeth behind the fangs. "Next to the cobra, the crait is credited with killing more human beings in India than any other snake" (Lydekker, *op. cit.*, p. 222). Of all the proteroglyphs perhaps the most famous is the cobra, of which the giant cobra ranges up to 13 feet in length. All the cobras have long ribs in the neck which can be spread widely apart when the snake is excited, forming a "hood." The spectacle-like mark on the back of this hood may be useful as a recognition mark between members of the same species; or possibly it may add to the terrifying appearance of the hood. It is not found complete in all cobras but is variously fragmented and may be entirely absent.

The mechanism for striking with the poison fangs in the cobras and their allies is essentially similar to that in the vipers, as described below.

The very poisonous sea-snakes (Hydrophiinae) are an aquatic branch of this proteroglyph section.

The vipers, family Viparidae, are the first of the Solenoglypha (tubular grooves). They have the maxillary bones of the upper jaw movably attached above to the downwardly projecting prefrontal bones at their anterior ends. Several successional pairs of poison-fangs (pl. 12.VII) are based upon the maxillae and when not in use are turned backward, lodging in a depression in the roof of the mouth. When the lower jaws are depressed by the contraction of the depressor mandibulae and other muscles the head is raised and the rod-like ectopterygoid and pterygoid bones are pushed forward against the rear of the maxillae. This causes the poison-fangs to swing forward into the striking position. With incredible speed the snake strikes downward and forward with its neck and head. The contraction of part of the jaw muscles squeezes the large poison-glands in the cheeks and squirts the poison-bearing saliva, through the canal in the fangs, into the wound.

Among the more famous of the Old World vipers are: the common viper (*Vipera verus*) of Europe and Asia, which is the only poisonous snake in Britain, the southern viper or asp (*Vipera aspis*) and the horned viper (*Cerastes*) of northern Africa and Arabia, bearing two short vertical horns above the eyes (Lydekker, *op. cit.*, pp. 229–236).

Among the spitting cobras of Africa and Asia (including some species of *Naja* as well as *Haemachatus*), there is a small hole on the lower front surface of each fang (fig. 12.51C), due to the failure of the tooth-germ to wrap itself completely around the poison duct. If the snake rears up when disturbed by man or beast, he may merely squirt poison droplets straight forward through these holes. If the poison strikes the eye of the enemy, it causes intense irritation and pain. This action may well frighten off a large animal such as an eland and save the snake from the danger of being stepped on as it lurks in the grass (Bogert, 1943).

The solenoglyph family of the pit-vipers (Crotalidae), which are common to Asia and America, are so called from the presence of two pits on the face between the nostrils and the eyes. These pits are now known to contain sensory organs that are stimulated by contact with the warm-blooded skin and body of a mammal and they give the signal to strike. This is another unique emergent organ, which is found only in the rattlesnakes. The rattle is a by-product of the exoskeleton and consists of jointed horny or quill-like segments which are budded off between the nail-like horn or "button" at the end of the tail in young snakes and the scaly portion of the tail. Possibly the beginnings of this organ are suggested in the tail of the spine-tailed Death Adder (*Acanthopsis antarcticus*), an Australian proteroglyph in which there is a paired series of large obliquely overlapping scales near the end of the tail (Lydekker, *op. cit.*, pp. 25, 26).

NEW ZEALAND'S PRIDE—THE TUATARA (*SPHENODON*)

Last Stand of a Famous "Living Fossil."—On several rocky islets off the northeast coast of New Zealand live the last survivors of an ancient reptilian stock that was related on the one hand to the lizards and on the other, to the crocodilians and dinosaurs. Tuatara, the Maoris called it, meaning spiny, in reference to the spiny crest along its neck and back (pl. 12.III); but all zoologists and students of evolution know it as *Sphenodon* (wedge-tooth). Fortunately for *Sphenodon* and for science, its few living representatives are vigilantly guarded and protected, under heavy penalties, by the New Zealand government and it is a high privilege even for a scientist to obtain permission to land among the jagged volcanic rocks where the tuataras make their home. In the daytime, as the government naturalist explains, most of them are asleep in their shady burrows; but when awakened they show no fear and allow themselves to be picked up gently without attempting to bite. The largest ones are about two feet long and look not unlike large iguana lizards, not only in general body-form and posture but in their spiny dorsal crest and the details of their five-toed hands and feet. Nor does their dark greenish yellow ground color seem inappropriate for a lizard, although the small yellowish dots, to which the specific name *punctatus* refers, add a distinctive note. The very large nocturnal eyes are another striking feature; but on the whole it would seem unreasonable to exclude *Sphenodon* from the lizard order Squamata if it were not for its differential characters, many of them in the outer and inner skeletons.

Its Primitive Skeleton (fig. 12.47A).—In the vertebral column of *Sphenodon* the centra are amphicoelous, that is, concave at both ends, and thus differ from those of lizards, which are procoelous, that is, with a concavity only at the front end and a ball or condyle on the rear end. The biconcave ends of the vertebrae indicate the presence of considerable remnants of the notochord, which is still well developed in the foetal stages figured by Schauinsland (1900).

The lower part of the neural arches bear only moderately developed anterior and posterior zygapophysial facets, with no trace of the accessory bracing facets called zygosphenes and zygantra, which are characteristic of the monitors and snakes. Consequently the vertebral column of *Sphenodon* is not as flexible nor are its units as firmly linked together as they are in the monitors and snakes. And there is no indication of convenient breaking planes in the proximal caudal vertebrae such as there is in many lizards. The intercentra are barely indicated in the region of the neck and thorax but in the tail they are represented by haemal rods as in reptiles generally. The atlas-axis section of the column is likewise extremely primitive, retaining the archetypal pattern for all higher vertebrates, including mammals, as will be noted later.

The girdles and limbs (figs. 12.47A, 48A) likewise are but little modified beyond the captorhinomorph stage, although the cleithrum has been lost and the anterior extension of the coracoid is perhaps secondarily cartilaginous. The scapulo-coracoid arch (fig. 12.46B) differs from the typical lizard type (C) in lacking fenestrae or windows; nor are the clavicles fenestrated. The scapula and suprascapula are not so widely expanded dorsally as they are in monitors, the musculature of the shoulder (fig. 12.19a) is quite primitive; the neck is shorter and the head larger than in typical lizards.

The dorsal ribs (fig. 12.48) are described as holocephalous (Williston, 1925, pp. 112, 213), that is, they have an undivided proximal end as in primitive reptiles, the inferior projection or head being in contact with the intercentral notch, the more dorsal part, corresponding to the tubercle, curving beneath the zygapophysial processes, whereas in lizards the single-headed ribs articulate well down on the sides of the centra. The cartilaginous ribs (fig. 12.46B) extend down to the sternum but are not secondarily prolonged below the abdomen as they are in some lizards.

Extremely primitive is the parasternum or "abdominal rib" system of *Sphenodon,* consisting of an elaborate arrangement of exoskeletal polyisomeres in the skin of the abdomen (fig. 12.46B). These extend all the way from the sternum to the pubis and they retain the basic features of the parasternum of the Permo-Carboniferous amphibians (fig. 12.45B, C) in so far as they include a median V-shaped series and rib-like lateral sets. The "rectus superficialis" and associated abdominal muscles in *Sphenodon* have connections with the parasternum much as they do in autarchoglossid lizards (Camp, 1923, pp. 378, 379); but in contrast with some of the latter, *Sphenodon* seems to use only its strong well-clawed limbs in digging.

The pelvis of *Sphenodon* (fig. 14.10B) has advanced somewhat beyond that of *Captorhinus* (fig. 12.8) in the fenestration of the puboischiadic plates and in the development of a large "pectineal" process in its front lower border, to or near which are fastened both the ambiens muscle of the thigh and the obliquus externus of the abdominal wall. The musculature of both the pectoral and pelvic limbs is, for a modern reptile, extremely primitive (cf. Fürbringer, 1873, 1900; Gregory and Camp, 1918; Miner, 1925) and the same is true of all the limb bones. The numbers of the phalanges on each of the digits also follow the primitive reptilian formula (2.3.4.5.3–4), as it is also in lizards.

The Skull and Teeth of *Sphenodon*.—The skull of *Sphenodon* (fig. 12.40) differs conspicuously from that of lizards (figs. 12.39C; 12.51A) and is much more primitive in the presence of a complete lower temporal bar formed of the quadratojugal and jugal bones. This bar, it will be recalled, was early absorbed in the lizards (fig. 12.39C); but the equivalent of the lateral embayment of lizards is bounded below in *Sphenodon* by the lower temporal bar and is there known as the lateral or lower temporal fossa (or fenestra). In a dissected *Sphenodon* the superficial portion of the temporal muscle mass, called capitimandibularis superficialis by L. A. Adams (1919), is seen to lie immediately beneath the lateral temporal opening (fig. 12.26C), and when the jaws are shut, the convex surface of this muscle may be seen filling the middle part of the fossa. The attachment of the muscle is mostly to the strengthened frame around the lower temporal window, namely, to the postorbital and squamosal above, the jugal anteriorly and inferiorly, and to the quadratojugal and squamosal posteriorly. The upper temporal fenestra of *Sphenodon* (fig. 12.41aA) is bounded by the postfrontal, postorbital, squamosal and parietal bones and differs from that of primitive reptiles and lizards (B) in the lack of the supratemporal ("tabulare," Camp). In this respect *Sphenodon* is more specialized than the primitive lizards, as it is also in the loss of the lacrimal bone and in the development of an ascending branch of the maxillary bones just behind the anterior narial opening (fig. 12.40C).

As viewed from above (fig. 12.41A), the skull of *Sphenodon* retains the basic pattern of the captorhinomorph reptiles but is much more specialized, especially in: (1) the pronounced enlargement of the orbits and consequent narrowing of the frontals; (2) the perforation of the old temporal roof, with the opening up of the very large supratemporal fenestra; (3) this opening up and absorption process has left only a narrow framework around the supratemporal fossa formed of the narrow, stiffened strips of the postfrontal, postorbital, squamosal and parietals; (4) the supratemporal (tabular) has been eliminated; (5) the pineal opening, in which lodges the reduced median or parietal eye, is squeezed between the temporal muscle crests on the parietal bones but is still present, as it is in typical lizards.

As compared with the side and top view of the very primitive two-arched *Youngina* skull (fig. 12.39B) from the Upper Permian of South Africa, the skull of *Sphenodon* is basically similar but more advanced: (a) in the enlargement of both the lateral and the dorsal temporal fenestrae; (b) the loss of the lacrimal; (c) in the development of a strong dorsal process in the maxilla; and (d) more especially in the dentition.

The downwardly pointed, upper front teeth of *Sphenodon* (fig. 12.40C) somewhat recall those of captorhinomorphs (fig. 12.29A) in so far as they overlap the forwardly inclined lower front teeth, but less sharply. These upper front teeth are separate in the embryo (Howes and Swinnerton) but later fuse into a tranversely set pair, which, with the lower front teeth, form effective nippers. The cheek teeth (fig. 12.40C2) are arranged on either side in two long parallel rows, the outer on the maxillae, the inner and shorter row on the palatines. The long groove between them receives the single row on the dentary, while the short, somewhat caniniform tooth on each side near the front of the lower jaw is received into a notch behind the prominent lateral upper front tooth. This makes an effective biting organ and is powered by the very large jaw muscles of the temporal fossa (fig. 12.26C). The cheek teeth are acrodont, that is, fastened to the edges of the jaws, and not thecodont, or set in sockets. In all these features *Sphenodon* is specialized well beyond the captorhinomorphs and even considerably beyond *Youngina.*

Notwithstanding these specializations, *Sphenodon* retains all five branches of the primitive pterygoid bones (fig. 12.40C2) as in captorhinomorphs; but its stapes has become more slender and rod-like and the same is true of its epipterygoid. In the lower jaw (C) the dentary has developed a long posterior process, which on the outer side has extended beneath the reduced articular.

All such osteological details add up to the conclusion that, as recognized by leading palaeontologists, *Sphenodon,* apart from a few of its own specializations such as large eyes and peculiar dentition, retains a long list of structural patterns in the masculature, vertebral column, pectoral and pelvic girdles and limbs, which have been inherited directly from the eosuchians and captorhinomorphs, forerunners of the later two-arched reptiles. Thus this New Zealand relict is a living witness of a critical stage in the evolution of the skeleton leading from the earliest cold-blooded quadruped respectively to the crocodilians, flying reptiles (pterosaurs), dinosaurs and birds.

Sphenodon is undoubtedly related more remotely to the lizards but as to its nearer relatives, it may well be a direct descendant of *Homoeosaurus* of the Jurassic of Europe, which is less specialized in the skull and dentition. This branch seems to be connected with the Sphenodonts by certain rare fragments of jaws and teeth from the Upper Jurassic (Morrison) of Wyoming (G. G. Simpson, 1926). The numerous and widely distributed Triassic rhynchosaurs (fig. 12.40B-B2) agree with *Sphenodon* in having on each side two parallel rows of upper teeth, separated by a groove into which fitted the equally straight single row of lower teeth; but the front of the jaws of rhynchosaurs convergently suggest those of a parrot and were probably covered especially on the outside with a horny beak. These may have pierced the shells of snails and mussels with their strong beaks and broken the pieces with their inner row of upper teeth and the straight edge of the lower beak.

CHAPTER XIII

BACK TO THE WATERS AND DOWN TO THE SEA

Contents

CHAPTER XIII

BACK TO THE WATERS AND DOWN TO THE SEA

FROM LIMBS TO PADDLES IN THE ICHTHYOSAURS AND PLESIOSAURS

IN CHAPTER XI it was noted that after some of the lobe-finned air-breathing fishes were transformed into well-limbed, mainly land-living labyrinthodonts, microsaurs and other amphibians, certain lines gradually reverted to water-living habits, as in typical frogs and some of the eel-like members of the newts and salamanders. Among the reptiles, Darwin's marine lizard (fig. 13.1) is famous for its semi-aquatic habits (p. 279) but its limbs have not been changed into paddles as they have in the extinct mosasaurs (fig. 13.2), ichthyosaurs and plesiosaurs. The mosasaurs have been discussed in Chapter XII. The main purposes of the present chapter are to develop the resemblances and differences between ichthyosaurs and plesiosaurs in their general mode of life and main skeletal features; and to indicate their quite different relationships with other reptilian orders.

ICHTHYOSAURS AND PLESIOSAURS: THEIR PARALLEL AND DIVERGENT FEATURES

The frontispiece to the "Book of the Great Sea Dragons" by Thomas Hawkins (1840) vividly portrays these monsters on the shore of a rocky islet on a stormy moonlit night. The center of interest is a giant *Ichthyosaurus* with huge gleaming eyes. His long pointed jaws threaten a recumbent *Plesiosaurus* with a swan-like neck. Several pelican-like winged dragons are hopefully snapping their long bills. The most valuable features of this folio are the many lithographic plates with good figures of well preserved skeletons of ichthyosaurs and plesiosaurs from the Jurassic strata of England.

The ichthyosaurs and the plesiosaurs (s.l.) ran through nearly parallel fossil histories, beginning in the Triassic and branching into many long-persisting lines, the last few ending in the Upper Cretaceous. The ichthyosaurs ranged in the seas from Spitsbergen in the north to South America, Queensland and New Zealand, and from Europe to California; they were usually accompanied by the plesiosaurs. However, these two well separated orders of reptiles were not close competitors, as their methods of pursuing and snatching fish were quite different.

Contrasts in Body-form and Locomotion.—The ichthyosaurs were superficially fish-like in body-form (fig. 13.6) and it is known from several extremely well preserved ones from the Lower Jurassic of Germany that the streamlined back was surmounted by a large dorsal fin, which, however, was made from the skin alone, like a whale's flukes. There was also a large vertically crescentic dermal tail fin, which was supported by the downwardly turned caudal section of the vertebral column. The vertebral column was composed of very many secondary polyisomeres with crowded neural arches and checker-like centra (fig. 13.8B). The ribs were numerous, slender and very long, especially behind the pectoral girdle to the middle of the convex ventral border. Behind this point the ribs rapidly diminished in length, arching well above the small ilium (fig. 13.8C2) and continuing into the short chevrons of the caudal region. The upper ends of the dorsal ribs were vertically bifid and articulated solely with the sides of the centra, not with long transverse processes from the sides of the neural arches, as they did in the plesiosaurs. All this indicates that in the short-necked ichthyosaurs the mid-body section was relatively narrow and that fast swimming was effected chiefly by lateral undulations of the body, the relatively small paddles being used in turning and in leisurely movements; whereas in the plesiosaurs (figs. 13.11; 13.16), which were not at all fish-like but always very saurian in appearance, the lungs were probably higher up, under the raised arches of the ribs, as in crocodilians, the neck was long to very long, the paddles very large, the tail short and without a caudal fin.

In ichthyosaurs the head (figs. 13.6; 13.7) was very large, and the jaws bill-like; the eyes enormous, whereas in the typical plesiosaurs (fig. 13.11C16) the head and the eyes were very small; but in the related pliosaurs the snout became long and heavy, somewhat like that of a long-snouted crocodilian, with large conical teeth. The long-necked plesiosaurs probably paddled slowly and quietly until near the prey, which was snatched by very swift turning movements of the body and side swings of the neck and head. The relatively short-necked pliosaurs may have made a sudden dash after the prey. The ichthyosaurs were already ichthyosaurs in all essential respects when they first appeared in the Middle Trias of Spitsbergen and North America (fig. 13.6B) but, as will be noted below, all ichthyosaurs retained indubitable marks of remote derivation from some sort of land-living, quadrupedal reptiles of Permian age. The plesiosaurs, on the other hand, were closely related to the nothosaurs (fig. 13.10A, B) and placodonts (fig. 13.18), in which the limbs were much less specialized for swimming, and clearly indicate that in their undiscovered ancestors of Permian age the limbs were no doubt well able to support the body on land.

Contrasts in Paddles, Limbs and Girdles.— The main features in which the ichthyosaurs and the plesiosaurs resemble each other are the flexible paddles, which in both orders are composed, beyond the humerus, of a multitude of bony secondary polyisomeres; but, whereas in the typical ichthyosaurs (fig. 13.9B) even the radius (R) and ulna (U) convergently resemble the bony polygons of the carpus, metacarpus and digits, in the plesiosaurs (figs. 13.11C; 13.16) the radius and ulna are somewhat less specialized and the phalanges, although multiplied, retain somewhat more of their constricted cylindrical form. Moreover, the most advanced ichthyosaurs have budded off several supranumerary digit-like rows of small nodules along the margins of the hand. By that time all these secondarily simplified polyisomeres were no doubt embedded in a continuous cartilaginous or connective matrix and covered with a leathery skin. And in all these features the paddles of both ichthyosaurs and plesiosaurs convergently foreshadowed those of the whales and dolphins of later geologic periods.

The highly specialized paddles of ichthyosaurs retain little or no suggestion as to any particular earlier type from which they may have been derived, and the same is true of their humeri; but the pectoral girdle of ichthyosaurs (fig. 13.8C, C1) is of relatively primitive reptilian type and has inherited the interclavicles, clavicles, coracoids and scapulae from such primitive reptiles as cotylosaurs, with minor modifications in proportion. Especially in typical plesiosaurs, on the other hand, the pectoral girdle was greatly modified as follows: (1) The coracoid plates, which are very stout even in the nothosaurs (fig. 13.11B), become greatly expanded on the ventral surface, forming a firm mid-ventral symphysis and serving as a wide base for the fan-like coracoid and pectoral muscles which pulled the humeri inward and backward (Watson, 1924). (2) The scapulae, shortening their ascending blades, sent forward strong "procoracoid" or proscapular processes which curled inward and met in the mid-ventral line beneath (3) the clavicles, which meanwhile had been reduced to small thin splints.

This newly emerging mechanism in the plesiosaurs, together with a less marked but similar expansion of their pubic plates, equipped the plesiosaurs with two very large pairs of distally flexible sweeps, which, powered by the enlarged limb muscles, made it possible to turn the body very quickly to right or left, thus bringing the head and neck within striking distance of the prey. The earlier stages of this line of evolution were retained in the Triassic nothosaurs, in which the coracoids, although stout and pillar-like, had not yet expanded into huge pectoral plates. The digits, at least in *Lariosaurus* (fig. 13.11A), were spreading and webbed, and the skull was less specialized in details than those of plesiosaurs.

SHELL-CRUSHING PLACODONTS

Another early offshoot of the plesiosaur stem were the Triassic placodonts (fig. 13.18). These had become specialized in developing large pebble-like teeth on the palate and lower jaw for smashing the shells of molluscs; and their plump rotund bodies were propelled more by the large tail than by the small paddles. One family, however, culminating in *Henodus chelyops* von Huene, was enclosed in a wide flattened turtle-like and leathery shell (fig. 13.22A) in which was embedded a mosaic of polygonal ossicles not unlike those of the leathery sea turtle Dermochelys. Moreover, only the rear pair of pebble-like teeth were retained (A1) and there was a wide flat, toothless bill. *Henodus* thus showed convergent resemblances to several different kinds of turtles, but its older heritage characters were indubitably nothosaurian.

CONTRASTS IN SKULL STRUCTURE

The temporal region of the skull in typical ichthyosaurs (fig. 13.7C) has been greatly crowded by the huge development of the outwardly facing eyes, so that the stout postorbital rim is squeezed far backward. The squamosal, supratemporal and postfrontal form a strong dorso-lateral bar bracing the orbit and the top of the skull. The jaw muscles have opened up a dorsal or supratemporal fenestra (*stf*).

In most ichthyosaurs the long tapering jaws have quite small pointed teeth but in the California Triassic *Omphalosaurus* the small rounded tooth crowns may have served in the seizing and breaking of small mollusc shells The anterior nares of ichthyosaurs were not a the front end of the bill but originally abov and behind it (*B, ant. nar.*). They led down ward into the internal nares (*C, int. nar.*), whic in the palatal view are bounded by the pre maxillae, prevomers (vomer), maxillaries an palatines, as in various other reptilian skulls The pterygoids were large and had fairly nor mal contacts with the prevomers, palatines basisphenoid and quadrates, but, possibly t provide added space for the pterygoid muscles which aid the temporals in overcoming the ad verse leverage of the long beak, the pterygoid extended back to the level of the occipital con dyle.

In the Triassic ichthyosaurs of Californi (fig. 13.7B) the eyes were neither enlarged no pushed far backward and the whole skull is les peculiar than it was in the later ichthyosaurs Nevertheless its primitive features are share with other reptiles; but the bones bounding it temporal fenestrae present a unique pattern c contacts which may possibly have been derive from the arrangement in the Permian capto rhinomorphs (fig. 12.29); the palatal patter (fig. 13.7C) has been derived from a simila source (A). The shoulder-girdle is relativel primitive and could also have been derive from that of a captorhinomorph type. The verte brae (fig. 13.8B) and limbs (fig. 13.9) are s transformed that they retain no clear evidenc of near relationships with earlier groups.

In the plesiosaurs (figs. 13.11C, 12 C, 14C the relatively small eyes look upward and ar entirely in front of the supratemporal fenes trae. The supratemporal bone, if present, is nc seen as such, and the postorbital and postfront remain near the posterior borders of the orbi These conditions are nearly foreshadowed i the lizard-like *Araeoscelis* (fig. 13.14A) of th American Permian, which, together with th South African Triassic *Protorosaurus* (fi 12.38a) and the American Triassic *Trilopho saurus* (fig. 12.38b), are regarded by J. T Gregory as possibly related to the stem of th plesiosaurs (fig. 12.1). All these reptiles ar therefore tentatively grouped by Romer (1945 in the subclass Synaptosauria, saurians with arches (upper and lower) together.

PERMIAN MESOSAURS

The Permian *Mesosaurus* (fig. 13.6A) and its allies, of Brazil and South Africa, were small semi-aquatic reptiles of inland waters, with a very long beak bristling with long curved, slender teeth. The body was lizard-like, with a long tail; but the ribs were relatively heavy and thick, possibly useful as ballast. The humerus and femur were fairly long and slender, the five-rayed hands and feet were probably webbed but otherwise not specialized, the hind limbs, however, being considerably larger than the fore limbs. The shoulder girdle, although small, was of fairly primitive reptilian type (fig. 13.8A1, A2) with stem-like interclavicle, moderately developed clavicles but no cleithra. The scapula was vertically low. The mesosaur pelvis was of quite primitive reptilian type. There was a well developed ventral armature of transverse rows of gastralia or "abdominal ribs," as in Carboniferous microsaurs and primitive reptiles (fig. 12.45, 46). The vertebrae (fig. 13.8A) with their swollen shoulders and low neural arches, suggest those of cotylosaurs but had no accessory zygapophyses.

This combination of characters would seem to be a favorable starting-point for the more advanced aquatic and fish-catching adaptations of the ichthyosaurs. And it could be argued in detail that there is no other known Permian reptile which appears to be so well suited to be an ancestor of the ichthyosaurs. Von Huene, however, after years of research, succeeded in securing *Mesosaurus* skulls in which the temporal region of the skull (fig. 13.7aA) was sufficiently well preserved for him to infer that the temporal opening was really a lateral temporal fenestra, as it was in the Permian pelycosaurs and mammal-like reptiles, namely, beneath the squamosal-postorbital, supratemporal and postfrontal, and not above those elements as it is in ichthyosaurs. Therefore in his later classification of the reptiles he removes the mesosaurs completely from the neighborhood of the ichthyosaurs and put them as an isolated side-branch near the stem of the mammal-like reptiles. The constriction of the temporal opening may, however, be due in part to the necessity for bracing the fulcrum of the excessively long jaws and to permit them to open widely. Moreover, none of the pelycosaurs or other known mammal-like reptiles seem to suggest real relationship to *Mesosaurus*. Possibly the mesosaurs may even illustrate in some way how the ichthyosaurs really began, without being a real founder of their line, which is given subclass rank by Romer (1945, p. 544) in recognition of their remoteness from other reptiles.

CHAPTER XIV

THE REIGN OF THE ARCHOSAURIANS (THECODONTS, CROCODILIANS AND DINOSAURS)

Contents

CHAPTER XIV

THE REIGN OF THE ARCHOSAURIANS (THECODONTS, CROCODILIANS AND DINOSAURS)

DOMINANT REPTILES WITH TWO TEMPORAL ARCHES

THE PRESENCE of two temporal arches on each side of the skull (fig. 12.39B) was used by Osborn (1903) as the leading character of his "Subclass Diapsida" (double arches) and he included in it not only the crocodilians, dinosaurs and their extinct relatives but also *Sphenodon* and the lizards and snakes (cf. fig. 12.1). Cope, however, had already (1891, p. 35) brought together the same orders under the superorder or subclass Archosauria. This name, which has clear priority, has been adopted by Romer (1933) and others and translated as "Ruling Reptiles."

THE PRIMITIVE TRIASSIC THECODONT *EUPARKERIA* AND OTHERS

The crocodilians and the dinosaurs are in some respects parallel offshoots (fig. 12.1) of the Triassic order Thecodontia. These were somewhat lizard-like in body-form (fig. 14.2a) but their caniniform cheek teeth were set in distinct sockets, whereas in the true lizards the marginal teeth are typically attached either to the inner side or to the distal edge of the jaws. One of the most primitive members of the order (figs. 14.3A, 6A) was *Euparkeria capensis* Broom from the Lower Trias of South Africa. Its upper and lower temporal arches were bounded by the same bones as in the Permian *Youngina* (fig. 12.39B) of South Africa, which is referred by Romer to the lepidosaurian subclass (including *inter alia* the lizards, snakes and *Sphenodon*); but *Euparkeria* had indeed advanced well beyond *Youngina* in the possession of a large preorbital fenestra bounded by the maxilla, lacrimal and jugal bones. Its function possibly may have been to lodge a large maxillary gland, but more probably it afforded attachment on its margins to the anterior branch of the pterygoid muscle, which in crocodilians lies just beneath the preorbital region of the bony face (fig. 14.7B). In *Ornithosuchus,* a relative of *Euparkeria* from the Upper Triassic of Europe, this preorbital fenestra is even larger and surrounded by a stiffened rim (fig. 14.6B). Moreover in the palatal view of the skull, the pterygoids of *Ornithosuchus* (fig. 14.27A) are prolonged backward as if to afford a firm base for the pterygoid muscles.

The lower jaw also in both genera, as well as in saurischian dinosaurs, bears an elongate oval "mandibular foramen," or more properly a fenestra (fig. 14.6A, B) similar to that which in the crocodilians (C) marks the middle part of the insertion area of the inner branch of the pterygoid muscle complex (figs. 14.7B). Indeed, a comparative review of the evidence indicates that the suborbital or anterior branch of the pterygoid muscle-mass was already in existence in the crossopterygian fishes, labyrinthodont amphibians, captorhinomorph reptiles and all their diversified relatives and descendants but that the Triassic thecodonts were among the earliest forms to develop a preorbital fenestra around the area of origin of this muscle and a "mandibular foramen" within its insertion area.

The resultant action of the opposite pterygoid and temporal muscles in the Thecodontia, as in the crocodilians (fig. 14.7B) and dinosaurs (fig. 15.18A), was in a vertical plane, so that the jaws formed a vertical spring-trap with sharply toothed edges. Probably the extremely wide open position of the jaws was achieved not only by the contraction of the relatively small depressor mandibulae muscles behind the quadrate-articular fulcrum, but also by the backward and downward pull of longitudinal throat muscles that ran from the anterior border of the sternum to the front end of the mandible.

Another indication of the ability to open the jaws very widely in these early thecodonts was the somewhat backward inclination of the quadrate-quadratojugal suspensorium and the position of the quadrate fulcra for the mandibles, in transverse line with, or even behind, the occipital condyle (figs. 14.6; 14.7). This arrangement, as in the crocodilians and other reptiles, also facilitated the upturning of the head upon the neck. Moreover, a line drawn along the alveoli of the upper cheek teeth (fig. 14.6) passes through or near the quadrate-articular joint, as is frequently the case in carnivorous reptiles, and this arrangement gives a scizzors-like effect to the cutting action.

In the top view, the *Euparkeria* skull was acutely triangular (fig. 14.3), with pointed nose, fairly large backwardly placed orbits, in these features agreeing in essentials with the eosuchian *Youngina* (fig. 12.39B) and contrasting with *Sphenodon* (fig. 12.40C1). The pineal foramen is conspicuously absent, in contrast with both *Youngina* and *Sphenodon,* but in agreement with crocodilians and dinosaurs.

In the occipital view, the strong transverse processes of the opisthotic bones held the quadrates firmly in place on the inner sides and resisted the medial tension of the pterygoid muscles, which would, if not so braced, constrict the gullet at the point where it needed ample room for the passage of the prey. These transverse opisthotic or paroccipital braces, already established in *Youngina,* were inherited by crocodilians (fig. 14.7B1), dinosaurs and related groups.

The general body-form of *Euparkeria* was not unlike that of a lizard. Of much larger bulk, but also somewhat lizard-like, was *Erythrosuchus* (fig. 14.2b), a contemporary and relative of *Euparkeria* from the Lower Triassic of South Africa. In this heavy-bodied, carnivorous saurian the limbs were very stout but, unlike many of its smaller relatives (fig. 14.2b), the thick-muscled fore-limbs were not much smaller than the hind-limbs, so that the gait must have been bent-limbed and clumsy. Its large skull, however, was essentially similar to those of other primitive thecodonts; that is, it was triangular as seen from above, with a pair of well developed preorbital fenestrae, relatively small supratemporal, and large lateral temporal fenestrae; and in side view the skull was quite triangular, high in the rear and sloping down to the rounded nose.

A related genus, *Chasmatosaurus,* had a long low skull (fig. 12.40A–A2), with many small teeth in its long jaws. Its rostrum, bearing eight teeth on each side, curved downward beyond the end of the lower jaw. Its temporal arches and preorbital fenestra and palate agreed in essentials with those of other thecodonts.

Bipedalism in Thecodonts.—In Aëtosaurus of the Upper Triassic of Europe the hind limbs (fig. 14.14) were somewhat longer than the forelimbs and it is not unlikely that the reptile walked on all fours with his hips elevated and his forelimbs bent sharply at the elbows. The acutely triangular skull was essentially like that of *Euparkeria.*

In another thecodont, *Saltoposuchus,* von Huene found that the forelimbs were very small as compared with the very long hind limbs and it is highly probable that this reptile (fig. 14.15a), which was about a yard long, could rear up and run on its hind legs, like some of the lizards (p. 272). Possibly it could also hop somewhat as a frog does. Its pelvis was better adapted for bipedal running than is that of the lizards by the somewhat helmet-like shape of the ilium, implying firm fixation at the sacrum, and by the great elongation downward and forward of the pubes, which in the dried skeleton were widely separated from the stout backwardly directed ischia by large pubo-ischiadic fenestrae. No doubt strong connective tissue or ligaments ran along the ventral surface in or on either side of the midline, connecting the pubes with the ischia. These ligaments completed the pivot-like function of the pelvis, which resisted the following tensions: (1) from the abdomen in front of it; (2) from the long tail behind it; (3) from the adductor and related ventral muscles of the thigh; (4) from the muscles running from the ilium to the cylindrical femur.

This type of pelvis, called triradiate, or more accurately triramous (three-branched) was prerequisite for that of the saurischian dinosaurs (fig. 14.17F-I). In *Saltoposuchus* and other bipedal thecodonts the body as a whole was pivoted on the pelvis (fig. 14.15a) at the opposite acetabula or sockets for the femora, and in rearing up and returning to the standing position the pelvis swung up and down like the beam of a steelyard or the walking beam of an older type of steamboat. The helmet-like expansion of the ilium afforded ample space, on the inner side, for the attachment to the vertebral column through the sacral ribs, on the top for the long axial muscles, on the outer side for the muscles that pass outward and downward from the crest and outer surface of the ilium to the thigh and leg. These muscles help to tilt the opposite ilia alternately toward and away from the midline as the backwardly directed limb is lifted from the ground and swung forward. The relatively massive tail, even in the small Triassic forerunners of the bipedal dinosaurs, not only assisted in pulling the right and left femora alternately backward but also in throwing the tail forcefully from one side to the other to aid in the sinuous forward motion and to prevent falling to one side or the other.

The ventral surface from the pubes to the sternum was protected in the thecodont reptiles by parasternum (fig. 14.15a) consisting of median and lateral rows of "gastralia" or abdominal ribs, essentially as in *Sphenodon* (fig. 12.46). This parasternum was transmitted to the giant bipedal carnivorous saurischia of later ages.

The pectoral girdle and limbs of the bipedal thecodonts (fig. 14.15a), although already reduced in size, had lost only the cleithrum of the primitive tetrapod stage. As in dinosaurs (fig. 14.8), the coracoid was single and plate-like and there was no bony anterior coracoid.

Along the mid-dorsal line in *Saltoposuchus,* from the occiput to the tip of the tail (fig. 14.15a) there was a series of horny plates and bony scutes. In the Upper Triassic *Aëtosaurus* of Europe the body was covered with an armor of rectangular bony plates (fig. 14.14A), including a median dorsal series and paired costal plates. In the American Triassic *Episcoposaurus* (*Desmatosuchus*) each of the lateral plates on the second to fifth cervical vertebrae bore a pair of outwardly directed bony horns increasing in length posteriorly. The presence of defensive armor suggests attacks from other fast-running reptiles, such as the carnivorous dinosaurs, which were already numerous and diversified.

All the thecodont reptiles noted above, as well as many others, are referred by Romer (1933, p. 597) to the suborder Pseudosuchia (false crocodiles). As in the lizards, this suborder includes ambulatory, bipedal, and possibly digging and water-living types; but limbless, serpent-like forms were conspicuously absent.

THE AQUATIC PHYTOSAURS

A definitely aquatic branch of the quadrupedal pseudosuchian stock was the suborder Phytosauria, which was formerly called Parasuchia (beside the crocodiles). It includes many genera from the Lower and Upper Trias, chiefly of Europe and North America. These saurians were very like the recent gavials in general form and appearance (fig. 14.2C) but they differed radically from all crocodilians in that the anterior nares were on top of the skull just in front of the orbits (fig. 14.4D), and that the long bony snout was prolonged in front of the nares, whereas in the true crocodilians, including the gavials, the nares are at the front end of the snout. Moreover, in the phytosaurs the "secondary palate" (of bones that have shelf-like processes that eventually close over the naso-pharyngeal duct) included only the premaxillae, maxillae, palatines and ectopterygoids but not the pterygoids (fig. 14.5B). In their very long, many-toothed jaws the phytosaurs convergently resembled other fish-catching vertebrates, such as the narrow-nosed garpike (fig. 9.18B), certain long and narrow-snouted labyrinthodonts (fig. 11.28), and especially the long-snouted mesosuchian crocodilians (fig. 14.7C) which replaced the phytosaurs in Jurassic times.

Among the phytosaurs, regional differentiation of the cervical, dorsal, lumbar and caudal regions was on the whole quite gradual, and the high, large neural arches (fig. 14.2C) indicate strong axial muscles and sufficient flexibility for lateral undulation in swimming, somewhat as they do in cetaceans. The double-headed ribs in the dorsal region were attached high up, that is, on the sides of the neural arches as in crocodilians and dinosaurs, indicating capacious lungs, as in other aquatic animals.

The carpus and tarsus of phytosaurs were cartilaginous, as sometimes happens in aquatic vertebrates (e.g. *Cryptobranchus*) but so far as known, the digits were not highly modified as paddles. The high blade of the scapula (figs. 14.2C, 8E) suggests those of dinosaurs but the coracoids were emarginate anteriorly and clavicles and interclavicles were present. The pelvis (fig. 14.2C) retained the plate-like pubes and ischia of primitive reptiles and was not very unlike that of certain saurischian dinosaurs (fig. 14.17I). Thus the phytosaurs, while retaining a basic thecodont heritage, resembled the long-snouted crocodilians in general appearance and habits, in the formation of a secondary palate, in the form and implantation of the teeth and other features; but agreed with the dinosaurs in having a preorbital fenestra, in the form of the scapula and of the pubi-ischiadic plate. Such incidental parallels with the crocodilians and dinosaurs illustrate the principle that parallelism frequently occurs in otherwise divergent branches of a common stock.

ORIGIN OF THE CROCODILIANS

The order Crocodilia apparently began in the Upper Triassic *Protosuchus* (fig. 14.14B) of North America. This small offshoot of the primitive Pseudosuchia was armored with rectangular bony plates, presumably covered in life with a horny surface, arranged in a pattern strikingly like that of its contemporary, the European *Aëtosaurus* (fig. 14.14A). But it was also distinctly crocodilian in the flattening of the occipital roof (fig. 14.7B1) and especially in the fact that the pubis had almost completely withdrawn from the lower border of the acetabulum and had been largely replaced there by the forwardly-grown acetabular process of the ischium. The crocodilians have carried to the extreme (fig. 14.10C) this process of replacement, in which one part retires from a given function and an adjacent part takes over that function; for in them the pubis has entirely withdrawn from the acetabulum and has devoted itself to its function of tying in the strongly built parasternum to the pelvis. But the genus *Protosuchus* of Barnum Brown and its allies differ enough from the Pseudosuchia on the one hand and from their more advanced relatives on the other, to be placed in a primitive suborder, Protosuchia.

EVOLUTION OF THE FISH-SNATCHING MESOSUCHIANS

The next major stage in the evolution of the crocodilians is represented by the numerous genera and species of Jurassic and Cretaceous teleosaurs (family Teleosauridae) of the suborder Mesosuchia (middle crocodilians). This suborder is distinguished from the higher Crocodilia (suborder Eusuchia) by the facts that the bony bridge under the naso-pharyngeal canal includes only the maxillae and the palatines, not the pterygoid bones, and that the internal narial opening lies in front of the transverse processes of the pterygoid bones, whereas in the Eusuchia the narial bridge has been prolonged much further backward by curved flanges from the pterygoid bones and the narial opening lies below the basioccipital (fig. 14.5C). So the bone is formed around the narial tube.

The typical mesosuchians were long-snouted, many-toothed fish-snatchers superficially resembling *Mystriosuchus* and other phytosaurs (fig. 14.5B) in this and related habitus features, but perfectly distinguished by the subordinal heritage characters of the palate, noted above. Some of them (*Pholidosaurus*) were fluviatile and estuarine in habit, like the existing crocodile (Romer, 1945, p. 221). Although the typical teleosaurs were marine in habit, their extremities were essentially like those of crocodiles, i.e. not greatly modified for swimming and probably able to support the body on land. However, the Upper Jurassic European Metriorhynchidae (including among other genera the misleadingly named *Geosaurus* (meaning earth saurian) were found in marine deposits and their hands and limbs and feet were markedly specialized as paddles (fig. 14.13), the bones on the preaxial border being curved in a forwardly convex series and enlarged to support the anterior margin of the paddles. The long hind legs of Geosaurus may have been useful in a sort of punting action in shallow water. Apparently in all the mesosuchians swimming was mainly by lateral undulations of the long tail, as already noted of the recent crocodilians.

The extremely long snouts, which were the rule in the Mesosuchia, were often associated with large supratemporal fenestrae and a long sagittal crest, implying powerful temporal muscles (fig. 14.7C); just as the large and strongly braced transverse wings of the pterygoid bones imply long and strong pterygoid muscles, both being necessary to overcome the adverse leverage of the very long jaws. In the Upper Jurassic *Alligatorellis* (which is classified by Romer among the Mesosuchia), however, the preorbital face is very short and the maximum width across the skull nearly equals its length (Williston, 1925, fig. 68C). The supratemporal fossae in this form are both short and narrow, the orbits relatively very large and the cranial table or occipital roof, wide and flat. These may possibly be infantile features and merely by retaining them in the adult, a narrow pointed triangular skull of primitive pseudosuchian type might be transformed into a wide short-faced skull with a flat occipital table and reduced supratemporal opening, as in *Brachygnathosuchus* and *Caiman.*

THE TYPICAL CROCODILIANS (EUSUCHIANS)

The reduction of the supratemporal fenestrae in the modern alligator (fig. 14.4E) and their secondary roofing over in *Caiman* by extension of the dorsal margins of the surrounding postorbitals and parietals, do not, however, imply that the temporal muscles themselves have disappeared but probably only that they have contracted ventro-dorsally and widened anteroposteriorly, as in the recent alligator, where the superficial branch inserts into the inner side of the mandible, medial to the large mandibular fenestra (Adams, 1919, p. 94). The eusuchian palate (fig. 14.5C), was already completed in the ? Upper Jurassic and Lower Cretaceous European *Hylaeochampsa* (Williston, 1925, fig. 68E, p. 84, Romer, 1935, p. 598).

This perfected bony naso-pharyngeal duct is associated, at least in recent crocodilians, with a unique branching of the membranous Eustachian or tubotympanal canal, which pierces the basioccipital and enters the chamber of the inner ear (fig. 14.6). Possibly this apparatus may convey to the central nervous system stimuli caused by pulsations due to varying pressures upon the tympanic membrane on the outside and by the air in the tubotympanal canal on the inside. For example, when a crocodilian suddenly dives into murky water he might hit his snout on a rock and block the bony tube leading from the nose to the trachea. Possibly this tubotympanal system may act on the radar principle, as do the sensitive nasal appendages of bats. In the gavial there are paired bony swellings or sacs protruding from either side of the palato-pterygoid tunnel or naso-pharyngeal canal. These sacs or bullae may perhaps act as resonating chambers for the loud sounds emitted at times by the larynx, or as pressure chambers which in some way regulate the pressure in the tubotympanal system. In this subject, observation and experiment on living crocodilians are obviously needed.

The crocodiles, the central family of this suborder, range from the Upper Cretaceous to the Recent periods and vary in one direction toward the wide-faced alligators and caimans, in the other, through *Tomistoma,* toward the very long-snouted gavials; these are more or less marine in habit and, as already noted, strongly resemble the long-snouted teleosaurs except that they have attained the Eusuchian stage in the bridging of the internal nares. One representative of the true crocodiles in the Upper Cretaceous of North America was the monster named *Phobosuchus* (fright crocodile) by Barnum Brown. Although only fragments of the skull and jaws are known, they are quite sufficient to suggest that this titanic crocodile was big enough to kill and devour some of his dinosaurian contemporaries, such as the duck-billed dinosaurs.

THE CROCODILIAN WITH A DINOSAUR FACE (SEBECOSUCHIANS)

In the Eocene of South America, G. G. Simpson (1937) discovered a strange crocodilian which he named *Sebecus,* whose high compressed face and sharp recurved teeth suggest those of carnivorous dinosaurs, but which was shown by various other parts of the skull to belong to the order Crocodilia (Colbert, 1946b). On account of its unique combination of characters it was excluded from any of the other crocodilian suborders and it is at present the sole known member of the suborder Sebecosuchia. The strikingly close resemblance of certain parts of the *Sebecus* skull to those of modern crocodilians, together with its relatively late position in time, suggest that it is an extinct side branch from some Cretaceous crocodilian which had retained the compressed triangular skull of the pseudosuchian stock, but had already attained the eusuchian grade in the temporal and pterygoid regions.

Thus the entire history of the orders Thecodontia and Crocodilia suggests that, among the descendants of an ancient stock, cumulative lines of anisomerism, that is, of diverse emphases in rates of cell division in different parts, were necessary pre-conditions for the vast genetic deployment of reptilian species, genera, families, suborders, orders and subclasses.

ORIGIN OF THE DINOSAURS

The Jurassic and later crocodilians on the whole represent a major amphibious-to-marine branch of the Archosauria; but their older Triassic relatives the Connecticut dinosaurs, together with the European relatives of the latter, arose from small swift-running bipedal forms and only in a later stage were such amphibious leviathans as the *Brontosaurus* evolved. The very name *dinosaur* (terrible reptile) has come to signify a gigantic and terrifying saurian, but some of the earliest and least specialized Upper Triassic dinosaurs (e.g. *Procompsognathus*) were not much larger than roosters, except for their long lizard-like tails (fig. 15.3C). But whether large or small the skeletons of the earlier dinosaurs combined numerous adaptations both for fast bipedal running and for insuring lightness and strength.

Remarkable Vertebrae in Coelurosaurians.— In the vertebrae of the smaller bipedal Triassic dinosaurs the principle of fenestration, which has been so often cited in previous pages, has operated intensively to combine lightness and strength. Comparison with the vertebrae of birds suggests that in the dinosaurs also there was an extensive branching system of air sacs connected with the lungs and that such air sacs, or possibly branches of the lungs themselves, were pressed against the sides of the vertebrae. There is convincing evidence from comparative anatomy that when a gland, a muscle, or a nerve, or any kind of visceral sac presses against a bony surface, the bone will gradually make room for its encroaching neighbor, sometimes sinking in the middle and forming rims around the invader as in the temporal fossae described above. In such dinosaur vertebrae the sides became infolded, leaving a strong median vertical partition between two large cavities (cf.

fig. 14.31a). This feature is referred to in the name Coelurosauria (hollow saurians), including the most primitive known group of Triassic dinosaurs and their descendants in the Jurassic and Cretaceous periods.

Signs of Bipedalism.—Preadaptations for swift-running were already beginning, or perhaps seen in an arrested primitive stage in the five-toed hind feet of the South African Triassic *Gryponyx* (fig. 14.24bA). In the front view of this foot the main axis of weight passed through the middle metatarsal (III), the second and fourth being subequal, on either side of it. The fifth metatarsal was very small and almost functionless; the first metatarsal, short and wide, supported the large-clawed hallux, which was raised well above the ground in running. The metatarsals of digits II, III, IV, were widened at the upper end, the second overlapping the third and the third overlapping the fourth. The main transverse joint was "mesotarsal" that is, across the middle of the tarsus and it was nearly hinge-like, permitting only a fore-and-aft movement of the foot upon the leg.

The phalangeal formula or sequence of numbers of the phalanges in digits I to V (2, 3, 4, 5, 1) very nearly conformed to the type preserved in primitive reptiles and birds. In the numerous genera of smaller coelurosaurs the hind foot had advanced further toward the extremely bird-like foot of the Upper Cretaceous *Ornithomimus* and *Struthiomimus* (fig. 14.24b C), in which metatarsals II, III, IV, were consolidated into a strong cannon bone, convergently resembling those of swift-running birds.

SUBORDINAL NAMES BASED ON FOOT-STRUCTURE

In spite of these now very evident bird-like characters of the feet of the swift-running carnivorous dinosaurs, Marsh invented for them the subordinal group names (a) Theropoda (meaning wild beast foot), in contrast with (b) Sauropoda (lizard foot), for the colossal amphibious *Brontosaurus, Diplodocus, et al.,* and (c) Ornithopoda (bird foot), for the herbivorous dinosaurs. Possibly to Marsh at that time the sharply-clawed digitigrade feet of his *Allosaurus* (fig. 14.24bB) may have suggested such a foot as that of the lion among mammals. Hardly more appropriate was the name Sauropoda, based upon the hind foot of *Brontosaurus* (E), which is not lizard-like as the name seems to imply, but superficially rather tortoise-like; while the hind feet (fig. 14.36a) of Marsh's "Ornithopoda," at least as represented in the horned dinosaurs (Ceratopsia) are superficially not as much bird-like as rhinoceros-like. Nevertheless, on the ground of priority and long custom, as well as because a name is only an accepted label for a concept, Marsh's names are retained by leading authorities.

THE SUBORDERS OF SAURISCHIAN DINOSAURS

As a result of vast labors by Marsh, Cope, von Huene and many others, von Huene has divided Marsh's suborder Theropoda into three main series (called infraorders by Romer, 1935, pp. 598–600), as follows:

1) Coelurosauria (Marsh), beginning as small, extremely light-limbed and bipedal Triassic and Jurassic (fig. 14.15b) dinosaurs; with small skulls, long necks and very long arms; culminating in the bird-like *Ornithomimus* and *Struthiomimus* (fig. 14.16) of the Upper Cretaceous.

2) Carnosauria (von Huene), bipedal, progressing toward the massive *Tyrannosaurus* carnivorous type (fig. 14.20a); with large skulls, large flesh-eating teeth and shorter necks.

3) Prosauropoda (von Huene), progressing toward the sauropod type; with relatively small skulls, small teeth, very long necks, short backs, massive limbs and short feet. *Plateosaurus,* the central form of this (fig. 14.29a) group, was much smaller and less specialized than the gigantic Sauropoda of the Jurassic and Cretaceous; it could evidently rear up on its hind legs or let its well developed forelimbs down in the secondarily quadrupedal posture.

4) Sauropoda. As noted above, the Jurassic Sauropoda (figs. 14.29b; 14.30a, b) had advanced well beyond the Prosauropoda in all parts of the skeleton and the group is accord-

ingly given subordinal rank, coördinate with the Theropoda. Taken together, the two suborders Theropoda and Sauropoda constitute the order Saurischia, with triramous pelvis (fig. 14.17C-I) in contrast with the order Ornithischia or herbivorous dinosaurs, in which the pelvis was typically quadriramous (fig. 14.18).

With this necessarily brief characterization of the main branches of the saurischian dinosaurs, we may now outline some of the outstanding features of the skeleton, with special reference to the body-form and other mechanical adaptations of one primitive, one intermediate, and one advanced type in each infraorder or suborder.

EVOLUTION OF THE COELUROSAURIA

The smaller Triassic coelurosaurians were probably related to such lightly-built pseudosuchians as *Saltoposuchus* (fig. 14.15aA), but they had acquired paired concavities in the vertebral centra. The forelimbs, especially the hands (fig. 14.24a), of what may be called the main line grew very long and slender as in the Jurassic *Ornitholestes* (B), culminating in the long, but stouter three-fingered hands of the Upper Cretaceous *Struthiomimus* (C). The latter, with their well developed claws and strong, semi-opposable thumb, seem to be adapted for grasping and climbing, while the long cylindrical femur, slender tibiae, long narrow feet with closely appressed metatarsals, all indicate fast-running, combined perhaps with climbing ability. The lightly built skull (fig. 15.18A) of *Ornitholestes* (fig. 14.15b), was relatively longer, as compared with the neck, than in *Struthiomimus* (fig. 14.16). In the latter the skull was without teeth and probably had a horny bill which may have been used in snapping up small animals, eggs, etc.

THE CARNOSAURIA, CULMINATING IN *TYRANNOSAURUS*, AND THEIR BIPEDAL ADAPTATIONS

In the infraorder Carnosauria, even the early members, from the Upper Trias of Europe and South Africa, were definitely carnivorous, the hands with sharp eagle-like claws and a large, partly opposable hook-like thumb (fig. 14.24a A). In the middle stage, represented by the Upper Jurassic *Allosaurus*, the skull was still well fenestrated and rather lightly built. In the Upper Cretaceous *Tyrannosaurus*, which in the erect bipedal position could raise its head more than 18 feet above the ground (fig. 14.21a), the skull had become massive (fig. 14.25), with much thickened bars and reduced fenestrae, the carnivorous teeth fewer but much larger. The fore-limbs in these giants (fig. 14.20aA) were reduced almost to vestiges, ending in a partly divergent thumb and slender second digit, the remaining digits (III-V) having been greatly reduced or eliminated. In these massive carnosaurs the ilium (figs. 14.21b, 23) was expanded and helmet-like, its dorsal crest concealing the neural spines of the sacral vertebrae in the side view and affording a fulcrum for huge iliofemoral muscles. When the animal rested on the ground, most of the weight of the body was supported by the pillar-like shafts and expanded lower ends of the pubes (fig. 14.21b).

THE PROSAUROPODA, BECOMING SECONDARILY QUADRUPEDAL

In the Prosauropoda, the primitive Upper Triassic *Yaleosaurus* had relatively stout hind limbs with five-toed, incipiently mesaxonic feet, as described above. The triangular skull (fig. 14.26A) may have been derived directly from the primitive *Euparkeria*-like thecodont skull (fig. 14.25B). *Plateosaurus* (fig. 14.29a) of the Trias of Europe differed from *Allosaurus* (fig. 14.19) in its much smaller head, smaller teeth and longer neck. With its moderate length of about 20 feet (Romer, 1935, p. 239), it was only about one-fourth as long as the very long and slender sauropod *Diplodocus* (fig. 14.30a) and it consequently needed far less specialization for supporting the body in the standing posture.

THE GIGANTIC SAUROPODA AND THE MARVELLOUS CONSTRUCTION OF THEIR VERTEBRAL COLUMNS

The suborder Sauropoda of the Jurassic and Cretaceous periods seems to be a direct outgrowth from the Triassic Prosauropoda and, as noted above, far exceeded that grade in its many specializations for securing lightness and strength in the skeleton.

The cavities in the sides of the vertebral centra, which were first seen in the coelurosaurs, were relatively smaller in the Carnosauria (fig. 14.21b) but were emphasized in the Prosauropoda and became very large in the Sauropoda (figs. 14.29b–31b). The latter further complicated their vertebrae by developing a complex system of vertical and oblique ridges on the sides of the tall neural arches (fig. 14.31a), which at the same time lightened and greatly strengthened each vertebra as a whole.

These crests and ridges were no doubt in part automatic growth responses to the oblique stresses induced by the great weight of the body, which probably exceeded 30 tons in the 75-foot *Brontosaurus*. Such ridges may also have arisen as partitions between adjacent divisions of the axial muscles. These axial muscles must have been highly developed and differentiated, more so than those of the crocodilians, otherwise it would have been impossible to support the enormous body weight, or to undulate the long, heavy neck and tail, or to raise the forepart of the body upon the femora, as it pivoted at the top of the femora, or to help the glutaeal muscles rock the body upon one side as the opposite leg was raised and swung forward.

The vertebrae of the sauropods, from the first cervical vertebra to the end of the tail, were regionally differentiated as follows:

1) The atlas, or first cervical vertebra, like that of other archosaurian reptiles, presented a concave ventral rim formed from the inner sides of its low neural arches and the anterior surface of its intercentrum. This rim received most of the ball-like occipital condyle protruding from the skull base (fig. 14.27D, E). The synovial membrane between the ball and the cup may have been derived in long past ages from an outfold of the notochordal sheath. The center of the atlas was filled by a protruding unit called the odontoid, which in mammals usually becomes fused to the second cervical vertebra or axis, but in reptiles still functions as the centrum (pleurocentrum) of the atlas (Williston, 1925, pp. 102–104). The odontoid thus received the thrust of the middle part of the occipital condyle. This first neck joint between the ball-like occipital condyle and the cup-like atlas provided for vertical, lateral and twisting oblique movements of the head.

2) The axis or second cervical vertebra provided a strong fulcrum both for the atlas and the skull. As in primitive reptiles (fig. 12.42) its neural arch and sides were the base for dorsal muscles running to the occiput and raising the head; from its ventral surface passed forward muscles running to the under side of the skull base and pulling the head down on the atlas.

3) The remaining cervical vertebrae of Sauropods constituted collectively a flexible or linked drawbridge (figs. 14.29a–30b), which could raise or lower the head over wide arcs, or swing it to one side or the other, or undulate the neck. The neck in turn was pivoted upon the thorax, but the effective length of the neck had been increased by partly cervicalizing, that is, permitting greater freedom of movement, to two or three of the anterior dorsal vertebrae. The very long posterior cervical vertebrae (fig. 14.31bB) with their horizontally long slender ribs, convergently suggest the neck vertebrae of swans (fig. 15.12C).

4) In the thoracic or dorsal vertebrae, freedom of movement was progressively restricted as the lengths of the ribs increased; this was partly because these vertebrae had to resist successively the cumulatively great adverse leverages transmitted to them from the very long and heavy neck; partly because, when the animal walked on land, the dorsal vertebrae had to support the weight of the heart, lungs, blood, and other parts directly beneath them, as well as part of the weight of the food-filled abdomen, which was behind and below them.

5) In the rear part of the relatively short thorax, the neural spines of the dorso-lumbar vertebrae increased in width (cf. fig. 14.31aA1) and decreased in height with the transverse section of the dorsal axial muscles as the load increased cumulatively to the sacrum.

6) What may be called the sacral strong-box or main fulcrum of the vertebral column consisted (fig. 14.32a, b) of: (a) several closely appressed or coalesced vertebral centra; (b) their massive neural arches and transverse processes; (c) the outwardly directed sacral ribs, which are intimately connected and fused with the transverse processes and coöperate with the latter in transmitting the loads, stresses and strains, passing both ways between the sacral strong-box, the limbs below, the tail behind and the body in front.

7) In the caudal vertebrae the neural arches became lower toward the rear (figs. 14.29b; 14.30a, b) and the V-shaped chevrons beneath the intercentral junctions are reduced and eliminated as the tail tapered to a long lash, consisting chiefly of diminishing cylindrical segments.

The foregoing brief sketch of the outstanding general features of the vertebral column in sauropods could be taken as the starting-point for a more detailed morphological and functional study of the vertebral skeleton and musculature of a lizard and a crocodilian in comparison with such a sauropod as *Diplodocus*. In the latter the details of the vertebrae have been rather fully recorded by Hatcher (1901) but apparently little has been done to discover their meaning, especially in comparison with the very systematic and constructive studies of Romer (1922, 1923), and of Miner (1925) on the limbs and girdles.

THE PILLAR-LIKE LIMBS AND SHORT TOES OF SAUROPODS

In the pillar-like, somewhat elephantine hind limbs of sauropods (figs. 14.30a, b) the femora in some cases reached a length of six feet, eight inches and were doubtless set in the middle of enormous, well rounded thighs not unlike those of an alligator (fig. 14.12C1). In standing, the knee was bent very little (figs. 14.29a–30b), so that the weight passed downward almost in a vertical line to the very short-toed, thickly padded feet. As in crocodilians and *Tyrannosaurus* (fig. 14.23), enormous adductor muscles passed outward from the great pubic plates to the femur, while the backwardly directed ischia supported the rear end of the reproductive and alimentary systems and gave origin to the ischiocaudalis and to part of the muscles running to the femur. The ends of the long bones were covered by growing cartilaginous caps not unlike those of large crocodilians or sea-turtles, but relatively thicker.

Could the Sauropods Walk on Land?—In the opinion of W. D. Matthew (1910) and others, the sauropods could not walk about on land because their limbs, it was thought, would collapse under the enormous weight unless they were supported by water pressure. Thus the sauropods were conceived solely as waders, punting, as it were, with their long hind limbs and rearing up in deeper water with their eyes at the surface. No one doubts that they could do this, but that they could not walk at all unless the body weight was eliminated by the pressure of the water was an arbitrary assumption and it was made without citation of any supporting evidence, such as conceivably might have been computed from civil engineers' laboratory tests, giving the loads under which hollow cylindrical columns of oak, subjected to pressure at both ends, would break down. Moreover, Matthew also assumed that some of the larger mammoths, such as *Elephas ganesa,* were near the danger line of collapse on account of their great weight, but he cited no cases of broken mammoth femora.

A superb series of fossil footprints of a sauropod walking on all fours were excavated by Barnum Brown and his assistants and are on exhibition in the American Museum of Natural History, New York. There is the record of a long trackway made by all four feet in walking across a flat, muddy surface. This quadrupedal trackway was closely paralleled by the bipedal trackway of a carnivorous dinosaur. Both animals were walking, the sauropod perhaps heading for the water and the theropod perhaps in pursuit. If the body of the sauropod had been even partly supported by water, it seems unlikely that such deep and regularly repeated imprints of the forefeet would have been made.

That the sauropods were only semiaquatic like hippopotami and that they had not given up walking on land is clearly indicated by the immensely strong but light construction of the vertebral column and especially by the sacral strong-box described above; all this was indeed the very opposite of the weakened or obliterated sacrum and the feeble ilia of such very large aquatic reptiles as the elasmosaurs (fig. 13.17), which depended solely upon the water for the support of the body.

AMAZINGLY SMALL SKULLS OF SAUROPODS

The skulls of sauropods were all basically similar derivatives of the *Plateosaurus* type (fig. 14.26B), but in *Morosaurus* (D) and allied genera the jaws were short and the face high; the skull also was relatively large in proportion to the length of the neck, and the teeth were about as large as teaspoons; while in *Diplodocus* the upper and lower jaws (E) were moderately prolonged downward and forward, the teeth had become pencil-like, somewhat procumbent and confined near the front end of the jaws. But the whole skull of *Diplodocus* was minute in relation to the enormously long neck and tail (fig. 14.30a).

It has usually been assumed that these colossal creatures could pluck up enough long-stemmed water plants with their tiny heads to feed their huge bodies and to build up their enormous skeletons, which required great quantities of lime; but from the structure of their jaws and teeth it seems more likely that they retained enough of their flesh-loving heritage to crave also some fleshy food combined with sufficient calcium for their skeletons. *Morosaurus,* for example, may have waited with camouflaged jaws for fresh-water Jurassic ganoid fishes (e.g., amioids) or plucked up thick-shelled *Unios*. *Diplodocus* may have browsed upon the fresh-water mussels, beds of which were once found near a *Diplodocus* skeleton (Holland, 1906).

CONTRASTING TYPES OF PELVIS IN SAURISCHIAN AND ORNITHISCHIAN DINOSAURS

The name "dinosaur" commonly covers two widely different orders (Saurischia, Ornithischia) of archosaurian reptiles, between which there are no known connecting links. The saurischians, as noted above (fig. 14.17), had a triramous pelvis in which the pubis extended downward and forward and the ilium was somewhat helmet-shaped with an expanded vertical blade. All the saurischians (including coelurosaurs, carnosaurs, prosauropods and sauropods) agreed in this basic plan of the pelvis, no matter how diverse were their body-forms and modes of locomotion and feeding. The typical ornithischians, on the contrary, had a quadriramous pelvis (fig. 14.18) in which the pubis was turned backward, while a fourth branch grew forward as the anterior process of the pubis. Both pubis and ischium in the most primitive ornithischians were prolonged downward and backward as slender parallel rods. The ilium was not at all helmet-shaped but was dorsoventrally narrow and prolonged anteriorly into a long curving anterior process. There was also a posterior process of the ilium, varying from long to short.

An equally wide contrast between these two orders is seen in the construction of the jaws and teeth, which in the primitive coelurosaurs (fig. 14.15B) and carnosaurs (fig. 14.25) were progressively carnivorous, and in the prosauropods (fig. 14.26B) and sauropods (C, D, E), retrogressively carnivorous; whereas in the ornithischians the teeth and jaws (figs. 14.37, 38) were specialized for slicing vegetation.

BIPEDAL *HYPSILOPHODON*, IGUANODONTS AND DUCKBILLS

As noted by Romer (1935, p. 248), *Hypsilophodon* of the Lower Cretaceous Wealden beds of Europe is among the most primitive of known ornithopod dinosaurs. It was quite small, scarcely more than a yard in length, with relatively slender and primitive extremities. In the hind foot, the middle or third metatarsal and its phalanges were the longest, with subequal II and IV on either side. The first metatarsal was fairly short, with two divergent phalanges. The fifth was vestigial. The tarsal joint passed across the middle of the tarsus. The hand of *Hypsilophodon* had a good, slightly divergent thumb, the second and third digits increasing in length, the fourth short, the fifth vestigial. These patterns of foot and hand are in line with the stages from Permian captorhinomorph to primitive Triassic archosaurian and would be a favorable preadaptive stage for the secondarily widened extremities of the duckbills and horned dinosaurs (figs. 1436b, 43b, 44).

Hypsilophodon retained some teeth in the premaxillae, which are lost in the fully beaked camptosaurs and duckbills (figs. 14.38a, b; 39).

Relatively primitive and well known ornithischians were the Upper Jurassic *Camptosaurus* of North America and the related *Iguanodon* of the Lower Cretaceous of Europe. *Camptosaurus,* although fully bipedal, contrasts with the earlier saurischians, which were very lightly built for swift running, in its slender pelvis, thick femur and tibia, short, heavy metatarsals and short toes with small blunt claws. The humerus, though short, was also stocky and the wide spreading hands, although much smaller than the hind feet, were used in the secondarily quadrupedal posture. Apparently the early ornithischians could run rather ponderously on their hind legs but preferred to browse leisurely on all fours.

The name *Iguanodon* refers to the superficial resemblance of the compressed, somewhat trifid tooth crowns to those of the lizard *Iguana*; but this was merely a convergence, as was the arrangement of the functional teeth in a straight row. In contrast with primitive saurischian skulls, the quadrate-articular joint of the iguanodonts was depressed below the level of the cheek teeth (fig. 14.38), an arrangement which permitted an obliquely vertical slicing movement of the outer sides of the lower teeth across the inner side of the upper teeth (fig. 14.37E, F).

The front part of the upper and lower jaws ends in curved, sharp-edged rims (fig. 14.39), not unlike those of turtles and adapted for the support and growth of a horny beak, which, however, is never preserved in the fossils.

In the "duck-billed" descendants of the camptosaurs, including the Cretaceous hadrosaurs or trachodonts (fig. 14.37F), the continuous vertical budding of new dental polyisomeres in interlocking and alternating, closely packed horizontal rows, has given rise to a huge "dental mill" of four great and heavy masses, two above and two below, comprising some 2200 individual teeth (B. Brown). Meanwhile the genetic and selective factors of evolution have also coöperated in (1) increasing the vertical height of the temporal muscles and (2) of the quadrate suspensor; (3) enlarging the vertical or coronoid arm on the dentary, thus further bettering the leverage of the temporal muscle; (4) enlarging the jugal as a brace for the quadrate suspensor. An incidental result has been the elevation of the temporal and occipital region and the downward and forward slope of the skull roof and forehead. (5) Meanwhile the facial bones supporting the beak have also grown downward and forward so that the mouth opening is well below the cheek teeth, an arrangement suitable for grubbing in muddy water with the wide duck-like beak. (6) The orbits, retaining their position above the jugals, thus find themselves far above the beak and in a good position for keeping watch for carnivorous enemies.

Supraorbital Nasal Tubes and Crests in Duckbills.—The nasal tubes in duckbills (fig. 14.38b) began to lengthen, extending downward with the beak and pushing backward and upward over the eyes, carrying with them special branches of the nasals and premaxillae. These became genetically dominant and diversified features, forming a great compressed helmet-like crest in the cassowary dinosaurs (*Corythosaurus*) or a long median beam, extending upward and backward far behind the occiput in *Saurolophus*. These varied duckbills could walk upright or on all-fours (fig. 14.38a), on land or in water, or swim well with their flattened hands and feet (fig. 14.36a). Their dorsal axial muscles ended in long ossified tendons, which formed intersecting grids on the sides of the high neural arches. In the bipedal posture, they would assist in holding up the back and in swimming they may have helped in sinuous movements of the back and tail.

Titanic Footprints.—An average-sized duckbill was about 30 feet long (Romer, 1935, p. 250) and in the bipedal erect position it could possibly raise its head about 15 feet above ground; but Barnum Brown discovered the three-toed footprints of a colossal duckbill in the Upper Cretaceous of Montana, each one 34 inches long and 34 inches wide, the stride being 15 feet 2 inches long and the total height estimated at 30 to 35 feet. The thigh bones that held up this huge mass are unknown. But even the footprints refute Matthew's idea that the limbs of the larger dinosaurs would collapse unless the weight of the body were neutralized by water pressure.

MONGOLIAN FORERUNNERS AND THE SECONDARILY QUADRUPEDAL HORNED DINOSAURS

From the Middle Cretaceous of Mongolia (fig. 14.43a) came the two small well preserved skeletons of *Psittacosaurus* (parrot saurian) and *Protiguanodon* described by Osborn (1924). These had compressed high, parrot-like beaks (fig. 14.45B) but their cheek teeth (fig. 14.37A) were essentially like those of *Iguanodon,* although much smaller. Their skeletons, adapted for both bipedal and quadrupedal progression, were remarkably generalized and would require only minor emphases to transform them into the *Protoceraptops* type (pl. 14.I), which in turn was definitely ancestral to the great horned dinosaurs (ceratopsians) of the Upper Cretaceous. An exceptionally primitive feature was the retention of well developed clavicles and an interclavicle, which were lost by all duckbills and horned dinosaurs.

Protoceratops, from the Middle Cretaceous of Mongolia, had an enormous skull in proportion to the size of its small, rather lightly built body (pl. 14.I). The parietal bones, growing backward and outward, fanned out into a great spreading crest (fig. 14.45C) which was braced on either side by outward-diverging bars from the top of the squamosals and in the midline by a stiff sagittal crest. The areas between the sagittal crest and the squamosal bars were plainly occupied by thin and widened extensions of the temporal muscles, which converged downward and forward to be inserted on the vertical coronoid processes of the lower jaw, as in other ornithischians. Large paired vacuities or fenestrae, probably covered by membrane, were on either side of the sagittal crest and in front of the thick terminal rim of the crest. In the top view (fig. 14.46C), the skull was triangular, with the large orbits about midway between the narrow beak and the spreading crest. In some of the smaller skulls one or two teeth were retained behind the beak in the premaxillae. This type of skull has evidently been derived from one like that of *Psittacosaurus* with a high beak but no backward extension of the supratemporal fenestrae and crests. The dental battery of *Protoceratops* (fig. 14.37C) was less massive than that of the duckbills but essentially like that of the horned dinosaurs (D). The postcranial skeleton (pl. 14.I) was strictly intermediate between the prevailingly bipedal type of *Camptosaurus,* with rather small forelimbs and the secondarily quadrupedal form, with massive forelimbs, of the horned dinosaurs (fig. 14.44). Possible swimming habits are suggested by the relatively long hands and feet and by the dorsal elongation of the spines of the caudal vertebrae.

The horned dinosaurs (Ceratopsia) were gigantic, fully quadrupedal derivatives of *Protoceratops,* acquiring massive bodies that were supported by nearly straight hind limbs and huge sharply bent forelimbs. The many different genera emphasized various features. *Monoclonius* (fig. 14.44) developed a tall conical horn above the nose; *Triceratops* (fig. 14.45) also had a pair above the eyes; other genera had smaller horn-like projections on the margins of the huge bony frill; in *Torosaurus* the very large frill was surrounded by large vacuities, etc. The huge head was directed downward and its horn or horns could be thrust upward against an attacking tyrannosaur.

All were land-living and quadrupedal, with broad four-toed hind feet and still wider five-toed forefeet. The back was arched upward above the laterally everted rim of the long ilium, which overhung the top of the femur. Over the arch of the back ran a network of ossified tendons.

STEGOSAURUS AND ITS "SACRAL BRAIN"

An early secondarily quadrupedal branch of primitive camptosaur-like ancestry were the stegosaurs, chiefly from the Lower, Middle, and Upper Jurassic. *Stegosaurus* itself (fig. 14.40) was alleged to have in the spinal canal above the hind limbs a far larger "brain" than the one in its microcephalic head, but it may be suspected that much of the room in the expanded spinal cavity was filled with meningeal membrane and vascular tissue, and may have served possibly as a storehouse of vessels containing oxygen. The highly arched back of *Stegosaurus* was surmounted by a series of thinly compressed bony plates, placed vertically, one on either side of the midline. They increased in size from small ones above the neck to huge ones above the hind limbs. The plates of opposite sides alternated, their tips directed upward and backward. They diminished in height over the proximal two-thirds of the tail. Behind them was a short interval, after which came two pairs of thin and sharply pointed long spikes.

BONE-HEADED DINOSAURS ("TROODONTS")

An exceptionally queer by-product of the ornithopod stock was a small Upper Cretaceous dinosaur named *Troödon* (fig. 14.42a). Its body-form as restored by Gilmore (1924b, 1931) was adapted for bipedal and quadrupedal progression, and was not unusual; but its short beakless skull was surmounted by a tall median swelling, which, together with the brain-case, consisted of dense, almost ivory-like solid bone with spikes on the surface. Related genera indicate that *Troödon* was not even the most specialized member of the family. But how could they have supported this dead weight on top of a small bipedal skeleton? Since such a dense skull would not be easily broken, or badly bitten, it may be imagined that *Troödon* scraped a shallow burrow for his body and plugged the entrance with his rock-like skull. Nor is it incredible that he might crash into anthills or termite nests with his head and lap up the insects with his thick tongue.

HEAVILY ARMORED ANKYLOSAURS

Less enigmatic were the ankylosaurs (fig. 14.42b), resembling horned dinosaurs in their highly arched, broad back, straight hind limbs and crooked forelimbs, but with very small heads and a thick armor consisting of small bony polygons arranged in transverse rows, covering the body and forming rings or bony spikes on the tail. Thus they were strikingly similar to glyptodonts among mammals and no doubt the bony scutes were likewise covered with a horny layer as in tortoises. The short wide triangular skulls also were protected by a thick casque of derm bones, produced at the corners into spikes. In one genus the end of the tail was enclosed in an enormous paired bony swelling. This heavy weapon could apparently be snapped quickly from one side to the other. It was supported by specially constructed vertebrae, with long overlapping zygapophyseal processes which would give both strength and resilience to the caudal axis

CONTRASTS, PARALLELISMS AND CONVERGENCES IN THE SAURISCHIA AND ORNITHISCHIA

In conclusion, the ornithischian dinosaurs seem to be a completely different order from the Saurischia, having little in common with them except (1) parallel tendencies to evolve bipedal and secondarily quadrupedal body-forms, together with (2) the frequent occurrence of gigantism, with consequent convergence in adaptations, especially in the limbs and vertebrae, for carrying great body weight.

The saurischians, as we have seen, may be traced quite directly from the primitive, prevailingly bipedal, thecodonts to the coelurosaurs and thence to the carnosaurs, prosauropods and sauropods. Nearly all of this great array were flesh-eaters and only the sauropods have been thought, and that too on rather doubtful evidence, to have evolved incipiently herbivorous habits.

The ornithischians, on the contrary, were specialized herbivores, diversified to live on the abounding floras of the Mesozoic era. Their ultimate derivation from a primitive archosaurian stock is sufficiently clear in the construction of their entire skeleton; but intermediate stages are lacking or barely suggested.

ROLES OF POLYISOMERISM AND ANISOMERISM IN SKELETAL TRANSFORMATIONS

As in all other groups, polyisomerism and anisomerism have played their usual important and complementary roles in evolution; but anisomerism is much more frequent. In the coelurosaurians, for example, the little air sacs, or offshoots from the lungs which filled the cavities on either side of the vertebrae, were polyisomeres and so were the 2200 teeth forming the great dental mill of the duckbill, or the multitude of polygonal bony scutes in the armor of the ankylosaurs. But the majority of the diagnostic characters that distinguish one genus or one family from another, are anisomeres, such as the extreme reduction in the size of the head in *Diplodocus* and *Stegosaurus,* or its great enlargement in *Protoceratops* and the later ceratopsians. Emergent diagnostic characters which are new or newly altered features may be either polyisomeres or anisomeres.

WHY WERE THE DINOSAUR HOSTS CUT OFF?

During the long millions of years of the Mesozoic era, dinosaurs of both orders spread widely on nearly all land masses and into many ways of life. Adaptations for bipedal walking and running, or for secondarily quadrupedal progression, are seen in many lines; but secondary adaptations for swimming are never gained by reducing the strong sacral box or changing the limbs into paddles. Leaping dinosaurs are unknown in spite of the greater size of hind legs over fore legs. Nor did any dinosaurs eliminate the limbs and take on the serpentine method of progression. Climbing may have been possible in the ornithopod *Hypsilophodon* and possibly in the slender-limbed *Ornitholestes* and *Struthiomimus* (fig. 14.16), but again with the possible exception of the last two, no dinosaur was provided with a patagium for sliding through the air. And only the earliest coelurosaurs could conceivably have given rise either to the flying reptiles (pterosaurs) or to the birds.

As the Mesozoic drew slowly to its close, the diversity of the dinosaurs increased. But then came a relatively brief period in the very latest stages of the Cretaceous epoch when dinosaurs of all kinds became extinct, all over the world; and not even in Madagascar or South America are they ever found in Paleocene, Eocene or later beds. One can readily account for the total elimination of the larger dinosaurs but this wholesale extinction of both orders could scarcely be due wholly to changes of climate and elevation; because turtles, lizards, snakes, *Sphenodon* and many of the crocodilians survived such changes and came through safely to later ages.

The famous hypothesis that the dinosaurs were exterminated by small egg-loving mammals is hardly convincing; for both egg-laying dinosaurs and small sharp-toothed mammals flourished together in the Middle Cretaceous of Mongolia, as they did also in the Jurassic of England. And not all the diversified small mammals evolved in the Tertiary period have exterminated the egg-laying turtles, lizards and crocodilians.

CHAPTER XV

BAT-WINGED REPTILES (PTEROSAURS) AND THE FEATHER-WINGED FRAME OF BIRDS

Contents

CHAPTER XV

BAT-WINGED REPTILES (PTEROSAURS) AND THE FEATHER-WINGED FRAME OF BIRDS

THE PTEROSAURS, PIONEER REPTILIAN FLYERS

DURING THE Jurassic and Cretaceous periods, that is, for about two-thirds of the Age of Reptiles, the bat-winged pterodactyls skimmed or flitted about above the heads of the sauropods and theropods, of the duckbills, horned dinosaurs and armored ankylosaurs; or swooped down just above the stream-haunting crocodilians and phytosaurs. The larger and later ones, like great soaring birds with a twenty-to-twenty-five-foot wing-spread, circled well off shore to snatch fish in front of the mosasaurs, plesiosaurs and ichthyosaurs and other marine competitors. And perhaps among the rocks and bushes on shore they laid their eggs, in spite of marauding monitor lizards and snakes.

Their Skinny Wings.—Their wings were of course the outstanding feature and very bat-like membranes they were (fig. 15.1), but attached only to the back of the enormously long fourth finger and not spread between the second, third, fourth and fifth fingers as in bats. The fifth finger in pterosaurs had probably been lost or greatly reduced before the pterosaurs began to fly. The small first, second and third fingers were free to be used as grappling hooks in climbing up on the rocks. The wing membrane on the fourth digit was continuous with the patagium or fold of skin extending along the sides and probably back to the hind limbs; in the long-tailed rhamphorhynchids the membrane was probably continued between the legs to the root of the tail, while the end of the long tail bore a flat pear-shaped horizontal rudder (fig. 15.1B).

Wing Base.—The long wings were based upon a wide shield-like sternum or breast-bone; their downward thrusts were delivered to it through the columnar coracoid bones. The upward thrusts came through the rod-like scapulae. In the largest pterosaurs (*Pteranodon,* fig. 15.2) the scapulae were movably connected at their upper ends with a socket on either face of a strong median ridge called the notarium, a new and emergent bone, probably in part the product of the connective tissue sheaths and dermbone scutes along the midline of the back, and in part due to the fusion of the tops of the neural spines. In the smaller pterosaurs (figs. 15.1, 3B), before the notarium took over the suspensory function, the scapulae, coracoids and breast-bone formed a sort of cradle upon which the body rested in flight. The relatively large humeri (fig. 15.4D), although provided with a flat, narrow crest for the large pectoral and deltoid muscles, was lightly but strongly built, the interior being more or less hollow.

Hind Limbs and Pelvis.—The hind limbs even in the most primitive pterosaurs were relatively short, with everted knees, and they doubtless served to stretch the gliding membrane. The five-toed hind feet were like delicate clawed hands—not at all adapted for running but probably useful in climbing about the steep cliffs of the seashore. The pelvis was relatively small but compact (fig. 15.1B), fastened to the vertebral column by transverse processes of the sacral vertebrae. Especially in the giant *Pteranodon,* the ilia (fig. 15.2) were prolonged forward and formed extended sacral contacts with seven vertebrae; they were also fastened posteriorly to some of the anterior caudal vertebrae.

Beneath the cup or acetabulum for the femur, the pelvis extended downward on each side as a strong plate, affording a base for relatively large muscles that ran out to the under side of the femora, while from the forwardly directed bars of the ilia, muscles converged toward the upper surface of the femur. In flying, these and other muscles were used to stretch the flying membrane outward and backward, and in landing it was probably turned downward to act as a brake and to tilt the wings and head upward. Attached to the lower edge of the ventral plates were a pair of flat bones, homologized by Williston with the "prepubes" of crocodilians, but which are nevertheless true pubes, as shown in Brown's *Protosuchus* (fig. 14.14B). In that case the whole of the posteriorly directed ventral plates are formed from the ischia, which served as an anchor for the tail muscles, while the pubes, displaced downward and outward, tied the parasternum to the pelvis as they do in crocodilians.

Thorax.—The thorax, especially in the larger pterosaurs (fig. 15.2), was expanded in front, doubtless to give room for the very large heart and lungs, but narrowed to the relatively narrow pelvic outlet, so that either the eggs or the newborn young must have been relatively smaller than those of crocodilians.

Skull and Jaws.—The jaws of pterosaurs (figs. 15.1, 2, 3B) were prolonged into a snapping beak, with slender teeth in the earlier forms, but in *Pteranodon* toothless and probably covered with a pointed horny bill. In *Pteranodon* (fig. 15.2) there was also an extremely long, backwardly directed sagittal crest, overhanging the neck and back, possibly assisting in balancing the wind pressure on the stout neck or perhaps aiding in sudden diving after fish.

PROBABLE ORIGIN OF PTEROSAURS FROM PRIMITIVE THECODONTS

The two-arched temporal region and the preorbital fenestrae (figs. 15.1–3) clearly indicate that the pterosaurs are disguised archosaurians, highly adapted for flight. But from which Triassic archosaurs were they derived? In his diagram, "The phylogeny of the archosaurian reptiles," Romer (1935, p. 210) connects the pterosaurs by a broken line with the thecodont common stock that gave rise to all the archosaurs; and this line comes out above between the central pseudosuchians and the phytosaurs. And indeed if we subtract the obvious specializations connected with flight, the residue of pterosaur characters is not inconsistent with Broom's hypothesis that such Triassic thecodonts as *Euparkeria* (fig. 14.3) and *Ornithosuchus* (B) would have made an excellent starting-point for the pterosaurs as well as for all the older archosaurian orders (fig. 12.1).

Even the Triassic phytosaurs, in spite of their pronounced specializations for aquatic life, retained many features which must have been present in the ancestors of the pterosaurs, including the long fish-catching jaws, the dorsal position of the nares, the forward and downward inclination of the quadrate and the upward and backward slope of the postorbito-squamosal bar (fig. 14.2C). The phytosaur vertebrae, girdles, limbs and feet were all far less specialized than those of pterosaurs.

The stretching of the skin on the sides of the neck, body and tail, extending outward to the hands and feet, has been paralleled in several orders of mammals (flying marsupials, flying "lemurs," flying squirrels, etc.), but so far as known it has always been preceded by very active, speedy locomotion combined with good climbing power.

The great elongation of the hands in pterosaurs is paralleled in the fast-running and possibly climbing coelurosaurs, *Ornitholestes* (fig. 14.15B) and *Struthiomimus* (fig. 14.16) but had not begun in the older and functionally bipedal *Procompsognathus* (fig. 15.3C). The relatively small size and poor walking powers of the hind feet of later pterosaurs (fig. 15.2, 3B) are evidently a secondary result of the over-development and great functional value of the flying membrane and wings. So the hind limbs, once the main running organs, became mere accessory hold-fasts.

PTEROSAURS AND BIRDS

The evidence indicates that the pterosaurs represent an older, less perfect series of experiments in transforming primitive, fast-running archosaurs into climbing, skimming and flying pterosaurs. The birds are the results of a somewhat later and far more elaborate and successful series, starting from simpler beginnings but ending to date in albatrosses, penguins, ostriches, pigeons, humming-birds, and thousands of other severally unique types.

The greatest difference between the primitive Upper Jurassic birds *Archaeopteryx* and *Archaeornis* (fig. 15.9), and their contemporary, the primitive pterosaur *Rhamphorhynchus* (fig. 15.2B), was that the birds' wings were already provided with long flight feathers (fig. 15.7), while the pterosaurs' wings were only lateral extensions of the skimming membrane fastened to the rear border of the greatly enlarged fourth digit of each hand.

THE FEATHER AND ITS LONG EVOLUTION

The emergence of feathers was one of the most far-reaching events in the entire history of vertebrate animals, because it made possible the vast system of branching species, genera, families and orders which constitutes the class of birds. In order to see this critical event in its proper perspective, we must revert for the moment to the oldest known chordates, which, it will be recalled, were the shell-skinned ostracoderms. The outer shell or exoskeleton of the cephalaspid division of this group was already highly complex in structure (fig. 6.17) and bone-cells had invaded the stratified basal layer, the vascular middle layer and the cosmine layer of formative cells. The outermost layer in the fossils was of enamel-like hardness and brilliance and devoid of bone-cells.

The evidence clearly indicates that the complex shell of ostracoderms became subdivided on the one hand into the shagreen denticles of sharks, and on the other into the ganoid scales of the older bony fishes. In the embryo fish each scale develops as a single productive organ, comparable with a tooth germ, or a hair follicle, or a feather papilla. Among surviving ganoid fishes the enamel or ganoin layer is secreted by the epidermis. In later fishes the bony base was gradually eliminated and the enamel layer gave place to a horny surface layer growing with concentric rings.

In brief, the shell of adult primitive vertebrates consisted of an epidermal outer layer capable in different forms of producing either enamel or keratin, and a complex bony base well supplied with blood vessels. These carried raw materials into the pulp cavities of the cosmine cells, which in turn secreted the inner and outer layers.

In the earliest known reptiles, the exoskeleton consisted of bony scutes channeled by vascular grooves and capped by horny surface layers, and these two constituents are well represented in the modern crocodilians and some lizards. Both layers were present in the primitive thecodonts and persisted in most of the archosaurs, including many dinosaurs. In the ancestors of the pterosaurs the scales seem to have been eliminated or at least greatly reduced, leaving a naked skin. In the earliest birds the bony base had been entirely eliminated, leaving only the horny epidermal layers. Nevertheless the embryonic feather papilla with its pulp or pith must represent the reptilian scale as a whole.

The Development of the Feather.—This embryonic papilla (fig. 15.6A, B) sinks into the skin from which it arose, forming a follicle or walled pit (D) around it. The minute core of the papilla buds off a vertical series of caps, forming the pith. The horny material forming the feather is secreted by the papilla into the space between the papilla and the follicle. The outermost layer of the papilla forms a thin temporary envelope around the growing spine-like "pin-feather." The shaft or quill of the formative papilla gives off on one side a branch called the after-shaft and a nearby tuft of minute downy branches (E); then it goes on to sprout off on either side a diverging set of barbs (F), which in turn subdivide on the edges into minute hair-like or hook-like barbules (G, J). These elaborate polyanisomeres are indeed more complex in pattern than the feather-like antennae of moths, or the midrib and veins of an ovate leaf. Antennae and leaves, however, are living and growing organs; but feathers, like the shells of molluscs, are dead by-products of daily metabolism.

Feathers and Mollusc Shells as Excretions.—Thus the bird, like the fish or the mollusc, is enclosed in an envelope of its own dead secretions or excretions, which are periodically cast off like deciduous teeth to make way for fresh replacements.

Successive Moults.—The successive crops of feathers are variously adapted as a downy covering for the nestling, as scales on the feet and tarsi, as primary and secondary wing feathers, wing coverts, tail coverts, tail feathers, etc., etc. Nor should one omit reference to the elaborate patches of color on certain feathers, which culminate in such complex and curious patterns as the ocelli or "eyes" of *Argus* or of peacock feathers.

In short, the feathery exoskeleton of birds, evolved ultimately from the many-layered shell of ostracoderms and more nearly from the much simpler reptilian bone-based scales, became streamlined externally, highly differentiated locally on the head and body and varied in accordance with the mode of life, phyletic relationships, species, subspecies, age and sex of the individual. The elimination of bony scutes and the emphasis of horny feathers not only lightened the body-covering but provided an insulation against external cold and heat and an effective blanket for the conservation of internal heat.

The Feathery Covering as a Heat Regulator.—Moreover, owing to the presence of surface muscles and to the overlapping of the feathers on the body, if the internal temperature became uncomfortably high, the bird could raise its feathers, somewhat as the slats can be opened in a venetian blind, to let some of the excessive heat escape.

THE BONY WING-FRAME IN PTEROSAURS AND BIRDS

To continue the contrast between the oldest Jurassic bird and its contemporary pterosaur, the possession of long flight feathers in the bird greatly increased the length and area of the supporting wing surface and made it necessary to provide the wing with a strong bony frame. The strengthening of the wing frame had barely begun in *Archaeopteryx* and *Archaeornis* (fig. 15C), whose long hands were basically identical with those of the running dinosaurs, *Ornitholestes* (fig. 14.15b) and *Struthiomimus* (fig. 15.8B). In the typical bird wing (D) are three distinct fingers, the first, second and third; the first is slightly divergent, the second much longer, the third shorter than the second. The middle of the third metacarpal is bowed a little away from the second, but coalesced with it at the upper or proximal end. No doubt the digits in *Archaeornis* were tied together by muscles, tendons, ligaments and skin, as in reptiles. The impressions of the long wing feathers (fig. 15.7) preserved in *Archaeornis* show that they were attached to the back of the hand and forearm, probably by ligament, as in recent birds. In the latter, strengthening of the wing bones (fig. 15.8D) is effected (a) by fusion of metacarpals II and III, (b) by elimination of one or two of the small phalanges, and (c) by marked enlargement of the proximal and distal phalanx of digit 2. The small vestige of the metacarpal of digit 1 supports the alula (little wing) or bastard wing. Thus *Archaeornis* retained the essential features of the three-toed grasping or climbing hands of the coelurosaurian dinosaurs, but in modern birds (except the hoatzin) only the thumb supports a claw and a little wing, while the second and third digits are immovably attached to each other and embedded in the base of the wing.

In correlation with these features, the feathered wing skeleton of a pigeon (fig. 15.9), representing a rather central type of bird, is not nearly as long, in comparison with the distance between the sockets for the pectoral and pelvic appendages as was the skin-wing of the primitive pterosaur *Rhamphorhynchus* (fig. 15.1), which was stretched on the back of the huge fourth digit and on the side of the body. But even more important than the differences in length of the wing was the fact that in the pterosaur the main joint in the wing was between the enormous first phalanx of the fourth digit and the fourth metacarpal, whereas, at least in the modern bird and apparently also in *Archaeornis,* the main joint was between the proximal end of the fused carpo-metacarpus and the radius and ulna.

Contrasts in Wing Folding.—Thus in walking or resting the bird-wing can be folded up on its N-shaped jointed frame and tucked in on the sides and back (fig. 15.9); whereas the main part of the pterosaur wing could only be bent sharply upward and backward at the base of the fourth digit (fig. 15.1), and if progression on the ground was at all possible, the elbows must have been bent outward and the forearms and metacarpus used as crutches, the wings either arched above the back or trailing in very exposed positions. Possibly the pterosaurs rested mostly on rocky surfaces, with the grasshopper-like elbows and knees bent and ready for a sudden spring into the air. The typical bird, on the contrary, depends only on the hind limbs for walking, running, hopping and for the initial leap into the air (fig. 15.5B).

Contrasts in Breast-bone and Wing Stroke.—The bird is also far superior to the pterosaur in the development of an enormous, deeply keeled breast-bone (figs. 15.9, 11), extending backward and downward beneath the capacious lung, chest and abdomen and serving as a base for the huge flying muscles, the pillar-like coracoids and the blade-like scapulae.

Especially noteworthy features are: (1) the "pneumatic foramen" by which the humeral branch of the air sacs enters the cellular interior of that bone; (2) the pulley-like "foramen triosseum," between the scapula, the coracoid and the humerus (fig. 15.11A2), which permits the tendon from the pectoralis minor muscle to pass upward and around to be inserted on the dorso-posterior surface of the upper end of the humerus; this arrangement gives propulsive power to the upstroke of the wing (fig. 15.5C), the downstroke being effected by the pectoralis major, attached to the large delto-pectoral crest (fig. 15.11).

SACRAL STRONG-BOX AND BENT-SPRING HIND LIMBS

In standing (fig. 15.9) or running the bird sternum and pectoral limbs are suspended by the ribs from the vertebral column; this in turn transmits its load to the hind limbs through the very extensive ilio-sacral strong-box and (figs. 15.11, 13) to the cup-like acetabula, which are the sole pivots, as in all bipeds. In all this region the typical bird far excels the pterosaurs and even the coelurosaurians in adaptations for combining lightness with sufficient strength, with the addition of ample room for the much larger load of food and eggs.

The bird pelvis (fig. 15.15C, D, F) recalls that of the ornithischian dinosaurs (E) in the backward turning of the long rod-like pelvis, which is parallel to and extends behind the anterior border of the ischium. But this is probably only a parallelism with the dinosaurs. More in detail, the strong, bent-spring hindlimbs include: (1) the cylindrical hollow femur; (2) the firm hinge-like knee (fig. 15.9B), well tied with ligaments; (3) the beam-like tibio-tarsus, triangular at its upper end to support the femur, columnar in the middle to resist bending, and hinge-like at the lower end, where it is united with the pulley-like upper half of the old tarsus; (4) the straight cannon-bone, of three coalesced metatarsals (II, III, IV); (5) the backwardly directed first digit, which acts as the rear fork in grasping and perching and as a spring-brace in landing and hopping; (6) the diverging digits 2, 3, 4, which afford needed friction against the substratum in perching, walking and hopping. Giant bipeds (ostrich, moa) reduce wings, lengthen tibio-tarsus.

COMPLEX VERTEBRAL COLUMN

The single ball-like occipital condyle of birds (fig. 15.20) is essentially reptilian in origin, as are also the atlas-axis complex. The remaining neck vertebrae bear saddle-shaped articular surfaces and appropriately placed zygapophyses, permitting wide ranges in twisting and curving the neck with minimal danger of self-dislocation (figs. 15.12C1, C2). Air sacs from the lungs, running along the neck region, help to mold the sides of the neck vertebrae, while the short cervical ribs, transverse processes and low neural arches give insertion to complex zigzagging muscles and tendons. Behind the neck the progressive immobilization of the thoracic vertebrae culminates in the very long lumbo-sacral tube or synsacrum (fig. 15.14) to which the sacroiliac flanges of the pelvics are attached.

The essentially reptilian caudal vertebrae in *Archaeopteryx* and *Archaeornis* were very numerous and tapered gradually to the end of the long tail, to which were attached the stems of the large, outward-and-backwardly divergent tail feathers.

MANY-JOINTED TAIL *VERSUS* TELESCOPED FAN-TAIL

In flying straight forward, the long and wide tail of *Archaeornis* (fig. 15.9A) offered additional support to the body and it may have been useful in effecting a gentle landing; but the muscles operating this tail must have been slender and weak and it must have been much less easy to manage than the "fan-tail" of modern birds (fig. 15.5B1). In the latter, the strong fan-like flight-feathers are firmly mounted on the compact, wedge-like ploughshare bone (fig. 15.13), which has been molded from the coalescence of a number of caudal vertebrae and by the outgrowth of a dorsal and lateral ridges. By this arrangement the fan-tail can be quickly drawn downward and forward in landing (fig. 15.5B2) or sharply raised, as in the display antics of "prairie chickens"*. Wing downstroke also eases shock of landing; tail downstroke adds to upthrust.

PRIMITIVE BIRD-SKULL COMPARED WITH BASIC ARCHOSAURIAN TYPE

The skull of *Archaeornis* (fig. 15.9A), as restored by Heilmann (1927), was essentially like those of such basic archosaurians as *Euparkeria* and *Ornithosuchus* (figs. 14, 3, 4), except that: (1) the eyes were larger, (2) the supratemporal opening smaller, (3) the preorbital face shorter, the nasal region being more delicate, (4) the teeth were smaller and did not extend so far back on the jaws, (5) the front half of the jaws was more slender, (6) the braincase was wider.

THE MODERNIZED BIRD-SKULL

The skull of typical modern birds (figs. 15.18B, B1), such as the fowl and pigeon, have advanced beyond *Archaeornis* in: (1) the great expansion of the orbits and braincase; (2) reduction or elimination of the olfactory parts of the skull and brain (fig. 15.22); (3) loss of the postorbito-squamosal bar and consequent merging of the upper and lower temporal openings; (4) loss of teeth and concomitant development of a horny beak.

PALATAL TYPES

The palatal region of *Archaeornis* is not known but presumably it would have retained a less specialized and more archosaurian arrangement of the vomer, palatines and pterygoids than those of modern birds. Even in the supposedly older or palaeognathous palates (fig. 15.20bD-F, 20cG, H) of ratite birds (emu, cassowary, moa, kiwi, ostrich, tinamou) the large internal nares were located behind the large beak and bounded dorsally by backward prolongations from the vomer, laterally by posterior extensions of the palatines. The anterior processes of the pterygoids had been excluded from the inner borders of the nares by the vomers and palatines; they had also lost their transverse or lateral processes and the ectopterygoid bones had disappeared; however, the pterygoids retained their contacts with the prominent basipterygoid processes and their bars that ran backward and outward to brace the quadrates were well developed. In the neognathous or modernized palates (fig. 15.20cI, J, K) the pterygoid bars articulate somewhat movably with oval facets on the basisphenoid. In the schizognathous subtype (I), found in most birds, the maxillo-palatine plates do not unite in front across the midline, as they do in the desmognathous palate of storks, ducks and geese (J), etc.; in the aegithognathous type of passerine birds the broad vomer is truncated in front. In the highly specialized palate of the parrots (K), derived from the schizognathous condition, the huge upper beak can be pushed upward (fig. 15.21B) by thrusts received from the palatines and rod-like pterygoids, which in turn are pushed forward by the quadrates when the jaw is depressed by the depressor muscles (fig. 15.8C, C1).

Modern birds have delicate stapedial rods and well developed drums and they are well known to have a good sense of hearing. Nor should the complicated voice organ or syrinx at the lower end of the trachea pass wholly unnoticed here, although it is only a by-product of the tracheal system and hardly to be considered in connection with skeletal structure. However, it is an "emergent" or once new organ, highly characteristic of birds and it is almost as important in their lives as the larynx is in man. By whatever steps its complex vibrating membranes and musculature may have arisen, or exactly how they function in detail, it is clear that the voice of a bird, like that of frogs and tree toads, responds to the bird's varying feelings.

SYNOPSIS OF THE ORDERS AND SUBORDERS OF BIRDS

The protean modifications of body-form, beaks, wings, feet, and plumage, soft anatomy and skeletal details in birds are treated in the vast literature of ornithology; but here it may be appropriate to refer very briefly to representatives of the main orders and suborders and to their possible relationships, under the following headings (fig. 15.25):

1) Cretaceous "toothed birds" (ODONTORNITHES). The famous *Hesperornis* was a highly specialized marine bird, with divergent, frog-like hind limbs and many close resemblances in skull and skeleton to the modern divers (Heilmann). The flat, crestless sternum and vestigial wings are merely convergent to the ratites (No. 2, below). Teeth retained from older Jurassic stage. Another Cretaceous bird, *Ichthyornis,* had a high keeled sternum, unreduced, complete wings, a generally tern-like form and is rightly made the type of another order (ICHTHYORNITHES). The toothed jaws may possibly not belong here but even if so, the bird may be related to the terns and gulls.

2) Ratite birds (RATITAE), including a wide range of flat-breasted birds with wings in various stages of reduction and usually with powerful hind limbs. All have the "palaeognathous" palate (fig. 15.20aD-F, 20bG, H). The wings are least reduced in the tinamous, which are still able to fly and may be related remotely to the fowls. At the other extreme, the kiwi is a specialized long-billed, worm-eating, pygmy survivor of the giant moas and lays relatively huge eggs, as they did. Resemblances between moas and running dinosaurs, cited by Lowe (1928), are assuredly convergent.

3) Archaic water-birds, including divers or loons (COLYMBI) and grebes (PODICIPEDES), superficially duck-like, but with lobed, not webbed, toes and schizognathous palate.

4) Tube-nosed marine birds (TUBINARES), including petrels and albatrosses (fig. 15.21A), the latter famous for their very wide wing-spread and soaring flight.

5) Fishing birds, with four long, webbed toes (STEGANOPODES), including (a) tropic birds (*Phaëthon*), with very long wings and two extremely long central tail quills. They capture fish by plunging into the sea, often diving from a great height. (b) Gannets, boobies, cormorants, frigate birds; (c) pelicans; (d) darters or snake birds. The palate is desmognathous.

6) Penguins (SPHENISCHI). Extremely spe

cialized for flying underwater, the strong wings changed into narrow pointed paddles covered with secondarily scale-like feathers. High keel on sternum. Excessively short thick metatarsals tending to become separate and converging toward those of carnivorous dinosaurs. Skull completely avian, schizognathous. Vertebrae equally avian, tail with ploughshare bone (pygostyle).

7) Waders (GRESSORES). Herons, storks, ibises, spoonbills, flamingoes. Long-necked, long-legged waders, typically walking about in swamps and marshes, never swimming or diving. Three front toes partly webbed, except in herons; rear toe long in herons, vestigial in flamingoes (PHOENICOPTERI). The latter related on the one hand to screamers, geese and ducks, and on the other to the eagles and vultures.

8) Geese, ducks, mergansers (ANSERES). Short-legged, more or less web-footed; the geese more terrestrial, the ducks thoroughly aquatic and good divers and swimmers. Desmognathous.

9) Screamers (PALAMEDEAE). Large, strong-winged, somewhat goose-like South American birds, with two strong spurs on each wing. Living in swamps and feeding on leaves and seeds of aquatic plants. Toes long, not webbed. Palate desmognathous. Possibly representing a fairly primitive, moderately long-legged stock, from which may have been derived the birds of prey on the one hand, and the ducks and geese on the other. Desmognathous.

10) Falcons, eagles, vultures (ACCIPITRES). Mostly soaring, raptorial birds with strongly grasping feet; legs typically short but long in the secretary bird. Possibly derived from the stem of the screamers. Desmognathous.

11) Fowls (*Gallus*), pheasants (*Phasianus*), grouse (*Tetrao*), curassows (*Crax*), megapodes, peacocks (*Pavo*), guinea fowls (*Numida*), turkeys (*Meleagris*), etc. (GALLI). Mostly ground-living birds, able to spring up in a short flight but not typically strong fliers. Males often pugnacious and striking with their long tarsal spines. Pheasants and peacocks noted for highly elaborate color patterns in males. Palate schizognathus. The arboreal hoatzin (*Opisthocomus*), famous for having claws on its thumb and index finger in the nestling, is probably related to the fowls, although often assigned to a distinct order (OPISTHOCOMI). Its adult sternum is excavated in front to accommodate the greatly enlarged gizzard.

12) Bustards (OTIDES), button-quails (Hemipodii or TURNICES), cranes and trumpeters (GRUES), rails (RALLI), sun-bitterns, etc. This large assemblage (Order GRUES) is connected with the next group, comprising

13) the plovers and sandpipers (CHARADRII), auks and murres (ALCI), gulls and terns (LARI) of the order LAROLIMICOLAE. The limbs are usually long or of moderate length, the feet usually not webbed (except in auks, gulls and terns). All are schizognathous. The giant extinct *Diatryma* from the Lower Eocene of North America possibly belongs in this assemblage. It convergently resembled the ratite birds in the great enlargement of the legs and vestigial condition of the wing.

14) Pigeons, doves, crowned pigeons, sandgrouse (Pteroclidae), dodo (*Didus*), etc. (COLUMBAE). The pigeons are very central birds and may be near the beginning of the picarian and perching birds. The palate is schizognathous.

15) The cuckoos (CUCULI) and plantain-eaters or turacos (MUSOPHAGI) seem to stand near the base of the picarian assemblages. In the typical cuckoos digit I can be placed wholly under, and digit IV wholly or partly under, a branch, and digits II and III on top of it. The road-runner (*Geoccocyx*), although running well on the ground, retains this perching adaptation in its feet. Bill, typically rather slender, becomes short, high and almost parrot-like in the ani (*Crotophaga*), long and almost toucan-like in the channel-bill (*Scythrops*). The brilliantly colored plantain-eaters have arched, serrated bills and erectile crests.

16) The parrots and macaws (PSITTACI), with about 500 species, have huge and excessively specialized movable beaks (fig. 15.21A) and grasping feet with two toes before and two behind.

17) The colies or mouse-birds (COLII), take their name not only from their ashy plumage but also from their quick, mouse-like movements on the branches of trees. All four toes are directed forward and they can hang head downward. They have strong, short beaks and erectile crests on their heads.

18) The night-jars, night-hawks, frogmouths, goatsuckers and whippoorwills (CAPRIMULGI) are crepuscular birds, very active flyers, with strong wings; they capture moths and other insects on the wing; and like swifts, they have small and weak legs. The middle toe is very long, with a serrated comb on its inner edge. The gape is extremely wide and owl-like, and fringed with hair-like feathers.

19) The owls (STRIGES) are justly famous as night hunters; they are well camouflaged in the moonlight, their great eyes are sensitive to the slightest movement and their ears are attuned to the faint rustlings of a mouse. With their soft downy plumage, they can either fly away or alight in silence, and even their toes are feathered. Nor do they fail to strike at the right second with their sharp beaks. Thus in a single night they often make a big haul of small mammals, reptiles, birds and large insects. The owls used to be classed with the birds of prey (No. 10) but they resemble the night-hawks (No. 18) and allied forms in other features. Among their own noteworthy specializations are the bell-like form of the eye and the marked contractility of the pupil, and the development of cavities between the inner and outer bony layers of the skull, which communicate with the inner ear and act as resonators.

20) The swifts (CYPSELI), like the swallows (Alaudidae) of the passerine order, often have long pointed wings and forked tails and are extremely quick in catching insects in the air. Moreover, they have very small beaks, wide gapes and weak feet. The "chimney swallow" of the United States is really a swift. The swifts, however, use the saliva from their enlarged salivary glands in constructing pocket-like, cradle-like or pouch-like nests; while the swallows make nests of mud, straw and feathers, or dig holes in banks.

21) The typical humming-birds (TROCHILI) include the smallest of all birds, one species, *Calypte helenae* of Cuba, being only 2¼ inches long; the largest is *Patagona gigas* of the Andes, 8½ inches long. Humming-birds excel all others in the extreme rapidity of their wing strokes, which enables them to hover in the air while they probe the flowers with their needle-like bills. Or in an instant they dart away and disappear. Their tiny skeleton bears a relatively enormous pectoral crest, the base for their thick wing-muscles; yet their minute feet, with three toes in front and one behind, easily support them when they rest on a twig. Nor do they fear to dart at any invader when defending their nests and young. In the jewel-like brilliancy of their plumage they excel even the birds of paradise. They are classified with the swifts in the order Micropodiformes but formerly both were referred to the order Passeres (perching birds).

22) The brilliant trogons (TROGONES), including the famous quetzal of Central and South America, and many others, have grasping feet, digits I, II, being opposable to III and IV as in the toucans.

23) The puff-birds and toucans, jacamars (GALBULAE) and woodpeckers (PICI), notwithstanding wide differences in appearance and habits, are referred to a single order (PICIFORMES). The woodpeckers alone include some 350 species. The habit of using their sharp bill with incredibly rapid strokes to pierce the bark and wood of trees has resulted in the strengthening of the neck and skull, and in the development of a bony curved spring from a branch of the hyoid which is connected with the tongue.

24) The kingfishers (Alcedinidae), todies (Todidae), motmots (Motmotidae), bee-eaters (Meropidae), the Coraciidae or rollers (so named from their rolling and tumbling flight), and possibly the hornbills (Bucerotidae) are all referred to the order CORACIIFORMES.

But if the picarian assemblage (Nos. 15–24) includes such an immense array of diversely specialized birds, what shall be said of the far vaster order PASSERES with its great multitude of families and thousands of species? The group is defined by technical characters, including the precise arrangement of the flexor and extensor tendons of the four toes (three in front and one behind, and all on the same level), which, as in most birds, automatically grasp the perch when the bird bends the ankle in settling down. The palate is aegithognathous (p. 318); the sternum has only a single rear notch on each side. Other characters include the number of secondaries (6+) in the wings and of retrices (12) in the tail, the absence of certain muscles, etc., and the helpless state of the young. The lowest family of the passeriform order, namely, the sparrow-like broadbills or Eurylaemidae, retain primitive features from some earlier "picarian" group, and the swifts and humming-birds also in some ways approach closely to the passeriform grade.

Here belong the great majority of birds in northern Europe and North America. Ornithologists have divided them provisionally, on the basis of the structure of the syrinx, into: SUBCLAMATORES, including only the broadbills; CLAMATORES, including the pittas, tyrants, manakins and cotingas, wood-hewers, ovenbirds and ant-thrushes; SUBOSCINES, including only the lyre-birds and the OSCINES or song birds, including (a) American songbirds; (b) Old World insect-eaters; (c) Old World finches, and (d) crows, orioles and birds of paradise.

THE FOSSIL RECORD OF BIRD LIFE

In conclusion, the fossil record of the evolution of birds is very incomplete and fragmentary as compared with that of reptiles and mammals; but it appears to warrant the following outline.

Among the thousands of footprints in the Triassic strata of the Connecticut valley, true bird tracks are conspicuously absent, nor are there any known traces of Triassic birds in any part of the world. But numerous bird-like preadaptations were then being evolved, especially in the skull, vertebral column, hands and feet of coelurosaurian dinosaurs and in the pelvis and hind limbs of ornithischian dinosaurs.

The Upper Jurassic *Archaeopteryx* and *Archaeornis* retained many basic archosaurian features but had reached the avian grade in possessing feathers. These complex exoskeletal polyanisomeres are a unique creation in the class Aves and they have predetermined innumerable changes in the skeleton and entire economy of birds.

The length and close appression of the metacarpals II, III, IV in *Archaeornis* and the presence of a mesotarsal, hinge-like joint were clear marks of not too distant relationship with the Triassic coelurosaurs and plainly indicate that the ancestral birds were functionally bipedal runners on the ground before they climbed up into the trees with their well clawed hands and feet and began to skim down from the trees by means of their very large flight- and tail-feathers.

The long-winged, lizard-tailed Jurassic birds would seem to have been devoid of special adaptations for swimming and would have been an easy prey for the mosasaurs, crocodilians and other hungry hunters. But some of their descendants or relatives began to shorten up the tail and develop a ploughshare bone and to strengthen the hind limbs for swimming. The culmination of such a specialized aquatic line is seen in the loon-like *Hesperornis* of the Upper Cretaceous, in which the long but vestigial humerus and even the very large, although keelless, sternum, indicate that its ancestors once had long wings, while the strong spreading hind limbs and lobate feet are distinct adaptations for aquatic life. But again, the coalesced metatarsals and long mesotarsal joint are souvenirs of primitive archosaurian runners and the socketed teeth even recall still more remote thecodont ancestors.

The Cretaceous *Ichthyornis* has evolved far in the opposite direction, attaining a high-keeled sternum and tern-like wings.

From some toothed Cretaceous stock related to *Hesperornis* but much less specialized may have been derived the archaic water-birds (loons, grebes) of the present time; while other forms, allied with *Ichthyornis*, would seem to be good structural ancestors for the gulls, which are known to date from the Eocene.

The earlier history of the ratites is lacking, except for one record in the Upper Cretaceous (*Chaenagnathus*). The reduction of the wings and of the keel on the sternum has been independently acquired in flightless cormorants, flightless rails and others; even the retention of the basipterygoid processes, the rod-like form and convergence of the opposite pterygoids and some of the other features of the palaeognathous palate may have also been acquired independently, as between the ostriches and the cassowaries; but other skeletal and anatomical features taken as a whole suggest that there is more than mere parallelism in the resemblances of the emus with the cassowaries, of the kiwi with the moas, of the tinamous with the fowls.

Romer's Vertebrate Paleontology (1935, pp. 602–610) provides the fossil records cited below.

Relatives of the cormorant-pelican order date from the Eocene (*Prophaeton, Protopelecanus*). The penguins date from the Oligocene and Miocene of South America and one of the tube-nosed birds is recorded from the Oligocene of Europe. Ancestral herons and storks also occur in the Eocene; while forerunners of the flamingos appear in the Upper Cretaceous. Such facts indicate that the adaptive branching of the marine and shore birds was well under way before the close of the Cretaceous and continued in Eocene and later times. The same is true as to the birds of prey, the fowls and pheasants, the cranes and rails and the plover group.

The earlier history of the pigeons is not known. But the owls date from the Eocene, the woodpeckers possibly from the Eocene, the cuckoos are recorded from the Oligocene, the swifts from the Oligocene, and the parrots from the Miocene. But even the last 25 million years may have given time enough for the differentiation of a good many of the 500 species of parrots. In the order Passeres, many families were differentiated in the Miocene and this seems to be the latest (early and mid-Tertiary) large-scale branching in the class Aves. Thus in the later Mesozoic the birds were in the midst of their long period of expansion, while the dinosaurs were nearing their relatively sudden extinction.

In conclusion, the basic bird heritage favored small size, and even "giant" birds were minute compared with thirty ton dinosaurs or fifty ton whales.

PART FOUR

(*Chapters XVI–XXII*)

THE RISE AND BRANCHING OF THE MAMMALS

CHAPTER XVI

FROM SLINKING REPTILES TO RUNNING MAMMALS: A CRITICAL STAGE IN EVOLUTION

Contents

CHAPTER XVI

FROM SLINKING REPTILES TO RUNNING MAMMALS: A CRITICAL STAGE IN EVOLUTION

DIVERGING ORIGINS OF BIRDS AND MAMMALS

A BIRD might be defined as a feather-winged biped, except that a few birds have lost their wings. And a mammal might be defined as a warm blooded vertebrate animal with a hairy skin, except for the whales which have almost completely lost the hairs. Such scant definitions give but a poor suggestion of the things they define. Both birds and mammals have a history and neither can be grasped without reference to their respective origins and diverse evolution. Both birds and mammals are warm blooded animals and have been derived from cold blooded lizard-like reptiles; so that there has been considerable parallelism in end results, which have been achieved by similar means. However, the older heritage and derivation of birds and mammals were widely different. The immediate ancestors of the birds were "two-arched" or archosaurian reptiles related to the crocodilians and dinosaurs; the immediate ancestors of the mammals were "single arched" or therapsid reptiles, whose only modern representatives are the mammals themselves. Both bird-ancestors and mammal-ancestors lived in the Permian and Triassic periods and in all probability belonged to quite different families or suborders of the "stem reptiles," or captorhinomorphs (fig. 12.1).

THE THRONGING EVOLUTIONARY LINES OF MAMMAL-LIKE REPTILES

Cross sections of the thronging evolutionary lines of mammal-like reptiles and their predecessors are known from: (1) various levels of the Permo-Carboniferous "Red Beds" of Texas and adjacent states (fig. 16.2), (2) the long sequence of fossil bearing horizons in the Karroo series (fig. 16.7a, b) of South Africa, (3) a similar series in Russia, and (4) a less complete sequence in South America.

THE CRAWLING PELYCOSAURS

The oldest known stage is abundantly represented among the pelycosaurs (figs. 16.1, 2) of the American Permo-Carboniferous (*cf.* the definitive monograph by Romer and Price 1940). These pelycosaurs range from fairly large lizard-like forms about four feet in length up to the eleven foot predator named *Dimetrodon* and the giant herbivore *Cotylorhynchus* estimated to weigh about a third of a ton (Romer, 1935, p. 276).

Primitive Skull Type. — *Varanosaurus* (fig. 16.2) about five feet long, had a narrow skull triangular as seen from above, with a large long face and nearly straight jaw lines. The marginal teeth were socketed and numerous, compressed with sharp edges, and but little regional differentiation, except for the larger size and canine-like form of the eighth and ninth upper cheek teeth—the forerunners of the greatly enlarged canines of the larger carnivorous pelycosaurs and therapsids. The orbits were fairly large their upper borders but slightly raised above the dorsal contour of the nasal region. Such a description would apply to many other reptilian skulls of different orders, but here the decisiv feature was the presence on each side of a single temporal opening, equal to the lower one of the two arched reptiles and lying beneath the postorbitosquamoid bar (fig. 16.4bB). The quadratojugal is less conspicuous than in many dinosaur skulls and is largely covered laterally by the squamosal and jugal (fig. 16.9A). The quadrate-articular joint lies nearly in line with the cheek teeth, as in other carnivorous reptiles; this permitted the head to be raised while the lower jaw was depressed, thus greatly increasing the gape. In *Ophiacodon* (fig. 16.2A) the skull was essentially similar to that of *Varanosaurus* but the single temporal opening was smaller, and the upper jaw vertically higher, the upper tooth row curved downward and the lower one upward. Thus to some extent, *Ophiacodon* and allied forms were intermediate between the straight-jawed *Varanosaurus* and the higher structural stage *Dimetrodon*.

The skull of *Dimetrodon* (fig. 16.9A) differed from that of *Varanosaurus* chiefly in the marked emphasis of the height of the upper jawbone (maxilla), in the festooning or downward curvature of the upper row of cheek teeth, in the development of large canine-like teeth in the premaxillae and anterior part of the maxillae. Similarly the lower tooth row was curved downward, which further improved the ability to grasp and tear a large victim. Partly as a result of the downward dip of the dentary-bone its rear upper border was curved upward and had begun to develop an "ascending process," partly overlapping the surangular (figs. 16.9A, 36A). This was also a preadaptation for the further evolution of this part in the later mammal-like reptiles and mammals. Romer and Price (1940, pp. 88–93) give clear evidence that: (1) in the process of tooth replacement, the pelycosaurs agreed with the Crocodilia, each new tooth-germ arising in the resorbed pocket on the inner side of the tooth then in place; (2) that there were two parallel series of teeth, one outer, one inner, arranged in alternating rows, each tooth of both rows being replaced by one of the same vertical set; (3) that there were several successive waves of replacement, passing from the rear forward.

The pterygoids (fig. 16.3B) of pelycosaurs retained all five branches of the captorhinomorph archetype (fig. 12.29). The stout transverse bars of the pterygoids bore a row of small, sharp teeth which must have been useful in piercing and breaking the prey. The rear branches of the pterygoid diverging outward to the quadrates were exceptionally strong and formed paired thrust-beams of the whole bony palate.

The vertebrae of the more primitive short-spined pelycosaurs (*e.g.*, *Varanops*, fig. 16.1A, and *Ophiacodon*, fig. 15.5E, E1) were very central and primitive in general type. In the neck the atlas and axis of *Dimetrodon* (figs. 16.5A; 16.6aA) were ideally primitive, with the centra deeply cupped by the notochord and the pleuro centrum (*plc*) of the atlas still unreduced and functioning as such; whereas in the mammal-like reptiles and mammals (fig. 16.6B) it tends more and more to coalesce with the centrum of the axis, while the neural arches of the atlas grow downward and join below in the mid-ventral line. Moreover there were small but well developed intercentra on all the neck vertebrae and the proatlas (fig. 16.6aA), possibly representing the small neural arch and spine of the "occipital vertebra," was still present.

In the larger dimetrodonts, attaining the body size of large alligators, the successive neural spines rise on a steep gradient (fig. 16.1C) from the second cervical to the mid-dorsal vertebra, where they may be about 27 inches high. Behind this they diminish in a descending curve to the anterior caudal region. This structure much exceeds in relative height, as compared with total length of body, the dermal crest on the back of the iguanid lizard *Basilosaurus americanus*, or the one on the tail of the "sail-tailed" aguanid lizard *Lophurus amboinensis*. In both these reptiles the crest is supported by the elongated spines of the vertebrae.

Various suggestions have been made as to the function of the bony crests in the pelycosaurs; but in the lizards, at least, the crest may perhaps serve as specific recognition marks or even for concealment. The bony spines of pelycosaurs imply an exuberant calcium metabolism rate and must have been covered with dermal tissue, as they doubtless were also in the carnivorous dinosaur *Spinosaurus*.

In another family of pelycosaurs, the Edaphosauridae, which likewise date back to the Pennsylvanian (Upper Carboniferous) the thick spines bear on both sides short transverse sprouts with rounded tips, all of which may have supported a mass of collagenous tissue enclosed in a tough leathery hide. This structure also indicates very prolonged and localized secretion of calcium somewhat as in spinose sea shells. As Romer intimates (1940, p. 276), the huge expanse of this membrane-covered "sail" greatly increased the area that, in unequal degrees, absorbed and radiated the sun's heat and the body heat. Agamid lizards, at least, are known to be subject to heat-stroke, and such a huge heat-absorber and radiator as that of the larger pelycosaurs must have been a responsibility as well as an asset. If so, they would probably react accordingly, by turning the minimum absorbing surface to the sun when their body temperature was too high or perhaps by digging themselves into the warm desert sand at night. It is also noteworthy that in the supposedly aquatic pelycosaur *Casea* the neural spines were extremely short. Such "speculations" may be worthwhile if they lead to further insight into the numerous factors of calcium metabolism in relation to local swellings or spines either in the internal skeleton or its outer covering.

Edaphosaurs with Cutting and Crushing Dentition.—The skulls of edaphosaurs were very small in proportion to their great rotund bodies. The skull was basically pelycosaurian, but on the strongly braced palatine and pterygoid bones were too large oval plates (fig. 16.3C) inclined obliquely downward, studded with small conical teeth. They were opposed by similar but inwardly inclined plates on the inner sides of the coronoid bones of the lower jaw. These crushing plates at first sight suggest those of the shell-crushing sparid fishes as do also the robust construction of the temporal region and lower jaw. Williston (1925, p. 233) accordingly regarded the edaphosaurs as "sub-aquatic or terrestrial invertebrate feeding reptiles"; but Romer (1935, pp. 275–276), perhaps because of the oblique position of the oval plates, calls them "masticating plates" and speaks of *Edaphosaurus* and *Casea* as "apparently herbivorus."

In *Casea* the very short round skull was close to that of the edaphosaur *Cotylorhynchus* (fig. 16.2M) with few, thick, very marginal teeth; there was a wide spread of small teeth on all the bones of the palate and a very narrow dental strip on the coronoid of the lower jaw (Williston, 1925, p. 51, fig. 42). The pineal foramen was exceptionally large and the spines in the neck and back very short; the ribs were very long and outwardly curved forming a huge thorax. The ilium had a helmet-shaped crest directed forward to give support to the muscles of the very thick thighs.

Representatives of the three suborders of pelycosaurs, namely of the carnivorous ophiacodonts and sphenacodonts and of the herbivorous edaphosaurs were already in Europe and North America by the Upper Carboniferous times (Romer, 1935, p. 600). They were abundant and diversified in the Lower (and Middle) Permian of North America and to some extent of Europe.

THE ANCIENT KARROO OF SOUTH AFRICA AND ITS VAST FOSSIL RECORD

Before tracing the adaptive and genetic branching of the extinct South African mammal-like reptiles and their varying progress toward the mammalian stage, let us first try to set forth a general geological picture of the Karroo region in which lie buried countless thousands of fossil records of reptilian life.

The "Seven Wonders of South Africa" are well advertised to all foreign visitors and tourists. But by far the greatest wonder, beside which the others shrink into minor features, is unknown to the tourist bureaus and therefore unknown to the tourists. So far as the travelling public is informed it is almost in vain that for a century past South African geologists and palaeontologists have devoted their long and productive lives to exploring the endless but fascinating problems relating to the sequence of ancient world events in South Africa, for no tourist guide tells his passengers the following story.

Table Mountain is a welcome and impressive sight as the traveller's ship at last comes to rest in the harbor of Cape Town. Its steep escarpment shows that it is a very hard rock, which is resisting well the eroding tooth of time and the encroachment of the ocean, which has partly cut it off from the mainland. When we take the train for the Transvaal, running from Cape Town northeastward, we first pass by the remnants of granitic intrusions (the "younger granites" of the geologists), which are now rounded but fairly steep, rather small mountains. Soon we are out on fertile lowlands, resting on the hard floor of the southwestern extension of the very ancient Transvaal-Nama rock system (fig. 16.7). Abutting against and resting on the eastern wall of this extension is the western end of the enormous boot-shaped or L-shaped Cape System. Table Mountain itself is a separate outlying fragment of the Cape System. As the train travels northeastward, it passes by successive levels of the Table Mountain series. The latter lies at the base of the Cape System and comprises about 5,000 feet of quartizitic sandstones and shales, with a tillite band. The tillite is composed of mudstone containing scattered pebbles and boulders of various kinds of rock, some of which are striated after the fashion of glaciated stone; these, together with other evidence, indicate the presence of moving ice (Union of South Africa Official Year Book, 1937, no. 18).

Resting upon the Table Mountain series are the 2500 feet of the Bokkeveld series of shales and sandstones. This contains a fairly abundant marine fauna of invertebrates which are closely related to the Lower and Middle Devonian faunas of South and North America and less closely related to the Lower Devonian of Europe (*op. cit.*, p. 13). Here then is a datum plane in time of an estimated age of over 300,000,000 years.

The older, or pre-Cape series of rocks, are classified in ten major series; these stretch downward, as it were, through the endless reaches of pre-Devonian time and their maximum thickness, ranging from 500 to 25,000 feet, if totalled would exceed 140,000 feet. The oldest of them, in Swaziland and elsewhere, may date far back to Archaean (Archaeozoic) time (*op. cit.*, chapt. i, p. 8).

As we travel northeastward around and across the mountains of the Cape System, passing through the gorge of the Hex River, we come to the outer edge of the Karroo System about one hundred and twenty miles northeast of Cape Town. Here we swing more to the east and slowly climb to Laingsburg, passing along the thin rim of the Dwyka series which forms the base of the Karroo.

The Karroo System, comprising more than 28,000 feet in maximum thickness of sedimentary rocks, has been warped into a huge oval geosyncline, or shallow basin (fig. 16.7). For many millions of years this basin was slowly sinking and at the same time its mountains and hills were being worn down by an enormous river system and redeposited as muds and sands; these in turn were eventually pressed into rock, but not before they had entombed the skeletons of multitudes of strange and now wholly extinct animals.

Stretching about 820 miles from southwest to northeast, the Karroo System is subdivided into four major stratified series (fig. 16.7a), ranging from 800 to 7,000 feet in maximum thickness. The entire section of the Karroo is nowhere seen at any single place but the sequence of its overlapping parts is recorded in many fine photographs in the Geologic Survey at Cape Town.

Near the base of the Dwyka, following the lowest, oldest shales, were the Dwyka tillites, which still bear the polished surfaces and straight grooves made by the grinding forces of the glaciers that pushed across them. But these glaciers, instead of being of equal age with those that entombed the mammoths of Europe, represented a far earlier period of low temperature and accumulation of snow and ice.

Unfortunately the fossil record of animal life of the Dwyka and succeeding Ecca formations is almost a blank but it is probable that the

Dwyka beds were deposited in Middle Permian times (Broom, 1932, pp. 9, 10). Beginning at the top of the Ecca and in the overlying Beaufort and Stormberg series, there is an astonishing display of nature's luxuriance.

Beyond Laingsburg we traverse the Ecca series, barren of fossils, and then rising somewhat we enter The Gouph and cross the wide exposure of the *Tapinocephalus* zone, where pareiasaurs and dinocephalians (fig. 16.8) once flourished. At Beaufort West we are on the edge of the great *Endothiodon* zone. The top of the plateau is reached between Beaufort West and Victoria West. Here we are in the *Cistecephalus* zone of the Beaufort series (fig. 16.7b). Much of the Karroo area has been shot through with sills of dolerite, which was once liquid crystalline rock; due to greater resistance to erosion, the remains of such dykes now stand up above the present general level like huge stockades.

To reach the *Lystrosaurus* zone, we must leave the main line at De Aar and go southeast to Middleburg. Here again we may change trains and go southwest to Graaf Reinet in the *Endothiodon* zone, or to the northeast, crossing the *Cynognathus* zone at Burgersdorp, eventually reaching Aliwal North at the edge of the Stormberg series at the top of the Karroo system.

These formations and their included fossils for a century past have been actively studied by geologists and palaeontologists: Andrew Geddes Bain, Rogers, Du Toit and Haughton, have described the Karroo formation; Sir Richard Owen, Harry Govier Seeley, David Meredith Sears Watson, and above all, Robert Broom, are the chief contributors to the classification and morphology of Karroo fossils; Sidney H. Haughton, and more recently Lieuwe D. Boonstra, have published valuable descriptions of the collections in the South African Museum; while in later years Friedrich von Huene, Elmer C. Case, Helga Pearson, Ferdinand Broili and J. Schröder, F. R. Parrington, Alfred S. Romer, Charles L. Camp, Everett C. Olson, have published noteworthy descriptions either of particular genera or of the collections which they have made there. The greatest work on the mammal-like reptiles of South Africa is that by Dr. Robert Broom, who, beginning in 1897, has produced a multitude of important papers and memoirs on Karroo fossils and on the problems of the origin of mammals, the morphology of the mammalian skull and the origin of man.

The roll-call of the Karroo animal fossils includes a crustacean, *Notocaris* Broom (from the Dwyka beds), various kinds of fresh-water fishes and stegocephalian amphibians, a riot of ponderous and bizarre pareiasaurs, a few generalized "two-arched" (diapsid) reptiles, and hundreds of described species of mammal-like reptiles (therapsids) distributed under nine suborders.

In Russia during approximately the same period we find a similar reckless opulence of life and similar titanic experiments upon the hardihood and adaptability of its vertebrate animals. But not a single one of the South African or Russian therapsids survives as a "living fossil" today. So that for the great series of Synapsida, or reptiles with one temporal arch on each side, the only representatives today are the highly evolved mammals, whereas for the Diapsida (with two temporal arches), including also the ancestral birds, we have as persistent primitives the tuatara (*Sphenodon*) and the crocodilians.

PROGRESSIVELY MAMMAL-LIKE REPTILES (THERAPSIDA) FROM SOUTH AFRICA AND RUSSIA

In the Middle Permian of South Africa were several genera, especially Broom's *Galepus* (fig. 16.12A) and *Galechirus* (fig. 16.11aA), which are tentatively referred by Romer (1945, pp. 278, 279, 600) to the Pelycosauria and which also seem to be in a transitional stage to their contemporaries and successors of the order Therapsida. In this great order, as in most others, the limits of our present book impose corresponding restrictions in the number of known fossil forms which may be cited effectively as supplying documentary evidence on the history of the entire series. Fortunately the main outlines of the evolution of the numerous and diversified mammal-like reptiles, and their gradual approach along partly parallel paths, to the mammalian stage, have been fairly well known for several decades past, thanks to the outstanding labors of Owen, Seeley, Broom, Watson and many later palaeontologists.

Therapsids were excessively abundant in the Middle and Upper Permian of South Africa, where they were represented by three main suborders (fig. 16.17), including six infraorders, many families, perhaps two hundred or more "genera," an unknown number of species, thousands of museum specimens and, according to Broom, countless thousands still in the rock. In the Lower Triassic of South Africa three of the infraorders (Titanosuchia, Tapinocephalia, Gorgonopsia) have disappeared; the Anomodonts are represented only by one family, the Therocephalians have nearly disappeared, except for the progressive Bauriidae. The Cynodontia, foreshadowed in the Upper Permian of Russia and South Africa, are diversified in the Triassic of South Africa and in South America are represented by a single advanced family (Traversodontidae). In the Upper Triassic of South Africa, possible derivatives of the smaller procynodonts are represented by the order Ictidosauria, some of which have almost crossed the indefinite zone between mammals and reptiles.

The skeletons of the older pelycosaurs, especially their vertebrae, girdles and limbs while extremely primitive, were also prerequisite and preadaptive to the diversified skeletons of therapsids and mammals. The side branches evolved many curious specializations but there was also frequent parallelism both in lateral specializations and in "main line" advances.

HIPPOPOTAMUS-LIKE DINOCEPHALIANS

The Permian Dinocephalia were mostly huge somewhat hippopotamus-like beasts. Some with long low skulls (Titanosuchids) are regarded as carnivores, others with swollen crania (fig. 16.8) and short rounded jaws are called herbivores. In spite of their great weight their feet retained all the bones of their pelycosaurian ancestors. In their hind feet, for example, the large thick disc-like astragalus never became closely united to the tibia as the corresponding bone (intermedium) did in the dinosaurs and birds, where a hinge-like joint was formed across the middle of the tarsus. On the contrary, in pelycosaurs (fig. 16.25B) and dinocephalians the entire hind foot, retaining considerable "give" in all its segments, converged obliquely toward the fibula (fig. 16.8). The lower end of the tibia must have been set in a cartilaginous piece, not preserved in the fossils, which in turn was movable articulated with a rounded prominence on the antero-internal part of the astragulus (fig. 16.26B, C). This arrangement permitted the tibia and fibula to be inclined upon and partly rotated on the upper end of the tarsus (B. Schaeffer, 1941). In the wrist also none of the elements were over specialized. The hands and feet were probably supported by large pads, the metacarpals and digits were short, ending in blunt tortoise-like claws.

The phalangeal formula in pelycosaurs (fig. 16.24B) retained the primitive reptilian numbers (2, 3, 4, 5, 4-3) as it did in the gorgonopsians. But in some of the latter (fig. 16.24C) the second phalanx of digit III and the second and third phalanges of digit IV were shortened proximo-distally. In certain specimens according to Broom these small phalanges became very small, almost reducing the formula to the mammalian stage (2.3.3.3.3), which is also recorded in the dinocephalian *Moschoides romeri* (Byrne, 1937, 1940).

The limb bones of the pelycosaurs were adapted to the bent-legged posture in walking (fig. 16.1), with the elbows and knees about on a level with the glenoids and acetabula; but among the therapsids (fig. 16.13A), even in the massive dinocephalians (fig. 16.8), the elbows and knees began to be opened up although the trackway was much wider than in typical mammals. In a skeleton of a lightly built gorgonopsian (*Lycaenops,* fig. 16.13A), as worked out by Colbert (1948), the hind legs are longer than the fore legs, the hip joint much higher than the shoulder joint, the back curved downward and forward, somewhat as in a rapidly striding alligator.

Increased efficiency in running was also indicated in the tarsus of the progressive Triassic therapsid *Bauria* (fig. 16.25F) in which the tibio-astragular contact was widened forming a transverse tibio-tarsal joint and the astragulus overlapped anterolaterally on the calcaneum, the latter having developed a posterior process or heel.

EVOLUTION OF THE PECTORAL GIRDLE AND LIMBS IN THERAPSIDS

The pectoral and pelvic girdles in the pelycosaurs and earlier therapsids retained many primitive reptilian features. On the front border of the scapula many retained a splint-like bone, the homologue of the cleithrum (fig. 16.20D *clt*) of older amphibians (C) and crossopterygians (A). The ancient division between the muscles of the neck and throat on the one hand and of the flanks and breast on the other was continued down the cleithrum to the spoon shaped clavicle and thence downward to the spade-like inter-clavicle (figs. 12.8; 12.46). The junction of the lower end of the cleithrum and the clavicle on each side was supported by a short anterior process (acromion) from the scapula, which becomes prominent in the anomodonts and cynodonts (fig. 16.20E) and was the forerunner and prerequisite of the acromion (G) of the mammalian scapula.

In primitive pelycosaurs the scapulo-coracoid (fig. 16.20D), derived remotely from the internal shoulder girdle of crossopterygian fishes (A), included (D) a large, broad scapular blade and a wide anteriorly curved coracoid plate divided by a longitudinal suture into a large pro-coracoid (*co1*), below the scapular blade, and a smaller coracoid (*co2*). On the lateral aspect of the (not shown) scapulocoracoid the anteroposteriorly elongated glenoid cavity, for the head of the humerus, was borne conjointly by the scapula and both coracoids. The anterior coracoid (*co1*), near its posterior end, bore the anteroventral corner of the glenoid facet and beneath this was the coracoid foramen for the passage of the supracoracoid nerve running outward to the supracoracoid and other muscles that pull the humerus forward.

Essentially the same arrangement of the entire shoulder girdle was retained in the Dinocephalians (fig. 16.8), Gorgonopsians (fig. 16.13A) and Therocephalians (fig. 16.13B); but in the anomodonts (fig. 16.11bC) and still more in the cynodonts (figs. 16.20E, 18.16A) the bony part of the pro-coracoid (fig. 16.20E, *co1*) diminished leaving a round cartilaginous border, in front of which a supracoracoid fenestra began to open between the clavicle (*clv*) and the procoracoid (*co1*). The glenoid facet shortened anteroposteriorly and became larger dorsoventrally (fig. 16.20E) as the dorsal side of the head of the humerus became globose. Meanwhile the anterior part of the coracoid muscle extended upward through the above-noted supracoracoid fenestra, working its way up near the anterior border of the scapula which was left vacant as the cleithrum dwindled and disappeared. Thus arose the supraspinatus muscle of mammals. The infraspinatus was similarly formed from the posterior part of the coracoid muscle mass. The spine or crest of the mammalian scapula, lying between the anterior and posterior spinatus muscles, seems to represent the outwardly turned anterior border of the cynodont scapula (figs. 16.14; 18.16A-C), the supraspinatus area being a new or mammalian crest between the supraspinatus and the subscapularis.

In earlier reptiles the scapulae had been nearly fixed in position by the attachment of the coracoid plates and interclavicles to the sternum (figs. 12.8; 12.46). In the duckbill platypus (*Ornithorhynchus*) by a further development of this stage the opposite coracoid plates can slide back and forth a little (fig. 18.16C) but the anteroposterior motion is limited by the firm attachment of the clavicles to the fixed interclavicles (cf. E).

The last step was taken by the typical mammals, perhaps in Jurassic and Cretaceous times: (a) the head of the humerus became almost spherical (fig. 16.21D); (b) the glenoid facet was turned downward rather than outward (fig. 16.20G); (c) the anterior coracoid dwindled and the posterior coracoid remained only as a small recurved hook. As all this happened the scapulae became free to slide forward and backward, while retaining their pivots on the sternum through the clavicles; the elbows were drawn in toward the flanks and the humeri came to lie mostly beneath the scapulae rather than lateral to them (fig. 18.22). The transformation of the shoulder-girdle from the typical reptilian to the mammalian stage was visualized in its broader outlines by Wilson and M'Kay (1893) in their study of the homologies of the borders and surfaces of the scapula in monotremes. It has since been carefully traced by Gregory and Camp (1918), Romer (1922a, b).

As to the parasternum (fig. 12.46 *abd. r*), in one of the older pelycosaurs (*Theropleura retroversa*) Williston shows a well developed ventral basket, consisting of closely set chevrons of gastralia (abdominal ribs) extending from the pectoral to the pelvic girdle. Little or nothing is known of the parasternum in therapsids and it may be assumed provisionally that it was thinned out and disappeared, except in so far as it may be represented by the lineae albae, which in man and other mammals divide the rectus abdominis muscle transversely into a closely linked series of muscular segments. In the gorgonopsian *Lycaenops* the strong anteroventral bar of the pubis (fig. 16.13a), which rather suggests the pubis of the crocodilians, must have given attachment either to the abdominal musculature or to remnants of the parasternum. In *Cynognathus* (fig. 16.14) the flat anteroventral brim of the pubis looks sufficiently like that of marsupials to suggest that it may have given attachment to a pair of broad epipubic bones.

EVOLUTION OF THE PELVIS AND HIND LIMBS

As to the evolution of the pelvis, the pubiischiadic plate in the older carnivorous pelycosaurs was not yet fenestrated (fig. 16.22A). However, its anterior pubic bar was emphasized, its posterodorsal ischial ridge more or less defined, and the middle portion of the intervening area was fairly thin. The acetabulum was obliquely ovate with the narrow end pointing downward and forward; the basin and rim of the acetabulum were formed almost equally by the ilium, pubis and ischium, the three bones meeting in a triradiate suture (fig. 16.22). To the narrow blade of the ilium, directed upward and backward, were attached the caudal muscles posteriorly and the iliocostales anteriorly. The herbivorous edaphosaurs, a side branch described by Romer and Price (1940), retained the plate-like unfenestrated pubi-ischiadic plate, but widened the iliac blade, developing its anterior process, which in *Casea* became helmet-shaped.

Among the therapsids the ponderous "herbivorous" dinocephalians (fig. 16.8) with their very wide thorax and abdomen, partly paralleled the aberrant pelycosaur *Casea* in everting the anterior blade of the ilium, but they also retained a rather stout posterior process; thus there was a strong base for the massive gluteal and other extensor muscles of the thigh and leg. The pubi-ischiadic plate was quite short anteroposteriorly in proportion to the total height of the pelvis. The round-oval acetabulur basin was very large and surrounded in part by a high rim. The strong anterior border of the pubis was sharply everted; and the anteroventral part was much smaller than the large postero-inferiorly directed ischium. There was no pubi-ischiadic fenestra and the foramen for the obturator nerve was small.

In the gorgonopsians *Lycaenops* (fig. 16.13aA) (*cf.* Colbert, 1948, p. 384) the ischium was very much larger than the pubis, extending downward below it and separated from it by a large ventrolateral gap. Comparison with the well preserved pelvis of *Moschops* (A.M.N.H. no. 5533 "C") suggests that the above mentioned gap would be bridged by thin bone or by cartilage and symphysial ligament, which would ventrally bound a true obturator fenestra, as suggested in Boonstra's *Aelurognathus* and in the remarkably well preserved pubiischiadic plate of *Cynognathus crateronotus* Seeley (fig. 16.14). The large ischium of

Lycaenops was doubtless the base for thick adductors and obturators of the femur, in this slinking speedy carnivorous reptile.

The opposite ischia in pelycosaurs and gorgonopsians in the rear view formed a sharp V which would permit only small eggs to pass through the pelvic outlet. But in *Cynognathus* the V seems to have been more obtuse, that is wider across the top, in accord with the relatively greater width across the lower part of the sacral ribs. Even in the egg-laying mammal *Ornithorhynchus* the maximum transverse diameter across the pelvic outlet was little if any wider than the distance across the transverse processes of the postsacral vertebrae.

In *Cynognathus* the anterior blade of the ilium (figs. 16.14, 16.22B), becoming large and helmet-shaped, increased the area for the extensors of the thigh and limb, which were needed to tilt the sacral region toward one side and hold it while the opposite leg swung forward.

EVOLUTION OF THE VERTEBRAL COLUMN

The centra of the dorsal vertebrae of the more primitive pelycosaurs (fig. 16.5E, E1) as noted above consisted essentially of short, modified cylinders with somewhat V-shaped cross section in the middle (Romer and Price, 1940, fig. 17, p. 98) and with concave anterior and posterior ends retaining traces of the embryonic notochordal tunnel in the middle of the centra. Small crescentic intercentra were present especially in the neck (fig. 16.5A, *ic1, ic2*) and back, becoming reduced in size toward the tail where they are represented by the large chevrons (fig. 16.1). The arrangement of the anterior and posterior zygapophyses (fig. 16.5E, E1) permitted little twisting but moderate dorsoventral flexure. The neural spines even in the relatively primitive ophiacodonts were fairly high, although not greatly prolonged upward as they were in the more advanced pelycosaurs (fig. 16.1C). In the dinocephalians (fig. 16.8) the dorsal vertebrae, apart from their large size and relative shortness did not differ essentially from the primitive pelycosaurian type, except that the neural arches were not unduly long and that the "sail" or dorsal crest was absent.

Among the dinocephalians the great body weight made it necessary for the sacral ribs to be very large and for the transverse processes on several presacral and post-sacral vertebrae to take part of the load. The top view shows indeed that there was a sort of synsacral box, somewhat similar to that of sauropod dinosaurs (fig. 14.32) but much smaller. Among the gorgonopsians (fig. 16.13a) the vertebrae of *Lycaenops* showed a very gradual change in passing from the cervical to the dorsal and lumbar regions. The first sacral vertebra had large transverse processes, forming a firm attachment for the pelvis, but the second sacral vertebra had a much smaller transverse process. The neural spines in *Lycaenops* were remarkably low in comparison with those of several other gorgonopsians (Colbert, *ibid.,* p. 378). The ribs as in pelycosaurs, were two-headed, except possibly in the lumbar region, the tubercle articulating with the transverse process, the long head extending downward and inward to the junction between the vertebrae.

POSSIBLE BEGINNING OF A DIAPHRAGM

In *Cynognathus* (fig. 16.14) and other cynodonts there was a marked regional differentiation of the presacral vertebrae. Beginning at the thirteenth dorsal the ribs rapidly shortened and the proximal parts were widened; each one had a triangular anterior process which extended under a similar process on the posterior margin of the rib in front of it. These overlapping relations were peculiar to the cynodonts. But they may also have been prerequisite for the apparently simplified stage of the typical mammals in which there is a fairly rapid differentiation in passing from the dorsal ribs to the lumbar transverse processes. The comparative anatomical and embryological evidence, ably summarized by Hyman (1942, pp. 254–255) indicates that the mammalian diaphragm is a composite structure derived from part of the transverse septum (between the pericardial and pleuroperitoneal cavities of fishes). The urodele, turtle and mammalian embryos supply intermediate stages between the fish and the mammal stages. The liver and its coronary ligament are enclosed in a sac from the same transverse septum.

"The phrenic nerves, which innervate the diaphragm, are derived from the more anterior spinal nerves entering the brachial plexus. This indicates that the myotomes which contribute to the diaphragm are of cervical origin and have migrated posteriorly" (*op. cit.*, p. 439).

This is partly because the future heart, lungs and liver in early embryos lie beneath the ganglia of the third, fourth and fifth cervical nerves, their twigs entering the "nephric fold" which forms part of the diaphragm (*cf.* Goddrich, E. S., 1930, pp. 650–651). These twigs combine to form the phrenic nerve and as the neck and thorax grow away from the abdomen the phrenic nerve is prolonged and the distal ends of the cervical myotomes stay with the remnants of the transverse septum and form the diaphragm.

The peculiar lumbar rib processes of cynodonts (fig. 16.14) were probably located in the intersection of the intercostal and perhaps iliocostal muscles, while the pointed outer ends of the ribs would probably be connected with part of the oblique abdominal system. All these muscles may have assisted to some extent in transverse wriggling movements, but not improbably they were in some way connected also with the emergence of the mammalian diaphragm, which seems to have been pushed forward by the large liver and stomach into the rear end of the thorax.

EVOLUTION OF THE GORGONOPSIAN SKULL, JAWS AND TEETH

The skull of the earlier therapsids including the dinocephalians (figs. 16.9 and 16.3D) and the gorgonopsians inherited the basic construction of the Dimetrodont pelycosaurs (figs. 16.9A, 16.3B); but like every other group they started their own specializations, some of which were in the direction of the mammalian stage. Among the pelycosaurs (fig. 16.2) the highly carnivorous ophiacodonts and sphenacodonts (including *Dimetrodon*) represent a pre-therapsid stage and among the therapsids the gorgonopsians inherited many basic pelycosaurian features but were specialized in the overdevelopment of the upper caniniform (fig. 16.18) teeth and in the variability and reduction in size of the cheek teeth.

The sabre-like upper canines of the Gorgonopsidae were the dominant feature of the skull and in *Lycaenops* (fig. 16.13aA) and related genera they were protected medially by a downward projection of the lower jaw, heightening the general resemblance to a sabre-tooth tiger (*cf.* Colbert, 1948). With such an equipment they could swing the lower jaw downward, clearing the upper canine, at the same time throwing the snout upward, then striking downward, piercing with the large incisors and shearing with the canine.

The gorgonopsian skull was well built to withstand the wrenching stresses experienced in killing and hauling the prey. The whole preorbital face formed at the top a strong, convexly arched brace (fig. 16.18), stiffened dorsally by the premaxillae, septomaxillae, nasals, prefrontals and frontals (fig. 16.4aC) which in turn transmitted part of their thrusts outward and backward to the strong orbital rims, to the occiput and to the temporal arches. The caniniform "incisors" (fig. 16.18) four in the premaxilla and one in the maxilla, of each side, were firmly seated in the transversely arched premaxillae and transmitted their thrusts upward and backward through the ascending processes of the premaxillae and the septomaxillae to the roof and sides of the face. The high side walls of the upper jaw consisted mostly of the vertical plates of the maxillae which contained the long curved upper part of the sabre-like canines (fig. 16.15bA) and the tooth germs of the large replacing canines. The opposite side walls of the skull were tied together below by the transversely arched bony palate (fig. 16.3E), including the ectopterygoids, palatines and transverse bars of the pterygoid. The circumorbital region (fig. 16.15bA) was braced in front by the lacrimal (*la*) and prefrontal (*prf*), above by the frontals and postfrontals (*pof*) behind by the postorbitals (*poorb*) and below by the jugals (*ju*). This circumorbital rim had to resist upward thrusts from in front, caused by the upward swing of the lower jaw, and by pressure from the rear, caused by strong temporal and pterygoid muscles. In the top view the strong centerpiece of the interorbital bridge typical of gorgonopsians (fig. 16.15bC) (connecting the bony face with the cranial vault) was the preparietal bone (*prpa*). In pelycosaurs Romer and Price (1940, p. 57) record the presence of paired grooves near the midline on the underside of the surface bones, commencing in the nasals and running back along the frontals to a point between the orbits; that is, to the site of the preparietals of therapsids. Beneath the frontals of pelycosaurs the middle part of the surface was in contact with the dorsal surface of the presphenoid and the interparietal ossification if present was covered by adjacent edge of the frontals and parietals. This median, unpaired parietal element is thus very probably due to a combination of a dermal bone on the surface (analogous with a Wormian bone) plus the fused vertical plates of the trough-like sphenethmoid, which in primitive tetrapods (figs. 11.23, 11.22b, and 12.13C) forms the side wall of the preorbital, or trabercular, parts of the cranium.

Lateral buckling of the gorgonopsian temporal region was stopped by the strong outwardly bowed longitudinal arch formed by the jugal and zygomatic process of the squamosal (fig. 16.15C). The temporal muscles, seated firmly in the temporal fossa, were tied in through their tough coverings: above by the strong postorbito-squamosal arch and parietal roof (fig. 16.4aC), on the inner side by the epipterygoid plate (fig. 16.33F) and the lateral wall of the inner ear (fig. 16.32D *pro*), to the rear by the strongly built squamosal bone, beneath by the rigid base of the cranium and by the quadrate branches of the pterygoid (fig. 16.15aA). These quadrate branches of the gorgonopsian pterygoids (fig. 16.3E), differ from the strong, straight, thick-rimmed ones of the pelycosaurs (fig. 16.3B) in that they have been squeezed in to form long secondary contacts with the basi- and parasphenoids, evidently due to the later expansion of the opposite pterygoid muscles. The transverse pterygoid processes (2) in the pelycosaurs (fig. 16.3B) were separated by a deep median interpterygoid cleft but in the gorgonopsians they meet in the middle and convergently resemble the transverse palatine bars of the mammals (fig. 16.72G).

The rear part of the inner braincase was based upon the basisphenoid and basioccipital, which transmitted their thrusts through the ball-like occipital condyle to the vertebrae and muscles of the neck and shoulder girdle. The rear wall of the braincase in the earlier therapsids (*e.g.* fig. 16.4bJ) was formed by the exoccipitals (*exoc*), squamosals (*sq*), tabulars (*tab*), interparietals (dermosupraoccipitals, *dsoc*). The membranous tubes of the semicircular canals and pouch containing the nerve of the inner ear plus part of the main line of the seventh cranial nerve, were lodged in the pro-otic-opisthotic bones (fig. 16.54B). These formed thick, transverse arches (fig. 16.4J, *opo*) which took up inward thrusts from the lower jaw and quadrates and transmitted them to the bony base of the cranium.

The lower jaw was of the usual complex, reptilian type, but in *Lycaenops* (fig. 16.13a) the dentary with its ascending branch already took up about $^{11}/_{14}$ of the total length. The surangular formed a strong dorsally convex arch, resting on the blade of the angular and inwardly on the prearticular. This surangular-angular prearticular arch formed part of the insertion area of the temporal muscles (including the outer layer or masseter) and helped brace the dentary in front and the articular behind. As seen from below the opposite halves of the mandible converged in front where they braced each other, transversely meeting in enlarged opposing surfaces which fuse together at an early age, forming an almost chin-like strong symphysis.

Thus the gorgonopsian skull, or indeed almost any skull of any vertebrate animal, consisted of a complex of arches, triangular trusses and straight beams, which are evident in the top, side, front, rear and underside views, as well as in sections in the longitudinal, transverse and horizontal planes.

Many of the foregoing and other features of the gorgonopsian skull not only served well the needs of their sabre-toothed owners, but were also prerequisite (except the reduction of the cheek teeth) for the various types of skull that are found in the collateral branches (dinocephalians, fig. 16.9B-F) and anomodonts (fig. 16.12C, D) as well as in the almost or quite direct ancestors of the mammals, namely the therocephalians, bauriamorphs, cynodonts and ictidosaurs (fig. 16.18). A brief review not only of the other main skull types, but also of some of the curious side branches will bring the transformation at least up to the beginnings of the mammalian class in the Uppermost Triassic and Lower Jurassic.

THE DINOCEPHALIAN SKULL

The dinocephalians were dominant therapsids of the *Tapinocephalus* beds (fig. 16.7) of South Africa and of beds of nearly equivalent Middle Permian age in Russia; although a few of their contemporaries were gorgonopsians and therocephalians. The long skulled dinocephalians called titanosuchids (fig. 16.9E) suggest essentially overgrown gorgonopsians except that they have less reduced cheek teeth. However Boonstra (1936) has cited about a dozen peculiar osteological details which, according to his view, could not have been inherited by dinocephalians from gorgonopsians. And indeed it is conceivable that the titanosuchids and the gorgonopsians of the Middle Permian of South Africa and Russia were derived independently of each other (fig. 16.17) from different genera of a basic therapsid stock of Lower Permian age, which would be near to or identical with the sphenacodont pelycosaurs (fig. 16.9A).

Among the short faced "herbivorous" branch of the dinocephalians, the skull (fig. 16.9C, F) parallels those of anomodonts (fig. 16.12) in the shortening and downbending of the face and the forward swing of the suspensors (quadrate). But, instead of acquiring a beak *Moschops* and others increased the size of the incisors (figs. 16.8; 16.9C, D). These were inclined forward and had very long roots (fig. 16.10) and short, stubby, heavily enamelled, oval crowns with steep inner slopes and blunt tips. These teeth were arranged in two parallel rows, partly like Bolk's exostichos (outer) and endostichos (inner) row. In the outer row which suggests the deciduous set of mammals, the teeth decreased in size from the anterior incisor to the last cheek tooth. The second or inner set, suggesting the permanent dentition of mammals, were formed in deep levels (fig. 16.10*p*) of the upper and lower jaws and as the first set were worn down by attrition against tough resistant food, the second set came up beneath them, resorbing the inner sides of the crowns of their predecessors, so that in old individuals (fig. 16.10B1) the worn down stumps of the "deciduous" set fitted like caps upon the high pointed crowns of the incoming teeth. Conceivably there may also have been a "prelacteal" set in very young individuals, but in large specimens there seems to be only two sets. In this respect the dentition of the dinocephalians would seem to have foreshadowed that of the mammals. Indeed from the great height of the teeth in both sets in the type (A1) of *Taurops macrodon* Broom it is evident that in a very young animal the tips of the teeth in either set would form first, and then as the jaws grew larger, the tooth pulps, growing away from the tips of the crowns, would slowly enlarge the cross section of the teeth, while adding new basal rims and pushing the tips up toward the surface.

A similar growth in cross section and height occurs for example in the marsupial *Phascolomys* whose cheek teeth in the very young individual grow larger as their growing pulps add to the proximal end of the crowns. For such reasons it may perhaps not be necessary to apply to the dinocephalians Bolk's complicated theory of the origin of the mammalian dentition. But we may provisionally adopt Broom's (1913) interpretation of the dental succession of both dinocephalians and cynodonts as being directly antecedent to the mammalian stage. In that case, the gigantic dinocephalians had progressed in the mammalian direction and away from the pelycosaurs, which according to Romer and Price (1940, p. 92) had a succession in alternating rows. According to Parrington (1936) the gorgonopsian upper canines belonged to a single replacing series of at least three teeth and thus retained one of the essentials of the typical reptilian stage.

Among the most striking features of the herbivorous dinocephalians is the immense thickening of the outer skull bones, especially on the forehead, and around the orbits (fig. 16.9C); the squamoso-jugal arches also were extremely large and thick and in some genera the backwardly prolonged nasals and prefrontals bear a huge rounded buffer above and in front of the eyes (fig. 16.9F). The surface of all these swollen bones is rough and they must have been covered with a very thick integument. Possibly these saurians lowered their heads and butted each other or their remote relatives the carnivorous dinocephalians.

The median foramen for the pineal eye was unusually large and in *Delphinognathus* and *Moschops* (fig. 16.8) it was set in the midst of a circular tumescence. Exceptionally large pineal eye-holes were also present in the supposedly aquatic pelycosaur *Casea* and in the thickset cotylosaurs *Diadectes* and *Chilonyx*.

The bones of the underside of the dinocephalian skulls while not swollen were very stoutly built (fig. 16.3D). The internal nares were exceptionally large, elongate ovals; the long prevomers (vomers) braced the premaxillae and were connected posteriorly with the wide anterior branches of the pterygoids. The latter were overlapped by large palatal lobes of the palatines, which lay on either side of the incipient trough leading from the anterior nares to the rear of the palate. The large pterygoids retained large transverse and quadrate processes (fig. 16.3D); these gave rise to the thick pterygoid muscles which extended outward and downward to the lower side of the articular bone of the mandible. The occipital condyle was single and convex as in primitive reptiles.

THE TURTLE-BEAKED ANOMODONTS

Another very curious side branch of the therapsid order was the suborder Dicynodontia (figs. 16.11a, b; 16.12), commonly called anomodonts, and thronged with genera chiefly from the Middle and Upper Permian of South Africa and Russia, but with later (Triassic) representatives in South Africa, Europe, Asia, North and South America. All the anomodonts were provided with turtle-like beaks, but most of the dicynodont family also had a pair of large curved upper canine teeth, at least in the males. Broom showed that the skulls called *Ondenodon* were toothless females of *Dicynodon*. One of the most curious anomodont skulls was that of the Triassic *Lystrosaurus* (fig. 16.11aB; 16.12D) which somewhat resembled a walrus skull in the marked shortening of the preorbital face and straight downward prolongation of the very long tusks. The Endothiodont family (e.g., fig. 16.11BC) had some small teeth on the palate. Exceptionally ample space for the jaw muscles was attained in the following ways: (1) by greatly widening the transverse diameters across the occiput and the squamosal-jugal arch (fig. 16.12E, H, I); (2) by elongation of the temporal fossa (G); (3) commonly by prolongation of the squamosal bar backward and upward (D, *sq*) and occasionally; (4) by pushing the orbit forward (H); (5) by developing a long ridge or strip on the outer face of the squamosal cheek-bar (figs. 16.11aB; 11bC) from which the superficial layer of the temporal muscle mass (*i.e.,* the masseter muscle) passed down to be inserted on the lateral face of the large articular bone. On the other hand the adverse leverages occasioned by the beak on the front end of the upper and lower jaws and by the nearly distal position of the canine tusks, were reduced by swinging the quadrate obliquely forward and downward and by bending the entire bony face sharply downward (D). The skulls of some anomodonts were not much larger than those of pond turtles. At the other extreme were the lumbering *Kannemeyria,* which paralleled the pariesaurs in its long ribs, curved back and short-toed hands and feet, the large *Aulacocephalus* with its wide snout (fig. 16.12I) and beyond all these the enormous skull of *Placerias* from the Triassic of New Mexico.

In their pectoral girdles the anomodonts were quite progressive toward the mammalian stage seen in the egg-laying *Ornithorhynchus* (p. 364). In the pelvis the helmet-shaped ilium (fig. 16.11aB) suggested that of the cynodonts and the wide ischial plate was produced downward below the level of the outwardly directed anterior bar of the pubis, but somewhat less than in the gorgonopsian *Lycaenops* (fig. 16.13a).

THE THEROCEPHALIANS AND THEIR SKULLS

The older therocephalians (fig. 16.19A, E, F) from the Middle Permian of South Africa somewhat resembled the contemporary gorgonopsians (fig. 16.17) in their predaceous features, but had longer skulls less turned down in the rear. The large upper canine tusks frequently occurred in pairs (fig. 16.15B), the temporal region (B, D) was elongate anteroposteriorly and the postorbital bar was short and widely separated from the squamosal by the parietal. The latter was a distinctly premammalian feature, and so was the reduction of the postorbital process of the postorbital bone in some of the smaller therocephalians (fig. 16.19B, D), so that the orbit opened into the temporal fossa. Equally important was the prolongation of the ascending blade of the dentary bone (fig. 16.36B), which gave rise later to the ascending branch of the mammalian lower jaw (figs. 16.36C, D; 37F, G). In the palatal view (fig. 16.15aB, C) the therocephalian skull differs from the gorgonopsian (A) in the presence of relatively long paired oval fenestrae between the pterygoid (*pter*) and ectopterygoid (*ect*) bones, the palatines (*pal*), the transverse processes of the pterygoids and the base of the cranium. The parasphenoid (*pas,* Broom's vomer) has a posterior expansion covering the basisphenoid—a primitive reptilian heritage.

The therocephalians branched (fig. 16.19) during the Middle and Upper Permian of South Africa into a large number of genera, which have been referred to five families. Of these, the Pristerognathidae were the oldest and on the whole least specialized. The Alopecopsidae were more or less in between the primitives and the more specialized branches. Of the Whaitsiidae, *Whaitsia* itself bears a very large preorbital basin which apparently lodged a large facial gland of unknown function. The Scaloposauridae (A-D) had small, very long slenderly built skulls, often with incomplete postorbital bar. They also had long slender limbs (fig. 16.13B) and probably lived on insects or other small creatures. The Bauriidae (fig. 16.19B), while essentially therocephalian, had advanced toward the mammalian grade in the development of secondary palatal plates on the maxillae and partly on the palatines which partly bridged over (fig. 16.30B) the primary internal nares (*int. nar*), without wholly concealing them in the palatal view (A, B). The quadrates, as in the Gorgonopsians (fig. 16.15A *qu*) were much reduced in size as compared with more primitive reptiles, and with them the quadrate branches of the pterygoids (*pt*). The opisthotics (*opo*), containing the posterior parts of the inner ears, were extended ventrolaterally, supporting the squamosals and affording insertion areas for some of the neck muscles that raise and turn the head (cf. fig. 16.6aB, B1).

THE ALMOST MAMMALIAN FOOT OF *BAURIA*

In the hind foot of *Bauria* (fig. 16.25F) the astragalus (*i*) partly overlapped the calcaneum (*fb*), which showed the beginning of a heel. The astragalus was developing a prominent transverse convexity for the tibia. Thus the movement of the foot was becoming more hinge-like at the tibio-tarsal joint as in mammals; whereas in primitive reptiles (B) the calcaneum (*fb*) was more disc-like and the flexure of the foot was more in the middle of the tarsus (Schaeffer, 1941a).

Thus *Bauria* and the cynodonts were independently evolving pre-mammalian features.

THE CYNODONTS EVOLVE TOWARD THE MAMMALS

A decided advance toward the mammalian grade was made by the cynodonts, which had forerunners in the Upper Permian of Europe, East Africa and South Africa (fig. 16.19N) but were found mostly in the South African Triassic (*O-Q*), with several related forms (fig. 16.44H) in the Triassic of South America. The cynodonts (figs. 16.40G, 43A, C), like the therocephalians, had lost the upper postorbito-squamosal contact, although they retained in the single temporal arch the contacts of the lateral processes of the postorbital and of the zygomatic process of the jugal, as well as the long bar of the jugal (or malar) extending beneath the squamosal. But whereas the entire temporal arch in the typical therocephalians was slender, that of the cynodonts was quite stout. This was probably correlated with the fact that most of the support for the large lower jaw had shifted from the greatly reduced quadrate (fig. 16.54B) to the much enlarged squamosal, and secondly, because the ascending ramus of the dentary (fig. 16.36C) had now become very large, indicating a corresponding increase in the volume of the temporal muscle (fig. 16.38), while the superficial or masseter layer of the old temporal muscle-mass had also increased in strength, pulling the jaw upward and straining the jugal bar downward (fig. 16.28A, A1).

The dentitions of the later cynodonts (fig. 16.40) exhibit progressive anisomerism or regional emphasis and differentiation into incisors, canines, premolars and molars, again foreshadowing the mammals. The upper molars have compressed crowns, recurved tips and jagged or cuspidate posterior borders; the roots were implanted in sockets. In the carnivorous cynodonts, the motion of the jaw was vertical and adapted for shearing, not grinding (fig. 16.28).

The secondary bony palate consists of wide thin shelves (figs. 16.30D, 31A) from the maxillae and short ones from the palatines, completely bridging the secondary internal nares, but its posterior edge was still in front of the last few upper molars, nor was it bordered by a raised transverse ridge as it is in typical mammals (fig. 16.31B). The reptilian "prevomers," fused to form the median vomer, were pushed backward along with the posterior border of the internal nares (A); laterally they were in contact with the remnants of the palatines (*pl*) that were not curved around to form the new bony palatal arch beneath the narial tunnels. The pterygoids (*pt*) still retained flaring transverse processes, although they bore no teeth and were thinner than those of gorgonopsians, another step in the direction of the mammals. The ectopterygoids were still present though small. With the reduction of the quadrates, the quadrate branches of the pterygoids had become quite delicate (fig. 16.31A) and served to support the thin, anteroposteriorly extended flanges of the epipterygoids (figs. 16.32, 33D, F). The latter (figs. 16.32, 33C, D *ept*) formed a secondary outer wall of the braincase, to which part of the pterygoid muscles (fig. 16.28A1) was attached, corresponding to the alisphenoid bone of mammals (fig. 16.32 *As*). The ancient parasphenoid (fig. 16.30C, *pas*) remained as a ventral cover of the basisphenoid, while its anterior process retained contact with the inwardly pinched pterygoids. A close approach toward the mammalian grade was the transverse spreading of the originally single occipital condyle into a double condyle (figs. 16.31A; 16.34C). This was probably in correlation with the lateral spreading of the atlas, which forms a circular cup in birds and typical reptiles. The rod-like stapes (figs. 16.53E, 16.54B), with its foot-plate in the fenestra ovalis of the inner ear, probably extended outward to touch the small quadrate. The quadratojugal was reduced to a small lateral annex of the quadrate (figs. 16.54B *qj*; 16.4bK).

In the top view, the skull of *Cynognathus* and related genera (fig. 16.4aE) retained the primitive triangular outline, but the temporal openings which originated in the pelycosaurs (fig. 16.9A) on the sides of the flat temporal region had extended their upward limits until they pressed the occipital roof into a high sagittal crest, so that they face more dorsally than laterally (fig. 16.4aE) and, being bounded medially by the parietals (*pa*), anteriorly by the postorbitals (*poorb*), postero-laterally by the squamosals (*sq*), they bear a deceptive but quite convergent resemblance to the true dorsal temporal fenestrae of *Sphenodon* (fig. 12.41aA1) and the other two-arched reptiles. In *Cynognathus* (fig. 16.28A), however, the dorsal extension of the temporal muscle masses was no doubt in correlation with the progressive dominance of the ascending branch of the dentary bone of the lower jaw, whereas in *Sphenodon* the large size of the upper temporal fenestra is due to the need to overcome the adverse leverage of strongly biting incisors and parallel rows of cheek teeth. In the lower jaw, the predominance of the ascending branch of the dentary noted above was in correlation with the compensatory reduction of the bones behind the dentary (fig. 16.36C), namely, the coronoid (*cor*), surangular (*sur*), prearticular (*prart*) and angular (*ang*). The reduction of these elements on both halves of the lower jaw was also in correlation with the reduction of the quadrate branch of the pterygoid (fig. 16.31 *pt*).

The suborder Cynodontia included a fairly wide range in size, from the small, presumably insect-eating procyonodonts to *Cynognathus crateronotus,* whose skull was about 18 inches long. All had relatively small pointed incisors (fig. 16.40), (typically four on each side in the upper and lower jaws) and moderately large canines, but the cheek teeth ranged from numerous simple pegs (fig. 16.40A) to compressed shearing molars (G). In *Tribolodon* (fig. 16.41), which is provisionally referred to the Cynodontia, each upper molar had three cusps in line, suggesting the triconodonts among Mesozoic mammals (fig. 17.11C), but not necessarily related to them.

Among the diademodont family of the same suborder, the upper cheek teeth crowns were successively wide ovals with single low cross-crests (fig. 16.44B1); the lower molars (D) bore cross-crests which in occlusion (B2) fitted between the crests of the upper molars. Two sets of teeth, apparently corresponding to the deciduous and permanent teeth of mammals, were found by Broom (1913) in a young jaw of *Diademodon* (D). Cynodonts, including diademodonts, also foreshadowed the typical mammalian dentitions in the facts: (1) that in occlusion (B2) the upper teeth projected beyond the lowers, and (2) that except at the posterior end of the tooth row, one lower tooth articulated with two uppers. Moreover, when, as in the South American traversodonts (H), the crowns of the upper teeth are more or less wedge-shaped, the inner tips of the upper molars articulate in the space between the rear of the corresponding lower and the front of the next lower molar.

ICTIDOSAURS AND OTHER TRIASSIC NEAR-MAMMALS

A still nearer approach to the mammalian stage was effected by the as yet little known order called by Broom Ictidosauria (fig. 16.19M) of the Triassic of South Africa. This includes a wide variety of fossil fragments (fig. 16.44G, E-E3) chiefly of jaws and teeth, some of which may eventually be referred to other groups. In a Triassic coal mine in North Carolina, two almost microscopic lower jaws (fig. 16.42) were found, which were described by Emmons more than a century ago under the names *Dromatherium* and *Microconodon*. Osborn (1888) classified them as primitive mammals and erected for them the order Protodonta, but Simpson (1928), with better comparative material available, referred them to the therapsid order Cynodontia. Romer (1935, p. 291) tentatively refers them to the order Octidosauria. The large ascending branch of the dentary sloped gently upward and backward, much as in the cynodonts, and the jaw elements behind the dentary must have been quite small. The dentition is well differentiated into incisors, canines, premolars and molars, the latter showing an incipiently triconodont stage, suggesting the beginning of the molar pattern of the Jurassic triconodont mammals (fig. 17.7E, D). In the type fossils of the Ictidosaurian group, called by Broom Ictidosaurians A and B (fig. 16.19M), the skull, as described by him, was almost mammal-like, having lost the postorbital bar and with very small elements behind the dominant dentary (fig. 16.36D).

CHANGES IN THE BRAIN-CAST AND BRAIN

There are two classes of evidence relating to the brain characters of primitive reptiles, of the mammal-like reptiles and of extinct mammalian types of the early Tertiary age. The first is the indirect evidence afforded by the general configuration of the parts of the skull surrounding the braincase, the second is the more direct evidence preserved by natural or artificial casts of the interior of the brain cavity, which show the position of the bony tunnels of the cranial nerves and the general location and relative development of the main divisions of the brain. It must be remembered, however, that the brains of certain reptiles, *e.g., Sphenodon, Chelone* (fig. 12.32) do not fit tightly into the braincase but are surrounded by thick envelopes of fibrous and vascular protective tissue which obscure more or less the outline of the brain itself in the cast of the brain cavity; whereas the brain of mammals fills almost the whole of the brain cavity, the meningeal wrappings being relatively thin, while even the blood vessels and venous sinuses which surround the brain are pressed on the inner side of the brain-wall and help by their imprints on the latter to locate the principal divisions of the brain.

Comparative anatomical data on the skulls and brains of fossil and recent amphibia and reptiles suggest that the brains of primitive reptiles were not unlike the brains of a turtle or of *Sphenodon* or of a lizard in general form, except that the pineal eye and its stalk, together with the corresponding parts of the brain, would be less reduced; that is, it would be in its pristine condition of complete functional development. The olfactory parts of the brain would be unreduced, the optic lobes rather small, the cerebellum very small; the right and left halves of the cerebrum long and but little swollen, largely olfactory in function, with hardly a beginning of the outgrowth of the neopallium, or higher brain. These inferences are fully confirmed by the endocranial casts, which are known in *Eryops,* a primitive amphibian (fig. 16.73B) and *Diadectes* (C), one of the Permian stem-reptiles.

As to the endocranial casts of the South African mammal-like reptiles, Watson (1913) has shown that in the case of *Diademodon,* an advanced cynodont, the cerebellum was relatively enlarged (fig. 16.73D), suggesting the known superior running and balancing power of the locomotor skeleton, while the forebrain had not yet put forth the neopallial outgrowths, at least to any marked extent. An ironstone cast of the interior of the nasal cavity of a small cynodont also described by Watson shows that in these animals the turbinate scrolls, which are characteristic of the mammals, were already beginning to be visible. This implies that both the olfactory and respiratory parts of the brain were approaching the mammalian condition, the advance in the respiratory adaptations being indicated by the presence in cynodonts of a separate naso-pharyngeal duct (fig. 16.30D); also by the relatively high regional differentiation of the neck, back and loins (fig. 16.14), which suggests the beginnings of a diaphragm.

More recently, Olson (1944, pp. 100–110) has carefully reconstructed the braincast from serial sections of the skulls of two or more anomodonts, gorgonopsians, therocephalians and cynodonts. Owing chiefly to the lack of ossification of the anterolateral walls of the braincase, he recovered but little direct evidence as to the size and appearance of the anterior half of the brain. On the other hand, his material afforded excellent impressions of the floor and sides of the brain stem and of the otic and pituitary regions. In general, the cerebellum was large and had well developed lateral lobes and flocculi, a primitive mammalian character. The pons was well developed, especially in cynodonts. The fenestra ovalis of the inner ear, which is very large in primitive reptiles, had not undergone the extreme reduction seen in mammals, as the stapes had not shrunk to its almost microscopic mammalian proportions. The sella turcica was very large in the gorgonopsians and therocephalians studied; it projected well below the general floor of the brain. It was very shallow in anomodonts, and in the cynodonts there was apparently a marked reduction in the size of the hypophysis and infundibulum, which moved closer to the ventral structures of the brain, as the mammalian grade was approached.

ORIGIN OF THE MAMMALS

All the facts cited in the foregoing discussion indicate that various descendants of the pelycosaur-therapsid stock evolved more or less independently toward the mammalian grade, and that, of the few which were not too heavily penalized by their own specializations, the ictidosaurs (fig. 16.19M), *Bienotherium* (fig. 17.1) and the Triassic microcleptids came very near the mammalian class on one side; while the cynodonts, including *Tribolodon* (fig. 16.41), seem to have had the prerequisite characters to lead to the Jurassic insectivorous mammals (figs. 17.8–17.18).

Cumulative evidence from palaeontology and comparative anatomy thus proves that the mammalian class as a whole sprang from the therapsid order of reptiles. Very probably that source is best represented collectively by the cynodonts and the ictidosaurs.

CONTRASTING EVOLUTION OF BIRDS AND MAMMALS

Locomotor System.—In review, birds and mammals may well have had a remote, broadly common source in the most primitive order (captorhinomorph cotylosaurs) of quadrupedal crawling reptiles (figs. 12.1, 12.8). But the swiftly running ancestors of the birds (cf. figs. 15.3C; 9, 24) very early became bipeds, probably using their enlarged forelimbs at first for climbing and later for flying. Much later a few of them at different times spent more time on the ground and reduced or lost their wings. A few took to the water and changed air-wings to water-wings and running feet to webbed feet; but none ever reverted, at least in the adult stage, to the quadrupedal mode of locomotion on the ground, of their remote lizard-like ancestors. During this transformation they made various improvements in the heart, lungs, etc., and in their insulating feather-coat and thus became "hot-blooded." On the other hand, they retained outstanding reptilian features in the cranium, jaws, feet and even in the brain.

The mammals, on the contrary, starting from a quite different order of cold-blooded reptiles, long retained the quadrupedal mode of progression and only took to climbing and bipedal walking at far later dates. Consequently the difference in general appearance between the skeletons of a running dinosaur (fig. 15.3C) and a pigeon (fig. 15.9B) is much more striking than that between the skeletons of a Permian lizard-like pelycosaur (fig. 16.1A) and a primitive mammal (opossum, fig. 18.22).

Skull and Jaws.—In passing from the typical two-arched reptile (fig. 15.18A) to the typical bird skull (B), the transformation in the jaws and skull greatly expanded the brain and braincase and substituted a beak for teeth, but had not wiped out the single ball-like occipital condyle (fig. 15.20aA, D–K), a basic reptilian feature, nor had it greatly reduced the quadrate and the post-dentary half of the jaw; whereas in the pelycosaur-mammal sequence the single basioccipital condyle (fig. 16.3B) had become double (F) and the bones behind the dentary greatly reduced (fig. 16.36). Meanwhile the dentary, having become dominant, established a new jaw-joint with the squamosal (fig. 16.39). Nor should we omit the differentiation of the teeth or the remodelling of their crowns in the mammalian direction (figs. 16.40, 44).

The birds, for greater power and skill in flying, transformed the bones of the three-fingered hand (fig. 15.8aC, D) to support the feathered wing. They developed a high keel on the breastbone (figs. 15.9B; 15.11) to support the flight muscles, a long synsacrum (figs. 15.13; 15.14) to stiffen the back and a retractile fan-shaped tail supported on a plough-shaped bone to act as a break in landing (fig. 15.5). The opossum (fig. 18.22), by comparison, retained quite primitive five-toed hands and feet and a long, almost reptilian tail and only moderate changes were made in the limbs in passing from crawling to climbing habits, except that the scapula was radically altered, the coracoid reduced and freed from the sternum, the helmet-like ilium had become a trihedral rod and the tibio-tarsal joint had been further developed.

Birds and mammals may also both be considered as high-powered living engines with many self-regulatory systems for maintaining a high and relatively stable body temperature. These long established hereditary adaptations have been obviously prerequisite for the immense deployment of both classes into orders, families, genera and species.

However, within the class of birds, contrasts in body form, skin, epidermal covering and skeleton (as between diving penguins, bipedal moas and quick-darting humming-birds) do not seem to be nearly as radical as those between whales, men and leaf-nose bats among mammals.

EVOLUTION OF MAMMALIAN SKIN, HAIR AND EARS

Hair and Skin.—The mammalian hairy coat and soft skin were parts of a new, emergent and unique organ-system of glandular, hair-coated skin not found in any other vertebrate class but to some extent paralleled by the feathery coat and pliable skin of birds. It will be recalled that in the very remote ancestors of all vertebrates, as represented by the Ordovician, Silurian and Devonian class of ostracoderms, the surface of the head and body was protected by a shell-like crust of very complex, many-layered construction (Chapter VI), subdivided into small scale-like plates on the body and into a shield on the head and thorax. In the ancestors of the sharks (Chapter VIII) this complex armor broke up into a leathery skin, bearing shagreen denticles. In the armored ganoids (Chapter IX) much of the primitive armor remained, but in the bony fishes it thinned down into derm-bones on the head and horny scales on the body. The earlier crossopterygians (Chapter X) also retained the essential features of the primitive armor and so did their land-living descendants, the labyrinthodonts (Chapter XI).

Some of the later amphibians (Chapter XI), including the frogs and salamanders, thinned the body-envelope into a glandular skin. Among the earlier reptiles (Chapter XII) certain of the diadectid cotylosaurs developed bony and presumably horn-covered armor on top of the head. Among the turtles (Chapter XII) similar elements gave rise to the complex carapace and plastron. In the crocodilians and some dinosaurs (Chapter XIV) the armor consisted of bony plates or osteoderms covered with horn, a combination seen also in some lizards (Chapter XII). In many other reptiles the osteoderms disappeared, leaving only the horny scales. In the fossil mammal-like reptiles (Chapter XVI) there is no known evidence of osteoderms but in the larger ones, such as the dinocephalians, there was probably a thick leathery skin. In the large skull of *Cynognathus* (fig. 16.40G1) the rough surface of the preorbital face also suggests a thick skin, but in the small cynodonts (fig. 16.43A) the smooth facial bones are much like those of mammals. In the egg-laying duck-bill platypus (Chapter XVIII) the smooth bony snout is covered with a pliable, hairless, leathery skin, while the smooth top of the skull is covered by a furry skin. Hence smoothness of bone surface in itself gives no hint whether the skin did or did not bear hair.

The cynodonts, though advanced toward the mammalian grade in many parts of the skeleton, could scarcely have had a complete diaphragm (see p. 334 above) and the cast of the interior of the braincase of *Diademodon* (fig. 16.73D) figured by Watson, is far closer to the reptilian than to the lower mammalian grade. Hence it is unlikely that the cynodonts had attained that high level of metabolic activity which is characteristic of the hairy mammals as compared with reptiles.

The evolution of hairs was apparently only one aspect of the complex process of transforming sluggish "cold-blooded" reptiles, in which the body temperature varied with that of the surroundings, into the mammals, in which the body temperature can long be maintained at a relatively stable level in spite of severe changes in the surroundings. In short, in typical mammals (setting aside whales with their thick, hairless insulating skin) the appearance of hair was probably concomitant with the evolution of a richly vascular and glandular skin, including sebaceous or oil glands and sudoriparous or sweat glands, the former tending to conserve body heat, the latter to cool the surface and lower the body temperature.

Hair is made of a horn-like material and in the embryo the formative hair-papilla is basically equivalent to the feather-germ of birds. And hairs, like birds' quills, have a stiff, horny outer layer and a cellular, air-filled pith. The arrangement of hairs in little groups of five, three, etc., in different mammals recalls the arrangement of scales in "geodesic" rows, like the scales of fishes, noted in part by DeMeijere, in part by D'Arcy Thompson and more fully by Breder (1947).

In the scaly anteaters (*Manis*) such hair groups may have been fused into scales (fig. 19.38D) which cover the entire body and are remarkably like those of the scale-tailed lizard (*Zonurus,* fig. 12.52b, Z). But as the rest of the anatomy of the scaly anteaters is completely mammalian, and as there is a wide range of epidermal and dermal structures in every class of vertebrates, the scaly armor of the manids can hardly have been inherited directly from reptilian scales but may rather be regarded as convergent imitations of them. Scale-like structures also occur on the tails of *Echinosorex* (*Gymnura*), an insectivore, and other small mammals, but as they are often associated with hairs, their status as direct derivatives of reptilian scales is doubtful.

In a hair of circular cross-section the papilla, originally cone-like, has pushed the horny tip upward and radially to form a cone-in-cone system. Flat scales on the tail might arise from cessation of upward growth and acceleration of radial growth, with resulting polygons due to mutual pressure or accommodation. Thomson (cited by Cunningham, 1903, p. 733) suggested that curly hairs may have arisen in the soft stage by being pulled by their erector muscles against the resistant convex surface of the sebaceous glands; human nails are molded in a fold of growing epidermis arched around the bony ungual phalanx. The strong flexors of the hand exert pressure on the volar surfaces, which are wide and flat; but if their bases are very narrow, the nails are also narrow, or vice versa, and form either claws or hoofs.

The little muscles of the skin which cause hairs to "stand on end" convergently recall the muscles that move the fins of fishes or those which fluff out the feathers of birds. In the porcupines the skin muscles are strong enough to erect the long quills; and in the Australian spiny "anteater" (*"Echidna"*) short moving quills appear to assist the animal in digging itself into dry ground. Some large hairs, such as the vibrissae of cats and rats are exceedingly sensitive to touch, which means that they are connected with cutaneous twigs of the labial branch of the infraorbital or superior maxillary nerve (V1).

The armor of armadillos and glypodonts among the American edentates (Chapter XIX) consists of bony dermal ossicles (fig. 19.30) covered with horny plaques which are analogous with the osteoderms and horny plaques of some lizards and dinosaurs; but in the edentate mammals the pieces of the armor are pierced by hairs. Vestiges of bony plaques (fig. 19.38) persist in the thick leathery skin of the ground sloths and giant anteaters (*Myrmecophaga*). In view of the lack of traces of such structures in the numerous remains of mammal-like reptiles, as well as in the several known orders of Jurassic mammals, the available evidence gives no ground for the assumption that the exoskeletal developments in the edentates are directly inherited from reptilian ancestors; that is, the rather striking resemblance between the carapace of ankylosaurian dinosaurs and those of edentates may be regarded as purely convergent.

What the later mammals did inherit from the mammal-like reptiles was not so much particular forms of scales as a potentially diverse and complex skin, which by further emphasis of its different constituents has produced hair, vibrissae, quills, horns, nails, hoofs, etc., from the epidermis; sweat, milk, oil, musk, malodorous or poisonous secretions, from its cutaneous glands. The skin as a whole is usually soft and pliable but it may become stiff or leathery or horny, or it may give rise to blubber or to bony plaques or to cartilaginous supports, to fans and ear trumpets as well as to radar-like receivers (in bats).

Ear.—The essential base of the inner ear consists of a pouch from the embryonic ectoderm (fig. 16.49F1), lined with sensitive cells filled with endolymph and supplied with nerves (fig. 16.49A–F) from the "acoustico-lateral" system. The inner ear is a greatly enlarged and differentiated unit in a series of minute sensory pits arranged along the sides of the head and body in fishes (fig. 7.21). The primitive embryonic ear-pouch (fig. 16.48B), filled with a fluid called endolymph, divides into three parts and thus gives rise dorsally to the utriculus and the semicircular canals, and ventrally to the sacculus. In the bony fishes (fig. 16.48C1) the cavities of the utriculus, sacculus and lagena contain stony accretions from the endolymph called otoliths. The otoliths of opposite sides are attached to end-branches of the eighth or acoustic cranial nerve (fig. 16.49B). When the fish bends over on either side, the otoliths presumably press harder against the nerves of that side or vice versa, and when the liquid in the cavities moves, it causes differential pressures on the nerves. This system is also sensitive to vibrations produced by striking objects under water, or by the grinding together of fishes' teeth, etc. Airborne sound waves may even be felt under water by certain fishes. When the air-breathing fishes came up on land, they retained the essential parts of the inner ear but got rid of the otoliths and received the airborne vibrations through the middle ear. This consisted chiefly of the stapes and drum membrane of amphibians (figs. 16.51, 52).

A local offshoot of the sacculus called the lagena or cochlea was begun in the lizards and crocodilians (fig. 16.49D) and further developed in the birds. In the mammals, the lagena (E, F) becomes wound around like a snail-shell (cochlea). In mammals the cochlea contains the newest and most highly differentiated units, sensitive to vibrations that are set up in the endolymph fluid by the oscillations of the tympanic membrane, which in turn receives them through the angulated chain of the malleus, incus and stapes (fig. 16.55). In the pelycosaurs the inner ear-cavities were basic to those of mammals but the cochlea left no definite trace (Romer and Price, 1940, pp. 73, 74). However, it was indicated in the higher mammal-like reptiles (*ibid.*). The cavity of the middle ear of mammals is located in a dorsal extension of the Eustachian tube (fig. 16.61B), that is, it was ultimately derived from the hyoid division of the oralo-branchial chamber of fishes.

As to the origin of the mammalian auditory ossicles, the stapes is the direct derivative of the reptilian bone of the same name (figs. 16.52D–F *stp;* 53–63) which in turn was derived through the amphibians (fig. 16.52B) from the upper end of the hyoid arch of fishes (fig. 16.52A). The perforation of the stapes by the stapedial artery and nerve, when present (fig. 16.53E), was likewise inherited from the reptile, amphibian and fish hyomandibular; but in some mammals the stapes has become rod-like, secondarily converging toward those of turtles, lizards (fig. 16.54A), snakes and birds.

The mammalian incus (fig. 16.56C), as shown by Reichert, Gaupp and their successors, is the direct derivative of the reptilian quadrate (A) but reduced to almost microscopic dimensions.

The mammalian malleus (figs. 16.59; 16.60B1) represents a combination of the reptilian articular, which was the proximal end of Meckel's cartilage plus a derm bone (goniale), identified by Gaupp (1911) as the homologue of the prearticular, but by Olson (1944) as more probably the dorsal branch of the angular. The saddle-like or cog-tooth joint between the malleus and the incus (fig. 16.59B, C) corresponds perfectly to the joint between the quadrate and the articular of reptiles (A). The tympanic bone apparently represents the angular (fig. 16.63A).

The tympanum or drum-membrane seems to be a complex from two widely different sources: (1) the small triangular dorsal part (membrana flaccida, figs. 16.55A3; 16.63 *RTY*) is both the homologue of the reptilian tympanum and the last vestige of the fish-operculum (fig. 16.51A, B, C); (2) the "new" or mammalian tympanum (figs. 16.58A; 59B, C; 62; 63 *MTYD*) is believed to have slowly evolved as the mandibular recess (*REC.M*), a pouch from the reptilian Eustachian tube (fig. 16.61 *EUST*) pressed against the inner side of the angular bone of the mandible (cf. figs. 16.59A, B; 63 *ANG, TY*).

As noted before (pp. 336, 341) the pelycosaurs and therapsids supply several stages in the reduction in size of all the postdentary elements, and in the ictidosaurians and *Bienotherium* (figs. 17.2B; 17.4B) they were probably small enough to vibrate with the tympanic membrane, which from good evidence (fig. 16.54A) was attached in part to the reduced quadrate and articular. Thus, according to cumulative evidence, the jaws of the ancestral mammal-like reptiles were gradualy differentiated (by anisomerism or emphasis of the dentary bone) into an anterior or dentary part, concerned exclusively with the procuring and masticatiôn of food, and a posterior part concerned less and less with the support of the dentary and more and more with the transmission of airborne vibrations, from the outer to the inner ear.

The outer ear of mammals is another emergent and unique class character, though it may be foreshadowed to some extent by the folds of skin that surround the tympanum in some lizards. Its rise was probably correlated with the differentiation of slips of the constrictor colli profundus muscle, which pushed their way forward around the ear opening and over the bony face beneath the skin. The ear flap may have once served partly as a movable lid or operculum, as it does still in the duckbill platypus (fig. 18.1A), developing later into the long ear trumpets of the anteater *Orycteropus* (fig. 19.24), or the enormous ear flaps of the African elephant, or the sensitive ear-folds of some bats (fig. 19.20). The cartilage support of the ear seems to be another emergent mammalian character, but it may represent the upper part of the hyoid cartilage of reptiles.

THE TRANSFORMATION OF REPTILES INTO MAMMALS: A SUMMARY

After reviewing the rise of submammalian characters in the fossil South African reptiles, we may now summarize and picture the contrasts between the primitive stem reptile and a typical mammal, which measure the numerous advances made by the mammals over their lizard-like ancestors.

PRIMITIVE REPTILE	TYPICAL MAMMAL
1) Variable body temperature, able to make only narrow and imperfect adjustments to changing external temperature. This fundamental physiological difference is partly dependent upon many of the detailed anatomical differences noted below	1) Relatively stable body temperature, able to make adjustments to a wide range of heat and cold and to delay fatal over-heating or freezing for a long time
2) Body covered with scales (poor heat insulators)	2) Body covered with hair, including fur, the latter an excellent heat insulator
3) Skin poor in glands	3) Skin rich in sweat glands and sebaceous glands
4) Only a transverse septum behind the chamber that encloses the heart	4) A muscular-tendinous dome or diaphragm which acts as a force pump to draw air into the lungs; it also accelerates the circulation by acting as a piston on the abdominal viscera
5) The nasal cavity imperfectly separated from the food passage	5) A separate tunnel for the inspired air, the nasal cavity being separated by both a bony and a soft palate from the food cavity
6) A complicated lower jaw made of many pieces	6) A greatly simplified and effective lower jaw made of only two pieces, the opposite dentary bones
7) Only a primary jaw-joint between the articular and the malleus	7) A secondary jaw-joint at the new junction of the dentary with the squamosal

PRIMITIVE REPTILE (continued)	TYPICAL MAMMAL (continued)
8) The primary jaw-joint serving primarily the masticatory function	8) The primary jaw-joint transformed into the cog-tooth articulation between the malleus and the incus
9) The hinder parts of the primary upper and lower jaws of large size, serving as the quadrate and articular	9) The hinder parts of the primary upper and lower jaws greatly reduced in size and serving as the incus and the malleus
10) The cavity of the middle ear not enclosing the quadrate and articular	10) The membranous lining of the cavity of the middle ear wrapped around the reduced quadrate (incus) and articular (malleus), which appear at first to be within this cavity but remain outside of its membrane
11) The angular bone of the lower jaw functioning as such	11) The angular bone probably transformed into the tympanic bone
12) A median occipital condyle on the basioccipital bone	12) Paired occipital condyles on the exoccipital bones
13) The locomotor skeleton essentially like that of a lizard, adapted chiefly for sprawling, with occasional very swift movements	13) The locomotor skeleton adapted for sustained and rapid movement, the fore and hind feet brought well under the body in standing and running
14) The pectoral girdle adapted for sprawling movements of the limbs, the coracoid plate large, the scapula relatively small, a large interclavicle being present	14) The pectoral girdle adapted for the mammalian ways of running and standing, the coracoid very small, the scapula large, provided with an out-turned spine and a distinct depression for the supraspinatus muscle, the interclavicle absent
15) The humerus very broad, the shaft short and the wide opposite ends sharply twisted on the shaft	15) The humerus relatively slender, the shaft long, the ends "untwisted"
16) The "reptilian formula" of the digital phalanges of the fore feet (2.3.4.5.3.)	16) The "mammalian formula" of the digital phalanges of the fore feet (2.3.3.3.3.)
17) The pelvis with large unperforated flat pubo-ischiadic plate, extending well in front of the socket for the femur	17) The pelvis with large perforation in the pubo-ischiadic plate, the latter extended well behind the socket for the femur
18) The ilium short and primitively extending behind the socket for the femur	18) The ilium long, extended well in front of the socket for the femur
19) The thigh-bone (femur) short, with its oval flattened head on top of the shaft, and without a neck	19) The femur long and cylindrical with a nearly spherical head protruding from the inner side of the upper end of the shaft and supported by a distinct "neck"
20) The outer border of the upper part of the shaft of the femur bearing a low oval swelling (for the attachment of the deep gluteal muscles coming from the ilium)	20) The upper end of the femur bearing a high projecting "great trochanter"
21) The under side of the shaft of the femur bearing a very large proximal depression	21) The under side of the shaft of the femur bearing a small but deep proximal depression
22) The middle part of the under side of the shaft of the femur bearing a high and thick "adductor crest"	22) The middle part of the under side of the shaft of the femur bearing a much reduced, inconspicuous adductor crest
23) The two main ankle bones like thick pancakes, with a notch between them	23) The two main ankle bones highly differentiated, forming an "astragalus" and a "calcaneum"; the astragalus (talus) mounted on top of the calcaneum with a more or less double-crested upper surface for the tibia and a globular head for the lower ankle bone; the calcaneum with a backwardly-directed lever for the attachment of the tendon of the calf muscles, an upwardly-directed socket and ridge for the reception of the astragalus and a flattened distal end directed downward and

PRIMITIVE REPTILE (continued)	TYPICAL MAMMAL (continued)
	forward and resting on one of the lower ankle bones (cuboid)
24) The feet plantigrade	24) The feet normally digitigrade
25) The reptilian formula of the digital phalanges of the hind feet (2.3.4.5.4.)	25) The mammalian formula of the digital phalanges of the hind feet (2.3.3.3.3.)

Every one of these marked differences noted above (pp. 347–49) between a primitive reptile and a typical mammal is related to the leading physiological difference, namely, that the reptile is "poekilothermal" or with the body temperature variable, that is, with inferior resistance to changes in the temperature of the environment, whereas the typical mammal is "homeothermal," that is, with a normally "constant" body temperature, which means in practice an ability to make a fairly wide range of adjustments to outside temperatures so that the body temperatures vary only within relatively narrow limits. Most of the remaining differences between a primitive reptile and a typical mammal also relate directly or indirectly to this fundamental difference, a difference which apparently determined the ultimate dominance of the mammals in their age-long struggle for supremacy over their more primitive cousins, the contemporary reptiles. Thus the oldest of the mammal-like reptiles of South Africa were definitely reptiles, while the later cynodonts and ictidosaurians were almost mammals.

WAS IT BY FOREORDINATION OR BY SURVIVAL OF THE FITTEST?

Philosophers who hold that the course of evolution from fish to man was completely planned before it started will see in the successive emergent stages from cold-blooded reptiles to warm-blooded mammals and in the numerous and undeniable adaptations, another example of teleology and/or aristogenesis. Experimental biologists and palaeontologists on the other side must continue to deny foreordination in view of the following classes of evidence: (1) that any given stage is merely prerequisite to all its diversified descendants; (2) that in evolution, as we see it taking place, changing parts are always acquiring new values and potentialities to the individual and to the species; (3) that progressive integrations, syntheses, incorporations, of structures and/or values, along lines of parallelism, divergence, mimicry, concealment, convergence, all lend an illusory appearance of intelligent guidance and far-seeing strategies. But from an experimental or pragmatic viewpoint, the possibilities for progressive divergent and convergent paths of evolution spring from the fact that similar repetitive events, situations and problems are constantly being met automatically by old and new solutions, as a result of: (a) the Malthusian struggle for existence; (b) both the conservative and the disruptive effects of the mechanism of heredity; (c) the selection over long periods of the better adapted averages.

CHAPTER XVII

THE DARK AGES OF MAMMALIAN HISTORY

(*MESOZOIC MAMMALS*)

Contents

CHAPTER XVII

THE DARK AGES OF MAMMALIAN HISTORY

THE CHINESE TRIASSIC NEAR-MAMMAL *BIENOTHERIUM*

To THE therapsid order Ictidosauria is very doubtfully referred the remarkably well preserved skull, jaws and other skeletal parts of *Bienotherium* (fig. 17.1–5), described by C. C. Young (1947) from the Triassic of China. This was at the opposite adaptive extreme from the shear-toothed cynodonts, since its molars were crushers; the uppers (fig. 17.4B) having roundly four-sided crowns, bearing three parallel rows of low tubercles on the outer, middle and inner sides respectively (fig. 17.5A2). The lower molars (fig. 17.5A1, A3) were narrower transversely and bore two parallel rows of tubercles which pressed into the middle valley and along the inner borders of the uppers. There were no canine-like teeth but at the front end of the premaxillae there was a well-separated pair of stout, downwardly directed, slightly recurved, obtusely conical incisors (figs. 17.1A, 4B), probably inc. 2. Similarly, in the front end of the lower jaw there was a pair of somewhat forwardly inclined stout teeth. Behind these teeth in both jaws there were large gaps or diastemata bounded by concave bony borders and well adapted for holding large round objects. Such a dentition, which was only superficially rodent-like, seems well adapted for seizing and piercing hard-shelled fruits and crushing the seeds. It evidently required exceptional size and strength of the jaw-muscles and of the supporting parts of the skull.

Fortunately, the skull itself of *Bienotherium* shows that this was indeed a fact. Thus the ascending ramus of the dentary was of enormous size and its high front border was about at right angles with the line of the cheek teeth and, as in mammals with crushing molars, the articulation of the jaw was well above the general plane of the cheek teeth. This enabled the whole premolar-molar series to be in occlusion at the same time, and it stands in wide contrast with the scizzors-like dentition of *Cynognathus* (fig. 16.40G1), in which the cutting power was concentrated first on the rear part of the tooth line and then passed forward, ending with the snapping shut of the incisors. The cheek arch of *Bienotherium* was likewise immense and arched upward, as it had to withstand the upward pull of the temporal and masseter muscles. As seen from below (fig. 17.4B), the massive cheek-arches are bowed outward to afford a wide cross-section and great power to the jaw muscles. For similar reasons the skull-top (fig. 17.3C) was produced into a very thick sagittal crest and the occiput was wide and strongly braced.

The ascending ramus (fig. 17.1A) of the *Bienotherium* dentary was produced backward to a blunt point, which may have already made contact with the transverse branch of the squamosal to form the beginning of the temporomandibular articulation, commonly known as the mammalian jaw-joint. For in typical reptiles (figs. 12.40A, C; 16.54aA) the joint between the skull and the lower jaw is located between the lower end of the quadrate and the upper surface of the articular bone, whereas in mammals (fig. 16.54bC) the similarly functioning but quite different joint lies between the transverse bar of the squamosal and the dentary bone.

On the inner side of the ascending ramus (fig. 17.2C) of the *Bienotherium* dentary there was a long, sharply defined crest, running obliquely upward and backward toward the squamosal, and below this crest was the remnant of a narrow bony rod, possibly representing the articular and prearticular bar of the reptilian jaw. But it is evident that these slender strips could have at best only served to tie the tooth-bearing part of the lower jaw to the skull and there was no room for, and no trace of the large surangular and angular which form the main rear part of the jaw in the lower therapsids (fig. 16.36A, B).

Thus *Beinotherium* had almost crossed the indefinite zone between the mammal-like reptiles and the mammals and whether it be classified as a very progressive reptile or a very early and specialized side branch of the mammals is partly a matter of definition.

The evidently adaptive but very wide differences in teeth, jaws and skull that separate *Bienotherium* from *Cynognathus* are comparable with those that separate the herbivorous *Diprotodon* (fig. 18.34G) from the carnivorous polyprotodont marsupial *Thylacinus* (fig. 18.25D; but there is a much closer resemblance between the molar teeth of *Bienotherium* (fig. 17.5) and those of the hitherto puzzling *Microcleptes* (*"Microlestes"*) and allied genera of the Upper Triassic of Europe, which were formerly referred to the mammalian order Multituberculata.

Closely related to the Chinese Triassic *Bienotherium* was the front part of the skull named *Tritylodon* (fig. 17.1B) by Seeley, from the Upper Triassic of South Africa. The systematic position of *Tritylodon* was long uncertain, some authors regarding it as a forerunner of the Jurassic multituberculate mammals, others referring it to the therapsid order of the Reptilia. Simpson (1928), while recognizing its relationship with the microcleptids, separated both from the true multituberculates.

SEED-CRACKING ALLOTHERIANS (MULTITUBERCULATES)

As noted above, the Upper Triassic Chinese *Bienotherium* (fig. 17.1–5) and its South African relative *Tritylodon* (fig. 17.1B) were already almost or quite mammals; but they may have belonged to a peculiar fruit or seed-crushing side branch, the microcleptids, characterized by squarish, flat-topped upper molars with three parallel rows of tubercles on the upper jaws and two in the lowers. These microcleptids were only superficially rodent-like but they were more probably nearer to the true Allotheria or multituberculates. Already in the Jurassic the Allotheria had developed much enlarged, compressed lower posterior premolars, with long parallel oblique grooves on the sides, while the lower molars had become quite small. Whether or not the Jurassic and later Allotheria were related to the microcleptids, their dentition (fig. 17.6) was more highly differentiated and presumably more specialized. In a much later allotherian, *Taeniolabes* (*"Polymastodon"*) from the Paleocene of North America, the grooved anterior lower premolar had become very small and the molars were very much larger. The general appearance of the skull in side view (fig. 17.6) is more rodent-like than that of other allotherians; but after intensive and extensive comparisons of all known allotherian material, Simpson (1933a) sets forth cogent evidence that the Allotheria are only pseudo-rodents, ancestral neither to the rodents nor to the marsupials nor to the monotremes, and that they were representatives of a very early and extinct side branch of the mammals.

JURASSIC EXPERIMENTS IN DENTAL MECHANISMS

An Oxford Student Starts a Controversy.—We are now considering a period which we may call the Dark Ages of mammalian history because of the relative paucity of the known mammalian fossils during the vast Mesozoic Era, or Age of Reptiles. Indeed all the known Jurassic and Cretaceous mammalian fossils in the world could be put into a small box, although, owing to the mouse-like size of most of these broken jaws and teeth, a great many different kinds would be crowded together in that box. With few exceptions, the Mesozoic mammals are known only from teeth or jaws, which were minute, imperfect, and very difficult to free from the rock in which they are imbedded. Nevertheless there is an astonishingly large number of species, genera and families among them, representing six orders and several subclasses of mammals. They have therefore been eagerly studied by a succession of eminent palaeontologists, who have realized the immense importance of these documents as bearing on the earliest ages of mammalian history.

The story of the gradual discovery and appreciation of the Mesozoic mammals is fully and graphically told by George Gaylord Simpson (1928, pp. 3–7; 1929, pp. 1–5). As early as 1764 a fossil lower jaw about an inch and a half long was collected by someone in the Stonesfield Slate at Stonesfield, Oxfordshire. But this specimen (type of *Amphilestes broderipii*) was not figured until 1871. In 1812 or perhaps 1814, a young man named William John Broderip was studying law at Oxford and at the same time studying natural history and making a collection of his own. One day, as he later wrote, "An ancient stone-mason . . . who used to collect for me made his appearance in my rooms at Oxford with two specimens of the lower jaws of mammiferous animals, imbedded in Stonesfield slate, fresh from the quarry." Broderip and his friend Professor Buckland (who purchased the other specimen) were convinced of the mammalian nature of these specimens (fig. 17.9–1, 3) but as it had long been held that no mammals were found in formations of the Secondary (Mesozoic) Age, they discreetly waited for several years until Cuvier, the illustrious founder of vertebrate palaeontology, made a visit to Oxford. He assured Buckland that they were indeed mammalian and not unlike the jaws of an opossum.

Following Buckland's brief announcement of the discovery in 1824, the first detailed description and figure of Broderip's specimen was published in 1825 by Prévost, who correctly identified it as mammalian and pointed out its "analogy" with the marsupials. De Blainville, however, in 1838 published a controversial paper in which he cast doubt on the mammalian nature of the two jaws and suggested that they were reptilian. Richard Owen finally settled the controversy in 1838 and 1842 in favor of their mammalian nature. Owen also published the first systematic revision and thorough study of all the material then available (1871).

These English fossils had come from two horizons, a few from the Stonesfield Slate of the middle Jurassic and the rest from the Purbeck of the uppermost Jurassic. From western North America (Wyoming), Marsh (1880–1894) described numerous minute jaws and teeth mostly representing the same families or closely allied ones as those of the English Jurassic. These were all from the Morrison formation (Uppermost Jurassic), which has also yielded abundant remains of the giant sauropod dinosaurs. The English Jurassic mammals were also the contemporaries of their dinosaurs.

The great importance of the Mesozoic mammals in the problem of the origin of diverse patterns of the mammalian molar teeth led Osborn to restudy and figure them in his monograph of 1888 and upon them he based a part of the famous Cope-Osborn theory of Trituberculy, dealing with the evolution of mammalian molar teeth. Later both Marsh (1889, 1891, 1892) and Osborn (1891, 1893) described a few very small teeth from the Upper Cretaceous of Wyoming which had been collected by ants and carried into the anthills. Some of these priceless little teeth belonged to relatives of existing marsupials (fig. 17.23B–D). Others represent placental insectivores (fig. 17.19D).

Mesozoic mammals have also been found, but rarely, in other parts of the world: in the Rhaetic or uppermost Triassic of Germany, England, and South Africa; in the uppermost Jurassic of Tendaguru, Tanganyika Territory, in East Africa; in the Upper Cretaceous of Mongolia (p. 357), and in the latest Cretaceous (Lance and Hell Creek) of Wyoming, Montana, South Dakota.

After years of laborious studies of this subject by earlier investigators there were many doubts, obscurities, ambiguities and erroneous conclusions, due in part to the imperfection of the material, in part to human errors of judgment. In the definitive monographs on the Mesozoic Mammals by Simpson (1928, 1929) many of these errors and obscurities were exposed and cleared away and the irreducible truths set forth in their natural order and beauty.

Diverse Insect-hunting Mammals of the Jurassic Period.—Three other Mesozoic mammalian orders are represented by very small fossil jaws from the Jurassic of England and North America: (1) the Triconodonta (fig. 17.7A–D) whose compressed molar teeth in the more typical genera each bore three cusps in a fore-and-aft line; (2) the Symmetrodonta (figs. 17.7E, F; 17.13; 17.14A, B) with V-like, cuspidate lower and upper molars; and (3) the Pantotheria, in which the lower molar crowns of the more primitive genus *Amphitherium* (figs. 17.15A; 17.16) had a high tricuspid, triangular, elevated "trigonid," from the posterior base of which projected a small, basin-like swelling (talonid). Unfortunately the upper teeth of this genus are not known but in related genera (fig. 17.20B) the talonid of the lower molar occludes with the internal tip of the upper molars, while the lower molar trigonid fits into the spaces between the upper molars (fig. 17.17B). This is a basic spatial relation in the teeth of many of the relatively primitive mammals of later ages.

Most of the other genera of pantotheres (fig. 17.18C–I) had transversely widened upper molars, each with a prominent internal cusp and a larger or more prominent outer cusp, together with oblique shearing edges; the very small lower molars had a V-shaped elevated trigonid and very small talonid (fig. 17.20B).

The docodonts (fig. 17.20, 21) are a highly specialized branch of the Pantotheria with large internal cusps on their upper molars and complex lower molars with large heels. All the typical dryolestids and the docodonts probably became extinct and so far as known have no representatives in the later mammals. Gidley (1906, p. 105) suggested that the complex molars of "*Dicrocynodon*" (*Docodon*) were apparently derived from the simple reptilian cone independently of the molars of *Dryolestes,* but Simpson (1929, p. 85) showed that the *Docodon* molar patterns were homologous with, but more differentiated than, those of less peculiar pantotheres (fig. 17.20B) and that the docodonts could be connected with the latter by means (fig. 17.21B) of the European genus *Peraiocynodon* (*idem,* 1928, pp. 109, 112, 125).

It is difficult to homologize with certainty the cusps of the upper molars of the dryolestids and docodonts (figs. 17.18C–I, 17.21) with those of later mammals (figs. 17.19, 17.22); but a tentative solution is proposed in figs. 17.18, 19. The problem is discussed in Gregory, 1934, pp. 191, fig. 10, 247–251, 305; and Butler, 1939, 1941, p. 435 *et seq.*; but it cannot be said that a final decision acceptable to all authorities has yet been reached.

The typical pantotherians have eight molars on each side, above and below, whereas later mammals have in primitive forms either 4/4 molars in the marsupials or 3/3 in the placentals. One family of pantotherians comprising the very minute forms of the family Paurodontidae has a reduced number of cheek teeth, including four to two premolars and four molars (Simpson, 1928). The dental formula of the lower jaw of the most primitive known paurodont, *Peramus tenuirostris* Owen, according to Simpson (1928, p. 122), is I_1 C_1 P_4 M_4. He writes as follows:

> "This formula" (he notes) "is of the greatest interest and outstanding importance for it is the only known occurrence of the cheek-tooth formula that must be considered ancestral for both placentals and marsupials. This takes an added significance when one recalls that *Peramus* is the most generalized of Purbeck pantotheres in its molar structure."

The material of these Jurassic orders is scarcely sufficient to decide whether the known genera all belong to wholly extinct subclasses or whether, for example, *Amphitherium* (fig. 17.15) and *Paurodon* may be at least structurally ancestral respectively to the insectivorous marsupials and the Insectivora of the placentals.

The general form of the skull in *Priacodon* (fig. 17.8), one of the triconodonts, as reconstructed by Simpson is rather like that of an opossum. The pantotheres must have had fairly long, tapering faces and low braincases not unlike that of *"Gymnura"* (*Echinosorex*, fig. 19.9A) among recent insectivores.

Three humeri of small Jurassic mammals studied by Simpson (1928, pp. 155–159) were of primitive mammalian type, i.e. intermediate in details between humeri of cynodonts and that of the opossum and less specialized than that of the duckbill platypus.

On the whole, the very small Jurassic mammals, many smaller than moles and none larger than kittens, were primarily active seekers after insect food and were evidently not endangered much by their huge contemporaries the dinosaurs. They are separated by a great gap in time from the better known mammals of Paleocene and later ages; but some of them assuredly foreshadowed the insectivorous and smaller carnivorous mammals of both the marsupial (fig. 17.19A, B) and the placental (fig. 17.19C, D) series. The lower dentition, especially that of *Amphitherium* (fig. 17.21) seems indeed to be an ideal stage in the evolution of the lower molar teeth toward the central tuberculo-sectorial type (figs. 17.25; 18.27A) of later mammals.

CRETACEOUS PIONEERS OF THE MARSUPIALS AND PLACENTALS

There is a long hiatus in the fossil mammalian record, covering the interval between the basal Cretaceous and the lower part of the upper Cretaceous; but we are extremely fortunate in having at least a few mammalian fossils from upper Cretaceous horizons, chiefly in western North America and Mongolia.

In 1889 Prof. O. C. Marsh described some minute isolated teeth from the "Laramie" (Lance) formation (Upper Cretaceous) chiefly of Wyoming. He noted the resemblance of some of them to opossum teeth but as they differed in several important details from that type and since they were millions of years older than the modern opossum, he proposed a "new family," which he named Cimolestidae. Eventually the news leaked out that these minute fossil teeth had been obtained by sifting the contents of ant hills, the ants having for some reason thrown them out during their extensive mining operations near the surface of an Upper Cretaceous country. Marsh's great rival, Cope, also sent collectors to rob the ants and they found the jaw which he named (1892) *Thlaeodon padanicus*. This was a much larger animal (fig. 17.23D), in which the last upper and lower premolar crowns formed very large bluntly conical cusps, adapted for breaking shells or small bones, while the relatively small molars retained their didelphiid heritage. Osborn (1893) described and figured another small lot of these Upper Cretaceous mammal teeth. He applied Marsh's name to some of them but gave several new names to certain ones that seemed to deserve them.

The most definitive discovery in this field was made by Dr. Barnum Brown in the course of his numerous excavations for dinosaurs in beds of Upper Cretaceous age on the Belly River in Alberta, Canada. Immediately underneath the bones of a trachodont dinosaur he found in place and in the same matrix fragments of the skull and a well preserved jaw of a small mammal (fig. 17.23A), which, after being carefully studied by Dr. W. D. Matthew (1916), proved to be a member of the American opossum family (Didelphiidae). Its lower teeth rather closely resembled those of the Virginia opossum except for minor differences. The special importance of Brown's discovery was that for the first time the skull fragments belonging with the jaws and teeth admitted no doubt that the direct ancestors of the American opossum were present in North America in Upper Cretaceous times before the dinosaurs became extinct.

Thus the humble American opossum has the distinction of representing the oldest known American family, its ancestors antedating the "first families of Virginia" by about seventy or eighty million years. It is also the most archaic "living fossil" among all the mammals of the northern world and with few exceptions its general skeletal and dental characters are so primitive that it has been admitted provisionally to the exhibit, in the American Museum of Natural History, entitled "The Skeleton from Fish to Man," as a representative of a primitive tree-living stage preceding the Primates (fig. 24.2a, bE).

This is not meant to imply, however, that the placental mammals, including man's ancestors, passed through a definitely marsupial or pouched stage, although, for reasons to be noted below, that possibility can hardly be considered as excluded.

While the subsequent history of the opossum group will be considered in the next chapter, let us return to the Upper Cretaceous world of dinosaurs and ancestral mammals and look at an equally important discovery, that of ancestral placental mammals in the Upper Cretaceous of Mongolia.

Near the Flaming Cliffs of Shabarakh Usu in the Gobi Desert the palaeontologists of Dr. Roy C. Andrews' expedition of 1923 discovered the dinosaur eggs which soon gained world-wide fame. From the same Upper Cretaceous beds they took out a superb series of skulls and skeletons of the dinosaurs that laid the eggs. But a still greater discovery of theirs was the finding of seven small skulls or parts of skulls, some of them with associated lower jaws, of "Mesozoic mammals" in the same formation (fig. 17.24). After being very carefully and skilfully extracted from the matrix by Mr. Albert Thomson, these precious relics were studied and described, at first by Gregory and Simpson (1926) and later by Simpson (1928). One of the skulls belonged to a multituberculate and helped to tie in one of the later families of North American multituberculates with its Jurassic predecessors, the ptilodonts. The others, when viewed in the long perspective of time and evolution which is set forth in this book, seem to warrant the following broad conclusions:

1) They belonged to placental, not marsupial, mammals and therefore at least the initial separation of the placentals and marsupials had taken place at some earlier period, presumably the mid-Cretaceous.

2) They were the predecessors of the placental insectivores and creodonts of the Paleocene and Eocene of North America and Europe.

3) They represented two families, four genera and four species and therefore indicate that at least one center of the evolution of the placental mammals was somewhere in central Asia or at any rate in the northern land mass.

4) One of these Gobi forms, named *Zalambdalestes* by Gregory and Simpson, was probably related to the stem of the early Tertiary insectivores of the family, Leptictidae (fig. 17.25A, B).

5) In the other main genus, named *Deltatheridium* (meaning, little beast with delta-shaped molars, like *Deltatherium* of the North American Paleocene), the upper molar crown patterns (C) were of the general type which had been predicated as preceding the "tritubercular" stage of the Paleocene and Eocene. The species was therefore named *"pretrituberculare."* This form or something very close to it may have been near the common ancestors of the centetoid insectivores (E) and of oxyclaenid creodonts or carnivores (F) of the Paleocene and Eocene epochs of the Tertiary period.

6) In general the discovery of these Upper Cretaceous placental mammals of Mongolia brought definite evidence in favor of the view of Huxley (1880) and Osborn (1888) that the varied mammalian types of molar teeth had all been derived by divergent evolution from a central insectivorous type which Cope and Osborn had named "tritubercular" in the upper, and "tuberculo-sectorial" in the lower molars (figs. 17.10, *4*).

In the Upper Cretaceous of North America, besides the marsupial teeth mentioned above, there were two or three isolated teeth, named *Gypsonictops* (fig. 17.19D), which have been carefully studied by Simpson (1929, pp. 137, 138) and referred to the leptictid family of placental insectivores. Although mere fragments, these specimens are of real importance because they extend the known range of this primitive insectivore family well down into the Upper Cretaceous.

THE PASSING OF THE GREATEST RACE

As the Mesozoic Era drew slowly to its close (a million years being less than $1/_{150}$ part of its length) the dinosaurs of both orders as well as pterosaurs, ichthyosaurs, plesiosaurs and other reptilian groups reached the climax of their size and diversity. By this time the reptilian hordes had spread over the entire accessible, habitable earth and were so well adapted to their environments that successful competition with them, much less their complete extermination, would have then seemed highly improbable. But the twilight of the gods of the reptilian world came with unexpected suddenness when the pent-up earth forces were let loose.

In North America and South America the elevation of the Rocky Mountain system brought in great changes, converting a low-lying land not dissimilar to modern Florida into a high mountainous and plateau country with scant rainfall and consequent drying-up of the swamps and jungles loved by the dinosaurs. In far-off Australia and New Zealand, as well as in North America and Europe, the Upper Cretaceous seas supported the great marine reptiles, which are never found in Tertiary horizons. Therefore the Upper Cretaceous was a disastrous time for the marine reptiles no less than for the dinosaurs. And since this period also marked the uplifting of great land areas and the draining of their shallow epicontinental seas, we may assume that the geologic changes were the prime factors in this wholesale extermination.

CHAPTER XVIII

BRANCHING EVOLUTION IN EGG-LAYING AND MARSUPIAL MAMMALS

Contents

CHAPTER XVIII

BRANCHING EVOLUTION IN EGG-LAYING AND MARSUPIAL MAMMALS

THE PLATYPUS NOT A "LINK BETWEEN BIRDS AND MAMMALS"

WHEN DR. ROY C. ANDREWS found fossil dinosaur eggs in Mongolia the discovery was "news" all over the world and so millions of people became more or less vaguely aware for the moment that dinosaurs laid eggs. People generally also know that chicks come out of eggs because eggs are sizable and conspicuous objects which are produced in millions and because pictures of fluffy chicks coming out of the egg appear frequently in the "funnies," on Easter cards and subway advertisements. But such microscopic objects as the eggs of dogs, cats and man, never being seen by the vast majority of people, seem to have a purely academic interest if indeed the public is at all aware of their existence. Even the Australians, who are at least as justly proud of their native fauna as the people of southern California are of their climate, do not mention the eggs of their famous duckbill platypus (fig. 18.1A) in their most ornate and conspicuous official publication on Australia and its wonders. To make matters worse, the platypus is usually spoken of as "a strange sort of link between the bird and the animal" (meaning *mammal*), which it assuredly is not. So far as the public is concerned, it was in vain that Dr. Caldwell in 1888 sent his triumphant cablegram to the British Association for the Advancement of Science: "Monotremes oviparous, eggs meroblastic." On the other hand, it is gratifying to note that it was assuredly not in vain that Mr. Harry Burrell (1927) stood for untold hours in the cold streams in the highlands of New South Wales, observing the living platypus, gently trapping them for his ingenious "platypusary" and finally being rewarded with fresh platypus eggs. For due largely to his persistent efforts not only to discover the facts but to popularize them, many highly instructive articles on the subject have appeared in Australian newspapers and magazines and strong efforts are being made to provide sanctuaries for this and other relics of Australia's glorious zoological past.

The main reasons why the erroneous idea that the "Duckbill Platypus" is a "sort of link between birds and animals" ever got abroad may be as follows: (1) that the writers of books and articles on popular natural history have followed each other like sheep in repeating a misleading half-truth that was common among the natural history books in Europe about the time Australia was settled by white people. (2) The platypus itself seems at first sight to give irrefutable evidence for the statement quoted, because its bill indeed looks somewhat like that of a duck and it has webbed feet which are used in swimming. The fact that the female duckbill lays eggs, seems to clinch the argument, for do not birds lay eggs? (3) Sometimes it is noted that the platypus bears a pair of cock's spurs (fig. 18.20C1) on its ankles. (4) On the other hand, the platypus evidently differs from birds in that it has no wings but has four legs "like an animal" and is covered with a sleek fur, like a seal. So, "why isn't it a link between birds and animals, after all?"

The answer must be divided into several parts:

1) Three of the leading bird-like features of the platypus, namely, (a) the "duckbill"

(fig. 18.1A), (b) the webbed feet, and (c) the tarsal spurs, are purely convergent or habitus characters.

a) The "bill" only superficially resembles the bill of a duck. In the living animal it is not at all horny, like the bill of a duck, but soft like kid-skin leather. It seems to represent a huge expansion of the muffle (fig. 18.10B, B1) found in some of the marsupials, e.g., the koala and *Notoryctes* (H. C. Raven).

b) Webbed feet, of course, are by no means confined to birds and the web of the feet of the platypus is not basically different from, though larger than, those which are between the bases of human fingers.

c) The spur (fig. 18.20C1) is entirely unlike a cock's spur in so far as it is a hollow tube, connected with a gland (?modified sebaceous) which gives out a poisonous secretion; the latter is described by Martin and Tidswell as "practically a solution of proteids, chiefly albumen but also proteose." Wood Jones (1924, p. 37) states that this secretion possessed a decidedly poisonous character and that its composition and its effects are probably akin to those of snake venom. H. C. Raven suggested to the writer that it probably grew out of a callosity that was originally associated with a tarsal skin gland such as occurs in some other mammals.

2) The egg-laying character is a primitive heritage from ancestral reptiles, as is also the complete separation of the right and left oviducts, while, on the other hand, the presence of true hair, of milk glands, a diaphragm, and many other anatomical characters, demonstrate that the platypus is exclusively a mammal.

3) At the time when the platypus became known to European naturalists the principle of convergent evolution had not yet been discovered, but the existence of "annectent forms" between different groups was widely recognized. The naturalists of that time saw nothing incongruous in the idea of a link between birds and "mammiferous quadrupeds" because they knew nothing that was relevant of the fossil history of the vertebrates and because the science of comparative anatomy had been but recently founded and the significance of its data was but poorly understood. Moreover the authors of English natural history books of that period were ultra-conservative and would have scarcely understood or appreciated De Blainville's classification of 1839–1864, in which he grouped the platypus with the spiny anteater (echidna) of Australia in a grand division or subclass of the mammals under the name "Ornithodelphes" (bird wombs); this was later changed by him to "Monotrèmes" (single opening) in allusion to the presence of a common opening or cloaca into which the reproductive, the excretory, and the digestive tracts all opened.

4) At that time it was not known that both birds and mammals are divergent branches from widely different parts of the primitive reptilian class; still less was it known either that the birds are "glorified reptiles" derived from the stem of the most typical reptiles, which are the diapsid or "two-arched" series, or that the mammals (including the platypus) are the transformed descendants of a quite different side branch of the reptilian group, the synapsid, or "single-arched" series.

In evolutional science, as in all other progressive sciences, every question in the course of being answered gives rise to other questions that extend like an infinite pattern of subdividing and often cross-linked branches. So then, if the duckbill is not a link between birds and mammals: (1) How nearly is it related to other recent mammals? (2) From which one, if from any, of the known orders of Mesozoic mammals has it been derived? (3) Does the anatomy of the duckbill and its allies throw any light upon the origin of mammals? (4) Did the monotremes give rise to the marsupials and placentals, or are the three divisions equal and coördinate? (5) To what extent can we draw from the anatomy of the duckbill illustrations of the anatomy of man's remote early mammalian ancestors?

HABITUS AND HERITAGE IN PLATYPUS AND ECHIDNA

As to the degree of the duckbill's relationships with other mammals, its nearest relationships are obviously with the spiny anteater (fig. 18.1B), *Tachyglossus* (echidna) of Australia (including Tasmania) and the long-billed anteater (*Zaglossus*) of New Guinea (fig. 18.1C). The last named is merely a specialized echidna with a long curved snout and the ability to arch his back and walk high on his post-like legs, which makes him look like a pygmy caricature of an elephant with very small ears. We may broadly contrast some of the habitus features of the platypus and the echidna, as follows:

	PLATYPUS	ECHIDNA
General habitat	Streams and pools	Swamps to deserts in any soil suitable for digging
Locomotion	Swimming by paddling and undulation of body; diving with broad tail; digging long tunnels with narrow claws (fig. 18.1A)	Slow, rocking amble; digging by means of broad strong hands and feet tipped with large curved claws
Skeleton	Relatively slender (fig. 18.17)	Extremely robust
Tail	Very large, beaver-like (fig. 18.1A)	Vestigial
Limbs (proximal segments)	Very short; held sharply akimbo	Very short; held sharply akimbo (fig. 18.1B) or elevated (C)
Feet	Changed to long paddles, with digging claws (fig. 18.1A)	Changed to broad digging organs (fig. 18.1B)
Integument	Sleek fur, like sealskin when in water	Very harsh fur, with quills (fig. 18.1B)
Diet	Insect larvae, worms, aquatic plants	Worms, insects
Bill	Flattened, duck-like	Tubular, snipe-like (fig. 18.5C)
Teeth	Present in young as vestigial molars (fig. 18.9B); replaced by hardened gumplates (figs. 18.5B; 18.6, 7)	Completely absent (fig. 18.5C)
Tongue	Flattened to work against palate	Tubular, sticky
Lower jaw	Well developed to support wide lower beak and oval gum-plates (fig. 18.6B)	Reduced to very slender rods
Smelling organs	Moderate or small (fig. 18.12B)	Greatly enlarged (fig. 19.36D)
Eyes	Small (fig. 18.1A)	Very small (fig. 18.1B)
External ears	Tubular opening just behind eye (fig. 18.1A)	Well developed pinna concealed by hair
Eardrum	Very small (fig. 18.14 *cav. tym.*)	Large (fig. 18.15B)
Brain	Relatively small	Large

Beneath these and many other conspicuous habitus differences there is a marked uniformity in subclass and ordinal heritage characters. Both animals lay sizable eggs (fig. 18.2), both have similar localized glandular areas where milk is secreted, in both the right and left oviducts of the females are completely separate as in reptiles, and so forth (cf. Gregory, 1910, pp. 158, 159). Nor do the skeletons, in spite of conspicuous habitus differences, fail to reveal numerous monotreme heritage features. We may safely conclude then that the duckbill is rather closely related to the echidna (*Tachyglossus*) and the proechidna (*Zaglossus*), and that together they form a closely compact mammalian order commonly called Monotremata. This group must, however, be an old one for various reasons, among them being the relatively wide structural gap between the two extremes and the absence of known intermediates either fossil or recent.

The monotremes retained an essentially reptilian separateness of the right and left oviducts, down to the urogenital sinus, and therefore they lack a true uterus. Moreover, they are the only adult mammals in which the opposite coracoids are large enough to reach the sternum (fig. 18.16E) or which retain an intact interclavicle (episternum). As noted above, they display several curiously bird-like features, such as the habits of laying eggs and of making nests for the young (the latter, however, being paralleled mong some small mammals), the presence of pair of horny spurs on the ankle as in fighting-ocks, but these spurs are grooved for the pas-age of a secretion from a poisonous crural land. Moreover, the nipples are represented nly by a depressed glandular area, which is cked, not sucked, by the young, and the brood ouch is said to be temporary, not permanent.

Joined with these and other highly excep-onal or submammalian features, are many thers which establish the status of these ani-ials as well within the mammalian class. Here elong the completely mammalian hair and dia-hragm, the general construction of the skull figs. 18.4–12) and indeed of the skeleton as a hole, which is much closer to the marsupial rade than to the therapsid reptilian (figs. 8.17–19) grade.

The monotremes differ sharply from the arsupials in the facts that they lay eggs, that e right and left oviducts are completely sepa-ited down to the cloaca, that they have an most completely reptilian shoulder-girdle (fig. 8.16) with interclavicles, clavicles, large pre-racoids and coracoids, whereas the adult arsupials (B) have lost the interclavicles and ecoracoids, while their coracoids are reduced small coracoid processes which become fused ith the scapula. Moreover the scapula of mo-otremes, like that of the cynodonts (fig. 8.16A), has at most an incipient stage of a praspinatus fossa, while that fossa in the arsupials is well developed (fig. 18.22). In e skull structure the two families of mono-emes agree with each other and contrast with e marsupials in a great many important fea-res to such an extent that while all parts of e marsupial skull are easy to identify by com-rison with the skulls of man and other pla-ntals, the correct identification of the corre-onding parts in the monotremes affords an possible task for the beginner and even led e experts of earlier days to several misidentifi-tions and misinterpretation of the parts. oreover, the construction of the brain in both not only mammalian but essentially mar-pial.

The long accepted explanation of the seemingly anomalous mixture of premammalian and mammalian characters in the duckbill and "anteater" has been that, although they preserve today some outstanding premammalian features, which all the ancestors of the higher mammals must also have had, yet they are the sole surviving representatives of an otherwise unknown extinct subclass of mammals which branched off at or near the beginning of the mammalian class.

THE PALIMPSEST THEORY OF THE ORIGIN OF THE MONOTREMES

A recent and quite different interpretation (Gregory, 1947) is that the order Monotremata is really much closer in origin to the marsupials than it is to the Triassic pro-mammals, and that some of the apparently "reptilian" features may be due in part to the now well known process of neoteny or paedogenesis (pp. 97, 253). According to this principle, certain ancient embryonic features which are ordinarly passed through before the adult stage are sometimes retained in the adult by the lagging of later growth stages either in vigor or in timing. In other words, the right and left oviducts, which in vertebrates generally are completely separate in the embryonic stages, may, in the monotremes, have simply failed to unite in the midline and thus made possible a seeming reversion to a reptilian stage. This initial check in development then permitted (according to this hypothesis) a partial and incomplete resurgence of the latent embryonic forces which in earlier ages would have produced a premammalian, not a mammalian, adult. The strength of this hypothesis, as applied to the case of the monotremes, lies chiefly in the great number and variety of the points of close agreement between the adult monotremes and the marsupials in such features as the arrangement of the cranial foramina (figs. 18.14), the mode of development and adult construction of the auditory ossicles (fig. 18.15) and internal ear, the close general resemblance of foetal monotremes (fig. 18.10) to foetal marsupials, rather than to embryo reptiles, the wholly marsupial stage of the brain, etc., as more fully set forth in the work cited above.

The duckbill lives in the pools of streams in eastern Australia and Tasmania and makes its long burrows and nests in their banks. In swimming, the large pectoral and pelvic paddles and large, somewhat beaver-like tail are all used for quick turning and diving. The skeleton and general anatomy, sense organs, brain and outer covering, are adapted for this double way of life. Its sleek, somewhat otter-like fur slips easily through the water and soon dries on land. The depressed, spatulate muzzle is covered by hairless, pliable skin which is uniquely extended but arises in the young embryo (fig. 18.10B1) from a patch of skin on the end of the muzzle, much like those of the embryo wombat and koala among marsupials. The muscles around the eyes, ears, forehead and neck are derivatives of the sphincter colli profundus and are innervated by the facialis branches of the seventh cranial nerves as in other mammals (Huber, 1930). When the animal dives, a part of this system pulls the skin together around the eyes and small ear holes. The outer ear tubes are prolonged forward outside the cheek arches to the rear border of the small round orbits (figs. 18.4B; 18.6B)—a unique specialization from a fully mammalian beginning. The hearing of the platypus is remarkably keen and its auditory ossicles (fig. 18.15) are completely mammalian in type and neither more nor less reptilian in their embryology than those of other mammals (De Beer, 1937).

The front ends of the bony jaws of the platypus (fig. 18.4B) enclosed in the spatulate bill, convergently suggest those of the duck-bill dinosaurs (C). The lower jaws (figs. 18.6; 18.9B) are composed solely of the dentary bones, as in mammals, the post-dentary elements having long since become minute and transformed into the malleus and tympanic ring (fig. 18.15C).

The young stages of platypus jaws contain the calcified germs or crowns of small teeth, comprising so-called incisors (lower), canines, premolars and molars, of which at least one upper and one lower premolar have replacing germs beneath their own crowns (Green, 1937). The irregularly cuspidate and distorted-looking low-crowned molars (fig. 18.8B, B1) have n[o] close resemblance with those of any othe[r] known animal. They are clearly in a degenerat[e] condition and are gradually pushed out of th[e] jaws by growing horny plates (fig. 18.9B) formed in the gums around the true teeth. These high-rimmed (figs. 18.5B; 18.9B) rounded basins take the place of teeth and ar[e] presumably more efficient in holding an[d] crushing the relatively enormous amount o[f] worms, insect larvae, etc., consumed by thi[s] very active animal, whose body temperatur[e] (Wood Jones, 1923) is only a little lower an[d] less stable than that of ordinary mammals.

The duckbill is a very efficient, mole-lik[e] burrower and makes long tunnels leading gentl[y] upward from the submerged openings in th[e] banks of the stream; local loops contain ampl[e] chambers for the nest and young, which ar[e] safely walled in temporarily while the parent [is] out searching for food, including worms an[d] perhaps subterranean and subaqueous insec[t] larvae as well as small "shrimps" (?branchio[-] pods, ?ostracods).

The five-toed hands and feet and indeed th[e] entire girdles and limbs (fig. 18.15B1) are we[ll] adapted both for digging and swimming. Th[e] web of skin on the large hands, which form fan[-] like, extensible paddles, can be folded unde[r] exposing the claws for digging.

In the pectoral girdle the exceptionally larg[e] coracoid and pectoral muscles are based partl[y] upon the expanded epicoracoid (precoracoid) plates and strong obliquely placed, pillar-lik[e] coracoids (fig. 18.16C); the latter in turn trans[-] mit the thrusts from the powerful forearms (fig. 18.17) obliquely inward and backward to th[e] strongly built sternum (fig. 18.16E). Th[e] strong bony sternum of monotremes is brace[d] in front by the wide flat episternum (*epist*) which has long, transversely extended bars sup[-] porting the clavicles (*clav*). To the outer en[d] of this episternal-clavicular crossbar the acro[-] mial processes of the scapulae are tied by liga[-] ments. Thus in ventral view the construction o[f] the well braced, triangular pectoral girdle of th[e] monotremes is in principle somewhat like th[at] f the earlier nothosaurs (fig. 13.10B); but in his case the enlargement of the coracoid bones s plainly correlated with the combination of addling and digging habits, and it must have trengthened the earlier more primitively reptilian episternum and clavicles. The climbing halangers, on the other hand, greatly reduced he coracoids in the adult; but in the foetus Broom, 1899) they retain the connection of he coracoids with the sternum (fig. 18.16D). This inference seems much less hypothetical han would be the case if we assumed that the houlder girdle of the platypus had acquired a arge interclavicle secondarily.

In the front view, the wide pectoral girdle of monotremes is seen to form a cradle for the fore art of the body, which is suspended from the pposite scapulae and is based on the wide episternum, epicoracoids and coracoids. In side iew, the blades of the scapulae (fig. 18.17) are nclined forward and to them are attached muscles that run forward to the sides of the large cciput, others that fan upward toward the vertebral column, the serrati that fan outward to he ribs, and the subscapulars that run down to he humeri. The platypus pectoral girdle as a whole retains clear traces of remote derivation rom cynodont-like ancestors (fig. 18.16A).

The pelvis of the platypus in the side view fig. 18.19) is remarkably small; in the ventral iew is much narrower than the pectoral girdle; his brings the thrusts from the opposite femora earer to the plane of the vertebral column and he femora are therefore efficiently placed for elivering larger forward thrusts in both swimming and digging. The large, flat, divergent epipubic bones (fig. 18.19) help to tie in the abdominal wall to the pelvis, as they do in marsupials.

Further comparisons of the skull, jaws and locomotor skeletons of both the duckbill and he echidna with those of cynodonts on the one and and marsupials on the other are set forth n the paper noted above (Gregory, 1947). Here it may suffice to note that the entire skeleton of echidna suggests derivation from a duckbill-like ancestor through the following outstanding changes in function and structure:

1) Retreat from water-living and great emphasis of digging power, involving loss of paddle-like adaptations in limbs and shortening of tail, with corresponding strengthening of the vertebral column as well as of the pectoral girdle and forelimbs, feet (fig. 18.20D, D1), sacrum, pelvis and hind limbs. Simultaneously the fine fur of the duckbill type was gradually replaced by horny spines of varying thickness, no doubt with corresponding increase in the size and strength of the muscles of the skin. In digging into hard ground or in forcing its way out of a box-cage, an echidna displays astonishing strength.

2) The bill of an early foetal platypus (fig. 18.10B) is relatively far shorter than that of the adult. To transform this snout into the narrow pointed bill of the echidna it would only be necessary to greatly increase its length (fig. 18.5C) and to eliminate the horny beak. Meanwhile all traces of teeth have disappeared, as has happened elsewhere among edentates and other anteating mammals.

In brief, the main points of the "palimpsest theory" of the origin of the monotremes are as follows:

1) The living platypus abounds in functional adjustments to a very ancient combination of: (a) digging long tunnels underground, perhaps primarily in pursuit of insect larvae, and (b) swimming and diving to reach the submerged entrance to its tunnel and possibly to dig for insect larvae in the muddy bottom.

2) Apart from its primitive mammalian characters noted above, its skeleton as a whole suggests derivation from some semiaquatic and fossorial branch of the Australian diprotodont marsupials, which may have been related to the phalangers and wombats. Some of the main items of evidence for these conclusions are depicted in figures 18.4–20.

3) Many of the differences between a young wombat skull and an adult platypus skull (fig. 18.6) are associated with the elongation of the muzzle, the reduction of the teeth and the related changes in the jaw muscles of the platypus. New polyisomeres are the horny gum plates which crowd out the retrogressive cheek

teeth. Somewhat similar changes would be needed to convert the primitive proboscidean (fig. 18.7A) into the ancestral mastodont (B).

4) Although the cheek teeth of the platypus (fig. 18.8B, B1) are quite retrogressive, the first and second molars are divided into anterior and posterior moieties by deep transverse folds, suggesting possible derivation from some form with W-shaped molar crowns such as those of the very young wombat (fig. 18.8A, A1). The loss of the dentition in the adult platypus indicates a profound change in its feeding habits, as an earlier set of specializations have been overlaid and largely obliterated by a new set.

5) In spite of many profound changes in habitus features, the skull of the platypus retains many basic heritage features in common with those of marsupials (figs. 18.4–14).

6) Although the monotremes share many significant heritage features with marsupials (Gregory, *op. cit.*), yet they may be more primitive than the latter in retaining an essentially reptilian shoulder girdle, with large coracoids in contact with the sternum (fig. 18.16E) as in the foetal marsupial (D). The platypus is also more primitive, more reptilian, than either marsupials or placentals, especially in the basic patterns of its reproductive system in both sexes, as well as in its retention of egg-laying habits.

DIVERSELY ADAPTED SKELETONS OF MARSUPIALS

The skeletons of the egg-laying mammals of Australia, while preserving some primitive mammalian features, have on the whole specialized themselves away from the central or main line of mammalian evolution by over-emphasizing the swimming and digging features in the platypus and the digging adaptations in the echidna. There is abundant evidence for the theory of Dollo (1899) that the nearer common ancestors of all the existing marsupials were tree-living forms not unlike the American opossum (fig. 18.22); but the pre-marsupial stage may well have been ground-living, egg-laying mammals related to the monotremes, but much less specialized and probably related also to some of the small Jurassic mammals (fig. 17.15A) other than allotherians.

The Virginia Opossum and Its Archetypal Skeleton.—The Virginia opossum is well adapted for climbing, having strong grasping power in its hind feet, in which the large first or great toe is directed inward and can extend beneath a branch, while the four remaining toes press on the top and side of it (cf. fig. 18.33A). The five-toed hands can also grasp the food (fig. 18.22), consisting of young birds, eggs, mice, frogs, fish and insects, fresh green shoots, fruits, berries, cherries, ripe beechnuts, acorns and formerly chestnuts, according to the season (Goodwin, 1935). It is a night prowler but is itself preyed upon by great horned owls, wildcats and foxes, men and dogs (*ibid.*, pp. 23, 24). It can inflict a severe bite with its sabre-like canines (fig. 18.24A); its sharp cusped lower molar trigonids (fig. 18.27A) shear into the spaces between its obliquely triangular upper molars, while the inner cusps of the latter fit into the talonid basins of the lowers. Thus it has preserved and developed the basic features of *Amphitherium* (fig. 17.16), its very primitive Jurassic prototype.

The opossum skull (fig. 18.24A) in most respects is far less specialized than are those of the monotremes (figs. 18.4–14), and it is far more easily compared with the skull of small cynodont reptiles (fig. 16.67–70), from which indeed it has inherited its basic features. Comparative anatomical and palaeontological data indicate that its Jurassic and later ancestors have gradually approached its present structural patterns by making the following alterations in the cynodont skull:

1) eliminating all traces of the prefrontals, postfrontals, postorbitals (fig. 16.69E, F), septomaxillae (fig. 16.67 *smx*) and ascending processes of the premaxillae (idem.);

2) prolonging the secondary palate backward (figs. 16.30–31);

3) supplying it with a new transverse bar from the palatines, which simulates the old transverse palatal bar of the pterygoids (fig. 16.31);

4) reducing the pterygoids to thin slips or "hamular processes" (fig. 16.31B);

5) developing a new pair of pterygoid-like flanges by drawing downward a pair of ridges from the old epipterygoids to form the side walls of the bony naso-pharyngeal duct behind the palate (figs. 16.31, 16.33F, 16.34A1, B1);

6) remoulding the old longitudinal spheneth-moid brain-trough (fig. 16.33C *ls*) to form the transversely extended orbitosphenoid (fig. 16.34B *Os*);

7) fusing the latter below with the basi-sphenoid;

8) expanding the old epipterygoids to form the alisphenoids (fig. 16.33D, D1, E) and fusing them below with the sides of the basisphenoid (fig. 16.34A);

9) fusing the fish-to-reptile proötic and opisthotic (figs. 16.48C; 12.37C) to form the mammalian petrosal (periotic) (fig. 16.54D), containing the inner ear;

10) uniting the petrosal with the squamosal to form the mammalian temporal bone (fig. 16.4L);

11) completing the new or mammalian jaw-joint between the squamosal and the dentary (fig. 16.39C);

12) greatly reducing the already small post-dentary elements (fig. 16.39B, C);

13) shoving them a little to the rear into the middle of the expanded Eustachian tube (figs. 16.59; 16.61; 16.63);

14) protecting the latter by developing another down-growing process from the sphenoid complex called the "alisphenoid bulla" (figs. 16.31; 34B1 *As*);

15) forcing a membranous, air-filled bubble from the embryonic Eustachian tube (fig. 16.61) to impinge against the remnant of the angular bone (figs. 16.59B; 16.63), thereby starting the fenestration of the angular bone; this seems to have been transformed into

16) the tympanic bone which supports the drum-membrane;

17) developing the outer ear from the rim of skin, connective tissue and cartilage that surrounds the upper end of the hyoid arch and the blind end of the Eustachian tube;

18) fastening this ear-trumpet (fig. 16.55 aA) to the outer side of the tympanic ring bone (fig. 16.54D *ectotymp*)

19) supplying it with slips from the facial muscles so as to turn it in all directions;

20) widening the occipital condyles (fig. 16.31);

21) fusing the tabulars with the surface of the supraoccipital (fig. 16.4K, L);

22) developing on their inner surface a bony transverse septum (interparietal) between the parietal lobes of the cerebrum and the cerebellum.

EVOLUTION OF THE MAMMALIAN VERTEBRAL COLUMN, RIBS, DIAPHRAGM, ETC., FROM THE CYNODONT STAGE

In the regionally differentiated vertebral column of the Virginia opossum (fig. 18.22) the separateness of the intercentrum of the atlas, and the sutural separation from the axis of the centrum of the atlas, forming the odontoid, are premammalian features. The column is further anisomerized or differentiated regionally as compared with that of *Cynognathus* (fig. 18.18) by: (1) the marked enlargement of the neural spines of the axis and third to seventh cervical vertebrae inclusive; (2) the transformation of the cervical ribs (fig. 18.18) into lateral processes for the attachment of the cervical muscles (fig. 18.22); (3) the sudden lengthening of the ribs in passing from the seventh cervical to the first dorsal; (4) the change in the direction and shape of the neural arches at the anticlinal vertebra (cf. fig. 18.26) at the summit of the dorsal curve and near the junction of the dorsal and lumbar regions; (5) the loss of ribs in the lumbar region (fig. 18.22); and (6) the substitution of transverse processes for them (cf. fig. 18.26); (7) the development of metapophyses (cf. fig. 18.19 *mtph*) or knobs above the anterior zygapophyses, which not only give attachment to some of the spinal muscles and tendons but also strengthen the column against wrenching and dislocation; (8) the segmentation of the sternum (fig. 18.22) into a series of pieces, sternebrae, which being connected by cartilaginous ribs with the true ribs, allow for respiratory movements of the diaphragm and ribs; (9) the forward doming of the diaphragm and abdominal cavity beneath the posterior part of the thorax, causing or influencing nos. 5 to 8.

The evolution of the diaphragm, which may have been well under way in the cynodonts, was complete in the opossum; it was a leading factor in many of these advances and was itself part of the general transformation from a variable (poecilothermal) to a more stable (homoeothermal) body temperature.

10) The two primary sacral ribs, remaining separate in very young marsupial skeletons, rapidly fused with their vertebrae and came to resemble the transverse sacral processes from the centra, which eventually replaced them.

11) Several coccygeal vertebrae also developed transverse processes, which reached outward to or near the ilia; (12) the true caudal vertebrae, although retaining the reptilian intercentra in the form of chevron bones (fig. 18.22), were more slender than those of therapsid reptiles, so that there is a sudden constriction behind the rump, whereas in the duckbill, as in typical reptiles, the body tapers gradually toward the tail. The tail of the Virginia opossum, notwithstanding its slenderness, is strongly prehensile and capable of supporting the weight of the body.

These changes and others were brought about by the interaction of cumulative hereditary variability or secular adjustment to changing mechanical and chemical conditions, such as: (1) differential increase in parts of the brain, tending to push the braincase from within; (2) changes in the jaw muscles and feeding methods; (3) changes in posture and mode of locomotion, with (4) resultant changes in the positions and angular relations of one part to another; (5) changes in the environment affecting metabolism and survival; (6) changes in the coöperative, competitive, or destructive relations of the organism to others.

SOUTH AMERICAN POLYPROTODONTS

Opossums of Central and South America.—The recent South American opossums are subdivided into numerous genera which are prevailingly arboreal, although one species (*"Didelphys" americana*) of Brazil is described as very small and shrew-like and without a prehensile tail, while the yapok (*Chironectes*) of Panama is a forest-living form of partly aquatic habits and webbed hind feet. It scrapes burrows under the stream banks and lives chiefly on crawfish and fish (Beebe, 1924). Some species of opossums depend largely upon fruit, another feeds chiefly upon crustacea (*"Didelphys" cancrivora*), which it pursues in the swamps.

The fossil record of the opossums (Romer, 1935, p. 611) gives fragments of a long history, beginning in the Upper Cretaceous of North America and continuing in the Paleocene and later periods of North and South America and Europe. One of the Upper Cretaceous members of this group, *Eodelphis,* was amazingly like a modern opossum in essential features of the skull and dentition (fig. 16.23A); another, named *Thlaeodon* (D), had a very stout lower jaw, supporting a massive blunt cheek tooth adapted for smashing something hard, but the dentitions of the other genera are very conservative and basically opossum-like.

Borhyaenids.—In the Miocene and later deposits of South America, the borhyaenids (fig. 18.21, 28A, B) so closely resembled the Australian *Thylacinus* (fig. 18.25D) and the dasyures in skull, teeth and skeleton that an equally close relationship was long assumed; but Simpson, after careful re-analysis (1941), concluded that this is a case of close parallelism rather than near relationship, and that the South American borhyaenids and the Australian *Thylacinus-Dasyurus* group are more probably independent derivatives of the once very widespread opossum family. But still further study of the facts seems desirable.

One of the branches of the South American borhyaenid stock culminated in *Thylacosmilus* (fig. 18.28C), which, as its name implies, was a marsupial (*Thylaco-*) that resembled a sabretooth tiger (*smilus*), since its skull possessed an essentially marsupial "heritage" beneath a sabre-tooth-like habitus.

Caenolestids.—Another primary superfamily of the marsupials is represented by a few small and inconspicuous Andean marsupials named *Caenolestes obscurus* (figs. 18.21, 28E) with allied species. These are the last survivors of a once thriving and diversified family that lived in Patagonia. They all have or had a pair of enlarged procumbent lower incisors (D, E) like those of Australian phalangers and their last lower premolars have conspicuous pointed crowns often grooved in part. Their upper molars (E1) also are not unlike those of phalangers. Such resemblances led to their being classed as diprotodonts, but their narrow feet are not "syndactylous" and the curious mingling of poly- and diprotodont features in their soft anatomy (Osgood, 1921) indicates that they are an independent South American branch, probably from some of the ancestral opossums (e.g. fig. 18.23B–B2) of Upper Cretaceous age (Simpson, 1945). In one branch of this caenolestoid group, the Polydolopidae, the crowns of the molars have budded additional conical cusps, so that they are literally "multituberculate," but the vague resemblance to the true multituberculates is wholly convergent (Gregory, 1910; Simpson, 1928).

AUSTRALIAN POLYPROTODONTS

From Mouse-like to Wolf-like Forms.—The American opossums, as already noted, are practically archetypal to all other marsupials, but as yet no really intermediate fossils have linked them with their distant relatives, the Australian dasyurids (fig. 18.21). These range from the mouse-like *Phascogale* to the rather civet-like dasyures and thick-skulled Tasmanian devil (*Sarcophilus*); their molar teeth (fig. 18.27) vary from small, sharp-cusped cutting triangles (B) to shearing blades (D). They are all well-clawed, slinking, ground-living animals without grasping power in the great toe.

The Marsupial Jumping "Mice" and "Banded Anteater".—The marsupial jumping "mice" (fig. 18.29, 18.37C–F) in their excessively long hopping limbs resemble the jerboas of Africa and the so-called kangaroo rats of North America, which are both placental rodents.

The "banded anteater" (*Myrmecobius,* figs. 18.21, 25E, 26) has a partly degenerate dentition, with secondary increase in the number of cheek teeth and compressed cuspidate molars (fig. 18.31F, 18.32F) which vaguely suggest those of Jurassic triconodonts (fig. 17.7C). But comparative study of the skull and entire skeleton indicates that *Myrmecobius* is really a small disguised relative of the "marsupial wolf" (*Thylacinus*) and of the dasyures.

The Marsupial False Mole.—A much more specialized side branch is the "marsupial mole" (*Notoryctes,* fig. 18.30B), which simulates the African Golden moles (A) in its body-form. It has two extremely large compressed claws on the strong forelimbs, for digging in hard ground; the small hind feet are adapted for scraping and throwing the dirt backward, the head is obtusely conical, the nose blunt, with a wide epidermal nasal shield; the cheek teeth (figs. 18.31G; 18.32G) with pricking triangles are superficially like those of Cape Golden moles (fig. 19.6J). The deeper features of the skeleton and anatomy are without exception marsupial and the resemblances to other mole-like mammals wholly convergent (Leche, 1907).

The Australian Bandicoots.—In this diversified Australian family (Peramelidae, fig. 18.21) the skull (fig. 18.25G), jaws and teeth (figs. 18.31H–K, 32H–K) do not show any convincing evidence of close relationship with the phalangeroids or herbivorous diprotodont marsupials, and the bandicoots seem to have been derived independently (fig. 18.21) from the ancestral insectivorous-omnivorous stock of the dasyuroids (Bensley, 1903).

AUSTRALIAN HERBIVOROUS MARSUPIALS (DIPROTODONTS)

The herbivorous branches (fig. 18.21) of the Australian marsupial fauna include the phalangers (wrongly called opossums), the koalas (*Phascolarctos*), the kangaroos and wallabies

(Macropodidae), the wombats (*Phascolomys*), and the huge extinct herbivores named *Diprotodon* and its congeners. All have a pair of large procumbent lower incisors (fig. 18.34) and the cheek teeth are variously adapted for cutting and crushing leaves or other herbage. The upper molars in phalangers (figs. 18.31PQR) have either two low conical outer tubercles, each with a low cross-crest, or four V-shaped crests (S, T); in kangaroos (M, N, O) and *Diprotodon* (U) there are paired cross-crests on both upper and lower molars; in the wombats, the upper and lower molars (V) have very high, curved crowns, growing from persistent pulps, for chewing tough roots. The wombats also have remarkably rodent-like front teeth. The jaws and skull (fig. 18.34F) are also superficially rodent-like, yet the wombat is assuredly only a ground-living and digging phalanger. The large, compressed, sharply pointed upper and lower front teeth of the extinct *Thylacoleo* (C), the "marsupial lion," were somewhat like the beak of a parrot, and below the stout cheek arches were a pair of long shearing blades. It was, therefore, supposed that *Thylacoleo* was a ferocious beast of prey; but later studies suggest that it was derived from the essentially herbivorous phalanger stock and that its teeth were well adapted for piercing and slicing large fruits with tough rinds.

The locomotor skeletons of the diprotodont Australian marsupials are adapted variously, but primarily for tree-climbing, as in the phalangers. In these the hind feet (fig. 18.37S) have the first digit large and sharply divergent, the second and third small, appressed and often conjoined by a web, the fourth enlarged, the fifth slightly divergent. From this central type the tiny musk-kangaroos (fig. 18.37X) lengthen the whole foot, especially the fourth metatarsal and phalanges; but they retain a small divergent first digit (hallux). In the most advanced kangaroos (Y) the first digit has disappeared, the second and third form paired vestigial remnants adapated for combing the skin, the fourth is very large and lies in the main axis of weight, which passes through the enlarged fourth metatarsal; the fifth metatarsal shares in the elongation of the fourth but remains slender, the fifth digit is also much smaller than the fourth. In the tree-kangaroo (Z) the feet are basically the same as in kangaroos but they have become secondarily shortened and widened; this not only gives better friction on the branches but actually shortens the leap and decreases the danger of falling. This is a pretty clear example of a shift in the direction of evolution, for the nearest relatives of the tree-kangaroos, namely, the rock wallabies (*Petrogale,* fig. 18.38), have narrow feet and toes and relatively longer fourth metatarsals, i.e., they are more like ordinary leaping kangaroos.

The extreme leaping adaptation in the feet of the kangaroos (fig. 18.37X, Y) differs basically from that of the small leaping marsupial "mice" (figs. 18.29C, D; 37C–F), and is nearer to that of the bandicoots or native "rabbits" (fig. 18.37N–R).

CROSS-CORRELATIONS OF DENTAL AND LOCOMOTOR TYPES

Within the limits of a family, there is often an invariable correlation between certain types of dentition and certain types of feet: e.g., in the phalangerid family, of diprotodont incisors (fig. 18.34B, D, E) with syndactylous feet (figs. 18.37S, T, U); but when the whole marsupial group is reviewed, similar dentitions may be found in combination with several widely different types of feet; and conversely similar types of feet may be associated with widely different types of dentition. For example, the dentitions of American opossums (fig. 18.27A) and Australian dasyures (B) are on the whole remarkably similar and yet the former have the hind feet well adapted for climbing (fig. 18.37A), while the latter have them adapted for running on the ground (B, I, J, K, M). Again, the dentition of the South American caenolestids (fig. 18.28D, E) is in general similar to that of some of the Australian phalangers, but the hind feet of the caenolestids are not syndactylous, while those of the phalangers are (fig. 18.37S, T, U). Likewise the cross-crested molars of the kangaroos (fig. 18.31

M–O) are similar to those of the extinct *Diprotodon* (U); but the feet of the kangaroos (fig. 18.37X, Y, Z) look extremely different from those of *Diprotodon* (W).

Conversely, superficially similar types of feet adapted for leaping are found in the small jumping "mice" (*Sminthopsis*), in the bandicoots (*Chaeropus*), and in the small musk kangaroos (*Hypsiprimnodon*), but the dentitions are widely different; and again, hind feet with a large, strongly developed, divergent great toe adapted for climbing are found in the American opossums (fig. 18.37A) and in the Australian phalangers (B), but the dentitions are completely different. Thus if the group is viewed as a whole, there is not an invariable correlation between a given type of teeth and a given form of feet.

When too much reliance was placed upon either teeth (Polyprotodontia, Diprotoontia) or feet (Diadactyla, Syndactyla) as exclusive criteria of classification, it sometimes led to widely unrelated animals being bracketed in the same partly artificial subordinal group, as when caenolestids were classed with "Diprotodontia" or peramelids in "Syndactyla"; but when adequate comparisons were made between several parts in relation to the evidence as a whole, it became evident that in making adjustments to similar needs, there had been parallelism in teeth or in feet between members of different families.

"MARSUPIAL HERITAGE" CHARACTERS

There is a set of characters which distinguish the marsupials from the higher or placental mammals and are seldom lost even in highly specialized forms. They include among others the following: (1) the angular process of the lower jaw is inflected (figs. 18.23, 24–26, 28, 33–34); (2) the jugal (malar) extends backward beneath the zygomatic process of the squamosal to the glenoid articular surface, where it often helps to form the anterior boundary for the movement of the condyle of the lower jaw; (3) the bony palate (fig. 18.24C1) is usually fenestrated; (4) there is usually a more or less enlarged "alisphenoid bulla" (fig. 18.24 *ty. As.*); (5) the basisphenoid is pierced by the paired foramina for the internal carotid arteries (fig. 18.13A, B *f. int. car.*); (6) the optic foramen of each side is fused with the foramen lacerum anterius (fig. 18.13A *f.l.a* + *f.opt*); (7) the proximal end of the nasals, as seen from above, are spreading (fig. 18.14A); (8) the adult dental formula in insectivorous-carnivorous forms is typically, inc. $\frac{5}{4}$ can. $\frac{1}{1}$ pms. $\frac{3}{3}$ ms. $\frac{4}{4}$ or a reduction of it (figs. 18.25, 27, 28); it is never inc. $\frac{3}{3}$ can. $\frac{1}{1}$ pms. $\frac{4}{4}$ ms. $\frac{3}{3}$ or a reduction of it as in placentals; (9) with few and doubtful exceptions only the posterior upper and lower premolars have deciduous predecessors; (10) in the vertebral column the atlas intercentrum of mammal-like reptiles (fig. 16.5AB*ic2*) is retained, while the centrum of the atlas, forming the epistropheus or odontoid of the axis, usually remains suturally separate from the axis; (11) the inferior arch of the atlas is often incomplete or cartilaginous in the midline; (12) the epipubic bones are normally present (figs. 18.22, 26, 35, 36); (but they are vestigial in *Thylacinus*); (13) the fibula sometimes bears a dorso-posterior process or "flabellum" (fig. 18.35); (14) the lower end or head of the astragalus is usually small or narrow, not expanded transversely as in many placentals.

Some of these and similar characters may not be very important functionally, but collectively they indicate a lower series of mammals; that is, the marsupial stock is probably older than that of the placental mammals.

CONVERGENCE BETWEEN MARSUPIALS AND PLACENTALS

Several striking cases of convergence arose between certain marsupials and merely analogus placental mammals, which has deceived the English-speaking Australians into calling Australian marsupial mammals "mice," "cats," "tigers," "moles," "rabbits," "bears," "badgers," "squirrels," etc.—all names of animals that belong to the subclass of placental mammals, which are dominant in the northern hemisphere as the marsupials are in Australia.

WHICH CAME FIRST, MONOTREMES, MARSUPIALS OR PLACENTALS?

The question (pp. 366, 367) to what extent the basic features of the anatomy of man's remote early mammalian ancestors may have been retained in the existing monotremes and marsupials has been answered to some extent in the preceding sections. That is, in the reproductive system and to a considerable extent in the brain, the monotremes seem to have retained very ancient mammalian characters; also in the underlying features of their chondrocranium, the principal features of the pectoral and pelvic girdles, and the basic characters of the limbs; but not in any of their specialized habitus features.

The further question as to whether the marsupials and placentals are coördinate derivatives of the "prototherians," or ancestral monotremes or whether and in what particulars one is more primitive than the other, may now be considered briefly.

That the placental mammals have been derived ultimately from animals that laid sizable eggs, covered with an eggshell or at least a tough membrane, is an *à priori* deduction from the generalization, based on other grounds, that all mammals have been derived from early reptiles. And, as this generalization is well supported from so many directions, the conclusion would be highly probable even if the marsupials were not known. The excessively small size of the eggs of the higher placental mammals, together with their loss of yolk and the development of a morula, is universally regarded by embryologists as indicative of a high grade of evolution.

The general marsupial method of reproduction is manifested today by so many highly specialized representatives of the group (kangaroos, koalas, wombats, etc.) that it is natural that their reproductive methods should be viewed with suspicion as being far from primitive. Moreover the presence of a vaginal sac or tube, median in position but of paired origin, in addition to paired lateral vaginal tubes which loop over the ureters, are usually regarded as irrevocable specializations which exclude the marsupial stock from ancestry to the placentals. Nevertheless the chief essentials of the *typical* marsupial method of development remain as follows: (1) the presence of a nipple-bearing area on the abdominal surface, with or without ridges or a fold of skin around it; (2) the presence of epipubic or marsupial bones in both sexes; (3) the presence of a yolk-sack placenta; (4) the absence or small development of an allantoic placenta as well as its variable nature; (5) the birth of the young at a very early stage, as compared with placentals; (6) the larval equipment of the newborn, which enables it to find its own way from the cloaca to the nipple; (7) the supposed presence of forced or automatic feeding of the young. But most or all of these may have been antecedents to the newer specializations in the placentals. Forceful feeding, for example, may have been prerequisite to the evolution of elaborate sucking mechanisms in the cheeks and throat of the young in the placentals. The presence of a well developed yolk-sac placenta in the more primitive marsupials may have been a prelude to the development of an incipient allantoic placenta, now beginning in *Phascolarctus,* but which was carried to such an extent in the placentals that it enabled many of them to give up the yolk-sac placenta and to become true placentals.

Since Dollo's "Law of the Irreversibility of Evolution" has been universally accepted by zoologists and palaeontologists, it seems usually to have been taken for granted that divergence in specialization is invariably the sign of divergence in phylogeny. In reference to the marsupials, "specializations" of their reproductive methods are numerous and obvious. Therefore, it is generally assumed, not even the most ancient marsupials can be ancestral to the placentals! But what if some of these marsupial specializations served merely as the approach to the placental specializations and were later replaced by them? The whole field of mammalian embryology might be reviewed profitably under this hypothesis.*

The several skeletal patterns of the highly specialized kangaroos, koalas, wombats, etc., are not to be considered as structurally ancestral to the primitive placental type. Nevertheless the skeleton of the more primitive insectivorous-carnivorous marsupials is not lacking in characters which seem to be on the whole more primitive than anything that is known in the placentals. The skull of the American opossum is indeed nearer to the construction of the skull of the mammal-like reptiles than is that of any placental mammal. The foregoing list (p. 371) of "marsupial heritage characters" severally and collectively distinguishes a marsupial from a placental skeleton. Some of these marsupial characters are more cynodont-like than the corresponding placental opposites; others are not. For example, the "alisphenoid bulla" of the marsupials (fig. 18.24 *ty. As*), which is a downward flange from the posterior rim of the alisphenoid bone, indicates that the tympanic chamber is medial to the squamoso-dentary joint (p. 367) and that it is in about the same position in relation to the quadrate (=incus) as it was in the mammal-like reptiles (fig. 16.34B1, A1). In placentals the alisphenoid part of the bulla is retained only in a few insectivores (fig. 19.8), but the position of the tympanic cavity is about the same.

The inflection of the angular region of the dentary (lower jaw) in marsupials is not present or recorded in the cynodont reptiles (fig. 16.31A), nor in the Mesozoic pantotherians (fig. 17.21A) nor the early placentals. It seems to be a response to the shifting of the internal pterygoid muscle from the angular bone to the dentary (fig. 16.38). In the base of the cranium, especially in the perforation of the basisphenoid by the pair of branches of the internal carotid artery, the marsupials (fig. 18.24C1 *f. car*) appear to be more primitive than the typical placentals, and the same seems to be true in such characters as: (a) the posterior spreading of the nasals, (b) the contact of the nasals with the lacrymals, (c) the exposure of the lacrymals in front of the orbit (perhaps secondarily increased in the larger forms, e.g., *Thylacinus*), (d) the contact of the mandibular condyle with the malar (jugal). The fenestration of the bony palate of typical marsupials has always been regarded as a marsupial specialization and it may perhaps be such.

The postcranial skeleton of marsupials is not lacking in features which we may reasonably regard as primitive mammalian characters preceding the further advances of the placentals. Here belongs with considerable probability:

1) The persistence of the separation of the odontoid (epistropheus) or centrum of the atlas vertebra, which in placentals very early fuses with the axis, but in cynodonts, monotremes and marsupials remains distinct.

2) The patella of placentals is a sesamoid or ligamentary bone which is not present in reptiles and is usually but poorly developed in marsupials. It is, however, present in monotremes, so that its exact status is doubtful.

3) The monotremes have a flabellum or projection at the upper end of the fibula which is occasionally present in marsupials but is not characteristic of placentals.

4) The number of presacral vertebrae is variable in both reptiles and mammals but in *Cynognathus* it is 28, in monotremes, 26 or 27, in marsupials, 26; in some placentals it is 26 but in others, 27; it rises to 37 in *Hyrax* and falls to 21 in certain armadillos. But because the sacral attachment may shift forward or backward, the exceptional numbers are found only in aberrant forms and 26–27 is probably primitive for both marsupials and placentals (Gregory, 1947).

5) In the sacrum such relatively primitive marsupials as the opossums and the Tasmanian devil (*Sarcophilus*) have distinct sacral ribs in

* This passage was written long before I had read the work of Edward McCrady, Jr. (1938), in which an essentially similar view is expressed. Pearson (1944) however has shown that the median part of the three-way vagina of some marsupials has developed from the union of a pair of kinks, or sharp bends, in the tubes, which originally served as receptacula seminis, and that the median opening into the cloaca, when it occurs, is secondary. In existing placentals there seems to be no trace of these sharp bends or diverticula and the median vaginal tube may represent only the fusion of the primary pair.

young stages, but in the placentals the sacral ribs are found free only in very young stages and are usually braced or replaced by transverse processes from the sacral vertebrae.

6) Marsupial or epipubic bones are well developed in monotremes, and to judge from the flattened anterior border of the pubes in *Cynognathus* they may have been present in cynodonts. The supposed traces of these structures in placentals (Weber, 1904, p. 109) are obscure at best, but since epipubic bones are associated with a generally lower status, they may well have been lost in even the earliest known placentals, and indeed they are already reduced to vestiges in the marsupial *Thylacinus*.

7) The pelvic inlet and outlet of typical marsupials are smaller than those of placentals, and Wortman (1901, p. 32 [424]) called attention to their small size in *Dromocyon,* one of the mesonychid creodonts, as an indication that the young of these early placentals were born in a relatively imperfect stage. Unfortunately this important observation was made on a highly specialized member of an early placental group and has not been extended to such truly primitive placentals as the Paleocene oxyclaenids on account of lack of material.

8) The hind feet of most marsupials are peculiarly specialized in various ways, those of the bandicoots, phalangers, kangaroos, wombats and related families all showing syndactylism, or the tendency toward pairing and reduction of the third and fourth digits, together with strong divergence of the first digit if present. This special adaptation for ridding the fur of parasites is deeply implanted in the heritage of the diprotodonts, as well as of the bandicoots, but is secondarily modified and disguised in the highly specialized wombats. Nevertheless it is probably not to be imputed to the more ancient marsupial stem because no traces of it have been detected in the families of the opossums, phascogales, dasyures, thylacines, marsupial moles, nor in any of the fossil South American carnivorous marsupials, nor in the caenolestids. Hence it is not to be expected in the stem placentals, nor is it found in any of their earliest known representatives.

9) The non-syndactylous marsupials, on the other hand, retain several primitive characters in both the hands and the feet which may reasonably be imputed to the stem placentals. Thus the typical opossum is pentadactylate (with five digits in the hand and the foot), unguiculate (clawed), plantigrade, the digits spreading and adapted for grasping the limbs of trees. As we shall see below (page 378) the late Dr. W. D. Matthew (1904) predicated exactly these characters for the stem placentals and pointed to the numerous apparent remains of this condition in the feet of the earliest known placentals of the Paleocene and Eocene ages.

10) The astragalus of the opossum with its incomplete lower process or head and its unkeeled upper surface has every appearance of being more reptilian, less highly evolved, than that of the earliest Eocene placentals, and the same is true of (11) the calcaneum with its short heel.

12) The semi-plantigradism of the primitive opossums appears to be really primitive and not at all the same as the alleged plantigradism of bears and other modern placental carnivores, which there is good reason to regard, along with the plantigradism of man, as quite secondary (H. C. Raven, 1936).

Huxley, Dollo, Bensley and their successors, including the present writer, have awarded to the common or Virginia opossum the honor of preserving most nearly the probable heritage of the stem marsupial. In view of what has been said above, it becomes increasingly probable that the oposum well deserves its place in our exhibit, "The Skeleton from Fish to Man," where it stands (figs. 24.1, 2) between the cynodont reptile and the basal primate.

CHAPTER XIX

RISE OF THE PLACENTAL MAMMALS

(THE OUTER FORM AND INNER FRAME OF INSECTIVORES, TREE-SHREWS, BATS, EDENTATES, RODENTS)

Contents

CHAPTER XIX

RISE OF THE PLACENTAL MAMMALS

PALEOCENE AND EOCENE EPOCHS, "DAWN OF THE RECENT" MAMMALS

THE STUDENT who enters vertebrate palaeontology by that branch which deals with the mammals and who by his own field experiences and reading gains a deep respect for the vast ages represented by many successive thousands of feet of Tertiary deposits, is apt to think of the Basal Eocene (Paleocene) faunas of New Mexico as being almost at the bottom of mammalian history (fig. 19.1). But the student of invertebrate palaeontology, approaching the Tertiary period through the incredibly long reaches of the Palaeozoic and Mesozoic eras, thinks of the Eocene epoch in its etymological sense, "Dawn of the Recent," which was given to it by invertebrate palaeontologists in recognition of the essentially modernized composition of its marine faunas. The old term, "Age of Mammals," which looks as if it had been invented before any fossil mammals had been found in Secondary (Mesozoic) rocks, persists not because it is accurate but because it conveys some sort of meaning in plain English, whereas "Caenozoic" or "Tertiary" sounds too learned to be given a chance by the newspaper writers who set limits to the public's vocabulary. In the present book we have approached the Age of Mammals from below and we therefore see it for what it is—namely, the last fifty million years of the one hundred and fifty odd million years of mammalian history.

Although two-thirds of the recorded history of the mammals had elapsed before the beginning of the so-called "Age of Mammals," yet the last third is comparable with the history of Europe since the Reformation and no less full of events that determined the present patterns in the Kaleidoscope of Evolution.

The late Dr. W. D. Matthew in his memoir on the Paleocene mammalian fauna of New Mexico (which was published in 1937, seven years after his death), sums up the relations of the "ancients" and "moderns" of the Paleocene and Eocene epochs in a diagram (fig. 19.1), which might be taken to represent the interior of a vast sinkhole that extends downward into the records of past time, passing down through the rock strata of Lower Eocene and Paleocene and reaching the underlying Cretaceous. The groups whose names are on the outer zones are the "moderns," while those on the inner zones are the "ancients," arranged in the order of the known or inferred antiquity of their respective branches. Of course the distinction between "ancients" and "moderns" was not absolute; nevertheless in the Paleocene (Puerco, Torrejon, Tiffany) there was a predominance of groups that were hold-overs from the Age of Reptiles and that were nearing the dates of their respective exits from the record, while in the Eocene proper we find that the balance has shifted and that there is a majority of orders and families that are still flourishing, or at least still in existence.

When the world was at last rid of the dinosaurs, the ancestral placentals could either stay on the ground and engage in the age-long war of the strong against the weak, or they could spend more time in the trees and develop more pronounced specializations for arboreal life.

THE THREE SURVIVING SUBCLASSES OF MAMMALS

(Monotremes, Marsupials, Placentals)

All surviving mammals are divided into three main groups: egg-layers, marsupials and placentals, and it is most unfortunate that this far-reaching fact, which in essentials has been grasped by zoologists for more than a century, should still be totally unknown to the world at large. This is partly because among zoologists the technical name Placentalia has long been discarded in favor of "Monodelphia" (single womb) or Eutheria (true beasts); another reason is that the placentals are not the only possessors of a placenta; for an organ having the same name and functions has been developed by convergence among the lizards in a species of Australian skinks (Flynn, 1923), and by parallelism among certain marsupials (J. P. Hill, 1898, 1899). But in no other group than the placental mammals does this organ maintain so high a rate of metabolic exchange between the mother and the foetus, nowhere else has it become so diversified, as in the placentals. In short, a placenta was one of the main organs of mammalian advance and it was only begun in the marsupials; for their young are born in a very imperfectly developed stage, while among the placentals the newborn young are usually much further advanced.

Such familiar animals as rats, squirrels, rabbits, cats, dogs, horses, cows, pigs, elephants, bats, monkeys and man are all placentals and indeed, except for the family of opossums, the marsupials are not now represented in the northern hemisphere.

THE ORDERS OF PLACENTAL MAMMALS

The subclass Placentalia (Eutheria) is divided into many fossil and recent orders, including among others the following:

1. INSECTIVORA
   tenrecs, moles, shrews, hedgehogs, etc.
2. EDENTATA
   armadillos, sloths, etc.
3. PHOLIDOTA
   scaly anteaters
4. RODENTIA
   rodents

4a. LAGOMORPHA
   picas, hares, rabbits

5. DERMOPTERA
   cobegos or "flying lemurs"
6. CHIROPTERA
   fruit bats, small bats
7. CARNIVORA
   creodonts, civets, hyaenas, cats, dogs, bears, mustelines, raccoons, pandas, sea lions, walruses, seals
8. CETACEA
   Toothed whales, whalebone whales
9. AMBLYPODA

9a. DINOCERATA
   extinct short-footed, hoofed mammals

10. HYRACOIDEA
    dassies
11. PROBOSCIDEA
    elephants, mastodons
12. SIRENIA
    manatees, dugongs
13. TUBULIDENTATA
    aardvarks
14. CONDYLARTHRA
    Eocene phenacodonts and others
15. LITOPTERNA
    an extinct South American order
16. NOTOUNGULATA
    mostly short-footed extinct South American ungulates
17. PERISSODACTYLA
    horses, tapirs, titanotheres, chalicotheres, rhinoceroses
18. ARTIODACTYLA
    pigs, hippopotami, oreodonts, camels, deer, antelopes, sheep, oxen
19. PRIMATES
    tree shrews, lemurs, tarsiers, monkeys, apes, man

The earliest known placentals, from the Upper Cretaceous of Mongolia (fig. 17.24), were very small Insectivora, known only from skulls and jaws. The predecessors, but not many of the direct ancestors, of the existing placental orders, appear in the Paleocene (Puerco, Torrejon, etc.) of North America (fig. 19.2), Europe and South America. Nearly all the orders surviving today, but few if any existing families, were represented in the Lower Eocene (Wasatch) fauna, about 50 million years ago. The first adaptive branching of the main orders of placentals appears to have started sometime before the Middle Cretaceous; some of these older stocks then branched and rebranched again through the long and continuous Upper Cretaceous, Paleocene, Eocene, Oligocene, Miocene, Pliocene and Pleistocene epochs, down to the exceedingly brief Holocene or Recent, which began perhaps not more than 25,000 years ago (Schuchert). Hundreds of genera and thousands of fossil and recent species of placental mammals have been discovered and named; untold thousands of details about their structure are in the records, including unique characters which distinguish each one from related forms. Moreover, the fossil histories are already sufficiently documented to establish the main outlines of evolution in many families. But the limitations of the present work make it necessary to restrict our account of the evolution of the skeleton to relatively few representative cases.

WERE THE ANCESTORS OF THE PLACENTAL MAMMALS ARBOREAL?

In 1899 Dollo showed that in the hind feet of tree-living marsupials (fig. 18.37), including American opossums and Australian phalangers, the first digit (hallux) was sharply set off from the others so that the foot could be used to grasp the limbs and branches of trees; and that even among marsupial families which now live on the ground some of the more primitive members, such as the musk-kangaroo and certain bandicoots, show clear traces of former arboreal adaptations (Chapter XVIII). Dollo's conclusion was adopted after careful consideration of the evidence by Bensley (1903), Gregory (1910) and later authors. In 1904 Matthew extended Dollo's theory to include the placental mammals and pointed out that among the older Eocene and Paleocene placental families, the more primitive forms (e.g., fig. 19.2) had radiating five-toed hands and feet with well developed claws or nails and at least moderately divergent first digits, apparently well able to grasp the limbs of trees; also that as diverse adaptations for living on the ground (as for running, leaping, etc.) became more complete, the divergent first digits usually were reduced and the feet lost their grasping power. Present evidence, however, suggests that the remote and still largely unknown Cretaceous ancestors of the placental mammals were not as much specialized for arboreal life as were the ancestral marsupials; but that they were able to scratch, dig, grasp or climb.

VORACIOUS PLACENTAL INSECT HUNTERS (INSECTIVORA)

All signs indicate that some of the small Jurassic "insectivores" such as *Amphitherium* (fig. 17.16), gave rise to *Deltatheridium* (fig. 17.24A) and allied genera in the Middle Cretaceous of Mongolia, which in turn may have led on the one hand to the existing *Echinosorex* (*"Gymnura,"* fig. 19.9A) and the hedgehogs (*Erinaceus,* B), and on the other to the tenrec of Madagascar (*Centetes,* fig. 19.3B) and its allies of the suborder Zalambdodonta.

Pretritubercular Molars. — *Deltatheridium* is known from several fossil skulls, some with well preserved teeth. The best skull (fig. 17.24), about 1.75 inches long, suggests that of a tiny flesh-eater with cat-like canines, obliquely triangular upper molar crowns with shearing edges and large interdental spaces; lower molar triangles, tricuspid with longer antero-external and shorter transverse sides, fitting into upper interdental spaces; narrow talonids or heels fitting between the inner and the two closely appressed outer cusps of the upper molars; premolars three, above and below, the first simple, the last with beginnings of the internal cusp on the upper and of the posterior cusp on the lower. The delicate sharp-cusped crowns of the cheek teeth, as well as the vertical movement of the jaw in *Deltatheridium,* indicate but little attrition. This suggests as the main food living insects and insect larvae rather than an omnivorous diet dealing with such tough materials as bones or nuts, which require chewing and more or less obliquely transverse rubbing movements of the jaw.

The generic name *Deltatheridium* (meaning little beast with V's) and the specific name *pretrituberculare* were given because the cheek teeth (fig. 17.24A2, A3) appear to supply an ideal intermediate between the wedge-like (tribosphenic), interlocking upper and lower molars (fig. 17.16) of the Jurassic "insectivores" (Pantotheria) and the highly diversified molar crowns of later placental mammals, which started from the basic "tritubercular" upper and "tuberculosectorial" lower molar patterns of the Cope-Osborn theory of trituberculy (figs. 20.7; 17.10).

The Very Primitive Skull of *Deltatheridium.* —The Upper Cretaceous skull of *Deltatherid-*

ium (fig. 17.24) has advanced beyond the Jurassic *Triconodon* (*Priacodon*) stage (fig.17.8) toward that of later insectivores, especially in the slenderization of the long zygomatic or cheek arch, which in the side view is nearly horizontal, instead of arching upward as it does in primitive marsupials. This may well be, partly because the masseter muscle had extended its origin upward beyond the level of the zygoma, arising from the strong outer fascia of the temporal muscle; partly because the thrust from the lower jaw was more backward than upward, against the squamosal, so that there was more tension than bending stress exerted upon the zygoma, as in the modern *Erinaceus* and other insectivores (cf. Dobson, 1882–1883, plates). A probably primitive placental skull character in *Deltatheridium* was the fact that, as evidenced by the low position of the jaw condyle, the preorbital face was only slightly bent downward upon the base of the braincase, that is, much as it was in the primitive carnivores (fig. 20.4A, B), insectivores (figs. 19.4, 13) and other mammals which direct their mouth and nose toward the ground in the search for food.

Unfortunately no limb bones or vertebrae were found with the Mongolian insectivore skulls, but in such recent insectivores as *Echinosorex* (*Gymnura*) a basically similar skull (fig. 19.13) is associated with rather short rat-like limbs adapted for scampering about in the underbrush (fig. 19.9).

"LES PETITS CARNASSIERS" (INSECTIVORA)

The order Insectivora includes the smallest known mammals (*Sorex parvus,* about two inches long), and few insectivores exceed the size of an ordinary rat. Exceptions, however, are the long-tailed African water-shrew (*Potamogale,* fig. 19.3C), which is about 17 inches long, and the large water mole (fig. 19.9) commonly called *Myogale moschata,* about 16 inches long. These water-living forms probably pursue crayfish and small fish. Some of the larger centetoids (fig. 19.5), with a skull 4½ inches long and relatively large canine teeth (Dobson, 1882, p. 74) would well deserve the name of *"les petits carnassiers"* applied to the shrews and their allies by 18th century French naturalists. But for the most part, the land-living insectivores are, as it were, kept small by the necessity of hunting such small prey as ground-living insects, larvae and worms.

THE MADAGASCAR INSECTIVORES AND THEIR PALEOCENE FORERUNNERS WITH V-SHAPED MOLARS

The North American Paleocene genus *Palaeoryctes* (ancient digger) of Matthew in its skull (fig. 19.4A) and teeth (fig. 19.6B) has advanced far beyond *Deltatheridium* (A) toward the modern centetoid insectivores; these survive chiefly in Madagascar, but with one outlying form (*Potamogale,* D) in Africa and another in the West Indies (*Solenodon,* I). All these, together with *Deltatherium* and certain possible successors in the Paleocene and Eocene of North America, are referred to the suborder Zalambdodonta. This name, meaning very lambda-like teeth, was long ago given in allusion to the V-like (inverted lambda) crowns of the cheek teeth in both the upper (fig. 19.6) and lower jaws. However, the simple V of the *Centetes* upper molar (E) seems to have been derived by the main high cusp having shifted inward toward the midline, the low internal cusp receding and shrinking to a slight ridge (cf. B, D, E). Meanwhile the posterior spur or talonid of the lower molars has greatly diminished in *Centetes* (fig. 19.4G) and almost or quite disappeared in the Cape Golden mole *Chrysochloris* (I). According to this view, the most primitive surviving molar type is that of the African water-shrew (figs. 19.6D, 4D), which in turn was compared by Matthew (1913) with the Paleocene genus *Palaeoryctes* (A) and the Upper Cretaceous *Deltatheridium.*

The centetoids (fig. 19.3) exhibit considerable adaptive branching from the minute, very long-tailed *Microgale* (A) to the stout tenrec (*Centetes,* B), which is several times larger and has a very short tail, a long rooting snout and is covered with sharp spine-like hairs not unlike those of the European hedgehogs (*Erinaceus,* fig. 19.9B). This, however, is certainly a convergence, as these two animals belong in widely different suborders. The tenrec and its congeners are mostly ambling, ground-living forms with strong claws for scratching but not overspecialized for digging. Their long sabre-like canines are effectively used against snakes and their tough spines make it difficult for a snake either to stab them or to swallow them. Their vertebrae, limbs and girdles are fairly primitive (fig. 19.5) but their skulls (fig. 19.4) are in many features peculiarly specialized.

The Skull of *Centetes.*—In *Centetes* and its allies (fig. 19.7A), for example, the opposite nasal bones are fused, which stiffens the boar-like snout on the dorsal side. Internally it is filled with elaborately infolded and branched turbinate bones, indicating high smelling power. The eyes and orbits are very small. The jugal (malar) has disappeared, leaving a gap in the dried skull between the stump-like zygomatic process of the squamosal and that of the maxilla. The somewhat tubular braincase is surmounted by a sagittal crest for the relatively thick temporal muscle and there is a sharp transverse occipital crest for the insertion of the strong neck muscles. The oblique cavity of the middle ear (fig. 19.8) is surrounded partly by the tympanic bone and by flanges from the squamosal, alisphenoid, basisphenoid and petrosal bones. This is a fine example of the power of such membranous bubbles or diverticula of the Eustachian tube to mold whatever bones they may happen to press against.

The African Water-shrew (*Potamogale*).— At the other extreme of body form, the long and supple African water-shrew (*Potamogale,* fig. 19.3C) has a very long, large tail, oval in cross-section, with thick muscles running longitudinally and separated into paired compartments both by septa and by oblique bony processes on either side of the neural arch. The hind feet are longer than the fore feet but not webbed, the main swimming power being in the tail. The clavicles, which were retained in primitive zalambdodonts, were lost in *Potamogale,* doubtless in connection with its newt-like swimming habits.

The West Indian *Solenodon.*—This zalambdodont (fig. 19.3D) uses its strong claws to tear open rotten tree trunks in search of grubs. It has an enlarged, sharp-tipped pair of upper incisors (fig. 19.4C), compressed and directed straight downward, which are probably used in pulling apart the rotten wood. Its shorter, partly procumbent, second lower incisors are deeply channeled on the inner side, whence the name *Solenodon* (channel tooth). Possibly these teeth may be merely folded over to increase their strength; they may also convey mucous secretions from the sublingual glands to the food.

The postcranial skeleton of *Solenodon* is stoutly built, adapted for moving about on the ground, for scratching and in some degree for digging or tearing, but not for tunneling.

Referred to the same family, Solenodontidae, are the North American Oligocene genus *Apternodus* (fig. 19.4B) and allied forms. In *Apternodus* the temporal region was strongly braced by a large flat, bony plate on the squamosal. The short lower jaw bore a very large, forwardly inclined ascending ramus by which the large temporal and masseter muscles could exert powerful cutting and breaking thrusts between the interlocking upper and lower molars. The front teeth were nipper-like but not large. By comparison with *Chrysochloris* (fig. 19.12H and *Myogale* (Dobson, 1882, pls. 14, 18), it seems highly probable that this crest in *Apternodus* was developed between: (a) the forwardly extended neck and limb muscles (e.g., trachelo-mastoid, spleniuscapitis, trapezius), (b) the zygomaticus, on the anterior edge of the crest and (c) the temporalis on the inner side of the crest.

The Cape Golden-moles (Chrysochlorids) and Their Highly Specialized Skeleton.—Another outlying branch of the zalambdodonts has given rise to the chrysochlorids or Cape Golden moles of Africa. These are highly specialized for digging in hard ground and they resemble *Noto-*

ryctes (fig. 18.30A), the Australian marsupial "mole" so closely (B) that it was thought by Cope that the two families were nearly related. But it was shown by Leche (1907) that this resemblance was another case of convergence, because *Notoryctes* still retains a substantial number of marsupial heritage or palaeotelic characters, while *Chrysochloris* retains equally significant zalambdodont (fig. 19.7D) heritage characters (Gregory, 1910, pp. 255–259). Broom (1916) found that in *Chrysochloris* the Jacobson's organ (a longitudinally placed membranous scroll, enrolled in cartilage, located on either side, near the floor of the nasal chamber) was much larger and better developed than it is in other zalambdodonts, and on the basis of this single difference he proposed to set up a separate order for the chrysochloroids. But notwithstanding their extreme specialization for tunnelling in dry ground, the golden moles retain some important characters that tend to connect them with the centetoids.

Among the outstanding digging adaptations in the Golden mole family were: (1) the bullet-like body-form (fig. 18.30B) with short conical head and cylindrical, tailless body; (2) the huge size of the olecranon process of the ulna and of the triceps muscle attached at right angles to it; the olecranon is about as long as the very short radius and this gives great power in pushing the huge claw on the third digit into the ground and in throwing the dirt backward; (3) the presence of a central ossified tendon in the powerful flexor digitorum profundus; in the dried skeleton it looks like a third arm bone; (4) the large twisted humerus (fig. 19.12E, E1) with its long entocondylar process giving rise to the flexors of the forearm and to the latissimus dorsi, which has shifted down the shaft to get a better pull on the forearm (Dobson, 1882, pl. XII); (5) the sharp eversion of the knees, giving right-angled insertion to the adductors and hamstring muscles on the tibia; (6) wide separation of opposite pubes at distal end, with consequent absence of pubic symphysis; (7) reduction of obturator fenestra to two small foramina and close union of short but strong pubis and ischium; (8) consequent transverse thickening of iliac bar; (9) strengthening of sacral attachment against adverse leverage at lower end of pubis.

No. 6 is closely correlated with No. 5 and so are Nos. 7 and 8. No. 6 also widens the pelvic exit and gives space for relatively large anal glands. But none of these features seem to be inconsistent with the derivation of *Chrysochloris* from zalambdodont ancestors (figs. 19.4, 6) related to *Palaeoryctes* and *Apternodus*.

UPPER CRETACEOUS FORERUNNER OF THE HEDGEHOGS

Another Upper Cretaceous Mongolian insectivore was named *Zalambdalestes* (fig. 19.13A) because its cheek teeth also were somewhat V-like. But a better preserved specimen (fig. 19.14A) showed that in the *Zalambdalestes* upper molars there were two main cusps near the outer border of the crown and that the crown as a whole was U-shaped rather than V-shaped, due to the fullness of the inner cusp. As the interdental spaces were narrower, the crowns were larger and better adapted for chewing and crushing than the pointed cusped crowns of *Deltateridium* (fig. 19.6A). Nor was there any inward displacement of the outer cusps as in *Deltatheridium,* or any enlarged upper or lower canines (fig. 19.13A), but a fairly large, somewhat canine-like, lateral upper incisor, overhanging a pair of procumbent lower incisors. In short, these teeth and jaws were apparently adapted for picking up and grinding very small objects rather than for fighting and for shearing flesh. The skull of *Zalambdalestes* (fig. 19.13A) was long and low, with fairly long tubular bony snout, small, highly placed orbits and slender, laterally outbowed zygomatic arches. The lower jaw, long and slender, the coronoid process delicate and inclined backward. Comparison of this skull and dentition with those of other insectivores indicates that it is basically like those of the family Leptictidae (fig. 19.13B, C) which ranged in North America from the Upper Cretaceous and Paleocene upward. A branch of this family gave rise to the living *Echinosorex* (fig. 19.9A) and the hedgehogs (Erinaceidae, figs. 19.9B; 19.13 D–G) (Simpson, 1928b).

THE CENTRAL POSITION OF THE HAIRY-TAILED HEDGEHOGS (ECHINOSORICINAE) AMONG PLACENTAL MAMMALS

Among surviving mammals, the hairy-tailed hedgehogs of East India, *Hylomys* and *Echinosorex* (*"Gymnura"*), are on the whole almost as central and primitive in comparison with other insectivores and indeed with all other placental orders, as the American opossum is in comparison with all other marsupials. The central position of *Echinosorex* (*"Gymnura"*) among the living Eutheria was clearly recognized by Huxley (1880, p. 657) and by Dobson (1882, p. iii). Of course the living *Hylomys* and *Echinosorex* are not as primitive as their Paleocene ancestors and relatives were, but they are known from complete skulls (fig. 19.13D, E) and associated skeletons, while few if any limb bones are definitely associated with the fossils named above.

In the upper molars of "Gymnura" (fig. 19.14L), to the primitive triad of cusps (protocone on the inner side; paracone, antero-external, metacone, postero-external) is added the new upgrowth from the basal ledge or cingulum, which is termed the hypocone, directed inward and backward. In the European and Asiatic spiny hedgehogs (O) the upper molar crowns develop two small cross-crests connecting the four main cusps; in the lower molars the posterior projections or talonids (fig. 19.13G) have become basin-shaped and the crown bears four main cusps, except on m_3 where there is a prominent median cusp (hypoconulid) on the rear end of the talonid. Meanwhile the skull has become short and wide, the malar (jugal) has been reduced to a small road (fig. 19.13G).

The spiny hedgehogs (fig. 19.9B) may well owe their survival partly to their prickly coats and to their ability to roll themselves up into a ball, which dogs are loath to bite. A similar adaptation was developed independently: in *Hemicentetes* (fig. 19.3B) and others among the centetoids, in the egg-laying *Tachyglossus* ("Echidna") among monotremes, and, to a greater degree, in the porcupines among rodents. These marked specializations have been avoided by *Echinosorex* (*"Gymnura"*), or possibly it has replaced spines by long hairs, as the Tasmanian species of echidnas have almost done.

LITTLE FURIES (SHREWS)

Among the most specialized of living insectivores are the shrews (Soricidae, fig. 19.9C, D) and moles (Talpidae, E, F), constituting the superfamily Soricoidea of the Suborder Dilambdodonta. The subordinal name, meaning, practically, double V's or inverted lambdas, was given in reference to the presence of two sharp V-shaped outer cusps on the upper molars of the shrews (fig. 19.10E, F1). Similar but inverted V's, also with sharp cutting edges, are borne by the lower molars and they shear into the notches between the upper V's. The upper incisors (fig. 19.10B, D) are relatively very large and directed forward; they look like the serrately edged mandibles of some insects and when opposed by the large procumbent pair of lower incisors, they serve to grasp, nip and cut insects, larvae, worms and other shrews. These tiny mammals are exceedingly active and pugnacious and one of them has been known to attack and kill a mouse twice its size (Goodwin, 1935, pp. 31, 41). The least shrew (*Sorex* (*Cryptotis*) *parvus*) is only about two inches long and is one of the smallest of all mammals. On account of their small size and intense activity, shrews radiate a large amount of heat in proportion to their mass; consequently they have colossal appetites and consume a relatively enormous quantity of food (*ibid.*, p. 35).

The skeleton of typical shrews is very lightly built, but in such small mammals the strength of bone is relatively high in proportion to the total weight. The smallest shrews seem to have been bred down almost to the same general order of magnitude as the insects which they pursue; but in power they surpass them as hummingbirds surpass bees. The water-shrew (*Sorex palustris*) has long hind feet fringed with short stiff hairs and it is an excellent swimmer. It can also dive, float, or run along the bottom of shallow pools or streams. If the pool is perfectly still, the surface tension is sufficient to allow it to run on the surface, rearing up and using the air bubbles held in the stiff hairs on the feet to cushion the weight (ibid., p. 37), somewhat as the water-striders do among insects, or as the jackana does among birds.

Unique Vertebrae of the Hero-shrew.—One of the largest shrews is the "Hero-shrew" (*Scutixorex congicus* Thomas) of the Belgian Congo, which is somewhat like a rat. This amazing creature, if stepped on by a barefooted native, is not killed, apparently because of the unique construction of its vertebral column (fig. 19.11). As described and figured by J. A. Allen (1917), the centra of the neck, back and loins, are very large and wide and their sides are covered with many small interlocking, bony zigzags which permit very little movement but protect the spinal nerve cord and its branches. The rest of the skeleton (fig. 19.11B), including the skull, girdles and limbs, conforms to the general shrew type (A) but is more stoutly built. At first sight the vertebrae suggest arthritic deformity, but though they are anomalous, they are not covered with irregular or spotty exostoses but with a beautifully formed, elaborately articulating system which is known only in members of this species. In this case a strikingly new skeletal system of secondary polyisomeres may have appeared suddenly in one isolated, inbred population (related perhaps to *Crocidura nyansae kivu*), which came to be dominant within a relatively small area; no intermediate forms are known to connect it with closely related species.

The Enigmatical Nesophontes.—In the West Indies the fossil *Nesophontes* was a small mammal whose skull (fig. 19.10A, A1) presents a puzzling mixture of shrew-like and zalambdodont characters. It seems, however, to be on the whole nearer to the shrews (H. E. Anthony, 1916, 1925).

TUNNELLING MOLES WITH HUGE HAND-SCOOPS

The typical moles (Talpidae) are highly adapted for tunnelling in soft ground (fig. 19.9F), chiefly by means of their relatively enormous five-fingered hands armed with large nails (fig. 19.12A). The width of the hands is even augmented by the presence of a sickle-shaped sesamoid bone in addition to the regular five fingers. To support and operate these huge scoops profound changes in the shoulder girdle and sternum have been necessary. In normal mammals the clavicles are curved rods by which the opposite scapulae are tied to the sternum; but in the moles (D) the clavicles (*clv*) have been greatly shortened, thickened, widened and adapted to withstand great pressure and bending stresses as well as tension. In effect they form two short blocks, which receive the thrusts from the humeri and transmit them inward and downward to the sternum, the anterior segment of which has been markedly prolonged forward and stiffened by a median crest. This combination has pulled the shoulder joints inward near to the midline and brought them in front of the thorax, so that the hands can reach forward in front of the nose. Thus the minimum width of the tunnel is determined not by the width across the shoulders as it would be, for example, in man, but by the width of the neck and forwardly directed limbs beneath it. By twisting and turning the mole would further reduce the effective width and practically screw itself forward into the ground.

In their new positions the clavicles (D) somewhat resemble the coracoids of the "spiny anteater" (Echidna) with which Vialleton (1924) proposed to homologize them. Parker and Haswell (vol. II, pp. 549, 643) interpret the clavicle of the mole as a compound element formed of the true dermal clavicle in front and of a mass of cartilage which is regarded as a procoracoid. But the latter may well be a neomorph, like the huge cartilaginous epiphysis on top of the young stages of the humeri of ungulates. The real coracoids of the mole are small processes on the glenoid surfaces of the scapulae in their normal mammalian relations, as well shown by Kistin (1929). Moreover, the clavicles of the mole are inescapably homologous with those of the Desman ("Myogale"), which are rod-like (fig. 19.12C) and have the right relations for mammalian clavicles, with the acromion of the scapula and with the sternum.

The sternum of the mole is strongly built and its anterior segment is prolonged forward (D) and has a low median crest for an anterior extension of the pectoral muscles. The strong thrusts which it receives from the humeri and clavicles are partly cushioned by the fan-like muscles around it, which act as living springs and tend to prevent overstraining of any one part or of one side. This fan-like arrangement (fig. 19.12B) also makes possible a more continuous and effective pull, as the angle of the moving lever changes with reference to the positions of the fulcrum and the load. Nor should it be forgotten that a fan is a limiting case of a flattened cone, or that ultimately all limb muscles have been derived from cone-in-cone extensions from the axial myomeres.

The scapulae of the mole are long and slender (D, *sc*), although stiffened with ridges. They give a relatively long range to the muscles that draw them upward, forward or backward, but they do not need to be thick and massive as do the muscles that pull the shovels sharply backward and inward in tunnelling, which is a kind of swimming in a resistant medium.

The humeri of the mole (G, G1) are very short, wide bones, which have been curved and twisted so much that the identity of their crests and tuberosities is difficult to recognize. From the wide rear surface (G1) of the humeri arises the massive triceps, which extends the forearm, thereby pushing the earth outward and backward. In the front view (G) the entocondylar crest and adjacent areas give rise to the strong pronators and flexors of the hand, which help to bend and twist the shovels as they dig. The very wide hand, in addition to its five strong fingers, has a long curved extra bone, the os falciforme, which may perhaps help in breaking off the roots of grass.

The rest of the mole skeleton is less specialized, the small, short hind feet being used partly for kicking the earth backward. The pelvis, however, is very long and narrow and the four to six unborn young must be extremely small.

The cranium of ordinary moles forms a thin bony envelope around the brain, with at most a slight sagittal crest. The eyes being very small, the orbits are practically absent and the external ears are minute and hidden under the fur. However, the otic region of the skull is well developed and the mole may hear by bone conduction. The snout is fairly long and conical; in the star-nosed mole (*Condylura,* B) it ends in a pair of radiating projections, one around each nostril, probably sensory in function. The incisors are small and peg-like, not enlarged or nipper-like as in the shrews; the cutting V's of the molars are sharply defined in the upper and lower jaws and there is less grinding action than in the shrews.

Not all moles have great shovel-like hands, and some Asiatic moles have small hands and feet and resemble shrews, but their teeth are mole-like.

Water-moles.—The Desmans (*"Myogale"*) or water-moles of Europe and Asia have the dentition and skull mole-like but, as noted above, their clavicles are simple rods, which apparently never became shortened and block-like, nor are the hands and feet nearly as specialized as those of moles.

SUMMARY ON INSECTIVORES

The hairy-tailed hedgehogs (*Echinosorex, "Gymnura"*) and the nearly allied *Hylomys* have on the whole a remarkably primitive skeleton and they seem to represent more nearly than any other surviving group the central type for all placental mammals, just as the opossum represents the central type of all marsupial mammals. The approximate remote ancestor of the *Gymnura* group in Upper Cretaceous times seems to be represented by the skull and jaws named *Zalambdalestes* from the Upper Cretaceous of Mongolia, by the teeth named *Gypsonictis* in the Upper Cretaceous of North America, by rather numerous genera founded mostly on teeth from the Paleocene of North America and by *Adapisorex* from the Paleocene of Europe. Unfortunately very little is known of the limbs of these fossil forms, but the living *Gymnura* preserves a very primitive type of limbs and five-rayed, well clawed hands and feet adapted for running about on the ground and scratching up insect food. This series (Zalambdolestidae, Leptictidae) leads to the rather varied family of the hedgehogs (Erinaceidae), of which the Gymnurinae represent persistently primitive forms.

Another Upper Cretaceous insectivore family, represented by *Deltatheridium* (the little beast with delta-like teeth), seems to stand near to the stem of the centetoid insectivores. These are now confined to Madagascar, West Africa and the West Indies, but somewhat more primitive forerunners are known from the Oligocene of North America. The centetoids, as we have just seen, are specialized away from the generalized "gymnurine" type, especially in their cheek teeth, skulls, limbs and feet. This group culminates in the excessively specialized Cape Golden moles, which some authors regard as a separate suborder. The shrews and moles are closely related but oppositely specialized families, forming one of the main divisions of the superfamily Soricoidea of the Insectivora.

As to the possible relationships of the insectivores to other orders of mammals, it is not improbable that the stem of the Leptictidae (figs. 19.13, 14) gave rise to the tree shrews (Tupaiidae, fig. 19.16) and thus to the lemuroids (Chapter XXIII) and higher Primates. The very primitive Upper Cretaceous Deltatheridiidae, so far as their teeth and skulls indicate, would seem to foreshadow such primitive Paleocene Carnivora as the Oxyclaenidae (fig. 20.11). In another direction *Deltatheridium* seems to be related to the Didelphodontinae (fig. 24.13C) of the Paleocene and Eocene (Simpson, 1945, p. 48), which were at least paralleling the true Carnivora.

THE TREE-SHREWS AND THEIR ALLIES (TUPAIOIDEA)

Of the existing tree-shrews of Indo-Malaysia, the squirrel-like *Tupaia* (fig. 19.15B) and closely related genera include many species. Some live on the ground and among bushes, with rather narrow feet; but the pen-tailed tree-shrew (C) of Borneo (*Ptilocercus lowei*) has diverging digits and is thoroughly arboreal.

The tree-shrews (Tupaiidae) of the Netherlands East Indies were formerly classified as Insectivora and future discoveries may indeed confirm their derivation from insectivorous Mesozoic forerunners. Haeckel (1866), probably following the observation of Peters that in the elephant-shrews the caecum was retained, while in other insectivores it was lacking, proposed the division Menotyphla (*remain,* blind [sac]), to include the elephant-shrews and the tree-shrews, while for the remaining insectivores he invented the term Lipotyphla, *concealed sac.* Van Kampen (1905, p. 447) showed that in the tree-shrews the tympanic bone took the form of a ring, which was enclosed by an auditory bulla made from the so-called entotympanic element (fig. 19.17C2). Carlsson (1909, 1922), on anatomical grounds, widely separated the tree-shrews from the elephant-shrews, regarding the former as related to the lemuroids, the latter to the erinaceoids. Gregory (1910) cited a long series of skeletal characters in which *Tupaia* differed widely from other insectivores and approached the lemurs. Wood Jones (1929) classed it with the lemurs in a group, Strepsirhini, coördinate with the true Primates. Matthew (1917), in describing the jaws, dentition and incomplete skeleton of *Nothodectes* (*Plesiadapis,* figs. 19.17B) of the Upper Paleocene, noted the fundamental resemblances on the one hand to the tree-shrews and on the other in certain features to the Eocene Notharctidae, which were primitive lemurs. Simpson (1935, 1945), after describing many plesiadapids, tarsioids and lemuroids from the Paleocene and Lower Eocene, treats the Tupaioidea as a superfamily of the lemuroid Primates. Enough has been said to indicate that in many respects the Tupaiidae represent a structurally intermediate group which has not yet fully attained the grade of the Primates but retains many insectivore features.

In general, the vertebral column, girdles, limb bones, hands and feet (fig. 19.16A) of the tree-shrews (including *Ptilocercus*) display no conspicuous terrestrial specializations and abound in primitive arboreal characters. They are indeed more primitive than the typical lemurs in the insectivorous modification of the incisors, in the presence of a pronounced W on the upper molars (B–1) and in the far less marked modifications of the hind foot for grasping the branches of trees (fig. 19.15C).

The next problem, from which of the known insectivore groups did the tupaiids come, leads apparently to elimination of the zalambdodonts, soricoids, and most of the erinaceomorphs. There remains, however, the probability that the plesiadapids, the tree-shrews and with them the ancestral primates, were derived from insectivores of Upper Cretaceous age related to *Zalambdalestes* (figs. 19.17A).

Although the tree-shrews have relatively large eyes, their brain has retained a fairly primitive stamp, with comparatively little development of the neopallium, or upper part of the forebrain. For this reason and from the fact that there are no known intermediate stages between them and the basic Insectivora, we are considering them as archaic relicts from Mesozoic times.

From the Oligocene of Mongolia, Simpson (1931) has described the skull (fig. 19.16C, C1) dentition, and parts of the skeleton of *Anagale*. This form is essentially a tupaioid but, like the plesiadapids, has some important characters tending to connect the tupaioids with the notharctid stem of the lemurs.

The elephant-shrews (figs. 19.15D; 19.16D) of Africa (Macroscelididae) have long hind legs, a long pointed muzzle, fairly large eyes and a relatively large brain. Their upper molars are divided by a transverse crease into two moieties, each containing more or less conic cusps. Apparently the least specialized member of the family is *Rhynchocyon* (fig. 19.15D), which convergently resembles an Australian bandicoot in its long pointed muzzle, long hind limbs and long narrow feet. It was formerly supposed to hop like a kangaroo but, according to Herbert Lang, who studied it in the field, it merely runs, like a tiny musk deer. At the other end of the family, *Elephantulus* has a delicate muzzle and a short wide braincase. Frechkop (1931), noting a few resemblances of the elephant-shrews to the Artiodactyla, was inclined to remove them to that order. As a result of an extended and systematic comparative analysis of the osteological and dental characters of the Macroscelididae, Gaynor Evans (1942) greatly strengthened the case for their relationship to the tree-shrews.

"FLYING LEMURS" (DERMOPTERA)

The so-called flying lemurs (colugos) of the East Indies are not lemurs and really do not fly but simply swoop or glide (fig. 19.15E) from one tree to another, or from tree to ground, like flying squirrels. As leaf- and fruit-eaters with extremely specialized dentition (fig. 19.18), the "flying lemurs" are widely different from primitive bats. However, they do show one strikingly bat-like character, as noted by Leche (1886), namely, that the patagium, or gliding membrane that extends along the sides of the body and out to the ends of the hands and feet, in this case forms webs between the long, well-clawed fingers, which it does not do in other gliding mammals (flying phalangers, flying squirrels, flying anomalurid rodents). Leche regarded this as a prerequisite character for the evolution of the bats. The slender long-bones of the limbs serve to stretch the membrane somewhat as the small, obliquely directed spar or sprit extends the sail of a boat. The patagium has been independently evolved in all the forms noted above, as it was also among the pterosaurs and birds and it is conspicuously absent in animals that live only on the ground, where it would be a hindrance. From the close relationship of the flying phalangers to ordinary arboreal phalangers, and of flying squirrels to true squirrels, it is evident that a gliding membrane has been evolved only among animals that leap about among the trees and that reach out their hands and feet to act as buffers in landing. The patagium of the "flying lemur" is doubtless stretched in part by the normal extensors of the limbs and spreaders of the fingers. As shown by Leche (1886) the true skin muscles are also joined with slips from the latissimus dorsi and superficial coracoid muscles and form fibres that branch out to the margin of the patagium.

The dentition of the "flying lemurs" is highly specialized in several ways. Two pairs of procumbent lower incisors have transversely widened crowns which are subdivided into combs (fig. 19.18A2) and used, according to H. C. Raven, in scraping the green coloring matter out of leaves. The canines, lateral upper incisor and anterior premolar crowns are compressed, with serrated edges (A). The upper molars (A1) each have a pair of high, sharply cuspidate outer V's, a pair of nipple-like intermediate cusps and a large, low internal tubercle or protocone. Each lower molar bears two narrow high V's, which fit neatly between the larger ones in the upper molars, while the basin-like talonid receives the pestle-like protocone. The motion of the jaw must be nearly vertical and quite unlike the obliquely lateral chewing movements of typical herbivorous mammals. Thus the molars of the "flying lemurs" function as do those of the dilambdodont insectivores, especially the mole-shrew group; but this is not improbably due to parallelism or convergence, especially as the skull of the "flying lemur" is extremely different from those of moles and shrews. It is, however, less different from the skull of the Oriental tree shrews (fig. 19.16A) which are almost lemurs, and it shows other points of resemblance to *Anagale* (C), an Oligocene side branch of the tree shrews, and some others to *Rhynchocyon* (D) of the family of the elephant-shrews (Macroscelidae).

Plagiomene, a probable forerunner of the colugos in the Lower Eocene of North America, had two triads of cusps on the outer side of the first and second upper molars (fig. 19.18B), one pair of large intermediate cones and one very large inner tubercle. Its lower molars (B1, B2) were also basic in cusp patterns to those of the colugo.

But even in the Upper Paleocene *Planetetherium* the teeth of this very ancient family were already rather specialized and do not give any indubitable clue to its origin except that they were primitive placentals of some sort (Simpson, 1945, p. 179).

THE HAND-WINGED BATS (CHIROPTERA) AND THEIR DIVERSIFIED SKELETONS

The Nyctitheriidae (meaning bat-beasts), from the Paleocene and Eocene of North America, are known from minute teeth and jaws, which suggest bat-teeth on the one hand and mole teeth on the other. So far as the teeth indicate, they would make good structural intermediates between the bats and earlier insectivores, but as yet not enough is known about the rest of the skeleton to be sure about their relationships.

The recent bats are divided into two suborders, the Megachiroptera, including the fruit bats, and the Microchiroptera, all the others. Bridging some of the differences between these two suborders is the beautifully preserved bat skeleton discovered by Jepsen.

The wings of bats (fig. 19.19) are merely enormously overgrown arms and hands, somewhat like those of pterosaurs (fig. 15.1), but with thin webs of skin stretched between the second, third, fourth and fifth metacarpals and digits. The hand has been flattened into a huge folding fan, while the supporting area has been further increased by the rest of the gliding membrane, which extends on the side of the neck and from behind the wings to the hind feet and between the legs and tail. The relatively short, well-clawed thumb (fig. 19.19d1) acts as a hook for climbing and clinging. The radius, a long slender, slightly curved cylinder, is strongly built and it is braced by the ulna, which is fused with it distally. Both are based on the cylindrical humerus. The hollow shaft of this bone, as it nears the distal end, is twisted slightly outward; this helps to direct the wings outward.

The humerus in its turn transmits the thrust of the wings to the wide scapula, which can move freely above the back, except that it is tied to the strong curved clavicle, and the latter to the transverse process of the sternum. The three bones, humerus, scapula, clavicle, are also tied together by ligaments and muscles and meet in an open pyramid at the base of the neck, where they also form the usual sling or movable cradle for the body. There is a rather small median process on the first joint of the sternum for the pectoral muscles but in wide contrast to the conditions in birds, the chief power for the wings comes not from the pectoral muscles arising in the keel of the sternum, but from all the muscles that are based on or are inserted in the scapula and clavicle. As the pivot for the clavicles, the anterior segment of the sternum is strongly built but not very large. In the smaller bats it tends to unite with the large first ribs and thus to form also a firm base for the thick neck muscles. The humerus in some of the small insectivorous bats has a triple articulation with the scapula, the main one at the head of the humerus head being flanked by accessory tubercles called trochin and trochiter (Miller, 1907). The neck vertebrae, especially in the fruit bats, are large and well muscled to support the very large head and jaws. The head in the fruit bats may be as long as the thorax. The hind limbs are small, as in pterosaurs, and the five-toed, well-clawed hind feet are well adapted for clinging. The pelvis is very small and narrow, and so also are the one or more young at birth.

Dental Crushers and Lancets.—In the more primitive insectivorous bats (*Rhinolophus, Vespertilio*) the upper and lower molar crowns (fig. 19.21A, H) are descriptively dilambdodont, that is, basically like those of moles and shrews. In the leaf-nosed bats (Phyllostomidae) the upper molars in some tend to become relatively very large (C, D), with enormous hypocones in the upper, and huge talonid basins in the lower molars (*Artibeus*). In another line, the cheek teeth (E, F) become narrower (*Hemiderma, Erophylla*). In the blood-sucking vampire, the second upper incisors and upper canines form great pointed blades (G*i2c*) for puncturing the skin and cutting the capillary vessels of the host, but the upper cheek teeth (*p*4, *m*1) have been greatly reduced.

The Tongue and Hard Palate.—In correlation with the shape of the tongue and the teeth, the palate of the Microchiroptera is transformed from the quite primitive triangular upper dental arch of *Otopterus* (fig. 19.23A) to the flattened arch of *Centurio* (G), which is actually much wider than it is long.

Skull Forms.—The side view of the skulls of the small bats (fig. 19.22) exhibit an amazing array of specializations. Perhaps the least specialized is that of the common Brown bat (*Myotis*). This skull (G) is a little like that of a young dog, except that the orbits are minute, the cheek arches slender and the muzzle coarse. In *Chilonatalis,* one of the Natalidae, the forehead is domed (I) and the muzzle low and slightly duck-like but with teeth. In *Desmodus* the cranium (F) is rounded, the bony face very short, with conspicuous protruding shear teeth. Among the Phyllostomoidea or New World bats, the skull of *Vampyrus* (A) suggests that of a wolf or bear on a minute scale, while that of *Leptonycteris,* with its long tapering muzzle (C) and slender jaws, recalls some of the smaller insectivores such as *Hylomys* (fig. 19.13D). *Centurio* is hyperbrachycephalic (fig. 19.22D) beyond any Mongol or other roundhead known to anthropologists. *Mormyrops* (E) combines the high cranium of a mastodon with the muzzle of a bulldog.

Cutaneous Radars.—But however strange each bat skull may appear to the layman, we may be sure that it is just as exactly adapted to its owner's use as are the still more bizarre noseleafs and other foliaceous appendages of the face (fig. 19.20). These have been shown to serve as inerrant radar-like recorders of pressure waves propagated in the air by the flying and screeching bat and received on the rebound from surrounding objects.

In some of the presumably more primitive South American vampire bats of the family Phyllostomidae, the nose-leaf takes the form of an erect spearhead (H). This median "antenna" may perhaps also serve to deflect vibrations coming head on into two streams, which would be further felt by the erect ears and antitragi. The probable usefulness of "shadow patterns," changing with the angle of incidence may also be noted.

Thus the flying bat reacts continuously and with great agility to an ever changing, three-dimensional "picture" of the angular relations and distances of his nose-leaf to approaching objects; much as a flying bird derives a comparable picture chiefly from his eyes.

From the immobile dermal crust of the basal ostracoderms to the pliable sensitive skin of mammals and thence to the nose-leaf of bats, evolution was always "emergent" in the sense that not only was each older stage a prerequisite for all the later ones but also that each advance opened up new opportunities for further advances, often in quite new directions.

EDENTATES, ARMORED OR HAIRY

Under this name was formerly classified an array of strange animals, living and fossil: sloths, ground sloths, American anteaters, armadillos, glyptodonts, scaly anteaters (manids), aardvarks (*Orycteropus*), and the fossil orders of tillodonts, taeniodonts, palaeanodonts. Of these, the first five and the last constitute the order Edentata in the restricted sense. The others are either definitely not related to the edentates (e.g. aardvarks) or at best only distantly related to them (e.g. manids). The very name Edentata (toothless) was never descriptively appropriate except for the American anteaters. But it has priority over the others and, according to the Rules of Nomenclature, it is the official designation of the order (Simpson, 1945, pp. 190–195).

Tillodonts and Taeniodonts. — The tillodonts range from the Upper Paleocene (fig. 19.1) to the Middle Eocene of North America. The upper molars, like those of many other Eocene mammals, have three main cusps arranged in a triangle, plus the beginning of a hypocone or upgrowth of the cingulum. The large, well preserved skull of *Tillotherium* from the Middle Eocene was superficially rodent-like, with a pair of quite large, procumbent and almost chisel-like incisors; but this sort of adaptation has been evolved many times independently in different orders and the rest of the skull is quite un-rodent-like. The dentition on the whole may well have been derived from a primitive Mesozoic insectivore stock (Simpson, 1945, p. 189).

Among the taeniodonts (figs. 19.26, 27) likewise of Paleocene and Eocene age, the older and less specialized small forms (*Onychodectes*, fig. 19.26A) had low-crowned tritubercular upper molars and elongate temporal fossae, implying strong biting and crushing power. In the most specialized ones the enlarged lower canines had long, curved crowns growing from persistent pulps (fig. 19.21D) and suggesting in the side view the beak of a parrot (hence, *"Psittacotherium,"* parrot beast). Stiffening enamel was retained as a prominent band (taenia) on the front border of the incisors and reduced in the rear, as in the incisors of rodents. The cheek teeth formed high, compressed columns (whence the name *Stylinodon,* E) and the very massive lower jaw was convergently rodent-like. Associated limb bones were very strongly built and the very large claws and wide-spreading extremities (fig. 19.31A, B) indicate powerful digging ability. Some features of the dentition and limbs were formerly regarded as indicating relationships with the true edentates but the North American taeniodonts are regarded as an independent series which merely paralleled the true edentates, some of which are recorded from the Eocene of South America (fig. 19.25).

AMERICAN EDENTATES (XENARTHRA) AND THEIR UNIQUE VERTEBRAL COLUMN

True edentates are divided by Simpson (1945, p. 69) into two suborders, Palaeanodonta and Xenarthra. The palaeanodonts (fig. 19.29) were armadillo-like in their skeletons, except that so far as known, they had no bony armor and their vertebrae lacked the accessory or xenarthral articulations (fig. 19.28B) which are so characteristic of the order Xenarthra. From one to three pairs of these extra-zygapophysial processes and articulations are located at both ends of the posterior dorsal and lumbar vertebrae. Like the regular inferior and posterior zygapophyses which they supplement, they are secondary polyisomeres and form a system of tongue-and-groove sliding joints. They may have been initiated in embryonic stages by extra creases and folds in the septa between the vertebrae. Their smooth surfaces show that they are covered with synovial membrane and no doubt they are tied in by strong ligaments. Thus they afford additional protection against dislocation, but at the same time restrict the movements to dorso-ventral extension and flexion. They must also assist the column in resisting the wrenching strains put upon it by the limbs and their muscles, and in the gigantic ground sloths such strains must have been relatively enormous.

The teeth of recent edentates have lost the enamel layer but the height of their crowns is often lengthened and especially in the cheek teeth the roots may be delayed in forming, so that the crown may long continue to grow and compensate for the attrition of these enamelless columns of dentine. The diet is omnivorous.

The Armadillos, Relatively Primitive. — Although the armadillos (figs. 19.24, 30A) are covered with an armor of horn-covered bony plaques on top of the head and over the back and sides, their skull, teeth and limb bones seem in many points to be less specialized than those of any other group of edentates. The armadillos (Dasypoda) are a very ancient group (fig. 19.25), dating back in South America to the Paleocene, at which time some of them (*Utaetus*) retained traces of enamel on the molar crowns (Simpson). Their powerful forelimbs are specialized for digging but not as much as those of the ground sloths. For example, the scapula is large (fig. 19.30A), with large acromial process, the clavicle strong, the humeri are very stout, with a twisted shaft and a wide distal end; the claws large, especially on digits II, III, IV (fig. 19.31D). The sacrum is extensive and the hind limbs strong.

Armadillos retain a more or less flexible carapace, with two to thirteen movable transverse bands (figs. 19.24, 30A) forming a girdle, so that they can roll up into a ball. In *Chlamyphorus,* the pichiciego, there is a massive circular convex shield on the rump, which is based upon the strongly braced ischium and postsacrum.

Tortoise-like Glyptodonts. — In the extinct glyptodonts (fig. 19.30B) the somewhat tortoise-like carapace formed a fixed shield composed of polygonal bony plates convergently recalling those of armored dinosaurs. The dorsal and lumbar vertebrae were immobilized by the armor. The neck vertebrae are reassembled into three movable segments which could be straightened out when the small head was extended, or strongly bent upon each other as the head was withdrawn, thus increasing the convergent similarity with the tortoises (M. Weber, 1928, vol. II, p. 196). The tail was ringed with bony plates and in *Doedicurus* and its congeners the distal end was fused into a massive club. The muzzle was short and high, the lower jaw stout with very large areas for the jaw muscles, the condyle of the lower jaw was raised very high above the level of the cheek teeth, as is frequently the case in mammals with grinding molars. The front teeth were reduced but in the lower jaw there was a large median channel for the protrusile tongue. The cheek teeth (fig. 19.33E), eight in number on each side above and below, were arranged in straight flat rows and had very high, three-lobed columnar crowns. From the malar arch (fig. 19.32E) on each side descended a long process for the origin of the masseter, which was inserted on the large curved rear border of the jaw.

The Mighty Ground Sloths and Their Enfeebled Kin, the Tree Sloths. — The larger ground sloths (fig. 19.34B), especially *Megatherium,* were beasts of mastodonic mass, able to rear up and pull down stout trees with their colossal arms and hands and to dig with the huge curved claws on all their feet. Their posterior dorsal and lumbar vertebrae were well braced by the primary and accessory articulations and tied in by strong ligaments and axial muscles. The transversely expanded blades of the ilia were not unlike those of mastodons but more robust. Their gigantic femora passed the huge weight down to the thick tibiae and strong fibulae but the hind feet with great curved claws rested chiefly on their outer sides. Apparently the remote ancestors of the later ground sloths combined digging and tree climbing. Some of the smaller ground sloths, not unlike *Nothrotherium* (C), must have shortened their caudal vertebrae, greatly lengthened their limbs, reduced the hands and feet to hooks, widened the cranium and shortened the face, and thus were changed into leaf-eating sloths (D). The evidence for this conclusion abounds in the entire skeleton of the sloths, which exhibits a pervasive heritage of ground sloth characters in the skull, vertebrae, girdles and limbs. This is of course somewhat disguised by later imposed habitus features, assumed in adjustment to their upside-down position when hanging in the trees. In short the tree sloths or Tardigrada (figs. 19.24, 25, 34D, 37C) are the last poor remnants of the mighty South American ground sloth fauna, left hiding in the forests; much as in New Zealand the kiwi is the last side branch of the giant moas; but in both cases a later overlying habitus has delayed the general recognition of the true relationship which is revealed by the heritage. In the case of the sloths, this was seen clearly by W. D. Matthew (1911, 1912) and is plainly suggested in G. G. Simpson's phylogenetic diagram (fig. 19.35).

The Hairy Anteaters.—The anteaters (Myrmecophagidae, Vermilingua, figs. 19.24, 25, 36C, 37E) represent another and quite different branch from the same basic stock, but they, following the lead of the long-skulled ground sloth *Scelidotherium* (fig. 19.36B), greatly lengthened the snout to form a curved tube; later they lost all the teeth, leaving the lower jaw as a pair of narrow curved sticks supporting the great protrusile tongue. The little anteaters, *Cyclopes,* then shortened the muzzle and became thoroughly arboreal.

SCALY "EDENTATES" (PHOLIDOTA)

The scaly anteaters (Manidae, figs. 19.24, 38B) of Africa, Asia and the East Indies, have slight claims to be classed with the real edentates. They resemble the anteaters especially in: the loss of teeth, the reduction of the lower jaws to small splint bones, the protrusile tongue, the cylindrical form of the cranium and the backward prolongation of the bony palate (fig. 19.39B). These are obviously convergent habitus features, because in the general anatomical and skeletal patterns they differ widely from the true or xenarthrous anteaters, not only in the absence of accessory zygapophyses in the vertebrae, but in the reduction of the acromial and coracoid processes of the scapula and in the absence of the clavicle (Weber, 1928, vol. II, p. 178).

Among the more noteworthy new or peculiar features are the T-shaped ungual phalanges of the embryonic hand and foot and, in *Manis javanica,* the narrow scoop-like form of the xiphisternum, which is surrounded by the sternoglossal muscles. In *M. tricuspis* this is prolonged into a J-shaped cartilaginous sheath reaching back to the pelvis. Inside this sheath is a peritoneal sac containing a loop of the muscles that move the extremely protrusile tongue (Weber, *op. cit.,* p. 177). The general resemblance of the horny imbricating epidermal scales of *Manis* (fig. 19.38D) to those of *Zonurus* among lizards has already been cited (p. 345) as an example of convergence.

HIGH-POWERED AARDVARKS (TUBULIDENTATA)

The aardvarks (*Orycteropus,* fig. 19.24) were also formerly regarded as edentates, but they resemble the latter mainly in habitus features, such as the presence of a long, somewhat cylindrical snout, a protrusile tongue, a backwardly prolonged bony palate and complexly folded ethmoid scrolls bearing olfactory epithelium (fig. 19.39A). Moreover, the vertebrae, girdles and limbs are powerfully built and when the creature digs furiously into termite hills, with his strong claws he kicks the earth far behind him. If a large termite hill were to fall down upon the digging aardvark, the massiveness and great strength of the lower back, pelvis and hind limbs would be useful in pushing his way out. The trumpet-like ears are somewhat like those of rabbits and antelopes.

The most peculiar feature of the aardvark is the construction of the cheek teeth, consisting of very many and very high parallel tooth prisms with persistent pulp canals, enclosed in an envelope of cement. These secondary polyisomeres may have evolved from the many small tubercles on the surface of a polybunous crown somewhat like those of warthogs. The tooth crowns are arranged in straight parallel series and, as in many other forms with a triturating dentition, the ascending ramus of the mandible is high, with its condyle high above the plane of the teeth, to insure simultaneous opposition of the entire set, at least of one side. The locomotor skeleton is much more primitive than those of edentates but in many points suggests derivation from some of the primitive Paleocene proto-ungulates or condylarths (Colbert, 1941).

SUMMARY ON EDENTATES

In conclusion, the Paleocene and Eocene tillodonts and taeniodonts may not be closely connected with the true edentates but they do indicate a transition from tritubercular, tribosphenic molars derived ultimately from those of primitive insectivorous placentals of late Mesozoic age to high-crowned, prismatic enamel-less crowns adapted for omnivorous-to-herbivorous habits, as in the true edentates (Xenarthra). Among these animals the extra zygapophysial facets on the posterior dorsal and lumbar vertebrae are associated with great strength in the limbs for digging, and they culminated in the ground sloths, some of which had the largest and strongest grasping hands of all known vertebrates. Protection is secured either by a reversion to reptilian-like osteoderms and horny scutes, as in the armadillos and glyptodonts, or by means of a tough thick leathery skin, containing small osteoderms but also producing coarse hair, as in the ground sloths.

The great anteaters and tree sloths are here regarded as dwarfed leftovers of different families of ground sloths. Fast running, aquatic and gliding adaptations in this order are conspicuously absent.

The scaly anteaters and aardvarks have merely a somewhat edentate-like habitus concealing a widely different ordinal heritage, the one of unsettled origin, the other possibly an offshoot of the early Paleocene hoofed mammals.

To return to the center of placental branching, one branch of the insectivores which ultimately produced the hedgehogs, seems to have given off near its base a series which led through the tree-shrews to the lemurs and thence by successive branching to the higher primates, culminating in the apes and man. But for the purpose of the present work it may be more useful and practical to defer the primates to the end of the systematic or taxonomic review of the mammalian orders (Chapter XXIII). At this point we may take up another side branch of the insectivore stock which led to the rodents.

THE SWARMING TRIBES OF RODENTS (RODENTIA, GLIRES)

The Order Rodentia was conveniently divided by Brandt (1855) into four suborders (fig. 19.40):

Sciuromorpha: squirrels, beavers, gophers, sewellels (mountain beavers) and many others;

Myomorpha: voles, lemmings, muskrats, rats and mice, dormice, jumping mice, jerboas, etc.;

Hystricomorpha: porcupines, guinea pigs, agutis, chinchillas, coypus, etc.;

Lagomorpha: pikas, hares and rabbits.

Fossil rodents are found from the Upper

Paleocene onward and in some groups tolerably clear outline histories are known through the Tertiary. The sciuromorphs begin with the relatively primitive, somewhat squirrel-like Ischyromyidae and branch toward the sewellels (Aplodontidae), squirrels, chipmunks, marmots, prairie dogs, flying squirrels, gophers, kangaroo rats and pocket mice.

A more specialized dwarf branch of the ischyromyoid stock gave rise to the swarming hosts of the myomorphs, especially the fossil and recent voles.

The Hystricomorpha date mainly from the Oligocene of Europe and the Miocene of South America. Although having the typical "hystricomorph" arrangement of the cheek arches (p. 393), they may be a somewhat artificial group, including South American and African subgroups of different origin (Simpson, 1945, pp. 209–213).

The Lagomorpha, foreshadowed in the Paleocene of Mongolia, and known from the Oligocene onward, are excluded from the true rodents and form a quite different order, possibly related to the ruminants among hoofed mammals.

The more or less rat-like, gnawing incisors (figs. 19.41, 42) are the leading feature of this great order. The single pair of enlarged upper incisors grow from persistent pulps and have an enamel band on the convex front surface and chisel-like edges, overhanging a similar but upwardly curved pair in the lower jaw. When the latter is pulled forward (chiefly by the external pterygoid muscles) the chisel-like edges of the lower incisors can be opposed to those of the upper and they cut the wood or other tough object between them. No trace of the upper and lower canines remains and there is a wide gap or diastema between the incisors and the horizontal rows of cheek teeth.

More or less similar adaptations for gnawing with the upper and lower incisors have been evolved independently in several other orders of mammals, e.g., in an allotherian, *Taeniolabis* (fig. 17.6), in the marsupial wombats (fig. 18.34), in the Eocene tillodonts; but such animals may always be distinguished from the rodents by marked differences in the jaws and cheek teeth.

Diverse Gnawing Mechanisms.—The suborders of Brandt were defined by Tullberg (1899) chiefly by characteristic differences in the spatial relations of two main layers of the masseter to certain parts of the skull. In the existing sewellels, Mountain "beavers" (fig. 19.41B) of the Rocky Mountains, as well as in their Eocene ancestors, the Ischyromyidae (A), the anterior lateral slip of the masseter (*mla*) ended on the anterior end of the lower border of the zygoma or cheek arch. In the beaver this muscle (D) has worked its way forward and upward in front of the cheek arch on to the side of the maxilla and it leaves its imprint on the skull in the form of a flat bony plate below and in front of the eye.

This muscle is pear-like with the neck of the pear extending up on to the flat plate noted above.

In the rats, representing the Myomorpha (fig. 19.42A, B), the outer slip of the masseter (*mla*) is fastened to the maxilla below and a little in front of the eye, but the inner slip (*mma*) has pushed its way upward and forward, medial to the eye, and it has begun to open a cleft between the anterior border of the orbit and the maxilla. In the hystriocomorphs, including the porcupines, the coypu (D) and others, the inner or deep slip (*mma*) becomes dominant, opening up a large bony arcade and extending widely on the side of the bony face. Meanwhile the outer anterior layer of the masseter (*mla*) remains attached to the lower border of the cheek arch. Transitional stages between the typical sciuromorph and the myomorph patterns occur in some fossil rodents (Albert Wood, 1936, 1937d, 1947).

The incisors of the true rodents (after the exclusion of the rabbits and their allies) remain fairly uniform in basal structure, differing chiefly in curvature and length, but the molar teeth become widely diversified in the different families. The Eocene Ischyromyidae have the roundly quadrangular upper molar crowns, surmounted by low tubercles and incipient cross-crests. This general pattern seems to be archetypal to all the more specialized ones (figs. 19.44–48).

New Polyisomeres in the Molar Crowns.—In many families the molar crowns become very high, with elaborately folded crests, as in the beavers (fig. 19.44D); in others there is a row of tall V-like prisms, as in the muskrats (fig. 19.48H), culminating in the anteroposteriorly elongate, many-plated molars of the capybara (K).

Rodent Skulls and Jaw Movements.—In accordance with their wood-gnawing dentition, all rodent skulls have the condyle of the lower jaw and the root of the zygomatic arch elevated well above the plane of the cheek teeth (figs. 19.41, 42), and, since the lower jaw must be drawn forward to bring the lower incisors into opposition with the uppers, the glenoid articular facet on the squamosal forms an anteroposteriorly directed groove. Usually the last premolar is more or less molariform, which gives the animal at least four teeth functioning as molars on each side above and below (figs. 19.44, 45, 47). The premaxillae are exceptionally large, to support the long curving upper incisors, and are securely sutured to the well braced maxillae and wide frontals. Except in some very small rodents with inflated skulls, the occiput is usually wide and the neck thick to support the relatively great strains upon the braincase from the teeth.

When the cheek teeth have high crowns, as in beavers and others, the alveolar pouches of the maxillae are correspondingly high and in some forms the opposite rows of molars may be squeezed inward, near to the midline (*Bathyergus*). They are well braced internally by the palatines. The long curved lower incisor crowns extend backward beneath the molars, sometimes almost to the base of the mandibular condyles. These latter form elongate ovals to fit into the anteroposteriorly elongate glenoid sockets of the squamosal. The angular process varies in breadth with the size and position of the posterior portion of the masseter on the outside and of the internal pterygoid muscle on the inside. The temporal complex (periotic, mastoid and tympanic) is small or moderately developed in primitive rodents, but becomes greatly inflated transversely in the small desert-living jerboas and kangaroo rats.

Thus, while the gnawing incisors exhibit only minor differences in the different families, chiefly in length and curvature, the range of structure in the molars is much greater, with corresponding differences in the parts that support them. The skulls, except in the parts directly related to the jaws and teeth, are all basically similar except in proportions, depending partly upon mass of body and mode of locomotion. Rodent skulls do not include nearly as many extremely aberrant and peculiar forms as are found in the orders of bats, edentates, whales, carnivores, ungulates and primates. Extremely elongate, toothless, cylindrical skulls adapted for ant-eating, such as occur among the edentates, are conspicuously absent, perhaps because the rodents are primarily gnawers of woody tissue and vegetation rather than insectivorous-omnivorous in diet.

Body-forms and Locomotor Habitus. — The body-form and skeleton of rodents, so far as their fossil history indicates, begins with the arboreal or semi-arboreal Ischyromyidae (fig. 19.40). The squirrels (C) and flying squirrels (E) emphasize the arboreal adaptations; the marmots (D) are essentially fat-bodied, digging squirrels. The gophers (E) are more specialized for digging and the related heteromyid rats for leaping. Kangaroo-like habits and long, slender hind limbs and feet have been developed independently in several families (fig. 19.46) of the sciuroid, muroid and dipodoid superfamilies (Hatt, 1932). The limbs, hands and feet of rodents are of course modified in accordance with the medium in which the animal lives and the mode of locomotion, as in gophers (Geomyinae) and the "mole-rats" (Bathyergidae), water-rats (Hydromyinae), tree mice (Dendromurinae), capybara (Hydrochoerinae), flying squirrels (Pteromyinae), African flying squirrels (Anomaluridae), dormice

(Gliridae), porcupines (Hystricidae, Erethizontidae). Usually the limbs are fairly short, with spreading hands and feet, but with no such bizarre distortions as occur in moles, golden moles, armadillos, ground sloths, anteaters and tree sloths.

The vertebral column among the jerboa-like families develops localized enlargements and fusions in the neck vertebrae to check the forward thrust of the relatively large head in landing (Hatt, 1932), but otherwise differences in the vertebrae in the different families are not very conspicuous.

SUMMARY

On the whole, the rodents are more or less defenseless, timorous and mostly small scampering mammals, the prey of snakes, owls, dogs, cats, minks and other carnivorous animals. Some of the smaller mice and voles could almost compete with the shrews as the smallest living mammals. The largest known rodents (*Castoroides, Hydrochoerus*) might compare in weight with a wild boar. But real giants, such as ground sloths or mastodons, are wholly foreign to the order.

THE HARES AND RABBITS (LAGOMORPHA)

None of the foregoing remarks apply to the hares, rabbits and pikas, which almost certainly belong in an order by themselves; for apart from their superficially rodent-like front teeth, they differ from the true rodents throughout the dentition, jaws, skull and vertebrae, girdles and limbs, as shown by Gidley (1912). In several ways they converge toward the caenotheres, a family of Eocene and Oligocene artiodactyls, but Hürzeler (1936), in a fine monograph on the caenotheres, considers these resemblances to be purely convergent. This leaves the hares and rabbits among the several "orphan groups" among placental orders; but for that matter, the same is true of the Rodent Order, which on anatomical grounds may have been derived from an as yet unknown or unrecognized family of Cretaceous or early Paleocene placentals with procumbent incisors and a taste for seeds, fruits and vegetation.

CHAPTER XX

ARCHAIC AND MODERNIZED CARNIVORA, INCLUDING SEALS (PINNIPEDIA)

Contents

CHAPTER XX

ARCHAIC AND MODERNIZED CARNIVORA, INCLUDING SEALS (PINNIPEDIA)

LION *VERSUS* OX: A STUDY OF OPPOSITES

THE STORY of the lion that leaped over the wall of the kraal, seized an ox in his strong jaws and then leaped back over the wall with his prize in his mouth, may not be literally true but it will serve to introduce the subject of the present and the next chapters, which is a brief review of the numerous kinds of flesh-eating placental mammals and of their far more diversified prey, the herbivorous placentals.

Of the many contrasts between the lion and the ox, the most central is that which involves their respective food habits. For in the lion's food, flesh with its proteins predominates, while the ox is limited to plants, especially grasses, composed mainly of carbohydrates. Thus the lion's food is on the whole more highly organized and less bulky. Its stomach remains simple and the intestine short. Its gastric juice contains sufficient hydrochloric acid to hasten the breaking down and digestion of the bones and connective tissue. The food of the ox is bulky, abounding in chlorophyll and woody tissue; its large stomach is divided into sacs or chambers, each with a different function; its intestine is very long. Digestion is effected by means of a great quantity of water and saliva, aided by countless millions of anaerobic bacteria and infusoria, which cause fermentation and produce a great quantity of carbonic acid gas. Relatively few mammals attain either of these extremely divergent specializations, the majority being able to subsist on a more or less mixed diet. In all cases digestion is essentially a process of hydrolysis, the food being reduced to finer and finer particles until it can be absorbed directly by the cells of the digestive tract.

Differences in Jaws, Teeth and Skull.—The differences in diet between the lion and the ox are likewise reflected in the differences in their dental mechanism and of the associated parts of the skull (fig. 20.2), as set forth in the following table.

	LION	OX
Jaws	short, powerful, wide, for strong vertical movements	long, slender, narrow for oblique side swing
Incisor teeth	present in both jaws	absent in upper jaw, replaced by pad
Lower incisors	sharp, for piercing, holding and tearing flesh	blunt, opposed to pad, for cropping grass
Canines	prominent, for killing and dragging the prey	upper canines absent; lower canines incisor-like, for cropping grass
Crowns of premolars and molars	compressed, blade-like, for shearing flesh and cutting bones	crescentic, long-crowned, for chewing the cud
Articular condyle of lower jaw	placed far down, on level with teeth, to produce a scissors effect, the rear teeth engaging first	placed far above the level of teeth, to bring all the cheek teeth on one side into play at once
Ascending branch of lower jaw	Very large, for attachment of powerful temporal muscle	slender, for small temporal muscle

	LION	OX
Body of jaw	massive, for attachment of powerful masseter muscle	slender, for attachment of slender masseter muscle
Method of cutting up food	into large chunks, with a few powerful bites	into many very small bits, with many strokes of the jaw

The general form of the skeleton in the lion enables it to make great leaps, to strike down its prey with its huge sharp clawed paws and to pull and haul the prey about with its strongly braced canine teeth. In the ox the bony frame resembles in essentials a cantilever bridge adapted for standing but also for running; the cropping machine can be let down or raised up on the end of a long jointed lever, the neck.

We shall presently see why we may infer that these profound contrasts have not existed from an initial moment of creation, but are due to the gradual summation of small differences and that the lion and the ox are the highly diversified descendants of primitive insectivorous placental mammals of the Mesozoic era.

MAIN CLASSIFICATION OF CARNIVORES

Unfortunately there is no really common English name for the mammalian Order Carnivora and even the word "order" in its zoological meaning is quite unknown to the public. Linnaeus in 1735 included in his Order "Ferae" (wild beasts) not only the true flesh-eaters but also the mole and hedgehog and bat; but later he removed these "little flesh-eaters," as the French called them, to the order "Bestiae," which also included the boar, the armadillo and the opossum. Through much later work it was gradually realized that the recent and fossil families included in the Order Carnivora (in the restricted sense) are related by descent from part of the extinct suborder Creodonta (flesh teeth). Their relationship with the modern Insectivora is probably remote but both the Orders Carnivora and Insectivora may have been derived from the still earlier (Jurassic) Order Pantotheria (Chap. XVII).

The existing flesh-eaters, exclusive of the seals and their allies, are divided into three main groups or superfamilies: (1) the Feloidea (Aeluroidea), including the civets, hyaenas, cats, etc. (figs. 20.1, 16–31); (2) the Canoidea (Arctoidea), including (figs. 20.32–38) the dogs, bears, raccoons, pandas and mustelines (martens, minks, skunks, otters, etc.). These collectively constitute the suborder Fissipeda (with cleft feet) in opposition to (3) the suborder Pinnipedia (with flippers), including the sea-lions, walruses and earless seals (figs. 20.39–41).

ADAPTIVE DIVERGENCE IN BODY-FORM, VERTEBRAE, AND LIMBS

The more typical Paleocene and Lower Eocene creodonts were thick-limbed beasts (fig. 20.3A) with rather spreading five-toed hands and feet, armed with blunt nail-claws. They may have been able to climb, swim, dig or run, but were not highly specialized for any one line. The more lightly built *Sinopa* (B), a Middle Eocene hyaenodont, had narrower feet adapted for grasping and running. It also had a highly arched back with an anticlinal region, where the direction of the neural arches was rather suddenly reversed. The Lower Oligocene *Hyaenodon* (C) was somewhat more dog-like, with compressed but not elongated hands and feet.

In the typical creodonts, the scaphoid, lunar and centrale of the carpus (fig. 20.10) were separate but rather closely appressed, but in the early fissipeds they were consolidated into a single bone, the scapho-lunar-centrale, which was retained by all families both of the fissiped and pinniped Carnivora.

DIGITIGRADISM AND PLANTIGRADISM

It used to be thought that plantigrade mammals were more primitive than digitigrade forms, such as dogs, cats and hyaenas (fig. 20.1A), in which both the wrist and the heel bones in standing were held high off the ground. But there are excellent reasons for inferring that plantigradism, at least in carnivores, is a secondary derivative of digitigradism and is associated either with tree-climbing, as in the palm-civets, or with the habit of rearing upward, as in the bears. Raven (1936) indeed showed that even the bears in walking keep their heels well off the ground and are only "plantigrade" on the front half of the foot. Digitigradism is often correlated with a sharp angulation of the second and first row of phalanges (fig. 20.32) and with the presence of thick dermal cushions under the metacarpals, metatarsals and phalanges. In the cats, digitigradism is not inconsistent with retractility of the claws. This is achieved by making the joint between the ungual and the second phalanx asymmetrical, so that by over-extension the claw can be folded back over the second phalanx. In the hind foot, digitigradism was attained as far back as among the bauriomorph therapsids (fig. 18.20A), which carried the heel well raised off the ground.

RUNNING, CLIMBING, DIGGING AND SWIMMING TYPES

Fast-running adaptations reach their climax in the very long-limbed cheetah among cats and in the greyhound among dogs. The greyhound progresses by making long leaps on all fours (fig. 20.32C), the vertebral column undulating up and down and slightly from side to side.

Climbing adaptations are especially evident in the wide hands of the giant panda, which has a functional "prepollex." It is highly improbable that this structure is homologous with or derived from the same named structure of the frog and other amphibians (Noble, 1931, pp. 108, 111, 117). It seems more likely that the prepollex of *Aeluropoda* is more comparable with the falciform radial sesamoid of the mole (fig. 19.12A).

Among the mustelines the body is primitively long and slender, culminating in the short-legged weasels and minks, in which it is almost snake-like. These "vermin," as they were formerly called, have probably been dwarfed to enable them to pursue the rats and mice of the underbrush into their holes in the ground. Some of the minks catch fish in the streams (Goodwin, 1935, p. 68). In the badgers the body is thick and the limbs strong and well clawed for digging. The sea otters converge toward the seals in their sleek coats and paddle-like hind limbs.

None of the known carnivores have adopted bipedal or leaping habits, and, although some are arboreal, none have become gliders.

EVOLUTION OF THE SHEARING TEETH AND MOLARS IN THE DIFFERENT FAMILIES

Flesh-eating is the outstanding feature of the central forms of this order and it is consistent with a variety of dental equipment, body-form and locomotor methods. The most primitive suborder, Creodonta, is represented by several families and many genera in the Paleocene and Lower Eocene of North America. Their jaws and dentitions vary greatly in detail but always with large, dog-like canines (figs. 20.3–7). The molars of creodonts were wedge-shaped (tribosphenic, "tritubercular") in the upper jaw and wedge- and basin-shaped ("tuberculo-sectorial") in the lower jaw (fig. 20.7).

Thus the dentition of the earlier Carnivora was basically identical with that of the older Insectivora, but in the Carnivora the premaxillae were shorter and rounder (fig. 20.4B1), the lower incisors more vertical (fig. 20.4B), less procumbent and the upper incisors were often set in a more transverse, less anteroposteriorly prolonged series (fig. 20.19).

In the apparently most primitive genera of creodonts, all three main cusps of the upper molars (proto-, para-, metacones) were conical to V-shaped, with two small accessory cusps (proto- and metaconules) lying between the outer and inner cusps (figs. 20.7; 20.11).

In the Lower and Middle Paleocene Oxyclaenidae (Arctocyonidae) the first and second upper molar crowns were like V's, but with a rounded inner cusp (fig. 20.11a). In the lower molar crowns the V's were reversed (11b), the tips being on the outer side. The anterior limb of the upper V ended in a low ridge-cusp called parastyle (fig. 20.7A1 *pas*); the external basal ridge or cingulum ended posteriorly in a low angular projection called the metastyle (*mts*). The upper and lower molars fitted together (A1) like the teeth and grooves in a cog-mechanism essentially as they did in Jurassic pantotheres (fig. 17.16) and in Upper Cretaceous insectivores (fig. 17.24a).

In the more typical flesh-eaters (fig. 20.7A1C) the metastyle (*mts*) was connected by an oblique crest with the metacone (*me*) and as this crest elongated and gave rise to an upper shearing blade, its antagonists, the paraconid (*pad*) and protoconid (*prd*) of the lower molar, also became flattened into a large blade divided into two parts by a deep notch (figs. 20.5, 6). This happened independently in different families. In the Oxyaenidae the longest upper shearing blade (fig. 20.9B) was developed on the first upper molar (m^1) and second (fig. 20.5) lower molar (m_2). In the Hyaenodontidae metastyle blades of increasing length (fig. 20.9F, G) were developed on p^4, m^1, m^2, and on m_1, m_2, m_3 of the lower series, but those on m^2 and m_3 were the longest. In the Miacidae (fig. 20.13), some of which were ancestral to the Fissipeda, the growth gradients from p^1 backward reached their peaks on p^4 and m_1, which thus became the "carnassial" or chief shear teeth. In such cases p^4 on the inner side (fig. 20.8) was separated from m^1 by an inverted V-shaped notch, into which the enlarged paraprotoconid blade of m^1 fitted; the talonid basin of m_1 received the protocone (*pr*) or internal cusp of m^1. In extreme shearing forms (such as *Patriofelis* and *Hyaenodon* among the creodonts, and the cats and some mustelines among fissipeds) the talonid of whichever molar was the main lower shear-tooth became reduced (fig. 20.5), finally disappearing in the ancestry of the Fissipeda. This had the advantage of bringing the cutting blades and cusps back nearer to the jaw muscles and thus increased their penetrating and shearing power.

The Paleocene and Eocene Miacidae (figs. 20.13–15) were classified with the creodonts by Cope and by Matthew but these authorities also saw that miacids were the source (fig. 20.16) of the feloids (civets, hyaenas, cats) and canoids (dogs, bears, raccoons, mustelines) (Simpson, 1945). Among the civets (Viverridae) the more primitive forms (fig. 20.25B) are nearer to the shearing than to the crushing-omnivorous dentition; the shearing type culminated in *Cryptoprocta* of Madagascar, which may be a survivor of the common Eocene stem that gave rise to both the viverrids and the true cats (figs. 20.16, 17G; 20.18D, 19D). The hyaenas are in a sense gigantic civets (figs. 20.1A, 18E, 19E) with shear teeth so large that they serve for breaking bones as well as shearing flesh. The shear on p^4 became reduced in the palm civets (figs. 20.17D, 21B-D) and their molar cusps became secondarily low and blunt. Another Madagascar viverrid, *Eupleres* (fig. 20.22D), has converged toward the anteaters and its cheek teeth have become very small (figs. 20.23E, 24E). In some of the smaller mongooses (fig. 20.26F) p^4, m^1, m^2 become secondarily widened transversely (fig. 20.28) and suggest the V-shaped molars of the zalambdodonts (fig. 19.6).

Among Paleocene creodonts, especially the arctocyonoids, the raised margin or cingulum at the postero-internal corner of the upper molar crowns (fig. 20.11a) gave rise to a cusp, the hypocone, which as it became larger eventually changed the outline of the crown from a triangular to a quadrangular contour. In general, quadrangular or "quadritubercular" upper molars are associated with elongated lower molars (fig. 20.11b) with four main cusps (protoconid, metaconid, hypoconid, entoconid) and a median posterior cusp, the hypoconulid (cf. fig. 20.7C). These are variously adapted to an omnivorous diet (bears) and sometimes to an herbivorous diet (e.g., giant panda, fig. 20.34V).

The more primitive dogs (Canidae) which are the central family of the Canoidea ("Arctoidea") retained strongly developed shear-teeth (fig. 20.34A, B) in p^4 and m_1 but did not neglect or sacrifice their remaining molars (m^1, m^2; m_1, m_2, m_3), emphasizing the protocone and hypocone-cingulum on the upper molars and the hypoconid and hypoconulid on the lower molars, especially m_3 (fig. **23.36A, B**).

In the Miocene *Ursavus* (20.34I), p^4 has lost its dominance and m^1, m^2 are markedly elongate anteroposteriorly, especially on the inner side, with bluntly conical cusps. Thus arose the peculiar dentition of the bears (J, K), adapted for crushing a wide variety of food.

A partly similar but quite independent transformation culminated in the pandas (L-V), which finally have enormous tuberculated molars, adapted for grinding bamboo stems. The giant panda (*Aeluropoda,* fig. 20.35F), although now exclusively herbivorous, was evidently derived by way of the small panda (*Aelurus,* E) from the older members of the raccoon family (Procyonidae, fig. 20.34T) which had an omnivorous dentition (Gregory, 1939).

Among the fur-bearers (mustelines) the more primitive forms (pl. 20.1J), including the pine martens (*Mustela*) had well developed shear teeth with long blades on p^4 and m_1. Their first upper molar was pestle-like in outline, with an expanded inner part and a sharp crosscrest. In the wolverine (*Gulo*) the shearing function was stressed but m^1 remained small (K). In the European badgers, m^1 became secondarily elongated, especially on the inner side and as the crushing function predominated, the shearing p^4 became shorter (A).

The less specialized ancestors of the otters (FF) had good shearing blades on p^4 and m_1 and only moderately large crushing molars, m^1 and m_2; but in the sea otters, p^4, m^1, and m_1, m_2 have become very large, with very low conical cusps, apparently adapted for crushing mollusc shells (N).

Thus the diverse cheek teeth of carnivores may all be traced back to those of the Paleocene creodonts of the family Oxyclaenidae (Arctocyonidae), which in turn were derived from the basic tribosphenic ("tritubercular") upper, tuberculo-sectorial lower molars of Upper Cretaceous insectivores and Jurassic pantotheres.

Interlocking Parts and Growth Gradients.—In the case of a cog-mechanism (figs. 20.7, 38), it is obvious that if the length of one arm of the series of V's or U's be changed, the shape of the notches between the cogs must be changed and the shape of the engaging cogs on the opposing surface must likewise be changed at the same time and in complementary ways. From comparative studies on the evolution of the teeth in mammals, it is evident that in such primitive mammals as the opossum (fig. 18.23), *Echinosorex* (*Gymnura,* fig. 19.13) and the oxyclaenid creodonts (fig. 20.7), there is a progressive widening of the crown as we pass backward from p^1 to m^1 or m^2; behind m^2 there is a sharp down slope to m^3. In other words, there has been a marked transverse growth gradient or anisomerization along this line, in which first the transverse diameter and then the oblique length of the protocone-metastyle diameter have been increased. Later growth gradients have left their record on the series leading from simple V-shaped molars to elongate crushing molars, as in the bears (fig. 20.34H-K) and pandas (T-V).

Since the interlocking parts of the upper and lower jaw must always change together, there must be correlated genetic predeterminants for the upper teeth which have inverse or mirror-image relationships to those of the lower teeth (cf. fig. 20.7). The correlation factors may be located in the stomadeal ingrowth and, earlier, in the blastopore.

PRINCIPAL FACTORS IN DIVERSIFYING THE SKULLS OF CARNIVORES

As in all other mammals, the principal factors in molding and diversifying the skulls of carnivores are: (1) the varying growth rates and form of the deciduous and permanent teeth and consequently of the tooth-bearing or alveolar portions of the bony face; (2) the height of the condyle of the lower jaw above the plane of the cheek teeth; (3) the size and directions of the temporal, masseter, external and internal pterygoid and digastric muscles, as well as of the muscles of the neck; (4) the size and position of the eyes; (5) the size and positions of the ethmo- and maxillo-turbinal scrolls (for smelling and for warming the air), as well as of the frontal sinuses; (6) the size, shape and position of the external, middle and internal ears and their accessory cavities; (7) the degree of inclination of the facial to the basicranial axis; (8) the height, length and breadth of all parts of the brain; (9) the posture of the head in relation to (a) the medium and (b) the locomotor methods and apparatus; (10) the form and nature of the dermal covering; (11) the restraining influence of adjacent parts, as of ligaments on their joints, which limit movement within certain paths; (12) the size, position, direction and form of the glenoid cavities and related parts that connect the jaw with the skull; (13) not the least of these formative skull factors would be the over-all mass of the organism and the proportionate size of the head to the rest of the body.

Somewhat similar lists could be prepared for the factors influencing the evolution of the vertebral column, ribs, girdles, limbs, hands and feet. Besides all these and similar habitus features, there are the family heritage features which probably make it easier for genetically predetermined changes to displace the norm of any particular dimension in one direction rather than another.

SUMMARY: CORRELATED EVOLUTION OF SKULLS, JAWS AND TEETH

After due study of the material, much more could be learned by comparative methods as to the form, functions and evolution of every part of the skull; but for the present it may suffice to review some of the outstanding contrasts between a few primitive carnivore skulls and their extreme derivatives.

1) In the more primitive creodonts (Oxyclaenidae) the skull (fig. 20.4) was moderately long, with but little down-bending of the bony face upon the basicranium and with a prominent sagittal crest. As seen from above, the bony muzzle was large and coarse, there was a sharp postorbital constriction with a short tubular region, followed by a low unexpanded braincase. The zygomatic arches were well developed and markedly bowed outward. The lower jaws were fairly long and capable of being opened very widely. The dentition was of primitive carnivorous type with tribosphenic molars and orthal (vertical) jaw movement (fig. 20.7).

1a) In *Patriofelis*, a cat-like derivative, the preorbital face was shortened, the jaws and zygomatic arches massive (fig. 20.3), with extreme shears on m_1 (fig. 20.9C) and m_2 (fig. 20.6E).

1b) Among the Middle and Upper Eocene mesonychid creodonts of North America the skulls ranged in size up to those of the larger bears, but the giant *Andrewsarchus* of the Upper Eocene of Mongolia had a skull nearly a yard long, supplied with large canines and blunt, low-cusped molars (fig. 20.12B).

2) In *Viverravus* and *Vulpavus*, representing the miacid ancestors (figs. 20.13, 14) of the Fissipeda, the skull was fundamentally as in (1), but the principal shears were on p^4 and m_1 and the gape of the mouth was a little shorter.

2a) In the modern less specialized civets, the skull (fig. 20.18B) differs from (2) chiefly in its larger braincase. The inner wall bears the imprint of four gyri, concentric around the fissure of Sylvius, which are characteristic of the brain of fissipeds. The auditory bulla (fig. 20.19B) is expanded and overlapped posteriorly by the paroccipital process of the exoccipital.

2b) In *Cryptoprocta* the skull (fig. 20.18D) has been derived substantially from (2a) by marked shortening of the face, widening of the zygomatic arches and braincase and emphasis of the shearing dentition (fig. 20.19D).

2c) In the sabre-tooth cats (fig. 20.31) the upper canines become enormous and sabre-like; the lower jaw develops a flange, perhaps for a muscular fold of the lower lip. The shear teeth are very large.

3) The skulls of the Herpestidae or mongoose family (figs. 20.26B-F, 27B-G) have evidently been derived from the ancestors of (2a). Some remain relatively long but in the mierkat (*Suricata*) the braincase (fig. 20.28E) is very wide, the face short with wide, anteroposteriorly narrow molars.

4) In *Eupleres* (fig. 20.22D), a structural derivative of the viverrid *Hemigalus* (B), which in turn is not far from (2a), the teeth are very small and the skull is assuming some of the anteater habitus (fig. 20.23E, 24E).

5) The dogs are widely diversified in skull and dentition, some (e.g., *Simocyon, Icticyon*) simulating the cats in shortening the face and stressing the shearing blades, others (Amphicyoninae) mostly with long face, enlarging the molars. Among domestic dogs, the extreme differences between the long narrow skull of the borzoi hound (fig. 20.32B) and the excessively short-faced, wide skull of the bulldog are associated with hereditary differences in the endocrine glands (Stockard, 1941). The King Charles spaniel's skull, with bulging, high forehead and short upturned jaws, is due to the persistence of foetal characters in the adult.

6) The skulls of bears, from a general morphologic-palaeontologic viewpoint, seem to be essentially those of giant dogs, with a secondary elongation of their molars. The polar-bear's skull is relatively long and narrow (fig. 20.33J); those of *Arctodus* and *Arctotherium* (I) short and wide.

7) The living procyonid skulls exhibit a wide range, from the very primitive civet-like *Bassariscus* (B) to such widely diverse end-forms as (a) the kinkajou (*Potos,* E), with a very short face, rounded braincase and dwarfed, flat-crowned molars; (b) the panda (*Aelurus,* F) and giant panda (*Aeluropoda,* G). The latter, as noted above, is strictly herbivorous and has wide grinding molars (fig. 23.34V). Yet in view of much morphologic and considerable palaeontologic evidence, the procyonid family seems to be of fairly unified origin.

8) Already in the more primitive mustelines or fur-bearers the bony face (pl. 20.I) is moderately short and the braincase both long and wide; the shear teeth relatively large, the first upper molar small. The skull of the sea-otter is widened, strengthened to support the huge crushing molars. Secondary increase in the size of the first molars is also notable among the badgers and related lines (A-D).

THE SEA-LIONS, WALRUSES AND EARLESS SEALS (PINNIPEDIA)

The otters and sea-otters have to some extent converged toward the seals (Pinnipedia), but in the latter the teeth, skulls, feet and many other features indicate that they have originated independently of the otters and probably from some other family of fissiped Carnivora.

Even the least specialized family of recent seals, namely, the eared seals or otaries, including the various northern (fig. 20.39A) and southern "sea-lions," are so highly specialized for marine habits that their habitus almost but not quite conceals their more remote superfamily heritage. In swimming, the eared seals practically fly under water with their very large pectoral "wings," somewhat after the manner of penguins, using their hind flippers in turning or in leaping up on a rock. The principal bend of the preaxial border of the pectoral flipper is near the lower end of the radius just above the wrist (fig. 20.40A). The thumb and its metacarpal have been secondarily enlarged and form the convex border of the distal segment of the paddle. The consolidated scapho-lunar-centrale bone supplies a firm base for the movements of the first, second and third metacarpal and digits and transmits their thrust to the concave socket on the wide distal end of the radius, while the narrow, block-like cuneiform (triquetrum) does the same for the smaller fourth and fifth metacarpals and digits, and transmits thei thrusts to the ulna. The humerus is rather shor and in side view somewhat S-shaped, with prominent delto-pectoral crest, large great tuberosity, convex head, prominent extensor-supinator crest and spiral channel for the brachialis muscle. The humerus is concealed beneath the thick skin of the body. The olecranon is very large, implying a massive triceps. The very large rounded scapula gives an ample base for the powerful muscles of its outer and inner surfaces and margins. The great cylindrical neck of the eared seals grades into the back and the first two dorsal vertebrae are turned upward to support and function with the neck. Even the scapulae extend up into the base of the neck. Whe moving about on land (fig. 20.39A), the head and neck can be elevated high above the pectoral limbs, which are widely separated to afford a firm base for this moving watch-tower. The greatest girth of the body is in a transverse plane through the glenoid articulations for the humeri. Behind this, the back slopes down rather steeply to the low hips and then more quickly to the stump of the tail. The ilia are very short, with crests turned outward, the pubes and ischia prolonged downward and backward between the short femora (fig. 20.40A). The rather long hind feet can be turned forward under the body, the large first metatarsal and digit forming the slightly curved anterior border. In the vertebral column the neural spines are relatively low and thick and the articulations such as to permit marked flexibility of the cervical and lumbar regions.

The skull of eared seals (Otariidae) shows the following conspicuous features:

(1) It is wide posteriorly, with ample braincase, which (2) is connected by a tubular constriction with (3) the wide, short fronto-nasal region. (4) On the inner walls of the braincase are imprints of the four concentric gyri around the Sylvian fissure. (5) The entrance to the nasal chamber is blocked by the large and highly folded mass of maxillo-turbinals, for heating the inspired air.

(6) Behind this, a second and smaller similar mass of ethmo-turbinal scrolls end posteriorly in the large cribriform plate of the olfactory fossa in the braincase.

(7) The periotic in ventral view is large and triangular, its anterior entotympanic wall inflated and undistinguishably joined with the tympanic bone.

(8) A large foramen for the internal carotid artery pierces the bulla.

(9) The jaw muscles are not large and there is but little sagittal crest.

(10) The teeth (fig. 20.41A) are usually "simple" pegs, with a basal cingulum, a small basal cusp and conical, slightly recurved crowns adapted for fish-catching. Very probably this is a retrogressive, secondarily simplified dentition.

(11) There are usually three incisors, one canine, four premolars and one molar on each side in the upper and lower jaws.

(12) There are two sets of teeth, all the permanent teeth except the true molars having deciduous predecessors.

These features, especially the coössified scapho-lunar-centrale, appear to be wholly consistent with the view that the eared seals are transformed fissipeds of some sort, and nos. (5) (6) (7) (8) suggest that they are modified relatives of the Mustelidae.

The skull of a baby walrus, showing the sutures and deciduous teeth, permits close comparison with that of a young eared seal and supports the accepted conclusion that the walruses (Odobaenidae) are nothing more than gigantic and highly specialized members of the otarioid series. In the adult male walrus (fig. 20.39B) the permanent upper canines grow to an enormous size and, pointing straight downward, they may serve as levers for uprooting bivalves as well as for fighting other males. The alveolar portions of the maxilla and indeed the entire skull are strengthened to meet the strains exerted by the tusks and by the corresponding power of the massive muscles of the neck and jaws. The alveoli of the canines have been folded around outside the anterior cheek teeth, which consist of five fairly long cylinders on each side above and below. The massive jaw, narrowing to a deep symphysis, smashes the shells with the impact of its cylindrical lower cheek teeth, the small lower canines being in line with the cheek teeth and closing inside the upper canines. From the Lower Miocene of Maryland the type and only known lower jaw of *Prorosmarus* (fig. 20.41C, C1) was that of an early walrus which, with the Pliocene *Alachtherium* (B, B1) of Europe, affords at least a structural link with the eared seals.

Moreover, the walrus can still bring his hind feet forward under the massive body and the construction of his girdles, limb bones, hands and feet are essentially that of the eared seals as modified by the enormous mass and somewhat different habits.

THE EARLESS SEALS (PHOCIDS)

The earless seals (fig. 20.39C) or Phocidae are so widely different from the eared seals that it is not certain that they have arisen from exactly the same ancestral stock. They have all advanced beyond the otariids in keeping their hind legs permanently directed backward and when they are on land, being unable to pull the hind limbs forward, they jerk forward on their bellies, bending their necks, backs, and hind limbs sharply upward and then rolling and lurching forward. Even the sea-elephants (D) progress on land in this unique way. The chief propelling force probably comes from a sudden forward thrust and backward jerk of the enormous neck, shoulders and back, aided by the pectoral limbs, the rear part of the body being sharply raised to reduce friction on the ground. Thus the vertebral column and its axial muscles, including flexors, extensors, abductors and adductors, play a dominant part in the forward locomotion of the earless seals, both on land and in the water, whereas in the eared seals (A) the backbone is held more stiffly and its muscles collectively reinforce or steady the thrusts of the large wing-like paddles against the water.

In the typical harbor seals (C) the hands are much smaller than the feet and are used rather in steering and slow paddling than in swimming forward; which is effected by undulations of the supple body transmitted to the large fan-like hind feet. When the opposite fans are held vertically and pressed together, they form a sculling organ, or if used separately, they act as breaks and rudders. Their fan-like shape is attained by lengthening and enlarging both the first and the fifth metatarsals and digits.

The cheek teeth of the living earless seals (fig. 20.42) comprise three principal types and many intergrading conditions: (1) compressed, serrated and pointed crowns (E, E1), with oval bases, which are set obliquely to the long axis of the jaw and thus partly overlap each other; (2) triconodont crowns (F), with three sharp cusps in line, in the Antarctic leopard seals, and (3) degenerate small pegs, in the hooded seal (G) and the sea-elephant (I).

The food of seals consists chiefly of fish, squids, crustaceans, but the leopard seals catch penguins, using their triconodont teeth as shears. The harbor seals are known to use their rather thick, blunt canines to bite blowholes in the ice forming above them and they keep the holes open by whirling themselves around vertically, using the canines as scrapers.

POSSIBLE ORIGINS OF THE SEALS

The skulls of the earless seals are basically so similar to those of the eared seals that pure convergence is highly improbable, the chief differences being in the jaws and dentitions. And both types of skull approach the musteline type, and especially those of the badgers and otters.

Simpson (1945, pp. 232–233) notes the occurrence of a Lower Pliocene Siberian fossil (*Semantor macrurus,* described by Orlov (1933)), which "barely possibly" may be "a pinniped-like lutrine (or other mustelid) rather than a lutrine-like pinniped." The presence of a long slim tail in itself is of course no evidence against pinniped relationship, as not a few short-tailed animals are closely related to long-tailed ones (e.g., the Manx cat to the ordinary cat, the uakari monkey to *Pithecia*). The limbs of *Semantor* combine otter-like and phocid-like characters. The skull, jaws and teeth are unfortunately unknown but so far as this fossil goes, it tends to support the view that the seals are aquatic relatives of the mustelid family. Close relationship to the otters is at first contra-indicated by the enormous crushing molars in the sea-otters and by the complete absence of such teeth in the known seals. But it is highly probable that, whichever family was ancestral to the seals, there were radical transformations in the cheek teeth to produce, on the one hand, evidently simplified pegs and, on the other, large triconodont-like shearing teeth. And there is no lack of cases that indicate a radical and relatively sudden change in the cheek teeth in connection with changes in food habits: e.g., the greatly enfeebled teeth of *Proteles* as compared with the large teeth of its relative, the hyaena; the minute teeth of the ant-eating *Myrmecobius,* as compared with those of its relatives among the Dasyuridae. In spite of inferred changes in the form of the teeth, in passing from terrestrial musteloids to the seals, the number and different kinds of teeth in the latter have suffered only minor changes. Thus Weber (vol. II, p. 347) notes that the dental formulae of pinnipeds center around that of the sea-lions:

$$I\frac{1.\ 2.\ 3}{0.\ 2.\ 3};\ C\frac{1}{1};\ P\frac{1.\ 2.\ 3.\ 4}{1.\ 2.\ 3.\ 4};\ M\frac{1.\ 2}{1.\ 2}$$

This differs from that of *Martes* of the Mustelidae chiefly in the loss of one lower incisor and the retention of the second upper molar.

On the whole the resemblances in the skull between both branches of the seals, especially those in the auditory and olfactory regions, suggest that: (1) both main branches of the pinnipeds have descended at least from a single ancestral superfamily, the Musteloidea; (2) that they may have come from some early Lower Oligocene mustelid which was already a good swimmer, and (3) that the lines leading to the seals were paralleled to some extent by the otters, which, however, ultimately evolved large crushing teeth. The very round, sleek heads, short jaws, large necks, short legs and extremely supple bodies and spreading hands and feet of the minks (*Mustela*), especially the extinct "sea-mink" (*Mustela macrodon*) (Goodwin, 1935, pl. IV), may give us an approximate picture of the general body-form of an ancestral pinniped of either main group. Such high-powered, hot-blooded mustelines, with their sleek insulating coats, would be preadapted to hunt fish in northern streams and gradually spread to the ocean shore and beyond. They appear to be far more completely preadapted for such a life than any known type of dogs, which are on the whole excellent runners but slow swimmers. As for bears (which are secondarily plantigrade, tailless, giant dogs) the polar bear, in spite of his aquatic habits, seems to be far nearer in all essentials to his land-living relatives than the seals are to theirs.

CHAPTER XXI

OUTER FORM AND INNER FRAME AMONG THE HOOFED MAMMALS (UNGULATA)

Contents

CHAPTER XXI

OUTER FORM AND INNER FRAME AMONG THE HOOFED MAMMALS (UNGULATA)

SIZE DIFFERENCES BETWEEN UNGULATES AND RODENTS

IN THE present book it will be necessary to deal rather briefly with the enormous assemblage of "hoofed mammals," which with few exceptions are provided with hoofs and are all herbivorous.

The ungulates parallel the rodents in their herbivorous diet but usually far surpass them in bulk. The horse's world is consequently very different from the rat's and all parts of their respective skeletons are adjusted to these contrasts. In general among animals, the size factor, although fully as important in their evolutionary history as in their individual lives, is far surpassed in importance by their total hereditary differences in habitus plus heritage. Some of the smaller ungulates, for example, such as the tragulines and the dik-diks, are hardly larger than jack rabbits and live in somewhat similar environments, yet perhaps every single bone in their respective skeletons would be sufficient to identify them either as artiodactyl ungulates or as lagomorph rodents. On the other hand, parallelism between different ungulates of similar locomotor or feeding habits is frequent, but convergence between widely removed families never extends to all parts of the skeleton and may be detected by a comparative analysis of the respective "habitus" and "heritage" of the forms.

TYPES OF LOCOMOTOR HABITUS IN UNGULATES

The locomotor habitus, or totality of special adaptations of the skeleton and related parts as adapted to different sizes, postures and gaits, may be designated under the following terms:

A. Ambulatory.—Relatively primitive adaptations for walking or running about, mostly found in Paleocene and Eocene protungulates of small or moderate size. Pentadactylate (with five digits), with somewhat divergent thumb (pollex) and great toe (hallux) and none of the digits greatly enlarged; e.g., the small Paleocene periptychids (p. 411) and the modern *Hyrax* (fig. 21.25A).

B. Mediportal.—Body-form stout, elbows and knees bent, metapodials not elongate but moderately wide, none greatly widened; hoofs small or moderate; e.g., *Ectoconus* (fig. 21.2) and modern tapir.

C. Graviportal (orthograde).—Very large, heavy-bodied ungulates with post-like limbs and more open angles at the elbows and knees; metacarpals (fig. 21.10C) and metatarsals (fig. 21.11B) very wide and flattened, astragalus very wide with keels flattened or absent; adapted for striding, not for galloping; e.g., *Dinoceras* (fig. 21.9) and modern elephant.

D. Saltatorial.—With long slender limbs, sharply bent at elbows and knees, and well arched backs; adapted for leaping; e.g., small antelopes.

E. Cursorial.—Completely unguligrade, with long, transversely rolled up or cylindrical hands and feet and marked reduction of digits I and V, emphasis of either metacarpal (or metatarsal) III in perissodactyls (fig. 21.48) or III and IV in artiodactyls (fig. 21.100).

F. Semifossorial.—"Pseudo-unguiculate," with hoofs pointed and recurved like claws, adapted for scratching or digging water-holes; e.g., chalicotheres (figs. 21.65, 66).

G. Semiarboreal.—With spreading hands and feet, more or less plantigrade, with friction pads on palms and soles; e.g., *Dendrohyrax* (modern tree cony, fig. 21.25A).

H. Semiaquatic.—With massive body, thick skin, very short limbs; e.g., hippopotamus (fig. 21.93).

I. Aquatic.—Body-form streamlined, fore-limbs changed into paddles; hind limbs lost; e.g., Sirenia (fig. 21.35).

Between these main types of locomotor habitus there are numerous intermediate stages and occasional shifts from one category to another, as from the small cursorial ancestral Lower Eocene perissodactyls to the much larger mediportal Middle Eocene titanotheres, and finally to the graviportal Lower Oligocene titanotheres (figs. 21.54, 55), or from the normal hoofed perissodactyls to the pseudo-unguiculate or "clawed" chalicotheres (fig. 21.66).

"KNUCKLE-JOINTED" PROTUNGULATES

Most of the genera and species of this extremely primitive yet diverse assemblage of Paleocene and Lower Eocene forerunners of the hoofed mammals are known only from fragmentary upper and lower jaws with teeth, but some have associated limb and foot bones, and three of them, viz., *Phenacodus* (fig. 21.12), *Ectoconus* (fig. 21.2), *Meniscotherium,* are represented by nearly complete skeletons. Collectively the condylarths are of the greatest value in the study of the early stages of evolution of the hoofed animals and of their mechanisms for locomotion and feeding.

The oldest and most primitive ungulates of the Paleocene and Eocene order Condylarthra (figs. 21.2–13) are rather closely related to their contemporaries, the most primitive members of the Creodonta (figs. 20.4A, 9A), or flesh-eaters. Simpson (1945, pp. 105, 216) therefore combines the flesh-eating and hoofed placental superorders and orders into a new Cohort, or grand division, of the Infraclass Eutheria, or Placentalia. To this Cohort he has given the appropriate name Ferungulata (from Ferae, Linnaeus' name for the Carnivora, plus Ungulata, hoofed mammals).

The smaller condylarths, e.g. *Hyopsodus* (fig. 21.13D), were perhaps about the size of squirrels, the larger ones, e.g. *Periptychus* (fig. 21.4A), would be nearer to large bulldogs in bulk, the largest, e.g. *Ectoconus* (fig. 21.2), might compare with an immature black bear. Rather spreading five-toed, semi-digitigrade hands and semi-plantigrade hind feet, with fairly short metacarpals and metatarsals (fig. 21.11D), so far as known, were characteristic of the hyopsodonts and the periptychids; the latter had more or less flattened nails, rather burly limbs and probably an ambling or usually slow gait. In the American Museum mounted skeleton of the large periptychid *Ectoconus* (fig. 21.2) the head is rather small, the back well arched, the tail long, the body stout, the limbs very muscular and the hands and feet short, as noted above. Well developed clavicles were retained; they are lost in all modernized ungulates; the radius could be twisted around the ulna, so as to supinate the hand (although perhaps not as completely as in man); the scaphoid, lunar and centrale bones of the wrist (fig. 21.10B) were not coalesced (contrast fissiped carnivora) and the lunar rested about equally on the magnum (capitatum) and unciform, as in other primitive placentals. The girdles and hind limbs also were exceedingly primitive and fundamentally like those of creodonts. The astragalus was pierced by the astragalar foramen (probably for a deep branch of the tibial artery); there were no keels on the upper end of the astragalus, while its convex distal end fitted into the concave socket of the upper surface of the navicular, so that there were rather free twisting movements at the ankle. Except for the presence of some extra cuspules on the marginal cingulum of the molars (fig. 21.6H, H1), the skeleton of *Ectoconus* as a whole is basically near to those of primitive creodonts.

The Famous Phenacodus.—The chief characteristics of the phenacodont family of the order

Condylarthra are beautifully displayed in the type of *Phenacodus primaevus* Cope from the Lower Eocene, which is a well preserved skeleton (fig. 21.12). It is about as large as a wolfhound, with five-toed hands and feet in an early running stage. The hands (fig. 21.10D) were less spreading than those of the periptychid *Ectoconus* (B). The wrists were in an early stage of being rolled around transversely, so that the middle digit (III) was becoming the central axis of the hand, while digits I, II, were being displaced to the inner posterior side and IV and V to the outer posterior side. Moreover, the hands were elongating and digits I–V were shortened, so that they no longer reached the ground in running; the ungual phalanx of the middle digit was large and incipiently spatulate, with a small distal cleft, indicating that it supported a thick nail or incipient hoof. The ungual phalanges of the other digits were of the same type but asymmetrical and not as wide.

The five-toed hind feet (fig. 21.11E) of *Phenacodus* were longer than the forefeet, with slender, short first and fifth digits. The astragalus had fairly well raised, rounded internal and external trochlear keels; the head or distal end of the astragalus was ball-like. Professor Cope, noting that the molar teeth of *Phenacodus* (fig. 21.5F) indicated herbivorous diet but that the astragalus had a ball-like head or condyle like those of the primitive carnivores, invented the name Condylarthra (knuckle-joint) for the extinct order of which *Phenacodus* was the leading representative. The generic name *Phenadocus* meant plain tip tooth and referred to the distinctness of the conical cusps composing the surface of the molar crowns. This was in contrast with the more complex molar patterns of modern ungulates, in which the primary cusps are often obscured by later crests, folds or accessory cuspules. Noting also the larger size of the middle or third digit of the hands and feet in this skeleton and realizing that this is the digit which has become the main axis of the feet in the horse (figs. 21.48, 49), Professor Cope thought that he had discovered the "five-toed atavus" of the horse whose existence had been inferred by Huxley.

Then he went further and inferred that, since *Phenacodus* was archetypal to the horse in respect to the number of its digits and the predominance of the middle digit, it must also be more primitive than the horse in the construction of its wrist (carpus).

"Serial" *versus* "Displaced" Bones of Wrist and Ankle.—Again, Cope noted that the wrist-bones of the horse (fig. 21.48F) in front view look somewhat like bricks and that the vertical joints of the upper row alternate with those of the lower row, after the manner of well laid bricks; but that in *Phenacodus* (A) the vertical joints between the wrist bones of the upper row seemed to be in line with those of the lower row, so that the upper and lower ones did not overlap but were like "unstruck bricks" (i.e., not alternating or interlocking). Upon this, Professor Cope based his idea that the "serial carpus," i.e., one in which the upper and lower rows were in vertical series, not alternating, was primitive for all ungulates. Some years later, however, Dr. W. D. Matthew (1897) described the carpus of *Euprotogonia* (*Protogonodon*), a direct predecessor of *Phenacodus,* from the basal Eocene (now called *Paleocene*) of New Mexico; noting that in this older, smaller and, in respect to the pattern of its molar teeth, still more primitive stage, the carpus was not "serial" but "displaced," like the carpus of creodonts. Palaeontologists also began to realize that *Phenacodus* with its "serial" carpus was not in the direct line of ancestry of the horse, but was a contemporary of *Eohippus,* the true "dawn horse," in which the carpus was fully "displaced." Later it was fully shown (Gregory, 1910) that although carpal bones may in the front view (fig. 21.10D) appear to be "serial," yet in the back view they are "displaced," partly because they are all primitively irregular or distorted polyhedrons, not bricks (fig. 21.66D). The pressures to which the carpal bones of hoofed mammals are subjected in standing, walking or running include both vertical and oblique thrusts; in the lunar or middle bone of the upper row (fig. 21.55) the vertical thrusts divide into two main streams, one passing more directly downward from the lunar to the magnum (fig. 21.66D), the other obliquely downward and to the outer side through the unciform. The matter was complicated by the presence of a separate small centrale bone in the primitive mammalian carpus (fig. 21.10A-C *ce*), where it was a remnant of the reptilian carpal pattern (fig. 16.24). This bone transmits some of the vertical thrusts of the lunar inward to the trapezoid. In the very primitive *Ectoconus* (fig. 21.10B) the centrale was still separate, as it was also in *Protogonodon* (fide Matthew), but in *Phenacodus* it had apparently fused with the scaphoid (D), as it had also in *Eohippus*. In the front view of the foot of *Eohippus* (fig. 21.48B) the scapho-centrale tended, as it were, to push the lunar off the magnum wholly on to the unciform, while the latter at the same time widened its contact with the lunar.

In *Phenacodus* (fig. 21.10), as in the creodonts and later ungulates, the upper ends of the metacarpals overlapped each other from the inner side outward. Thus the proximal end of the second metacarpal (mtc. II) overlapped on mtc. III and abutted above against the trapezium, trapezoid and magnum; that of mtc. III supported the magnum and thrust upward and outward against the unciform; that of mtc. IV supported the unciform and thrust upward and outward against mtc. V.

Even in the standing pose all this complex system of interlocking joints would have collapsed instantly under the weight of the body if it had not been tied together by strong connective tissue and ligaments, which resisted the tensions tending to disrupt the system, as the bony facets and processes resisted the pressures. Moreover, it was of course only by means of the tension-resisting half of the system that the various carpal bones were not dislocated, when, for example, the animal reached forward with its right front paw and attempted to get up from the ground, flexing the hand upon the forearm, opening up all the carpal joints in front and squeezing together those in the rear. In this position of extreme flexion, the processes on the rear or underside of the carpal bones (fig. 21.66D) come closer together, the magnum being the centerpiece upon and around which the others slide. Obviously they could not slide upon one another if it were not for their smooth glistening surfaces and for the lubricating synovial fluid surrounding them. Thus each carpal bone moves, as it were, along on its own orbit, impelled by the forces around it and restrained by its fellows and by its own bands.

The hands of *Phenacodus,* with their reduced thumbs and fifth digits, retained but little of the grasping power of the flexible fan-like hands of the primitive carnivores, and in so far as they were already partly rolled up into cylindrical supporting props, they foreshadowed the hands of all later ungulates. Even the carpus of *Phenacodus* was on the whole quite primitive and, as long ago observed by Osborn (1889), it would easily have been modified into the carpus of the elephant, chiefly by the differential widening of some of its pieces.

The hind-foot of *Phenacodus* in all its parts was far more primitive than that of the horse and it still retained much of the freely movable ankle of its creodont ancestors. Although *Phenacodus* itself was not the direct ancestor of the horse, its hind-foot, in order to be transformed into that of the horse, needed chiefly to be greatly lengthened, with elimination of digits I and V, great enlargement of metatarsal III and its digit, marked vertical flattening of the carpus and rolling up at its sides to form a cylindrical column.

The *Phenacodus* hind-foot could easily be transformed into that of the elephant chiefly by the great widening and flattening of all its parts and by differential reduction of the ungual phalanges. Thus *Phenacodus,* although apparently the terminal member of its own side branch, was not far removed from the common ungulate stock which gave rise to both the widely contrasting extremes represented by the fast-running or cursorial horse and the massive striding elephant. We shall indeed have frequent occasion in this chapter to refer to the leading contrasts in proportional measurements (anisomerism) between ambulatory, mediportal (ancestral or intermediate), cursorial and graviportal ungulates, such as were evolved by parallelism in related orders or families, or by convergence between families or orders of placental mammals which were of widely different derivation. Within the superorder Protungulata these main locomotor types were diversely represented in the orders Condylarthra, Litopterna, Notoungulata, Astrapotheria, Pantodonta, Dinocerata and others.

Adaptive Branching of the Condylarths.— Among the order Condylarthra the mioclaenids, periptychids and hyopsodonts were mostly small primitive forms with ambulatory limbs and short, five-toed hands and feet, associated with diverse, primitive to moderately specialized, dentitions. *Ectoconus* of the periptychids, as we have seen, was more massive than *Phenacodus* and retained very primitive ambulatory to mediportal features in the pectoral and pelvic girdles, limbs, hands and feet. *Phenacodus,* although within the mediportal group, had entered upon the path leading to cursorialism.

Although very diverse in generic characters, the upper molars of condylarths (fig. 21.5B-G) suggest herbivorous modifications beyond the sharp-cusped insectivore type to the primitive crushing stage, with three either rounded, conical or partly V-shaped main cusps. A cingulum-hypocone, proto- and metaconules, and usually parastyles and metastyles, are present. In the Lower Eocene *Hyopsodus powellianus* upper molars (fig. 21.14A) of this general type were associated with five-toed feet (fig. 21.11D) which were moderately short and spreading and provided with flat nails or incipient hoofs. Thus this rather small mammal represented the ambulatory or slow-running stage.

In *Hyopsodus paulus* the skull (fig. 21.13D) was fairly long and low, with the canines not enlarged but basically like the simple upper anterior premolars. The adult dental formula was in the primitive placental stage:

$$I\frac{3}{3}\ C\frac{1}{1}\ P\frac{4}{4}\ M\frac{3}{3}$$

A mesostyle or fold was arising at the junction of the external cingulum and the W-shaped crest borne by the large para- and metacones (fig. 21.14A); m^1, m^2 of the upper molars had well developed hypocones; m^1, m^2, m^3, with conspicuous intermediate conules, the protoconule (*pl*) forming with the protocone a low oblique crest, the metaconule (*ml*) remaining separate from the hypocone. The lower molars (A1) had blunt cusps, the outer row (protoconid, hypoconid) incipiently U-shaped; all three lower molars with hypoconulids (*hld*); hypoconulid of m_3 large. The last upper premolars (p^4) were transversely bicuspid, the last lower premolars compressed, with high incipient trigonid and low talonid.

In general the skulls, jaws and dentitions of the mioclaenids, hyopsodontids and phenacodontids were archetypal to the vast diversity of those of the later ungulates.

THE SLENDER LITOPTERNS OF SOUTH AMERICA

The order Litopterna was an exclusively South American branch from the condylarths and in the patterns of its cheek teeth some of its earlier members are with difficulty distinguished from the South American condylarth *Didolodus* (fig. 21.15A). One main branch, culminating in *Macrauchenia* of the Pampean Pleistocene, was superficially somewhat like a gigantic llama in locomotor adaptations (fig. 21.16A); but their teeth (fig. 21.15E) and skulls were radically different from those of camels. Although their lower limb bones were long and moderately slender, their hands and feet were not long and their usual gait may have been striding or walking rather than galloping. They would therefore be called mediportal (p. 410).

The ordinal name, Litopterna (smooth heel) was given in reference to the presence of a small, smoothly convex, knuckle-like ridge borne on the outer side of the calcaneum (fig. 21.16A) and serving for the support of the distal end of the fibula. But although chosen for emphasis in the name of the order, this fibulo-calcaneal facet is prominent also in the toxodonts, protypotheres and all other extinct South American ungulates except *Astrapotherium* (Scott, 1937, p. 539). The fibulo-calcaneal facet was also present in the small and very horse-like litopterns (figs. 21.16B; 21.49F, G) of the Patagonian Miocene. In one of these highly cursorial forms (*Diadiaphorus*) the middle digits (III) were enlarged and the lateral ones (II, IV) reduced, so that the hands and feet resembled those of the three-toed horses. In the other (*Thoatherium*) reduction of the lateral digits had gone so far that both fore and hind feet were functionally monodactyl, that is, with the weight passing through the enlarged middle toe as in the later horses. Their molar teeth, with curved cutting crests (fig. 21.15D, E), also suggest those of the three-toed horses of the northern hemisphere, but they differ from them in significant features and their basic pattern is definitely of litoptern origin. The same is true of the detailed construction of their limbs, which, although convergently horse-like in habitus, are litoptern in heritage.

THE SOUTH AMERICAN EXTINCT ORDER OF NOTOUNGULATA

For many aeons, that is, during the entire Caenozoic era, estimated at 60 million years, the order Litopterna evolved in South America in the same faunal regions along with another major constellation of the superorder Protungulata, namely, the vast order Notoungulata (southern ungulates). This order is divided by Simpson into the suborders Notioprogonia (with three families), Toxodonta (eight families), Typotheria (two families), Hegetotheria (one family). These diverse beasts were no doubt closely adapted to their respective environments on their own continent; which was isolated from North America from the Paleocene until Lower Pliocene times, when the first later immigrants from the north began to find their way in. Even the South American mammalian faunas themselves, it is now generally inferred, had been derived ultimately (that is, possibly in Upper Cretaceous times) from stray immigrants from the north, which included primitive insectivorous marsupials and primitive insectivorous placentals. After South America became isolated, perhaps early in late Cretaceous and early Paleocene times, the primitive insectivorous marsupials started on their long evolution which produced the somewhat wolf-like borhyaenids (fig. 18.28aB, C) and finally the "marsupial sabre-tooth" (E). But these marsupials (to judge from their distant relatives in North America and Australia) retained an essentially marsupial brain, with no corpus callosum and a limited development of the higher centers, and it is certain that even today marsupials as a whole are not as intelligent as the higher placentals. Consequently the South American herbivores were left free to cope only with relatively slow and unintelligent enemies, the native carnivorous marsupials, and thanks to the isolation of their continent, they were long saved from the ravages of placental creodonts, wolves and larger cats which harried the ungulates of the northern world. Accordingly only a few of the litopterns evolved swift horse-like forms and only a few of the notoungulates produced swift hare-like animals; but the great majority of notoungulates were relatively slow, peaceful herbivores, and very few even acquired horns for fighting among themselves.

The notoungulates are so diverse in habitus features that there are but few characters that hold good for all of them, but Roth noted as an ordinal character the presence of a cavity in the temporal region above the internal ear, which was possibly lined with a membranous pouch from the Eustachian canal. In the lower molar teeth of the more primitive ones (figs. 21.19G, H, I; 21.20A, A1) the postero-internal cusp or entoconid remained separate from the long crest extending backward from the protoconid; traces of this condition may be seen even in highly specialized derivative patterns (J, K). The upper molars (D, E) often have high crowns with oblique folds and crests a little like those of rhinoceroses, but in the less specialized ones (C, B) there are unmistakable traces of the tribosphenic or so-called tritubercular ground plan.

The least specialized families of the Notoungulata, namely, the Arctostylopidae, the Henricosbornidae and the Notostylopidae, have been brought together by Simpson (1934) to form the suborder Notioprogonia. These are known mostly from small jaws, teeth and some skulls. The upper cheek teeth of the Henricosbornidae, from the Paleocene and Lower Eocene of South America, were remarkably primitive (C), with low cusps and ridges, all clearly derived from the tribosphenic basal type. The upper molars of *Notostylops* (B) were more progressive, with moderately high crowns and flattened outer wall, oblique crests and conspicuous, sharply defined depressions or "islands" left between the crests and ridges. The skull, lower jaw and dentition of some notostylopids (fig. 21.21A) were somewhat like those of the existing "dassies" or hyracoids of Africa, but their respective ordinal heritage features were quite distinct. The Arctostylopidae (fig. 21.20A, A1) are known only from one small lower jaw fragment, found in the Lower Eocene of North America, and another found (fig. 21.19A) in the Paleocene of Mongolia, but both bearing the clear notoungulate stamp. It is uncertain whether these forms were immigrants from South America, via the Andean-Rocky Mountain path to North America, or left-overs, on the way from Asia to South America.

The central suborder of the notoungulate order was the Toxodonta, ranging from small dassie-like or rabbit-like forms to the giant *Xotodon,* with a skull somewhat like that of a very large rhinoceros. *Toxodon,* discovered by Darwin in the Pleistocene of South America, was a massive beast (fig. 21.18A) about nine feet long and about four and one-half feet high, with a huge body supported by very thick lower limbs, and extremely wide short hands and feet. The long neural spines of the second to fifth dorsal vertebrae collectively formed a high hump somewhat like that of a bison, and functioning for the support of the great, short-necked skull; the latter with large downwardly curved upper incisors (whence the name *Toxodon,* bow tooth). These met the forwardly inclined lower incisors to form the front end of a large cropping or grazing machine, the tall-crowned cheek teeth cutting the tough fodder with their sharp edges. The ilia, like those of ground sloths and mastodons, were greatly expanded laterally. The great gluteal muscles doubtless were needed to prevent collapse when the opposite foot was lifted off the ground; the abdominal wall was tied to the strong pubis, the thick hamstrings and part of the adductors to the ischium. The extremely short three-toed feet were doubtless inserted into the front part of a very thick cushioned column, as in the elephant and the rhinoceros.

Almost at the other extreme from *Toxodon* were the small Notohippidae, with skulls (fig. 21.21C) a little like those of three-toed horses, but the cheek teeth related to those of other toxodonts; also the small hyrax-like forms called *Archaeohyrax* by Ameghino. The feet of *Rhynchippus* were functionally three-toed but without reduction of the lateral metacarpals. The ungual phalanges were deeply cleft, implying large possibly somewhat claw-like hoofs.

The toxodont family Isotemnidae is represented in the American Museum of Natural History by a mounted skeleton of *Thomashuxleya* (fig. 21.17A) from the Lower Eocene of South America. As shown by Simpson (1935, 1936b), it was a moderate, middle-sized animal with a relatively very large head, stout limbs and fairly short, spreading hands. The upper canines were well defined, the molars with cutting crests. In the same Museum the family Leontiniidae is represented by a large slab containing the crushed skeletons of *Scarrittia,* also described by Simpson. This was a fairly large and heavy animal, with long, somewhat columnar limbs, very short-toed hands and feet and high-crowned, obliquely crested molars.

Homalodotherium (fig. 21.17B) from the Lower Miocene of Patagonia was so named because of the even, uninterrupted series of high crowned teeth in both the upper and the lower jaws, without any space or diastema behind the canines and with only moderate differences in patterns of adjacent crowns, starting from the central incisors, around through the canines and premolars, to the last or third molar. In the ancestral protoungulates (figs. 21.3, 21.6, 21.13B), as in all other primitive mammals, regional differentiation of the tooth crowns into incisors, canines, premolars and molars was well marked. Thus in *Homalodotherium* as in *Rhynchippus* (fig. 21.21C) and others the gradual transition in appearance from incisors to molars, together with the folded patterns of the molars, suggests a secondary lessening of the differences or *dedifferentiation* between the formerly well differentiated growth forces that build up the different sets of crown-patterns. However, the molar patterns have not been dedifferentiated; but the premolars have gradually become more like the molars (cf. figs. 21.19D, E; 21.21C, D), the incisors and canines have gradually become more like premolars.

In the ancestry of the horse the same process of molarization finally produced premolar patterns which are almost or quite indistinguishable from those of the adjacent molars (fig. 21.46). In *Homalodotherium,* however, as in other notoungulates (fig. 21.19B, D), the molarization of the premolars was not quite complete; that is, the crown of the last upper premolar (p^4) was not as large as that of m^1 and there was only an incipient division of the inner and outer margins of the crown into anterior and posterior moieties. The incisor and canine teeth were gradually approaching the premolar stage by the upgrowth of the inner basal cingulum and the heightening of the entire crown.

The limbs of *Homalodotherium* were adapted for taking long strides (fig. 21.17B) and supporting this herbivore as it cropped or grazed. But it had the advantage over its swarming competitors in that its terminal phalanges were cleft, presumably for the support of fairly large claws, with which it might scratch for succulent roots or scrape its way through the earth at the bottom of a much used waterhole. In these features *Homalodotherium* converged somewhat toward the extinct "clawed ungulates" or chalicotheres of North America (fig. 21.65). Before the frequency and extent of convergence and parallelism were universally recognized, these resemblances between *Homalodotherium* and the chalicotheres misled the eminent South American palaeontologist F. Ameghino into regarding the South American ungulates with clawed feet as ancestral to the North American chalicotheres. But the latter were closely related to other perissodactyls (p. 430) and were of northern (Palaearctic) origin and distribution, while *Homalodotherium* and its allies were distinctly of notoungulate, South American, origin and distribution.

The earlier members of the suborder Typotheria were small Eocene ungulates with a rather deep chin region (fig. 21.21B). Ameghino thought they were ancestral apes and accordingly named them *Notopithecus*; but they were later shown to be related to *Protypotherium*; this Upper Oligocene to Middle Miocene genus foreshadowed the later typotheres, which culminated in the Pleistocene. In *Protypotherium* all the teeth, from the first incisor to the last molar, had tall, sharp-edged crowns, crowded close together, the upper ones slanting somewhat backward, the lowers, forward. The masseteric or angular region of the lower jaw was expanded and the zygomatic arch consisted of three superposed bars contributed by the maxilla, jugal (malar) and squamosal. The auditory bullae were inflated. The orbits were rather large and the nasal chamber ample. The animal evidently had keen senses to warn it of danger and its locomotor equipment included moderately long, bent hind limbs, lumbar region with well developed parapophyses, long pelvis; thorax small with feeble neural arches and slender, not long, ribs; hind feet four-toed with small narrow claws, digits slender, metatarsals not shortened nor widened, tarsus narrow, lower leg slender; forelimbs shorter than hind limbs; hands four-fingered, rather delicate, second metacarpal and digit slightly enlarged, small claws; scapula of moderate size, no clavicles. Thus the skull superficially resembled the *Hyrax* (dassie) type (fig. 21.23B) and Ameghino indeed named one of the protypotheres *Argyrohyrax* ("silver," from Argentine), but the locotomor skeleton is rather cat-like and very unlike that of the dassie.

In *Mesotherium* (*"Typotherium"*) of the Pampean (Pleistocene) of South America the skull was basically like that of *Protypotherium,* except that it was much larger and the median pairs of upper and lower incisors were somewhat rodent-like.

In the suborder Hegetotheria (Simpson, 1945, p. 130) the skull and dentition were basically like those of the protypotheres but with a marked emphasis of the rodent-like features of the incisors. *Pachyrukhos* Ameghino of this suborder was also somewhat rabbit-like, especially in its long hind legs and short tail. But it is very likely that restorations of the animal in life have overemphasized its rabbit-like habitus features.

PARADOXICAL ASTRAPOTHERES

The skeleton (fig. 21.18) of *Astrapotherium* (lightning beast) at first sight looks as if it had been put together from parts of widely different fossil and recent animals; but this unique assemblage of parts has been well established by the discovery of an almost complete skeleton, which was later mounted in the Chicago (Field) Museum of Natural History, and described by Riggs (1935) and Scott (1937). The upper canines are prolonged downward into long curved tusks diverging laterally beyond the large erect lower canines (fig. 21.21D). Above the root of the tusks was the high, transversely narrow nasal channel, with the nasal bones reduced to short stubs. Comparisons with the tapir skull show that, starting from the retracted nasals, a long flexible nose must have curved forward and downward, carrying with it the narrow upper lip, which was fastened to the small toothless, downwardly turned premaxillae. These nubbin-like premaxillae are very different from those of ruminant artiodactyls, which support a flat, spreading pad opposing the lower incisors. They may rather have served for some of the upper lip muscles that would pull the proboscis tip toward the mouth. *Astrapotherium* cannot be correctly restored with a wholly elephant-like trunk because its skull differs from that of the elephant in the marked transverse narrowness of the narial channel and the nose-lip complex was probably much more tapir-like than elephant-like. It may indeed have been inflatable, like that of the elephant-seal, as well as extensile and prehensile. As if to make room for an enlarged buccinator pouch behind the tusks, the first and second upper premolars are absent and the third reduced and crowded. There is a very long diastema between the lower canines and the cheek teeth, a long lingual channel extending to the incisors. The latter were slightly procumbent and arranged in a spatulate row, somewhat as in the giraffe, and it is likely that they assisted the extensile tongue in cropping the herbage. The molars have high sharp edges suggestive of those of the North American Oligocene *Metamynodon,* an aquatic rhinoceros. The frontal region behind the small, slightly upcurved nasals was inflated. The orbits were rather small and far below the level of the nasals. In front of the lower rim of the orbit was an oblique, upward and forwardly directed wide groove, which may have lodged a lateral diverticulum of the nasal passage as in the tapir. The anterior superficial branch of the masseter muscle may have been attached to the low protuberance on the zygomatic process of the maxillary bone, the chief muscles of the snout arising from the prominent and oblique lower border of the orbit.

The cervical vertebrae of *Astrapotherium* are moderately well developed, as they should be to support this very large head, but the dorsal spines are exceptionally small, as are also all the dorsal and lumbar vertebrae (fig. 21.18B). The thorax is long, the lumbar region long, the tail slender and short. The heavy head must have been partly suspended by the serratus and subscapular muscles from the tall scapulae and robust fore-limbs. The scapula was pointed on top, with a high crest or spine and a very large wide acromion, which slightly overlapped the stout humerus. The hands were very small and spreading, with very short digits; the radius and ulna moderate. The ilium was widely expanded, the femur long, columnar, the tibia stout, not long, the spreading hind feet remarkably small. This assemblage of features suggests that the animal lived alongside quiet pools and lakes, feeding in the reeds and wading about, with its weight supported mainly by the water. The flexible backbone and short spreading feet would be useful in swimming or wading and the expanded ilia would assist it in rearing up while in the water or even when reaching upward for leaves.

The resemblances of the astrapotheres in their jaws and teeth to the North American *Metamynodon* is mainly convergent; but their relationships, even with the notoungulates, is also one of partial convergence (Simpson, 1945, p. 238). The skull of *Trigonostylops* of the Eocene of South America, which was formerly thought to be a small primitive astrapothere, yielded many detailed features in which it differs from astrapotheres. On the basis of his intensive study of the morphology of its skull, Simpson concludes that *Trigonostylops* and the astrapotheres ". . . represent anciently divergent lines probably related near their base, derived from some litoptern-like and condylarth-like ancestry, within the South American complex but quite distinct from the Notoungulata as such" (*op. cit.*). Not many of the more conspicuous habitus features of *Astrapotherium* seem to have been developed among the true notoungulates; yet there are notoungulates which have characters that may be hypothetically ascribed to an early pre-*Astrapotherium* ancestor. For example, in the Lower Eocene isotemnid toxodont *Thomashuxleya* (fig. 21.17A), as described by Simpson, the skull was already large, the upper canines conspicuous, the lower jaw long, the forelimbs stout with short spreading hands; the neural spines of the dorsal and lumber vertebrae remarkably low, the tail short, the blade of the ilium flaring partly outward. But there was no retraction of the nasals, or long postcanine diastema, or crowding and reduction of the premolars; the molars also had less tall crowns and were relatively less dominant. Nor were the hind limbs post-like or the fingers excessively short. The last few items would be expected also in some degree in any fairly primitive litoptern, such as *Henricosbornia* (fig. 21.19C) or in a small condylarth such as *Hyopsodus* (fig. 21.13D). In short this seems to be one of numerous cases where the doubt as to the precise line of ascent does not preclude a comparative morphologic analysis of the probable successive stages that led to the observed terminal specializations. Thus it seems that the remote condylarth-litoptern ancestors of *Astrapotherium* had to pass through a stage that paralleled the notoungulate *Thomashuxleya* in the features listed above before they began to evolve the end specializations of *Astrapotherium.*

FALSE ELEPHANTS (PYROTHERIA)

In *Pyrotherium* from the Oligocene of Patagonia the extreme retraction of the nasals (fig. 21.22) and the backward displacement of the narial opening at first suggest relationship with the elephants and mastodons; so also do the cheek teeth, which bear sharp anterior and posterior cross-crests not unlike those of *Dinotherium*. The enlarged procumbent upper and lower incisors, the massiveness of the zygomatic arch, and some other features, all augment the resemblance. Nevertheless it would be unsafe to assume that *Pyrotherium* is a genuine proboscidean, not a spurious and very imperfect imitation. The possession of an anterior and a posterior cross-crest on each molar tooth is no reliable sign of relationship with the Proboscidea, as it has often been evolved independently in other orders, viz.: among the marsupials in the diprotodonts and kangaroos; among the placentals of the order Perissodactyla, in the helaletids, lophiodonts and tapirs; among the Artiodactyla in the peccaries and certain pigs (*Lestriodon*), as well as in the Eocene dichobunid *Tapirulus*. In the molarization of its premolars *Pyrotherium* recalls the litoptern *Macrauchenia* and in the relatively large size of the procumbent pincers-like incisors it far surpasses the proboscidean *Moeritherium*.

The presence of these stout incisive tusks and of so many double-crested molars, with possibly very resistant woody food, evidently required immense jaw muscles, as evidenced by: (a) the greatly expanded angular region of the lower jaw, as well as of the coronoid region; (b) the extremely massive and strongly braced zygomatic arch, which is directed sharply inward and backward, as in *Homalodotherium*; (c) the great size and thickness of the opposite premaxillae and maxillae, which are fused into a long rostrum comparable with that of the litoptern *Macrauchenia*; (d) the extreme retraction of the narial passage is also a point of similarity with *Macrauchenia*. In short, although the remote ancestry of *Pyrotherium* is unknown, the resemblances of its teeth and skull to those of *Moeritherium* appear to be purely convergent, while its points of agreement with the litopterns suggest parallelism and common origin from a remote pre-litoptern stock.

ENVIRONMENT, ISOLATION AND EVOLUTION

In concluding this very brief summary of the evolution of the skeleton of the "South American complex," we may well pause here to ask why, both in North and South America, the molars of most hoofed animals of many different families should have independently increased in height and complexity of pattern, and why the premolars should have followed suit.

The high open country along the Andean-Rocky Mountain axis, on the plains of which lived many of the ungulates of both North and South America, was frequently covered with volcanic ashes and dust, and in this material many of their skeletons were entombed and later fossilized. In herbivorous animals, forced to browse or graze on gritty herbage, low-crowned molar teeth would soon wear down and in so doing they would predetermine a relatively short life span and a still shorter reproductive period. Thus the intricate processes of natural selection, operating upon hereditary variations in such an environment, would, so to speak, pay good dividends on higher and stronger, i.e., more complexly folded, tooth crowns having ample, long active tooth-pulps, strong dentine and very hard, raised edges of enamel coating. The spread of siliceous grasses, which presumably followed progressive elevation of the region, further increased the rate of wear on the tooth crowns of the cropping incisors and of all the cheek teeth, demanding higher and higher limits of resistance. Such factors affecting tooth structure, plus the constants of gravitation, of the principles of the lever, of the balance of forces, etc., in addition to competition of large populations—all operated on animals of various sizes and diversity of heritage. Thus arose, on one hand, the numerous convergences between animals of similar habitus in North and South America, and on the other, the accumulation, in each continent, of unique heritage features which serve to distinguish the ungulates that were evolved on opposite sides of the former gap between them.

Following Simpson's classification (1945) of the hoofed mammals, we come next to the Superorder Paenungulata, including the Orders Pantodonta, Dinocerata and Pyrotheria, plus the assemblage formerly called Subungulata (Proboscidea, Hyracoidea, Sirenia).

STUMP-FOOTED PANTODONTA AND DINOCERATA

The members of the Order Pantodonta (figs. 21.8; 21.4B, C), from the Paleocene and Eocene of North America and Europe, ranged from the size of pigs (*Pantolambda*) to that of hippopotami, that is, from medi- to subgraviportal semiaquatic stages. *Pantolambda,* the apparent patriarch of the family Coryphodontidae, had two V-shaped or inverted lambda-like cusps (para- and metacones) on the outer sides of its upper molar teeth (fig. 21.7A, G), with a large rounded low cusp (protocone) on the inner side. The premolars had a single V on the outer side and a low internal cusp. The opposing V's on the lower molars, much as in *Coryphodon* (E1), had low or reduced front legs and high-crested rear ones. The anterior lower V fitted into the inverted triangular space between two upper molars, the posterior lower V fitted on the crown of its opponent between the para- and metacone tips of the outer V's. The basic plan of the *Pantolambda* teeth had all been inherited from Upper Cretaceous insectivorous-carnivorous ancestors, but in *Pantolambda* it was presumably being adapted for the cutting of vegetation. In *Coryphodon* (fig. 21.8B) the strong vertebral column was arched upward, the limbs short and massive, the five-toed hands and feet very short, with flat small nails (cf. figs. 21.10C; 21.11B).

The coryphodonts, ranging from the Upper Paleocene of North America to the Middle Oligocene of Mongolia, culminated in animals with large, somewhat hippopotamus-like skulls (fig. 21.8B). Another family, the Barylambdidae, had surprisingly small heads, large rounded bodies, and very massive limbs and very short feet (fig. 21.8A).

The members of this order (fig. 21.9) formed their cheek teeth in a different way (fig. 21.7J, K) from that followed by the Pantodonta (G, H), in the end (K, F) acquiring functionally bilophodont molars, that is, with two cross-crests on each upper and lower molar. They ended as graviportal giants in the Middle to Upper Eocene of North America and Mongolia. These excessively low-brained beasts stood (fig. 21.9) with nearly vertical columnar limbs and very short hands and feet, which were doubtless supported by thick enlarged pads like those of elephants. When enraged they took long strides forward, doubtless lowering their long narrow heads and trying to ram the enemy with their horns or to rip him with their sabre-like tusks. These "horns" were conical, club-like bony outgrowths from the top of the skull, probably covered with tough hide. They were arranged in two or three pairs over the nose, eyes and occiput. Their vertebrae, girdles and limb bones so strikingly recalled those of mastodons and other proboscideans that, when first discovered, they were referred to that order; but their skulls and dentitions could hardly be more widely different and there can be no doubt that the resemblances between the Dinocerata and the Proboscidea are largely convergent.

MULTIPLE EVOLUTION OF THE ELEPHANT ORDER (PROBOSCIDEA)

A brief description of the skeleton of the American mastodon may be a convenient starting-point for an outline of the general evolution of this proboscis-bearing order. Although the trunk is not preserved, it was evidently very large, as shown by the massiveness of the nasal region. It was not only heavy in itself but to judge from çomparison with that of recent elephants, it could be used in conjunction with the tusks to wrap around and carry very heavy objects, such as large teakwood logs. Thus the trunk and large tusks together could exert enormous adverse leverage against the rest of the body, which therefore had to be massive enough to afford an effective base and anchor for this peculiar application of the principles of the drawbridge and derrick. That is evidently why the scapulae, humerus and forearms of the Mastodonts are so enormous, why the carpals are flattened, the metacarpals widened. The hoofs were mere nubbins because the feet were inserted into great cushions of connective tissue. To the very large triangular scapulae were attached not only the great triceps and other muscles of the arm, but also certain of the lateral neck muscles which held up the wide pillars of the occiput; while the entire thorax was supported by the scapulae, chiefly through the fan-like serrati muscles coming up from the ribs, and indirectly by the deltoid on the outside and the subscapularis on the inside, which helped the arm muscles to keep the scapulae upright and slightly inclined inward. The enormous widespread ilia and column-like hind limb held up the abdominal wall below and the sacral region above. The neural spines formed only a moderate hump above the upper ends of the scapulae, but were collectively strong enough to resist the tensions from the thick ligament of the neck, which was inserted in a deep depression in the midline of the occiput.

While the mastodon skull is only moderately specialized, the skulls of stegodonts, mammoths (fig. 21.28) and elephants are further modified in correlation with: (a) the enormous size of the tusks, whose sockets run far upward and backward into the correspondingly braced maxillae (fig. 21.32B1); (b) the reduction of the premolars; (c) the anteroposterior multiplication and vertical heightening of the transverse crests of the molars (fig. 21.34F′-H′), each crest being composed of small columns derived from the tubercles on the surface of the crowns. Thus a massive dental mill is formed; although these grindstones, which in old animals are reduced on each side to one immense molar above and one below (fig. 21.28E), do not rotate upon each other like man-made millstones but at each stroke of the jaw the lower pair push the food upward and forward and rub it against the upper pair.

Among the combined effects of these factors are that: (a) the upper and lower jaws are greatly shortened, thickened and deepened; (b) the glenoid sockets are elevated very far above the level of the grinding surface of the molars; (c) the external auditory meatus becomes a small pore or tube bounded below by flanges from the squamosal and mastoid bones and located above the condyle of the lower jaw; (d) the zygomatic arches are greatly strengthened and bent downward and forward; (e) the orbits move forward and downward in line with the sockets of the tusks; (f) the remnant of the nasals is pushed upward and backward; (g) the areas for the fastening of the massive trunk in front and of the neck muscles in the rear are greatly increased by the expansion of the sponge-like, air-filled bony cells lying between the progressively separating, inner and outer tables of bone (fig. 21.32B1); (h) on the inside of the skull (fig. 21.33) the basi- and orbito-sphenoid bones form the central keystone for a system of intersecting arches in the anteroposterior, transverse and horizontal planes, which prevent the crushing of the brain by the powerful forces around it (Gregory, 1903).*

By what cumulative steps did this amazing complex of new and unexcelled natural mechanisms emerge from less specialized ancestral stages? The branching and sub-branching of the proboscidean hosts, especially during the second half of the Caenozoic era, have been so luxurious that the explorers and systematists, whose patient labors have brought the vast material to light, have felt compelled by the facts to give names to several hundred species of fossil Proboscidea. After years of labor, Osborn (1936) succeeded in arranging the material under five suborders, eight families, twenty-one subfamilies and forty-three genera. By further research, Simpson (1945), while adopting the main features of Osborn's classification, has been able to clarify many puzzling problems, to eliminate some divisions and to reduce the system to three suborders, five families and twenty-two genera.

Four great structural stages in the evolution of the Proboscidea were clearly recognized and pithily described by C. W. Andrews (1906). Subsequent material and research have confirmed his insight. They may be outlined as follows:

Stage I. *Moeritherium.* Upper Eocene and early Oligocene of Egypt. Skull (figs. 21.28B, 29B) small, long, narrow, low, not inflated; small orbits, near anterior end, above premolars. Free end of nasals short, above proximal end of short rostrum. Proboscis not yet formed. Second upper incisors moderately enlarged and curved downward, opposing forwardly inclined lower pair (i_2). Upper canines vestigial, lower

* Among the abbreviations used in fig. 21.31 are the following:

Pmx, Premaxillary
Mx, Maxillary
Mx p, Maxillary pouch (for molars)
Ma, Malar
Po f, Postorbital ridge of frontal
pal Mx, Palatine ledge of maxillary
a na, Anterior nares
ty p, Anterior process of tympanic
tp h, Tympanohyal
eu, Eustachian opening of tympanic
a p f, Anterior palatine foramina (canals)
i^2, Tusk
Pl, Palatine
Ps, Presphenoid
Bs, Basisphenoid
Bo, Basioccipital
p As, Pterygoid wing of alisphenoid
Pt, Pterygoid
p n, Posterior (internal) nares
i o f, Infraorbital foramen
c a s, Alisphenoid canal
f l m, Foramen lacerum medius
f ov, Foramen ovale (confluent with *f l m*)
Sq, Squamosal
$p^3(dm^2)$, Third premolar (or second deciduous molar of authors)
Ex o, Exoccipital
pg, Postglenoid ledge of squamosal
p ty, Posttympanic ledge of squamosal, which with *pg* forms a secondary external auditory meatus (*e a m*)
Ty, Tympanic bulla
Vo, Vomer
i c c, Canal for internal carotid artery
f st m, Stylomastoid foramen
f l p, Foramen lacerum posterius
c f, Notch, a vestige of condylar foramen (?) (confluent with *f l p*)
$p^4(dm^3)$, Fourth premolar (or third deciduous molar of authors)

canines absent. Upper premolars 3 and 4 (fig. 21.34B) rounded, not molariform; three upper molars oblong, with four low cones in two transverse pairs; basal cingulum thick, beaded; lower molars narrower, with two transverse pairs of low cusps and incipient posterior cingulum ridge. Lower jaw (fig. 21.28B) with expanded angle and high condyle, located near the rear end and beneath small auditory meatus. Zygoma stout, malar thick, extending beneath squamosal to glenoid.

Stage II. *Phiomia* and *Palaeomastodon.* Lower Oligocene of Egypt. Skull (C) short, wide, high, somewhat inflated; orbits larger, above second molars. Nasals very short, pointed, a proboscis evidently present, supported by the narrow premaxillo-maxillary rostrum, which bears a single pair of short, downwardly curved tusks. Upper tusks divergent passing outside of lower tusks. Upper and lower molars bi- to trilophodont, posterior premolars submolariform (fig. 21.34C). Lower jaw with lower incisor tusks directed forward (fig. 21.28C). Malar short, stout, abutting against stout zygomatic process of maxilla, extending backward beneath zygomatic process of squamosal to glenoid.

Stage III. *Gomphotherium* (*Trilophodon*). Miocene of Europe and North America. Skull (fig. 21.29D) basically as in II but with further forward and transverse growth of rostrum and of divergent upper tusks. Molars essentially trilophodont (fig. 21.34F′) but with fourth ridge beginning. Long lower incisor tusks, mostly buried in alveoli beneath long channel for tongue (cf. fig. 21.30E).

Stage IV. Stegodonts, Mammoths and Elephants. Lower Pliocene to Recent. Skull extremely short and wide, greatly inflated (figs. 21.28D, E, 29E, F). Upper tusks becoming very large, often curving upward and inward; lower tusks absent. Upper and lower jaws greatly shortened, the rostral part of lower jaw vestigial or absent (fig. 21.32A, B), together with the rostral part of the tongue. Molars with many plates (fig. 21.34H′), becoming extremely high, ultimately reduced in old animals to a single pair (m3) above and below.

The *Moeritherium* skull, jaw and dentition differ so widely in general appearance (figs. 21.28B, 29B) even from those of *Phiomia* and *Palaeomastodon* (C) that Osborn (1936) put it in a separate suborder, Moeritherioidea, and regarded it as an early semiaquatic side branch of the proboscidean order, having possible relationships also with the Sirenia. While *Moeritherium* itself could at best be only a great granduncle, rather than a great grandfather of the later Proboscidea, its skull nevertheless retains a protoungulate basic pattern, combined with the features recognized by C. W. Andrews to be prerequisite in pre-proboscideans. The stout malar extends back to the glenoid, underlying the zygomatic process of the squamosal; the small external auditory meatus is more or less circular and nearly enclosed by contact of the postglenoid and posttympanic processes; the small orbits are immediately above the third and fourth upper premolars, as they are in *Dinotherium,* whereas in other proboscideans with a well developed trunk the orbit is pushed back above m^2. m^3; the dental formula $\left(\frac{3.\ 1.\ 3.\ 3}{2.\ 0.\ 3.\ 3}\right)$ is intermediate between that of the protoungulate (condylarth) $\left(\frac{3.\ 1.\ 4.\ 3}{3.\ 1.\ 4.\ 3}\right)$ and of *Phiomia* $\left(\frac{1.\ 0.\ 3.\ 3}{1.\ 0.\ 3.\ 3}\right)$. The molars of *Moeritherium* (figs. 21.34, B, E′) had already acquired the bilophodont ground-plan and even the beginnings of the trefoil pattern on m_1, which was further evolved in *Phiomia.*

In *Moeritherium,* the nasals (figs. 21.28B, 29B) are but slightly displaced posteriorly, although already shortened in front, and the notch between the nasal and the premaxilla is neither large enough nor deep enough to afford room for a flexible trunk like that of the tapir. It is very likely that the marked enlargement of the second upper incisor and the procumbency of its mate in the lower jaw were at least begun before the short fleshy muzzle began to evolve into a trunk. *Moeritherium* retained all its adult cheek teeth in a horizontal row and so do the palaeomastodonts, whereas in later Proboscidea (except *Dinotherium*) the premolars were gradually eliminated and the times of eruption of the molars retarded toward the rear; thus an originally vertical to oblique pattern of replacement finally gave rise to a succession from the rear forward along downwardly curved arcs (fig. 21.32B1) in upper and lower jaws.

There is a wide range of variation in the lower tusks and the front half of the lower jaw, from the long narrow tusks and symphysial region of *Palaeomastodon* to (a) the very wide "shovel" made from the lower tusks and their supporting alveolar bone in *Platybelodon* (fig. 21.30E); (b) the downturned lower tusks of *Rhynchotherium dinotherioides* and of Dinotherium; (c) the progressively shortening and dwindling tusks of the true mastodons; (e) the greatly shortened and tuskless symphysial region of the stegodonts and elephants.

The postcranial skeleton of *Moeritherium*, although imperfectly known, had apparently not yet reached the graviportal stage with widely expanded ilia. The flattening of the carpals and tarsals, the widening of the metacarpals, the marked shortening of the digits and the opening up of the angles at the elbows and knees, which were characteristic of all known proboscideans beyond the *Moeritherium* stage, were also characteristic of other graviportal ungulates, such as the Dinocerata (fig. 21.9), Astrapotheria (fig. 21.18B), Embrithopoda (fig. 21.25B) and to a somewhat less extent in the secondarily graviportal titanotheres (fig. 21.54B) among perissodactyls. There are indeed considerable differences in the heights of the spines on the dorsal and lumbar vertebrae as between different genera of mastodons and mammoths, which may be associated with differences in the habitual posture of the skull, whether directed downward toward the ground or with the neck elevated, as in the mammoth; but as a whole (except for *Moeritherium*) the locomotor portions of the skeletons of the proboscideans are restricted to the orthograde or graviportal type and show a far narrower range of adaptive branching compared with those of the order Notoungulata; these include such widely diverse forms as: (a) the narrow-clawed, rather cat-like skeleton of *Protypotherium,* (b) the secondarily clawed, scratching feet and strong limbs of *Homalodotherium* (fig. 21.17B), (c) the somewhat rabbit-like *Pachyrukhos* skeleton; (d) the massive, high-humped frame of *Toxodon* (fig. 21.18A), with its abnormally squat legs and very short three-toed hands and feet; (e) the pseudo-graviportal, perhaps semi-aquatic skeleton of *Astrapotherium* (fig. 21.18B).

THE ROCK-DWELLING "CONIES" (HYRACOIDEA)

The damans, dassies, or "conies" of the Old Testament (*Procavia syriaca*), are plump, somewhat marmot-like mammals living among the rocks in Syria, Palestine and Arabia. Others live in the semi-arid Karroo of South Africa (*Procavia capensis*); still others climb up trees, chiefly in Central and East Africa (e.g. *Dendrohyrax arboreus,* etc.). They live in small isolated colonies and are variable in so many details of the skeleton, soft anatomy and pelage that Oldfield Thomas united them all in a single genus (*Procavia*), but later systematists find it more convenient to group them into three genera. These are the sole survivors of a once extensive African fauna, and fossil remains of hyracoids have been found in the Lower Oligocene of Egypt, the Lower Miocene of East Africa, and the Lower Pliocene of Greece.

The skeleton of the recent *Procavia* (*"Hyrax"*) has a highly arched back (fig. 21.25A), a very long lumbar region and very short tail, as indicated in the following comparison (data mostly from Weber, 1928):

	DORSALS	SACRALS	CAUDALS
Phenacodus primaevus	20	? 3	?
Recent hyracoids	27–30	5–7	4– 8
Recent perissodactyls	22–24	5–6	17–22
Recent artiodactyls	19	5–7	6–24
Recent proboscideans	23	4	31

In the side view, the very long lumbar region and short tail suggest that as the back was arched upward, the sacral attachment may have shifted backward from one vertebra to the next, the form of the vertebrae being readjusted accordingly, somewhat as among the toothed whales, the number of functionally lumbar vertebrae increased as the ilia receded. However the high number of thoraco-lumbar vertebrae in the hyrax may have been obtained, it probably compensates for the shortness of the legs and aids in climbing; for they move quickly about among the rocks and their plantigrade hind feet bear large cushions which are said to exert a suction on smooth surfaces.

As in the perissodactyls, the recent hyracoids have four toes in the forefoot, of which the metacarpals of digits II, III, IV are subequal, the fifth being smaller. The carpals are block-like and arranged in vertical series, as in the condylarth *Phenacodus*. The centrale is separate from the scaphoid, a primitive placental character. The flat nails on the three middle digits suggest hoofs. In the functionally three-toed hind foot, the hallux (I) is vestigial and digits II, III, IV are subequal, III being slightly longer, so that the axis of weight passes through the middle digit as in the perissodactyls. The second digit of the hind foot bears a peculiar upwardly-curved, pointed nail, which is used for scratching and combing the coarse fur.

The large astragalus in the hyrax has a raised trochlear pulley for the tibia and fibula, the latter also touching a rounded eminence on the outer rim of the calcaneum and recalling a similar but better defined facet in the litopterns and many notoungulates. There is a sharp step-like facet on the mesial side of the astragalus which articulates with the internal malleolus of the tibia and probably transmits part of the upward thrust from the foot to the tibia, especially when the sole of the foot is turned inward, as in climbing. The distal end or head of the astragalus is about as wide as the proximal end and bears an incipiently concavo-convex facet for the equally wide navicular upon which it mainly rests in *Procavia capensis*. But, according to Osborn (1889), in the tree hyrax (*Dendrohyrax arboreus*) the astragalus also has a broad facet for the cuboid, which is the main support for the calcaneum. Thus, as well noted by Weber and Abel (1928, p. 431), one species of hyrax is no longer taxeopodal (serial) but has become diplarthral (displaced) as these terms were used by Cope (see p. 412).

Indeed the alternative whether there shall be or shall not be an oblique astragalo-cuboid facet and thus whether a given animal shall be a "diplarth" or a "taxeopod" depends upon the relative rates of increase (growth in the individual, or anisomerism in phylogeny) of all the bones of the tarsus in any or all the three planes of space; also upon the angular direction of the long axis of the astragalus. If the astragalus becomes very wide, much wider than the navicular, it will overlap on the cuboid, as in the Dinocerata; but if the navicular has also greatly widened, it may take over the cuboid contact and separate the astragalus widely from the small cuboid, as in *Mastodon*. Again, the lower part or distal exremity of the calcaneum may be long and the whole astragalus short, then the lower end of the astragalus will be widely separated from the cuboid, as in *Protypotherium*. In the recent hyracoids the distal ends of the astragalus and calcaneum are nearly on the same transverse plane and a slight lengthening of the former would bring it in contact with the cuboid. It should be noted, however, that the contacts seen in the front view may not fully represent the pertinent facts. Among the primitive Eocene titanotheres, for example, the front view of the foot indicates a very small contact between the astragalus and the cuboid, but this contact widens in the deeper planes and as time went on, it widened further until it became conspicuous even in the front view (fig. 21.55H).

The skull (fig. 21.23B), jaws and teeth (fig. 21.24D) of *Procavia* present a puzzling combination of habitus features, vaguely recalling those of small palaeotheres, rhinoceroses, and protypotherian notoungulates. In the tree-living hyraces with short-crowned cheek teeth, the upper molar crown patterns approach the bunoselenodont type of the extinct titanotheres and palaeotheres, that is, there is a low conical cusp on the inner side and two crescent-like ones on the outer side; but in the high-crowned rock-living hyraces the molars have oblique cutting outer walls and cross-crests somewhat as in the rhinoceroses. The premolars p^1 to p^4 are progressively molariform. The central upper incisors are enlarged and sharply pointed, triangular in cross-section. The opposing two lower pairs are procumbent, their chisel-like crowns subdivided by two vertical grooves, which may perhaps serve in combing the skin. Canines, absent in adults, are vestigial in the deciduous set. Inspection of the jaws and teeth (figs. 21,23B; 21.24D) indicates that in chewing the jaws move obliquely, so that the entire lower premolar-molar row of one side sweeps across the upper row from the outer side inward, as in perissodactyls. In view of the relatively large grinding area of the twelve more or less molariform teeth on each side, it is not surprising that the angular region of the lower jaw for the insertion of the masseter muscle is greatly expanded, or that secondary postorbital processes are formed by the frontal and malar (jugal) to brace the orbit posteriorly against the thrusts of the temporal and masseter muscles.

The long skull of *Geniohyus* (fig. 21.23) from the Lower Oligocene of Egypt is much more primitive than that of the modern *Hyrax*. In the greater expansion of the angular region of the lower jaw and the incipient enlargement of the central incisors, it differs from that of *Eohippus* among the perissodactyls; and it contrasts widely with that of the notoungulate *Protypotherium* in its much less hypsodont teeth and in all its crown patterns, slender zygomatic arch and much smaller orbit. The feet of recent hyracoids, however, agree with those of protypotheres in having four functional digits in the fore feet and three in the hind feet; but as these numbers also hold good in the earlier perissodactyls of different families, they do not in this case supply decisive evidence of ordinal relationship. Moreover, the presence of four digits in the hand and three in the hind foot in these representatives of three different orders is consistent with the fact that the fore foot needs a wider surface than does the hind foot, because it catches hold of the ground in front of the weight and drags the body toward it, whereas the hind foot after it passes the vertical, has a greater effect in pushing the body from the rear.

In short, both the recent and the extinct hyracoids may be remotely related to the perissodactyls only by way of derivation from a common ancestral protoungulate (condylarth) stock, and they seem also to have mainly convergent resemblances to certain extinct South American ungulates (fig. 21.24E).

COLOSSAL ARSINOITHERES (EMBRITHOPODA)

The giant *Arsinoitherium* from the Lower Oligocene of Egypt, a contemporary of the hyracoids, was a huge beast about 11 feet long (fig. 21.25B). On top of its massive skull was a pair of enormous horn cores, united near the base and curving forward and upward. They were probably covered by still longer true horns. At the rear base of the principal pair of bony horns there was a second pair, but of minute size and conical form. The cheek teeth are all hypsodont (i.e. with high crowns) and the upper molars have the outer borders deeply infolded toward the flat inner sides; the premolars are submolariform but not infolded.

The postcranial skeleton was of the graviportal type like that of the Proboscidea and, as noted by C. W. Andrews (1906), the individual limb bones were much like those of *Mastodon*. But in view of the profound differences in all parts of the skull and dentition from the corresponding parts in the Proboscidea, no close relationship between them seems possible and the graviportal resemblances were evidently convergent. Indeed, Andrews held that the nearer relationship of this, so far as known, isolated genus was probably with the Hyracoidea (cf. fig. 21.26). All three orders are regarded as probable survivors of a once widespread and diversified, distinctly African fauna (Osborn, 1900) which evolved its new types during aeons of geographic isolation and inbreeding.

GROTESQUE "SIRENS" (SIRENIA)

The sea-cows, including the manatees (fig. 21.35A) and the dugongs, appear to be a whale-like branch of the primitive pre-proboscidean stem, somewhat as the true whales seem to be a derivative of a far older primitive placental carnivorous stem (Chapter XXII). The entire skeleton of the sea-cow has been transformed from a quadrupedal to a completely aquatic stage and the streamlined cylindrical body is enclosed in a thick, pliable leathery skin. This bears only a few sparsely distributed hairs in the embryo, except for the bristles around the thick lips. Locomotion is chiefly by means of the transversely expanded convex or forked tail, which, as in the whales, is composed only of tough skin and is without bony support except for the small posterior caudal vertebrae. The pectoral appendages are externally paddle-like although supported by a somewhat flattened but otherwise little modified bony limb and hand. The hind limbs are represented by vestiges of the femora and in the dugong by a pair of rod-like innominate bones. The latter in the manatee are vertically short, with ventrally flaring, small pubo-ischiadic plates. The opposite iliac processes are loosely attached by ligaments to the outer ends of the wide transverse processes of the single sacral vertebra, but ventrally the opposite innominates are well separated, being connected only by a transverse ligament.

The 17–19 pairs of ribs are excessively thick and heavy. They serve as ballast to steady the animal when it is feeding. The neck is very short. Mammals generally, except the recent sloths, have seven cervical vertebrae but the manatee has lost one and the dugong has seven or eight. These vertebrae, while much shortened anteroposteriorly and closely appressed to each other, are not fused as they are in some whales.

The neural arches of all the vertebrae are stout but low, from the beginning of the dorsals to near the end of the bony tail. Behind the ribs there are wide transverse processes from the first lumbar, diminishing in width caudally. Beginning with the second postsacral vertebra, there are stout median ventral processes or haemapophyses. In cross-section, behind the sacral vertebrae (fig. 21.35A2) the powerful longitudinal axial musculature is divided into two dorsal and two ventral sets, which can raise or lower the horizontally expanded tail or move it laterally, or combine the two movements in sculling.

The bony pectoral extremities are enclosed in flat leathery paddles, but in the manatee there are traces of nails on the edges of the paddles, on the distal ends of the second, third and fourth digits. The first metacarpal is present but the phalanges of this digit are reduced. There is no multiplication of the phalanges as in whales and little if any shortening of the forearm and humerus. The fifth metacarpal is the largest and it is spread somewhat apart from the fourth, convergently resembling the same feature in some aquatic reptiles. The carpals are moderately flattened and fused to form only two or even one proximal piece. There are no clavicles but the large scapulae retain a vertically shortened spine and laterally projecting acromion.

The skull of the dugong (fig. 21.36H) retains a pair of downward and forwardly directed tusks, which are absent in the manatee (C). These small tusks are implanted in relatively huge alveolar portions of the premaxillae and the latter also support a horny pad on their postero-ventral faces; the pad opposes a similar one borne by the swollen, down-turned anterior end of the mandible. The bony rostrum supports the immense, thick muscular lips. In the manatee these lips are folded in on the sides and can be protruded or retracted, while the stout bristles on them help to pull in the eel-grass upon which the animal feeds. The cheek teeth of the manatee (fig. 21.37G) have increased in number, up to twenty or more in each of the four rows (two above and two below), but of these only about six in each row are in place at any one time. This high number is attained partly by combining the deciduous and the permanent teeth in a single succession from front to rear. The teeth are formed in the rear of the alveolar pouches and they move forward gradually, finally dropping out at the front end of the alveolar groove, somewhat as they do in the elephant. The crowns of the cheek teeth are more or less bilophodont, that is, each with two cross-crests. These are not sharply edged but are made up of numerous tubercles which wear down into a continuous ridge.

The manatee, having lost the upper tusks, has only a feebly developed rostrum; but the dugong has retained more of the massive decurved rostrum which was present in some of the fossil Sirenia (fig. 21.36) that have been found mostly in the marine Pliocene, Miocene and Oligocene of Europe. The oldest known sirenians come from the Eocene of the West Indies and of Egypt, and their skulls were already essentially sirenian. In one of them, however (*Protosiren*), the pelvis was much less specialized than the rod-like pelvic bones of the dugong, and since it retained a normal-looking acetabulum, it is probable that a small hind limb was still present.

In *Desmostylus* from the Miocene of California and Japan, each of the molar tooth crowns (fig. 21.37B) consisted of a cluster of small, parallel cylindrical columns, arranged in three successive pairs with a median posterior one. The bony rostrum (fig. 21.38A) was very large and included one upper pair of long upper tusks curved slightly downward, and two pairs of smaller lower tusks set in a shovel-like lower jaw (A2). Preliminary reports of various skeletal parts found in the Japanese marine Miocene indicate the existence of a large quadrupedal mammal with massive limbs, which may possibly belong with the skull of Desmostylus. The almost hippopotamus-like lower tusks and massive many-columned molars of *Desmostylus* would seem to be by no means inconsistent either with semiaquatic rather than marine habits, or with remote derivation from such a primitive proboscidean as *Moeritherium* (fig. 21.36A).

From what precedes it seems very likely that the order Sirenia, like the order Cetacea, as we know them, are merely aquatic branches of some larger orders whose earlier members had been diversely adapted to many other ways of life.

Recent sirenians share with the elephants some significant common characters, especially in the soft anatomy, and C. W. Andrews, Matsumoto and others have supported earlier authorities in regarding the orders Proboscidea, Hyracoidea and Sirenia as offshoots of an as yet otherwise undiscovered or unrecognized common stock. Such a common ancestral source may even be structurally represented among Eocene condylarths by *Hyopsodus* (fi 21.13D), with its gently procumbent fro teeth, primitive cheek teeth and short, sprea ing five-toed hands and feet (fig. 21.11D); b intermediate stages that would connect it wit the subungulates are wanting.

Ameghino and others assumed that at th base of the superorder Subungulata (Paenungu lata Simpson in part) there was a real relation ship between the hyracoids and some of th extinct South American ungulates, but renewe study of the evidence makes it seem more likel that both are to some extent similar but ind pendent offshoots of different protoungulates c condylarths, possibly of early Paleocene ag and wide northern distribution. Although th Sirenia may well have begun to diverge fro this ancestral stock in early Paleocene time their ultimate derivation from quadrupedal lan mammals is incontestably indicated by the re tention of a reduced pelvis and vestigial fem even in the modern sirenians.

FROM SWIFT-LIMBED COURSERS TO PONDEROUS GIANTS, AMONG THE HORSE RHINOCEROSES AND OTHERS (PERISSODACTYLA)

In the "Odd-toed" ungulates, including th tapirs, rhinoceroses, horses and their numero fossil allies, the symmetrical middle or thi digits and their metapodials (i.e. metacarpa or metatarsals) are enlarged so that the ma axis of weight passes through them. The earl est known perissodactyls of the Lower Eoce of Europe and North America had already lo the thumb in the hands (fig. 21.48B) and digi I and V in the hind-feet (fig. 21.49B). Bo hands and feet had lost the grasping power their remote five-toed ancestors and were begi ning to be rolled into cylindrical columns, res ing chiefly on the enlarged hoof of the midd digit and thus were unguligrade rather tha digitigrade. The metapodials of even the mo primitive known forms were already quite lo in proportion to their width, so that the wri and heel were raised high off the ground (fi 21.50). Thus the hands and feet of these ear perissodactyls were already functioning as stilts and the animal had to balance itself on top of four bent-lever springs (fig. 21.39A). The pectoral levers have nine successive segments and eight joints, the pelvic levers, ten segments and ten joints in each vertical series. Of course there were the same numbers of segments and joints in such primitive mammals as the opossum, but in the latter the spreading hands and feet gave much wider support and reduced the adverse leverages and the dangers of falling and dislocation. So great are the strengths of bones and ligaments that in the line leading from *Eohippus* to the horse the lengths of the middle metapodials steadily increased, reaching its climax, however, in the small, light-limbed *Neohipparion* (fig. 21.27B) and in the young foals of modern horses, in which the limbs are relatively longer than in the adults. But in the arger and heavier horses the width of the middle digits and metapodials increased more apidly than their lengths (fig. 21.48F).

The "cannon-bones" or middle metapodials of horses must resist vertical crushing strains as do the pillars of a temple, and transverse breaking strains as does a baseball bat when it trikes the ball. In the cursorial method of locomotion adopted by the early perissodactyls and perfected in the horse (fig. 21.39B), the hind eet in running really bat the ground backward, while the front feet pull it toward the animal. f the ground is not too slippery, its reaction ends the animal bounding forward.

Two important characters of the astragalus of cursorial perissodactyls, as compared with hat of cursorial artiodactyls (figs. 21.83B, 04F) are: (1) that in the former (fig. 21.49B) here is a wider contrast in general appearance etween the proximal and distal ends of the stragalus than there is in the artiodactyls (fig. 1.104F); for in the perissodactyl the pair of eels or carinae on the proximal end of the stragalus, as well as the groove between them, vere set obliquely to the general axis of the canon-bone in the front view, whereas in the rtiodactyls the carinae and grooves were usully more nearly in line with the long axis of the annon-bones; and (2) that in the perissodactyls the distal end of the astragalus is primitively narrower and rather asymmetrical, with a small cuboid facet and a wide navicular facet, whereas in the artiodactyls the distal end is usually as wide as the proximal end, the cuboid facet is well developed, and the whole bone tends to be symmetrical with its long axis in line with that of the metapodials. In secondarily graviportal perissodactyls, especially in the later titanotheres (fig. 21.55H), both ends of the astragalus widen and a good cuboid facet develops, but the distal end never becomes as pulley-like, or as much like the proximal end, as it does in the artiodactyls.

The increase in speed which characterized the evolution of most lines of the horse family was probably favored by a corresponding increase in the speed of wolves and other wild dogs that hunted them down (fig. 21.1); and their quickness in leaping to one side and in bucking must have been a considerable factor in escape after sudden attacks by lions and other cats. But just as every man has the defects of his qualities, so increased speed in the horses, which was dependent in part upon an increase in length of limbs, in turn required thicker muscles, increased energy-output, more food-fuel, bigger bellies, and so forth. Since cross-sections increase with the squares of linear dimensions, and mass with their cubes, there evidently comes a time in the evolution of any large-bodied race when the added protection gained from increased bulk is offset by the effort and expense in terms of food consumed. But such relations involve complex statistical problems with many fortuitously varying environmental factors.

In some orders of vertebrates, certain phyletic lines seem to have given rise to individuals of diminishing size, until dwarfed or minute forms have become normal. A fairly clear case of this principle is that of the sea-horses (*Hippocampus*) cited by Breder (1942). One may suspect that similar selective factors have produced the "least shrew," the weasels, which go into the burrows of rats and moles, and possibly the dik-diks and duikers among antelopes. Among the perissodactyls, dwarf forms (e.g. *Nanippus*) may have been derived from larger breeds and dwarfing may have produced the very small cursorial Eocene rhinoceros named *Triplopus* and the little Mongolian titanothere *Microtitan* (Granger and Gregory, 1943).

In certain families of perissodactyls the cursorial primitives of the Lower Eocene, gave rise to lines which, as their bulk increased (figs. 21.53, 54) widened their metapodials (fig. 21.55) and slowed down the increase of length in their limb bones. Among the titanotheres, the Lower Eocene *Lambdotherium* (fig. 21.56A) was a small cursorial form that was not very different from the ancestors of the horse (*Eohippus*) except for certain differences in the upper molar crowns. In the first real titanothere, *Eotitanops* (B), also of the Lower Eocene, the hands and feet were relatively wider (21.55). In later stages, as the body increased in size, the hands and feet became wider and relatively shorter, so that in the latest American titanotheres of the Lower Oligocene, the hands and feet were becoming secondarily graviportal, with short free digits and wide metapodials. Thus the direction of evolution in the titanotheres shifted from the cursorial to the secondarily graviportal type, as it did also among the later and heavier rhinoceroses. Among the latter, one Asiatic branch of the subfamily Paraceratheriinae of Osborn passed from the cursorial to the super-gigantic stage and ended in *Baluchitherium* of the Lower Oligocene of central Asia, the largest of all known land mammals. In this colossal beast (fig. 21.80B), which is estimated to have attained a height of 17 to 19 feet at the shoulders, the middle digit and metacarpal of the hand (A) continued to bear the chief weight and were markedly increased in width. The neck bones of the largest *Baluchitherium* were about three times as high as those of an adult white rhinoceros, and are comparable in size with the neck vertebrae of the larger dinosaurs. Its skull, which was about four and one half feet long, was small in proportion to the enormous weight. No other known proboscidean or ground sloth came anywhere near *Baluchitherium* in size and weight (Granger and Gregory, 1935, 1936), and it affords a clear example of the great strength of bones and ligaments in large, land-living mammals.

Another branch of rhinoceroses led to *Teleoceras,* with excessively short legs and huge thorax.

A quite different family were the amynodonts (fig. 21.76A) semi-aquatic forms with wide bodies, bent limbs, short wide muzzles and high-crowned crested molars.

In the chalicotheres (figs. 21.65, 66) the very large ungual phalanges were deeply cleft and curved upward, so that they suggested those of a gigantic carnivore with huge claws. Nevertheless these beasts belonged in the order Perissodactyla, as shown by the plans of their molar teeth, which resembled those of titanotheres and of the skull, which was somewhat horse-like. They represent a shift from the primitive cursorial to a long-limbed, mediportal, scratching and digging type somewhat analogous with the secondarily clawed notoungulate *Homalodotherium* (fig. 21.17B).

The adaptive branching of the dentition and skull of some typical perissodactyls was very briefly as follows:

Hyracotherium (*Eohippus,* fig. 21.42A) retained the primitive placental dental formula $\frac{3.\ 1.\ 4.\ 3}{3.\ 1.\ 4.\ 3}$ which was originally adapted for an insectivorous-omnivorous diet; but its nearer ancestors had no doubt gradually become herbivorous, although retaining small canines in the upper and lower jaws. Its first upper premolar already separated by a diastema from the second, was destined to be retained through the ages and to give rise to the variable "wolf tooth" in the modern horse. The second, third and fourth upper premolars of *Eohippus* (fig. 21.46bB) increased successively in width, but p[4] had only three main cusps, including a rounded single internal cusp, in contrast with the four main cusps and quadrate contour of the low-crowned molars. The protoconule and metaconule of the molars each bore a small crest directed outward and forward and lying between the two main outer and the two main inner cusps. In succeeding stages (*Orohippus* to *Equus*) the molar crowns increased in height and more slowly in anteroposterior length, the two outer cusps (*pa, me*) became large and U-shaped, forming an external fold or mesostyle (*ms*) between them. The protocone (*pr*) eventually became flattened (J) on the inner side. The two small oblique ridges on the proto- (*pl*) and metaconule (*ml*), becoming larger, developed into somewhat crescentic crests with wrinkled borders; they tended to enclose two valleys, the anterior and posterior "lakes," which eventually were filled with dental cement. In the lower molars the crowns increased their height but became long and narrow (fig. 21.47aH) the enlarged proto- (*prd*) and hypoconids (*hyd*) alternating with the enlarged and subdivided metaconid (*med*) and the single entoconid. The hypoconulid (*hld*) on m_3 remained prominent. The premolars increased in complexity, p^4, p^3, p^2, in the order named, but in area p^2 finally became the largest of the six grinding teeth on each side (figs. 21.46aJ). There were closely correlated changes in the lower premolars (fig. 21.47aH), the result being that the later horses (fig. 21.42D) carried a dental battery of twenty-four grinding teeth, all with high crowns.

In the completed deciduous dentition (fig. 21.45aA1) there are three molariform teeth on each side, above and below. In a foetus of the fifth to sixth month (A) the germs of the crowns of the milk teeth, especially the molars, are already growing and the last or fourth deciduous molar is forming in a bony pocket immediately beneath the eye. At birth (A1) the milk molars are all in place and the germs of the first permanent molars are under way, the upper one in a pocket under the eye (m1), the lower in a pocket in front of the ascending branch of the mandible. These teeth gradually increase in height and move downward and forward in the upper jaw, upward and forward in the lower.

The succeeding molar teeth (A3, A4) are formed in the same regions and then move respectively downward and forward in the upper jaw and upward and forward in the lower; the bony tissue is absorbed in front of them and reformed behind them. The roots long remain widely open to permit ample blood supply to the deepening crowns, but gradually they become long tapering as the crowns are ground down (A6). Meanwhile due to the forward growth of the jaws the true molars move forward, until in a horse 39 years old m^1, which began to form below the eye, lies far in front of it.

The great increase in height of the cheek teeth has caused a corresponding increase in the height and length of the maxillae and of the alveolar region of the lower jaw. The complex folding of the tooth crowns has greatly increased the areas of friction between the food and the grinding surface of the teeth and this has required a corresponding expansion of the angular or masseteric area toward the rear end of the lower jaw (figs. 21.42, 45). Again, simultaneous contact of the entire upper and lower tooth rows of one side, as in other cases already noted, is achieved by elevating the condyle of the jaw far above the level of the cheek teeth, while the expansion of the masseter has evoked a long horizontal ridge on the side of the bony face. The orbits in the horse family have as it were retreated upward and backward and a secondary postorbital process has been developed, coming down from the frontal and serving to brace the shortened, strong zygomatic process of the squamosal (fig. 21.42A–D). In the interior of the skull, the delicate scrolls of the ethmo-turbinals are very complex and extensive and are the basis for an acute olfactory sense. The lesser wings of the sphenoid (orbitosphenoids) are large and well developed and they assist in supporting the frontal roof, giving rise in rare cases to a pair of humps called "horns" (Chubb, 1934). The preorbital part of the face, especially in the three-toed horses, bears one or more deep depressions (fig. 21.43A), which are believed to have lodged lateral diverticula of the nasal passage (B1). In the modern horses and zebras these are reduced to vestiges (B) and the depressions have nearly or quite disappeared (Gregory, 1920).

In the titanotheres (fig. 21.53), the skull in *Lambdotherium* (fig. 21.56A1) was essentially like that of *Eohippus,* that is, with a relatively long face and slender jaws. As time went on, one branch of the titanotheres lengthened the nasal region (fig. 21.62D) and cranium and culminated in the Upper Eocene *Dolichorhinus* and allied genera. Another Eocene branch (*Manteoceras,* C) widened the skull and zygomatic arches and developed bony eminences above the orbits. In one end stage (*Brontotherium*) the huge bodies (fig. 21.53) became secondarily graviportal, the skull roof was curved upward, the bony "horns" developed into enormous Y-shaped, transversely flattened outgrowths (fig. 21.58bJ) which seem to have been covered by a callous leathery skin. The zygomatic arches were expanded laterally into huge bony swellings (I1), which probably often received the impact of a charging rival. The occiput was greatly widened to support the enormously thick neck muscles and the neural spines above the scapula grew upward to form a huge dorsal hump (fig. 21.54D), from which the ligaments of the neck radiated downward on to the cervical vertebrae and the occiput.

The upper molars of the titanotheres (figs. 21.57; 21.60) are called buno-selenodont because they have a large conical cusp (protocone) on the inner side and two large V's or crescents on the outer side. The proto- and metaconules (figs. 21.60A, B, *pl, ml*) were abortive and never gave rise to large cross-crests as they did in the horses. Similarly in the lower molars the crowns remained W-shaped (A1, C1). The molarization of the premolars was delayed (fig. 21.56B1) and never completed (fig. 21.60B).

The tapirs started in the Lower Eocene from small forms (*Homogalax,* fig. 21.72A) which were closely related to *Eohippus,* but soon developed the small crests on the proto- and metaconules into two large cross-crests (B–G) which were sheared by similar crests on the lower molars. P2–4 were molarized early.

The skull of tapirs is chiefly remarkable for the features associated with the proboscis (fig. 21.70A), namely, the retraction of the nasals and the development of two curved notches on either side of the nasals in which the cartilaginous proximal end of the proboscis fits. The muscles that operate the proboscis are located on or near the rim of bone in front of each orbit. The proximal end of the nasal chamber is divided into right and left halves by a cartilaginous or bony median partition analogous with the cartilaginous nasal septum of man.

In the early hornless Eocene rhinoceros *Hyrachyus* and in related genera the skull (fig. 21.74C) was about twice as long as that of *Eohippus,* with the orbit about in the middle, and a fairly large bony face. The incisors and canines were not specialized but the molars (fig. 21.77A) already had obliquely placed, flat outer crests (ectoloph) and two oblique cross-crests. The premolars were becoming molariform. The lower molars (B) also bore two narrow oblique cross-crests which sheared across the three crests of the uppers (C1). In subsequent stages of the typical rhinoceroses the crowns of one pair of upper incisors (i^1) became elongate compressed, with a median cutting ridge (fig. 21.79D), one pair of lower incisors (i_2) became enlarged and procumbent, their inner ends shearing across the blades of the upper pair. In both the "black" (*bicornis*) and the "white" (*simus*) rhinoceros of Africa the incisors are absent. The upper molars increased the height of the crowns and eventually developed small accessory folds (fig. 21.78 A–C) called crista, crotchet and antecrochet. In the Pleistocene *Elasmotherium* (D, D1) of Siberia the enamel edges of these main crests became elaborately crimped, thus increasing the grinding surface and converging superficially toward the molar patterns of later horses (G).

The earlier rhinoceroses were hornless (fig. 21.75A, B). The Oligocene and later diceratheres had a pair of small bony swellings on the slightly upturned ends of the nasals. The horns of recent rhinoceroses (fig. 21.81A–D) are composed of agglutinated hairs. The *Elasmotherium* (E) had a huge bony swelling on top of its head, which probably bore an enormous horny boss.

The rhinoceroses thus exhibit a branching of skeletal types which was more numerous and more widespread than that of the horses and palaeotheres; starting with small, moderately swift cursorial forms (fig. 21.75A), one dwarf branch ends in the very slender-limbed *Triplopus,* another in the semi-aquatic, almost hippopotamus-like amynodonts (fig. 21.76A); others end in the large upland rhinoceroses of Asia and Africa, while one branch culminated in the supergiant *Baluchitherium* (fig. 21.80).

In short, the unguligrade perissodactyls, branching from small and primitive cursorial forms, gave rise to: (a) extreme cursorial, (b) mediportal, (c) secondarily graviportal, (d) secondarily clawed and (e) semi-aquatic types; but arboreal, gliding, fossorial and aquatic forms are conspicuously absent.

DOUBLE-JOINTED ANKLES AND OTHER MARVELS AMONG THE SWINE, CAMELS, DEER AND OXEN (ARTIODACTYLA)

In the "even-toed" ungulates (fig. 21.82) the second and third metapodials and digits are subequal and the main axis of weight passes down between them. The astragalus has a double pulley, one at the proximal, one at the distal end, for articulation respectively with the fused tibia and fibula above and with the navicular and cuboid below. The ruminant artiodactyls (including deer, giraffes, antelopes, sheep, oxen, etc.) have the stomach divided into four parts, each with its special functions, which collectively make it possible for the animal to store away sufficient quantities of food and then to regurgitate parts of it and "chew the cud." In the pig this specialization is less advanced and these divisions are less sharply marked.

Many of the differences between artiodactyls and perissodactyls in the teeth, jaws, skulls and locomotor parts of the skeleton are associated with the foregoing outstanding ordinal differences.

The artiodactyl ankle-joint arose when an astragalus of primitive creodont type (fig. 21.83A), such as that of *Claenodon,* widened its distal end and developed a wide facet for the cuboid (p. 425), at the same time emphasizing the trochlear keels for the conjoined lower ends of the tibia and fibula and changing the transverse axes of the several bones to secure a double hinge (Schaeffer, 1947). This double ankle-joint (H–H3) permitted an extreme flexure of the ankle and an emphasis of the bent-lever action of the primitively long hind limbs. However, the hands and feet themselves (fig. 21.82), so far as known in the less specialized early Tertiary artiodactyls, were not as narrow as they were in the primitive Eocene perissodactyls *Eohippus* and *Heptodon,* and not nearly as long and narrow as they became in later artiodactyls (figs. 21.100, 104).

Some artiodactyl lines long retained the lateral metapodials and digits (II, V) and in the case of the semi-aquatic hippopotami (fig. 21.95bI, I1) all the metapodials were widened, as well as the wrist and ankle.

The ungual phalanges and hoofs on the second and third digits of artiodactyls usually look like mirror images of each other and are often in contact on either side of a narrow cleft (figs. 21.95, 100, 104). Taken together, they suggest the single undivided hoof of the horse and the adjectives used by Aristotle in describing the feet of animals, translated by early English writers as multifid, bifid, solidungulate, fissiped, seem to imply the idea that the "cloven foot" was somehow due to the partial cleavage of a solid foot.

Origin of the "Cloven Hoof."—As to the remote origin of the "cloven-hoofed" stage, in the feet of the very primitive Eocene condylarth *Hyopsodus* (fig. 21.11D) the third and fourth metapodials were subequal, as they continue to be in the Oligocene *Agriochoerus* (fig. 21.82A, A1) and others, which, however, show some indication of the mirror-image relationship between the ungual phalanges of these digits. The gradual coalescence of the metapodials in the ruminants (fig. 21.104) forced the terminal or ungual phalagnes together and made the outer side of the third digit face the inner surface of the fourth.

Adaptive Divergence of the Hands and Feet.—One early Eocene side branch, the anoplotheres, apparently became secondarily semiaquatic, spreading the second metapodial of the hand (fig. 21.82F) and foot (F1) widely apart from the third and fourth, apparently to support a web. At the other extreme, some of the giraffe-like camels (fig. 21.100D, D1) of Miocene and Pliocene times greatly lengthened the remaining metapodials and digits (III, IV) and developed soft pads beneath the phalanges, which enabled them to walk swiftly on sandy ground (fig. 21.101). The earlier tragulines (fig. 21.103), deer and antelopes were small and slender, able to make sudden leaps on their spring-like limbs and feet. But in some lines (e.g., entelodonts, fig. 21.89A), as the mass increased, the limb bones and extremities widened but without losing their essentially cursorial features. The surviving giraffes, in spite of increasing bulk, even surpassed the giraffe-camels noted above in retaining very long stilt-like limbs and a modified cursorial gait.

The contrast between artiodactyls and perissodactyls is conspicuous in the hands and feet and is maintained, but less sharply, in the fan-like scapula, in the femur, which usually lacks a third trochanter, and in the ilium, which even in very heavy-bodied forms does not expand laterally as much as it does in secondarily graviportal perissodactyls.

The Vertebral Column.—The variable appearance of the vertebral column in the side view in the artiodactyls is the result of the following factors and others:

1) If the hind limbs are much longer than the fore limbs, even if they are much bent at the knees and ankles, they will tend to raise the pelvis and acetabulum so high that the column will dip downward from the top of the lumbar arch, as in the small primitive cursorials *Caenotherium* (fig. 21.96), *Leptomeryx* (fig. 21.103) and *Blastomeryx*.

2) If the fore limbs are much longer than the hind limbs, the column will slope upward, as in the giraffe.

3) When the neural spines of the more anterior thoracic vertebrae increase rapidly in height, as in *Dinohyus* (fig. 21.89) and *Bison*, forming a high hump, they serve to support the ligaments of the neck and of the large or heavy skull.

The strongly arched vertebral column of certain primitive artiodactyls (fig. 21.96B) enables the animal to make sudden leaps. There are, however, normally only 19 thoraco-lumbar vertebrae, with typically four sacrals. While this limits the amount of curvature of the lumbar region, it also stiffens the bow-like action of the vertebral column in leaping.

PRIMITIVE EOCENE ARTIODACTYLS

The adaptive branching in this order is extremely extensive and only a few outstanding lines of the evolution of the skeleton may now be noted. The order was probably derived from some as yet undiscovered condylarth in Paleocene times, the chief ordinal character, the double-jointed astragalus, having arisen in the manner noted above (p. 433).

In the Lower Eocene, members of the artiodactyl dichobunids (figs. 21.84B-D, 87) and their relatives were already present in both Europe and North America; while in the Upper Eocene and Lower Oligocene, forerunners of the pigs, entelodonts, anoplotheres, anthracotheres, oreodonts, camels, tragulines and many extinct lines were differentiated. The earlier deers appeared in the Lower Miocene, as well as the forerunners of the giraffes. The antilocaprids or American antelopes have their forerunners in the Middle Miocene, as do also some of the true antelopes, but the great bulk of the known antelopes, oxen, sheep, goats, etc., date from the Pliocene and Pleistocene.

The oldest and most primitive Eocene artiodactyls belong to the suborder Suiformes (or Non-ruminantia) and to the infraorder Palaeodonta, including among others, the Lower to Upper Eocene Dichobunidae and allied families (Simpson, 1945).

In *Diacodexis,* which is referred to this group, the triangular upper molar crowns (fig. 21.84B) bore three large conical main cusps, besides small oblique crests on the proto- and metaconules and well developed external, anterior and postero-internal cingula. In *Dichobune* (D) and related genera, all three primary cusps were conical to V-shaped, the cingulum-hypocone was well developed and the proto- and meta-conules were more or less crescentic, the fourth upper premolars bicuspid (with one main outer and one lower inner cusp), the other premolars compressed. The canines, at least in some genera, were small and all the teeth, in both upper and lower jaws, from the first incisor to the third upper molar, were in a continuous row without interspaces (fig. 21.87B). The bony face and jaws were fairly long, the eyes moderately large, the postorbital process prominent but not closing the orbit posteriorly.

In the Middle Eocene *Homacodon* (fig. 21.84C) the first and second upper molars had six somewhat pyramidal cusps and the narrower four-cusped lower molars (C1) had the two outer cusps crescentic. The last upper premolar was bicuspid, the last lower premolar compressed, the third, second and first premolars simple, compressed-conical. The molars and premolars were brachyodont (with low crowns). *Homacodon* was a small animal, with a fairly large, well-convoluted brain; the orbits were rather large (fig. 21.87A) and said to be without postorbital processes—an unusual feature in an artiodactyl.

Emphasis of the upper canines and great elongation of the bony jaws gave rise to *Achaenodon* (fig. 21.88A) and the entelodonts or elotheres (B), which looked like gigantic pigs but they had a very highly spined dorsal hump (fig. 21.89) and narrow two-toed feet, like buffaloes. These died out in the Lower Miocene.

In the Cebochoerids (fig. 21.99A, B), derivatives of the dichobunoid stem, the four main cusps of the upper molars became crescentic and the angular region of the jaws very deep, to afford space for enlarged masseters, while the bony face was short, producing a superficial suggestion of the skulls of the spider-monkeys.

PIGS AND PECCARIES

The ancestors of the peccaries (fig. 21.89B) and pigs, which evolved from another branch of the dichobunoid stock, after giving rise to upper molars with two pairs of cones arranged transversely (fig. 21.91D), began to evolve subsidiary tubercles (E-G) on the surface of the antero-posteriorly elongate crowns, with extra cingulum-ridges and cusps on the growing rear end. This line culminated in the wart-hog (*Phacochoerus,* G), in which the third molars consist of clusters of tall parallel cylinders, somewhat analogous with those in the aberrant sirenian *Desmostylus* but much more numerous. The orbits in the wart-hogs (fig. 21.90C) have moved upward and backward almost to the rear upper corner of the skull, while the preorbital face is bent sharply down upon the braincase.

The wild pigs are distinguished for using their snouts and tusks as levers to turn up the ground. Accordingly their long bony snouts are braced appropriately and the distal end of the snout is strengthened with a large prenasal ossicle, which prevents collapse of their nasal passages. Their upper canine tusks, acquiring an outward curl, wear obliquely on the upwardly curved lowers (B). Finally in the babirusa of Celebes the maxillary alveolar pouches are folded completely over so that the upper tusk grows directly upward and then curves over, growing spirally downward. The lower tusks follow suit but at a distance.

HIPPOPOTAMI NOT SUINA

The hippopotami (fig. 21.93C) are usually classified with the pigs under the Suina, but Colbert (1935a) has adduced strong evidence that they are highly specialized, semiaquatic derivatives of the diversified family of anthracotheres (A, B), which flourished in the Oligocene, Miocene and Pliocene of North America, Europe and Asia. This was the leading family of the Infraorder Ancodonta. The anthracotheres were rather pig-like animals, with large incisors and canines, but their molar cusps were crescentic. In *Ancodus leptorhynchus* of this family the bony muzzle became very long and tapering (*Bothriodon*) (fig. 21.94A). The hippopotami, in becoming semiaquatic, have widened their feet (fig. 21.95I, I1), emphasized the canine tusks (fig. 21.94C) and changed their paired molar cusps (fig. 21.95F, F2) into opposed "trefoils" (converging toward those of early mastodonts). These are obviously habitus features, like the elevation of the nasal openings and eyes to the upper surface of the skull, but all the older heritage features tend to ally the hippopotami with the anthracotheres.

ANOPLOTHERES, CAINOTHERES

The family of anoplotheres, from the Upper Eocene of Europe, are related to the anthracotheres but were small animals with small canines and low-crowned teeth in a continuous series (fig. 21.92A). Their peculiar feet have been noted above (pp. 433, 434).

Another branch of the Ancodonta were the cainotheres (fig. 21.96) of the Upper Eocene and Oligocene of Europe, which, as noted above, were remarkably hare-like in general habitus features but differed in retaining basically dichobunoid skulls and dentitions.

"RUMINATING HOGS" (OREODONTS)

The oreodonts (figs. 21.97, 98), ranging from the Upper Eocene to the Middle Pliocene of North America, were once called ruminating hogs, because their molar crowns bore two pairs of crescents (fig. 21.99C, D) like those of ruminants, while their general appearance was somewhat more hog-like on account of their usually short legs, short necks and fairly large skull. But their extremities were much more primitive than those of hogs (fig. 21.95), retaining four separate metapodials and, in the hand, a small but perfect thumb (fig. 21.82A, B). In the *Promerycochoerus* line of oreodonts (figs. 21.97, 98), culminating in *"Pronomotherium,"* the nasals became quite short and the narial opening moved backward and upward, so that the skull looked somewhat tapir-like and there was evidently a proboscis.

In the closely related agriochoerids the terminal phalanges of the hands and feet were compressed and pointed and evidently equipped with sharp claw-like hoofs (fig. 21.82A, A1). The skull of *Agriochoerus* (fig. 21.98A) was of primitive artiodactyl type, except for the emphasis of the canines.

XIPHODONTS AND CAMELS

The camels, of the suborder Tylopoda, were preceded by the Xiphodonts from the Upper Eocene and Lower Oligocene of Europe, which resembled the camels in their very long slim limbs and functionally two-toed feet, compressed premolars and selenodont (crescent toothed) molars.

In the true camels (Camelidae) (beginning with the North American, Upper Eocene *Protylopus*) the molar crowns became higher as the animals ate tougher and more gritty herbage. In the narrow, two-toed feet (fig. 21.100) the long metapodials were closely appressed and ultimately fused, but externally the paired digits, each resting on its own pad, do not give the impression of a "divided hoof" as they do in the oxen. The earlier skulls (fig. 102A), with moderate changes, retained the basal Eocene anoplothere (fig. 21.92) and dichobunid (fig. 21.87) heritage. The diversified camels are divided into six subfamilies and many genera. They range from the small *Stenomylus* to the mighty *Titanotylopus*; but all had cushioned, distally plantigrade toes adapted to traversing the deserts, not for leaping and bounding among the rocks and forests.

TRAGULOIDS

The forerunners of the suborder Ruminantia are included in the infraorder Tragulina, ranging from the Upper Eocene to the present time. This suborder embraces three superfamilies, five families and about thirty-one genera (Simpson, 1945, p. 151), all extinct except the living water chevrotain (*Hyaemoschus*) of Africa and the chevrotains (*Tragulus*) of Asia. These small hornless forms more nearly resemble the Eocene dichobunids (fig. 21.87) and anoplotheres (fig. 21.92) than do the existing pigs (fig. 21.90) and peccaries (fig. 21.89B), which have large tusks and specialized skulls.

The molar patterns of the traguloids are completely selenodont, that is, with two crescents on the outer cusps (para- and metacones) and two on the inner; but the postero-internal cusp was not a hypocone (fig. 21.85B_2-D_2), as it appears to be, but an enlarged metaconule (*ml*), which takes the place of the vanished hypocone. Among the earlier Tragulina, at least one genus, *Archaeomeryx* (fig. 21.105A) from the Eocene of Mongolia, has retained three upper incisors on each side. These are lost in the recent tragulines (fig. 21.107) and in the Pecora (deer, etc.), in which they are replaced by the spatulate end of the palate supported by the flattened premaxillae. The small lower canine in this genus was procumbent (fig. 21.105A2) and in line with the three lower incisors, another prerequisite for the Pecora stage. *Archaeomeryx* was accordingly regarded by Matthew and Granger (1925) and by Colbert (1941) as the forerunner of the true Pecora.

The living tragulines chew the cud but their stomach is nearly but not quite as complex as in typical ruminants. As they are completely hornless, their only weapons are the enlarged, downwardly curved, almost sabre-like canines (fig. 21.107B, C). They have curved backs and long bent limbs adapted for quickly bounding away. The African water chevrotain, *Hyaemoschus,* has narrow but moderately well preserved side toes (II, V), which are reduced in *Tragulus* (fig. 21.104D, E).

THE WELL-HORNED PROTOCERATIDS

A very peculiar and aberrant family, doubtfully assigned to the Tragulina, were the **Protoceratidae** (fig. 21.106), ranging from the Lower Oligocene to the Lower Pliocene. Here hornlike swellings appear, at least in the males, apparently for the first time among artiodactyls. In *Protoceras celer* Marsh there are paired protuberances (A) from the sides of the maxillae, extending upward well above the level of the nasal channel; the nasals themselves end above and in front of the eyes. Just in front of the eyes there was a second pair of very small conical protuberances, while directly above and slightly behind the eyes, the third and largest pair were directed upward and outward. Apparently none of these protuberances bore either bony antlers or true horns, but they must have been covered with skin, like the similar protuberances in the giraffes. In *Syndyoceras* (B) of the Lower Miocene, the successor of *Protoceras,* the maxillary protuberances have grown upward, uniting at the base and then diverging outward and backward. The long supraorbital protuberances curve upward and inward. In *Synthetoceras* and *Prosynthetoceras* (C) from the Lower Pliocene, the supramaxillary pair are much longer, uniting in an antler-like, median beam until near the top, when they fork into two prongs. The large supraorbitals now diverge like the true horns of longhorned cattle.

The forefoot in *Protoceras* (fig. 21.100A) retained four separate metacarpals; in the rather camel-like hind feet (A1) two middle metatarsals were elongated but separate and those of digits II and V were slender splints without phalanges (Scott, 1940, p. 592). The navicular, cuboid and ectocuneiform bones of the tarsus remained separate (*idem*), whereas in typical ruminants they become ankylosed in adults and form a solid base for the hinge-like tarsus above and for the metatarsals below.

THE PECORA

The infraorder Pecora, containing the deer, giraffes, antelopes, oxen, etc., practically begins with the little hornless deer *Blastomeryx* and related forms (fig. 21.109A) of the family Cervidae from the Lower Miocene to the Middle Pliocene of North America, with a possible forerunner (*Eumeryx*) of the Lower Oligocene of Mongolia. A notable feature is the pair of sabre-like canine tusks, like those of the existing hornless musk-deer (*Moschus*) and the Chinese water deer (*Hydropotes*).

Aletomeryx, a primitive deer from the Lower Miocene of North America, had a pair of low bony "pedicles" projecting from the frontals above the orbits (fig. 21.109B), each supporting a small antler with several radially spreading tines (Frick, 1937). In another genus of this subfamily (*Cranioceras*), a long median pedicle curved upward from the top of the occiput, while above the orbits a pair of pedicles grew straight upward. Such long pedicles in the recent muntjak deer of Asia support simple forked antlers (E). In the more typical deer (fig. 21.108) the antlers, which are first developed as a simple spike, become progressively forked; finally in the moose and in the extinct Irish deer the antlers become palmated or webbed.

Development of Antlers.—The development of antlers in the individual is connected with the maturation of the gonads and, like the plumage and spurs of the cock, they are apparently a sex-linked male character. In the deer the warm thick skin known as the "velvet" first appears on the surface of a pair of tumor-like swellings on the frontal bones and as these grow outward, the "velvet" secretes the antler as its core, at the same time stimulating or drawing out a supporting pedicle from the frontal bone. Thus the antler-producing organs act somewhat like the large glands on the face in front of the orbit, but instead of secreting a waxy material, they deposit solid bone. And in order to do this, they must, like tooth germs, have a strong attraction for osteoblasts floating in the blood stream. During the period of outgrowth the antler-forming organ gives rise to folds at the base, called the burr, where the beam joins the pedicle. When the circulatory system stops supplying new material to the antler-organ, it begins to shrivel up and finally gets frayed and worn off, leaving the dense bony antler exposed.

The seasonal development of antlers precedes the fighting and rutting season and is no doubt correlated with seasonal activities of the thyroids, parathyroids, pituitary, adrenal, interstitial and other glands. Thus the antler-glands are seen to be dermal polyisomeres arranged in three pairs in *Protoceras* but limited to a single pair in Pecora (except *Tetracerus*).

The antler-glands are also special by-products of the annual endocrine growth and reproductive cycle, which also accelerates the shedding and renewal of hair and the development and replacement of teeth. The branching of the antlers either radially, as in the extinct polyclad deer (fig. 21.108I), or on one side or another of the beam, is obviously due to the branching of the antler-forming organ, the antler being merely the inert core. The presence of the antlers and their use as spears and levers impose new loads on the skull, which is specially braced either with new trabeculae or with dense bone wherever the strains are greatest. Thus the frontals, which carry the antlers, extend upward and backward. Meanwhile the parietals withdraw from the skull roof to form side walls for the frontals.

As the deer are mostly forest-living browsers (i.e., feeding more on bushes than on grass), with a moderately long and flexible neck, their muzzle is not as sharply bent down upon the base of the braincase as it is in some of the antelopes. Their molar teeth, with the usual four crescents in the upper (fig. 21.112) and two in the lower rows, have only moderately high crowns, giving a relatively large total area for grinding comparatively soft herbage.

ANCIENT AND MODERN GIRAFFES

The modern giraffes (figs. 21.110, 111) may be regarded as abortive and highly specialized side branches of the early deer, in which the horn-forming organ stimulated the outgrowth of pedicles from the frontal bone which became suturally separate from it. In the modern okapi (fig. 21.111B) of the forests of the Belgian Congo the antlers of the deer seem to be represented only by a small bony nubbin on top of the bony, skin-covered "horns," and in the living giraffes even this nubbin is absent. In some of the Miocene and Pliocene giraffes the rear pair of bony outgrowths (fig. 21.110B) attained immense size, taking the form of warped and partly branching horns (h1), but again no antlers were present. The typical giraffes, browsing on the leaves of fairly tall trees, retained large molar cusps with only moderately high crowns. Their necks became inordinately long and their limbs stilt-like, the spines of several of the more anterior dorsal vertebrae being elongated to support the enormous crane-like neck.

EVOLUTION OF TRUE HORNS IN PRONGHORNS, ANTELOPES, OXEN, SHEEP AND GOATS

Another cluster of side branches gave rise to the swarming American antelopes (Antilocaprinae) of the North American Miocene and Pliocene, of which the pronghorn (fig. 21.114E) is the sole survivor. Here the bony pedicles took many forms, some variously forked (fig. 21.113A, C), others twisted into a doubly helicoid column (E). But the epithelial layers of the formative skin gave rise to a deciduous horny covering, not to a bony antler (Frick, 1937). The molars of the antilocaprids had very high and narrow crowns, enabling them to browse upon the siliceous herbage.

The climax in the diversity of true horns rather than antlers occurred in the Old World antelopes, oxen, sheep and goats. Here also the bony outgrowth from the frontal (os cornu) is covered by concentric horny layers (fig. 21.115). Such horns may be vertically screw-shaped as in the kudu (fig. 21.116aG), very long, tapering and slightly recurved as in the gemsbok (*Oryx*, K), lyrate or sharply bent in the middle with upwardly turned tips as in the hartebeests (116bP), turned downward and then upward to form hooks as in the gnu (R), arching widely outward and then upward as in some of the oxen (B), or spirally wound outward as in the bighorn sheep (fig. 21.117C). Such horns often conform more or less to the formulae for the generation of solid curves either by the rotation of a plane figure around an axis or by the radial growth of vectors along a certain path (D'Arcy Thompson, 1942). Some horns resemble the shells of molluscs: thus the screw-shaped horn of the kudu suggests the younger part of the shell in *Vermetus,* the long tapering, curved cylinder of the oryx recalls the *Dentalium* shell, the horns of the bighorn sheep suggest certain helicoid ammonites.

CORRELATION OF HORNS AND SKULL

The skulls and other parts of the skeleton of the Pecora are appropriately modified, often in accordance with the way in which the horns are used. For example, in the hartebeest the facial part of the skull is bent sharply downward, practically at right angles upon the braincase, while the occipital condyles and atlas are modified so that the top of the horns, which are directed upward and backward in the standing pose, can be turned directly forward.

DOMINANCE OF THE ARTIODACTYL ORDER

Thus the Artiodactyla, endowed with a highly specialized and efficient stomach and an equally adaptable locomotor system, are and have long been the dominant ungulates especially of the northern continents and Africa. Simpson (1945, p. 35) gives the following numbers for the totals of fossil and recent genera regarded as valid:

Artiodactyla	419	Herbivorous placental mammals (ungulates, rodents, lagomorphs)	1505
Perissodactyla	158	Carnivora (except pinnipeds)	337
Notoungulata	105		
All other ungulates	171		
All ungulates	853		
Rodentia	619		
Lagomorpha	33		

Thus in number of genera the Artiodactyla are the most diversified of all ungulates, and among herbivorous placentals, they are outnumbered only by the rodents.

CROSS CORRELATIONS OF LOCOMOTOR AND DENTITIONAL TYPES AMONG UNGULATES

Within this great Cohort Ferungulata almost any given type of locomotor habitus as listed above may be found in combination with either of two or more quite different habitus patterns of the dentition. For example, generally similar skeletons of cursorial habitus among the older perissodactyls were combined with different habitus patterns of the dentition that were characteristic of different families, including Equidae (horses), Tapiridae (tapirs), Hyrachyidae. Hyracodontidae (hornless rhinoceroses). Again, some early perissodactyls and artiodactyls, although differing in certain skeletal features, were cursorial and not dissimilar in general appearance, but their dental patterns were widely different. In some cases, on the other hand, rather similar dental patterns may be seen in two forms of widely different locomotor habitus. For example, the cheek teeth of *Moropus* are rather similar to those of *Dolichorhinus,* yet the former is a chalicothere (a peculiar family of perissodactyls in which the hoofs have been transformed into huge claws), while the latter belongs to a quite different perissodactyl family (the titanotheres) with short hoofs and mediportal or graviportal habitus. Many examples of such convergent habitus and divergent heritage have been cited in the previous pages.

CHAPTER XXII

FISH-LIKE BY CONVERGENCE (CETACEA)

Contents

CHAPTER XXII

FISH-LIKE BY CONVERGENCE (CETACEA)

ARCHAIC WHALES (ARCHAEOCETI)

A "Vulgar Error" About Whales.—The fish-like habitus of modern whales is a sufficiently good disguise to have fooled the majority of mankind for ages, and that whales are fishes is one of the thousand or more "vulgar errors" which Sir Thomas Browne (1646) failed to refute. The first known step toward the recognition of the true status of whales was taken by Aristotle, who listed animals of the whale kind among warm-blooded vivipara. Linnaeus, who had a genius for looking beneath the surface of things, boldly put together his "Class Mammalia," animals with breasts, to include among others the order Cete, for the whales, porpoises, dolphins and their supposed allies, the sea cows.

The Earliest Forerunners.—The direct fossil ancestors of the various families of modern whales are unknown below the Lower Oligocene. The Archaeoceti, which were the earliest known members of the order, ranging from the Middle Eocene to the Upper Oligocene, were already highly specialized for aquatic life and most of them belong to extinct side branches. Nevertheless their skull structure was far less specialized than that in modern whales.

The typical archaeocetes are found in the Upper Eocene of Alabama and adjacent states, and in the Upper Eocene of Egypt; they include the so-called *Basilosaurus* ("*Zeuglodon*"), which in its general form (fig. 22.1A) is one of the most amazing of all vertebrates. The "*Zeuglodon*" (yoke-tooth) was discovered in Louisiana in 1832, and was made famous in 1845 by an enterprising German collector, Dr. Albert Koch. He put together the fossil remains which he had dug up into a mounted skeleton called Koch's *Hydrarchos* (water chief) and exhibited it in many parts of Europe and North America. The most critical and conscientious studies of Johannes Müller (1849) and of Remington Kellogg (1928) have shown that apart from minor errors in Koch's mounted skeleton, the principal one was that he had put into it the remains of the vertebral columns of two animals, claiming a total length of 114 feet, thus producing a "sea serpent" of amazing length. But even if Koch's animal had been cut down by nearly half its length, there still remained an unassailable length of fifty-five feet, as shown in the United States National Museum skeleton, which was most accurately mounted under the direction of the late Dr. J. W. Gidley (1913).

The greater part of the length of this animal (fig. 22.1A) is due to two factors: (1) the increase in number of the centra behind the ribs (as compared with any relatively primitive land mammal), and (2) the increasing length of the individual centra from the mid-dorsals caudally, except near the end of the tail. These numerous centra were evidently secondary polyisomeres and they gave great flexibility to the vertebral column of these very swift, somewhat moray-like predators. However, according to Kellogg (1936), in the genus *Zygorhiza* (fig. 22.1B), also from the Upper Eocene, the homologous centra were but little elongated and much less specialized; but the skull of this genus, apart from size differences, is practically indistinguishable from that of *Basilosaurus*.

The extreme specialization of what is possibly the male zeuglodont or *Basilosaurus* seems to have influenced Gidley and Miller to treat the Archaeoceti as if they were a hopelessly specialized group which could all be safely eliminated from the ancestry of the true whales and whose whale-like features might be all set aside as due to parallelism or convergence. But after long study, the evidence still seems to me to support the view of C. W. Andrews (1906) and O. Abel (1902), who pointed out essential features in which the skull of *Zeuglodon* (fig. 22.2B) foreshadowed those of the odontocetes (toothed whales). Probably the oldest and seemingly the most primitive of all known archaeocetes is represented by portions of two lower jaws described by C. W. Andrews (1920) from a clay formation in Southern Nigeria, of possibly Lower Eocene age. In these jaws, named *Pappocetus lugardi* by Andrews, the cheek teeth retained certain characters that recall those of early land-living carnivores (Kellogg, 1928, pp. 38, 39).

The next oldest and most primitive known representative of the archaeocetes is *Protocetus* Fraas from the Middle Eocene of Egypt. It is known only from the skull but it was no doubt already a whale, at least in the making, and is rightly assigned to the suborder Archaeoceti. Its numerous whale-like or dolphin-like features include: its long pointed rostrum, the dorsal position of its nostrils, the general character of its dentition, the inflation and involution of its tympanic bulla, the breadth of the occipital condyles, the rearward prolongation of the bony palate, the sharp projection of the supraorbital flange of the frontals.

Among the more primitive pre-cetacean characters of *Protocetus* are the dental formula (? I^3 C^1 Pm^4 M^3), the differentiation of the cheek teeth into premolars and molars. The upper premolars and molars somewhat suggest those of hyaenodonts in their compressed, more or less shearing para- and metacones, but it had advanced beyond the hyaenodonts in the reduction of the internal bud or protocone of the premolars and molars and in the remodelling of the outer side of their crowns into a single large triangle with the paracone as its tip. All these features as well as the inflation of the bulla would be readily derived from the conditions known in *Hyaenodon*.

Hence Fraas did not hesitate to conclude that *Protocetus* was intermediate between the creodonts and the cetaceans, and despite subsequent objections relating to the archaeocetes which will presently be considered, this conclusion still seems to be morphologically correct, even though the actual splitting off of the archaeocete stock from land-living creodonts may well have occurred in an earlier horizon than that of *Protocetus*. On the whole the archaeocetes were remarkably uniform in skull and dental structure, especially in view of the astonishingly great difference in the vertebrae between *Zeuglodon* (*Basilosaurus*) and *Zygorhiza* (fig. 22.1B), the latter being far less unlike those of typical whales.

Kellogg's invaluable tables of the geological and geographical distribution of whales (1928, pp. 32–36) indicate the oldest archaeocetes of the family Protocetidae possibly in the Lower Eocene and certainly in the Middle and Upper Eocene of Africa; these were followed by the Dorudontidae (including *Zygorhiza* and others) in the Upper Eocene of North America and in the Lower Oligocene of Europe and New Zealand. *Basilosaurus* (assigned to its own family) appears in the Middle Eocene of Africa and persisted in the Upper Eocene of Africa, North America and Europe.

TOOTHED WHALES (ODONTOCETES)

In the oldest supposed odontocetes (of the family Agorophiidae) the skull is more or less intermediate between the archaeocete and odontocete stages; their age is Upper Eocene. Then come the squalodonts (fig. 22.2D), very rare in the Oligocene but abundant in the Lower Miocene. In the latter period there were long-snouted dolphins and ziphiids. In the Middle and Upper Miocene marine formations these families were very numerous. The Pliocene marks the climax of the dolphins and sperm-whales (physeterids). The whalebone whales (Mysticeti) as such are unknown before the Middle Oligocene; among them the cetotheres are abundant in the Middle Miocene, the rorquals and balaenids in the Pliocene. Thus so far as the fossil record shows, the archaeocetes were evolved first; then came the first odontocetes and mysticetes, but the expansion of the odontocetes preceded that of the mysticetes.

Adaptive Branching of the Teeth.—Apart from the Upper Eocene agorophiids, which some authors include in the archaeocetes, the oldest of the odontocetes were the squalodonts (fig. 22.2D) mainly of Lower Miocene age, with a forerunner (*Microcetus*) in the Upper Oligocene. The teeth of squalodonts (fig. 22.4C), like those of other families of odontocetes, underwent a period of rapid multiplication, the total number rising from 56 in *Prosqualodon* eventually to 180± in *Eudelphinus*. Meanwhile these teeth lost all trace of derivation not only from a more remote tritubercular ground plan but even from a nearer squalodont pattern (fig. 22.4C), and they became simple pegs, often with conical tips (D) set in long rows on the elongate fish-catching jaws; thus they converged toward the teeth of fish-catching ichthyosaurs, plesiosaurs, mesosuchians, etc. In certain lines of dolphins the base and root of each of the cheek teeth became much compressed transversely but elongate anteriorly (L), while the tip remained pointed, so that the teeth consisted of a series of "battle axes" with their bases set lengthwise or obliquely in the jaws. Traces of this condition are seen in the sperm whales (G) and in some ziphioid whales; in some of the latter the teeth are reduced to a single pair, sometimes erect and tusk-like in the middle of the jaw (J). In another direction the true dolphins, originally with many teeth (E), give rise on the one hand to the strong-jawed killers (F, F1), on the other to the single spirally wound left tusk of the male narwhal (K); the male also retains another but vestigial tusk on the right side, while the female may have two atrophied tusks.

Skull Structure.—In the oldest known toothed whales of several families the rostrum is very long (fig. 22.5C). In such cases the long facial portion of the skull is strongly articulated with and based upon the equally strong braincase by a system of anteroposterior and transverse arches and braces, the latter including the backwardly extended processes of the premaxillae and maxillae. The entire skull indeed is composed of a system of structural opposites or counterparts that maintain the equilibrium of forces around the junction between the skull and the neck, but the primitive symmetry of the system (fig. 22.5D-G) is disturbed in the more specialized forms by the twisting of the bony supports of the respiratory sacks or nostrils and their associated parts (figs. 22.5, 6).

Inspection of Kellogg's text and illustrations (1928) shows that a secondary shortening of the rostrum with widening across its base has occurred independently among several lines of odontocetes; e.g., among squalodonts, in *Prosqualodon*; among the true porpoises (Delphinidae), in *Delphinodon, Globiocephalus, Monodon* (narwhal) and *"Orca"* (*Orcinus*), the killer whale; among the ziphioids especially in the bottle-nosed whale (*Hyperoödon*); this transformation is very marked in the Physeteridae (fig. 22.5), culminating in the sperm whale (*Physeter,* F) and the pygmy sperm whale (*Kogia,* G).

The Sperm Whale and Its Colossal Nose.—The sperm whale is so named because a large one may yield 80 barrels, or about three tons of sperm oil. A large part of this yield comes from the interior of a huge rounded protuberance at the front end of the head called the case (figs. 22.6A-A2), which may be about twenty feet long (fig. 22.7F). Whalers have given special names to some of the gross parts of this complex, anatomists (e.g., Pouchet and Beauregard, 1885; Schulte, 1918) have given accurate descriptions of its many parts; but it remained for H. C. Raven, partly as a result of his dissection of a baby sperm whale, partly as a result of E. Huber's (1934) experiments upon the blowhole region of living dolphins, to work out a theory as to the way in which the parts worked in the sperm whale, and for Gregory and Raven (1933) to recognize the whole thing as a supergigantic transformed nose (figs. 22.6, 22.7), a nose that has lost its smelling power, but retained and specialized its breathing functions, developing new and complex accessories and becoming by far the greatest known nose of all time. Moreover, the spermaceti oil in the interior of the "case" is doubtless part of the whale's store of extra fuel or food material which is accumulated when and where food is abundant and drawn off in times of scarcity (cf. Hjort, J., 1937, p. 25). Thus the spermaceti organ in the interior of the case has also acquired some of the function of the camel's hump.

The titanic nose of the sperm whale rests upon the top of the bony rostrum formed by the premaxillae and maxillae but it is so immense that the rostrum practically forms only a stiff floor and keel for it. It evidently begins to develop early in the foetus and the ascending processes of the maxillae grow around its convex base and are therefore concave dorsally. Similarly at its rear end it pushes the bones of the forehead backward and moulds them around itself into a concave basin with a high rim (fig. 22.5F). At the bottom of this nasal basin are two circular openings, like drain pipes, the left one small, the right several times larger (fig. 22.6A, A3). These lead down to the glottis, or upper end of the windpipe, which can be inserted into their lower ends just behind the long narrow bony palate.

The Spermaceti Organ.—The spermaceti organ itself is a huge hollow cylindrical bag (fig. 22.6A1) with its interior filled with waxy spermaceti oil and its surface surrounded with strong circular and longitudinal muscles. Its long axis lies above the rostrum (A), resting on the thick oily tissue called the "junk." The posterior end of the cylindrical organ is produced into two plugs, a large one which can be pushed into the left narial tube, and a small one for the small right narial tube (A5). These bony tubes (fig. 22.5F) are connected anteriorly with two main membranous tubes, the left and right nasal passages, which in turn have accessory pouches or diverticula (fig. 22.6A4). The left nasal passage (fig. 22.6A) passes from the bony narial tube obliquely upward and forward around the outer side of the spermaceti organ but beneath the blubber, into a bag which lies near the top of the nose, beneath the blowhole. The right nasal passage, immediately after it leaves the small right bony narial tube, gives off the enormous frontal sack (fig. 22.6A, A3), expanded transversely but thin anteroposteriorly, which covers the forehead and forms a thin layer between the forehead and the rear end of the spermaceti organ. After giving off this frontal sack, the right nasal tube passes forward underneath the spermaceti organ and then, turning upward, opens by a wide tightlipped valve (called by sailors the "monkey's muzzle" or *museau du singe* (A4)) into the "distal sack" beneath the blowhole. These two nasal passages plus the sack beneath the blowhole correspond as a whole to the naso-pharyngeal passage of the human nose and the spermaceti organ is a new organ which has developed deep in the ethmoid region of the nose. Gregory and Raven (1933) concluded that whatever other functions it might have, its muscular coats controlled the plugs that fit into the right and left bony narial tubes and therefore would have to do with the regulation of air pressure in diving. The highly complex adjustments to pressure during prolonged submergence in order to escape suffocation and to avoid "diver's sickness" through excess of nitrogen are ably summarized by Kellogg (1928, pp. 672–675).

Thus equipped the sperm whale can dive to great depths, perhaps in pursuit of the swift squids upon which it feeds. Occasionally it encounters giant squids (*Archaeteuthis*) with immense tentacles and sucking discs like saucers, which leave their imprints on the blubber skin of the sperm whale's nose (Murray and Hjort, 1912). In some sperm whales the mandible (figs. 22.1D; 22.4G) may be upward of twenty feet long. It carries a row of rounded teeth, anteroposteriorly oval in cross section, set in deep sockets. No teeth are present in the upper jaw.

The locomotor skeleton (fig. 22.1D) of the sperm whale is fundamentally the same as that of other toothed whales (fig. 22.1C).

In the pygmy sperm whale (*Kogia breviceps*) the rostrum has been greatly shortened and the cranium widened, so that the spermaceti organ (which is essentially the same as in *Physeter*) is contained in its rear by a high amphitheatre formed chiefly by expanded flanges of the premaxillae and maxillae (fig. 22.5G). Less developed stages of the entire nose are seen in the narwhal (fig. 22.6B) and in ordinary dolphins.

WHALEBONE WHALES (MYSTICETI)

Aquatic Adaptations of the Greatest Living Animals.—A man on a beach, viewing a stranded blue or sulphur-bottom whale (fig. 22.8A) at close range, might well wonder how such a colossal mass could be pushed through the water headfirst. But during the long ages in which successive living whale models were being tried out, natural selection gradually modelled the mass into an amazingly fish-like form, with well rounded, bluntly pointed head, the body tapering toward the rear and suddenly expanding into the flukes.

The very large size of the head (fig. 22.9) as a whole is chiefly due to the enormous size of the mouth and throat in which the whale engulfs its food, consisting of schools of herring and millions of shrimp-like copepods. If this huge head were attached to a slender neck of normal proportions, the adverse leverage against the thorax would be irresistible, while the concavity on either side between the head and the thorax would set up seriously retarding whirlpools. Efficient streamlining has been attained by greatly shortening the neck, widening the occipital region, filling the space between head and thorax with muscles running from the thorax to the occiput and from the shoulder-girdle to the sides of the head. At the same time the telescoped and coalesced cervical vertebrae form a strong pivot upon which the head may move in front and the rest of the body behind (fig. 22.9A).

The rounded wedge (fig. 22.10A) formed by the curved bony rostrum and the converging halves of the lower jaw, somewhat as in a fish, forms the prow, which is driven forward by thrusts from the rear delivered mostly through the vertebral column (fig. 22.9A). The latter serves somewhat the same function as the keel of a ship but instead of being an almost rigid rod on the lower side it lies along the back and is divided by the individual vertebrae into movable segments. The ribs serve much the same purpose as do their namesakes in a ship but again they are movably articulated with the segmented axis (fig. 22.9A). Beneath the concavity of the ribs, next to the backbone, on each side is housed a great cellular mass, the lungs (fig. 22.12A3); these are represented in a ship partly by the ventilating system and in a submarine also by the air tanks. The very large main pump (heart) is located on the underside, near the middle of the ribbed section (A3). The body cavity behind and beneath the ribs contains the complex stomach and intestine. These are collectively analogous in part with the power plant of a ship. The intestine is essentially a very long, much coiled cylinder, the inner walls of which digest the raw fuel material, transforming it into diverse fluid products: some of the latter, bearing oxygen-hungry haemoglobin, are carried through the circulatory tubes to the oxygen tanks (lungs); others, laden with fuel-food, are carried to the motors (muscles).

Since a continuous wheel-like revolution of one part upon another would be stopped by the twisting of the attached ligaments, the whaleship uses for propulsion a basically simple up-and-down and oblique sculling system. The motors consist of vast multitudes of contractile muscle fibres arranged in opposing sets above and below and on opposite sides of the segmented axis. The dorsal part of the axial musculature is continued forward above the backbone (fig. 22.12A2) and is fastened to a huge bony plate borne on the back of the occiput (fig. 22.10A).

This occipital plate is inclined forward and upward to the top of the skull behind the blowhole (figs. 22.9A1; 22.3D). The word "telescoping" was applied by G. S. Miller (1923) to this overlapping on the top of the skull, as a result of which the supraoccipital ultimately creeps over the parietals and frontals and in extreme cases (fig. 22.9) almost or quite gains contact with the nasals, premaxillae and maxillae. Meanwhile the opposite parietals (*pa*) are squeezed out on either side so that they no longer meet on top of the skull; but progressive and opposite overlapping would perhaps better describe the evolution of this condition. Miller (1923, p. 76) attributes the "telescoping" to the squeezing effect of water pressure from in front and the thrusts of the moving body from the rear; doubtless, however, there was little if any slipping or sliding but merely slow overlap and overgrowth of bones in response to enlarging nuchal muscles and mouth cavity in baleen whales, or to nuchal muscles and backwardly shifting nasal complex in odontocetes.

Each muscle fibre has its own nerve-wire and these wires, compounded into increasing twigs, branches and trunks, finally reach the control system (spinal cord and brain), where their activities are regulated by a corresponding ascending series which represents the numerous recording instruments in a modern ship and is known collectively as the sensory system.

All this assemblage, of infinite intricacy in detail but thoroughly mammalian in general plan, is enclosed and held together by an extremely thick and strong streamlined integument, commonly known as blubber, probably because it blobs or oozes oil. On either side of the caudal vertebrae the integument is built out into two huge and fairly stiff flukes (fig. 22.8), which serve mainly for propulsion but probably to a certain extent for steering, according to the selective actions of the paired dorsal and ventral sets of muscles surrounding the caudal segment of the vertebral column (fig. 22.12A2). The pectoral paddles (fig. 22.8) can be turned in different planes and, being near or at the widest part of the body, are doubtless important for steering.

The pelvic limbs of the ancestral land mammals have disappeared entirely from the surface but a vestigial tibia, femur and pubi-ischiadic rod persist on each side (fig. 22.9A). Equally noteworthy is the fact that since the pelvis is no longer used to support the backbone, the ilia have disappeared, the normal attachment of the ilium at the sacrum is lost and the sacral region has been desacralized, the transition from the lumbar to the caudal centra being quite gradual (fig. 22.9). All this adds considerably to the functional length of the tail.

The Right Whale's Food Filter System.—The right whale (fig. 22.9A), like all its fellows of the whalebone order (Mysticeti) has the whole upper facial portion of the skull curved upward into a strong median arch from which depends on each side a series of numerous transversely widened, anteroposteriorly flattened sheets of "whalebone," or baleen (fig. 22.11). It took the genius of a Darwin to recognize that these must have been derived by extreme emphasis from the transverse palatal ridges of ordinary mammals. They are made of horny, agglutinated hair-like elements which fray out at the lower end to form collectively a sieve for the minute food organisms (copepods, etc.); these are taken into the cavernous mouth and swept down by the enormous muscular tongue (fig. 22.11A).

Transitory Teeth.—In an embryo of *Balaenoptera musculus,* 123 cm. long, Kükenthal (1890, Taf. XXV) figures a row of 53 little conical tooth crowns along the premaxillo-maxillary alveolar tract, possibly corresponding to the deciduous dentition of normal mammals. A few of the peg-like crown tips were double. According to Weber (1928, p. 378), these vestigial teeth were later absorbed *in utero* before the foetus had reached half its full length. All this suggests that the remote ancestors of the mysticetes were fish-catching odontocetes with elongate jaws and numerous teeth, and that as the body size became gigantic, the animal came to engulf schools of fish and crustaceans rather than individuals; also that during the transition stages the tongue increased in functional importance *pari passu* with the horny palatal plates.

The upper jawbones (premaxillae, maxillae), being curved up into the arch that supports the "whalebone sieve," and the lower jaws being curved outward beyond the sieve, there can be no shutting of the upper and lower bony jaws. In a large whale the length of the lower jaw may be thirteen feet. Consequently when the mouth is open, the adverse leverage at the front end of the jaw from the pressure of the water as the great mass moves forward must be immense. It is not surprising therefore that the zygomatic portion of the squamosal should be very strongly built (fig. 22.9B) or that the temporal muscle should fill the temporal fossa above and behind the orbit and be inserted by a thick tendon into the coronoid process (Carte and Macalister, 1869, p. 222). The main part of the masseter, although limited to the rear end of the cheek, was short and thick (fig. 22.12A *mass.*). On the other hand, when the mouth was closed the great power necessary to depress the mandible was supplied partly by the large "depressor mandibulae" (or posterior belly of the digastric, fig. 22.12A2), partly by the very long sterno-mandibularis (A1).

With the relative immobility of the lips is associated the reduction of the buccinator and orbicularis oris muscles.

Function of the Submaxillary Pouch.—Between the distended sides of the mandible was the huge "submaxillary pouch." Carte and Macalister (1869, p. 204) note that in the piked whale (*Balaenoptera acuto-rostrata,* fig. 22.8B) "the mouth and fauces were lined by mucous membrane . . ., which was arranged in longitudinal folds similar to those on the geniothoracic region of the animal; this membrane, together with the mylohyoid muscle and skin, formed the wall of the great submaxillary pouch [cf. fig. 22.12A1, *cavum ventrale*], which from the elastic nature of its component parts was capable of great distention; this enabled the animal to take into its mouth a large quantity of water containing such substances as constituted its food; the latter was retained in the mouth by the straining mechanisms of the baleen plates, while the superfluous water was expelled by the contraction of the pouch." Doubtless the huge tongue also took part in this sweeping action as the minute crustaceans were swept off the baleen plates. The skin on the under side of the throat pouch bears a system of longitudinal grooves which extend in diminishing transverse rows far backward along the ventral surface (fig. 22.8A). Carte and Macalister (*ibid.,* p. 203) note that "the subcutaneous tissue [of these folds] being highly elastic, rendered the skin in this region very distensible." It would not be surprising if these whales, after closing the mouth, could force the water in the throat pouch and the food in it up into the mouth cavity and ingest the food at leisure.

Sense Organs, Skull and Brain.—As a result of extreme aquatic specialization, the main sense organs of a whalebone whale, inherited originally from primitive land-living mammals, have undergone profound modifications, which are clearly set forth by Kellogg (1928, pp. 191–199). The olfactory bulb of the brain in the little piked whale although small is still present, as well as three very small ethmoturbinal plates (Weber, 1928, p. 372); but the maxilloturbinal system, which is largely developed in all normal mammals, is greatly reduced and the forepart of the nasal chamber forms a large tubular bony anterior nares leading to the pouch beneath the blowhole. The long right and left narial tunnels (fig. 22.10B) lead obliquely downward and backward to the posterior narial openings at the rear borders of the palatine and pterygoid bones (B1); into these fits an intranarial prolongation of the larynx. Thus the inspired air passes directly into the trachea and to the lungs, instead of being discharged first into the back part of the pharynx.

The olfactory sense being reduced, one might expect the eyes to be remarkably large, as they were in the ichthyosaurs and giant squids; but they are relatively small, at least in proportion to the huge size of the skull. However, they have large and well adapted lenses (Kellogg, *ibid.,* p. 195). The frontal roof of each orbit in whalebone whales (fig. 22.10A) is produced laterally into a long supraorbital (*s.orb.fr.*) half-funnel covering the optic nerves and eye muscles. Especially in the right whale (fig. 22.9A) this tube is prolonged outward and downward, so that the eye lies far below the top of the head at the angle of the mouth. With the eyes in this position, near or at the widest part of the body, the whale may be able to sense the approach of other large objects from either side and at any angle to his own line of advance. Thus he may see the shadows of schools of small fish or his fellows who are travelling with him, or the reflection of the submerged portion of an iceberg, or the swift rush of an attacking *Orca*.

Whales have inherited from primitive land mammals the elaborate apparatus of the outer, middle and inner ear but have made more extensive modifications in it than have less thoroughly aquatic mammals. Even the seals began to inflate the tympanic bone but the whales have gone much further in the same direction, fusing the massive periotic and tympanic, closing the external ear tube, immobilizing the chain of ear ossicles (malleus, incus, stapes) and apparently receiving water-borne vibrations through bone conduction to the inner ear (cf. Kellogg, 1928, pp. 202–206).

The cutaceous sensory system of whales is very highly developed and with it are those parts of the cortex which receive and coördinate cutaneous sensations with motor response. The cortex of whales is indeed greatly expanded and convoluted, as is also the cerebellum. These general features are reflected to some extent in the global form of the braincase.

Reproduction, Growth and Migration.—Some of the most amazing of all features of whales relate to their modes of mating, reproduction, growth and migration. Mating in the manner of terrestrial quadrupeds being impossible, an intromittant organ of great length has developed in the male, operated by the huge ischio-cavernosus muscle, which is tied posteriorly to the pubi-ischiadic rods. According to Johan Hjort (1937, p. 22), the period of gestation in the sulphur bottom or blue whale is seven months and the newborn babe has a length of 21 feet and weighs two tons, or as much as 20 men. The animals consume an enormous amount of food and grow fast, putting on 200 pounds in 24 hours. In males, sexual maturity is attained at 22.6 months, in females at 23.7 months.

In regard to migration, the great whales of both orders annually travel over long distances between the cold waters of north and south seas, often following regular routes, timed with the periods of greatest food supplies in given regions.

"If a blue whale puts on 10 knots speed," Hjort (*ibid.*, p. 22) writes, "it develops about 47 horsepowers. For hours on end I have watched my friends the whalers trying unsuccessfully to overtake a blue whale in the Antarctic waters. Hour after hour the impatient gunner would call for full speed, but though he could steam at 13 or 14 knots he often had to give up, while I felt a thoroughly unbusinesslike admiration for the powers of the whale and marveled at the wonderful machinery that nature has constructed out of flesh and blood."

Almost any modern whalebone whale would conform in a general way to the foregoing very brief outline; among the widely ranging differences between the genera and species we may note: (1) size (from about 14 feet in the pygmy piked-whale, *Balaenoptera acuto-rostrata,* to 90 feet or more in the blue whale, *Sibbaldus*); (2) length of longest whalebone plates from about 12–18 inches in the California gray whale (*Rhachianectes*) (True, 1904, p. 290) to a maximum of 10–12 feet in the Greenland, Arctic or bowhead whale (*Balaena mysticetus*) (Encycl. Brit. 14 Ed., article on Cetacea, vol 5, p. 168).

Ancestry of the Mysticetes.—*Cetotherium,* the Miocene ancestor of the modern whalebone whales, was somewhat less specialized than the latter in its skull (fig. 22.3D) in that the supraoccipital projection had not yet made contact with the ascending processus of the premaxillae and maxillae and its baleen plates were probably short; it is, however, much less specialized than any known dolphin or toothed whale in the failure of its maxillae to overspread the supraorbital plates of the frontal (D, E). The skull of *Patriocetus* from the Upper Oligocene of Austria, according to Abel, was at least structurally ancestral to that of the cetotheres but it retained functional teeth. According to Abel (1913), *Patriocetus* affords a connecting link with the basal odontocete. Kellogg, however, feels cautious about the possibility of a common origin of the two orders and Gerritt S. Miller opposed it chiefly for the following reasons: (1) in the mysticetes the greater part of the maxilla ends posteriorly in a process in front of the orbit; its ascending process is narrow and does not expand transversely so as to cover the roof of the supraorbital process of the frontal (cf. fig. 22.3D; 22.10A), whereas in the toothed whales (Odontoceti) the dorsal or ascending process of the maxilla (fig. 22.5) is hugely developed, extending backward and upward and expanding laterally so as to form a secondary roof covering the frontals; (2) in the telescoping of the bones of the skull roof among the mysticetes (fig. 22.9A1) the ascending processes of the maxillae and premaxillae may both gain narrow contacts with the supraoccipital, which has been expanded and pushed forward until it forms both the roof and the nuchal surface of most of the braincase; whereas in the odontocetes the nuchal plate of the supraoccipital expands transversely but remains more nearly above the occipital condyles (fig. 22.5).

In the mysticetes the small size of an ascending frontal process of the maxilla and the presence of its normal ventral process are both connected with the modification of the rostrum for carrying the baleen plates. In the odontocetes the lack of a ventral process and the great overdevelopment of the ascending flange of the maxilla are both connected with the profound transformation of the nasal complex as described above. And it seems indeed that the skulls of the archaeocetes (fig. 22.5B) form an ideal structural ground plan for the divergent paths leading perhaps through *Archaeodelphis* (fig. 22.3C) on the one hand to the mysticetes (D) and on the other to the odontocetes (E).

PART FIVE

(*Chapters XXIII, XXIV*)

THE DEVIOUS PATHS TO MAN

CHAPTER XXIII

ORIGIN, RISE AND DEPLOYMENT OF THE PRIMATES

Contents

CHAPTER XXIII

ORIGIN, RISE AND DEPLOYMENT OF THE PRIMATES

EVOLUTION AND THE PRIMATES

FROM A PALAEONTOLOGIST'S viewpoint, one of the greatest discoveries ever made by man was that the earth is not the center of the solar system, for this discovery tended to undermine the entrenched philosophy of western European civilization, which was that the earth and the heavens were created for man. The second discovery was formulated by Linnaeus (1758), who recognized and named the class Mammalia and classified man as a Mammal and a member of an order to which he gave the name Primates, meaning chiefs or heads of the animal kingdom.

Even in this, Linnaeus could not quite free himself from the anthropocentric philosophy, for he followed tradition in conceiving a graded hierarchy of beings from the "lowest" to the "highest." However, this old idea of the "stairway of beings" prepared the minds of many philosophers for the acceptance, a full century later, of the doctrine of the origin of "higher" from "lower" types by transformation rather than by special creation of each grade (Daudin, 1926; Torrey and Felin, 1937).

Darwin's monumental investigations of the fact of evolution and of its mechanisms, including the principles of hereditary variation and of natural selection, caused a revolution in the thinking of many people regarding the nature, origin and destiny of man. As a result of the popularization of evolution led by Huxley and Haeckel, man was for a time seen by part of the public, in his natural perspective: on the one hand as a made-over animal with all the frailties of his animal origin, on the other hand as a unique product of evolution which can look inward upon himself and utter his reflections in articulate speech.

The next stage was the patient opening up of the wonders of the fossil records of past ages and the development of the methods of comparative anatomy, including embryology. From such sources cumulative evidence has shown that evolution is a fact. Finally came the epochal discoveries in genetics which have again established the principle of the natural selection and linking of new fortuitous mutations as the method by which transformations are effected. A mutation may of course be large or small, useful or lethal. It may be made permanent through the segregation and inbreeding of relatively small populations which include a considerable number of individuals that possess it. Most mutations are either not useful or associated with lethal hereditary characters. In nature, selection for survival of the species over long periods may rarely operate long on single characters but must rest on the linking and resultant of the selective value of many fortuitously varying characters.

Having accepted evolution as a fact, biologists went on to search for the details of the mechanisms of heredity, while many palaeontologists became buried in their technical problems.

Meanwhile various self-perpetuating organizations that seem to be dependent for their very existence upon the continuing acceptance by their public of their essentially anthropocentric traditions and dualistic philosophy, have gradually regained their ascendency. Their policies have been first to ignore evolution, then, if forced to notice it, to laugh it off, to cast scorn and ridicule upon prominent evolutionists, or to prove that evolution cannot be because it is logically impossible. Signs are not wanting, however, that even in great centers of traditional learning which have been hostile to Darwinian evolution, there are alert, honest and relatively impartial minds who have become convinced by evidence that the age-long transformation of organic beings is a fact of nature. Such authors, while standing firmly on the doctrine of the dual nature of man, seem to be willing to admit that the body of man might have been created through long ages of evolution, but from some always "unknown" lower form whose existence must first be "proved" by evidence acceptable to their infallible system.

Persistent partisans in some states have succeeded in making it illegal to teach evolution or even to refer to it except as an "hypothesis." But since intensive personal investigations of material facts of evolution, rather than quotations from outworn controversial texts, are prerequisite for a first-hand knowledge of the subject, the teaching of evolution in the schools by those who have not had such training is of doubtful educational value.

Far more damaging to the intelligence of the public is the widespread habit of most newspapers and magazines of ignoring the vital evolutionary background of human affairs as if it had no public importance or no application to the news of the day. How many people realize that a great part of the wickedness and troubles of man spring from the fact that he is still a scion of the apes, self-centered, capricious, jealous and vengeful, but with his powers for good and evil infinitely multiplied?

How many university professors and high dignitaries of church and state humbly accept the fact that they are vertebrate animals? that their very jaws and teeth testify to their membership in the mammalian Order of Primates?

Such deplorable ignorance has grown partly because of the intellectual blinders that have been fastened by custom and tradition upon the minds of young and old. This method discourages and diverts effective criticism of traditional teaching as to the origin and nature of man and as to the true source of man's virtues and of his wickedness. Accordingly a large part of the public, which is unconscious of its own ignorance, prefers astrology, tradition and racism to evolutionary science. At the other extreme, an infamous perversion of Darwinian theory was taught and millions of allegedly inferior people were exterminated to insure the dominance of the "master race." Thus a superficial knowledge of evolutionary science may be far worse than none at all.

In view of the foregoing considerations the author of the present work is distinctly not attempting to "prove evolution," since competence to appreciate the cumulative evidence for evolution can be gradually developed, only among those who work long and earnestly to acquire the necessary technical knowledge: in genetics, general morphology, physiology, comparative anatomy, palaeontology, odontology, geology, etc. The present chapter is mainly an attempt to summarize the origin and branching of the skeletal patterns of primates, including man, in so far as that outline has gradually been pieced together, like a vast picture-puzzle, by the patient labors of hundreds of investigators.

TREE-SHREWS (TUPAIOIDS) AND THE DAWN OF THE PRIMATES

In Chapter XIX it was shown that among the placental insectivores, dating at least from Upper Cretaceous times, there was a wide adaptive branching. Most of the branches, including the zalambdodonts (centetoids), the European hedgehogs (erinaceoids), moles and shrews (soricoids), finally became extremely specialized for life on or under the ground. But some of the earlier forerunners of the insectivores must have preserved a relatively primitive skeleton (cf. fig. 19.16A) with five-toed hands and feet capable of grasping the branches of trees, with freedom of movement of the radius and ulna to permit full supination and pronation. The large hands were used not only to grasp branches but to seize insects, berries and perhaps small birds' eggs and nestlings. The cheek teeth inherited the basic patterns of the tritubercular upper molars and tuberculo-sectorial lower molars, which are found in the earliest known members of the insectivores, creodonts and other primitive placental orders. Active and free movements in the trees encouraged keen sight, large eyes, an alert sense of balance and a large, quickly adaptable brain. Such were and are the tree-shrews (tupaioids), whose modern survivors have been classed either as progressive insectivores (Gregory, 1910, 1916), or as primitive lemuroids (Wood Jones, 1929, Simpson, 1945).

EOCENE AND LIVING LEMURS (LEMUROIDS)

Notharctus, Ideal Structural Prototypal Primate. —The existing lemurs of Madagascar (fig. 23.2) were rightly regarded by earlier authorities as an essentially archaic group of Primates, which, aside from certain peculiar specializations in their incisors and lower canines, have remained on a relatively primitive primate stage in the general characters of their skull, brain and locomotor skeleton. Remote ancestors or collateral ancestors of the lemurs have been found in the Lower and Middle Eocene of North America (Notharctinae) and in the Upper Eocene of Europe (Adapinae).

The skeleton of *Notharctus* (fig. 23.1A) is fundamentally identical with that of the modern *Lemur* (B), especially in its hands and feet; but it is less specialized than *Lemur* in retaining more normally arranged incisors and canines, as noted below (fig. 23.10A, E). The humerus of *Notharctus* (fig. 23.4E) is seen to be intermediate between those of more primitive mammals (A-D) and those of the existing Primates (F-I).

The earlier notharctine mandibles (fig. 23.5D, D1) were essentially like those of the earlier creodonts, and their cheek teeth (figs. 23.5D, 7A) were adapted chiefly for vertical movements of the mandible. In the later notharctids (fig. 23.10B) and still more in the adapids (fig. 23.10C) the areas for the masseter in the angular region and on the zygomatic arch were much strengthened and the same is true of the area for the internal pterygoid on the inner side of angle of the mandible, while the elevation of the mandibular condyle gave room for increased length of the temporal, masseter and pterygoid muscles. Meanwhile the opposite halves of the mandible, originally connected only by ligament (fig. 23.5D), became fused and thereby withstood better the twisting stresses generated by oblique movements of the mandible. These tendencies were carried to an extreme in the Recent Indrisidae (fig. 23.15D-F), which have hugely expanded angular regions and very stout zygomatic arches; similar, if less extreme, adaptations have been developed also in both New World and Old World monkeys.

The nasal chamber of *Notharctus* (fig. 23.10A) was about as large as it is in *Lemur* (E), indicating an acute sense of smell. The orbits were also well developed, but relatively smaller than in a *Lemur* skull of the same length. They look forward, upward and outward but, due to the fact that in recent lemurs the eyeballs protrude, especially on the outer side, the direction of the eyes of *Notharctus* was very probably more forward and less outward than would be suggested by the orbits themselves. The formation of a secondary postorbital brace from the frontal and the malar in these early primates paralleled that in other mammals (e.g., ruminants, horses) and was conditioned in part by increasing stresses of the temporal, masseter and pterygoid muscles.

In the region of the middle ear of *Notharctus* the construction and relations of the ento- and ectotympanic (fig. 23.11A) to each other and to the cochlea of the internal ear differed only in minor details from that described by Stehlin in *Adapis* (B), even the tube for the internal stapedial branch of the carotid artery (*int. car. can.*) being in the same position. As this whole complex is found intact in the modern lemurs and indrisids, it constitutes an important item of evidence that these forms inherited it from the primitive lemuroid stock and that *Notharctus* was an early branch of that stock.

The evolution of the Notharctinae of the North American Eocene has been traced, largely as a result of the field work of Matthew and Granger (1915) from the lower levels of the Lower Eocene of Wyoming up through ascending levels to the top of the Bridger (Middle Eocene), above which it is not certainly recorded. Osborn (1902), Matthew (1915), and the present writer (1920, 1937) have all studied the closely graded sequence of forms as represented mostly by jaws and dentitions (fig. 23.6A-L) and occasionally by incomplete skulls (figs. 23.9; 10A, B). The systematic revision of the species was most judiciously and carefully done by Matthew and Granger and later the species of *Notharctus* were again revised by Granger and Gregory (1917). The main results may be briefly stated:

1) In the oldest horizon (Clark's Fork) of the Paleocene, the upper molars (fig. 23.6A) were roundly triangular with three conical main cusps, besides a protoconule (*pl*) and a metaconule (*ml*). In these oldest known stages of the family the subconic form of the outer cusps, in combination with a shearing oblique protocone-protoconule crest, suggest an insectivorous-omnivorous-frugivorous diet, perhaps not unlike that of a raccoon. The older species (A-D), originally with more triangular upper molar crowns, were grouped in the genus *Pelycodus* (A-F).

2) As we pass upward through successive horizons (A-L) the protocone (*pr*) tends to bud off a hypocone or "pseudohypocone" (*pshy*) which was so called because the true hypocone, formed from the basal cingulum, was not distinct; as this pseudohypocone becomes larger it tends to fill out the postero-internal corner of the crown (F-L), thus reducing the "interdental embrasure" or space into which the trigonid of the lower molar formerly fitted. The anteroposterior diameters of the upper molar crowns, especially on the first and second upper molars, were increasing somewhat faster than the transverse diameters, and the para- and metacones (*pa, me*) were also increasing faster in this direction. Thus the upper molar crowns were transformed from tritubercular (A), roundly triangular, to quadritubercular subquadrate contours (L). Meanwhile also a small bead-like swelling called the mesostyle (*ms*), which appears on the middle of the outer side of the crown at the junction of the para- and metacones (A, B), becomes more and more pronounced (D-L). The fairly progressive Lower Middle Eocene species, with subquadrate upper molars and well developed mesostyles, were grouped in the genus *Notharctus* (I-K). Thus in the latest species (*Telmatolestes crassus,* L) the outer wall of the crown presents two rounded crescents and the inner half of the tooth bears large rounded proto- and pseudohypocones, the protoconule (*pl*) well ridged, the conical metaconule (*ml*) isolated. Thus the general pattern superficially suggests the buno-lopho-selenodont type of primitive condylarths (fig. 21.14), a partly convergent resemblance.

3) The lower molars (fig. 23.7) went through corresponding but less conspicuous changes, involving the reduction and disappearance of the paraconid (*pad*) and the development of crescents on the trigonid and talonid (B-D).

4) While all this was going on in the molars, the third and fourth upper premolars were (fig. 23.6) making a feeble attempt, so to speak, to become molarized, the end result being that their main outer cusp became divided into two cusps (G-L), corresponding to the para- and metacones of the molars, and their main inner cusp was just beginning to develop a low swelling in the position of the "pseudohypocone" (*ps. hy*). In the corresponding lower premolars (fig. 23.7) the hypoconids (*hyd*), which sweep across the isthmus and down the valley between the para- and metacones, were also developing. But so far as the record indicates, the race became extinct before this process could reach the limit of its inherent possibilities, in contrast with the accelerated process of molarization in the premolars of the horse family (figs. 21.45; 21.46).

5) As the rates of evolution in size (fig. 23.8) varied in the different genetic lines, it happened that in the Bridger formation (Middle Eocene), where these rare fossils are more numerous, a few relatively small and more primitive forms referred to *Notharctus matthewi, N. osborni* were contemporaneous or nearly so

with those of a larger, more advanced genus (*Telmatolestes*).

The foregoing results are important in any general survey of the evolution of mammals because they afford a pretty clear-cut case, based on excellent material of the considerable amount of transformation of the premolar and molar patterns effected during ten million odd years of the Lower and Middle Eocene.

The generic name *Notharctus*, literally "bastard dog," refers to a supposed mixture of carnivorous and "pachyderm" (herbivorous) characters. This name was given in 1870 by Joseph Leidy, one of the great pioneers of North American vertebrate palaeontology, who in his monograph (1873) on "The Extinct Vertebrate Fauna of the Western Territories" penned the following morphologic analysis of the type and then only known lower jaw of *Notharctus tenebrosus*:

"In many respects the lower jaw of Notharctus resembles that of some of the existing American monkeys quite as much as it does that of any of the living pachyderms. Notharctus agrees with most of the American monkeys in the union of the rami of the jaw at the symphysis, in the small size of the condyle, in the crowded condition of the teeth, and in the number of incisors, canines and true molars, which are also nearly alike in constitution [fig. 23.13]. Notharctus possesses one more premolar and the others have a pair of fangs. The resemblance is so close that but little change would be necessary to evolve from the jaws and teeth of Notharctus that of a modern monkey. The same condition which would lead to the suppression of a first premolar, in continuance would reduce the fangs of the other premolars to a single one. This change, with a concomitant shortening and increase of depth of the jaw, would give the characters of a living Cebus. A further reduction of a single premolar would give rise to the condition of the jaw in the Old World apes and man" (1873, pp. 89, 90).

These were, however, merely the *possibilities* of *Notharctus;* whether or not any of these possibilities were actually realized will be discussed during the following general comparative sketch of the dentition, skull and postcranial skeleton of *Notharctus* which is based upon the materials monographed by Gregory (1920).

The dental formula $\text{Inc}\frac{2}{2}\text{C}\frac{1}{1}\text{Pm}\frac{3}{3}\text{M}\frac{3}{3}$ was the same as in the Upper Eocene Adapinae (fig. 23.10C) and was well suited to give rise by reduction to the dental formulae of the Lemuridae, Lorisidae, Indrisidae and other lemuroids, just as well as it may have been reduced to give rise to those of the New World and Old World primates. But this does not necessarily imply that *Notharctus* itself was the direct ancestor of any of the later families. All that it may mean is that *Notharctus* inherited this primitive dental formula from still older members of its own family.

The median upper incisors of *Notharctus* (fig. 23.9A1) were basically like those of the modern lemuroids, especially *Propithecus* (fig. 23.15), but with a more elongate compressed tip. Quite possibly this sort of incisor, instead of being evolved from one with a transverse cutting edge, as in monkeys, had been reduced from a more insectivorous type of incisor, namely, one with a compressed crown and prominent tip, essentially like that of certain insectivores (*e.g.*, fig. 19.17B) but not so large. This hypothesis would fit in with the numerous evidence of relationship between the tupaioids or Menotyphla (p. 385), including the Plesiadapidae and the true lemuroids, including the earlier Notharctinae. The early (Paleocene) enlargement of the median upper incisors and the acceleration of growth-force toward this tooth may have been one of the factors involved in the absence of the third upper incisor in primates. The lateral upper incisors (i^2) of *Notharctus* (fig. 23.9A1) were quite small but functional teeth, with slightly compressed blunt tip; this tooth had developed only a slight suggestion of the nipping form of its larger neighbor.

The upper canines of *Notharctus* in some skulls were enlarged and laniary (fig. 23.9A1), in others feeble (23.10A) and less curved. In later primates similar differences denote the males and females, and in *Notharctus* the supposed males also have higher sagittal and lambdoidal crests (fig. 23.9A), larger zygomatic arches and more strongly braced mandibles. The two lower incisors (i_1, i_2) on each side were known in but few specimens (fig. 23.9B); in occlusion they fitted on the inner side of the uppers and had already begun to develop a little of the chisel-edge form of their homologues in later primates. Although slightly procumbent, they had not assumed the specialized compressed form seen in modern lemurs (fig. 23.5C); nor had the lower canines, which were somewhat like those of a dog, begun to be transformed into the incisor series. The somewhat dog-like appearance of the male lower and upper canines and incisors of *Notharctus* (fig. 23.9) was assuredly not inherited from a primitive carnivorous placental mammal, such as the creodonts or small condylarths, because there is much evidence to indicate that the primates never went through a creodont stage, but branched off from early tree-shrews, the latter in turn probably originating in unknown Cretaceous leptictoid insectivores (Simpson). The absence of procumbency of the incisors and reduction of the lower canines led both Wortman (1903, p. 172) and Gidley (1923) to infer that because the Eocene *Notharctus* did not display the peculiar specialization of modern lemurs, it could not be a lemuroid. Through the unfortunate choice of staking everything on certain arbitrarily chosen diagnostic characters and of trusting blindly to the principle of the irreversibility of evolution, they apparently expected to find the procumbent lemuroid incisors and canines of modern lemurs already developed in the lower and middle Eocene ancestors of lemuroids; similarly they failed to note the significance of the fact that even the upper Eocene Adapidae (fig. 23.10C), which were closely allied in many features to the Lemuridae, had not yet acquired procumbency of the front teeth. In the majority of its characters *Notharctus* was indeed a lemuroid but they dismissed such characters as being "merely primitive."

Nevertheless, the present author never at any time suggested that *Notharctus* was directly ancestral to the modern lemuroids, as he specifically referred to the evidence that the European Adapinae were at least near to such ancestors; what he did then (1920) and does still affirm is that the earliest notharctids, represented by the small *Pelycodus ralstoni* (fig. 23.6A), *Pelycodus trigonodus* (B) and others were among the most primitive of all then known lemuroids; and that even *Notharctus* retained a long series of primitive primate characters, many of which are still preserved in one or more of the modern lemuroids. We may also note here that a close study of the interlocking relations of the upper and lower cheek teeth (fig. 23.12b) suggests that the later notharctids (B1), like the adapids, had made considerable progress away from a primitive insectivorous stage with merely vertical movements of the mandible (fig. 23.5D), toward a real grinding movement from side to side, and that this tends to explain the resemblance of the later notharctine molar patterns to those of primitive Eocene ungulates (figs. 21.5; 14).

Fossil and Recent Lemurs of Madagascar.— This great island, the home of other "living fossils," is also the last refuge of the typical lemurs, the most primitive of the existing primates; these have come down to us with remarkably few modifications of their skeletons from their remote ancestors of the Paleocene and Eocene epochs. As seen in a zoological park, the lemur has a fox-like face (fig. 23.2A), with pointed muzzle and long, sharp, upper canine teeth. Its eyes are quite large, as they often are in nocturnal animals. Its hairy coat bears contrasting light and dark areas and its tail is long and thick, probably a cryptic pattern for an animal that crouches in the trees. If the keeper will let us in to the monkey-house on a moonlight night, we may hear low growling and sharp KWOK, KWOK sounds and see their eyes gleaming as they turn toward us. Then we understand why the natives of Madagascar are said to have called these animals "ghosts," which was translated by classical scholars into *Lemures,* the household ghosts of the Romans.

At first sight there seems to be nothing especially monkey-like in a lemur and it is not until we catch sight of its grasping hands and feet and slender fingers (fig. 23.2A) that we recognize a simian feature. For such reasons the German word *Halbaffen* happens to be very apt for a popular name, the lemurs being more or less intermediate between tree shrews and monkeys.

Since the dentition is the first clue to the general food habits, we note that in the typical lemurs the three lower incisor-like teeth on each side are small compressed rods, sharply procumbent and with pointed tips (fig. 23.5C), and that the outer units of this series are the lower canines, which have been reduced and taken over into the incisor series, a convergence toward the ruminants. It was supposed that these little incisors and canines were used as combs for the fur, but direct observation of living lemurs by M. Russell Stein (1936) indicates that the fur is not combed, but licked by the large tongue. Other observations on African bush-babies (*Galago*) by Lowther (1939) show that their incisors and canines are used as sharp picks, as in removing excrescences on the skin and probably in dismembering insects. The median upper incisors are small, compressed, with low elongate crowns and longitudinal crests, and in general plan not unlike the longitudinal cutting incisors of *Notharctus* (fig. 23.11A). The upper canines in the males (fig. 23.10E) are "laniary," that is, they are long curved daggers with sharp cutting edges. The females have small weak canines with less curved crowns. The anterior lower premolars (p_2) are compressed, pointed, with cutting edges (fig. 23.10E); the upper fourth premolar (fig. 23.12D) has a cutting outer edge and a blunt internal cusp. The upper molars (D) are roundly triangular with somewhat cutting outer cusps, a large swollen protocone and a hypocone cingulum. All this suggests a mixed diet of fruit, leaves, insects and perhaps nestling birds. The relatively slender mandible (fig. 23.10E) and zygomatic arches of *Lemur* are, in the light of evidence cited (p. 456), assuredly retrogressive.

The muzzle is large, the tip of the nose vertically divided with a black glandular area, not dissimilar to that of a dog and quite unlike the nose of monkeys and apes. Internally the nas chamber is capacious, with unreduced olfac tory scrolls; the olfactory lobes and relate parts of the brain are also well developed, espe cially by contrast with the reduction of thes parts in the monkeys and apes. The unreduce bony face (fig. 23.10E) is but little inclined t the base of the cranium, much less than in monkey.

The brain of *Lemur* is fairly large, muc more so than in primitive carnivores, and th braincase (fig. 23.10E) swollen and capaciou The large orbits are supported posteriorly b conjoined processes from the frontal and mal bones. Hearing must be keen, as indicated b the well inflated entotympanic bulla (fi 23.11C), one on each side. Inside of the bul is the ring-like tympanic bone, supporting th tympanum, or drum-membrane. This contras with the condition in the monkeys (fig. 23.35B in which the tympanic (*ectotym*) bones are n enclosed within the entotympanic bullae.

The most conspicuous feature of the lemur limbs is the foot (fig. 23.3B1), in which th very large hallux or "great toe" diverges sharp from the rest of the foot and is flattened distal into a rounded disc, bearing a flat nail. Th large metatarsal of this digit has at its upper en a conspicuous olecranon or elbow-like proce into which the tendon of the peroneus long muscle, on the back of the lower leg, is inserte This enlarged hallux is one limb of a biramo grasping organ, the other being formed by th rest of the digits and their supporting pad Moreover when the large hallux is fully ab ducted, part of the olecranon-like projection its upper end fits into a slot formed between th surrounding bones of the tarsus. In this positio a continuous effort to keep the hallux fully ab ducted (as in grasping a large limb) may b unnecessary.

After a careful review of this and other ev dence, it was concluded that the foregoing typ of foot, which was present in all known pr mates of the Eocene epoch, was the basic pa tern for the foot structure of all primat (Gregory, 1920) and that it permits a usef combination of firm grasping, climbing, leapin nd safe landing. But if we were to accept the postulates of those who seem to hold that a pecialization once gained can never be lost, hen the foregoing specializations of the lemur oot, even though dating back to the Lower Eocene, would exclude all the lemurs that possess it from the ancestry of the higher primates.

In the hands the pollex (thumb) of *Lemur* fig. 23.3B) is less divergent, more like the ther digits, than is the hallux, another basic rimate feature, the highly divergent thumb of ibbon and man being a later specialization.

The arched vertebral column of *Lemur* (fig. 3.1B) is well differentiated into cervical, thoracic, lumbar, sacral, coccygeal and caudal reions. In general it is a flexible column, with ong lumbar parapophyses indicating leaping abits. The caudal vertebrae are elongate and apering distally, with gradually dwindling ransverse processes; this is in accord with the on-prehensile character of the tail.

The range in body-form of the still living Lemuridae is considerable. At one extreme tands the very small and active *Microcebus* nd *Chirogale* (fig. 23.23), apparently nocturnal forms with very large eyes, narrow and hort muzzle. In the tarsus, according to De Blainville, the navicular is more or less elongate proximo-distally and so is the distal porion of the calcaneum (fig. 23.24B). This is a eaping adaptation which has been developed o an extreme in the galagos (C) and in the tarsioids (fig. 23.26)—a good example of parallelism, or the independent evolution of similar djustments in related families. A peculiar member of the lemur family is *Myoxicebus* (fig. 3.15C), in which the rounded skull and short wide muzzle superficially suggest a sloth and he cheek teeth bear low obtusely conical rowns, probably indicating frugivorous habits.

Among the fossil lemurs of Madagascar, dug rom old bogs and probably not very ancient, was a giant lemur, *Megaladapis,* with a skull fig. 23.10D) about as large as that of a black ear, with a stumpy tail, relatively short limbs for a lemur) and a stout backbone. This form, although a specialized side branch, is not far rom the border line between the true lemurs and the Adapidae, fossil lemuroids of the Upper Eocene and Lower Oligocene of Europe. As shown in Stehlin's monograph (1912), *Adapis* agrees with *Lemur* in all fundamental characters, such as the inclusion of the ring-like tympanic bone within the inflated entotympanic (fig. 23.11B); while from its cheek teeth (fig. 23.12C), with low conical protocones and somewhat cutting para- and metacones, those of *Lemur* (D) could most easily be derived. But *Adapis* (fig. 23.10C) was far less specialized than the modern lemurs in that its lower incisors and canines were but slightly procumbent, and not reduced, while its lacrimal bones (*la*) were not secondarily enlarged in front of the orbits as they are in *Lemur* (E). For such reasons those who expected remote Eocene ancestors to foreshadow their specialized modern descendants in all characters were slow to recognize that the Adapidae were only primitive lemurs

Camouflaged Night Prowlers (Indrisids).—A second family of modern lemuroids in Madagascar is the Indrisidae, which were described and figured by Milne Edwards (in Grandidier's monograph, 1875). These animals (figs. 23.14–19) are more highly specialized for leaping in the trees and grasping the branches. In *Indris,* the most peculiar member of the family, all four limbs are extremely long and strongly angulated (fig. 23.16). In the quadrupedal standing pose the vertebral column slopes upward toward the elevated rump and stump-like tail. The hands and feet are extremely long, the foot biramous with very large hallux and very long remaining metatarsals and digits; in the hand, the index finger often diverges from the others and tends to follow the thumb, so that there is a cleft between these two and the rest—recalling the zygodactyl forefoot of the potto (p. 463). The muzzle of *Indris* is large (fig. 23.15E) and the cheek teeth have low conical cusps (fig. 23.18D), all frugivorous characters. The lower canine is procumbent and incisiform. The erect lower tooth which suggests a canine is really a premolar (p_3). All these genera parallel the Old World monkeys in the loss of p^1,

p^2. A large, brilliantly colored inverted V stands out above and around the tail stump against the very dark color of the back (fig. 23.14B). This may be a "follow me" sign, like the "flag" of the white-tailed deer.

In *Avahis,* another member of this family, the general appearance (C) is almost monkey-like, the tail being long and the face much shortened; but the hands and feet are indrisine in character. *Propithecus* (fig. 23.14A) and *Mesopropithecus* (fig. 23.17A) are probably the least specialized of the family. They have a moderately developed muzzle and the first and second upper molars bear four V's (fig. 23.18B), simulating those of other leaf-eating mammals (e.g., primitive ruminants and *Phascolarctus*). These molars are also elongated anteroposteriorly as in other leaf-eaters.

A fossil member of the Indrisidae inappropriately named *Palaeopropithecus* has a far larger massive skull (fig. 23.17B) with hugely expanded masseteric and internal pterygoid areas on the mandible, massive zygomatic arches and reduced entotympanic bullae. Its upper molar patterns (fig. 23.18C) are coarse, blunted derivatives of the *Propithecus* type.

The adaptive branching among the indrisioid lemurs was thus fairly pronounced, but it is not nearly as wide as it is among the phalangeroid marsupials. Single features, such as the diprotodont modification of the lower incisors, the selenoid cusps of the molars, the enlargement and grasping adaptations of the feet and hands may be matched in the two groups (e.g., *Mesopithecus* and *Phascolarctus*) but there are few closely convergent pairs and the indrisioids are lacking in such extremely diverse specializations as were achieved by *Thylacoleo,* the wombat, the kangaroos or the great diprotodontids.

The Indrisidae are evidently related to the Lemuridae and may perhaps have been derived from some such primitive adapid as *Pronycticebus* (fig. 23.18A) of the Upper Eocene of Europe, as the true lemurs were from some other adapids.

The Monkey-like Archaeolemur and Its Allies.—Another member of this strange extinct Malagasy fauna was *Archaeolemur* (fig. 23.17C) with a somewhat monkey-like skull (fig 23.17C), bilophodont molars, trenchant pre molars, procumbent lower incisors and canines one of the anterior premolars (p_3) simulates short lower canine.

Even more monkey-like was the skull c *Hadropithecus* (fig. 23.17D). In the exceller plates in the memoir by Lamberton (1938), w note that: (1) the posterior upper premolar (p^4) of *Hadropithecus* were wider than m although not fully bilophodont; (2) in the an terior lower premolars (p_2) the crown wa compressed, with a pointed tip and obliqu anterior face that sheared behind the posterc internal face of the small upper canine—some what as in Old World monkeys; (3) the las deciduous molar (dm_3) was large and bi- t tri-lophodont, that is, it was more molariforr than its replacing tooth (p_4), as is usually th case in placental mammals; (4) the anteric lower deciduous cheek-tooth was small an crowded, partly simulating a canine; (5) th permanent lower canine (regarded as an incisc by authors) was small, procumbent and assc ciated with the single procumbent incisor, as i Indrisidae.

Comparative study of these extinct Malagas lemuroids in the British Museum (Natural Hi tory) led to the conclusion (Gregory, 191 pp. 435–441) that their simian features ha been brought about from a lemuroid base an were merely convergent toward the higher pr mates; this is also the conclusion of Lamberto (1938, p. 43).

The Incomparable Aye-aye.—By far the mo bizarre of the Malagasy lemuroids is the ay aye (*Daubentonia*), a round-faced mamm (fig. 23.19) with great compressed, paired u per and lower front teeth, degenerate minu cheek teeth, large ears and an amazingly lor slender fourth digit of the hand, tipped with large curved claw. According to the reports a good observer to Sir Richard Owen (1866 the aye-aye moves slowly along the branches certain trees, listening for the slight soun made by boring grubs beneath the bark. Havir located a promising spot, the aye-aye with his chisel-like incisors and strong jaws cuts a deep slot in the wood, opening into the tunnel of the grubs. He then inserts his elongate probe-like finger, hooks the grub with the claw and pulls it forth through the slot up to his waiting tongue. No better example could be found of the complete coöperation of different parts of one organism, doubtless controlled by the motor cortex of the forebrain.

There appears to be no feature in the dentition, skull or postcranial skeleton of the aye-aye which seems inconsistent with remote derivation from an indrisine stock. But its incisors, jaws and molars more or less closely resemble those of certain fossils described by Stehlin (1912, pp. 1435–1447, 1490–1494) from the Upper Eocene of Europe; he suggests (pp. 1474–1489) that the aye-aye may be connected with the Plesiadapidae, a specialized branch of the tree-shrews. Since the Plesiadapidae and with them the tree-shrews are probably near the ultimate source for all the lemuroids, the aye-aye may conceivably represent a separate branch from the same source. But it closely resembles the Indrisidae in so many features that a separate derivation seems unlikely (Simpson, 1945, p. 181).

Madagascar may have received its strange mammalian fauna, including its zalambdodont insectivores (p. 379), its viverrids (p. 401), and its varied lemuroids, from the northwest, perhaps in mid-Tertiary times; because primitive relatives of all these groups are known in the old Tertiary of Europe and of North America. The existing lemuroids of the continent of Africa—all of the family Lorisidae (Nycticebidae), also have a supposed ancestor (*Pronycticebus*) in the same European Eocene center, as well as close relatives in the late Tertiary (*Indraloris* Lewis) and Recent fauna of India.

The Leaping Galagos and the Clinging Lorises.—The family Lorisidae includes two subfamilies of widely different locomotor habits: the first subfamily, including the bush-babies (Galaginae) have long slender hind limbs and can leap up in the air and catch a flying insect (fig. 23.23C, D); the second subfamily (Lorisinae) includes the extremely short-footed lorises (A, B), which move stealthily about, clinging to the limbs with their large hands and spreading feet or rolling themselves up to sleep with their heads tucked under their backs and their hands and feet clutching the branch on which they rest. In accordance with these contrasts, two of the ankle bones of the galagos, namely, the navicular and calcaneum, are greatly elongated (fig. 23.24C), while in the lorises they are very short (fig. 23.21). The hands of the galagos are true hands, but in the potto, the most specialized of the lorises, they become biramous clamps, in which the second digit is reduced or lost.

The differences in the dentition (fig. 23.23) and skulls (fig. 23.22) between these two subfamilies are less pronounced than those between the feet. Both have the procumbent lower incisors and canines, together with minor modifications of the tri-to-quadritubercular upper molar plan (fig. 23.23). Both also contrast with the Lemuridae and Indrisidae in having the entotympanic bullae less extended outward, leaving the tympanic ring exposed (van Kampen, 1905, p. 665). But in the galagos the muzzle tends to be very long and pointed, while in the lorises it is short; the eyes of the galagos are larger and so also are their brains. The gap between them must be very ancient.

The Upper Eocene genus *Pronycticebus,* known from a well preserved skull (fig. 23.22B) and dentition (fig. 23.29bB), was referred to the Lorisidae by Grandidier, but to the Adapidae by Le Gros Clark (1934b). So far as one can tell from the available evidence, it may be a transitional form between the Adapidae and the Lorisidae. On the other hand, the striking resemblances in the skulls between *Galago* (E) and the tarsioid *Necrolemur* (Gregory, 1922, p. 208) may possibly be due to close relationship rather than to parallelism or convergence; but even in that case the characters that tie the Lorisinae and Galaginae together (Weber, 1928, p. 735) may not all be parallelisms. According to Tate Regan (1930), the enamel prisms of the Lorisidae agree in structure with those of tarsioids and platyrrhines and differ from those of Adapidae, Madagascar lemuroids and catarrhine monkeys, but there are reasons (see p. 472) for doubting the importance of this item of evidence.

All the foregoing lemuroids belong to terminal family branches. The Lemuridae and Indrisidae have every mark of derivation from the Adapidae; the archaeolemurs seem to be merely a monkey-like branch of the indrisid stock; the aye-aye may be a highly specialized branch from the archaeolemurs. In many characters the Lorisinae are nearer to the Lemuridae than to the others and were probably also derived from the adapid stem.

WEIRD HOMUNCULI (TARSIOIDS)

The second great group of Eocene primates is the group of the tarsioids, but before considering them let us sketch their sole existing descendant, the Spectral Tarsier (figs. 23.25, 26) of Indonesia.

Except for its very long thin leaping legs and long slender tail, one could enclose the body of this elf in the hollow of his hand. Its head is round and seems to be far too big for its body. The greater part of the face is taken up by the huge eyeballs; between them the nose is diminutive. The ears are wide and can be "reefed" like a sail. As a whole, the face is owl-like. The jaws are extremely short (fig. 23.27), but the mouth bristles with sharp-cusped teeth. The upper molar crowns (fig. 23.29E) have low conical cusps in a rounded tritubercular pattern. The neck is short and the creature can swivel its head around quickly. The long slender digits end in subcircular discs like those on the toes of certain arboreal tree frogs; the great toe is relatively large, sharply divergent and of grasping type; the navicular and calcaneum of the tarsus (fig. 23.26) are greatly elongated (whence the name tarsier) as in the galagos (fig. 23.24C); the tibia is very long, rod-like, the femur also long and strongly angulated on the tibia. Sudden leaping from branch to branch and the catching of insects on the wing are thus suggested, and in view of the very small total weight, relatively enormous leaps among the branches may be predicated.

The owl-like globose head of *Tarsius* (fig. 23.26) is very large as compared with the thorax, larger than in the marmosets (fig. 23.37) and only the small total mass and the high strength of bony and muscular tissue in relation to mass makes it possible for such a huge globe to be propelled on top of such a small backbone. Again, the same conditions have made it possible for the limbs to be so long and for a long extra segment (naviculo-calcaneal) to be intercalated in the foot, which adds greatly to the catapult-throwing-stick mechanism (fig. 23.26). The load which this mechanism throws consists of two main parts: (1) the head and (2) the rest of the body. The leg-springs and vertebral column are so arranged that both in starting and in landing the head and vital organs are cushioned against the shock involved in such sudden accelerations and decelerations. The long delicate digits, each with its round terminal disc, all controlled by relatively strong muscles and tendons, contribute greatly to the cushion-action, as does also the curvature of the neck. Perhaps for these and like reasons the consolidation and great strengthening of some neck vertebrae, which is characteristic of the long-leaping jerboas, is conspicuously absent in *Tarsius*. Here the great globular occiput is supported by a wide but slender atlas, the spine of the axis and succeeding vertebrae being small or moderate. Nor are the neural spines of the second to fourth dorsal vertebrae much enlarged for the attachment of a great ligament of the neck, as they are in certain ungulates. Probably the tendency for the large head to pitch forward in landing is controlled in part by the extensors of the back, also by the trapezius and other muscles that spread from the shoulder blade to the occiput.

Although the lower jaw (fig. 23.27C) is strongly built, it is not deep, its angular region being but little expanded; the zygomatic arches are slender and there is no prominent masseteric process beneath the malar. All these are signs that the masseter is either not very voluminous or that it is spread over the deep fascia on the side of the face, somewhat as it is in zalambdodont insectivores. The temporal muscle is evidently spread over the latteral surface of the occiput but is not strong enough to require either a thick braincase or high sagittal and nuchal crests. The huge brain must have had the leading part in moulding the cranium around it, while the enormous eyeballs have chiefly moulded the bones around the orbits.

The relatively huge brain—again compared with the mass as a whole—is superficially like an owl's brain—a remarkable convergence, since in the owl it is the thalamic floor, in *Tarsius,* the roof, of the forebrain that has been greatly developed; but in both cases it is the great nocturnal eyes, high balancing power and precise coördination of sensory stimuli and motor response that have contributed to the convergence. The brain of *Tarsius* is predominantly an "eye-brain" with prominent projection of the optic parts on to the occipital pole, which is accordingly conspicuous (Elliot Smith, 1924, fig. 15).

The lower part of the forebrain is moulded around the great backwardly converging cones formed by the eyeballs and by their muscles. The optic nerves, as it were, tie the eye-cones on to the under side of the forebrain. The orbito-sphenoid bone forms a thin wrapping around the optic nerves and adjacent parts.

All the olfactory parts of both brain and skull are reduced, as indicated by the narrowly restricted olfactory chamber. Partly as a result of this, the upper end of the lacrymal duct, with its bone and tunnel, are pushed out by the huge eyeballs, away from the surface of the narrow nasal chamber. The reduction of the dorsal, olfactory part of the nasal chamber leaves the breathing tubes, or nostrils, projecting laterally to form a very short platyrrhine-like nose.

As if to compensate for the reduction of the sense of smell, the auditory parts of the skull and brain are highly developed. Thus the opposite entotympanic chambers (fig. 23.28C) are greatly inflated and prolonged forward and inward until they nearly meet across the mid-line immediately behind the pterygoid fossae. The ectotympanic is also inflated and forms on each side a sort of saucer with a small tube on its outer surface for the attachment of the funnel leading to the large outer ear. The internal carotid artery passes through a conspicuous tunnel (*car. for.*) in the middle of each bulla.

Tarsius is the sole survivor of a group of primates, first recorded in the Paleocene of North America, which flowered out in the lower Middle Eocene and apparently came to an end (except for the line leading to *Tarsius*) in the Upper Eocene and Lower Oligocene of Europe. They range from very minute, insectivore-like forms in the Paleocene (Gidley, 1923; Simpson, 1937) to the Upper Eocene *Necrolemur* (fig. 23.28B), which was about the size of a newborn kitten. The anterior upper and lower incisor teeth vary widely: in many of the genera they were enlarged and somewhat rodent-like (Matthew, 1915); but in a few cases were small and more "normal" in appearance (*i.e.,* with low tips). It would usually be taken for granted that the small incisors are primitive, but since the tree shrews already had the median upper incisors considerably enlarged, compressed and with blunt hook-like tips as in the Tupaiidae (fig. 19.16) and Plesiadapidae (fig. 19.17B); the same may have been the case in the ancestral tarsioids, that is both tarsioids and lemuroids may have stemmed from primitive tree-shrews with procumbent upper and lower incisors, the large second upper incisors (I^2) bearing a compressed crown and downwardly curved hook-like tip.

Teilhard de Chardin (1916–1921, p. 7) figured an upper dentition of an Upper Eocene tarsioid, *Pseudoloris parvulus,* in which there was a minute alveolus for the true median incisor, medial to the alveolus for the enlarged I^2. The incisor dental formula of this genus was $I\frac{3}{?}$, whereas in perhaps all other known Primates it was normally $I\frac{2}{2}$ or less. In 1927 the same author described a very small primate jaw from the Lower Eocene of Belgium with possibly three lower incisors. Simpson (1940, p. 190), who appropriately named this specimen *Teilhardina,* writes of it as follows:

"*Teilhardina* is of extraordinary interest not only as the oldest European anaptomorphid but also because it is, in the known parts, the most nearly generalized and the most primitive of all known tarsioids, or even of all known primates (if the tupaioids be excluded from the Order, contrary to my present opinion). Despite its close resemblance to *Omomys,* it differs from that likewise primitive genus in numerous slight particulars throughout the dentition, and every difference seems to stamp *Teilhardina* as less distant from truly prototypal primate morphology."

In correlation with the enlargement of the second incisors and the development of a partial diastema in these short-faced forms, the anterior premolars were often crowded. The more even spacing and lack of crowding in the surviving *Tarsius,* associated as it is with the enfeebled jaws, may all be part of a drift away from the primitive insectivorous dentition.

Pseudoloris (fig. 23.29D) of the Upper Eocene of France (cf. Teilhard de Chardin, 1916–1921, pl. I) seems also to be a direct ancestor to *Tarsius* and connects it with the Eocene tarsioids, which were formerly placed in a separate family, Anaptomorphidae.

Cope's *Anaptomorphus homunculus* skull (figs. 23.27A; 23.28A) from the Lower Eocene of North America approached that of *Tarsius* in the enlargement of the eyes, of the brain, and of the auditory bullae, as well as in the dentition. Matthew (1915) discovered that Cope had first applied the generic term *Anaptomorphus* to the species *aemulus,* based on a lower jaw in which the median incisors were not enlarged, whereas they were much enlarged in the jaw of "*Anaptomorphus*" *homunculus;* he therefore referred Cope's species *homunculus* (based on the skull) to a new genus which he named *Tetonius homunculus,* in reference to the town in New Mexico near which it was found. Thus the evidence of the skulls of *Pseudoloris* and *Tetonius* establishes the existence of ancestral tarsioids far back in the Lower Eocene, while later discoveries push them well down into the Paleocene (Gidley, 1923; Simpson, 1937). Cope in his brilliant, if often hasty, way intimated that his *homunculus* might be a remote ancestor of man, and Wood Jones (1929), Woollard (1925), Le Gros Clark (1934) and others have cited many anatomical characters in which the modern *Tarsius* resembles the higher primates, especially man, and differs from the lemuroids.

On the other hand Woollard (1925, p. 1182) concludes that:

"There is no need to analyse the many divergent views put forth about the zoological affinities and the significance of *Tarsius.* With all the evidence of its anatomy marshalled, there can be no doubt that *Tarsius* is a Lemur of the Lemurs and is annectant to the early Eocene primitive placentals, and that standing at the base of the Primate stem it reaches forth to the simian forms and is annectant to the Anthropoidea."

The structure of the tarsus and of the hallux is known in *Necrolemur* (Schlosser, Stehlin) and in a partially preserved foot that was provisionally referred by Matthew to the middle Eocene tarsioid *Hemiacodon*; it is fundamentally identical with that in *Tarsius* (fig. 23.26), that is, both the navicular and the calcaneum were elongate (much less so, however, in *Hemiacodon* than in *Tarsius*) and the hallux is large and sharply divergent. Thus, so far as the evidence indicates, the primitive Eocene tarsioids were already advanced, specialized arboreal leapers with large and strongly divergent grasping hallux. This last character is in full accord with the evidence derived from the presence of just such a grasping hallux in the hind feet of the Eocene lemuroid *Notharctus* (fig. 23.3) and in all the existing Lemuroidea and Tarsioidea, as well as in the monkeys and apes (fig. 24.11); these facts further indicate that this very deeply stamped structural pattern would date back, at least in its preparatory stages, to the Paleocene and even earlier insectivorous ancestors of the primates. And indeed even *Ptilocercus* of the modern tree shrews (fig. 19.17B, C), presumably a survivor from the dawn of the primates, does show this divergent hallux, although possibly a little retrogressive. It suggests also that the premium on large eyes and sure-footed leaping was already leading to large "eye-brains" capable eventually of visualizing past events and literally foreseeing (imagining) future events.

NEW WORLD MONKEYS (CEBOIDS) AND THE RISE OF BRAINS

If the prize of world dominance had been awarded by natural selection solely on the criterion of the size of the brain in relation to the total body weight, then the New World monkeys, or some of them, might well have overspread the habitable earth. In course of geologic time they might conceivably have risen to high levels of intelligence. In that case the history of the primates might even have been investigated and recorded by some big brained descendant of the Capuchins, rather than by distant cousins of the rhesus monkeys and baboons.

The factual basis for the foregoing fancy is that in the South American squirrel-monkey (*Saimiri* or *Chrysothrix sciureus,* fig. 23.34K, K1) and some of its relatives the brain weight is $\frac{1}{17}$ of the total body weight (Max Weber, 1928, p. 791), whereas in man it is about $\frac{1}{35}$.

In the forests of Brazil and Central America this expanded brain is busied chiefly in keeping the feeble monkey folk up in the trees and out of harm's way, but long ago, in the African uplands, monkeys with not too large brains and sabre-like canines came down from the trees and roamed in marauding bands, eventually becoming baboons.

In short a wider dispersal of the New World monkeys seems to have been limited rather by their genetic failure to develop new adaptations for ground-living. In the spider monkeys and the marmosets we behold the extremes, on the one hand, of free swinging and, on the other, of timorous clinging, but always in the trees that shelter and feed them.

To the North American public of an earlier day the Capuchin monkey (fig. 23.31) was a not infrequent sight, because members of this or allied species were imported in considerable numbers from Central America and carried around by itinerant organ-grinders. With moving appeal these quaint beings would hold out their hands for pennies and catch them too. I remember that on one occasion, when pennies were being tossed to him, a *Cebus* reached up and caught them in the air with remarkable accuracy. Here at last is a true hand that is able to do things requiring a high degree of coördination, such as putting a cork back in a bottle (Thomas Belt, The Naturalist in Nicaragua, p. 93), and here is a brain that can teach itself to use longer and longer rakes to reach a gradually receding prize until the eleventh rake is many times longer than the monkey's own rake, i.e., his arm and hand (Warden and Galt, Journ. Phychol., 1941, vol. 11, pp. 3–21). Tilney (1928, pp. 245, 284) states that the brain of the howler monkey (*Alouatta*), a relative of *Cebus,* has advanced far beyond that of *Tarsius* in the development of complex interconnecting systems in many parts of the brain stem and of the neopallium.

Shall we regard the tarsioids as "too specialized" to give rise to *Cebus*? We note, for example, that *Cebus* (fig. 23.31) lacks the great elongation of the navicular and calcaneum that is characteristic of *Tarsius* (fig. 23.26). No doubt that is true, but if the tarsus had been no longer than it is in the Eocene tarsioid *Hemiacodon,* would the "principle of the irreversibility of evolution" permit such a tarsus to shorten up again? All we can say now is that in the later proboscideans the principle of irreversibility did not prevent the secondary shortening of the formerly long bony rostrum of the mandible, which their forefathers had gradually lengthened (fig. 21.28). It did not prevent the long narrow hand of the primitive perissodactyls from giving rise to the short wide hand of the later titanotheres (fig. 21.55). It did not prevent the slender leaping foot of primitive kangaroos from being transformed into the relatively short wide foot of the tree kangaroos (fig. 18.37Z). In short we may suspect that the idea of irreversibility has sometimes been invoked to deny that certain transformations which required marked changes in relative lengths and breadths actually did occur and that this idea has long delayed the correct interpretation of the evidence.

The evolution of a strong prehensile tail (figs. 23.30, 32) among the primitive Cebidae may have compensated for some reduction of their leaping habits and a secondary shortening of the tarsus. They still do leap, but they have also become adept at hanging from the branches and in swinging from branch to branch, as well as in running and leaping on the tops of the branches.

The skulls of *Cebus* and other platyrhines (figs. 23.34*a, b, c*) have advanced well beyond those of tarsioids (fig. 23.27), as have also their brains. The voluminous cebid brain has moulded the cranium around it and the large eye-cones have impressed upon the surrounding malar, frontals and sphenoid several diagnostic characters; for example: (1) the malars (fig. 23.34c) have spread upward and backward, lateral to the frontals, until they have come in contact with the parietals; (2) the malars have also crept inward between the eye-cones and the temporal muscle until they have nearly shut off the interior of the orbits from the temporal fossae; (3) meanwhile the parietals, sharing as it were the coronal increase of the parietal lobes or association areas of the brain, as well as the forward growth of the temporal lobes, have crept forward above the alisphenoids to gain contact with the orbital plates of the malars; (4) the marked expansion of the brain in a postero-lateral direction has filled out the base, sides and top (fig. 23.34c) of the occipital region; (5) the transverse expansion of the brain has also moved the ectotympanic rings outward (fig. 23.35B). (6) The entotympanic bullae, however, remain anchored to the pterygoid region, as in tarsioids (fig. 23.28), nearer the mid-point of the brain-floor and beneath the periotic portion of the temporal complex. There is nothing in all this that seems incompatible with derivation from a primitive Eocene lemuroid stock.

The dentition of cebids (fig. 23.36H) differs markedly from that of *Tarsius* (fig. 23.29E) and only a remote connection seems at first to be indicated. For in cebids (fig. 23.35B) we meet for the first time what may be called a subhuman type of upper incisors, in which the median pair has transversely wide edges (figs. 23.34c, 35B) suitable for cutting the stems of fruits and leaves, and is opposed by somewhat similar teeth in the lower jaw (fig. 23.34b). The lateral pair of upper incisors is less sharply trenchant, with rounder tips, those of the lower jaw being intermediate in form between the median incisors and the canines. This kind of incisor is superficially quite unlike the nipping incisors of plesiadapids and notharctids (figs. 19.17B, 23.9A1), but may nevertheless have been derived from them in connection with the retraction and widening of the muzzle, mainly by twisting the crown on the root so that its posteroexternal border was shifted laterad.

The upper canines of the male *Cebus* are set vertically, with rather straight, bluntly-pointed tips. The upper dental arch, as viewed from below (fig. 23.35B) is often wider across the canines than it is across the third molars. The premolars of cebids (fig. 23.36) are becoming the dominant polyisomeres in the dentition. Their crowns are transversely oval, bicuspid, all three very similar. The first and second upper molars are subquadrate, with low conical cusps tending to be modelled into two low blunt cross-crests. The third molar is relatively small. The lower premolars are likewise three in number on each side (fig. 23.13B1), there being no trace of the first premolar of primitive Eocene lemuroids. The lower molars often bear low transverse ridges on the trigonid and talonid but these cross ridges are not as strongly emphasized as they are among the Old World monkeys. The lower jaws (fig. 23.34) are stoutly built, the opposite halves braced by early union at the symphysis, the areas for the masseter, pterygoid and temporal muscles all well developed. On the whole, these jaws and dentitions should be adapted to a mixed diet.

Cebus (figs. 23.30–37) and its allies agree with the short faced lemuroids and tarsioids in respect to the location of the large forwardly-directed eyes, in the shortening of the muzzle and in the separation of the right and left nostrils, which are directed laterally (hence the group name Simiae platyrrhinae or Platyrrhinae, often applied to the New World monkeys).

Family Tree of the South American Monkeys.—Probably the most primitive living cebids are the little douroucoullis (figs. 23.31A; 23.34A), or Owl monkeys (*Aotus, "Nyctipithecus"*) and the true titi, *Callicebus* (fig. 23.34aB). The skull of *Aotus* (A) somewhat resembles the *Tarsius* type (fig. 23.27C) in its very large eyes and reduced nose, but differs widely in its diagnostic platyrrhine characters.

In *Callicebus* (fig. 23.34aB) there is a great expansion of the angular region of the mandible (indicating an expansion of the masseter) and the skull is robustly built. In these smaller forms the upper molars (fig. 23.36A, B) are more primitive than those of *Cebus* (H), with clearer traces of the tritubercular pattern.

Perhaps from an ancestor of *Callicebus,* with its expanded mandible, arose the woolly monkeys (*Lagothrix,* fig. 23.34bG) and the two genera of spider monkeys, *Bractyteles* (H), *Ateles* (I), both very specialized in their very long limbs and highly prehensile tails (fig. 23.32). The spider monkeys literally use the curly tip of the tail as a hook to reach things with, as well as a holdfast and anchor in swinging beneath the branches. Their mandibles (fig. 23.34bH, I) are greatly expanded posteriorly and their dentition (fig. 23.36E) is essentially as in *Callicebus* (B).

The saki *Pithecia* (fig. 23.33A) and the ouakari cacajao (C), may represent an early side branch in which the molar cusps began to be less pronounced and to flatten down as the crown became higher (fig. 23.36D). Especially in the cacajao (fig. 23.34aD) the upper and lower incisors were becoming procumbent, so that a pseudo-squirrel habitus is now being assumed, presumably with a shift from leaves to harder food, such as nuts.

The howler monkey (*Alouatta,* fig. 23.33B) seems to be related to the spider monkeys but has become very specialized: first in the huge expansion of the mandible (fig. 23.34E) and the development of more or less W-shaped crests on the anteroposteriorly elongate molars (fig. 23.36C). The resemblance to the molar patterns of the Eocene lemuroid *Notharctus* (fig. 23.13A) may be partly convergent. The skull of *Alouatta* (fig. 23.34aE) has been greatly strengthened to support its huge jaws. In the side view the jaw conceals the enormous laryngeal sac which is inside the inflated basihyal. The orbits slope upward, even more than in other cebids.

Cebus itself (fig. 23.34bJ) may likewise represent a branch from the primitive *Callicebus* stock (fig. 23.34aB), but its mandible is less expanded posteriorly (fig. 23.34bJ). The squirrel monkey (*Saimiri* or *Chrysothrix,* fig. 23.34K, K_1) seems to be a dwarfed relative of *Cebus* with a huge brain (see above, p. 467). Its dentition and jaws are small and suited possibly rather for soft-bodied larvae and insects than for the tough rinds of fruits and nuts.

The Timorous Marmosets.—The marmosets (Callitrichidae, "Hapalidae") are timorous, rather short-legged pygmies (fig. 23.37) which cling tightly to the trees by means of their digital integument and claw-like nails. Some of them rival *Pithecia* (fig. 23.33A) in the length of their hairy coats and manes; these may be useful in shedding the rain, in concealment, and in occasional displays of their none-too-amiable dispositions. Most of their digits are provided with more or less compressed folded nails, which function as claws (as in the tree-shrews) and are indeed regarded as true intermediates between claws and nails by Le Gros Clark (1936). Their hallux, however, is still strongly divergent and has the flat nail of the cebids. The three premolars (fig. 23.36) have now become larger than the molars and the third upper molar has been lost.

The genus *Callimico* (fig. 23.36H) affords a structural link tending to connect the marmosets with the smaller Cebidae, since it retains the third upper molar, although small. Moreover, in view of the evident agreement of the marmosets with the cebids, in the skull (fig. 23.34a, b, c) and other characters, it is difficult to conceive that their clawed feet are not in specialized association with their great reduction in body size and with the assumption of clinging habits.

The palaeontology of the New World monkeys is regrettably brief. They first appear, in the Lower Miocene of Patagonia, as a small imperfect skull named *Homunculus* by Ameghino, who regarded it as an ancestor of man. But careful studies by Bluntschli (1931) and Stirton (in press) show that *Homunculus* is closely related to *Aotus* and other cebids. Apparently the monkeys of Central America have come up from the south, after the uplift of a bridge between North and South America, in Lower Pliocene times. Thus there is a sharp break in the record below this point; that is, there are no further known fossil links that definitely connect the New World monkeys with either the tarsioids (Wortman, 1904) or the adapids and notharctids (Wortman, 1903, p. 174; Gidley, 1923). Although these alternative hypotheses have been noted above in other connections, we may here attempt to indicate the main merits and difficulties in each of them.

With regard to the possibility that the New World Primates were derived from *Notharctus* or its relatives, it is literally true that there is no feature known to the present writer in the entire dentition, skull and skeleton of *Notharctus* which definitely eliminates it from ancestry to the New World Primates; that is, the skeletal characteristics of *Notharctus* are demonstrably more primitive, nearer the tree shrew ancestors, than are the corresponding characters of any of the New World monkeys. The morphological difficulties in accepting *Notharctus* or any of the earlier members of its family as ancestors of the New World monkeys are: first, that the notharctids are closely related to the adapids and the adapids to the lemuroids; second, that if we take *Aotus* and *Homunculus* to represent the primitive cebids, these forms superficially resemble *Notharctus* or any lemuroid much less than they resemble the tarsioids. However, the tarsioids' main claim for ancestry to the cebids is not that some of the tarsioids approach *Aotus* and *Homunculus* in the enlargement of the eyes, reduction of the nasal chamber, etc. (which is open to suspicion as convergence), but that even the recent *Tarsius,* in spite of many millions of years of genetic separation from the ancestors of the higher primates, still agrees with the latter in the many deep-seated characters of the brain and other anatomical structures, including the mode of placentation (Elliot Smith, Tilney, Wood Jones, Woolard, Le Gros Clark).

The advanced specializations of the early tarsioids for arboreal nocturnal habits, reduction of the olfactory sense, great enlargement of eyes, increasing ability to make huge leaps and grasp the tree in landing, all placed the premium of survival upon a rapidly increasing brain; this in turn may have been the focal point of selection in the post-tarsioid stages that are preserved in the New World monkeys.

The zoogeographic and paleogeographic evidence bearing on the phylogenetic relations of the monkeys of the New and Old Worlds to each other and to the lemuroids and tarsioids is complex and indecisive. *Homunculus,* the earliest known South American monkey, was part of the extensive Santa Cruz Lower Miocene fauna, of which the varied ungulates, edentates and most of the rodents seem to have been derived remotely from North American ancestors of Paleocene or Lower Eocene age. The supposed African relationships of the Santa Cruz "hystricomorph" rodents are now ascribed to parallelism by competent authorities; but the South American and African characin and cichlid fishes are not so easily separated (p. 165). In short, it seems that the South American *Homunculus* may, according to present evidence, have been derived either from such a very primitive Paleocene North American tarsioid as *Paromomys* (fig. 23.29A), or from a relatively advanced Eocene lemuroid such as *Pronycticebus.*

OLD WORLD MONKEYS (CERCOPITHECOIDS): A PRELUDE TO BAD MANNERS

One of the numerous inhibitive results of tradition in some essentially Fundamentalist zoological gardens is that the curators in their manifold responsibility to trustees, city officials nd meticulous specialists, are very careful to have their animals correctly identified in the labels, even down to subspecies, but very seldom make any attempt to bring out the broader contrasts between larger groups. In such zoos Old World monkeys (figs. 23.38, 39) are not even contrasted with New World monkeys (figs. 23.30–33), a distinction which an intelligent child of six or seven years could easily apply once it was pointed out to him, or an older child could discover for himself if the incentive were strong enough.

Contrasts with New World Monkeys.—What then are some of these significant differences which divide the monkeys of the worlds Old and New into two profoundly separated groups? In the first place, the naturalists of the eighteenth and early nineteenth centuries noticed that in the typical New World monkeys the nose (fig. 23.31) was rather flat and bulbous, with the rounded nostrils well separated from each other and opening at the sides of the base of the nose; whereas in the Old World monkeys (fig. 23.38) the lower end of the nose was V-shaped, with the slit-like nostrils nearly meeting beneath the nose and opening more or less downward. On this account De Blainville (1834) named the New World monkeys *"singes à narines éloignées,"* while the Old World types were denominated as having *"narines raprochées."* From this it was an easy step for Sir Richard Owen (1868) to call the New World monkeys "Simiae platyrrhinae," and the Old World types "Simiae catarrhinae." But long before Owen's use of these terms, Hemprich (1820), in a textbook on natural history, had divided the monkeys into Platyrrhina and Catarrhina (Simpson, 1945, pp. 64, 66).

Another curious difference is seen in the high specialization of the strongly prehensile tail among the American monkeys and its at most incipient development among the Old World forms. Although not all American monkeys have a prehensile tail (the tail is bushy and nonprehensile in a few possibly specialized types), whenever it is present it plays an important part in climbing and can even serve in the grasping of objects not within reach of the hands. The muscles of the lower back and tail are highly developed, while the tail bones bear easily recognizable marks. Even certain centers in the brain of *Alouatta* (*Mycetes*), according to Tilney (1928, pp. 257, 258), reflect the functional importance of this "fifth hand." Widely different is the non-prehensile tail of the typical Old World monkeys, which hangs down behind them; when they leap it swings like an inert pendulum; yet it is sometimes possibly of some use as a drag or balance. But Herbert Lang observed that, in two species of mangabey monkeys in the forests of the Belgian Congo, although the tail is not technically prehensile, "having no hairless tactile pads on its under side as in some South American primates, it is often wrapped around branches in a semi-spiral manner to assist in securing a firmer hold in certain positions. The hair of the tail in young specimens is quite smooth but it is relatively heavy, —an indication of the increase of the muscular and sinewy attachments along the vertebrae which are used for climbing in later life." (Herbert Lang, in J. A. Allen, 1925, legend of Pl. LXXXVI.)

The shortening of the face and the formation of a postorbital partition, noted above (p. 468) is common to the two groups, but accompanied by various and significant differences: (1) In the side view of the cranium of the New World type (fig. 23.40A) we find that the expanded malar (*ml*) or cheek bone is wrapped around the back of the eye to such an extent that it has attained a sutural contact with the forwardly extending parietal bone, whereas in the typical Old World monkeys (B) the malar does not meet the parietal, from which it is separated by the greater wing of the sphenoid bone (A, *als*). (2) In the New World monkeys the tympanic bone (fig. 23.41A) forms a large open ring surrounding the eardrum and from the lower border of this ring projects forward and inward a large concave shell of bone (entotympanic) which is connected with the periotic or bone of the innear ear; whereas in the Old World monkeys the tympanic bone forms a long spout or funnel (B, C), narrow at the outer end and expanding suddenly on the inner end, where it is fused with the underlying periotic bone. This spout covers the under side of the long tube, the bony ear tube, the eardrum being fastened near its inner end. The bony ear tube has indeed a double function: on the outer side it serves for the attachment of the pinna or external ear; on the inner, for the attachment of the tympanum or drum membrane.

Thus the form of the tympanic in the Old World monkeys differs widely from that of their New World relatives and there is no direct fossil evidence that it has been derived therefrom; but in foetal stages of both the New World and the Old World series the tympanic bone goes through a stage in which it is a simple large ring incomplete at the upper end, surrounding the drum membrane. Essentially similar conditions are figured by van Kampen (1905, p. 673) for the young *Tarsius,* so that possibly the starting-point for both New World and Old World tympanic patterns may be found in different genera of Eocene tarsioids; especially since the adult *Necrolemur,* a European Eocene tarsioid, has an expanded tympanic bone bearing a short tube or spout (*ext. aud. meat*) on its outer side (fig. 23.28B), which would appear to be a suitable starting-point for the long gutter of the Old World type. On the other hand, in some other genus of tarsioids a continued lateral shortening of the tympanic spout, together with an expansion of its lower part, might easily result in the New World type.

Evidence from Enamel Structure.—The probability that the New World and Old World monkeys have been derived independently from different families of Eocene primates appears to be strengthened by the fact, stressed by Tate Regan (1930, pp. 383–392), who stated that "Thornton Carter's researches on the microstructure of the enamel of the teeth of the Primates show that in the Eocene Adapidae and the Madagascar Lemurs (Lemuriformes), and in the Catarrhina, the enamel prisms have straight edges and are separated by a small amount of interstitial substance, but that in the Asiatic and African Lemurs (Lorisidae), the Eocene *Hemiacodon, Tarsius* and the Platyrrhines, the prisms have wavy edges and are separated by a larger amount of interstitial substance." Tate Regan considered that "this investigation is decisive," and accordingly he worked out a new classification of the Primates and a diagram "showing relationships of the suborders and families of the Primates, in which the Lorisoids, Tarsioids and Platyrrhines are closely related by virtue of having enamel prisms with wavy edges, while the Lemuroids and Catarrhina (Old World monkeys, anthropoids, man) represent a cluster of families held together chiefly by the common possession of enamel prisms with straight edges."

The obviously weak points of this arrangement are that it leans too heavily on a single difference, which may have been acquired independently in different groups, and secondly, that it separates the Lorisidae too widely from the lemuroids, with which they agree in many important characters of the nose, hands, feet, external genitalia, dentition, etc. Thirdly, the diagram in question ignores the morphological difficulties involved in separating the Eocene "Omomyidae" from the Eocene Necrolemuridae and the Recent *Tarsius.* Nevertheless it can not be denied that the common possession of enamel prisms with wavy edges by some Eocene tarsioids (*Hemiacodon*) and by American monkeys adds somewhat to the probability of the inference that the New World monkeys arose in the western continent from some of the highly diversified tarsioid stock.

In evaluating the evidence from enamel structure on the problem in hand, we may well keep in mind the following thought. It is characteristic of one modern school of systematists that in classifying fishes and mammals they seek to discover one decisive character which appears to be constant in a wide diversity of forms and which they then choose to be the leading one in their analytical finding keys. But the experience of the majority of modern vertebrate palaeontologists, who are in closest contact with real phylogenetic divergence, indicates that single characters may be stable for long periods and then suddenly begin to vary; also that single characters may hold true throughout a large series of forms and then become variable and of little systematic value in others. The names of the larger groups are usually based on some single leading character; but the practice of neglecting to make a wide and strictly comparative analysis of all available characters bearing on the phylogenetic relations of the forms classified would, if it became general, tend to set back the science of classification to a pre-Linnaean stage in which classifications were merely arbitrary and unnatural assemblages.

Tate Regan's classification and phylogeny of the Primates (1930) was rejected by J. P. Hill (1932), who held that the resemblance in the straight-edged enamel prisms between the catarrhines and the true lemuriform stock does not seem to be enough to warrant the direct derivation of the former from the latter, as in the phylogenetic diagram of Tate Regan. Hill's memoir on "The Developmental History of the Primates" adds weight to the evidence for his conclusion, which is as follows (p. 162):

"The fundamental agreements that exist in the details of the early developmental processes in the existing Platyrrhine and Catarrhine Monkeys justify the postulation of a common ancestral stock from which both took origin. For this hypothetical stock I have adopted the designation Pithecoid. The resemblances in question lend no support whatever to the idea that these two groups of Monkeys are of diphyletic origin and were evolved quite independently, the Platyrrhines from a Lorisiform stock in N. America and the Catarrhines from Lemuriform ancestors in the Old World. Such an origin takes no cognisance of the enormous developmental hiatus that exists between the Lemuroid and the Pithecoid and implies developmental parallelism of an unprecedented order and, moreover, for such to occur in two groups so closely allied by their major morphological characters that systematists by common consent place them together in the same sub-order, seems to me in the highest degree improbable."

Taking the investigations of J. P. Hill, G. B. Wislocki, Elliot Smith, Tate Regan and others into consideration, it seems probable that the New World and Old World monkeys represent divergent offshoots of an earlier primate stock, or stocks; that the New World monkeys have been derived from New World Eocene tarsioids (in the widest sense), the Old World monkeys from Old World tarsioids, which were distantly related to but not of identical genera with their New World cousins. So far as known records indicate (1948), the common stock source of all primates, including New World and Old World monkeys, apes, man and recent lemuroids and *Tarsius,* is represented collectively by the known Paleocene forerunners of the tree shrews, plesiadapids, notharctids, adapids and tarsioids.

The known fossil and recent Old World monkeys show a comparatively restricted range in their dentitions (figs. 23.40B, C; 23.44–47C): all retain the same dental formula $\left(I\frac{2}{2}, C\frac{1}{1}, P\frac{2}{2}, M\frac{3}{3}\right)$, all tend to have more or less sabre-like canines, at least in the males, with a shearing surface on the antero-external face of the anterior lower premolar; nearly all have bilophodont upper and lower molars, the single partial exception being noted below. Moreover, all have "cheek pouches" (due to expansible buccinator muscles) and flat sitting-pads or ischial callosities (fig. 23.39), attached to the expanded lower ends of the ischia. Accordingly there is but one family, Cercopithecidae (Lasiopygidae), but there are two sharply marked subfamilies, (1) the Cercopithicinae, including among many others the macaques and baboons; and (2) the Semnopithecinae (Presbytinae), including the frugivorous and leaf-eating langurs and related Asiatic monkeys, which have a large sacculated stomach. The body-form varies from the small slender arboreal *Miopithecus talapoin* (fig. 23.42) to the burly dog-like baboons (fig. 23.43).

From Short-faced Macaques to Long-faced Baboons.—Among the macaques and baboons there are many existing stages (fig. 23.44) from the primitive short-faced macaques (A) to the excessively specialized long-faced mandrill (F). Many correlated changes can be followed in this transformation, among which we may note:

(1) the marked anteroposterior elongation of the upper and lower jaws, of the palate and molar teeth; (2) the consequent forward and downward slope of the ascending ramus of the mandible; (3) the increase in size of the jaw muscles and bony crests; (4) the development in the males of great manes or ruffs around the shoulders and head and (5) of longitudinal skin folds of bright red and blue along the sides of the muzzle and (6) of long curved upper canine tusks. Evidently this group is now undergoing a period of rapid speciation, or mutation, the mutations involving body size as well as many other characters. The existence of structural intergrades between the short-faced macaque (A) and the extremely long-faced baboon (F) does not in itself prove that the series was orthogenetic (undeviating) rather than branching. However, the skull and dentition of the short-faced macaque (A) are far less different from the generalized Old World monkey type (fig. 23.42) than are those of the dog-faced baboons (F).

In some of the macaques the lengthening of the face after reaching a certain point has been overtaken and surpassed by an increase in the height of the face. Thus apparently has been produced the strange face of the Gelada baboon (*Theropithecus*) of West Africa (cf. fig. 23.39B). This beast specializes in tantrums and intimidations. Bouncing up and down and making truly horrible grimaces and threats, he probably succeeds in bluffing all but the most desperately hungry leopards in his native home. The limbs and backbone of the baboons (fig. 23.43) in comparison with those of the primitive talapoin (fig. 23.42) have been modified for secondarily quadrupedal locomotion on rough ground and rocks. Especially in the larger baboons (fig. 23.39B), the limbs have been shortened and remodelled so that the general contour is somewhat dog-like. But the hands and feet retain clear evidence of derivation from a climbing stage. Thus the great toe, though somewhat reduced, is still divergent and grasping, while the hands, though beginning to suggest the paws of running mammals, retain their grasping power.

Even among the macaques the beginnings of social life, including mutual assistance and accommodation, have been discovered by the psychologists. In the rough school of life the young macaque slowly climbs the grades of social education, especially in robbery, intimidation, favoritism and unearned privilege. The mores of the baboons, as recorded by Zuckerman (1932), foreshadow the worst systems of human ethics.

Herbivorous Old World Monkeys.—The herbivorous monkeys (Colobinae or Semnopithecinae), with sacculated stomachs, are largely arboreal. Their skulls vary from the small, almost globe-like cranium and short jaws of *Pygathrix rubicunda* (fig. 23.46D), though the more massive cranium, obliquely-deepened face and robust jaws of *Colobus satanus* (B) and culminate in the almost orang-like rounded cranium and deep heavy jaws of *Rhinopithecus bieti* (H). One of the strangest-looking (fig. 23.38D) of all the higher primates belongs in this division, namely, the *Nasalis,* or proboscis monkey, whose females and young have turned-up noses.

With their large brains, strong curiosity and piercing eyes, monkeys evidently scrutinize the faces which interest them. Their diversified and conspicuous color patterns, their whiskers, crests and other facial adornments (cf. Elliot, 1912, vol. 2) may be useful partly for concealment in the light and shade of the forest, but also as tribal marks and individual features.

The Puzzling Oreopithecus.—Most of the fossil members of the Old World or catarrhine monkeys date from the later Tertiary and of these only one (*Oreopithecus*), from the Pliocene of Italy, shows a marked departure in its molar patterns (fig. 23.47B2, B) from the general catarrhine type. As described and figured by Schwalbe (1915), the upper molar crowns (B2) of *Oreopithecus* bear a pattern of four roundly conical "mountains" arranged in two transverse pairs but with a sharp "crista obliqua" running obliquely outward from the protocone to the metacone (B2, *cr. obl.*), and a short "crista posterior" (B2, *cr. post.*) running from the hypocone (*hy*) obliquely forward and outward to the crista obliqua. Similarly the main lower molar cusps (B) were arranged in two transverse pairs, but with a long "crista obliqua" running obliquely from the hypoconid (3) to the metaconid (2), and a short "crista anterior" running inward and backward from the protoconid (1) to the base of the crista obliqua. The third lower molar bore a hypoconulid (cusp. 5). Both upper (p^3, p^4) and both lower (p_3, p_4) premolars were roundly bicuspid, the anterior lower premolar crown being neither compressed nor shearing. The upper canines (*c*) were small and could not have been tusk-like; there were apparently no large diastemata in either the upper or the lower jaws. The anteroposterior length of the second deciduous molar (*dm* 2) equalled that of the two permanent premolars.

On account of the mixed resemblances of the *Oreopithecus* teeth, on the one hand to those of Old World monkeys (C), on the other to those of anthropoid apes (D), Schwalbe (*op. cit.*, p. 219) made for it a new family, "Oreopitheciden," inserted below the "Hylobatiden" or gibbons. A new and direct comparison of Schwalbe's figures and casts of *Oreopithecus,* received from the Paris Museum, with the Siwalik anthropoids shows (fig. 23.47) that Schwalbe correctly identified the parts of the upper and lower molar patterns with those of apes and monkeys; but, notwithstanding the further agreement in dental formula, analogous molar parts are found also among such Eocene bunodont artiodactyls as *Cebochoerus* (fig. 21.99A), which are, however, far more primitive in retaining p^2 and p_2. To sum up, *Oreopithecus* may well be a unique survivor from a basal catarrhine stock, related on one hand to the herbivorous monkeys, on the other to the tailless or anthropoid apes; but there is a bare possibility that it may be a monkey-like descendant of some short-faced Eocene artiodactyl such as *Cebochoerus.*

Postscript, June, 1950. A recent paper by Hürzeler (1947), describing all known material of *Oreopithecus,* adds further evidence for Schwalbe's conclusion. This genus tends to lessen the gap between the cercopithecoid and hominoid stocks.

TAILLESS OR ANTHROPOID APES: OUR SIMIAN COUSINS (PONGIDS)

A remarkable combination of characters is displayed in an incomplete lower jaw from the Upper Eocene of Burma, named *Amphipithecus mogaungensis* by Colbert (1937). The generic name *Amphipithecus,* meaning virtually "both ways ape," is appropriate, for if classified only by reason of the presence of three lower premolars on each side, this little jaw would be referred to the New World or platyrrhine stock; but when judged by the totality of its known characters, it seems to be a true anthropoid (in the wide sense); but *Amphipithecus* alone among Old World forms, has retained p_2 of the typical placental dentition, p_3 and p_4 having given rise to the so-called "p_1" and "p_2" of all other known Old World monkeys, apes and man.

In the Lower Oligocene of Egypt was found a minute jaw, the molars of which bore conical cusps arranged virtually in pairs (fig. 23.47A). This was named *Apidium phiomense* by Osborn (1908), who suggested that it might be a primitive stage in the ancestry of the catarrhine monkeys. Its real relationships, however, are uncertain.

The Gibbons and Their Forerunners.—From the same locality and horizon came two other very important and famous small lower jaws, one named *Parapithecus* (fig. 23.48A, A1), the other, *Propliopithecus* (B, B1), by Schlosser (1911). In *Parapithecus* the jaw was wide in the rear, pointed in front, and generally suggestive of a tarsioid type. The dental formula was $I\frac{}{2}\,C\frac{}{1}\,P\frac{}{2}\,M\frac{}{3}$, as in Old World monkeys and apes (fig. 23.58). The lower incisors (A) were slightly procumbent but not at all rodent-like—an important difference from the majority of the tarsioids; the lower canines were larger than either of the premolars; the premolars had roundly conical tips and barely incipient indications of trigonid and talonid. In the lower molars were four main cusps (1, 3, 2, 4) and a low inconspicuous median posterior cusp, or hypoconulid (5). By pairing the four cusps transversely, these molars could conceivably give rise to the Old World bilophodont monkey type, or, by emphasizing the hypoconulid (5) and abandoning the tendency to develop cross-crests from the opposite pairs, one could derive from the *Parapithecus* molar the anthropoid ape pattern (B). As far as known, *Parapithecus* is a far better link between the whole Old World monkey group (including even the apes and man) and the tarsioid stock than any other known member of the Old World monkeys, which have all become diversely specialized away from a primitive *Parapithecus* type. The sabre-like upper canines in male Old World monkeys (figs. 23.42–46), the compressed and shearing surface of their anterior lower premolar (= p_3 of *Notharctus*), and the sharply bilophodont character of the molars, are found in almost all the known catarrhines, except *Oreopithecus* (fig. 23.47B). In view of the probable origin of the Primates as a whole from plesiadapid-like ancestors (fig. 19.17B) with somewhat enlarged median procumbent lower incisors and reduced canines, it seems not improbable that the sabre-like form of the canines in male catarrhine monkeys, as well as the shear-like surface of the anterior lower premolars (really p_3 of the primitive dentition), are later specializations which had evidently not yet been acquired by *Parapithecus*.

The second little fossil jaw (fig. 23.48B) from the Lower Oligocene of Egypt was named *Propliopithecus* by Schlosser because it resembled *Pliopithecus* (C), the "ancestral gibbon" of the Miocene and Pliocene of Europe. The jaw itself was remarkably large in proportion to the small size of the teeth, the horizontal ramus or body of the mandible being both deep and robust, the "angular" region of the mandible expanded and the condyle elevated high above the cheek teeth. All this clearly implies relatively large, strong muscles in proportion to the cross-section area of the teeth. Hence the penetrating and crushing power were relatively great and this little ape could probably crack and grind quite tough objects. Similar conditions in the New World monkeys produced the relatively very large deep mandible of *Callicebus* (fig. 23.34aB), but with somewhat different molar patterns. This great power of the jaws was of enduring value to most of the later apes, in which the excess of power to tooth size was maintained as high as it needed to be in view of their rapid rise to gigantism.

Apparently the hypoconulid, or median posterior cusp (fig. 23.58B, 5), had developed at an earlier stage from the basal cingulum. It was already present on all three molars of *Propliopithecus*. It later aligned itself (5) with the proto- and hypoconids (1, 3), so that there came to be three main cusps (1, 3, 5) on the outer, or buccal, side of the lower molars and two (2, 4) on the inner. Thus the two-and-two symmetry of the molar cross-crests of Old World monkeys was never attained in the anthropoid apes, and though there were low transverse crests on the meta- and entoconids (cusps 2, 4), the proto- and hypoconids (1, 3) failed to develop strong cross-crests (C-F). This was undoubtedly an expression of one of the very many genetic differences that became cumulative after the separation of the two groups (monkeys and apes) possibly in Middle or Upper Eocene time. And since *Amphipithecus, Parapithecus, Propliopithecus* and their immediate successors were all found in Asia, Africa or Europe, there is good evidence for the conclusion that the monkeys and anthropoid apes of the Old World originated there, possibly from some Upper Eocene tarsioid allied with *Parapithecus*.

The next clear records of anthropoid apes are from the Miocene of the Vienna basin and from the Lower Miocene of Tanganyika, East Africa. In the former were numerous fragmentary fossils named *Pliopithecus* (figs. 23.48C, C1; 58C), which have been monographed by Hoffman (1893). In the general plan and details of its dentition this ape was ancestral to the modern gibbons and at the same time not distantly related to *Proconsul* (fig. 23.57A) and the Miocene ancestors of the chimpanzee. But we shall return to these genera presently.

With the gibbons (fig. 23.51) of Southeastern Asia and the East Indies we come to a new and revolutionary departure from the types produced by the lower suborders and families of primates. For the gibbons are masters of the "flying trapeze" and of the method of progression called "brachiation." Living in tall trees with long swaying branches, they run out to near the end of a branch and as it bends and swings back, they leap forward and catch with their long hands and arms on the branches of the next tree. Perhaps because of their great daring and selfconfidence and of the necessity for instantaneous action in landing, the gibbons are subject to great risks from falling. A. H. Schultz (1944, p. 124) found that many gibbons in their native forests carried the marks of former injuries, especially of broken limbs.

When on the ground the gibbons run erect, with their long arms raised like balancing poles, their knees flexed but swinging outward and their strong great toes directed inward. When resting, they sit upright, although their ischial callosities are reduced. They can hold a banana in one hand, using the thumb and index finger of the other hand to strip it.

The gibbons are, on the whole, the first of the man-like apes, but they show a mixture of relatively primitive with specialized features. To begin with the latter, the gibbons have paralleled the baboons and other Old World monkeys in developing long upper canine teeth (fig. 23.53A), especially in the males, in which these teeth are compressed with sharp-cutting edges. The blade of the upper canine overlaps the shearing surface on the antero-external part of the crown of the anterior lower premolar (fig. 23.48aD), but all this is probably only a parallelism with the baboons, because the very primitive *Propliopithecus* (fig. 23.48A) had very little indication of this specialization. The lower anterior premolars of the gibbon are compressed and rather elongate anteroposteriorly. All the lower molar teeth have very low crowns (fig. 23.48D) without cingula and as compared with those of *Pliopithecus* (C) they give the impression of being less omnivorous, more adapted for mashing fruits. The upper molars (fig. 23.57D) have three main low conical cusps and a low hypocone, with a rounded central basin bounded laterally by the internal slopes of the para- and metacones. As is often the case when sabre-like upper canines are present, the horizontal part or body of the lower jaw (fig. 23.48D) is stout but not deep, in contrast with the deeper jaw in the primitive *Propliopithecus.*

The cranium of the gibbon (fig. 23.51) is moulded around the voluminous and well-developed brain and the muscular crests are not strongly emphasized. The orbits (fig. 23.53A) are large but the orbitotemporal fissure is less narrow than it is in the higher apes. The interorbital constriction of the nasal chamber is much less than it is in typical catarrhines. The nostrils, although V-shaped, or catarrhine (fig. 23.49A), show an early stage in the lateral expansion, which reaches its climax in the gorilla. The gibbon also retains a conspicuous, relatively primitive feature on the inner wall of the cranium: a large subarcuate fossa (fig. 23.63A) on the mesial surface of the periotic. This fossa is present in at least the great majority of mammals of many orders and it is retained in all the lower primates, including the lemuroids, tarsioids, platyrrhines and catarrhines, up to and including the gibbon; but in the higher anthropoids (fig. 23.63B) and man (C) it is much reduced or wanting.

The postcranial skeleton (fig. 23.51) of the gibbons abounds in characters which are both well-suited to its particular habits and prophetic of the conditions in the later anthropoids and man. For example, the chest is wide and rounded in section to afford room for the large lungs and heart, while the abdomen is slender or of moderate girth. The clavicles are retained and form strong rods connecting the arms with the sternum, which are useful for the muscles that raise and swing the very long arms; thus the shoulders are far apart, in contrast with the narrow width across the shoulders in fast-running carnivores and ungulates. The lower arms can be freely pronated and supinated and the hands, although serving as suspension hooks in swinging, can also be used as true hands. The arrangement of several of the arm muscles, according to Strauss (1941), is highly specialized in connection with brachiating habits.

The centra of the lumbar vertebrae are wider in proportion to their length than are the corresponding parts in animals that run fast on all fours. The paraphophyses of the lumbars, which are prolonged in fast-running quadrupeds, are here rather short. The sacral region is short, the first sacral vertebra being moderately expanded transversely. The coccygeal vertebrae are fused into a coccyx. The lack of a tail, which distinguishes the anthropoid apes and man from most of the lower primates, may well be due to a mutation which occurred very early in the history of the anthropoid group.

In the pelvis of the gibbon the ilium (figs. 24.6, 7) has already begun to lose its rod-like or trihedral section and to widen the areas for the iliacus and deep gluteal muscles. The ischial callosities, although retained as vestiges in the skin, make little or no imprint on the lower end of the ischium. The long hind limb of the gibbon is on the whole fairly man-like except for the sharp divergence of the hallux and the failure to form a bridge, or instep of the foot.

The still existing gibbons include three genera (*Symphalangus, Hylobates, Brachytanytes*), the skeletons of which have been most carefully measured by Schultz (1933, 1944).

Fossil Apes of East Africa and Asia (Dryopithecinae). — From the Lower Miocene of Kenya, Dr. Hopwood (1933) of the British Museum (Natural History) has reported upon a collection of fossil mammals which includes teeth and incomplete jaws referred to several peculiar genera of apes. All the apes are probably related to *Pliopithecus,* but the most noteworthy is *Proconsul* (fig. 23.50A), so named because of the resemblance in its jaws and dentition to "Consul," a chimpanzee of the London Zoological Gardens. Additional discoveries of *Proconsul* material were later made by and under the direction of Dr. L. S. B. Leakey and described by D. G. MacInnes (1943) and by Leakey. Of these the most important were (fig. 23.50A): (1) the front part of the bony face with the lower rim of the orbit; it included also the nasals, maxillae, premaxillae, left upper canine, all the cheek and the greater part of the bony palate and upper dental arch; (2) the lower jaw, with a nearly complete lower dental arch and symphyseal region. Much of the left ascending ramus, including the mandibular condyle, was preserved. As a whole, the dentition of *Proconsul* (figs. 23.50A, 57A), besides being probably ancestral to that of the chimpanzee (fig. 23.57C), retains many basic features seen in *Pliopithecus* (fig. 23.58C). In another direction it is plainly more primitive than that of *Sivapithecus* (B) and related fossil genera. The high ascending ramus of the mandible (fig. 23.50A) is inclined more or less backward as in baboons and some orangs; together with the long nasal bones it indicates a high, long face somewhat like that of the cercopithecid monkey *Erythrocebus* (fig. 23.45A).

Most authors would probably assume that *Proconsul* could not be ancestral to man because it displays so few unequivocally human characteristics; but that may well be because, at the relatively remote period (Lower Miocene) in which *Proconsul* lived, distinctively human characters had not begun to be differentiated from primitive ape characters. In the upper molars of a related genus (*Xenopithecus*) from the same formation, the internal cingulum is emphasized beyond the condition in *Proconsul* and the molar cusps are bunoid or low and conical. This condition may represent a relatively late stage in the transformation from a small sharp-cusped, insectivorous molar to one with a low and wide crown with a more or less wrinkled surface. The presence of either internal or external cingula seems to mark a relatively rapid peripheral growth of the base of the crown and a delay in the growth and calcification of the cusps.

Dryopithecus (tree ape) includes several species of fossil apes (figs. 23.55; 23.56A-A3) from the Middle Miocene and Lower Pliocene of Europe (Spain, France, the Rhine and the Vienna basin). A closely related genus, *Sivapithecus* (C-C3) occurs in Lower Pliocene of the Siwalik Hills in India. These and other named extinct genera of India are all based on fossil teeth and a few imperfect jaws, but the patterns of all the teeth are so characteristic and distinctive that they constitute important documents of the evolution of the anthropoid apes and man. We may for brevity's sake describe them collectively, noting the minor details and technical differences only when essential to the larger story.

The incisors of the subfamily Dryopithecinae, although seldom preserved (fig. 23.55A), seem to have conformed to the general type that is common both to anthropoids and man: that is, the central upper incisors had a transverse cutting edge (fig. 23.56C1); in the lateral upper incisors the cutting edge if developed was narrower; both the lower incisors had rather narrow cutting crowns and were slightly procumbent. The upper canines were large in the males (fig. 23.55A), much smaller in supposed females (fig. 23.56C3). The larger male canines were not as recurved or sabre-like as in the gibbons but more robust, with a straight vertical groove on the anterior edge and a basal swelling on the inner side, all much as in *Proconsul.* In one supposed female canine (figs. 23.56C1, C3; 73B) the tip was low and there was a strong basal swelling and internal cingulum or basal tubercle (*b.t.*); this tooth was therefore described as sub-premolariform (Gregory, Hellman and Lewis, 1938). The lower canines were smaller and had lower crowns than the uppers. When the lower jaw was closed (fig. 23.56C3), the tips of the lower canines fitted into the paired diastemata between the lateral incisors and the upper canines. Both upper premolars (C1) were bicuspid with transversely oval crowns and indications of anterior and posterior cingula. Both had three roots (fig. 23.74H), two on the outer, one on the inner side (Pilgrim, 1915). The anterior lower premolar (fig. 23.55) was set obliquely behind the stout subvertical root of the lower canine. Its anterolateral face bore an oblique shearing surface which articulated with the posterointernal face of the upper canine. This part bore the enlarged protoconid (*prd*) and was supported by a large anteroexternal root. The posterior, sloping part of the crown, corresponding to the talonid, was directed more or less obliquely inward and was supported by a root. Within the genus *Sivatherium,* one species, *S. pilgrimi,* has the axis of the crown of this tooth (as seen from above) less oblique (fig. 23.56C2) and the shearing face suggests that of an Old World monkey; but, in another case (C5) referred to, *S. indicus,* the axis of the crown is more oblique, the shearing surface is becoming swollen, the posterointernal part of the crown tends to form only an internal spur. This type definitely suggests the "bicuspid" form of lower anterior premolar which is occasionally attained in the chimpanzee (fig. 23.59D) and has become further modified in man (H).

The first and second upper molars (fig. 23.56A, C) of the Dryopithecinae were quadritubercular with rather low conical cusps, the surface varying from sharper ridges and crests in *Dryopithecus fontani* to a flatter but more wrinkled surface in *Palaeosimia.* The lower molars (figs. 23.55A1; 56A2-A4; 58D; 60C) inherited the basal plan of three cusps: protoconid (1), hypoconid (3) and hypoconulid (5) on the outer side, and two (metaconid (2) and entoconid (4)) on the inner side. For convenience and brevity, Gregory and Hellman (1926, pl. X) gave numbers to these cusps, 1, 3, 5, on the outer, 2, 4, on the inner side of the crown.

Between the five cusps was a system of oblique and transverse grooves in the enamel surface (really fissures between the areas of the growing cusps) arranged around a central Y (figs. 23.56A3; 58D; 60A-C). This entire complex of five main cusps with their bounding grooves was designated as the *"Dryopithecus* pattern" by the writer in 1916. By comparison of this lower molar pattern with those of older Primates (figs. 23.58A-C) it was shown that the trigonid of the primitive mammalian molar is represented in *Dryopithecus* only by cusps 1 (protoconid) and 2 (metaconid) and by the groove at the front end ("fovea anterior"), which is a trace of the trigonid basin. The talonid basin of the primitive molar pattern is represented by the "central depression," which is surrounded by cusps 2, 3, 5, and 4, and into which the protocone of the upper molar fits.

Hrdlička (1924) proposed a special terminology for the lower molar patterns of *Dryopithecus* and other apes. He invented the terms *precuspidal fossa* for the "fovea anterior" of authors and *postcuspidal fossa* for the central basin of the lower molar.

The lower jaw of *Dryopithecus* and its allies is remarkably stout and deep (fig. 23.55A), with a massive sloping symphysis and broad ascending ramus. Such a jaw requires a strongly-braced skull with large projecting malars, heavy zygomatic arches and supraorbital braces, and a strongly-built cranium, but not necessarily a high sagittal crest.

Considered as a whole, the Dryopithecinae range from small forms (*Ramapithecus* Lewis) about the size of a gibbon, to *Sivapithecus indicus* (fig. 23.58C), of which some jaws exceed those of an adult male chimpanzee. There are minor differences in the form of the molar cusps, the height of the molar crowns and the degree of wrinkling of the enamel; and a rather wide variation in: (1) the shape of the upper canine and of the anterior lower premolars; (2) the varying relative breadth of the lower molars, the lowest breadth index of m_2 being 78.6 in *Sugrivapithecus gregoryi,* the highest, 106, in *Bramapithecus thorpei* (Gregory, Hellman and Lewis, 1938, p. 8); (3) the third upper molars also vary, especially in the degree of development of the hypocone; (4) the relative depth of the lower jaw in comparison with the length of the molars.

The *Dryopithecus* group was revised by Pilgrim (1915), Gregory and Hellman (1926), and Gregory, Hellman and Lewis (1938), who made extensive comparisons with the teeth of recent anthropoids and of fossil and recent man. They concluded that different members of the *Dryopithecus* group gave rise at different times to the orang, chimpanzee, gorilla and man. The views of Weidenreich and Broom on this subject will be considered later (pp. 488, 490).

Very little is known as to the limbs of the Dryopithecinae. A humerus referred to *Dryopithecus fontani* and a femur (*Paidopithex*) suggest relationships with gibbon and all the other apes, and man. The ulna from the Miocene of Austria named by Ehrenberg (1938) *Austriacopithecus* combines ape-like and baboon-like features (cf. *Oreopithecus,* pp. 474, 475). The humerus from the same region in measurements comes close to the chimpanzee (*ibid.*, p. 95).

The Deliberate Orang.—A peculiar side branch from the Dryopithecinae of India led to the orang (fig. 23.62), which lives in the rain-forests of Sumatra and Borneo, where their long rusty-colored or dark orange hair sheds the rain and protects them like a cloak. We may infer that with the advent of giantism the orangs reduced the relative length of their leaps from tree to tree and relied more on their increasingly long reach and the progressively hook-like form of their hands. As their habitus became more sloth-like, even their hind limbs became more and more specialized for grasping, tension and suspension, and in climbing they are able to reach remarkably far forward and upward with their feet. The end result was that their hind limbs have become poorly adapted for supporting their increasing weight on the ground and that they use their very long forearms as crutches (fig. 23.52).

The whole head of the orang has become highly specialized. In the shape of the cranium (fig. 23.50) the orang suggests a King Charles spaniel, because in both cases antero-posterior cranial growth has been retarded at an early stage and the round infantile dome is retained in the adult. In the adult orang, however, the cranium is greatly widened at the base (fig. 23.54B) and is supported by stout transverse arches through the malars and orbits and across the occiput (Bluntschli, 1929). The large orbits or eye-cones (fig. 23.49B) are shifted so directly forward that sometimes between them the upper part of the nasal chamber is almost squeezed out of existence. In typical Old World monkeys and gibbons, partly as a result of the forward direction of the eyes, the supraorbital ridges are well developed; but in the orangs (fig. 23.50), due to the backward and upward growth of the cranium, the supraorbital ridges have retreated and no longer project much above the orbits, which are directed somewhat upward. The lack of a sagittal crest may be associated with the great increase in the size of the masseter muscles, as indicated by the massiveness of the malar arches and the expansion of the masseteric surface on the angular region of the mandible. The bony face is massive and produced downward and forward, while the lips can be protruded far forward. The auditory opening, and with it the postglenoid process, have been displaced backward and upward. This has been associated with a lengthening of the temporal fossa and an increase in the cross-section area of the temporal and internal pterygoid muscles. The upper dental arches of the orang vary considerably in form (Hellman, 1918) but are often well rounded and wider across m^2 than across the canine (fig. 23.54A).

Meanwhile the front of the jaws, the lips, and the median upper incisors (fig. 23.54A) have become very wide. In the males the canine tusks are powerful weapons and are also useful in cracking open the thick rinds of the durian and other fruits upon which the orang feeds. The mandible of the orang retains the primitive depth and has even increased its thickness. The molars are massive, with rather high crowns and flattened surfaces covered with enamel wrinkles which largely conceal the four primary cusps (fig. 23.54A).

The laryngeal sacs become enormously inflated, especially in old males. Escaping through the embryonic gill-clefts in the hyoid complex, they extend downward beneath the skin, covering the front of the chest, reaching into the pits of the arms and swelling out the cheeks into great subcircular discs (fig. 23.49B). These new polyisomeres, which are so conspicuous in the orang, are also well developed in the gorilla (H. C. Raven) and traces of them are recorded in man.

The brain of the orang, except for its shortness and roundness, in general resembles those of the chimpanzee and gorilla. Its convolutions are more complex than those of the gibbon and especially in certain regions are approaching a human pattern. In the temporal region of the brain Fr. Connelly (1936) records in certain orang skulls an evident tendency toward folding over to form "opercula" around the insula of Reil. In man this folding over is believed to be connected with the correlation of visual and auditory association centers.

Captive orangs in zoological gardens often convey an impression of keen awareness of the things that interest them. I once saw a certain orang in the London Zoo, who took a long barley straw and flicked the bearded end over the baseboard of his cage into an open leader or gutter with water in it; immediately he jerked back the lower half of the straw like a fishing-pole, grabbed the swinging wet end with his other hand and drew it between his lips. Several times he cast his line in this way, bringing up a little water each time and solemnly passing the straw through his lips. Another very large old orang in the Pretoria, South Africa, zoo was begging, with his hand shoved far out between the bars of his cage. He was seated on a piece of burlap bag and when he arose to go to the other side of the cage, he took the burlap with him, using one hind foot as a hand to drag it along, and swiftly threw it into place as he sat down. In observing these and other living apes we recognize in them the kind of behavior which is ordinarily regarded as evidence of "mind." On another day I saw the same old orang rolling something around and when he opened his mouth I was alarmed to see in it a jagged piece of a small mirror. Evidently it pleased him to do this as he made no effort to remove the object or to spit out the sharp triangular broken piece of it. I at once sought out the keeper and told him what the orang was doing. He replied that the animal had done that before and when told to give up the mirror had held it out and then quickly put it back in his mouth. If the philosopher Descartes could have observed the deliberate behavior of this ape, he might have hesitated to classify all animals as mindless automata. Or else he might have agreed with the Malays in regarding the orang-utan as a wild man of the forests.

Orangs, as well as chimpanzees, assuredly

enjoy themselves even by doing "fool things." Perhaps it would be more accurate to say that, with their keen memories for things that interest or amuse them, they not only respond quickly in simple sensory-motor sequences but have the ability (a) to compare a present sensory pattern with their memories of similar past experiences, (b) to recognize the common terms between complex present and past situations and (c) to respond in ways that have pleased them in the past. They also respond to sudden psychosomatic impulses, such as to get up, move about and sit down somewhere else, as if tired of doing nothing.

The Lively Chimpanzee.—The most lively and adventurous of the existing great apes is the chimpanzee, who can run very quickly on the ground and is fully at home in the trees. His arms are shorter than those of the orang and his hands, though long, are wider in proportion to their length (fig. 24.10a). When on the ground the chimpanzee normally stands and runs on all fours (fig. 23.49C1), the long forearms used somewhat as crutches. The muscular strength per pound of body weight of a certain chimpanzee was more than twice as great as that of an average male student in an agricultural college (Bauman, 1923). This great margin of strength over weight, together with the long arms and short legs, enables the ape to climb rapidly and hurl himself about with surprising quickness. With this great strength goes a keen realization of the spatial relations of all parts of the body to gravitation and to surrounding objects. Fortunately the chimpanzee preserves for us what appears to be a fairly central anthropoid type of body, which avoids the excessively long forearms and very long hands of the orang and the gigantism of the gorilla. The chimpanzees, and indeed all the great apes, seem to be relatively heavy-bodied derivatives of the small primitive *Propliopithecus* type; but it also seems probable that the very small stem-anthropoid had less elongate and slender limbs and hands than those of the gibbons (fig. 23.51). Increasing bulk has brought increasing width of some parts, as in the ilia of gorillas (fig. 24.7C), and a relative shortening of others, as in the hind limbs of gorillas (fig. 23.61).

That the chimpanzee has been derived from a semi-bipedal, gibbon-like ancestor is suggested by the fact that the outside of the hand and the knuckles, rather than the palms of the hands, are used for supporting the weight of the body; in other words, the hands are being secondarily used as feet rather than as grasping and climbing organs (fig. 23.49C1). In spite of the length of its fingers and the relative shortness of its thumb, the chimpanzee can learn to pick up a needle, using the thumb and index finger as a pair of pincers. In disproof of statements in the literature implying that the thumb of the chimpanzee is not fully opposable, H. C. Raven took photographs of his chimpanzee grasping a cylinder and indisputably opposing the thumb to the tip of the finger. With her eyes able to focus on an object held in the hand and with a "mind" determined to overcome obstacles, Mr. Raven's "Meshie" learned to untie knots and to pass herself through many loops tied around her body. In climbing up a pole she showed the greatest skill in flipping upward the chain by which she was tied. In brief, the manual dexterity and quickness of this chimpanzee was amazing.

Meanwhile the experimental psychologists, especially Mme. Kohts (1935), Professor Yerkes (1916–1948) and his colleagues, have given a great diversity of tests to chimpanzees, which are usually highly coöperative and eager not only to secure the rewards but to be doing something. With great zest they keep at the business of the moment, whether it be stacking boxes or listening through their ear-phones for the signal to press the button, or choosing the right token, which they may insert in the "chimpomat" to secure different kinds of prizes.

The dentition of the chimpanzee (figs. 23.50B; 53; 54; 57; 58) is derived immediately from the *Pliopithecus-Dryopithecus* type (especially *Proconsul*) described above. The width across the upper canines in adult males (23.53B) is so pronounced that the upper dental arch (fig. 23.54) often diverges from the molars toward the canines. These canines are strong tusks and with them the animal is ready at any time either to bite savagely or to use them as picks for tearing things apart. The central upper incisors are large with sharp incisive edges and basal swellings, the lateral incisors smaller, more pointed; the lower incisors have narrow high crowns. The anterior lower premolar varies from the more compressed and shearing type to the sub-bicuspid stage (fig. 23.59D). The lower molars retain the *Dryopithecus* pattern (figs. 23.58E; 60E) but are typically smaller than those of orangs of equal skull length; the crowns are very low and the cusps rather delicate. The shape of the ascending ramus and coronoid process of the mandible and the position and height of the condyle above the cheek teeth vary considerably (G. S. Miller, 1915, pl. 5).

On the whole, the chimpanzee is the least modified of the modern descendants of the Dryopithecinae, which extended from Spain through Egypt and East Africa to India. Its immediate ancestor may well be *Proconsul* (fig. 23.50A), as suggested by Hopwood (1933).

The Massive Gorilla.—Rather nearly related to the chimpanzee but more narrowly limited to the African rain forest is the gorilla, the giant among recent anthropoids. A male of the Highland gorilla (*Gorilla beringei,* fig. 23.49aD) of the Belgian Congo, weighing 460 pounds, was by no means the largest (H. C. Raven, 1931), while a certain West African gorilla (*Gorilla gorilla*), which had lived long in the San Diego, California, Zoo, attained an abnormal weight of over 600 pounds. With such immense weight and strength in proportion, adult male gorillas need to fear nothing but man on the ground and they guard their young ones against the leopards, which are the only large carnivores that inhabit the forests. At night the females and young make their beds of branches in the trees, while the big male makes his bed at the foot of the tree by seizing first one armful of underbrush after another and sitting down upon it. In the daytime they take alarm at the slightest noise made by man and go bounding away, the big ones easily forcing a pathway through the underbrush. The male acting as rear guard occasionally rears up, beats his mighty chest with a reverberating rattle of strokes and utters a terrifying scream of defiance. If followed persistently, he may occasionally charge, hurling himself at the hunter. On such an occasion H. C. Raven (Gregory and Raven, 1936, p. 269) stopped a charging gorilla with a shot in the face, the animal dropping not more than fifteen feet away from him.

These massive beasts tear up the shoots of certain bushes and draw them through their mouths, stripping off the green surface with their canine teeth and throwing away the white stems. They are very fond of sugarcane, cutting off the tender buds and chewing the stalks. They invade the vegetable gardens of the natives and easily push over the banana trees to get at the tender leaves at the center. The abdomen of the male gorilla is very large and the stomach and intestines can carry an immense quantity of vegetable matter in solution. In a certain Highland gorilla which was being preserved for anatomical study by an expedition from Columbia University and the American Museum of Natural History, the stomach and intestine in spite of their great size were essentially human in type, the stomach being neither subdivided into large compartments as in the ruminants, nor sacculated as in the leaf- and fruit-eating members of the Old World monkeys. The vermiform appendix of this specimen was essentially similar to one figured by Huntington (1903) and by Lorin-Epstein (1932) as a variant of the human appendix. This illustrates the curious contrast between the somewhat baboon-like habitus, including the general contour and appearance of the gorilla, especially when on all fours, and the numerous skeletal and other characters in which both the gorilla and the chimpanzee are structurally nearer to the ancestral source of man than they are to the baboon.

In the skull of the male adult gorilla the bony face is long, being produced downward and forward in front of the eyes, which look straight forward beneath the prominent supraorbital torus (fig. 23.50D). The upper nasal chamber is not restricted and the lower nasal chamber bears the same number and essentially the same arrangement of maxillo-turbinals as in the chimpanzee and man. There is a well developed frontal sinus, and maxillary antra and sphenoidal sinuses are capacious. The nasal bones are much less reduced than they are in the orang and there is often a low but distinct nasal bridge. The nostrils are excessively wide with flaring rims (fig. 23.49D), so that the nasal V of the Old World monkeys has had its wings greatly expanded. The rim of the nostrils is strengthened by alar cartilages as in man. The facial muscles of the gorilla and chimpanzee are advanced beyond the monkey type and approach the human in several characters (E. Huber, 1930).

The brain of the gorilla as described by Tilney (1928) and others is, so to speak, "almost human," not only in its general construction but in numerous special features. Thus Le Gros Clark (1927, p. 475), in describing the brain of the young gorilla "John Daniels II," gives the following summary:

> "Attention has been especially drawn to the following points: the relative complexity of the parietal lobes in contrast with the frontal and occipital lobes; the remarkable extent to which the insula is concealed by surrounding opercula, so that the insula on the right side is completely buried from the surface, thus reproducing the condition to be found in the human brain; an asymmetry of the brain and venous sinuses which corresponds to the typical asymmetry of the human brain."

In adult males the upper canines are tusklike and the palatal arch has nearly parallel sides (fig. 23.69), due to the anteroposterior lengthening of the molars and doubtless also to the shape of the tongue, which is longer and relatively narrower than in man.

The upper and lower incisors and canines (fig. 23.69) are of the anthropoid ape type, the canines very large in the male, much smaller and with a low crown in the female. The upper premolars are supported by two outer and one inner roots (fig. 23.74I). The molars, elongated anteroposteriorly, bear sharp-edged cusps with a minimum of enamel wrinkling (fig. 23.70). The lower anterior premolar approaches the sub-bicuspid type (fig. 23.59E). In correlation with the deepening of the bony face below the orbit, the ascending rami and condyles of the mandible (fig. 23.50D) are very high, which permits nearly simultaneous opposition of all the cheek teeth. In order to operate this great dental mill, the temporal, masseter and internal pterygoid muscles and their crests and surfaces are all strongly developed (fig. 23.62bA).

In response to the powerful stresses initiated not only by the heavy jaw muscles but by the immense extensor muscles of the neck which are inserted on the nuchal platform (figs. 62b; 50D), the cranium of the gorilla (and to a less extent that of the chimpanzee) is strengthened with various bony arches and tracts, the whole forming systems of balanced trusses and forces around the mandibular and occipital condyles as pivots (figs. 23.54; 64). As a part of these stiffening systems the squamous (*sq*) branch of the temporal bone (fig. 23.62b) usually overlaps the frontal at the pterion, thus excluding the alisphenoid (*als*) from contact with the parietal (*pa*). On the inner side of the braincase (fig. 23.64A) the opposite frontals (*fr*) send down processes which in adults tend to cover up the contact between the mesethmoid and the sphenoid and to afford firm bases for the immense supraorbital ridge. The presence of stiffening crests and tracts of bone around the periphery of the jaw-muscles makes possible the thinning of parts of the skull roof, where, however, the bone although thin is exceedingly tough. These and similar habitus characters are usually found also in the chimpanzee and are plainly structural adjustments to the anthropoid ape way of living or *vice versa*; but they are cited by Wood Jones (1929, pp. 338–340) as tending to show that these apes are not related to man. However, these descending processes from the frontals in anthropoids merely overlap the ethmoid bones but they do not disrupt the old contact between the sphenoid and the ethmoid (fig. 23.64A) which lies beneath them (Gregory, 1928).

The vertebrae, limbs and girdles of the gorilla

reflect its near relationship to the chimpanzee but many of the bones have been transversely widened, due in part to the great total weight, in part to the emphasis of infantile characters, as in the width of the ilium. The ilia are expanded transversely (figs. 24.6, 7), for the wider abdomen and the gluteal muscles; the femur is short and relatively wide, for the huge muscles of the thigh. The spines of several of the cervical vertebrae are enlarged for the great muscles of the neck, which are inserted on the nuchal platform (fig. 23.61). The gorilla's hands (fig. 24.10a), including the digits, especially in comparison with those of the light-bodied gibbon, are widened in proportion to their length, and in the adult gorilla carpus the centrale is fused with the scaphoid. The sole of the foot (fig. 24.11a) is very large, wide, and secondarily plantigrade (H. C. Raven, 1936). In walking, the great toe is directed inward as in all anthropoid apes. Nevertheless the muscles of the foot (fig. 24.12) are clearly homologous with those of man, differing chiefly in their relative sizes and directions.

The detailed comparative studies by Wislocki (1932) of the placentation and female reproductive tract of man, anthropoids and lower primates lead him to make the following general statement: "The study of the reproductive tract of man places him undoubtedly among, but not equally close to, all of the Simiae. From the standpoint of reproduction, man shows the closest anatomical relationship to the anthropoid apes and not to the catarrhine or platyrrhine monkeys. The present findings amply support this statement, in spite of the contrary findings of Bischoff and Bolk upon the external genitalia. It is true, nevertheless, that in regard to the topography of the external genitalia, man does not closely resemble the great apes, especially in post-natal ontogenetic stages. But, when considered from the standpoint of the internal reproductive tract and placentation especially, his close kinship to the great apes can not be denied or dismissed" (p. 196).

In the male gorilla the os penis is present but it is reduced to the merest vestige (Weinert, 1932, p. 174). In man it is entirely absent. Thus the gorilla has nearly lost one of its supposed distinctions from man.

EXTINCT APE-MEN FROM SOUTH AFRICA (AUSTRALOPITHECINAE)

In 1925 Professor Raymond A. Dart of the University of the Witwatersrand, Johannesburg, South Africa, announced in *Nature* the discovery of a fossil skull, containing a natural endocranial cast of a new type of anthropoid, which he named *Australopithecus africanus* (African southern ape). This uniquely important fossil (figs. 23.65A; 66A) had been found in a quarry of "desert limestone" near Taungs, on the eastern edge of the Kalahari desert in Bechuanaland, about 220 miles southwest of Johannesburg. It was seen in the house of a farmer by a student of Professor Young, and the latter obtained the skull and handed it over to Professor Dart for study and description. It was a small skull, not wholly unlike that of a chimpanzee child, with all the deciduous teeth in place as well as the first permanent upper and lower molars. But it differed from young chimpanzee skulls in the larger brain with more fully developed parietal area. The deciduous molars (fig. 23.67) were more like those of man and the first permanent upper and lower molars much larger than those of chimpanzees, relatively wider than those of gorillas, with more man-like cusps. Professor Dart further noted that this ape lived more than a thousand miles south of the known ranges of the chimpanzees and gorillas. Although the occipital condyles were missing in the specimen, he concluded from the conformation of the surrounding bones that the head was held more erect than in the chimpanzee and that the creature was more bipedal.

There were many fossil baboon skulls in the cave from which the type had come and from a close study of the ways in which they had been broken before fossilization, Dart inferred that the adult Australopitheci had broken the baboon skulls open to get at the brains. He also pointed out that the discovery of this skull tended to support Darwin's inference that Africa, which is the home of man's nearest existing relatives, the chimpanzee and gorilla, may also have been the scene of the great transformation from ancestral anthropoid to man.

Dart's epochal discovery was received with scant enthusiasm by many of his colleagues. Dr. Schwartz, a mammalogist, after comparing Dart's description with skulls of young gorillas, published a brief note in *Nature* (1936, p. 969) suggesting that the new form was a gorilla, possibly a new species. Dr. Hrdlička (1925) said in effect that Dart's specimen was inadequate as a basis for any satisfactory conclusions because, being that of a young animal it was not comparable with the type skulls of *Pithecanthropus* and Neanderthal, which belonged to old individuals! Wolfgang Abel (1931) came to the conclusion that its resemblances to human dentitions were mere parallelisms and that the principle of irreversibility estopped us from considering *Australopithecus* as anything but another end form!

However, Sollas (1925), Elliot Smith (1925), Broom and the present writer (1933, 1938, 1939, 1940) supported Dart, after having studied the excellent casts and photographs of the original which he had placed at our disposal.

In 1936 Dr. Robert Broom of the Transvaal Museum, Pretoria, the veteran South African palaeontologist to whom the mostly unknowing world is indebted for the discovery of many important stages in the evolution of the mammal-like reptiles, published the first note on "A new fossil anthropoid skull from South Africa," describing parts of the skull (fig. 23.65B) which had been obtained from another "limestone quarry" at Sterkfontein, not far from Krugersdorp, about twenty miles northwest of Johannesburg. The fossils had been blown out of the solid "bone breccia" and after locating the spot from which they had been blasted, Dr. Broom obtained valuable data. After several years' collecting in this quarry, he had secured most of the skull, including the upper jaw with alveoli of the incisors and canines and the crowns of the premolars and molars; a second smaller specimen included half of the upper dental arch, with well preserved lateral incisor canine, anterior premolar and first upper molar also the ascending branch of the maxilla just in front of the orbit. Additional specimens include the front part of the lower jaw and a well preserved lower canine. Several other bones including a wrist bone and a toe bone were found later. These have all been described by Broom in *Nature,* and finally in his memoir on "The South African Man Apes" (1946).

In 1938 through the courteous invitation of Drs. Broom and Dart and their respective institutions, the writer and his colleague Dr. Milo Hellman visited South Africa and made an independent study of Dart's and Broom's material, securing casts, photographs, measurements and observations. From this material we made provisional reconstructions of the upper and lower dentitions and dental arches (fig. 23.70) of Broom's *Plesianthropus,* applying the principles and procedures which had been worked out in our previous studies on the occlusal relations of the upper and lower teeth in man and other mammals (Hellman, 1919a, b; Gregory and Hellman, 1926; Gregory, Hellman and Lewis, 1938).

At that time Dr. Broom had recently discovered in another locality (Kromdraai) about two miles distant from Sterkfontein the incomplete skull and dentition which he made the type of *Paranthropus robustus* (fig. 23.65C).

Since then he has discovered many important parts of these ape-men, especially: (1) a well preserved skull of an old female *Plesianthropus*; (2) some parts of two amazingly large jaws, the teeth of which are appreciably larger than those of the huge fossil jaw from Java, named *Meganthropus* by von Koenigswald; (3) a small pelvis of human or subhuman type, with transversely widened ilia; (4) several other parts of skeletons that collectively indicate an erect posture. Meanwhile, Dart, working in another locality (Makapansgat) has secured beautiful dentitions and jaws of his *Australopithecus prometheus,* so named because of its supposed fire-making habit. The several braincasts range from about 450 cc. in *Plesianthropus,* to 650 cc. in *Paranthropus*

(Schepers, 1946); that is, they are within or near the size range for gorillas.

In the course of time various other types of fossil mammals have been found in the bone breccia of the Sterkfontein and Kromdraai caves and have been described by Broom. The geologist Cooke (1938) who made a careful study of the present Sterkfontein cave, infers that it represents the remains of an earlier cave, the roof of which had collapsed, and that the bone breccia was contained in a landslide which had filled the cave. This occurred during a late "pluvial" period in the sequence of "dry" and "wet" phases recorded in this region.

Extensive excavations by the University of California, South African expedition in 1947, 1948 showed that there were four main horizons or phases: (1) the gray breccia, with remains of horse and antelope but without baboons; (2) the pink breccia, containing baboon skulls of the type found at Sterkfontein and Makaspansgat; (3) open caves containing Middle Stone Age implements sealed in the floor; (4) others with more recent artifacts, showing Bushman techniques (C. L. Camp, Science, 1948, p. 550). At present writing (April, 1949), further studies of the still accumulating material bearing on the geologic ages of the deposits at Taungs, Makapansgat, Sterkfontein, Kromdraai and other quarries are being made by Dart, Broom, Camp, Barbour, Cooke, Van Riet Low and others. Meanwhile the question whether the South African ape-men are the direct ancestors or merely the granduncles of modern man is relatively unimportant. Certain it is that they combined brains of only ape-like bulk with teeth of human or near-human patterns. Their nearest known fossil relatives seem to be the "ape-men" (*Meganthropus* and *Pithecanthropus*) of Java and the Peking man (*Sinanthropus*) of China; the latter may, however, be of somewhat later age (Middle Pleistocene).

The outstanding characteristic of the adult male *Pleisanthropus* and *Paranthropus* was that they had massive jaws (fig. 23.66) and cheek teeth, the latter much larger than those of the Heidelberg man (fig. 23.77). This falls in line with other discoveries, as noted by Weidenreich (1945), tending to show that in early races of man the males had massive skulls and jaws, although they may not have been proportionately tall. The supraorbital torus (ridge), well preserved in the type of *Plesianthropus transvaalensis* (fig. 23.66B), was well developed and the cranium rounded above, without median crest. The braincast (dotted outline) was not longer than that of a large gorilla but was relatively wider (fig. 23.64a), indicating an advance beyond the gorilla stage. The lateral upper incisor (preserved in the female topotype, fig. 23.66B) had a relatively small crown of primitive human form tending toward the "medium shovel-shape" of Hrdlička's classification (1920); the median incisors are not preserved but from the space left for them their crowns were evidently wider than the lateral incisors—a feature common to apes and men. The upper canine crown, also preserved in the female topotype (fig. 23.66B), is quite small as compared with those of male apes, and in its shape also the crown is of primitive human type (fig. 23.73C); nevertheless it is readily derivable from the "subpremolariform" female canine (B) known in certain members of the *Dryopithecus* group (Gregory and Hellman, 1939, p. 345). From the form and position of the incisors and the character of their worn tips it is plain that there was an edge-to-edge bite of the upper and lower incisors and canines such as occurs in some aged apes and was normal at least in some primitive men. The lower canine, represented by a beautifully preserved left lower crown (fig. 23.68F), was found at the type locality in the matrix in juxtaposition with a symphysial fragment. The unworn crown is basically like that of certain female gorillas and orangs (E) but is much smaller and has a lower tip. Similar but still smaller lower canines (G) are found in certain females of *Sinanthropus* (Weidenreich).

There is decisive evidence that in *Plesianthropus* the diameter of the upper dental arch across the canines (figs. 23.68A, 70B) was considerably less than the width across the opposite m^2. The form of the upper dental arch was thus approaching the human stage.

Among the many bits of evidence tending to show that apes and men are the diverse derivatives of a common stock are the following facts: (1) that in both groups the number of permanent premolars is the same, namely, two on each side of the upper and lower jaws. These premolars correspond to p^3, p^4, and p_3, p_4, respectively, of the dentition of primitive placental mammals, although they are designated as p^1, p^2, and p_1, p_2 by most anthropologists; (2) that in apes and primitive men the crowns of "p^1," "p^2" are bicuspid (fig. 23.69A, B) and so similar to each other except in minor details that in some cases close scrutiny is required to distinguish either between p^1 and p^2 or between those of apes and men.

The upper premolars of *Pleisianthropus* (figs. 23.68A, 70B) with regard to size come nearest to those of the larger *Sinanthropus* males; in the reduction of buccal roots to one (Broom, 1936, p. 719) they also attain the *Sinanthropus* stage, but they are more primitive and ape-like in retaining unfused, widely divergent buccal and lingual roots; the outer or buccal face of p^1 (fig. 23.66B) lacks the asymmetry which is characteristic of its homologue in typical apes (fig. 23.74H); on the other hand p^1 is more ape-like than human in so far as the mesial and distal borders of the buccal face of the crown diverge more sharply from each other than in most men; again, in the occlusal aspect of the upper premolar crowns the elevation of the buccal above the lingual cusps is slight and the crowns wear into gently convex surfaces which can be nearly matched in certain human palates. Thus the upper premolars contribute their share in tending to show that *Plesianthropus* presents a unique mixture of ape and human or subhuman characteristics.

With regard to the lower premolars (figs. 23.68A1, B1; 74A-G), Weidenreich, for reasons that are fully set forth in his memoir (1937, pp. 53–60), concludes that "the first lower premolar [of the common ancestors of ape and man] has never been of a pronounced sectorial type but it was always more molariform, with a buccal and lingual main cusp and a distinct talonid." He also concludes (p. 168):

"Although the canine of *Sinanthropus* has already undergone a reduction. yet it cannot be surmised that it had at any time such a tusk-like structure as is true in anthropoids; the premolar likewise probably never had such a sectorial character as that found in anthropoids. These teeth apparently were the centre of special differentiations in anthropoids which were already effective in *Dryopithecus*. Hence, the hominids must have branched off from the common stem of the anthropoid stock before *Dryopithecus* and the other fossil members of this group developed. The stem was thus divided into two branches, one with a more homomorphic canine group leading to or represented by the hominids, and the other with a heteromorphic canine group leading to *Dryopithecus*, its relatives and its descendants, that is to say, to the anthropoids in the strict sense of the term."

Gregory and Hellman dissented from these conclusions, at least in part, and referred to their papers already published (1916; 1926, p. 94; 1938, p. 24) for detailed evidence in support of the following summary.

The oldest of the diversified and widely distributed dryopithecine group is the genus *Proconsul* (fig. 23.50A) of Hopwood from the Lower Miocene of East Africa. Although considered by Hopwood as an ancestor only of the chimpanzee, we regard it also as morphologically near to the stem of the entire ape-man stock. The crown of the first lower premolar in the presumably male jaw of *Proconsul* (Hopwood, 1933, pl. 6) has been badly worn by the upper canine and suggests the shearing type of other anthropoids. In the referred jaw described by MacInnes (1943) the excellent cast sent by Dr. Leakey shows that the first lower premolar (fig. 23.74A), although somewhat worn and on the whole larger than p_2, yet provides the essential basis for both shearing (B-D) and bicuspid stages (E-G). The crown of p_2 in *Proconsul* is incipiently molariform, with parts recognizably homologous with the trigonid and talonid of more primitive mammalian molars. Both p_1 and p_2 are supported by two well separated roots; one anterior (mesial), the other posterior (distal) (Hopwood, 1923, pl. 6, fig. 8). The crown and roots of p_1 are set obliquely, so that the mesial root projects more on the buccal, the distal root more on the lingual side. Even in *Dryopithecus fontani* (fig. 23.55) of the late Tertiary of Europe, and in *Sivapithecus* (fig. 23.74B) of India, although the pointed tip of the crown is pronounced, both crown and roots of p_1 are rotated somewhat farther outward, the former shearing face of the crown has become more convex, the metaconid becomes enlarged and in a certain specimen of *Sivapithecus* (C) the tooth begins to approach the bicuspid pattern; the latter is carried further in certain chimpanzees (fig. 23.59D) and nearly perfected in man (E-G) (*cf.* Gregory and Hellman, 1926, pp. 94, 99).

An almost perfectly preserved second left lower premolar (fig. 23.68A1) from Sterkfontein is provisionally referred to *Plesianthropus*. It has a high trigonid, a lower talonid, and in these basic features it is reminiscent of the remote insectivorous-omnivorous lemuroid ancestors of the higher primates. It has advanced beyond the primitive ape condition of *Proconsul* in that its crown does not extend postero-internally into an angulate corner but is much rounded, with a thick, heavily beaded posterior (distal) basal ridge. It is more primitive than p_2 of *Sinanthropus* in having a wider crown and larger talonid. P_2 in *Sinanthropus,* however, varies considerably from a more asymmetrical contour with pronounced talonid to a symmetrical oval bicuspid with reduced talonid (Weidenreich, 1937, pl. XXVIII, figs. 273b, c, d).

The roots of this tooth (p_2) in *Plesianthropus* consist of a mesial and a distal root closely appressed. In general these roots are strictly comparable with those of primitive apes but are more appressed mesiodistally and thus intermediate between the conditions in apes and in man. The same is true with regard to the position of these roots in *Plesianthropus,* for in apes the mesial root of p_2 extends more buccally than the distal root and clear traces of this condition are retained in *Plesianthropus.*

P_2 in *Paranthropus* is essentially similar to that of *Plesianthropus*. Its crown (fig. 23.68B1) projects buccally into a well rounded protuberance; its antero-internal corner is rounded and it has a high metaconid (*med*). As compared with p_2 of *Sinanthropus,* its contour is more or less between Figures 273*b* and *d* of Weidenreich's Plate XXVIII.

The roots of p_1 and p_2 in *Paranthropus,* although imperfectly preserved, definitely correspond to those of p_2 in *Plesianthropus.* The most careful comparisons show that they are progressive beyond the dryopithecine stage and are transitional to those of primitive man.

Dr. von Koenigswald has kindly informed me that in *Pithecanthropus robustus* two specimens of the first upper premolar (fig. 23.74J) show all three roots of typical anthropoids (H, I); also that De Terra (1905, pp. 240, 241) and Busch (1896, pp. 164–174) have reported this condition as a rather rare variant in man.

From the foregoing it will be evident that both *Plesianthropus* and *Paranthropus* with their small front teeth and advanced premolars had progressed far in the human direction.

The fundamental identity in general plan between the molars of fossil and/or recent apes and of primitive men has been noted by many authors: e.g., Osborn (1892, 1907); Gaudry (1901); Röse (in Selenka, 1898–1903); Remane (1922); Gregory (1916); Gregory and Hellman (1926); but its full significance has been missed by several anatomists and anthropologists, notably Wood Jones, Straus and Broom. The molars of *Plesianthropus* and *Paranthropus* fall well within the limits of the general anthropoid-primitive human plan (figs. 23.70, 71, 72). They inherited from primitive apes the very large size of the third upper and lower molars; their first and second upper molars retained large undiminished hypocones (fig. 23.71); in the lower molars (fig. 23.72) the five main cusps and the sixth accessory cusp of the "*Dryopithecus* pattern" were all present; but the pattern as a whole was in a transitional stage (D, E) pointing toward the cruciform or plus pattern of human molars.

In the lower jaw the corpus or horizontal ramus (fig. 23.66) was very massive, as in apes. The symphyseal region was also very massive but lacked the "simian shelf" (Broom, 1947), which was also absent in the Dryopithecinae (Gregory and Hellman, 1926, figs. 4, 16, 17).

The most outstanding feature of *Paranthropus* was the unique flattening of the maxilla below and in front of the nasal floor (fig. 23.65C1). We carefully examined the original and could see no evidence of *post mortem* crushing or distortion. The surface is very smooth and the shape suggests that there must have been a great transverse dilatation of the nostril sacks and upper lip. An associated feature is the extreme forward position and expansion of the root of the malar bones, the opposite ones projecting slightly beyond the plane of the concave nose bridge. The *Paranthropus* face, as indicated by the type fossil, thus appears to be highly peculiar, exceeding Wadjak I (as represented in Damon's excellent cast) in the flattening of the face, reduction of the canine and forward position of the spreading cheek bones. Its cheek teeth, on the contrary, are very large, while those of Wadjak I are reduced and have finely wrinkled enamel crowns. A remarkably human-looking character noted by Broom (1938, p. 379) is seen in the large semicircular bony auditory meatus which lies between the reduced postglenoid and posttympanic processes and is neither modified into a spout nor jammed against the postglenoid process as it is in the gorilla. However, in this feature the adult *Paranthropus* has merely retained a character which may be seen in infant skulls of orangs, chimpanzees and gorillas. In short, the upper dental arch (*Plesianthropus*) was of the convergent or paraboloid type (fig. 23.70), essentially as in man. The canines, at least in females, were small and more or less transitional. The anterior lower premolar (*Paranthropus*) was in general almost human. The upper premolars retained the primitive buccal and lingual roots, which are more or less fused in man, as did the lowers. The molars were very large, far exceeding those of Heidelberg and approaching those of gorilla in size but subhuman in the modification of the "*Dryopithecus* pattern" toward the plus or cruciform human type. The jaws attained enormous size and the skulls were massive, except perhaps in small females, the range in size being remarkably wide. In respect to the volume of the brain, however, the *Plesianthropus* type fell within the range of male gorillas.

The "child" skull, Dart's *Australopithecus,* reveals the deciduous dentition of a member of this subfamily. It has extremely small canines (even for milk teeth) and incisors (fig. 23.67); but its deciduous upper and lower molars are much more human than ape-like in their patterns; its permanent first molars are very large but approach those of *Plesianthropus* except in minor details.

Broom (1946, pp. 110, 132) rejects *Dryopithecus* as an ancestor of *Plesianthropus* and man, partly because, on the crown of the first lower deciduous molar of the gorilla, chimpanzee and orang (and thus presumably of their common ancestor) there is but one main cusp, whereas in baboons, *Australopithecus* and man the corresponding crown bears four cusps. Thus Broom, relying on this difference, discounts or ignores the many fundamental agreements in the deciduous and permanent dentition as a whole that are common even to modern apes and primitive man.

Both Weidenreich and Broom apparently expect to find diagnostic human characters in the canines and lower premolars at a period preceding the shortening of the face and transformation of the upper and lower dental arches (see p. 494); whereas Gregory, Hellman and Lewis (1938) and Weinert (1932), accepting the evidence of the anthropoid brain, reproductive system, blood precipitin tests, dentition and skeleton as a whole, regard the Australopithicinae and mankind as of probable Lower Pliocene or Upper Miocene derivation from some one of the several genera of the Dryopithecinae (but not necessarily from any of the Indian genera).

COLOSSAL FOSSIL HOMINOID FROM CHINA (GIGANTOPITHECUS)

In 1935 G. H. R. von Koenigswald, to whom the world owes also the discovery of *Pithecanthropus* skulls II, III, IV, reported that among fossils that he had purchased in a Hong Kong dispensary was a huge third lower molar (fig. 23.75F) exceeding the largest gorilla's m_3 in length and evidently representing a new genus and species, which he named *Gigantopithecus blacki.* Later he obtained another very large specimen, a worn second upper molar (L), which apparently belonged to the same genus and species, and still later, another third lower molar. From the color and other features of the attached matrix and of the fossils, he thinks it probable that they came from the "Yellow earth," or Sino-Malayan fauna of China, of Middle Pleistocene age (Weidenreich, 1946, p. 64). The marked elongation or narrowness of m_3 is normal in the Dryopithecinae and recent anthropoids, but not in ancient man, while the well worn condition of the crown suggests that this tooth erupted early as in anthropoids and that its eruption was not delayed nearly until maturity as in modern man. In its general pattern (fig. 23.75F) the crown pattern of the third lower molar of *Gigantopithecus* is plainly a derivative of the *Dryopithecus* pattern (D) and its main parts are clearly homologous with those of *Plesianthropus* and *Paranthropus.* Weidenreich (1945, pp. 77–89) regards *Gigantopithecus* as the largest and most primitive known member of the human group.

THICK-SKULLED OGRES FROM JAVA, CHINA, AND EUROPE

Human Titan from Java (Meganthropus).—In 1941, Dr. von Koenigswald published the discovery in the Lower Pleistocene of Java of the front part of an enormous lower jaw containing part of the symphysis, the alveoli of the incisors and canines and the well preserved premolars (Weidenreich, 1945, p. 15). This fragment (fig. 23.78D), which von Koenigswald named *Meganthropus paleojavanicus,* has been very carefully studied by von Koenigswald (1942), Teilhard de Chardin and Weidenreich (1945, 1946). It completely dwarfs all other then known human jaws, although it was itself eclipsed by the colossal *Gigantopithecus* (fig. 23.75F). The canine, as indicated by its alveolus, was relatively small and erect; the anterior lower premolar was of primitive human type in so far as its antero-buccal face was not flattened for shearing against the upper canine. Its crown and roots, however, are strongly asymmetrical, the large lingual root being obliquely postero-internal to the smaller antero-internal root which supports the "fovea anterior," or trigonid fossa. Although somewhat more advanced than the anterior lower premolar of *Paranthropus* (fig. 23.74E), all parts of p_1 of *Meganthropus* can still be definitely homologized with those of female chimpanzees and gorillas, especially with those in which the antero-buccal shearing face is but little enlarged or is rounded rather than flat (A, B).

Extensive comparative study by Weidenreich of braincasts, skulls, jaws and dentitions of all known forms of early Pleistocene man has led him to the far-reaching conclusion (1945, p. 124; 1946) that *Homo sapiens,* instead of being the descendant of pygmy ancestors, is really a diminished derivative of such supergiants as *Gigantopithecus* and *Meganthropus* and that, while the stature of later races has diminished, the relatively large brain of *Homo sapiens* has been inherited in part from the absolutely large brains of the Pleistocene supergiants. Further evidence in favor of his conclusion has been supplied by the gigantic jaws of *Paranthropus crassidens* from Swartkrans, South Africa, described by Broom. All signs now (1949) point to a rather close relationship between the South African Australopithecinae and the East Asiatic-Malayan *Pithecanthropus-Meganthropus* group. Definite associations between very large skulls or jaws and large limb bones are lacking. In South Africa some of the jaws and teeth of the Australopithecinae far surpass those of *Homo sapiens* in size, but the several and only known femora, perhaps belonging to small females, indicate the proportions of pygmies. Possibly both physical extremes (giants and pygmies) were present in the Pleistocene of South Africa, Java and China, as they were later among the Africans, on Mt. Carmel (Palestine), and in Southern Europe (Cro-Magnon, Grimaldi).

The Java Ape-man and the Peking Man.—The stories of *Pithecanthropus,* the "Java ape-man" and of *Sinanthropus,* the "Peking man," have been told so often and so well (*e.g.,* Weidenreich, 1946) that we may condense them here to the merest outline, dealing only with the questions: Whence came these palaeolithic low-brows, who were their nearest relatives, and to whom did they give rise?

Even when only the original skull-top of *Pithecanthropus* was available, and before the Peking remains had been discovered, careful comparative studies had convinced most anthropologists that *Pithecanthropus* belonged within the limits of the human family. Dubois (1897, 1923a, 1923b, 1934), however, who had worked out a formula designed to bring out the correlation between stature and brain mass, found that the ratio of brain mass to length of femur in *Pithecanthropus* was exactly what it would be if a gibbon's cranium had been enlarged to the size of that of *Pithecanthropus.* He regarded *Pithecanthropus* not as a direct ancestor of man, but as an independent line. He called it a "giant gibbon," a descendant of the common ancestral man-ape stock, which he conceived as having had a gibbon-like skull (Weidenreich, 1940, pp. 33, 34).

On the other hand, other anthropologists and students of the brain noted that the cast of the interior of the skull cap (fig. 23.64) of *Pithecanthropus,* which practically revealed the shape of the brain in top view, needed chiefly to be filled out a little in front and made somewhat fuller in the region of the parietal eminences in order to duplicate the cast of the Neanderthal, to which its general contours conformed rather closely but with somewhat smaller over-all dimensions. Much later, when good crania of the Peking skulls were available, it was seen by Davidson Black (1931) that *Sinanthropus* (fig. 23.7) was closely related to *Pithecanthropus*; indeed Weinert (1928, 1931) proposed to abandon the genus *Sinanthropus* and refer the species *pekinensis* to the genus *Pithecanthropus.* Later, three new crania, an upper jaw and a lower jaw of *Pithecanthropus* were discovered in Java by Dr. G. H. R. von Koenigswald (1942), who made a new reconstruction of the cranium. Fr. Teilhard (1943), the eminent French palaeontologist, carefully checked the evidence and confirmed the accuracy of von Koenigswald's results, which had been severely criticized by Dubois. Weidenreich (1945), after an intensive study of all the material, supervised the making of a new reconstruction of the entire skull of the larger form, which he named *Pithecanthropus robustus* (fig. 23.76).

The skull of *Pithecanthropus* as compared with that of an average white man was far more massive. In side view the cranium was not elevated but "bun-shaped," that is, long and low tapering in the rear. The frontal region (preserved in *P. erectus*), instead of forming a high nearly vertical, well rounded forehead, sloped sharply backward above and behind the orbits. Above the ear holes the cranium was relatively low, its sides sloping upward to the strong median ridge (A1). Thus the cross-section above the ear holes was strongly trussed, which would protect the brain against heavy strokes from above. The occipital surface was strengthened by massive marginal swellings. Thus the suboccipital muscles must have been correspondingl thick and powerful. In front view (fig 23.76A2) as restored by Weidenreich the uppe face was very wide, bounded above by the mas sive supraorbital ridges and at the sides by th wide and strong cheek bones. Unfortunatel the ascending or frontal rami of the maxilla o upper jaw were not preserved, so that the exac distance from the nose bridge to the tip of th premaxilla is unknown, but the known superio maxillary region of *Pithecanthropus robustu* was relatively wider and lower than it is i large male gorillas. The width across the cheek must have been large to accommodate the larg masseter and buccinator muscles which are im plied by the form of the lower jaw, which i much more massive than in modern man. Th very large upper dental arch, as partly pre served in *Pithecanthropus robustus,* wa rounded in front, with the canine not tusk-lik as in male gorillas, and the diameter across th canines was less than that across m²; *Pithecan thropus* was also much more human than apelike in the crown patterns of the canines, premolars and molars.

The beautifully preserved crania of the Solo men of Java collected by von Koenigswald and described in detail by Weidenreich and Koenigswald (1950), although distinctly later (Upper Pleistocene), and nearer to *Homo sapiens* in some features, also retained many primitive details inherited from the *Pithecanthropus* stage (Weidenreich). Unfortunately the jaws and teeth are unknown.

Weidenreich's intensive studies again reveal the fact that human skulls differ from those of apes not in basal morphological patterns but in the emphasis of some features and in the displacement or reduction of others (anisomerism).

In three skulls of *Pithecanthropus* the brain sizes range from 775 cc. in a well preserved normal adult, to a little over 900 cc. (Weidenreich, 1945, p. 93). The smaller capacity, Weidenreich notes, is not larger than that of a modern human infant of about eleven to twelve months of age. It is about double the size of the brain of an adult chimpanzee, but only one-quarter larger than the largest known ape brain. The La Chapelle-aux-Saints Neanderthal braincast measured about 1620 cc., while that of a typical modern man is about 1320 cc.

Thus if brain volumes were reliable criteria of intelligence (which they are not), the smallest-brained *Pithecanthropus* would have been only a little more intelligent than a male gorilla; while the large-brained Neanderthals would have been considerably more intelligent than an average modern European. But whatever the inherent possibilities of the Javan and Peking people's brains may have been, their bank of learning and tradition was still in a relatively early stage of accumulating a favorable balance. In Mid-Pleistocene times their capacities both for deliberate evil and for good were limited. Their rage may have been fearful but after the killing came the genial relaxation of the feast.

When *Pithecanthropus* is given the benefit of every doubt, his ogre-like features (fig. 23.76) can hardly be overlooked; he would therefore be a veritable bogey to all true Fundamentalists and anathema to those learned pundits who so long reviled and rejected him.

The Peking man was named *Sinanthropus pekinensis* originally on the basis of a single lower molar tooth, in which Davidson Black (1927), with almost prophetic insight, saw a new type of Palaeanthropic man, lower than the Heidelberg or Neanderthal species. The long continued and systematic explorations of the Geological Survey of China, sponsored by the Rockefeller Foundation and carried on under the successive direction of Davidson Black and Franz Weidenreich, have yielded a rich harvest of evidence concerning the Peking man and the kind of world in which he lived. The extremely thick skull bones, strong brow ridges (fig. 23.77A), massive jaws (B) and efficient grinding dentition of the *Sinanthropus* people were probably of high survival value in their savage world, in which the rule of *Eat or be eaten* came first. Anthropologists suspect that the Peking man and indeed all the earlier cave dwellers preferred the brains and the marrow of the long bones of men, cave bears, etc. Even the relatively small-toothed Eskimos perform wonders in cutting, tearing and chewing tough carcasses; while to the pygmies of the Congo forest and to the wild aborigines of Australia practically nothing that once lived is not toothsome. Gregory and Hellman in their comparative studies on the dentitions of anthropoid apes and man, noting that early man must have come upon the half eaten carcasses left by the lion, hyaena and cave bear, concluded that *Sinanthropus* and Neanderthal were opportunists who did not hesitate to be carrion feeders or cannibals, as well as vegetarians.

In Weidenreich's splendid memoirs (1936, 1937) on the mandibles and dentition of *Sinanthropus* is set forth the evidence for the wide variability in the details of the lower jaws and of the permanent and deciduous dentitions. The lower teeth of *Sinanthropus* (figs. 23.68G, 23.71D, D1; 72K, 75G) in many respects foreshadow those of the Heidelberg and Neanderthal men, but the dentition as a whole is definitely more primitive. Some of the lower canines (fig. 23.68G), for instance, retain clear traces of identity in plan with the lower canine of a certain orang, while others lose these traces and approach the low-crowned, less differentiated form of the Neanderthal and later types. In the unworn upper molar crowns (figs. 23.75K; 71D, D1) the primary grooves branch extensively, producing a more or less wrinkled effect; but the wrinkles are not as numerous or as delicate as in the orang (fig. 23.75C). In the lower molars (fig. 23.75G) the *Dryopithecus* pattern is disguised but not entirely obliterated by the wrinkling, although sometimes approaching the cruciform or plus pattern (Weidenreich, 1937, pl. XXX, fig. 296). The radiographs (skiagrams) of the lower molars (Weidenreich, 1937, pl. XXXIII, fig. 314) show that the so-called "taurodontism" due to the depression of the point of bifurcation of the roots, was somewhat less than in the Heidelberg lower molars but more pronounced than in modern man.

The anterior lower premolars (Weidenreich, 1937, pl. XXVIII, figs. 272, 273) are completely bicuspid with no unquestionable traces of a former shearing face (fig. 23.74F), so that, as noted above (p. 488), Dr. Weidenreich disbelieves that they were derived from such a stage and cites many other characters tending, in his opinion, to show that the pre-Peking stage was a still undiscovered type of anthropoid which had retained more "primitive" characters than had any of the *Dryopithecus* group.

Weidenreich in many papers cited strong evidence for the origin of the human brain, skull, erect posture and foot (1921–22, 1923) from a primitive anthropoid stage, yet so far as I know he never accepted the evidence cited by Remane (1921) and by Gregory and Hellman (1926, pp. 40, 94–101) as tending to show that when the entire bony muzzle was shortened and the canines were reduced, the lower canine slipped behind the upper, while the anterior lower premolar crown was twisted, so that its talonid and roots were turned inward while its roots tended to be reduced and fused into one. In brief, to reject the entire dryopithecine group as ancestral to man and apes and further to assume the existence of wholly unknown types of anthropoid apes with essentially human characters of the canines and anterior lower premolars, seems to be equivalent to evoking the unknown to explain the known, especially when the profound transformation of many other parts of the teeth and skeleton are freely admitted.

Weidenreich showed that several features of the Peking jaws and dentition are retained in jaws of North China people, including an obliquely longitudinal swelling torus mandibularis on the lingual surface of the jaw. There are also traces among them of the primitive median longitudinal ridge on top of the skull and of other *Sinanthropus* characters which collectively suggest derivation from *Sinanthropus,* at least in part.

Over fifty species of animals, mostly mammals, lived in the vicinity of Chou Kou Tien, the home of *Sinanthropus*. Their fossil relics, identified by Fr. Teilhard de Chardin (1929, 1934, 1936), agree closely with those of the Middle Pleistocene of Europe.

The Geological Survey of China used to call this "Lower Pleistocene" [of China] but Fr. Teilhard, Dr. De Terra, Dr. Weidenreich and others concur that it is about equivalent to the Middle Pleistocene of Europe. Extensive field studies in Java by Dubois, Selenka, von Koenigswald, De Terra and Teilhard de Chardin have afforded the adequate data for equating the *Pithecanthropus erectus* level of Java with the earlier *Sinanthropus* levels of China. If we use the last five thousand years of human civilization as a unit of measurement, the age of the *Pithecanthropus* and Peking fossils would be possibly one hundred times as great. But that would be only about one-half of the Pleistocene, which is conservatively estimated at one million years (Zeuner, 1946). And this staggering figure in turn is but one five-hundredth part of the estimated age from the beginning of the Palaeozoic era (see p. 9) to the present. Time moves in a deliberate way its wonders to unfold.

As to the possible relationships of *Pithecan*

thropus and *Sinanthropus,* Weidenreich (1945) showed that from the colossal *Meganthropus* (fig. 23.78D) downward there is a morphological series, including among others *Pithecanthropus robustus, Pithecanthropus erectus, Sinanthropus* (from large to small), Rhodesian man, Neanderthal, Predmost mammoth hunters, to Aurignacian man (*Homo sapiens*). McCown and Keith (1939) have shown that the fossil people of Palestine included both large and small individuals, the latter with a mixture of Neanderthal and *sapiens* characters. Nevertheless, Dr. Weidenreich's challenging thesis that the Javan supergiants may well represent the primitive stature of mankind and that modern races have been reduced in stature will doubtless be a major incentive for further exploration and discovery by anthropologists for several decades to come.

HEIDELBERG AND NEANDERTHAL

The Heidelberg jaw (figs. 23.77bC, 79A) was discovered in a sandpit at Mauer near Heidelberg, Germany, in 1907. It lay buried beneath some 79 feet of gravel, clay, silt and loess (wind-blown sand) and its discoverer, Dr. Otto Schoetensack of the University of Heidelberg, took all due care to have it photographed *in situ* before removing it from the matrix. According to good evidence, the deposit in which it lay was laid down during the long second interglacial stage of Mid-Pleistocene. In strength and massiveness this jaw bone surpasses those of the vast majority of mankind. The region of the chin does not protrude, as in modern man, but differs little from that of *Plesianthropus* (fig. 23.77bA). The longitudinal section of the symphysis (C), while basically ape-like, bears a small elevation; this appears to represent the "genial tubercles" of man (A. S. Woodward), to which was attached the tendon of the fan-like genioglossus muscle of the tongue. The exceedingly broad ascending ramus affords insertion for powerful jaw muscles. The teeth and the U-shaped dental arch are definitely but primitively human. The teeth, though small compared to the size of the jaw bone, are actually of robust size; the molar pulp cavities are incipiently deepened, prefiguring the specialized condition of Neanderthal teeth.

Although the Heidelberg jaw is gigantic in comparison with those of most modern men, it is not as large as that of *Pithecanthropus robustus* (fig. 23.76A) and quite small beside that of *Meganthropus javanicus* (fig. 23.78D). Its molar teeth are not much larger than those of some modern men and almost pygmy-like compared with those of *Gigantopithecus* (fig. 23.75L, F).

The Heidelberg jaw (fig. 23.77bC) is more primitive than that of modern man in lacking a chin, in the great thickness of its corpus mandibulae, and in the width of the ascending ramus. The patterns of all its tooth crowns can be nearly matched in modern man. The pulp cavity of the molars was less expanded ("taurodont") than those of typical Neanderthals, but more so than in typical modern man. Most of the features, in fact, indicate a structural stage intermediate between *Sinanthropus* and modern man (fig. 23.79C).

According to Osborn (1943), remains of Neanderthal man seem to have been found as early as 1700, but the first important specimen was obtained from a quarry at Gibraltar in 1848, and its significance was not understood for many years.

"In 1864, William King created the name *Homo neanderthalensis* for a fragmentary fossil skeleton found in 1856 in a cave deposit in the Neander valley (German: *thal* or *tal*) near Düsseldorf, Germany. But that a distinctive and archaic species of mankind had been discovered was not generally recognized until after the excavation of two skeletons, in 1886, near Spy, Belgium, together with numerous flint implements and fossil remains of mammoth, woolly rhinoceros, and other extinct mammals. To date, fossil specimens representing nearly a hundred individuals of Neanderthal man have been recovered, scattered from Palestine and Crimea to Germany and Spain. Most of them have been excavated, in a more or less fragmentary condition, from consolidated débris or as burials in the floors of caves and rock shelters.

"Various other scientific names have been used to refer either to Neanderthal man collectively or to individual specimens. Among these are: *Homo* *primigenius, H. mousteriensis, H. antiquus, H. europaeus, H. krapinensis, H. calficus,* and *Palaeoanthropus neanderthalensis.* According to international rules of priority, *neanderthalensis* is the proper specific name. . . ." *

According to Weidenreich (1946, p. 93), the Chapelle-aux-Saints Neanderthal skull (fig. 23.79B) described by Boule (1913) has a larger brain capacity (1625 cc.) than that of *Sinanthropus,* which ranges from 915 cc. to 1225 cc.; this in turn is considerably greater than that of *Pithecanthropus,* whose capacity ranged from 775 cc. to a little over 900 cc. In modern European man it ranges from 910 cc. to 2100 cc.

Most of the teeth in the aged Chapelle-aux-Saints skull had been lost during life, but in Le Moustier (fig. 23.80B, B1) the teeth are quite well preserved and retained much more primitive ape-like features than in typical modern man (C, C1). Thus the Neanderthal stage of the upper and lower cheek teeth was the next to the latest one in the structural series from primitive Mesozoic insectivorous pantotherians (fig. 24.13A) to man (G).

The fifth to seventh cervical and anterior thoracic vertebrae of the Chapelle-aux-Saints man had relatively long neural arches as compared with those of typical modern man, which was not surprising in view of the large size of the occipital plate, implying equally robust suboccipital muscles. The proportions of the body differed from those of a modern Australian aboriginal in having the skull and jaws larger, the braincase lower, the arms and legs shorter. Boule noted certain primitive ape-like features in the astragalus (1913) but he was opposed to the derivation of man from apes and regarded modern man, Neanderthal man and apes as independent lines.

PILTDOWN HIGHBROW

In 1911, a workman who was mending a road near Piltdown Common, Fletching, in Sussex, England, crashed his pick into a fossil human skull and threw the pieces into the spoil heap. Soon afterward, Charles Dawson, an English lawyer whose hobby was archaeology and who had long been keeping on the lookout for relics of early man in that vicinity, rediscovered these sad evidences of human ignorance. Carried to the British Museum (Natural History), the pieces were most carefully studied by the eminent palaeontologist, Arthur Smith Woodward, who patiently tried to reconstruct the skull, including the lower jaw, from the quite good pieces that were left. In his report to the Geological Society of London (1913, read December 18, 1912), he showed that while the lower jaw was almost indistinguishable from that of an ape, and the walls of the skull were excessively thick, the skull showed no trace of supraorbital ridges, the forehead was rounded and many of the details of the bones were remarkably like those of modern human skulls.

This skull was unquestionably found in the Piltdown formation, a gravel deposit made during the Pleistocene when the River Ouse was still far above its present trench and on a level of the "high terrace" on which Piltdown Commons lies. A considerable collection of fragmentary mammalian fossils was taken out from the same place, including, among others, teeth of rhinoceros, hippopotamus, horse, beaver.

Except for the human remains, these all belonged to extinct species of animals whose presence in deposits of the Pleistocene epoch had long been known from various localities in England. Among them was a piece of mammoth tusk which it was thought had been shaped by human hands but possibly owes its roughhewn end to the gnawing of beavers or porcupines; there was also a piece of a molar tooth of an extinct stegodont elephant, identified by Osborn (1929; 1942, p. 966; 1943) as *Archidiscodon planifrons*; several eoliths and a flint implement of Chellean type similar to those found in other interglacial deposits both in England and on the continent. The waterworn fragments of the mastodon and rhinoceros teeth, according to Woodward, had slumped and drifted down from levels above the Piltown Commons that were there when the River Ouse was only beginning to cut down its valley.

* *The Age of Man,* Amer. Mus. Nat. Hist., 1944.

In course of time, Smith Woodward's reconstruction of the Piltdown skull fragments was most critically and closely studied by many other scientists, including Keith, Elliot Smith, Pycraft, Gerrit S. Miller, McGregor and others. To make a long story unduly short, in the original reconstruction the midline was probably not exactly identified, the forehead was too low and the braincast too small. All this was corrected in subsequent reconstructions by McGregor and others. Whatever the precise figure for the brain capacity may have been (McGregor estimates it as about 1300 cc.), the fact remains that on the whole most of the bones of the cranium, apart from their great thickness, could be nearly matched among modern human skulls. On the other hand, the lower jaw and remaining molar teeth were so perfectly ape-like that several anthropologists have decided that it belonged to an entirely different sort of animal from that which owned the skull; it has accordingly been made the type of a new genus and species of fossil ape—the sole one known in England.

A canine tooth was found later at the original site by Fr. Teilhard de Chardin. Whether it is a right lower or a left upper is less important for the moment than the fact that it recalls the canines of a certain female orang (Gregory, 1914, p. 195); but if it belongs with the skull, it is certainly human or subhuman. The lower molar teeth, although much worn, are of the general *Dryopithecus* type, with five cusps and a certain arrangement of grooves (p. 479) which is common to anthropoid apes and primitive men. In particular, they rather closely resemble the worn crowns of certain orang molars.

Some time later, Mr. Dawson found in another locality three small pieces of a fossil skull of a "second Eoanthropus," showing part of the supraorbital bone (again without a supraorbital torus), a bit of the occiput and a single lower molar tooth of the opposite side from those of the type. The English authorities took this to be a complete confirmation of the association of the original ape-like jaw with the Piltdown skull, but to Hrdlička (1922, 1923), Weidenreich (1946, p. 23) and others the evidence was unconvincing, seeing that the new lower molar might well have belonged to the opposite side of the original jaw. But whatever the final verdict may be with regard to the lower jaw, it is certain that at the time when the Ouse was on a level far higher than it is at present, there was a thick-skulled, large-brained man present in England, the contemporary of many now extinct mammals of Pleistocene age.

The Javan fossil skull named *Homo modjokertensis* by von Koenigswald was that of a young infant with a smooth cranial dome. It was found in a relatively low level assigned to the Lower Pleistocene and may be the young of *Pithecanthropus robustus*. The subglobose form of cranium which is found in infant anthropoids was already becoming evident in several adult human skulls in early or Mid-Pleistocene times and especially in adult modern men it is to some extent a retained infantile feature. Hence it is possible that too high a systematic value has been put upon the presence or absence of the retreating forehead and supraorbital torus in man, even though adult males of catarrhine monkeys and of the anthropoid-human group (except *Proconsul*, orang and modernized man) have this torus.

HOMO SAPIENS

The Swanscombe (Kent) skull was found in the Middle Gravels of the "Barnfield pit" at a depth of 24 feet (7 metres) from the surface (Report on the Swanscombe Skull, Jour. Roy. Anthropol. Inst. of Great Britain and Ireland, 1938). These gravels are an alluvium of the Pleistocene Thames and belong to the interglacial period of the Middle Pleistocene. "The Committee [making the Report] is unanimous in accepting the Swanscombe skull as an indigenous fossil of the 100-foot Terrace of the Lower Thames." The skull was found *in situ*, June 29, 1935, by Mr. A. T. Marston, L.D.S., of Clapham, who had for some years visited the pit at regular intervals as a collector of Paleolithic implements. The associated flint implements are of Middle Acheulian type. "The associated fossil fauna (mammalian and molluscan) is of a

Middle Pleistocene interglacial type, testifying to a 'parkland' biotope and to a climate that was probably slightly warmer than that of the present day." . . . "The Barnfield Middle Gravels would probably be accounted by most Continental authorities, on the available palaeontological and archaeological evidence, as belonging to a late stage of the Great Interglacial Mindel-Riss period" (*op. cit.,* p. 97).

Although the occipital and left parietal bones are the only parts of this human skull recovered, both these are nearly complete; they show little sign of wear and they articulate perfectly. They yielded an endocranial cast of 1325 cc. estimated capacity. The cephalic index must have been close to 78. The chief exceptional features of this modern-looking skull were the unusual breadth of the occipital bone and the exceptional thickness of both bones. The skull was probably that of a female who died in her early twenties (*op. cit.,* pp. 97, 98).

Thus the Swanscombe skull of Mid-Pleistocene age, like the Piltdown of a possibly earlier horizon, was of the smooth, well-rounded type, with a relatively large, well-convoluted brain.

That *Homo sapiens* dates back at least to the Mindel-Riss interglacial period, of Middle Pleistocene Age, would be no surprise to vertebrate palaeontologists, who are familiar with the evidence that the latest, or almost latest, stages of evolution of many mammals had been reached in the late Pliocene or Pleistocene epochs. But, as Weidenreich (1937, 1946) has so fully shown, the fact that the earlier representatives of *Homo sapiens* were contemporaneous with the Neanderthal race does not neutralize the evidence of comparative osteology that the Neanderthal stage retained many relatively primitive features which were gradually lost in the more advanced members of *Homo sapiens.*

CHAPTER XXIV

MAN'S DEBT TO THE PAST

Contents

CHAPTER XXIV

MAN'S DEBT TO THE PAST

INTRODUCTION: SOLAR ENERGY AND LIFE (cf. Chapter I)

THE PREHISTORY of man has been discovered in part by archaeologists, who have dug into man's past from above downward, starting with the written history of Egypt, for example, and working back to the earliest known Egyptians. But the geologists and paleontologists, reading the "sermons in stones and books in running brooks," have brought to light records of the Procession of the Vertebrates extending from the Middle Ordovician period to the present time, an estimated span of perhaps four hundred million years. Using the combined results of geology, palaeontology and comparative anatomy, we have endeavored in the preceding chapters to locate the origin of this myriad host of the vertebrates and to follow its main branchings and deployments; but now it seems worth while to bring together and summarize the parts of this longer story which relate especially to the ancestry of man. We shall therefore start as near to the origin of the procession as we can and trace the successive branchings and turnings of the long road that finally led to the peculiar side path on which man is now travelling.

Marcus Aurelius Antoninus in his Meditations gives thanks to his parents and elders, from whom in the formative days of his childhood and youth he had received the guiding principles of his career. We too may well pay the tribute of acknowledgment to our remote ancestors, both for our successive patterns of life and for the physical equipment for living. Nor did they fail to add to the heritage of the race at every stage in the long and tortuous road of our ascent.

In order to understand more clearly the main structural stages which finally led to man, we shall also do well to take account of those which branched off in other directions; because even so they illustrate the general principles of evolution and especially do they indicate what paths of adaptation had to be either avoided or abandoned early by the ancestors of the vertebrates from ostracoderms to man.

1. FROM HYDROCARBONS TO PROTOPLASM

There was apparently no definite beginning in man's longer life history. Although the record is broken, many stages still exist from the lowest living things upward; and the chemists and biologists have amassed evidence that there has been a progressive up-building from the inorganic to the organic, from the simpler hydrocarbons to highly complex mixtures of the proteins and other substances with long chain molecules that constitute protoplasm.

Protoplasm is an extremely complex solar engine and the stages by which it was built up are not yet clear. But chemists have succeeded in analyzing the green and yellow pigment matter of plants and have concluded "that chlorophyll is derived in nature from the sugars which it is itself designed to produce through photosynthesis" (Encycl. Brit., 14th ed., vol. 5, p. 611). Other evidence also shows that the chloroplasts or chlorophyll-producing units under the influence of sunlight produce carbohydrates, whereas the nucleus has to do with the elaboration of proteins. Thus from very early times primitive plants possessed these two highly constructive kinds of organs, chloroplasts and nuclei.

There are annectant stages between chlorophyll-producing, one-celled plants (Protophyta) and one-celled animals (Protozoa). In the latter, the chlorophyll-producing elements are reduced ultimately to zero and only the nucleus remains. Hence not only the Protozoa but all types of animals, as well as parasitic plants and bacteria, depend either directly or indirectly upon the green plants, for these store up part of the vital energy of the sun's rays in their carbohydrate food stuffs. Hence also it is easy to see why all animals are equipped to take their food stuffs either from the plants or from other animals, and why Natural Selection has so often favored the evolution of aggressive predators, from amoeba to man.

Both plants and animals use the process of hydrolysis or solution, in preparing their food material for integration and incorporation into living tissue. Merely by increase in bulk the physical conditions of the organism's surface become different from those of its interior: layers and membranes form, differential osmotic pressures develop, while certain substances, tending to pass from within outward, may either be excreted or form a tough rind, bark or exoskeleton.

From the evidence of still existing primitive plants, botanists infer that the plant kingdom started its career in fresh-water and that the ability to live in salt water came later, as did also the invasion of the land. In a brief but very clear summary of the Evolution of the Plant Kingdom, Dr. Merle C. Coulter (in "The Nature of the World and of Man," Univ. Chicago Press, 1926, pp. 216–239) traces the succession from primitive water plants (thallophytes) to amphibious plants (bryophytes) to the first woody plants (pteridophytes).

At each stage the plants were accompanied by their remote relatives and enemies, the animals, which likewise became progressively complex but in different ways. Plants, having to conserve and utilize their chlorophyll, built up their leaves, stems and roots to support and protect it and to supply it with raw materials. Hence the land plants mostly became fixed to the land by their roots and as they grew they turned their leaves toward the sunlight. But they had also adopted the bisexual mode of reproduction and the single-celled spores of primitive water plants still swim by means of flagella in the manner of some one-celled animals. This method in both groups, as well as the alternation of generations, insured a wide dispersal of the seed.

There was indeed a good deal of parallelism and convergence between the earlier plants and animals: in the development of free-floating or actively swimming spores or larvae; in the development of either attached or stalked and radiate adult body forms. For example, there are stalked or plant-like forms among Protozoa (*Vorticella, Zoothamnion, Dendrosoma*); among Coelenterata (hydroids, hydroid corals, graptolites, sea-anemones, sea-fans, corals); Echinodermata (sea-lilies), Annelida (tubiculous annelids). But the animals, being relieved of the limitations and responsibilities imposed by the making of chlorophyll, were thereby free to specialize in methods of pursuit, capture and ingestion of their food and prey. Fast-moving, bilateral predators with segmented exo- or endo-skeletons are thus conspicuously absent in plants, and even among animals are found chiefly in the Arthropoda and Chordata.

As long as both plants and animals were minute, single-celled organisms, floating in water or propelling themselves by cilia or flagella, they required but little adjustment against the downward pull of gravitation or against the weight of one part against the other. But when increasing growth pressure from the assimilation of food resulted in cell division through many generations, and the daughter cells remained clustered around the mother cell, increasing mass posed many new problems.

In many-celled plants the outer cell layers tended to screen the inner ones from the sun's rays, with consequent slowing down of the making of chlorophyll and of food materials. In plant masses left by the tide in mud flats the outer layers would dry up into a tough film and thus tend to conserve the vital moisture in the interior (Coulter, *op. cit.,* p. 227). In such primitive amphibious plants as the mosses, the outer sheet-like layer forms a protective transparent cuticle beneath which are spaces housing the food-making chloroplasts. Carbon dioxide enters the chambers containing the chloroplasts by means of pores in the outer cuticle. Root-like processes from the bottom layers absorb water from the moist substratum.

As adaptation for land-living progressed, water, containing dissolved nitrates, phosphates, etc., was conducted upward from the roots to the food-making leaves by means of a system of tubes whose thick walls were strengthened by woody tissue. Another set of tubes ("phloem") conducted the food downward from the food-forming leaves to the trunk and roots.

Growth is checked and stopped by fall and winter, but in the spring a new layer or ring forms on and near the outer surface. The young plant can rear itself upward against the downward pull of gravitation by hydraulic growth pressure. As growth continues, the outer layers are supported by the inner layers, forming a cone-in-cone-like system which subdivides itself at the top into leaves and branches, at the bottom into roots.

2. ONE-CELLED PROTISTA

To his excessively remote single-celled ancestors of Precambrian age man owes the basic properties of every one of the billions of cells that collectively form his body. We may perhaps gain some idea of the probable character of these Precambrian cells from the living *Euglena,* which has not yet wholly lost the power to build up chlorophyll and is thus functionally intermediate between plants and animals. Each of these single cells, with its cell-stuff (cytoplasm), nucleus, limiting membrane and vibratile flagellum, was a little world in itself and in it the principles of polyisomerism (repetition of similar parts) and anisomerism (emphasis of some parts and reduction of others) ruled just as universally as they do in all the growth stages of man and in all the products of human activity.

3. ONE-CELLED ANIMALS

Through a long series of intermediate stages man has inherited the predominant selfishness of his nature from those far-off, first animal cells which ceased to produce chlorophyll and began to obtain the prize of stored-up radiant energy by devouring other cells. But the corrective principle of exchange of products and services also began to operate as soon as it became necessary for two parental cells to unite to form a zygote in order to renew the waning power of cell division. Thus Sex makes its entry near the beginning of Act I. No wonder it has proved to be a difficult force to restrict within properly sanctified channels!

4. MANY-CELLED ANIMALS

Whenever the daughter cells of a single zygote failed to separate completely and began to form aggregations, the possibilities were opened for further polyisomerism and anisomerism leading to local differentiation, the division of labor and the coöperation of parts in maintaining the whole organization. Thus arose the diverse phyla of Metazoa, including the first two-layered, bell-like coelenterates with their sack-like gut, sensitive nerve ring, reproductive cells, stinging cells, supporting tissue and contractile neuro-muscular strands.

5. SPONGES

Among many Protozoa (Chapter I) the flagella were usually used for driving the cells through the water, but in the sponges they serve to draw the food-bearing water into the interior, which consists of a more or less complex system of canals leading to surface pores. This system thus serves both for aëration and nutrition. The digestive cells lining the interior of the canals bear protoplasmic collars in which the food particles adhere. The digestive or endoderm cells are enclosed externally by a layer of small cells, the ectoderm, while a third layer, the mesogloea or mesoderm, secretes supporting spicules between adjacent tubes.

The sponges illustrate several grades in the evolution of complex corporations or organisms, from relatively simple polyanisomeres. In the less complex sponges the combined circulatory and digestive system is exceedingly simple. There is but little sensitivity and the skeletal parts are inert spicules excreted between the sponge tubes and not organized into an articulated skeleton.

Even the simplest sponges are usually supposed to be excluded from the line of ascent to higher animals, partly because the adults are attached rather than free-swimming animals. But it is still possible that even in higher groups the attached adult stage is older than the free-floating or free-swimming animals (cf. p. 27). The simplest sponges agree with many higher groups in having free-floating ciliated embryos; also the embryo changes into a double-walled cup or gastrula by the overgrowth of the outer layer.

6. COELENTERATES

The coelenterates (Chapter I) are definitely more advanced than the sponges in that they have developed a pouch-like or tubular digestive sack without side pores. Some of their digestive cells bear a flagellum but are without protoplasmic collars. The large ectoderm cells are produced into "muscular processes," which serve especially for contraction of the tentacles and body wall. The attached hydroid *Bougainvillea ramosa* is convergently tree-like, but it buds off free-swimming, cup-like medusae. In different coelenterates the gonads may be borne by either the fixed or the medusa stage and there may be an alteration of sexual and asexual generations, as in early plants.

The coelenterates are typically radiate in structure, the mouth and gut being in the center, surrounded by the contractile bell and its tentacles. The latter bear thread-cells or nematocysts, which serve to numb or poison small fishes, which are then absorbed by the jellyfish. The nematocyst contains a coil of hollow thread which is shot outward upon contact with a foreign substance by the contraction of the sensitive tentacle. These nematocysts, which are characteristic of the coelenterates, are unknown in higher groups but are somewhat like the similar-acting trichocysts of certain ciliate Protozoa (Parker and Haswell, I, 1897, p. 123). The higher coelenterates exhibit further stages in the integration of diverse polyanisomeres into a complex organism, as in *Physalia,* the "Portuguese man-of-war," with its float, stinging tentacles, nutritive zooids and reproductive zooids.

Many other aspects of "evolution emerging" are illustrated in the flower-like anemones and multiform corals—all sessile or attached forms which fasten the primitive gut to the body-wall with complex radiating septa or mesenteries and excrete highly diversified bases. The mesenteries are moved by a strong muscular system which is said to contain "both ectodermal and endodermal fibres and endodermal muscle processes" (Parker and Haswell, vol. I, p. 181).

But with all their diversity, the coelenterates never develop a real mesodermal muscle system or a segmented locomotor apparatus. Nevertheless the coelenterates show us some very early stages in the evolution of digestive, locomotor and control systems.

The typical comb-jellies (Ctenophora) are pear-shaped or ovoid relatives of the coelenterates, in which comb-like swimming plates are arranged in eight vertical series around the outside. A pair of long retractile tentacles protrude from deep sheaths on opposite sides, the equatorial section being thus bilateral. Nematocysts are replaced by adhesive cells. On top of the pear-shaped body is a peculiar apparatus containing a mass of calcareous particles supported by four curved "legs," formed from groups of large cilia. This organ is connected by ciliated grooves which pass outward and downward to the series of swimming-plates. Apparently it acts "as a kind of steering-gear or apparatus for the maintenance of equilibrium" (*op. cit.,* p. 203).

Ctenoplana, a relative of the ctenophores,

has flattened down on the bottom and "creeps on its ventral surface like a worm" (*op. cit.*, p. 212). It has retained the dorsal organ and the ciliated comb-plates in modified form. Thus its present habitus is somewhat like that of a flatworm, but its heritage is coelenterate.

7. FLATWORMS AND OTHERS

The typical triclad flatworms (Chapter I) have taken a long step toward the higher types of invertebrates. They are extended bilaterally, with head and tail ends and creeping habitus. They have a true mesoderm which gives rise to deep dorso-ventral and longitudinal, as well as to superficial circular muscles; by means of these the body is alternately squeezed up into a hump and thrust forward. The central intestine is represented by a vacuolated nucleated mass without lumen, but often with very numerous lateral diverticula. The mouth is ventral, leading into a muscular, often protrusile pharynx, but there are no jaws. The nervous system is sharply differentiated and includes bilateral and radiating nerve cords, converging toward a relatively large "brain," upon which rests a pair of eyes of varying complexity. The paired otocysts are sacs containing otoliths of carbonate of lime, suggesting the so-named structures in fishes and possibly concerned with the reception of water-borne vibrations. A vascular system of water-vessels performs an excretory function and the ciliary flame-cells suggest the nephridial excretory organs of annelid worms. The reproductive organs are complex and both sexes are developed in the same individual. The impregnated ovum divides into four small dorsal "micromeres" and four large ventral "megameres." The micromeres give rise to the ectoderm, the megameres to the meso- and endoderm. In general this cleavage pattern stands between those of the ctenophores and the annelids.

From the flatworms evolution emerging has given rise to a wide diversity of intestinal parasites (flukes, tapeworms, etc.). During their life history the flukes undergo profound transformations as they are passed from the sheep-host to the snail-host and back to another sheep by way of the feces and infected water or grass. Some of the cestodes produce hooked embryos (cysticercoids) which in turn give rise to cysts containing clusters of daughter cysts and hooked heads or scolices. In the human tapeworm the scolex may bud off as many as 850 segments, forming a narrow ribbon several yards in length (*op. cit.*, pp. 231, 233). Thus they live unconscious in a bath of food and produce innumerable offspring.

The less specialized roundworms (nematodes) live freely in fresh or salt water, damp earth, decaying matter, etc.; but the dreaded *Trichina spiralis* is wholly parasitic. The thread-worms (gordioids) are free-living only in the sexual stage and become parasitic in the asexual stage.

Doubtfully referred to the roundworm phylum (Nemathelminths) are the "hook-headed worms" (Acanthocephala) parasitic in the intestines of vertebrates of all classes. The extraordinary diversity of these animal parasites, the "devilish ingenuity" and complexity of their life histories and adaptations, have not failed to excite the amazement of biologists and philosophers. Although their phyletic origins are often obscure (Baylis, 1938), these pests owe their existence neither to the workings of a malevolent Designer nor to the guidance of special tutelary deities (p. 5); they have, on the contrary, evolved under the automatic, cumulative and statistical operation of the Principles of Transformation (pp. 14, 15) extended through countless aeons.

Widely contrasting with these internal parasites are the free-swimming arrow-worms (*Sagitta*). Their general habitus is fish-like, with head, trunk, tail and paired bilateral fin-folds. But they seem to be allied with free-swimming namatodes, especially in their muscular system (Parker and Haswell, vol. I, p. 293).

Another free-swimming marine group are the nemertean worms (Nemertinea). These have an elongate cylindrical or flattened body, which is not segmented and is devoid of appendages.

The entire surface of the body is covered with vibratile cilia (as in typical flatworms); the mouth is anterior and from it may be protruded a very long muscular proboscis, which is well supplied with nerves and is probably a tactile organ. In those nemerteans which go through a metamorphosis there is a helmet-shaped, free-swimming ciliated larva, or pilidium, which is not dissimilar to the free-swimming trochosphere larvae of marine annelid worms.

The "worms" noted above, both parasitic and free-living, belong to at least two great phyla, Platyhelminthes (flatworms) and Nemathelminthes (roundworms). These, according to the researches of embryologists, belong to two major divisions or superphyla of the animal kingdom (cf. Allee, 1926, p. 263). The ctenophores and flatworms lie near the base of the "Arthropod" series, of which the more primitive marine members have "trochosphere," helmet-like, free-swimming larvae. The mesoderm is typically derived from a pair of "pole cells"; these are buds from the "second somatoblast" of an early cleavage stage (fig. 5.2A). This series also includes the nemertines, rotifers, annelids, molluscs, trilobites, crustaceans, arachnids and insects.

In this "teloblastic" or arthropod group of phyla, if one of the earlier cleavage cells is separated from its fellows, it is incapable of giving rise to an entire embryo, but can only produce a particular part. This type of cleavage is called "determinate."

In the opposite group of phyla, the mesoderm cells merely bud from the endoderm cells and in later embryonic stages form either a lamina or pouches on either side of the embryonic gut. This "chordate series" includes the roundworms (Nemathelminthes), the arrowworms (*Sagitta*), the starfishes, etc. (Echinoderms), the prochordates (*Balanoglossus, Amphioxus,* tunicates) and the vertebrates. It is further characterized by the fact that if in an early cleavage stage one of the daughter cells is separated from the rest it may grow into an entire embryo. This type of cleavage is called "indeterminate."

8. ROTIFERS

The rotifers or "wheel animalcules" are microscopic relatives of the worms. They are abundant in ponds and ditches; several of them have a convergent resemblance to some of the infusoria among the Protozoa; but, as remarked by Parker and Haswell (vol. I, p. 299), they are multicellular, bilateral animals with all the essential parts of other worm-like forms. Some are fixed in the adult state, attached by a bag-like test from which may protrude a cluster of ciliated lobes or tentacles. Others have a distally forked tail by which they can attach themselves; or they may let go and swim with the aid of cilia. Some have one or more pairs of appendages provided with long muscle bands and ending in long setae, by which the animal can skip through the water. In the pharynx of rotifers there is a large muscular chamber called the mastax, containing a complex of pieces, named incus and malleus, which serves to break up the food. The larvae are well ciliated, free-swimming trochospheres not unlike those of primitive marine annelids, to which the rotifers may be remotely related.

9. ANNELIDS

The typical marine annelid worms (Chapter I), although the most primitive of the "arthropod series," are much more advanced than the marine flatworms, especially in the development of a jaw-bearing head and of paired locomotor-respiratory appendages (parapodia). The body in the adult is divided into metameric segments or polyanisomeres, each one when fully developed containing: (1) a chamber of the body cavity; (2) a section of the alimentary canal and other organs and (3) a pair of parapodia.

Again among the annelids we have the question, which came first, the attached or the free-swimming stage, and again, in all those which undergo an indirect development, we have free-swimming trochosphere larvae. In the earthworms, however, which are highly specialized for digging, the eggs are enclosed in a capsule containing food material and the embryos develop directly into complete little worms. Some of the marine worms develop ciliated tufts or branchiae, often forming tree-like tentacles. Another highly specialized annelid is the "parchment worm," which lives buried in a U-shaped, parchment-like tube. Some of its appendages are modified into sensitive flaps or valves, which by their rhythmic movements draw in the food-bearing current through one aperture of the tube and push it out of the other.

Thus the annelids no less than other phyla illustrate the principle that evolution emerging is a veritable *dea ex machina*. She is an opportunist who fashions her creatures automatically, selecting and building up their differences, adapting them respectively to the diverse opportunities, limitations, exemptions and advantages of their environment and period. In so far as different places, requirements, opportunities and inherited features respectively resemble each other, she repeats similar features among widely different phyletic heritages.

On the other hand, her new and unique assemblages and highly individual patterns are due to: (1) equally unique and unpredicted combinations of genic factors; (2) new features in the biotic or physical environment, which in turn open new and unpredicted opportunities and impose new limitations and restrictions upon others, following Osborn's principle of "action, reaction and interaction."

10. MOLLUSCS
(cf. Chapter II)

The long continued survival of early standardized types, the profuse production of new types and many other principles of evolution emerging are beautifully illustrated in the evidence bearing on the history of the molluscs (Chapter II). It was long customary to regard the oval-shelled limpets that cling tightly to the rocks between tide limits as near the ancestral molluscs; but the available evidence suggests that the limpet habitus is an end stage of a long series of prior adaptations. According to the hypothesis tentatively put forth in Chapter II, the remoter pre-Cambrian ancestors of the molluscs were probably attached marine relatives of the annelids and rotifers. They may have lived in trumpet-like shells secreted by a mantle or surface envelope and basically similar to those of the conulariids. These creatures had a U-shaped or looped gut, several protruding tentacles, but little if any head and only a rudimentary nervous system. Behind the tentacles was probably a calcareous operculum, to which were attached the muscles that quickly drew the animal back into its shell and shut the lid.

The nearest surviving relatives of these ancestral molluscs may be the elephant-tusk shells (*Dentalium*), which at first sight seem to be aberrantly specialized animals. But Lacaze Duthiers long ago cited strong anatomical evidence for their connection on the one hand with the bivalves and on the other hand with the snails.

In many of the early palaeozoic orthoceratid ancestors of the nautiloids, squids and cuttlefishes, the shell was long and straight or but slightly curved. It consisted of very many cone-in-cone segments with an inner axial tube or phragmacone, marking successive budding and growth of the shell, which may have originally been attached to the sea floor at the tip of the cone, like a cup coral.

The dominant creative organ of all known molluscs is the mantle, a filmy glandular skin that envelopes the growing animal and produces the shell, the breathing organs and other vital parts. This molluscan mantle itself became a sort of creative agent or teleological machine. In response to unfolding opportunities it produced the widest extremes in molluscan shell forms. The resulting vast hosts of orders, families, genera and species of snails and bivalves are confusing and baffling to all but seasoned conchologists, because of endless convergence and parallelism combined with unique individual shell patterns. The unique features of every shell pattern suggest the wide creative possibilities of evolution emerging, while the frequency of parallelism and convergence define its boundaries and limitations.

When the primitive trumpet-shell mollusc was left stranded on its side at low tide, many evolutionary possibilities were opened to it. It could continue to shut itself in by means of its strong opercular muscle or it could squirm around a little and try to bury itself in the sand. In the latter case, as the selective value of such reactions increased, the muscular sheath around the intestine became strengthened on the ventral surface and a creeping foot was evolved, so that the animal could drag about its long conical shell, much as the augur snails do today.

The presumably older method of shell growth, namely, by means of a series of periodically produced concentric hollow cones and a central stalk, was retained and further developed among the cephalopod molluscs. By faster growth of the entire shell system on the dorsal surface, curved and, later, bilateral spiral shells were evolved, as in the typical nautiloids. In certain lines as the animal moved forward to secrete its latest chamber, some of the empty chambers that were left behind became filled with gas and tended to float, thus lightening the heavy load that the animal carried on its back. The next step in certain other lines was to reduce the weight of the shell by simplifying its complex foldings, as in *Nautilus* and *Argonauta*. With the further increase of aggressive organs, the shell became small or vestigial and the animals took to piracy on the open seas (cuttles and squids) or to sudden assault and robbery (devilfishes).

In general form the symmetrical tusk-shaped pteropods seem to lead, through *Atlanta,* to the more normal gastropods, although the wing-like flaps or anterior foot lobes of the pteropods may well be their own peculiar line of specialization. But the real branching off of all the major molluscan classes from a proto-molluscan stock must have occurred in Pre-Cambrian times, partly because the Cambrian *Palaeacmaea* already possessed a very limpet-like, bilaterally symmetrical, conical shell and doubtless a strong oval sucker-like foot for clinging to the rocks. Thus the apparent series of stages (Chapt. 2, p. 37) which lead from normal helicoid shells to *Hipponyx*-like forms and thence to the limpets may be only deceptive convergences.

In the early Palaeozoic bellerophonts the shell was symmetrically coiled with a rapidly expanding oval aperture, and there seem to be many intermediate forms leading to asymmetry through faster growth on the right or left side of the mantle, and thus to the normally helicoid snail-shells.

Among those older molluscs which are symmetrical are the Amphineura, or chitons, which compete with the limpets in clinging to the rocks between tidewater. But their oval shells are divided transversely into articulated valves, which collectively suggest symmetrical segmentation or metamerism. Internally, however, there is but little suggestion of metamerism or body segments. It is true that some of the chiton's relatives, the shell-less Aplacophora, have certain striking anatomical characters which led Thiele and others to refer them to a separate class, Solenogastres, regarded as connecting the chitons directly with the flatworms. While there are not wanting other signs of remote relationship between molluscs and worms, including the possession of trochophore-like larvae, yet it seems more likely on the basis of present evidence that the metamerism of the chitons is quite superficial and that the naked Aplacophora are secondarily worm-like. The divided dorsal shell of chitons may even be a result of the serial budding of the depressor muscles.

By whatever path the remote proto-gastropod stock was evolved, it eventually gave rise to an enormous diversity of shell forms which even surpasses that of the cephalopods. Obviously if one side of a bilaterally symmetrical plane figure grows faster than the other, the contour will curve toward the slower-growing side, and if growth be faster on the back than on the belly, a dorso-ventral curve results. The concurrence of both conditions in a solid body will lead to a warped or helicoid surface. By appropriate adjustments of the growth rates in time and in the three dimensions of space, the mathematicians could conceivably construct machines which would generate many of the standard forms of gastropod shells, as brilliantly set forth by D'Arcy Thompson. A first stage in the development of such a machine, although as yet operating only in two dimensions, has indeed been invented by Breder (1947). But his machine shows that in many kinds of curves there are critical points beyond which further increase in a given growth rate or dimension results in a great and sudden change in the form of the curve traced by the recording needle. In other words, these transformations are so radical that after a given critical point has been passed the recorded curves may bear but little visual similarity to their pre-critical predecessors. Such conditions may indeed set a limit to the increase of certain growth rates in shells, e.g., the downward pitch of the screw. A partly analogous case occurs during the development of the worm shells (vermetids, etc.), in which the earlier coils are normally helicoid; but then there is a sudden crisis and the tube twists first one way and then another, finally producing an irregular tangled mass of contorted tubing.

Such unexplained crises in growth factors may have been partly responsible for the riotous profusion of irregular and unique color patterns, especially in *Turbo petholatus,* as well as among the volutes, cones and many other families. For in general the color patterns are influenced not only by all the factors which determine the form of the shell itself, but also by the many variables in the growth and behavior of the pigment-forming organs. For example, when forward growth temporarily ceases along the advancing outer rim of the aperture, the individual pigment cells of this rim may either stop work altogether, leaving a white streak, or they may give rise to spots. If there are many adjacent spots, a line may be formed at right angles to the "revolving lines." But sometimes some of the pigment makers seem, as it were, to go on strike and to lose their sense of time, of teamwork and of direction. Single pigment streaks then appear to be growing in random directions and an orderly rhythmic pattern breaks up into erratic patches and meanderings. In such cases Chance has apparently triumphed over Law. Even these "random" lines, however (as on the shell of *Cymbium aethiopicum*), often tend to follow the curvature of the shell along a given section. In reticulated patterns (as in *Conus reticulatus*) the diamond-shaped spots look as if they were formed at the intersection of successive series of "geodesic" bands that were wrapped obliquely around the shell. Perhaps the V's of olives and *Conus porphyrio* may have arisen at the intersection of oblique geodesic pigment lines rather than of bands. The ordinary "revolving" lines or spots all wind around the shell more nearly at right angles to its axis.

Not less astonishing than the forms and color patterns of gastropods are their pharyngeal odontophores, bearing flexible rasps or radulae studded with thousands of minute teeth arranged in transverse rows on a movable strap. By means of this apparatus most sea snails can bore circular holes in the shells of their victims. Like other molluscan marvels, these delicately adjustable and complex machines are often taken for granted by malacologists. But how did the odontophore itself arise? Perhaps the first part of the answer is that the odontophore of gastropods is part of a muscular protrusile organ called the introvert, consisting of part of the pharynx; second, that it is analogous with a somewhat similar protrusile pharyngeal organ, called the proboscis, in marine annelids. These likewise bear chitinous denticles and jaw plates, all being operated by numerous muscle slips. Another partly analogous pharyngeal organ is the mastax of rotifers. The comb-like denticles of the molluscan radula are to some extent suggested by the hook-like or comb-like setae borne by the parapodia of certain annelids. Another form which has a protrusile pharyngeal sack armed with chitinous papillae is the gephyrean worm *Sipunculus.* Some of the minute "jaws" and denticles called conodonts from Silurian and Devonian horizons, closely resemble those of certain annelids, but others suggest the teeth of *Peripatus.* They were also compared by Loomis with the radular denticles of gastropods. In either case it remains probable that the molluscan odontophore is a further development of a protrusile denticulated pharynx of the general type seen in annelids.

The usefulness of the radular dentitions of gastropods to the taxonomist in supplying apparently reliable criteria for natural orders may eventually aid the evolutionist in his search for stages leading backward to less highly developed ancestral pharyngeal mechanisms below the level of existing orders. The stiletto-like poison fangs of the cones are specialized away from radulae with numerous small teeth. The long narrow radulae of limpets, which scrape the algae or bryozoans off the rocks, may be more primitive than those of predaceous sea-snails, which use their radulae to bore holes in the thick shells of clams. Presumably the protomolluscs must have fed on small forms such as Bryozoa, before attacking large well armored ones. The radulae of elephant-tusk shells (Scaphopoda), chitons, snails and sepiid cephalopods all agree in having central, marginal and lateral denticles arranged symmetrically on one side and the other of a central tooth (Pelseneer, 1906, pp. 6, 7).

Taxonomic works often put the bivalves (Pelecypoda) before the snails (Gastropoda), and the latter ahead of the cephalopods. The cephalopods do indeed attain the highest grades among all molluscs in their locomotor apparatus and brains. Moreover, the class of pelecypods was quite distinct from all others even in Cambrian times, and in view of present evidence their bilaterality may well be regarded as a heritage from the proto-Mollusca, shared with the primitive cephalopods and scaphopods. The same may be true of the complete absence of a head and the very small size of the cerebropleural ganglion, which is possibly homologous with and at least analogous to the cerebral and pleural ganglia of gastropods (cf. Parker and Haswell, vol. I, pp. 641, 642, 679).

The majority of pelecypods dig in the sand or mud and feed upon microscopic organic particles. These are drawn through the siphon into the capacious gill chamber between the relatively very large gill-lamellae bearing cilia. The gills differ widely in details in the different orders, ranging from simple plume-like ctenidia (cf. Gastropoda) to long double folds with cross-bars, convergently suggesting those of *Amphioxus*. The digging habit is made possible by the shell, which is divided into right and left valves, movably articulated by the hinge and operated by the reaction between the one or two adductor muscles and the resilifer, or elastic ligament, the latter tending to open the shell. The shell protects the delicate internal organs from pressure and serves as a fulcrum for the muscular foot. The latter pushes the sand aside and creates a depression into which the shell sinks. This entire digging habitus, including the bivalve shell with its hinge, the adductor muscles, the digging foot and the prolongation of the mantle into a siphon, are probably not to be looked for in the remote undiscovered protomolluscan.

Despite the very low grade and size of their central nervous system, the cumulative operation of natural selection upon "heritable variation" (mutation) continued through long ages on a statistical basis has produced many lines of adaptive advance for the pelecypodan hosts, as indicated by the large number of long enduring structural variations in shells, hinges, adductor muscles, gills, etc. The locomotor apparatus consists primarily of the muscular foot, probably in coöperation with the adductor muscles; but in the pectens the adductors and the resilifer (elastic ligament) merely by clapping the valves open and shut provide a simple means for sudden excursions in the water, and in the file shells (*Lima*) perhaps also for aeration of the plume-like functional gills.

Each and all of these once new "basic patents" among the molluscs is worthy of extended consideration, but we may here choose but one. Let it be granted that one can follow divergent evolutionary paths outward from the simple cog-like polyisomeres of the primitive taxodonts (fig. 2.47) to the varied and highly anisomerous hinges of *Megalodon*, *Spondylus*, the venerid clams, the oysters, the pholids, etc.; but how do the convexities on one valve come to fit so perfectly into the concavities on its opposite, so that one forms the mirror image or mould of the other? This question applies to hundreds of like cases throughout the animal kingdom: e.g., to the "lock-and-key" relations of accessory reproductive organs in opposite sexes, to the interlocking relations of the bones of the carpus in mammals, to the occlusal relations of upper and lower teeth, to the correlated relations of zygapophysial processes in vertebral columns, etc., etc. Unfortunately, little if anything is said about such problems in works on embryology, but we may here hazard the suggestion that such lock-and-key relations are preceded in phylogeny and probably in ontogeny by the buckling up, folding, invagination, evagination, etc., of an originally extended tissue layer or layers. For example, if the original layers were composed of bilaterally compressed, conical polyisomeres, all evenly spaced, the results of such right-left folding would bring the convexities on the right-facing surfaces of the *left* side into contact with the depressions between adjacent cones of the *right* side. Such may be the beginning of the simple cog-and-ratchet relation of the right and left "teeth" in the "taxodont" hinge of primitive ark shells (fig. 2.49), from which by anisomerism and selection in different directions the diversified heterodont, teleodont and certain other types of hinge may have been derived.

11. ONYCHOPHORA (PERIPATUS)

The caterpillar-like *Peripatus* (Chapter III) of South Africa, Australia, India, South America, has one of the most ancient known direct lineages among all the vast hosts of jointed animals. For its unmistakably direct ancestor *Ayesheyia* was discovered by Walcott in the Mid-Cambrian of British Columbia. As noted in Chapter III, *Peripatus* is basically annelid in general plan, but in some features has advanced in the direction of the arthropods. It has two pairs of jaws, bearing rows of denticles in succession and suggesting those of conodonts, which may indeed be allied with or ancestral to *Peripatus*.

Peripatus appears also to represent a persistently primitive stage in the evolution of the insects, which it resembles in the possession of respiratory tracheae and in its well-jointed hooked appendages. It is more primitive than insects and agrees with the centipedes in that the polyanisomeres which constitute its body segments have not yet become differentiated into a tripartite thorax. Thus its generalized morphologic pattern has changed but little in some 400 million years; and it is the more precious to zoologists because it still stands at an early milepost on the road between the annelids and the insects.

But *why did Peripatus stop where it was,* while its collateral descendants the later centipedes and insects went on so much further along their diverse paths? The easy answer, "We do not know," would hardly be fair either to the facts so thoroughly established by the geneticists or to the mathematical analyses based on long continued experiments upon populations of fruit-flies by Sewell Wright and his predecessors and co-workers. *Peripatus* has remained substantially the same for more than 400 million years *because:* (1) its surviving populations are those remnants of the old stock which escaped admixture with the later mutations and have somehow survived the calamities introduced by lethal combinations of genes; (2) the surviving populations of *Peripatus* in all their spreading and migrations (presumably from the northern to southern lands) have stayed within a relatively narrow ecologic zone and within viable climates, while those who did not do so were either destroyed or went on to new adaptations and lost much of their *Peripatus* heritage.

12. ARTHROPODS

The very primitive Lower Cambrian trilobites (Chapter III) had a large and well developed head, large paired eyes, numerous body segments each with its articulated paired appendages; the latter were typically divided into a postero-ventro-medial gill-bearing branch and an antero-lateral leg-like branch. The mouth was provided with one pair of antennae and fcur pairs of appendages, of which the basal pieces functioned as jaws. Collectively these paired appendages were homologous with the parapodia of annelids but were composed of articulated segments. The dorsum presented one median and opposite lateral rows of exoskeletal plates. In these and many other features the typical trilobites contrasted widely with the molluscs, to which they could be only very remotely related by common ancestry from a Pre-Cambrian proto-annelid stock. Probably convergent therefore is the curious resemblance between the paired shells of Paleozoic entomostracan crustaceans (*Palaeocypris, Estheria,* etc.) and such small bivalve molluscs as *Posidonia, Posidonomya* (Boule and Piveteau, 1935, pp. 157, 136).

Leif Störmer (1939, 1944) shows that the gill-bearing branch of the appendages of trilobites differs in location and serial homology from those of crustaceans, and for this and other reasons he derives the trilobites and crustaceans as parallel offshoots of a primitive Pre-Cambrian annelid stock. The molluscs may be regarded as a peculiarly specialized side branch of the primitive annelid stock, while its progressive descendants were the trilobites and other arthropods.

In contrast with the typical molluscs, the primitive arthropods developed a chitinous instead of calcareous exoskeleton and they swam with their articulated appendages by waving them rhythmically along the sides of the body. The appendages of trilobites, as in the various types of Cambrian entomostracan crustaceans, spring off from the ventral surface and their presence there has doubtless conditioned the development of the ventral nerve chain and its ganglia. The median row of dorsal transverse plates probably protected the straight gut and the segmental heart; each of its plates also served as a center-piece for the transverse arch that was the fulcrum for one pair of appendages. These appendages appear to be homologous with the parapodia of annelids, while the trilobite head-shield seems to be a complex corresponding with the annelid median praestomium (bearing the anterior tentacles and eyes) plus peristomium (bearing its clustered tentacles laterally and the mouth ventrally). The minute embryonic stages of Cambrian brachiopods foreshadow first, the trilobite head-shield, next, the first and second transverse segments, and later the tail shield or pygidium. Thus they are definitely trilobitic in type, quite different (except in basic features) from the embryos and Nauplius larvae of living crustaceans.

The trilobite exoskeleton (including the locomotor appendages), or something quite close to it in basic features, gave rise to those of the merostomes (including the eurypterids) and the scorpions. In the living "king crabs" (*Limulus*) the huge head-shield (cephalothorax) and the abdominal carapace, with the ventral median endosternite, together form a very strong base for the segmented appendages. The more distal caudal segments in some trilobites are covered with a sizable median dorsal plate, the pygidium. As such it may have been useful either for protection or perhaps as a minor lever. In the Cambrian merostome *Sidneyia inexpectata* of Walcott the terminal caudal piece took the form of a transversely placed fan-like tail fin, suggesting the tail of a crayfish; but in another merostome (*Emeraldella*) there was a long median caudal spine like those of many eurypterids.

The crustaceans and many insects pass through successive moults as the body increases in size, and they cast off the old skin and emerge with a new one. This also contrasts with the molluscs, which simply secreted a new, enlarged extension of the shell as they grew forward, without throwing off the old shell. These and many other factors enable the arthropods as a whole to move much more rapidly than the molluscs, except the squids. The limuloids, eurypterids and scorpions represent successive advances beyond a primitive arthropod stage, the scorpions acquiring relatively huge pincer-claws to aid them in killing insects. These groups very early gave up the simple exposed gills of trilobites and developed covered book-gills, which were especially useful for respiration on land. Meanwhile the distal segments of certain paired appendages were changed into powerful claws, while four other pairs of appendages became walking limbs. The collective fulcrum and sockets for the claw-bearing and walking appendages were supplied by the stout ventral "sternal" segments, which were crowded obliquely together near the mid-ventral line of the thorax. The caudal fin-spine of primitive merostomes is represented in the scorpions by the curved caudal spine, which is provided with a large poison gland.

The spiders seem to be a highly specialized side branch of the scorpions, greatly shortened anteroposteriorly and widened transversely, like the larvae of certain eurypterids. Their most marvellous possession is the web-spinning organ, including the spinnerets, and the ability of certain kinds to construct orbicular webs of relatively enormous size. Spiders are often credited with exceptional "intelligence," but it is highly probable that for their seemingly complex performances in spinning an orbicular web they depend not upon "intelligence" but upon a no less marvellous inherited "wisdom of the body," which, for example, makes them always turn through a constant angle after completing each connecting thread between the main cables of a large web (Savory, 1928). Even more wonderful is the Bolas spider (Gertsch, 1947), which hurls at its prey a whirling thread tipped with a drop of viscous silk.

The trilobite stock was formerly regarded as ancestral also to the insects but Störmer and others regard it as much more probable that the insects have been derived remotely from primitive Pre-Cambrian annelids and that their nearer ancestors were at least related to the centipedes and millipedes (Chilopoda). In either case the typical adult insects have probably condensed and consolidated the numerous segments of the primitive arthropod into: (1) a complex head, (2) a tripartite articulated thorax supporting three pairs of walking legs as well as one or two pairs of wings, and (3) a segmented abdomen not bearing legs. The chitinous exoskeleton of the essentially cylindrical thorax is evidently stiff enough to withstand the varied stresses coming from the head, wings, legs and abdomen.

According to Graber and J. Demoor (cited in Encycl. Brit., Ed. 14, vol. 13, p. 420), the legs of insects are usually moved in two sets of three, the first and third legs of one side moving with the second leg of the other. One tripod thus affords a firm base of support while the legs of the other tripod are brought forward to their new positions. Here is truly a "basic patent" developed within the class and distributed very widely within its limits.

"The wings of insects are in all cases developed after hatching, the younger stages being wingless and often unlike the parents in other respects. In such cases the development of wings and the attainment of the adult form depend upon a more or less profound transformation or metamorphosis." (Encycl. Brit., Ed. 14, vol. 13, p. 418.) Thus wings also appear to have arisen within the insect classes and Snodgrass (1935, p. 212) has shown that they probably arose through change of function of two pairs of expanded dorsal integumentary plates. It may also be suspected that originally they served as accessory respiratory lamellae and that their strengthening "veins" are remnants of respiratory tubes.

The wings, primitively in two pairs, may become massive and leathery, as in the elytra or anterior pair of beetles. The second pair may become vestigial, as in mosquitos, or vanish completely, as in some other Diptera.

The paired appendages of the primitive arthropod head give rise to the antennae, mandibles and first and second maxillae of insects. The mandibles and maxillae, primitively functioning as biting jaws, are often modified into sucking tubes, as in mosquitos, bugs and butterflies. Diversified also are the modifications of the integument around the cloacal exit into ovipositors, which may be thin and blade-like as in katydids, or very long and needle-like as in ichneumons.

In brief, it may be stated that entomologists have recorded, described and named many hundreds of unique organs or parts of organs: all adaptive and "preadaptive" and all products of emergent evolution, among the hundreds of thousands of species of insects.

However many and marvellous the mechanisms of insects may be, they are but the prelude to the wonders of their instinctive habits, culminating in the increasingly autonomous colonies and societies of the wasps, termites, ants and bees. While the genetic mechanisms behind these intricate phenotypic and social patterns are as yet imperfectly known, the fact that there are gradations leading, for example, from relatively simple conditions in the ponerine ants to the most complex social system of the higher ants, adds to the cumulative evidence for Darwin's conclusion: namely, that evolution proceeds by the age-long application of Natural Selection to heritable variation. And in this inexorable system the Malthusian forces of population pressure and limited food supply have normally continued to direct the trend of selection in any given line toward the survival of those non-lethal heritable variations which could be and were successfully incorporated into an already complicated type of organism.

Thus the many orders, the very many families, the thousands of genera and the hundreds of thousands of species of insects have evolved since the relatively primitive Palaeodictyoptera of Carboniferous times.

13. BRACHIOPODS

Brachiopod shells (Chapter IV) look somewhat like those of molluscan bivalves, but their valves cover the dorsal and ventral surfaces, the hinge being transverse, while those of the pelecypod molluscs cover the right and left sides, with the hinge along the back. This contrast was well established in the Cambrian and the superficial resemblances in shell form between the two types are surely convergent. The phylogenetic gap between the brachiopods and the arthropod series appears indeed to be very deep.

Reduced to its simplest terms, a brachiopod may be conceived as essentially a stalked, three-layered sack, in which the opposite corners of the capacious mouth have been produced into spirally wound "arms" or tentacles furnished with numerous vibratory filaments which draw the food-bearing water toward the mouth. An enveloping mantle, analogous with that of molluscs, secretes the dorsal and ventral valves of the shell; while complex opposing sets of muscles operate the valves. In the Inarticulata there is no hinge and the upper and lower valves are independently connected with the stalk; in the Articulata they are articulated with each other.

The brachia or "arms" assume many characteristic forms; their skeletal supports (secreted by the mantle) are often preserved in the fossils and are used in classifying the families. Brachiopod shells differ widely among themselves in the relations of length, width and thickness, as well as in degree of fluting and surface ornamentation.

The adult brachiopods, like the pelecypod molluscs, are quite headless; but as in similar cases, this headlessness is in part evidently concomitant with the great enlargement of the food-catching tentacles and their enclosure in a huge shell-covered preoral cavity; also with the complete absence of adult locomotor organs.

The larval brachiopods are free-swimming and broadly suggest the free-swimming larvae of the annelid-arthropod series, but with some characteristic features peculiar to trilobites. According to Parker and Haswell (vol. I, p. 345):

> "The setae of Brachiopods, sunk in muscular sacs, are marks of annulate affinities, since such organs are found elsewhere only among Chaetopoda and Gephyrea. The form of the larva tells in the same direction, the eye-bearing head region or prostomium and the provisional setae being very striking characters. But the segmentation of the Brachiopod is quite different from that of the annulate larva, in which new segments are always added behind those previously formed, and in which metamerism always affects the mesoderm."

In some ways the brachiopods recall the rotifers, which are also bilateral with dorsoventral differentiation, often stalked, with elaborate bilobed ciliated tentacular folds around the mouth and with a ciliated trochosphere. But all this gives but a meager answer to the question: From what sort of earlier molluscoids were the brachiopods derived? Probably the best answer to the problem of the origin of the brachiopods is that given by Allee (1927, pp. 229–230):

"In the Brachiopoda we find an indeterminate type of cleavage and the coelom arises as an outpocketing from the primitive gut; these characteristics we associate with the chordate series. The development of a nerve plate crowned with a long tuft of cilia, together with the position of the mouth and the presence of annelid-like segments and appendages seem to relate them to an annelid-like trochophore. Perhaps they belong to neither line, but should be represented as constituting another branch from the ancestral coelenterate stock which separated from the others before the arthropod line had developed determinate cleavage."

14. POLYZOA (BRYOZOA)

The typical sea mats (Polyzoa or Bryozoa) or corallines (Chapter IV) contrast with their relatives the brachiopods in their high specialization for colonial life. In the simplest terms, a typical polyzoan zooecium or zooid (of *Bugula*) consists of a chitinized cylindrical stalk containing a U-shaped digestive tract. The mouth, surrounded by a horseshoe-shaped band of tentacles, can be protruded beyond the stalk or quickly retracted by a long muscle attached to the bottom of the cylinder. The little cylinder is continuous below with a hollow creeping stolon, simple or branched, and the colony increases by continuous budding. To the sides of the cylindrical stalk are attached minute avicularia, resembling birds' heads and equipped with snapping jaws. These often seize minute worms or crustaceans and are believed to have a defensive value. They are regarded as modified zooids and in some types take the form of long whip-like appendages which enable the colony to perform creeping movements. There is nothing remotely suggesting these avicularia in the brachiopods or phoronids, their nearest known relatives. But they are somewhat analogous with the pedicellariae of sea-urchins (Parker and Haswell, vol. I, pp. 363, 389). Both are probably neomorphs within their respective groups, in which tegumentary spines moved by muscles may be replaced by minute appendages with snapping jaws. Another not necessarily alternative hypothesis is suggested above (Chapter IV, p. 79): namely, that the snapping jaws of avicularia may represent the two valves of the shell of a diminutive brachiopod with their adductor muscles.

In some polyzoa the tentacle-bearing, food-collecting zooid may be withdrawn like a "Jack-in-the box," into a basal chamber. As the retractor muscles pull it into the box, the tentacles draw together and water is squeezed out of the "compensating sack" to make room for it, while a double lid closes down over the pores or holes for the zooid and its sack.

Phoronis is a slender cylindrical marine animal which resembles some of the Polyzoa and brachiopods in having a horseshoe-shaped lophophore with its arms spirally wound. The animals live in associations, each one enclosed in a simple membranous or leathery tube. Apart from the lophophore, there are none of the amazing special features of either the brachiopods or the Polyzoa. The free-swimming, ciliated larva, called an Actinotrocha, slightly suggests the early larvae of a brachiopod and the endoderm sack likewise gives rise to lateral pouches of mesoderm. In view of their almost naked state, it is not likely that they are less specialized, except by degeneration, than the brachiopods.

15. ECHINODERMS

In spite of the extreme dissimilarity between echinoderms (Chapter IV) and all other invertebrate phyla, they do possess this in common with the brachiopods, namely, that they develop a body cavity (coelome) out of lateral mesodermic pouches from the primitive gut. In the brachiopods these coelomic sacs, as in other phyla, form respectively the mesoderm of the body wall and the peritoneal covering of the viscera. In the echinoderms, on the other hand, the left enterocoelic pouch gives rise to a secondary outgrowth, the hydrocoele, which in turn develops into the "ambulacral" system. The latter is essentially a hydrostatic mechanism for the operation of the very numerous suckers or tube feet. It consists mainly of five radiating water-canals, one running out into each of the five arms of a starfish and giving off many little canals on either side. Each little canal (or radial vessel) communicates with a single tube-foot. Each tube-foot bears a little sucker at the lower end and its tube opens above into a small ampulla or pouch. The latter, when alternately compressed and released, probably acts like a syringe in forcing the water into the tube-foot and withdrawing it. Each of these multitudinous tube-feet is a sucker of small strength, but collectively and in time they can overcome the contractile force of an oyster's large retractor muscle. When they have pulled open the shell enough, the starfish pushes in part of its eversible stomach and begins to ingest its victim.

This amazing application of hydraulic power to predatory habits has become possible through the essentially jointed construction of the radiating arms of the starfish. The entire exoskeleton consists of a tough, hard integument, in which are embedded numerous plates of calcareous material. In the fresh condition this exoskeleton presents a limited degree of flexibility (Parker and Haswell, vol. I, p. 346). On the functionally ventral side of each arm is a prominent "ambulacral" groove between the rows of tube-feet. The covered gallery to which the word "ambulacrum" refers contains the "ambulacral ossicles" that brace against the suction of the tube-feet and ampullae. In pulling open a bivalve shell, one or three of the five arms take hold of the rock, while the others are applied to the oyster shell.

This unique mechanism apparently had its beginnings in such relatively primitive echinoderms as the early palaeozoic carpoid cystids. The most primitive of the carpoids were pear-shaped or globose sacs, sometimes stalked. They contained a U-shaped digestive tract with the mouth and anus openings near together at the top of the sac. A mosaic of small tube-bearing plates covered the surface and on some of them there were branching clusters of grooves presumably ciliated, converging toward the mouth.

In the apparently more advanced Palaeozoic cystoids there were five large ambulacral grooves, foreshadowing the arms of sea-lilies and starfishes. Starfishes (asteroids) appear to be highly modified carpoid cystoids which have lost their stalk and twisted themselves over so that their mouth and ambulacral grooves face downward rather than upward. Their tube-feet are evaginated outgrowths of the hydrocoele, which was in turn derived from the mesenteric pouches (Parker and Haswell, vol. I, p. 361).

Another line led to the more or less global to flattened sea-urchins (echinoids), some of which have tube-feet as well as movably articulated spines. The mouth of sea-urchins faces downward, as in starfishes. It is surrounded by an excessively complex dental mechanism known as "Aristotle's lantern." This comprises five vertically suspended, radially arranged teeth with sharply pointed or chisel-like tips, all suspended in a framework of numerous pieces and operated by five elaborate sets of muscles. Presumably Aristotle's lantern can cut the stems and pluck up small attached animals such as hydroids and bryozoa. Prerequisite adaptations or preadaptations no doubt included an eversible mouth and pharynx, somewhat analogous to those of sea-snails but surrounded by muscularly controlled dermal plates.

This mechanism affords a good illustration of the simultaneous control by the body as a whole of mechanisms made up of numerous coöperating parts. And yet the chief parts of the epidermal nervous system of the sea-urchin consist of: (1) a nerve ring surrounding the mouth and probably operating Aristotle's lantern; (2) a set of five radial nerves, running upward along the inner surface of the five main ambulacral sectors of the shell; they connect at the upper end with (3) the five rudimentary eyes, which are located on the ring of ocular plates, on the upper surface of the body. The deep and coelomic parts of the nervous system which are present in the starfish are here only feebly developed (Parker and Haswell, vol. I, p. 367) and there is no cerebral ganglion. Thus the sea-urchin is a remarkably intricate but largely automatic organism.

16. PROCHORDATES

The typical sea-squirts, ascidians or tunicates (Chapter V) doubtless form a great specialized side branch from the undiscovered or unrecognized chordate stem, nor can they be securely linked with any of the invertebrate phyla. They belong on the three-layered level, distinctly above the two-layered coelenterates. But they contrast sharply with the annelid-arthropod, or teloblastic series in both their developmental and adult stages. They do not have ciliated, trochophore larvae and their free-swimming, "tadpole" larvae differ from those of invertebrates in possessing a notochord and a dorsal nerve.

After the tadpole larva becomes attached by its adhesive papillae, it undergoes a retrogressive metamorphosis, losing its larval tail, tail muscles, notochord, nerve-cord, eye and otolith, and developing an enormous pharynx or branchial chamber; the latter is provided with a grid of circularly disposed main vessels and very numerous cross-connecting vessels. Meanwhile the integument has developed a thick test or tunic, consisting largely of a substance related to the cellulose of plants.

Some of the fixed ascidians give rise to colonies of zooids embedded in gelatinous material without separate tests. The "Larvacea," which seem to be neotenic larvae, expand the test and float away in it, using their large vibratile tails as propellers. The Thaliacea expand into free-swimming or fusiform bodies provided with muscular hoops. The pyrosomids form hollow banded casks, lined with zooids; the reproductive zooids form stolons, which bud off new units. The latter derive nourishment from the oviduct of the parent by means of a vascular disc called a placenta.

In conclusion, the long and thinly muscled vibratile tail of the larval simple ascidians suggests (perhaps deceptively) an intermediate stage between the jointed stalk of such food-sifting echinoderms as *Mitrocystites* (fig. 4.6B) and the strongly muscled tails of *Amphioxus* and the ostracoderms. But the possession of this tail was not sufficient to insure further progress from the tunicate toward the vertebrate stage. The expanded, many-grilled pharynx of fixed tunicates became one of the dominant organs leading to a retrogressive metamorphosis with marked reduction of the invertebrate-like larval brain and complete resorption of the larval notochord and caudal nerve cord.

The next group still surviving, which retains some seemingly "invertebrate" features combined with others suggesting the vertebrates, is the Hemichorda, including, among others, the acorn-worm (*Balanoglossus*) and related genera. The adult *Balanoglossus* is worm-like, living in the mud of the crowded tidal zone, slowly oozing through the mud, engulfing it and extracting small food particles. Despite its sluggishness under laboratory conditions, it would seem highly probable that it uses its acorn-like "proboscis" and large "collar" as a hydraulic ram to push itself into the mud, part of the mud going into its slit-like mouth. Its capacious pharynx, although tongueless, may be adapted for squeezing or in some way conditioning or sorting the mud. Perhaps some of the mud is voided through the gill slits. Slow movements of the body are performed by its ciliated surface and by the muscular body wall, but the vigorous musculature of fast-digging organisms is lacking.

Despite the presence of a small so-called notochord at the root of the proboscis and despite the "fundamentally chordate" plan of ventral gut, axial notochord and dorsal nerve, the adult acorn-worm seems to have diverged very far from *Amphioxus,* to which it appears to be remotely allied by its embryology. The free-floating, ciliated "Tornaria" larva of some species of *Balanoglossus* suggests the larvae of some echinoderms and this fact inspired Garstang with the stimulating theory that the vertebrates arose from some pre-echinoderm form by "neoteny" in the larval stages, gradually losing the old adult body form and transforming the larval Tornaria into an adult primitive chordate. Although Garstang's theory lacks direct proof, there are indications that the great advance toward the vertebrate type, which is exemplified in the locomotor system of *Amphioxus,* was indeed attained by: (1) steadily improving the locomotor system of the motile larva; (2) retaining this active locomotor system until maturity; and (3) gradually delaying the ancient sessile or sedentary stage until older life stages, so that it was finally eliminated entirely.

In *Amphioxus* possible souvenirs of an earlier, less active, bottom-living adult stage may include: (1) its enormous pharynx, provided with a fine respiratory grill and recalling that of the sessile ascidian; (2) its ciliated endostyle or ventral groove; (3) its habit of darting into the sand by means of its spear-like rostrum (supported by an abnormally long and large notochord) and of poking its head out and remaining stationary with its oral cirri spread out in the manner of a stalked annelid or rotifer; (4) its circlet of sensitive oral cirri, analogous to tentacles; (5) its retractile velum, which can probably assist in preventing unwelcome objects from entering the pharynx; (6) the extremely small size and poor development of its adult brain; (7) its lack of a true heart, the function of which is performed by the numerous aortal bulbs at the base of each main branchial artery. Moreover in its embryology the swift-darting, fish-like *Amphioxus* agrees with the sluggish, worm-like *Balanoglossus* in several features mentioned above (p. 87; fig. 5.6) that are possibly indicative of remote relationship.

17. ORIGIN OF THE VERTEBRATES *

According to the illuminating suggestions of Percy Raymond and of Homer Smith, the need for protection was not the primary reason for the high development of the exoskeleton in many invertebrates and earlier vertebrates. At least the initial stage appears to have been due to the deposition of insoluble material from some of the secretory tubes that pierced the basal layers of the integument.

The nakedness of *Amphioxus* was formerly assumed to be a primitive, not a secondary character, so that zoologists did not expect the earliest forerunners of the vertebrates to be covered with armor; therefore they were slow to recognize the relative primitiveness of the Devonian and Silurian ostracoderms (potsherd skins). Recently, thanks largely to the labors of Stensiö, it has come to be realized by some zoologists that, considered as a class, the ostracoderms (Chapter VI) were the lowest and oldest of the vertebrates.

The diversified ostracoderms of the Upper Silurian and Devonian were nearing the end of their history as a class and we do not know their direct ancestors among the invertebrates; while even their relationships to the various living prochordates are not clear. Conceivably the ostracoderm class was more primitive than *Amphioxus* not only in possessing a well protected exoskeleton but even in the basic features of the pharynx, the latter being provided with an essentially fish-like system of gills and gill-chambers, in contrast with the expanded, grill-like branchial sieve of *Amphioxus* (Chapter V); just as the gill system and gill-slits of an ordinary shark are far less complex, less advanced, than the elaborate, finely-meshed branchial sieve of the whale shark (Chapter VIII), which feeds on very small organisms.

Although the existing *Amphioxus* may, as compared with the ostracoderms, be degenerately specialized in some features, it yet seems probable that, taking the prochordates collectively, they do represent a sort of intermediate grade tending to connect the vertebrate stock with that major division of the Metazoa which includes the echinoderms and perhaps the brachiopods and is described as enterocoelic.

17a. Contrasting Types of Mesoderm.—At this point it seems advisable to recall the evidence of embryology (Chapter VI) concerning the rise of the mesoderm, to which the vertebrates owed their locomotor and skeletal systems.

The development of a third cell layer, the mesoderm, covering the body-cavity with a thin veil and giving rise to both muscular and supporting tissues, immensely widened the opportunities for diverse complications in the systems for locating, pursuing, capturing, ingesting and digesting living prey.

The phyla of triploblastic, or three-layered,

* Cf. Chapters VI, VII.

animals began very early to separate into two major groups, according to the way in which the mesoderm was formed. In the more primitive members of the "teloblastic" series the mesoderm was produced by the multiplication of a single cell (the "second somatoblast") which was budded off at an early stage of the subdivision of the fertilized egg. This teloblastic series included the annelid worms, crustaceans, insects, molluscs and related phyla. On the other hand, the vertebrates, including eventually man and other mammals, were apparently not derived from any of the teloblastic series but owed to them only stimulating competition and opportunities to eat or be eaten.

In the second, or "enterocoelic" series, the mesodermal tissue was produced by the multiplication of mesoderm cells on either side of the archenteron, or primitive gut, as in the flatworms, echinoderms, brachiopods, *Balanoglossus, Amphioxus* and the vertebrates.

In the older remote relatives of the vertebrates, namely, brachiopods, echinoderms, balanoglossids, the larvae were free-floating or free-swimming, moving by the action of bands of lashing cilia, but the adults were fixed or sluggish creatures originally sifting minute food-particles from the sea water.

According to the theory developed in this work (pp. 98, 99; fig. 5.20), the still undiscovered adult ancestors of the vertebrates were fixed or sluggish "food-filterers," but their larvae, like the larvae of ascidians and *Amphioxus,* swam rapidly by the vibration of a long tail supported by a notochord. As the locomotor and food-getting adaptations of these late larval stages were improved, the old adult conditions of sluggishness or attachment were shortened and eventually eliminated. In this way, through the improvement of the larval stages, a motile adult stage may have evolved out of an older fixed food-filtering stage.

In short, the major theses that emerge from this and other evidence are: (1) that slow-moving, food-sifting invertebrates and prochordates were the remote forerunners and at least collateral ancestors of the basically predatory vertebrates, from fish to man; (2) that the shift from the "lower" to the "higher" grade was effected by improving the locomotor apparatus of the quick-darting larvae and by accelerating the maturation of the germ cells, so that the transformation into a sedentary adult was finally eliminated, i.e., by neoteny (Chapter V). None of this seems inconsistent with the possibility that, although at present naked, *Amphioxus* may have been derived from creatures that were by no means naked but provided with some sort of overlapping tegumental plates, which served on the one hand as a protective covering and on the other as fulcra for the attachment of the more superficial layers of the muscles of the head and body.

In general, the mesoderm, whether arising from the lower side of one of the primary endoderm cells (as in annelids and other members of the annelid-arthropod series), or from the "mesoderm crescent" around the blastopore at the junction of the ecto- and mesoderm, as in *Amphioxus* (Conklin), always comes to lie between the ectoderm and the endoderm. Collectively it is the protean servant and partner of both older germ layers as well as of the body as a whole.

Even in tunicates, part of the mesoderm, which lay in a pair of clusters between the primitive gut (endoderm) and the sensitive surface (ectoderm) had been detached from the gut to serve as the active partner of the locomotor apparatus, while the long turgid notochord furnished the recoil for the undulating tail.

One of the remote primary functions of the mesoderm had been to assist the digestive tube in getting in contact with food. Its contractile elements, the future muscles of the mouth, throat and cheek, helped to squeeze forward the mouth and lips, or to protrude the eversible "introvert" (extrovert), as in annelids and snails. When the ancestors of the vertebrates shifted gradually toward predatory habits, the mesoderm coöperated with the ectoderm in producing some sort of prickles or teeth to assist either the mouth or the introvert (pharynx) in seizing and eventually cutting the prey. Nor was the mesoderm found wanting in furnishing any other necessary parts of the mouth and pharynx with appropriate muscles. Meanwhile the splanchnic layer of the mesoderm, which covered the digestive tube, also supplied the necessary muscles for peristaltic or other movements of the stomach and intestines. In vertebrates these visceral muscles mostly remained "unstriped" and were called "involuntary" muscles in contrast with the striped muscles of the jaws and locomotor system.

Raymond Dart (1924a, b, c) in three very significant essays showed that progressively, as we ascend from the lower invertebrates to the vertebrates, the "autonomic nervous system" (which has to do with the action and reaction of the "involuntary" or unstriped muscles of the nutritive, vascular and reproductive systems) comes increasingly under the control of the cerebrospinal or central nervous system, which was originally concerned chiefly with the "voluntary" striped muscles of the locomotor system and with the guidance of the body as a whole.

In the ascidians, as we have seen (Chapter V), the notochord of the free-swimming larva is confined to the tail, as are also its muscles; but in the *Amphioxus* embryo the notochord, arising as a small cluster of cells near the blastopore, rapidly grows forward along the back, finally pushing itself out far beyond the mouth. Meanwhile a series of dorsal buds of the mesoderm give rise to the "protovertebrae" or mesodermal segments; these later differentiate into the V-shaped myomeres, or muscle segments, each with its own sensory and motor nerves. The dorsal muscle segments later grow downward, around the sides of the enlarged pharynx and abdomen, and upward, around the notochord and spinal nerve cord, increasing the vertical diameter of the body. This relatively great muscle power, imparting a high speed to the darting lancelet, requires increased oxidation and increased food, which are secured by the greatly enlarged pharynx.

18. JAWLESS OSTRACODERMS

The oldest known fossil chordates or primitive vertebrates were the ostracoderms, which are first recognizable in small fragments from the Ordovician age but flourished in the Silurian and Devonian ages. As a class they were already basic vertebrates and were far removed from any known group of invertebrates except in so far as they were probably members of the enterocoelic series, with distant connections with the echinoderms and brachiopods. They had indeed already undergone a profound transformation, since they possessed a dorsal nerve cord lying above a notochord, which in turn ran just above the tubular gut. On each side of the gut there was a long row of V-shaped muscle segments, while the brain was dorsal to the common cavity of the mouth and gill-arches. These are basic vertebrate characters, completely realized in the prochordate *Amphioxus* but not in any known invertebrate. They were transmitted intact to all the later vertebrates, including man.

The ostracoderms were bilaterally symmetrical, streamlined and tadpole-like, pushing their large cephalo-thorax by lateral undulations caused by segmental muscles on either side of the notochord. They were also strongly differentiated dorsoventrally and before they acquired swift swimming habits they probably rested on the sandy bottom or clung to submerged rocks. They resembled *Amphioxus* and the larval lampreys in having a large chamber for the mouth and gills beneath the braincase and probably also in drawing in small food particles either by the lashing of ciliated tracts on the floor of the pharynx or by rhythmic pumping action of the floor of the throat. They had a row of pouches on each side containing the gills, much as in the larval lamprey. These pouches were separated by partitions that contained internal skeletal tissue and it was from these partitions that the cartilages of the jaws, hyoid arch and gill arches of fishes were formed.

Thus man owes to the ostracoderms the "visceral arches" of his embryonic life, from which the cores of his jaws, hyoids and vocal apparatus were formed. In the ostracoderms, however, the oral pouches and their supporting arches were in series with the gill-supports and were not yet specialized to serve as jaws.

The ostracoderms (Chapter VI) are indeed often classified as a superclass (Agnatha), including only the ostracoderms and their specialized surviving derivatives, the cyclostomes (lampreys and hagfishes), in contrast with the Gnathostomata (vertebrates with compound jaws), including the classes Placoderms, Sharks, Bony Fishes (Ganoids, Teleosts), Dipnoans, Amphibians, Reptiles, Birds and Mammals.

In the light of the comparative embryology of recent prochordates and cyclostomes (Chapter V) we can now realize that the mesoderm, to which the Palaeozoic ostracoderms doubtless also owed their well developed locomotor apparatus, had further endowed them with a complex cranium and an "oralobranchial chamber" of the most primitive known type.

The ostracoderm head included an unpaired naso-hypophysial tube, a pineal eye, paired eyes, paired internal ears and lateral line organs, arranged in the same order as they or their representatives are in the embryos of all vertebrates, including man. Their brains also were of basic vertebrate type and the same is true of their paired cranial and spinal nerves, by means of which the sensory and motor systems coöperated to produce reflex actions, responses to stimuli, much as they do in later vertebrates. In their circulatory system, notochord, median fins and in some cases even the beginnings of paired fins, they also clearly foreshadowed the fishes and higher vertebrates.

From the behavior of their descendants we may safely infer that in addition to merely reflex actions they had a certain rudimentary sense of the body as a whole, in which the neural counterparts of memories of similar situations in their past experience conspired, either with physiological need (desire) or with fear (dislike), to stimulate appropriate action toward or away from the focus of interest.

19. PLACODERMS WITH COMPLEX JAWS

The placoderms (Chapter VII) were a widely varying group of fish-like vertebrates of Devonian age, most of them with a well armored head and thoracic shield. Possibly not one of them, so far as known, may be considered to be a direct ancestor to the higher classes of vertebrates which culminated in the mammals, including man. Yet viewed as a class they showed several advances above the ostracoderms and their adaptations proved to be prerequisite for further promotion to the higher grades.

1) The ostracoderms are classed as "jawless" (Agnatha) even though some of them had movable plates around the margins of the mouth; for so far as known such surface plates were not supported by "visceral arches" of inner skeletal origin. But the placoderms, so far as known, had complex jaw arches, which foreshadowed those of higher vertebrates; that is, there were movable, articulated endoskeletal upper and lower jaws which supported the surface jaw plates; these in turn bore dental tubercles.

In the ostracoderms the endoskeletal cores of the future upper and lower jaws were presumably in series with the endoskeletal ridges between the hyoid and branchial arches, and all these arches were immovable and continuous with the mesodermal endoskeletal tissue that surrounded the brain and paired sense organs. But in the placoderms, and still more in the fishes and higher vertebrates, the "visceral arches" (jaws plus hyobranchial arches) became enlarged and bent up into segments which were movably articulated with each other and with the braincase (Romer, 1937). Embryologists (e.g., Reichert, Gaupp) have long since recognized the significance of the "visceral arches" in the embryos of man and other vertebrates, but it remained for palaeontologists (especially Stensiö, Watson, Romer) to trace these visceral arches to their origin in the ostracoderms and placoderms (cf. Chapter VII).

2) The second, or hyoid, arch, at least in the acanthodians (which are classed as placoderms by Moy-Thomas and others) originally bore half-gills on both its upper and lower segments, exactly like the branchial arches. In the sharks, ganoids and higher fishes the upper segment of the hyoid arch lost most of its gill and became modified into the hyomandibular,

which assists the oral arch in the support of the lower jaw; meanwhile the lower half of the hyoid arch, which in the acanthodians (Chapter VII) retained all its gill-bearing function and connections, lost its gill-slit and half gill and gave rise to the hyoid bones of higher vertebrates. Thus man owes the foundations of his vocal apparatus to these very remote ostracoderms and placoderms and, as shown in Chapter XVI, he is indebted to the same source for the parts that finally were transformed into the auditory ossicles of his middle ear (stapes, incus, malleus).

3) The placoderms also foreshadowed the higher vertebrates in so far as the bony shell which surrounded their thoracic region was eventually shortened anteroposteriorly and modified into a supporting structure for the pectoral fins; and this thoracic girdle has every appearance of being homologous with the dermal portion of the pectoral girdle of higher vertebrates (Chapter VII).

4) The deep or "endoskeletal" pectoral girdle of ostracoderms, according to Stensiö, is homologous with the coraco-scapular arch of higher vertebrates. Of this complex pectoral girdle the higher mammals, including man, retain only (a) the scapulocoracoid, representing the "endoskeletal shoulder-girdle" and (b) the clavicles, representing the dermal pectoral girdle (fig. 24.3).

5) The several surface bones of the skull of placoderms were grouped around the brain-trough and the sense-organs; they covered the musculature of the jaws and branchial arches and afforded bases for the neck and jaw muscles, as they do in the fishes, but with differences in the number and contacts of the individual pieces. As shown in Chapters IX, X, land-living vertebrates reduced and eliminated many of the individual skull bones of the earlier fishes and placoderms, so that the human skull and its appendages are composed of but few (28) individual bones, while in the primitive fishes and placoderms the skull as a whole, including the parts of the "visceral" or oralo-branchial basket, comprised a large number of bones. Nevertheless it may truly be said that man owes the basic plan and general arrangement of his skull bones not only to his nearer ancestors but to his very remote ancestors of the ostracoderm and placoderm grades.

We may pass by the sharks and other cartilage fishes (Chapter VIII) which are not as near to the direct line of ascent toward man as were the basal placoderms (such as *Macropetalichthys*); because sharks very early modified the exoskeleton of the head and thorax, reducing it to the state of small shagreen denticles, and because they and their allies sacrificed the bone cells (which were already developed in the early placoderms) and retained in the adults the cartilaginous endoskeletal tissue of their larval stages; whereas the placoderms and still more the true fishes and higher vertebrates developed the bony exoskeleton and almost completely ossified their endoskeletons, retaining the larval cartilage stage merely as a matrix for the deposition of bone.

We must also pass by the very ancient Devonian ganoid *Cheirolepis* and other palaeoniscoids, which reduced the muscular core of their paired pectoral and pelvic fins, developed the scale-rays (lepidotrichs) and became the ancestors, not of the higher vertebrates, but of the ganoids and bony fishes (Chapter IX).

Similarly we must set aside the dipnoans or lung-fishes (Chapter X), which, although possessing lungs, very early acquired a peculiar fan-like dentition on their bony palates and inner sides of their lower jaw. This fan-like dentition may be traced upward directly into the modern lung-fishes; but was one of many curious specializations which led the lung-fishes away from the line of ascent toward man.

20. LOBE-FINNED FISHES

Thus in our quest for man's ancestors we come directly to the lobe-finned or crossopterygian fishes (Chapters X, XI), which also date back to the Middle Devonian. Of these *Eusthenopteron* appears to stand quite near to the ancestors of the tetrapods or land-living vertebrates. The earliest known ancestors or near-ancestors which we can thank for the beginnings of our lungs were these Devonian lobe-fins, which were related to, but more primitive than, the dipnoans, and which, like them, had internal as well as external nares or nostrils similar to those of their collateral descendants, the earliest amphibians.

The lobe-fins still passed most of their lives in the water, as shown by their undiminished gill-arches and opercular bones; but they had strong muscular lobes on their pectoral fins and since they lived in the same swampy environment with the lung-fishes, it is highly probable that during dry seasons they moved about in the drying pools and even wriggled from one mud hole to the next, both to feed on the fishes that were dying there and to moisten their gills, which they still retained.

The lobe-fins foreshadowed their successors the amphibians in their skulls, vertebral columns and muscular pectoral and pelvic limbs; but they had only begun to be transformed into amphibians and still retained the full complement of fish bones in the skull, jaws and branchial arches; even their paired fins bore a large web of scale-rays or lepidotrichs and they possessed well developed paddle-like dorsal and anal fins and a large triple-pointed caudal fin. The dipnoans, which in some ways were also evolving toward the amphibians, eventually reduced their dorsal and anal fins and finally eliminated their caudal fins. There is indirect evidence that the branch of the lobe-fins which gave rise to the amphibians also reduced and finally eliminated the dorsal, anal and caudal fins, together with the scale-rays of the paired fins, while enlarging the muscular bases of the pectoral and pelvic paddles.

The bony skeleton of their paired fins was laid down in the core of a cone-in-cone system of muscular layers and septa, which had pushed outward from the W-shaped muscle segments or myomeres of the sides of the body. The outer parts of this cone-in-cone system produced the more distal "mesomeres", as well as the "radials," the proximal parts of the system produced the "basipterygials" and eventually the scapulo-coracoids of the pectoral fins and the "pelvis" of the pelvic fins. The ways in which these paired paddles were turned, bent and twisted as they were transformed from paddles into limbs have been discussed in Chapter XI.

In short, man has to thank the crossopterygian or lobe-finned fishes not only for transmitting to him all the deeper vertebrate adaptations which they had inherited from the placoderms and ostracoderms, but also for taking the early steps which led to the profound transformation of water-living fishes into land-living tetrapods.

21. EMERGENCE OF AMPHIBIANS

The evolution of all lines of vertebrates has constantly required the reduction and loss of old specializations as new specializations came up, and the systematists who habitually assume that newly discovered forms are at once "too specialized" to be derived from any known earlier form and "too specialized" to be ancestral to any known later forms, are reducing the picture of evolution to an indefinite number of disconnected fragments. For example, it was formerly assumed that the known crossopterygians were "too specialized" to be ancestral to the amphibians: (a) because their median fins were too well developed and their paired fins not developed enough; (b) because they had a sort of movable joint across their foreheads by means of which the front part of the head could be tilted upward on the rear part, and because no trace of this joint is discernible in the skulls of the oldest amphibians; (c) because there were difficulties in homologizing the so-called "frontals" of crossopterygians with the true frontals of early amphibians. But Westoll (fig. 11.19) has described a skull of a lobe-fin from the Upper Devonian of Quebec, named by him *Elpistostege,* in which the skull roof has advanced beyond the *Eusthenopteron* stage toward the amphibian, and in which the arrangement of the bones of the skull-roof is such as to suggest that the so-called "frontals" of *Eusthenopteron* gave rise to the parietals of the amphibians, while the true frontals (hitherto called "nasals") of *Eusthenopteron* gave rise to the frontals of the amphibians (fig. 11.20). Moreover the studies of Watson and Romer indicate that the front part or ethmosphenotic division of the braincase of the crossopterygians merely extended backward and became attached to the rear or otic-occipital division. At the same time the snout or preorbital part of the skull in the earliest amphibians became longer and there was a marked shortening of the otic-occipital division and of its covering bones.

The transformation of lobe-fins into primitive amphibians, as noted in Chapter XI, involved the reduction and loss of the bony operculum, which seems to have been replaced by a membranous cover to the gill-chamber, as the latter gave rise to both the tympanic cavity and the Eustachian or tubo-tympanal canal. Meanwhile the hyomandibular was reduced and changed into the stapes of Amphibia. In the crossopterygians the row of lateral gulars was formed on each side in the lower part of the opercular fold and these are homologous with the branchiostegal bones of ganoid fishes. Some of these bones gave rise to the anterior and posterior splenials and angulars of both crossopterygians and early amphibians. The teeth of crossopterygians, which had complexly infolded or labyrinthine bases, were inherited with but little change by the oldest amphibians, which are called labyrinthodonts in allusion to this feature.

The complex upper and lower jaws of lobe-fins, with their inner core forming the palato-quadrates and Meckel's cartilage, and their sheathing bones forming the premaxillae, maxillae, dentaries, coronoids, etc., gave rise directly to those of the early amphibians. Nor was there at first any profound change in the vertebral column, except that as the thrusts of the limbs and the weight of the body on land caused greater stresses on the column, its neural arches grew larger and developed strong pre- and post-zygapophyseal articulations, the bony centra (both inter- and pleurocentra) increased and the notochordal space at the core of the centra diminished. Meanwhile with the development of walking limbs, the shoulder-girdle was profoundly transformed; the posttemporal bone, which in fishes connects the girdle with the rear corners of the skull, was eliminated (except in one known early amphibian), the cleithrum was reduced and the scapulo-coracoid greatly enlarged.

The pelvic rods (basipterygia) of the lobe-fins developed on each side a dorsal process, the ilium which by growing upward began to gain contact with the sacral ribs and thus changed into the amphibian pelvis.

In the nervous system the change from water-living to land-living resulted especially in the reduction of the lateral line organs, although grooves for them are retained in the skulls of the early amphibians.

The breeding habits of the lobe-fins are unknown to us but with very few exceptions fertilization in fishes is "external," that is, it is effected merely by the simultaneous discharge of eggs and sperm from the parents. This method persists in the modern frogs and since the gilled forms, which have been called branchiosaurs, have been shown by Romer to be in many cases only the young of labyrinthodonts, it seems very probable that in the latter a frog-like method of reproduction, including external fertilization, was still practised.

In short, the amphibians as a whole are transitional between fishes and tetrapods, having acquired tetrapod limbs in the adult stage but retaining fish-like methods of mating and going through conspicuously fish-like aquatic stages in their embryology. Most of the amphibians, however, drifted off into many curiously specialized side branches represented by the salamanders, frogs and caecilians of the present day, and only the earliest and most primitive amphibians gave rise to the reptiles and higher vertebrates. But all the changes noted above and many others, between the crossopterygian and early labyrinthodont stages, were prerequisite and "preadaptational" for further advancement toward any of the higher vertebrates, including man.

22. STEM REPTILES

The early amphibians bred in the water and gave forth eggs with a watery envelope, which then developed into fish-like limbless larvae that later sprouted limbs and came up on land. But the reptilian class very early began to lay eggs on land and, partly by means of the amniotic fold, they provided inside a tough shell a watery environment for the embryo. Recent turtles, *Sphenodon,* lizards, snakes and crocodilians usually lay eggs on land or, in some cases, bring forth their young alive. It is highly unlikely that the latter method had been evolved in the cotylosaurian stem reptiles, which very probably laid eggs with shells and scraped nests or depressions in which the eggs were covered with earth or sand. The practice of internal fertilization, with or without accessory intromittent organs in the males, is universal among existing reptiles, birds and mammals and may safely be predicated of the stem reptiles.

These cotylosaurs differed from the typical amphibians in avoiding the flattening of the skull and the shortening of the ribs; they retained the basic features of the earliest amphibians in most parts of the skeleton. Thus the skull inherited the fairly high temporal region of the earliest amphibians and the ribs remained long and curved. On the other hand, the typical cotylosaurs had advanced beyond the earliest amphibians in several features, for example, the intertemporal bones of the skull were lost (except in *Seymouria*) and the supratemporals greatly reduced or eliminated. In the smaller cotylosaurs the tabulars and postparietals (dermosupraoccipitals) were turned down sharply on to the nuchal surface of the skull and the posttemporal fossae of later reptiles were opened up above the paroccipitals (opisthotics). The occipital condyle, which was tripartite in the earliest amphibians, was becoming circularly convex to bean-shaped through the protrusion of the basioccipital portion. The intercentra were reduced or vestigial and the centra were becoming "holocentrous," that is, not subdivided as they were in the earlier amphibians, but made up chiefly of the enlarged paired pleurocentra, which formed a complete ring. The pectoral arch was reducing the cleithrum and the pelvis acquired a firm contact with one or two pairs of enlarged sacral ribs. The limbs of the larger cotylosaurs were very stout and retained the sharply bent form of the early amphibians but in the smaller forms they were less stubby and more lizard-like.

Seymouria, from the Permo-Carboniferous Red Beds of Texas, is in many respects intermediate between the primitive labyrinthodont amphibians and the cotylosaurs but it was probably not very nearly related to the smaller cotylosaurs or captorhinids which were more or less directly ancestral to all the reptiles and their varied descendants.

Thus man owes to the cotylosaurian stages of his evolution the emancipation from the semi-aquatic life and developmental habits of the ancestral amphibians, which opened the way to the conquest of the land by the ancestral reptiles. These in turn exploited all opportunities for crawling, creeping, running, digging, climbing and flying. Some of them became aquatic and even took on various deceptively fish-like habitus features. In other words, the cotylosaurs stood just below a great fork (fig. 12.1) in the line of advance: on the left, hosts of their descendants deployed into the soon crowded ranks of reptiles and birds; on the right, the path led to the pelycosaurs, the mammal-like reptiles, the mammals and man.

Thus it happens that the crow, representing a high type of bird, and man, stand on small twigs at the opposite sides of this great development. If crows philosophized, it would be natural for a loquacious old crow to assert that Crowland was the center of the universe and that he was the center of Crowland. He would most likely resent the suggestion that his own remote ancestors had been not crows at all but squirmy things with no wings! And not many of the human race would welcome the thought that their remote ancestors avoided the path that ended logically in dinosaurs and birds only by turning into the long detour through the stem pelycosaurs to the mammal-like reptiles, to primitive mammals, monkeys, apes, ape-men and brutal cave man.

Thanks to Darwin, the outlines of that vast array are clearing.

23. PRIMITIVE PELYCOSAURS

The typical pelycosaurs of the Permo-Carboniferous of Texas (Chapter XVI) were large reptiles which may be definitely excluded from the line of march toward man by their possession of a great crest or frill on the back, which was supported by the elongated neural spines of the vertebrae. But in the smaller pelycosaurs such as *Varanosaurus* these neural spines were not elongated and the skeleton as a whole was not unlike that of a monitor lizard but much more primitive in details. In all the pelycosaurs the skull abounded in features which were primitive for those of the mammal-like reptiles and hence ultimately ancestral to the mammals, including man. For example, as seen from above the skull was more or less like an inverted V, and as seen from the side its face was long and fairly high and its temporal region high.

In the cotylosaurs, no less than in the ancestral amphibians and lobe-fins, the temporal muscles of the jaws were covered externally by a continuous shell of bone formed on each side by the cheek plate or squamosal. But in the pelycosaurs we find this region on each side perforated by a round opening called the lateral temporal fossa; this was transmitted directly to the mammal-like reptiles and thence with some modifications to the mammals and man.

The perforation of the temporal region of the skull was due to the strength and activity of the temporal muscles and affords an example of the "principle of fenestration" of Gregory and Adams (1915), according to which the activity of muscles tends to strengthen the periphery and to excavate the middle of their areas on the bones to which they are fastened. Numerous examples of this principle are seen especially in the skull of reptiles (pp. 268, 273, 296), in the shoulder-girdle of lizards, in the pelvic girdle of dinosaurs and of mammals. The lateral temporal fossa or fenestra of the pelycosaurs was bounded above by the postorbital-squamosal bar, below by the jugal-quadratojugal, and posteriorly by the squamosal. The bar-like or arch-like form of these bones was directly due to the fact that bone strengthens itself by means of bony trabeculae along the lines of greatest stress and thins out along the areas of less stress. Although these principles were already in operation in parts of the skeletons of placoderms, labyrinthodonts and cotylosaurs, none of those forms had developed temporal fossae, their temporal regions being covered with a bony shell formed by the squamosal bones. Thus the lateral temporal openings were developed first in the pelycosaurs.

The pelycosaurs also developed further the paired five-branched pterygoid braces of the palato-pterygo-quadrate arches, which they had inherited from the cotylosaurs and primitive amphibians. The pterygoid bones were important in strengthening the upper jaw of these carnivorous beasts and their importance preceded the reduction of the pterygoids in mammals.

The teeth of pelycosaurs were set in sockets as in mammals, instead of being merely attached by the base of the crowns as in lobe-fins and labyrinthodonts.

The pelycosaurs retained reduced intercentra beneath some of their vertebrae and the centra were still pierced by the notochord. Their pectoral girdles foreshadowed that of the earlier mammal-like reptiles and the same is true of their pelvis and the general pattern of their limbs. Most of them were predatory animals, as were all of man's ancestors above the grade of ostracoderms and up to the frugivorous anthropoids. Their brains, to judge from the interior of their braincases, were essentially similar to those of turtles, *Sphenodon,* lizards and crocodiles: that is, they had well developed olfactory, optic, otic, balancing and reflex circuits, but they had not yet begun to develop the higher parts of the forebrain.

24. MAMMAL-LIKE REPTILES (Chapter XVI)

In the extinct mammal-like reptiles of South Africa and Russia the earlier forms, from the Middle Permian age, had advanced considerably beyond the pelycosaurs, especially in so far as their limbs were longer and better adapted for running. The dinocephalians, although retaining many relatively primitive features, early wandered away from the main line by becoming gigantic and by thickening the bones on top of the skull. Moreover they developed a large pineal eye (Chapter XII), which was set in a protruding boss on top of the forehead; but since they had also perfectly good paired eyes, we may suspect that the pineal eye had some other function besides ordinary vision; perhaps it may have specialized in catching some sort of actinic rays as a substitute for deficient antirhachitic vitamins. Thus even though these near-cyclopean monsters were not in the direct line of ancestry to man, they may help in solving the riddle of the function of the pineal eye, which was present in the ostracoderms, placoderms, crossopterygians, labyrinthodonts, cotylosaurs, pelycosaurs and mammal-like therapsid reptiles—in other words, in the line leading toward the mammals and man.

Another group of mammal-like reptiles which definitely removed themselves from the line of human ascent was the anomodonts (Chapter XVI), of which the typical members had lost all their teeth except the canines, while their gums had hardened into a strong parrot-like beak. This induced marked changes in the temporal region of the skull, which further removed the anomodonts from the line of human ascent.

The more central groups of the mammal-like reptiles were the gorgonopsians and therocephalians. These had quite similar skulls except for minor differences and had inherited their basic features from the pelycosaur stem. The earlier gorgonopsians were more primitive than the therocephalians in that the lateral temporal fossa still lay entirely below the postorbito-squamosal bar, whereas in the therocephalians it extended dorsally, disrupting the postorbito-squamosal contact and spreading up on to the sides of the parietals, which were thus often produced dorsally into a sagittal crest. Continuation of the same process of expansion of the temporal muscles in later therocephalians reduced the posttemporal-jugal bar to a thin ring; in the most advanced therocephalians this bar was incomplete or eliminated, thus leaving no complete bony rim between the orbits and the temporal fossae, as in the earlier mammals.

The pterygoid region of therocephalians was also more advanced than that of gorgonopsians in that it had developed fossae or circular openings between the ectopterygoid and the palatine bones.

Both gorgonopsians and therocephalians enjoyed a wide range in size, from almost rat-like little beasts to super-tigers. The relative lengths and widths of their skulls varied, as did the number of their feeble cheek teeth. In certain therocephalians, however, we find an early stage in the process of developing accessory cusps on the sides of the main cusps of the molar teeth. This process of cuspidation was carried further in the cynodonts and was an important theme in the prelude to the origin of the mammals.

In *Bauria,* from the Triassic of South Africa, the hind foot had made a notable advance toward the primitive mammalian stage in so far as its calcaneum had developed a distinct tuber or heel bone and was extending laterally beneath the proximal part of the astragalus. In continuation with this process, the astragalus in mammals (Chapters XVI, XXIV), including man, rides on top of the calcaneum, so that the powerful muscles of the calf, inserted on the tuber or heel, exert a strong lifting action on the weight coming down through the lower leg. The phalangeal formula (I2 II3 III3 IV3 V3) was already that of the mammals.

The cynodonts, from the Upper Permian and Triassic of South Africa and the Upper Permian of Russia, carried the line of advance well forward toward the mammals. They were evidently ravenous beasts, probably slinking through the underbrush and making a sudden dash at their prey. They ranged in size from weasels to crocodiles, but *Cynognathus crateronotus,* whose wolf-like skull base was about 13 inches long, had quite small legs. The small cynodonts, however, had relatively longer limbs and their skulls were strikingly mammal-like. The lumbar ribs of the cynodonts came straight out from the stout vertebrae and were then expanded into V-shaped distal ends, each of which was inclined so as to pass under the following rib.

Thus the movements of the vertebral column were probably more lateral than vertical. Even the very small cynodonts had similar but somewhat less expanded lumbar ribs. This peculiar specialization may have preceded the far more pronounced modification of the entire lumbar and sacral regions in mammals.

The shoulder girdle had already lost the cleithrum, but it retained the median interclavicle, paired clavicles and scapulo-coracoids of primitive reptiles. The coracoids, however, were much reduced in size as compared with those of primitive reptiles, thus foreshadowing the mammalian stage.

Many parts of the skull were very progressive toward the mammalian grade:

1) Swelling of the temporal muscles caused the development of a high sagittal crest, which appeared to squeeze the diminishing foramen for the pineal eye almost out of existence. The sides of the occipital region, compressed between the enlarging temporal and nuchal muscles, were continued dorsally as lambdoid crests.

2) The ascending ramus of the dentary of the lower jaw, which was already present but not large in the gorgonopsians and therocephalians, attained immense size (at least in *Cynognathus crateronotus*), while the bones behind the dentary (mainly coronoid, surangular, angular and articular, the latter representing Meckel's cartilage), were correspondingly reduced.

3) The pterygoid bones, which had formerly formed a very large part of the brace for the upper jaw, began to reduce the branch running to the diminishing quadrate and, doubtless due to the increasing size of the internal pterygoid muscles, were squeezed together against the high median keel of the parasphenoid.

4) At the same time the palatal shelves of the maxillary and palatine bones, which were barely beginning in the earlier mammal-like reptiles, grew toward each other until they met in the midline, arching over the backwardly prolonged internal nares and forming a bony secondary palate of mammalian type.

5) The stapes, inherited from earlier stages, extended laterally to the small quadrate, which was destined to become the incus of mammals, while the articular and angular bones already appear to have begun to take on the functions of the mammalian incus and tympanic bones.

6) The occipital condyle was spreading laterally on to the exoccipital bones and was indeed more mammalian than reptilian in appearance.

7) The atlas and axis also were submammalian in stage, as were also the remaining vertebrae.

8) The dentition in *Cynognathus* was also of submammalian type, with differentiated incisors and canines and cuspidate premolars and molars. In the related genera *Gomphognathus* and *Diademodon* the crowns of the upper cheek teeth were markedly expanded transversely so as to form ovals which were surmounted by low ridges, so as to suggest the beginning of a crested or subherbivorous type.

9) The great enlargement of the ascending ramus of the dentary brought its posterior corner very close to the squamosal bone and prepared the way for further advances in this region in the mammals.

10) On the sides of the braincase the fan-like epipterygoids were beginning to take on the functions of their derivatives, the alisphenoids of mammals, while the sphenethmoid or primitive brain trough was becoming differentiated into ethmoid, presphenoid and basisphenoid.

11) The brain, to judge from the casts of the interior of the brain chamber of *Diademodon,* figured by Watson, was still in an essentially reptilian stage. That is, the olfactory portions of the forebrain were still dominant, the brain was very narrow transversely except for a slight expansion of the medulla and there was not even a beginning of the expansion of the neopallium or higher part of the forebrain.

Two quite small skulls from the Triassic of South Africa have been referred to a new suborder Ictidosauria by Broom and, according to his descriptions, they had almost reached the mammalian grade. The ascending ramus of the dentary was large and sharply elevated, the bones behind the dentary being small and nearing their reduced condition in the mammals.

The secondary palate was well developed and the orbits were not separated posteriorly from the temporal fossae.

25. ORIGIN OF THE MAMMALS

In concluding his valuable memoir on the structure of the braincase of therapsid reptiles, Olson (1944) notes the lack of further connecting links between the mammal-like reptiles and the mammals and, observing that there was considerable parallelism in the advancement of the therocephalians and cynodonts, suggests that the chief mammalian characters may have arisen independently in different phyla and that the mammalian grade may have been reached more than once and by somewhat different routes. It would not be surprising if this suggestion were confirmed, at least to a certain extent, by future discovery, as it already seems probable that the tritylodonts, Triassic forerunners of the multituberculate mammals, may have been derived from the Ictidosauria (Chapters XVI, XVII) and the remaining mammals seem to be widely removed from the multituberculates. However, as noted elsewhere in the present work (pp. 347, 366, 373), all the surviving major groups of mammals (monotremes, marsupials and placentals) are tied together by the possession of so many common mammalian characters, that plurality of origin (from different suborders of mammal-like reptiles) does not seem to be a necessary hypothesis to account for their differences.

The later orders of mammals may conceivably have been derived either from the therocephalians or from the bauriamorphs or from the cynodonts, since these three suborders were rather closely related; but the fact remains that of these three, the small cynodonts had attained a very marked approach to primitive mammals in the entire construction of their skulls, jaws and dentitions, vertebral column, pectoral and pelvic girdles and limbs; although they were somewhat less advanced in their limbs and girdles than in their skulls. At all events the record is clear that no known group of vertebrates except the mammal-like reptiles was advancing in the mammalian direction, and that the cynodonts and ictidosaurians were nearing the mammalian grade.

The rather daring experiments of the cynodonts in enlarging the dentary bone of the lower jaw and reducing its postdentary elements were "preadaptations," that is, they were prerequisite, to the final revolutionary steps of the abandonment of the old postdentary jaw elements as such, and of their transformation into ear bones. Moreover the acquirement of the new dentary-squamosal joint between the lower jaw and the skull, together with the growing power of cuspidation and variability in the shape of the molar crowns, opened up all the immensely varied possibilities of diverse apparatus for chewing different kinds of food, which were so fully exploited by the mammals.

Man therefore owes to these early predators the first steps in the emancipation of his line from the cold-blooded reptilian grade with highly variable body temperature to the mammalian grade with more stable body temperature.

26. DEPLOYMENT OF THE MESOZOIC MAMMALS (CHAPTER XVII)

At the top of the Triassic the record of advance toward the higher mammals is broken and is not resumed until the Middle and Upper Jurassic. It is true that *Tritylodon,* the oldest known mammal, occurs in the Upper Triassic of South Africa and China and that its skull and other parts of the skeleton of a related form (*Bienotherium*) have been described by Young (1947) from remarkable material found in China; but *Tritylodon* was the earliest known multituberculate and, as most clearly shown by Simpson (1928, 1929), the multituberculates were a separate division of the mammals, which did not give rise to the higher mammals. It is even conceivable that the multituberculates may have come from a form like *Bauria,* while the higher mammals may have been derived from the smaller cynodonts. In either case the multituberculates had long since taken the road toward the Left and were already highly specialized for seed-grinding and nut-cracking, while the direct ancestors of mammals, retaining their predatory habits, had fought their way up from the reptilian lowlands to the mammalian highlands. During this transitional period reptilian scales were probably being replaced by a later product of the many-layered skin, namely, hairs, each with a basal pulp and a rapidly growing tip of horny material.

The glands of the deep layers of the skin became differentiated into: (a) sudoriparous or sweat glands, secreting a slightly saline fluid which by evaporation tended to lower the surface temperature; (b) sebaceous or oil glands, which helped to keep the skin pliable. On the ventral surface of breeding females some of the sebaceous glands became greatly hypertrophied, giving rise to (c) lacteal glands, the milky secretion of which was at first licked by the newborn young.

The young were probably still hatched from small but not microscopic eggs, which were incubated by the mother. In the living monotremes (Chapter XVIII), or egg-laying mammals of Australia, the newborn young, like those of many reptiles, are not like larvae, which are wholly different in appearance from their parents, but more like young animals of the same species, which after a short period of maternal care will be able to shift for themselves. Not improbably the newborn young of the first mammals inherited some of this reptilian precociousness but in the advance toward the later mammals they became specialized for longer periods of nursing and parental care.

When the known fossil record begins again in the Middle and Upper Triassic, four groups of mammals were already represented (Chapter XVII): multituberculates, triconodonts, symmetrodonts and pantotherians, all of which have been most clearly revised and figured in the memoirs by G. G. Simpson. Of these, the triconodonts retain the most likeness to the small cynodonts of an earlier period in their skulls, jaws and teeth, and may be provisionally regarded as the most primitive of the true mammals. In *Triconodon* and allied genera the upper and lower molars each bore three compressed cusps arranged in a fore-and-aft line. This may well be too advanced in its own line to have given rise to the teeth of later mammals, but another genus, *Amphilestes,* although no earlier in time than *Triconodon,* seems to be more primitive in having less difference between its premolars and molars and less developed accessory cusps on its molars.

It was formerly thought (by Cope and Osborn) that the laterally compressed molars of *Amphilestes* gave rise directly to the triangular molars of the typical mammals by the rotation or circumduction of the small accessory cusps toward the outer side in the upper, and toward the inner side in the lower molars, so as to form "reversed triangles" of the upper and lower molar crowns; but direct palaeontologic evidence of this change is still wanting and it seems more probable that the compressed teeth of *Amphilestes* type experienced a rapid transverse growth of the central cusp of the molar crowns, the direction of faster growth being inward in the upper, outward in the lower teeth. Various apparent stages in this process are found among the next order, Symmetrodonta, of which the upper molar teeth of *Amphidon* and *Eurylambda* are compressed, with but slight internal buds, while those of *Spalacotherium* are much wider transversely and have high internally placed tips or principal cusps.

In the third order, Pantotheria, this apparent process seems to be further accelerated, until in *Dicrolestes* the axis of fastest growth has shifted far to the inner or lingual side of the upper molars and toward the outer side of the lower molars. Whether or not any of these genera were in the direct line of ascent to the insectivores and other later orders, they were at least conducting early experiments in the occlusal relations of the upper and lower cheek teeth which have long been of interest to the student of the human dentition and its evolution. In any case it is certain that one of the pantotherians, named *Amphitherium,* had a remarkably generalized dentition which seems to connect the stem of the lines leading to the insectivores and higher mammals to the Jurassic order Pantotheria. Moreover, this very small jaw from the Middle Jurassic near Oxford, England, so much resembles that of a small marsupial that it was determined by Cuvier to be that of a small opossum, although it differs from all marsupials in the greater number of its teeth and in the non-inflection of the small finger-like angular process of the lower jaw. These were both probably primitive characters as compared with the fewer cheek teeth and inflected angle of the opossum. In the other direction, *Amphitherium* seems to be more primitive than any of the known placental mammals.

A skull and upper teeth that would fit on this jaw must have combined many features seen in both primitive marsupials and primitive placentals, such as: face long and but little bent down upon the cranial base; olfactory parts of skull elongate, eyes small or moderate, temporal fossae elongate anteroposteriorly and not separated by a secondary postorbital bar from the orbits, masseter muscle not much expanded but extending a little posteriorly on to the small angle; dentition insectivorous with slightly procumbent lower incisors, subcaniniform canines; anterior premolars simple, posterior premolars but little advanced toward the molar type, but with an internal bud on the upper and an incipient talonid on the lower; upper molars "tritubercular," but with well developed external cingulum cusps, no hypocone, the crown resting on three roots, one internal and two external; lower molars "tuberculo-sectorial" (fig. 18.27A), with a high trigonid and depressed talonid, the crowns supported by two main roots in tandem, one anterior, the other posterior. Such a generalized picture is based on numerous forms of known primitive mammals of Mesozoic and Caenozoic times and it is extremely probable that the portrait just sketched would accurately fit our own ancestors at a period when the main line was beginning to fork toward the marsupials on the Left and toward the placentals on the Right.

27. THE THREE MAJOR GRADES OF MODERN MAMMALS

The egg-laying mammals (Chapter XVIII) are represented in the existing fauna by very differently specialized animals, the semiaquatic duckbilled Platypus, and the spiny anteater, which nevertheless retain an almost reptilian grade of construction in the anatomy and mode of function of their reproductive systems. In the female the right and left oviducts, instead of fusing to form a median uterus as in higher mammals, remain separate as far down as the cloaca, as in reptiles, while the intromittent organ of the male differs from that of higher mammals in that it does not serve as a common exit for both genital and excretory products, in which respect it has also retained certain reptilian features. Moreover the fact that both these animals lay sizable eggs is assuredly a pre-mammalian, and therefore a pre-human, feature.

In the marsupials (Chapter XVIII) the newborn young have very large pectoral limbs, by means of which they are able to cling to the ventral surface of the mother and to crawl into her pouch, where they seize one of the teats in their mouths and become attached. The mother squirts milk into their mouths, but they are not choked by it because their epiglottis is produced upward into the posterior narial opening so that they can breathe without getting milk in their windpipes. It would be supposed by some authors that this is an extreme specialization which would prevent the marsupial stem from being ancestral to the placental stem; but it may well be that the forceful feeding by the mother has prepared the way for the active sucking by the young which is highly developed among the placentals. This higher development in turn may have been conditioned by a precocious development of the orbicularis oris, buccinator and throat muscles in the young placental mammals.

Meanwhile the development of pectoral limbs in the early placentals may have been retarded to later stages and the whole intrauterine period greatly lengthened by the increase and improvement of the allantoic placenta (p. 372), of which only beginning stages have been evolved among the modern marsupials. Thus it may well be that the monotremes, marsupials and placentals, instead of being completely divergent groups, are connected by their stem forms in an ascending series, Prototheria, Metatheria, Eutheria, as recognized by T. H. Huxley (1880). (See Gregory, 1947, p. 46.)

28. BASAL PLACENTALS *

After another long break in the fossil record between the Upper Jurassic and the Upper Cretaceous, we find that in the latter period basic marsupials and very primitive placentals were both present. The marsupials, although extremely primitive, were already tending toward the Left and have been dealt with in Chapter XVIII. The Upper Cretaceous placentals, on the contrary (Chapter XVII), are of capital importance in the present chapter, because they included several representatives of the Order Insectivora, which was regarded by Huxley and Osborn as the parent group for the entire placental series. Among them was a very small beast named *Deltatheridium,* which seems to be on or near the border between insectivores and carnivores, while another, named *Zalambdalestes,* was probably on the way toward *Ictops* and some of the later insectivores. Its possible relations to the tree-shrews and primates are considered below.

These Upper Cretaceous forerunners of the placentals show what was happening in the teeth, jaws and skull: (1) The dental formula was reduced from the higher numbers in most of the Mesozoic mammals to Incisors $\frac{3}{3}$, Canines $\frac{1}{1}$, Premolars $\frac{3}{3}$, Molars $\frac{3}{3}$. (2) The deciduous teeth of these earliest forms are not known but their modern descendants have deciduous and permanent incisors, canines and premolars, whereas in the typical marsupials only the posterior premolars were replaced by successors. The true molars in both marsupials and placentals begin to come in with the last deciduous molar and are usually regarded as being delayed members of the deciduous series. (3) The upper and lower molar crowns exhibited the "reversal of the triangles" which was basic for all mammals: that is, the tip of the "trigon" was on the inner side of the upper molars, while that of the "trigonid" was on the outer side of the lower molars. (4) The lower molars were of the "tuberculo-sectorial" type that is, they had both a trigonid and a posterior heel or talonid which received the internal tip of the upper molars. (5) The trigonid of each lower molar fitted between the trigons of two upper molars and was received either into the "interdental embrasures" or on the marginal portions of the upper molars.

* Cf. Chapter XIX.

All these relations were transmitted to the early tree-shrews (Chapter XIX) and to the primates (Chapter XXIII), including man.

(6) In *Deltatheridium* the cheek teeth are presumably ancestral to the V-shaped teeth of the earlier creodonts, which had definite metastyle blades; but in *Zalambdalestes* the upper molars are transversely widened with two conical cusps near the outer border, as would be expected in an ancestor of *Ictops* and its allies of the hedgehog group (Erinaceoidea), and probably also in the ancestry of the primates.

(7) In both types the last upper premolar was less advanced in structure than the molars and the latter were still transversely wide V's with no beginning of a posterointernal cusp or hypocone.

(8) In its jaws and skull *Deltatheridium* likewise seems to foreshadow the creodonts and their relatives of later ages, while the *Zalambdolestes* skull was long and low, with long muzzle, reduced canines and procumbent upper and lower incisors as in typical insectivores.

(9) Neither *Zalambdolestes* nor *Deltatheridium* had begun to develop a secondary postorbital-malar bar, which was later developed in some of the primates.

(10) *Zalambdolestes* was both older and more primitive than the Paleocene tree-shrews (Plesiadapidae, Chapter XIX) or their modern relatives the Tupaidae and it may have been near their line of ancestry as well as to the *Ictops*-erinaceoid group. It was also far more primitive than either the Paleocene tarsioids or the Eocene lemuroids, so that it may well be the nearest known Upper Cretaceous forerunner of the primates, including man.

29. RISE OF THE PRIMATES

The modern tree-shrews throughout the dentition, skull and skeleton have retained extremely primitive characters which have been gradually lost in the lemuroids, tarsioids, monkeys, apes and man, as set forth in Chapter XXIII. No doubt each successive stage left the older stock free from the newer features, partly by moving to other localities, partly by inbreeding with its near congeners and by developing antipathies to all others.

The pen-tailed tree-shrew (*Ptilocercus*) of Borneo has retained on the whole a very primitive skeleton (Chapter XIX), including a few moderate adaptations for arboreal life. Thus it has spreading hands and feet but not such marked adaptations for climbing as are found in the lemurs. Its dentition is fairly primitive and consistent with derivation from Paleocene plesiadapids: mainly by retrogression of their enlarged hooked median incisors. All the tupaids agree with the lemurs in having the tympanic ring enclosed by the expanded tympanic bulla.

The postcranial skeleton of the Paleocene Plesiadapidae so far as known is consistent, on the one hand, with their ancestral position to the modern tree-shrews and, on the other hand, with a fairly close relationship with both the Paleocene tarsioids and the Lower Eocene lemuroids (see pp. 385, 458, 465).

Both the tarsioids and the lemuroids very early evolved a strongly grasping type of hind foot with large, markedly divergent great toe, or hallux. This biramous, grasping foot, together with a hand with long fingers and partly opposable thumb, may reasonably be imputed to all the early primates.

Most of the tarsioids emphasized the procumbent lower incisors, reduced the canines and developed very large orbits and an expanded brain. These animals were of very small size and nocturnal, insectivorous habits. The typical lemuroids (pp. 459, 460, 461) branched widely in habits, shape of skull and dentition, but did not give rise to the Old World monkeys, apes and man. The New World monkeys may possibly have sprung either from Paleocene forerunners of the Notharctidae of the lemuroid suborder; or from some unknown tarsioid-like forerunners of the Old World series.

By Lower Eocene times our own simian forebears were old settlers in the trees, finding security and abundant food there. Some of the brethren of the loris tribe fell into a day lethargy and crept about timidly at night, searching for grubs and fat moths. But our tribe with eager hands grabbed the berries, seeds and insects and stuffed them into their cheeks. Their upper molar teeth, which in the earlier lemuroids retained the rounded, three-sided pattern, developed good hypocones on the postero-internal corners of m^1, m^2, and thus changed the three-sided crown into a more quadrangular contour. In the lines that led to the Old World, or catarrhine monkeys, the four main cusps became associated in two transverse pairs, as in the macaques and baboons, both in the upper and lower molars. The males of this tribe developed large canines and elongate oblique shearing surfaces on the opposing antero-external face of the "first" lower premolar.

The Old World monkeys, with few exceptions, retained a long tail; but perhaps in late Oligocene times some of our own ape ancestors lost their long tails; but they could still sit up on their callous haunches and chatter.

The apes at first retained a more triangular arrangement of the three main cusps in the upper molars and they never yoked the proto- and paracone, the hypo- and metacone, respectively, into a pair of cross-crests. In the lower molars they long retained a relatively primitive pattern with the opposite cusps, protoconid-metaconid, hypoconid-entoconid, not yoked into two transverse crests. Moreover the upgrowth of the hypoconulid (cusp 5) near the outer to mid-axis of the crown was conspicuous in the apes, but very rare in the Old World monkeys.

A very small lower jaw, named *Parapithecus*, from the Lower Oligocene of Egypt, has slightly procumbent lower incisors, small canines and low-crowned premolars and molars with conical cusps. It may be related to the tarsioid stock but it also appears to be a forerunner of the Old World monkeys, anthropoid apes and man. It lacked the first and second lower premolar on each side, as do all the known Old World monkeys (except *Amphipithecus* Colbert from the Upper Eocene of Burma), the apes and man; thus the two remaining premolars were really p_3, p_4.

The earliest known anthropoid ape is *Propliopithecus* from the Oligocene of Egypt, also represented by a quite small lower jaw. The face was evidently rather short, with full cheek bones, as indicated by the expanded angular area for the masseter. The lower molar crowns bear three low conical cusps on the outer and two on the inner side, an arrangement that foreshadows the "*Dryopithecus* pattern" of the lower molars of the anthropoid ape stock. As indicated by its name, *Propliopithecus* preceded *Pliopithecus* of the Lower Miocene of Austria, which in turn appears to be related on the one hand with the oriental gibbons and on the other with *Proconsul* and allied genera of the Lower Miocene of East Africa.

In the gibbons the bulk of the body never became very great and these apes can run swiftly along the boughs, rearing up on their hind legs, springing upward to catch the branches and hurling themselves recklessly from one tree to another. The living gibbons indeed excelled all other animals in brachiation —the practice of swinging and leaping from branch to branch with the arms directed upward and the body suspended from the hands. With increasing bulk of body, such reckless leaping proved too costly in terms of broken limbs and ribs. The heavy-bodied orangs solved the problem by swinging leisurely from branch to branch with their long arms; but they remained in the forests. The chimpanzee, gorilla and ape-man spent more time on the ground and explored new possibilities in locomotion and feeding.

Primitive anthropoid apes are represented by several Upper Miocene and Pliocene genera of the subfamily Dryopithecinae, of Europe, East Africa and India. Different genera of this subfamily gave rise to: (1) the orang, with extreme specialization of the limbs for arboreal life; (2) the chimpanzee, an excellent climber, but progressing rapidly on all fours on the ground, using the knuckles of the hands to support the weight of the fore part of the body; (3) the giant gorillas, which are becoming too heavy for safety in the trees and spend most of their waking hours ambling through the forests; (4) the Pleistocene South African man-apes (Australopithecinae), which agree with the human stock in their wider frontal lobes of the brain and in having small canines and bicuspid anterior lower premolars; the pattern of their lower molars is transitional to that of primitive man. Although the known Australopithecinae may be the great-uncles rather than the greatgrandfathers of man, they are structurally intermediate between the older anthropoid stock and the subhuman types represented by *Meganthropus* (p. 491), *Pithecanthropus* and *Sinanthropus*.

30. APE INTO MAN

The English anatomist Tyson (1689), after making careful comparative dissections of a chimpanzee (his "pygmy"), very clearly stated that in many ways the anatomy of the "pygmy" was more man-like than monkey-like. Linnaeus, Lamarck, Darwin, Haeckel, Huxley and their successors all saw and proclaimed the same fact and did not fear to make the proper inferences therefrom; but all classes of civilized society were shocked by this revolutionary idea and, as Pennant (1781) naively remarked, "My vanity will not suffer me to rank mankind with *Apes, Monkeys, Maucaucos* [Lemurs] and *Bats,* the companions Linnaeus has allotted us even in his last system." The same false pride, and that kind of exclusive dualism which is taken for granted by most of the white race, still prevent the vast majority even of college graduates from making any effective personal application of the facts: (a) that man is a made-over ape and (b) that man is a vertebrate animal, with all the capacity for selfishness and predation that constitutes "original sin"; from which the race is always needing redemption. The emergence of *Homo sapiens* is an event of such enormous import that many anthropologists have taken too close a view of it and have indeed failed to take any effective interest in the long series of ancestral stages that preceded and were prerequisite for the emergence of man.

That there has been a relatively rapid transformation in the case of man seems never to have been discovered by those writers who have assumed that even remote ancestors to be admitted as such must already exhibit at least a "tendency" or "trend" to develop all the very specializations which now distinguish man from all the apes. For example, Cope once suggested that the tiny Eocene tarsioid *Anaptomorphus,* with its relatively huge brain and eyes and short face, would prove to be a sort of Homunculus! Here was a foreshadowing of Wood Jones's theory that the ancestors of man were related to the tarsioids and that they never passed through an anthropoid ape stage. But evidence from thousands of cases proves that: during the aeons of geologic time the kaleidoscopic interactions of variation, heredity and selection have cumulatively produced profound transformations, both in structure and function. That is why, as we follow the taxonomic grades upward from individuals to populations, subspecies and higher categories the differences increase in number and magnitude.

Among the numerous major transformations, dealt with in previous chapters, we may recall: the evolution of lobe-finned fishes into the first amphibians, of basal amphibians into stem reptiles, of certain small running diapsid reptiles into flying reptiles, and of others into birds, of flying birds into penguins, of mammal-like reptiles into mammals, of primitive insectivorous-carnivorous placental mammals into whales, of primitive protungulates into elephants, of ground-living insectivores into moles, of tree-shrews into "flying lemurs" and bats.

A few of the major changes that were effected during the transformation from primitive anthropoid ape to modern man were as follows:

1) Gibbons, chimpanzees and gorillas can all rear up and walk on their hind legs, at least for short distances. The gibbon balances with his long fore arms and all apes when rearing up can waddle with bent knees. As Cunningham long since showed, the "lumbar curve," which is brought to perfection in man and enables the head to be balanced above the femora, is beginning in the chimpanzee and the gorilla. According to the interpretation developed in an earlier paper (Gregory, 1934), this double curve in the backbone was in its later stages accompanied by: (a) a *secondary lengthening* of the presacral region, brought about by the backward growth of the posterosuperior spine of the ilia, (b) a vertical shortening and (c) marked transverse spreading across the sacrum, which thus tended to shift caudally, freeing the former lumbo-sacral vertebra to the lumbar series, so that man as a rule has one more presacral vertebra than the gorilla. But this condition has been mistakenly regarded as more primitive than the shorter lumbar region of the gorilla. As a result of the wide transverse spreading and vertical shortening of the ilia, the great sciatic notch has arisen merely by the narrowing of the concavity beneath the spreading ilium, which is present in the gorilla.

2) By lengthening of the thigh and lower leg, the upright gait was encouraged, speed in running was increased and an alternately retarded and accelerated pendulum swing was attained in walking.

3) Many authors have noted that when the gorilla walks his great toes point inward (fig. 24.11a), thereby giving a wide base for his waddling, bent-kneed gait (fig. 23.61); but when man walks the great toe is drawn in beside the other toes, the legs straightened, and a pendulum-like and speedy movement of the limbs is attained. An inwardly-directed, divergent great toe has indeed been characteristic of known infra-human primates from the Eocene lemurs (fig. 24.9B) and tarsioids (*Hemiacodon, Necrolemur*) onward; it thus seemed reasonable to infer (Gregory, 1916, 1920) that the human stage, associated as it is with the fully upright gait, much enlarged brain, greatly shortened face (fig. 24.14H) and highly modified teeth (fig. 24.13G) was the derived and not the primitive stage. These results have been contested by those scientists who rely upon the

assumed applicability of the "principle of irreversibility" to prevent an allegedly "more specialized" condition from giving rise to an allegedly "less specialized" one. But according to Weidenreich (1913, 1921, 1923), Gregory (1916, 1920, 1936), Morton (1927), and Elftman and Manter (1935), the human foot bears evidence of its origin from an ape-like foot, which when the animal came down from the trees was still held in the grasping position, with a high arch; whereas in the chimpanzee and gorilla the foot has been allowed to slump into a flat-footed position. Meanwhile the great toe has been adducted toward the other digits, while the strong transverse fascial band beneath the skin (fig. 24.9) has developed into the hallucial portion of the transverse metatarsal ligament (H. C. Raven, 1936). The shortening of the four outer toes decreased their adverse leverage in raising the weight on to the toes. The arrangements (fig. 24.12) of the tendons and muscles of the foot in man and the anthropoids are quite consistent with the foregoing interpretation (Raven, 1936).

4) The shoulder girdle (fig. 24.3), arm (fig. 24.4) and hand (fig. 24.10a; pl. 24.III) required only changes in proportions and angles and so did the pelvis (figs. 24.6, 7), hind limb (fig. 24.8) and foot (figs. 24.9; 11a, 12). By appropriate shifts of certain muscles (Raven), the fingers attained more individual freedom as the hands gave up their locomotor functions and improved their manual skill.

5) As the upright gait was developed, the skull became balanced on top of the doubly curved erect column (fig. 24.1), instead of being tied on to a horizontal column, and there were many correlated changes. As the occipital condyles moved forward in relation to the backwardly extending occiput, the shapes of the atlas and axis (fig. 16.6aD) and the position of their ligaments were correspondingly altered, while the main ligament of the neck was much reduced.

6) Meanwhile the rapidly growing brain (fig. 24.14H) increased the cranial flexure in the plastic foetal stages, the cranial vault was expanded above the level of the temporal muscles and as the jaws shortened, the temporal muscles diminished, as did also the nuchal muscles. Further growth of the occipital poles of the brain (containing the optic projection fibres) lengthened the cranium anteroposteriorly; its rapid transverse expansion, coming later, brought in the high round-heads.

In the final human stage (fig. 23.66C) the massive, heavy brain is curved anteroposteriorly above the atlas; the basicranial axis extends gently upward; the height of the brain and braincase has greatly increased and the bony face has been withdrawn largely beneath the braincase. Due to the pronounced radial expansion (fig. 23.64B) from the pituitary fossae outward, the frontal lobes have pushed the forehead forward, the temporal lobes have pushed the alisphenoids forward, the occipital lobes have extended the occipital wall far behind the foramen magnum.

That the canines of man have been secondarily reduced and the anterior lower premolars remodeled from the compressed to the bicuspid form has been maintained by Remane (1924a, b) and the present author (cf. pp. 488, 489) against the verdict of most later anthropologists, who suggest that the anthropoid ancestor of man branched off far below the level of the Dryopithecinae. Their case, on the surface of things, has been strengthened by the finding in the Pleistocene of Java part of a gigantic jaw (fig. 23.78D), named *Meganthropus palaeojavanicus* by von Koenigswald, which has small canines and bicuspid anterior lower premolars, as was also the case in the Australopithecinae, in *Pithecanthropus robustus, Sinanthropus* (Weidenreich, 1937), Heidelberg, Neanderthal and modern man.

In support of the conclusions of Remane (*ibid*) and of Gregory, Hellman and Lewis (1938), the shortening and reduction of the canine crowns in a certain female dryopithecine (fig. 23.73D) may be again cited; and the apparently gradual stages of the transformation of the anterior lower premolar crown (fig. 23.74). Moreover, a pre-dryopithecine derivation of man seems inconsistent with: (a) the occurrence of the modified "*Dryopithecus* pattern" in the lower molars of some men (figs. 23.60; 23.67C; 72) and (b) the basic resemblances, joined with some differences in proportional emphasis of certain parts between the brains and reproductive organs respectively of the modern anthropoid apes, especially the gorilla, and of man.

The long dominant misapplication of Dollo's principle of irreversibility has conditioned many investigators to assume without evidence that there never was a reduction of canines or a transformation of the lower premolars and that the remote Lower Miocene ancestors of man must already have had small canines and essentially human anterior lower premolars. Thus in the memoir by Broom and Schepers (1946) all known Miocene and Pliocene apes (including the Dryopithecinae) are arbitrarily rejected as ancestors of man, on the basis of single differences, such as the size of the male canines or the lack of a shearing blade on the anterior lower premolar in Australopithecinae and man. Other investigators also have failed to realize that the transformation of the lower canines and anterior premolars from the primitive *"Dryopithecus"* to the human stage was evidently only a small part of the reduction and retraction of the upper and lower jaws and was in turn associated with such far-reaching changes as: (a) acceleration or retardation of growth rates in various parts of the cranium; (b) shifting genetic factors; (c) changing environmental pressures and opportunities.

In short, Natural Selection, operating upon the products of secular genetic changes, resulted in a profound reorganization of the skull, backbone, limbs, hands and feet, etc., which culminated in the upright posture, the bipedal gait and the bimanual perfection of man. But here it should be noted again that Sewell Wright and others have shown that there is a well founded statistical basis for the assumption that when, in the history of any given evolutionary series, environmental opportunities and penalties happen to favor certain combinations of desirable improvements (as in brain, teeth, limbs), such combinations do occur and continue over long periods of time.

Thus, in spite of the profound contrasts between modern man and ape in many habitus features, they still possess a deep-seated anatomical common heritage, which is merely masked by their divergent specializations.

31. FROM TROPISM TO HUMAN BEHAVIOR

The principle of functional integration, or acting as a whole, which is universal in living things, must begin on the inorganic level and be primarily a result of homogeneity of composition. In other words, polyisomeres of any given kind and generation naturally act alike and behave in concert; they are attracted by and toward each other and toward certain forces and repelled by others. Moreover, they soon discover, as it were, the principle called *e pluribus unum;* that is, they resist disruption and scattering better if they stick together.

Not even viruses are completely homogeneous and ordinary protoplasm includes a great mixture of different complex chemical substances. The different collocations of these elements, in reference to the environment and to the center of mass of the organism, would give rise to local differentiation and physiological division of function, such as found in the dorsal and ventral, anterior and posterior, regions of vertebrates.

Since dividing polyanisomeres tend to produce offspring which are like the parents, at least in general plan, the species tends to persist and its organic pattern becomes extended in time and space. But given a limited available area and a limited food supply, the Malthusian principle of increasing competition comes in and hence the living world is thronged with competing hereditary organizations. For each sentient species the perpetuation is insured in the long run, first, because every one of its members is motivated by self-centered desires for pleasure and fears of pain, and secondly, because each species is geared to various recurrent environmental cycles which give it suitable opportunities for feeding and reproduction. In a relatively stable or balanced environment, adherence to the hereditary pattern and the same environment yields the greatest number of survivors; but in an environment which is undergoing a severe secular shift, if the species is to survive, part of it must either emigrate in unconscious search for the old life conditions or meet the present crisis with appropriate adjustments.

As there is a hereditary pattern for the organism as a whole, so there are many contributory patterns for the special organs or organ systems. Experimental embryologists have shown that such an embryonic organ as the optic bud or a limb bud has its own "field of force" which evokes and controls the appearance and growth of the complete organ. Hence arises the struggle for existence between competing organs, as when, for example, in the ancestry of the elephants, as one pair of upper and one pair of lower incisors became enlarged into tusks while the other incisors dwindled and were eliminated; later the large lower incisors got in the way of the enlarging proboscis and were also cast out.

In ciliate Protozoa each cell is a unitary organization, acting as a whole, but it has its major constituents, such as the nucleus, and its organelles, such as flagella, cilia, etc. Even a protozoan behaves somewhat like a higher animal, receiving signals which tell it of either the proximity of food or the presence of a distasteful environment. Because each protozoan of a given species is a sort of standardized polyanisomere which is bound to react between fairly uniform limits to such recurrent variables as light, temperature, acidity, etc., definite tropisms arise and are the basis for later behavior patterns.

Nevertheless, repeated stimuli of the same kind may leave some chemical trace in the protoplasm and at the next experiment the animal may show what may be called signs of fatigue, satiety and disgust, or of eagerness and pleasure. Thus the beginnings of a learning process based on rewards and punishment date far down on the scale of life.

In such a relatively low vertebrate as a shark the sensory and neuro-muscular apparatus has already become excessively complicated. The organization is still functioning as a whole, but it has, so to speak, to watch a very great number of sensory records and to set in motion innumerable motor responses. Its task is simplified by numerous reflex arcs, in which a given stimulus automatically starts a given response; but at any moment suspicion and fear may arise (or their psychosomatic equivalents in the shark); then through some unknown method of struggle between competing stimuli, the former motor response may be first arrested and then thrown into reverse.

Presumably even a shark may experience something like hesitation, in which two competing stimuli, such as hunger and fear (whether based on memory or on preadaptive behavioral patterns), may balance each other until other memories or new stimuli upset the balance. Thus hesitation may look like deliberate choice, but "intelligence" may have begun merely with a restraining interval in which present stimuli are required to register in the presence of memories or chemical traces of past stimuli and of associations evoked by the results of yielding to them. The trigger-action of the last impulse to break the deadlock often starts a rapid kaleidoscopic change of value-patterns and in man it may be one of the reasons for the illusion of free will.

But if the whole process in the shark is far too complicated for any simple explanation on the bases of simple tropisms and reflexes (C. J. Herrick, 1944), how infinitely more complex it must be in man! For in man the primary parts of the nervous system have passed under control of a gigantic incubus, the neopallium, within which the swirling kaleidoscopic streams of memories are constantly urging, Do this, Don't do that, often in the crazy fashion of dreams. Why, then, do we not all become insane criminals? One answer, of course, is the inherited "wisdom of the body," which saves us many disasters, plus our cumulative commonsense; this is gained by experience, primarily from the memory of the associated results of our own acts, and secondarily, from the memory and example of what we have learned from others.

Compared with lower vertebrates, an adult chimpanzee has an enormous and highly convoluted brain; he can learn a great deal, both from his own experience and from what he is taught by his human master. He discovers for himself the principle of trial and error and has amazingly versatile command of his locomotor system. Moreover, with the coöperation of fellow chimpanzees, he discovers for himself important social principles, such as "One for all and all for one." By means of certain noises and gestures, he does to a limited extent advertise his own feelings.

But all this and much more collectively is far overweighed by the simple fact that chimpanzees have never invented any kind of articulate system of communication, in which signals have been transformed into a connected series of verbal symbols for states of feeling, actions, things, spatial relationships, or values to the speaker; nor do they use a language in which the meaning of the whole message is conditioned by the sequence and juxtaposition of its units or by rising or falling inflections or emphases.

Yet there is strong evidence that an anthropoid ape brain was eventually transformed into a human brain. The increase of the brain beyond the norm of existing anthropoids is illustrated in the increasing range of variability and generally rising cranial capacity of *Australopithecus, Plesianthropus, Paranthropus, Pithecanthropus, Sinanthropus,* Neanderthal and *Homo sapiens* (Broom and Schepers, 1946). Articulate speech was very probably near its beginning in *Australopithecus.* Meanwhile there was an increasing synthesis of sensory and motor parts of the forebrain, as well as of their connections in the cerebellum and brain stem. Doubtless there was long continued and severe selection for those subhuman strains which could produce larger and better brains and more and more effective speakers. Speech itself eventually came to be valued and the tribes of leaders and prophets, of teachers and grammarians, arose to expand, record and regulate it. Thus thinking became verbalized and by this new shorthand method of reasoning men blundered into new stupidities and worked out new triumphs. With the progress of printing, the power of the exceptionally eloquent individuals in some cases became multiplied a millionfold and when their peculiar doctrines became part of a powerful tradition, they were able to impose them upon countless followers and to exert enormous influence on the course of civilization.

Effective comparison and contrast are the basis for description and enumeration and the measure of intelligence. Some philologists and ethnologists have said, in effect, that all known human languages, either directly or by circumlocution, can contrast *me* and *you, here* and *there, inside* and *outside, now* and *then, today* and *tomorrow, mother* and *offspring, "father"* (in different senses), *friends* and *enemies,* and so on for untold thousands of ideas. By means of this immeasurably potent apparatus of articulate speech man collectively has transformed himself and created many worlds: such as codes of behavior, civilizations, philosophies, mythologies, religions, sciences, inventions, systems of education, the fine arts, the game of chess and thousands of others. But it is no wonder that to the man on the street the idea that man could develop out of any kind of speechless ape might seem merely funny. Not a few philosophers and mathematicians, with no real knowledge of the evidence that they themselves and their systems are by-products of the "procession of the vertebrates," have been inclined to set man on a lonely pedestal and to treat such a naively human personification as "mind" as if it were truly an objective individual entity instead of an artificial summation for billions of different psychosomatic responses by human and other sentient beings.

Speculation as to the steps by which the ape brain and mind were expanded into the human brain and mind will be deemed worthless by scientists of the "École des Faits"; nevertheless it may be worth while to brave their scorn and try what can be done in view of: (a) the transitional stage of the extinct Australopithecinae, (b) the wide range in intelligence among living adults, and (c) the observed transformation undergone by each person from birth to death. We ourselves can perhaps remember enough along the last named line to admit that there has been an enormous growth in conditioned and association systems of ideas between the infant and the adult. In a thesis by H. G. Wells (1942) he exposed the "illusion of personality" and showed that each person is not one and indivisible but a compromise and composite of many association systems which may even split up and re-form into multiple personalities. Indeed, we can easily see (in others) the loosening effects of alcohol and anaesthetics upon otherwise firmly regulated personalities.

A human "mind" may be crudely likened to a forest, or to a jungle. It is certain that what is there now has been planted there in times past and that what will be there in the future depends in part at least upon what is there now. Whatever special habits or skills I may now have were begun long ago and have grown by repetition. Thousands of ways in which I now behave have grown from the traditions, commandments, taboos and customs impressed upon me by my parents, teachers and public opinion. Then there is the constructive, self-conscious personality which I have built up of my own repetitive choice and which doubtless modifies the other factors in varying degrees, encouraging some and restraining others. In short, the principle of integration and organization has been working among my memories and habits until it has produced a more or less adaptable but (I hope) fairly well oriented composite personality. The principle of struggle for survival among the parts of an organism has also operated among my memories; but, parenthetically, I wish I knew how to scotch the inhibitor that blocks the memory paths when I suddenly try to remember the names of persons and authors whom I have known well for years past.

In the light of the foregoing we find new meaning in the verse, "As a man thinketh in his heart, so is he," and "the mind's the standard of the man." So it is but natural that an author's writings usually contain some record, it may even be in disguised form, of his personality, while an accurate biography is also a valuable record of an episode in evolution.

Many transformations in the sound of spoken words are recorded in the rise and differentiation of the Romance languages from Latin, in the shift of the vowels and consonants in the Aryan group, in the composite origin and history of the English language and in the wide differentiation of the American Indian languages. The stabilizing influence of the scribes on the development of language is evident in the comparative study of Sumerian and Assyrian inscriptions, of hieratic and demotic Egyptian and Old, Middle and Modern English.

In brief, words, phrases, sentences, ideas and general association systems have themselves had their origins and their transformations, like the songs of birds. But in man custom and tradition have played the part of heredity and the principles of evolution, such as integration, the struggle for existence, polyisomerism and anisomerism, have constantly operated in the origin, obsolescense and extinction of words and, indeed, of whole languages.

32. THE ROLES OF INSTINCT, LEARNING AND TRADITION

In man, instincts or inherited predispositions to react in certain patterns have been reduced to a minimum, and the neopallium, or "new brain," has largely taken over the care and general management of its owner. Some steadiness and continuity of direction in human evolution have been supplied by the conflict of individual needs and desires with the restraining effects of learning and tradition. The bank of learning, including science as one of its branches, has provided innumerable "basic patents" and tables indicating the limits of safety in building or bridge designs of a given type. An engineer may imagine that he has designed the entire structure of a bridge but really he has only adapted it by evaluating certain variables and by creating some new adaptive features. The designers of the complex calculating machines used by astronomers undoubtedly know that, as in the writing of a detective story, they have only worked backward from an assumed end. They appreciate the fact that the very equations and solutions which the machine is designed to handle are the products of centuries of trial and error methods, not wholly unlike those of Natural Selection itself.

Much of what passes for original design among men is due in part to man's faculty of imitation, which has developed in adaptation to a social life, in part to tradition, which has taken over some of the functions of heredity. Since man-made machines and Nature's own machines have to work in the same physical world and use the principles of the lever, of the spring, of the crusher, etc., it is not surprising to evolutionists to note very numerous cases of convergence between them, as in the body-forms of some whales and submarines. But though natural machines have many advantages over human machines, such as self-repair, self-feeding, self-regeneration, they also suffer many limitations: in land animals the limit of height has rarely exceeded twenty feet, whereas the Empire State Building, from subbasement to flagpole, is more than 1,000 feet high. Certain whales may reach a length of 90 or 100 feet, but some ocean liners exceed 850 feet in length. The principle of the wheel is unknown in animals, whereas it is essential in millions of man-made machines. For armor, natural machines rely on chitin, horn or bone, whereas battleships have huge thick plates of steel.

In conclusion, the classical argument that because a watch is "designed," so is any natural mechanism, compares two things which are merely analogous, not homologous, and which differ profoundly in their modes of origin.

Under present systems of education and journalism the public as a whole will never be able to grasp the cumulative effect of Natural Selection in building up such adaptive marvels as the Weberian apparatus of carps and catfishes; but for students who can appreciate it the evidence does show that nervous systems, like all others, have evolved through the ages on a competitive basis, and that in the case of human evolution high survival values have accrued in the long run and, with many exceptions, to the owners of the best brains—best, that is, not only in the repair of past misfortunes but in adapting their owners to a variable future.

33. BRAINS AS ORGANS OF FUTURITY

Sensation of some sort is admittedly a property of animals. Plants also, as botanists have shown, behave *as if* their responses to sunlight and water were due to feelings. The lower limit of sensation in an encysted stage of a parasitic animal may be near zero, but extreme sensitivity to faint stimuli of different orders doubtless occurs in the antennae of ants, the noseleaf of bats, and the retina of certain abyssal fishes. Vibrations of some kind, of a wide range in amplitude and rate, are undoubtedly essential both in the giver and the receiver of stimuli.

In a not-living object, such as a rock, there are no special organs either for receiving or resisting outside forces; but if the weight of mountains be pressing against the rock floor of a tunnel at the bottom of a deep mine it may buckle up and be squeezed together until it is strong enough to resist the thrusts from the sides. Here is an example of an adjustment without benefit of a nervous system. Rocks, metals, wood, eto., react to impact, pressure, heat and other forces in such a way that the effects of repeated or increasing stimuli of the same kind may be cumulative until a critical point is reached, when there is often a sudden breakdown and a rearrangement of material. Up to a certain point the tensile strength of iron cables may even be increased by long continued stress, and it is said that a razor blade may recover part of its sharpness if given a sufficient rest.

Living matter, although far more complex, reacts to outside forces in more or less the same ways. Repeated pressure or abrasion on the skin in a mammal, for example, may stimulate the multiplication of epidermal cells that form callous areas; and it is well known that exercise of limb muscles increases their strength. But living things far excel inert objects in their adaptive responses to varied stimuli. Nevertheless living organisms are synthesized from not living materials, and their superior adaptability has been improved, along with the complexity of their organization, through the ages. Adaptability is shown in every organ system of the body and especially in the body as a whole; but one of the principal agents in adaptability, at least in many-celled animals, is the nervous system. The foregoing statements may all seem elementary or trite to specialists, but they lead to rather surprising conclusions.

What a nervous system is and exactly how it works sooner or later brings up the question, what is it that receives, perceives, understands, knows the sensations? The answer of the inquirer is inevitably determined in part by his own conditioned responses, including his education and philosophy. Here it must suffice to state that the present author is conditioned to accept without any mental reservation the far-reaching conclusion that "mental" activities are limited to nervous systems and that the abstraction "mind" is chiefly a distillate from the behavioral activities of the higher organisms. If an amoeba can learn to avoid dangerous or unpleasant stimuli, that does not mean necessarily that an amoeba is directed by a "mind" but that stimuli that have led to a partial destruction of tissue have thereby changed the chemical and physical configuration of parts and induced a reversal of the reaction. Nevertheless able investigators have found that when all due allowances are made for simple conditioned responses, ants and bees have a faculty corresponding to memory, while man's nearest relatives, the apes and monkeys, even show the earlier stages of reasoning. In short, experimental and comparative evidence indicate that past reactions of nervous systems leave physical traces which are the basis of conditioned responses, of memory and, later, of reasoning. The effects of a stimulus may and do often continue after the original stimulus itself has ceased, as in "after images" and the continued pain following an injury. Sensory memories seem indeed to be somewhat like automatically recorded echoes and images of earlier stimuli-and-response complexes. Many of the activities of the body, especially the nervous system, thus have to do with the recovery of equilibrium, with recuperation, healing of wounds, and other adjustments to past events.

The rhythmic or cyclical repetition of certain environmental events, including seasons, days and nights, high tides and low tides, induce corresponding alternations and reversals in the responses of living things. Sunrise is the time for diurnal animals to get up and for nocturnal animals to retire. But a given cyclical or repetitive event may occur gradually, as when the shadows lengthen and the sun goes down, so that the creatures come to associate these small signals with memories of the completed event. Thus arises a vast world of anticipative reactions. Even the repair of past damage has a high value for the future, but distinctly anticipatory responses to events that have not yet arrived have progressively increasing values, both as insurance and as investment (Gregory, 1924).

One of the pitfalls is that the signals of the new event may not agree completely with those of its predecessor; thus the urge of circumstance may set off an inappropriate and wrong response; while delayed response for verification of signals may also lead to disaster. Hence arise errors in reasoning, including mistaken identification and the vast realms of falsification; hence also the premiums, both in human and extra-human affairs, for progressive improvements in accuracy of recording and in reliability of estimating wholes from samples of parts.

34. *HOMO SAPIENS*, PRIMATE EXTRAORDINARY

Perhaps during early Pliocene times, before the mountain ice sheets of Europe had spread down into the plains, our alert prehuman ancestors were already running upon their hind legs and steadily improving their hands, feet and brains, in a country of open plains and sparse forests. Somewhat later the South African ape-men or some nearly related northern forms gave rise to the *Pithecanthropus robustus* and other thick-skulled, big-brained giants of the Pleistocene ice age. Thence, according to Weidenreich, by diminishing stature and increasing alertness, the giants shrank into the babbling, gesticulating, finally articulate savages of Heidelberg and Neanderthal; still later arose their tall, highly intelligent successors, the Aurignacians and Cro-Magnons.

When man learned to produce an abundant artificial food supply, it became possible to feed warriors and captains, priests and kings, and to rob other tribes on an ever-widening scale. Much later the philosophers were evolved, and they began to inquire persistently into the nature and meaning even of man himself. In the long run we became partly self-conscious and imagined ourselves to be the master race of all time. Thus we strutted to the center of the stage and calmly assumed that around our world and man's fate the rest of the universe turned.

THE HUMAN FRAME AND THE SCIENCE OF BONES

A human skeleton with a scythe over his shoulder has long been a symbol of Death. Skulls were formerly often figured on tombstones, "poor Yorick's" skull was immortalized by Shakespeare; Apothecaries and pirates gave wide publicity to the skull-and-crossbones symbol, but of course for quite different purposes. Nurses, doctors, surgeons and the makers of artificial limbs mostly take only a practical interest in bones and but few young biologists develop enthusiasm for the skeletons of the frog and the cat.

If present popular interest in the science of osteology could be measured and plotted in such units as thirty lines of newstype or two minutes of radio time, then the score of osteology, including such topics as the growth of bone, the evolution of different types of skeletons, etc., for the past few years would be almost zero. The score for geology would hardly be better, unless economic geology were included; new finds of dinosaurs in the Southwest and recent discoveries of fossil ape-men in South Africa and Java would lift the curve a little for palaeontology and anthropology; but all these put together would make only a very low hump in comparison with the scores for politics, theatre, sports, which would rise in mounting millions. In spite of this enormous handicap of negative publicity, it will presently be shown that the study of skeletons of all ages has already yielded much of potential interest to both laymen and scientists.

When human bones are buried in the ground, they acquire a mysterious fascination to diggers, young and old, including archaeologists both amateur and professional. After all, a human skeleton or even an odd limb bone from an ancient cemetery represents a once living and breathing human being and one's imagination is free to reconstruct all its entire life story.

Bones, recent or fossil, are seldom used to illustrate any general principles, nor are general principles much used (with certain noteworthy exceptions) to explain either the forms of bones or the general construction of the skeleton as a whole. In other words, osteology as usually taught is still in the descriptive stage and should, for the most part, be called osteography. Books on anatomy usually treat each part of the human skeleton as if it were a separate *"Ding an sich."* In Cunningham's Anatomy (1903), for example, the item by item description of the paired temporal bones fills eight pages; the description of the entire skeleton, including its joints, requires no less than two hundred and thirty-seven large octavo pages. This is of course quite necessary and useful for descriptive purposes and for the precise knowledge required by surgeons. Medical students may well be appalled by the task of learning the names, locations and boundaries of the bones of the human skull, together with the names and chief features of all its ridges, angles, eminences, processes, depressions, fossae, grooves, canals, foramina, etc., etc. That, however, is the hard, static way of learning about things which are parts of a growing whole and are more easily comprehended if seen in their natural associations and with reference to their functions or use in the entire organism.

Nevertheless there has long been and there still is a real science of osteology, and a tiny fragment of it may be found in some dictionaries under the word *biceps*. This is illustrated by a neat little cut, taken in part from T. H. Huxley's Physiology and showing that when the biceps muscle pulls up the forearm it is acting like the power in a leverage system of the third order, since it is inserted on the lever (radius) between the fulcrum (humerus) and the weight (W) held in the hand. Indeed, the conception of the shoulder, arm and hand, acting as a lever, or really as a complex system of levers, leads to a better understanding of the skeleton as a whole than that which is usually conveyed by the formal definitions in dictionaries.

Surgeons and doctors have long known that the ribs and rib muscles, the muscular diaphragm and the abdominal muscles may act separately or together not only as a great pump for the lungs but also in movements of contraction and relaxation of the abdominal walls. However, it required the ingenuity of Sir Arthur Keith, in his little book, Engines of the Human Body, to make the action of the diaphragm clear to intelligent children.

In the same author's illuminating lectures on man's postural disorders (reported in the British Medical Journal in March and April, 1923), he showed in detail how such painful troubles as the slipping forward of the viscera, dislocations of the lower lumbar vertebrae and their discs, the undue stretching of the sacral ligaments, the collapse of the arches of the foot and similar ills, are often the result in part of imperfect or insufficient adaptations to man's upright posture. Here, in a word, was an invaluable application to the study of the skeleton of the main facts and principles of human origin and early evolution. This was welcome to those palaeontologists and comparative anatomists who studied the skeletons of fossil and recent animals as adaptive mechanisms rather than as "dry bones."

The human skeleton has often been defined as the "bony framework of the body." Framework, however, seems to imply a stiff, well-joined frame like that of a wooden house, whereas the skeleton as a whole combines: (1) a fairly rigid but slowly growing skull; (2) many well-jointed vertebral segments that permit sliding, bending or twisting movements of one part upon another; (3) the upper and lower appendages with all their highly mobile parts.

In general, the skeleton of vertebrates from fish to man consists of: (1) an articulated axial series, namely the braincase, vertebrae and ribs; (2) an oralo-branchial series, comprising the jaws and gill arches; and (3) an appendicular series, including the bony girdles and fins or limbs. Each segment may serve either as a support for an adjacent part or as a lever for moving other parts or outside objects.

The human skeleton becomes far more comprehensible if we consider it in the light of its long evolution. We have therefore begun with the skeletal or hard parts of the simplest organisms and then have worked outward along the branching lines of evolution that led to such amazing creatures as squids, humming-birds, elephants, whales and men.

In previous chapters it has been shown that the skeletal system of the lower animals often comprises two parts of quite different origin, namely: (1) the outer skeleton (exoskeleton), which is especially well developed in crustaceans, insects, shell-bearing molluscs and primitive vertebrates; and (2) the inner skeleton (endoskeleton), which finally becomes predominant in higher vertebrates, including man.

There is evidence that in the remote predecessors of the vertebrate animals the exoskeleton was predominant and the endoskeleton only beginning; whereas in man, the endoskeleton having become the sole support of the body, the exoskeleton is represented chiefly by the teeth, nails, hair and skin.

PART SIX

EVOLUTION EMERGING: RETROSPECT AND PROSPECT

CHAPTER XXV

LAW AND CHANCE

Contents

CHAPTER XXV

LAW AND CHANCE

REDUPLICATION AND EMPHASIS IN EVOLUTION

A REGRETTABLE effect of the apparent conflict between Chaos and Order is that the easiest way to reduce the kingdom of Chaos and advance that of Order is to specialize in some small department. Whether we are investigating some of the thousands of problems relating to the heredity of fruitflies or whether our speciality happens to be that of second assistant trouble-shooter in a wayside garage, we gradually learn how to apply standard symbols in our particular trade and proceed to view outsiders, especially those who give advice unasked, with more or less suspicion. Such, we regret to state, is often the attitude of scientists of the *"école des faits"* toward "theorists" of the *"école des idées." "Besser ein Steinchen der Wahrheit als ein ganzer Schwindelbau"* must have been said by a very superior factualist who in his youth had failed in Philosophy I. As we ourselves happen to be a sort of educational hybrid between Science and Philosophy, we have never been loath to attempt to reconcile our so often estranged parents. Up to date, however, neither side has paid much attention to us. But we intend to continue our well meant efforts.

In earlier passages we noted that the concept of evolution must take into account the relatively unchanging side of nature as well as its measurable or describable transformations; that we should study the relatively fixed background as well as the moving or changing object. The chief characteristics and principal transformations among the graptolites, a particular type of invertebrate animals, were shown to be reducible to the same general terms which had been listed by a certain philosopher (Woodbridge) as applicable to all sorts of reality. In the present chapter we shall develop the theme further by showing how it is possible to pass downward from the abstract qualities of Individuality, Continuity, Potentiality, Chance, Purpose, to general recurrences or principles that are more nearly recognizable as falling within the domain of the biological sciences.

We have seen that, considering a single strip of graptolites (Chapter I), any two adjacent individuals and the niches that they lived in were "as like as two peas in a pod." This similarity of adjacent individuals of the same general derivation occurs so often in the mineral, plant and animal kingdoms that it seems as if there must be some widely recognized general term to describe this relationship. I found that there were several terms that described particular instances of such a situation but I did not succeed at that time in finding any single term which seemed to cover all such homogeneous units. I called them *polyisomeres* (many, equal, parts) and I continue to use this term even though (as I afterward found) the chemists had already invented the term *polyisomer* for a particular application of the same idea. Polyisomeres may be, for example, a row of graptolites, all the leaves on a tree, or other biologic units; they may be a row of cog-teeth on a zipper, the pebbles in a stratum of conglomerate, or a string of musical notes or drum beats. In so far as they are repetitive units they are called *polyisomeres* and the condition of being alike in this way is called *polyisomerism.*

A given lot of polyisomeres may be produced

simultaneously, as when a parental mass explodes, so to speak, into units of the next generation, or they may be produced successively by recurrent or rhythmic activities involving a stop-and-go or faster and slower alternation, as in normal growth increments, the egg-capsules of whelks, and the like. In either case polyisomeres reflect the fundamental quality of Continuity.

But each polyisomere has its own Individuality, for even if they look exactly alike they will have different histories in space and time, as in the case of different leaves on a tree or different ones of the paired appendages of Crustacea, or different generations (F_1, F_2, etc.).

In so far as different polyisomeres show difference in sizes or in emphasis of any given character, they are termed *anisomeres* (unequal parts).

Polyisomeres (implying repetition, units, serial numbers) and anisomeres (implying either positive or negative emphasis, degrees) form an associative or correlative pair of categories of wide application in nature and even in the special worlds of man. After we have once learned to know them, we are forced to wonder why others, especially students of the natural and social sciences, have not commonly recognized them as such. The answer is that the differences between polyisomeres and polyanisomeres of different kinds and origins are so conspicuous that each kind has received its own name, and since most of our thinking is verbal, we often fail to recognize that specific concrete things may usually be grouped in some larger categories. Daisies, flies, sand grains, chocolate bars, votes, vertebrae and mosquitos, we know separately and we are so busy in dealing effectively with each new kind of polyisomeres or polyanisomeres that, especialy if we are "practical" men, we may fail completely to recognize the utility of lumping them all together under a single class.

It is not until we ask ourselves, why is the world crowded with different kinds of polyisomeres and anisomeres, that we come to the principle of progressive divergence between originally homogeneous polyisomeres. In cases where sufficient evidence is available, it becomes highly probable that in any given group, such as graptolites, the remotely ancestral life forms were all more or less alike. The very existence of great numbers of species of any particular genus, family, or higher groups, indicates progressive anisomerous differentiation from an ancestral common stock of polyisomeres.

In nature, any given series of polyisomeres or polyanisomeres are more or less alike because:

1) they have been derived by budding, etc., from one source and are made of the same material throughout;

2) they express the tendency of living units to reproduce themselves by growth (anisomerism) and by repetition of parts (e.g., *Salpa* chains);

3) there may be significant economies in repeating the same product;

4) the materials used in preadaptation and adaptation are polyisomerous and anisomerous;

5) the inorganic environment is also organized into polyisomeres, anisomeres and polyanisomeres, and each type of organic polyanisomeres lives in or is associated with an environmental polyanisomeric system, the parts of the former being adjusted to the parts of the latter, often in a "lock-and-key" relationship.

So too the works of mankind tend to be polyisomerous:

1) because man like nature found that polyisomerism may be advantageous for the sake of economy, efficiency, the relative ease of manufacture, storage, handling;

2) because of the essentially polyisomerous nature of the natural materials used by man and found by him ready to be conditioned (anisomerized) in certain features;

3) because a series of polyisomeres applied in the same direction may have a cumulative effect as in a pile driver or if spread in different directions a mass effect, as in methods of drilling, sowing, or killing by machine guns.

Since the world is crowded with polyisomeres and since these are of multitudinous kinds, it should be worth while for us to inquire whether there may be some way of classifying polyisomeres which will be helpful in the study of evolution. A row of peas in a pod may be taken as an example of *simple* polyisomeres. To this division we might refer the holes in a simple sponge, the simplest kind of sponge spicules, the spikes of a spiny-oyster (*Spondylus*), the spots of pigment on a tiger-cowry, the "teeth" along the outer border of the aperture of a cowry shell, the curved rows of sand left by the retreating waves on a beach. As already noted, simple polyisomeres of the same kind may arise either through simultaneous subdivision or successive budding, as rhythmic growth of a parent mass; they may be either the mass itself or some imprint or other record of the presence, movement or growth of the mass, as in cuneiform inscriptions, the thecae or niches of any single strip of graptolites, the holes in the shell of *Haliotis* (abalone), the hexagonal prisms of a beehive.

In *subdivided polyisomeres,* every polyisomere of any given set may be incised or flounced, or folded in the same way. Such are the sutures of ammonites, the elaborately plicated folds on certain *Murex* shells, the stellate, cruciform and other many-rayed spicules of sponges, the rows of gill-combs on the appendages of Crustacea.

The distinction between simple and subdivided polyisomeres will of course be elusive whenever a simple polyisomere begins to subdivide.

In *aggregate polyisomeres,* individual polyisomeres, in so far as they are conceived or seen as simple units, are grouped, combined or connected in some regularly repeated design or pattern set off against a background, as soldiers moving in squads, companies and regiments, all seen from above. Here belong repetitive clusters of any kind, like the tetrads of the graptolite *Tetragraptus,* bunches of grapes seen as bunches, a grove of pine trees seen against the sky.

Again, the distinction between subdivided and aggregate polyisomeres may be difficult to determine either because of the occurrence of transitional and mixed conditions or because the mode of origin of the aggregation is not fully known.

Up to this point each set of polyisomeres which we have considered, whether *simple, subdivided* or *aggregate,* has been conceived, at least for the moment, as made up of pure or unmixed elements, that is, of one kind only. But in the organic world perhaps the greater number of polyisomeres are composed of more than one kind of material and may therefore be called mixed or *heterogeneous* polyisomeres or polyanisomeres. These may be of varying degrees of complexity according to the number of different kinds of units entering into each larger polyisomere. An example would be the row of eye-spots along the edge of the mantle in scallops (*Pecten*). Although all the eyes of a given *Pecten* look alike, they are each composed of several different tissues. A similar example would be a bunch of orchids plucked from the same plant, the buds from a single *Salpa,* the different members of a single colony of corals, and in still greater degree, the assemblage of differentiated castes and individuals which make up one ant-state of a given species. That is, the ant-state itself as compared with others of the same species would be a heterogeneous polyisomere, or giant polyanisomere. The zygotes of any given species may be regarded as polyanisomeres because each zygote contains hereditary material from both its parents. The chromosomes of any given individual would be polyanisomeres because they are severally composed of more than one kind of heredity-bearing material. In the chromosome map of a given individual strain the genes are presumably polyanisomeres of a lower order of complexity.

Under the same general heading may be classed secondary polyisomeres, or *pseudopolyisomeres,* in which the units compared look alike but are of different origin.

The interrelations between the constitutent phenomena of organic and cultural evolution are so close that the refusal of some cultural anthropologists to continue the search for the evolution or history of ethnic traits tends toward particularism and isolation. As an example of the fruitfulness of close coöperation between cultural and physical anthropologists and other scientists may be cited the monographs of the Bishop Museum (Honolulu) bearing on the origin and migrations of the Polynesians. The combined evidence indicates that the Polynesians as a whole were derived from some of the early Indonesians (already a mixed race) who in their great sailing catamarans followed the coasts of Papua and adjacent islands and thereby picked up a significant amount of Negrito blood and certain cultural features. Venturing farther and farther outward into the Pacific, they eventually established one or more centers in the mid-Pacific and from thence extended northward to Hawaii, eastward to Easter Island and southwestward to New Zealand. The early settlers of New Zealand (Morioris) were but little different from the Maoris. The latter, under the guidance of their chiefs, retained a completely Polynesian culture, including the language, but living in isolation for many centuries, they skillfully adapted parts of their old culture to changed conditions. Meanwhile they developed a copious folklore and a complex system of *tabus* for the regulation of daily life, hunting, war and peace. Thus man behaves as a natural organism and though he has often done his best to conceal the fact and to act as if he were superior to nature, in the long run even his thinking processes may yield to analysis in evolutional terms.

Let us turn now to a consideration of the differences between individual polyisomeres of the same sort. As already noted, we find that, at least in the organic world, however closely two polyisomeres of the same set may resemble each other, they yet differ not only in some measurable way but also in respect to their differences of position, that is, their numerical order. The different peas in a single row will also differ measurably in their diameters, as do the peas of successive generations bred under certain conditions. In large collections of *Tellina* shells there are numerous intergrades between purple, red and yellow. Among fusinid shells (fig. 2.25A-F) one finds at one extreme very slender tall shells with high spines and prolonged siphon tubes, and at the other extreme, very low wide shells with almost no spire and greatly reduced siphon tubes. The progressive reduction in spire and siphon is due to arrested or negative growth gradients, the increase in transverse diameters to positive growth gradients and the phenomenon of change in dimensions has been called anisomerism (unequalization of parts).

All adult human heads are polyanisomeres of a single species, which are subject to wide anisomerism or variation in respect to proportional width, height, breadth of face, etc. In fact every change or transformation implies change of emphasis or anisomerism. Anisomerism was formerly imperfectly recognized in part under the terms growth, differentiation, allometry, heterogony.

In short we may sum up as follows:

- I. Polyisomeres
  1. Homogeneous (with only one kind of unit)
     - Simple
     - Subdivided
     - Aggregate
  2. Heterogeneous (with more than one kind of unit)
- II. Anisomeres (differentiated polyisomeres)
- III. Polyanisomeres (combinations of I-1, I-2, and II)

We are now ready to return to the question: How did the world come to be so full of polyisomeres and polyanisomeres? The first answer seems to be that at least in the solar system, and probably everywhere else, the polyisomerous intra-atomic forces are ever acting, that is, the "prime mover" of the entire system never stops. This is no less true of the radiant activity of the sun. This, however, although never failing, is subject to periodic and apparently variable outbursts, and, owing to the rotation of the earth, most points on the earth's surface, except at the poles, are subject to daily increase and decrease of exposure to the sun's rays. Thus the immediate source of terrestrial life is both polyisomerous and anisomerous in character. Next, the chlorophyll of the plants forms a self-renewing but varying treasury of solar energy, which is the basal stuff of life. The physical background is also subject to recurrent but variable forces. Hence it is no wonder that all the products of the interplay of cosmic forces and terrestrial environment should bear the stamp of their genesis, namely, repetition and varying emphasis.

As soon as we realize that any sort of recurrences are in themselves polyisomeres, it follows that the latter must be abundant in a world in which every sunrise starts and every sunset stops or slows down the basal process of photosynthesis in plants; that at every low tide billions of limpets, barnacles and other creatures clamp down on the rocks and remain quiet, without new food until the next tide reaches them.

If all recurrences were simple and sharply defined and all environments constant, the world would contain only uniformly patterned polyisomeres and polyanisomeres. But in a world where there are so many variable forces and so many chance meetings and unions of forces and of values, we find a corresponding diversity of these categories. For example, the cumulative force of gravitation, when the sun and moon are both in line with the earth, produces extra high tides; these reach far up the strand and bring death or derangement to many, but to the waiting schools of silver fish (*Menidia*) they bring the only opportunity of the season for breeding. It is of course the seasonal succession that causes the corresponding growth in the woody tissue of trees to be recorded as rings (typical and simple polyisomeres) but it is the seasonal variations in moisture that cause the anisomeric record of yearly thickness, and, as we have seen in a previous chapter, as the number of recurrences and emphases increases, the possibilities of new combinations and unexpected results go up in geometric ratio.

In such a world of both simple and complex recurrences and unexpected conjunctions and emphases it is small wonder that animals, if they are to exist at all, must have various ways of adjusting themselves to recurrent changes in their environment.

And here we come upon another at first surprising and far-reaching result. It will be to the advantage of an animal if he can anticipate danger and seek safety in one way or another before the trouble descends upon him in full force. The range cattle, it is said, do not wait for the storm to break but start for shelter as soon as they sense its approach. Hence many adaptational systems have an anticipatory function. Moreover, if one gets, say, the first few symptoms of an attack, it is better to make full preparations as if the attack were already here. But either desire or fear predispose us to the logical fallacy *ex unum disce omnes* (the great source of most mistakes), which being translated means, don't wait for the full signals, *act now*. So here we have, as one result of the recognition of the far-reaching character of polyisomeres and anisomeres, a clue to the basal reasons for our behavior in the world of nerve-signals and association-values. Moreover, one of the reasons why the neopallium is so useful is that it provides a background of varied experiences and memories upon which the present stimulus-pattern may be projected. At least in simple cases the deliberate judgment, yes or no, stop or go, is likely to be more abidingly useful than the uninhibited momentary impulse.

E PLURIBUS UNUM

Perhaps the greater part of the causes of evolution lies deep within the nature of the hereditary mechanism and of its reactions to Natural and Artificial Selection; but on the phenotypic plane of observation, although the facts of evolution are infinitely varied, many of the leading principles of evolution when once grasped by the observer appear to operate at all levels of organization, from inorganic material and forces outward along the diverging paths of descent with modifications.

E pluribus unum is the principle of progressive increase, organization and integration. During the course of evolution as well as of individual development, new characters and units of structure appear at given levels and these units are subsequently combined and made to work together either in further extensions of old organs and functions or in entirely new forms and functions. The organism or line of organisms may, as it were, squeeze several units together and unite them into one organ, as in the separate metatarsals in reptiles, which were united into a single metatarsus (or tarso-metatarsus) in birds. The agents in carrying out such operations during individual development have been called "organizers," or "fields of force," and the evidence indicates that though invisible these organizing fields of force are just as real as the animals themselves; also that they in turn result from the interaction of specific chemical configurations with the materials of growth. Undoubtedly the individual animal receives through the interaction of heredity and environment some sort of regulating or adjusting system whereby the dangers which its kind have been accustomed to meet are usually rendered nugatory.

From such facts and considerations has arisen the doctrine of holism (Smuts, 1931), which stresses the wholeness, the self-defensive, self-perpetuating reactions of organic beings. The fact that the whole is in one sense really greater than the sum of its parts has been discovered independently by several authors, who have used such words as Élan de Vie, Vitalism, Emergence (Lloyd Morgan), Wisdom of the Body, "Creative Evolution," "Aristogenesis" (Osborn). Bateson intended to brush this off as fatuous optimism in his scornful allusion to Dr. Pangloss, who maintained that everything happens for the best in the best of all possible worlds. Nevertheless it is a fact that the properties of water are very different from those of H_2 and O. Here is the "creative" aspect of all reality but when traditional words and phrases like "Creation," "Adaptation" are avoided, it seems plausible that if the simple principle of the composition of forces holds on atomic levels, then all the properties of H_2O would presumably result from the *combination* (which is something more than simple addition) of H_2 and O *plus* the energy necessary to combine them.

Curiously enough, the author (Osborn) who most stoutly defended the creative aspect of reality would not realize that this creativeness gave rise to a certain unpredictability or *Chance.*

These fields of force have themselves developed in the individual from the interaction of *their* predecessors, which leads back to the genetic constitution of the parent zygote. But the oak itself is not in the acorn; all that is in the acorn is a set of chemicals which will react to the energy of the sun and the material from the soil in such a way as to produce eventually a seedling; but this seedling will react only in such ways as to produce an oak; probably because it has a very specific constitution and is packed with a specific set of materials and fields of force.

The concept of holism correctly emphasizes that the living body as a whole is greater and different from the mere summation of its parts; but the body includes many automatic mechanisms which tend to correct any deviation beyond safe limits; this may only show that the interaction of genetic and environmentally selective factors over long periods has been sufficient to produce the abstraction that there is some sort of superior field of force which regulates all the others.

CHANGE OF FUNCTION AND EVOLVING PANELS OF HERITAGE, HABITUS AND ENVIRONMENT

All the great changes in pattern, as from hands to wings (Chap. XV), have plainly been accompanied by equal changes in function, as from grasping to flying organs. But to effect such major changes it has usually required at least tens of millions of years. Therefore there may be long sequences of generations in which both the functional values and their corresponding patterns appear to be stable. The individual animal, when confronted with changed values in his ordinary movements, for example, as when a dog swims, may succeed as best he can in adapting his hereditary cursorial patterns to the new medium. But doubtless there are some breeds of dogs which make better swimmers than others, not so much through a radical difference in physical build as in their superior tolerance to and persistence in a medium to which they are but poorly adapted. A sea-lion, on the contrary, although derived ultimately from more or less cursorial carnivores, virtually flies through the water and his entire body-form and functions have been made over in adaptation to a once new and strange medium. The respective anatomical patterns of dog and sea-lion are both strongly hereditary and, in essentially similar cases, it has been abundantly indicated that no improvement by practice or exercise which either of them could achieve as individuals would be transmitted to their offspring. Nevertheless there is clear anatomical evidence that even by the extremely slow effects of Natural Selection upon chance mutations, the locomotor patterns of the fast-moving terrestrial carnivores were gradually overlaid and replaced by the later swimming patterns of the ancestors of the sea-lion. Thus the habitus of the remote ancestor becomes a diminishing part of the successive heritages of its highly modified descendants.

Much of this has come about through the divergent pressure of Natural Selection, acting through competition and the environment, upon each animal. These divergent pressures took part in opening the paths leading from primitive land mammal respectively to dog and sea-lion.

LIFE ZONES, ADAPTIVE BRANCHING, PARALLELISM, CONVERGENCE

Whenever the records are ample, innumerable intergrades and combinations are found between even the most diverse habitus patterns as well as between the different types of environment. In most cases also the environment of the embryos and larvae contrast widely with those of the adults of the same species, as do their structural patterns, food and habits.

The kinds of environment in which a single subspecies can adapt itself are usually few and not very unlike, but as we ascend the taxonomic scale the range of the environments often increases with the number of genera, subfamilies, families, etc., until we arrive at the class. The class, unless it is represented only by a lone survivor from extremely ancient times, such as *Peripatus,* usually has representatives in many different environments and its members fill innumerable special adaptive niches and inter-connectives, as in insects (cf. Wheeler in Social Insects). Hence it is that more or less similar main adaptive types of any one large systematic group may be found in one or several other large groups, as between mammals and reptiles (Table II).

TABLE II

CONVERGENCE IN LOCOMOTOR HABITUS AND FOOD HABITUS BETWEEN REPTILES AND MAMMALS

LOCOMOTOR HABITUS	REPTILES	MAMMALS
Crawling, with everted limbs	† Cotylosaurs	Platypus
Quadrupedal running	† Higher mammal-like reptiles	Primitive mammals (e.g., opossum)
Graviportal (with massive limbs and short feet)	† Sauropod dinosaurs	Elephants
Bipedal hopping	† Saltopus (small dinosaurs)	Jerboa
Flying	† Pterosaurs	Bats
Crawling, with small limbs and long body	Certain lizards (scincs)	Weasels
Digging, with clawed hands and feet	Tortoises	Armadillos
Swimming, hands paddle-like	† Ichthyosaurs	Dolphins
FOOD HABITUS		
Predaceous, with shearing upper teeth	† *Cynognathus*	Tasmanian pouched "wolf" (*Thylacinus*)
Herbivorous, with grinding teeth	† Duckbill dinosaur (*Hadrosaurus*)	Elephants
Fish-catching, with long, many-toothed rostrum	† Phytosaurs	† Archaeocetes
Blunt-toothed, for crushing	† Placodonts	Sea otters

† Extinct.

TABLE III

EVOLVING PANELS OF HERITAGE, HABITUS AND ENVIRONMENT

	HABITUS (of adult)		HERITAGE	ENVIRONMENT			
Example	*Food habits*	*Locomotion*	*Class*	*Climate*	*Milieu or terrain*	*Botanic*	*Social grade*
Peripatus	Carnivorous	Caterpillar-like Many parapodia	Onychophora	Temperate to subtrop.	Rotting wood	Fallen trees	Solitary
Periplaneta	Omnivorous	Flying	Hexapoda	Temp.-trop.	Crevices	Varied	Solitary
Culex	Sanguinivorous	Flying	Hexapoda	Temp.-trop.	Air	Trees, bushes	Swarms
Termites	Vegetarian	Flying or flight-less	Hexapoda	Tropical	Earth and woody tissue	Varied	Socialized
Wild honey-bees	Nectar gathering	Flying	Hexapoda	Temp.-trop.	Air; holes in trees	Varied	Socialized
Amphioxus	Food-sifting for diatoms, etc. Jawless	Darting by myomeres	Prochordata	Moderate	Salt water	Diatoms Desmids	Solitary to schooling
† *Cephalaspis*	? Pumping in small food	Head shield and slow swimming	Ostracodermi	Moderate	Fresh water	....	? Solitary
Petromyzon larva	Food-sifting	Undulating by myomeres	Cyclostomata	Moderate	Fresh water		
adult	{Piscivorous Rasping lips				Fresh and salt water	....	Solitary
Polypterus	Piscivorous Predatory	Undulating by myomeres, plus paddles	Osteichthyes	Tropical	Fresh water	....	Solitary
† *Eryops*	Carnivorous Peg teeth	Primitive tetrapod	Amphibia	? Moderate	Swampy	Primitive forest	? Solitary
Frog	Fly-catching	Leaping, swimming	Amphibia	Temp.-trop.	Fresh water, moist land	Varied	Solitary
† *Cynognathus*	Carnivorous	Fast-crawling	Reptilia	? Semi-arid	....		Solitary
Opossum (*Didelphis*)	Insectivorous-carnivorous	Climbing	Mammalia	Moderate	Trees	Varied	Solitary
Lemur	Insectivorous-frugivorous	Climbing	Mammalia	Tropical	Trees	Trees	Transitional
Gorilla	Frugivorous	Climbing and on ground	Mammalia	Tropical	Trees	Trees	In families and small bands
Modern man	Omnivorous	Bipedal	Mammalia	Widely varied	Varied	Varied	Socialized

† Extinct.

"PREADAPTATION" *VERSUS* "DESIGN"

Through the age-long sifting action of Natural Selection many animals inherit such a configuration of the nervous system that they instinctively prepare far in advance for the oncoming of winter, certainly without knowing why they do it; they are merely "wound up" to do it. During the course of individual development, inherited automatic and unconscious futuristic or anticipatory values may be everywhere evident to the observer but not to the reactor. The egg can hardly know why its germinal disc is subdividing at an increasingly high speed; nor do the fields of force around the blastopore know why they are endowing different cells with different sets of potentialities or that these differences in tempo are of critical importance for the young bird that will shortly break through the shell and be fed by its parents, until its wings are ready.

All this has been inherited from the past but it works out as if all were under the control of a central office.

In brief, the poet who said, "I looked around to find my past, but lo, it had gone before," clearly anticipated both Cuénot and Davenport, who grasped the meaning of structural preadaptations. Indeed, the present writer regards both man-made and natural designs as parallel or convergent results, (a) of the preadaptive nervous system of man, and (b) of the cumulatively adaptive patterns of nature.

"LAW" *VERSUS* "CHANCE"

When a resting sting-ray (*Dasybatus hastatus*) is stepped on, it swings it long lash-like tail, which bears a poisonous spine consisting of a long brittle calcareous spike, bordered with two rows of sharp little recurved hooks. An incautious human bather stepping on a sting-ray concealed in the muddy sand may receive an extremely painful wound. Certain kinds of sand-sharks (*Charcharias*), and the hammer-head (*Sphyrna*) swoop down and grab the sting-rays and thus may get their mouths or cheeks pierced by the stinging spine. Dr. Gudger (1932) reports several specimens which had been pierced by many spines, showing that the poisonous effects, if any, upon the shark in earlier encounters with a sting-ray did not prevent him from making later attacks.

These facts supply a good example of several general principles relating to the ways of evolution:

1) The poisonous spine itself has evidently arisen by long continued anisomerism or emphasis of the spine of the second dorsal fin.

2) The poisoning of the wounds in fish or other creatures which were unfortunate enough to come in contact with this weapon, was partly due to the presence on the rear borders of the spike of a series of secondary polyisomeres, i.e., the little hooks mentioned above, which cause severe lacerations.

3) The poisonous effects upon man of wounds made by the sting-ray's spine, are due to the introduction into the human blood stream of the secretions of certain "poison glands" at the base of the ray's sting.

4) The evolution of the sting-ray's weapon was probably well under way in Cretaceous times, when the ancestors of man were arboreal, insectivorous tree shrews, living in the forests well away from the bays in which the sting-rays lurked.

5) Thus the present poisonous effects on man are due to the then long distant and unpredictable intersection or coincidence in later time and space of a then nonexistent human foot and a then incomplete sting-ray's sting.

6) It would be quite arbitrary to assume that the evolution of the sting-ray's weapon was influenced appreciably, if at all, by the presence of contemporary arboreal and insectivorous-frugivorous tree-shrews, or by the then very remote possibility that much later some tree-shrews eventually gave rise to the branching lines leading to lemurs, monkeys, apes and ultimately to bipedal wading man; especially since this sting is proved to be not always effective in protecting the ray against its immediate devourers, the prowling sand sharks.

7) No doubt a practically infinite number of prerequisite events and conditions took place

in geologic time, leading respectively on one side to the evolution of the sting rays and their stings, and on the other to the evolution of early man and of modern incautious bipedal bathers. And on both sides each event in its turn had required a similar spreading network of prerequisite antecedents, and a similar intersection of networks at an incalculable number of times and places.

8) Nevertheless and in spite of the unpredictable course of many random events at a given time and place, yet in the long run the respective evolution on the one hand of the sting-ray's sting and on the other of the incautious bather's foot have been respectively preconditioned by the cumulative effect of repetitive or polyisomerous situations, which have developed more or less rhythmically, as in events determined by recurring sunlight and darkness, by winter and summer, by seasonal times for breeding and not breeding, or for growth and arrest of growth, or for seeking this type of environment or that, and by thousands of others.

9) In so far as similar events occur rhythmically, they increase their chances of meeting *similarly* rhythmical series with which they can cross or intersect; e.g., the more incautious bathers and the more sting-rays there may be in a given locality, the greater the chance for collisions between them. That is, the more often differently conditioned series exist near each other, the greater are the chances for meetings or combinations between them. This is part of the "kaleidoscope theory" of evolution proposed elsewhere by the writer.

10) The net result of these considerations is to suggest that newly emergent or creative evolutionary events have not been foreordained but have been preconditioned.

11) Natural "laws" (not humanly conceived commands or prohibitions) are recognized by man through their effect in causing repetitive, recurrent phenomena. "Chance" is the name for random or unexpected events, due to new or newly observed intersection or collision of different natural laws at a given time or space. Hence "Law" and "Chance" are not mutually exclusive but complementary aspects of natural events or phenomena.

12) "Chance" and "Law" are correlative terms, designating opposite attributes of events, between which are intermediate or mixed events or attributes.

Thanks to the enormous labors of geneticists, working with plants, fruitflies, rats, etc., and taxonomists, studying the exact geographic boundaries of related subspecies respectively of plants, insects, fishes, snakes, small mammals, etc., it has been shown that when two nearly related species overlap they often interbreed and produce new subspecies. If any one of the latter become isolated and establish an inbreeding population, certain new features, such as intensified color or an extra spot, or longer tails, may eventually become conspicuous. New characters in the animals, either young or old, apart from differences due to nutrition, altitude and other physical factors, are often caused by changes in the grouping or combinations of the chromosomes. These are the bearers of the genes which control hereditary features. These genes have been likened to an invisibly minute package of chemicals, differing in qualities and amounts, which affect the rates of growth of different parts of the forming embryo. During the process of cell division the chromosomes and with them the genes form opposing lines and are fused, redivided and otherwise rearranged. Usually the combinations are such that when the individual grows up its measurements, etc., fall within the norm of its group. But sometimes, when the genes get too badly mixed, a new character suddenly appears, such as eyelessness in flies, or loss of pigmentation in the skin of mammals.

Very often these new characters are said to be "lethal," because they are the outward signs of a weak constitution which either fails to survive to adulthood or is gravely handicapped in life's struggles. But out of many such abnormal occurrences one new or intensified character, such as increased size or an extra vertebra in the lumbar region, may prove to be either innocuous or useful in the animal's continual efforts to get food, to resist harsh features of the environment, to escape enemies, to capture mates.

By artificial selection, different strains, bearing different special features, may be repeatedly crossed and by long continued selective breeding such strange creatures as lion-headed and telescope carps have been derived from ordinary carps; bulldogs, King Charles spaniels and borzoi hounds from wolf-like dogs, etc. Within the lifetime of one experimentalist (Stockard) several quite new varieties of dogs have been started but not fixed. On the other hand, the experience of palaeontologists suggests that the evolution of new subspecies may require several thousand years, new species, tens or hundreds of thousands, new genera, millions, new families, scores of millions, new orders, one to three hundred millions, new classes, from 300 million years upward.

Turning to the environment, every part of the land is slowly being changed by the forces of heat and cold, water and ice, volcanic eruptions, etc., etc. In a given spot, changes may be almost imperceptible for many years, but in certain regions sudden and unpredictable tidal waves or flooded rivers may inflict widespread destruction. Over long periods the uplift of new mountain-chains may introduce far-reaching transformations of the climate, terrain and flora.

It is evident that even if changes in the genes may be initiated by cosmic rays, yet the complex shuffling, combination and resorting of the genes in the cell divisions of a fertilized egg are determined in large part by its own set of rules or laws and that these laws operate within limits that are directly determined by the internal configurations of the parts and their reactions to outside forces; but that Natural Selection, as exerted by the relatively sudden uplift of a mountain chain or the periodic flooding of the ancestral home can only propose new tests of viability and adaptability to the developing organism, and that it can act only as a screen or sifter for those best equipped to survive the test.

To sum up, within the developing individual there are many causal chains and networks which by reaction with the immediate environment determine his individual form and appearance; and there are many other quite independently working factors which have produced the continuous environmental test of his adaptiveness and viability. Each causal series or network has its own contributary series and networks, which widen out indefinitely as we look backward through the ages. Whether a given horse will win a race depends *inter alia* upon his hereditary stamina and swiftness, upon whether the track is soggy or hard, and upon the skill of his jockey. Such a combination of factors is called, in barber shops, luck and in laboratories randomness. Ten decades of study of fossil and recent animals have convinced geneticists and palaeontologists that in the sense used above chance has played an important part in the origin and evolution of each individual, population, variety, subspecies, etc. (cf. Simpson).

"IRREVOCABILITY" AND THE "KALEIDOSCOPE THEORY"

In remote Pre-Cambrian times, more than five hundred millions of years ago, the endless clash of positive and negative forces within the atom, as described by Schrödinger (1946), had already produced a statistically operating system which made possible the summation, integration and differentiation of chemical and physical reactions and interactions. Thus the system of emerging evolution was "by necessity" building itself cumulatively and in a branching way, even as it was constantly being pruned and guided by natural selection in adaptation to internal and external environment.

The changing patterns of a kaleidoscope afford various similes of the ways of evolution. (1) A slight and gradual shift of the lever (representing Natural Selection) will normally produce only a small change in the hereditary pattern. (2) The persistent elements of the kaleidoscope patterns are collectively analogous to the older "heritage" of the organism, the changed details correspond to the "habitus." (3) If the next few moves of the lever be small and in the same direction, the "heritage" pattern will slowly be modified by the "habitus" responses. But if the lever be shifted through wider arcs, the changes in pattern will be more radical and the older features will be wiped out accordingly. Thus small shifts of the lever in the same direction may induce progressive "adaptive" changes in the pattern, analogous to those that have occurred in "orthogenetic" series, as when the low-crowned, relatively simple molars of *Eohippus* changed gradually into the extremely high-crowned and complex molars of the horse. (4) But sometimes the lever (Natural Selection) moves in the opposite direction, as when long wings in ground-living insects on small islets became disadvantageous, so that Selection favored shortening the wings. Indeed many cases indicate that whenever a given type passes from one medium to another, as from fresh to salt water, from water-living to land-living, from ground-living to flying or the reverse, radical changes in survival values occur, especially in all parts of the locomotor organs (fig. 14.1); so that profound changes in proportion as well as increase or decrease in number usually supervene (e.g., reptiles to birds, lizard-like reptiles to ichthyosaurs and plesiosaurs).

It is true that the results of the past are irrevocable, but the direction or trend of evolution is not always "irreversible"; in some cases (as when lower tusks in proboscideans were first enlarged and then rapidly reduced) a specialization can be reduced and lost; and after a reduction in number of parts, there may be a secondary increase in number or a reversal in proportions among the surviving units (pp. 150, 424, 430).

ENVOY: THE COSMIC KALEIDOSCOPE OF EVOLUTION

Throughout the extensive series of transformations which have been all too briefly envisioned in this book, including the "main line" (from the anthropocentric viewpoint), every earlier ancestral stage was obviously prerequisite or "preadaptive" as a point of departure for the later one. As the next stage was reached, new opportunities arose for the utilization of many old attributes and for the further modification of others. Thus some characters that had been gained in an earlier habitus were being shifted into the heritage column, and other old habitus and heritage characters were gradually being cancelled as they were displaced by new ones. Thus *Homo sapiens* is in very truth the heir of all the ages and the net product of an incalculable series of generations from "Amoeba to Man."

The endlessly branching history of the outer form and inner frame is due in part to the complex genetic mechanism which in reaction with environmental stimuli was always producing new accelerations or retardations of growth. These were manifested either as new or additional polyisomeres or as newly emphasized anisomeres.

Finally, geneticists (Sewell Wright, 1932) and palaeontologists (Simpson, 1947) are finding that the genetic mechanism, in reaction with the environment, sometimes permits and even favors, on a purely statistical basis, (1) the persistence of possibly advantageous new features; (2) the correlation or combination of advantageous features in different parts of the organism with each other; (3) of the organism with the environment. Thus the value of the whole has always been greater than that of the sum of its parts.

And so the cosmic kaleidoscope keeps turning round and round, slowly but endlessly dissolving old combinations while creating new patterns, new values, new opportunities.

BIBLIOGRAPHY

(For Bibliography Table of Contents See pp. xxiii ff.)

PART I

TIME AND EVOLUTION—THE LOWER INVERTEBRATES

(Cf. Vol. I, Chapters I–III)

Cosmic Background of Life: Light, Space-Time, Nuclear Physics

BRAGG, W. 1933. The universe of light. New York.

EINSTEIN, A. 1929. Field theories, old and new. New York Times, Feb. 3, Sec. 9. [General and special theories of relativity.]—1946. The meaning of relativity. Princeton Univ. Press.—1950. The meaning of relativity. 3d edition, including the generalized theory of gravitation. Princeton Univ. Press.

JEANS, J. 1929. The universe around us. New York.—1939. The expanding universe and the origin of the great nebulae. Nature, 158.—1944. The galactic system. Ibid., 41.

KARLSON, P. 1936. The world around us: A modern guide to physics. New York.

LECOMTE DU NOÜY, P. 1937. Biological time. New York.

MARSHAK, R. E. and H. A. BETHE. 1941. The sources of stellar energy. Trans. N. Y. Acad. Sci., 74.

MILNE, E. A. 1949. Origin of the universe. Nature, 855. [Critical review of L' hypothèse de l'atome primitif: Essai de cosmogonie par Prof. Georges Lemaître. London. Trenchant criticism of "curved space" and of "lack of a deeper theory of gravitation."]

RUSSELL, H. N. 1937. The fuel of the stars. Sci. Amer., 12.

RUTHERFORD, E. 1937. The newer alchemy. New York.

SCHRODINGER, E. 1946. What is life? Cambridge and New York.

SLACK, C. M. 1937. Nuclear physics chart. New York Times, July.

Geologic Time

ASHLEY, G. H. 1932. Geologic time and the rock record. Bull. Geol. Soc. Amer., 477.

BARRELL, J. 1917. Rhythms and the measurements of geologic time. Bull. Geol. Soc. Amer., 745.

BULLARD, E. C. 1945. Geological time. Mem. and Proc. Manchester Lit. Phil. Soc. (1943–45), 55. [Revised time estimates.]

GOODMAN, C. and R. D. EVANS. 1941. Age measurements by radioactivity. Bull. Geol. Soc. Amer., 491.

HEVESY, G. VON. 1930. The age of the earth. Science, 509.

HOLMES, A. 1927. The age of the earth. London.—1946. An estimate of the age of the earth. Nature, 680.—1947. A revised estimate of the age of the earth. Ibid., 127.

KNOPF, ADOLPH et al. 1931. Physics of the earth. IV. The age of the earth. Council Nat. Acad. Sci., 80.

LOUDERBACK, G. D. et al. 1935. The geologic and the cosmic age scales. Science, 51.

SCHUCHERT, C. 1927. The earth and its rhythms. New York and London.

ZEUNER, F. E. 1946. Dating the past. London.

The Solar System: Origin of Planets (including Earth)

ALFVEN, H. 1943. Non-solar planets and the origin of the solar system. Nature, 721.

BANERJI, A. C. 1944. Non-solar planetary systems. Nature, 779.

BETHE, H. A. 1939. [Possible relations of earth to sun.] Physical Review, 434.

CHAMBERLIN, T. C. and R. D. SALISBURY. 1906. Geology. Vol. II. [Planetesimal hypothesis, 38.] New York.

EDGEWORTH, K. E. 1944. Origin of the solar system. Nature, 140.

GAMOW, G. 1948. Biography of the earth: Its past, present and future. New York.

HUNTER, A. 1944. Origin of the solar system. Nature, 255.

JEANS, J. 1936. Evolution of the solar system. Nature, 532.—1943. Nonsolar planetary systems. Nature, 721.

JEFFREYS, H. 1944. Origin of the solar system. Nature, 592.

LYTTLETON, R. A. 1944. Origin of the planets. Nature, 592.

McCREA, W. H. 1945. New views of the origin of the solar system. Nature, 466.

MOULTON, F. R. 1928. The planetesimal hypothesis. Science, 549.—1935. Review of The architecture of the universe by W. F. G. Swann. Science, 47. [Defence of Chamberlin and Moulton's planetesimal theory.]

RUSSELL, H. N. 1940a. A famous theory weakens. Sci. Amer., 140. [Encounter theory of the origin of planets.]—1940b. The origin of the earth. Sci. Amer., 266. [Summary of H. A. Bethe on origin of sun's energy in formation of planets.]

SCHMIDT, O. J. 1944. A meteoric theory of the origin of the earth and planets. Comptes Rendus (Doklady) de l'Acad. Sci. de l'URSS, no. 6.

WEIZSÄCKER, C. F. VON. 1944. Ueber die Entgegensplanetensystems. Zeits, f. Astrophysik.

WHIPPLE, F. L. 1948. The dust cloud hypothesis. Scientific American, May, 35.

Earth History: Historical Geology

CHAMBERLIN, T. C. and R. D. SALISBURY. 1904, 1906. Geology. Vols. I–III. New York.—1909. A college textbook of geology. New York.

GEIKIE, A. 1903. Textbook of geology. Vols. I–IV. London.

GRABAU, A. W. 1913. Principles of stratigraphy. New York.—1921. A text-book of geology. New York.—1940. The rhythm of the ages: Earth history in the light of the pulsation and polar control theories. Peking.

SCHUCHERT, C. 1931. Outlines of historical geology (2nd ed.). New York.

SCHUCHERT, C. and C. O. DUNBAR. 1941. A textbook of geology. New York.

Lithosphere, Hydrosphere

ADAMS, L. H. 1937. The earth's interior, its nature and composition. Sci. Month., 199.

DALY, R. A. 1933. The depths of the earth. Science, 95.

MACELWANE, J. B. 1946. The interior of the earth. Amer. Scientist, 177.

Origin of Life: Pre-Cambrian Life

FLETT, J. S. 1929. Marble. Encycl. Brit. (14th ed.). [Mineral nature of "Eozoön."]

HALDANE, J. B. S. 1944. Radioactivity and the origin of life in Milne's cosmology. Nature, 555.

KIRKPATRICK, R. 1912. On the stromatoporoids and Eozoön. Ann. Mag. Nat. Hist., 341. [Note on Eozoön canadense, oldest known organism from Lower Laurentian sea-bottom. Considered to be a foramenifer "nearly related to Labechia and Beatricia."]

KLIGLER, I. J. 1918. The evolution of bacteria. Science, 320.

LIPMAN, C. B. 1932. Are there living bacteria in stony meteorites? Amer. Mus. Novitates, no. 588.

OSBORN, H. F. 1917. The origin and evolution of life. New York.

RAYMOND, P. E. 1935. Pre-Cambrian life. Bull. Geol. Soc. Amer., no. 3.

WALCOTT, C. D. 1899. Pre-Cambrian fossiliferous formations. Bull. Geol. Soc. Amer., 199.

Life Processes: Growth and Form, Metabolism, Food, Excretion, Photochemistry, Photosynthesis

HOPKINS, F. G. 1933. Some chemical aspects of life. Science, 219.

HUXLEY, J. S. 1932. Problems of relative growth. New York.

RUSSELL, E. S. 1916. Form and function: A contribution to the history of animal morphology. London.

SHEPPARD, W. E. 1929. Article on Photochemistry. Encycl. Brit. (14th ed.).

SHERMAN, H. L. 1925. The chemistry of food and nutrition. New York.

SMITH, H. W. 1939. Studies on the physiology of the kidney. Univ. Kansas School of Medicine.—1943. Lectures on the kidney. Ibid. [Origin of vertebrate skeleton.]

THOMPSON, D'ARCY W. 1942. On growth and form (2nd ed.). New York.

Evolution, General:

Modes, Principles, Factors, Genetics, Speciation, Principles of Classification, Epistemology, Adaptation, Chance versus "Law," Selection, Environments, Camouflage and Mimicry

BATESON, W. 1894. Materials for the study of variation . . . London.—1914. Address of the President, Brit. Assoc. Adv. Sci., Science, 287, 319.

BAYLIS, H. A. 1938. Helminths and evolution. In: Evolution: Essays . . . presented to Prof. E. S. Goodrich . . . Oxford.

BEER, G. R. DE. 1938. Embryology and evolution. In: Evolution: Essays . . . presented to Prof. E. S. Goodrich . . . Oxford.

BERG, L. S. 1926. Nomogenesis or evolution determined by law. London.

BOGERT, C. M. et al. 1943. Criteria for vertebrate subspecies, species and genera. Ann. N. Y. Acad. Sci., 105.

BOYDEN, A. 1934. Precipitins and phylogeny in animals. Amer. Nat., 516.—1935. Genetics and homology. Quart. Rev. Biol., 448.—1942. Systematic serology . . . Physiol. Zool., 109.—1943a. Homology and analogy . . . Quart. Rev. Biol., 228.—1943b. Serology and animal systematics. Amer. Nat., 234.

BREDER, C. M. JR. 1946. Analysis of the deceptive resemblances of fishes to plant parts . . . Bull. Bingham Oceanogr. Coll., Yale Univ.—1947. Analysis of geometry of symmetry. Bull. Amer. Mus. Nat. Hist., 327.

BROOM, R. 1932. Evolution: Design or accident. In: Our changing world-view. Johannesburg. Univ. Witwatersrand Press.

CLARK, A. H. 1928. Animal evolution. Quart. Rev. Biol., 523.—1937. Eogenesis, the origin of animal forms. Acta Biotheoretica, 181.

COLLINS, F. H. 1889. An epitome of the synthetic philosophy . . . [of Herbert Spencer]. New York.

COPE, E. D. 1871. The method of creation of organic forms. Proc. Amer. Philos. Soc., 229. [Anteroposterior repetitive acceleration.]—1887. Origin of the fittest: Essays on evolution. New York.

CRISSMAN, P. 1945. Causation, chance, determinism and freedom in nature. Sci. Monthly, 455.

CUÉNOT, L. 1909. Le peuplement des places vides dan la nature et o'origine des adaptations. Rev. gén. d. Sci., no. 1, 5.—1914. Théorie de la préadaptation. Scientia, 60.—1925. L'Adaptation. Encycl. Scient . . . (Dr. Toulouse). Paris. [Mutation and preadaptation, 135; co-adaptations, 265.]

DARWIN, C. 1859. The origin of species by means of natural selection. . . . London.—1868. The variation of animals and plants under domestication. 1st ed. London.—1901. Journal of researches . . . during the voyage . . . of H. M. S. "Beagle." New ed. London.

DARWIN, C. G. 1931. The uncertainty principle. Science, 653.

DAUDIN, H. Undated. I. De Linné à Jussieu: Méthodes de la classification et idée de série en botanique et en zoologie (1740–1790). Études d'hist. des sci. nat. Paris.—1926. II. Cuvier et Lamarck: Les classes zoologiques. Ibid.

DAVENPORT, C. B. 1903. The animal ecology of the Cold Spring Sand-spit, with remarks on the theory of adaptation. Decen. Publ., Univ. Chicago. [Theory of preadaptation.]—1930. The mechanism of organic evolution. Jour. Washington Acad. Sci., 317.—1933. The crural index. Amer. Jour. Phys. Anthrop., 333. [Structure determines function, 352.]

DEAN, B. 1908. Accidental resemblance among animals: A chapter in un-natural history. Pop. Sci. Monthly, 304. [Catfish with "crucifix."]

DOBZHANSKY, T. 1941. Genetics and the origin of species. 2nd ed. Columbia Univ. Biol. Ser. New York.

DOHRN, A. 1876. Der Ursprung der Wirbelthiere und das Princip des Functionswechsels: Genealogische Skizzen. Leipzig.

DOLLO, L. 1893. Les lois de l'évolution (Résumé). Bull. Soc. Belge d. Géol., 164.

EIMER, G. H. T. 1889–1895. Die Artbildung und Verwandtshaft bei den Schmetterlingen . . . Leipzig.—1890. Organic evolution as the result of inheritance of acquired characters according to the laws of organic growth. New York.—1897. Orthogenesis der Schmetterlinge . . . Leipzig.

EMERSON, A. E. 1945. Taxonomic categories and population genetics. Entomol. News, 14.

ERRINGTON, P. L. 1943. Analysis of mink predation upon muskrats . . . Agric. Exp. Station, Iowa State Coll. Agric. Research Bull., 798. [Constructive synthesis of field studies in Natural Selection.]—1946. Predation and vertebrate populations. Quart. Rev. Biol., 144.

FRAIPONT, C. and S. LECLERQ. 1932. L'Evolution: Adaptations et mutations-berceaux et migrations. Paris.

GAUSE, G. F. 1947. Problems of evolution. Trans. Connecticut Acad. Arts and Sci., 17. [Experimental evidence for organic selection.]

GOLDSCHMIDT, R. B. 1938. Physiological genetics. New York. [Mimicry in butterflies.]—1940. The material basis of evolution. Yale Univ. Press, New Haven.—1945. Mimetic polymorphism, a controversial chapter of Darwinism. I. Quart. Rev. Biol., 147.—II. Ibid., 205.

GREGORY, W. K. 1914. Locomotive adaptations in fishes illustrating "Habitus" and "Heritage." Ann. N. Y. Acad. Sci. (for 1913), 267.—1924. On design in nature. Yale Review, 334.—1933. Basic patents in nature. Science, 561.—1934. Polyisomerism in cranial and dental evolution among vertebrates. Proc. Nat. Acad. Sci., vol. 20, 1.—1935a. On the evolution of the skulls of vertebrates with special reference to heritable changes in proportional diameters (anisomerism). Ibid., vol. 21, 1.—1935b. Reduplication in evolution. Quart. Rev. Biol., 272.—1936. The transformation of organic designs: A review of the origin and deployment of the earlier vertebrates. Biol. Reviews, Cambridge Philos. Soc., 311.

GUNDERSON, A. and G. T. HASTINGS. 1944. The interdependence of plant and animal evolution. Scient. Monthly, 63.

HAECKEL, E. 1866. Generelle Morphologie der Organismen . . . Bd. I. Allgem. Anat. . . . Bd. II. Allgem. Entwick. . . . Berlin.—1903a. Anthropogenie oder Entwickelungsgeschichte des Menschen. 2nd ed. Leipzig. [Haeckel's "Urwirbeltier," I, 270.]—1903b. Die Welträthsel . . . Volks-Ausgabe, 48, bis 67. Bonn.—1909. Natürliche Schöpfungs-Geschichte: . . . Vorträge über die Entwickelungslehre. Berlin.

HEISENBERG, W. 1935. Wandlungen in den Grundlagen der Naturwissenschaft. Leipzig.

HENDERSON, L. J. 1913. The fitness of the environment . . . New York.

HOLMES, S. J., 1944. Recapitulation and its supposed causes. Quart. Rev. Biol., 319.

HOBBES, T. 1651. The Leviathan; or the matter, form and power of a commonwealth, ecclesiastical and civil (Ed. Michael Oakeshoot). Oxford. 1946.

HUBBS, C. L. 1940. Speciation of fishes. Amer. Nat., 198.—1943. Criteria for subspecies, species and genera . . . Ann. N .Y. Acad. Sci., 109. —1944. Concepts of homology and analogy. Amer. Nat., 289.

HUXLEY, J. S. 1927. Developmental rates and genetic factors . . . Brit. Jour. Exp. Biol., 917.—1931. Notes on differential growth. Amer. Nat., 289.—1932. Problems of relative growth. London.

JEPSEN, G. L. 1948. Genetics, paleontology and evolution. Princeton Univ. Bicent. Conf. Princeton.

KINSEY, A. C. 1937. Supra-specific variation in nature and in classification . . . Amer. Nat., 206.

LAMARCK, J. B. P. A. DE M. 1809. Philosophie zoologique . . . Paris. [Inheritance of acquired characters, 233 et seq.; cf. Simpson, 1947, 483.]

LEATHES, J. B. 1926. Function and design. Science, 387.

LILLIE, F. R. 1913. The mechanism of fertilization. Science, 524.—1927a. Physical indeterminism and vital action. Ibid., 139.—1927b. The gene and the ontogenetic process. Ibid., 361.—1929. Embryonic segregation and its role in the life history. Wilhelm Roux' Archiv f. Entwicklsmech. der Organismen, 499.

LIVINGSTON, B. E. 1934. Environments. Science, 569.

MALTHUS, T. R. 1798. Essay on the principle of population . . . London.—1926. First essay on population, 1798 . . . Reprint. Roy. Econ. Soc. London.

LINNAEUS, C. 1758. Systema Naturae . . . Editio Decima. Holmiae.

MAYR, E. 1942. Systematics and the origin of species . . . Columbia Biol. Ser. New York.

MAYR, E., G. L. STEBBINS, JR. and G. G. SIMPSON. 1945. Symposium on age of the distribution pattern of the gene arrangements in Drosophila pseudo-obscura. Lloydia, 69.

MORGAN, T. H. 1910. Chance or purpose in the origin and evolution of adaptation. Science, 201.—1917. The theory of the gene. Amer. Nat., 513.—1925. Evolution and genetics. Princeton Univ. Press.—1932. The scientific basis of evolution. New York.

MULLER, H. J. 1916. The mechanism of crossing over. Amer. Nat., 193.—1938. Bearings of the Drosophila work on problems of systematics. Proc. Zool. Soc. London, 55.—1942. Isolating mechanisms, evolution and temperature. Biol. Symposia, 71.

NOBLE, G. K. and H. T. BRADLEY. 1933. The mating behavior of lizards; its bearing on the theory of sexual selection. Ann. N. Y. Acad. Sci., 25.

NOPCSA, F. 1923. Reversible and irreversible evolution: A study based on reptiles. Proc. Zool. Soc. London, 1045.—1926. Heredity and evolution. Ibid., part 2, 633.

OSBORN, H. F. 1892. The Cartwright Lectures [on evolution] for 1892. Medical Record, 197, 253, 449, 533.—1894. From the Greeks to Darwin. . . . Columbia Univ. Biol. Ser. 2nd ed., 1928. New York.—1897. The limits of organic selection. Amer. Nat., 944.—1898. Modification and variation, and the limits of organic selection. Abstr., Proc. Amer. Assoc. Adv. Sci., 239.—1902a. Homoplasy as a law of latent or potential homology. Amer. Nat., 259.—1902b. The law of adaptive radiation. Ibid., 353.—1906. The causes of extinction of Mammalia. Ibid., 769, 829.—1908a. The four inseparable factors of evolution . . . Science, 148.—1908b. Coincident evolution through rectigradations. Ibid., 749.—1912. The continuous origin of certain unit characters as observed by a palaeontologist. Harvey Soc. Vol., 7th ser., 153.—1915. Origin of single characters as observed in fossil and living animals and plants. Amer. Nat., 193.—1916a. The origin and evolution of life upon the earth. Sci. Monthly, 5, 170, 289, 313, 502, 601.—1916b. The origin and evolution of life. New York.—1917. Application of the laws of action, reaction and interaction in life evolution. Proc. Nat. Acad. Sci., 7.—1926. Evolution, III—the palaeontological aspect. Encycl. Brit. (13th ed.), 1073.—1929. The titanotheres . . . Chapter XI. Causes of the evolution and extinction of the titanotheres. Monogr. 55, U. S. Geol. Surv., II, 805.—1933. Aristogenesis, the observed order of biomechanical evolution. Proc. Nat. Acad. Sci., 699.—1934. Aristogenesis, the creative principle in the origin of species. Amer. Nat., 193.

PAINTER, T. S. 1934a. A new method for the study of chromosome aberrations and the plotting of chromosome maps in Drosophila melanogaster. Genetics, 175.—1934b. Salivary chromosomes and the attack on the gene. Jour. Heredity, 465.

PARKER, G. H. 1924. Organic determinism. Science, 518.

PARR, A. E. 1926. Adaptiogenese und phylogenese . . . Abhandl. z. Theorie der Organischen Entwicklung. Heft 1. Berlin.

PLOUGH, H. H. 1942. Temperature and evolution . . . Biol. Symposia, 3.

RIDDLE, O. 1947. Endocrines and constitution in doves and pigeons. Carnegie Inst. Washington Publ. 572.

RITTER, W. E. 1932. Why Aristotle invented the word entelecheia. I. Quart. Rev. Biol., 377.—1934. Part II. Ibid., 1.

SCHRODINGER, E. 1946. What is life? The physical aspect of the living cell. Cambridge and New York. [Indeterminism of alternate plus or minus intra-atomic charges a basic factor for evolution of living matter. Cumulative changes in growth or organic plan dependent upon large populations and statistical majorities among atoms, all subject to similar intra-atomic indeterminate conditions.]

SHULL, A. F. 1936. Evolution. New York.—1942. Two decades of evolution theory. Amer. Nat., 171.

SIMPSON, G. G. 1937. Patterns of phyletic evolution. Bull. Geol. Soc. Amer., 303.—1941. The role of the individual in evolution. Jour. Washington Acad. Sci., 1.—1944. Tempo and mode in evolution. Columbia Univ. Biol. Ser. New York.—1945. Evidence from fossils and from the application of evolutionary rate distributions. Lloydia, vol. 8, 103.—1947. The problem of plan and purpose in nature. Sci. Monthly, 481.

SINGER, C. 1925. The evolution of anatomy . . . New York.—Undated. The story of living things: A short account of the evolution of the biological sciences. New York.

SMUTS, J. C. 1931. The scientific world-picture of today. Science, 297. [Holism.]

SPENCER, H. 1864a. The classification of the sciences. New York.—1864b. First principles of a new system of philosophy. New York.—1867–1872. The principles of biology. Vols. I, II. New York. [Homogeneity to heterogeneity, vol. I, 160; evolution and dissolution, vol. II, 4.]

STROMER VON REICHENBACH, E. 1912. Lehrbuch der Palaozoologie. Bd. II. Schlussbetrachtungen. Leipzig. [First statement of the principle of the reduction of skull elements in passing from lower to higher vertebrates.]

STURTEVANT, A. H. 1937. Essays on evolution. I. On the effects of selection on mutation rate. Quart. Rev. Biol., vol. 12, 464.—1938a.—II. On the effects of selection on social insects. Ibid., vol. 13, 74.—1938b. III. On the origin of interspecific sterility. Ibid., vol. 13, 333.

SUMNER, F. B. 1919. Adaptation and the problem of "organic purposefulness." Amer. Nat., 193. —1923. . . . Origin and inheritance of specific characters. Ibid., 238.—1929. Is evolution a continuous or discontinuous process? Sci. Monthly, 72.

SWINNERTON, H. H. 1939. Paleontology and the mechanics of evolution. Quart. Jour. Geol. Soc. London, 33.

THOMPSON, D'ARCY W. 1911. Magnalia Naturae; or, The greater problems of biology. Science, 417.—1913. On Aristotle as a biologist, with a prooemion on Herbert Spencer. Oxford.—1942.

On growth and form. Cambridge Univ. Press, First ed., 1917.

TORREY, H. B. and F. FELIN. 1937. Was Aristotle an evolutionist? Quart. Rev. Biol., 1.

VIALLETON, L. 1924. Morphologie générale. Membres et ceintures des vertébrés tétrapodes . . . Paris. [His thesis is to disprove transformism (evolution).]

WATSON, D. M. S. 1929. Adaptation. Brit. Assoc. Adv. Sci., Sect. D. Zool., 1.

WEIDENREICH, F. 1931. Über Umkehrbarkeit der Entwicklung. Palaeont. Zeitschr., 177.

WHEELER, W. M. 1922–1923. Social life among the insects. Scient. Monthly, vol. 14, 497; ibid., vol. 15, 67, 119, 235, 320, 385, 527; ibid., vol. 5, 16, 160, 312.—1926. Emergent evolution and the social. Science, 433.—1928. Emergent evolution and the development of societies. New Sci. Ser. New York.

WILLEY, A. 1936. Reductions and reversions. Trans. Roy. Soc. Canada, 115.

WOODBRIDGE, F. J. E. 1903. The problem of metaphysics. Philos. Rev., 367.—1937. Nature and mind. Selected essays . . . Columbia Univ. Press. New York.

WOODWARD, A. S. 1938. Palaeontology and the Linnaean classification. Proc. Linnean Soc. London, 238.

WRIGHT, S. 1932. The roles of mutation, inbreeding, crossbreeding, and selection in evolution. Proc. Internat. Congress of Genetics, 356.—1945. The differential equation of the distribution of gene frequencies. Proc. Nat. Acad. Sci., 382.

Palaeontology: General Works
(including both Invertebrates and Vertebrates)

BERRY, E. W. 1929. Paleontology. New York.

BOULE, M. and J. PIVETEAU. 1935. Les fossiles . . . Paris.

GOLDRING, W. 1931. Handbook of paleontology . . . Part I. The fossils. Part II. The formations. N. Y. State Mus. Handb., nos. 9, 10.

RAYMOND, P. E. 1939. Prehistoric life. Harvard Univ. Press.

WOODWARD, A. S. 1898. Outlines of vertebrate palaeontology . . . Cambridge.

ZITTEL, K. A. VON. 1883, 1887. Traité de Paléontologie. Paris, Munich, Leipzig.—1918. Grundzüge der Paläontologie (Paläozoologie). Revised by F. Broili and M. Schlosser. II. Abt.: Vertebrata. Munich and Berlin.

Palaeogeography, General:
Land Bridges, Isostasy versus Drift of Continents, Geographical Distribution

BLACK, D. 1931. Palaeogeography and polar shift . . . Bull. Geol. Soc. China, 105.

DARLINGTON, P. J. The geographical distribution of cold-blooded vertebrates. I. Quart. Rev. Biol., March, no. 1, 1. II. Ibid., June, no. 2, 105.

DAVID, T. W. E. 1914. Geology of the Commonwealth. Federal Handb. of Australia, VII.

GRABAU, A. W. 1940. The rhythm of the ages: Earth history in the light of the pulsation and polar control theories. Peking.

HUENE, F. VON. 1929. Versuch einer Skizze der Paläogeograph. Beziehungen Südamerikas. Geol. Rundschau, 81.—1933. Die südamerikanische Gondwanafauna. Forschungen und Fortschritte, vol. 9, no. 9.

MATTHEW, W. D. 1915. Climate and evolution. Ann. N. Y. Acad. Sci., 171.—1939. Idem, 2nd ed. Special Publ., N. Y. Acad. Sci.

SCHMIDT, K. P. 1943. Corollary and commentary for "Climate and Evolution." Amer. Midland Naturalist, 241.

SCHUCHERT, C. 1910. Palaeogeography of North America. Bull. Geol. Soc. Amer., 427.—1915. Climates of geologic time. Carnegie Inst. Washington Publ. no. 192, 263.—1928a. The hypothesis of continental displacement. In: The Theory of Continental Drift by W. A. Van der Gracht et al, 104.—1928b. Review of the late Paleozoic formations and faunas with special reference to the ice-age of Middle Permian time. Bull. Geol. Soc. Amer., 769.—1929. Geological history of the Antillean region. Ibid., 337.—1932. Gondwana land bridges. Ibid., 875.

SIMPSON, G. G. 1946. Tertiary land bridges. Trans. N. Y. Acad. Sci., 255.

TRELEASE, W. 1918. Bearing of the distribution of the existing flora of Central America and the Antilles on former land connections. Bull. Geol. Soc. Amer., 649.

WATTS, W. W. 1935. Form, drift and rhythm of the continents. Science, 203.

WILLIS, B. 1932. Isthmian links. Bull. Geol. Soc. Amer., 917.

ZEUNER, F. E. 1936. Palaeobiology and climates of the past. Problems of Paleontology, vol. 1, 199.

Paleobotany

BERRY, E. W. 1920. Paleobotany: A sketch of the origin and evolution of floras. Smiths. Inst., Ann. Rept. for 1918, 289.—1945. The beginnings and history of land plants. Johns Hopkins Univ. Studies in Geol., no. 14, 1.

GUNDERSON, A. 1928. The story of plant evolution. Brooklyn Botanic Garden Leaflets (16), no. 1. [Interdependence of plants and animals.]

GUNDERSON, A. and G. T. HASTINGS. 1944. Interdependence in plant and animal evolution. Scient. Monthly, 63.

SEWARD, A. C. 1898–1919. Fossil plants . . . Vols. 1–4. Cambridge Univ. Press.

Invertebrates (Fossil and Recent), General Works, Evolution of-, Textbooks

ALLEE, W. C. 1927. The evolution of the invertebrates. In: Nature of the World and of Man. Univ. of Chicago Press.

BRONN, H. G. 1880–1940. Klassen und Ordungen des Thier-Reichs wissenschaftlich dargestellt in Wort und Bild. [Vols. I–V on invertebrates.] Leipzig.

BROOKS, W. K. 1890. Handbook of invertebrate zoology. Boston.

BUCKSBAUM, R. 1938. Animals without backbones. Univ. of Chicago Press.

GOLDRING, W. 1929, 1931. Handbook of paleontology . . . Parts I, II. New York State Mus. nos. 9, 10. Albany.

GRABAU, A. and H. W. SHIMER. 1909. North American index fossils: Invertebrates. Vol. I. New York.

HYMAN, L. H. 1940. The invertebrates: Protozoa through Ctenophora. New York.

MAYER, A. G. 1905. Sea-shore life: The invertebrates of the New York coast. New York Aquarium Nature Ser.

NEUMAYR, M. 1889. Die Stämme des Thierreiches: Wirbellose Thiere. Wien und Prag.

PARKER, T. J. and W. A. HASWELL. 1897. A textbook of zoology. Vol. I. London.

SCHINDEWOLF, O. H. 1938. Handbuch der Paläozoologie. Bd. 1–5, Lief. 1–2. Berlin.

SCHUCHERT, C. 1921. Methods of determining the relationships of marine invertebrate fossil faunas. Bull. Geol. Soc. Amer., 339.

SHIMER, H. W. 1914. An introduction to the study of fossils. New York.

STORER, T. I. 1943. General zoology. New York.

TWENHOFEL, W. H. and R. S. SCHROCK. 1935. Invertebrate paleontology. New York.

WOODS, H. 1937. Palaeontology. 7th ed. Cambridge Biol. Ser.

ZITTEL, K. A. VON. 1924. Grundzüge der Paläontologie (Paläozoologie). I. Abteilung: Invertebrata. Revised by F. Broili. Munich and Berlin.

Protista, Protophyta, Protozoa, Sponges, Coelenterates, Conularids, Graptolites

BIDDER, G. P. 1929. Sponges. Encycl. Brit., 14th ed. Vol. 21, 254.

BOUČEK, B. 1939. Conularida. In: Handbuch der Paläozoologie (Ed. O. H. Schindewolf), Bd. 2A, Lief. 5. Berlin. [True conularians treated as a separate order of coelenterates allied with hydroids and scyphozoans.]

BULMAN, O. M. B. 1936. On the graptolites prepared by Holm. Arkiv f. Zoologi, Bd. 28A, no. 17.—1938. Graptolithina. In: Handbuch der Paläozoologie (Ed. O. H. Schindewolf). Berlin. [For affinities of the graptolites, see pp. 62–64.]

CALKINS, G. N. 1933. The biology of the Protozoa. 2nd ed. Philadelphia.

FRITSCH, F. E. 1929. Protophyta. Encycl. Brit., 14th ed. Vol. 18, 615.

HYMAN, L. H. 1940. The invertebrates: Protozoa through Ctenophora. New York.

JAHN, T. L. 1946. The euglenoid flagellates. Quart. Rev. Biol., 246.

KIDERLEN, H. 1937. Die Conulariden. Über Bau und Leben der ersten Scyphozoa. Neues Jahrb. f. Mineral. etc. Stuttgart.

KIESLINGER, A. 1939. Scyphozoa. In: Handbuch der Paläozoologie (Ed. O. H. Schindewolf), 69. Berlin.

KLIGLER, I. J. 1917. The evolution and relationships of the great groups of bacteria. Jour. Bacteriol., March.

KNIGHT, J. B. 1937. Conchopeltis Walcott, an Ordovician genus of the Conularida. Jour. Paleont., 186.

KUHN, O. 1939. Hydrozoa. In: Handbuch der Paläozoologie (Ed. O. H. Schindewolf). Berlin.

RUEDEMANN, R. 1932. Guide to the fossil exhibits of the New York State Museum. Circular N. Y. State Mus. Albany. [Graptolites.]

WILLEY, A. 1897. On Ctenoplana. Quart. Jour. Micros. Sci., 323.

Mollusca

BRONN, H. G. 1896–1907. Klassen und Ordnungen des Tier-Reichs . . . [See Simroth, H.]

CALMAN, W. T. 1929. Teredo. Encycl. Brit., 14th ed., vol. 21, 946.

COSSMANN, M. and G. P. 1910–1913. Iconographie complète des coquilles fossiles de l'Éocène des environs de Paris. Tome 2.

CRAMPTON, H. E. 1916, 1925a, 1932. Studies on the variation, distribution, and evolution of the genus Partula . . . Carnegie Inst. Washington, Publs. 228, 228A, 410.—1925b. Contemporaneous organic differentiation in the species of Partula living in Moorea, Society Islands. Amer. Nat., 5.

GABRIEL, C. J. 1936. Victorian sea shells . . . Field Naturalists' Club of Victoria.

GARDNER, J. 1945. Mollusca of the Tertiary formations of northeastern Mexico. Geol. Soc. Amer., Mem. no. 11.

GRABAU, A. W. 1902. Studies of Gastropoda. Amer. Nat., 917.—1935. Studies of Gastropoda. Nat. Univ. Peking.

HAAS, F. and W. WENZ (Editors). 1922. Archiv f. Molluskenkunde. Frankfurt.

HIRASE, S. 1938. A collection of Japanese shells . . . 6th ed. Tokyo.

JACKSON, R. T. 1886–1893. Pelecypoda: the Aviculidae and their allies. Mem. Boston Soc. Nat. Hist., vol. 4, 277.

KNIGHT, J. B. 1933. The gastropods of the St. Louis, Missouri, Pennsylvanian Outlier: VI. The Neritidae. Jour. Paleont., 359.

LINDEN, M. VON. 1896. Die Entwicklung der Skulptur und der Zeichnung bei Gehäuseschnecken des Meeres. Zeits. f. Wiss. Zool., 261.—1898a. Unabhängige Entwicklungsgleichheit bei Schneckengehäusen. Biol. Centralbl. 697.—1898b. Unabhängige Entwicklungsgleichheit (Homoögenesis) bei Schneckengehäusen. Zeitschr. f. Wiss. Zool., 708.

NAEF, A. 1913. Studien zur generellen Morphologie der Mollusken. 1. Teil: Über Torsion und Asymmetrie der Gastropoden. Ergeb. u. Fortschr. d. Zool., 73.

PELSENEER, P. 1906. Mollusca. In: A treatise on zoology by E. Ray Lankester, Part V. London.

PERRY, L. M. 1940. Marine shells of the southwest coast of Florida. Bull. Amer. Paleont., no. 95, Paleont. Research Inst.

POWELL, A. W. B. 1937. The shellfish of New Zealand . . . Auckland.

PRASHAD, B. 1932. Pila (the apple-snail). Indian Zool. Mem. Lucknow.

ROBSON, G. C. 1929. Mollusca. Encycl. Brit., 14th ed. Vol. 15, 674.

SIMROTH, H. 1896–1907. Mollusca, II. Abt.: Gastropoda prosobranchia. In: Klassen u. Ordn. des Thier-Reichs . . . (Ed. H. G. Bronn).

SMITH, M. 1937. East coast marine shells . . . Ann Arbor, Michigan. 4to.—1938. A catalog of the recent species of the Family Muricidae. Part I. Tropical Photographic Laboratory, Lantana, Florida.—1940. World-wide sea shells. Ibid.—1942. Review of the Volutidae . . . Ibid.—1944. Panamic marine shells . . . Ibid, 4to.

SÖDERSTRÖM, A. 1925. Die Verwandtschaftsbeziehungen der Mollusken. Upsala.

THIELE, J. 1925a. Solenogastres. In: Handbuch der Zoologie (Ed. Kükenthal and Krumbach).—1925b. Mollusca: Unterklasse der Gastropoda: Prosobranchia . . . Ibid., 40.

TRUEMAN, E. R. 1942. The structure and deposition of the shell of Tellina tenuis. Jour. Roy. Micros. Soc. London, 69.

WEBB, W. F. 1936. Handbook for shell collectors. . . . 4th ed. Rochester, N. Y.

WENZ, W. 1938–1940. Gastropoda: Prosobranchia. In: Schindewolf's Handbuch der Paläozoologie, Bd. VI, Teil 5.

WRIGLEY, A. 1948. The color patterns and sculpture of molluscan shells. Proc. Malacolog. Soc. London, 31 May, 206.

ZITTEL, K. A. VON. 1924. Grundzüge der Paläontologie (Paläozoologie). I. Abt. Invertebrata. Munich and Berlin.

Annelids and Arthropods: Trilobites, Eurypteroids, Limuloids, Crustaceans, Peripatus, Insects

BEECHER, C. E. 1895. The larval states of trilobites. Amer. Geol., 166.

CALMAN, W. T. 1935. The meaning of biological classification. Proc. Linnean Soc. London, 145. [Phylogeny of the Crustacea, 154–158.]

CLARKE, J. M. and R. RUEDEMANN. 1912. The Eurypterida of New York. Mem. N. Y. State Mus., no. 14.

CRANE, J. 1941. Crabs of the genus Uca from the west coast of Central America. Zoologica (N. Y.) 145.

DAVID, T. E. and R. J. TILLYARD. 1936. Memoir on fossils of the late Pre-Cambrian (Newer Proterozoic) from the Adelaide Series, South Australia. Sydney. 8vo. [Supposed primitive arthropods.]

EMERSON, A. E. 1941. Taxonomy and ecology. Ecology, vol. 22, 213.

GARSTANG, W. and R. GURNEY. 1938. The descent of Crustacea from trilobites and their larval relations. In: Essays on Aspects of Evolutionary Biology (Ed. G. R. de Beer), 271.

GERTSCH, W. J. 1947. Spiders that lasso their prey. Natural History, 152.

HANDLIRSCH, A. 1906. A revision of American Paleozoic insects. Proc. U. S. Nat. Mus., 661. —1908. Die fossilen Insekten und die Phylogenie der rezenten Formen. Leipzig. Vols. I, II. —1926–1927. Arthropoda. In: Handbuch d. Zoologie (Eds. Kükenthal and Krumbach), Bd. 3.

IMMS, A. D. 1937. Recent advances in entomology. 2nd ed. Philadelphia.

LANG, A. 1921. Handbuch des Morphologie des Wirbellosen Tiere. Continued by Earl Hescheler. 4 Bd. Arthropoda. Jena.

LUTZ, F. E. 1918, 1935. Field book of insects. New York.—1931a. Insects vs. the people . . . Natural History, vol. 31, no. 1, 49.—1931b. In defense of insects. Scient. Monthly, 367.—1941. A lot of insects. New York.

O'CONNELL, M. 1916. The habitat of the Eurypterida. Bull. Buffalo Soc. Nat. Sci., 1.

RAASCH, G. O. 1939. Cambrian Merostomata. Geol. Soc. Amer., Special Paper no. 19.

RAYMOND, P. E. 1920. The appendages, anatomy and relationships of trilobites. Mem. Connecticut Acad. Arts and Sci.

SAVORY, T. H. 1928. The biology of spiders. London. [Making an orb-web, p. 147, fig. 65.]

SCHNEIDER, A. 1885. Die einzelligen Drüsen und die Stellung von Peripatus im System. Zool. Beiträge, 116.

SNODGRASS, R. E. 1935. Principles of insect morphology. New York.—1938. Evolution of the Annelida, Onychophora and Arthropoda. Smithsonian Miscell. Coll., vol. 97, no. 6.

STORMER, L. 1933. Are the trilobites related to the arachnids? Amer. Jour. Sci., 147.—1939. Studies on trilobite morphology. Part I. Thoracic appendages and their phylogenetic significance. Norsk geol. tidsskr., 143.—1942. Part II. Larval development, segmentation and sutures, and their bearing on trilobite classification. Ibid., 49.—1944. On the relationships and phylogeny of fossil and recent Arachnomorpha . . . Skr. Norske Vidensk. Akad. i Oslo.

SÜFFERT, F. 1929. Morphol. Erscheinungsgruppen in der Flügel-zeichnung der Schmetterlinge . . . Archiv. f. Entwickl. mechanik der Organismen (Festschrift f. Hans Spemann), 299.

WALCOTT, C. D. 1884. The Cambrian faunas of North America. U. S. Geol. Surv., Bull. 287.—1886. Studies in the Cambrian faunas of North America. Ibid., no. 30.—1890. The fauna of the Lower Cambrian or Olenellus zone. U. S. Geol. Surv., 10th Ann. Rept., 509.

WHEELER, W. M. 1910. Ants: Their structure, development and behavior. Columbia Univ. Biol. Ser., no. 9.—1933. Colony-founding among ants, with an account of some primitive Australian species. Harvard Univ. Press.

PART II

EMERGENCE OF THE VERTEBRATES

(Cf. Vol. I, Chapters IV–IX)

Brachiopods, Bryozoa (Polyzoa), Echinoderms

ABEL, O. 1920. Lehrbuch der Paläozoologie. Jena. [Descriptions of carpoid echinoderms, pp. 280–281.]

AIYAR, R. G. 1938. Salmacis (the Indian sea-urchin). Indian Zool. Mem. no. 7.

BATHER, F. A. 1928. Echinoderms. Encycl. Brit., 14th ed. Vol. 7, 895.

BATHER, F. A., J. W. GREGORY and E. S. GOODRICH. 1900. The echinoderms. In: A Treatise on Zoology (Ed. E. Ray Lankester). London.

BEECHER, C. E. and J. M. CLARKE. 1889. Development of some Silurian Brachiopoda. N. Y. State Mus. Mem. 1.

FENTON, C. L. 1931. Studies of evolution in the genus Spirifer. Publ. Wagner Free Inst. Sci. Philadelphia.

GRABAU, A. W. 1917. [Evolution in Spirifer.] See Origin and Evolution of Life by H. F. Osborn, 1917, fig. 36.

HARMER, S. F. 1929. Polyzoa. Encycl. Brit., 14th ed. Vol. 18, 196.

LANG, A. 1894. Lehrbuch der Vergleichenden Anatomie der Wirbellosen Thiere. 4 Theil., Kap. 8, 9. Echinodermen und Enteropneusten. Jena.

SWINNERTON, H. H. 1923. Outlines of palaeontology. London. [Aristocystis, p. 118; food grooves, pp. 120–129; stalked condition of echinoderms, pp. 119–129.]

WALCOTT, C. D. 1912. Cambrian Brachiopoda. U. S. Geol. Surv., Monogr. 51.

Prochordates and Origin of Vertebrates: Amphioxus, Ascidians, Balanoglossids

ALLEE, W. C. 1927. The evolution of the invertebrates. In: "The Nature of the world and of man" (Ed. H. H. Newman). Univ. of Chicago Press. [Relations of invertebrates and vertebrates, 260.]

BAER, K. E. 1828. Ueber Entwicklungsgeschichte der Thiere. Koenigsberg.

BALFOUR, F. M. 1875. Comparison of the early stages in the development of vertebrates. Quart. Jour. Micros. Sci., 207.—1880–1881. A treatise on comparative embryology. Vols. I, II. London.

BATESON, W. 1884. The early stages in the development of Balanoglossus . . . Quart. Jour. Micros. Sci., vol. 24, 207.—1885. The later stages in the development of Balanoglossus . . . Ibid., vol. 25, 81.—1886. Continued account of the later stages in the development of Balanoglossus . . . Ibid., vol. 26, 512.

BEARD, J. 1886. Some annelidan affinities in the ontogeny of the vertebrate nervous system. Nature, 259.—1888. The old mouth and the new: A study in vertebrate morphology. Anat. Anz., 15.

BEER, G. R. DE. 1928. Vertebrate zoology . . . New York.—1930. Embryology and evolution. Oxford.—1938. Embryology and evolution. In: Evolution: Essays . . . presented to Prof. E. S. Goodrich . . . (Ed. G. R. de Beer). Oxford.—1940. Embryos and ancestors. In: Monographs on Animal Biology (Ed. G. R. de Beer). Oxford.

BROOKS, W. K. 1899. The foundations of zoology. Columbia Univ. Biol. Ser. London. [Tunicata and origin of the vertebrates.]

BULLOCK, T. H. 1940. The functional organization of the nervous system of Enteropneusta. Biol. Bull., vol. 79, 91.

CONKLIN, E. G. 1929. Problems of development. Amer. Nat., 5.—1932. The embryology of Amphioxus. Jour. Morph., 69.—1933. The development of isolated and partially separated blastomeres of Amphioxus. Jour. Exp. Zool., 303.

COPE, E. D. 1885. The position of Pterichthys in the system. Amer. Nat., 289. [Proposes Antiarcha as an order of Tunicata, p. 291.]

CUVIER, G. L. C. F. D. 1817. Le règne animal distribué d'après son organisation. Tome I. Paris.

DELAGE, Y. and E. HÉROUARD. 1898. Les Procordés. In: Traité de Zoologie Concrète by Delage and Hérouard. Tome 8. Paris.

DOHRN, A. 1876. Der Ursprung der Wirbelthiere und das Princip des Functionswechsels. Leipzig.

FELL, H. B. 1948. Echinoderm embryology and the origin of chordates. Biol. Reviews, vol. 23 (1), 81. Cambridge.

GARSTANG, W. 1894. Preliminary note on a new theory of the phylogeny of the Chordata. Zool. Anz., 122.—1922. The theory of recapitulation: A critical restatement of the biogenetic law. Jour. Linnean Soc. London (Zool.), 81.

GASKELL, W. H. 1896. The origin of vertebrates. Proc. Cambridge Philos. Soc. 19.—1898–1906. On the origin of vertebrates deduced from the study of ammocoetes. Jour. Anat. Physiol. London.—1908. The origin of vertebrates. London.

GEOFFROY SAINT-HILAIRE, É. 1818. Philosophie anatomique. Tome I. Paris. [Comparison of invertebrates and vertebrates.] 1830. Principes de philosophie zoologique, discutés en Mars 1830, au sein de l'Academie Royale des Sciences. Paris.

GEOFFROY SAINT-HILAIRE, I. 1847. Vie, travaux et doctrine scientifique d'Étienne Geoffroy Saint-Hilaire. Paris.

GISLÉN, T. 1930. Affinities between the Echinodermata, Enteropneusta and Chordonia. Zool. Bidrag f. Uppsala, vol. 12, 199.

GOODRICH, E. S. 1895. On the coelom, genital ducts and nephridia. Quart. Jour. Micros. Sci., vol. 37, 477.—1909. Vertebrata craniata (First fascicle: cyclostomes and fishes). In: A Treatise on Zoology (Ed. E. Ray Lankester), part 9. London.—1930. Studies on the structure and development of vertebrates. London.—1934. The early development of the nephridia in Amphioxus. Parts I, II. Quart. Jour. Micros. Sci., vol. 76, 499, 655.

GREGORY, W. K. 1935. Reduplication in evolution. Quart. Rev. Biol., 272.—1936. The transformation of organic designs: A review of the origin and deployment of the earlier vertebrates. Biol. Reviews (Cambridge), 311. [Amphioxus ? from ostracoderms.]—1946. The roles of motile larvae and fixed adults in the origin of the vertebrates. Quart. Rev. Biol., 348.

HARMER, S. F. et al. 1910. Fishes and ascidians. Cambridge Nat. Hist., vol. 7.

HUBRECHT, A. A. W. 1883. On the ancestral forms of the Chordata. Quart. Jour. Micros. Sci., 349. —1887. The relations of the Nemertea to the Vertebrata. Ibid., 605.

JAECKEL, O. 1918. Über fragliche Tunicaten aus dem Perm Siciliens. Palaeont. Zeitschr., II Bd., 66.

KOVALEVSKY, A. 1866. Entwickelungsgeschichte der einfachen Ascidien. Mem. Acad. Sci. Saint-Petersb.—1871. Weitere Studien über die Entwickelung der einfachen Ascidien. Arch. mikr. Anat., 101.

KUPFER, C. 1870. Die Stammverwandtschaft zwischen Ascidien und. Wirbelthieren. Arch. mikr. Anat., 115.

LANG, A. 1894. Lehrbuch der Vergl. Anat. der Wirbellosen Thiere. 4 Theil. Echinodermen und Enteropneusten. Jena.

LEACH, W. J. 1944a. The archetypal position of Amphioxus and Ammocoetes and the role of endocrines in chordate evolution. Amer. Nat., 341.—1944b. Some deficiencies in Amphioxus as an ancestral type . . . Turtox News, vol. 22, no. 4.

MCBRIDE, E. W. 1896. Note on the formation of the germinal layers in Amphioxus. Proc. Philos. Soc. Cambridge, 151.—1898. The early development of Amphioxus. Quart. Jour. Micros. Sci., 589.—1914. Text-book of embryology. Vol. I. Invertebrata. London.

MASTERMAN, A. T. 1897a. On the Diplochorda. Quart. Jour. Micros. Sci., 281. [Bateson's genetic distinction between "segmented" Invertebrata and segmented Vertebrata, p. 330.]—1897b. On the "notochord" of Cephalodiscus. Zool. Anz. vol. 20, 443.

METSCHNIKOFF, E. 1866. Ueber eine Larve von Balanoglossus. Arch. Anat. Physiol., Jahrg. 1866, 592.—1870. Untersuchungen über die Metamorphose einiger Seethiere. 1. Ueber Tornaria. Zeitschr. f. Wiss. Zool., 131.—1881. Über die systematische Stellung von Balanoglossus. Zool. Anz., 139, 153. [Unites enteropneusts and echinoderms into one group, Ambulacraria.]

MINOT, C. S. 1897. Cephalic homologies: A contribution to the determination of the ancestry of vertebrates. Amer. Nat., 927.

MORGAN, T. H. 1891. Growth and development of Tornaria. Jour. Morph. 407.—1894. The development of Balanoglossus. Ibid., 1.

OWEN, R. 1883. Essays on the conario-hypophysial tract and the aspects of the body in vertebrate and invertebrate animals. London.

PACKARD, A. S. 1884. Aspects of the body in vertebrates and arthropods. Amer. Nat., 855. [Cf. The Microscope, October, 1884, pp. 225–227. Article under same title, signed N.]

PATTEN, W. 1884. The development of phryganids, with a preliminary note on the development of Blatta germanica. Quart. Jour. Micros. Sci., 549.—1891. On the origin of vertebrates from arachnids. Ibid., 317.—1902. On the structure and classification of the Tremataspidae. Amer. Nat., 379.—1903. On the structure of the Pteraspidae and Cephalaspidae. Amer. Nat. 827.—1912. The evolution of the vertebrates and their kin. Philadelphia.

SEDGWICK, A. 1884. On the origin of metameric segmentation and some other morphological questions. Quart. Jour. Micros. Sci., 43.

SEMPER, C. 1875–1876. Die Verwandtschaftsbeziehungen der gegliederten Thiere. Arbeit. zool.-zoot. Inst. Wurzburg.

SPENCER, W. K. 1938. Some aspects of evolution in Echinodermata. In: Evolution: Essays . . . presented to Prof. E. S. Goodrich . . . Oxford.

STORMER, L. 1933. Are the trilobites related to the arachnids? Amer. Jour. Sci., 147. [Triarthrus, pp. 147–150.]

WILHELMI, R. W. 1942. The application of the precipitin technique to theories concerning the origin of vertebrates. Biol. Bull., 179.

WILLEY, A. 1894. Amphioxus and the ancestry of the vertebrates. Columbia Univ. Biol. Ser. II.

WILSON, E. B. 1892. The cell-lineage of Nereis: A contribution to the cytogeny of the annelid body. Jour. Morph., 361.

Vertebrates, Fossil and Recent: General (Textbooks, Classification, Phylogenetics, Comparative Anatomy)

ABEL, O. 1912. Grundzüge der Palaeobiologie der Wirbeltiere. Stuttgart.—1920. Lehrbuch der Palaozoologie. Jena.

BEER, G. R. DE. 1928. Vertebrate zoology: An introduction to the comparative anatomy, embryology, and evolution of chordate animals. New York.

BRONN, H. G. 1880–1940. Klassen und Ordnungen des Thier-Reichs . . . [Vol. 6, Parts 1–5, Vertebrata.] Leipzig.

COLE, F. J. 1944. A history of comparative anatomy. London.

COPE, E. D. 1898. Syllabus of lectures on the Vertebrata. Philadelphia.

GEGENBAUR, C. 1898–1901. Vergl. Anat. der Wirbelthiere mit Berüchsichtigung der Wirbellosen. I, II. Leipzig.

GOODRICH, E. S. 1930. Studies on the structure and development of vertebrates. New York.

GREGORY, W. K. 1936. The transformation of organic designs: A review of the origin and deployment of the earlier vertebrates. Biol. Reviews (Cambridge), vol. 11, 311.

HUENE, F. VON. 1936. Kurze Übersicht über die Geschichte der Vertebraten . . . Palaeont. Zeitschr., 198.—1937. Das Problem der Phylogenie. Problems of Paleontology, vols. 2, 3, 615.—1940. Die stammesgesch. Gestalt der Wirbeltiere . . . Palaeont. Zeitschr., 55.

HUXLEY, T. H. 1872. A manual of the anatomy of vertebrated animals. New York.

HYMAN, L. H. 1942. Comparative vertebrate anatomy. Univ. Chicago Press.

JEPSEN, G. L. 1944. Phylogenetic trees. Trans. N. Y. Acad. Sci., 81.

LANKESTER, E. R. (Editor). 1900–1909. A treatise on zoology. Parts 1–9. London.

LYDEKKER, R. Undated. The new natural history. Vols. I–V. New York.

NAEF, A. 1926. Studien zur systemat. Morphol. und Stammesgesch. der Wirbeltiere. Ergebn. u. Fortschr. d. Zool. 7. Bd.

OWEN, R. 1840–1845. Odontography . . . Vol. I. Text. Vol. II. Atlas. London.—1866–1868. On the anatomy of vertebrates. Vols. I–III. London.

PARKER, T. J. and W. A. HASWELL. 1940. A textbook of zoology. Vol. II. 6th edition, revised by C. Forster-Cooper. London.

ROMER, A. S. 1945. Vertebrate Paleontology. 2nd ed. Univ. Chicago Press. [First edition, 1933.]

SCHMALHAUSEN, J. J. Fundamentals of the comparative anatomy of vertebrates. Moscow and Petersburg. Government Publ.

VALLOIS, H. V. 1922. Les transformations de la musculature de l'épisome chez les vertébrés. Arch. d. Morphol. Gén. et Expér. Paris.

WIEDERSHEIM, R. E. E. 1906. Vergleichende Anatomie der Wirbeltiere. Jena.—1907. Einführung in die Vergleichende Anatomie der Wirbeltiere. Jena.

WOODWARD, A. S. 1898. Outlines of vertebrate palaeontology for students of zoology. Cambridge Nat. Sci. Manuals, Biol. Ser.

ZITTEL, K. A. VON. 1902. Text-book of palaeontology. Translated and edited by Charles R. Eastman. Vol. II.—1923. Grundzüge der Paläontologie (Paläozoologie). II. Abt.: Vertebrata. Munich and Berlin.—1932. Text-book of Palaeontology. Translated and edited by C. R. Eastman. 2nd English ed. by A. S. Woodward. London.

Vertebrates, Embryology, General

BALFOUR, F. M. 1875. A comparison of the early stages in the development of vertebrates. Quart. Jour. Micros. Sci., 207.—1880–1881. A treatise on comparative embryology. Vols. I, II. London.

BEER, G. R. DE. 1930. Embryology and evolution. Oxford.—1940. Embryos and ancestors: Monographs on animal biology (G. R. de Beer, Ed.). Oxford.

DETWILER, S. R. and H. B. ADELMANN. 1928. Morgan on Entwicklungsmechanik. Quart. Rev. Biol., 419.

GEOFFROY SAINT-HILAIRE, E. 1822. Philosophie anatomique des monstruosités humaines . . . Tome 2nd. Paris.

HUXLEY, J. 1924. Early embryonic differentiation. Nature, 276.

KERR, J. G. 1929. Vertebrate embryology. Encycl. Brit., 14th ed., vol. 23, 98.

LANKESTER, E. R. 1877. Notes on embryology and classification. Quart. Jour. Micros. Sci., 399.

PATTEN, B. 1927. The embryology of the pig. Philadelphia.

ROMER, A. S. 1942. Cartilage an embryonic adaptation. Amer. Nat., 394.

SPEMANN, H. 1927. Neue Arbeiten uber Organisatoren in der tierischen Entwicklung. Die Naturw. Wochenschr. f. d. Fortschr. d. Reinen u. d. Angewandten Naturw., 946.

Vertebrate Skeleton, General: Cartilage and Bone, Origin and Growth, Fenestration, Vertebrae

BRASH, J. C. 1928. Growth of the alveolar bone. Internat. Jour. Orthodontics, 196.—1934. Some problems in the growth and developmental mechanics of bone. Edinburgh Med. Jour., 305.

EVANS, F. G. 1939. The morphology and functional evolution of the atlas-axis complex from fish to mammals. Ann. N. Y. Acad. Sci., 29.

GADOW, H. 1933. The evolution of the vertebral column. Cambridge Univ. Press.

GOODRICH, E. S. 1930. Studies on the structure and development of vertebrates. London.

GREGORY, W. K. 1937. The bridge-that-walks. Natural History, 33.

GUDGER, E. W. 1929. Some early and late illustrations of comparative osteology. Ann. Med. History, 334.

HUXLEY, T. H. 1872. A manual of the anatomy of the vertebrated animals. New York.

OWEN, R. 1848. On the archetype and homologies of the vertebrate skeleton. London.

ROCKWELL, H., F. G. EVANS and H. C. PHEASANT. 1938. The comparative morphology of the vertebrate spinal column: Its form as related to function. Jour. Morph., 67.

ROMER, A. S. 1942. Cartilage an embryonic adaptation. Amer. Nat., 394.

SLIPJER, E. J. 1947. De Voortbewegingsorganen. Leerboek d. Vergel. Ontleedkunde der Vertebraten.

SMITH, H. W. 1939. Studies in the physiology of the kidney. Univ. Kansas. [Suggests mode of origin of the vertebrate skeleton.]

STENSIO, E. A. 1927. The Downtonian and Devonian vertebrates of Spitsbergen. Part I. Family Cephalaspidae. Skrift. om Svalbard og Nordishavet . . . [Bone before cartilage, pp. 30–32, 333–334, 374.]

STROMER VON REICHENBACH, E. 1912. Lehrbuch d. Paläozoologie. II. Teil: Wirbeltiere. Leipzig und Berlin. [Progressive reduction of skeletal elements from lower to higher forms in vertebrates and invertebrates, p. 282. Cf. Williston, 1925; Gregory, Roigneau et al, 1935.]

Vertebrate Skull, General: Visceral Arches and their Muscles

ADAMS, L. A. 1919. A memoir on the phylogeny of the jaw muscles in recent and fossil vertebrates. N. Y. Acad. Sci., 51.

BEER, G. R. DE. 1937. The development of the vertebrate skull. Oxford.

EDGEWORTH, F. H. 1935. The cranial muscles of vertebrates. Cambridge.

GOODRICH, E. S. 1930. Studies on the structure and development of vertebrates. London.

GREGORY, W. K. 1920. Evolution of the lacrymal bone of vertebrates. Bull. Amer. Mus. Nat. Hist., 95.—1927, 1929. The palaeomorphology of the human head: Ten structural stages from fish to man. Part I. Quart. Rev. Biol., 267. Part II. Ibid., 233.—1931. Certain critical stages in the evolution of the vertebrate jaws. Internat. Jour. Orthodontia, Oral Surgery and Radiography, 1138.—1934. Polyisomerism and anisomerism in cranial and dental evolution among vertebrates. Proc. Nat. Acad. Sci., 1.—1935. On the evolution of the skulls of vertebrates with special reference to heritable changes in proportional diameters (anisomerism). Ibid., 1.

GREGORY, W. K. and L. A. ADAMS. 1915. The temporal fossae of vertebrates in relation to the jaw muscles. Science, 763.

GREGORY, W. K. and M. ROIGNEAU, et al. 1935. "Williston's Law" relating to the evolution of skull bones in the vertebrates. Amer. Jour. Phys. Anthrop., 123. [Cf. Stromer von Reichenbach, 1912. Williston, 1925.]

IHLE, J. E. W., P. N. VAN KAMPEN, H. F. NIERSTRASZ, and J. VERSLUYS. 1927. Vergleich. Anat. der Wirbeltiere. Berlin.

JAEKEL, O. 1927. Der Kopf der Wirbeltiere. Ergebn. d. Anat. u. Entwickl., 815.

KLAAUW, C. J. VAN DER. 1948. Size and position of the functional components of the skull: A contribution to the architecture of the skull . . . Arch. Neerlandaises de Zool., tome IX, le et 2e livr.

LOCY, W. A. 1895. Contribution to the structure and development of the vertebrate head. Jour. Morph., 497.

RETZIUS, M. G. 1881. Das Gehörorgan der Wirbelthiere. Vol. I. Das Gehörorgan der Fische und Amphibien. Stockholm.

SLIPJER, E. J. 1947. De Musculatuur van den Kop. Leerb. d. Vergel. Ontleedk. der Vert., 309.

STROMER VON REICHENBACH, E. 1912. Lehrbuch der Paläozoologie. Bd. II. Leipzig. [First statement in regard to reduction of skull elements in passing from fish to man. Cf. Gregory, Roigneau, et al, 1935; also Williston, 1925.]

WILLISTON, S. W. 1925. Osteology of the reptiles (Ed. W. K. Gregory). Harvard Univ. Press.

Vertebrates, General, including Fishes: Sense Organs, Nervous System, Brain, Psychology

BLACK, D. 1917–1921. The motor nuclei of the cerebral nerves in phylogeny: A study of the phenomena of neurobiotaxis. Jour. Comp. Neurol., vol. 27, 467; vol. 28, 379; vol. 32, 61; vol. 34, 233.

DETWILER, S. R. 1923. Experiments on the transplantation of the spinal cord in Amblystoma . . . Jour. Exp. Zool., 339.—1927a. The effects of extensive muscle loss upon the development of spinal ganglia in Amblystoma. Ibid., vol. 48, 1.—1927b. Die Morphogenese des peripheren und centralen Nervensystems der Amphibien in Licht experimenteller Forschungen. Die Naturw., 873, 895.

EDINGER, T. 1929. Die fossilen Gehirne. Ergeb. d. Anat. u. Entwickl., 1.—1937. Paläoneurologie. Fortschr. d. Paläont., 235.

HERRICK, C. J. 1943. The cranial nerves: A review of fifty years. Denison Univ. Bull., 41.

KAPPERS, C. U. A. 1907. Die phylogenetischen Verlagerungen der motorischen Oblongata-Kerne, ihre Ursache und ihre Bedeutung. Neurolog. Centralb., 834.—1908. Weitere Mitteilungen bezüglich der phylogenetischen Verlagerung der motorischen Hirnnervenkerne: Der Bau des autonomen Systems. Folia Neurobiolog., 157.—1929. The evolution of the nervous system in invertebrates, vertebrates and man. Haarlem.

KAPPERS, C. U. A., C. C. HUBER and E. C. CROSBY. 1936. The comparative anatomy of the nervous system of vertebrates, including man. Vols. I, II. New York.

KAPPERS, C. U. A. and W. F. THEUNISSEN. 1908. Die Phylogenese des Rhinencephalons des Corpus Striatum und der Vorderhirn-Commisuren. Folio Neurobiolog., 173.

KUPFFER, C. W. VON. 1906. Die Morphologenie des Centralnervensystems (Bdellostoma, Petromyzon, Elasmobranchier, Ganoiden, Teleostier). In: Hertwig's Handbuch der vergl. u. exper. Entwick. der Wirbeltiere. Jena.

PAPEZ, J. W. 1929. Comparative neurology. New York.

RETZIUS, M. G. 1871. Das Gehörlabyrinth der Knochenfische. In: Anat. Untersuch. 1. Lief.

SCHEPERS, G. W. H. 1948. Evolution of the forebrain: The fundamental anatomy of the telencephalon. Capetown, South Africa. [Evolution of forebrain from prevertebrate stages to man.]

STENSIÖ, E. A. 1927. The Downtonian and Devonian vertebrates of Spitzbergen. Part I. Family Cephalaspidae. Skrift. om Svalbard og Nordishavet, no. 12.—1932. The cephalaspids of Great Britain. London. [Brain, cranial nerves and cranial vessels.]

WALIS, G. L. 1942. The vertebrate eye and its adaptive radiation. Bull. Cranbrook Inst. Sci., Bloomfield Hills, Michigan.

WIMAN, C. 1918. Über Gehirn und Sinnesorgane bei Tremataspis. Geol. Inst. Bull., Upsala.

Fishes (Fossil and Recent), General:

Textbooks, Classification, Catalogues, Phylogeny

AGASSIZ, J. L. R. 1833–1843. Recherches sur les poissons fossiles. Vols. 1–5. Folio. Neuchatel.

BERG, L. S. 1940. Classification of fishes, both recent and fossil. Trav. de l'Inst. Zool. de l'Acad. d. Sci. d. l'URSS, vol. 5, 87. English translation, p. 346.

BOULENGER, G. A. 1910. Fishes (Systematic account of Teleostei). In: The Cambridge Nat. Hist., 541. London.

BRIDGE, T. W. 1904. Fishes (exclusive of the systematic account of Teleostei). Ibid., vol. 7, 139. London.

CUVIER, G. and A. VALENCIENNES. 1828–1849. Histoire naturelle des poissons. 22 vols. Paris.

DAY, F. 1880–1884. The fishes of Great Britain and Ireland. 8 vols. London and Edinburgh.

DEAN, B. 1895. Fishes, living and fossil. Columbia Univ. Biol. Ser. New York.

GILL, T. N. 1893. Families and subfamilies of fishes. Mem. Nat. Acad. Sci., 127.

GOODE, G. B. and T. H. BEAN. 1896. Oceanic ichthyology . . . Mem. Mus. Comp. Zool., Harvard Coll. [Special Bull. U. S. Nat. Mus., 1895.]

GOODRICH, E. S. 1909. Vertebrata craniata (First Fascicle: Cyclostomes and fishes). In: A treatise on zoology (Ed. Sir E. Ray Lankester). London.

GREGORY, W. K. and F. LAMONTE. 1947. The world of fishes . . . New ed. Amer. Mus. Nat. Hist. New York.

HAY, O. P. 1902. Bibliography and catalogue of the fossil Vertebrata of North America. Bull. U. S. Geol. Surv., no. 179. [Classification of fishes, pp. 252–255.]—1929. Second bibliography and catalogue of the fossil Vertebrata of North America. Vols. I, II. Carnegie Inst. of Washington. [Classification of fishes, pp. 514–519.]

HUBBS, C. L. 1923. A classification of fishes including families and genera . . . Science, 181.

JORDAN, D. S. 1923. A classification of fishes, including families and genera as far as known. Stanford Univ. Publ.

JORDAN, D. S. and B. W. EVERMANN. 1896. The fishes of North and Middle America . . . Bull. U. S. Nat. Mus., vol. 47.

KYLE, H. M., 1926. The biology of fishes. New York.

LYDEKKER, R. 1903. Fishes. In: The New Natural History, vol. 5, 314. New York.

MOY-THOMAS, J. A. 1939. Palaeozoic fishes. New York.

NEWBERRY, J. S. "1889" [1890]. The Palaeozoic fishes of North America. Monogr. U. S. Geol. Surv.

NORMAN, J. R. 1931. A history of fishes. New York.

PANDER, C. H. 1856. Fossilen Fische des Silurischen Systems der Russisch-Baltischen Gouvernements. Atlas, 4to. St. Petersburg.—1860. Ueber die Saurodipteren, Dendrodonten, Glyptolepiden und Cheirolepiden des Devonischen Systems. 4to. St. Petersburg.

REGAN, C. TATE. 1906. Pisces. In: Biologia Centrali-Americana (1906–1908.)—1929. Article on fishes. Encycl. Brit., 14th ed.

ROMER, A. S. 1946. The early evolutio of fishes. Quart. Rev. Biol., 33.

ROULE, L. 1926. Les poissons et le monde vivant des eaux . . . Paris.

STOYE, F. H. 1935. Tropical fishes for the home . . . New York.

TRAQUAIR, R. H. 1900. The evolution of fishes. Rept. Brit. Assoc. Adv. Sci. (1899): Zoology, 768.

WOODWARD, A. S. 1889–1901. Catalogue of the fossil fishes in the British Museum (Natural History). 4 vols. Vol. I. Elasmobranchii. Vol. II. Elasmobranchii to Chondrostei. Vol. III. Chondrostei (concl.) to Isospondyli. Vol. IV. Isospondyli (concl.) to Anacanthini.—1906. The study of fossil fishes. Proc. Geol. Assoc. London, 266. [Evolution and succession from crossopterygian to teleost.]—1915. The use of fossil fishes in stratigraphical geology. Ibid., 52. [Succession of Ostracodermi, Elasmobranchii, Holocephali, Dipnoi, Crossopterygii, Actinopterygii.]

Fishes:

Locomotion, Body Form, Growth, Symmetry, Skeleton, Body Musculature, Squamation

ABEL, O. 1922. Ueber den Wiederholten Wechsel der Körperformen im Laufe der Stammesgeschichte der Teleostomen. Bijd. tot de Dierkunde. Kon. Zool. Genootschap. Natura Artis Magistra te Amsterdam, 73.

BREDER, C. M. JR. 1924. Respiration as a factor in locomotion of fishes. Amer. Nat., 145.—1926. The locomotion of fishes. Zoologica (New York), 159.—1947. An analysis of the geometry of symmetry with especial reference to the squamation of fishes. Bull. Amer. Mus. Nat. Hist., 321.

GOODRICH, E. S. 1908. On the scales of fish, living and extinct, and their importance in classification. Proc. Zool. Soc. London (for 1907), 751.

GRAY, J. 1933. Studies in animal locomotion. I. The movement of fish, with special reference to the eel. Jour. Exper. Biol. [A geometrical analysis of vectors and components involved.]

GREENE, C. W. and C. H. GREENE. 1914. The skeletal musculature of the king salmon. Bull. Bureau Fisheries (for 1913), 21.

GREGORY, W. K. 1914. Locomotive adaptations in fishes illustrating "habitus" and "heritage." Ann. N. Y. Acad. Sci. (for 1913), 267.—1928a. Studies on the body-forms of fishes. Part I. The body-forms of fishes and their inscribed rectilinear lines. Palaeobiologica, vol. 1, 93.

HEINTZ, A. 1935. How the fishes learned to swim. Smithsonian Rept. (for 1934), 223.

HOUSSAY, F. 1900. La forme et la vie: Essai de la methode mecanique en zoologie. Paris.—1912. Forme, puissance et stabilite des poissons. Paris.—1914. The effect of water pressure upon the form of fishes: A study of evolution of form resulting from conditions of life and habits. Scient. American, Suppl., 376.

HUBBS, C. L. 1926. The structural consequences of modification of the developmental rate in fishes . . . Amer. Nat., 57.

HUBBS, C. L. and L. C. HUBBS. 1945. Bilateral asymmetry and bilateral variation in fishes. Michigan Acad. Sci., Arts and Letters, 1944 (1945), 229.

HUXLEY, J. S. 1931. Notes on differential growth. Amer. Nat., 289.

LOCHHEAD, J. H. 1942. Control of swimming position by mechanical factors and proprioception. Quart. Rev. Biol., 12.

MOODIE, R. L. 1922. The influence of the lateral-line system on the peripheral osseous elements of fishes and Amphibia. Jour. Comp. Neurol., 319.

ROULE, L. 1926. Les poissons et le monde vivant

des eaux . . . Tome I. Les formes et les attitudes. Paris.

Rugh, R. et al. 1947. The mechanics of development. Ann. N. Y. Acad. Sci.

Shapiro, S. 1938. A study of proportional changes during the post-larval growth of the blue marlin (Makaira nigricans ampla Poey). Amer. Mus. Novitates, no. 995.—1943. The relationship between weight and body-form in various species of scombroid fishes. Zoologica (New York), 87.

Starks, E. C. 1901. Synonomy of the fish skeleton. Proc. Washington Acad. Sci., 507.

Thompson, D'A. W. 1942. On growth and form. New York. [1st ed., 1917.]

Fishes: Giants and Pygmies

Gudger, E. W. 1926. A study of the smallest shark-suckers (Echeneididae) on record . . . Amer. Mus. Novitates, no. 234.—1928. The smallest known specimens of the sucking-fishes . . . Amer. Mus. Novitates, no. 294.—1933. Photographs of the whale shark, the greatest of all the sharks. Scient. Monthly, 273.—1934. The largest fresh-water fishes. Nat. Hist. Mag., 282.—1937. Bathytoshia, the giant stingaree of Australia . . . Australian Mus. Mag., 205.—1941a. The quest for the smallest fish. Nat. Hist. Mag., 216.—1941b. The whale shark unafraid . . . Amer. Nat., 550.—1943. The giant fresh-water fishes of South America. Scient. Monthly, 500.—1944. The giant freshwater perch of Africa. Ibid., 269.—1945a. The giant freshwater fishes of Asia. Jour. Bombay Nat. Hist. Soc., 374.—1945b. Giant freshwater fish of Europe. The Field, Aug. 4.

Herre, A. W. 1929. The smallest living vertebrate. Science, 329.

Te Winkel, L. E. 1935. A study of Mistichthys luzonensis with special reference to conditions correlated with reduced size. Jour. Morph., 463.

Fishes:

Head, Lateral Line Canals, Skull, Brain, Cranial Nerves

[For diagnostic skull characters, see fish orders and families below.]

Allis, E. P. Jr. 1888. The anatomy and development of the lateral line system in Amia calva. Jour. Morph., 463.—1895. The cranial muscles and cranial and first spinal nerves in Amia calva. [I.] Ibid., 485.—1897. [II.] Ibid., 487.—1898. On the morphology of certain of the bones of the cheek and snout of Amia calva. Ibid., 425.—1899. On certain homologies of the squamosal, intercalar, exoccipitale and extrascapular bones of Amia calva. Anat. Anz., 49.—1900. The premaxillary and maxillary bones, and the maxillary and mandibular breathing valves of Polypterus bichir. Ibid., 257.—1903. The skull and the cranial and first spinal muscles and nerves in Scomber scomber. Jour. Morph., 45.—1905. The laterosensory canals and related bones in fishes. Intern. Monats. Anat. Physiol., 401.—1909. The cranial anatomy of the mail-cheeked fishes. Zoologica, Bd. 22.—1913. The homologies of the ethmoidal region of the selachian skull. Anat. Anz., 322.—1914. The pituitary fossa and trigeminofacialis chamber in selachians. Ibid., 225.—1917. The lips and the nasal apertures in the gnathostome fishes and their homologues in the higher vertebrates. Proc. Nat. Acad. Sci., 73.—1919a. The lips and the nasal apertures in the gnathostome fishes. Jour. Morph., 145.—1919b. The myodome and trigemino-facialis chamber of fishes and the corresponding cavities in higher vertebrates. Ibid., 207.—1919c. The homologies of the maxillary and vomer bones of Polypterus. Amer. Jour. Anat., 349.—1922. The cranial anatomy of Polypterus . . . Jour. Anat., 189.—1928. Concerning the pituitary fossa, the myodome and the trigemino-facialis chamber in recent gnathostome fishes. Ibid., 95.—1935. On a general pattern of arrangement of the cranial roofing bones in fishes. Ibid., 233.

Beer, G. R. de. 1924. Studies on the vertebrate head. Part I. Fish. Quart. Jour. Micros. Sci., 287.—1925. Contributions to the development of the skull in sturgeons. Ibid., 671.

Bridge, T. W. 1877. The cranial osteology of Amia calva. Jour. Anat. Physiol., 605.

Cuvier, G. and A. Valenciennes. 1828. Histoire naturelle des poissons. Vol. I. Paris. [Early history of research on comparative anatomy of the skull of vertebrates.]

Derscheid, J. M. 1923. Contributions à la morphologie cephalique des vertébrés. A. Structure de l'organe olfactif chez les poissons. I. Osteichthyes, Teleostei, Malacopterygii. Ann. Soc. Roy. Zool. d. Belgique, 79.

De Villiers, C. 1948. Recherches sur le crane dermique des teleosteans. Ann. Paleont. (1947–1948), 3.

EDGEWORTH, F. H. 1925–1926. On the hyomandibula of Selachii, Teleostomi and Ceratodus. Jour. Anat., 173.

GREGORY, W. K. 1933. Fish skulls: A study of the evolution of natural mechanisms. Trans. Amer. Philos. Soc., 75. [With bibliography of circa 350 references, chiefly on fish skulls.]—1935a. On the evolution of the skulls of vertebrates with special reference to heritable changes in proportional diameters (anisomerism). Proc. Nat. Acad. Sci., 1.—1935b. "Williston's Law" relating to the evolution of skull bones in the vertebrates (with Marcelle Roigneau et al). Amer. Jour. Phys. Anthrop., 123. [Williston was antedated by Stromer (1912) in the statement of this principle.]

HERRICK, C. J. 1899. The cranial and first spinal nerves of Menidia: A contribution upon the nerve components of the bony fishes. Jour. Comp. Neurol., 153.

HOLMGREN, N. 1942. General morphology of the lateral sensory line system of the head in fish. Kungl. Svenska Vetenskapsakad. Handl., 1.

HOLMGREN, N. and E. STENSIÖ. 1936. Kranium und Visceralskelett der Akranier, Cyclostomen und Fische. Handb. d. vergleich. Anat., Bd. 4, 233.

HUBBS, C. L. 1919. A comparative study of the bones forming the opercular series of fishes. Jour. Morph., 61.

JAEKEL, O. 1927. Der Kopf der Wirbeltiere. Ergebn. Anat. u. Entwick., 815.

KAPPERS, C. U. A. and E. B. D. FORTUYN. 1921. Vergl. Anat. des Nervensystem. 2 vols. Haarlem.

KESTEVEN, H. L. 1925. A third contribution on the homologies of the parasphenoid, ectopterygoid and pterygoid bones and of the metapterygoid. Jour. and Proc. Roy. Soc. New South Wales, 41.—1925–1931. Contributions to the cranial osteology of the fishes. Parts I–VII. Records of the Australian Museum: I, vol. 14, 271 [Tandanus]; II, vol. 15, 132 [maxillae in eels]; III, vol. "14" (16), 201 [nomenclature]; IV, vol. "14" (16), 208 [Platycephalus]; V, vol. "14" (16), 233 [maxillo-ethmoid articulations]; VI, vol. 16, 316 [Pagrosomus]; VII, vol. 18, 236 [Neoceratodus].

KINGSBURY, B. F. 1926. Branchiomerism and the theory of head segmentation. Jour. Morph., 83

MOY-THOMAS, J. A. 1938. The problem of the evolution of the dermal bones in fishes. In: Essays on Evolution, 305. Oxford.—1941. Development of the frontal bones of the rainbow trout. Nature, 681 [cf. Westoll, 1941].

OMARKHAN, M. 1948. The morphology of the chondrocranium of Gymnarchus niloticus. Jour. Linnean Soc. London, June, 452.

PARKER, G. H. 1903. The sense of hearing in fishes. Amer. Nat., 185.

PARKER, W. K. 1873. On the structure and development of the skull in the salmon (Salmo salar, L.). Philos. Trans. Roy. Soc. London, 95.—1881. On the structure and development of the skull in sturgeons . . . Proc. Roy. Soc. London, 142.—1883. Idem. Philos. Trans. Roy. Soc. London, 139.—1882. On the development of the skull in Lepidosteus osseus. Proc. Roy. Soc. London, 107.—1883. Idem. Philos. Trans. Roy. Soc. London, 443.

PARRINGTON, F. R. 1949. A theory of the relations of lateral lines to dermal bones. Proc. Zool. Soc., vol. 119, part I, 65.

PEHRSON, T. 1940. The development of dermal bones in the skull of Amia calva. Acta Zool., 1.

SEWERTZOFF, A. N. 1899. Die Entwickelung des Selachierschadels: Ein Beitrag zur Theorie der Korrelativen Entwickelung. Festschr. für Carl von Kupffer, 281.—1928. The head skeleton and muscles of Acipenser ruthenus. Acta Zool., 193.

STARKS, E. C. 1901. Synonomy of the fish skeleton. Proc. Washington Acad. Sci., 507.—1908. On the orbitosphenoid in some fishes. Science, 413.—1926. Bones of the ethmoid region of the fish skull. Stanford Univ. Publ., Biol. Sci., vol. 4, 139.

STENSIÖ, E. A. 1947. The sensory lines and dermal bones of the cheek in fishes and amphibians. Kungl. Svenska Vetenskapsakad. Handl., no. 3.

SWINNERTON, H. H. 1902. A contribution to the morphology of the teleostean head skeleton . . . Quart. Jour. Micros. Sci., 503.

WESTOLL, T. S. 1941. Latero-sensory canals and dermal bones. Nature, 168. [Cf. Moy-Thomas, 1941.]

Fishes: Inner Ears, Otoliths

FROST, G. A. 1925–1930. A comparative study of the otoliths of the neopterygian fishes. Ann. Mag. Nat. Hist. (9), vols. 14, 15, 16, 17, 18, 19, 20; (10), vols. 1, 2, 4, 5.

Fishes: Habits, Behavior, Response

BREDER, C. M., JR. 1929. Report on synentognath habits and development. Carnegie Inst. Year Book (1928–1929), 279.—1932a. An interesting scorpion fish [Zebra-fish, Pterois]. Bull. N.Y. Zool. Soc., 30.—1932b. On the habits and development of certain Atlantic Synentognathi. Carnegie Inst. Washington Publ. No. 435.—1934. An experimental study of the reproductive habits and life history of the cichlid fish, Aequidens latifrons (Steindachner). Zoologica (N. Y.), 1.—1935a. The reproductive habits of the common catfish, Ameiurus nebulosus (Le Sueur) . . . Ibid., 143.—1935b. Sex recognition in the guppy, Lebistes reticulatus Peters. Ibid., 187.

BREDER, C. M., JR. and F. HALPERN. 1946. Innate and acquired behavior affecting the aggregation of fishes. Physiol. Zoology, 154.

MORROW, J. E. 1948. Schooling behavior in fishes. Quart. Rev. Biol., 27.

PARR, A. E. 1927. A contribution to the theoretical analysis of the schooling behavior among fishes. Occasional Papers, Bingham Oceanogr. Coll., 1. —1931. Sex dimorphism and schooling behavior among fishes. Amer. Nat., 173.

Fishes:

Visceral Arches, Gills and Accessory Respiratory Organs, Hyoid and Opercular Region, Jaws and Teeth, Musculature

ADAMS, L. A. 1919. A memoir on the phylogeny of the jaw muscles in recent and fossil vertebrates. Ann. N. Y. Acad. Sci., 51.

ALLIS, E. P., JR. 1895. The cranial muscles and cranial and first spinal nerves in Amia calva. I. Jour. Morph., 485.—1897. II. Ibid., 487.—1915. The homologies of the hyomandibula of the gnathostome fishes. Ibid., 563.—1916. The so-called labial cartilages of Raia clavata. Quart. Jour. Micros. Soc., 95.—1917a. The homologies of the muscles related to the visceral arches of the gnathostome fishes. Ibid., 303.—1917b. The lips and the nasal apertures in the gnathostome fishes and their homologues in the higher vertebrates. Proc. Nat. Acad. Sci., 73.—1918. On the origin of the hyomandibula of the Teleostomi. Anat. Rec., 73.—1919. On the homologies of the squamosal bone of fishes. Ibid., no. 2.—1920. The constrictor muscles of the branchial arches in Acanthias blainvillii. Jour. Anat., 222.—1925. On the origin of the V-shaped branchial arch in the Teleostomi. Proc. Zool. Soc. London, 75.—1928. Concerning the homologies of the hyomandibula and preoperculum. Jour. Anat., 198.

CARTER, G. S. 1935. Respiratory adaptations of the fishes of the forest waters . . . Linnean Soc. London, Jour. Zool., 219. [Accessory respiratory organs of electric "eel," Gymnotus, and Plecostomus.]

EATON, T. H., JR. 1939. Suggestions on the evolution of the operculum of fishes. Copeia, 42.

GAUPP, E. 1904. Das Hyobranchialskelet der Wirbeltiere. Ergebn. d. Anat. u. Entwick. (1905), 808.

HOFER, H. 1945. Zur Kenntnis der Suspensionsformen des Kieferbogens . . . bei den Knochenfischen. Zool. Jahrb. (Anat.), 321.

HOLMQUIST, O. 1910. Der Musculus Protractor Hyoidei (Geniohyoideus Auctt.) und der Senkungsmechanismus des Unterkiefers bei den Knochenfischen. Lunds Univ. Arsskrift, Bd. 6, Nr. 6.

HUBBS, C. L. 1919. A comparative study of the bones forming the opercular series of fishes. Jour. Morph., 61.

LUBOSCH, W. 1923. Der Kieferapparat der Scariden und die Frage der Streptognathie. Anat. Anz., 10.

STARKS, E. C. 1916. The sesamoid articular: A bone in the mandible of fishes. Leland Stanford Junior Univ. Pub.

TRETJAKOFF, D. 1926. Das Skelett und die Musculatur im Kopf des Flussneunauges. Zeitsch. Wiss. Zool., 267.

VETTER, B. 1874. Untersuch. zur vergl. Anat. der Kiemen-und-Kiefermusculatur der Fische. I. Jena. Zeitschr. f. Naturw., N. F., vol. 1, 405.—1878. II. Ibid., vol. 5, 431.

Fishes:

Aerating Organs, Gills, Swim-bladder, Suprabranchial Organs, Oral Valves

AKITA, Y. K. 1936. Studies on the physiology of swimbladder. Jour. Faculty of Science, Tokyo Imp. Univ., 111.

GUDGER, E. W. 1946. Oral breathing valves in fishes. Jour. Morph., 263.

JAEKEL, O. 1927. Über die Atemorgane der Wirbeltiere. Palaeont. Zeitsch., 250.

MUIR, E. H. 1925. A contribution to the anatomy and physiology of the air-bladder and Weberian ossicles in Cyprinidae. Prof. Roy. Soc. London, 545.

Fishes: Vertebrae

GABRIEL, M. L. 1944. Factors affecting the number and form of vertebrae in Fundulus heteroclitus. Jour. Exp. Zool., 105.

GADOW, H. F. 1933. The evolution of the vertebral column . . . Cambridge Univ. Press.

FORD, E. 1937. Vertebral variation in teleostean fishes. Jour. Marine Biol. Assoc. United Kingdom, 1.

JORDAN, D. S. 1919. Temperature and vertebrae in fishes. Science, 336.

Fishes: Fins, Median and Paired, Girdles

BRAUS, H. 1901. Die Muskeln und Nerven der Ceratodusflosse. Zool. Forsch., Denkschr. d. Med. Nat. Gesell. Hena, Bd. 4.

DEAN, B. 1896. The fin-fold origin of the paired limbs, in the light of the ptychopterygia of Palaeozoic sharks. Anat. Anz., 673.—1902a. Biometric evidence in the problem of the paired limbs of the vertebrates. Amer. Nat., 837.—1902b. Historical evidence as to the origin of the paired limbs of vertebrates. Ibid., 767.—1906. Notes on the living specimens of the Australian lung-fish, Ceratodus forsteri, in the Zoological Society's collection. Proc. Zool. Soc. London, 168. [Postures and movements of paired fins.]

EATON, T. H. 1945. Skeletal supports of the median fins of fishes. Jour. Morph., 193.

FÜRBRINGER, M. 1873. Zur vergl. Anat. der Schultermuskeln. Jenaische Zeitschr., 237.—1900. Zur vergl. Anat. der Brustschulter Apparates und der Schultermuskeln. Ibid., 215.

GEGENBAUR, C. 1864–1872. Untersuch. zur vergl. Anat. der Wirbelthiere. Heft 2. Schultergürtel der Wirbelthiere. Brustflosse der Fische. (1865)

GOODRICH, E. S. 1906. Notes on the development, structure and origin of the median and paired fins of fish. Quart. Jour. Micros. Sci., 333.

GREGORY, W. K. and H. C. RAVEN. 1941. Studies on the origin and early evolution of paired fins and limbs. Parts I–IV. Ann. N. Y. Acad. Sci., 273.

GRENHOLM, A. 1923. Studien über die Flossenmuskulatur der teleostier. Uppsala Univ. Arsskrift.

HOLLISTER, G. 1936. Caudal skeleton of Bermuda shallow water fishes. I. Zoologica (N. Y.), vol. 21, 257.—1937a. Part II. Ibid., vol. 22, 265.—1937b. Part III. Ibid., 385.

HOWELL, A. B. 1933. Homology of the paired fins in fishes. Jour. Morph., 451.

KRYZANOVSKY, S. 1927. Die Entwicklung der Paarigen Flossen bei Acipenser, Amia und Lepidosteus. Acta. Zool., 277.

MAYER, P. 1886. Die unpaaren Flossen der Selachier. Mitt. Zool. Stat. Neapel., 217.

MIVART, ST. G. J. 1879. Notes on the fins of elasmobranchs, with considerations on the nature and homologues of vertebrate limbs. Trans. Zool. Soc. London, 439.

REGAN, C. TATE. 1904. The phylogeny of the Teleostomi. Ann. Mag. Nat. Hist., 329. [Evolution of median and paired fins.]

SCHMALHAUSEN, J. J. 1912. Zur Morphologie der unpaaren Flossen. I. Die Entwicklung des Skelettes und der Muskulatur der unpaaren Flossen der Fische. Zeitschr. f. wissenschaftl. Zool., Bd. 100, 509.—1913. II. Ibid., Bd. 104, 1.

SEWERTZOFF, A. N. 1926. Die Morphologie der Brustflossen der Fische. Jena. Zeitschr. f. Naturw., 343.—1934. Evolution der Bauchflossen der Fische. Zool. Jahrb., 415.

STARKS, E. C. 1930. The primary shoulder girdle of the bony fishes. Stanford Univ. Publ., 149.

SWINNERTON, H. H. 1905. A contribution to the morphology and development of the pectoral skeleton of teleosteans. Quart. Jour. Micros. Sci., 363.

THACHER, J. K. 1876. Median and paired fins, a contribution to the history of vertebrate limbs. Trans. Connecticut Acad. Sci., 281.

Fishes: Coloration

BEEBE, W. 1943. Pattern and color in the cichlid fish, Aequidens tetramerus. Zoologica (N. Y.), 13.

BEEBE, W. and JOHN TEE-VAN. 1932. New Bermuda fish . . . Zoologica (N. Y.), 109. [Coloration.]

COTT, H. B. 1941. Adaptive coloration in animals. Oxford Univ. Press.

DALTON, H. C. and H. B. GOODRICH. 1937. Chromatophore reactions in the normal and albino paradise fish, Macropodus opercularis L. Biol. Bull., 535.

GOODRICH, H. B. 1929. Mendelian inheritance in fish. Quart. Rev. Biol., 83.—1935. The development of hereditary color patterns in fish. Amer. Nat., 267.—1939. Chromatophores in relation to genetic and specific distinctions. Ibid., 198.

GOODRICH, H. B. et al. 1934. Germ cells and sex differentiation in Lebistes reticulatus. Biol. Bull., 33.

GOODRICH, H. B. and M. HEDENBURG. 1941. The cellular basis of colors in some Bermuda parrot fish . . . Jour. Morph., 493.

GOODRICH, H. B., G. A. HILL and M. S. ARRICK. 1941. The chemical identification of gene-controlled pigments in Platypoecilus and Xiphophorus and comparisons with other tropical fish. Genetics, no. 6, 573.

GOODRICH, H. B. and R. NICHOLS. 1931. The development and the regeneration of the color pattern in Brachydanio rerio. Jour. Morph., 513.

GOODRICH, H. B. and M. A. SMITH. 1937. Genetics and histology of the color pattern in the normal and albino paradise fish, Macropodus opercularis L. Biol. Bull., 527.

GORDON, M. 1927. The genetics of a viviparous top-minnow, Platypoecilus; the inheritance of two kinds of melanophores. Genetics, 253.—1928. Pigment inheritance in the Mexican killifish: Interaction of factors in Platypoecilus maculatus. Jour. Heredity, 551.—1931a. Morphology of the heritable color patterns in the Mexican killifish, Platypoecilus. Amer. Jour. Cancer, 732.—1931b. Hereditary basis of melanosis in hybrid fishes. Ibid., 1495.

LONGLEY, W. H. and S. F. HILDEBRAND. 1941. Systematic catalogue of the fishes of Tortugas, Florida; with observations on color, habits and local distribution. Carnegie Inst. Washington Publ. 535.

PARKER, G. H. 1943a. Animal color changes and their neurohumors. Quart. Rev. Biol., 205.—1943b. Coloration of animals and their ability to change their tints. Scient. Monthly, 197.

SCHMIDT, J. 1919. Racial studies in fishes. II. Experimental investigations with Lebistes reticulatus (Peters) Regan. Jour. Genetics, 147.

STOCKARD, C. R. 1915. A study of wandering mesenchymal cells on the living yolk-sac and their developmental products: chromatophores, vascular endothelium and blood cells [Studies on Fundulus]. Amer. Jour. Anat., 525.

TOWNSEND, C. H. 1929. Records of changes in color among fishes. Zoologica (N. Y.), 321.

Fishes: Luminescence

BOULENGER, G. A. 1910. Fishes (Systematic account of Teleostei). Cambridge Nat. Hist. London. [Luminous organs, p. 624.]

BRIDGE, T. W. 1910. Fishes (exclusive of the systematic account of Teleostei). Cambridge Nat. Hist. London. [Luminous organs, pp. 178–181; relation of chromatophores to scales, pp. 166–170; spines and scales, p. 183 ff.]

HARVEY, E. N. 1915. Studies on light production by luminous bacteria. Amer. Jour. Physiol., 230. —1920. The nature of animal light. Philadelphia.—1940. Living light. Princeton Univ. Press.

HARVEY, E. N. and others. 1948. Bioluminescence. Ann. N. Y. Acad. Sci., 329.

Deep-Sea (Abyssal) Fishes

BEEBE, W. 1933. Deep-sea stomiatoid fishes . . . Copeia, no. 4, 160.—1934. Deep-sea fishes of the Bermuda Oceanographic Expedition. Family Idiacanthidae. Zoologica (N. Y.), 149.

BRAUER, A. 1908. Die Tiefseefische. In: Wiss. Ergebn. d. Deutsch. Tiefsee-Exped. a.d. Dampfer "Valdivia" (Ed. Carl Chun), vol. 15, 1.

DOLLO, L. 1904. Poissons. Expédition Antarctique Belge. Zoologie. Anvers. 4to.

GARSTANG, W. 1932. The phyletic classification of the Teleostei. Proc. Leeds Philos. Lit. Soc., 240. [Unites stomiatoids and scopeloids; = Lampadephori.]

GILBERT, C. H. 1905. The deep-sea fishes of the Hawaiian Islands. Bull. U. S. Fish Commission, 1903 (1905), 575.

GOODE, G. B. and T. H. BEAN. 1895. Oceanic ichthyology: A treatise on the deep-sea and pelagic fishes of the world. . . . Special Bull. U. S. Nat. Mus.; Mem. Mus. Comp. Zool. Harvard Coll. (1896).

GREGORY, W. K. and G. M. CONRAD. 1936. Pictorial phylogenies of deep-sea Isospondyli and Iniomi. Copeia, no. 1, 21.

MURRAY, SIR J. and J. HJORT. 1912. The depths of the ocean. London.

NORMAN, J. R. 1929. The teleostean fishes of the Family Chiasmodontidae. Ann. Mag. Nat. Hist., 529.

NUSBAUM, H. J. 1923. Études d'anat. comp. sur les Poissons provenant des campagnes scientifique de S.A.S. le Prince de Monaco. Vol. 65, 1.

PARR, A. E. 1927a. Ceratioidea. Bull. Bingham Oceanogr. Coll., vol. 3, art. 1.—1927b. The stomiatoid fishes of the suborder Gymnophotodermi . . . Ibid., art. 2.—1929. A contribution to the osteology and classification of the Orders Iniomi and Xenoberyces . . . Occ. Papers, Bingham Oceanogr. Coll., Peabody Mus. Nat. Hist., Yale Univ., no. 2.—1930a. On the osteology and classification of the pediculate fishes . . . Ibid., no. 3.—1930b. On the probable identity, life-history and anatomy of the free-living and attached males of the ceratioid fishes. Copeia, 129.—1930c. A note on the classification of the stomiatoid fishes. Ibid., 136.

PARKER, T. J. 1886. Studies in New Zealand ichthyology. I. On the skeleton of Regalecus argenteus. Trans. Zool. Soc. London, 5.

REGAN, C. T. 1902. On the classification of the fishes of the suborder Plectognathi . . . Proc. Zool. Soc. London, 284.—1907. On the anatomy, classification and systematic position of the teleostean fishes of the suborder Allotriognathi. Ibid., 634.—1911. The anatomy and classification of the teleostean fishes of the orders Berycomorphi and Xenoberyces. Ann. Mag. Nat. Hist., vol. 7, 1.—1912. The anatomy and classification of the teleostean fishes of the Order Lyomeri. Ibid., vol. 10, 347.—1923. The classification of the stomiatoid fishes. Ibid., vol. 11, 612. [Likeness between Photichthys and Elops, p. 613.]—1924. The morphology of a rare oceanic fish, Stylophorus chordatus Shaw . . . Proc. Roy. Soc. London, 193.—1925. New ceratioid fishes from the North Atlantic, the Caribbean Sea and the Gulf of Panama collected by the "Dana." Ann. Mag. Nat. Hist., 561.—1926. The pediculate fishes of the suborder Ceratioidea. Oceanogr. Repts. edited by the "Dana" Committee, no. 2.

REGAN, C. T. and E. TREWAVAS. 1929. The fishes of the families Astronesthidae and Chauliodontidae. Oceanogr. Repts. edited by the "Dana" Committee, no. 5.—1930. The fishes of the families Stomiatidae and Malacosteidae. Ibid., no. 6.—1932. Deep-sea angler fishes (Ceratioidea). Carlsberg Foundation's Oceanogr. Exped. . . . Copenhagen.

TCHERNAVIN, V. 1946. The ventral part of the hyoid gill-slit and a mandibular "operculum" in some bony fishes. Nature, 303. Special characteristics of the Lyomeri and other abyssal fishes.—1947. Six specimens of Lyomeri in the British Museum (with notes on the skeleton of Lyomeri). Jour. Linnean Soc. London, 287.

TREWAVAS, E. 1933. On the structure of two oceanic fishes, Cyema atrum Gunther and Opisthoproctus soleatus Vaillant. Proc. Zool. Soc. London, part 3, 601.

TROTTER, E. S. 1926. Brotulid fishes from the Arcturus Oceanographic Expedition. Zoologica (N. Y.), 107.

WOODWARD, A. S. 1898. The antiquity of the deep-sea fauna. Natural Science, 257.

ZUGMAYER, E. 1911. Poissons provenant des campagnes du Yacht Princess Alice (1901, 1910). Résult. Campagn. Scient., Monaco, vol. 35.

Blind Fishes

BREDER, C. M. 1944. Ocular anatomy and light sensitivity studies on the blind fish from Cueva de los Sabinos, Mexico. Zoologica (N. Y.), 131.

Electric Fishes

BALLOWITZ, E. 1899. Das elektrische Organ des afrikanischen Zitterweises (Malapterurus electricus). Jena.

COATES, C. W. 1946. Electrical characteristics of electric tissue. Ann. N. Y. Acad. Sci., art. 4.

COATES, C. W. and R. T. COX. 1936. Preliminary note on the nature of the electrical discharges of the electric eel, Electrophorus electricus (Linnaeus). Zoologica (N. Y.), vol. 21, part 11, 125.

COATES, C. W., R. T. COX and L. P. GRANATH. 1937. The electric discharge of the electric eel, Electrophorus electricus (Linnaeus), Ibid., vol. 22, part 1, 1.

DAHLGREN, U. 1906. The electric organ of the stargazer, Astroscopus (Brevoort). Anat. Anz., 387.

ELLIS, M. M. 1913. The gymnotid eels of Tropical America. Mem. Carnegie Mus., vol. 6, no. 3, 109.

EWART, J. C. 1888. The electric organ of the skate . . . Raia radiata. Philos. Trans. Roy. Soc. London, 539.

FRITSCH, G. T. 1887, 1890. Die elektrischen Fische. 1. Malapterurus electricus. 2. Die Torpedinideen. Leipzig. Folio.

NACHMANSOHN, D., R. T. COX, C. W. COATES and A. L. MACHADO. 1942. Action potential and enzyme activity in the electric organ of Electrophorus electricus (Linnaeus). Jour. Neurophysiol., 499.

WHITE, E. G. 1918. The origin of the electric organs in Astroscopus guttatus. Carnegie Inst. Washington Publ. no. 252, 139.

Fishes: Faunal Works (Recent and Fossil)

ANDREWS, C. W. 1906. A descriptive catalogue of the Tertiary Vertebrata of the Fayum, Egypt. London. 4to. [Catfishes.]

BEAN, T. H. 1903. Catalogue of the fishes of New York. N. Y. State Mus., Bull. 60. [Amia, pp. 73–76.]

BEEBE, W. 1926. The Arcturus adventure: An account of the New York Zoological Society's First Oceanographic Expedition. New York.—1942. Atlantic and Pacific fishes of the genus Dixonina. Zoologica (N. Y.), vol. 27, no. 8, 43.

BEEBE, W. and J. TEE-VAN. 1928. The fishes of Port-au-Prince Bay, Haiti . . . Zoologica (N. Y.), vol. 10, no. 1, 1.—1933. Field book of the shore fishes of Bermuda. New York.

BIGELOW, H. B. and W. W. WELSH. 1925. Fishes of the Gulf of Maine. Bull. U. S. Bureau of Fisheries (for 1924).

BOULANGER, G. A. 1909–1916. Catalogue of the fresh-water fishes of Africa in the British Museum (Natural History). Vols. 1–4. London.

BREDER, C. M., JR. 1929. Field book of marine fishes of the Atlantic Coast. New York and London.

COPE, E. D. 1884. Vertebrata of the Tertiary formations of the West. Book I. Rept. U. S. Geol. Surv. of the Territories, III. [Fishes, 1–100.]

CROOK, A. R. 1892. Ueber einige fossile Knochenfische aus der mittleren Kreide von Kansas. Palaeontographica, 108.

DAVID, L. R. 1943. Miocene fishes of Southern California. Geol. Soc. Amer. Special Papers, no. 43.

DAY, F. 1865. The fishes of Malabar. London.—1875–1878. The fishes of India . . . London. 4to.—1880–1884. The fishes of Great Britain and Ireland. London and Edinburgh.

DOLLO, L. and R. STORMS. 1888. Sur les téléostéens du Rupélieu. Zool. Anz., vol. 11, 265.

EVERMANN, B. W. 1902a. [Fishes of Porto Rico.] Bull. U. S. Fish Comm. 1900 (1902), vol. 20, part. 1.—1902b. Summary of the scientific results of the Fish Commission Expedition to Porto Rico. Ibid.

FOWLER, H. W. 1936. The marine fishes of West Africa . . . Bull. Amer. Mus. Nat. Hist., vol. 70, part 2, 607.—1937. Zool. results Third De Schauensee Siamese Expedition. Part VIII.—Fishes . . . Proc. Acad. Nat. Sci. Philadelphia, 125.

GARMAN, S. 1899. Repts. on an exploration of the West Coast of Mexico, Central and South America, and off the Galapagos Islands, in charge of Alexander Agassiz . . . [Part I] The fishes. Mem. Mus. Comp. Zool. Harvard Coll., vol. 24.

GOODE, G. and T. H. BEAN. 1896. Oceanic ichthyology . . . Ibid., vol. 22.

GREGORY, W. K. 1923. A Jurassic fish fauna from western Cuba, with an arrangement of the families of holostean ganoid fishes. Bull. Amer. Mus. Nat. Hist., 223.

HAY, O. P. 1903a. On certain genera and species of North American Cretaceous mactinopterous fishes. Bull. Amer. Mus. Nat. Hist., vol. 19, 1.—1903b. On a collection of Upper Cretaceous fishes from Mount Lebanon, Syria. Ibid., vol. 19, 395.

JORDAN, D. S. and B. W. EVERMANN. 1896–1900. The fishes of North and Middle America. . . . Bull. U. S. Nat. Mus., vol. 47, parts 1–4.—1905. The aquatic resources of the Hawaiian Islands. Part I. The shore fishes. Bull. U. S. Fish Comm., 1903 (1905).

LONGLEY, W. H. and S. F. HILDEBRAND. 1941. Systematic catalogue of the fishes of Tortugas, Florida; with observations on color, habits and local distribution. Carnegie Inst. Washington Publ. 535.

MCCULLOCH, A. R. 1921. Check-list of the fish and fish-like animals of New South Wales. Australian Zoologist, 24.—1922. Ibid., 86.—1934. Check-list of the fish and fish-like animals of New South Wales. Parts 1–3. 3rd ed.

MEEK, S. E. and S. F. HILDEBRAND. 1923. The marine fishes of Panama. Part I. Field Mus. Nat. Hist., Publ. 215.—1925. Part II. Ibid., Publ. 249.—Part III. Ibid., Publ. 709.

NICHOLS, J. T. 1918. Fishes of the vicinity of New York City. Amer. Mus. Nat. Hist., Handbook ser. no. 7.—1943. The fresh-water fishes of China. Natural History of Central Asia, vol. 9, Amer. Mus. Nat. Hist.

NICHOLAS, J. T. and C. M. BREDER, JR. 1927. The marine fishes of New York and southern New England. Zoologica (N. Y.).

NICHOLS, J. T. and L. GRISCOM. 1917. Fresh-water fishes of the Congo Basin . . . Bull. Amer. Mus. Nat. Hist., 653.

REGAN, C. T. 1914. The Antarctic fishes of the Scottish National Antarctic Expedition. Trans. Roy. Soc. Edinburgh, 229.

ROUGHLEY, T. C. 1916. Fishes of Australia and their technology. Sydney.

SAINT-SEINE, P. 1949. Les poissons des calcaires lithographique de Cerin (Ain). Nouv. Arch. du Mus. d'Histoire nat. de Lyon, fasc. 2.

SCHLAIKJER, E. M. 1937. New fishes from the Continental Tertiary of Alaska. Bull. Amer. Mus. Nat. Hist., 1. [Phylogeny of the Centrarchidae.]

SCHRENKEISEN, R. 1938. Field book of fresh-water fishes of North America north of Mexico. New York.

SCHULTZ, L. P. 1943. Fishes of the Phoenix and Samoan Islands . . . Smiths. Inst., U. S. Nat. Mus. Bull.

STENSIÖ, E. A. 1921. Triassic fishes from Spitzbergen. Part I.—1925. Part II. Kungl. Svenska Vetenskaps-akad. Handl. (3).—1931. Upper Devonian vertebrates from East Greeland . . . Meddel. om Groenland, Bd. 86.

WEBER, M. and L. F. DE BEAUFORT. 1911–1936. The fishes of the Indo-Australian Archipelago. Vols. I–VII. Leiden. Vol. VIII, 1940, by L. F. de Beaufort.

WOODWARD, A. S. 1895. On the fossil fishes of the Upper Lias of Whitby. Proc. Yorks. Geol. Polytech. Soc. [Leptolepis, a primitive teleost.] —1902–1912. The fossil fishes of the English Chalk. Parts I–VII. Monogr. Palaeontographical Soc., vol. 56, 57, 61, 62, 63, 64, 65.—1916–1918. The fossil fishes of the English Wealden and Purbeck formations. Parts I–III. Ibid., vols. 69, 70, 71.—1925. . . . The fossil fish-fauna of the English Purbeck beds. Geol. Mag., vol. 2, 145.—1926. The fossil fishes of the Old Red Sandstone of the Shetland Islands. Trans. Roy. Soc. Edinburgh, vol. 54, 567.—1942. Some new and little-known Upper Cretaceous fishes from Mount Lebanon. Ann. Mag. Nat. Hist., 537.

? Fishes: Conodonts

AMSDEM, T. W. and A. K. MILLER. 1942. Ordovician conodonts from the Bighorn Mountains of Wyoming. Jour. Paleont., vol. 16, no. 3, 301.

ELLISON, S. 1944. The composition of conodonts. Jour. Paleont., vol. 18, no. 2, 133. [Bearing on classification; may be fish.]

HASS, W. H. 1941. Morphology of conodonts. Jour. Paleont., vol. 15, no. 1, 71.

HOLMES, G. B. 1928. A bibliography of the conodonts with descriptions of early Mississippian species. Proc. U. S. Nat. Mus., vol. 72, art. 5.

SCHMIDT, H. 1934. Conodonten-Funde in ursprünglichen Zusammenhang. Palaeont. Zeitschr., Bd. 16, 76. [Regards conodonts as denticulations on the gill and hyoid arches of certain primitive fishes . . .]

SCOTT, H. W. 1934. The zoological relationships of the conodonts. Jour. Paleont., vol. 8, no. 4, 448. [Relationships of conodonts still uncertain.]—1942. Conodont assemblages from the Heath formation, Montana. Jour. Paleont., vol. 16, no. 3, 293. [Jaw apparatus could operate equally well for fish or annelids.]

ULRICH, E. O. and R. S. BASSLER. 1926 (1927). A classification of the tooth-like fossils, conodonts . . . Proc. U. S. Nat. Mus., vol. 68, 1.

Agnatha: Ostracoderms

BROILI, F. 1933. Die Gattung Pteraspis in den Hunsrückschiefern. Sitz. d. Bayerisch. Akad. d. Wissensch., 1.

BROTZEN, F. 1933. Weigeltaspis nov. gen. und die Phylogenie der panzertragenden Heterostraci. Central. f. Min., etc., 648.—1935. Beiträge zur Vertebratenfauna des westpodolischen Silurs und Devons . . . Arkiv f. Zoologi, Bd. 28 A, no. 22. [Protaspis, a very primitive pteraspid; Brachipteraspis suggests connection between pteraspids and drepanaspids.]

BRYANT, W. L. 1926. On the structure of Palaeaspis . . . Proc. Amer. Philos. Soc., 256.—1932. Lower Devonian fishes of Bear Tooth Butte, Wyoming. Ibid., 225.—1933–1934. The fish fauna of Bear Tooth Butte, Wyoming. Parts I–III. Ibid., vols. 72, 73. [Ostracoderms, placoderms of Lower Devonian age.]—1935. Cryptaspis and other Lower Devonian fossil fishes from Beartooth Butte, Wyoming. Ibid., vol. 75, 111.—1936. A study of the oldest known vertebrates, Astraspis and Eriptychius. Ibid., vol. 76, 409.

BULMAN, O. M. B. 1930. On the general morphology of the anaspid Lasanius Traquair. Mag. Nat. Hist., 354.

EASTMAN, C. R. 1917. Fossil fishes in the collection of the United States National Museum. Proc. U. S. Nat. Mus., 235. [Astraspis desiderata Walcott, p. 238.]

GREGORY, W. K. 1935. On the evolution of the skulls of vertebrates with special reference to heritable changes in proportional diameters (anisomerism). Proc. Nat. Acad. Sci., vol. 21, 1. Ostracoderms.

HEINTZ, A. 1933. Neuer Fund von Archegonaspis in einem obersilurischen Geschiebe. Zeitsch. f. Geschiebeforschung, 123.—1938. Über die ältesten bekannten Wirbeltiere. Die Naturwissenschaften, vol. 26, 49.—1939. Cephalaspids from Downtonian of Norway. Det. Norske Videnskaps-Akad. i Oslo. I, no. 5, 1.

HOLMGREN, N. 1943. Agnather och Gnathostomer. Kungl. Svenska Vetenskaps-Akad. Arsbok, 337.

KIAER, J. 1924. The Downtonian fauna of Norway. I. Anaspida. Det. Norske Vidensk.-Akad. i Oslo. I, no. 6, 1.—1928. The structure of the mouth of the oldest known vertebrates, pteraspids and cephalaspids. Palaeobiologica, 117.—1930. Ctenaspis: a new genus of cyathaspidian fishes . . . Skrift. om Svalbard og Ishavet Undersök., no. 33, 1.—1932a. New coelolepids from the Upper Silurian of Oesel (Esthonia). Archiv f. d. Naturkunde Estlands, X Bd., 3 Lief. —1932b. The Downtonian and Devonian vertebrates of Spitsbergen. IV. Suborder Cyathaspida . . . Skrift. om Svalbard og Ishavet Undersök., no. 52, 1.

KIAER, J. and A. HEINTZ. 1935. The Downtonian and Devonian vertebrates of Spitsbergen. V. Suborder Cyathaspida. Part I. Tribe Poraspidei. Ibid., no. 40, 1.

LANKESTER, E. R. 1868. The Cephalaspidae. In: A monograph on the fishes of the Old Red Sandstone of Britain (Powrie and Lankester). Part I. Palaeontograph. Soc. 4to.

PANDER, C. H. 1856. Monographie der Fossilen Fische des Silurischen Systems der Russisch-Baltischen Gouvernements. Text (4to) and atlas (folio). St. Petersbourg.

PATTEN, W. 1903. On the structure and classification of the Tremataspidae. Mem. de l'Academie Imp. des Sciences de St. Petersbourg, vol. 13, no. 5.—1931. New ostracoderms from Oesel. Science, vol. 73.—1935a. The ostracoderm genus Dartmuthia Patten. Amer. Jour. Sci., 323. —1935b. Oeselaspis, a new genus of ostracoderm. Ibid., 453.—1935c. The ostracoderm Order Osteostraci. Science, 282.—1938. The Tremataspidae. Parts I, II. Amer. Jour. Sci., 172.

ROHON, J. F. 1889. Ueber Unter-Silurische Fische. Mélanges Geol.-Paléont., Acad. Imp. Sci., St. Petersbourg, vol. 33, 7.—1892, 1893. Die Obersilurische Fische von Oesel. I. Theil. Thyestidae und Tremataspidae. Mem. Acad. Imp. d. Sci. d. St. Petersbourg. II. Theil. Selachii, Dipnoi, Ganoidei, Pteraspidae u. Cephalaspidae. Ibid.

STENSIÖ, E. A. 1927. The Downtonian and Devonian vertebrates of Spitsbergen. Part I. Family Cephalaspidae. Skrift. om Svalbard . . . , no. 12, 1. [Bone primary, cartilaginous skeleton secondary, pp. 30, 333, 334, 374. Relationship between Osteostraci and Anaspida, cephalaspids and lampreys, 343–373.]—1932. The cephalaspids of Great Britain. Brit. Mus. (Nat. Hist.). London.—1939. A new anaspid from the Upper Devonian of Scaumenac Bay in Canada . . . Kungl. Svenska Vetenskaps.-Akad. Handl., Bd. 18, no. 1, 1. [Endeiolepis, n.g., fig. 7.]

STETSON, H. C. 1927. Lasanius and the problem of vertebrate origin. Jour. Geol., 247.—1928a. A restoration of the anaspid Birkenia elegans Traquair. Ibid., 458.—1928b. A new American Thelodus. Amer. Jour. Sci., 221.

TRAQUAIR, R. H. 1898a. On Thelodus pagei, Powrie. Trans. Roy. Soc. Edinburgh, vol. 39, 595. —1898b. Report on fossil fishes collected . . . in the Silurian . . . of Scotland. Ibid., 827. [Thelodus, Ateleaspis, Anaspida.]—1905. Supplementary report on Silurian fossil fishes. Ibid., 879. [Thelodus, Lanarkia, Ateleaspis.]

ULRICH, E. O. 1911. Revision of the Palaeozoic system. Bull. Geol. Soc. Amer., 281.

WALCOTT, C. D. 1892. Preliminary notes on the discovery of a vertebrate fauna in Silurian (Ordovician) strata. Bull. Geol. Soc. Amer., 153. Astraspis desiderata Walcott, p. 166.

WESTOLL, T. S. 1945. A new cephalaspid fish from the Downtonian of Scotland, with notes on the structure and classification of ostracoderms. Trans. Roy. Soc. Edinburgh, 341.

WHITE, E. I. 1935. The ostracoderm Pteraspis Kner and the relationships of the agnathous vertebrates. Philos. Trans. Roy. Soc. London, 381.—1946. Jamoytius kerwoodi, a new chordate from the Silurian of Lanarkshire. Geol. Mag., 89. [Type of new genus, family, order: Euphanerida, of ostracoderms, without armor and with well developed fin-folds.]

WOODWARD, A. S. 1920. On certain groups of fossil fishes. Proc. Linnean Soc., London (1919–1920), 25. [Anaspida, Heterostraci, Osteostraci, Antiarchi.]

Agnatha: Cyclostomes

COPE, E. D. 1889. Synopsis of the families of Vertebrata. Amer. Nat., 849. [Characteristics of Class Agnatha (pp. 852–853); Ostracodermi with cyclostomes = Agnatha.]

DEAN, B. 1899. On the embryology of Bdellostoma stouti . . . In: Festschr. Carl von Kupffer, 221. Jena.—1900. The Devonian "lamprey," Palaeospondylus gunni Traquair, with notes on the systematic arrangement of the fish-like vertebrates. Mem. N. Y. Acad. Sci., vol. 2, 1. [Type of "Class Cycliae" Dean; but more probably a small aphetohyoidean related to Acanthodii.—C. Forster Cooper, in Parker and Haswell, vol. II, 170.]

HEINTZ, A. 1931. A new reconstruction of Dinichthys. Amer. Mus. Novitates, no. 457.

HOLMGREN, N. and E. A. STENSIÖ. 1936. Kranium und Visceralskelett der Akranier, Cyclostomen und Fische. Hand. d. vergleich. Anat., Bd. 4, 233.

HUSSAKOF, L. 1906. Studies on the Arthrodira. Mem. Amer. Mus. Nat. Hist., 105.

JOHNSTON, J. B. 1902. The brain of Petromyzon. Jour. Comp. Neurol., 1.—1905. The cranial nerve components of Petromyzon. Morph. Jahrb., 149.—1908. Additional notes on the cranial nerves of petromyzonts. Jour. Comp. Neurol., 569.

LEACH, W. J. 1946. Oxygen consumption of lampreys, with special reference to metamorphosis and phylogenetic position. Physiol. Zool., 365.

PARKER, W. K. 1884. On the skeleton of the marsipobranch fishes. Part I. The myxinoids. Part II. The lamprey. Philos. Trans. Roy. Soc. London, 373, 411.

REYNOLDS, T. E. 1931. Hydrostatics of the suctorial mouth of the lamprey. Univ. California Publ. Zool., 15.

SCOTT, W. B. 1882. Beiträge zur Entwickl. der Petromyzonten. Morph. Jahrb. 101.—1887. The embryology of Petromyzon. Jour. Morph., 253.

STOCKARD, C. R. 1906. The development of the mouth and gills in Bdellostoma stouti. Jour. Anat., 481.

Placoderms:

Antiarchi (Bothriolepids, Phyllolepids), Arthrodira, Acanthaspids (Dolichothoraci), Brachythoraci (Arthrodira sensu stricto), Ptyctodontida, Petalichthyida, Rhenanida

BROILI, F. 1929. Acanthaspiden aus dem rheinischen Unterdevon. Sitz. d. Bayerischen Akad. d. Wissensch., 143.—1930a. Neue Beobachtungen an Lunaspis. Ibid., 47.—1930b. Über Gemündina Stürtzi Traquair. Abh. d. Bayerischen Akad. d. Wissensch., 1. 4to.—1933a. Weitere Fischreste aus den Hunsrückschiefern. Sitz. d. Bayerischen Akad. d. Wissensch., 269.—1933b. Ein Macropetalichthyide aus den Hunsrückschiefern. Ibid., 417.

BRYANT, W. L. 1934. The fish fauna of Beartooth Butte, Wyoming. Part II. Pisces, Order Arthrodira, Family Acanthaspida. Proc. Amer. Philos. Soc., 127.

COPE, E. D. 1885. The position of Pterichthys in the system. Amer. Nat. 289. [Proposes Antiarcha as an order of Tunicata, p. 291.]

DEAN, B. 1896. On the vertebral column, fins and ventral armoring of Dinichthys. Trans. N. Y. Acad. Sci., 157.—1901. Palaeontological notes. I. On two new arthrodires from the Cleveland Shale of Ohio. Mem. N. Y. Acad. Sci., 85. II. On the characters of Mylostoma Newberry. Ibid., 101. III. Further notes on the relationships of the Arthrognathi. Ibid., 110.

DENISON, R. H. 1940. [Reconstruction of soft anatomy of a Devonian armored fish, Bothriolepis . . .] Dartmouth Coll. Mus. Rept., 13.—1941. The soft anatomy of Bothriolepis. Jour. Paleont., 553.

DUNKLE, D. H. and P. A. BUNGART. 1940. On one of the least known of the Cleveland Shale Arthrodira. Scient. Publ. Cleveland Mus. Nat. Hist., 29.—1942a. The infero-gnathal plates of Titanichthys. Ibid., 49.—1942b. A new genus and species of Arthrodira from the Cleveland Shale. Ibid., 65.

GROSS, W. 1931. Asterolepis ornata Eichw. und das Antiarchi-Problem. Palaeontographica: Beitr. Naturgesch. der Vorzeit. [Movements of pectoral appendages, pp. 39–46.]—1934. Der histologische Aufbau des Phyllolepidenpanzers. Centralb. f. Min., etc.—1941. Die Bothriolepis-Arten der Cellulosa-Mergel Lettlands. Kungl. Svenska Vetenskaps-Akad. Handl. (3).

HEINTZ, A. 1928. Einige Bemerkungen über den

Panzerbau bei Homosteus und Heterosteus. Det Norske Videnskaps-Akad. i Oslo, 1.—1929a. Die Downtonischen und Devonischen Vertebraten von Spitsbergen. II. Acanthaspida. Skrift. om Svalbard og Ishavet, no. 22, 1.—1929b. Die Downtonischen und Devonischen Vertebraten von Spitsbergen. III. Acanthaspida. Nachtrag. Ibid., no. 23, 1.—1931a. A new reconstruction of Dinichthys. Amer. Mus. Novitates, no. 457. —1931b. Untersuchungen über den bau der Arthrodira. Acta Zool., 225.—1931c. Revision of the structure of Coccosteus decipiens Ag. Norsk. geol. tidsskr., (B), 291.—1932a. The structure of Dinichthys: A contribution to our knowledge of the Arthrodira. The Bashford Dean Memorial Vol., art. 4, 113. Amer. Mus. Nat. Hist. New York. 4to.—1932b. Beitrag zur Kenntnis der Devonischen Fischfauna Ost-Gronlands. Skrift. om Svalbard og Ishavet, no. 42, 1.—1932c. Über einige Fischreste aus dem Hunsrück-Schiefer. Centralb. f. Min., etc., 572. —1933? 1934? Revision of the Estonian Arthrodira. Part I. Family Homostiidae Jaekel. Arkiv. f. d. Naturkunde Estlands, X Bd., 4 Lief., 1.—1933. Some remarks about the structure of Phlyctaenaspis acadica Whiteaves. Norsk. geol. tidsskr., 127.—1937. Die Downtonischen und Devonischen Vertebraten von Spitsbergen. VI. Lunaspis-Arten . . . Skrift. om Svalbard og Ishavet, no. 72, 1.—1938. Notes on Arthrodira. Norsk. geol. tidsskr., no. 18.

HUSSAKOF, L. 1906. Studies on the Arthrodira. Mem. Amer. Mus. Nat. Hist., 105.

JAEKEL, O. 1902. Ueber Coccosteus u. die Beurtheiling der Placodermen. Sitz.-Bericht. d. Gesell. Naturforsch. Freunde, 103.—1907. Über Pholidosteus n. g. die Mündbildung u. Korperform der Placodermen. Ibid., 4.—1919. Die Mündbildung d. Placodermen. Ibid., 63.—1927. Untersuchungen über die Fischfauna von Wildungen [Arthrodira]. Palaeont. Zeitschr., 329.

M'COY, F. 1848. On some new fossil fish of the Carboniferous period. Ann. Mag. Nat. Hist., 1, 115. [Coelacanths, placoderms, cestracionts and hybodonts. Proposal of Placodermi as group of fishes, p. 6.]

MILLER, H. 1852. The Old Red Sandstone . . . 5th ed. Edinburgh. [Pterichthys, pls. 1, 2.]

NEWBERRY, J. S. 1875. The structure and relations of Dinichthys, with descriptions of some other new fossil fishes. Rept. Geol. Surv. Ohio, 54.

PANDER, C. H. 1857. Ueber die Placodermen des Devonischer Systems. St. Petersbourg.

PATTEN, W. 1912. The evolution of the vertebrates and their kin. Philadelphia. [Reconstruction of Bothriolepis, figs. 247, 248; supposed relations to arachnids and vertebrates, fig. 258.]

STENSIÖ, E. A. 1925. On the head of the macropetalichthyids, with certain remarks on the head of the other arthrodires. Field Mus. Nat. Hist., Publ. 232, 87. [Bone precedes cartilage, p. 186.] —1931. Upper Devonian vertebrates from East Greenland. Meddel. om Gronland, 1. [Chiefly on the Antiarcha.]—1934a. On the heads of certain arthrodires. I. Pholidosteus, Leiosteus, and acanthaspids. Kungl. Svenska Vetenskaps-Akad. Handl., 1. [Bone to cartilage, retrogressive series, p. 66 ff.]—1934b. On the Placodermi of the Upper Devonian of East Greenland. I. Phyllolepida and Arthrodira. Meddel. om Gronland, 1. [Relationships and systematic position of the phyllolepids, pp. 32–34.]—1936. On the Placodermi of the Upper Devonian of East Greenland. Suppl., Part I. Ibid., 1. [Phyllolepida an order of placoderms related to typical Arthrodira.]—1939. On the Placodermi of the Upper Devonian of East Greenland. 2nd Suppl., Part I. Ibid., 5. [Species of phyllolepids.]—1940. Über die Fische des Devons von Ostgrönland. Mitteil. d. Naturf. Gesell. Schaffhausen (Schweiz), 132.—1942. On the snout of arthrodires. Kungl. Svenska Vetenskaps-Akad. Handl., 1. Kujdanowiaspis, a primitive acanthaspid (Dolichothoraci).—1944. Contributions to . . . vertebrate fauna of the Silurian and Devonian of Western Podolia. II. Notes on two arthrodires from the Downtonian of Podolia. Arkiv. f. Zool., 1.—1945a. On the heads of certain arthrodires. II. On the cranium and cervical joint of the Dolichothoraci (Acanthaspida). Kungl. Svenska Vetenskaps-Akad. Handl., Bd. 22, 1. [Skull of Kujdanowiaspis compared with that of Coccosteus, Eusthenopteron, Amia—a fundamental monograph.]—1945b. On the Placodermi of the Upper Devonian of East Greenland. Part II. Antiarchi: Subfamily Bothriolepinae. Palaeozoologica Groenlandica, vol. 2. Copenhagen.

STENSIÖ, E. A. and G. SÄVE-SÖDERBERGH. 1938. Middle Devonian vertebrates from Canning Land and Wegener Peninsula (East Greenland.) Meddel. on Groenland, Bd. 96, no. 6, 1. [Asterolepis an antiarch.]

TRAQUAIR, R. H. 1904. A monograph on the fishes of the Old Red Sandstone. Part 2, no. 2. The Asterolepidae. Palaeontographical Soc., 91. London.

WATSON, D. M. S. 1934. The interpretation of arthrodires. Proc. Zool. Soc. London, 437.—1938. On Rhamphodopsis, a ptyctodont from the Middle Old Red Sandstone of Scotland. Trans. Roy. Soc. Edinburgh, vol. 59, 397.

WESTOLL, T. S. 1945. The paired fins of placoderms. Trans. Roy. Soc. Edinburgh, 381.

WOODWARD, A. S. 1941. The head shield of a new macropetalichthyid fish (Notopetalichthys hillsi, gen. et sp. nov.) from the Middle Devonian of Australia. Ann. Mag. Nat. Hist., 91.

Palaeospondylia

BULMAN, O. M. B. 1931. Note on Palaeospondylus gunni, Traquair. Ann. Mag. Nat. Hist., 179. [Affinities discussed, pp. 18, 190.]

COOPER, C. F. 1940. Editor: A textbook of zoology, by Parker and Haswell. 6th ed. Vol. II. London. [Palaeospondylus probably a small aphetohyoidean related to Acanthodii.—C.F.C., pp. 170–172.]

DEAN, B. 1900. The Devonian "lamprey," Palaeospondylus gunni Traquair, with notes on the systematic arrangement of the fish-like vertebrates. Mem. N. Y. Acad. Sci., 1.

MOY-THOMAS, J. A. 1939. Palaeozoic fishes. London.

NEWBERRY, J. S. 1890. The Palaeozoic fishes of North America. Monogr. U. S. Geol. Surv. (1889).

SOLLAS, W. J. and I. SOLLAS. 1903. Palaeospondylus gunni. Philos. Trans. Roy. Soc. London, 267.

STENSIÖ, E. A. 1927. The Downtonian and Devonian vertebrates of Spitsbergen. Part I. Family Cephalaspidae. Skrift. om Svalbard og Nordishavet, no. 12. [Palaeospondylus, pp. 374–378.]

TRAQUAIR, R. H. 1893. A further description of Palaeospondylus gunni, Traquair. Proc. Roy. Physical Soc. Edinburgh, 87.

Acanthodii (Aphetohyoidea, in part)

DEAN, B. 1907. Notes on acanthodian sharks. Amer. Jour. Anat., 209.

MOY-THOMAS, J. A. 1935. Notes on the types of fossil fishes in the Leeds City Museum. Proc. Leeds Philos. Soc., vol. 2, 451; vol. 3, 111.

WATSON, D. M. S. 1937. The acanthodian fishes. Philos. Trans. Roy. Soc. London, 49. [Gemündina, p. 138.]

WOODWARD, A. S. 1906. On a Carboniferous fish fauna from the Mansfield district, Victoria. Mem. Nat. Mus. Melbourne, no. 1. [Acanthodii, pp. 1–14.]—1935. The affinities of the acanthodian and arthrodiran fishes. Ann. Mag. Nat. Hist., 392.

"Cartilage Fishes": Sharks, Rays, Chimaeroids

ALLIS, E. P. 1923. The cranial anatomy of Chlamydoselachus anguineus. Acta Zoologica, 123.

BALFOUR, F. M. 1878. A monograph on the development of elasmobranch fishes. London.—1881. On the development of . . . the paired fins of the Elasmobranchii . . . its bearings on the nature of the limbs of the Vertebrata. Proc. Zool. Soc. London, 656.

BEEBE, W. and J. TEE-VAN. 1941. Fishes from the Tropical Eastern Pacific. Part 2. Sharks. Zoologica (N. Y.), 93.

BEER, G. R. DE and J. A. MOY-THOMAS. 1935. On the skull of Holocephali. Philos. Trans. Roy. Soc. London, 287.

BREDER, C. M. JR. 1929. Field book of marine fishes of the Atlantic Coast from Labrador to Texas . . . New York.

BROWN, C. 1900. Ueber das Genus Hybodus und seine systematische Stellung. Palaeontographica, 149.

CLAYPOLE, E. W. 1893. The cladodont sharks of the Cleveland Shale. Amer. Geologist, 325.—1895a. The cladodonts of the Upper Devonian of Ohio. Geol. Mag., 473.—1895b. Recent contributions to our knowledge of the cladodont sharks. Amer. Geologist, 363.

COPE, E. D. 1884a. A Carboniferous genus of sharks still living. Science, 275. [Refers to Samuel Garman's Chlamydoselachus.]—1884b. The skull of a still living shark of the Coal Measures. Amer. Nat., 412. [Refers Chlamydoselachus anguineus to Didymodus.]—1884c. Pleuracanthus and Didymodus [in their relation to Chlamydoselachus]. Science, 645.

DANIEL, J. F. 1928. The elasmobranch fishes. 2nd ed. Univ. California Press.

DAVIS, J. W. 1885. Note on Chlamydoselachus anguineus Garman. Proc. Yorkshire Geol. Soc., 98. [Study of teeth of Chlamydoselachus and Pleuracanthus.]—1892. On the fossil fish-remains of the Coal Measures of the British

Islands. Part I. Pleuracanthidae. Scient. Trans. Roy. Soc. Dublin, 703.

DEAN, B. 1894. Contributions to the morphology of Cladoselache (Cladodus). Jour. Morph., 85. —1904. Notes on the long-snouted chimaeroid of Japan, Rhinochimaera (Harriotta) pacifica (Garman) Mitsukuri. Jour. Coll. Sci. Imperial Univ. Tokyo, Japan, vol. 19, art. 4.—1906. Chimaeroid fishes and their development. Carnegie Inst. Washington Publ. 32.—1909. Studies on fossil fishes (sharks, chimaeroids, and arthrodires). Mem. Amer. Mus. Nat. Hist., vol. 9, 211. [Cladoselache and origin of paired fins.]

DENISON, R. H. 1937. Anatomy of the head and pelvic fin of the whale shark, Rhineodon. Bull. Amer. Mus. Nat. Hist., 477.

DOHRN, A. 1884. Studien zur Urgeschichte des Wirbelthierkörpers. IV. Die Entwicklung und Differenzierung der Kiemenbogen der Selachier. Mitt. Zool. Stat. Neapel, Bd. 5, 102.

EASTMAN, C. R. 1897. Tamiobatis vetustus, a new form of fossil skate. Amer. Jour. Sci., 85.

FRITSCH, A. J. 1883–1895. Fauna der Gaskohle und der Kalksteine der Permformation Böhmens. Vols. I–III. Prag. [Pleuracanthus, Xenacanthus.]

GARMAN, S. 1885. Chlamydoselachus anguineus Garm., a living species of cladodont shark. Bull. Mus. Comp. Zool. Harvard Coll.—1913. The Plagiostomia (sharks, skates and rays). Mem. Mus. Comp. Zool. Harvard Coll. Vols. I, II. 4to.

GEGENBAUR, C. 1872. Untersuchungen zur Vergleichenden Anatomie der Wirbelthiere. Heft 3. Das Kopfskelet der Selachier . . . Leipzig.

GOODRICH, E. S. 1909. Vertebrata craniata (First fascicle: Cyclostomes and Fishes). London.—1930. Studies on the structure and development of vertebrates. London.

GREGORY, W. K. 1935. Winged sharks. Bull. N. Y. Zool. Soc., 129.

GUDGER, E. W. 1915. Natural history of the whale shark, Rhineodon typicus Smith. Zoologica (N. Y.), 349.—1932. Cannibalism among the sharks and rays. Scient. Monthly, May, 403.—1935. The geographical distribution of the whale shark (Rhineodon typus). Proc. Zool. Soc. London, 863.—1940. The breeding habits, reproductive organs and external embryonic development of Chlamydoselachus . . . Bashford Dean Memorial Vol., 523.—1941. The feeding organs of the whale shark, Rhineodon typus. Jour. Morph., 81.

GUDGER, E. W. and B. G. SMITH. 1933. The natural history of the frilled shark Chlamydoselachus anguineus. Bashford Dean Memorial Vol., 245.

HARRIS, J. E. 1938. The dorsal spine of Cladoselache. Scient. Publ. Cleveland Mus. Nat. Hist., 1.

HASSE, J. C. F. 1879–1885. Das natürliche System der Elasmobranchier auf Grundlage des Baues und der Entwicklung ihrer Wirbelsäule . . . Parts I, II and Suppl. Jena. [Vertebrae of elasmobranchs; classification by vertebral characters.]

HAWKES, O. A. 1906. The cranial and spinal nerves of Chlamydoselachus anguineus Gar. Proc. Zool. Soc. London, 959.

HAY, O. P. 1900. The chronological distribution of the elasmobranchs. Trans Amer. Philos. Soc., 63.

HOLMGREN, N. 1940. Studies on the head in fishes. Part I. Development of the skull in sharks and rays. Acta Zoologica, 51.

JAEKEL, O. 1906. Neue Rekonstruktionen von Pleuracanthus sessilis und von Polyacrodus (Hybodus) Hauffianus. Sitz. Ber.-naturf. Freunde, Berlin, no. 6, 155.

KOKEN, E. 1907. Ueber Hybodus. Geol. u. Pal. Abhandl., 261.

MOY-THOMAS, J. A. 1935. The structure and affinities of Chondrenchelys problematica Tr. Proc. Zool. Soc. London, 391.—1936a. The structure and affinities of the fossil elasmobranch fishes from the Lower Carboniferous rocks of Glencartholm, Eskdale. Proc. Zool. Soc. London, 761.—1936b. On the structure and affinities of the Carboniferous cochliodont Helodus simplex. Geol. Mag., 488.—1939. The early evolution and relationships of the elasmobranchs. Biol. Rev., 1.

MOY-THOMAS, J. A. and E. I. WHITE. 1939. On the palatoquadrate and hyomandibula of Pleuracanthus sessilis Jordan. Geol. Mag., vol. 76, 459.

MÜLLER, J. and F. G. J. HENLE. 1838–1841. Systematische Beschreibung der Plagiostomen. Berlin. Folio.

MURPHY, R. C. and J. T. NICHOLS. 1916. The shark situation in the waters about New York. Brooklyn Mus. Quarterly, 145.

NICHOLS, J. T. and R. C. MURPHY. 1916. Long Island fauna. IV. The sharks (Order Selachii). Brooklyn Mus. Sci. Bull., 1.

NIELSEN, E. 1932. Permo-carboniferous fishes from East Greenland. Meddel. om Groenland, vol. 86. Kobenhavn.

PARKER, W. K. 1879. On the structure and development of the skull in sharks and skates. Trans. Zool. Soc. London, 189.

PRASHAD, R. R. 1945. A study of the succession of teeth in elasmobranchs. Jour. Madras Univ., 25.

REGAN, C. T. 1906. Classification of the selachian fishes. Proc. Zool. Soc. London, 722.

ROMER, A. S. 1942. Cartilage an embryonic adaptation. Amer. Nat., 394.

SEWERTZOFF, A. N. 1899. Die Entwickelung des Selachierschädels. Festschr. z. Carl von Kupffer, 281. Jena.

SMITH, A. 1829. Contributions to the natural history of South Africa . . . Zool. Jour., vol. 4. [Rhineodon typus, pp. 443–444.]—1849. Pisces (Vol. 4 of his illustrations of the zoology of South Africa). London. [Rhinodon typus, pl. 26.]

SMITH, B. G. 1937. The anatomy of the frilled shark Chlamydoselachus anguineus Garman. Bashford Dean Memorial Vol., 333.—1942. The heterodontid sharks: Their natural history and the external development of Heterodontus japonicus . . . Ibid., 649.

SMITH, H. M. 1925. The whale-shark (Rhineodon) in the Gulf of Siam. Science, 438.

STENSIÖ, E. A. 1925. On the head of the macropetalichthyids with certain remarks on the head of the other arthrodires. Field Mus. Nat. Hist., Publ. 232, 87. [Cartilaginous skeleton in sharks secondary, pp. 160–164.]

THACHER, J. K. 1876. Median and paired fins, a contribution to the history of vertebrate limbs. Trans. Connecticut Acad. Sci., 281.

WHITE, E. G. 1930. The whale-shark, Rhineodon typus. Description of skeletal parts and classification . . . Bull. Amer. Mus. Nat. Hist., 129.—1936. A classification and phylogeny of the elasmobranch fishes. Amer. Mus. Novitates, no. 837.—1937. Interrelationships of the elasmobranchs with a key to the Order Galea. Bull. Amer. Mus. Nat. Hist., 25. [A notable and successful effort to clarify and simplify the major classification of elasmobranchs.]

WHITLEY, G. P. 1926. Sharks. Australian Mus. Mag., 13.—1940. The fishes of Australia. Part I. Sharks, rays, devil-fish and other primitive fishes of Australia and New Zealand. Roy. Zool. Soc. New South Wales.

WOODWARD, A. S. 1889–1901. Catalogue of the fossil fishes in the British Museum (Natural History). 4 vols. London. Part I . . . Elasmobranchii. Part II . . . Elasmobranchii (Acanthodii [concluded], Holocephali, etc. [Introduction traces evolution of fishes in geologic time.]—1906. On a Carboniferous fish fauna from the Mansfield district, Victoria. Mem. Nat. Mus. Melbourne [Acanthodian].—1916. On a new species of Edestus from the Upper Carboniferous of Yorkshire. Quart. Jour. Geol. Soc., 1.—1919. On two new elasmobranch fishes (Crossorhinus jurassicus, sp. nov., and Protospinax annectans) from the Upper Jurassic Lithographic Stone of Bavaria. Proc. Zool. Soc. London (1918), 231.—1920. On the dentition of the petalodont shark, Climaxodus. Quart. Jour. Geol. Soc., 1.—1921. Observations on some extinct elasmobranch fishes. Proc. Linnean Soc. London, 29.—1924. Un nouvel Elasmobranche (Cratoselache Pruvosti, gen. et sp. nov.) du calcaire carbonifère de Denée. Libre Jubil. Cinquant fond. Soc. Geol. Belgique, 59. —1928. President's Address, Dorset Natural History and Antiquarian Field Club. Dorchester. [Note restorations of Hybodus, Chlamydoselache and Jurassic ganoids.]—1940. The affinities of the Palaeozoic pleuracanth sharks. Ann. Mag. Nat. Hist., 323.

Bony Fishes (sensu lato): Osteichthyes, Actinopterygii, Origins, Phylogeny, Classification (General) (See also Fish Skull)

BOULENGER, G. A. 1910. Fishes (Systematic account of Teleostei). In: The Cambridge Natural History, 541. London.

BROUGH, J. 1936. On the evolution of bony fishes during the Triassic period. Biol. Rev., 385 (Cambridge).

GARSTANG, W. 1931. The phyletic classification of the Teleostei. Proc. Leeds Philos. and Lit. Soc., 240.

GREGORY, W. K. 1907. The orders of teleostomous fishes: A preliminary review of the broader features of their evolution and taxonomy. Ann. N. Y. Acad. Sci., 437.

REGAN, C. T. 1904. The phylogeny of the Teleostomi. Ann. Mag. Nat. Hist., (7), vol. 13, 329. —1909. The classification of teleostean fishes. Ibid., (8), vol. 3, 75.—1910. Notes on the classification of the teleostean fishes. Proc. Seventh Internat. Zool. Congress, Boston (1907).—1929. Article on Fishes. Encycl. Brit., 14th ed. [Classification of Osteichthyes.]

SCHAEFFER, B. 1947. Cretaceous and Tertiary actinopterygian fishes from Brazil. Bull. Amer. Mus. Nat. Hist., 1.

WOODWARD, A. S. 1942. The beginning of the teleostean fishes. Ann. Mag. Nat. Hist., (11), vol. 9, 902.

Ganoid Fishes (except Crossopterygians): Palaeoniscoids, Sturgeons (Chondrostei), Holostei, Subholostei, Amioids

ALDINGER, H. 1937. Permische Ganoidfische aus Ostgroenland. Meddel. Groenland, vol. 102, 1.

BROUGH, J. 1931. On fossil fishes from the Karroo system and some general considerations on the bony fishes of the Triassic period. Proc. Zool. Soc. London, 235.—1936. On the evolution of bony fishes during the Triassic period. Biol. Reviews, vol. 11, 385.—1939. The Triassic fishes of Besano, Lombardy. Brit. Mus. (Nat. Hist.). London.

EATON, T. H. JR. 1939. A paleoniscid brain case. Jour. Washington Acad. Sci., 441.

GILL, E. L. 1923. The Permian fishes of the genus Acentrophorus. Proc. Zool. Soc. London, 19.—1925. The Permian fish Dorypterus. Trans. Roy. Soc. Edinburgh, 643.

LEHMAN, J. P. 1947. Description de quelques exemplaires de Cheirolepis canadensis (Whiteaves). Kungl. Svenska Vetenskapsakad. Handl., vol. 24, no. 4.

MAYHEW, R. L. 1924. The skull of Lepidosteus platostomus. Jour. Morph., 315.

MÜLLER, J. 1844. Über den Bau und die Grenzen der Ganoiden und über das natürliche System der Fische. Abh. Akad. Wiss. Berlin, 117.

NIELSEN, E. 1936. Some few preliminary remarks on Triassic fishes from East Greenland. Meddel. om Groenland, Bd. 112, no. 3.

RAYNOR, D. H. 1941. The structure and evolution of the holostean fishes. Biol. Reviews, 218.

REGAN, C. T. 1923. The skeleton of Lepidosteus, with remarks on the origin and evolution of the lower neopterygian fishes. Proc. Zool. Soc. London, 445.

SEWERTZOFF, A. N. 1923. The place of the cartilaginous ganoids in the system and the evolution of the osteichthyes. Jour. Morph., 105.—1928. The head skeleton and muscles of Acipenser ruthenus. Acta Zool., 193.

STENSIÖ, E. A. 1921. Triassic fishes from Spitsbergen. Vienna.—1925. Part II. Kungl. Svenska Vetenskapsakad. Handl., Bd. 2, 1.—1932. Triassic fishes from East Greenland collected by the Danish Expeditions in 1929–1931. Meddel. om Groenland, Bd. 83, 1.

STOLLEY, E. 1920. Beiträge zur Kenntnis der Ganoiden des deutschen Muschelkalks. Palaeontographica (1919–1921), 25.

TRAQUAIR, R. H. 1875. On the structure and systematic position of the genus Cheirolepis. Ann. Mag. Nat. Hist., 237.—1877–1914. The ganoid fishes of the British Carboniferous formations: Palaeoniscidae. Parts 1–7. Monogr. Palaeontographical Soc., vols. 31 (1877), 55 (1901), 61 (1907), 63 (1909), 64 (1911), 65 (1912), 67 (1914).—1879. On the structure and affinities of the Platysomidae. Trans. Roy. Soc. Edinburgh, 343. 4to.

WADE, R. T. 1935. The Triassic fishes of Brookvale, New South Wales. Brit. Mus. (Nat. Hist.). London.

WATSON, D. M. S. 1925. The structure of certain palaeoniscids and the relationships of that group with other bony fish. Proc. Zool. Soc. London, 815. [Cheirolepis.]—1928. On some points in the structure of palaeoniscid and allied fish. Proc. Zool. Soc. London, 49.

WESTOLL, T. S. 1944. The Haplolepidae, a new family of late Carboniferous bony fishes . . . Bull. Amer. Mus. Nat. Hist., 1.

WHITE, E. I. 1939. A new type of palaeoniscoid fish, with remarks on the evolution of the actinopterygian pectoral fins. Proc. Zool. Soc. London, 41. [Cornuboniscus budensis, gen. et sp. n.]

WOODWARD, A. S. 1895a. Catalogue of the fossil fishes in the British Museum (Natural History). Part III. Containing the actinopterygian Teleostomi of the orders Chondrostei (concluded), Protospondyli, Aetheospondyli and Isospondyli (in part). London.—1895b. A contribution to the knowledge of the fossil fish fauna of the English Purbeck beds. Geol. Mag., 145.—1897. A contribution to the osteology of the Mesozoic amioid fishes, Caturus and Osteorachis. Ann. Mag. Nat. Hist., 292.—1901. Catalogue of the fossil fishes in the British Museum (Natural History), Part IV. Containing the actinopterygian Teleostomi of the suborders Isospondyli (in part), Ostariophysi, Apodes, Percesoces, Hemibranchii, Acanthopterygii and Anacanthini. London [Introduction gives outlines of evolution of teleosts].—1906. The study of fossil fishes. Presidential address. Proc. Geol. Assoc. London, 266. [Restoration of Cheirolepis.]—1924. The animals of the Carboniferous period, with special reference to discoveries in Yorkshire. The Naturalist, 105.—1934. Notes on some recently discovered Palaeozoic fishes. Ann. Mag. Nat. Hist., 526.—1939. The affinities of the pycnodont ganoid fishes. Ibid., 607. —1942. The beginning of the teleostean fishes. Ibid., 902.

Polypterus a Ganoid
(See also Fish Skull)

ALLIS, E. P., JR. 1922. The cranial anatomy of Polypterus . . . Jour. Anat., 189.

BUDGETT, J. S. 1901. On some points in the anatomy of Polypterus. Trans. Zool. Soc. London, 323.—1902. On the structure of the larval Polypterus. Ibid., 315.

GOODRICH, E. S. 1908. On the systematic position of Polypterus. Rept. Brit. Assoc. Adv. Sci. (1907), 545.—1928. Polypterus a palaeoniscid ? Palaeobiologica, 87.

JARVIK, E. 1947. Notes on the pit-lines and dermal bones of the head in Polypterus. Festskr. Prof. Nils von Hofsten. Zool. Bidrag f. Uppsala, 600.

POLLARD, H. B. 1892. On the anatomy and phylogenetic position of Polypterus Zool. Jahrb., 387.

Lower Teleost Fishes: Order Isospondyli

BOLIN, R. L. 1939. A review of the myctophid fishes of the Pacific Coast of the United States and of Lower California. Stanford Ichthyological Bull., 89.

BRIDGE, T. W. 1895. On certain features in the skull of Osteoglossum formosum. Proc. Zool. Soc. London, 302.—1900. The air-bladder and its connection with the auditory organ in Notopterus borneensis. Jour. Linnean Soc. London: Zool., 503.

COCKERELL, T. D. A. 1925. The affinities of the fish Lycoptera middendorffi. Bull. Amer. Mus. Nat. Hist., 313.

CROOK, A. J. 1892. Ueber einige fossile Knochenfische von den mittleren Kreide von Kansas. Palaeontographica. Portheus.

ERDL, M. P. 1847. Beschreibung des Skeletes von Gymnarchus niloticus . . . Abh. k. Bayer. Akad. Wiss., 209.

OSBORN, H. F. 1904. The great Cretaceous Fish Portheus molossus Cope. Bull. Amer. Mus. Nat. Hist., 377.

PARKER, W. K. 1873. On the structure and development of the skull of the salmon (Salmo salar, L.). Philos. Trans. Roy. Soc. London, 95.

REGAN, C. T. 1916. The British fishes of the subfamily Clupeinae and related species in other seas. Ann. Mag. Nat. Hist., 1.

RIDEWOOD, W. G. 1904a. On the cranial osteology of the fishes of the families Elopidae and Albulidae . . . Proc. Zool. Soc. London, 35.—1904b. On the cranial osteology of the clupeoid fishes. Ibid., 448.—1904c. On the cranial osteology of the fishes of the families Mormyridae, Notopteridae and Hyodontidae. Jour. Linnean Soc. London: Zool. (1903–1906), 188.—1905a. On the cranial osteology of the fishes of the families Osteoglossidae, Pantodontidae and Phractolaemidae. Ibid., 252.—1905b. On the skull of Gonorhynchus Greyi. Ann. Mag. Nat. Hist., 361.

SWINNERTON, H. H. 1903. Osteology of Cromeria nilotica and Galaxias attenuatus. Zool. Jahrb., 58.

WHITE, T. E. 1942. A new leptolepid fish from the Jurassic of Cuba. Proc. New England Zool. Club, 97.

WOODWARD, A. S. 1896. On some extinct fishes of the teleostean family Gonorhynchidae. Proc. Zool. Soc. London, 500.—1901. Catalogue of the fossil fishes in the British Museum (Natural History). Part IV. Containing the actinopterygian Teleostomi of the suborders Isospondyli (in part), Ostariophysi, Apodes, Percesoces, Hemibranchii, Acanthopterygii and Anacanthini. Brit. Mus. (Nat. Hist.). London.

Order Ostariophysi: Weberian Apparatus

ADAMS, L. A. 1940. Some characteristic otoliths of American Ostariophysi. Jour. Morph., May 1, 497.

ANON. 1929. Article on Loach. Encycl. Brit., 14th ed.

BRIDGE, T. W. and A. C. HADDON. 1889. Contributions to the anatomy of fishes. I. The air-bladder and Weberian ossicles in the Siluridae. Proc. Roy. Soc. London, 309.—1893. Contributions to the anatomy of fishes. II. The air-bladder and Weberian ossicles in the Siluroid fishes. Philos. Trans. Roy. Soc. London, 65.

COCKERELL, T. D. A. 1925. The affinities of the

fish Lycoptera middendorffi. Bull. Amer. Mus. Nat. Hist., 313.

EASTMAN, C. R. 1917. Dentition of Hydrocyon and its supposed fossil allies. Bull. Amer. Mus. Nat. Hist., 757.

EDWARDS, L. F. 1926. The protractile apparatus of the mouth of the catostomid fishes. Anat. Rec., 257.

EIGENMANN, C. H. 1917–1927. The American Characidae. Parts I–IV. Mem. Mus. Comp. Zool. Harvard Coll., vol. 43, pp. 1, 103, 209, 311.

EIGENMANN, C. H. and G. S. MYERS. 1929. The American Characidae. Part V. Mem. Mus. Comp. Zool. Harvard Coll., 429.

ELLIS, M. M. 1913. The gymnotid eels of tropical America. Mem. Carnegie Mus., 109.

GRABER, V. 1886. Die Äussern Mechanischen Werkzeuge der Tiere. I. Wirbeltiere. Leipzig. [Mechanism of tongue, p. 114; carp skull, pp. 79–83, figs. 44, 45.]

GREGORY, W. K. and G. M. CONRAD. 1937. The structure and development of the complex symphysical hinge-joint in the mandible of Hydrocyon lineatus Bleeker, a characin fish. Proc. Zool. Soc. London (for 1936), 975.—1938. The phylogeny of the characin fishes. Zoologica (N. Y.), 319.

GUDGER, E. W. 1925. The crucifix in the catfish skull. Nat. Hist. Mag., 371.

KINDRED, J. E. 1919. The skull of Amiurus. Illinois Biol. Monogr.

KRUMHOLZ, L. A. 1943. A comparative study of the Weberian ossicles in North American ostariophysine fishes. Copeia, no. 1, 33.

MUIR, E. H. 1925. A contribution to the anatomy and physiology of the air-bladder and Weberian ossicles in Cyprinidae. Proc. Roy. Soc. London, 545. [Apparatus perceives wave-pressures impinging on the body. Air-bladder divided into two by a sphincter.]

MYERS, G. S. 1943. Review of Fishes of Western South America by C. H. Eigenmann and W. R. Allen. Copeia, no. 1, 60. [Discussion of Phylogeny of the Characins by W. K. Gregory and G. M. Conrad.]

NELSON, E. M. 1948. The comparative morphology of the Weberian apparatus of the Catostomidae and its significance in systematics. Jour. Morph., Sept., 225.

NICHOLS, J. T. 1930. Speculation on the history of the Ostariophysi. Copeia, no. 3, 148.

REGAN, C. T. 1911a. The classification of the teleostean fishes of the order Ostariophysi. 1.—Cyprinoidea. Ann. Mag. Nat. Hist., 13.—1911b. The classification of the teleostean fishes of the order Ostariophysi.—Siluroidea. Ibid., 553.—1912. A revision of the South American characid fishes of the genera Chalceus, Pyrrhulina, Copeina and Pogonocharax. Ibid., 387.—1929. Article on Cat-fish. Encycl. Brit., 14th ed., vol. 5.

SAGEMEHL, M. 1885. Beiträge zur vergleichenden Anatomie der Fische. III. Das Cranium der Characiniden nebst allgemeinen Bemerkungen über die mit einem Weber'schen Apparat versehenen Physostomen-familien. Morph. Jahrb., 1.—1891. Beiträge zur vergleichenden Anatomie der Fische. IV. Das Cranium der Cyprinoiden. Ibid., 489.

SHELDEN, F. F. 1937. Osteology, myology and probable evolution of the nematognath pelvic girdle. Ann. N. Y. Acad. Sci., 1.

WRIGHT, R. R. 1886. On the skull and auditory organ of the siluroid hypophthalmus. Proc. and Trans. Roy. Soc. Canada (for 1885), 107

Order Apodes

NORMAN, J. R. 1926. The development of the chondrocranium of the eel (Anguilla vulgaris). Philos. Trans. Roy. Soc. London, 369.

REGAN, C. T. 1912. The osteology and classification of the teleostean fishes of the order Apodes. Ann. Mag. Nat. Hist., 377.

Order Mesichthyes:
Haplomi, Iniomi (except Deep-sea Forms), Microcyprini, Synentognathi, Thoracostei, Salmopercae

BREDER, C. M., JR. 1929a. Field observations on flying fishes: A suggestion of methods. Zoologica (N. Y.), 295.—1929b. Report on synentognath habits and development. Carnegie Inst., Year Book (1928–1929), 279.—1932. On the habits and development of certain Atlantic Synentognathi. Papers from Tortugas Laboratory, vol. 28, 1.—1937. The perennial flying-fish controversy. Science, 420.—1938. A contribution to the life histories of Atlantic Ocean flying-fishes. Bull. Bingham Oceanogr. Coll., Peabody Mus. Nat. Hist., Yale Univ., 1.

BREDER, C. M., JR. and H. E. EDGERTON. 1942. An analysis of the locomotion of the seahorse,

Hippocampus, by means of high speed cinematography. Ann. N. Y. Acad. Sci., 145.

DOLLO, L. 1909. Les poissons voiliers. Zool. Jahrb., 419.

HUBBS, C. L. 1924–1943. Studies of the fishes of the order Cyprinodontes. I, II, III, IV, Univ. Mich. Mus. Zool. Misc. Publ., no. 13 (1924);—VI, ibid., no. 16 (1926);—XVI, ibid., no. 42 (1939);—V, Occ. Papers, Mus. Zool. Univ. Mich., no. 148 (1924);—VIII, ibid., no. 198 (1929);—IX, ibid., no. 230 (1931);—X, ibid., no. 231 (1931);—XII, ibid., no. 252 (1932);—XIII, ibid., no. 301 (1934);—XIV, ibid., no. 302 (1935);—XV, ibid., no. 339 (1936);—XVII, ibid., no. 433 (1941);—XVIII, ibid., no. 458 (1942);—VII, Copeia, no. 164 (1927);—XI, ibid., no. 2 (1932);—XIX, ibid., no. 1 (1943).—1933. Observations on the flight of fishes, with a statistical study of the flight of the Cypselurinae and remarks on the evolution of the flight of fishes. Papers Michigan Acad. Sci., Arts and Letters, vol. 17 (for 1932), 575.—1937. Further observations and statistics on the flight of fishes. Ibid., vol. 22, 641.

JUNGERSEN, H. F. E. 1908. Ichthyotomical contributions. I. The structure of the genera Amphisile and Centriscus. Danske Vidensk. Skrift. Naturv., 41.

REGAN, C. T. 1911a. The anatomy and classification of the teleostean fishes of the order Iniomi. Ann. Mag. Nat. Hist., 120.—1911b. The anatomy and classification of the teleostean fishes of the order Salmopercae. Ibid., 294.—1911c. The osteology and classification of the teleostean fishes of the order Microcyprini. Ibid., 320.—1911d. The classification of the teleostean fishes of the order Synentognathi. Ibid., 327.

RIDEWOOD, W. G. 1913. Notes on the South American freshwater flying-fish Gastropelecus, and the common flying-fish, Exocoetus. Ann. Mag. Nat. Hist., 544.

ROBERTSON, G. M. 1943. Fundulus sternbergi, a Pliocene fish from Kansas. Jour. Paleont., 305.

STARKS, E. C. 1902. The shoulder girdle and characteristic osteology of the hemibranchiate fishes. Proc. U. S. Nat. Mus., 619.—1904a. A synopsis of characters of some fishes belonging to the order Haplomi. Biol. Bull., 254.—1904b. The osteology of Dallia pectoralis. Zool. Jahrb., 249.

STOYE, F. H. 1947–1948. The fishes of the order Cyprinodontes. Parts 1–3. Aquarium Jour., Feb., May, July–August, Oct., Nov., Dec., 1947, Jan., 1948.

TCHERNAVIN, V. V. 1947. Further notes on the structure of the bony fishes of the order Lyomeri (Eurypharynx). Jour. Linnean Soc. London, 377.

Spiny-finned Fishes (Acanthopterygii, sensu lato): Berycoids, Percoids, Labroids, Labyrinthici, Chaetodonts, Mail-cheeked Fishes (Scleroparei or Scorpaenoids), Jugulares, Gobioids, Xenopterygii, Trachinoids, Zeoids

BOLIN, R. L. 1944. A review of the marine cottid fishes of California. Stanford Ichthyol. Bull., 1.—1947. The evolution of the marine Cottidae of California with a discussion of the genus as a systematic category. Ibid., 153.

BOULENGER, G. A. 1895. Catalogue of the perciform fishes in the British Museum. 2nd ed. Vol. I. Centrarchidae, Percidae and Serranidae (part). Brit. Mus. (Nat. Hist.). London.

DAY, A. L. 1914. The osseous system of Ophiocephalus striatus Bloch. Philippine Jour. Sci., 19.

HERRE, A. W. and H. R. MONTALBAN. 1928. The Philippine siganids. Philippine Jour. Sci., 151.

REGAN, C. T. 1911a. The anatomy and classification of the teleostean fishes of the orders Berycomorphi and Zenoberyces. Ann. Mag. Nat. Hist., 1.—1911b. On the cirrhitiform percoids. Ibid., 259.—1911c. The osteology and classification of the gobioid fishes. Ibid., 729.—1913a. The osteology and classification of the teleostean fishes of the order Scleroparei. Ibid., 169.—1913b. Classification of the percoid fishes. Ibid., 111.

RENDAHL, H. 1930. Pegasiden-Studien. Arkiv f. Zool. (1929–1930).

SMITH, H. M. 1936. The archer fish. Nat. Hist. Mag., no. 6, 2.

STARKS, E. C. 1898. The osteology and relationships of the family Zeidae. Proc. U. S. Nat. Mus., 469.—1899. The osteology and relationships of the percoidean fish, Dinolestes lewini. Ibid., 113.—1902. The relationship and osteology of the caproid fishes or Antigoniidae. Ibid., 565.—1904. The osteology of some berycoid fishes. Ibid., 601.—1905. The osteology of Caularchus maeandricus (Girard). Biol. Bull., 292.—1907. On the relationship of the fishes of the family Siganidae. Biol. Bull. (Woods Hole), 211.—1923. The osteology and relationships of the uranoscopoid fishes, with notes on other fishes with jugular ventrals. Stanford Univ. Publ. Biol. Sci., 261.

SCHULTZ, L. P. 1944. A revision of the American clingfishes, Family Gobiesocidae, with descriptions of new genera and forms. Proc. U. S. Nat. Mus., 47.

UHLMANN, E. 1921. Studien zur Kenntnis des Schädels von Cyclopterus lumpus L. 1 Teil. Morphogenese des Schädels. 2 Teil. Enstehung des Schädel-knochen. Jena. Zeitschr. f. Naturw., 275.

WILBY, G. V. 1936. On the gobiesocid genus Rimicola. Copeia, 116.

Allotriognathi

GREGORY, W. K. 1935. Nature's sea serpent. Nat. Hist. Mag., 431. [On the ribbon-fish, Regalecus, and other trachypterids.]

HANCOCK, A. and D. EMBLETON. 1849. Account of a ribbon-fish (Gymnetrus) taken off the coast of Northumberland. Ann. Mag. Nat. Hist., 1.

JONES, F. W. 1929. On Trachypterus (Regalecus). Extract from Journal (1906) quoted by Weber and De Beaufort in The fishes of the Indo-Australian Archipelago, vol. 5, 92. Leiden.

PARKER, T. J. 1884. On a specimen of the great ribbon fish (Regalecus argenteus, n. sp.) lately obtained at Moeraki, Otago. Trans. and Proc. New Zealand Inst. (1883), 284.—1886. Studies in New Zealand Ichthyology.—I. On the skeleton of Regalecus argenteus. Trans. Zool. Soc. London, 5.

REGAN, C. T. 1907a. Descriptions of the teleostean fish Velifer hypselopterus and of a new species of the genus Velifer. Proc. Zool. Soc. London, 633.—1907b. On the anatomy, classification and systematic position of the teleostean fishes of the suborder Allotriognathi. Ibid., 634. [Velifer related to Regalecus.]

WEBER, M. and L. F. DE BEAUFORT. 1929. The fishes of the Indo-Australian Archipelago. Vol. V. Leiden.

WHITLEY, G. 1933. Studies in ichthyology. Records Australian Mus. [Regalecus, excellent figure of head and mouth.]

Percesoces

DOLLO, L. 1909. Les Téléostéens à ventrales abdominales secondaires. Verhandl. d. K. K. zool.-bot. Gesellsch. in Wien, 135.

GUDGER, E. W. 1918. Sphyraena barracuda: Its morphology, habits and history. Carnegie Inst. Washington Publ. no. 252, 53.

GUDGER, E. W. and C. M. BREDER. 1928. The barracuda (Sphyraena) dangerous to man. Jour. Amer. Med. Assoc., 1938.

JORDAN, D. S. and C. L. HUBBS. 1919. Studies in ichthyology: A monographic review of the family of Atherinidae or silversides. Leland Stanford Junior Univ. Publ., 1.

STARKS, E. C. 1899. The osteological characters of the fishes of the suborder Percesoces. Proc. U. S. Nat. Mus., 1.

Phallostethi

BAILEY, R. J. 1936. The osteology and relationships of the phallostethid fishes. Jour. Morph., 453.

HERRE, A. W. C. T. 1939. The genera of Phallostethidae. Proc. Biol. Soc. Washington, 139.—1942. New and little known phallostethids, with keys to the genera and Philippine species. Stanford Ichthyol. Bull., 137.

MYERS, G. S. 1928. The systematic position of the phallostethid fishes with diagnosis of a new genus from Siam. Amer. Mus. Novitates, no. 295.—1935. A new phallostethid fish from Palawan. Proc. Biol. Soc. Washington, 5.

TE WINKEL, L. E. 1939. The internal anatomy of two phallostethid fishes. Biol. Bull., 59.

Scombroids (sensu lato):
Marlins, Swordfishes, Carangoids, Luvarus, Stromateids

BEEBE, W. 1941. A study of a young sailfish (Istiophorus). Zoologica (N. Y.), 209.

CONRAD, G. M. 1937a. The brain of the swordfish (Xiphias gladius). Amer. Mus. Novitates, no. 900.—1937b. The nasal bone and sword of the swordfish (Xiphias gladius). Ibid. no. 968.

GREGORY, W. K. and G. M. CONRAD. 1937. The comparative osteology of the swordfish (Xiphias) and the sailfish (Istiophorus). Amer. Mus. Novitates, no. 952.—1943. The osteology of Luvarus imperialis, a scombroid fish: A study in adaptive evolution. Bull. Amer. Mus. Nat. Hist., 225.

GUDGER, E. W. 1940. The alleged pugnacity of the swordfish and the spearfishes as shown by their attacks on vessels. Mem. Roy. Asiatic Soc. Bengal, 215.

JOHNSEN, S. 1918–1919. Notes on Luvarus imperialis Raf., a fish new to the fauna of Norway. Bergens Mus. Aarbok (1918–1919), 1.

KISHINOUYE, K. 1923. Contributions to the comparative study of the so-called scombroid fishes. Jour. Coll. Agric., Imp. Univ. Tokyo, 293.

LAMONTE, F. 1945. North American game fishes. Garden City, New York.

LAMONTE, F. and D. E. MARCY. 1941. Swordfish, sailfish, marlin, and spearfish. Ichthyol. Contrib. Int. Game Fish Assoc., 3.

REGAN, C. T. 1902. A revision of the fishes of the family Stromateidae. Ann. Mag. Nat. Hist., 115.

ROULE, L. 1924. Étude sur l'ontogénese et la croissance avec hypermétamorphose de Luvarus imperialis Rafinesque . . . Ann. de l'Inst. Oceanogr., I, 119.—1929. Présentation d'un squelette de Lampris luna. Bull. Mus. d'Hist. Nat. Paris.

STARKS, E. C. 1909. The scombroid fishes. Science, 572.—1910. The osteology and mutual relationships of the fishes belonging to the family Scombridae. Jour. Morph., 77.—1911. Osteology of certain scombroid fishes. Leland Stanford Junior Univ. Publ., 1.

WAITE, E. R. 1902. Skeleton of Luvarus imperialis, Rafinesque . . . Rec. Australian Mus., 292.

Plectognaths

BREDER, C. M. and E. CLARK. 1947. A contribution to the visceral anatomy, development and relationships of the Plectognathi. Bull. Amer. Mus. Nat. Hist. 287.

GREGORY, W. K. and H. C. RAVEN. 1934. Notes on the anatomy and relationships of the ocean sunfish (Mola mola). Copeia, 145.

LÜTKEN, C. F. 1864. Om en ved Sevedöi Begyndelsen 1862 opdreven "Kaempe-Klumpfisk" (Mola nasus Raf.). Vidensk. Meddel. Naturh. Foren. Kjobenhavn, 379.

PARR, A. E. 1927. On the functions and morphology of the postclavicular apparatus in Spheroides and Chilomycterus. Zoologica (N. Y.), 245.

RAVEN, H. C. 1939a. Notes on the anatomy of Ranzania truncata, a plectognath fish. Amer. Mus. Novitates, no. 1038.—1939b. On the anatomy and evolution of the locomotor apparatus of the nipple-tailed ocean sunfish (Masturus lanceolatus). Bull. Amer. Mus. Nat. Hist., 143.

REGAN, C. T. 1902. On the classification of the fishes of the suborder Plectognathi . . . Proc. Zool. Soc. London, 284.

STEENSTRUP, J. and C. LÜTKEN. 1898–1901. Spolia Atlantica. Bidrag til Kundskab om Klumpeller Maane-fiskene (Molidae). Det Kongelige Danske Vidensk. Selsk. Skrift, 1.

THILO, O. 1899. Die Entstehung der Luftsäcke bei den Kugelfischen. Anat. Anz., 73.

Discocephali

GUDGER, E. W. 1926. A study of the smallest shark-suckers (Echeneididae) on record, with special reference to metamorphosis. Amer. Mus. Novitates, no. 234.—1930. Some old-time figures of the shipholder, Echeneis or Remora, holding the ship. Isis, 340.

REGAN, C. T. 1912. The anatomy and classification of the teleostean fishes of the order Discocephali. Ann. Mag. Nat. Hist., 634.

STORMS, R. 1888. The adhesive disk of Echeneis. Ann. Mag. Nat. Hist., 67.

Heterosomata

CHABANAUD, P. 1934. Hétérogénéité des téléostéens dyssymétriques. Bull. Soc. zool. de France, 275.—1936. Le neurocrane osseux des téléostéens dyssymétriques après la métamorphose. Ann. de l'Inst. Océanogr., 223.—1938. Contribution à la morphologie et à la systématique des téléostéens dyssymétriques. Arch. d. Mus. Nat. d'Hist. Nat., 59.

COLE, F. J. and J. JOHNSTONE. 1901. Pleuronectes (the plaice). Proc. and Trans. Liverpool Biol. Soc. (1901–1902), 145.

HUBBS, C. L. 1945. Phylogenetic position of the Citharidae, a family of flatfishes. Miscell. Publ. Mus. Zool., Univ. Michigan, no. 63.

NORMAN, J. R. 1934. A systematic monograph of the flatfishes (Heterosomata). Vol. I. London, 4to.

WILLIAMS, S. R. 1902. Changes accompanying the migration of the eye, and observations on the tractus opticus and tectum opticum in Pseudopleuronectes americanus. Bull. Mus. Comp. Zool. Harvard Coll., 1.

Blennies, Brotulids, Anacanths, Symbranchoids

ADAMS, L. A. 1908. Description of the skull and separate cranial bones of the wolf-eel (Anarrhichthys ocellatus). Kansas Univ. Sci. Bull., 331.

COCKERELL, T. D. A. 1916. The scales of the brotulid fishes. Ann. Mag. Nat. Hist., 317.

EMERY, C. 1880. Le specie del genere Fierasfer nel Golfo di Napoli e regioni limitrofe. In: Fauna und flora des Golfes von Neapel . . . Leipzig.

PARR, A. E. 1946. The Macrouridae of the western North Atlantic and Central American seas. Bull. Bingham Oceanogr. Coll., Peabody Mus. Nat. Hist., Yale Univ., 1.

REGAN, C. T. 1903. On the systematic position and classification of the gadoid or anacanthine fishes. Ann. Mag. Nat. Hist., 459.—1912a. The osteology of the teleostean fishes of the order Opisthomi. Ibid., 217.—1912b. The anatomy and classification of the symbranchoid eels. Ibid., 387.—1912c. The classification of the blennioid fishes. Ibid., 265.

TROTTER, E. S. 1926. Brotulid fishes from the Arcturus Oceanographic Expedition. Zoologica (N. Y.), 107.

VAILLANT, L. 1905. Le genre Alabes de Cuvier. Comptes Rendu, 1713. [On the affinities of Alabes with blennioids rather than with symbranchioid eels.]

Pediculati

BEEBE, W. and J. CRANE. 1947. Eastern Pacific Expeditions of the New York Zoological Society. 37. Deep-sea ceratioid fishes. Zoologica (N. Y.), 151.

CUVIER, G. 1817. Sur le genre Chironectes Cuv. (Antennarius Commers.). Mém. Mus. Nat. d'Hist. Nat., III, 418.

GREGORY, W. K. and G. M. CONRAD. 1936. The evolution of the pediculate fishes. Amer. Nat., 193.

REGAN, C. T. 1912. The classification of the teleostean fishes of the order Pediculati. Ann. Mag. Nat. Hist., 277.

REGAN, C. T. and E. TREWAVAS. 1932. Deep-sea angler-fishes (Ceratioidea). Carlsberg Foundation's Oceanogr. Exped. . . . Rept. No. 2, 1.

WATERMAN, T. H. 1939a. Studies on deep-sea angler-fishes (Ceratioidea). I. . . . Bull. Mus. Comp. Zool. Harvard Coll., 65.—1939b. II. Ibid., 82.—1948. III. The comparative anatomy of Gigantactis longicirra Waterman. Jour. Morph., 81.

PART III

AIR-BREATHING FISHES, AMPHIBIANS, REPTILES, BIRDS

(Cf. Vol. I, Chapters X–XV)

Crossopterygii (Tassel-fins, Lobe-fins) From Paddles to Paired Limbs

(See also Tetrapoda)

BRYANT, W. L. 1919. On the structure of Eusthenopteron. Bull. Buffalo Soc. Nat. Sci., 1.

BYSTROW, A. P. 1939. Zahnstruktur der Crossopterygier. Acta Zool., 283.

EATON, T. H., JR., 1939. The crossopterygian hyomandibular and the tetrapod stapes. Jour. Washington Acad. Sci., 109.

FRITSCH, A. 1895. Fauna der Gaskohle und der Kalksteine der Permformation Böhmens. Vol. 1. Prag. [Encl. Megalichthys.]

GOODRICH, E. S. 1902. On the pelvic girdle and fin of Eusthenopteron. Quart. Jour. Micros. Soc., 311.—1919. Restorations of the head of Osteolepis. Jour. Linnean Soc.: Zoology, 181.

GREGORY, W. K. and H. C. RAVEN. 1941. Studies on the origin and early evolution of paired fins and limbs. Part II. A new restoration of the skeleton of Eusthenopteron. Ann. N. Y. Acad. Sci., 293.

GROSS, W. 1936a. Beiträge zur Osteologie baltischer und rheinischer Devon-Crossopterygier. Palaeont. Zeitschr., 129.—1936b. Neue Crossopterygier aus dem baltischen Oberdevon. Zentralb. f. Min., Geol. u. Pal., Abt. B, 69.

HILLS, E. S. 1943. The ancestry of the Choanichthyes. Australian Jour. Sci., 21.

HUSSAKOF, L. 1908. Catalogue . . . of fossil vertebrates in Amer. Mus. Nat. Hist. Part I. Fishes. Bull. Amer. Mus. Nat. Hist., 1. [Type of Sauripterus taylori, fig. 28.]—1912. Notes on Devonic fishes from Scaumenac Bay, Quebec. New York State Mus. Bull., 127. [Eusthenopteron, p. 131.] —1918. Catalog . . . of fossil fishes in Mus. Buffalo Soc. Nat. Sci. Bull. Buffalo Soc. Nat. Sci., 1. [Eusthenopteron, pp. 176–178, pl. 70, fig. 2.]

JARVIK, E. 1937. On the species of Eusthenopteron found in Russia and the Baltic States. Bull. Geol. Inst. Upsala, 63.—1942. On the structure of the snout of crossopterygians and lower gnathostomes in general. Zool. Bidrag fran Uppsala (1941–1942), 235.—1944. On the exoskeletal shoulder-girdle of teleostean fishes with special reference to Eusthenopteron foordi Whiteaves. Kungl. Svenska Vetenskapsakad. Handl., 1.—1948. On the morphology and taxonomy of the Middle Devonian osteolepid fishes of Scotland. Ibid., 1.

LANKESTER, E. R. 1908. Guide to the Gallery of Fishes in the . . . British Museum (Natural History). London. [Restoration of Eusthenopteron, fig. 37, p. 66.]

PANDER, C. H. 1856. Die fossilen Fische des silurischen Systems der Russich-Baltischen Gouvernements. (Ueber die Saurodipterinen . . .) St. Petersburg. 4to. [Osteolepis, pp. 8–21, pl. C.]

PARKER, T. J. and W. A. HASWELL. 1940. A textbook of zoology. Vol. II, 6th ed., revised by C. Forster Cooper. London and New York. [Choanichthyes, a term invented by Romer to include lung-bearing fishes, i.e. Crossopterygii plus Dipnoi, p. 284.]

ROMER, A. S. 1937. The braincase of the Carboniferous crossopterygian Megalichthys nitidus. Bull. Mus. Comp. Zool. Harvard Coll., 1.—1941. Notes on the crossopterygian hyomandibular and braincase. Jour. Morph., 141.

SÄVE-SÖDERBERGH, G. 1933. The dermal bones of the head and the lateral line system of Osteolepis macrolepidotus Ag . . . Nova Acta Regiae Soc. Sci. Upsaliensis.—1941. Notes on the dermal bones of the head in Osteolepis macrolepidotus Ag . . . Zool. Bidrag fran Uppsala, 523.

SCHAEFFER, B. 1941. A revision of Coelacanthus newarki and notes on the evolution of the girdles and basal plates of the median fins in the Coelacanthini. Amer. Mus. Novitates, no. 1110.—1948. A study of Diplurus longicaudatus with notes on the body form and locomotion of the Coelacanthini. Ibid., no. 1378.

SMITH, J. L. B. 1940. A living coelacanthid fish from South Africa. Trans. Roy. Soc. South Africa, 1.

STENSIÖ, E. A. 1922. Notes on certain crossopterygians. Proc. Zool. Soc. London, 1241.—1925. Note on the caudal fin of Eusthenopteron. Arkiv f. Zool., no. 11.—1931. Upper Devonian vertebrates from East Greenland . . . Meddel. om Groenland, no. 1.—1932. Triassic fishes from East Greenland. Ibid., no. 3. [Laugia, g.n., pp. 46–48.]—1937. On Devonian coelacanthids of Germany . . . Kungl. Svenska Vetenskapsakad. Handl., 1.—1940. Über die Fische des Devons von Ostgrönland. Mitt. Naturf. Ges. Schaffhausen, 132.

TRAQUAIR, R. H. 1876. On the structure and affinities of Tristichopterus alatus, Egerton. Trans. Roy. Soc. Edinburgh, 383.—1911. Les poissons Wealdiens de Bernissart. Mem. Mus. Roy. d'Hist. nat. de Belgique (1910), 1. [Coelacanths.]

WATSON, D. M. S. 1921. On the coelacanth fish. Ann. Mag. Nat. Hist., 320.

WATSON, D. M. S. and H. DAY. 1916. Notes on some Palaeozoic fishes. Manchester Mem. [Crossopterigii.]

WELLBURN, E. D. 1902. On the genus Megalichthys, Agassiz . . . Proc. Yorkshire Geol. and Polytechnic Soc. (1900–1902), 52.

WESTOLL, T. S. 1936. On the structure of the dermal ethmoid shield of Osteolepis. Geol. Mag., 157.—1937. On a specimen of Eusthenopteron from the Old Red Sandstone of Scotland. Ibid., 507.—1939. On Spermatodus pustulosus Cope, a coelacanth from the "Permian" of Texas. Amer. Mus. Novitates, no. 1017.—1940. New Scottish material of Eusthenopteron. Geol. Mag. 65. [Elpistostege, gen. nov.; frontal of fishes homologous with parietal of tetrapods, pp. 71, 72.]

WHITEAVES, J. F. 1887. Illustrations of the fossil fishes of the Devonian Rocks of Canada. Part I. Trans. Roy. Soc. Canada (1886), 101. [Eusthenopteron.]—1889. Part II. Ibid. (1888), 77. [Eusthenopteron, pls. 6, 7, 10.]

WOODWARD, A. S. 1922. Observations on crossopterygian and arthrodiran fishes. Presidential address. Proc. Linnean Soc., 27.

Dipnoi

BRAUS, H. 1901. Die Muskeln und Nerven der Ceratodusflosse . . . In: R. Semon, Zool. Forsch., vol. 1, 137.

BROILI, F. 1933. Weitere Fischreste aus den Hunsrückschiefern. Sitz. d. Bayer. Akad. d. Wissensch., 269. [Skull roof of dipnoans compared with that of ostracoderms, pp. 300–303, fig. 12.]

BYSTROW, A. P. 1944. On the dentition of Fleurantia denticulata. Comptes rendus (Doklady) d. l'Acad. d. Sci. d. l'U.R.S.S., 31.

COOPER, C. F. 1937. The Middle Devonian fish fauna of Achanarras. Trans. Roy. Soc. Edinburgh, 223. [Dipterus.]

DEAN, B. 1906. Notes on the living specimens of the Australian lungfish, Ceratodus forsteri, in the Zoological Society's collection. Proc. Zool. Soc. London, 168.

DOLLO, L. 1895. Sur la phylogénie des Dipneustes. Bull. Soc. Belge d. Geol., d. Paleont. et d'Hydrol., 79.

FRITSCH, A. 1889. Fauna der Gaskohle und der Kalksteine der Permformation Böhmens. Vol. 2. Stegocephali (Schluss).—Dipnoi, Selachii (Anfang). Prag.

GOODRICH, E. S. 1909. Vertebrata Craniata (First Fascicle: Cyclostomes and fishes). In: A treatise on zoology (Lankester), Part 9. London. [Lungs of Dipnoi and Tetrapoda, pp. 223–227, fig. 217.] —1925. Cranial roofing-bones in Dipnoi. Jour. Linnean Soc. London, 79.

GRAHAM-SMITH, W. and T. S. WESTOLL. 1937. On a new long-headed dipnoan fish from the Upper Devonian of Scaumenac Bay, P. Q., Canada. Trans. Roy. Soc. Edinburgh (1936–1937), 241.

GÜNTHER, A. 1872. Description of Ceratodus, a genus of ganoids from Queensland, Australia. London. 4to.

KELLICOTT, W. E. 1905. The development of the vascular and respiratory systems of Ceratodus. Mem. N. Y. Acad. Sci., vol. 2, part 4.

PANDER, C. H. 1858. Über die Ctenodipterinen des devonischen Systems. St. Petersburg. 4to. Dipnoi: Dipterus, Cheirodus, Holodus, etc.

PARKER, W. N. 1892. On the anatomy and physiology of Protopterus annectens. Trans. Roy. Irish Acad. Dublin, 109.

ROMER, A. S. 1936. The dipnoan cranial roof. Amer. Jour. Sci., 241.

ROMER, A. S. and H. J. SMITH. 1934. American Carboniferous dipnoans. Jour. Geol., 700.

SÄVE-SÖDERBERGH, G. 1937. On Rhynchodipterus elginensis n.g., n.sp., representing a new group of dipnoan-like Choanata from the Upper Devonian of East Greenland and Scotland. Arkiv. f. Zool.

SCHNEIDER, A. F. 1886. Über die Flossen der Dipnoi und die Systematik von Lepidosiren und Protopterus. Zool. Anz., 521.

SEMON, R. 1898. Die Entwickelung der paarigen Flossen des Ceratodus forsteri. In: Zool. Forschungsr. in Australien . . ., vol. 1, part 2, 59.—1901. Die Zahnentwickelung des Ceratodus forsteri. Denkschr. Med. Nat. Ges. Jena, vol. 4.

SMITH, H. W. 1930. Lung-fish. Scient. Monthly, 467. [Protopterus.]—1932. Kamongo. New York. [Protopterus ethiopicus will drown in water if prevented from breathing free air.]

TRAQUAIR, R. H. 1890. Notes on the Devonian fishes of Scaumenac Bay and Campbelltown in Canada. Geol. Mag. N. S., 15.

WATSON, D. M. S. and E. L. GILL. 1923. The structure of certain Palaeozoic Dipnoi. Jour. Linnean Soc. London, 163.

Tetrapoda, General: Origin, Major Classification, Earliest Tetrapods

(See also Vertebrates, General, Tetrapoda, Paired Limbs, etc.)

BROILI, F. 1913. Unser Wissen über die ältesten Tetrapoden. Fortschr. d. Naturw. Forsch., 51.

CASE, E. C. 1915. The Permo-Carboniferous Red Beds of North America and their vertebrate fauna. Carnegie Inst. Washington, Publ. no. 207.

CASE, E. C., S. W. WILLISTON, and M. G. MEHL. 1913. Permo-Carboniferous vertebrates from New Mexico. Ibid., no. 181.

COPE, E. D. 1880. Second contribution to the history of the Vertebrata of the Permian formation of Texas. Proc. Amer. Philos. Soc., 38. [Eryops megacephalus, pls. 1–4.]

CREDNER, H. 1881–1893. Die Stegocephalen und Saurier aus dem Rothliegenden des Plauen'schen Grundes bei Dresden. I–X. Zeitschr. d. deutsch. geol. Gesellsch., vols. 33–36, 37, 38, 42.

DAWSON, J. W. 1863. Air-breathers of the Coal Period . . . Montreal.—1895. Synopsis of the air-breathers of the Paleozoic of Canada. Proc. and Trans. Roy. Soc. Canada.

EFREMOV, J. A. 1939a. First representative of Siberian early Tetrapoda. Comptes rendus (Doklady) d. l'Acad. d. Sci. d. l'U.R.S.S., 105. —1939b. On the significance of the Upper Palaeozoic continental basins of Siberia in the palaeontology of the early Tetrapoda. Ibid., 248. [Summary in English, 270.]—1939c. On the Permian Tetrapoda-fauna of the U.S.S.R. and stratigraphical subdivision of the continental Permian. Bull. d. l'Acad. d. Sci. d. l'U.R.S.S., 272. [Summary in English, 287.]—1940. Preliminary description of the new Permian and Triassic Tetrapoda from U.S.S.R. Travaux d. l'Inst. Paleontol., vol. 10. [Summary in English, 93.]

GREGORY, J. T. 1948. A new limbless vertebrate from the Pennsylvanian of Mazon Creek, Illinois. Amer. Jour. Sci., 636. [Aistopoda regarded as snake-like derivative of basic reptilian stock: Infraclass Captorhina, order Microsauria.]

GREGORY, W. K. 1913. Crossopterygian ancestry of the Amphibia. Science, 806.—1915. Present status of the problem of the origin of the Tetrapoda, with special reference to the skull and paired limbs. Ann. N. Y. Acad. Sci., 317.

GREGORY, W. K. and H. C. RAVEN. 1941. Studies on the origin and early evolution of paired fins and limbs. Ann. N. Y. Acad. Sci., 273. [Bearing on origin of tetrapods.]

HUENE, F. VON. 1943a. Die Korrelation in phyletischen Auftreten der Pflanzen und der Tetrapoden. Neuen Jahrb. f. Min., etc., 26.—1943b. Grundsätzliches über die Entfaltung der früher Landwirbeltiere in der Erdgeschichte. Naturw. Monatsschr. aus der Heimat, 70.—1944a. Die Verwandtschaft einiger früher Tetrapoden-Gruppen. Palaeont. Zeitschr., 410.—1944b. Palaontologische Grundzüge des Stammesgeschichte der früher Tetrapodenzweige. Biol. Zentrabl., vol. 65, 268.—1948. Short review of the lower tetrapods. Robert Broom Commemorative Vol., 65. [A brief review of wide interest.]

PARKER, T. J. and W. A. HASWELL. 1940. A textbook of zoology. Vol. 2. 6th ed., revised by C. Forster-Cooper. London. [Relations of Tetrapoda to fishes.]

ROMER, A. S. 1936. Studies on American Permo-Carboniferous tetrapods. Publ. Lab. Pal., U.S.S.R., vol. 1, 85.—1941a. The first land animals. Nat. Hist. Mag., 236.—1941b. Earliest land vertebrates of this continent. Science, 279. —1945. Vertebrate paleontology. 2nd ed. Univ. Chicago Press.

SÄVE-SÖDERBERGH, G. 1934. Some points of view concerning the evolution of the vertebrates and the classification of this group. Arkiv f. Zool., 1.

SEWERZOFF, A. N. 1926. Der Ursprung der Quadrupeda. Palaeont. Zeitschr., 75.

WATSON, D. M. S. 1917. Sketch classification of the Pre-Jurassic tetrapod vertebrates. Proc. Zool. Soc. London, 167.—1926. The evolution and origin of the Amphibia. Croonian Lecture. Philos. Trans. Roy. Soc. London, 189. [Eogyrinus, figs. 18–25; Megalichthys, figs. 32, 33, 36–39; Eusthenopteron, figs. 34, 35.]—1942. On Permian and Triassic Tetrapoda. Geol. Mag., 81.

WESTOLL, T. S. 1938a. Ancestry of the tetrapods. Nature, 127. [Elpistostege, Devonian crossopterygian; skull roof transitional from crossopterygian to labyrinthodont.]—1938b. The origin of the tetrapods and their relation to the bony fishes. Jour. Brit. Assoc. Adv. Sci., 59.—1942. Relationships of some primitive tetrapods. Nature, 121.—1943. The origin of the tetrapods. Biol. Rev., 78.

WILLISTON, S. W. 1916a. The osteology of some American Permian vertebrates. Contrib. Walker Mus., Univ. Chicago, 165.—1916b. Synopsis of the American Permocarboniferous Tetrapoda. Ibid., 193.

Tetrapoda: Skull and Jaws, Jaw Muscles, Visceral Arches, Hyobranchial Region

ADAMS, L. A. 1919. A memoir on the phylogeny of the jaw muscles in recent and fossil vertebrates. Ann. N. Y. Acad. Sci., 51.

ALLIS, E. P. 1936. Comparison of the laterosensory lines, the snout and the cranial roofing bones of the Stegocephali with those in fishes. Jour. Anat., vol. 70, 293.

BROOM, R. 1913. On the structure of the mandible in the Stegocephalia. Anat. Anz., 73.

FÜRBRINGER, M. 1922. Das Zungenbein der Wirbeltiere insbesondere der Reptilien und Vögel . . . Abh. d. Heidelberger Akad. Wiss.

GADOW, H. 1888. On the modifications of the first and second visceral arches with special reference to the homologies of the auditory ossicles. Philos. Trans. Roy. Soc. London, 451.

GAUPP, E. 1899. Ontogenese und Phylogenese des schalleitenden Apparates bei den Wirbeltieren. Ergeb. d. Anat. u. Entwickelungsges. (1898), 990.—1902. Über die Ala temporalis des Säugerschädels und die Regio orbitalis einiger anderer Wirbeltierschädel. Anat. Heft. (Merkel und Bonnet), no. 61, 161. [Homologizes mammalian ala temporalis with basipterygoid of lizard.]—1911. Beiträge zur Kenntnis des Unterkiefers der Wirbeltiere. II. Die Zusammensetzung des Unterkiefers des Quadrupeden. Anat. Anz., 433.—1913. Die Reichertsche Theorie (Hammer-, Amboss, und Kieferfrage). Archiv f. Anat. u. Entwickelungsges. (1912), Supplement-Band.

GOODRICH, E. S. 1915. The chorda tympani and middle ear in reptiles, birds and mammals. Quart. Jour. Micros. Sci., 137.

GREGORY, W. K. 1917. Second report of the committee on the nomenclature of the cranial elements in the Permian Tetrapoda. Bull. Geol. Soc. Amer., 973.—1920. Studies in comparative myology and osteology. No. IV. A review of the evolution of the lacrymal bone of vertebrates with special reference to that of mammals. Bull. Amer. Mus. Nat. Hist., 95.

HUENE, F. VON. 1913. The skull elements of the Permian Tetrapoda . . . Bull. Amer. Mus. Nat. Hist., 315.

LUBOSCH, W. 1915. Vergleich. Anat. d. Kaumuskeln der Wirbeltiere. 1. Die Kaumuskeln der Amphibien. Zeitschr. Naturw., 51. Jena.

MOODIE, R. L. 1908. The lateral line system in extinct Amphibia. Jour. Morph., 511.

ROMER, A. S. 1937. The braincase of the Carboniferous crossopterygian Megalichthys nitidus. Bull. Mus. Comp. Zool. Harvard Coll., 1. —1940. Mirror image comparison of upper and lower jaws in primitive tetrapods. Anat. Rec., 175.—1941. Notes on the crossopterygian hyomandibular and braincase. Jour. Morph., 141.

SÄVE-SÖDERBERGH, G. 1935. On the dermal bones of the head in labyrinthodont stegocephalians and primitive Reptilia . . . Meddel. om Groenland . . .—1944. New data on the endocranium of Triassic Labyrinthodontia. Arkiv. f. Zool. 1. —1945. Notes on the trigeminal musculature in non-mammalian tetrapods. Nova Acta Regiae Soc. Sci. Upsaliensis, 1.

SAWIN, H. J. 1941. The cranial anatomy of Eryops megacephalus. Bull. Mus. Comp. Zool. Harvard Coll., 407.

STADTMÜLLER, F. 1931. Über eine Cartilago Pararticularis am Kopfskelet von Bombinator und die Schmalhausensche Theorie zum Problem der Gehörknöchelchen. Zeitschr. f. Anat. u. Entwickl., 792. [Extracolumella = hyomandibular; parartic. cart. of Bombinator = vestige of hyomand., p. 800.]

SUSHKIN, P. P. 1927. On the modifications of the mandibular and hyoid arches and their relations to the brain-case in the early Tetrapoda. Palaeont. Zeitschr., 263. [Diadectes, figs. 19–22; Kotlassia, fig. 39.]

THYNG, F. W. 1906. Squamosal bone in tetrapodous Vertebrata. Proc. Boston Soc. Nat. Hist. (1904–1906), 387.

WATSON, D. M. S. 1916. On the structure of the brain-case in certain Lower Permian tetrapods. Bull. Amer. Mus. Nat. Hist., 611.

WESTOLL, T. S. 1940. New Scottish material of Eusthenopteron. Geol. Mag., 65. ["Frontal" of fishes homologous with parietal of tetrapods, pp. 71, 72.]—1943. The hyomandibular of Eusthenopteron and the tetrapod middle ear. Proc. Roy. Soc. B, 393.

WILLISTON, S. R. 1913. The primitive structure of the mandible in amphibians and reptiles. Jour. Geol., 625.

Tetrapoda: Locomotor System, Girdles and Paired Limbs, Origin, Footprints

BRAUS, H. 1901. Die Muskeln und Nerven der Ceratodusflosse: Ein Beitrag zur vergleichenden Morphologie der freien Gliedmassen bei niederen Fischen und zur Archipterygiumtheorie. In: R. Semon, Zool. Forschungsr. in Australien . . . vol. 1, 137.

BROOM, R. 1913. On the origin of the cheiropterygium. Bull. Amer. Mus. Nat. Hist., 385.

CARMEN, J. E. 1927. Fossil footprints from the Pennsylvanian system in Ohio. Bull. Geol. Soc. Amer., 385.

CHEN, H. K. 1935. Development of the pectoral limb of Necturus maculosus. Univ. Ill. Biol. Monogr., vol. 14, no. 1.

COPE, E. D. 1891. On the characters of some Paleozoic fishes. Proc. U. S. Nat. Mus., 447. [Paired fins of Megalichthys nitidus Cope, pp. 457, 458, pl. 32.]

EVANS, F. G. 1946. The anatomy and function of the foreleg in salamander locomotion. Anat. Rec., 257.

GEGENBAUR, C. 1868. La torsion de l'humerus. Ann. d. Sci. Nat., Zool., 55.

GILMORE, C. W. 1926. Fossil footprints from the Grand Canyon. I. Smiths. Miscell. Coll., vol. 77, no. 9.—1927. II. Ibid., vol. 80, no. 3.—1928a. III. Ibid., vol. 80, no. 8.—1928b. Fossil footprints from the Fort Union (Paleocene) of Montana. Proc. U. S. Nat. Mus., vol. 74, art. 5.

GRAY, J. 1944. Studies in the mechanics of the tetrapod skeleton. Jour. Exper. Biol., 88.

GREGORY, W. K. 1911. The limbs of Eryops and the origin of paired limbs from fins. Science, 508.—1918. Note on the origin and evolution of certain adaptations for forward locomotion in the pectoral and pelvic girdles of reptiles and mammals. In: Studies in comparative myology and osteology, no. III, by W. K. Gregory and C. L. Camp. Bull. Amer. Mus. Nat. Hist., 515. —1935. Further observations on the pectoral girdle and fin of Sauripterus taylori Hall, a crossopterygian fish from the Upper Devonian of Pennsylvania, with special reference to the origin of the pentadactylate extremities of Tetrapoda. Proc. Amer. Philos. Soc., 673.—1937. The bridge-that-walks . . . Nat. Hist. Mag., 33.

GREGORY, W. K., R. W. MINER and G. K. NOBLE.

1923. The carpus of Eryops and the structure of the primitive chiropterygium. Bull. Amer. Mus. Nat. Hist., 279.

GREGORY, W. K. and H. C. RAVEN. 1941. Studies on the origin and early evolution of paired fins and limbs. Parts I–IV. Ann. N. Y. Acad. Sci., 273.

HAINES, R. W. 1938. The primitive form of epiphysis in the long bones of tetrapods. Jour. Anat., 323.—1939. A revision of the extensor muscles of the forearm in tetrapods. Ibid., 211.

HALL, J. 1843. Geology of New York. Part IV. In: Natural History of New York. New York, Boston, Albany. [Sauripterus, figs. 1–3, pl. 3.]

HOLMGREN, N. 1933. On the origin of the tetrapod limb. Acta Zool., 185.—1939. Contribution to the question of the origin of the tetrapod limb. Ibid., 89.

HOWELL, A. B. 1933a. Homology of the paired fins in fishes. Jour. Morph., 451.—1933b. Morphogenesis of the shoulder architecture. Quart. Rev. Biol., 247.—1935a. The primitive carpus. Jour. Morph., 105. Sauripterus.—1935b. Morphogenesis of the shoulder architecture. Part III. Amphibia. Quart. Rev. Biol., 397.—1936. Phylogeny of the distal musculature of the pectoral appendage. Jour. Morph., 287.

KLAATSCH, H. 1896. Die Brustflosse der Crossopterygier. Ein Beitrag zur Anwendung der Archipterygium-Theorie auf die Gliedmassen der Landwirbelthiere. Festschr. von Carl Gegenbaur, 259. [Polypterus, musculature of pectoral limb, pls. 2, 3.]

MINER, R. W. 1925. The pectoral limb of Eryops and other primitive tetrapods. Bull. Amer. Mus. Nat. Hist., 145.

MORTON, D. J. 1926. Notes on the footprint of Thinopus antiquus. Amer. Jour. Sci., 409.

OLSON, E. C. 1936. The ilio-sacral attachment of Eryops. Jour. Paleont., 648.

PETRONIEVICS, B. 1918. Note on the pectoral fin of Eusthenopteron. Ann. Mag. Nat. Hist., 471.

RABL, C. 1901. Gedanken und Studien über den Ursprung der Extremitäten. Zeits. f. Wissensch. Zool., 474.

ROMER, A. S. 1924. Pectoral limb musculature and shoulder-girdle structure in fish and tetrapods. Anat. Rec., 119.—1942. The development of tetrapod limb musculature—the thigh of Lacerta. Jour. Morph., 251.—1944. The development of tetrapod limb musculature—the shoulder region of Lacerta. Ibid., vol. 74, 1.

ROMER, A. S. and F. BYRNE. 1931. The pes of Diadectes: Notes on the primitive tetrapod limb. Paleobiologica, 25. [Relations of pre- and postaxial borders between fish and tetrapod.]

SCHAEFFER, B. 1941. The morphological and functional evolution of the tarsus in amphibians and reptiles. Bull. Amer. Mus. Nat. Hist., 395.

SCHMALHAUSEN, J. J. 1915. The development of the extremities in amphibians and their significance in the question of the origin of the extremities of terrestrial vertebrates. Moscou.—1917a. On the dermal bones of the shoulder-girdle of the Amphibia. Revue Zool. Russe, II. Moscou.—1917b. On the extremities of Ranidens sibiricus Kessl. Revue Zool. Russe, 129.

STEINER, H. 1921. Hand und Fuss der Amphibien: Ein Beitrag zur Extremitätenfrage. Anat. Anz., 515.—1935. Beiträge zur Gliedmassentheorie: Die Entwicklung des Chiropterygium aus den Ichthyopterygium. Revue Suisse de Zool., 715.

VIALLETON, L. 1924. Morphologie générale. Membres et ceintures des vertébres tétrapodes: Critique morphologique du Transformisme. Paris. [Excellent figures of girdles and limbs.]

WATSON, D. M. S. 1913. On the primitive tetrapod limb. Anat. Anz., 24.—1914. The cheirotherium. Geol. Mag., 395. [Tetrapod footprints.] —1917. The evolution of the tetrapod shoulder-girdle and fore-limb. Jour. Anat., 1.

WESTENHÖFER, M. 1926. Vergleichend-morphologische Betrachtungen über die Entstehung der Ferse und des Sprunggelenkes der Landwirbeltiere mit besonderer Beziehung auf den Menschen. Arkiv f. Frauenkunde und Konstitutionsforschung, Berlin.

WESTOLL, T. S. 1943. The origin of the primitive tetrapod limb. Proc. Roy. Soc., 373.

WILDER, H. H. 1919. The appendicular muscles of Necturus maculosus. Zool. Jahrb. Suppl., 15 (Festschr. f. J. W. Spengel, vol. 2).

WILLARD, B. 1935. Chemung tracks and trails from Pennsylvania. Jour. Paleont., 43. [Supposed origin of Amphibia, p. 52; Paramphibius, pls. 10, 11; later shown to be tracks of a limuloid.]

Tetrapoda: Vertebrae

EVANS, F. G. 1939. The morphology and functional evolution of the atlas-axis complex from fish to mammals. Ann. N. Y. Acad. Sci., 29.

GADOW, H. 1933. The evolution of the vertebral column: A contribution to the study of vertebrate phylogeny. Cambridge Univ. Press.

GREGORY, W. K., H. ROCKWELL and F. G. EVANS. 1939. Structure of the vertebral column in Eusthenopteron foordi Whiteaves. Jour. Paleont., 126.

HUENE, F. VON. 1926. Zur Frage der phylogenetischen Bedeutung des Wirbelbaues der Tetrapoden. Paleont. Zeitschr., vol. 7, 260.—1942. Die Wirbelstrukturen der Tetrapoden und ihre stammesgeschichtliche Wichtigkeit. Ibid., vol. 23, 219.

NOPCSA, F. 1930. Über die Orientierung konvexokonkaver Gelenkflächen. Anat. Anz., 401.

OLSON, E. C. 1936. The dorsal axial musculature of certain primitive Permian tetrapods. Jour. Morph., 265.

ROCKWELL, H., F. G. EVANS and H. C. PHEASANT. 1938. The comparative morphology of the vertebrate spinal column: Its form as related to function. Jour. Morph., 87.

WILLISTON, S. W. 1918. I. The evolution of vertebrae. II. The osteology of some American Permian vertebrates. Contrib. Walker Mus., 75.

Amphibia: General

GADOW, H. 1909. Amphibia and Reptiles. In: Cambridge Natural History, vol. 8.

NOBLE, G. K. 1931. The biology of the Amphibia. New York.

Amphibia: Brain and Nervous System

BLACK, D. 1917. The motor nuclei of the cerebral nerves in phylogeny: A study of the phenomena of neurobiotaxis. Part II. Amphibia. Jour. Comp. Neurol., 379.

KINGSBURY, B. F. 1903. Columella auris and nervus facialis in the Urodela. Jour. Comp. Neurol., 313.

KINGSBURY, B. F. and H. D. REED. 1908. The columella auris in Amphibia. Anat. Rec., 81.

KINGSLEY, J. S. 1902. The cranial nerves of Amphiuma. Tufts Coll. Studies, no. 7, 293.

Amphibia, Upper Devonian to Triassic Labyrinthodonts (Stegocephalia), Lepospondyls, etc.

AMALITSKY, V. P. 1921. Dvinosauridae: Excavations in the North Dvinsk region by Professor V. P. Amalitsky. Petrograd.—1924a. On the Dvinosauridae, a family of labyrinthodonts from the Permian of North Russia. Ann. Mag. Nat. Hist., 50.—1924b. On a new Cotylosauria of the family Seymouridae from the Permian of North Russia. Ibid., 64. [Kotlassia.]

AUGUSTA, J. 1936. Ein Stegocephalen-Bauchpanzer aus dem mährischen Perm . . . Zentralbl. f. Min., etc., 453.—1937. Bemerkungen zu ben Stegocephalen Melanerpeton pusillum Fr. und Branchiosaurus umbrosus Fr. aus dem böhmischen Perm. Věstnik Královské České Společnosti Nauk, Praha.

BAUR, G. 1896. The Stegocephali, a phylogenetic study. Anat. Anz., 657.

BROILI, F. 1904. Permische Stegocephalen und Reptilien aus Texas. Palaeontographica, 1. 4to. [Diplocaulus, Trimerorachis, Dissorophus, Labidosaurus, etc.]

BROILI, F. and J. SCHRÖDER. 1937. Beobachtungen an Wirbeltieren der Karrooformation. XXV. Über Micropholis Huxley. XXV. Über Lydekkerina Broom. Sitz. d. Bayer. Akad. d. Wissensch. 19.

BROOM, R. 1913. Studies on the Permian temnospondylous stegocephalians of North America. Bull. Amer. Mus. Nat. Hist., 563.—1915. The Triassic stegocephalians, Brachyops, Bothriceps and Lydekkerina. Proc. Zool. Soc. London, 363.

BULMAN, O. M. B. and W. F. WHITTARD. 1926. On Branchiosaurus and allied genera (Amphibia). Proc. Zool. Soc. London, 533.

BYSTROW, A. P. 1935. Morphologische Untersuchungen der Deckknochen des Schädels der Wirbeltiere. I. Mitteilung. Schädel der Stegocephalen. Acta Zool., 65.—1938a. Dvinosaurus als Neotenische Form der Stegocephalen. Acta Zool., 209.—1938b. Zahnstruktur der Labyrinthodonten. Ibid., 387.—1939. Blutgefässystem der Labyrinthodonten (Gefässe des Kopfes). Ibid., 125.

BYSTROW, A. P. and J. A. EFREMOV. 1940. Benthosuchus sushkini Efr., a labyrinthodont from the Eotriassic of Sharzhenga River [Ang. Sharjenga]. Travaux de l'Inst. Paléont. Acad. Sci. d. l'U.R.S.S.

CASE, E. C. 1917. The environment of the amphibian fauna at Linton, Ohio. Amer. Jour. Sci., 123.—1933. Progressive chondrification in the Stegocephalia. Proc. Amer. Philos. Soc., 265.—1946. A census of the determinable genera of the Stegocephlia. Trans. Amer. Philos. Soc., 325.

COLBERT, E. H. and B. SCHAEFFER. 1947. Some Mississippian footprints from Indiana. Amer. Jour. Sci., Oct., 614.

COPE, E. D. 1871. Observations on the extinct

batrachian fauna of the Carboniferous of Linton, Ohio. Proc. Amer. Philos, Soc., 177.—1873. On some new Batrachia and fishes from the Coal Measures of Linton, Ohio. Proc. Acad. Nat. Sci., Philadelphia, 340.—1874. Supplement to the extinct Batrachia, Reptilia and Aves of North America. I. Catalogue of the air-breathing Vertebrata from the Coal Measures of Linton, Ohio. Trans. Amer. Philos. Soc., 261.

COPE, E. D. and W. D. MATTHEW. 1915. Hitherto unpublished plates of Tertiary Mammalia and Permain Vertebrata. U. S. Geol. Surv. Monogr. [Eryops, pls. X–XIII; Megalichthys, pl. I.]

CREDNER, H. 1881–1893. Die Stegocephalen und Saurier aus dem Rothliegenden des Plauen'schen Grundes bei Dresden. I–X. Zeitschr. d. deutsch. geol. Gesellsch., vols. 33–35, 37, 38, 42.

DAWSON, J. W. 1863. Air-breathers of the Coal Period: A descriptive account of the remains of land animals found in the Coal Formation of Nova Scotia. Montreal.—1870. Notes on some new amphibian remains from the Carboniferous and Devonian of Canada. Quart. Jour. Geol. Soc. London, 166.—1895. Synopsis of the air-breathers of the Paleozoic of Canada. Proc. and Trans. Roy. Soc. Canada, vol. 12, sect. 4, 1.

DOUTHITT, H. 1917. The structure and relationships of Diplocaulus. Contrib. Walker Mus., 1. Univ. Chicago.

EFREMOV, J. A. 1837. Notes on the Permian Tetrapoda and the localities of their remains. On the labyrinthodonts from U.S.S.R. . . . Travaux d. l'Inst. Paléont. Acad. Sci. d. l'U.R.S.S. [English summary, pp. 23–27.]

EMBLETON, D. and T. H. ATTHEY. 1874. On the skull and other bones of Loxomma. Ann. Mag. Nat. Hist., 38.

FRAAS, E. 1888. Die Labyrinthodonten der Schwäbischen Trias. Stuttgart, I. Theil.—1889. II. Palaeontographica, vol. 36.—1913. Neue Labyrinthodonten der Schwäbischen Trias. Mitteil a. d. Kgl. Naturalien-Kabinett, 275. Stuttgart. 4to.

FRITSCH, A. 1883. Fauna der Gaskohle und der Kalksteine der Permformation Böhmens. Prag. 4to. [Branchiosauridae, Apateonidae, Aistopoda, Nectridea, Limnerpetidae, Hylonomidae, Microbrachidae.]

GADOW, H. 1909. Amphibians and reptiles. In: Cambridge Natural History, vol. 8. ["Proreptilia" founded on Eryops et al, p. 285.]

HUENE, F. VON. 1912. Der Unterkiefer von Diplocaulus. Anat. Anz., 472.—1922. Beiträge zur Kenntnis der Organisation einiger Stegocephalen der Schwäbischen Trias. Acta Zool., 395.

HUXLEY, T. H. 1867. On a collection of fossil vertebrates from the Jarrow Colliery, County Kilkenny, Ireland. Trans. Roy. Irish Acad., 351.

JARVIK, E. 1948. Note on the Upper Devonian vertebrate fauna of East Greenland and on the age of the ichthyostegid stegocephalians. Arkiv f. Zool. K. Svenska Vetenskapsakad., Bd. 41A, no. 13, 1.

KUSMIN, T. M. 1938. On the primitive features in the skull structure of the late stegocephalians. Problems of Paleontology, vol. 4, 1. [English translation, p. 30.]

MEHL, M. G. and E. B. BRANSON. 1928. Auditory organs of some labyrinthodonts. Bull. Geol. Soc. Amer., 485.

MOODIE, R. L. 1911. The temnospondylous Amphibia and a new species of Eryops from the Permian of Oklahoma. Kansas Univ. Sci. Bull., 235.—1916. The Coal Measures Amphibia of North America. Publ. Carnegie Inst. Washington, no. 238.

NILSSON, T. 1937. Ein Plagiosauride aus dem Rhät Schonens: Beiträge zur Kenntnis der Organisation der Stegocephalengruppe Brachyopoidei. Lunds Geol.-Min. Inst., 1.—1939. Cleithrum und humerus der Stegocephalen und rezenten Amphibien; auf grund neuer funde von Plagiosaurus depressus Jaekel. Lunds Univ. Ärsskr., N. F., vol. 35, no. 10.—1943a. Über einige postkraniale Skelettreste der triassischer Stegocephalen Spitsbergens. Bull. Geol. Instit. Upsala, 227.—1943b, 1944. On the morphology of the lower jaw of Stegocephalia . . . Kungl. Svenska Vetenskapsakad. Handl., vols. 20, 21.—1946a. On the genus Peltostega Wiman and the classification of the Triassic stegocephalians. Ibid., vol. 23.—1946b. A new find of Gerrothorax rhaeticus Nilsson, a plagiosaurid from the Rhaetic of Scania. Lunds Geol.-Min. Inst., no. 109, 1.

PARRINGTON, F. R. 1948. Labyrinthodonts from South Africa. Proc. Zool. Soc. London, 426.

ROMER, A. S. 1930. The Pennsylvanian tetrapods of Linton, Ohio. Bull. Amer. Mus. Nat. Hist., 77.—1936. Studies on American Permo-Carboniferous tetrapods. Problems of Paleontology, 85.—1939. Notes on branchiosaurs. Amer. Jour. Sci., 748.—1941. Evolution of the Amphibia (Abstract). Bull. Geol. Soc. Amer., 1931.—1945. The late Carboniferous vertebrate fauna of Kounova (Bohemia) compared with that of the Texas Redbeds. Amer. Jour. Sci., 417.—1947. Review of the Labyrinthodontia. Bull. Mus. Comp. Zool. Harvard Coll, 368.

ROMER, A. S. and R. V. WITTER. 1942. Edops, a primitive rhachitomous amphibian from the Texas Redbeds. Jour. Geol., 925.

SÄVE-SÖDERBERGH, G. 1932. Preliminary note on Devonian stegocephalians from East Greenland. Meddel. om Groenland, vol. 94, no. 7. [Skulls of "Ichthyostegalia."]—1936. On the morphology of Triassic stegocephalians from Spitsbergen, and the interpretation of the endocranium in the Labyrinthodontia. Kungl. Svenska Vetenskapsakad. Handl., vol. 16, no. 1. —1937. On the dermal skulls of . . . labyrinthodonts from the Triassic of Spitsbergen and North Russia. Bull. Geol. Instit. Upsala, 189.

STEEN, M. C. 1931. The British Museum collection of Amphibia from the Middle Coal Measures of Linton, Ohio. Proc. Zool. Soc. London (1930), 849.—1934. The amphibian fauna from the South Joggins, Nova Scotia. Ibid., 465. —1937. On Acanthostoma vorax Credner. Ibid., 491.—1938. On the fossil Amphibia from the Gas Coal of Nýřany and other deposits in Czechoslovakia. Ibid., 205.

SUSHKIN, P. P. 1936. Notes on the Pre-Jurassic Tetrapoda from U.S.S.R. III. Dvinosaurus amalitzki, a perennibranchiate stegocephalian . . . Edited by J. A. Efremov. Acad. Sci. U.S.S.R., 43.

WATSON, D. M. S. 1912. The larger Coal Measures Amphibia. Proc. Manchester Lit. Philos. Soc., 1. [Embolomerous vertebrae of Loxomma and Peteroplax more primitive than rachitomous type.]—1913. Batrachiderpeton lineatum Hancock and Atthey, a Coal-Measure stegocephalian. Proc. Zool. Soc. London, 949.—1919. The structure, evolution and origin of the Amphibia.—The "Orders" Rachitomi and Stereospondyli. Philos. Trans. Roy. Soc. London, vol. 209, 1.—1921. On Eugyrinus wildi (A.S.W.), a branchiosaur from the Lancashire Coal Measures. Geol. Mag., 70.—1926a. The Carboniferous Amphibia of Scotland. Palaeontologia Hungarica (1921–1923) (1926), 221.—1926b. The evolution and origin of the Amphibia. Croonian Lecture. Philos. Trans. Roy. Soc. London, 189. —1940. The origin of frogs. Trans. Roy. Soc. Edinburgh (1939–1940), 195.—1942. On Permian and Triassic tetrapods. Geol. Mag., 81.

WILLISTON, S. W. 1909. New or little-known Permian vertebrates: Trematops, new genus. Jour. Geol., 636 [Pelvic limb.]—1910. Cacops, Desmospondylus: New genera of Permian vertebrates. Bull. Geol. Soc. Amer., 249.—1915. Trimerorhachis, a Permian temnospondyl amphibian. Jour. Geol., 246.—1916. The skeleton of Trimerorhachis. Ibid., 291.

WILSON, J. A. 1941. An interpretation of the skull of Buettneria, with special reference to the cartilages and soft parts. Contrib. Mus. Paleont. Univ. Michigan, 71.

Amphibia, Jurassic to Recent

BRONN, H. G. 1873–1878. Klassen und Ordnungen der Amphibien . . . In: Die Klassen und Ordnungen des Thier-Reichs . . . vol. 6. Leipzig und Heidelberg.

DUNN, E. R. 1923. The geographical distribution of amphibians. Amer. Nat., 129.

EVANS, F. G. 1944. The morphological status of the modern Amphibia among the Tetrapoda. Jour. Morph., 43.

GADOW, H. 1901. Amphibians and reptiles. In: Cambridge Natural History, vol. 8.

KLAAUW, C. J. VAN DER. 1924. Bau und Entwickelung der Gehörknöchelchen: Literatur, 1899–1923. Erkeb. d. Anat. u. Entwick., 565.

NOBLE, G. K. 1931. The biology of the Amphibia. New York and London.

REED, H. D. 1909. The columella auris in Amphibia. Jour. Morph., 549.

ROMER, A. S. and T. EDINGER. 1942. Endocranial casts and brains of living and fossil Amphibia. Jour. Comp. Neurol., 355.

Urodela (Caudata)

FRANCIS, E. T. B. 1934. The anatomy of the salamander. Oxford.

MOODIE, R. L. The ancestry of the caudate Amphibia. Amer. Nat., 361.

MOOKERJEE, H. K. 1930. On the development of the vertebral column of Urodela. Philos. Trans. Roy. Soc. London, 415.

NOBLE, G. K. 1927. The plethodontid salamanders; some aspects of their evolution. Amer. Mus. Novitates, no. 249.

PARKER, W. K. 1876. On the structure and development of the skull in the urodelous Amphibia. Part I. Philos. Trans. Roy. Soc. London, 529.

PIATT, J. 1935. A comparative study of the hyobranchial apparatus and throat musculature in the Plethodontidae. Jour. Morph., 213.

SMITH, B. G. 1912 a, b. The embryology of Cryptobranchus allegheniensis. I. Jour. Morph., vol. 23, no. 1; II. Ibid, no. 3, 455.—1916. III. Ibid., vol. 42, no. 1, 197.

VERSLUYS, J. 1909. Die Salamander und die Ursprünglichsten vierbeiningen Landwirbeltiere. Naturw. Wochenschr., vol. 8, no. 3.

Anura (Salientia)

BADENHORST, C. E., J. D. MASS, W. A. MAREE, G. K. SLABBERT. 1945. Cranial morphology of Anura. Five papers in Ann. Univ. Stellenbosch, vol. 23, sect. A, nos. 2–6.

EATON, T. H., JR. 1939. Development of the frontoparietal bones in frogs. Copeia, 95.—1942. Are "frontoparietal" bones in frogs actually frontals? Jour. Washington Acad. Sci., 151.

GAUPP, E. 1896–1904. A. Eckers und R. Wiedersheims Anatomie des Frosches. Auf Grund eigener Untersuchungen durchaus neu bearbeitet. Part I, 1896; Part II, 1899; Part III, 1904. Braunschweig.

MOOKERJEE, H. K. 1931. On the development of the vertebral column of Anura. Philos. Trans. Roy. Soc. London, 165.

NOBLE, G. K. 1922. The phylogeny of the Salientia. I. The osteology and the thigh musculature; their bearing on classification and phylogeny. Bull. Amer. Mus. Nat. Hist., 3.

NOBLE, G. K. and M. E. JAECKLE. 1928. The digital pads of the tree frogs: A study of the phylogenesis of an adaptive structure. Jour. Morph. and Physiol., 259.

PUSEY, H. K. 1943. On the head of the liopelmid frog, Ascaphus truei. I. The chondrocranium, jaws, arches, and muscles of a partly-grown larva. Quart. Jour. Micros. Sci., 105. [Supports Noble's view that Liopelma is a quite primitive frog; opposes Säve-Söderbergh.]

RAMASWAMI, L. S. 1934. Contributions to our knowledge of the cranial morphology of some ranid genera of frogs. Part I. Proc. Indian Acad. Sci., vol. I, no. 2, 80.—1935a. The cranial morphology of some examples of Pelobatidae (Anura). Anat. Anz., 65.—1935b. Contributions to our knowledge of the cranial morphology of some ranid genera of frogs. Part II. Proc. Indian Acad. Sci., vol. 2, no. 1.—1937. The morphology of the bufonid head. Proc. Zool. Soc. London (1936), 1157.—1939. Some aspects of the anatomy of Anura (Amphibia): A review. Proc. Indian Acad. Sci., 41.

STADTMÜLLER, F. 1931. Zur Entwicklungsgeschichte der cartilago supraorbitalis bei Alytes obstetricans. Anat. Anz., 241. [Chondrocranium of Amphibia.]

WATSON, D. M. S. 1940. The origin of frogs. Trans. Roy. Soc. Edinburgh (1939–1940), 195.

Caecilia (Apoda)

DUNN, E. R. 1942. The American caecilians. Bull. Mus. Comp. Zool. Harvard Coll., 439.

SARASIN, P. and F. SARASIN. 1887–1890. Ergebnisse Naturwiss. Forsch. auf Ceylon. II. Band. Zur Entwickl. und Anat. der ceylonesischen Blindwühle Ichthyophis glutinosus, L. Wiesbaden, 4to.

Reptilia, General:
Textbooks, Major Classification, Phylogeny
(See also Vertebrata, Tetrapoda)

BAUR, G. 1887. On the phylogenetic arrangement of the Sauropsida. Jour. Morph., 93. [Stereosternum, Proganosauria, p. 103; turtles related to stem of plesiosaurs, pp. 97–99.]

BROOM, R. 1924. On the classification of the reptiles. Bull. Amer. Mus. Nat. Hist., 39.

DITMARS, R. L. 1922. Reptiles of the world: tortoises and turtles, crocodilians, lizards and snakes of the Eastern and Western Hemispheres. New York.

FÜRBRINGER, M. 1900. Beitrag zur Systematic und Genealogie der Reptilien. Jenaisch. Zeits. f. Naturw.

GREGORY, J. T. 1948. The structure of Cephalerpeton and affinities of the Microsauria. Amer. Jour. Sci., Sept., 559. [Microsaurs regarded as an order of the Infraclass Captorhina, Subclass Eureptilia. See also Tetrapoda, General, above.]

HOFSTEN, N. VON. 1941. On the phylogeny of the Reptilia. Zool. Bidrag fran Uppsala. Festskr. Prof. S. Ekman, 501.

HUENE, F. VON. 1914. Beiträge zur Geschichte der Archosaurier. Geol. u. Paläontol. Abh., 3.—1925. Stammlinien der Reptilien. Centralb. f. Min., etc., 229.—1944. Die Zweiteilung des Reptilstammes. Neuen Jahrb. f. Min., etc., 427.

LYDEKKER, R. and others. Undated. Reptiles. In: The New Natural History, vol. 5, sect. 9. New York.

NOPCSA, F. 1923a. Reversible and irreversible evolution: A study based on reptiles. Proc. Zool.

Soc. London, 1045.—1923b. Die Familien der Reptilien. Fortschr. d. Geol. u. Palaeontol. Berlin.—1928. The genera of reptiles. Palaeobiologica, vol. 1, 163.

OLSON, E. C. 1947. The family Diadectidae and its bearing on the classification of reptiles. Fieldiana: Geol., vol. 11, no. 1. [Divides Class Reptilia into two subclasses: Parareptilia (including Seymouria, Diadectes, Kotlassia) and Eureptilia (true reptiles, including captorhinomorphs).]

OSBORN, H. F. 1903. The reptilian subclasses Diapsida and Synapsida and the early history of the Diaptosauria. Mem. Amer. Mus. Nat. Hist., 1.

SEELEY, H. G. 1887–1895. Researches on the structure, organization and classification of the fossil Reptilia. Parts I–IX. Philos. Trans. Roy. Soc. London, vols. 178–186.

WILLISTON, S. W. 1917. The phylogeny and classification of reptiles. Jour. Geol., 411.—1925. The osteology of the reptiles. Harvard Univ. Press. Cambridge.

ZITTEL, K. A. VON. 1932. Text-book of palaeontology. 2nd English ed. London.

Reptilia, Fossil and Recent; Miscellaneous
(See also Tetrapoda)

DITMARS, R. L. 1905. The reptiles of the vicinity of New York City. Amer. Mus. Jour., 93.

EFREMOV, J. A. 1940a. New discoveries of Permian terrestrial vertebrates in Bashkiria and the Tchkalov Province. Comptes Rendus (Doklady) de l'Acad. d. Sci. d. l'U.R.S.S., 412.—1940b. Die Mesen-Fauna der permischen Reptilien. Neuen Jahrb. f. Min., etc., Suppl. to vol. 84, B, 379.

LULL, R. S. 1912. The life of the Connecticut Trias. Amer. Jour. Sci., 397.

MEYER, H. VON. 1856. Zur Fauna der Vorwelt. III. Saurier aus dem Kupferschiefer der Zechstein-Formation. Frankfurt. Folio.—1858. Reptilien aus der Steinkohlen-Formation in Deutschland. Cassel. Folio.—1860. Zur Fauna der Vorwelt. IV. Reptilien aus dem Lithographischen Schiefer des Jura in Deutschland und Frankreich. Frankfurt. Folio.

PEYER, B. 1944. Die Reptilien von Monte San Giorgio. Neujahrsblatt, Naturforsch. Gesellsch. in Zürich, 146. Stück.

SCHMIDT, K. P. 1919. Contribution to the herpetology of the Belgian Congo . . . Bull. Amer. Mus. Nat. Hist., 385.

STROMER, E. 1936. Ergebn. der Forschungsreisen Prof. E. Stromers in den Wüsten Ägyptens. VII. Abh. d. Bayer. Akad. d. Wissenschaften, N. F., Teil 33.

WILLISTON, S. W. 1911. American Permian vertebrates. Univ. Chicago Press.—1916. The osteology of some American Permian vertebrates. II. Contrib. Walker Mus., Univ. Chicago, 165.—1918. III. Ibid., 87.

Reptilia: Brain and Nervous System

BLACK, D. 1920. The motor nuclei of the cerebral nerves in phylogeny: A study of the phenomena of neurobiotaxis. Part III. Reptilia. Jour. Comp. Neurol., 61.

CRAIGIE, E. H. 1936. Notes on cytoarchitectural features of the lateral cortex and related parts of the cerebral hemisphere in a series of reptiles and birds. Trans. Roy. Soc. Canada, 87.

DENDY, A. 1910. On the structure, development and morphological interpretation of the pineal organs and adjacent parts of the brain in the tuatera (Sphenodon punctatus). Proc. Roy. Soc., 629.

EDINGER, L. 1896. Untersuch. über die vergl. Anat. des Gehirns. Neue Studien über des Vorderhirn der Reptilien. Sencken. naturf. Ges., 313.

EDINGER, T. 1927. Das Gehirn der Pterosaurier. Zeitschr. f. Anat. u. Entwickl., 105.—1941. The brain of Pterodactylus. Amer. Jour. Sci., 661.

HERMAN, W. 1925. The relation of the corpus striatum and the pallium in Varanus and a discussion of their bearing on birds, mammals and man. Brain, vol. 48, 362.

SCHEPERS, G. W. H. 1948. Evolution of the forebrain. The fundamental anatomy of the telencephalon, with special reference to that of Testudo geometrica. Cape Town, So. Africa. [Vertebrate homologies of the reptilian telencephalon.]

WARNER, F. J. 1931. The cell masses in the telencephalon and diencephalon of the rattlesnake, Crotalus atrox. Proc. K. Acad. Sci. Wetens. Amsterdam, 1156.—1935. The medulla of Crotalus atrox. Jour. Nervous and Mental Disease, 504.—1942. The development of the diencephalon of the American water snake (Natrix sipedon). Trans. Roy. Soc. Canada, 53.—1946. The fibre tracts of the forebrain of the American diamond-back rattlesnake (Crotalus adamanteus). Proc. Zool. Soc. London, 22.

Reptilia: Circulatory and Respiratory Systems
(See also Vertebrates, Comparative Anatomy)

BOGERT, C. M. 1939. Reptiles under the sun. Nat. Hist. Mag. vol. 44, no. 1, 26. [Adjustments to changing temperature.]

DOMBROWSKI, B. 1930. Zur Phylotektonik der respiratorischen Muskulatur der Reptilien und Säugetiere. Zeitschr. f. Anat. u. Entwicklungsgeschichte, vol. 93, 353.

HUNTINGTON, G. S. 1911. The development of the lymphatic system in the reptiles. Anat. Rec., 261.

MATHUR, P. N. 1944. The anatomy of the reptilian heart. Part I. Varanus monitor (Linné). Proc. Indian Acad. Sci., 1.

Reptilia: Skull, Temporal Arches, Hyoid System
(See also Vertebrate Skull, Tetrapod Skull and Jaws)

BAUR, G. 1894. Bemerkungen über die Osteologie der Schläfengegend der höheren Wirbeltiere. Anat. Anz., 315.

BROOM, R. 1922a. On the temporal arches of the Reptilia. Proc. Zool. Soc. London. 17.—1922b. On the persistence of the mesopterygoid in certain reptilian skulls. Ibid., 455.

FÜRBRINGER, M. 1922. Das Zungenbein der Wirbeltiere insbesondere der Reptilien und Vögel . . . Abh. d. Heidelberger Akad. d. Wiss., vol. 11.

GAUPP, E. 1900. Das Chondrocranium von Lacerta agilis . . . Anat. Hefte, 433.

GREGORY, W. K. 1913. Homology of the "lacrimal" and of the "alisphenoid" in recent and fossil reptiles. Bull. Geol. Soc. Amer., 241.—1920. . . . A review of the evolution of the lacrimal bone of vertebrates . . . Bull. Amer. Mus. Nat. Hist., 95.

GREGORY, W. K. and L. A. ADAMS. 1915. The temporal fossae of vertebrates. Science, May 21, 763.

HUENE, F. VON. 1923. Contribution to the vomer-parasphenoid question. Bull. Geol. Soc. Amer., 459.

PARRINGTON, F. R. 1937. A note on the supratemporal and tabular bones in reptiles. Ann. Mag. Nat. Hist., 69.

SEWERTZOFF, S. A. 1929. Zur Entwicklungsgeschichte der Zunge bei den Reptilien. Acta Zool., 231.

WILLISTON, S. W. 1904. The temporal arches of the Reptilia. Biol. Bull., 175.

Reptilia: Jaws and Teeth
(See also Vertebrate Skull, Fish Jaws and Teeth, Tetrapod Jaws, etc.)

BAUR, G. 1895. Ueber die Morphologie des Unterkiefers der Reptilien. Anat. Anz., vol. 11, no. 13.

BOLK, L. 1912. On the structure of the dental system of reptiles. Akad. van Wetenschappen te Amsterdam, 950.

PARRINGTON, F. R. 1936. Further notes on tooth-replacement. Ann. Mag. Nat. Hist., 109.

WATSON, D. M. S. 1912. On some reptilian lower jaws. Ann. Mag. Nat. Hist., 573.

Reptilia: Locomotor System, Skeleton, Girdles and Limbs, Limb Musculature.
(See also Tetrapod Locomotor System)

BROOM, R. 1920. On the structure of the reptilian tarsus. Proc. Zool. Soc. London, 143.

FÜRBRINGER, M. 1900. Zur vergl. Anat. des Brustschulterapparates und der Schultermuskeln. IV Teil. (Lacertilier, Rhynchocephalier und Crocodilier.) Jenaischen Zeitschr. f. Naturw., 215.

GADOW, H. 1882. Beiträge zur Myologie der hinteren Extremität der Reptilien. Morphol. Jahrb. (Gegenbaur), 329.

GREGORY, W. K. and C. L. CAMP, 1918. Studies in comparative myology and osteology. No. III. Bull. Amer. Mus. Nat. Hist., 447. Part 3. Note on the origin and evolution of certain adaptations for forward locomotion in the pectoral and pelvic girdles of reptiles and mammals. Ibid., 515. Part 4. Note on the morphology and evolution of the femoral trochanters in reptiles and mammals. Ibid., 528. Part 5. Second note on the evolution of the coracoid elements in reptiles and mammals. Ibid., 545. Summary and conclusions. Ibid., 553.

HOWELL, A. B. 1936. Morphogenesis of the shoulder architecture. Part IV. Reptilia. Quart. Rev. Biol., 183.

MIVART, ST. G. 1867. Notes on the myology of Iguana tuberculata. Proc. Zool. Soc. London, 766.

MOODIE, R. L. 1908. Reptilian epiphyses. Amer. Jour. Anat., 443.

ROMER, A. S. 1922. The locomotor apparatus of certain primitive and mammal-like reptiles. Bull. Amer. Mus. Nat. Hist., 517.

SCHAEFFER, B. 1941. The morphological and functional evolution of the tarsus in amphibians and reptiles. Bull. Amer. Mus. Nat. Hist., 395. [Bearing on classification and phylogeny of reptiles.]

Reptilia: Vertebrae, Ribs (See also Tetrapod Vertebrae)

HUENE, F. VON. 1940. Die Bedeutung der doppelköpfigen Caudalrippen. Zentralb. f. Min., etc., no. 4, 115.

NOPCSA, F. Über prozöle und opisthozöle Wirbel. Anat. Anz., 19.

WILLISTON, A. W. 1910. New Permian reptiles: rhachitomous vertebrae. Jour. Geol., 585.

Seymouriamorpha

AMALITZKY, V. P. 1924. On a new Cotylosauria of the family Seymouridae from the Permian of North Russia. Ann. Mag. Nat. Hist., 64 [Kotlassia.]

BROILI, F. 1904. Permische Stegocephalen und Reptilien aus Texas. Palaeontographica, 1. [Seymouria baylorensis.]—1927. Über den Zahnbau von Seymouria. Anat. Anz., 185.

BYSTROW, A. P. 1940. On the microscopical structure of the scales of the dorsal armour of Kotlassia prima Amal. Bull. d. l'Acad. d. Sci. d. l'U.R.S.S., 125.—1944. Kotlassia prima Amalitzky. Bull. Geol. Soc. Amer., 379.

COPE, E. D. 1896. The ancestry of the Testudinata. Amer. Nat., 398. [Conodectes, n.g.]

SUSHKIN, P. P. 1923. On the representatives of the Seymouriamorpha, supposed primitive reptiles, from the Upper Permian of Russia . . . Occ. Papers, Boston Soc. Nat. Hist., 179. [Kotlassia, figs. 1, 4–6.]—1926. Notes on the Pre-Jurassic Tetrapoda from Russia. I, II, III. Palaeontologia Hungarica, I, 323. [Seymouriamorpha, pp. 337–341, Kotlassia.]

WATSON, D. M. S. 1919. On Seymouria, the most primitive known reptile. Proc. Zool. Soc. London, 1918 (1919), 267.

WHITE, T. E. 1939. Osteology of Seymouria baylorensis Broili. Bull. Mus. Comp. Zool. Harvard Coll., 325.

WILLISTON, S. W. 1910. Cacops, Desmospondylus, new genera of Permian vertebrates. Bull. Geol. Soc. Amer., 249 [Desmospondylus related to Seymouria.]—1911. Restoration of Seymouria baylorensis Broili, an American cotylosaur. Jour. Geol., 232.

Cotylosauria (including Pareiasuria)

AMALITZKY, V. P. [Illustrations for a memoir on Pareiasaurus, Inostrancevia, Dicynodon and other Permian (?) reptiles and amphibians.] Unpublished.

BOONSTRA, L. D. 1929–1933. Pareiasaurian studies. [See also Haughton and Boonstra.] Ann. South African Mus., vol. 28. [Manus, 97; pes, 113; hind limb, 429; osteology and myology, fore limb, 437.] Vol. 31. [Cranial osteology, 1; dermal armour, 39; vertebral column and ribs, 49.]—1930. A contribution to the cranial osteology of Pareiasaurus serridens (Owen). Ann. Univ. Stellenbosch, 1.—1932a. A note on the hyoid apparatus of the Permian reptiles (Pareiasaurians). Anat. Anz., 65.—1932b. The phylogenesis of the Pareiasauridae: A study in evolution. South African Jour. Sci., 480.

BROILI, F. and J. SCHRÖDER. 1936. Beobachtungen an Wirbeltieren der Karrooformation. XXI. Über Procolophon Owen. Sitz. d. Bayer. Akad. d. Wiss., 230.

BROOM, R. 1913a. On the cotylosaurian genus Pantylus Cope. Bull. Amer. Mus. Nat. Hist., 527. [A specialized microsaurian with crushing teeth.—Romer, 1945.]—1913b. On the structure and affinities of Bolosaurus. Ibid., 509.—1914. Some points in the structure of the diadectid skull. Ibid., 109.—1930. On a new species of Anthodon (A. gregoryi). Amer. Mus. Novitates, no. 448.—1938. On a new type of primitive fossil reptile from the upper Permian of South Africa. Proc. Zool. Soc. London, 535. [Millerina, "between cotylosaurs and synapsids and diapsids"; see p. 541.]

CASE, E. C. 1905. The osteology of the Diadectidae and their relations to the Chelydosauria. Jour. Geol., 126.—1910. New or little known reptiles and amphibians from the Permian (?) of Texas. Bull. Amer. Mus. Nat. Hist., 163.—1911. A revision of the Cotylosauria of North America. Carnegie Inst. Washington Publ. no. 145.

CASE, E. C. and S. W. WILLISTON. 1912. A description of the skulls of Diadectes lentus and Animasaurus carinatus. Amer. Jour. Sci., 339.

COLBERT, E. H. 1946. Hypsognathus, a Triassic reptile from New Jersey. Bull. Amer. Mus. Nat. Hist., 225.

COPE, E. D. 1878. Descriptions of new Vertebrata from the Upper Tertiary formations of the West.

Proc. Amer. Philos. Soc., 219. [Diadectes molaris.]—1880. The skull of Empedocles. Amer. Nat., 304. [First use of name "Cotylosauria."]—1896a. The reptilian order Cotylosauria. Proc. Amer. Philos. Soc. (1895), 436.—1896b. Second contribution to the history of the Cotylosauria. Ibid., 122.—1898. Syllabus of lectures on the Vertebrata. Univ. Pennsylvania, Philadelphia. [Chelydosauria.]

HARTMANN-WEINBERG, A. 1929. Über Carpus und Tarsus der Pareiasauriden. Anat. Anz., 401.—1930. Zur Systematik der Nord-Düna-Pareiasauridae. Palaeont. Zeitschr., 47.—1937. Pareiasauriden als Leit-fossilien. Problems of Paleontology, vols. II, III, 649.

HAUGHTON, S. H. 1929–1933. Pareiasaurian studies [see also Haughton and Boonstra]. Ann. South African Mus., vol. 28. [Notes on some pareiasaurian braincases, 88.]

HAUGHTON, S. H. and L. D. BOONSTRA. 1929–1933. Pareiasaurian studies [see also L. D. Boonstra]. Ann. South African Mus., vol. 28. [Classification based on skull features, 79; mandible, 261; osteology and myology, hind limb, 297.]

HUENE, F. VON. 1912. Die Cotylosaurier der Trias. Palaeontographica, 69. [Procolophonia.]—1920a. Sclerosaurus und seine Beziehungen zu anderen Cotylosaurien und zu den Schildkröten. Zeitschr. f. induktive Abstammungs - Vererbungslehre, 163.—1920b. Ein Telerpeton mit gut erhaltenem Schädel. Centralb. f. Min., etc., nos. 11, 12, 189. [Procolophonia.]—1943. Zur Beurteilung der Procolophoniden. Neue Jahrb. f. Min., etc., 192.—1944. Pareiasaurierreste aus dem Ruhuhu-Gebiet. Palaeont. Zeitschr., 386.

OLSON, E. C. 1937. A mounted skeleton of Labidosaurus Cope. Jour. Geol., 95.—1947. The Family Diadectidae and its bearing on the classification of reptiles. Fieldiana: Geology, 1.

PRICE, L. I. 1935. Notes on the brain case of Captorhinus. Proc. Boston Soc. Nat. Hist., 377.

ROMER, A. S. 1943. Recent mounts of fossil reptiles and amphibians in the Museum of Comparative Zoology. Bull. Mus. Comp. Zool. Harvard Coll., 331. [Including Diadectes.]—1944. The Permian cotylosaur Diadectes tenuitectus. Amer. Jour. Sci., 139.—1946. The primitive reptile Limnoscelis restudied. Ibid., 149.

ROMER, A. S. and F. BYRNE. 1931. The pes of Diadectes: Notes on the primitive tetrapod limb. Paleobiologica, 25.

SEELEY, H. G. 1888. Researches on the structure, organization and classification of the fossil Reptilia. II. On Pareiasaurus . . . Philos. Trans. Roy. Soc. London, 59.—1892. VII. Further observations on Pareiasaurus. Ibid., 311.

SUSHKIN, P. P. 1928. Contributions to the cranial morphology of Captorhinus Cope . . . Palaeobiologica, 263.

WATSON, D. M. S. 1914a. On the skull of a pareiasaurian reptile, and on the relationship of that type. Proc. Zool. Soc. London, 155.—1914b. Procolophon trigoniceps, a cotylosaurian reptile from South Africa. Proc. Zool. Soc., London, 735.

WELLES, S. P. 1941. The mandible of a diadectid cotylosaur. Univ. California Publs., Bull. Dept. Geol. Sci., 423.

WESTOLL, T. S. 1942. Ancestry of captorhinomorph reptiles. Nature, 667.

WILLISTON, S. W. 1908. The Cotylosauria. Jour. Geol., 139.—1909. New or little known Permian vertebrates. Pariotichus. Biol. Bull., 241.—1910. The skull of Labidosaurus. Amer. Jour. Anat., 69.—1911. American Permian vertebrates. Univ. Chicago Press. [Limnoscelis, figs. 4, 13.]—1912. Restoration of Limnoscelis, a cotylosaur reptile from New Mexico. Amer. Jour. Sci., 357.—1917. Labidosaurus Cope, a Lower Permian cotylosaur reptile from Texas. Jour. Geol., 309.

Chelonia

BAUR, G. 1887. On the morphogeny of the carapace of the Testudinata. Amer. Nat., 89.—1893. Classification of Cryptodira. Amer. Nat., 671.—1896. Bemerkungen über die Phylogenie der Schildkröten. Anat. Anz., 561.

COPE, E. D. 1896. The ancestry of the Testudinata. Amer. Nat., 398. [Describes new family, Otocoelidae; new genera. Otocoelus, later shown to be a dissorophid, and Conodectes.]

GOETTE, A. 1899. Über die Entwicklung des Knöchernen Rückenschildes (Carapax) der Schildkröten. Zeitschr. f. Wiss. Zool., 407.

GREGORY, W. K. 1946. Pareiasaurs versus placodonts as near ancestors to the turtles. Bull. Amer. Mus. Nat. Hist., 275.

HAY, O. P. 1905. On the group of fossil turtles known as the Amphichelydia; with remarks on the origin and relationships of the suborders, superfamilies and families of testudines. Bull. Amer. Mus. Nat. Hist., 137.—1908. The fossil turtles of North America. Carnegie Inst. Washington Publ. no. 75.—1922. On the phylogeny of the shell of the Testudinata and the relationships of Dermochelys. Jour. Morph., 421.

HUENE, F. VON. 1928. Einige Schildkrötenreste aus der obersten Trias Württembergs. Centralbl. f. Min., etc., 509. [Humerus of primitive chelonian type. Concludes that in the Trias, besides terrestrial amphichelydians, there were also marine cryptodirans.]—1943. Bemerkungen über Valléns ausführung über den Schildkrötenpanzer. Neuen Jahrb. f. Min., etc., 198.

JAEKEL, O. 1914. Über die Wirbeltierfunde in der oberen Trias von Halberstadt. Palaeontol. Zeitschr., 155. [Plateosaurus, Stegochelys.]—1916. Die Wirbeltierfunde aus dem Keuper von Halberstadt. Serie II. Testudinata. Teil 1. Stegochelys dux, n.g., n. sp. Ibid., 88.—1918. Stegochelys dux.-Nachschrift "Triassochelys" pro "Stegochelys." Ibid., 251.

MOODIE, R. L. 1908. The relationship of the turtles and plesiosaurs. Kansas Univ. Sci. Bull., 319.

NILSSON, T. 1945. The structure of the cleithrum in plagiosaurids and the descent of Chelonia. Arkiv. f. Zool., 1.

OWEN, R. 1849. On the development and homologies of the carapace and plastron of the chelonian reptiles. Mem. Trans., Geol. Soc. London, 151.

RUCKES, H. 1929. Studies in chelonian osteology. I. Truss and arch analogies in chelonian pelves. Ann. N. Y. Acad. Sci., 31. II. The morphological relationships between the girdles, ribs and carapace. Ibid., 81.

SIMPSON, G. G. 1938. Crossochelys, Eocene horned turtle from Patagonia. Bull. Amer. Mus. Nat. Hist., 221.—1942. A Miocene tortoise from Patagonia. Amer. Mus. Novitates, no. 1209.—1943. Turtles and the origin of the fauna of Latin America. Amer. Jour. Sci., 413.

VERSLUYS, J. 1913. On the phylogeny of the carapace and on the affinities of the leathery turtle, Dermochelys coriacea. Brit. Assoc. Adv. Sci., 1913, 1.—1914. Über die Phylogenie des Panzers der Schildkröten und über die Verwandtschaft der Lederschildkröte (Dermochelys coriacea). Zeits. f. Palaeont., 321.

WATSON, D. M. S. 1914. Eunotosaurus africanus Seeley and the ancestry of the Chelonia. Proc. Zool. Soc. London, 1011.

WIELAND, G. R. 1906. The osteology of Protostega. Mem. Carnegie Mus., 279.—1909. Revision of the Protostegidae. Amer. Jour. Sci., 101.

WILLIAMS, E. E. 1950. Variation and selection in the cervical central articulations of living turtles. Bull. Amer. Mus. Nat. Hist.

ZANGERL, R. 1939. The homology of the shell elements in turtles. Jour. Morph., 383.

Marine Reptiles, General and Miscellaneous

ABEL, O. 1924. Die Eroberungszüge der Wirbeltiere in die Meere der Vorzeit. Jena.

ANDREWS, C. W. 1910–1913. A descriptive catalogue of marine reptiles of the Oxford Clay. Part I, 1910. Part II, 1913. London, 4to.

HAWKINS, T. 1834. Memoirs of Ichthyosauri and Plesiosauri. Extinct monsters of the ancient earth. London. Folio.—1840. Book of the great sea dragons: Ichthyosauri and Plesiosauri. London. Folio.

MEYER, H. VON. 1847–1855. Zur Fauna der Vorwelt. II. Die Saurier des Muschelkalkes mit Rücksicht auf die Saurier aus Buntem Sandstein und Keuper. Frankfurt am Main. Folio. [Nothosaurs.]

OSBURN, R. C. 1906. Adaptive modifications of the limb skeleton in aquatic reptiles and mammals. Ann. N. Y. Acad. Sci., 447.

PEYER, B. 1932. Die Triasfauna der Tessiner Kalkalpen. I. Einleitung. Abh. d. Schweis. Palaeont. Gesell. (1931), vol. 50.—1936–1937. X. Clarazia schinzi n.g., n.sp. XI. Hescheleria rübeli n.g., n.sp. XII. Macrocnemus bassanii Nopcsa. Ibid., vols. 57 (1936); 58, 59 (1937).

WILLISTON, S. W. 1914. Water reptiles of the past and present. Univ. Chicago Press. [Progressive reduction in cranial bones, pp. 3, 21–25.]

Marine Reptiles: Mesosaurs

BROOM, R. 1907. On some new fossil reptiles from the Karroo Beds of Victoria West, South Africa. Trans. South African Philos. Soc., 31. [Heleosaurus.]

COPE, E. D. 1885. A contribution to the vertebrate paleontology of Brasil. Proc. Amer. Philos. Soc., 1. [Stereosternum.]

GERVAIS, P. 1867–1869. Zoologie et Paléontologie Générales. Atlas. Paris. [Mesosaurus, pl. 42.]

HUENE, F. VON. 1940. Das unterpermische Alter aller Mesosaurier führenden Schichten. Zentr. f. Min., etc., Jahr., 200.

McGREGOR, J. H. 1908. On Mesosaurus brasiliensis . . . from the Permian of Brasil. [Stereosternum.]

Commissao dos Estudos das Minas de Carvao de Petra de Brasil, 302.

Marine Reptiles: Ichthyosauria

BROILI, F. 1916. Einige Bemerkungen über die Mixosauridae. Anat. Anz., 474.

HUENE, F. VON. 1922. Die Ichthyosaurier des Lias und ihre Zusammenhänge. Berlin. 4to.—1925. Einige Beobachtungen an Mixosaurus cornalianus (Bassani). Centralbl. f. Min. etc., 289.—1931. Neue Studien über Ichthyosaurier aus Holzmaden. Abh. Senckenberg, Naturf. Ges., 345.—1935. Neue Beobachtungen an Mixosaurus. Palaeontol. Zeitschr., 159.—1937. Die Frage nach der Herkunft der Ichthyosaurier. Bull. Geol. Instit. Upsala, 1. [Mesosaurus, fig. 1.] —1943. Bemerkungen über primitive Ichthyosaurier. Neues Jahrb. f. Min., etc., 154.

JAEKEL, O. 1904. Eine neue Darstellung von Ichthyosaurus. Marz-Protokoll d. Deutschen geol. Gesell., 26.

MERRIAM, J. C. 1908. Triassic Ichthyosauria with special reference to American forms. Mem. Univ. California, 1.

REPOSSI, E. 1902. Il Mixosauro degli Strati Triasici di Basano in Lombardia. Atti della Soc. Ital. di Sci. Nat., 1.

Marine Reptiles: Sauropterygia, Nothosaurs, Plesiosaurs, Placodonts

BROILI, F. 1912. Zur Osteologie des Schädels von Placodus. Palaeontographica, 147.

CASE, E. C. 1936. A nothosaur from the Triassic of Wyoming. Contrib. Mus. Paleont., Univ. Michigan.

DAMES, W. 1895. Die Plesiosaurier der süddeutschen Liasformation. Abh. d. Konigl. Preuss Akad. d. Wiss., Berlin.

DREVERMANN, F. 1924. Schädel und Unterkiefer von Cyamodus. Abh. d. Senckenberg-naturf. Gesell., 291.—1933. Die Placodontier. 3. Das Skelett von Placodus gigas Ag. Ibid., 323.

FRAAS, E. 1910. Plesiosaurier aus dem oberen Lias von Holzmaden. Palaeontographica.

HUENE, E. VON. 1944. Cymatosaurus und seine Beziehungen zu anderen Sauropterygiern. Neues Jahrb. f. Min., etc., 192.

HUENE, F. VON. 1931. Ergänzungen zur Kenntnis des Schädels von Placochelys und seiner bedeutung. Geol. Hungarica, ser. Palaeont., 1.—1933. Die Placodontier. 4. Zur Lebensweise und Verwandtschaft von Placodus. Abh. Senckenberg-naturf. Gesell., 365.—1936. Henodus chelyops, ein neuer Placodontier. Palaeontographica, 99.—1938. Der dritte Henodus. Ergänzungen zur Kenntnis des Placodontiers Henodus chelyops Huene. Ibid., 105.—1942. Pachypleurosauriden im süddeutschen obersten Muschelkalk. Zentralb. f. Min., etc., 290.

JAEKEL, O. 1902. Ueber Placochelys n.g. und ihre Bedeutung für die Stammesgeschichte der Schildkröten. Neues Jahrb. f. Min., etc., 127. [Turtles related to Sauropterygia.]—1905. Über den Schädelbau der Nothosauriden. Sitz.-Bericht. d. Gesell. naturf. Freunde, 60. Berlin. —1907. Placochelys placodonta aus der Obertrias des Bakony. Resultate d. Wiss. erforschung d. Balatonsees, vol. 1, part. 1.

MEYER, H. 1860. Zur Fauna der Vorwelt: Reptilien aus dem Lithographischen Schiefer des Jura in Deutschland und Frankreich. Frankfurt. Folio. [Nothosaurus.]

PEYER, B. 1931–1939. Die Triasfauna der Tessiner Kalkalpen. Abh. d. Schweiz. Palaeont. Gesell., vol. 51. III. Placodontia. IV. Ceresiosaurus . . . V. Pachypleurosaurus . . . Ibid., vol. 52. VII. Neubeschreibung der Saurier von Perledo. Ibid., vols. 53, 54. VIII. Weitere Placodontierfunde. Ibid., vol. 55. XIV. Paranothosaurus. . . . Ibid., vol. 62.

PRAVOSLAVLEFF, P. 1916. Restes d'un Elasmosaurus, trouvés dans le Crétacé superieur de la Province du Don. Travaux de la Soc. Imp. des Naturalistes de Petrograd, vol. 38, 153.

STROMER, E. 1935. Ergebn. d. Forschungsr. Prof. E. Stromers in den Wuesten Aegyptens. II. 15. Plesiosauria. Bayer. Akad. Wiss., 3.

WATSON, D. M. S. 1924. The elasmosaurid shoulder-girdle and fore-limb. Proc. Zool. Soc. London, 885.—1911a. A plesiosaurian pectoral girdle from the Lower Lias. Mem. and Proc. Manchester Lit. and Philos. Soc., vol. 55, part 2, no. 16.—1911b. The Upper Liassic Reptilia. Microcleidus . . . Ibid., vol. 55, part 2, no. 17.

WELLES, S. P. 1943. Elasmosaurid plesiosaurs, with description of new material from California and Colorado. Mem. Univ. California, 125.

WHITE, T. E. 1940. Holotype of Plesiosaurus longirostris Blake and classification of the plesiosaurs. Jour. Paleont., 451.

WILLISTON, S. W. 1906. North American plesiosaurs: Elasmosaurus, Cimoliasaurus and Polycotylus. Amer. Jour. Sci., 221.—1907. The skull of Brachauchenius, with observations on the relationships of the plesiosaurs. Proc. U. S. Nat. Mus., 477.

ZANGERL, R. 1935. B. Peyer, Die Triasfauna der Tessiner Kalkalpen. IX. Pachypleurosaurus . . . Osteologie, Variations-breite, Biologie. Abh. d. Schweiz. Palaeontol. Gesell., vol. 56, 1.

Trachelosauria

HUENE, F. VON. 1931. Über Tanystropheus und verwandte Formen. Neues. Jahrb. f. Min., etc., Beil-Band 67, B, 65.—1944. Über die systematische Stellung von Trachelosaurus aus dem Buntsandstein von Bernburg. Ibid., 170.

PEYER, B. 1931. Die Triasfauna der Tessiner Kalkalpen. II. Tanystropheus longobardicus Bass. sp. Abh. d. Schweiz. Palaeont. Gesell., vol. 50, 6. [Tribelesodon, supposed pterosaur, based on tail vertebrae of Tanystropheus.]

Protorosauria

GREGORY, J. T. 1945. Osteology and relationships of Trilophosaurus. Univ. Texas. Publ., 273.

HAUGHTON, S. H. 1924. Reptilian remains from the Karroo Beds of East Africa. Quart. Jour. Geol. Soc., 1. [Protorosauria.]—1929. Notes on the Karroo Reptilia from Madagascar. Trans. Roy. Soc. South Africa, 125. [Hovasaurus and Tangasaurus, Protorosauria.]

HUENE, F. VON. 1944. Beiträge zur Kenntnis der Protorosaurier. Neues. Jahrb. f. Min., etc., 120.

SEELEY, H. G. 1887. Researches on the structure, organization and classification of the fossil Reptilia. I. On Protorosaurus Speneri (von Meyer). Philos. Trans. Roy. Soc. London, 187.

WATSON, D. M. S. 1914. Broomia perplexa, gen. et. sp. n., a fossil reptile from South Africa. Proc. Zool. Soc. London, 995.

WILLISTON, S. W. 1913a. An ancestral lizard from the Permian of Texas. Science, 825. [Araeoscelis near to Protorosauria.]—1913b. The skulls of Araeoscelis and Casea, Permian reptiles. Jour. Geol., 743.—1914. The osteology of some American Permian vertebrates. I. Araeoscelis. Jour. Geol., 364. [Derivation of lizards from primitive Protorosauria, 392.]

Squamata (Lizards, Snakes)

BAHL, K. N. 1937. Skull of Varanus monitor (Linn.) Rec. Indian Mus., vol. 39, 133.

BEEBE, W. 1924. Galapagos, world's end. New York. [Habits of semi-aquatic and upland iguanoid lizards.]—1944. Field notes on the lizards of Kartabo, British Guiana, and Caripito, Venezuela. Part I. Gekkonidae. Zoologica (N. Y.), 145.

BELLAIRS, A. D'A. 1949. The anterior brain-case and interorbital septum of Sauropsida, with a consideration of the origin of snakes. Jour. Linn. Soc. London, June, 482.

BROILI, F. 1926. Ein neuer Fund von Pleurosaurus aus dem Malm Frankens. Abh. d. Bayer. Akad. d. Wiss., 1.

BROOM, R. 1913. On the squamosal and related bones in the mosasaurs and lizards. Bull. Amer. Mus. Nat. Hist., 507.—1925. On the origin of lizards. Proc. Zool. Soc. London, 1.—1935. On the structure of the temporal region in lizard skulls. Ann. Transvaal Mus., 13. [Quadratojugal in ancestors of lizards, pp. 14–16.]

CAMP, C. L. 1923. Classification of the lizards. Bull. Amer. Mus. Nat. Hist., 289.—1945. Prolacerta and the protorosaurian reptiles. Amer. Jour. Sci., vol. 243.

COPE, E. D. 1887. The origin of the fittest. New York. [Homoplasy; body forms of lizards.]—1900. Crocodilians, lizards and snakes of North America. Rept. U. S. Nat. Mus. (1898), 153.

FLYNN, T. T. 1923. On the occurrence of a true allantoplacenta of the conjoint type in an Australian lizard. Rec. Australian Mus., 72.

GAUPP, E. 1898. Zur Entwicklungsgeschichte des Eidechsenschädels. Ber. Naturf. Gesell. zu Freiburg i. B., 302.

GILMORE, C. W. 1928. Fossil lizards of North America. Mem. Nat. Acad. Sci., 1.—1938. Description of new and little-known fossil lizards from North America. Proc. U. S. Nat. Mus., 11. —1942. Paleocene faunas of the Polecat Bench Formation, Park County, Wyoming. Part II. Lizards. Proc. Amer. Philos. Soc., 159.

GILMORE, C. W. and G. L. JEPSEN. 1945. A new Eocene lizard from Wyoming. Jour. Paleont., 30.

GREGORY, W. K. 1934. Whence came the "Dragons of Komodo"? Bull. N. Y. Zool. Soc., 68. [Pictorial phylogeny of the Squamata.]

LYDEKKER, R. and others. Undated. The new natural history. Reptiles. Vol. V, sec. 9, New York.

MOODIE, R. L. 1907. The sacrum of the Lacertilia. Biol. Bull., 84.

MURPHY, R. C. 1940. The most amazing tongue in Nature. Nat. Hist. Mag., 260. [Mechanism of protrusile tongue of chamaeleons.]

NOBLE, G. K. and H. T. BRADLEY. 1933. The

mating behavior of lizards; its bearing on the theory of sexual selection. Ann. N. Y. Acad. Sci., 25.

PARKER, W. K. 1879. On the structure and development of the skull in the Lacertilia. Part I. On the skull of the common lizards. Philos. Trans. Roy. Soc. London, 595.—1881. On the structure of the skull in the chamaeleons. Trans. Zool. Soc. London, 77.

PARRINGTON, F. R. 1935. On Prolacerta broomi, gen. et. sp. n., and the origin of lizards. Ann. Mag. Nat. Hist., 197.

POPE, C. H. 1935. The reptiles of China: turtles, crocodilians, snakes, lizards. In: The natural history of Central Asia, vol. 10. Amer. Mus. Nat. Hist. 4to.

WATSON, D. M. S. 1914. Pleurosaurus and the homologies of the bones of the temporal region of the lizard's skull. Ann. Mag. Nat. Hist., 84.

ZANGERL, R. 1944. Contributions to the osteology of the skull of the Amphisbaenidae. Amer. Midland Naturalist, vol. 31, 417.—1945. Contributions to the osteology of the post-cranial skeleton of the Amphisbaenidae. Ibid., vol. 33, 764.

Mosasaurs (Pythonomorpha)

BAUR, G. 1892. On the morphology of the skull in the Mosasauridae. Jour. Morph., 1.

CAMP, C. L. 1923. Classification of the lizards. Bull. Amer. Mus. Nat. Hist., 228. [Mosasaurs related to varanoids, 323–325.]—1942. California mosasaurs. Mem. Univ. California, vol. 13, no. 1.

DOLLO, L. 1894. Nouvelle note sur l'osteologie des mosasauriens. Bull. Soc. Belge d. Géol. (1892), 219.—1904. Les mosasauriens de la Belgique. Ibid., vol. 18, 207.—1913. Globidens Fraasi, mosasaurien mylodonte nouveau du Maestrichtien (Crétacé supérieur) . . . Arch. d. Biol., vol. 28, 609.—1924. Globidens alabamaensis: Mosasaurien mylodonte américain retrouvé dans la Craie d'Obourg (Sénonien supérieur) . . . Ibid., vol. 34, 167.

GORJANOVIC-KRAMBERGER, K. 1892. Aigialosaurus, eine neue Eidechsen a. d. Kreideschiefern der Insel Lesina . . . Glasnik. Soc. Hist. Nat. Croatica, vol. 7, 74. [Connects mosasaurs with varanoid lizards.]

HUENE, F. VON. 1910. Ein Ganzes Tylosaurus-Skelet. Geol. u. Palaeontol. Abh., 297.—1911. Über einen Platecarpus in Tübingen. Neues Jahrb. f. Min., etc., 48.

OSBORN, H. F. 1899. A complete mosasaur skeleton, osseous and cartilaginous. Mem. Amer. Mus. Nat. Hist., I, 167.

WILLISTON, S. W. 1904. The relationships and habits of the mosasaurs. Jour. Geol., 43.

Serpentes

BOGERT, C. M. 1943. Dentitional phenomena in cobras and other elapids, with notes on adaptive modifications of fangs. Bull. Amer. Mus. Nat. Hist., 285.

CAMP, C. L. 1923. Classification of the lizards. Bull. Amer. Nat. Hist., 289. [Relations of snakes to anguimorphine lizards, p. 301.]

CLARK, P. J. and R. F. INGER. 1942a. Scale reduction in snakes. Copeia, no. 3, 163.—1942b. Scale reduction studies in certain non-colubrid snakes. Ibid., no. 4, 230. [Negative anisomerism.]

DUNN, E. R. 1942. Survival value of varietal characters in snakes. Amer. Nat., 104.

HAAS, G. 1930. Über das Kopfskelett und die Kaumusculature der Typhlopiden und Glauconiiden. Zool. Jahrb., 1.

PARKER, W. K. 1878. On the structure and development of the skull in the common snake (Tropidonotus natrix). Philos. Trans. Roy. Soc. London, 385.

SIMPSON, G. G. 1933. A new fossil snake from the Notostylops Beds of Patagonia. Bull. Amer. Mus. Nat. Hist., 1. A large booid.

Rhynchocephalia

BROWN, B. 1905. The osteology of Champsosaurus Cope. Mem. Amer. Mus. Nat. Hist., IX, 1.

CONRAD, G. M. 1940. By boat to the Age of Reptiles. Nat. Hist. Mag., no. 4, 224. [Chiefly on Sphenodon.]

COPE, E. D. 1876. On some extinct reptiles and Batrachia from the Judith River and Fox Hills Beds of Montana. Proc. Acad. Nat. Sci. Philadelphia, 340. [Choristodera, p. 350.]

DENDY, A. 1910. On the structure, development and morphological interpretation of the pineal organs and adjacent parts of the brain in the tuatara (Sphenodon punctatus). Proc. Roy. Soc. London, 629.

GILMORE, C. W. 1909. New rhynchocephalian reptiles from the Jurassic of Wyoming . . . Proc. U. S. Nat. Mus., 35.

HOWES, G. B. and H. H. SWINNERTON. 1901. On the development of the skeleton of the tuatara,

Sphenodon punctatus . . . Trans. Zool. Soc. London, 1.

HUENE, E. VON. 1935. Ein Rhynchocephale aus dem Rhät (Pachystropheus n.g.). Neues Jahrb. f. Min., etc., Beil.-Band, 74, 441.

HUENE, F. VON. 1910. Über einen echten Rhynchocephalen aus der Trias von Elgin, Brachyrhinodon Taylori. Neues Jahrb. f. Min., Geol. u. Palaont., 29. [Palaeohatteria, pp. 38–42.]—1929. Über Rhynchosaurier und andere Reptilien aus den Gondwana-Ablagerungen Südamerikas. Geol. u. Palaeont. Abh., 1.—1938. Stenaulorhynchus, ein Rhynchosauride der ostafrikanischen Obertrias. Nova Acta Leopoldina, 83.—1939a. Die Lebensweise der Rhynchosauriden. Palaeont. Zeitschr., 232.—1939b. Die Verwandtschafts geschichte der Rhynchosauriden des südamerikanischen Gondwanalandes. Physis (Revista de la Sociedad Argentina de Ciencia Naturales), 499.—1942. Ein Rhynchocephale aus mandschurischen Jura. Neues Jahrb. f. Min., etc., Beil.-Band, 244.

SÄVE-SÖDERBERGH, G. 1947. Notes on the braincase in Sphenodon and certain Lacertilia. Festskr. tillägnad Prof. Nils von Hofsten. Zool. Bidrag f. Uppsala, 389.

SCHAUINSLAND, H. 1900. Weitere Beiträge zur Entwickl. der Hatteria. Arkiv. f. mikroskop. Anat. u. Entwickl., vol. 56 (57), 747.

SIMPSON, G. G. 1926. The fauna of Quarry Nine, American terrestrial Rhynchocephalia. Amer. Jour. Sci., 1.

SLADDEN, B. and R. A. FALLA. 1928. Alderman Islands: A general description, with notes on the flora and fauna. New Zealand Jour. Sci. and Technol., 193. [Home of surviving tuataras.]

Theocodontia: Eosuchians, Pseudosuchians (Aëtosaurs), Phytosaurs, Pelycosimians

ANDERSON, H. T. 1936. The jaw musculature of the phytosaur, Machaeroprosopus. Jour. Morph., 549.

BROILI, F. and J. SCHRÖDER. 1934. Beobachtungen an Wirbeltieren der Karrooformation. V. Über Chasmatosaurus van hoepeni Haughton. Sitz. d. Bayer. Akad. d. Wiss., 225.—1936. Beobachtungen an Wirbeltieren der Karrooformation. XX. Beobachtungen an Erythrochampsa Haughton. Ibid., 229.

BROOM, R. 1913. On the South-African pseudosuchian Euparkeria and allied genera. Proc. Zool. Soc. London, 619. [Potential ancestor of dinosaurs, pterosaurs, birds, pp. 630–632; pseudosuchian pubis, p. 623.]—1922. An imperfect skeleton of Youngina capensis, Broom, in the collection of the Transvaal Museum. Ann. Transvaal Mus., 273. [Ancestry of two-arched reptiles, p. 275.]—1924. Further evidence on the structure of the Eosuchia. Bull. Amer. Mus. Nat. Hist., 67.—1932. On some South African pseudosuchians. Ann. Natal. Mus., 55.

CAMP, C. L. 1930. A study of the phytosaurs, with description of new material from western North America. Mem. Univ. California., 1.

COLBERT, E. H. 1947. Studies of the phytosaurs Machaeroprosopus and Rutiodon. Bull. Amer. Mus. Nat. Hist., 53.

FRAAS, O. 1877. Aëtosaurus ferratus Fr. Die gepanzerte Vogel-Eidechse aus dem Stubensandstein bei Stuttgart. Württemberg. naturwiss. Jahresh., 5.

GOODRICH, E. S. 1942. The hind foot of Youngina and fifth metatarsal in Reptilia. Jour. Anat., 308.

HAUGHTON, S. H. 1921. On the reptilian genera Euparkeria Broom and Mesosuchus Watson. Trans. Roy. Soc. South Africa, 81.

HUENE, F. VON. 1911a. Über Erythrosuchus Vertreter der neuen Reptil-Ordnung Pelycosimia. Geol. u. Palaeontol. Abh., 1.—1911b. Beiträge zur Kenntnis und Beurteilung der Parasuchier. Ibid., 1.—1913. A new phytosaur from the Palisades near New York. Bull. Amer. Mus. Nat. Hist., 275.—1915. On reptiles of the New Mexican Trias in the Cope Collection. Bull. Amer. Mus. Nat. Hist., 485. [Thecodonts.]—1920a. Osteologie von Aëtosaurus ferratus . . . Acta Zool., 465.—1920b. Stammesgeschichtl. Ergeb. . . . an Trias-Reptilien. Zeitschr. f. induktive Abstammungs-und Vererbungslehre, 159. [Restorations of Saltoposuchus and Procompsognathus; skull of Aëtosaurus, fig. 1.]—1921. Neue Pseudosuchier und Coelurosaurier aus dem Württemberg. Keuper. Acta Zool., 329. [Aëtosaurus, Saltoposuchus, Procompsognathus.]—1922a. Neue Beiträge zur Kenntnis der Parasuchier. Jahrb. d. Preussich. Geologisch. Landesanstalt, 1921 (1922), 59. [Origin of phytosaurs from pseudosuchians.]—1922b. Kurzer Überblick über die triassiche Reptil-Ordnung Thecodontia. Centralbl. f. Min., etc., 40. [Origin of phytosaurs from pseudosuchians.]—1926. Gondwana-Reptilien in Südamerika. Palaeont. Hungarica (1924–1926), 1. [Pelycosimia.]—1936a. Übersicht über Zusammensetzung und Bedeutung der Thecodontia. Centralbl. f. Min., etc., 162.—1936b. The constitution of the Thecodontia. Amer. Jour. Sci., 207.—1938. Ein grosser Stagonolepide aus der jüngeren Trias Ostafrikas. Neues. Jahrb. f. Min., etc., Beil.-Band, 80, 264.—1939. Ein kleiner Pseudosuchier und ein Saurischier aus den ostafrikanischen Mandaschichten. Ibid., 61.—1940. Eine Reptilfauna aus der Ältesten Trias Nordrusslands. Ibid., 84.

McGregor, J. H. 1906. The Phytosauria, with especial reference to Mystriosuchus and Rhytidodon. Mem. Amer. Mus. Nat. Hist., 27.

Mehl, M. G. 1915. The Phytosauria of the Trias. Jour. Geol., 129.—1928. The Phytosauria of the Wyoming Triassic. Denison Univ. Bull., 141.

Newton, E. T. 1894. Reptiles from the Elgin Sandstone. Philos. Trans. Roy. Soc. London, 573. [Ornithosuchus, pls. 54–56; affinities, pp. 598–601.]

Olson, E. C. 1936. Notes on the skull of Youngina capensis Broom. Jour. Geol., 523.

Olson, E. C. and R. Broom. 1937. New genera and species of tetrapods from the Karroo Beds of South Africa. Jour. Paleont., 613. [Youngoides, fig. 3.]

Sawin, H. J. 1947. The pseudosuchian reptile Typothorax meadei, new species. Jour. Paleont., 201.

Watson, D. M. S. 1912a. Mesosuchus Browni, gen. et spec. nov. Rec. Albany Mus., 296.—1912b. Eosuchus Colletti, gen. et spec. nov. Ibid., 298.

Woodward, A. S. 1907. On a new dinosaurian reptile (Scleromochlus). Quart. Jour. Geol. Soc. London, vol. 63.

Crocodilians

Broom, R. 1927. On Sphenosuchus and the origin of the crocodiles. Proc. Zool. Soc. London, 359.

Brown, B. 1933. An ancestral crocodile. Amer. Mus. Novitates, no. 638. [Protosuchus.]—1942. The largest known crocodile. Nat. Hist. Mag., 260. [Phobosuchus.]

Colbert, E. H. 1946a. The eustachian tubes in the Crocodilia. Copeia, no. 1, 12.—1946b. Sebecus, representative of a peculiar suborder of fossil Crocodilia from Patagonia. Bull. Amer. Mus. Nat. Hist., 217.

Colbert, E. H., R. B. Cowles and C. M. Bogert. 1946. Temperature tolerances in the American alligator and their bearing on the habits, evolution and extinction of the dinosaurs. Bull. Amer. Mus. Nat. Hist., 327.

Fraas, E. 1902. Meer-Crocodilier (Thallattosuchia) des oberen Jura unter specieller Berücksichtigung von Dacosaurus und Geosaurus. Palaeontographica, 1.

Gadow, H. 1882. Untersuchungen über die Bauchmuskeln der Krokodile, Eidechsen und Schildkröten. Morph. Jahrb., 57.

Gregory, W. K. and C. L. Camp. 1918. Studies in comparative myology and osteology. No. III. Parts I–VI. Part II. A comparison of the muscle areas of the pelvis of Alligator, Struthio, and Ornitholestes. Bull. Amer. Mus. Nat. Hist., 515.

Manter, J. T. 1940. The mechanics of swimming in the alligator. Jour. Exp. Zool., 345.

Mook, C. C. 1921a. Notes on the postcranial skeleton in the Crocodilia. The dermo-supraoccipital bone in the Crocodilia. Bull. Amer. Mus. Nat. Hist., 67.—1921b. Skull characters of recent Crocodilia wtih notes on the affinities of the recent genera. Bull. Amer. Mus. Nat. Hist., 123.—1925a. The ancestry of the alligators. Nat. Hist. Mag., 407.—1925b. A revision of the Mesozoic Crocodilia of North America. Bull. Amer. Mus. Nat. Hist., 319.—1934. The evolution and classification of the Crocodilia. Jour. Geol., no. 3.

Parker, W. K. 1883. On the structure and development of the skull in the Crocodilia. Philos. Trans. Roy. Soc. London, 263.

Romer, A. S. 1923. Crocodilian pelvic muscles and their avian and reptilian homologues. Bull. Amer. Mus. Nat. Hist., 533.

Simpson, G. G. 1935. A new crocodilian from the Notostylops Beds of Patagonia. Amer. Mus. Novitates, no. 623.—1937a. New reptiles from the Eocene of South America. Ibid., no. 927. [Sebecus icaeorhinus.]—1937b. An ancient Eusuchian crocodile from Patagonia. Ibid., no. 965.

Troxell, E. L. 1925. Mechanics of crocodile vertebrae. Bull. Geol. Soc. Amer., 605.

Dinosauria: General

Brown, B. 1935. Sinclair Dinosaur Expedition, 1934. Nat. Hist. Mag. vol. 36, 3.

Brown, B. and E. S. Schlaikjer. 1941. The rise and fall of the dinosaurs. Nat. Hist. Mag., vol. 48, no. 5.

Colbert, E. H. 1945. The dinosaur book: The ruling reptiles and their relatives. Man and

Nature Publ., Handbook no. 14, Amer. Mus. Nat. Hist.

LULL, R. S. 1915. Sauropoda and Stegosauria of the Morrison of North America compared with those of Europe and Western Africa. Bull. Geol. Soc. Amer., 323.—1917. On the functions of the "sacral brain" in dinosaurs. Amer. Jour. Sci., 471.

MARSH, O. C. 1877. Introduction and succession of vertebrate life in America. Popular Science Monthly, 515. [Relationship between crocodilians and dinosaurs, p. 526.]—1895. On the affinities and classification of the dinosaurian reptiles. Amer. Jour. Sci., 483. [Belodon, p. 485.]—1896. The dinosaurs of North America. 16th Ann. Rept. U. S. Geol. Surv. (1894–1895), 133.

ROMER, A. S. 1923. The ilium in dinosaurs and birds. Bull. Amer. Mus. Nat. Hist., 141.

SCHUCHERT, C. 1918. Age of the American Morrison and East African Tendaguru formations. Bull. Geol. Soc. Amer., 245.

SWINTON, W. E. 1934. The dinosaurs: The story of a group of extinct reptiles. London.

Dinosauria: Coelurosauria

HUENE, F. VON. 1910. Ein primitiver Dinosaurier aus der Mittleren Trias von Elgin. Geol. u. Palaeont. Abh., 25.—1925. Eine neue Rekonstruktion von Compsognathus. Centralbl. f. Min., etc., 157.—1934. Ein neuer Coelurosaurier in der thüringischen Trias. Palaeont. Zeitschr., 145. [Reconstruction of Halticosaurus.]

HUENE, F. VON and R. S. LULL. 1908. On the Triassic reptile, Hallopus. Amer. Jour. Sci., 113.

NOPCSA, F. 1903. Neues über Compsognathus. Neues Jahrb. f. Min., etc., Beil.-Band, 476.

OSBORN, H. F. 1903. Ornitholestes Hermanni, a new compsognathoid dinosaur from the Upper Jurassic. Bull. Amer. Mus. Nat. Hist., 459.

Dinosauria: Saurischia (Theropoda, Sauropoda)

BIEN, M. N. 1940. Discovery of Triassic saurischian and primitive mammalian remains at Lufeng, Yunnan. Bull. Geol. Soc. China, 225.

BIRD, R. T. 1939. Thunder in his footsteps. Nat. Hist. Mag., 254.—1941. A dinosaur walks in the Museum. Ibid., 74.

BROOM, R. 1911. On the dinosaurs of the Stormberg, South Africa. Ann. South African Mus., 291. [Massospondylus, pls. 15–17.]

BROWN, B. 1938. The mystery dinosaur. Nat. Hist. Mag., 190.

GILMORE, C. W. 1907. The type of the Jurassic reptile Morosaurus agilis redescribed . . . Proc. U. S. Nat. Mus., 151.—1920. Osteology of the carnivorous Dinosauria in the United States National Museum . . . Antrodemus (Allosaurus) and Ceratosaurus. U. S. Nat. Mus. Bull., 1.—1925. A nearly complete articulated skeleton of Camarasaurus, a saurischian dinosaur from the Dinosaur National Monument, Utah. Mem. Carnegie Mus., 347.—1932. On a newly mounted skeleton of Diplodocus . . . Proc. U. S. Nat. Mus., 1.—1936. Osteology of Apatosaurus . . . Mem. Carnegie Mus., 175.

GREGORY, W. K. 1905. The weight of the Brontosaurus. Science, 566.—1920. Restoration of Camarasaurus and life model. Proc. Nat. Acad. Sci., 16.

HATCHER, J. B. 1901. Diplodocus (Marsh): Its osteology, taxonomy and probable habits, with a restoration of the skeleton. Mem. Carnegie Mus., vol. 1, 1.—1903a. Osteology of Haplocanthosaurus . . . Ibid., vol. 2, 1.—1903b. Additional remarks on Diplodocus. Ibid., vol. 2, 72.

HAY, O. P. 1910. On the manner of locomotion of the dinosaurs, especially Diplodocus, with remarks on the origin of the birds. Proc. Washington Acad. Sci., 1.

HOLLAND, W. J. 1906. The osteology of Diplodocus Marsh . . . Mem. Carnegie Mus., 225.—1924. The skull of Diplodocus. Ibid., 379.

HUENE, F. VON. 1907–1908. Die dinosaurier der Europäischen Triasformation mit Berücksichtigung der Aussereuropäischen Vorkommnisse. Geol. u. Palaeont. Abh., Suppl.-Band 1.—1914. Nachträge zu meinen früheren Beschreibungen Triassischer Saurischia. Ibid., vol. 13, 68.—1915. Beiträge zur Kenntnis einiger Saurischier der schwäbischen Trias. Neues Jahrb. f. Min., etc., 1.—1926a. Vollständige Osteologie eines Plateosauriden aus dem schwäbischen Keuper. Geol. u. Palaeont. abh., 139.—1926b. The carnivorous Saurischia in the Jura and Cretaceous formations, principally in Europe. Revista del Museo de La Plata, 35.—1932. Die fossile Reptil-Ordnung Saurischia, ihre Entwicklung und Geschichte. Monogr. z. Geol. u. Palaeont. Text and atlas.

JANENSCH, W. 1914a. Übersicht über die Wirbeltierfauna der Tendaguru-Schichten . . . [Sauropoda.] Archiv f. Biontologie, 81. 4to.—1914b.

Die Gliederung der Tendaguruschichten im Tendagurugebiet und die Entstehung der Saurierlagerstätten. Ibid., 227. [Sauropods in salt-mud layers.]—1925. Die Coelurosaurier und Theropoden der Tendaguru-Schichten Deutsch-Ostafrikas. Palaeontographica, Suppl.-Band 7, 30. —1929a. Material und Formengehalt der Sauropoden in der Ausbeute der Tendaguru-Expedition. Ibid., 1.—1929b. Die Wirbelsäule der Gattung Dicraeosaurus. Ibid., 39. — 1929c. Magensteine bei Sauropoden der Tendaguru-Schichten. Ibid., 137.—1935. Die Schädel der Sauropoden Brachiosaurus, Barosaurus und Dicraeosaurus aus den Tendaguru-Schichten Deutsch-Ostafrikas. Ibid., 147.—1937. Skelettrekonstruktion von Brachiosaurus brancai aus den Tendaguru-Ostafrikas. Zeitschr. d. Deutschen Geol. Gesell., 550. — 1938a. Brachiosaurus, der grösste sauropode Dinosaurier aus dem oberen Jura von Deutsch-Ostafrika. Forschungen u. Fortschritte, 140.—1938b. Aufstellung der Skelettrekonstruktion des Brachiosaurus. Aus der Heimat, Naturw. Monats., 124.

LAMBE, L. M. 1903. The lower jaw of Dryptosaurus incrassatus (Cope). Ottawa Naturalist, 133.—1914. On a new genus and species of carnivorous dinosaur from the Belly River formation of Alberta . . . Ibid., vol. 28, 13.—1917. The Cretaceous theropodous dinosaur Gorgosaurus. Geol. Surv. Canada. Mem. 100, 1.

LULL, R. S. 1919. The sauropod dinosaur Barosaurus Marsh . . . Mem. Connecticut Acad. Arts and Sci., 1.

MARSH, O. C. 1884. Principal characters of American Jurassic dinosaurs. Part VIII. The order Theropoda. Amer. Jour. Sci., 329.

MATTHEW, W. D. 1905. The mounted skeleton of Brontosaurus in the American Museum of Natural History. Amer. Mus. Jour., 63.—1910. The pose of sauropodous dinosaurs. Amer. Nat., 547.

MOOK, C. C. 1914. The dorsal vertebrae of Camarasaurus Cope. Bull. Amer. Mus. Nat. Hist., 223.—1917a. Criteria for the determination of species in the Sauropoda, with description of a new species of Apatosaurus. Ibid., 355.—1917b. The fore and hind limbs of Diplodocus. Ibid., 815.—1918. The habitat of the sauropod dinosaurs. Jour. Geol., 459.

NOPCSA, F. 1930. Zur Systematik und Biologie der Sauropoden. Palaeobiologica, 40.

OSBORN, H. F. 1899a. A skeleton of Diplodocus. Mem. Amer. Mus. Nat. Hist., vol. 1, 191.—1899b. Fore and hind limbs of carnivorous and herbivorous dinosaurs from the Jurassic of Wyoming. Bull. Amer. Mus. Nat. Hist., 161.—1912. Crania of Tyrannosaurus and Allosaurus. Mem. Amer. Mus. Nat. Hist., 1.—1917. Skeletal adaptations of Ornitholestes, Struthiomimus, Tyrannosaurus. Bull. Amer. Mus. Nat. Hist., 733.—1920. Reconstruction of the skeleton of the sauropod dinosaur Camarasaurus Cope (Morosaurus Marsh). Proc. Nat. Acad. Sci., 15.

OSBORN, H. F. and C. C. MOOK. 1921. Camarasaurus, Amphicoelias, and other sauropods of Cope. Mem. Amer. Mus. Nat. Hist., 249.

RIGGS, E. S. 1903. Structure and relationships of opisthocoelian dinosaurs. Field Columbian Mus., 165.

ROMER, A. S. 1923. The pelvic musculature of saurischian dinosaurs. Bull. Amer. Mus. Nat. Hist., 605.

RUSSELL, L. S. 1948. The dentary of Troödon, a genus of theropod dinosaurs. Jour. Paleont., Sept., 625. [Not the same as Troödon Gilmore.]

TORNIER, G. 1909. Wie war der Diplodocus carnegii wirklich gebaut? Sitz. d. Gesell. naturf. Freunde zu Berlin, 193.

WATSON, D. M. S. 1914. The cheirotherium. Geol. Mag. 395.

WIMAN, C. 1929. Die Kreide-Dinosaurier aus Shantung. Palaeontologia Sinica.

Dinosauria: Ornithischia
Iguanodontia, Ceratopsia, Stegosauria, Ankylosauria, Troödontia

BROWN, B. 1908. The Ankylosauridae, a new family of armored dinosaurs from the Upper Cretaceous. Bull. Amer. Mus. Nat. Hist., 187. —1912a. The osteology of the manus in the Family Trachodontidae. Ibid., 105.—1912b. A crested dinosaur from the Edmonton Cretaceous. Ibid., 131.—1913. The skeleton of Saurolophus, a crested duck-billed dinosaur from the Edmonton Cretaceous. Ibid., 387.—1914a. Anchiceratops, a new genus of horned dinosaurs from the Edmonton Cretaceous of Alberta. With discussion of the origin of the ceratopsian crest and the brain casts of Anchiceratops and Trachodon. Ibid., 539.—1914b. A complete skull of Monoclonius, from the Belly River Cretaceous of Alberta. Ibid., 549.—1914c. Corythosaurus casuarius, a new dinosaur from the Belly

River Cretaceous, with provisional classification of the Family Trachodontidae. Ibid., 559.—1914d. Leptoceratops, a new genus of Ceratopsia from the Edmonton Cretaceous of Alberta. Ibid., 567.—1916. Corythosaurus casuarius: skeleton, musculature and epidermis. Ibid., 709.—1917. A complete skeleton of the horned dinosaur Monoclonius and description of a second skeleton showing skin impressions. Ibid., 281.

BROWN, B. and E. M. SCHLAIKJER. 1940a. The origin of ceratopsian horn-cores. Amer. Mus. Novitates, no. 1065.—1940b. The structure and relationships of Protoceratops. Ann. N. Y. Acad. Sci., 133.—1940c. A new element in the ceratopsian jaw, with additional notes on the mandible. Amer. Mus. Novitates, no. 1092.—1942. The skeleton of Leptoceratops, with the description of a new species. Ibid., no. 1169.—1943. A study of the troödont dinosaurs, with the description of a new genus and four new species. Bull. Amer. Mus. Nat. Hist., 115.

COLBERT, E. H. 1945. The hyoid bones in Protoceratops and in Psittacosaurus. Amer. Mus. Novitates, no. 1301.—1948. Evolution of the horned dinosaurs. Evolution, June, 145.

DOLLO, L. 1905. Les dinosauriens adaptés a la vie quadrupède secondaire. Bull. Soc. Belge de Geol., de Paleontol., etc. (Bruxelles), 441.

GILMORE, C. W. 1905. The mounted skeleton of Triceratops prorsus. Proc. U. S. Nat. Mus., 433.—1909. Osteology of the Jurassic reptile Camptosaurus . . . Ibid., 197.—1914. Osteology of the armored Dinosauria in the United States National Museum. . . . U. S. Nat. Mus. Bull., 1.—1915. Osteology of Thescelosaurus, an orthopodous dinosaur from the Lance Formation of Wyoming. Proc. U. S. Nat. Mus., 591.—1917. Brachyceratops: A ceratopsian dinosaur from the Two Medicine Formation of Montana . . . U. S. Geol. Surv., Professional Paper 103.—1918. A newly mounted skeleton of the armored dinosaur Stegosaurus stenops in the United States National Museum. Proc. U. S. Nat. Mus., 383.—1919. A new restoration of Triceratops, with notes on the osteology of the genus. Ibid., 97.—1923. A new species of Corythosaurus, with notes on other Belly River Dinosauria. Canadian Field Naturalist, 46.—1924a. On the genus Stephanosaurus, with a description of the type specimen of Lambeosaurus lambei Parks. Bull. Canada Geol. Surv., 29.—1924b. On Troödon validus, an orthopodous dinosaur from the Belly River Cretaceous of Alberta, Canada. Univ. Alberta, Dept. Geol. Bull., 1. [Cf. Russell, L. S. 1948, above.]—1925. Osteology of ornithopodous dinosaurs from the Dinosaur National Monument, Utah. Parts 1, 2, 3. Mem. Carnegie Mus., 385.—1931. A new species of troödont dinosaur from the Lance Formation of Wyoming. Proc. U. S. Nat. Mus., art. 9.—1937. On the detailed skull structure of a crested hadrosaurian dinosaur. Ibid., 481.

GREGORY, W. K. and W. GRANGER. 1923. Protoceratops andrewsi, a pre-ceratopsian dinosaur from Mongolia. With appendix on the structural relations of the Protoceratops Beds, by Charles P. Berkey. Amer. Mus. Novitates, no. 72.

GREGORY, W. K. and C. C. MOOK. On Protoceratops, a primitive ceratopsian dinosaur from the Lower Cretaceous of Mongolia. Amer. Mus. Novitates, no. 156.

HATCHER, J. B., O. C. MARSH, and R. S. LULL. 1907. The Ceratopsia . . . Monogr. U. S. Geol. Surv.

HUENE, F. VON. 1912. Beiträge zur Kenntnis des Ceratopsidenschädels. Neues Jahrb. f. Min., 146.

HUENE, F. VON and R. S. LULL. 1908. Neubeschreibung des Originals von Nanosaurus agilis Marsh. Neues. Jahrb. f. Min., etc., 134.

HUXLEY, T. H. 1869. On Hypsilophodon foxii, a new dinosaurian from the Wealden of the Isle of Wight. Quart. Jour. Geol. Soc. London, 3.

JANENSCH, W. 1925. Ein aufgestelltes Skelett des Stegosauriers Kentrurosaurus aethiopicus E. Hennig aus den Tendaguru-Schichten Deutsch-Ostafrikas. Palaeontographica, Suppl.-Band 7, 257.

KRIPP, D. 1933. Die Kaubewegung und Lebensweise von Edmontosaurus spec. auf Grund der mechanisch-konstruktiven Analyse. Palaeobiol., vol. 5, 409.

LAMBE, L. M. 1913. A new genus and species of Ceratopsia from the Belly River formation of Alberta. Ottawa Naturalist, 109.—1915. On Eoceratops canadensis, gen. nov., with remarks on other genera of Cretaceous horned dinosaurs. Canada Geol. Surv., Mus. Bull. no. 12.—1920. The hadrosaur Edmontosaurus from the Upper Cretaceous of Alberta. Canada Geol. Surv., Mem. 120.

LULL, R. S. 1903. Skull of Triceratops serratus. Bull. Amer. Mus. Nat. Hist., 685.—1908. The cranial musculature and the origin of the frill in the ceratopsian dinosaurs. Amer. Jour. Sci., 387.—1910a. The armor of Stegosaurus. Ibid., 201.—1910b. Stegosaurus ungulatus Marsh, recently mounted at the Peabody Museum of Yale University. Ibid., 361.—1910c. The evolution of the Ceratopsia. Proc. 7th Intern. Zool. Cong. (1907).—1921. The Cretaceous armored dinosaur, Nodosaurus textilis Marsh. Amer. Jour. Sci., 97.—1933. Revision of the Ceratopsia or horned dinosaurs. Mem. Peabody Mus. Nat. Hist., Yale Univ.

LULL, R. S. and N. E. WRIGHT. 1942. Hadrosaurian dinosaurs of North America. Geol. Soc. Amer., Special Papers, no. 40.

NOPCSA, F. 1902. Dinosaurierreste aus Siebenbürgen (Schädelreste von Mochlodon), mit einem Anhange: Zur Phylogenie der Ornithopodiden. Kaiserl. Akad. d. Wiss., vol. 72, 1.—1904. Dinosaurierreste aus Siebenbürgen. (Weitere Schädelreste von Mochlodon). Ibid., vol. 74, 1.—1911a. Omosaurus lennieri: un nouveau dinosaurien du Cap de la Hève. Bull. Soc. Géol. d. Normandie.—1911b. Note on British dinosaurs. IV. Stegosaurus priscus, sp. nov. Geol. Mag., 109.—1923. Part VI. Acanthopholis. Ibid., 193.—1928. Palaeontological notes on reptiles. Geol. Hungarica, Ser. Palaeont.—1929. Dinosaurierreste aus Siebenbürgen. V. Ibid. [Struthiosaurus transsylvanicus, n. sp.; pls. 1–4; Thyreophoroidea, p. 71.]

OSBORN, H. F. 1923. Two Lower Cretaceous dinosaurs of Mongolia. Amer. Mus. Novitates, no. 95. [Psittacosaurus, Protiguanodon]. — 1924. Psittacosaurus and Protiguanodon: two Lower Cretaceous iguanodonts from Mongolia. Ibid., no. 127.

OWEN, R. 1865. Monograph of the fossil Reptilia of the Liassic formations. . . . [Scelidosaurus harrisonii.]

PARKS, W. A. 1920. The osteology of the trachodont dinosaur Kritosaurus incurvimanus. Univ. Toronto Studies, Geol. Ser. 5.—1923. Corythosaurus intermedius, a new species of trachodont dinosaur. Ibid., 5.

POMPECKJ, J. F. 1920. Das angebliche Vorkommen und Wandern des Parietalforamens bei Dinosauriern. Sitz. d. Gesell. Naturf. Freunde, no. 3, 109.

ROMER, A. S. 1927. The pelvic musculature of ornithischian dinosaurs. Acta Zool., 225.

RUSSELL, L. S. 1932. On the occurrence and relationships of the dinosaur Troödon. Ann. Mag. Nat. Hist., 334.

STERNBERG, C. M. 1933. Relationships and habitat of Troödon and the nodosaurs. Ann. Mag. Nat. Hist., 231.—1940a. Ceratopsidae from Alberta. Jour. Paleont., 468.—1940b. Thescelosaurus edmontonensis, n. sp., and classification of the Hypsilophodontidae. Ibid., 481.—1942. New restoration of a hooded duck-billed dinosaur. Ibid. vol. 16, 133.—1945. Pachycephalosauridae proposed for dome-headed dinosaurs; Stegoceras lambei, n. sp., described. Ibid., vol. 19, 534.

VERSLUYS, J. 1923. Der Schädel des Skelettes von Trachodon annectens in Senckenberg-Museum. Frankfurt.

WIELAND, G. R. 1911. Notes on the armored Dinosauria. Amer. Jour. Sci., 112.—1912. On the dinosaur-turtle analogy. Mem. d. R. Accad. d. Sci. d. Inst. d. Bologna (1911–1912).

Pterosaurs

BROWN, B. 1943. Flying reptiles. Nat. Hist. Mag., No. 3, 104.

BROILI, F. 1925. Ein Pterodactylus mit Resten der Flughaut. Sitz. d. Bayer. Akad. d. Wiss., 23.—1927a. Ein Exemplar von Rhamphorhynchus mit Resten von Schwimmhaut. Ibid., 29.—1927b. Ein Rhamphorhynchus mit Spuren von Haarbedeckung. Ibid., 49.

EATON, G. F. 1910. Osteology of Pteranodon. Mem. Connecticut Acad. Arts and Sci.

EDINGER, T. 1941. The brain of Pterodactylus. Amer. Jour. Sci. 665.

HUENE, F. VON. 1914. Beiträge zur Kenntnis des Schädels einiger Pterosaurier. Geol. u. Palaeont. Abh., 57.

OWEN, R. 1860. On the vertebral characters of the order Pterosauria, as exemplified in the genera Pterodactylus (Cuvier) and Dimorphodon (Owen). Philos. Trans. Roy. Soc. London, 161.

SEELEY, H. G. 1901. Dragons of the air: An account of extinct flying reptiles. New York.

STROMER, E. 1913. Rekonstruktionen des Flugsauriers Rhamphorhynchus Gemmingi H. v. M. Neues Jahr. f. Min., etc., 49.

WATSON, D. M. S. and E. H. HANKIN. 1914. On the flight of pterodactyls. Aeronautical Jour., no. 72.

WILLISTON, S. W. 1897. Restoration of Ornithostoma (Pteranodon). Kansas Univ. Quart., 35.

—1911. The wing-finger of pterodactyls, with restoration of Nyctosaurus. Jour. Geol., 696.

WOODWARD, A. S. 1902. On two skulls of the ornithosaurian Rhamphorhynchus. Ann. Mag. Nat. Hist., 1.

Birds, General: General Classification

ABEL, O. 1912. Grundzüge der Palaobiologie der Wirbeltiere. Stuttgart. [Archaeopteryx flight, p. 353.]

BEDDARD, F. E. 1898. The structure and classification of birds. London.

BELON, P. 1555. L'Historie de la Nature des Oyseaux . . . Chap. 12. L'Anatomie des ossements des oyseaux, conferée avec celle des animaux terrestre et de l'homme. [Cited by Gudger, 1929.]

EVANS, A. H. 1909. Birds. In: Cambridge Natural History, vol. 9.

FRECHKOP, S. 1939. De la differenciation des oiseaux. Gerfaut, fasc. 2, 100.

FÜRBRINGER, M. 1888. Untersuchungen zur Morphologie und Systematik der Vögel . . . I, II. Folio. Jena.

GADOW, H. F. 1929. Article on Bird. In: Encycl. Brit., 14 ed.

GADOW, H. F. and E. SELENKA. 1891. Vögel. In: H. G. Bronn's Klassen und Ordnungen des Thier-Reichs . . . I. Anatomischer Teil, vol. 6.

GUDGER, E. W. 1929. Some early and late illustrations of comparative osteology. Ann. Medical Hist., May, 334.

HEADLEY, F. W. 1895. The structure and life of birds. London.

HUXLEY, T. H. 1867. On the classification of birds . . . Proc. Zool. Soc. London, 415.—1868a. On the animals which are most nearly intermediate between birds and reptiles. Geol. Mag., 357.—1868b. On the classification and distribution of the Alectoromorphae and Heteromorphae. Proc. Zool. Soc. London, 294.—1869. Further evidence of the affinity between the dinosaurian reptiles and birds. Quart. Geol. Soc. London, 12.

LAMBRECHT, K. 1933. Handbuch d. Palaeornithologie. Berlin.

LYDEKKER, R. and others. Birds. In: The New Natural History, vol. 3, sect. 6, 289; vol. 4, sect. 7, 1; ibid., sect. 8, 289.

NEWTON, A. and H. GADOW. 1893–1896. A dictionary of birds. London.

PYCRAFT, W. P. 1910. A history of birds. London.

SHUFELDT, R. W. 1909. Osteology of birds. N. Y. State Mus., Bull. 130.

WETMORE, A. 1930. A systematic classification for the birds of the world. Proc. U. S. Nat. Mus., 1.—1933. Development of our knowledge of fossil birds. In: Fifty Years' Progress of American Ornithology, 1883–1933, p. 231. Lancaster, Pa.

Birds: Miscellaneous, Systematic

CHAPIN, J. P. 1917. The classification of the weaver birds. Bull. Amer. Mus. Nat. Hist., 243. —1932. The birds of the Belgian Congo. Part I. Ibid., vol. 65. Part II (1939). Ibid., vol. 75.—1936. A new peacock-like bird from the Belgian Congo. Revue de Zool. et de Botanique Africaines, vol. 29, fasc. 1.

CHAPMAN, F. M. 1926. The distribution of bird-life in Ecuador. A contribution to the study of the origin of Andean bird-life. Bull. Amer. Mus. Nat. Hist., part 1, 1; part 2, 133.

DELACOUR, J. 1944. A revision of the Family Nectariniidae (sunbirds). Zoologica (N. Y.), vol. 29, 17.

ENGELS, W. L. 1940. Structural adaptations in thrashers (Mimidae: genus Toxostoma) with comments on interspecific relationships. Univ. California Publ., Zool., 341.

HOWARD, H. 1932. Eagles and eagle-like vultures of the Pleistocene of Rancho La Brea. Carnegie Inst. Washington, publ. no. 429.

LOWE, P. R. 1939. On the systematic position of the swifts (Suborder Cypseli) and humming birds (Suborder Trochili) . . . Trans. Zool. Soc. London.

Birds: Origin

BEEBE, W. 1915. A Tetrapteryx stage in the ancestry of birds. Zoologica (N. Y.) 39.—1938. Why do birds have scales on their legs? Bull. N. Y. Zool. Soc., 20.

BÖKER, H. 1927. Die biologische Anatomie der Flugarten der Vögel und ihre Phylogenie. Jour. f. Ornithologie, 305.—1930. Über das Verhaltnis der Dinosaurier zu den Vögeln. Morphol. Jahrb., 223.

BROOM, R. 1906. On the early development of the appendicular skeleton of the ostrich, with remarks on the origin of birds. Trans. South African Philos. Soc.. 355.—1913. On the

South-African pseudosuchian Euparkeria and allied genera. Proc. Zool. Soc. London, 619. [Origin of diapsids and birds.]

COPE, E. D. 1867. Account of extinct reptiles which approach birds. Proc. Acad. Nat. Sci. Philadelphia, 234.

GREGORY, W. K. 1916. Theories of the origin of birds. Ann. N. Y. Acad. Sci., 31.—1933–1934. Remarks on the origins of the ratites and penguins, with discussion by Robert C. Murphy. Proc. Linnean Soc. N. Y., nos. 45, 46.—1946. Some critical phylogenetic stages leading to the flight of birds. Ibid., nos. 54–57 (1941–1945).

HEILMANN, G. 1927. The origin of birds. New York.

HUXLEY, T. H. 1870. Further evidence of the affinity between the dinosaurian reptiles and birds. Quart. Jour. Geol. Soc. London.

MATTHEW, W. D. 1928. Outline and general principles of the history of life. Synopsis of lectures in Paleontology I. Univ. of California Press. [Origin of feathers as insulating structures, p. 144.]

NOPCSA, F. 1929. Noch einmal Proavis. Anat. Anz., 265.

OSBORN, H. F. 1900. Reconsideration of the evidence for a common dinosaur-avian stem in the Permian. Amer. Nat., 777.

PYCRAFT, W. P. 1906. The origin of birds. Knowledge and Scientific News, Sept., 531.

STRESEMANN, E. 1927–1934. Sauropsida: Aves. In: Handbuch der Zoologie . . . (Kükenthal and Krumbach), vol. 7. [Origin of birds, pp. 728–734.]

WILLISTON, S. W. 1915. Trimerorhachis . . . Jour. Geol., vol. 23, no. 3, 246. [Consolidation of instep bones in oldest birds indicates running habit, p. 250.]

Birds: Plumage, Color Patterns

BEEBE, W. 1908. Preliminary report on an investigation of the seasonal changes of color in birds. Amer. Nat., 34.

BIEDERMANN, W 1928. Das Federkleid der Vögel. Ergebn. der Biol., vol. 3. In: Vergl. Physiol. des Integuments der Wirbeltiere, II, 388. Berlin.

DARWIN, C. 1871. The descent of man . . . Part 2. London. [Ocelli or eye-like spots on the plumage of birds, 552 ff.]

WILLIER, B. H. 1941. An analysis of feather color pattern produced by grafting melanophores during embryonic development. Amer. Nat., 136.

Birds: Skull and Jaws, Hyoids
(See also Birds, General; Birds, Origin)

BROOM, R. and G. T. BROCK. 1931. On the vomerine bones in birds. Proc. Zool. Soc. London, 737.

KRIPP, D. V. 1933a. Der Oberschnabel-Mechanismus der Vögel . . . Morphol. Jahrb., vol. 71, 469.—1933b. Beiträge zur mechanischen Analyse des Schnabelmechanismus. Ibid., vol. 72, 541.—1935. Die mechanische Analyse der Schnabelkrümmung und ihre Bedeutung für die Anpassungsforschung. Ibid., vol. 76, 448.

McDOWELL, S. 1948. The bony palate of birds. Part I. The Palaeognathae. The Auk, Oct., 520. [A stimulating contribution. W. K. G.]

PARKER, W. K. 1866. On the structure and development of the skull in the ostrich tribe. Philos. Trans. Roy. Soc. London, 113.—1869. On the structure and development of the skull in the common fowl (Gallus domesticus). Ibid., 755. —1879. On the structure and development of the bird's skull. Part II. Trans. Linnean Soc. London, 99.

PYCRAFT, W. P. 1900. On the morphology and phylogeny of the Palaeognathae (Ratitae and Cypturi) and Neognathae (Carinatae). Trans. Zool. Soc. London, 149.—1901. Some points on the morphology of the palate of the Neognathae. Jour. Linnean Soc., Zool., 343.

REICHERT, C. 1837. Über die Visceralbogen der Wirbeltiere im Allgemeinen und deren Metamorphosen bei den Vögeln und Säugetieren. Archiv f. Anat., Physiol. u. wissensch. Medicin, 120.

SCHESTAKOWA, G. S. 1934. Zur Frage über die Homologie der Gehörknöchelchen; Die Entwicklung des schalleitenden Apparates der Vögel. Bull. Soc. Nat. Moscou, 225.

Birds: Locomotor System, Skeleton, Wings,
Feathers, Flight, Respiratory System

BÖKER, H. 1929. Flugvermögen und Kropf bei Opisthocomus cristatus und Stringops habroptilus. Morphol. Jahrb., 152.

BUNNELL, S. 1930. Aeronautics of bird flight. The Condor, Nov.–Dec., 269.

EWART, J. C. 1921. The nestling feathers of the mallard, with observations on the composition, origin and history of feathers. Proc. Zool. Soc. London, 609.

FISHER, H. L. 1945. Locomotion in the fossil vulture Teratornis. Amer. Midland Naturalist, 725.

FÜRBRINGER, M. 1902. Zur vergl. Anat. des Brustschulterapparates und der Schultermuskeln. V. Teil. Vögel. Jena. Zeitschr. f. Naturw., 289.

GADOW, H. 1912. On the origin of feathers. Archiv f. Naturges., vol. 78, part A. 209.

GILBERT, P. W. 1939. The avian lung and air-sac system. The Auk, 57.

GRIFFIN, D. R. 1944. The sensory basis of bird navigation. Quart. Rev. Biol., vol. 19, 15.

HEADLEY, F. W. 1912. The flight of birds. London.

HORTON-SMITH, C. 1938. The flight of birds. London.

HOWELL, A. B. 1937. Morphogenesis of the shoulder architecture: Aves. The Auk, July, 364.—1938. Muscles of the avian hip and thigh. Ibid., January, 71.

LEBEDINSKY, N. G. 1914. Über den Processus pectinealis des Straussen beckens und seine phylogenetische Bedeutung. Anat. Anz., 84. [1. Carinatae.—Proc. pectin. formed by ilium; 2. Apteryx casuarius.—P.p. formed by ilium and pubis; 3. Struthio.—P.p. formed by pubis only. Concludes that P.p. is a neomorph.]

MAREY, E. J. 1890. Le vol des oiseaux. Paris.

NOPCSA, F. 1918. Über den Längen-Breiten-Index des Vögelsternums. Anat. Anz., 510.

PARKER, W. K. 1888. On the structure and development of the wing in the common fowl. Philos. Trans. Roy. Soc. London, 385.

PYCRAFT, W. P. 1922. Birds in flight. London.

ROMER, A. S. 1923a. Crocodilian pelvic muscles and their avian and reptilian homologues. Bull. Amer. Mus. Nat. Hist., 533.—1923b. The ilium in dinosaurs and birds. Ibid., 141.—1927. The development of the thigh musculature of the chick. Jour. Morph., 347.

SHUFELDT, R. W. 1890. The myology of the raven . . . London.

STEINER, H. 1922. Die Ontogenetische und Phylogenetische Entwicklung des Vögelflügelskelettes. Acta Zool., 307. [Carpus of primitive tetrapods and birds, p. 340 et seq.]

STOLPE, M. and K. ZIMMER. 1939. Der Vögelflug. Seine anatomisch-physiologischen und physikalisch-aerodynamischen Grundlagen. Leipzig.

WARNER, L. H. 1931. Facts and theories of bird flight. Quart. Rev. Biol., 84.

Birds: Brain

BLACK, D. 1921. The motor nuclei of the cerebral nerves in phylogeny. A study of the phenomena of neurobiotaxis. Part IV. Aves. Jour. Comp. Neurol., 233.

CRAIGIE, E. H. 1928. Observations on the brain of the humming bird . . . Jour. Comp. Neurol., 377.—1929. The cerebral cortex of Apteryx. Anat. Anz., 97. [Evidence that the avian neocortex has been reduced from a multilaminar condition.]—1930. Studies on the brain of the kiwi (Apteryx australis). Jour. Comp. Neurol., 225.—1935a. The cerebral hemispheres of the kiwi and of the emu (Apteryx and Dromiceius). Jour. Anat., 380.—1935b. The hippocampal and parahippocampal cortex of the emu (Dromiceius). Jour. Comp. Neurol., 563.—1936a. Notes on cytoarchitectural features of the lateral cortex and related parts of the cerebral hemisphere in a series of reptiles and birds. Trans. Roy. Soc. Canada (3), 87.—1936b. The cerebral cortex of the ostrich (Struthio). Jour. Comp. Neurol., 389.—1939. The cerebral cortex of Rhea americana. Ibid., 331.

CRAIGIE, E. H. and R. M. BRICKNER. 1927. Structural parallelism in the midbrain and tweenbrain of teleosts and of birds. Proc. Kon. Akad. van Weten. te Amsterdam, 695.

EDINGER, T. 1926. The brain of Archaeopteryx. Ann. Mag. Nat. Hist., 151.

HUNTER, J. I. 1923. The fore-brain of Apteryx australis. Proc. Kon. Akad. van Weten. te Amsterdam, 807.

WIMAN, C. and T. EDINGER. 1941. Sur les crânes et les encéphales d'Aepyornis et de Mullerornis. Bull. de. l'Acad. Malgache.

Birds: Archaeopteryx, Archaeornis

DAMES, W. 1884. Ueber Archaeopteryx. Palaeont. Abh., 119.—1897. Über Brustbein, Schulter-und Beckengurtel der Archaeopteryx. Sitz. d. Köngl. Preuss. Akad. d. Wiss., Berlin, 817.

HUXLEY, T. H. 1868. Remarks upon Archaeopteryx lithographica. Proc. Roy. Soc. London, 243.

LOWE, P. R. 1935. On the relationship of the struthiones to the dinosaurs and to the rest of the avian class, with special reference to the position of Archaeopteryx. Ibis, April, 398.

OWEN, R. 1864. On the Archaeopteryx of von Meyer, with a description of the fossil remains of a long-tailed species, from the Lithographic Stone of Solenhofen. Philos. Trans. Roy. Soc. London (1863), 33.

PETRONIEVICS, B. 1925. Ueber die Berliner Archaeornis . . . Ann. Geol. Peninsule Balkanique, vol. 8, fasc. 1, 5.—1928. Zur Pubisfrage der Archaeornis. Anat. Anz., 342.

PETRONIEVICS, B. and A. S. WOODWARD. 1917. On the pectoral and pelvic arches of the British Museum specimen of Archaeopteryx. Proc. Zool. Soc. London, 1.

Birds: "Odontornithes"

MARSH, O. C. 1880. Odontornithes: A monograph of the extinct toothed birds of North America. U. S. Geol. Exploration of the 40th Parallel. Washington.

THOMPSON, D'A. W. 1890. On the systematic position of Hesperonis. Univ. Coll., Dundee, vol. 1, 97.

Birds: Penguins and Diving Birds

LOWE, P. R. 1933. On the primitive characters of the penguins and their bearing on the phylogeny of birds. Proc. Zool. Soc. London, part 2, 483.

MURPHY, R. C. and F. HARPER. 1921. A review of the diving petrels. Bull. Amer. Mus. Nat. Hist., vol. 44, 494.

PYCRAFT, W. P. 1899. Contribution to the osteology of birds. Part IV. Pygopodes. Proc. Zool. Soc. London, 1018.

SIMPSON, G. G. 1946. Fossil penguins. Bull. Amer. Mus. Nat. Hist., vol. 87, 1.

Birds: Ratites and Flightless Birds

ANDREWS, C. W. 1897. Note on a nearly complete skeleton of Aepyornis from Madagascar. Geol. Mag., 241.

ARCHEY, G. 1941. The moa: A study of the Dinornithiformes. Bull. Auckland Inst. and Mus.

BOAS, J. E. V. 1929. Biol.-Anat. Studien über den Hals der Vögel. Mem. d. l'Acad. Roy. d. Sci. et d. Lettres de Danemark, 105.

GADOW, H. 1880. Zur Vergl. Anat. der Muskulatur des Beckens und der hinteren Gliedmasse der Ratiten, Jena.

LOWE, P. R 1928. Studies and observations bearing on the phylogeny of the ostrich and its allies. Proc. Zool. Soc. London, 185.

MATTHEW, W. D. and W. GRANGER. 1917. The skeleton of Diatryma, a gigantic bird from the Lower Eocene of Wyoming. Bull. Amer. Mus. Nat. Hist., 307.

TROXELL, E. L. 1931. Diatryma, a colossal heron. Amer. Jour. Sci., 19.

PART IV

MAMMAL-LIKE REPTILES, ORIGIN, RISE AND BRANCHING OF MAMMALIAN ORDERS
(Cf. Vol. I, Chapters XVI–XXII)

Mammal-like Reptiles: Pelycosaurs and Their Horizons

BAUR, G. and E. C. CASE. 1897. On the morphology of the skull of the Pelycosauria and the origin of the mammals. Anat. Anz., 109.—1899. The history of the Pelycosauria, with a description of the genus Dimetrodon, Cope. Trans. Amer. Philos. Soc., 1.

BROOM, R. 1910. A comparison of the Permian reptiles of North America with those of South Africa. Bull. Amer. Mus. Nat. Hist., 197.—1914. A further comparison of the South African dinocephalians with the American pelycosaurs. Ibid., 135.

CASE, E. C. 1903. The structure and relationships of the American Pelycosauria. Amer. Nat., 85.—1904. The osteology of the skull of the pelycosaurian genus, Dimetrodon. Jour. Geol., 304.—1905. The morphology of the skull of the pelycosaurian genus, Dimetrodon. Trans. Amer. Philos. Soc., 5.—1907. Revision of the Pelycosauria of North America. Carnegie Inst. Washington Publ. 55.

GILMORE, C. W. 1919. A mounted skeleton of Dimetrodon gigas in the United States National Museum . . . Proc. U. S. Nat. Mus., 525.

HUENE, F. VON. 1905. Pelycosaurier in deutschen Muschelkalk. Neues Jahr. f. Min., etc., 321.—1913. Das Hinterhaupt von Dimetrodon. Anat. Anz., 429.—1925. Ein neuer Pelycosaurier aus der unteren Permformation Sachsens. Geol. Pal. Abh. Jena. 215.

OLSON, E. C. 1936. The dorsal axial musculature of certain primitive Permian tetrapods. Jour. Morph., 265.

ROMER, A. S. 1927. Notes on the Permo-Carboniferous reptile Dimetrodon. Jour. Geol., 673.

ROMER, A. S. and L. W. PRICE. 1940. Review of

the Pelycosauria. Geol. Soc. Amer., Special Papers, no. 28.

WATSON, D. M. S. 1916. Reconstructions of the skulls of three pelycosaurs in the American Museum of Natural History. Bull. Amer. Mus. Nat. Hist., 637.

WILLISTON, S. W. 1911. American Permian vertebrates. Univ. of Chicago Press.—1915. A new genus and species of American Theromorpha, Mycterosaurus longiceps. Jour. Geol., 554.—1916. The osteology of some American Permian vertebrates. II. Contrib. Walker Mus., 165.

Therapsid Faunal Horizons in South Africa, Russia, South America, North America

BAIN, A. G. 1852. On the geology of southern Africa. Trans. Geol. Soc., 175.

BROOM, R. 1909. An attempt to determine the horizons of the fossil vertebrates of the Karroo. Ann. S. African Mus., 285.

BYSTROW, A. P. 1935. Rekonstruktionsversuche einiger Vertreter der Nord-Dwina Fauna. Proc. Acad. Sci. U.S.S.R., 289. [Text in Russian with German translation.]

DU TOIT, A. L. 1926. The geology of South Africa. London. [Sequence of Karroo horizons.]

HAUGHTON, S. H. 1919. A review of the reptilian fauna of the Karroo system of South Africa. Trans. Geol. Soc. South Africa.—1924. The fauna and stratigraphy of the Stormberg Series. Ann. South African Mus., 323.

HUENE, F. VON. 1925. Die südafrikanische Karroo formation als geologisches u. faunistisches Lebenbild. Forschr. d. Geol. u. Palaeont., vol. 11, 12.—1926. Gondwana-Reptilien in Südamerika. Palaeontologia Hungarica.—1931. Beitrag zur Kenntniss der Fauna der südafrikanischen Karrooformation. Geol. u. Palaeontol. Abhandl.—1940. Die Saurier der Karroo-Gondwana und verwandten Ablagerungen in faunistischer, biologischer und phylogenetischer Hinsicht. Neues Jahr. f. Min. etc., Beil.-Band 83, 246.

ROGERS, A. W. 1937. Union of South Africa Official Year Book, No. 18. Revised by S. H. Haughton. Geol. Surv. South Africa. [Geological structure of the Union. Chapt. 1, with geological map.]

WATSON, D. M. S. 1913. The Beaufort Beds of the Karroo System of South Africa. Geol. Mag., 388.—1914. The zones of the Beaufort Beds and of the Karroo System in South Africa. Ibid., 203.

Therapsida (sensu stricto): General and Miscellaneous; Origin of Mammals; Major Classification

AMALITSKY, V. [Illustrations for a memoir on Pareiasaurus, Inostrancevia, Dicynodon and other Permian (?) reptiles and amphibians.] Unpublished. In Osborn Library of the American Museum of Natural History, New York.

BROILI, F. and J. SCHRÖDER. 1935. Beobachtungen an Wirbeltieren der Karrooformation. Sitz. d. Bayer. Akad. d. Wiss., Jahrg. 1935, 93. Parts 8–14. [On various therapsids. Good figures.]

BROOM, R. 1907. The origin of the mammal-like reptiles. Proc. Zool. Soc. London, 1047.—1910. A comparison of the Permian reptiles of North America with those of South Africa. Bull. Amer. Mus. Nat. Hist., 197.—1913. On the origin of the mammalian digital formula. Anat. Anz., 230.—1914. On the origin of mammals. Croonian Lecture. Philos Trans. Roy. Soc. London, vol. 206.—1915. Permian, Triassic and Jurassic reptiles of South Africa. Bull. Amer. Mus. Nat. Hist., 105.—1924. On the classification of the reptiles. Ibid., 39.—1932. The mammal-like reptiles of South Africa and the origin of mammals. London.—1937a. A few more new fossil reptiles from the Karroo. Ann. Transvaal Mus., 141.—1937b. A further contribution to our knowledge of the fossil reptiles of the Karroo. Proc. Zool. Soc. London, 299.—1938. On recent discoveries throwing light on the origin of the mammal-like reptiles. Ann. Transvaal Mus., 253.—1940a. Some new Karroo reptiles from the Graaff-Reinet district. Ibid., 71.—1940b. On some new genera and species of fossil reptiles from the Karroo Beds of Graaff-Reinet. Ibid., 157.—1941. Some new Karroo reptiles, with notes on a few others. Ibid., 193.—1943. Some new types of mammal-like reptiles. Proc. Zool. Soc. London, 17.—1948. A contribution to our knowledge of the vertebrates of Karroo Beds of South Africa. Trans. Roy. Soc. Edinburgh, part 2, no. 21, p. 86. [Many new forms: cotylosaurs, therocephalians, gorgonopsians, anomodonts, precynodonts.]

EFREMOV, J. A. 1940. Preliminary description of the new Permian and Triassic Tetrapoda from U.S.S.R. Travaux d. l'Inst. Palaeont., vol. 10, part 2. [English summary, pp. 93–140. Venjukovia, figs. 16–23, pl. 8; Inostrancevia, pl. 14.]

HAUGHTON, S. H. 1922. On some Upper Beaufort

Therapsida. Trans. Roy. Soc. South Africa, 299. —1932. On a collection of Karroo vertebrates from Tanganyika Territory. Quart. Jour. Geol. Soc. London, 634.

HUENE, F. VON. 1925. Stammlinien der Reptilien. Centralbl. f. Min., etc., 229.—1935–1942. Die fossilen Reptilien des südamerikanischen gondwanalandes . . . Munich. [Belesodon, pp. 98, 99; Traversodon, p. 138.]—1944. Paläontologische Grundzüge des Stammesgeschichte der frühen Tetrapodenzweige. Zeits. f. wiss. Zool., 259.

OLSON, E. C. 1944. Origin of mammals based upon cranial morphology of the therapsid suborders. Geol. Soc. Amer., Special Papers, no. 55.

OSBORN, H. F. 1898. The origin of the Mammalia. Amer. Nat., 309.

OWEN, R. 1876. Description of the fossil Reptilia of South Africa in the collection of the British Museum. Vol. I, text. Vol. 2, plates. London, 4to.

PARRINGTON, F. R. 1946. On the quadratojugal bone of synapsid reptiles. Ann. Mag. Nat. Hist., 780.

SEELEY, H. G. 1895. Researches on the structure, organization and classification of the fossil Reptilia. Part II, sect. 5. On the skeleton in new Cynodontia from the Karroo Rocks. Philos. Trans. Roy. Soc. London, 59.

WATSON, D. M. S. 1942. On Permian and Triassic tetrapods. Geol. Mag., 81.

WILLISTON, S. R. 1917. The phylogeny and classification of reptiles. Jour. Geol. 411.—1925. The osteology of the reptiles. Harvard Univ. Press. [Including therapsids.]

Dinocephalia

BOONSTRA, L. D. 1935. Voorhistoriese Diere. IV. Die Jongspan, vol. 1, no. 49. [Restoration of Moschops skeleton.]—1936a. Some features of the cranial morphology of the tapinocephalid deinocephalians. Bull. Amer. Mus. Nat. Hist., 75.—1936b. The cranial morphology of some titanosuchid deinocephalians. Ibid., 99.

BROILI, F. and J. SCHRÖDER. 1935. Beobachtungen an Wirbeltieren der Karrooformation. VIII. Ein Dinocephalen-Rest aus den unteren Beaufort-Schichten. Sitz. d. Bayer. Akad. d. Wiss., 93.

BROOM, R. 1914. A further comparison of the South African dinocephalians with the American pelycosaurs. Bull. Amer. Mus. Nat. Hist., 135.—1923. On the structure of the skull in the carnivorous dinocephalian reptiles. Proc. Zool. Soc. London, 661.—1928. On Tapinocephalus and two other dinocephalians. Ann. South African Mus., 427.—1929. On the carnivorous mammal-like reptiles of the family Titanosuchidae. Ann. Transvaal Mus., 9.—1936. On the structure of the skull in a new type of dinocephalian reptile. Proc. Zool. Soc. London, 733. [Eudinosuchus vorsteri.]

BYRNE, F. 1937. A preliminary report on a new mammal-like reptile from the Permian of South Africa. Trans. Kansas Acad. Sci., vol. 40, 221. [Moschoides romeri, pls. 1–3.]—1940. Notes on the evolution of the mammal-like reptiles. Ibid., 291. [Comparison of Moschoides romeri with other dinocephalians.]

EFREMOV, J. A. 1940. Ulemosaurus svijagensis Riab., ein Dinocephale aus den Ablagerungen des Perm der U.S.S.R. Nova Acta Leopoldina, N.F., 155.

GREGORY, W. K. 1926. The skeleton of Moschops capensis Broom, a dinocephalian reptile from the Permian of South Africa. Bull. Amer. Mus. Nat. Hist., 179.

RIABININ, A. N. 1938. Vertebrate fauna from the Upper Permian deposits of the Sviaga Basin. 1. A new dinocephalian, Ulemosaurus svijagensis n.g., n.sp. Ann. Central Geol. and Prospecting Scient. Research Mus. (Tschernyschew Museum). [English summary, pp. 40–52.]

WATSON, D. M. S. 1914. The Deinocephalia, an order of mammal-like reptiles. Proc. Zool. Soc. London, 749.

Anomodontia

ARAMBOURG, C. 1943. Un squelette de Lystrosaurus au Muséum national d'Histoire naturelle. Bull. du Mus., vol. 15, 351.

BROILI, F. and J. SCHRÖDER. 1935–1937. Beobachtungen an Wirbeltieren der Karrooformation. Sitz. d. Bayer. Akad. d. Wiss. VI. Über den Schädel von Cistecephalus Owen. Ibid., 1. XII. Über einige primitive Anomodontier-Schädel aus den unteren Beaufort-Schichten. Ibid., 223. —1936. XVI. Beobachtungen am Schädel von Emydochampsa Broom. Ibid., 21. XVII. Ein neuer Anomodontier aus der Cistecephalus-Zone. Ibid., 45.—1937. XXVIII. Über einige neue Anomodontier aus der Tapinocephalus-Zone. Ibid., 118.

BROOM, R. 1912a. On some points in the structure of the dicynodont skull. Ann. South African Mus. 337. [Oudenodon described and figured.]—1912b. On some new fossil reptiles from the Permian and Triassic beds of South Africa. Proc. Zool. Soc. London, 859. [Taurops, Scymnognathus, Galeops, Aelurosaurus, Pristernognathus, Ictidognathus, Dicynodon, etc.]—1915. On the anomodont genera, Pristerodon and Tropidostoma. Proc. Zool. Soc. London, 355.—1921. On some new genera and species of anomodont reptiles from the Karroo Beds of South Africa. Proc. Zool. Soc. London, 647. [Oudenodon confirmed as female of genus Dicynodon. See Broom, 1909.]—1935. A new genus and some new species of mammal-like reptiles. Ann. Transvaal Mus., 1. [Oudenodon, female of Dicynodon, pp. 8–12.]—1938. On two new anomodont genera. Ibid., 247.

CAMP, C. L. 1948. The dicynodont ear. In: Robert Broom Commemorative Volume, 109.

CASE, E. C. 1934. Description of a skull of Kannemeyeria erithrea Haughton. Contrib. Mus. Paleont. Univ. Michigan, 115.

HAUGHTON, S. H. 1918. Investigations in South African fossil reptiles and Amphibia. Part 10. Descriptive catalogue of the Anomodontia . . . Part I. Ann. South African Mus., 127. Part II. Some new carnivorous Therapsida, with notes upon the braincase in certain species. Ibid., part 6, 175.

HUENE, F. VON. 1922. Zur Osteologie des Dicynodon-Schädels. Palaeont. Zeits., 58.—1933. Die südamerikanische Gondwana-Fauna. Forsch. u. Fortschr., no. 9, Berlin. [Mounted skeleton of Stahleckeria and of Belesodon.]—1936. Ein Stahleckeria-Schädel. Zentr. f. Min., etc., 507.—1942. Die Anomodontier des Ruhuhu-Gebietes in der Tübinger Sammlung. Palaeontographica, 154.

JAEKEL, O. 1904. Ueber den Schädelbau der Dicynodonten. Sitz-Ber. d. Ges. Naturf. Freunde, 172.

OLSON, E. C. 1937. The skull structure of a new anomodont. Jour. Geol., 851.

OLSON, E. C. and F. BYRNE. 1938. The osteology of Aulacocephalodon peavoti Broom. Jour. Geol., 177.

PARRINGTON, F. R. 1945. On the middle ear of the Anomodontia. Ann. Mag. Nat. Hist., 625.

PEARSON, H. S. 1924a. The skull of the dicynodont reptile Kannemeyeria. Proc. Zool. Soc. London, 793.—1924b. A dicynodont reptile reconstructed. Proc. Zool. Soc. London, 827. [Kannemeyeria.]

ROMER, A. S. Stahleckeria lenzii, a giant Triassic Brazilian dicynodont. Bull. Mus. Comp. Zool., Harvard Coll., 465.

SEELEY, H. G. 1888–1894. Researches on the structure, organization and classification of the fossil Reptilia. V. On associated bones of a small anomodont reptile (Keirognathus cordylus (Seeley). Philos. Trans. Roy. Soc. London, vol. 179 B, 487.—1889. VI. On the anomodont Reptilia and their allies. Ibid., vol. 180, 215.—1894. VIII. Further evidences of the skeleton in Deuterosaurus and Rhopalodon, from the Permian Rocks of Russia. Ibid., vol. 185, 663.

SOLLAS, I. B. J. and W. J. SOLLAS. 1914. A study of the skull of a Dicynodon by means of serial sections. Philos. Trans. Roy. Soc. London (1913), 311.—1916. On the structure of the dicynodont skull. Ibid., 531.

SUSHKIN, P. P. 1926. Notes on the Pre-Jurassic Tetrapoda from Russia. I. Dicynodon amalitskii, n. sp. II. Contributions to the morphology and ethology of the Anomodontia. Palaeontologia Hungarica (1921–1925), 323. Budapest.

WATSON, D. M. S. 1912. The skeleton of Lystrosaurus. Records, Albany Mus., 287.—1913a. The limbs of Lystrosaurus. Geol. Mag., 256.—1913b. Some notes on the anomodont brain case. Anat. Anz., 210.

Gorgonopsia, Therocephalia, Bauriamorpha

BOONSTRA, L. D. 1934a. A contribution to the morphology of the Gorgonopsia. Ann. South African Mus., 137.—1934b. Additions to our knowledge of the South African Gorgonopsia, preserved in the British Museum (Natural History). Ibid., 175.—1934c. A contribution to the morphology of the mammal-like reptiles of the suborder Therocephalia. Ibid., 215. [Euchambersia mirabilis, p. 261, fig. 35.]—1934d. On an aberrant gorgonopsian, Burnetia mirabilis Broom. South African Jour. Sci., 462.—1938. On a South African mammal-like reptile, Bauria cynops. Palaeobiologica, 164.

BROILI, F. and J. SCHRODER. 1934–1936. Beobachtungen an Wirbeltieren der Karrooformation. Sitz. d. Bayer. Akad. d. Wiss.—1934. II. Über den Cynodontier Tribolodon frerensis Seeley.

Ibid., 163. III. Ein Gorgonopside aus den unteren Beaufort-Schichten, 179. IV. Ein neuer Gorgonopside aus den unteren Beaufort-Schichten, 209.—1935. VII. Ein neuer Bauriamorphe aus der Cynognathus-Zone. Ibid., 21. XIII. Über die Skelettereste eines Gorgonopsiers aus den unteren Beaufort-Schichten, 279. XIV. Ein neuer Vertreter der Gorgonopsiden-Gattung Aelurognathus, 331.—1936. XV. Ein Therocephalier aus den unteren Beaufort-Schichten. Ibid., 1. XVI. Beobachtungen am Schädel von Emydochampsa Broom, 21. XVII. Ein neuer Anomodontier aus der Cistecephalus-Zone, 45. XXIII. Ein weiterer Therocephalier aus den unteren Beaufort-Schichten, 283. XXIV. Über Theriodantier-Reste aus der Karrooformation Ostafrikas, 311.

BROOM, R. 1915. On some new carnivorous therapsids in the collection of the British Museum. Proc. Zool. Soc. London, 163.—1920. On some new therocephalian reptiles from the Karroo Beds of South Africa. Ibid., 343.—1927. On a new type of mammal-like reptile from the South African Karroo Beds (Anningia megalops). Ibid., 227.—1930. On the structure of the mammal-like reptiles of the sub-order Gorgonopsia. Philos. Trans. Roy. Soc. London, 345. [Lycaenops, pl. 27, fig. 1.]—1932. The mammal-like reptiles of South Africa and the origin of mammals. London. [Cynarioides, fig. 46.]—1936. On the structure of the skull in the mammal-like reptiles of the suborder Therocephalia. Philos. Trans. Roy. Soc. London, 1.—1937. On the palate, occiput and hind foot of Bauria cynops Broom. Amer. Mus. Novitates, no. 946.—1938a. On a new family of carnivorous therapsids from the Karroo Beds of South Africa. Proc. Zool. Soc. London, 527.—1938b. On a nearly complete therocephalian skeleton. Ann. Transvaal Museum, 257.

COLBERT, E. H. 1948. The mammal-like reptile Lycaenops. Bull. Amer. Mus. Nat. Hist., 353.

HAUGHTON, S. H. 1918. Investigations in South African fossil reptiles and Amphibia. (Part 11). Some new carnivorous therapsids, with notes upon the brain-case in certain species. Ann. South African Mus., 175.—1924. On some gorgonopsian skulls in the collection of the South African Museum. Ibid., 499.

HUENE, F. VON. 1938. Drei Theriodontier-Schädel aus Südafrika. Palaeont. Zeits., 297.

OLSON, E. C. 1937. The cranial morphology of a new gorgonopsian. Jour. Geol., 511.—1938a. The occipital, otic, basicranial, and pterygoid regions of the Gorgonopsia. Jour. Morph., 141. —1938b. Notes on the brain case of a therocephalian. Ibid., 75.—1944. Origin of mammals based upon the cranial morphology of the therapsid suborders. Geol. Soc. Amer., Special Papers, no. 55.

PARRINGTON, F. R. 1939. On the digital formulae of theriodont reptiles. Ann. Mag. Nat. Hist., 209.

PRAVOSLAVLEFF, P. A. 1927a. Gorgonopsidae (III). Akad. Nauk, Severo-Dvinskie raskopki, Prof. V. P. Amalitsky. Leningrad. 4to. [Inostrancevia, figs. 2–20, pls. 1–10.]—1927b. Gorgonopsidae (IV). Ibid.

SCHAEFFER, B. 1941. The pes of Bauria cynops Broom. Amer. Mus. Novitates, no. 1103.

SEELEY, H. G. 1888. Researches on the structure, organization and classification of the fossil Reptilia. III. On parts of the skeleton of a mammal from Triassic rocks of Klipfontein, Fraserberg, South Africa (Theriodesmus phylarchus, Seeley), illustrating the reptilian inheritance in the mammalian hand. Philos. Trans. Roy. Soc. London, 141.—IX. On the Therosuchia. Ibid. (1894), 987.

WATSON, D. M. S. 1913. On some features of the structure of the therocephalian skull. Ann. Mag. Nat. Hist., 65.—1914. Notes on some carnivorous therapsids. Proc. Zool. Soc. London, 1021. —1921. The bases of classification of the Theriodontia. Ibid., 35. [Changes required to transform Arctognathus into a cynodont, p. 79.]—1931. On the skeleton of a bauriamorph reptile. Ibid., 1163.

Cynodontia, Protodonta

BROILI, F. and J. SCHRÖDER. 1934–1936. Beobachtungen an Wirbeltieren der Karrooformation. Sitz. d. Bayer. Akad. d. Wiss.—Zur Osteologie des Kopfes von Cynognathus. Ibid., 95.—II. Über den Cynodontier Tribolodon frerensis Seeley. Ibid., 163.—1935. IX. Über den Schädel von Gomphognathus Seeley. Ibid., 115.—X. Über die Bezähnung von Trirachodon Seeley. Ibid., 189.—XI. Über den Schädel von Cynidiognathus Haughton. Ibid., 199.—1936. XVIII. Über Cynodontier-Wirbel. Ibid., 61.—XXIII.

Ein neuer Galesauride aus den unteren Beaufort-Schichten. Ibid., 283.

BROOM, R. 1909. On the shoulder girdle of Cynognathus. Ann. South African Mus., 283.—1911. On the structure of the skull in cynodont reptiles. Proc. Zool. Soc. London, 893.—1938a. On the structure of the skull of the cynodont, Thrinaxodon liorhinus, Seeley. Ann. Transvaal, Mus., 263.—1938b. The origin of the cynodonts. Ibid. 279. [Cynodonts evolved from therocephalians, p. 279.]

GREGORY, W. K. and C. L. CAMP. 1918. Studies in comparative myology and osteology. No. III. Part 1. A comparative review of the muscles of the shoulder-girdle and pelvis of reptiles and mammals, with an attempted reconstruction of these parts in Cynognathus, an extinct therapsid reptile. Bull. Amer. Mus. Nat. Hist., 447.

HAUGHTON, S. H. 1920. On the genus Ictidopsis. Ann. Durban Mus., 243.—1924. On Cynodontia from the Middle Beaufort Beds of Harrismith, Orange Free State. Ann. Transvaal Mus., 74. [Glochinodontoides, figs. 50–55.]

HUENE, F. VON. 1928. Ein Cynodontier aus der Trias Brasiliens. Centralb. f. Min., etc., 251.

PARRINGTON, F. R. 1933. On the cynodont reptile Thrinaxodon liorhinus Seeley. Ann. Mag. Nat. Hist., 16.—1934. On the cynodont genus Galesaurus, with a note on the functional significance of the changes in the evolution of the theriodont skull. Ann. Mag. Nat. Hist., 38—1935. A note on the parasphenoid of the cynodont Thrinaxodon liorhinus Seeley. Ann. Mag. Nat. Hist., 399.—1946. On the cranial anatomy of cynodonts. Proc. Zool. Soc. London, 181.

SEELEY, H. G. 1895. Researches on the structure, organization and classification of the fossil Reptilia. Philos. Trans. Roy. Soc. London. Part IX, sect. 4. On the Gomphodontia. Ibid., 1. Sect. 5. On the skeleton in new Cynodontia in the Karroo Rocks. Ibid., 59. [Cynognathus, figs. 1–32; Tribolodon frerensis, figs. 33, 34.] Sect. 6. Associated remains of two small skeletons from Klipfontein, Fraserburg. Ibid., 149.

SIMPSON, G. G. 1926. Are Dromatherium and Microconodon mammals? Science, 548.—1927. On the cynodont reptile Tribolodon frerensis, Seeley. Ann. Mag. Nat. Hist., 28.

SUSHKIN, P. P. 1929. Permocynodon, a cynodont reptile from the Upper Permian of Russia. Congrès internat. de Zoologie, 1927, 804.

WATSON, D. M. S. 1911. The skull of Diademodon, with notes on those of some other cynodonts. Ann. Mag. Nat. Hist., 293.—1913. Further notes on the skull, brain and organs of special sense of Diademodon. Ibid., 217.—1920. On the Cynodontia. Ibid., 506.

Mammals, General:
Major Source-books, Textbooks,
General Catalogues, Bibliographies

BEDDARD, F. E. 1902. Mammalia. In: Cambridge Natural History, vol. 10. London.

BLAINVILLE, H. M. D. DE. 1839–1864. Ostéographie ou description iconographique du squelette et du système dentaire des mammifères. Text, 4 vols. 4to. Atlas, 4 vols., folio. Paris.

BOULE, M. and J. PIVETEAU. 1935. Les fossiles: éléments de paléontologie. Paris.

CAMP, C. L. and V. L. VANDERHOOF. 1940. Bibliography of fossil vertebrates, 1928–1933. Geol. Soc. Amer., Special Papers, no. 27.

CAMP, C. L., D. N. TAYLOR and S. P. WELLES. 1942. Bibliography of fossil vertebrates, 1934–1938. Geol. Soc. Amer., Special Papers, no. 42.

CUVIER, G. L. C. F. D. 1812. Recherches sur les ossemens fossiles de quadrupèdes . . . Paris.

FLOWER, W. H. 1885. An introduction to the osteology of the Mammalia. 3rd ed. London.

FLOWER, W. H. and R. LYDEKKER. 1891. An introduction to the study of mammals, living and extinct. London.

GREGORY, W. K. 1910. The orders of mammals. Parts I, II. Bull. Amer. Mus. Nat. Hist.

HAY, O. P. 1902. Bibliography and catalogue of the fossil Vertebrata of North America. U. S. Geol. Surv., Washington.—1929–1930. Second bibliography and catalogue of the fossil Vertebrata of North America. Vol. I, II. Carnegie Inst. Washington.

LYDEKKER, R. 1887. Catalogue of the fossil Mammalia in the British Museum (Natural History). Part 5. Containing the group Tillodontia, the orders Sirenia, Cetacea, Edentata, Marsupialia, Monotremata, and Supplement. London.

OSBORN, H. F. 1910. The age of mammals in Europe, Asia and North America. New York.

PALMER, T. S. 1904. Index Generum Mammalium: A list of the genera and families of mammals. North American fauna, No. 23. U. S. Dept. Agriculture, Division Biol. Surv.

PARKER, T. J. and W. A. HASWELL. 1940. A textbook of zoology. Vol. II. 6th ed., revised by C. Forster-Cooper. London and New York.

ROMER, A. S. 1945. Vertebrate paleontology. 2nd ed. Univ. Chicago Press. [1st ed., 1933.]

SCHLOSSER, M. 1923. Mammalia. In: Karl von Zittel's Grundzüge der Paläontologie, vol. 2, Vertebrata. Munich and Berlin.

STROMER, E. 1912. Lehrbuch der Paläozoologie. II. Teil: Wirbeltiere. Leipzig.

TROUESSART, E. L. 1897–1905. Catalogus mammalium tam viventium quam fossilium. Nova editio (prima completa). Parts 1–6 and suppl. Berlin.

WEBER, M. 1904. Die Säugetiere. Einführung in die Anatomie und Systematik der recenten und fossilen Mammalia. Jena.

WEBER, M. and O. ABEL. 1928. Die Säugetiere . . . 2nd. ed. Bd. II: Systematischer Teil. Jena.

WEBER, M. and H. M. DE BURLET. 1927. Die Säugetiere. . . . 2nd ed. Bd. I. Anatomischer Teil. Jena.

WIEDERSHEIM, R. 1907. Einführung in die Vergleichende Anatomie der Wirbeltiere. Jena.

Mammals: Natural Histories

CABERA, A. 1925. Historia natural vida de los Animales in las Plantas y de la Tierra. Tome I. Zoologie (Vertebrados). Barcelona.

GESNER, C. 1551. Historiae Animalium. Lib. I. De Quadrupedibus viviparis. Tiguri. Folio. Lib. II. De quadrupedibus oviparis.

HORNADAY, W. T. 1904. The American natural history . . . New York.

LYDEKKER, R. [undated.] The new natural history. Vols. I–VI. New York.

Mammals: Origin
(See also Mammal-like Reptiles, above)

BROOM, R. 1914. On the origin of mammals. Croonian Lecture. Philos. Trans. Roy. Soc. London, vol. 206.—1932. The mammal-like reptiles of South Africa and the origin of mammals. London.

GREGORY, W. K. 1910. The orders of mammals. Bull. Amer. Mus. Nat. Hist. [Origin of mammals, pp. 113–141.]

MATTHEW, W. D. 1904. The arboreal ancestry of the Mammalia. Amer. Nat., 455, 811.

OLSON, E. C. 1944. Origin of mammals based upon cranial morphology of the therapsid suborders. Geol. Soc. Amer., Special Papers, no. 55.

OSBORN, H. F. 1898. The origin of the Mammalia. I. Amer. Nat., 309.—1899. II. Amer. Jour. Sci., 92.—1900. III. Occipital condyles of reptilian tripartite type. Amer. Nat., 943.

Mammals: Major Classification into Orders, Phylogeny

BLAINVILLE, H. M. D. DE. 1916. Prodrome d'une nouvelle distribution systématique du règne animal. Jour. de Physique, de Chimie, d'Hist. nat. et des Arts, 244.—1934. "Cours de la faculté des sciences, 1834." Fide Palmer, T. S., Index generum mammalium, 1904, p. 780.

GEOFFROY SAINT-HILAIRE, E. and G. CUVIER. 1795. Mémoire sur une nouvelle division des mammifères, et sur les principes qui doivent servir de base dans cette sorte de travail. Magasin Encyclopédique, 1re année, vol. 2, 164.

GREGORY, W. K. 1907. The place of Linnaeus in the unfolding of science; his views on the Class Mammalia. Popular Science Monthly, 121.—1908. Linnaeus as an intermediary between ancient and modern zoology; his views on the Class Mammalia. Ann. N. Y. Acad. Sci., 21.—1910. The orders of mammals. I. Typical stages in the history of the ordinal classification of mammals. II. Genetic relations of the mammalian orders . . . Bull. Amer. Mus. Nat. Hist., vol. 27.—1929. Article on Mammalia. Encycl. Brit., 14th ed.

GILL, T. 1870. On the relations of the orders of mammals. Proc. Amer. Assoc. Adv. Sci., 267.—1872. Arrangement of the families of mammals with analytical tables. Smiths. Miscell. Coll., no. 230.

HUXLEY, T. H. 1869. An introduction to the classification of animals. London. [Monodelphian orders, as in the classification of 1864, grouped in accordance with placental characters.]—1880. On the application of the laws of evolution to the arrangement of the Vertebrata and more particularly of the Mammalia. Proc. Zool. Soc. London, 649.

JEPSEN, G. L. 1944. Phylogenetic trees. Trans. N. Y. Acad. Sci., 81.

LINNAEUS, C. 1735. Systema naturae, sive regna tria naturae systematice proposita per classes, ordines, genera et species. Lugduni Batavorum. Folio.—1894. Systema naturae, regnum animale. 10th ed. (1758) revised. Lipsiae.

MATTHEW, W. D. 1928. The evolution of mammals in the Eocene. Proc. Zool. Soc. London (1927), 947.—1937. Paleocene faunas of the San Juan Basin, New Mexico. Trans. Amer.

Philos. Soc. [Relationships of earlier and later orders.]—1943. Relationships of the orders of mammals (Edited and annotated by G. G. Simpson). Jour. Mammal., 304.

OSBORN, H. F. 1910. The age of mammals in Europe, Asia and North America. New York. [Classification of mammals, pp. 513–563.]

OWEN, R. 1859. On the classification and geographical distribution of the Mammalia. London.

SIMPSON, G. G. 1931. A new classification of mammals. Bull. Amer. Mus. Nat. Hist., 259.—1945. The principles of classification and a classification of mammals. Ibid., vol. 85.

TROUESSART, E. L. 1897–1905. Catalogus mammalium tam viventium quam fossilium. Nova editio (prima completa) Parts 1–6 and suppl. Berlin.

WINGE, H. 1882. Om Pattedyrenes Tandskifte issaer med Hensyn til Taendernes Former. Vidensk. Meddel. fra den naturh. Foren, 15.—1941. The interrelationships of the mammalian genera. Vol. I. Monotremata, Marsupialia, Insectivora, Chiroptera, Edentata. Translated from the Danish by E. Deichmann and G. M. Allen. Kobenhavn.

Mammals: Evolution Factors:
Speciation, Genetic and Endocrine Factors, Skin Color and Color Patterns, "Habitus" (Adaptive) Factors, Ecologic Factors (as bearing on Selection)
(See also "Evolution," in Part I)

COTT, H. B. 1938. Concealing coloration in animals. Photogr. Jour., Roy. Photogr. Soc., 563.

EDINGER, T. 1942. The pituitary body in giant animals, fossil and living . . . Quart. Rev. Biol., 31. [Endocrine factor.]

ERRINGTON, P. L. 1943. An analysis of mink predation upon muskrats in North-Central United States. Iowa State Coll. Agriculture and Mechanic Arts. Experiment Station, Research Bull. no. 320. [Ecologic factors, selection.]

GREGORY, W. K. 1936. Habitus factors in the skeleton of fossil and recent mammals. Proc. Amer. Philos. Soc., 429.—1943. Environment and locomotion in mammals. Nat. Hist. Mag., 222.

HALL, E. R. 1943. Criteria for vertebrate subspecies, species and genera: The mammals. Ann. N. Y. Acad. Sci., 141. [Genetic factors]

HOWELL, A. B. 1944. Speed in animals: Their specializations for running and leaping. Univ. Chicago Press. [Adaptive factors.]

JONES, F. W. 1943. Habit and heritage. London.

MORGAN, T. H. 1911. The influence of heredity and of environment in determining the coat colors in mice. Ann. N. Y. Acad. Sci., 87.

OSBORN, H. F. 1915. Origin of single characters as observed in fossil and living animals and plants. Amer. Nat., 193.

PYCRAFT, W. P. 1935. Concerning cats and their coloration. Illust. London News, Nov. 2, 734.

ROOSEVELT, T. 1911. Revealing and concealing coloration in birds and mammals. Bull. Amer. Mus. Nat. Hist., 119.

SUMNER, F. B. 1934. A test of the possible effects of visual stimuli upon the hair color of mammals. Proc. Nat. Acad. Sci., 397.

SUMNER, F. B. and H. S. SWARTH. 1924. The supposed effects of the color tone of the background upon the coat color of mammals. Jour. Mammal., 81.

STOCKARD, C. R. and others. 1941. The genetic and endocrine basis for differences in form and behavior as elucidated by studies of contrasted pure-line dog breeds and their hybrids. Wister Inst. Anat. and Biol. Philadelphia.

Mammalian Embryology, Placentation
(See also Mammals: Textbooks, above; and Man: Embryology, below)

BEER, G. R. DE. 1930. Embryology and evolution. Oxford. [Clandestine evolution, p. 30.]—1940a. Embryos and ancestors. Oxford.—1940b. Embryology and taxonomy. In: The new systematics (Julian Huxley, editor). Oxford.

FLYNN, T. T. 1923. On the occurrence of a true allantoplacenta of the conjoint type in an Australian lizard. Rec. Australian Mus., 72.

FLYNN, T. T. and J. P. HILL. 1947. The development of the Monotremata. Part VI. The later stages of cleavage and the formation of the primary germ-layers. Trans. Zool. Soc. London, part 1, 1. [Close agreement between monotreme and marsupial methods, the latter a little more advanced toward the monodelphian stage.]

FRECHKOP, S. 1941. Le placenta du Daman et la valeur systématique de cet organe. Ann. Soc. Roy. Zool. de Belgique, 150.

MCCRADY, E. JR. 1938. The embryology of the opossum. Amer. Anat. Mem. Philadelphia.

WISLOCKI, G. B. 1930. On an unusual placental form in the Hyracoidea: Its bearing on the theory of the phylogeny of the placenta. Contrib. Embryol., no. 122, Carnegie Inst. Washington Publ. 407.

WISLOCKI, G. B. and O. P. VAN WESTHUYSEN. 1940. The placentation of Procavia capensis, with a discussion of the placental affinities of the Hyracoidea. Ibid., no. 171.

Mammalian Anatomy
(See also Mammalia: Textbooks, above)

CUVIER, G. L. C. F. D. 1800. Leçons d'anatomie comparée. Paris.

CUVIER, G. L. C. F. D. and C. L. LAURILLARD. 1849. Anatomie comparée recueil de planches de myologie . . . Folio. Paris.

ELLENBERGER, W. and H. BAUM. 1891. Systematische und topographische Anatomie des Hundes (Canis). Berlin.

ELLENBERGER, W., H. BAUM and H. DITTRICH. 1898. Handbuch der Anatomie der Tiere für Künstler. Leipzig.

GREENE, E. C. 1935. Anatomy of the rat. Trans. Amer. Philos. Soc., n.s., vol. 27. Philadelphia.

HOWELL, A. B. 1926. Anatomy of the wood rat: Comparative anatomy of the subgenera of the American wood rat (genus Neotoma). Baltimore.

KNIGHT, C. R. 1947. Animal anatomy and psychology for the artist and layman. New York.

MITCHELL, P. C. 1905. On the intestinal tract of mammals. Trans. Zool. Soc. London, 437.

PUTNAM, B. 1947. Animal X-rays: A skeleton key to comparative anatomy. New York.

REIGHARD, J. and H. S. JENNINGS. 1902. Anatomy of the cat. 2nd ed., revised. New York.

SCHMALTZ, R. 1909. Atlas der Anatomie des Pferdes. II. Teil. Topographische Myologie. Berlin.

SISSON, S. 1938. The anatomy of the domestic animals. 3rd ed. Philadelphia.

Mammalian Locomotor System (Skeleton and Musculature)
(See also Vertebrata, Tetrapoda and Mammalia, General, above; and Man, below)

BLAINVILLE, H. M. D. DE. 1839–1864. Ostéographie . . . (See Mammalia, General).

BRASH, J. C. 1928. Growth of the alveolar bone. Internat. Jour. Orthondontics, 196.—1934. Some problems in the growth and developmental mechanics of bone. Edinburgh Med. Jour., 305.

CHUBB, S. H. 1929. How animals run: Some interesting laws governing animal locomotion . . . Nat. Hist. Mag., 543.—1932. Vestigial clavicles and rudimentary sesamoids: Their development and functions in mammals. Amer. Nat.

CUVIER, G. L. C. F. D. 1812. Recherches sur les ossemens fossiles des quadrupèdes . . . Vols. I–X, with atlas (2 vols.) Folio. Paris.—1849. Anatomie comparée recueil de planches de myologie. Edited by M. Laurillard. Folio. Paris.

DAVENPORT, C. B. 1941. Responsive bone. Proc. Amer. Philos. Soc., 65. [Structure of bone, development of lines of bony rods, determined by stresses.]

ELFTMAN, H. O. 1929. Functional adaptations of the pelvis in marsupials. Bull. Amer. Mus. Nat. Hist., 18. [Musculature in relation to locomotion.]

EVANS, F. G. 1939. The morphology and functional evolution of the atlas-axis complex from fish to mammals. Ann. N. Y. Acad. Sci., 29. [Musclature and function.]

EVANS, F. G. and V. E. KRAHL. 1945. The torsion of the humerus: A phylogenetic survey from fish to man. Amer. Jour. Anat., 303.

FLOWER, W. H. 1885. An introduction to the osteology of the Mammalia. 3rd ed. New York.

GREGORY, W. K. 1935. The pelvis from fish to man: A study in paleomorphology. Amer. Nat., 193.—1936. Habitus factors in the skeleton of fossil and recent mammals. Proc. Amer. Philos. Soc., 429.—1943. Environment and locomotion in mammals. Nat. Hist. Mag., 222.—1949. The humerus from fish to man. Amer. Mus. Novitates, no. 1400.

HANSON, F. B. 1920. The problem of the coracoid. Anat. Rec., 327.

HOWELL, A. B. 1936. The phylogenetic arrangement of the muscular system. Anat. Rec., 295. —1937. Morphogenesis of the shoulder architecture. Part VI. Therian mammals. Quart. Rev. Biol., 440.—1941. The femoral trochanters. Field Mus. Nat. Hist., 279.—1944. Speed in animals: Their specializations for running and leaping. Univ. Chicago Press.

ROMER, A. S. 1922a. The locomotor apparatus of certain primitive and mammal-like reptiles. Bull. Amer. Mus. Nat. Hist., 517. [Correlation of reptilian and mammalian musculature.]—1922b. The comparison of mammalian and reptilian coracoids. Anat. Rec., 39.—1924. The lesser trochanter of the mammalian femur. Ibid., 95.

ROMER, A. S. and F. BYRNE. 1931. The pes of Diadectes. Notes on the primitive tetrapod limb. Palaeobiologica, 25.

SCHAEFFER, B. 1941. The pes of Bauria cynops Broom. Amer. Mus. Novitates, no. 1103.

Mammalian Vertebral Column
(See also Vertebrata, Tetrapoda, above; Man, below)

CAVE, A. J. E. 1934. On the vertebral epiphyses of Mammalia. Proc. Zool. Soc. London, 225.

HATT, R. T. 1932. The vertebral column of ricochetal rodents. Bull. Amer. Mus. Nat. Hist., 599. [Adaptive modifications of vertebral column.]

OLSON, E. C. 1936. The dorsal axial musculature of certain Permian tetrapods. Jour. Morph., 265.

SENSENIG, E. C. 1943. The origin of the vertebral column in the deer-mouse, Peromyscus maniculatus rufinus. Anat. Rec., 123.

SLIJPER, E. J. 1946a. Comparative biologic anatomical investigations on the vertebral column and spinal musculature of mammals. Verhand. d. Koninkl. Nederlandsche Akad. v. Wetens., no. 5.—1946b. Over de Wervelkolom van Onze Huisdieren. Tijdsch. voor Diergeneeskunde, 677.

TODD, T. W. 1922. Numerical significance in the thoracicolumbar vertebrae of the Mammalia. Anat. Rec., 261.

Mammalian Skull and Jaws
(See also Mammalia, General, above; and Man, below)

ADAMS, L. A. 1919. Memoir on the phylogeny of the jaw muscles in recent and fossil vertebrates. Ann. N. Y. Acad. Sci., 51.

BEER, G. R. DE. 1928. Vertebrate zoology: An introduction to the comparative anatomy, embryology, and evolution of chordate animals. New York.—1937. The development of the vertebrate skull. Oxford.

BROOM, R. 1895. On the homology of the palatine process of the mammalian premaxillary. Proc. Linnean Soc., N.S.W., 477. [Prevomer proposed for name of bone on median side of the cartilage of Jacobson, pp. 484, ff.]—1897. On the comparative anatomy of the marsupial Organ of Jacobson. Ibid.—1898. A contribution to the comparative anatomy of the mammalian Organ of Jacobson. Trans. Roy. Soc. Edinburgh. —1909a. Observations on the development of the marsupial skull. Proc. Linnean Soc, N.S.W., 195. [Origin of alisphenoid, pp. 211, ff.; monotreme skull, pp. 208, ff.]—1909b. On the Organ of Jacobson in Orycteropus. Proc. Zool. Soc. London, 680.—1927. Some further points on the structure of the mammalian basicranial axis. Ibid., 233.—1935. The vomer-parasphenoid question. Ann. Transvaal Mus., 23.

CASE, E. C. 1924. A possible explanation of fenestration in the primitive reptilian skull, with notes on the temporal region of the genus Dimetrodon. Contrib. Mus. Geol. Univ. Michigan.

EDGEWORTH, F. H. 1935. The cranial muscles of vertebrates. Cambridge Univ. Press.

GAUPP, E. 1898. Zur Entwickl. des Eidechsenschädels. Berichte d. Naturf. Gesell. zu Freiburg i. B., 302.—1900. Das Chondrocranium von Lacerta agilis. Ein Beitrag zum Verständnis des Amniotenschädels. Anat. Hefte, 433.—1901. Alte Probleme und neuere Arbeiten über den Wirbeltierschädel. II. Teil. Neuere Arbeiten über das Knorpelcranium. Ergeb. d. Anat. u. Entwickel. (1900), 847. ["Tropibasic" and "platybasic" skull types, 979, 995.]—1902. Über die Ala temporalis des Säugerschädels und die Regio orbitalis einiger anderer Wirbeltierschädel. Anat. Heft (Merkel und Bonnet), Heft 61, p. 161.

GREGORY, W. K. 1913. Critique of recent work on the morphology of the vertebrate skull, especially in relation to the origin of mammals. Jour. Morph., vol. 24.—1920. Studies in comparative myology and osteology. No. IV. A review of the evolution of the lacrimal bone of vertebrates with special reference to that of mammals. Bull. Amer. Mus. Nat. Hist., 95.—1927. The palaeomorphology of the human head. Ten structural stages from fish to man. Part I. The skull in norma lateralis. Quart. Rev. Biol., 267.—1929. Part II. The skull in norma basalis. Ibid., vol. 4, 233.—1936. Air-conditioning in nature. Nat. Hist. Mag., vol. 38, 382. [Mammalian skull, ethmoid region.]

GREGORY, W. K. and G. K. NOBLE. 1924. The origin of the mammalian alisphenoid bone. Jour. Morph. and Physiol., 435.

HUENE, F. VON. 1923. Contribution to the vomer-parasphenoid question. Bull. Geol. Soc. Amer., 459.

KESTEVEN, H. L. [?1917] The homology of the mammalian alisphenoid and of the echidna-pterygoid. I. The homology of the mammalian alisphenoid. Jour. Anat., vol. 52, 449.—1947. The evolution of the maxillo-palate. Proc. Linnean Soc., N.S.W. (1946), 73.

MEAD, C. S. 1909. The chondrocranium of an embryo pig, Sus scrofa . . . Amer. Jour. Anat., 167.

MULLER, J. 1934. The orbitotemporal region of the skull of the Mammalia. Wis-en Natuurkunde aan de Rijksuniversiteit te Leiden. [Extensive discussion of bone homologies and contacts, especially the lacrimal and alisphenoid.]

PARRINGTON, F. R. and T. S. WESTOLL. 1940. On the evolution of the mammalian palate. Philos. Trans. Roy. Soc. London, 305.

Mammalian Teeth: Origin and Evolution
(See also Man, below)

ADLOFF, P. 1917. Zur Entwickelungsgeschichte des Zahnsystems von Centetes ecaudatus nebst Bemerkungen zur Frage der Existens einer präpermanenten Dentition. Anat. Anz., 593.—1935. Über die Cope-Osborn'sche Trituberkulärtheorie und über eine neue Theorie der Differenzierung des Säugetiergebisses von M. Friant, Paris. Ibid., 81.

AMEGHINO, F. 1896. Sur l'evolution des dents des mammifères. Bol. Acad. Nac. Ciencias de Córdoba, 381.

BARDENFLETH, K. S. 1913. Notes on the form of the carnassial tooth of carnivorous mammals; with a critical sketch of the most important tooth-cusp-theories. Vidensk. Meddel. fra Dansk naturh. Foren., 67.

BOLK, L. 1915. On the relation between the dentition of marsupials and that of reptiles and monodelphians. Proc. Kon. Akad. v. Wet. te Amsterdam, nos. 4, 5.

BROOM, R. 1909. Some observations on the dentition of Chrysochloris and on the tritubercular theory. Ann. Natal. Government Mus., 129.—1913. On evidence of a mammal-like dental succession in the cynodont reptiles. Bull. Amer. Mus. Nat. Hist., 465.

BUTLER, P. M. 1937. Studies of the mammalian dentition. I. The teeth of Centetes ecaudatus and its allies. Proc. Zool. Soc. London, 103.—1939. Studies of the mammalian dentition. Differentiation of the post-canine dentition. Ibid., 1.—1941. A theory of the evolution of mammalian molar teeth. Amer. Jour. Sci., 421.—1946. The evolution of carnassial dentitions in the Mammalia. Proc. Zool. Soc. London, 198.

CHRISTIE-LINDE, A. A. 1912. On the development of the teeth of the Soricidae . . . Ann. Mag. Nat. Hist., 602.

COLBERT, E. H. 1939. Some studies of adaptations in dentitions of mammals, including man. Amer. Jour. Orthodontics and Oral Surgery, 952.

COLYER, F. 1945. Variations in number of teeth of animals. Dental Rec., 121.

COPE, E. D. 1873. On the types of molar teeth. Proc. Acad. Nat. Sci., Philadelphia, 371.—1874. On the homologies and origin of the types of molar teeth of Mammalia Educabilia. Jour. Acad. Nat. Sci., Philadelphia, 71.—1883a. The tritubercular type of superior molar tooth. Proc. Acad. Nat. Sci., Philadelphia, 56.—1883b. Note on the trituberculate type of superior molar and the origin of the quadrituberculate. Amer. Nat., 407.—1884. On the trituberculate type of molar tooth in the Mammalia. Proc. Amer. Philos. Soc. (1883), 324.

FRECHKOP, S. 1933. Notes sur les mammifères. XIII. Note préliminaire sur la similitude des molaires supérieures et inférieures. Bull. d. Mus. roy. d'Hist. nat. d. Belgique, no. 7.—1935a. Remarques sur l'évolution des dents molaires chez les mammifères, par E. Patte. Revue d. Questions scientifiques, July 20, 177.—1935b. Sur l'évolution de la dentition des mammifères. Ann. d. la Soc. roy. zool. de Belgique, 38.—1935c. Notes sur les mammifères. XVIII. Trituberculie, polyisomérisme et symmétrie des dents des mammifères. Bull. Mus. roy. d'Hist. nat. d. Belgique, vol. 11, no. 25, 1.

FRIANT, M. 1932a. L'influence de la grandeur du corps sur la morphologie dentaire chez les mammifères. Compes rendus de l'Acad. d. Sci., 482. —1932b. La théorie de la trituberculie et l'influence de la taille sur la forme des dents. Arch. Mus. nat. d'Hist. nat., Paris, 83.—1933. Contribution a l'étude de la différenciation des dents jugales chez les mammifères: Essai d'une théorie de la dentition. Publ. d. Mus. d'Hist. nat., no. 1.

GIDLEY, J. W. 1906. Evidence bearing on tooth-cusp development. Proc. Washington Acad. Sci., 91. [Supports premolar analogy theory.]

GOODRICH, E. S. 1894. On the fossil Mammalia from the Stonesfield Slate. Quart. Jour. Micros. Sci., 1. [Very primitive forms of mammalian dentition.]

GREGORY, W. K. 1922. The origin and evolution of the human dentition. A palaeontological review. Parts I–V. Baltimore.—1926. Palaeontology of the human dentition: Ten structural stages in the evolution of the cheek teeth. Amer. Jour. Phys. Anthropol., 401.—1934a. Polyisomerism and anisomerism in cranial and dental evolution among vertebrates. Proc. Nat. Acad. Sci., 1.—1934b. A half century of trituberculy. The Cope-Osborn theory of dental evolution, with a revised summary of molar evolution from fish to man. Proc. Amer. Philos. Soc., 169.—1935. Reduplication in evolution. Quart. Rev. Biol., 272.—1941. Evolution of dental occlusion from fish to man. In: Development of occlusion, by W. K. Gregory, B. H. Broadbent and M. Hellman. Univ. Press., Philadelphia.

HUXLEY, T. H. 1880. On the cranial and dental characters of the Canidae. Proc. Zool. Soc. London, 283–284. [Regards molar of Centetes as primitive for Carnivora. Interprets molar cusps by the "premolar analogy" theory.]

MAJOR, C. I. F. 1873. Nagerüberreste aus Bohnerzen Suddeutschlands und der Schweiz. Palaeontographica, 75. [Theory of polybuny.]—1893. On some Miocene squirrels, with remarks on the dentition and classification of the Sciurinae. Proc. Zool. Soc. London, 179. [Infers that primitive mammalian molars were polybunous.]

OSBORN, H. F. 1888a. The nomenclature of the mammalian molar cusps. Amer. Nat., 926. [Proposes and defines names in accordance with the Theory of Trituberculy.]—1888b. The evolution of mammalian molars to and from the tritubercular type. Ibid., 1067.—1907. Evolution of mammalian molar teeth to and from the triangular type. New York.

SCHLOSSER, M. 1892. Die Entwickelung der verschiedenen Säugethierzahnformen in Laufe der geologischen Perioden. Verhandl. deutsch. odontolog. Gesellsch., 203.

SCOTT, W. B. 1892. The evolution of the premolar teeth in the mammals. Proc. Acad. Nat. Sci., Philadelphia, 405. [Proposes and defines names for premolar cusps.]

SHAW, D. M. 1917. Form and function of teeth: A theory of "maximum shear." Jour. Anat. Physiol., 97.

SIMPSON, G. G. 1933a. The "plagiaulacoid" type of mammalian dentition. Jour. Mammal., 97.—1933b. Critique of a new theory of mammalian dental evolution. Jour. Dent. Research, 261. [Critique of Friant's theory.]—1936. Studies of the earliest mammalian dentitions. Dental Cosmos. [Origin of tribosphenic molars, p. 16.]—1947. Note on the measurement of variability and on relative variability of teeth of fossil mammals. Amer. Jour. Sci., 522.

TIMMS, M. 1903. The evolution of the teeth in the Mammalia. Jour. Anat. Physiol., 131.

VALLOIS, H. [?1937] Les théories sur l'origine des molaires. L'Orthodontie francaise, 185.

WINGE, H. 1882. Om Pattedyrenes Tandskifte isaer med Hensyn til Taendernes Former. Vidensk. Meddel. fra den naturh. Foren. [Winge's theory of origin and homologies of mammalian molar cusps.]

WOOD, A. E. and H. E. WOOD, II. 1933. The genetic and phylogenetic significance of the presence of a third upper molar in a modern dog. Amer. Midland Naturalist, 36.

WOODWARD, M. F. 1893. Contributions to the study of the mammalian dentition. Part I. On the development of the teeth of the Macropodidae. Proc. Zool. Soc. London, 450.—1896. Part II. On the teeth of certain Insectivora. Ibid., 557.

WORTMAN, J. L. 1886. The comparative anatomy of the teeth of the Vertebrata. Amer. System of Dentistry, 351.

Deciduous Dentition; Development and Replacement of Teeth
(See also Man, below)

ADLOFF, P. 1904. Ueber den Zahnwechsel von Cavia cobaya. Anat. Anz., 141.

BOLK, L. 1921, 1922. Odontological essays. I. On the development of the palate and alveolar ridge in man. Jour. Anat., vol. 55, parts 2, 3, p. 138—II. On the development of the enamel-germ. Ibid., 152.—III. On the tooth-glands in reptiles and their rudiments in mammals. Ibid., 219.—1922. IV. On the relation between reptilian and mammalian teeth. Ibid., 107.—V. On the relation between reptilian and mammalian dentition. Ibid., 55.

LECHE, W. 1909. Zur Frage nach der stammesgesch. Bedeutung des Milchgebisses bei den Säugetieren. I. Zool. Jahrb., 449.—1915. II. Viverridae, Hyaenidae, Felidae, Mustelidae, Creodonta. Ibid., 275.

PARRINGTON, F. R. 1936a. On the tooth-replacement in theriodont reptiles. Philos. Trans. Roy.

Soc. London, 121.—1936b. Further notes on tooth-replacement. Ann. Mag. Nat. Hist., 109.

WOODWARD, M. F. 1894. On the succession and genesis of the mammalian teeth. Science Progress, 438.

Mammalian Jaws and Jaw Muscles, Visceral Arches, Hyoid, Larynx
(See also Vertebrata, above, and Man, below)

ADAMS, L. A. 1919. Memoir on the phylogeny of the jaw muscles in recent and fossil vertebrates. Ann. N. Y. Acad. Sci., 51.

BLUNTSCHLI, H. and H. SCHREIBER. 1929. Anatomie: Über die Kaumuskulatur. Die Fortschritte der Zahnheilkunde, vol. 5, 1.

BROCK, G. T. 1938. The cranial muscles of the gecko. A general account, with a comparison of the muscles in other gnathostomes. Proc. Zool. Soc. London, 735.

EDGEWORTH, F. H. 1914. On the development and morphology of the mandibular and hyoid muscles of mammals. Quart. Jour. Micros. Sci., 573.

GAUPP, E. 1905. Die Nicht-Homologie des Unterkiefers in der Wirbeltierreihe. Verhandl. anat. Gesellsch., 125.—1911a. Beiträge zur Kenntnis des Unterkiefers der Wirbeltiere. II. Die Zusammensetzung des Unterkiefers der Quadrupeden. Anat. Anz., 433.—1911b. III. Das Problem der Entstehung eines "sekundaren" Kiefergelenkes bei den Säugern. Ibid., 609.

GREGORY, W. K. 1931. Certain critical stages in the evolution of the vertebrate jaws. Internat. Jour. Orthodontia, Oral Surgery and Radiography, 1138.—1943. The earliest known fossil stages in the evolution of the oral cavity and jaws. Amer. Jour. Orthodontics and Oral Surgery, 253.

HOLLISTER, N. 1918. [Differences in jaws and cheek-arches of wild and captive lions, due to differences in food.] In: East African mammals . . . U. S. Nat. Mus. Bull. 99, 11.

LIGHTOLLER, G. H. S. 1939. Probable homologues. A study of the comparative anatomy of the mandibular and hyoid arches and their musculature. Part I. Comparative myology. Trans. Zool. Soc. London, 349.

LUBOSCH, W. 1908. Das Kiefergelenk der Edentaten und Marsupialier. Nebst Mittheilungen über die Kaumuskulatur dieser Thiere. In: Semon, Zool. Forschungsr. in Australien . . . IV. Denkschr. d. Med.-Naturwiss. Gesellsch. Jena, 519.

PARRINGTON, F. R. and T. S. WESTOLL. 1940. On the evolution of the mammalian palate. Philos. Trans. Roy. Soc. London, 305. [Evolution of jaw muscles.]

SPRAGUE, J. M. 1943. The hyoid region of placental mammals with especial reference to the bats. Amer. Jour. Anat., 385.

SUSHKIN, P. P. 1927. On the modifications of the mandibular and hyoid arches and their relations to the brain-case in early Tetrapoda. Palaeontol. Zeitschr., 263.

Mammalian Auditory Ossicles, Tympanic Region, Inner Ear
(See also Vertebrata, above)

BONDY, G. 1907. Beiträge zur vergleichenden Anatomie des Gehörorgans der Säuger (Tympanicum, Membrana Shrapnelli und Chordaverlauf). Anat. Hefte, 293.

BROOM, R. 1912. On the structure of the internal ear and the relations of the basicranial nerves in Dicynodon and on the homology of the mammalian auditory ossicles. Proc. Zool. Soc. London, 419.

CARUS, C. G. 1818. Lehrbuch der Zootomie. Leipzig. [Homologizes incus of mammals with quadrate of reptiles.]

DORAN, A. H. G. 1879. Morphology of the mammalian ossicula auditus. Trans. Linnean Soc. London (1878), 371. [Arranged according to orders and families.]

EATON, T. H., JR. 1939. The crossopterygian mandibular and the tetrapod stapes. Jour. Washington Acad. Sci., 109.

GADOW, H. 1901. The evolution of the auditory ossicles. Anat. Anz., 436.

GAUPP, K. 1899. Ontogenese und Phylogenese des schalleitenden Apparates bei den Wirbeltieren. Ergeb. d. Anat. u. Entwickl., 990.—1911. Beiträge zur Kenntnis des Unterkiefers der Wirbeltiere. I. Der Processus anterior (Folii) des Hammers der Säuger und das Goniale der Nichtsäuger. Anat. Anz., 97.—1913. Die Reichertsche Theorie (Hammer-, Amboss- und Kieferfrage). Archiv f. Anat. u. Entwickl. (1912), Suppl.-Bd.

GREGORY, W. K. 1910. The orders of mammals. Part II. Genetic relations of the mammalian orders; with a discussion of the origin of the Mammalia and of the problem of the auditorry ossicles. Bull. Amer. Mus. Nat. Hist. [Quadrate = incus theory, pp. 125–143.]—1913. Critique of recent work on the morphology of the vertebrate skull, especially in relation to the origin of mammals. Jour. Morphol., vol. 24. [Origin of mammalian auditory ossicles.]

KAMPEN, P. N. VAN. 1905. Die Tympanalgegend des Säugetierschädels. Morphol. Jahrb., 321.

KLAAUW, C. J. VAN DER. 1922. Über die Entwickelung des Entotympanicums. Tijdschr. d. Ned. Dierk. Vereen, 135.—1923. Die Skelettstückchen in der Sehne des Musculus stapedius und nahe dem Ursprung der Chorda tympani. Zeitschr. f. Anat. u. Entwickl., 32.—1931. The auditory bulla in some fossil mammals, with a general introduction to this region of the skull. Bull. Amer. Mus. Nat. Hist., 1.

MCCLAIN, J. A. 1939. The development of the auditory ossicles of the opossum (Didelphys virginiana). Jour. Morph., 211.

MECKEL, J. F. 1820. Handbuch der menschlichen Anatomie. Bd. IV. Halle. [Recognized homology of mammalian malleus with reptilian articular, p. 47.]

OLSON, E. C. 1944. Origin of mammals based upon cranial morphology of the therapsid suborders. Geol. Soc. Amer., Special Papers, no. 55. [Homologies of reptilian jaw bones and mammalian auditory ossicles.]

PALMER, R. W. 1913. Notes on the lower jaw and ear ossicles of a foetal Perameles. Anat. Anz., 510.

REICHERT, C. 1837. Über die Visceralbogen der Wirbeltiere im Allgemeinen und deren Metamorphosen bei den Vögeln und Säugetieren. Archiv f. Anat. Physiol. u. Wiss. Medicin, 120.

RETZIUS, M. G. 1881. Das Gehörorgan der Wirbelthiere. II. Das Gehörorgan der Reptilien, der Vögel und der Säugethiere. Stockholm. 4to.

SHESTOKOVA, G. S. [?1947] On the homology of the auditory ossicles of mammals. [?] Proc. Acad. Sci. U.S.S.R., 91.

STREETER, G. L. 1907. On the development of the membranous labyrinth and the acoustic and facial nerves in the human embryo. Amer. Jour. Anat., 140.

WESTOLL, T. S. 1943. The hyomandibular of Eusthenopteron and the tetrapod middle ear. Proc. Roy. Soc. Edinburgh, 393.—1944. New light on the mammalian ear ossicles. Nature, 770, Dec. 16.—1945. The mammalian middle ear. Ibid., 114, Jan. 27.

Mammalian Face, Facial Skeleton and Musculature

(See also Vertebrata, above, and Man, below)

ANSON, B. J. 1929. The comparative anatomy of the lips and labial villi of vertebrates. Jour. Morph. and Physiol.

BENDER, O. 1907. Die Schleimhautnerven des Facialis, Glossopharyngeus und Vagus. Studien zur Morphologie des Mittelohres und der benachbarten Kopfregion der Wirbelthiere. In: R. Semon, Zool. Forsch . . . 4 Bd., 341. Denskchr. d. Med.-Naturwiss. Gesellsch. Jena.

GREGORY, W. K. 1929. Our face from fish to man. New York.

HILL, W. C. O. 1948. Rhinoglyphics: Epithelial sculpture of the mammalian rhinarium. Proc. Zool. Soc. London, vol. 118, 1.

HUBER, E. 1922. Über das Muskelgebiet des Nervus facialis beim Hund, nebst allgemeinen Betrachtungen über die Facialist-Muskulatur. I. Teil. Morphol. Jahrb., 1.—1923. II. Teil. Ibid., 354.—1930. Evolution of facial musculature and cutaneous field of trigeminus. Part I. Quart. Rev. Biol., 133. Part II. Ibid., 389.

RUGE, G. 1896. Über das peripherische Gebiet des N. facialis bei Wirbeltieren. Festschr. Carl Gegenbaur, vol. 3. Leipzig.

Mammalian Nervous System, Sense Organs, Brain, Brain Form and Skull Form

(For Mammalian Behavior, see Man: Comparative Psychology)

BEACH, F. A. 1942. Central nervous mechanisms involved in the reproductive behavior of vertebrates. Psychol. Bull., 200.

BURLET, M. DE. 1927. Sinnesorgane. 1. Hautsinnesorgane. In: Die Säugethiere, by Max Weber. 2d ed. Bd. I, 178

CLARK, W. E. LE GROS. 1926. The mammalian oculomotor nucleus. Jour. Anat., 426.—1933. The brain of the Insectivora. Proc. Zool. Soc. London (1932), 975.

EDINGER, T. 1948. Evolution of the horse brain. Geol. Soc. Amer., Mem. 25. [Eohippus brain of remarkably low type.]

FRIANT, M. 1947. Anatomie comparée du cerveau. Paris. Collection Orion.

JOHNSON, G. L. 1901. Contributions to the comparative anatomy of the mammalian eye. Philos. Trans. Roy. Soc., 1.

LARSELL, O. 1945. Comparative neurology and present knowledge of the cerebellum. Bull. Minnesota Med. Foundation, 73.

PAPEZ, J. 1929. Comparative neurology . . . New York.

Mesozoic Mammals and Near-Mammals (Protodonta, Microleptids, Allotheria, Triconodonta, Symmetrodonta, Pantotheria)

AMEGHINO, F. 1903. Los Diprotodontes del Orden de los Plagiaulacoideos y el origin de los Roedores y de los Polimastodontes. Ann. d. Mus. Nac. Buenos Aires, 81.

BLAINVILLE, H. M. D. DE 1838a. Doutes sur le prétendu Didelphe de Stonesfield. Comptes Rendus Acad. Sci. Paris, 402.—1838b. Nouveaux doutes sur le prétendu Didelphe de Stonesfield. Ibid., 727.

BROILI, F. and J. SCHRÖDER. 1936. Beobachtungen an Wirbeltieren der Karrooformation. XIX. Ein neuer Fund von Tritylodon Owen. Sitz. d. Bayer. Akad. d. Wiss., 187.

BROOM, R. 1910. On Tritylodon and on the relationships of the Multituberculata. Proc. Zool. Soc. London, 760.—1914. On the structure and affinities of the Multituberculata. Bull. Amer. Mus. Nat. Hist., 115. [Tritylodon not a marsupial, pp. 119, 120.]

BUCKLAND, W. 1824. Notice on Megalosaurus. Trans. Geol. Soc. London, 390. [First published notice of two Jurassic mammal jaws later described and named Amphitherium.]

BUTLER, P. M. 1939a. The post-canine teeth of Tritylodon longaevus Owen. Ann. Mag. Nat. Hist., 514.—1939b. The teeth of the Jurassic mammals. Proc. Zool. Soc. London, 329.

COPE, E. D. 1892. On a new form of Marsupialia from the Laramie formation. Proc. Amer. Soc. Adv. Sci., 177. [Describes genus Thlaeodon.]

GIDLEY, J. W. 1909. Notes on the fossil mammalian genus Ptilodus, with descriptions of new species. Proc. U. S. Nat. Mus., 611.

GOODRICH, E. S. 1894. On the fossil Mammalia from the Stonesfield Slate. Quart. Jour. Micros. Sci., 1.

GRANGER, W. and G. G. SIMPSON. 1928. Multituberculates in the Wasatch Formation. Amer. Mus. Novitates, no. 312.—1929. A revision of the Tertiary Multituberculata. Bull. Amer. Mus. Nat. Hist., 601.

KÜHNE, W. G. 1943. The dentary of Tritylodon and the systematic position of the Tritylodontidae. Ann. Mag. Nat. Hist., 589.—1947. [Report on Lower Liassic infillings of Carboniferous limestone fissures, Somerset, England, where bones were found of mammal-like reptiles (Tritylodontidae).] Quart. Jour. Geol. Soc. London (1946).

MARSH, O. C. 1880. Notice of Jurassic mammals representing two new orders. Amer. Jour. Sci., 235.—1887. American Jurassic mammals. Ibid., 326.—1889a. Discovery of Cretaceous Mammalia. I. Ibid., 81.—1889b. Part II. Ibid., 177.—1891a. Note on Mesozoic Mammalia. Proc. Acad. Nat. Sci. Philadelphia, 237.—1891b. On the Cretaceous mammals of North America. Rept. Brit. Assoc. Adv. Sci., Leeds (1890), 853.—1892. Discovery of Cretaceous Mammalia. Part III. Amer. Jour. Sci., 249.

OSBORN, H. F. 1888. On the structure and classification of the Mesozoic Mammalia. Jour. Acad. Nat. Sci. Philadelphia, 186.—1891. A review of the Cretaceous Mammalia. Proc. Acad. Nat. Sci. Philadelphia, 124.—1893. Fossil mammals of the Upper Cretaceous Beds. Bull. Amer. Mus. Nat. Hist., 311.

OWEN, R. 1838a. Observations on the fossils representing the Thylacotherium prevostii (Valenciennes) . . . and on Phascolotherium bucklandi. Trans. Geol. Soc. London, 47.—1838b. Description of the remains of marsupial Mammalia from the Stonesfield Slate. Proc. Geol. Soc. London, 17.—1871. Monograph of the fossil Mammalia of the Mesozoic Formations. Palaeontographical Soc., vol. 24 ("1870"), 1.—1884. On the skull and dentition of a Triassic mammal, Tritylodon longaevus (Owen) from South Africa. Quart. Jour. Geol. Soc., 146.

PARRINGTON, F. R. 1941. On two mammalian teeth from the Lower Rhaetic of Somerset. Ann. Mag. Nat. Hist., 140.—1947. On a collection of Rhaetic mammalian teeth. Proc. Zool. Soc. London, 707.

PREVOST, L. C. 1825. Observations sur les schistes calcaires oolitiques de Stonesfield en Angleterre, dans lesquels ont été trouvés plusieurs ossemens fossiles de mammifères. Ann. d. Sci. nat. Paris, 389.

SEELEY, H. G. 1895. Researches on the structure, organization and classification of the fossil Reptilia. Part IX. Sect. 2. The reputed mammals from the Karroo Formation of Cape Colony

(Tritylodon). Philos. Trans. Roy. Soc. London (1894), 1019.

SIMPSON, G. G. 1925–1928. Mesozoic Mammalia. I. American triconodonts. Part 1. Amer. Jour. Sci., vol. 10, 145. Part 2. Ibid., 334. II. Tinodon and its allies. Ibid., 451. III. A preliminary comparison of Jurassic mammals. Ibid., 559.—1926. IV. The multituberculates as living animals. Ibid., vol. 11, 228. V. Dromatherium and Microconodon. Ibid., vol. 12, 87.—1927. VI. Genera of Morrison pantotheres. Ibid., vol. 13, 409. VII. Taxonomy of Morrison multituberculates. Ibid., vol. 14, 36. VIII. Genera of Lance mammals other than multituberculates. Ibid., 121. IX. The brain of Jurassic mammals. Ibid., 259.—1928. X. Some Triassic mammals. Ibid., vol. 15, 154. XI. Brancatherulum tendagurense Dietrich. Ibid., 303. XII. The internal mandibular groove of Jurassic mammals. Ibid., 461. —1926. Are Dromatherium and Microconodon mammals? Science, 548.—1928. A catalogue of the Mesozoic Mammalia in the Geological Department of the British Museum. London.—1929. American Mesozoic Mammalia. Mem. Peabody Mus., Yale Univ.—1932. The supposed occurrence of Mesozoic mammals in South America. Amer. Mus. Novitates, no. 530. —1933a. The "plagiaulacoid" type of mammalian dentition: A study of convergence. Jour. Mammal., 97.—1933b. Paleobiology of Jurassic mammals. Palaeobiologica, 127.—1935. The first mammals. Quart. Rev. Biol., 154.—1937a. Skull structure of the Multituberculata. Bull. Amer. Mus. Nat. Hist., 727.—1937b. A new Jurassic mammal. Amer. Mus. Novitates, no. 943.

SIMPSON, G. G. and H. O. ELFTMAN. 1928. Hind limb musculature and habits of a Paleocene multituberculate. Amer. Mus. Novitates, no. 333.

YOUNG, C. C. 1940. Preliminary notes on the Mesozoic mammals of Lufeng, Yunnan. Bull. Geol. Soc. China, 93.—1947. Mammal-like reptiles from Lufeng, Yunnan, China. Proc. Zool. Soc. London, 537.

Land Bridges, Migration Routes, Dispersal, Zoogeographic Boundaries, Climate and Evolution

COLBERT, E. H. 1939. The migrations of Cenozoic mammals. N. Y. Acad. Sci., 89.

FORBES, H. O. 1893. Antarctica: A supposed former southern continent. Nat. Sci., 54.

MATTHEW, W. D. 1906. Hypothetical outlines of the continents in Tertiary times. Bull. Amer. Mus. Nat. Hist., 353.—1915. Climate and evolution. Ann. N. Y. Acad. Sci., 171.—1930. The dispersal of land animals. Scientia, July, 33.—1939. Climate and evolution. 2d ed., revised and enlarged. N. Y. Acad. Sci.

MAYR, E. 1944. Wallace's Line in the light of recent zoogeographic studies. Quart. Rev. Biol., 1.

RAVEN, H. C. 1935. Wallace's Line and the distribution of Indo-Australian mammals. Bull. Amer. Mus. Nat. Hist., 1.

SIMPSON, G. G. 1940. Antarctica as a faunal migration route. Proc. Sixth Pacific Sci. Congress (1939), 755.—1941. Range as a zoological character. Amer. Jour. Sci., 785.—1943. Mammals and the nature of continents. Ibid., 1.—1946. Tertiary land bridges. Abstr., Trans. N. Y. Acad. Sci., 255.—1947. Evolution, interchange, resemblance of the North American and Eurasian Cenozoic mammalian faunas. Evolution: Internat. Jour. Organic Evol., vol. 1, 218.

TOIT, A. L. DU. 1944. Tertiary mammals and continental drift. A rejoinder to George G. Simpson. Amer. Jour. Sci., 145.

Tertiary and Quaternary Mammals and Horizons (e.g. Osborn's Age of Mammals)

COPE, E. D. 1884. The Vertebrata of the Tertiary Formations of the West. Book I. Rept. U. S. Geol. Surv. of the Territories.

LEIDY, J. 1873. Contributions to the extinct vertebrate fauna of the Western Territories. Rept. U. S. Geol. Surv. of the Territories, vol. 1, 14.

MARSH, O. C. 1878. Introduction and succession of vertebrate life in America. Amer. Assoc. Adv. Sci. (1877). Popular Sci. Monthly, March-April.

OSBORN, H. F. 1894. The rise of the Mammalia in North America. Proc. Amer. Assoc. Adv. Sci. (1893), 188.—1900. Correlation between Tertiary mammal horizons of Europe and America. Ann. N. Y. Acad. Sci., 1.—1910. The age of mammals in Europe, Asia and North America. New York.

SIMPSON, G. G. 1933. Glossary and correlation charts of North American Tertiary mammal-bearing formations. Bull. Amer. Mus. Nat. Hist., 79.—1947a. Holarctic mammalian faunas and continental relationships during the Cenozoic. Bull. Geol. Soc. Amer., 613.—1947b. A continental Tertiary time chart. Jour. Paleont., 480.

STOCK, C. 1948. Pushing back the history of land mammals in Western North America. Bull. Geol. Soc. Amer., 327. [Correlation of Clarno with Bridger Eocene. Sespe from Upper Uinta to Lower Harrison; successive marine transgressions in the California region.]

WOOD, H. E. and others. 1941. Nomenclature and correlation of the North American Continental Tertiary. Bull. Geol. Soc. Amer., 1.

Mammals: Paleocene and Eocene of North America

COPE, E. D. 1888. Synopsis of the vertebrate fauna of the Puerco series. Trans. Amer. Philos. Soc., 298. [Loxolophus priscus (Cope), p. 337, fig. 6.]

DENISON, R. H. 1937. Early Lower Eocene mammals from the Wind River Basin, Wyoming. New England Zool. Club, 11.

JEPSEN, G. L. 1930a. New vertebrate fossils from the Lower Eocene of the Bighorn Basin, Wyoming. Proc. Amer. Philos. Soc., 117.—1930b. Stratigraphy and paleontology of the Paleocene of Northeastern Park County, Wyoming. Ibid., 463. [Phenacodaptes, pl. 9.]—1940. Paleocene faunas of the Polecat Bench Formation, Park County, Wyoming. Ibid., 217.

MATTHEW, W. D. 1897. A revision of the Puerco fauna. Bull. Amer. Mus. Nat. Hist., 259.—1909. The Carnivora and Insectivora of the Bridger Basin, Middle Eocene. Mem. Amer. Mus. Nat. Hist., 289.—1928. The evolution of mammals in the Eocene. Proc. Zool. Soc. London (1927), 947.—1937. Paleocene faunas of the San Juan Basin, New Mexico. Trans. Amer. Philos. Soc. 1.

MATTHEW, W. D. and W. GRANGER. 1915–1918. A revision of the Lower Eocene Wasatch and Wind River faunas. Parts I–V. Bull. Amer. Mus. Nat. Hist., vols. 34, 38. [See Creodonta, Condylarthra, Entelonychia, Primates, Insectivora, Glires, Edentata.]

OSBORN, H. F. and C. EARLE. 1895. Fossil mammals of the Puerco Beds . . . Bull. Amer. Mus. Nat. Hist. [Triisodon, fig. 7.]

SCOTT, W. B. 1889. The geological and faunal relations of the Uinta Formation. In: The Mammalia of the Uinta Formation, by W. B. Scott and H. F. Osborn. Trans. Amer. Philos. Soc., 461.—1945. The Mammalia of the Duchesne River Oligocene. Ibid., 209.

SCOTT, W. B. and H. F. OSBORN. 1890. The Mammalia of the Uinta Formation. Parts I–V. Trans. Amer. Philos. Soc., n.s., (1889), 461. [See Creodonta, Rodentia, Artiodactyla, Perissodactyla.]

SIMPSON, G. G. 1928. A new mammalian fauna from the Fort Union of southern Montana. Amer. Mus. Novitates, no. 297.—1932. A new Paleocene mammal from a deep well in Louisiana. Proc. U. S. Nat. Mus., 1.—1935a. The Tiffany fauna, Upper Paleocene. 1. Multituberculata, Marsupialia, Insectivora and (?)Chiroptera. Amer. Mus. Novitates, no. 795.—1935b. New Paleocene mammals from the Fort Union of Montana. Proc. U. S. Nat. Mus., 221.—1936a. Census of Paleocene mammals. Amer. Mus. Novitates, no. 848.—1936b. Additions to the Puerco fauna, Lower Paleocene. Ibid., no. 849.—1936c. A new fauna from the Fort Union of Montana. Ibid., no. 873.—1937. The Fort Union of the Crazy Mountain Field, Montana, and its mammalian faunas. Smiths. Inst., U. S. Nat. Mus., Bull. 169.—1946. The Duchesnean fauna and the Eocene-Oligocene boundary. Amer. Jour. Sci., 52.

VAN HOUTEN, F. B. 1945. Review of latest Paleocene and early Eocene mammalian faunas. Jour. Paleont., 421.

WORTMAN, J. L. 1901–1902. Studies of Eocene Mammalia in the Marsh Collection, Peabody Museum. Parts I, II. Amer. Jour. Sci., ser, 4, vols. 11–14. [See Carnivora, Primates. Origin of tritubercular molar, vol. 13, pp. 41–46.]

Mammals: Oligocene, Miocene, and Pliocene of North America

CLARK, J. 1937. The stratigraphy and paleontology of the Chadron Formation in the Big Badlands of South Dakota. Ann. Carnegie Mus., 261.

DOUGLASS, E. 1906. The Tertiary of Montana. Mem. Carnegie Mus., 203.

FURLONG, E. L. and others. 1946. Fossil vertebrates from western North America and Mexico. Contribs. to Paleontology, Carnegie Inst. Washington.

GREGORY, J. T. 1942. Pliocene vertebrates from Big Spring Canyon, South Dakota. Univ. Calif. Publ., Bull. Dept. Geol. Sci., 307.

MATTHEW, W. D. 1901. Fossil mammals of the

Tertiary of northeastern Colorado. Mem. Amer. Mus. Nat. Hist., I, 354.—1903. The fauna of the Titanotherium Beds at Pipestone Springs, Montana. Bull. Amer. Mus. Nat. Hist., 197.—1907. A Lower Miocene fauna from South Dakota. Ibid., 169.—1918. Contributions to the Snake Creek fauna . . . Bull. Amer. Mus. Nat. Hist., 183.—1924. Third contribution to the Snake Creek fauna. Ibid., vol. 50, 59.

MATTHEW, W. D. and C. C. MOOK. 1933. New fossil mammals from the Deep River Beds of Montana. Amer. Mus. Novitates, no. 601.

MCGREW, P. O. 1938. The Burge fauna: A Lower Pliocene mammalian assemblage from Nebraska. Univ. California Publ., Bull. Dept. Geol. Sci., 309.

SCHLAIKJER, E. M. 1933–1935. Contributions to the stratigraphy and palaeontology of the Goshen Hole area, Wyoming. Parts 1–4. Bull. Mus. Comp. Zool., Harvard Coll., vol. 76, nos. 1–4. [See Insectivora, Carnivora, Ungulata, Rodentia.]

SCOTT, W. B. and G. L. JEPSEN. 1936–1941. The mammalian fauna of the White River Oligocene, Parts 1–5. Trans. Amer. Philos. Soc., vol. 28. [See Insectivora, Carnivora, Rodentia, Lagomorpha, Artiodactyla, Perissodactyla.]

SIMPSON, G. G. 1932. Miocene land mammals from Florida. Florida State Geol. Surv., Bull. no. 10.—1943. The discovery of fossil vertebrates in North America. Jour. Paleont., 26.

STIRTON, R. A. 1939a. Cenozoic mammal remains from the San Francisco Bay region. Univ. California Publ., Bull. Dept. Geol. Sci., 339.—1939b. Methods and procedure in the Valentine question. Amer. Jour. Sci., 429.

STIRTON, R. A. and H. F. GOERTZ. 1942. Fossil vertebrates from the superjacent deposits near Knights Ferry, California. Univ. California, Bull. Geol. Sci., 447.

WALLACE, R. E. 1946. A Miocene mammalian fauna from Beatty Buttes, Oregon. Carnegie Inst. Washington, Publ. 551, 113.

WHITE, T. E. 1940. New Miocene vertebrates from Florida. Proc. New England Zool. Club, 31.—1941. Additions to the Miocene fauna of Florida. Ibid., 91. [Mephititaxus, pl. 14.]—1942. The Lower Miocene mammal fauna of Florida. Bull. Mus. Comp. Zool. Harvard Coll., 1.

WOOD, H. E. and A. E. WOOD. 1937. Mid-Tertiary vertebrates from the Texas Coastal Plain . . . Amer. Midland Naturalist, 129.

Mammals: Pleistocene and Recent of North America

ELLIOT, D. G. 1904. The land and sea mammals of Middle America and the West Indies. Field Columbian Mus., Zool. ser., vol. 4, 441.

GOODWIN, G. G. 1935. The mammals of Connecticut. State Geol. and Nat. Hist. Surv., Bull. no. 53.

HAY, O. P. 1912. The Pleistocene Period and Its Vertebrata. Dept. Geol. and Natural Resources of Indiana, 36th Ann. Rept. (1911), 540.

MERRIAM, J. C. and C. STOCK. 1925. Papers concerning the Palaeontology of the Pleistocene of California and the Tertiary of Oregon. Parts I–III. Contrib. to Palaeont., Carnegie Inst. Washington Publ. 347.

STOCK, C. 1930. Rancho La Brea: A record of Pleistocene life in California. Los Angeles Mus., Publ. no. 1.

Mammals: Tertiary to Recent of Old World; Correlation of Mammalian Faunal Horizons

ANDREWS, C. W. 1903. Notes on an expedition to the Fayûm, Egypt, with descriptions of some new mammals. Geol. Mag., 337.—1906. A descriptive catalogue of the Tertiary Vertebrata of the Fayûm, Egypt. British Museum (Natural History). London.—1914. On the Lower Miocene vertebrates from British East Africa . . . Quart. Jour. Geol. Soc., 163.

ARAMBOURG, C. and J. PIVETEAU. 1929. Les vertébrés du Pontien de Salonique. Ann. d. Paléont., 59. [Lower Pliocene.]

BEADNELL, H. J. L. 1901. The Fayûm depression: A preliminary notice of the district of Egypt containing a new Palaeogene fauna. Geol. Mag., 540.

COLBERT, E. H. 1935a. Distributional and phylogenetic studies on Indian fossil mammals. Amer. Mus. Novitates, no. 796.—1935b. Siwalik mammals in the American Museum of Natural History. Trans. Amer. Philos. Soc., vol. 26, 1.—1938. Fossil mammals from Burma in the American Museum of Natural History. Bull. Amer. Mus. Nat. Hist., 255. [Amphipithecus, figs. 13–15.]

DEPÉRET, C. 1917. Monographie de la faune de mammifères fossiles du Ludien Inférieur d'Euzet-les-Bains (Gard). Ann. d. l'Univ. d. Lyon, n.s., I, 288.

FILHOL, H. 1879. Étude des mammifères fossiles de Saint-Gérard le Puy (Allier). Part I. Biblioth. d. l'École des Hautes Études, Sci. nat., vol. 19, 1. Paris.—1884. Mémoires sur quelques mammifères fossiles des Phosphorites du Quercy. Toulouse (1882). [Fauna of Lower Oligocene.]

GRABAU, A. W. 1927. A summary of the Cenozoic and Psychozoic deposits with special reference to Asia. Bull. Geol. Soc. China, 151.

GREGORY, W. K. 1927. Mongolia, the new world. IV. The Mongolian age of mammals. Scientific Monthly, 337.

HOPWOOD, A. T. 1929. New and little-known mammals from the Miocene of Africa. Amer. Mus. Novitates, no. 344.—1940. Fossil mammals and Pleistocene correlation. Proc. Geol. Assoc., vol. 51, part 1, 79.

KOENIGSWALD, G. H. R. VON. 1935. Die fossilen Säugetierfaunen Javas. Proc. Kon. Akad. v. Wet. te Amsterdam, vol. 38, no. 2.

MATTHEW, W. D. 1929. Critical observations upon Siwalik mammals (exclusive of Proboscidea). Bull. Amer. Mus. Nat. Hist., 437.

MATTHEW, W. D. and W. GRANGER. 1923. New fossil mammals from the Pliocene of Sze-Chuan, China. Bull. Amer. Mus. Nat. Hist., 563.—1924. New insectivores and ruminants from the Tertiary of Mongolia, with remarks on the correlation. Amer. Mus. Novitates, no. 105.—1925. New ungulates from the Ardyn Obo Formation of Mongolia, with faunal list and remarks on correlation. Ibid., no. 195.

OSBORN, H. F. 1900. Correlation between Tertiary mammal horizons of Europe and America. Ann. N. Y. Acad. Sci.—1908. New fossil mammals from the Fayûm Oligocene, Egypt. Bull. Amer. Mus. Nat. Hist., 265.

OWEN, R. 1846. A history of British fossil mammals and birds. London.

PEI, W. C. 1936. On the mammalian remains from Locality 3 at Choukoutien. Palaeontologia Sinica, vol. 7, part 5. [Pleistocene mammals.]

SCHLOSSER, M. 1887–1890. Die Affen, Lemuren [etc.] des europäischen Tertiärs . . . Teil I–III. Wien. 4to. [Separat-Abdruck: Beitrage zur Palaontologie Österreich-Ungarns, vols. VI, VII, VIII.]—1924. Tertiary vertebrates from Mongolia. Palaeontologia Sinica, ser. C, vol. 1, no. 1.

STEHLIN, H. G. 1903–1916. Die Säugetiers des schweizerischen Eocaens. Parts I–VII. Abh. schweiz. paläont. Gesell., vols. 30–41. [See Ungulata, Primates.]

STROMER, E. 1926. Reste Land-und Süsswasser-Bewohnender Wirbeltiere aus den Diamantfeldern Deutsch-Südwestafrikas. In: Erich Kaiser, Die Diamantenwüste Südwestafrikas, vol. 2, 107.

TEILHARD DE CHARDIN, P. 1916–1922. Les mammifères de l'écocène inférieur francais et leurs gisements. Ann. d. Paléont., vol. 10 (1916–1921), 171; vol. 11 (1922).—1926. Description des mammifères tertiaires de Chine et de Mongolie. Ibid., vol. 15, 1. [See Ungulates, Perissodactyls, Rhinoceroses.]—1936. Fossil mammals from Locality 9 of Choukoutien. Palaeontologia Sinica, part 4. [Pleistocene.]—1938. The fossils from Locality 12 of Choukoutien. Ibid., no. 5. ["Late Pliocene."]

TEILHARD DE CHARDIN, P. and P. LEROY. 1942. Chinese fossil mammals. A complete bibliography, analysed, tabulated, annotated and indexed. Publ. no. 8, Inst. Géo-biologie, Pékin.

TEILHARD DE CHARDIN, P. and J. PIVETEAU. 1930. Les mammifères fossiles de Nihowan (Chine). Ann. d. Paléont., vol. 19. ["Late Pliocene."]

TEILHARD DE CHARDIN, P. and C. C. YOUNG. 1931. Fossil mammals from the late Cenozoic of northern China. Palaeontologia Sinica, vol. 9, part 1. ["Lower Pliocene" antelopes and Pleistocene rodents.]—1936. On the mammalian remains from the archaeological site of Anyang. Ibid., vol. 12, part 1. [Recent.]

VIRET, J. 1929. Les faunes de mammifères de l'Oligocène supérieur de la Limagne bourbonnaise. Ann. D. l'Univ. d. Lyon, n.s., I, part 47, 1.

YOUNG, C. C. 1944. Note on the first Eocene mammal from South China. Amer. Mus. Novitates, no. 1268.

ZDANSKY, O. 1930. Die alttertiären Säugetiere Chinas nebst stratigraphischen Bemerkungen. Palaeontologia Sinica. [Eocene and later horizons.]

Mammals: Tertiary to Recent of South America

AMEGHINO, F. 1887. Enumeracion sistemática de las especies de mamiferos fósiles coleccionados por Cárlos Ameghino en los terrenos eocenos de la Patagonia austral y depositados en el Museo la Plata.—1897. Mammifères crétacés de l'Argentine. Deuxième contribution à la connaissance de la faune mammalogique des couches à Pyrotherium. Bol. Inst. Geogr. Argentino,

Buenos Aires, vol. 18, 406. [Supposed Cretaceous age based partly on misleading evidence. See G. G. Simpson, 1932a.]—1906. Les formations sédimentaires du crétacé supérieur et du tertaire de la Patagonie avec un parallèle entre leurs faunes mammalogiques et celles de l'ancien continent. An. Mus. Nac. Hist. Nat., Buenos Aires, vol. 15.—1913–1936. Obras completas y correspondencia cientifica de Florentino Ameghino. 24 vols. La Plata, Argentina.

BERRY, E. W. 1932. Fossil plants from Chubut Territory collected by the Scarritt Patagonian expedition. Amer. Mus. Novitates, no. 563. [Bearing on Notoungulata and geological succession in Patagonia.]

BOULE, M. and A. THEVENIN. 1920. Mammifères fossiles de Tarija. Paris. 4to. [Rich Pleistocene fauna.]

BURMEISTER, H. 1854. Systematische Uebersicht der Thiere brasiliens . . . von Rio de Janeiro und Minas geraës. Theil I. Säugethiere (Mammalia). Berlin.—1879. Description physique de la République Argentine . . . Tome III. Animaux Vertébrés. Part 1. Mammifères vivante et étients. Buenos Aires.

GAUDRY, A. 1904–1909. Fossiles de Patagonie. No. 1. Dentition de quelques mammifères. Mém. Soc. Geol. France, vol. 12, no. 1.—1906a. No. 2. Étude sur une portion du Monde antarctique. Ann. d. Paléont., vol. 1, 101.—1906b. [No. 3.] Les attitudes de quelques animaux. Ibid., 1.

GERVAIS, F. L. P. 1855. Mammifères. Animaux nouveaux, ou rare, recueillis pendent l'expédition dans les parties centrales de l'Amérique du Sud. Paris.

GOODWIN, G. G. 1946. Mammals of Costa Rica. Bull. Amer. Mus. Nat. Hist., vol. 87, 271. [Excellent notes on food and locomotor habits.]

HATCHER, J. B. 1903. Reports of the Princeton University Expedition to Patagonia, 1896–1899. Vol. I. Narrative of the expedition. Geography of southern Patagonia. Princeton. 4to.

KRAGLIEVICH, L. 1930. La formación friaseana del Rio Frias . . . y su fauna de mamiferos. Physis, 127.

LOOMIS, F. B. 1914. The Deseado formation of Patagonia. Amherst College, 232.

LYDEKKER, R. 1894. Contributions to a knowledge of the fossil vertebrates of Argentina. An. Mus. La Plata, Paleontologica Argentina, vols. 2, 3. Folio.

ROTH, S. 1903a. Los ungulados sudamericanos. An. Mus. La Plata, Paleontologica Argentina, vol. 5. [Proposes order Notoungulata and recognizes its distinctness from North American and European orders.]—1903b. Noticias preliminarias sobre nuevos mamiferos fósiles del Cretáceo superior y Terciario inferior de la Patagonia. Rev. Mus. La Plata, 133.

SCHAUB, S. 1935. Säugetierfunde aus Venezuela und Trinidad. Abh. d. Schweiz. Palaeont. Gesell., vol. 55, 1.

SCOTT, W. B. 1903–1932. (Editor) Reports of the Princeton University Expedition to Patagonia, 1896–1899. Vols. [5, 6, 7, Palaeontology].—1937. A history of land mammals in the Western Hemisphere. Revised ed. New York.

SIMPSON, G. G. 1932a. The supposed association of dinosaurs with mammals of Tertiary type in Patagonia. Amer. Mus. Novitates, no. 566.—1932b. Some new or little-known mammals from the Colpodon beds of Patagonia. Ibid., no. 575.—1933. Stratigraphic nomenclature of the Early Tertiary of Central Patagonia. Ibid., no. 644.—1934a. Provisional classification of extinct South American hoofed mammals. Ibid., no. 750.—1934b. Attending marvels. A Patagonian Journal. New York.—1935a. Early and Middle Tertiary geology of the Caiman Region, Chubut, Argentina. Amer. Mus. Novitates, no. 775.—1935b. Descriptions of the oldest known South American mammals, from the Rio Chico Formation. Ibid., no. 793.—1935c. Occurrence and relationships of the Rio Chico fauna of Patagonia. Ibid., no. 818.—1940. Review of the mammal-bearing Tertiary of South America. Proc. Amer. Philos. Soc., 649.—1941. The Eogene of Patagonia. Amer. Mus. Novitates, no. 1120.—1942. Early Cenozoic mammals of South America. Proc. Eighth Amer. Congress Geol. Sci. (1940). Mesozoic and Cenozoic faunas and floras, 303.—1943a. Turtles and the origin of the fauna of Latin America. Amer. Jour. Sci., 413.—1943b. Notes on the mammal-bearing Tertiary of South America. Proc. Amer. Philos. Soc., 403.—1948. The beginning of the age of mammals in South America. Bull. Amer. Mus. Nat. Hist., 5. [Marsupials, Edentates, Condylarths, Litopterns, Notioprogonia.]

Mammalian Orders: Monotremes

BEER, G. R. DE and W. A. FELL. 1935. The development of the Monotremata. Part III. The development of the skull of Ornithorhynchus. Trans. Zool. Soc. London, 1.

BEMMELEN, J. F. VAN. 1899. . . . Concerning the palatine-, orbital-, and temporal regions of the monotreme skull. Proc. Kon. Akad. v. Wetensch. te Amsterdam, 81.—1901. Der Schädelbau der Monotremen. In: R. Semon, Zool. Forsch. in Australien . . . III. Bd. Monotremen und Marsupialier. II, IV. Lief. Denkschr. d. Med.-Naturwiss. Gesellsch. Jena, vol. 6, 729.

BRESSLAU, E. 1907. Die Entwickelung des Mammarapparates der Monotremen, Marsupialier und einiger Placentalier. Ein Beitrag zur Phylogenie der Säugethiere. I. Entwickelung und Ursprung des Mammarapparates bei Echidna. In: R. Semon, Zool. Forsch. in Australien . . . IV. Bd. Morphol. Verschiedener Wirbeltiere, 455.—1912. II. Der Mammarapparat des erwachsenen Echidna-Weibchens. Ibid., vol. 7, 627.

BURRELL, H. 1927. The platypus. Sydney.

DENKER, A. 1901. Zur Anatomie des Gehörorgans der Monotremata. In: R. Semon, Zool. Forsch. in Australien . . . III. Bd. Monotremen und Marsupialier. II, III. Lief., 635.

EGGELING, H. 1899. Ueber die Stellung der Milchdrüsen zu den übrigen Hautdrüsen. I. Mitth.: Die ausgebildeten Mammardrüsen der Monotremen und die Milchdrüsen der Edentaten . . . In: R. Semon, Zool. Forsch. in Australien . . . IV. Bd. Morphologie . . . 77.—1901. II. Mitth.: Die Entwickelung der Mammardrüsen, Entwickelung und Bau der übrigen Hautdrüsen der Monotremen. Ibid., vol. 7, 173.—1907. Nachtrag zur II. Mitth.: Neue Beobachtungen über die Mammardrüsenentwickelung bei Echidna. Ibid., vol. 7, 333.

FLEAY, D. 1944. The birth of a baby platypus. Animal Kingdom (N. Y. Zool. Soc.), 51.

FLYNN, T. T. and J. P. HILL. 1939. The development of the Monotremata. Part IV. Growth of the ovarian ovum, maturation, fertilisation, and early cleavage. Trans. Zool. Soc. London, 445.

GAUPP, E. 1908. Zur Entwickel. und vergl. Morphol. des Schädels von Echidna aculeata var. typica. In: R. Semon, Zool. Forsch. in Australien . . . III. Bd. Monotremen und Marsupialier. II. Teil. Denkschr. d. Med.-Naturwiss. Gesellsch. Jena. 539.

GÖPPERT, E. 1901. Beiträge zur vergleichenden Anatomie des Kehlkopfes und seiner Umgebung mit besonderer Berücksichtigung der Monotremen. In: R. Semon, Zool. Forsch. in Australien . . . II, IV. Lief. Denkschr. d. Med.-Naturwiss. Gesellsch. Jena. vol. 6, 533.

GÖSSNITZ, W. VON. 1901. Beitrag zur Diaphragmafrage. In: R. Semon, Zool. Forsch. in Australien . . . IV. Bd.: III Lief. Denkschr. d. Med.-Naturwiss. Gesellsch. Jena, vol. 7, 205.

GREENE, H. L. H. H. 1937. The development and morphology of the teeth of Ornithorhynchus. Philos. Trans. Roy. Soc. London, 367.

GREGORY, W. K. 1910. The orders of mammals. Part II. Bull. Amer. Mus. Nat. Hist., vol. 27. [Genetic relations of the Monotremata, pp. 144–162.]—1947. The monotremes and the palimpsest theory. Ibid., vol. 88, 1.

GREGORY, W. K. and C. L. CAMP. 1918. Studies in comparative myology and osteology. No. III. Bull. Amer. Mus. Nat. Hist., vol. 38, art. 15, part 4, 528. [Review of limb musculature of Ornithorhynchus.]

HILL, J. P. and J. B. GATENBY. 1926. The corpus luteum of the Monotremata. Proc. Zool. Soc. London, 715.

HOWELL, A. B. 1936. The musculature of antebrachium and manus in the platypus. Amer. Jour. Anat., 425.—1937a. Morphogenesis of the shoulder architecture. Part V. Monotremata. Quart. Rev. Biol., 191.—1937b. The swimming mechanism of the platypus. Jour. Mammal., 217.

JONES, F. W. 1923. The mammals of South Australia. Part I. The monotremes and the carnivorous marsupials. Handbooks of the Flora and Fauna of South Australia, British Science Guild (South Australian Branch). Adelaide.

KEIBEL, F. 1903. Zur Anatomie des Urogenitalkanals der Echidna aculeata var. typica. Anat. Anz., 301.

KESTEVEN, H. L. 1929. The skull of Ornithorhynchus: Its later development and adult features. Jour. Anat., 447.

McKAY, W. J. S. 1894. The morphology of the muscles of the shoulder-girdle in monotremes. Proc. Linnean Soc., N.S.W., 263.

MARTIN, C. J. and F. TIDSWELL. 1894. Observations on the femoral gland of Ornithorhynchus and its secretion; together with an experimental enquiry concerning its supposed toxic action. Proc. Linnean Soc., N.S.W., 471.

MATTHEW, W. D. 1915. Climate and evolution. Ann. N. Y. Acad. Sci., 171. [Monotremata, p. 270.] 2d ed., 1939.

MECKEL, J. F. 1826. Ornithorhynchi paradoxi descriptio anatomica. Folio. Leipzig.

MIVART, ST. G. 1866. On some points in the anatomy of Echidna hystrix. Trans. Linnean Soc. London, 379.

OWEN, R. 1884. Evidence of a large extinct monotreme (Echidna Ramsayi, Ow.) from the Wellington Breccia Cave, New South Wales. Philos. Trans. Roy. Soc. London, 273.

PEARSON, H. S. 1926. Pelvic and thigh muscles of Ornithorhynchus. Jour. Anat., 152.

POULTON, E. B. 1894. The structure of the bill and hairs of Ornithorhynchus paradoxus, with a discussion of the homologies and origin of mammalian hair. Quart. Jour. Micros. Sci., 143.

RÖMER, F. 1898. Studien über das Integument der Säugethiere. II. Das Integument der Monotremen. In: R. Semon, Zool. Forsch. in Australien . . . III. Bd. Monotremen und Marsupialier. II. 1 Teil. Denkschr. d. Med.-Naturwiss. Gesellsch. Jena, vol. 6, 189.

RUGE, G. 1895. Die Hautmuskulatur der Monotremen und ihre Beziehungen zum Marsupial-und Mammarapparate. In R. Semon, Zool. Forsch. in Australien . . . II. Bd. Monotremen und Marsupialier. Denkschr. d. Med.-Naturwiss. Gesellsch. Jena, vol. 5, 77.

SEMON, R. 1894a. Beobachtungen über die Lebensweise und Fortpflanzung der Monotremen nebst Notizen über ihre Körpertemperatur. In: R. Semon. Zool. Forsch. in Australien . . . Denkschr. d. Med.-Naturwiss. Gesellsch. Jena, vol. 5, 3.—1894b. Die Embryonalhüllen der Monotremen und Marsupialier. Ein vergleichende Studie über die Fötalanhänge der Amnioten. Ibid., vol. 5, 19.—1894c. Zur Entwickelungsgeschichte der Monotremen. Ibid., vol. 5, 61.

SIMPSON, G. G. 1928. A catalogue of the Mesozoic Mammalia in the geological department of the British Museum. London. [Relationships of monotremes, pp. 151, 154–158, 182, 183.]—1929. The dentition of Ornithorhynchus as evidence of its affinities. Amer. Mus. Novitates, no. 390.—1938. Osteography of the ear region in monotremes. Ibid., no. 978.

SPENCER, B. and G. SWEET. 1899. The structure and development of the hairs of monotremes and marsupials. Part I. Quart. Jour. Micros. Sci., 549.

WATSON, D. M. S. 1916. The monotreme skull: A contribution to mammalian morphogenesis. Philos. Trans. Roy. Soc. London, 311.

WEBER, M. 1904. Die Säugetiere: Einführung in die Anatomie und Systematik der recenten und fossilen Mammalia. Jena. [Sections on monotremes, 317–331.]—1927. Die Säugetiere. Bd. I. Anatomischer Teil. Ibid. [Fig. 10, after de Mayere; no. 11, arrangement of platypus hair into groups suggests that of beavers and otters.]

WESTLING, C. 1889. Anatomische Untersuchungen über Echidna. K. Svenska Vet.-Akad. Handl., 1.

WILSON, J. T. and W. J. S. MCKAY. 1893. On the homologies of the borders and surfaces of the scapula in monotremes. Proc. Linnean Soc., N.S.W., 377.

ZIEHEN, T. 1897. Das central nerven System der Monotremen und Marsupialier . . . I. Theil. Makroskopische Anatomie. In: R. Semon. Zool. Forsch. in Australien . . . III. Bd. II. 1 Lief. Denkschr. d. Med.-Naturwiss. Gesellsch. Jena, vol. 6, 1.

Marsupials: General

ABBIE, A. A. 1937. Some observations on the major subdivisions of the Marsupialia, with especial reference to the position of the Peramelidae and Caenolestidae. Jour. Anat., 429.—1941. Marsupials and the evolution of mammals. Australian Jour. Sci., 77.

ANDERSON, C. 1936. The origin of Australian mammals. Australian Mus. Mag., 133.

BENSLEY, B. A. 1903. On the evolution of the Australian Marsupialia; with remarks on the relationships of the marsupials in general. Trans. Linnean Soc. London, 83.

BERRY, E. W. 1919. Upper Cretaceous floras of the eastern Gulf Region in Tennessee, Mississippi, Alabama and Georgia. U. S. Geol. Surv., Professional Paper 112. [Extinct species of Eucalyptus found in North America, p. 126.]

BRESSLAU, E. 1912. Die Entwickelung des Mammarapparates der Monotremen, Marsupialier und einiger Placentalier. Ein Beitrag zur Phylogenie der Säugethiere. III. Entwickelung des Mammarapparates der Marsupialier, Insectivoren, Nagethiere, Carnivoren und Wiederkäuer. In: R. Semon, Zool. Forsch. in Australien . . . IV. Bd. Morphologie . . . Wirbeltiere. Denkschr. d. Med.-Naturwiss. Gesellsch. Jena, vol. 7, 647.

BROOM, R. 1899. On the development and morphology of the marsupial shoulder-girdle. Trans. Roy. Soc. Edinburgh, 749.

DOLLO, L. 1899. Les ancètres des marsupiaux étaient-ils arboricoles? Travaux Station zoologique de Wimereux, vol. 7, 588.

EGGLING, H. 1905. Ueber die Stellung der Milchdrüsen zu den übrigen Hautdrüsen. III. (letzte) Mittheilung: Die Milchdrüsen und Hautdrüsen der Marsupialier. In: R. Semon, Zool. Forsch. in Australien . . . IV. Bd. Denkschr. d. Med.-Naturwiss. Gesellsch. Jena, vol. 7, 299.

ELFTMAN, H. O. 1929. Functional adaptations of the pelvis in marsupials. Bull. Amer. Mus. Nat. Hist., 189.

FLYNN, T. T. 1922. The phylogenetic significance of the marsupial allantoplacenta. Proc. Linnean Soc., N.S.W., 541.

FRECHKOP, S. 1930. Notes sur les mammifères. III. Au sujet du nombre des chromosomes chez les marsupiaux. Bull. d. Mus. roy. d'Hist. nat. de Belgique, vol. 6, no. 18.

GREGORY, W. K. 1910. The orders of mammals. Bull. Amer. Mus. Nat. Hist., vol. 27. [Genetic relations of the Marsupialia, 197–231.]

PEARSON, J. 1945. The female urogenital system of the Marsupialia with special reference to the vaginal complex. Papers and Proc., Roy. Soc. Tasmania (1944), 71.—1946. Some problems of marsupial phylogeny. Rept. 25th Meeting, Australian and New Zealand Assoc. Adv. Sci. . . . , 72. Adelaide. [A basic analysis of relationships based on anatomy of urogenital system, vaginal complex and parturition, placentation, evolution of the mammalian placenta.]

THOMAS, O. 1888. Catalogue of the Marsupialia and Monotremata in the collecton of the British Museum. London.

DOLLO, L. 1900. Le pied du Diprotodon et l'origine arboricole des marsupiaux. Bull. Scient. d. la France et d. la Belgique, 278.

FLOWER, W. H. 1868. On the affinities and probable habits of the extinct Australian marsupial, Thylacoleo carnifex, Owen. Proc. Geol. Soc., London, 307.

FLYNN, T. T. 1911a. Notes on marsupialian anatomy. 1. On the condition of the median vaginal septum in the Trichosuridae. Papers and Proc., Roy. Soc. Tasmania.—1911b. Contributions to a knowledge of the anatomy and development of the Marsupialia. No. 1. The genitalia of Sarcophilus satanicus (♀). Proc. Linnean Soc., N.S.W. (1910), 873.

FRECHKOP, S. 1930a. Notes sur les mammifères. I. Sur certains caractères correlatifs chez les Peramelidae (Marsupialia). Bull. du Mus. roy. d'Hist. nat. de Belgique, vol. 6, no. 5.—1930b. II. Caractères distinctifs et phylogenie du wombat (Phascolomys) et du koala (Phascolarctus). Ibid., no. 12.

GREGORY, W. K. 1924. Australia, the land of living fossils. Nat. Hist. Mag., vol. 24, no. 4, 4.—1944. Australia—the story of a continent. Nat. Hist. Mag., vol. 53, no. 8, 360. [Family tree of marsupials.]

Marsupials: Australia, New Guinea

ABBIE, A. A. 1939a. A masticatory adaptation peculiar to some diprotodont marsupials. Proc. Zool. Soc. London, 261.—1939b. The mandibular meniscus in monotremes and marsupials. Australian Jour. Sci., 86.

BROOM, R. 1895. On a small fossil marsupial with large grooved premolars. Proc. Linnean Soc., N.S.W., 563. [Burramys parvus, n.g., n.sp.]—1938. On the affinities and habits of Thylacoleo. Proc. Linnean Soc., N.S.W., 57.

CARLSSON, A. 1914. Über Dendrolagus dorianus. Zool. Jahrb., 547.—1915. Zur Morphologie des Hypsiprymnodon moschatus. Kungl. Svenska Vetenskaps. Handl.

HILL, J. P. 1898. The placentation of Perameles. Quart. Jour. Micros. Sci., 384.—1899. Contributions to the embryology of the Marsupialia. II. On a further stage in the placentation of Perameles. Ibid., vol. 43, part 1.—1900. On the foetal membranes, placentation and parturition of the native cat (Dasyurus viverrinus). Anat. Anz., 364.—1910. The early development of the Marsupialia, with special reference to the native cat (Dasyurus viverrinus). Quart. Jour. Micros. Sci., 1.—1918. Some observations on the early development of Didelphys aurita. Ibid., 91.

JONES, F. W. 1924. The mammals of South Australia. Part II. The bandicoots and the herbivorous marsupials (the syndactylous Didelphia). Handbooks of the Flora and Fauna of South Australia, British Science Guild, 132. Adelaide. —1925. Part III. The Monodelphia. Ibid., 271.—1930. A re-examination of the skeletal characters of Wynyardia bassiana, an extinct Tasmanian marsupial. Roy. Soc. Tasmania: Papers and Proc., 96.

LONGMAN, H. A. 1921. A new genus of fossil marsupials. [Euryzygoma, a giant diprotodont.]

Mem. Queensland Mus., 65.—1934. Restoration of Euryzygoma dunense. Ibid., 201.

OWEN, R. 1845. Report on the extinct mammals of Australia . . . Brit. Assoc. Adv. Sci. (1844), 1. [Jaw of Diprotodon, mistaken for "pachyderm."] —1859–1876. On the fossil mammals of Australia. Philos. Trans. Roy. Soc. London. Parts I (1859); II (1866); III (1870), on Diprotodon; IV (1871), on Thylacoleo; V (1872), on Nototherium; VI, VII (1872), on Phascolomys; VIII (1874); IX (1874); X (1876), on Macropodidae.—1879. On Hypsiprymnodon Ramsay, a genus indicative of a distinct family (Pleopodidae) in the diprotodont section of the Marsupialia. Trans. Linnean Soc. London, 573.—1883a. On the affinities of Thylacoleo. Philos. Trans. Roy. Soc. London, part 2, 575.—1883b. Pelvic characters of Thylacoleo carnifex. Ibid., 639.—1884. Description of teeth of a large extinct (marsupial?) genus Sceparnodon, Ramsay. Ibid., 245.—1887. Additional evidence of the affinities of the extinct marsupial quadruped Thylacoleo carnifex (Owen). Ibid., vol. 178, 1.

PEARSON, J. 1944. The vaginal complex of the rat-kangaroos. Australian Jour. Sci., 80.

POCOCK, R. I. 1926. The external characters of Thylacinus, Sarcophilus, and some related marsupials. Proc. Zool. Soc. London, 1037.

RAVEN, H. C. 1929a. Strange animals of the island continent. I. Nat. Hist. Mag., 83.—1929b. II. Ibid.—1939. The identity of Captain Cook's kangaroo. Jour. Mammal., 50.—1942. Unique mammals of Australia. Fauna (Philadelphia), 104.

RAVEN, H. C. and W. K. GREGORY. 1946. Adaptive branching of the kangaroo family in relation to habitat. Amer. Mus. Novitates, no. 1309.

SELENKA, E. 1892. Studien über Entwick. der Tiere. I. Bd., V. Heft. Beutelfuchs und Känguruatte (Phalangista et Hypsiprymnus). Wiesbaden.

SPENCER, B. 1900. A description of Wynyardia bassiana, a fossil marsupial from the Tertiary Beds of Table Cape, Tasmania. Proc. Zool. Soc. London, 776.

STIRLING, E. C. 1907a. [Diprotodon australis. Note and five plates, showing mounted skeleton and restoration of D. australis.] Rept. Public Library, Museum and Art Gallery of South Australia (for the years 1905–1906).—1907b. Reconstruction of Diprotodon from the Callabonna deposits, South Australia. Nature, 543.

STIRLING, E. C. and A. H. C. ZIETZ. 1900. Fossil remains of Lake Callabonna. Part I. Description of the bones of the manus and pes of Diprotodon australis, Owen. Mem. Roy. Soc. South Australia, vol. 1, part 1.

TATE, G. H. H. 1945a. Results of the Archbold Expeditions. No. 52. The marsupial genus Phalanger. Amer. Mus. Novitates, no. 1283.—1945b. No. 54. The marsupial genus Pseudocheirus and its subgenera. Ibid., no. 1287.—1945c. No. 55. Notes on the squirrel-like and mouse-like opossums (Marsupialia). Ibid., no. 1305.—1947. No. 56. On the anatomy and classification of the Dasyuridae (Marsupialia). Bull. Amer. Mus. Nat. Hist., 97.—1948a. No. 59. Studies on the anatomy and phylogeny of the Macropodidae (Marsupialia). Bull. Amer. Mus. Nat. Hist., 233.—1948b. No. 60. Studies in the Peramelidae (Marsupialia). Ibid., 313.

Marsupials: Didelphyidae

COPE, E. D. 1884. The Tertiary Marsupialia. Amer. Nat., 686.—1892. On a new genus of Mammalia from the Laramie formation. Ibid., 758. [Describes Thlaeodon padanicus.]

HARTMAN, C. G. 1916. Studies in the development of the opossum Didelphys virginiana L. I. History of the early cleavage. II. Formation of the blastocyst. Jour. Morph., vol. 27, no. 1.—1920. V. The phenomena of parturition: The method of transfer of young to the pouch. Anat. Rec., vol. 19, no. 5.—1923. Breeding habits, development, and birth of the opossum. Smiths. Rept. for 1921, 347.—1928. The breeding season (Didelphis virginiana) and the rate of intrauterine and postnatal development. Jour. Morph. and Physiol., 143.

HILL, J. P. and E. A. FRASER. 1925. Some observations on the female uro-genital organs of the Didelphyidae. Proc. Zool. Soc. London, part 1, 189.

MATTHEW, W. D. 1916. A marsupial from the Belly River Cretaceous. With critical observations upon the affinities of the Cretaceous mammals. Bull. Amer. Mus. Nat. Hist., 477. [Eodelphis browni, n.g., n.sp.]

SELENKA, E. 1887. Studien über Entwickelungsgeschichte der Tiere. Bd. I. IV. Heft. Das Opossum (Didelphys virginiana). Wiesbaden.

TATE, G. H. H. 1933. A systematic revision of the marsupial genus Marmosa, with a discussion of the adaptive radiation of the murine opossums (Marmosa). Bull. Amer. Mus. Nat. Hist. 1.

Marsupials: South America

AMEGHINO, F. 1900. Mamiferos del cretáceo inferior de Patagonia (Formación de las areniscas abigarradas). Com. d. Mus. Nac. d. Buenos Aires, 197. [Proteodidelphys praecursor, p. 20.]

GREGORY, W. K. 1933. On the "habitus" and "heritage" of Caenolestes. Jour. Mammal., 106.

OBENCHAIN, J. B. 1925. The brains of the South American marsupials Caenolestes and Orolestes. Field Mus. Nat. Hist., publ. 224, 175.

OSGOOD, W. H. 1921. A monographic study of the American marsupial, Caenolestes; with a description of the brain of Caenolestes by C. Judson Herrick. Field Mus. Nat. Hist., publ. 207.

RIGGS, E. S. 1934. A new marsupial saber-tooth from the Pliocene of Argentina and its relationships to other South American predacious marsupials. Trans. Amer. Philos. Soc., vol. 24, part 1. [Thylacosmilus, a specialized borhyaenid.]

SIMPSON, G. G. 1928. Affinities of the Polydolopidae. Amer. Mus. Novitates, no. 323.—1941. The affinities of the Borhyaenidae. Ibid., no. 1118.

SINCLAIR, W. J. 1905. The marsupial fauna of the Santa Cruz Beds. Proc. Amer. Philos. Soc., 73. [Cladosictis lustratus, pl. 1; Prothylacynus patagonicus, pl. 2.]—1906. Mammalia of the Santa Cruz Beds. Part III. Marsupialia. Princeton Univ. Exped. to Patagonia . . . , vol. 4, 333.

THOMAS, O. 1895. On Caenolestes, a still existing survivor of the Epanorthidae of Ameghino, and the representative of a new family of recent marsupials. Proc. Zool. Soc. London, 870.

TOMES, R. F. 1863. Notice of a new American form of marsupial. Proc. Zool. Soc. London, 50.

WOOD, H. E. II. 1924. The position of the "sparassodonts"; with notes on the relationships and history of the Marsupialia. Bull. Amer. Mus. Nat. Hist., 77.

Placental Mammals: General

ABBIE, A. A. 1939. The ancestors of the Eutheria. Nature, 523.

CLARK, W. E. LE G. 1939. The ancestors of the Eutheria. Ibid., 600.

COPE, E. D. 1872. On the primitive types of the orders of Mammalia Educabilia. Amer. Philos. Soc., 1.

GIDLEY, J. W. 1919. Significance of divergence of the first digit in the primitive mammalian foot. Jour. Washington Acad. Sci., 273. [Critique of Matthew's theory of arboreal ancestry of placental Mammalia. Holds divergence of first digit merely a primitive character, as in Sphenodon.]

MATTHEW, W. D. 1904. The arboreal ancestry of the Mammalia. Amer. Nat., 811.

OSBORN, H. F. 1894. A division of the eutherian mammals into the Mesoplacentalia and Cenoplacentalia [terms subsequently altered to Meseutheria and Ceneutheria]. Trans. N. Y. Acad. Sci., 234.

Insectivora: General

BROOM, R. 1915. On the organ of Jacobson and its relations in the "Insectivora." Part 1. Tupaia and Gymnura. Part 2. Talpa, Centetes and Chrysochloris. Proc. Zool. Soc. London, 157, 347.

CLARK, W. E. LE G. 1933. The brain of the Insectivora. Proc. Zool. Soc. London (1932), 975.

DOBSON, G. E. 1882. A monograph of the Insectivora, systematic and anatomical. Parts 1–3. Including the families Erinaceidae, Centetidae and Solenodontidae. London. 4to.

FRECHKOP, S. 1932. Notes sur les mammifères. VIII. De la forme des molaires chez les Insectivores. Bull. d. Mus. roy. d'Hist. nat. d. Belgique, 1.

GREGORY, W. K. 1927. Mongolian mammals of the "age of reptiles." Scient. Monthly, 225.

GREGORY, W. K. and G. G. SIMPSON. 1926. Cretaceous mammal skulls from Mongolia. Amer. Mus. Novitates, no. 225.

MATTHEW, W. D. and W. GRANGER. 1915. A revision of the Lower Eocene Wasatch and Wind River faunas. Part IV. Entelonychia, Primates, Insectivora (Part). Bull. Amer. Mus. Nat. Hist., 429.—1918. Part V. Insectivora (continued), Glires, Edentata. By W. D. Matthew. Ibid., 565.

PARKER, W. K. 1885. The structure and development of the skull in the Mammalia. Part III. Insectivora. Philos. Trans. Roy. Soc., 121. [Erinaceus, pp. 124–159, pls. 16–22.]

PATTERSON, B. and P. O. MCGREW. 1937. A soricid and two erinaceids from the White River Oligocene. Field Mus. Nat. Hist., Geol. Ser., 245.

SCOTT, W. B. 1903. Mammalia of the Santa Cruz Beds. I. Edentata. II. Insectivora. III. Glires. Rept. Princeton Univ. Exp. to Patagonia, Vol. V. Princeton, 4to.

SCOTT, W. B. and G. L. JEPSEN. 1936. The mammalian fauna of the White River Oligocene. Part I. Insectivora and Carnivora. Trans. Amer. Philos. Soc., 1.

SIMPSON, G. G. 1928a. Further notes on Mongolian Cretaceous mammals. Amer. Mus. Novitates, no. 329. [Deltatheridiidae.]—1928b. Affinities of the Mongolian Cretaceous insectivores. Ibid., no. 330.

SMITH, G. E. 1902. Notes on the brain of Macroscelides and other Insectivora. Linnean Soc.'s Jour., Zool., 443.

SPRAGUE, J. M. 1944. The hyoid region in the Insectivora. Amer. Jour. Anat., 175.

WORTMAN, J. L. 1920. On some hitherto unrecognized reptilian characters in the skull of the Insectivora and other mammals. Proc. U. S. Nat. Mus., 1. [Pseudo-reptilian characters.]

WINGE, H. 1917. Udsigt over Insektaedernes indbyrdes Slaegtskab. Vidensk. Medd. fra Dansk naturh. Foren., Bd. 68, 83.

Zalambdodonts (Centetoids)

ALLEN, G. M. 1910. Solenodon paradoxus. Mem. Mus. Comp. Zool. Harvard Coll., 1.

ALLEN, J. A. 1908. Notes on Solenodon paradoxus Brandt. Bull. Amer. Mus. Nat. Hist., 505.

BROOM, R. 1916. On the structure of the skull in Chrysochloris. Proc. Zool. Soc. London, 449.

BUTLER, P. M. 1941. A comparison of the skulls and teeth of the two species of Hemicentetes. Jour. Mammal., 65.

COOPER, C. F. 1928. On the ear region of certain of the Chrysochloridae. Philos. Trans. Roy. Soc. London, 265.

LECHE, W. 1907. Zur Entwick. des Zahnsystems der Säugetiere . . . II. Teil: Phylogenie. II. Heft: Der Familien der Centetidae, Solenodontidae und Chrysochloridae. Zoologica, 1. Stuttgart. 4to.

MATTHEW, W. D. 1910. On the skull of Apternodus . . . Bull. Amer. Mus. Nat. Hist., 33.—1913. A zalambdodont insectivore from the Basal Eocene. Ibid., 307. [Palaeoryctes.]

MOHR, E. 1936–1938. Biologische Beobachtungen an Solenodon paradoxus Brandt in Gefangenschaft. I–IV. Zool. Anz., vols. 113–122.

ROUX, G. H. 1947. The cranial development of certain Ethiopian "Insectivores" and its bearing on the mutual affinities of the group. Acta Zoologica, 169. [Anatomy of the chondrocranium in a soricoid (Suncus), a tupaioid (Elephantulus), a chrysochloroid (Eremitalpa). Bearing on ordinal relationships discussed.]

SCHLAIKJER, E. M. 1933. Contributions to the stratigraphy and palaeontology of the Goshen Hole area, Wyoming. I . . . The structure and relationships of a new zalambdodont insectivore from the Middle Oligocene. Bull. Mus. Comp. Zool. Harvard Coll., 1.—1934. A new fossil zalambdodont insectivore. Amer. Mus. Novitates, no. 698.

VERRILL, A. H. 1907. Notes on the habits and external characters of the Solenodon of San Domingo (Solenodon paradoxus). Ann. Mag. Nat. Hist., 68.

Soricoids (Moles, Shrews)

ALLEN, J. A. 1917. The skeletal characters of Scutisorex Thomas. Bull. Amer. Mus. Nat. Hist., 769.

ANTHONY, H. E. 1916. Preliminary diagnosis of an apparently new family of insectivores (Nesophontes). Bull. Amer. Mus. Nat. Hist., 725.—1925. Mammals of Porto Rico, living and extinct: Chiroptera and Insectivora. N. Y. Acad. Sci., Sci. Surv. of Porto Rico and the Virgin Islands, 1. [Nesophontes.]

BEER, G. R. DE. 1929. The development of the skull of the shrew. Philos. Trans. Roy. Soc. London, 411.

GOODWIN, G. G. 1942. The taming of a shrew. Nat. Hist. Mag., 282.

KISTIN, A. D. 1929. The mole clavicle. Jour. Mammal., 305.

SCHREUDER, A. 1940. A revision of the fossil water-moles (Desmaninae). Archiv. Néerlandaises de Zool., 201.

Erinaceoids (Leptictids, Erinaceids, Dimylids)

BATE, D. M. A. 1932. A new fossil hedgehog from Palestine. Ann. Mag. Nat. Hist., 575.

BUTLER, P. M. 1947. On the evolution of the skull and teeth in the Erinaceidae, with special reference to fossil material in the British Museum. Proc. Zool. Soc., part II, 446. [A basic contribution.—W.K.G.]

FRIANT, M. 1943. Mammifères. Fasc. II. Insectivora. A. Insectivora vera. Sous-fasc. I.—Erinaceidae. Catalogue . . . d'ostéol. d. Mus. nat. d'Hist. nat. Paris.

HÜRZELER, J. 1944a. Über einen dimyloiden Erinaceiden (Dimylechinus nov. gen.) aus dem Aquitanien der Limagne. Eclogae geol. Helvetiae,

460.—1944b. Beiträge zur Kenntnis der Dimylidae. Schweiz. Palaeont. Abh., 1.

JEPSEN, G. L. 1930a. Stratigraphy and paleontology of the Paleocene of northeastern Park County, Wyoming. Proc. Amer. Philos. Soc., 435. [Insectivora.]

LECHE, W. 1902. Zahnsystems der Säugethiere . . . II. Theil: Phylogenie. I. Heft: Die Familie der Erinaceidae. Zoologica, 1.

MATTHEW, W. D. 1903. A fossil hedgehog from the American Oligocene. Bull. Amer. Mus. Nat. Hist., 227.—1929. A new and remarkable hedgehog from the later Tertiary of Nevada. Univ. California Publ., Bull. Dept. Geol. Sci., 93.

SIMPSON, G. G. 1927. Mammalian fauna of the Hell Creek formation of Montana. Amer. Mus. Novitates, no. 267. [Gypsonictis, primitive leptictid, p. 6.]

VIRET, J. 1938. Étude sur quelques Erinacéidés fossiles, spécialement sur le genre Palaerinaceus. Trav. Lab. géol. Sci., Lyon, 1.—1940. Étude sur quelques Erinacéidés fossiles (suite) genres Plesiosorex, Lanthanotherium. Ibid., 33.

Tupaioids (Tree Shrews, Jumping Shrews)
(See also Primates, Lemuroidea)
Dermoptera

LECHE, W. 1886. Über die Säugethiergattung Galeopithecus: eine morphologische Untersuchung. Kongl. Svenska Vetenskaps.—Akad. Handl., 1.

SHUFELDT, R. W. 1911. The skeleton in the flying lemurs, Galeopteridae. Philippine Jour. Sci., 139.

Bats (Chiroptera)

ALLEN, G. M. 1939. Bats. Harvard Univ. Press. Cambridge.

ALLEN, J. A., H. LANG and J. P. CHAPIN. 1917. The Amer. Mus. Nat. Hist. Congo Exped. Collection of Bats. Parts I, II, III. Bull. Amer. Mus. Nat. Hist., 405.

ANTHONY, H. E. 1925. Mammals of Porto Rico, living and extinct: Chiroptera and Insectivora. N. Y. Acad. Sci., Sci. Surv. of Porto Rico. . . .

GALAMBOS, R. and D. R. GRIFFIN. 1942. Obstacle avoidance by flying bats: The cries of bats. Jour. Exp. Zool., 475.

HARTRIDGE, H. 1945. Acoustic control in the flight of bats. Nature, 490.

MATTHEW, W. D. 1917. A paleocene bat. Bull. Amer. Mus. Nat. Hist., 569.

MILLER, G. S. 1907. The families and genera of bats. U. S. Nat. Mus., Bull., 57.

REVILLIOD, P. 1917. Fledermäuse aus der Braunkohle von Messel bei Darmstadt. Abh. Hessischen Geolog. Landesanstalt, Darmstadt, 158.

SPRAGUE, J. M. 1943. The hyoid region of placental mammals with especial reference to the bats. Amer. Jour, Anat., 385.

TATE, G. H. H. 1941a. Results of the Archbold Expeditions. No. 35. A review of the genus Hipposideros with special reference to Indo-Australian species. Bull. Amer. Mus. Nat. Hist., 353.—1941b. No. 36. Remarks on some Old World leaf-nosed bats. Amer. Mus. Novitates, no. 1140.—1941c. No. 38. Molossid bats of the Archbold collections. Ibid., no. 1142.—1941d. No. 39. Review of Myotis of Eurasia. Bull. Amer. Mus. Nat. Hist., 537.—1941e. No. 40. Notes on vespertilionid bats. Ibid., 567.—1942a. No. 47. Review of the vespertilionine bats, with special attention to genera and species of the Archbold collections. Ibid., 221.—1942b. No. 48. Pteropodidae (Chiroptera) of the Archbold collections. Ibid., 331.—1946. Geographical distribution of the bats in the Australasian archipelago. Amer. Mus. Novitates, no. 1323.

TATE, G. H. H. and R. ARCHBOLD. 1939. Results of the Archbold Expeditions. No. 23. A revision of the genus Emballonura (Chiroptera).

VESEY-FITZGERALD, B. 1947. The senses of bats. Smiths. Rept. for 1947, 317.

Rodents (except Lagomorphs)

ALLEN, J. A. 1922. Sciuridae, Anomaluridae and Idiuridae collected by the American Museum Congo Expedition. Bull. Amer. Mus. Nat. Hist., 39.

BRANDT, J. F. 1855. Beiträge zur nähern Kenntniss der Säugethiere Russlands. Mém. Acad. Imp. Sci. St. Petersbourg, 1.

CABRINI, R. and J. ERAUSQUIN. 1941. La articulacion temporomaxilar de la rata. Revista Odontológica de Buenos Aires, 1.

COOK, H. J. and J. T. GREGORY. 1941. Mesogaulus praecursor, a new rodent from the Miocene of Nebraska. Jour. Paleont., 549.

FRECHKOP, S. 1932a. Notes sur les mammifères. IX. De la forme des dents molaires des rongeurs sciuromorphes. Bull. d. Mus. roy. d'Hist. nat. de Belgique, no. 12, 1.—1932b. X. Contribution à la classification écureuils africains. Ibid., no. 19, 1.—1932c. XII. De l'évolution de la forme des molaires chez les rongeurs hystricomorphes. Ibid., no. 34, 1.—1936. XIX. Le hamster montrant la différence fundamentale entre les molaires des rongeurs et celes des ongulés. Ibid., no. 18, 1.

FRIANT, M. 1936. Recherches sur la morphologie dentaire et les affinités des rongeurs fossiles et actuels du groupe des Coendinés (Steiromys-Erethizon-Coendu). Proc. Zool. Soc. London, 725.—1945. Le télencéphale des Dasyproctidae, rongeurs américains. Bull. Soc. Zool. France, 15.

GIDLEY, J. W. 1907. A new horned rodent from the Miocene of Kansas. Proc. U. S. Nat. Mus., 627.

HATT, R. T. 1929. The red squirrel: Its life history and habits, . . . Bull. N. Y. State Coll. of Forestry, vol. 2, 7.—1930. The biology of the voles of New York. Ibid., vol. 5, 513.—1932. The vertebral column of ricochetal rodents. Bull. Amer. Mus. Nat. Hist., vol. 63, 599.—1940. Lagomorpha and Rodentia other than Sciuridae, Anomaluridae and Idiuridae, collected by the Amer. Mus. Congo Exped. Ibid., vol. 76, 457.

HERSHKOVITS, J. 1944. A systematic review of the neotropical water rats of the genus Nectomys (Cricetinae). Univ. Michigan, Miscell. Publ. Mus. Zool., no. 58, 1.

HILL, J. E. 1935. The cranial foramina in rodents. Jour. Mammal., vol. 16, no. 2.

HINTON, M. A. C. 1926. Monograph of the voles and lemmings (Microtinae), living and extinct. Brit. Mus. (Nat. Hist.).

HOLLISTER, N. 1919. East African mammals in the United States Museum. Part II. Rodentia, Lagomorpha, and Tubulidentata. U. S. Nat. Mus., Bull. no. 99, 1.

HOWELL, A. B. 1926. Voles of the genus Phenacomys. I. Revision of the genus Phenacomys. II. Life history of the red tree mouse (Phenacomys longicaudus). U. S. Dept. Agric., Bur. Biol. Surv., N. Amer. Fauna, no. 48, 1.—1927. Revision of the North American lemming mice. Ibid., no. 50, 1.—1932. The saltatorial rodent Dipodomys: The functional and comparative anatomy of its muscular and osseous systems. Proc. Amer. Acad. Arts and Sci., 377.

KRAGLIEVICH, L. 1930. Diagnosis osteologico-dentaria de los generos vivientes de la subfamilia "Caviinae." An. d. Mus. nac. de Hist. nat., 59.

LÖNNBERG, E. 1924. On a new fossil porcupine from Honan with some remarks about the development of the Hystricidae. Geol. Surv. China, Palaeontologia Sinica, 1.

MCGREW, P. O. 1941. The Aplodontoidea. Geol. Ser. Field Mus. Nat. Hist., 1. [Mylagaulus, figs. 2, 9–12.]

MAJOR, C. I. F. 1899. On fossil dormice. Geol. Mag., 492.

MATTHEW, W. D. 1902. A horned rodent from the Colorado Miocene. With a revision of the Mylagauli, beavers and hares of the American Tertiary. Bull. Amer. Mus. Nat. Hist., 291.—1905. Notice of two new genera of mammals from the Oligocene of South Dakota. [Eutypomys and Heteromeryx.] Ibid., vol. 21, 21.—1910. On the osteology and relationships of Paramys and the affinities of the Ischyromyidae. Ibid., vol. 28, 43.

MATTHEW, W. D. and J. W. GIDLEY. 1904. New or little known mammals from the Miocene of South Dakota, American Museum Expedition of 1903. II. Carnivora and Rodentia, by W. D. Matthew. Bull. Amer. Mus. Nat. Hist., vol. 20, 246. [Includes note on evolution of rodent teeth.]

MATTHEW, W. D. and W. GRANGER. 1925. Fauna and correlation of the Gashato formation of Mongolia. Amer. Mus. Novitates, no. 189.

MILLER, G. S., JR. 1927. The rodents of the genus Plagiodontia. Proc. U. S. Nat. Mus., art. 16, 1.

MILLER, G. S., JR. and J. W. GIDLEY. 1918. Synopsis of the supergeneric groups of rodents. Jour. Washington Acad. Sci., 433.

RIGGS, E. S. 1899. The Mylagaulidae: An extinct family of sciuromorph rodents. Field Columbian Mus., Geol. Ser., no. 1, 181.

RUSCONI, C. 1934. Apuntes sobre la evolución ontogénetica de los molares del género agouti (Rodentia). Revista Odontológica, vol. 22.

SCHAUB, S. 1925. Die hamsterartigen Nagetiere des Tertiärs und ihre lebenden Verwandten: Eine systematisch-odontologische Studie. Abh. Schweiz. Palaeontol. Gesellsch., vol. 45 (1921–1925), 1.—1928. Bemerkungen über Schädelbau, Gebiss und systematische Stellung des Genus Lophiomys. Eclogae geol. Helvetiae, 267. [Rodentia, Myomorpha.]—1930a. Fossile Sicistinae. Ibid., no. 2, 616.—1930b. Quartäre und jungtertiäre Hamster. Abh. Schweiz. Palaeontol. Gesell., vol. 49.—1938. Tertiäre und quartäre Murinae. Ibid., vol. 61, 1.

SCHWARZ, E. and H. K. SCHWARZ. 1943. The wild and commensal stocks of the house mouse, Mus musculus Linnaeus. Jour. Mammal., 59.

SCOTT, W. B. 1905. Glires of the Santa Cruz beds. Repts. Princeton Univ. Exped. Patagonia, 384.

SETON, E. T. 1928. Lives of game animals. Vol. IV. Rodents, etc., New York.

SIMPSON, G. G. 1941. A giant rodent from the Oligocene of South Dakota. Amer. Mus. Novitates, no. 1149.

STIRTON, R. A. 1934. A new species of Amblycastor from the Platybelodon beds, Tung Gur formation, of Mongolia. Amer. Mus. Novitates, no. 694.—1935. A review of the Tertiary beavers. Univ. California Publ., Bull. Dept. Geol. Sci., vol. 23, 391.

STOCK, C. 1935. New genus of rodent from the Sespe Eocene. Bull. Geol. Soc. Amer., 61. [Phylogeny of aplodontids.]

SUMNER, F. B. 1932. Genetic, distributional, and evolutionary studies of the subspecies of deer mice (Peromyscus). Bibliographia Genetica IX, 1. The Hague.

TATE, G. H. H. 1935. The taxonomy of the genera of neotropical hystricoid rodents. Bull. Amer. Mus. Nat. Hist., 295.

TEILHARD DE CHARDIN, P. 1942. New rodents of the Pliocene and Lower Pleistocene of North China. Inst. Géo-Biologie, Pekin, no. 9.

TULLBERG, T. 1899. Ueber das System der Nagethiere; Eine phylogenetische Studie. Königl. Gesellsch. d. Wiss. zu Upsala.

WOOD, A. E. 1931. Phylogeny of the heteromyid rodents. Amer. Mus. Novitates, no. 501.—1932. New heteromyid rodents from the Miocene of Florida. Florida State Geol. Surv., Bull. 10, 45. —1935. Evolution and relationship of the heteromyid rodents; with new forms from the Tertiary of western North America. Ann. Carnegie Mus., 73.—1936a. The cricetid rodents described by Leidy and Cope from the Tertiary of North America. Amer. Mus. Novitates, no. 822.—1936b. Geomyid rodents from the Middle Tertiary. Ibid., no. 866.—1936c. Fossil heteromyid rodents in the collections of the University of California. Amer. Jour. Sci., 112.—1937a. Additional material from the Tertiary of the Cuyama Basin of California. Ibid., 29.—1937b. Parallel radiation among the geomyoid rodents. Jour. Mammal., 171.—1937c. Fossil rodents from the Siwalik beds of India. Amer. Jour. Sci., 64.—1937d. Rodentia. In: The mammalian fauna of the White River Oligocene, by W. B. Scott and G. L. Jepsen. Part II. Trans. Amer. Philos. Soc., 157.—1947. Rodents—A study in evolution. Evolution: Internat. Jour. Organic Evol., no. 3, 154.

WOOD, A. E. and R. W. WILSON. 1936. A suggested nomenclature for the cusps of the cheek teeth of rodents. Jour. Paleont., 388.

YOUNG, F. W. 1937. Studies of osteology and myology of the beaver (Càstor canadensis). Michigan State Coll., Agr. Exp. Sta. Mem. no. 2.

Lagomorphs (Hares and Rabbits)

GIDLEY, J. W. 1912. The lagomorphs an independent order. Science, 285.

GREEN, M. 1942. A study of the Oligocene Leporidae in the Kansas Univ. Mus. of Vert. Pal. Trans. Kansas Acad. Sci., 229.

HÜRZELER, J. 1936. Osteologie und odontologie der Caenotheriden. Abh. d. Schweiz. Palaeont. Gesell., vol. 58–59. [Notes many resemblances between caenotheres and lagomorphs but ascribes them to convergence.]

LYON, M. W., JR. 1904. Classification of the hares and their allies. Smiths. Miscell. Coll., no. 1456, 322.

MAJOR, C. J. F. 1899. On fossil and recent Lagomorpha. Trans. Linnean Soc. London, 433.

WOOD, A. E. 1940. Lagomorpha. In: The mammalian fauna of the White River Oligocene by W. B. Scott and G. L. Jepsen. Part III. Trans. Amer. Philos. Soc., 271.—1942. Notes on the Paleocene lagomorph, Eurymylus. Amer. Mus. Novitates, no. 1162.

Edentates (Xenarthra)

CABRERA, A. 1944. Los gliptodontoideos del Araucaniano de Catamarca. Revista del Museo de La Plata, Sec. Paleont., vol. 3.

COLBERT, E. H. 1942. An edentate from the Oligocene of Wyoming. Notulae Naturae, no. 109.

HOLMES, W. W. and G. G. SIMPSON. 1931. Pleistocene exploration and fossil edentates in Florida. Bull. Amer. Mus. Nat. Hist., 383.

KLAAUW, C. J. VAN DER. 1930. On the tympanic region of the skull in the Megatherium. Proc. Zool. Soc. London, Part 1, 127.—1931. On the tympanic region of the skull in the Mylodontinae. Ibid., Part 3, 607.

LULL, R. S. 1915. A Pleistocene ground sloth, Mylodon harlani, from Rock Creek, Texas. Amer. Jour. Sci., 327.—1929. A remarkable ground sloth. Mem. Peabody Mus. Yale Univ., vol. 3, part 2.—1930. The ground sloth, Nothrotherium. Amer. Jour. Sci., 344.

LYDEKKER, R. 1894. Vertebrados fosiles de la Argentina. Contributions to knowledge of the fossil vertebrates of Argentina. 2. The extinct edentates of Argentina. An. Mus. La Plata, Paleontologica Argentina. Folio.

MATTHEW, W. D. 1911. The ground sloth group. Amer. Mus. Jour., vol. 11, 113.—1912. Ancestry of the edentates. Ibid., vol. 12, 300.—1931. Genera and new species of ground sloths from the Pleistocene of Cuba. Amer. Mus. Novitates, no. 511.

MILLER, R. A. 1935. Functional adaptations in the forelimb of the sloths. Jour. Mammal., 38.

OSBORN, H. F. 1904. An armadillo from the Middle Eocene (Bridger) of North America. Bull. Amer. Mus. Nat. Hist., 163. [Metacheiromys.]

PARKER, W. K. 1874–1885. On the structure and development of the skull in the Mammalia. Part II. Edentata. Philos. Trans. Roy. Soc. London.

SCOTT, W. B. 1903–1904. Edentata of the Santa Cruz beds. Rept. Princeton Univ. Exped. Patagonia, vol. 5.

SICHER, H. 1944. Masticatory apparatus of the sloths. Field Mus. Nat. Hist., Zool. Ser., 161.

SIMPSON, G. G. 1931. Metacheiromys and the Edentata. Bull. Amer. Mus. Nat. Hist., 295.—1932. Enamel on the teeth of an Eocene edentate. Amer. Mus. Novitates, no. 567.—1941. A Miocene sloth from southern Chile. Ibid., no. 1156.

STOCK, C. 1914a. The systematic position of the mylodont sloths of Rancho La Brea. Science, n.s., vol. 39, 761.—1914b. Skull and dentition of the mylodont sloths of Rancho La Brea. Univ. California Publ., Bull. Dept. Geol., vol. 8, no. 18.—1917a. Recent studies on the skull and dentition of Nothrotherium from Rancho La Brea. Ibid., vol. 10, no. 10.—1917b. Further observations on the skull structure of mylodont sloths from Rancho La Brea. Ibid., no. 11.—1917c. Structure of the pes in Mylodon harlani. Ibid., no. 16.—1925. Cenozoic gravigrade edentates of western North America; with special reference to the Pleistocene Megalonychinae and Mylodontidae of Rancho La Brea. Carnegie Inst. Washington, Publ. 133.

STRAUS, W. L., JR. and G. B. WISLOCKI. 1932. On certain similarities between sloths and slow lemurs. Bull. Mus. Comp. Zool. Harvard Coll., 45.

WHITE, T. E. 1942. The Lower Miocene mammal fauna of Florida. Bull. Mus. Comp. Zool. Harvard Coll.

WINGE, H. 1915. Jordfundne og nuvlevende Gumlere (Edentata) fra Lagoa Santa, Minas Geraes, Brasilien. E. Museo Lundii, vol. 3, 3.

WISLOCKI, G. B. 1928. Observations on the gross and microscopic anatomy of the sloths . . . Jour. Morphol. and Physiol., 317.

WORTMAN, J. L. 1897. The Ganodonta and their relationship to the Edentata. Bull. Amer. Mus. Nat. Hist., 59.

Manids (Pholidota)

BROOM, R. 1939. The organ of Jacobson in the scaly anteater. Ann. Transvaal Mus., 323.

FRECHKOP, S. 1931. Notes sur les mammifères. VI. Quelques observations sur la classification des Pangolins (Manidae). Bull. Mus. roy. d'Hist. nat. de Belgique., vol. 7, no. 22.

FRIANT, M. 1944. Le cerveau des Pangolins arboricoles d'Afrique. Rev. Zool. Bot. Africa, vol. 38, nos. 1–2.

HATT, R. T. 1934a. Pangolins. Nat. Hist. Mag., vol. 34, no. 6, 725.—1934b. The pangolins and aardvarks collected by the Amer. Mus. Congo Exp. Bull. Amer. Mus. Nat. Hist., 643.

Aardvarks (Tubulidentata)

ADLOFF, P. 1934. Über die Zähne von Orycteropus. Zeitschr. f. Anat. u. Entwick., 710.

ANTHONY, M. R. 1934. La dentition de l'oryctérope. Ann. d. Sci. nat., Zool., 290.

BROOM, R. 1909a. The organ of Jacobson in Orycteropus. Proc. Zool. Soc. London, 680.—1909b. On the milk dentition of Orycteropus. Ann. South African Mus., vol. 5, 381.

CLARK, W. E. LE G. and C. F. SONNTAG. 1926. A monograph of Orycteropus afer. III. The skull, by W. E. Le G. Clark. The skeleton of the trunk and limbs: General summary, by C. F. Sonntag. Proc. Zool. Soc. London, 445.

COLBERT, E. H. 1933. The presence of tubulidentates in the Middle Siwalik Beds of northern India. Amer. Mus. Novitates, no. 604.—1941. A study of Orycteropus gaudryi from the Island of Samos. Bull. Amer. Mus. Nat. Hist., 305.

FRECHKOP, S. 1937. Notes sur les mammifères. XXI. Sur les extrémités de l'oryctérope. Bull. Mus. roy. d'Hist. nat. de Belgique, vol. 13, no. 19.

HELBING, H. 1933. Ein Orycteropus-Fund aus dem Unteren Pliocaen des Roussillon. Schweiz. Paläont. Gesellsch., 256.

JEPSEN, G. L. 1932. Tubulodon taylori, a Wind River Eocene tubulidentate from Wyoming. Proc. Amer. Philos. Soc., 255.

SONNTAG, C. F. 1925. A monograph of Orycteropus afer. II. Nervous system, sense-organs and hairs. Proc. Zool, Soc. London, 1185.

Carnivora: Creodonta

COLBERT, E. H. 1933. The skull of Dissopsalis carnifex Pilgrim, a Miocene creodont from India. Amer. Mus. Novitates, no. 603.

COPE, E. D. 1873. On the flat-clawed Carnivora of the Eocene of Wyoming. Proc. Amer. Philos. Soc., vol. 13, 198.—1880. On the genera of the Creodonta. Ibid., vol. 19, 76.—1881. On some Mammalia of the lowest Eocene beds of New Mexico. Ibid., 484. [Protogonia.]—1884. The Creodonta. Amer. Nat. 255, 344, 478.

DENISON, R. H. 1938. The broad-skulled Pseudocreodi. Ann. N. Y. Acad. Sci., 163.

GIDLEY, J. W. 1919. New species of claenodonts from the Fort Union (Basal Eocene) of Montana. Bull. Amer. Mus. Nat. Hist., 541.

GAZIN, C. L. 1946. Machaeroides eothen Matthew, the saber-tooth creodont of the Bridger Eocene. Proc. U. S. Nat. Mus., 335.

MATTHEW, W. D. 1901. Additional observations on the Creodonta. Bull. Amer. Mus. Nat. Hist., 1.—1905. Notes on the osteology of Sinopa, a primitive member of the Hyaenodontidae. Proc. Amer. Philos. Soc. 69.—1906. The osteology of Sinopa, a creodont mammal of the Middle Eocene. Proc. U. S. Nat. Mus., 203.—1937. Paleocene faunas of the San Juan Basin, New Mexico. Trans. Amer. Philos. Soc., vol. 30. [Skeleton of primitive creodonts.]

MATTHEW, W. D. and W. GRANGER. 1915. A revision of the Lower Eocene Wasatch and Wind River faunas. Part I. Order Ferae (Carnivora), Suborder Creodonta, by W. D. Matthew. Bull. Amer. Mus. Nat. Hist., 1.

OSBORN, H. F. 1900. Oxyaena and Patriofelis restudied as terrestrial creodonts. Bull. Amer. Mus. Nat. Hist., 269.—1909. New carnivorous mammals from the Fayûm Oligocene, Egypt. Ibid., 415.—1924. Andrewsarchus, giant mesonychid of Mongolia. Amer. Mus. Novitates, no. 146.

SCOTT, W. B. 1890. The Mammalia of the Uinta Formation. Part II. The Creodonta, Rodentia and Artiodactyla. Trans. Amer. Philos. Soc. (1889), 461.—1895. The osteology of Hyaenodon. Jour. Acad. Nat. Sci. Philadelphia, 499.

SIMPSON, G. G. 1936. Additions to the Puerco fauna, Lower Paleocene. Amer. Mus. Novitates, no. 849. [Protogonodon grangeri, n.sp., fig. 4.]

WORTMAN, J. L. 1894. Osteology of Patriofelis, a Middle Eocene creodont. Bull. Amer. Mus. Nat. Hist., 129.—1899. Restoration of Oxyaena lupina Cope with description of certain new species of Eocene creodonts. Ibid., 139.—1901–1902. Studies of Eocene Mammalia in the Marsh Collection, Peabody Museum. Part I. Carnivora. Amer. Jour. Sci., vol. 11, 333. [Vulpavus.] Ibid., vol. 12, 281. [Characters distinguishing Creodonta from Carnassidentia, Subfamily Mesonychinae.] Ibid., 377. [Subfamily Mesonychinae: Dromocyon], 421. [Dromocyon.] Ibid., vol. 13 (1902), 39. [Mesonyx, Dissacus; origin of tritubercular molar.] Ibid., 115. [Family Oxyaenidae.] Ibid., 197. [Limnocyon.] Ibid., 433. [Family Hyaenodontidae.] Ibid., vol. 14, 17. [Creodonta concluded; Summary.]

Fissipedia: General and Miscellaneous

COLBERT, E. H. 1939. Carnivora of the Tung Gur Formation of Mongolia. Bull. Amer. Mus. Nat. Hist., 47.

COPE, E. D. 1875. On the homologies of the sectorial tooth of Carnivora. Proc. Acad. Nat. Sci. Philadelphia, 20.—1882. On the systematic relations of the Carnivora Fissipedia. Proc. Amer. Philos Soc., 471.—1888. The mechanical origin of the sectorial teeth of the Carnivora. Proc. Amer. Assoc. Adv. Sci., 36th Meeting, 254.

DAHR, E. 1942. Über die Variation der Hirnschale bei wilden und zähmen Caniden. Ein Beitrag zur Genealogie der Haushunde. Arkiv. f. Zool. Stockholm, no. 16. [Carnivora.]

HELBING, H. 1928a. Carnivoren aus der miocänen Molasse der Schweiz. Schweiz. Paläont. Gesell., 232.—1928b. Carnivoren aus dem Miocän von Ravensburg und Georgengsmünd. Ibid., 377.

LYDEKKER, R. Undated. The carnivores: Order Carnivora. In: New Natural History, vol. 1, sect. 2. New York.

MATTHEW, W. D. 1909. The Carnivora and Insectivora of the Bridger Basin, Middle Eocene. Mem. Amer. Mus. Nat. Hist., 291.—1912. Carnivora and Rodentia. Bull. Geol. Soc. Amer., 181.

MERRIAM, J. C. 1906. Carnivora from the Tertiary Formations of the John Day Region. Univ. California Bull. Geol., vol. 5, no. 1.

POCOCK, R. I. 1929. Carnivora. Encycl. Brit., 14th ed., vol. 4.

POHLE, H. 1920. Zur Kenntnis der Raubtiere. II. Die Stellung der Gattungen Amphictis und Nandinia. Sitz. d. Gesell. natur. Freunde, Berlin, no. 1, 48.

SCHLAIKJER, E. M. 1935. Contributions to the stratigraphy and palaeontology of the Goshen Hole area, Wyoming. III. A new basal Oligocene formation. Bull. Mus. Comp. Zool. Harvard Coll., 71. [Carnivora, Perissodactyla, Artiodactyla.]

SCOTT, W. B. and G. L. JEPSEN. 1936. The mammalian fauna of the White River Oligocene. Part I. Insectivora and Carnivora. Trans. Amer. Philos. Soc., part 1, 1.

TEILHARD DE CHARDIN, P. 1915. Les carnassiers des Phosphorites du Quercy. Ann. d. Paléont. (1914–1915), 103.

WORTMAN, J. L. 1901. Studies of Eocene Mammalia in the Marsh Collection, Peabody Museum. Part I. Carnivora (continued). Amer. Jour. Sci., vol. 12, 143 [Family Viverravidae], Ibid., 193 [Viverravidae, continued].

WORTMAN, J. L. and W. D. MATTHEW. 1899. The ancestry of certain members of the Canidae, the Viverridae and Procyonidae. Bull. Amer. Mus. Nat. Hist., 109.

ZDANSKY, O. 1924. Jungtertiäre Carnivoren Chinas. Palaeont. Sinica, vol. 2, 1.

Aeluroids: Viverridae, Hyaenidae, Felidae

ALLEN, J. A. 1924. Carnivora collected by the American Museum Congo expedition. Bull. Amer. Mus. Nat. Hist., 73.

CARLSSON, A. 1902. Ueber die systematische Stellung von Eupleres goudoti. Zool. Jahrb., vol. 16, 217.—1910. Die genetischen Beziehungen der madagassischen Raubtiergattung Galidia. Ibid., 559.—1911. Ueber Cryptoprocta ferox. Ibid., vol. 30, 419.—1920. Ueber Arctictis binturong. Acta Zool., 337.

GERVAIS, P. 1874. Dentition et squelette de l'Euplère de Goudot. Jour. de Zool., 237.

GRAY, J. E. 1870. Description of an adult skull of Eupleres goudoti. Proc. Zool. Soc. London, 824.

GREGORY, W. K. and M. HELLMAN. 1939. On the evolution and major classification of the civets (Viverridae) and allied fossil and recent Carnivora: A phylogenetic study of the skull and dentition. Proc. Amer. Philos. Soc., 309.

HELBING, H. 1925. Das Genus Hyaenaelurus Biedermann. Schweiz. Paläont. Gesell., Eclogae geol. Helvetiae, 214.

MARSH, O. C. 1872. Preliminary description of new Tertiary mammals. Part I. Amer. Jour. Sci., 122. [Viverravus gracilis.]

MATTHEW, W. D. 1910. The phylogeny of the Felidae. Bull. Amer. Mus. Nat. Hist., 288.

MERRIAM, J. C. and C. STOCK. 1932. The Felidae of Rancho La Brea. Carnegie Inst. Washington Publ. no. 422.

MILNE EDWARDS, A. and A. GRANDIDIER. 1867. Observations anatomique sur quelques mammifères de Madagascar. I. De l'organisation du Cryptoprocta ferox. Ann. d. Sci. nat., 314.

MIVART, St. G. 1882. On the classification and distribution of the Aeluroidea. Proc. Zool. Soc. London, 135.

ORLOV, J. A. 1936. Tertiäre Raubtiere des Westlichen Sibiriens. I. Machairodontinae. Traveaux d. l'Inst. Paléozool. d. l'Acad. d. Sci. d. l'U.R.S.S., vol. V, 111.

OSGOOD, W. H. 1932. Mammals of the Kelley-Roosevelts and Delacour Asiatic expeditions. Field Mus. Nat. Hist., Zool. Ser., 193. [Hemigalus, rare viverrid.]

POCOCK, R. I. 1915a. On some of the external characters of Cynogale bennettii Gray. Ann. Mag. Nat. Hist., vol. 15, 351.—1915b. On the species of the mascarene viverrid Galidictis, with the description of a new genus and a note on "Galidia elegans." Ibid., vol. 16, 113.—1915c. On some of the external characters of the palm-civet (Hemigalus derbyanus Gray) and its allies. Ibid., 153.—1915d. On the feet and glands and other external characters of the paradoxurine genera Paradoxurus, Arctictis, Arctogalidia and Nandinia. Proc. Zool. Soc. London, 387.—1915e. On some of the external characters of the genus Linsang, with notes upon the genera Poiana and Eupleres. Ann. Mag. Nat. Hist., 341.—1915f. On some external characters of Galidia, Galidictis, and related genera. Ibid., 351.—1916a. The tympanic bulla in hyaenas. Proc. Zool. Soc. London, 303.—1916b. On the external characters of the mongooses (Mungotidae). Ibid., 349.—1916c. On some of the external characters of Cryptoprocta. Ann. Mag. Nat. Hist., vol. 17, 413.—1916d. The alisphenoid canal in civets and hyaenas. Proc. Zool. Soc. London, 442.—1932. The marbled cat (Pardofelis marmorata) and some other Oriental species, with the definition of a new genus of the Felidae. Ibid., 741.—1933a. The civet-cats of Asia. Jour. Bombay Nat. Hist. Soc., 423. —1933b. The rarer genera of Oriental Viverridae. Proc. Zool. Soc. London, 969.—1937. The mongooses of British India, including Ceylon and Burma. Jour. Bombay Nat. Hist. Soc., 211.—1939–1941. Fauna of British India, including Ceylon and Burma. Mammalia. Vol. I. Primates and carnivores. Families Felidae and Viverridae. London.

SCHAUB, S. 1925. Ueber die Osteologie von Machaerodus cultridens Cuvier. Schweiz. Paläont. Gesell., 255.—1934. Observations critiques sur quelques Machairodontides. Eclogae geol. Helvetiae, 399.

SHORTRIDGE, G. C. 1934. The mammals of southwest Africa. Vol. I. London. [Genetta albiventris, opp. p. 115.]

SIMPSON, G. G. 1937. The Fort Union of the Crazy Mountain Field, Montana, and its mammalian faunas. U. S. Nat. Mus., Bull. 169. [Primitive miacid carnivore with shear tooth on p^4.]—1941. The species of Hoplophoneus. Amer. Mus. Novitates, no. 1123.

TERRY, R. J. 1917. The primordial cranium of the cat. Jour. Morph., 281.

Arctoids: Miacids, Canids, Ursids, Procyonids, Mustelids

ALLEN, G. M. 1929. Mustelids from the Asiatic expeditions. Amer. Mus. Novitates, no. 358.

BARDENFLETH, K. S. 1913. On the systematic position of Aeluropus melanoleucus. Mindeskr. Japetus Steenstrup, art. 17. Kobenhavn.

BREUER, R. 1933. Über das Vorkommen sogenannter keilförmiger Defekte an den Zähnen von Ursus spelaeus und deren Bedeutung für die Paläobiologie. Palaeobiologica, 103.

CABRERA, A. 1932. Sinopsis de los Canidos Argentinos. La Revista del Centro de Ing. Agronomos . . . no. 145. 489.

CARLSSON, A. 1905. Ist Otocyon caffer die Ausgangsform des Hundegeschlechts oder nicht? Zool. Jahrb., 717.—1925. Über Ailurus fulgens. Acta Zool., vol. 6, 269.

CHAPMAN, F. M. 1935. José: Two months from the life of a Barro Colorado coati. Nat. Hist. Mag., vol. 35, no. 4, 299.—1936. José, 1936. Ibid., vol. 36, 126.—1937. José, 1937. Ibid., vol. 37, 524.

CLARK, J. 1937. The stratigraphy and paleontology of the Chadron formation in the Big Badlands of South Dakota. Ann. Carnegie Mus., 261. [Mustelavus, ancestral mustelid.]

COLBERT, E. H. 1939. Wild dogs and tame—past and present. Nat. Hist. Mag., Feb., 90.

COPE, E. D. 1883. The extinct dogs of North America. Amer. Nat., 235.

DEGERBOL, M. 1930. Über prähistorische, dänische Hunde. Vidensk. Medd. f. Dansk. naturh. Foren., 17.

FISHER, E. M. 1939a. The sea otter, past and present. Proc. Sixth Pacific Sci. Congr., Oceanogr. and Marine Biol., 223.—1939b. The sea otter in California. Ibid., vol. 4, Zool., 231.—1939c. Habits of the southern sea otter. Jour. Mammal., 21.—1941. Notes on the teeth of the sea otter. Ibid., 428.

FRICK, C. 1926. The Hemicyoninae and an American Tertiary bear. Bull. Amer. Mus. Nat. Hist., 1.

FURLONG, E. L. 1932. A new genus of otter from the Pliocene of the Northern Great Basin Province. Carnegie Inst. Washington Publ. no. 418, 94.

GAZIN, C. L. 1936. A new mustelid carnivore from the Neocene beds of northwestern Nebraska. Jour. Washington Acad. Sci., 199.

GREGORY, W. K. 1933. Nature's wild dog show. Bull. N. Y. Zool. Soc., 83.—1936. On the phylogenetic relationships of the giant panda (Ailuropoda) to other arctoid Carnivora. Amer. Mus. Novitates, no. 878.

HALL, E. R. 1936. Mustelid mammals from the Pleistocene of North America, with systematic notes on some recent members of the genera Mustela, Taxidea and Mephitis. Carnegie Inst. Washington Publ. no. 473, 41.—1939. Remarks on the primitive structure of "Mustela stolzmanni" with a list of the South American species and subspecies of the genus "Mustela." Physis (Revista de la Soc. Argentina de Cienc. Nat.), vol. 16, Sec. Zool. (Vertebrados), 159.

HELBING, H. 1930. Zwei Oligocaene Musteliden (Plesictis genettoides Pomel, Palaeogale angustifrons Pomel). Schweiz. Palaeontol. Gesell., vol. 50, 1.—1932. Über einen Indarctos-Schädel aus dem Pontien der Insel Samos. Nebst einem Anhang: Hyaenarctos spec. aus dem Pliocaen von Vialette (Haute-Loire). Ibid., vol. 52, 1.—

1935. Cyrnaonyx antiqua (Blainv.), ein Lutrine aus dem ëuropaeischen Pleistocaen. Eclogae geol. Helvetiae, 563.—1936. Die Carnivoren des Steinheimer Beckens. A. Mustelidae. Palaeontographica, Suppl.-Bd. 8, 1. [Trochotherium, fig. 9, pls. 2, 3, and figs. 18–33 of pl. 4.]

HÜRZELER, J. 1944. Zur revision der europäischen Hemicyoniden. Verhandl. d. Naturf. Gesell. in Basel, 131.

JONES, F. W. 1921. The status of the dingo. Trans. Roy. Soc. S. Australia, 254.—1939a. The "thumb" of the giant panda. Nature, 157.—1939b. Forearm and manus of the giant panda. Proc. Zool. Soc. London, 113.

KLATT, B. 1928. Vergl. Untersuchungen an Caniden und Procyniden. Zool. Jahrb., 217.

KRAGLIEVICH, L. 1928. Contribución al Conocimiento de los Grandes Cánidos extinguidos de Sud América. Anales de la Soc. Cient. Argentina, 25.

LANKESTER, E. R. 1901. On the affinities of Aeluropus melanoleucus, A. Milne-Edwards. Trans. Linnean Soc. London (1900–1903), 163.

LONNBERG, E. 1928. Contributions to the biology and morphology of the badger, Meles taxus, and some other Carnivora. Arkiv f. Zool., vol. 19A, no. 26.

MATTHEW, W. D. 1902a. A skull of Dinocyon from the Miocene of Texas. Bull. Amer. Mus. Nat. Hist., 129.—1902b. On the skull of Bunaelurus, a musteline from the White River Oligocene. Ibid., 137.—1902c. New Canidae from the Miocene of Colorado. Ibid., 281. [Cynodictis viverrinus, a civet-like ancestral dog.]—1909. The Carnivora and Insectivora of the Bridger Basin, Middle Eocene. Mem. Amer. Mus. Nat. Hist., no. 9, 291.—1930. The phylogeny of dogs. Evolutionary phylogenies. Jour. Mammal., 117.

MATTHEW, W. D. and W. GRANGER. 1923. New fossil mammals from the Pliocene of Sze-Chuan, China. Bull. Amer. Mus. Nat. Hist., 563.

MATTHEW, W. D. and R. A. STIRTON. 1930. Osteology and affinities of Borophagus. Univ. California Publ., Bull. Dept. Geol. Sci., 171.

MCGRAW, P. O. 1935. A new Cynodesmus from the Lower Pliocene of Nebraska with notes on the phylogeny of the dogs. Univ. California Publ., Bull. Dept. Geol. Sci., 305.—1937. The genus Cynarctus. Jour. Paleont., 444.—1938. Dental morphology of the Procyonidae, with a description of Cynarctoides, gen. nov. Geol. Ser. Field Mus. Nat. Hist., 323.—1939. A new Amphicyon from the deep river Miocene. Ibid., no. 23, 341.—1941. A new procyonid from the Miocene of Nebraska. Ibid., no. 5, 33. [Upper Miocene Marsland; Phlaocyon marslandensis, n. sp.]—1944. The Aelurodon saevus group. Ibid., no. 13, 79.

MERRIAM, J. C. 1903. The Pliocene and Quaternary Canidae of the Great Valley of California. Univ. Calif. Bull. Geol., 277.—1913. Notes on the canid genus Tephrocyon. Ibid., 359.

MERRIAM, J. C., C. STOCK and C. L. MOODY. 1916. An American Pliocene bear. Univ. Calif. Bull. Geol., vol. 10, 87.

MERRIAM, J. C. and C. STOCK. 1925. Relationships and structure of the short-faced bear, Arctotherium, from the Pleistocene of California. Carnegie Inst. Washington Publ. 347.

PEI, W. C. 1934. On the Carnivora from locality of Choukoutien. Palaeont. Sinica, vol. 8, 1. Mustelidae.

PETERSON, O. A. 1910. Description of new carnivores from the Miocene of western Nebraska. Mem. Carnegie Mus., 205.

PILGRIM, G. E. 1933a. The genera Trochictis, Enhydrictis, and Trocharion, with remarks on the taxonomy of the Mustelidae. Proc. Zool. Soc. London (1932), 845.—1933b. A fossil skunk from Samos. Amer. Mus. Novitates, no. 663.

POCOCK, R. I. 1921a. The auditory bulla and other cranial characters in the Mustelidae. Proc. Zool. Soc. London, 473.—1921b. On the external characters and classification of the Mustelidae. Ibid., 803.—1929a. Some external characters of the giant panda (Ailuropoda melanoleuca). Ibid., 975.—1929b. The structure of the auditory bulla in the Procyonidae and the Ursidae, with a note on the bulla of Hyaena. Ibid. (1928), 963.—1932a. The black and brown bears of Europe and Asia. Part I. Jour. Bombay Nat. Hist. Soc., 771.—1932b. Part II. Ibid., 101.—1935. The races of Canis lupus. Proc. Zool. Soc. London, part 3, 647.—1936. The foxes of British India. Jour. Bombay Nat. Hist. Soc., vol. 39, 36.—1939. The prehensile paw of the giant panda. Nature, 206.

RAVEN, H. C. 1936. Notes on the anatomy of the viscera of the giant panda (Ailuropoda melanoleuca). Amer. Mus. Novitates, no. 877.

REVILLIOD, P. 1926. Étude critique sur les genres de Canides Quarternaires sud-americains et description d'un crâne de Palaeocyon. Mém. Soc. Paleont. Suisse, 114.

REYNOLDS, S. H. 1909. A monograph of the British Pleistocene Mammalia. The Canidae. Palaeontographical Soc., vol. 2, part 3.—1912. A monograph of the British Pleistocene Mammalia. The Mustelidae. Ibid., part 4.

RODE, K. 1935. Untersuchungen über das Gebiss der Bären. Monogr. z. Geol. u. Palaeont., part 7.

ROMER, A. S. and A. H. SUTTON. 1927. A new arctoid carnivore from the Lower Miocene. Amer. Jour. Sci., 459.

SCHLOSSER, M. 1899. Parailurus anglicus und Ursus Bockhi aus den Ligniten von Baróth-Köpecz. Mitth. Jahrb. Königl. Ungarisch. Geol. Anstalt., part 2.

SCOTT, W. B. 1898. Notes on the Canidae of the White River Oligocene. Trans. Amer. Philos. Soc., 325.

SEGALL, W. 1943. The auditory region of the arctoid carnivores. Field Mus. Nat. Hist., Zool. ser., vol. 29, no. 3, 33.

SICHER, H. 1944. Masticatory apparatus in the giant panda and the bears. Field Mus. Nat. Hist., Zool. ser., vol. 29, no. 4, 61.

SIMPSON, G. G. 1946. Palaeogale and allied early mustelids. Amer. Mus. Novitates, no. 1320.

SOWERBY, A. DE C. 1932. The pandas or cat-bears. China Jour., 296.

VANDERHOOF, V. L. and J. T. GREGORY. 1940. A review of the genus Aelurodon. Univ. Calif. Publ. Bull. Dept. Geol. Sci., 143.

VIRET, J. 1942. Observations sur les canides du genre Pseudamphicyon. Ann. d. l'Univ. d. Lyon (1941–1942), 1.

WORTMAN, J. L. 1901a. Studies of Eocene Mammalia in the Marsh Collection, Peabody Museum. Part I. Carnivora. Amer. Jour. Sci., 333. [Introduction; Family Canidae.]—1901b. Part I. Carnivora (continued). Ibid., 437. [Canidae.]

WORTMAN, J. L. and W. D. MATTHEW. 1899. The ancestry of certain members of the Canidae, the Viverridae and Procyonidae. Bull. Amer. Mus. Nat. Hist., 109.

WOODWARD, A. S. 1915. On the skull of an extinct mammal related to Aeluropus from a cave in the Ruby Mines at Mogok, Burma. Proc. Zool. Soc. London, 425.

Pinnipedia

BERGERSEN, B. 1931. Beiträge zur Kenntnis der Haut einiger Pinnipedien unter Besonderer Berücksichtigung der Haut der Phoca groenlandica. Det Norske Videnskaps-Akademi i Oslo. I, 1.

BERRY, E. W. and W. K. GREGORY. 1906. Prorosmarus alleni, a new genus and species of walrus from the Upper Miocene of Yorktown, Virginia. Amer. Jour. Sci., 444.

DAL PIAZ, G. B. 1929. I Mamiferi fossili e viventi delle Tre Venezie. Parte Sistematica, no. 4. Pinnipedia.

DOUTT, J. K. 1942. A review of the genus Phoca. Ann. Carnegie Mus., 61.

HUBER, E. 1934. Anatomical notes on Pinnipedia and Cetacea. Carnegie Inst. Washington Publ. no. 447, 105.

KELLOGG, R. 1922. Pinnipeds from Miocene and Pleistocene deposits of California . . . and a résumé of current theories regarding origin of Pinnipedia. Univ. Calif. Publ., Bull. Geol. Sci., 23.—1927. Fossil pinnipeds from California. Carnegie Inst. Washington Publ. no. 346, 25.

OSBURN, R. C. 1906. Adaptive modifications of the limb skeleton in aquatic reptiles and mammals. Ann. N. Y. Acad. Sci., 447.

VAN BENEDEN, P. J. 1877–1886. Description des ossements fossiles des environs d'Anvers. Ann. Mus. roy. d'Hist. nat. de Belgique, vols. 1, 4, 7, 9, 13, parts 1–5, text and plates in folio. [Cetacea, Pinnipedia.]

Ungulata: General
(See also Mammalian Skeleton)

COLBERT, E. H. 1935. Siwalik mammals in the American Museum of Natural History. Trans. Amer. Philos. Soc., vol. 26, 1.

FRECHKOP, S. 1936. Notes sur les Mammifères. XX.—Remarques sur la classification des Ongulés et sur la position systématique des Damans. Bull. Mus. roy. d'Hist. nat. de Belgique, no. 37, 1.

GREGORY, W. K. 1912. Notes on the principles of quadrupedal locomotion and on the mechanism of the limbs in hoofed animals. Ann. N. Y. Acad. Sci., 267.

GREGORY, W. K. and E. H. COLBERT. 1939. On certain principles of evolution illustrated in the mammalian orders Perissodactyla and Artiodactyla. Acad. d. Sci. d. l'U.R.S.S., Inst. A. N. Sewertzoff d. Morfologie évolutive. Mem. Vol. to A. N. Sewertzoff, I, 97. Moscou, Leningrad.

HOCHSTETTER, F. 1942. Ueber die Harte Hirnhaut

und ihre Fortsätze bei den Säugetieren . . . Akad. d. Wissensch. i Wien, vol. 106, part 2.

OSBORN, H. F. 1889. The evolution of the ungulate foot. In: The Mammalia of the Uinta formation by William B. Scott and Henry Fairfield Osborn. Part IV. Trans. Amer. Philos. Soc., 531.

SIMPSON, G. G. 1945. The principles of classification and a classification of mammals. Bull. Amer. Mus. Nat. Hist., vol. 85. [Classification of ungulates, pp. 123–162; 233–272.]

Amblypoda (sensu lato):
Taligrada, Pantodonta, Dinocerata

COPE, E. D. 1873. On the short-footed Ungulata of the Eocene of Wyoming. Proc. Amer. Philos. Soc., 38.—1884–1885. The Amblypoda. Amer. Nat. (1884), 1110, 1192.—Vol. 19 (1885), 40. [Teeth of coryphodonts and uintatheres, 1884, p. 1117.]

GRANGER, W. and W. K. GREGORY. 1934. An apparently new family of amblypod mammals from Mongolia. Amer. Mus. Novitates, no. 720.

MARSH, O. C. 1886. Dinocerata: A monograph of an extinct order of gigantic mammals. U. S. Geol. Surv., Washington.

OSBORN, H. F. 1898a. A complete skeleton of Coryphodon radians. Notes upon the locomotion of this animal. Bull. Amer. Mus. Nat. Hist., 81.—1898b. Evolution of the Amblypoda. Part I. Taligrada and Pantodonta. Ibid., 169. [Pantolambda, fig. 2, p. 172.]—1913. The skull of Bathyopsis, Wind River uintathere. Ibid., vol. 32, 417.

OSBORN, H. F. and W. GRANGER. 1931. Coryphodonts of Mongolia. Amer. Mus. Novitates, no. 459.

PATTERSON, B. 1933. A new species of the amblypod Titanoids from western Colorado. Amer. Jour. Sci., 415.—1934. A contribution to the osteology of Titanoides and the relationships of the Amblypoda. Proc. Amer. Philos. Soc., vol. 73, no. 2.—1935. Second contribution to the osteology and affinities of the Paleocene amblypod Titanoides. Ibid., 143.—1937. A new genus, Barylambda, for Titanoides faberi, Paleocene amblypod. Geol. Ser. Field Mus. Nat. Hist., 229.—1939a. A skeleton of Coryphodon. Proc. New England Zool. Club, 97.—1939b. New Pantodonta and Dinocerata from the Upper Paleocene of western Colorado. Geol. Ser. Field Mus. Nat. Hist., 351. [Skeleton of Barylambda; skull, teeth and incomplete skeleton of Haplolambda.]

SIMPSON, G. G. 1929. A new Paleocene unitathere and molar evolution in the Amblypoda. Amer. Mus. Novitates, no. 387.—1932. A new Paleocene mammal from a deep well in Louisiana. Proc. U. S. Nat. Mus., art. 2. [Anisonchus fortunatus.]

WOOD, H. E. II. 1923. The problem of the unitatharium molars. Bull. Amer. Mus. Nat. Hist., 599.

Condylarthra

COPE, E. D. 1874. Report on the vertebrate palaeontology of Colorado. In: Ann. Rept. U. S. Geol. and Geogr. Surv. of the Territories . . . (1873) by F. V. Hayden. Washington. [Phenacodus, p. 441.]—1882. On the brains of the Eocene Mammalia Phenacodus and Periptychus. Paleontological Bull., no. 36, 563. Washington.

JEPSEN, G. L. 1930. Stratigraphy and paleontology of the Paleocene of Northeastern Park County, Wyoming. Proc. Amer. Philos. Soc., 463. [Phenacodaptes, pl. 9 (figs. 1–5), pp. 517–520.]

MATTHEW, W. D. 1897. A revision of the Puerco Fauna. Bull. Amer. Mus. Nat. Hist., 259. [Sections on Euprotogonia and the Periptychidae.—1937. Paleocene faunas of the San Juan basin, New Mexico. Trans. Amer. Philos. Soc. [Condylarthra.]

MATTHEW, W. D. and W. GRANGER. 1915a. A revision of the Lower Eocene Wasatch and Wind River faunas. Part II. Order Condylarthra, Family Hyopsodontidae, by W. D. Matthew. Bull. Amer. Mus. Nat. Hist., vol. 34, 311.—1915b. A revision of the Lower Eocene Wasatch and Wind River faunas. Part III. Order Condylarthra, Families Phenacodontidae and Meniscotheriidae, by Walter Granger. Ibid., 329.

OSBORN, H. F. 1898. Remounted skeleton of Phenacodus primaevus . . . Bull. Amer. Mus. Nat. Hist., 159.

SIMPSON, G. G. 1933. Braincasts of Phenacodus, Notostylops and Rhyphodon. Amer. Mus. Novitates, no. 622.

TEILHARD DE CHARDIN, P. 1921. Note sur la presence dans le Tertiare inférieur de Belgique d'un condylarth appartenant au groupe des Hyopsodus. Bull. Acad. roy. de Belgique, no. 6, 357.

Hyracoidea

ANDREWS, C. W. 1906. A descriptive catalogue of the Tertiary Vertebrata of the Fayûm, Egypt. Brit. Mus. (Nat. Hist.). London. [Fossil hyracoids.]

FISCHER, E. 1903. Bau und Entwickelung des Carpus und Tarsus vom Hyrax. Jenaisch. Zeitsch. f. Naturw., 691.

MATSUMOTO, H. 1926. Contribution to the knowledge of the fossil Hyracoidea of the Fayûm, Egypt . . . Bull. Amer. Mus. Nat. Hist., 253.

OSBORN, H. F. 1898. On Pliohyrax kruppii Osborn, a fossil hyracoid from Samos, Lower Pliocene, in the Stuttgart Collection . . . Proc. Int. Cong. Zool., Sect. B, Vertebrata, 173. Cambridge.—1906. Milk dentition of the hyracoid Saghatherium from the Upper Eocene of Egypt. Bull. Amer. Mus. Nat. Hist., 263.

URSING, B. 1934. Untersuchungen über Entwicklung und Bau des Hand-und-fuss-Skeletts bei Mammalia. II. Procavia daemon. Lunds Univ. Arsskrift, Bd. 30, nr. 12. Kungl. Fysiografiska Sällskapets. Handl., Bd. 45, nr. 12.

WELLS, L. H. 1939. The endocranial cast in recent and fossil hyraces (Procaviidae). So. African Jour. Sci., 365.

WISLOCKI, G. B. 1930. On an unusual placental form in the Hyracoidea: Its bearing on the theory of the phylogeny of the placenta. Contrib. Embryology, Carnegie Inst. Washington, 83.

WISLOCKI, G. B. and O. P. VAN DER WESTHUYSEN. 1940. The placentation of Procavia capensis, with a discussion of the placental affinities of the Hyracoidea. Contrib. Embryology, Carnegie Inst. Washington, 65.

Embrithopoda: Arsinoitheres

ANDREWS, C. W. 1906. A descriptive catalogue of the Tertiary Vertebrata of the Fayûm, Egypt. Brit. Mus. (Nat. Hist.). London.

Proboscidea

ADAMS, A. L. 1877–1881. Monograph of British fossil elephants. Palaeont. Soc. London.

ANDREWS, C. W. 1903. On the evolution of the Proboscidea. Philos. Trans. Roy. Soc. London, 99.—1906. A descriptive catalogue of the Tertiary Vertebrata of the Fayûm, Egypt. Brit. Mus. (Nat. Hist.). London.—1908. On the skull, mandible, and milk dentition of Palaeomastodon, with some remarks on the tooth change in the Proboscidea in general. Philos. Trans. Roy. Soc. London, 393.

BOAS, J. E. V. and S. PAULLI. 1908. The elephant's head. Studies in the comparative anatomy of the organs of the head of the Indian elephant. Part 1. The facial muscles and the proboscis. Folio. Jena.

BORISSIAK, A. 1936. Mastodon atavus n.sp., der primitivste vertreter der gruppe M. angustidens. Travaux de l'Inst. Paleozool., 171.

COLBERT, E. H. 1942. The geologic succession of the Proboscidea. In: Proboscidea: A monograph . . . by Henry Fairfield Osborn, vol. II, 1421. American Museum of Natural History. 4to.

COOPER, C. F. 1924. On remains of extinct Proboscidea in the Museums of Geology and Zoology in the University of Cambridge. I. Elephas antiquus. Proc. Cambridge Philos. Soc., 108.

EALES, N. B. 1926. The anatomy of the head of a foetal African elephant, Elephas africanus (Loxodonta africana). Trans. Roy. Soc. Edinburgh, 491.

FALCONER, H. and P. T. CAUTLEY. 1846–1849. Fauna antiqua Sivalensis . . . I. Pachydermata: Elephas, Mastodon. London.

FRICK, C. 1926. Tooth sequence in certain trilophodont-tetrabelodont mastodons. Bull. Amer. Mus. Nat. Hist., 122.—1933. New remains of trilophodont-tetrabelodont mastodons. Ibid., 505.

GREGORY, J. T. 1945. An Amebelodon jaw from the Texas Panhandle. Univ. Texas Publ. 4401, 477.

GREGORY, W. K. 1903. Adaptive significance of the shortening of the elephant's skull. Bull. Amer. Mus. Nat. Hist., 387.—1934. Significance of the supra-symphysial depression and groove in the shovel-tusked mastodonts. Jour. Mammal., 4.

HOPWOOD, A. T. 1935. Fossil Proboscidea from China. Palaeont. Sinica, vol. 9, part 3.

LULL, R. S. 1909. The evolution of the elephant. Amer. Jour. Sci., 169.

MACINNES, D. G. 1942. Miocene and Post-Miocene Proboscidea from East Africa. Trans. Zool. Soc. London, 33.

MATSUMOTO, H. 1922. Revision of Palaeomastodon and Moeritherium, Palaeomastodon intermedius and Phiomia osborni, new species. Amer. Mus. Novitates, no. 51.—1923. A contribution to the knowledge of Moeritherium. Bull. Amer. Mus. Nat. Hist., 97.—1929a. On Loxodonta (Palaeoloxodon) namadica (Falconer and Cautley) in Japan. Sci. Rept. Tôhoku Imp. Univ., 2d ser. (Geology), vol. 13, no. 1.—1929b. On Loxodonta (Palaeoloxodon) tokunagai Matsumoto, with remarks on the descent of loxodontine elephants. Ibid., 7.—1929c. On Parastegodon Matsumoto and its bearing on the descent of earlier elephants. Ibid., 13.

OSBORN, H. F. 1900. The angulation of the limbs of Proboscidea, Dinocerata, and other quadrupeds in adaptation to weight. Amer. Nat., 89.—1921a. Evolution, phylogeny and classification of the Mastodontoidea. Bull. Geol. Soc. Amer., 327.—1921b. Adaptive radiation and classification of the Proboscidea. Proc. Nat. Acad. Sci., 231.—1925a. The elephants and mastodonts arrive in America. Nat. Hist. Mag., vol. 25, no. 1. —1925b. Final conclusions on the evolution, phylogeny, and the classification of the Proboscidea. Proc. Amer. Philos. Soc., 17.—1926. Mastodons and mammoths of North America. Amer. Mus. Nat. Hist., Guide Leaflet no. 62; Nat. Hist. Mag., vol. 23 (1923), 3, and vol. 25 (1925), 3.—1931. Palaeoloxodon antiquus italicus sp. nov., final stage in the "Elephas antiquus" phylum. Amer. Mus. Novitates, no. 460. —1935. The ancestral tree of the Proboscidea. . . . Proc. Nat. Acad. Sci., 404.—1936. Proboscidea: A monograph of the discovery, evolution, migration and extinction of the mastodonts and elephants of the world. Vol. I. Moeritherioidea, Deinotherioidea, Mastodontoidea. Edited by M. R. Percy. Monogr. Amer. Mus. Nat. Hist.—1942. Vol. II. Stegodontoidea, Elephantoidea. Ibid.

OSBORN, H. F. and W. GRANGER. 1932. Platybelodon grangeri, three growth stages, and a new serridentine from Mongolia. Amer. Mus. Novitates, no. 537.

SCHLESINGER, G. 1912. Studien über die Stammesgeschichte der Proboscidier. Jahrb. d. K. K. Geol. Reichsanstalt, part 1, 87.—1917. Die Mastodonten des K. K. Naturhistorischen Hofmuseums. Denks. d. K. K. Naturh. Hofmus., 1. 4to.

TAKAI, F. 1936. On a new fossil elephant from Okubo-mura . . . Japan. Proc. Imp. Acad. Tokyo, 19. [Parastegodon may be transitional between Stegodontinae and Elephantinae.]

TEILHARD DE CHARDIN, P. and M. TRASSAERT. 1937. The proboscidians of southeastern Shansi. Palaeontologia Sinica, part 1, 1.

TOKUNAGA, S. and F. TAKAI. 1938. Fossil elephants from Totigi Prefecture, Japan. Japanese Jour. Geol. and Geogr., nos. 1, 2.

WATSON, D. M. S. 1944. History of elephants. A review of "Proboscidea . . ." by Henry Fairfield Osborn. Nature, 5.

Sirenia

ABEL, O. 1912. Die eocänen Sirenen der Mittelmeerregion. I. Der Schädel von Eotherium aegyptiacum. Palaeontographica, 289.—1922. Desmostylus: Ein mariner Multituberculate aus dem Miocän der nordpazifischen Küstenregion. Akad. d. Wiss. in Wien, no. 14. [Wrongly refers Desmostylus to Multituberculata.]

DEPÉRET, C. and F. ROMAN. 1920. Le Felsinotherium serresi des sables Pliocènes de Montpellier et les Rameaux phylétiques des Siréniens fossiles de l'ancien monde. Arch. d. Mus. d'Hist. nat. de Lyon, 1.

EDINGER, T. 1933. Über Gehirne tertiärer Sirenia Ägyptens und Mitteleuropas sowie der rezenten Seekühe. Ergebn. der Forschungsr. Prof. E. Stromers in den Wüsten Ägyptens. V. Tertiäre Wirbeltiere. Abh. d. Bayerischen Akad. d. Wissensch. part 20.—1939. Two notes on the central nervous system of fossil Sirenia. Bull. Faculty Sci., no. 19, The Fouad I Univ., 43.

HATT, R. T. 1934. The American Museum Congo Expedition manatee and other recent manatees. Bull. Amer. Mus. Nat. Hist., 533.

HAY, O. P. 1915. A contribution to the knowledge of the extinct sirenian Desmostylus hesperus Marsh. Proc. U. S. Nat. Mus., vol. 49, 381.—1923. Characteristics of sundry fossil vertebrates. Pan-American Geologist, 101. [Proposes to divide Sirenia into two suborders: Desmostyliformes and Trichechiformes.]—1924. Notes on the osteology and dentition of the genera Desmostylus and Cornwallius. Proc. U. S. Nat. Mus., 1.

LEPSIUS, G. R. 1882. Halitherium schinzi die fossile Sirene des Mainzer Beckens. Abh. d. Mittelrhein. geol. Vereins, 1. 4to.

MATTHES, E. 1927. Zur Entwicklung des Kopfskelettes der Sirenen. II. Das Primordial-cranium von Halicore dugong. Zeits. f. Anat. Entwick., parts 1, 2.

MATTHEW, W. D. 1916. New sirenian from the Tertiary of Porto Rico, West Indies. Ann. N. Y. Acad. Sci., 23.

MERRIAM, J. C. 1906. On the occurrence of Desmostylus, Marsh. Science, no. 605, 151.—1911. Notes on the genus Desmostylus of Marsh. Univ. Calif. Bull. Geol., no. 18, 403.

MURIE, J. 1874. On the form and structure of the manatee (Manatus americanus). Trans. Zool. Soc. London, part 3 (1872), 127.—1885. Further observations on the manatee. Ibid., part 2 (1880), 19.

SICKENBERG, O. 1934. Beiträge zur Kenntnis Tertiärer Sirenen. I. Die Eozänen Sirenen des Mittelmeergebietes. II. Die Sirenen des Belgischen Tertiärs. Mém. d. Mus. roy. d'Hist. nat. d. Belgique, no. 63.—1938. Ist Desmostylus eine Sirene? Palaeobiologica, no. 2 (Abel Festschrift), 340.

SIMPSON, G. G. 1932. Fossil Sirenia of Florida and the evolution of the Sirenia. Bull. Amer. Mus. Nat. Hist., 419.

STROMER, E. 1921. Untersuchung der Huftbeine und Huftgelenke von Sirenia und Archaeoceti. Bayer. Akad. d. Wiss., 41.

TAKAI, F. 1939. The mammalian faunas of the Hiramakian and Togarian stages in the Japanese Miocene. Tôhoku Imp. Univ. Jubilee Publ., 189. [Distribution of Desmostylus in Japanese Miocene.]

TOKUNAGA, S. 1936. Desmostylus found near the town of Yumoto, Fukushima Prefecture. Jour. Geogr., Tokyo, no. 572, 473. [In Japanese, with English summary.]

VANDERHOOF, V. L. 1937. A study of the Miocene sirenian Desmostylus. Univ. Calif. Publ. Bull. Dept. Geol. Sci., no. 8, 169.—1942. An occurrence of the Tertiary marine mammal Cornwallius in Lower California. Amer. Jour. Sci., 298.

WISLOCKI, G. B. 1935. The placentation of the manatee (Trichechus latirostris). Mem. Mus. Comp. Zool. Harvard Coll., no. 3, 159.

WOODWARD, H. 1885. On an almost perfect skeleton of Rhytina gigas (Rhytina stelleri, "Steller's sea-cow") . . . from the Pleistocene peat-deposits on Behring's Island. Quart. Jour. Geol. Soc., 457.

YOSHIWARA, S. and J. IWASAKI. 1902. Notes on a new fossil mammal. Jour. Coll. Sci., Imp. Univ. Tokio, Japan, art. 6, 1. [Desmostylus.]

Early South American Ungulates (Pre-Pliocene)
Litopterna

(See also Mammals: Tertiary to Recent of South America, especially Ameghino, Lydekker, Roth, Scott, Simpson, et al.)

BURMEISTER, H. 1864. Beschreibung der Macrauchenia patachonica Owen (Opisthorhinus . . .) . . . Abh. d. Naturf. Gesell. zu Halle, vol. 9, 1.—[1864?] Descripcion de la Macrauchenia patachonica. Pp. 52–66. [No publisher's imprint.]—1885. Neue Beobachtungen an Macrauchenia patachonica. Nova Acta . . . vol. 47, no. 5, 239.

SCOTT, W. B. 1910. Mammalia of the Santa Cruz Beds. Part I. Litopterna. Repts. Princeton Univ. Exp. to Patagonia. Paleontology, vol. 7.

SIMPSON, G. G. 1935. Descriptions of the oldest known South American mammals, from the Rio Chico formation. Amer. Mus. Novitates, no. 793. [Ernestokokenia, figs. 5–7.]

Notoungulata:
Notioprogonia, Toxodonta, Typotheria, Hegetotheria

CABRERA, A. 1937. Notas sobre el suborden "Typotheria." I–IV. Notas del Museo de La Plata, II: Paleontologie, no. 8, 17.

LYDEKKER, R. 1894. Vertebrados fosiles de la Argentina [Contributions to a knowledge of the fossil vertebrates of Argentina]. 3. A study of extinct Argentine ungulates. An. Mus. La Plata, Paleontologica Argentina, vol. 2 (1893), part 3. Folio. [Typotherium, pls. 2, 3.]

MATTHEW, W. D. 1915. Entelonychia, Primates, Insectivora (part). Bull. Amer. Mus. Nat. Hist., vol. 34, 429. [Arctostylops, pp. 429–433 and pl. 15.]

MATTHEW, W. D. and W. GRANGER. 1925. Fauna and correlation of the Gashato formation of Mongolia. Amer. Mus. Novitates, no. 189. [. . . "Palaeostylops confirms the view that the South American Tertiary hoofed mammals were originally derived from the north, although undergoing a great secondary evolution in the Neotropical region."—W.D.M.]

OWEN, R. 1840. The zoology of the voyage of H.M.S. Beagle . . . Part I. Fossil Mammalia. London. [Toxodon and Macrauchenia.]

PATTERSON, B. 1932. The auditory region of the

Toxodontia. Field Mus. Nat. Hist., Geol. Ser. 6, no. 1.—1934a. The auditory region of an Upper Pliocene typotherid. Ibid., no. 5, 83.—1934b. Upper premolar-molar structure in the Notoungulata, with notes on taxonomy. Ibid., no. 6, 91.—1934c. Cranial characters of Homalodotherium. Ibid., No. 7, 113.—1936. The internal structure of the ear in some notoungulates. Ibid., no. 15, 199.—1937. Some notoungulate braincasts. Ibid., no. 19, 273.

RIGGS, E. S. 1935. Description of some notoungulates from the Casamayor (Notostylops) Beds of Patagonia. Proc. Amer. Philos. Soc., 163.

RUSCONI, C. 1936. La supuesta afinidad de Argyrolagus con los Typotheria. Bol. Acad. Nac. d. Ciencias en Cordoba, 173.

SCOTT, W. B. 1912. Mammalia of the Santa Cruz Beds. Part II. Toxodonta. Repts., Princeton Univ. Exp. Patagonia, vol. 6, 111.—1930. A partial skeleton of Homalodontotherium from the Santa Cruz Beds of Patagonia. Field Mus. Nat. Hist., Geol. Mem.—1932. Nature and origin of the Santa Cruz fauna. With additional notes on the Entelonychia and Astrapotheria. Repts. Princeton Univ. Exp. Patagonia: Palaeontology, vol. 7, part 3, 193.

SERRES, M. DE. 1867. De l'ostéographie du Mesotherium et de ses affinités zoologique. Compte rendus Ac. Sci., vol. 65; 6, 140, 273, 429, 593, 740, 841. [Typotherium.]

SIMPSON, G. G. 1932a. New or little-known ungulates from the Pyrotherium and Colpodon Beds of Patagonia. Amer. Mus. Novitates, no. 576.—1932b. Cochilius volvens from the Colpodon Beds of Patagonia. Ibid., no. 577.—1932c. Skulls and brains of some mammals from the Notostylops Beds of Patagonia. Ibid., no. 578.—1933. Braincasts of two typotheres and a litoptern. Ibid., no. 629. [Hegetotherium, fig. 1.]—1934. Provisional classification of extinct South American hoofed mammals. Ibid., no. 750. [Notioprogonia, pp. 9, 15, 16.]—1935. An animal from a lost world. Nat. Hist. Mag., vol. 35, no. 4, 316. [Restoration of skeleton of Thomashuxleya, a primitive notoungulate.]—1936a. Structure of a primitive notoungulate cranium. Amer. Mus. Novitates, no. 824.—1936b. Skeletal remains and restoration of Eocene Entelonychia from Patagonia. Ibid., no. 826. [Thomashuxleya, figs. 1, 2.]—1940. The names Mesotherium and Typotherium. Amer. Jour. Sci., 518. [Valid name of this genus is Mesotherium.]—1945. A Deseado hegetothere from Patagonia. Ibid., 550.

SINCLAIR, W. J. 1908. The Santa Cruz Typotheria. Proc. Amer. Philos. Soc., 64.—1909. Mammalia of the Santa Cruz Beds. Part I. Typotheria. Repts. Princeton Univ. Exp. Patagonia, vol. 6.

Astrapotheria

RIGGS, E. S. 1935. A skeleton of Astrapotherium. Geol. Ser., Field Mus. Nat. Hist., vol. 6, no. 13, 167.

SCOTT, W. B. 1909. Mammalia of the Santa Cruz Beds. Part IV. Astrapotheria. Repts. Princeton Univ. Exp. Patagonia, vol. 6: Palaeontology, 301.—1937. The Astrapotheria. Proc. Amer. Philos. Soc., 309.

SIMPSON, G. G. 1933. Structure and affinities of Trigonostylops. Amer. Mus. Novitates, no. 608.

Pyrotheria

GAUDRY, A. 1906. Fossiles de Patagonie. Étude sur une portion du monde antarctique. Ann. Paléont., vol. 1, 101.—1909. Fossiles de Patagonie. No. 4. Le Pyrotherium. Ibid., vol. 6, 1.

LOOMIS, F. B. 1914. The Deseado formation of Patagonia. Amherst College, 1. [Fine skull of Pyrotherium.]

Perissodactyla: General

COPE, E. D. 1881. The systematic arrangement of the order Perissodactyla. Proc. Amer. Philos. Soc., 377.

GIDLEY, J. W. 1912. Perissodactyla. Bull. Geol. Soc. Amer., 179.

OSBORN, H. F. 1889. The Perissodactyla. In: The Mammalia of the Uinta formation, by W. B. Scott and H. F. Osborn. Part III. Trans. Amer. Philos. Soc., vol. 16 (1889), 505.

PILGRIM, G. E. 1925. The Perissodactyla of the Eocene of Burma. Mem. Geol. Surv. India, Palaeontologia Indica, vol. 8, no. 3. 4to.

SCOTT, W. B. 1941. Perissodactyla. In: The mammalian fauna of the White River Oligocene, by W. B. Scott and G. L. Jepsen. Trans. Amer. Philos. Soc., vol. 28, part 5, 747.

WOOD, H. E., II. 1937. Perissodactyl suborders. Jour. Mammal., 106.

1926. The horse in the tiger's skin. Bull. N. Y. Zool. Soc., 111.

Eohippus to Equus

BURMEISTER, H. 1875. Die fossilen Pferde der Pampasformation, Buenos Aires. Folio. [Description of Hippidium and other extinct horses of South America.]

CABRERA, A. 1935. Estado actual de los Conocimientos sobre el origen del género "Equus." Revista de la Facultad de Agronomia y Veterinaria, I, vol. 8, 16.

CAMP, C. L. and N. SMITH. 1942. Phylogeny and function of the digital ligaments of the horse. Mem. Univ. Calif., no. 2, 69.

CHUBB, S. H. 1912. Notes on the trapezium in the Equidae. Bull. Amer. Mus. Nat. Hist., 113.—1934. Frontal protuberances in horses. An explanation of the so-called "horned horse." Amer. Mus. Novitates, no. 740.

COLBERT, E. H. 1935. Distributional and phylogenetic studies on Indian fossil mammals. II. The correlation of the Siwaliks of India as inferred by the migrations of Hipparion and Equus. Amer. Mus. Novitates, no. 797.

COOPER, C. F. 1932. The genus Hyracotherium. A revision and description of new specimens found in England. Philos. Trans. Roy. Soc. London, 431.

EDINGER, T. 1948. Evolution of the horse brain. Geol. Soc. Amer., Mem. 25. [Eohippus brain extremely primitive, almost reptilian.]

ELLENBERGER, W., H. BAUM and H. DITTRICH. 1898. Handbuch der Anatomie der Tiere für Künstler. Leipzig.

EWART, J. C. 1905. The multiple origin of horses and ponies. Smiths. Rept. (1905), 437. Washington.

GIDLEY, J. W. 1901. Tooth characters and revisions of the North American species of the genus Equus. Bull. Amer. Mus., Nat. Hist., 91.—1903. A new three-toed horse. Ibid., 465.—1907. Revision of the Miocene and Pliocene Equidae of North America. Ibid., 865. [Revision of species of later three-toed horses, with observations on the phylogeny.]

GRANGER, W. 1908. A revision of the American Eocene horses. Bull. Amer. Mus. Nat. Hist., 221.

GREGORY, W. K. 1920. Studies in comparative myology and osteology. No. V. On the anatomy of the preorbital fossae of Equidae and other ungulates. Bull. Amer. Mus. Nat. Hist., 265.—

KOVALEVSKY, W. 1873. Sur l'Anchitherium aurelianense Cuv. et sur l'Histoire paléontologique des chevaux. Mém. l'Acad. Imp. d. Sci. de St. Petersbourg, vol. 20, no. 5.

LEIDY, J. 1869. Extinct mammalian fauna of Dakota and Nebraska. Jour. Acad. Nat. Sci. Philadelphia, vol. 7. [Species of three-toed horses described.]

LOOMIS, F. B. 1926. The evolution of the horse. Boston.

MATTHEW, W. D. 1926. The evolution of the horse: A record and its interpretation. Quart. Rev. Biol., no. 2, 139.

MATTHEW, W. D. and S. H. CHUBB. 1932. Evolution of the horse. Part I. Evolution of the horse in nature, by W. D. Matthew. Part II. The horse under domestication: Its origin and the structure and growth of the teeth. 6th ed., revised. Amer. Mus. Nat. Hist., Guide Leaflet Ser., no. 36.

MATTHEW, W. D. and R. A. STIRTON. 1930. Equidae from the Pliocene of Texas. Univ. Calif. Publ., Bull. Dept. Geol. Sci., 349.

MERRIAM, J. C. 1915a. New species of the Hipparion group from the Pacific coast and Great Basin provinces of North America. Univ. Calif. Publ., Bull. Dept. Geol., vol. 9, no. 1—1915b. New horses from the Miocene and Pliocene of California. Ibid., no. 4, 49.—1919. Tertiary mammalian faunas of the Mohave Desert. Ibid., vol. 11, no. 5, 437. [Equidae, 548–568.]

OSBORN, H. F. 1905. Origin and history of the horse. Address before New York Farmers, Metropolitan Club, Dec. 19, 1905.—1907. Points of the skeleton of the Arab horse. Bull. Amer. Mus. Nat. Hist., 259.—1912. Craniometry of the Equidae. Mem. Amer. Mus. Nat. Hist., vol. 1, part 3, 57.—1918. Equidae of the Oligocene, Miocene, and Pliocene of North America. Iconographic type revision. Mem. Amer. Mus. Nat. Hist., N.S., vol. 2, part 1.

OWEN, R. 1841. On Hyracotherium leporinum. Trans. Geol. Soc., 203. [Earliest record of the four-toed horse, not then recognized as ancestral to the horse family.]—1842. Description of the fossil remains of a mammal (Hyracotherium leporinum) . . . from the London Clay. Trans. Geol. Soc. London, 203.—1858. Description of a small lophiodont mammal (Pliolophus vulpiceps, Owen) from the London Clay, near Harwich. Quart. Jour. Geol. Soc., 54.

PILGRIM, G. E. 1938. Are the Equidae reliable for the correlation of the Siwaliks with Caenozoic stages of North America? Rec. Geol. Surv. India, 457.

RIDGWAY, W. 1905. The origin and influence of the thoroughbred horse. Univ. Press. Cambridge.

SCHLAIKJER, E. M. 1932. The osteology of Mesohippus barbouri. Bull. Mus. Comp. Zool., Harvard Coll., vol. 72, 391.—1937. A study of Parahippus wyomingensis and a discussion of the phylogeny of the genus. Ibid., vol. 80, 255.

SCHMALTZ, R. 1909. Atlas der Anatomie des Pferdes. II. Teil: Topographische Myologie. Berlin.

SEFVE, I. 1912. Die fossilen Pferde Südamerikas. K. Svenska Vetensk. Handl., Stockholm, no. 6.

STEHLIN, H. G. 1929. Bemerkungen zu der Frage nach der unmittelbaren Ascendenz des Genus Equus. Eclogae geol. Helvetiae, no. 2, 186.

STEHLIN, H. G. and P. GRAZIOSI. 1935. Richerche sugli Asinidi fossili d'Europa. Mém. Soc. Paléont. Suisse, vol. 56.

STIRTON, R. A. 1940. Phylogeny of North American Equidae. Univ. Calif. Publ., Bull. Dept. Geol. Sci., no. 4, 165.—1941. Development of characters in horse teeth and the dental nomenclature. Jour. Mammal., no. 4, 434.

STIRTON, R. A. and W. CHAMBERLAIN. 1939. A cranium of Pliohippus fossulatus from the Clarendon Lower Pliocene fauna of Texas. Jour. Paleont., no. 3, 349.

WORTMAN, J. L. 1896. Species of Hyracotherium and allied perissodactyls from the Wahsatch and Wind River Beds of North America. Bull. Amer. Mus. Nat. Hist., 81.

Titanotheres

GRANGER, W. and W. K. GREGORY. 1943. A revision of the Mongolian titanotheres. Bull. Amer. Nat. Hist., 349.

GREGORY, W. K. 1929a. The muscular anatomy and the restoration of the titanotheres. In: Titanotheres of ancient Wyoming, Dakota and Nebraska, by H. F. Osborn. U. S. Geol. Surv., Monogr. 55, vol. 2, 703.—1929b. Principles of leverage and muscular action. Ibid., 727.—1929c. Summary of harmonic and differential allometrons in the skulls and feet, and an interpretation of the phylogeny of the titanotheres. Ibid., 828.

HERSH, A. H. 1934. Evolutionary relative growth in the titanotheres. Amer. Nat., 537.

OSBORN, H. F. 1896. The cranial evolution of Titanotherium. Bull. Amer. Mus. Nat. Hist., 157.—1902. The four phyla of Oligocene titanotheres. Bull. Amer. Mus. Nat. Hist., 91.—1908. The four inseparable factors of evolution. Theory of their distinct and combined action in the transformation of the titanotheres, an extinct family of hoofed animals in the order Perissodactyla. Science, 148. [Heredity, environment, ontogeny, selection. Relative positions in space and time not explicitly noted.].—1929. The titanotheres of ancient Wyoming, Dakota and Nebraska. Vols. I, II, U. S. Geol. Surv., Monogr., 55. 4to.

PETERSON, O. A. 1924. Osteology of Dolichorhinus longiceps Douglass, with a review of the species of Dolichorhinus in the order of their publication. Mem. Carnegie Mus., 405.

Chalicotheres

BORISSIAK, A. 1945. The chalicotheres as a biological type. Amer. Sci., 667.

COLBERT, E. H. 1934. Chalicotheres from Mongolia and China in the American Museum. Bull. Amer. Mus. Nat. Hist., 353.—1935. Distributional and phylogenetic studies on Indian fossil mammals. III. A classification of the Chalicotherioidea. Amer. Mus. Novitates, no. 798.

FILHOL, H. 1894. Observations concernant quelques mammifères fossiles nouveaux du Quercy. Ann. d. Sci. nat., Zoologie, 129. [Chalicotherium, p. 150.]

HOLLAND, W. J. and O. A. PETERSON. 1914. The osteology of the Chalicotherioidea . . . Mem. Carnegie Mus., 189.

OSBORN, H. F. 1913. Eomoropus, an American Eocene chalicothere. Bull. Amer. Mus. Nat. Hist., 261.

PETERSON, O. A. 1907. Preliminary notes on some American chalicotheres. Amer. Nat., 733.

Tapirs, Helaletids, Lophiodonts

DEPÉRET, C. 1903. Études paléontologiques sur les Lophiodon du Minervois . . . Arch. d. Mus. d'Hist. nat. de Lyon, 1.—1904. Sur les caractères et les affinités du genre Chasmotherium

Rütimayer. Bull. Soc. Géol. d. France, 569.—1910. Études sur la famille des Lophiodontidae. Ibid., 558.

FILHOL, H. 1888. Étude sur les vertébrés fossiles d'Issel (Aude). Mem. Soc. Géol. d. France, 1.

MURIE, J. 1872. On the Malayan tapir Rhinochoerus sumatranus (Gray). Jour. Anat. Physiol., 131. [Nasal diverticulum described and figured.]

PETERSON, O. A. 1919. [Rept. on helaletids, etc. from the Upper Eocene of the Uinta Basin.] Ann. Carnegie Mus., 127. [Helaletes, pls. 42, 43; Desmatotherium, pl. 44.]

SCHAUB, S. 1928. Der Tapirschädel von Haslen: Ein Beitrag zur Revision der oligocänen Tapiriden Europas. Schweiz, Palaeont. Gesell., 1.

SCHLAIKJER, E. M. 1937. A new tapir from the Lower Miocene of Wyoming. Bull. Mus. Comp. Zool. Harvard Coll., 231.

SIMPSON, G. G. 1945. Notes on Pleistocene and Recent tapirs. Bull. Amer. Mus. Nat. Hist., 33.

STEHLIN, H. G. 1903. Die Säugetiere des schweizerischen Eocaens; Critischer Catalog der Materialien. I. Chasmotherium—Lophiodon. Abh. Schweiz. Palaeont. Gesell., 1.—1904. II. Palaeotherium - Plagiolophus - Propalaeotherium. Ibid. (1905), 155.—1906. III. Lophiotherium-Anchilophus-Pachynolophus. Nachträge. Schlussbetrachtungen über die Perissodactylen. Ibid. (1905), 447.

TROXELL, E. L. 1922a. The status of Homogalax, with two new species. Amer. Jour. Sci., 288.—1922b. Helaletes redefined. Amer. Jour. Sci., 365.

Rhinoceroses

BEDDARD, F. E. and F. TREVES. 1889. On the anatomy of Rhinoceros sumatrensis. Proc. Zool. Soc. London. [Anatomy of the face.]

BORISSIAK, A. 1914. Concerning the dental apparatus of Elasmotherium caucasicum n. sp. Bull. Acad. Imp. Sci. St. Pétersbourg, 5551—1938. A new Dicerorhinus from the Middle Miocene of North Caucasus. Travaux d. l'Inst. Palaeozool., 7.

BRANDT, J. F. 1877. Versuch einer Monographie der Tichorhinen nashörner nebst Bemerkungen über Rhinoceros leptorhinus Cuv . . . Mém. d. l'Acad. Imp. d. Sci. de St. Pétersbourg, no. 4, 1.—1878a. Tentamen synopseos Rhinocerotidum viventium et fossilium. Ibid., no. 5, 1. 4to.—1878b. Mittheilungen ueber die Gattung Elasmotherium besonders den Schaedelbau derselben. Ibid., No. 6, 1.

CARTER, T. D. and J. E. HILL. 1942. Notes on the lesser one-horned rhinoceros, Rhinoceros sondaicus. 1. A skull of Rhinoceros sondaicus in the American Museum of Natural History. Amer. Mus. Novitates, no. 1206.

COLBERT, E. H. 1932. Aphelops from the Hawthorn formation of Florida. Florida State Geol. Surv. Bull., no. 10, 55.—1942. Notes on the lesser one-horned rhinoceros, Rhinoceros sondaicus. 2. The position of Rhinoceros sondaicus in the phylogeny of the genus Rhinoceros. Amer. Mus. Novitates, no. 1207.

COOPER, C. F. 1923. Baluchitherium osborni (?Syn. Indricotherium turgaicum, Borissiak). Philos. Trans. Roy. Soc. London, 35.—1934. The extinct rhinoceroses of Baluchistan. Ibid., 569.

DAL PIAZ, G. 1930. I. Mammiferi dell'Oligocene Veneto: Trigonias ombonii. Mem. dell' Inst. Geol. d. R. Universita di Padova, 1.

DIETRICH, W. O. 1935. Ueber M3 von Rhinoceros (Tichorhinus) antiquitatis Blumenbach. Sitz. d. Gesell, naturf. Freunde, June 28, 112.

GRANGER, W. and W. K. GREGORY. 1935. A revised restoration of the skeleton of Baluchitherium, gigantic fossil rhinoceros of Central Asia. Amer. Mus. Novitates, no. 787.—1936. Further notes on the gigantic extinct rhinoceros Baluchitherium from the Oligocene of Mongolia. Bull. Amer. Mus. Nat. Hist., 1.

GREGORY, W. K. 1935. Building a super-giant rhinoceros. Nat. Hist. Mag., no. 4, 340.

GREGORY, W. K. and H. J. COOK. 1928. New material for the study of evolution. A series of primitive rhinoceros skulls (Trigonias) from the Lower Oligocene of Colorado. Proc. Colorado Mus. Nat. Hist., no. 1.

MATTHEW, W. D. 1932. A review of the rhinoceroses with a description of Aphelops material from the Pliocene of Texas. Bull. Dept. Geol. Sci., Univ. Calif. Publ., no. 12, 411.

OSBORN, H. F. 1898a. The extinct rhinoceroses. Mem. Amer. Mus. Nat. Hist., no. 1, part 3, 75.—1898b. A complete skeleton of Teleoceras fossiger. Notes upon the growth and sexual characters of this species. Bull. Amer. Mus. Nat. Hist., 51.—1900. Phylogeny of the rhinoceroses of Europe. Ibid., 229.—1923a. Baluchitherium grangeri, a giant hornless rhinoceros from Mongolia. Amer. Mus. Novitates, no. 78.

—1923b. The extinct giant rhinoceros Baluchitherium of Western and Central Asia. Nat. Hist. Mag., no. 3, 208.—1936. Amynodon mongoliensis from the Upper Eocene of Mongolia. Amer. Mus. Novitates, no. 859.

PETERSON, O. A. 1920. The American diceratheres. Mem. Carnegie Mus., no. 6, 399.

STEHLIN, H. G. 1930. Bemerkungen zur Vordergebissformel der Rhinocerotiden. Schweiz. Paläont. Gesell., Eclogae geol. Helvetiae, no. 2, 644.

TROXELL, E. L. 1921a. New amynodonts in the Marsh Collection. Amer. Jour. Sci., 1.—1921b. New species of Hyracodon. Ibid., 34.—1921c. Caenopus, the ancestral rhinoceros. Ibid., 41.—1921d. A study of Diceratherium and the diceratheres. Ibid., 197.—1922. The genus Hyrachyus and its subgroups. Ibid., vol. 4, 38.

WOOD, H. E., II. 1926. Hyracodon petersoni, a new cursorial rhinoceos from the Lower Oligocene. Ann. Carnegie Mus., no. 2, 315.—1927. Some early Tertiary rhinoceroses and hyracodonts. Bull. Amer. Paleont., no. 50, 1.—1929a. American Oligocene rhinoceroses.—A postscript. Jour. Mammal., no. 1, 63.—1929b. Prohyracodon orientale Koch, the oldest known true rhinoceros. Amer. Mus. Novitates, no. 395.—1931. Lower Oligocene rhinoceroses of the genus Trigonias. Jour. Mammal., no. 414.—1932. Status of Epiaceratherium (Rhinocerotidae). Jour. Mammal., no. 2, 169.—1934. Revision of the Hyrachyidae. Bull. Amer. Mus. Nat. Hist., 181.—1938. Cooperia totadentata, a remarkable rhinoceros from the Eocene of Mongolia. Amer. Mus. Novitates, no. 1012.—1941. Trends in rhinoceros evolution. Trans. N. Y. Acad. Sci., no. 4, 83.

Artiodactyla: General

HOPWOOD, A. T. 1929. New and little-known mammals from the Miocene of Africa. Amer. Mus. Novitates, no. 344.

LOOMIS, F. B. 1925. Dentition of artiodactyls. Bull. Geol. Soc. Amer., 583.

LULL, R. S. 1920. New Tertiary artiodactyls. Amer. Jour. Sci., 83.

MATTHEW, W. D. 1929. Reclassification of the artiodactyl families. Bull. Geol. Soc. Amer., 403.—1934. A phylogenetic chart of the Artiodactyla. Jour. Mammal., 207.

PETERSON, O. A. 1912. Artiodactyla. Bull. Geol. Soc. Amer., 162.

PILGRIM, G. E. 1928. The Artiodactyla of the Eocene of Burma. Mem. Geol. Surv. India, Palaeontologia Indica, 1.—1941. The dispersal of the Artiodactyla. Biol. Rev., 134.

SCHAEFFER, B. 1947. Notes on the origin and function of the artiodactyl tarsus. Amer. Mus. Novitates, no. 1356.—1948. The origin of a mammalian ordinal character. Evolution, June, 164.

SCOTT, W. B. 1889. The Creodonta, Rodentia, and Artiodactyla. In: The Mammalia of the Uinta formation, by W. B. Scott and H. F. Osborn. Part II. Trans. Amer. Philos. Soc., part 3, 471.—1940. Artiodactyla. In: The mammalian fauna of the White River Oligocene, by W. B. Scott and G. L. Jepsen. Part IV. Ibid., 363.

STEHLIN, H. G. 1906. Die Säugetiere des schweizerischen Eocaens . . . IV. Dichobune- Mouillacitherium- Meniscodon- Oxacron. Abh. d. Schweiz. Paläont. Gesell., vol. 33, 597.—1908. V. Choeropotamus- Cebochoerus- Choeromorus- Haplobunodon- Rhagatherium- Mixtotherium. Ibid., 691.—1910. VI. Catadontherium- Dacrytherium- Leptotheridium- Anoplotherium- Diplobune- Xiphodon- Pseudamphimeryx- Amphimeryx- Dichodon- Haplomeryx- Tapirulus- Gelocus.- Nachträge.- Artiodactyla incertae sedis.- Schlussbetrachtungen über die Artiodactylen. Ibid., vol. 36 (1909, 1910), 839.

Entelodonts

KOWALEWSKY, W. 1873. Osteologie des Genus Entelodon Aym. Mém. d. l'Acad. Imp. d. Sci. d. St. Pétersbourg, 415.

MARINELLI, W. 1924. Untersuchungen über die Funktion des Gebisses der Entelodontiden. Palaeontol. Zeitschr., 25.

PETERSON, O. A. 1909. A revision of the Entelodontidae. Mem. Carnegie Mus., no. 3, 41. [Dinohyus, figs. 29–70, pls. 55–62.]

SCOTT, W. B. 1898. The osteology of Elotherium. Trans. Amer. Philos. Soc., 273.

SINCLAIR, W. J. 1922. The small entelodonts of the White River Oligocene. Hyracodons from the Big Badlands of South Dakota. Amer. Philos. Soc., no. 1, 53.

TROXELL, E. L. 1920. Entelodonts in the Marsh Collection. Amer. Jour. Sci., 243.

Suina (Peccaries and Pigs)

COLBERT, E. H. 1933a. The skull and mandible of Conohyus, a primitive suid from the Siwalik Beds of India. Amer. Mus. Novitates, no. 621.

—1933b. An Upper Tertiary peccary from India. Ibid., no. 635.—1935a. Distributional and phylogenetic studies on Indian fossil mammals. IV. The phylogeny of the Indian Suidae and the origin of the Hippopotamidae. Ibid., no. 799.—1935b. A new fossil peccary, Prosthennops niobrarensis, from Brown County, Nebraska. Nebraska State Mus., Bull. 44, 419.—1935c. Siwalik mammals in the Amer. Mus. Nat. Hist. Trans. Amer. Philos, Soc., vol. 26, 1.—1938a. Pliocene peccaries from the Pacific Coast region of North America. Carnegie Inst. Washington Publ. no. 487, 241.—1938b. Brachyhyops, a new bunodont artiodactyl from Beaver Divide, Wyoming. Ann. Carnegie Mus., 87.

GAZIN, C. L. 1938. Fossil peccary remains from the Upper Pliocene of Idaho. Jour. Washington Acad. Sci., no. 2, 41.

GIDLEY, J. W. 1920. Pleistocene peccaries from the Cumberland Cave deposit. Proc. U. S. Nat. Mus., 651.

HANSON, F. B. 1919a. The coracoid of Sus scrofa. Anat. Rec., vol. 16, 197.—1919b. Nerve foramina in the pig scapula. Ibid., no. 6, 289.—1919c. The development of the sternum in Sus scrofa. Ibid., no. 1, 1.—1920. The development of the shoulder girdle of Sus scrofa. Ibid.

LEAKEY, L. S. B. 1943. New fossil Suidae from Shunguru, Omo. Jour. East Africa Nat. Hist. Soc., no. 2, 45.

MEAD, C. S. 1909. The chondrocranium of an embryo pig, Sus scrofa. Amer. Jour. Anat., no. 2, 167.

OSBORN, H. F. 1883. On Achaenodon, an Eocene bunodont. Contrib. E. M. Mus. Geol. and Archaeol. Princeton Coll., Bull. no. 3, 23.

PEARSON, H. S. 1923. Some skulls of Perchoerus (Thinohyus) from the White River and John Day formations. Bull. Amer. Mus. Nat. Hist. 61.—1927. On the skulls of early Tertiary Suidae, together with an account of the otic region in some other primitive Artiodactyla. Philos. Trans. Roy. Soc. London, 389.—1928. Chinese fossil Suidae. Palaeontologia Sinica, part 5, 51.

PETERSON, O. A. 1906. New suilline remains from the Miocene of Nebraska. Mem. Carnegie Mus., no. 8, 305.—1914. A mounted skeleton of Platygonus leptorhinus in the Carnegie Museum. Ann. Carnegie Mus., nos. 1, 2; 114.

PILGRIM, G. E. 1926. The fossil Suidae of India. Mem. Geol. Surv. India, Paleontologia Indica, no. 4, 1.

PIRA, A. 1909. Studien zur Geschichte der Schweinerassen, insbesondere derjenigen Schwedens. Zool. Jahrb., suppl. 10, pt. 2, 16.

RUSCONI, C. 1930. Las Especies Fosiles Argentinas de Pecaries ("Tayassuidae") y sus Relationes con las del Brasil y Norte America. An. d. Museo Nac. de Hist. Nat., 121.

SHAW, J. C. M. 1938. The teeth of the South African fossil pig. (Notochoerus capensis syn. Meadowsi) and their geological significance. Trans. Roy. Soc. South Africa, part 1, 25.—1939. Growth changes and variations in wart hog third molars and their palaeontological importance. Ibid., part 1, 51.

SHAW, J. C. M. and H. B. S. COOKE. 1941. New fossil pig remains from the Vaal River gravels. Trans. Roy. Soc. South Africa, part 4, 293.

SINCLAIR, W. J. 1914. A revision of the bunodont Artiodactyla of the Middle and Lower Eocene of North America. Bull. Amer. Mus. Nat. Hist., 267. [Dichobunidae.]

STEHLIN, H. G. 1899 (1900). Ueber die Geschichte des Suiden-Gebisses. I. Teil. Abh. d. Schweiz. Paläontol. Gesell., 1.

STOCK, C. 1928. A peccary from the McKittrick Pleistocene, California. Contrib. Palaeont., Carnegie Inst. Washington Publ. no. 393, 23.

Ancodonts, Anthracotheres, Hippopotami

COLBERT, E. H. [On origin of hippopotami, see 1935a, 1935c above.]

DAL PIAZ, G. 1926. Osservazioni sulla formula dentaria del genere Anthracotherium. Atti della Accad. Scient. Veneto-Trentino-Istriana, 1.—1932. I. Mammiferi dell' Oligocene Veneto: Anthracotherium monsvialense. Mem. dell' Istituto Geol. della R. Univ. di Padova, 1.

KOENIGSWALD, G. H. R. VON. 1934. Die spezialisation des incisiven-gebisses bei den Javanischen Hippopotamidae. Proc. Kon. Akad. van Wet. te Amsterdam, no. 9, 1.

KOWALEWSKY, W. 1873a. Monographie der Gattung Anthracotherium Cuv. und Versuch einer natürlichen Classification der fossilen Hufthiere. Palaeontographica (1876), 133.—1873b. On the osteology of the Hyopotamidae. Mém. d. l'Acad. Imp. d. Sci. d. St. Pétersbourg, no. 5 et dernier, 19.

MATTHEW, W. D. 1909. Observations upon the genus Ancodon. Bull. Amer. Mus. Nat. Hist., 1.

REYNOLDS, S. H. 1920. A monograph of the British

Pleistocene Mammalia. Hippopotamus. Palaeontographical Soc., part 1, 1.

SCOTT, W. B. [1895?] The structure and relationships of Ancodus. Jour. Acad. Nat. Sci. Philadelphia, part 4, 461.

SCOTT, W. B. and G. L. JEPSEN. 1940. [See Artiodactyla above.]

SIEBER, R. 1936. Remarques sur les Anthracotherium de l'Oligocène français. Bull. Soc. l'Hist. nat. de Toulouse, no. 4, 1.

STEHLIN, H. G. 1929. Artiodactylen mit fünffingriger Vorderextremität aus dem europäischen Oligocän. Verhandl. d. Naturf. Gesell. in Basel, part 2, 599.

TROXELL, E. L. 1921. The American bothriodonts. I. Amer. Jour. Sci., 325.

Caenotheres

HÜRZELER, J. 1936. Osteologie und odontologie der Caenotheriden. Abh. d. Schweiz. Palaeont. Gesell., vol. 58–59.

Oreodonts

BLACK, D. 1920. Studies of endocranial anatomy. II. On the endocranial anatomy of Oreodon (Merycoidodon). Jour. Comp. Neurol., no. 3, 272.

COLBERT, E. H. 1943. A Miocene oreodont from Jackson Hole, Wyoming. Jour. Palaeont., vol. 17, no. 3, 298.

LOOMIS, F. B. 1933. Three oreodont skeletons from the Lower Miocene of the Great Plains. Bull. Geol. Soc. Amer., vol. 44, 723.

MATTHEW, W. D. 1932. A skeleton of Merycoidodon gracilis and its adaptive significance. Bull. Dept. Geol. Sci., Univ. Calif. Publ., no. 2, 13.

PETERSON, O. A. 1914. The osteology of Promerycochoerus. Ann. Carnegie. Mus., nos. 1, 2; 149. —1923. Restoration of Merychyus elegans subsp. minimus Peterson. Ann. Carnegie Mus., no. 1, 96.—1928. Osteology of Phenacocoelus typus Peterson. Mem. Carnegie Mus., no. 3, 131.—1931. Two new species of agriochoerids. Ann. Carnegie Mus., nos. 3, 4; 341.

SCHULTZ, C. B. and C. H. FALKENBACH. 1940. Merycochoerinae: A new subfamily of oreodonts. Bull. Amer. Mus. Nat. Hist., 213.—1941. Ticholeptinae: A new subfamily of oreodonts. Ibid., 1.—1947. Merychyinae, a subfamily of oreodonts. Ibid., 157.

SCOTT, W. B. 1894. Notes on the osteology of Agriochoerus, Leidy (Artionyx O. and W.). Proc. Amer. Philos. Soc., 243.

THORPE, M. R. 1937. The Merycoidodontidae: An extinct group of ruminant mammals. Mem. Peabody Mus. Nat. Hist., part 4.

WORTMAN, J. L. 1895. On the osteology of Agriochoerus. Bull. Amer. Mus. Nat. Hist., art. 4, 145.

Tylopoda (Camels)

BARBOUR, E. H. and C. B. SCHULTZ. 1939. A new giant camel, Gigantocamelus fricki, gen. et sp. nov. Bull. Univ. Nebraska State Mus., no. 2, 17.

CABRERA, A. 1935. Sobre la Osteologia de Paleolama. An. d. Mus. Argentino d. Ciencias Nat., 283.

KRAGLIEVICH, L. J. 1946. Sobre Camélidos Chapadmalenses. Notas del Museo de La Plata, Paleontologia, no. 93, 317.

LOOMIS, F. B. 1910. Osteology and affinities of the genus Stenomylus. Amer. Jour. Sci., 297.

MATTHEW, W. D. 1910. On the skull of Apternodus and the skeleton of a new artiodactyl. Bull. Amer. Mus. Nat. Hist., 33. [Eotylopus.]

PETERSON, O. A. 1904. Osteology of Oxydactylus. A new genus of camels from the Loup Fork of Nebraska . . . Ann. Carnegie Mus., no. 3, 434. —1908. Description of the type specimen of Stenomylus gracilis Peterson. Ibid., nos. 3, 4; 286.—1911a. A new camel from the Miocene of Western Nebraska. Ibid., no. 2, 260.—1911b. A mounted skeleton of Stenomylus . . . and remarks upon the affinities of the genus. Ibid., no. 2, 267.

RUSCONI, C. 1931. La Dentadura de Palyaeolama en relacion a la de otros Camelidos. Revista de Medicina Veterinaria, no. 6, 250.

SCOTT, W. B. 1891. On the osteology of Poebrotherium: A contribution to the phylogeny of the Tylopoda. Jour. Morphol. vol. 5, no. 1.— 1940. Artiodactyla. In: The mammalian fauna of the White River Oligocene by W. B. Scott and G. L. Jepsen, Part IV.

STOCK, C. 1928. Tanupolama, a new genus of llama from the Pleistocene of California. Contrib. to Palaeont., Carnegie Inst. Washington Publ. no. 393.

TEILHARD DE CHARDIN, P. and M. TRASSAERT. 1937. The Pliocene Camelidae, Giraffidae and Cervidae of Southeastern Shansi. Palaeontologia Sinica, Nat. Geol. Surv. China, no. 1.

Ruminantia
(Traguloids, Deer, Giraffes, Pronghorns, Antelopes, Goats, Sheep, Oxen)

ARAMBOURG, C. 1941. Antilopes nouvelles du Pleistocène ancien de l'Omo (Abyssinie). Bull. d. Mus. (Paris), no. 4, 339.

BARBOUR, E. H. and C. B. SCHULTZ. 1941. A new species of Sphenophalos from the upper Ogallala of Nebraska. Bull. Univ. Nebraska State Mus., no. 6, 59.

BOAS, J. E. V. 1934. Über die verwantschaftliche Stellung der Gattung Antilocapra und der Giraffiden zu den übrigen Wiederkäuern. Det. Kgl. Danske Vidensk. Selskab., Biol. Meddel. 11, no. 3.

BOHLIN, B. 1937. Einige Jungtertiare und Pleistozene Cavicornier aus Nord-China. Nova Acta Regiae Soc. Sci. Upsaliensis, part 1.—1939. Gazella (Protetraceros) Gaudryi (Schlosser) and Gazella dorcadoides Schlosser. Bull. Geol. Instit. Upsala, 79.

BROWN, B. 1926. A new deer from the Siwaliks. Amer. Mus. Novitates, no. 242.

CARLSSON, A. 1926. Über die Tragulidae und ihre Beziehungen zu den übrigen Artiodactyla. Acta Zool., 69.

COLBERT, E. H. 1933. A skull and mandible of Giraffokeryx punjabiensis Pilgrim. Amer. Mus. Novitates, no. 632.—1935. Distributional and phylogenetic studies on Indian fossil mammals. V. The classification and the phylogeny of the Giraffidae. Ibid., no. 800.—1936a. Tertiary deer discovered by the American Museum Asiatic expeditions. Ibid., no. 854.—1936b. Palaeotragus in the Tung Gur formation of Mongolia. Ibid., no. 874.—1938a. The giraffe and his living ancestor. Nat. Hist. Mag., no. 1, 46.—1938b. The relationships of the okapi. Jour. Mammal., no. 1, 47.—1940. Some cervid teeth from the Tung Gur formation of Mongolia, and additional notes on the genera Stephanocemas and Lagomeryx. Amer. Mus. Novitates, no. 1062.—1941. The osteology and relationships of Archaeomeryx, an ancestral ruminant. Ibid., no. 1135.

COLBERT, E. H. and R. G. CHAFFEE. 1939. A study of Tetrameryx and associated fossils from Papago Spring Cave, Sonoita, Arizona. Amer. Mus. Novitates, no. 1034.

DAWKINS, W. B. 1878. Contributions to the history of the deer of the European Miocene and Pliocene strata. Quart. Jour. Geol. Soc., May, 402.—1883. On the alleged existence of Ovibos moschatus in the forest-bed and its range in space and time. Ibid., November, 575.

DERSCHEID, J. M. and H. NEUVILLE. 1925. Recherches anatomiques sur l'Okapi, Okapia Johnstoni Scl. Revue Zool. Africaine, part 1.

FRECHKOP, S. 1946. Notes sur les mammifères. XXIX.—De l'Okapi et des affinitiés des Giraffidés avec les Antilopes. Bull. Mus. roy. d'Hist. nat. de Belgique. no. 1.

FRICK, C. 1937. Horned ruminants of North America. Bull. Amer. Mus. Nat. Hist., vol. 69.

FURLONG, E. L. 1931. Distribution and description of skull remains of the Pliocene antelope Sphenophalos from the Northern Great Basin Province. Carnegie Inst. Washington Publ. 418, 27.—1943. The Pleistocene antelope, Stockoceros conklingi, from San Josecito Cave, Mexico. Carnegie Inst. Washington Publ. Ibid., 551.

HARLÉ, E. and H. G. STEHLIN. 1914. Un capridé quaternaire de la Dordogne, voisin du thar actuel de l'Himalaya. Bull. Soc. géol. de France, vol. 13 (1913), 422.

HESSE, C. J. 1935. New evidence on the ancestry of Antilocapra americana. Jour. Mammal., no. 5, 307.

HUXLEY, J. S. 1931. The relative size of antlers in deer. Proc. Zool. Soc. London, part 3, 819.

JOLEAUD, L. 1916. Cervus (Megaceroides) algericus Lydekker. Recueil et Notices de la Societé Archeol. de Constantine, 1.—1917. Les gazelles pliocène et quaternaires de l'Algerie. Bull. d. la Soc. géol. de France, 208.

LANKESTER, E. R. 1902. On Okapia, a new genus of Giraffidae from Central Africa. Trans. Zool. Soc. London, part 6, 279.—1910. Monograph of the okapi. Brit. Mus. (Nat. Hist.) London. Atlas.

LÖNNBERG, E. 1933. Description of a fossil buffalo from East Africa. Arkiv. f. Zool., no. 17, 1.

LOOMIS, F. B. 1928. Phylogeny of the deer. Amer. Jour. Sci., 531.

LYDEKKER, R. 1898a. The deer of all lands: A history of the family Cervidae, living and extinct. London.—1898b. Wild oxen, sheep and goats of all lands, living and extinct. London.

MATSUMOTO, H. 1926. On some fossil cervids from Shantung, China. Sci. Rept. Tôhoku Imp. Univ., 2d ser. (geol.), no. 2, 27.

MATTHEW, W. D. 1902. The skull of Hypisodus, the smallest of the Artiodactyla, with a revision of the Hypertragulidae. Bull. Amer. Mus. Nat.

Hist., 311.—1904. A complete skeleton of Merycodus. Ibid., 101.—1908. Osteology of Blastomeryx and phylogeny of the American Cervidae. Ibid., 535.

MATTHEW, W. D. and W. GRANGER. 1925. New mammals from the Shara Murun Eocene of Mongolia. Amer. Mus. Novitates, no. 196. [Archaeomeryx optatus, n.g., n.sp.]

MILNE-EDWARDS, A. 1864. Recherches anatomiques, zoologiques et paléontologiques sur la famille des chevrotains. Ann. d. Sci. nat., 49.

NOBACK, C. V. and W. MODELL. 1930. Direct bone formation in the antler tines of two of the American Cervidae, Virginia deer (Odocoileus virginianus) and wapiti (Cervus canadensis); with an introduction on the gross structure of antlers. Zoologica (New York), no. 3, 10.

PAVLOW, M. W. 1927. Rangifer tarandus et formes rapprochées. Bull. Soc. d. Naturalistes d. Moscou, Sect. Géol., 137.

PILGRIM, G. E. 1911. The fossil Giraffidae of India. Mem. Geol. Surv. India, Palaeontologia Indica, Mem. 1.—1937. Siwalik antelopes and oxen in the American Museum of Natural History. Bull. Amer. Mus. Nat. Hist., 729.—1939. The fossil Bovidae of India. Mem. Geol. Surv. India, Palaeontologia India, Mem. 1.—1947. The evolution of the buffaloes, oxen, sheep and goats. Jour. Linnean Soc. London, no. 279, 272.

PILGRIM, G. E. and A. T. HOPWOOD. 1928. Catalogue of the Pontian Bovidae of Europe in the Department of Geology. Brit. Mus. (Nat. Hist.). London.

PILGRIM, G. E. and S. SCHAUB. 1939. Die schraubenhörnige Antilope des europäischen Oberpliocaens und ihre systematische Stellung. Abh. d. Schweiz. Palaeont. Gesell., 1.

POCOCK, R. I. 1933. The homologies between the branches of the antlers of the Cervidae based on the theory of dichotomous growth. Proc. Zool. Soc. London, part 2, 377.—1935. The incisiform teeth of European and Asiatic Cervidae. Ibid., part 1, 179.—1936. Preliminary note on a new point in the structure of the feet of the okapi. Ibid., part 2, 583.

REYNOLDS, S. H. 1929. A monograph on the British Pleistocene Mammalia.- The giant deer. Palaeontographical Soc., part 3, 1.—1933. A monograph on the British Pleistocene Mammalia.- The red deer, reindeer, and roe. Ibid., part 4, 1.—1934. A monograph on the British Pleistocene Mammalia.- Ovibos. Ibid., Suppl., 1.—1938. A monograph on the British Pleistocene Mammalia.- The Bovidae. Ibid., part 6, 1.

SCHAUB, S. 1928. Die Antilopen des Toskanischen Oberpliocäns, Eclogae geol. Helvetiae, no. 1, 260.

SCHLOSSER, M. 1924. Ueber die systematische Stellung jungtertiärer Cerviden. Centr. f. Min. etc., no. 20, 634.

SCHWARZ, E. 1937. Die fossilen Antilopen von Oldoway. Wiss. Ergeb. Oldoway-Exp., part 4, 7.

SCLATER, P. L. and O. THOMAS. 1894–1900. The book of antelopes. London.

SCOTT, W. B. 1895. The osteology and relations of Protoceras. Jour. Morph., 303.—1899. The selenodont artiodactyls of the Uinta Eocene. Trans. Wagner Free Inst. of Sci., 15.

SICKENBERG, O. 1933. Parurmitherium rugosifrons, ein neuer Bovide aus dem Unterpliozän von Samos. Palaeobiologica, 81.

STEHLIN, H. G. 1928a. Ueber die systematische Stellung des Genus Leptobos. Schweiz. Paläont. Gesell., Eclogae geol. Helvetiae, 217.—1928b. Bemerkungen über die Hirsche von Steinheim am Aalbuch. Ibid., 245. [Evolution of antlers of Dicroceros.]—1937. Bemerkungen über die miocaenen Hirschgenera Stephanocemas und Lagomeryx. Verhandl. d. Naturf. Gesell. in Basel, 193.—1939. Dicroceros elegans Lartet und sein Geweihwechsel. Ibid., no. 2, 162.

STIRTON, R. A. 1932. An association of horncores and upper molars of the antelope Sphenophalos nevadanus from the Lower Pliocene of Nevada. Amer. Jour. Sci., 46.—1944. Comments on the relationhips of the cervoid family Palaeomerycidae. Ibid., 633.

TEILHARD DE CHARDIN, P. 1939. The Miocene cervids from Shantung. Bull. Geol. Soc. China, no. 3, 269.

TEILHARD DE CHARDIN, P. and M. TRASSAERT. 1938. Cavicornia of southeastern Shansi. Palaeontologia Sinica, no. 6.

TOKUNAGA, S. and F. TAKAI. 1936. A new roedeer, Capreolus (Capreolina) mayai . . . from the Inland Sea of Japan. Jour. Geol. Soc. Japan, no. 515, 642.—1939. A study of Metacervulus astylodon (Matsumoto) from the Ryukyu Islands, Japan. Trans. Biogeogr. Soc. Japan, no. 2, 221.

TROXELL, E. L. 1920. A tiny Oligocene artiodactyl, Hypisodus alacer, sp. nov. Amer. Jour. Sci., 391.

YOUNG, C. C. 1936. New finds of fossil Bubalus in China. Bull. Geol. Soc. China, no. 4, 505.

Cetacea: General and Miscellaneous

ALLEN, G. M. 1921a. A new fossil cetacean. Bull. Mus. Comp. Zool. Harvard Coll., no. 1, 1.—1921b. Fossil cetaceans from the Florida phosphate beds. Jour. Mammal., no. 3, 144.

BRANDT, J. F. 1873. Untersuchungen über die fossilen und subfossilen Cetacean Europas. Mém. d. l'Acad. Imp. d. Sci. d. St.-Pétersbourg, no. 1.

DENKER, A. 1902. Zur Anatomie der Gehörorgans der Cetacea. Anat. Hefte, 421.

GAWN, R. W. L. 1948. Aspects of the locomotion of whales. Nature, vol. 161, no. 4080, 44.

GRAY, J. and D. A. PARRY. 1948. Aspects of the locomotion of whales. Ibid., no. 4084, 199.

HARMER, S. F. 1929. Article on Cetacea. Encycl. Brit., 14 ed.

HUBER, E. 1927. Comparative anatomical investigations on the facial musculature of the whale. Anat. Rec., no. 1, 41.—1934. Anatomical notes on Pinnipedia and Cetacea. Carnegie Inst. Washington Publ. no. 447, 105.

KELLOGG, R. 1928a. The history of the whales; their adaptation to life in the water. I. Quart. Rev. Biol., no. 1, 29.—1928b. II. Ibid., no. 2, 174.—1938. Adaptation of structure to function in whales. In: Coöperation in Research, Carnegie Inst. Washington Publ. no. 501, 649.—1942. Tertiary, Quaternary, and Recent marine mammals of South America and the West Indies. Proc. Eighth Amer. Scient. Cong. (1940): Zoology, 445.—1944. Fossil cetaceans from the Florida Tertiary. Bull. Mus. Comp. Zool., Harvard Coll., no. 9, 433.

KÜKENTHAL, W. 1889–1893. Vergleichend-Anatomische und Entwickelungsgeschichtliche Untersuchungen an Walthieren. I, II, III. Denkschr. d. med.-nat. Gesell., Jena.

MILLER, G. S., JR. 1925. The telescoping of the cetacean skull. Smiths. Miscell. Coll., no. 5, 1.

MURRAY, J. and J. HJORT. 1912. The depths of the ocean; . . . London.

RAVEN, H. C. 1942. Some morphological adaptations of cetaceans (whales and porpoises) for life in the water. Trans. N. Y. Acad. Sci., no. 2, 23.

SLIJPER, E. J. 1936. Die Cetaceen, vergleich. Anatomie und System. Capita Zoologica, part 1, 1. Univ. Utrecht. 4to.—1938. Die Sammlung rezenter Cetacea des Musée roy. d'Hist. nat. de Belgique. Mededeel. van het Kon. Natuurhist. Mus. van Belgie, no. 10, 1.

VAN BENEDEN, P. J. and P. GERVAIS. 1880. Ostéographie des Cétacés vivants et fossiles. comprenant la description et l'iconographie du squelette et du système dentaire de ces animaux . . . Paris. Folio.

WEBER, M. 1923. Die Cetaceen der Siboga-Expedition. Siboga- Exp., vol. 58. Leiden.

WINGE, H. 1921. A review of the interrelationships of the Cetacea. Smiths. Miscell. Coll., no. 8, 1.

Cetacea: Archaeoceti (Zeuglodonts)

ANDREWS, C. W. 1920. A description of new species of zeuglodont and of leathery turtle from the Eocene of southern Nigeria. Proc. Zool. Soc. London, no. 22, 309. [Pappocetus lugardi.]

DART, R. 1923. The brain of the Zeuglodontidae. Proc. Zool. Soc., 615; ibid., 652. London.

FRAAS, E. 1904. Neue Zeuglodonten aus dem unteren Mitteleocän vom Mokattam bei Kairo. Geol. u. Palaeont. Abh., part 3, 199. Jena. [Protocetus possibly derived from hyaenodont stem.]

GIDLEY, J. W. 1913. A recently mounted Zeuglodon skeleton in the United States National Museum. Proc. U. S. Nat. Mus., 649.

KELLOGG, R. 1936. A review of the Archaeoceti. Carnegie Inst. Washington Publ. 482.

MÜLLER, J. 1849. Über die fossilen Reste der Zeuglodonten von Nordamerica, mit Rücksicht auf die europäischen Reste aus dieser Familie. Berlin. Folio.

POMPECKJ, J. F. 1922. Das Ohrskelett von Zeuglodon. Senckenbergiana, parts 3, 4.

SMITH, G. E. [1903?] The brain of the Archaeoceti. Proc. Roy. Soc., vol. 71, 322.

STROMER, E. 1908. Die Archaeoceti des Ägyptischen Eozäns. Beiträge z. Palaont. u. Geol. Osterreich-Ungarns u. d. Orients, 106.

TRUE, F. W. 1908. The fossil cetacean, Dorudon serratus Gibbes. Bull. Mus. Comp. Zool. Harvard Coll., no. 4, 65. [Zygorhiza, n.g., p. 78.]

Cetacea: Odontoceti
(Toothed Whales: Dolphins, Xiphioids, Sperm Whales)

ABEL, O. 1901. Les dauphins longirostres du Boldérien (Miocène supérieur) des environs d'Anvers. Part I. Mém. Mus. roy. nat. de Belgique, vol. 1, 1.—1902a. Part II. Ibid., vol. 2, 101.—

1902b. Die Ursache der Asymmetrie des Zahnwalschädels. Sitz. k. Akad. Wiss., 510. Wien.—1905a. Les Odontocètes du Boldérien (Miocène supérieur) d'Anvers. Mem. Mus. roy. d'Hist. nat. de Belgique, 1.—1905b. Eine Stammtype der Delphiniden aus dem Miocän der Halbinsel Taman. Jahrb. k. k. geolog. Reichanstalt, Wien, part 2, 375.—1931. Das Skelett der Eurhinodelphiden aus dem Oberen Miozän von Antwerpen. III. Teil und Schluss . . . Mém. Mus. roy. d'Hist. nat. de Belgique, no. 48, 191.

ALLEN, G. M. 1926. Fossil mammals from South Carolina. Bull. Mus. Comp. Zool. Harvard Coll., no. 14, 447. [Primitive cetacean, Archaeodelphis.]—1941. A fossil river dolphin from Florida. Ibid., no. 1.

BENHAM, W. B. 1901a. On the larnyx of certain whales (Cogia, Balaenoptera and Ziphius). Proc. Zool. Soc. London, vol. 1, 278.—1901b. On the anatomy of Cogia breviceps. Ibid., vol. 2, 107.

CABRERA, A. 1926. Cetáceos fósiles del Museo de La Plata. Revista Mus. d. La Plata, 363.

DAL PIAZ, G. 1916a. Gli Odontoceti del Miocene Bellunese. Parte Terza: Squalodelphis fabianii. Mem. dell' Inst. Geol. della R. Univ. di Padova, vol. 5, 3.—1916b. Parte Quarta: Eoplatantista italica. Ibid., 3.

FLOWER, W. H. 1872. Description of the skeleton of the Chinese white dolphin (Delphinus sinensis, Osbeck). Trans. Zool. Soc. London (1870), part 2, 151.

GERVAIS, P. 1854–1855. Histoire naturelle des mammifères. Vol. II. Paris. [Mesoplodon, pp. 320, 321.]

KERNAN, J. D. and H. VON W. SCHULTE. 1918. Memoranda upon the anatomy of the respiratory tract, foregut, and thoracic viscera of a foetal Kogia breviceps. Bull. Amer. Mus. Nat. Hist. 231.

LE DANOIS, E. 1910. Recherches sur l'anatomie de le tête de Kogia breviceps (Blainv.). Arch. d. Zool. Exp. et Gen., vol. 6, 149.—1911. Recherches sur les visceres et la squelette de Kogia breviceps (Blainv.) avec un résumé de l'histoire de ce Cetace. Ibid., 465.

LYDEKKER, R. 1894. Contributions to a knowledge of the fossil vertebrates of Argentina. 2. Cetacean skulls from Patagonia. An. Mus. La Plata, Paleont. Argentina. vol. 2 (1893), 1.

MILLER, G. S., JR. 1918. A new river-dolphin from China. Smiths. Miscell. Coll., no. 9.

POUCHET, G. and H. BEAUREGARD. 1885. Note sur "l'Organe des Spermaceti." Comptes rendus hebdomadaires d. Séances et Mémoires de la Soc. de Biol., vol. 2, 342.—1892. Recherches sur le cachalot. Nouv. Arch. d. Mus. d'Hist, nat., vol. 4, 1.

RAVEN, H. C. 1937. Notes on the taxonomy and osteology of two species of Mesoplodon (M. europaeus Gervais, M. mirus True). Amer. Mus. Novitates, no. 905.—1942. On the structure of Mesoplodon densirostris, a rare beaked whale. Bull. Amer. Mus. Nat. Hist., 23.

RAVEN, H. C. and W. K. GREGORY. 1933. The spermaceti organ and nasal passages of the sperm whale (Physeter catodon) and other odontocetes. Amer. Mus. Novitates, no. 677.

RAY, J. 1671. An account of the dissection of a porpoise. Philos. Trans. Roy. Soc. London, vol. 6, 639. [One of the first to discover relationships of cetaceans with land animals.]

SCHULTE, H. VON W. 1917. The skull of Kogia breviceps Blainv. Bull. Amer. Mus. Nat. Hist., 361.

SCHULTE, H. VON W. and M. DE FOREST SMITH. 1918. The external characters, skeletal muscles, and peripheral nerves of Kogia breviceps (Blainville). Ibid., 7.

SLIJPER, E. J. 1939. Pseudorca crassidens (Owen), ein Beitrag zur Vergleichenden anatomie der cetaceen. Zool. Mededeelingen. vol. 21, 242.

VAN BENEDEN, P. J. 1864–1867. Recherches sur les Ossements provenant du Crag d'Anvers. Les Squalodons et Supplement. Mém. Acad. roy. de Belgique, vol. 35, 3; vol. 37, 3.

WISLOCKI, G. B. On the placentation of the harbor porpoise Phocaena phocaena (Linnaeus). Biol. Bull. no. 1, 80.

Cetacea: Mystaceti (Whalebone Whales)

ABEL, O. 1914. Die Vorfahren der Bartenwale. Denks. k. Akad. Wiss., 155.

ALLEN, G. M. 1916. The whalebone whales of New England. Mem. Boston Soc. Nat. Hist., no. 2, 107.

ANDREWS, R. C. 1914. Monographs of the Pacific Cetacea. I. The California gray whale (Rhachianectes glaucus Cope). Mem. Amer. Mus. Nat. Hist., part 5, 227.—1916. II. The sei whale (Balaenoptera borealis). Ibid., part 6, 289.

BEAUREGARD, H. 1882. Étude de l'articulation temporo-maxillaire chez les Balaenoptères. Jour. d. l'Anat. et d. la Physiol., Année 18, 16.

CARTE, A., and A. MACALISTER. 1869. On the anatomy of Balaenoptera rostrata. Philos. Trans. Roy. Soc. London (1868), no. 8, 201.

FLOWER, W. H. 1865. Observations upon a fin-whale (Physalus antiquorum Gray) recently stranded in Pevensey Bay. Proc. Zool. Soc. London, no. 11, 699.

FRECHKOP, S. 1944. Notes sur les Mammifères. XXVIII.- Essai d'interprétation biologique de la structure des Cétacés. Bull. Mus. roy. d'Hist. nat. de Belgique., no. 13, 1.

HJORT, J. 1937. The story of whaling: A parable of sociology. Scient. Monthly, 19.

MACKINTOSH, N. A. 1947. The natural history of whalebone whales. Smiths. Rept. (for 1946), 235.

PACKARD, E. L. and R. KELLOGG. 1934. A new cetothere from the Miocene Astoria formation of Newport, Oregon. Carnegie Inst. Washington Publ. 447, 1.

PETERS, N. 1939. Über Grobe, Wachstum und Alter des Blauwales [Balaenoptera musculus (L) und Finnwales (Balaenoptera physalus) (L)]. Zool. Anz., parts 7, 8; 193.

RIDEWOOD, W. G. 1922. Observations on the skull in foetal specimens of whales of the genera Megaptera and Baleanoptera. Philos. Trans. Roy. Soc. London, 209.

SCHULTE, H. VON W. 1916. Anatomy of a foetus of Balaenoptera borealis. Mem. Amer. Mus. Nat. Hist., part 6, 389.

TRUE, F. W. 1904. The whalebone whales of the western North Atlantic compared with those occurring in European waters, with some observations on the species of the North Pacific. Smiths. Contrib., vol. 33.

PART V

THE DEVIOUS PATHS TO MAN ANTHROPOLOGY: PRIMATES (Cf. Vol. I, Chapters XXIII–XXV)

Anthropology: General (See also Man)

BOAS, F. and others. 1938. General anthropology. Boston. New York.

DARWIN, C. 1874. The descent of man and selection in relation to sex. 2d ed. 1909. New York.

DENIKER, J. 1915. The races of man. New ed. New York.

HAECKEL, E. 1903. Anthropogenie oder Entwickelungsgeschichte des Menschen. Revised ed. 2 vols. 1st ed. 1874. Leipzig.—1908. Unsere Ahnenreihe . . . Jena.

HOOTON, E. A. 1942. Man's poor relations. New York.—1946. Up from the ape. Revised ed. New York.

HOWELLS, W. 1944. Mankind so far. Garden City, N. Y.

KEANE, A. H. 1900. Man, past and present. Cambridge Univ. Press.

KROBER, A. L. 1948. Anthropology. Race, language, culture, psychology, prehistory. New York.

MARTIN, R. 1928. Lehrbuch der Anthropologie. II. Bd. Kraniologie, Osteologie. 2d ed. Jena.

MCGREGOR, J. H. 1938. Human origins and early man. In: Boas, F. and others. General anthropology, 24. New York.

ROMER, A. S. 1933. Man and the vertebrates. 3d ed. 1941. Univ. Chicago Press.

WEIDENREICH, F. 1946. Apes, giants and man. Univ. Chicago Press.

Primates: General (Systematic Works on Fossil and Recent Primates)

BLAINVILLE, H. M. D. DE. 1834. [Classification of "les Primates" adopted in 1834, published by Gervais in 1836 in: Dictionaire pittoresque d'Histoire naturelle, vol. 4, p. 619.]—1839–1864. Ostéographie ou description inconographique comparée du squelette et du système dentaire des mammifères recent et fossiles. 4 vols. Text, 4to; atlas, folio. Paris.

CLARK, W. E. LE G. 1930. The classification of the primates. Nature, 236.—1936. The problem of the claw in primates. Proc. Zool. Soc. London, part 1, 1.—1949. History of the primates: An introduction to the study of fossil man. Brit. Mus. (Nat. Hist.). London.

ELLIOT, D. G. 1912. A review of the primates. Monogr. Amer. Mus. Nat. Hist. 4 vols. 4to.

HEMPRICH, W. 1820. Grundriss der Naturgeschichte für höhere Lehranstalten. Entworfen von Dr. W. Hemprich. Berlin and Vienna. [Division of monkeys into Platyrrhina and Catarrhina.]

HILL, J. P. 1932. The developmental history of the primates. Philos. Trans. Roy. Soc. London, 45.

OWEN, R. 1868. On the anatomy of vertebrates. Vol. 3. Mammals. London.

REGAN, C. T. 1930a. The evolution of the primates. Ann. Mag. Nat. Hist., 383.—1930b. The classification of the primates. Nature, 125. [Microstructure of enamel of teeth.]

STRAUS, W. L., JR. 1931. The form of the tracheal cartilages of primates, with remarks on supposed taxonomic importance. Jour. Mammal., 281.

WISLOCKI, G. B. 1929. On the placentation of primates, with a consideration of the phylogeny of the placenta. Carnegie Inst. Washington Publ. 394, 51.

Primates: Fossil
(See also Lemuroids, Tarsioids, etc.)

ABEL, O. 1931. Die Stellung des Menschen im Rahmen der Wirbeltiere. Jena.

GIDLEY, J. W. 1923. Paleocene primates of the Fort Union, with discussion of relationships of Eocene primates. Proc. U. S. Nat. Mus., art. 1, 1.

OSBORN, H. F. 1902. American Eocene primates and the supposed rodent family Mixodectidae. Bull. Amer. Mus. Nat. Hist., 169.

SCHLOSSER, M. 1887–1890. Die Affen, Lemuren, etc. des europäischen Tertiärs. Beiträge Palaeont. Oesterreich-Ungarns und des Orients, vol. 6, Part I, 1.—1888. Part II. Ibid., vol. 7, 225.—1890. Part III. Ibid., vol. 8, 387.

SIMPSON, G. G. 1930. Primates and pangolins from the Asiatic expeditions. Amer. Mus. Novitates, no. 429.—1935. The Tiffany fauna, Upper Paleocene. III. Primates, Carnivora, Condylarthra, and Amblypoda. Ibid., no. 817.—1940, Studies on the earliest primates. Bull. Amer. Mus. Nat. Hist., 185.

TEILHARD DE CHARDIN, P. 1916–1921. Sur quelques primates des Phosphorites du Quercy. Ann. d. Paléont., vol. 10, 1.

WORTMAN, J. L. 1903–1904. Studies of Eocene Mammalia in the Marsh Collection, Peabody Museum. Part II. Primates. [No. 1.] Amer. J. Sci., vol. 15, no. 87, 163. [No. 2.] Classification of the Primates, no. 89, 399. [No. 3.] Origin of the Primates, no. 90, 419.

Tupaioids

ALLEN, J. A. 1922. The American Museum Congo Expedition Collection of Insectivora. Bull. Amer. Mus. Nat. Hist., vol. 47, 1. [Tooth replacement in Rhynchocyon.]

CARLSSON, A. 1909. Die Macroscelididae und ihre Beziehungen zu den übrigen Insectivoren. Zool. Jahrb., part 4, 349.—1922. Über die Tupaiidae und ihre Beziehungen zu den Insectivora und den Prosimiae. Acta Zool., vol. 3, 227.

CLARK, W. E. LE G. 1924a. The myology of the tree-shrew (Tupaia minor). Proc. Zool. Soc. London, part 2, 461.—1924b. On the brain of the tree-shrew (Tupaia minor). Ibid., part 4, 1053.—1925. On the skull of Tupaia. Ibid., part 2, 559.—1926. On the anatomy of the pen-tailed tree-shrew (Ptilocercus lowii). Ibid., part 4, 1179.—1927. On the tree-shrew (Tupaia minor). Ibid., part 1, 254.—1928. On the brain of the Macroscelididae (Macroscelides and Elephantulus). Jour Anat. (London), vol. 62, part 3, 245.—1929. The thalamus of Tupaia minor. Ibid., vol. 63, part 2, 177.

DAVIS, D. D. 1938. Notes on the anatomy of the treeshrew Dendrogale. Field Mus. Nat. Hist., Zool. Ser., no. 30, 383.

EVANS, F. G. 1942. The osteology and relationships of the elephant shrews (Macroscelididae). Bull. Amer. Mus. Nat. Hist., 85.

FRECHKOP, S. 1930. Notes sur les mammifères. IV. Au sujet des crânes des Tupaiidae (Insectivores). Bull. d. Mus. roy. d'Hist. nat. d. Belgique, vol. 6, no. 21.—1931. Notes sur les mammifères. V. Note preliminaire sur la dentition et la position systematique des Macroscelididae. Ibid., vol. 7, no. 5.

HENCKEL, K. O. 1928. Das Primordialkranium von Tupaia und der Ursprung der Primaten. Zeits. f. Anat. u. Entw., Parts 1, 2, 204.

JEPSEN, G. L. 1934. Revision of the American Apatemyidae and the description of the new genus, Sinclairella, from the White River Oligocene of South Dakota. Proc. Amer. Philos. Soc., vol. 74, no. 4, 287.

KLAAUW, C. J. VAN DER. 1929. On the development of the tympanic region of the skull in the Macroscelididae. Proc. Zool. Soc. London, 491.

LYON, M. W., JR. 1913. Treeshrews: An account of the mammalian family Tupaiidae. Proc. U. S. Nat. Mus., 1.

MATTHEW, W. D. 1917. The dentition of Nothodectes. Bull. Amer. Mus. Nat. Hist., 831.

SIMPSON, G. G. 1931. A new insectivore from the Oligocene, Ulan Gochu horizon, of Mongolia. Amer. Mus. Novitates, no. 505. [Anagale tending to unite tupaioids with Eocene lemuroids.]

—1935. The Tiffany fauna, Upper Paleocene. II. Structure and relationships of Plesiadapis. Ibid. no. 816.

TROXELL, E. L. 1923. The Apatemyidae. Amer. Jour. Sci., 503.

Lemuroids

AFFOLTER. 1937. Les organes cutanes brachiaux d'Hapalemur griseus. L'Acad. Malgache, 1.

BLUNTSCHLI, H. 1938. Die Sublingua und Lyssa der Lemuriden-Zunge. Anat. Inst. Univ. Bern, part 2, 127.

CARLETON, A. 1936. The limb-bones and vertebrae of the extinct lemurs of Madagascar. Proc. Zool. Soc. London, part 1, 281.

CLARK, W. E. LE G. 1931. The brain of Microcebus murinus. Proc. Zool. Soc. London, part 2, 463.—1934. On the skull structure of Pronycticebus Gaudryi. Ibid., part 1, 19. [Pronycticebus commonly regarded as a lorisid, allied to Adapis and Notharctus.]—1945. Note on the palaeontology of the lemuroid brain. Jour. Anat., part 3, 123.

GRANDIDIER, A. 1875. Histoire, physique, naturelle et politique de Madagascar. Vol. IX. Histoire naturelle des mammifères, by M. Alph. Milne Edwards et Alf. Grandidier, tome 4. Paris.

GRANDIDIER, G. 1902. Observations sur les lémuriens disparus de Madagascar . . . Bull. Mus. Hist. nat., Paris, 497, 507.—1905. Recherches sur les lémuriens disparus et en particulier sur ceaux qui vivaient à Madagascar. Nouv. Archives du Mus., vol. 7.

GRANGER, W. and W. K. GREGORY. 1917. A revision of the Eocene primates of the genus Notharctus. Bull. Amer. Mus. Nat. Hist., 841.

GREGORY, W. K. 1915a. On the relationship of the Eocene lemur Notharctus to the Adapidae and to other primates. Bull. Geol. Soc. Amer., vol. 26, 419.—1915b. On the classification and phylogeny of the Lemuroidea. Ibid., 426.—1920. On the structure and relations of Notharctus, an American Eocene primate. Mem. Amer. Mus. Nat. Hist., part 2, 49.

HELLER, F. 1930. Die Säugetierfauna der mitteleozänen Braunkohle des Geiseltales bei Halle a. S. Jahrb. d. Halleschen Verbandes, 13.

HENCKEL, K. O. 1927. Das Primordialkranium der Halbaffen und die Abstammung der höheren Primaten. Verhandl. d. Anat. Gesellsch., 108.

HILL, W. C. O. 1936. The affinities of the lorisoids. Ceylon Jour. Sci., part 3, 287.

HÜRZELER, J. 1946a. Zur Charakteristik, systematischen Stellung. Phylogenese und Verbreitung der Necrolemuriden aus dem europäischen Eocaen (Vorläufige Mitteilung). Bericht d. Schweiz. Paläontol., Eclogae geol. Helvetiae, no. 2, 352.—1946b. Gesneropithex Peyeri nov. gen. nov. spec., ein neuer Primate aus dem Ludien von Gösgen (Solothurn). Ibid., 354.

INCE, F. E. 1937. A contribution to a normal table of the development of Loris lydekkerianus. Trans. Zool. Soc. London, part 2, 43.

LAMBERTON, C. 1937. Contribution à la connaissance de la faune subfossile de Madagascar. Note III. Les Hadropithèques. Bull. de l'Acad. Malgache, 1. [Archaeolemur, palate and dentition, pl. 3.]—1938. Dentition de lait de quelques lémuriens subfossiles malgaches. Lab. de Zool. des Mammifères, no. 2, 57.—1939. Contribution à la connaissance de la faune subfossile de Madagascar. Notes IV–VIII. Lémuriens et Cryptoproctes. Mém. de l'Acad. Malgache, part 27, 1.—1941. Note IX. Oreille osseuse des lémuriens. Ibid., 1.—1947. Note XVI. Bradytherium ou Palaeopropithecus? Bull. d. l'Acad. Malgache (1944–1945), 1.

LOWTHER, F. DE L. 1939. The feeding and grooming habits of the galago. Zoologica (N. Y.), part 4, 477.—1940. A study of the activities of a pair of Galago senegalensis moholi in captivity, including the birth and postnatal development of twins. Ibid, part 4, 433.

MAJOR, C. I. F. 1894. On "Megaladapis madagascariensis," an extinct gigantic lemuroid from Amholisatra. Philos. Trans. Roy. Soc. London, 15.

MILLER, R. A. 1943. Functional and morphological adaptations in the forelimbs of the slow lemurs. Amer. Jour. Anat., no. 2, 153.

MILNE EDWARDS, and A. GRANDIDIER. 1875. Histoire naturelle des mammifères. Ordre des Lémuriens, famille des Indrisinés. Histoire physique, naturelle et politique de Madagascar, publiée par Alfred Grandidier. Vol. I, Text, Atlas.

OWEN, R. 1866. On the aye-aye (Chiromys, Cuvier . . .). Trans. Zool. Soc. London, art, 2, 33.

POCOCK, R. I. 1918. On the external characters of the lemurs and of Tarsius. Proc. Zool. Soc. London, part 1, 19.

RAU, A. S. and N. S. SAHASRABUDHE and others. 1930–1931. Contributions to our knowledge of the anatomy of the Lemuroidea. III. The skull of Loris lydekkerianus. Jour. of Mysore Univ., vol. 4, no. 2, 1.—1931. IV. The vertebral column and the appendicular skeleton of Loris lydekkerianus. Ibid., vol. 5, no. 1. V. The orbit of Loris lydekkerianus. Ibid., no. 2.

SIMPSON, G. G. 1935. The Tiffany fauna, Upper Paleocene. II. Structure and relationships of Plesiadapis. Amer. Mus. Novitates, no. 816.

SMITH, G. E. 1903. Further notes on the lemurs, with especial reference to the brain. Linnean Soc.'s Jour.-Zool., 80.—1907. On the form of the brain in the extinct lemurs of Madagascar, with some remarks on the affinities of the Indrisinae. Trans. Zool. Soc. London, part 2, 163. [Appendix by H. F. Standing, 1908.]

STANDING, H. F. 1908. On recently discovered subfossil primates from Madagascar. Trans. Zool. Soc. London, part 2, 59.

STEHLIN, H. G. 1912. Die Säugetiere des schweizerischen Eocaens. VII. Teil. I. Halfte. Einleitende Bemerkungen zum Genus Adapis und zu der Gruppe des Adapis pariensis Blainville. Abh. d. schweiz. palaeont. Gesellsch., 1165.

STEIN, M. R. 1936. The myth of the lemur's comb. Amer. Nat., 19.

STRAUS, W. L., JR. and G. B. WISLOCKI. 1932. On certain similarities between sloths and slow lemurs. Bull. Mus. Comp. Zool. Harvard Coll., no. 3, 45.

WORTMAN, J. L. 1903. Studies of Eocene Mammalia in the Marsh Collection, Peabody Museum. Part II. Primates. [No. 4.] Suborder Cheiromyoidea. Amer. Jour. Sci., vol. 16, 345. ["Metacheiromys" supposed primate; later shown to be related to the edentates.]—1904. [No. 5.] Suborder Anthropoidea. Ibid., vol. 17, 219. [Notharctus and its relatives regarded as related to South American monkeys.]

Tarsioids

ANTHONY, R. and H. V. VALLOIS. 1926. Catalogue raisonné . . . d'Ostéologie . . . d. Mus. d'Hist. nat. Mammifères. Fasc. IV. Primates. Sous-Fasc. 1.- Tarsiidae. Paris. [Osteology of Tarsius, p. 23.]

COOPER, C. F. 1910. Microchoerus erinaceus (Wood). Ann. Mag. Nat. Hist., 39.

COPE, E. D. 1872. On a new vertebrate genus from the northern part of the Tertiary basin of Green River. Proc. Amer. Philos. Soc., 554. [Describes Anaptomorphus aemulus.]—1882a. An anthropomorphous lemur. Amer. Nat., 73.—1882b. Contributions to the history of the Vertebrata of the Lower Eocene of Wyoming and New Mexico . . . I, II. Proc. Amer. Philos. Soc., 139. [Anaptomorphus homunculus, 152.]

CLARK, W. E. LE G. 1930. The thalamus of Tarsius. Jour. Anat., (England), part 4, 371.

FULTON, J. F. 1939. A trip to Bohol in quest of Tarsius. Yale Jour. Biol. and Med., 561.

HENCKEL, K. O. 1930. Zur Entwickl. des Fussskeletts von Tarsius spectrum L. Morphol. Jahrb., parts 3, 4, 636.

HÜRZELER, J. 1948. Zur Stammesgeschichte der Necrolemuridae. Schweiz. Palaeont. Abh., Bd. 66, 1. [Concludes that Necrolemuridae are not properly classed as tarsioids, chiefly because the true tympanic is enclosed within the bulla.]

HUNTER, J. I. 1923. The oculomotor nucleus of Tarsius and Nycticebus. Brain, part 1, 2.

MATTHEW, W. D. and W. GRANGER. 1915. A revision of the Lower Eocene Wasatch and Wind River faunas. Part IV. Entelonychia, Primates, Insectivora (part), by W. D. Matthew. Bull. Amer. Mus. Nat. Hist., 429. [Tetonius (Anaptomorphus) homunculus, p. 456.]

SMITH, G. E. 1920. Discussion on the zoological position and affinities of Tarsius. Proc. Zool. Soc. London, 465.—1924. The evolution of man: Essays. London. [Brain of Tarsius, fig. 15.]

STEHLIN, H. G. 1916. Die Säugetiere des schweizerischen Eocaens. VII. Teil, II. Hälfte. Caenopithecus- Necrolemur- Microchoerus- Nannopithex- Anchomomys- Periconodon- Amphichiromys- Heterochiromys- Nachträge zu Adapis.- Schlussbetrachtungen zu den Primaten. Abh. d. schweiz. Palaeont. Gesellsch., 1299. [Mostly tarsioids and plesiadapids.]

TILNEY, F. 1927. The brain stem of Tarsius: A critical comparison with other primates. Jour. Comp. Neurol., no. 3, 371.

WOOLLARD, H. H. 1925. The anatomy of Tarsius spectrum. Proc. Zool. Soc. London, no. 70, part 3, 1071.

WORTMAN, J. L. 1904a. Studies of Eocene Mammalia in the Marsh Collection, Peabody Museum. Part II. Primates. [No. 6] Omomys pusillus Marsh. Amer. Jour. Sci., vol. 17, 133.—1904b. [No. 7] On the affinities of the Omomyinae. Ibid., 203.

New World Monkeys (Ceboids)

BARTH, M. 1918. Über die funktionele Struktur des oberkiefer-apparates bei Neuweltaffen. Anat. Heft., parts 168, 169, 173.

BEATTIE, J. 1927. The anatomy of the common marmoset (Hapale jacchus Kuhl). Proc. Zool. Soc. London, 593.

BELT, T. 1874. The naturalist in Nicaragua . . . Revised ed., 1888. London. [Intelligence and behavior of white-faced Cebus monkey, pp. 92, 93.]

BLUNTSCHLI, H. 1929. Ein eigenartiges an Prosimierbefunde erinnerndes Nagelverhalten am Fuss von platyrrhinen Affen. Zeitschr. f. wissensch. Biol., Festschr. für Hans Spemann, III. Teil, 1.—1931. Homunculus patagonicus und die ihm zügereiheten Fossilfunde aus den Santa-Cruz-Schichten Patagoniens. Morphol. Jahr., 811.

CABRERA, A. 1939. Los monos de la Argentina. Physis, Revista de la Soc. Argentino de Cienc. Nat., vol. 16, Sec. Zool. (Vertebrados), 1.

MONTAGU, F. M. A. 1929. The discovery of a new anthropoid ape in South America? Scient. Monthly, 275. [Supposed "anthropoid ape," probably a large spider monkey.]

POCOCK, R. I. 1925. Additional notes on the external characters of some playtrrhine monkeys. Proc. Zool. Soc. London, part 1, art. 3, 27.

RUSCONI, C. 1935. Las especies de primates del oligocene de Patagonia (gen. Homunculus). Parts I–VIII. Revista Argentina de Paleont. y Antropol., Ameghinia, vol. 1, no. 2, 39; Parts IX–XIV, ibid., no. 3, 71; Parts XV–XVII, ibid., no. 4, 103.

SCHULTZ, A. H. 1926. Studies on the variability of platyrrhine monkeys. Jour. Mammal., no. 4, 286.

Old World Monkeys (Cercopithecoids)

BROOM, R. 1940. The South African Pleistocene cercopithecid apes. Ann. Transvaal Mus., part 2, 89.

HARTMAN, C. G., G. W. CORNER and others. 1941. Embryology of the rhesus monkey (Macaca mulatta). Carnegie Inst. Washington Publ. 538.

HARTMAN, C. G. and W. L. STRAUSS, JR. (Editors) 1933. The anatomy of the rhesus monkey (Macaca mulatta). Baltimore.

LEAKEY, L. S. B. 1943. Notes on Simopithecus oswaldi Andrews from the type site. Jour. E. Africa Nat. Hist. Soc., nos. 1, 2; 39–44.

MOLLISON, T. 1924. Zur systematischen Stellung des Parapithecus fraasi Schlosser. Zeitchr. f. Morphol. u. Anthropol., 205.

OSBORN, H. F. 1908. New fossil mammals from the Fayûm Oligocene, Egypt. Bull. Amer. Mus. Nat. Hist., 265. [Apidium, possibly a primitive monkey.]

POCOCK, R. I. 1926. The external characters of the catarrhine monkeys and apes. Proc. Zool. Soc. London (1925), part 4, 1479.—1928. Langurs or leaf monkeys of British India. Jour. Bombay Nat. Hist. Soc., nos. 3, 4, 472.—1932. The rhesus macaques (Macaca mulatta). Ibid., no. 3, 530.

SCHULTZ, A. H. 1937. Fetal growth and development of the rhesus monkey. Carnegie Inst. Washington Publ. 479, 71.—1942. Growth and development of the proboscis monkey. Bull. Mus. Comp. Zool. Harvard Coll., no. 6, 279.

WASHBURN, S. L. 1944. The genera of Malaysian langurs. Jour. Mammal., no. 3, 289.

WISLOCKI, G. B. and G. L. STREETER. 1938. On the placentation of the macaque (Macaca mulatta) from the time of implantation until the formation of the definitive placenta. Carnegie Inst. Washington Publ. 496.

ZUCKERMAN, S. 1926. Growth-changes in the skull of the baboon, Papio porcarius. Proc. Zool. Soc. London, 843. [Transition from skull type of macaque to that of baboon conforms to double logarithmic growth curve.]

Anthropoids (Fossil and Recent Anthropoid Apes)

ALLEN, J. A. 1925. Primates collected by the American Museum of Natural History Congo Expedition. Bull. Amer. Mus. Nat. Hist., 283.

BABOR, J. F. and Z. FRANKENBERGER. 1930–1932. Studien zur Naturgeschichte des Gorillas. I. Teil. Biologia Generalis, vol. 6, part 4, 553. II. Teil. Ibid., vol. 7, part 3, 367. III. Teil. Beitrag zur Morphologie und Morphogenese des Grosshirns des Gorillas. Zeitschr. f. Anat. u. Ent., vol. 97, part 6, 780.

BAUMAN, J. E. 1923. The strength of the chimpanzee and orang. Scient. Monthly, no. 4, 432.

BINGHAM, H. C. 1929. Observations on growth and development of chimpanzees. Amer. Jour. Phys. Anthrop., no. 3.—1932. Gorillas in a native habitat. [Rept. of exped. from Yale Univ.

and Carnegie Inst. of Washington on Mountain Gorilla in Parc. Nat. Albert, Belgian Congo.]

BOLK, L. 1925. On the existence of a dolichocephalic race of Gorilla. Kon. Akad. van Wet. te Amsterdam, no. 2, 204.

BRANDES, R. 1931. Über den Kehlkopf des Orang-Utan in verschiedenen Altersstadien mit besonderer Berücksichtigung der Kehlsackfrage. Morphol. Jahrb., parts 1, 2.

BRUHN, J. M. and F. G. BENEDICT. 1936. The respiratory metabolism of the chimpanzee. Proc. Amer. Acad. Arts and Sci., vol. 71, no. 5, November.

CLARK, W. E. LE. G. 1948. Skull of Proconsul from Rusinga Island. Nature, Oct. 30.

COLBERT, E. H. 1937. A new primate from the Upper Eocene Pondaung formation of Burma. Amer. Mus. Novitates, no. 951. [Amphipithecus Colbert, a very primitive ape with three lower premolars.]

COMELLA, J. F. DE V. and M. C. PAIRO. 1947. Dos nuevos antropomorfos de Mioceno Español y su Situacion dentro de la moderna Sistemática de los Simidos. Notas y Comm. del Inst. Geol. y Min. de España, 91. [Upper Miocene Sivapithecus occidentalis; Villalta-Crusafont Pontian: Hispanopithecus laietanus; lower molars.]

COOLIDGE, H. J., JR. 1929. A revision of the genus Gorilla. Mem. Comp. Zool. Harvard Coll. no 4, 293.—1933. Pan paniscus; pigmy chimpanzee from south of the Congo River. Amer. Jour. Phys. Anthropol., no. 1.

DEPÉRET, C. 1911. Sur la decouverte d'un grand singe anthropoide du genre Dryopithecus dans le Miocène moyen de la Grive-Saint-Alban. Comptes rendus, vol. 153, 32.

EHRENBERG, K. 1938. Austriacopithecus, ein neuer menschenaffenartiger Primate aus dem Miozän von Klein-Hadersdorf bei Poysdorf in Niederösterreich (Nieder-Donau). Sitz. d. Akad. d. Wissenschaften in Wien, part 4, 71. [Humerus referred by Schlosser to Dryopithecus.]

FRECHKOP, S. 1935. Notes sur les mammifères. XVII.- À propos du chimpanzé de la rive gauche du Congo. Bull. d. Mus. roy. d'Hist. nat. d. Belgique, no. 2.

GAUDRY, A. 1890. Le Dryopithèque. Mém. Soc. Géol. France, Paléontol. I., part 1.

GREGORY, W. K. 1915. Is Sivapithecus Pilgrim an ancestor of man? Science, 341.—1927. Hesperopithecus apparently not an ape nor a man. Ibid., 579. [Much worn type tooth compares with less worn first upper premolar of a large peccary from the same locality and level.]

GREGORY, W. K., B. BROWN and M. HELLMAN. 1924. On three incomplete anthropoid jaws from the Siwaliks, India. Amer. Mus. Novitates, no. 130.

GREGORY, W. K. and M. HELLMAN. 1923a. Notes on the type of Hesperopithecus haroldcookii Osborn. Amer. Mus. Novitates, no. 53.—1923b. Further notes on the molars of Hesperopithecus and of Pithecanthropus. Bull. Amer. Mus. Nat. Hist., 509. [Cf. Gregory, 1927.]—1926. The dentition of Dryopithecus and the origin of man. Anthropol. Papers, Amer. Mus. Nat. Hist., part 1.

GREGORY, W. K., M. HELLMAN and G. E. LEWIS. 1938. Fossil anthropoids of the Yale-Cambridge India Expedition of 1935. Carnegie Inst. Washington Publ. 495.

GREGORY, W. K. and H. C. RAVEN. 1935–1936. In quest of gorillas. Scient. Monthly, November 1935–December 1936, vols. 41–43.—1937. In quest of gorillas. New Bedford, Mass.

GYLDENSTOLPE, N. 1928. Zoological results of the Swedish Expedition to Central Africa, 1921. Vertebrata. No. 5. Mammals from the Birunga volcanoes north of Lake Kivu. Arkiv. f. Zool., no. 4.

HOFMANN, A. 1893. Die Fauna von Göriach. Abh. d. K. K. geol. Reichsanstalt, 1. Wien. 4to. [Hylobates (Pliopithecus) antiquus.]

HOOIJER, D. A. 1948. Prehistoric teeth of man and of the orang-utan from Central Sumatra, with notes on the fossil orang-utan from Java and Southern China. Zool. Mededeel., vol. 29, 175.

HOPWOOD, A. T. 1933a. Miocene primates from British East Africa. Ann. Mag. Nat. Hist., 96. [Proconsul africanus, gen. et. sp. nov.]—1933b. Miocene primates from Kenya. Linnean Soc.'s Jour.-Zool., 437. [Proconsul, pl. 6.]

HÜRZELER, J. 1947. Alsaticopithecus Leemanni nov. gen. nov. spec., ein neuer Primate aus dem unteren Lutétien von Buchsweiler im Unterelsass. Ber. Schweiz. Pal. Ges., 26 Jahresversam. Eclogae geol. Helvetiae, vol. 40, no. 2, 343. [Upper and lower teeth possibly related to Propliopithecus and Parapithecus but left incertae sedis.]

KEITH, A. 1926. The gorilla and man as contrasted forms. The Lancet, 490.

KOENIGSWALD, G. H. R. VON. 1935. Eine fossile Säugetierfauna mit Simia aus Südchina. Proc. Kon. Akad. van Wet. te Amsterdam, no. 8, 872.

LARTET, E. 1837. Note . . . sur la découverte récente d'une machoire de singe fossile. Comptes rendus,, vol. 4, 85. Paris.—1856. Note sur un grand singe fossile qui se rattache au groupe des singes supérieurs. Ibid., vol. 43, 219.

LEAKEY, L. S. B. 1934. Adam's ancestors: An up-to-date outline of what is known about the origin of man. New York.—1943. A Miocene anthropoid mandible from Rusinga, Kenya. Nature, 319.—1946. Ape or primitive man? Interpretation of fossil finds in Kenya. The Times, London, Aug. 24.

LEWIS, G. E. 1934. Preliminary notice of new man-like apes from India. Amer. Jour. Sci., 161. —1936. A new species of Sugrivapithecus. Ibid., 450.—1937. A new Siwalik correlation. Ibid., 191.

MACINNES, D. G. 1943. Notes on the East African Miocene primates. Jour. East Africa and Uganda Nat. Hist. Soc., nos. 3-4 (77, 78), 141. [Proconsul, pls. 24–28.]

NOBACK, C. V. 1930. Growth of infant gorilla. Amer. Jour. Phys. Anthropol., no. 2, 165.

OSBORN, H. F. 1922a. Hesperopithecus, the first anthropoid primate found in America. Amer. Mus. Novitates, no. 37.—1922b. Hesperopithecus, the anthropoid primate of western Nebraska. Nature, 281. [Cf. Gregory, 1927.]

PILGRIM, G. E. 1915. New Siwalik primates and their bearing on the question of the evolution of man and the Anthropoidea. Rec. Geol. Surv. India, part 1. [Suggests man from Sivapithecus.] —1927. A Sivapithecus palate and other primate fossils from India. Mem. Geol. Surv. India, Palaeontologia Indica, 1.

PIRA, A. 1913. Beiträge zur Anatomie des Gorilla. (I) Morphol. Jahr., parts 1, 2, 309.—1914. (II) Ibid., part 2, 12.

PITMAN, C. R. S. 1935. The gorillas of the Kayonsa Region, western Kigezi, S. W. Uganda. Proc. Zool. Soc. London, part 3, 477.

POHLIG, H. 1895. Paidopithex rhenanus, n.g.n.sp., le singe anthropomorph du Pliocène rhénan. Bull. Soc. Belge d. Géol., 1.

RAVEN, H. C. 1931. Gorilla, greatest of all apes. Nat. Hist. Mag., no. 3, 231.

RAVEN, H. C. and others. 1950. The Henry Cushier Raven Memorial Volume on the anatomy of the gorilla. 4to.

SCHLOSSER, M. 1903. Antnropodus oder Neopithecus. Centralbl. f. Min., etc., 512. [Related to Dryopithecus.]—1911. Oligocänen Landsäugetiere aus dem Fayûm. Beiträge zur Paläont. u. Geol. Osterreich-Ungarns u. d. Orients., 51. [Parapithecus, Propliopithecus.]

SCHULTZ, A. H. 1927. Studies on the growth of gorilla and of other higher primates, with special reference to a fetus of gorilla, preserved in the Carnegie Museum. Mem. Carnegie Mus., no. 1.—1930. Notes on the growth of anthropoid apes. Rept. Laboratory and Museum of Comp. Pathology, Philadelphia, 34.—1933a. Observations on the growth, classification and evolutionary specialization of gibbons and siamangs. Human Biol., vol. 5, no. 2, 212.—1933b. Notes on the fetus of an orang-utan with some comparative observations. Rept. Laboratory and Museum of Comp. Pathology, Philadelphia.—1933c. Die Körperproportionen der erwachsenen catarrhinen Primaten, mit spezieller Berücksichtigun der Menschenaffen. Anthropol. Anz., parts 2, 3, 154.—1934. Some distinguishing characters of the mountain gorilla. Jour. Mammal., no. 1, 51.—1937. Die Körperproportionen der afrikanischen Menschenaffen im foetalen und im erwachsenen Zustand. Neue Forsch. in Tierzucht u. Abstammungslehre, 1. —1940. Growth and development of the chimpanzee. Carnegie Inst. Washington Publ. No. 518.—1941. Growth and development of the orang-utan. Ibid., no. 525.—1942. Morphological observations on a gorilla and an orang of closely known ages. Amer. Jour. Phys. Anthrop., no. 1, 1.—1944. Age changes and variability in gibbons. A morphological study on a population sample of a man-like ape. Ibid., no. 1.

SCHWALBE, G. 1915. Über den fossilen Affen Oreopithecus Bambolii: Zugleich ein Beitrag zur Morphologie der Zähne der Primaten. Zeitschr. f. Morphol. u. Anthropol., part 1, 149.

SELENKA, E. 1898. Studien über Entwicklungsgeschichte der Tiere. IV. Menschenaffen. I. Rassen, Schädel und Bezahnung des Orangutan. Wiesbaden.—1899. II. Schädel des Gorilla und Schimpanse. Ibid., 92.

SONNTAG, C. F. On the anatomy, physiology, and pathology of the orang-outan. Proc. Zool. Soc. London, part 2, no. 24, 349.

STRAUS, W. L., JR. 1936. The thoracic and abdominal viscera of primates with special reference to the orang-utan. Proc. Amer. Philos. Soc., no. 1, 1.—1942. The structure of the crown-pad of the gorilla and of the cheek-pad of the orang-utan. Jour. Mammal., no. 3, 276.

STROMER, E. 1928. Bemerkungen über den tertiären Gibbon Pliopithecus antiquus. Ber. d. Naturw. Vereins f. Schwaben u. Neuburg, 64.

TYSON, E. 1699. Orang-Outang, sive Homo Sylvestris, or, the anatomy of a pygmie compared with that of a monkey, an ape, and a man. Roy. Soc. London. [A young chimpanzee.]

WISLOCKI, G. B. 1932. On the female reproductive tract of the gorilla with a comparison of that of other primates. Carnegie Inst. Washington Publ. 433, 163.—1933. Gravid reproductive tract and placenta of the chimpanzee. Amer. Jour. Phys. Anthropol., no. 1, 81.—1936. The external genitalia of the simian primates. Human Biol., no. 3, 309.

WOODWARD, A. S. 1914. On the lower jaw of an anthropoid ape (Dryopithecus fontani) from the Upper Miocene of Lérida (Spain). Quart. Jour. Geol. Soc., vol. 70, 316.

YERKES, R. M. 1925. Almost human. New York.

YERKES, R. M. and A. W. YERKES. 1929. The great apes: A study of anthropoid life. Yale Univ. Press.

South African Man-apes (Australopithecinae)

ABEL, W. 1931. Kritische Untersuchungen über Australopithecus africanus Dart. Morphol. Jahrb., part 4, 539.

ADLOFF, P. 1932. Das Gebiss von Australopithecus africanus. Dart. Einige ergänzende Bemerkungen zum Eckzahnproblem. Zeits. f. d. ges. Anat., 145.—1939. Die südafrikanischen fossilen Menschenaffen und der Ursprung des menschlichen Gebisses. Anthropol. Anz., nos. 1, 2, 72. [Small canines of, cited for his theory of origin of man.]

ARAMBOURG, C. 1943. Sur les affinités de quelques Anthropoides fossiles d'Afrique et leurs relations avec la lignée humaine. Comptes rendus. . . . 593.

BROOM, R. 1925. On the newly discovered South African man-ape. Nat. Hist. Mag., no. 4, 409.—1929. Note on the milk dentition of Australopithecus. Proc. Zool. Soc. London, part 1, no. 11, 85.—1934. On the fossil remains associated with Australopithecus. South African Jour. Sci., 471.—1936a. A new fossil anthropoid skull from South Africa. Nature, 486.—1936b. The dentition of Australopithecus. Ibid., 719.—1937a. The Sterkfontein ape. Ibid., 326.—1937b. Discovery of a lower molar of Australopithecus. Ibid., 681.—1938a. The Pleistocene anthropoid apes of South Africa. Ibid., 377.—1938b. New light on the origin of man. Forum, nos. 22, 23.—1938c. Further evidence on the structure of the South African Pleistocene anthropoids. Nature, 897.—1939a. Les singes anthropoides fossiles de l'Afrique du Sud et leurs relations à l'homme. Revue Scient., 172.—1939b. The dentition of the Transvaal Pleistocene anthropoids, Plesianthropus and Paranthropus. Ann. Transvaal Mus., 303.—1939c. A restoration of the Kromdraai skull. Ibid., 327.—1939d. On the affinities of the South African Pleistocene anthropoids. South African Jour. Sci., 408.—1941a. Structure of the Sterkfontein ape. Nature, 86.—1941b. Mandible of a young Paranthropus child. Ibid., no. 3733.—1942. The hand of the ape-man, Paranthropus robustus. Ibid., 513.—1943a. South Africa's part in the solution of the problem of the origin of man. South African Jour. Sci., 68.—1943b. An ankle-bone of the ape-man, Paranthropus robustus. Nature, 689.—1947a. The upper milk molars of the ape-man, Plesianthropus. Ibid., 602.—1947b. Discovery of a new skull of the South African ape-man, Plesianthropus. Ibid., 672.—1947c. The mandible of the Sterkfontein ape-man. South African Science, no. 1, 14.

BROOM, R. and J. T. ROBINSON. 1947a. Jaw of the male Sterkfontein ape-man, Plesianthropus. Nature, 153.—1937b. Further remains of the Sterkfontein ape-man, Plesianthropus. Ibid., 430.—1948. Size of the brain in the ape-man, Plesianthropus. Ibid., 438.

BROOM, R. and G. W. H. SCHEPERS. 1946. The South African fossil ape-men: the Australopithecinae. Part I. The occurrence and general structure of the South African ape-men, by Robert Broom. Part II. The endocranial casts of the South African ape-men, by G. W. H. Schepers. Transvaal Mus. Mem., no. 2.

CLARK, W. E. LE G. 1946. Significance of the Australopithecinae. Nature, 863.—1947a. Observations on the anatomy of the fossil Australopithecinae. Jour. Anat., part 3, 300.—1947b. The importance of the fossil Australopithecinae in the study of human evolution. Science progress, no. 129, 377.

COOKE, H. B. S. 1938. The Sterkfontein bone breecia: A geological note. South African Jour. Sci., 204.

DART, R. A. 1925. Australopithecus africanus: the man-ape of South Africa. Nature, 195.—1929. A note on the Taungs skull. South African Jour. Sci., 648.—1934. The dentition of Australopithecus africanus. Folia Anatomica Japonica, part 4, 207.—1940. The status of Australopithecus. Amer. Jour. Phys. Anthropol., 167.—1948a. The infancy of Australopithecus. In: Robert Broom Commemorative Volume, 143.—1948b. The Makapansgat protohuman Australopithecus prometheus. Amer. Jour. Phys. Anthropol., Sept., 259.—1948c. The adolescent mandible of Australopithecus prometheus. Ibid., Dec., 391. —1949. The predatory implemental technique of Autralopithecus. Ibid., March, 1.

GREGORY, W. K. 1939. The bearing of Dr. Broom's and Dr. Dart's discoveries on the origin of man. Assoc. Sci. and Technical Socs. South Africa, Ann. Proceedings (1938–1939), 25.

GREGORY, W. K. and M. HELLMAN. 1938. Evidence of the australopithecine man-apes on the origin of man. Science, vol. 88, 615.—1939a. Fossil man-apes of South Africa. Nature, 25.—1939b. The South African fossil man-apes and the origin of the human dentition. Jour. Amer. Dental Assoc., 558.—1939c. The dentition of the extinct South African man-ape Australopithecus (Plesianthropus) transvaalensis Broom. A comparative and phylogenetic study. Ann. Transvaal Mus., part 4, 339.—1940. The upper dental arch of Plesianthropus transvaalensis Broom, and its relations to other parts of the skull. Amer. Jour. Phys. Anthropol., 211.—1945. Revised reconstruction of the skull of Plesianthropus transvaalensis Broom. Ibid., 267.

HAUGHTON, S. H. 1948. Notes on the australopithecine-bearing rocks of the Union of South Africa. Trans. and Proc. Geol. Soc. S. Africa (1947), 55.

HRDLIČKA, A. 1925. The Taungs ape. Amer. Jour. Phys. Anthropol., no. 4.

JONES, T. R. 1937. A new fossil primate from Sterkfontein, Krugersdorp, Transvaal. South African Jour. Sci., 709.

KOENIGSWALD, G. H. R. VON. 1942. The South African man-apes and Pithecanthropus. Carnegie Inst. Washington Publ. 530, 205.

MILLER, G. S., JR. 1930. Note on the milk dentition of Australopithecus. Jour. Mammal., no. 1, 92.

SCHEPERS, G. W. H. 1948. Problems in brain evolution. In: Robert Broom Commemorative Volume, 191. Sulcal homologies; origin of the pseudo-lunate sulcus; transformations of the parietal lobe; phylogenesis of Heschl's convolution; evolution of the frontal lobe.

SCHWARZ, E. 1936. The Sterkfontein ape. Nature, 969.

SCHWARZ, R. 1931. Das Abstammungsproblem des Menschen. Die Fortschritte d. Zahnheilkunde. . . . 753.

SENYÜREK, M. S. 1941. The dentition of Plesianthropus and Paranthropus. Ann. Transvaal Mus., part 3, 293.

SHAW, J. C. M. 1939. Further remains of a Sterkfontein ape. Nature, 117.—1940. Concerning some remains of a new Sterkfontein primate. Ann. Transvaal Mus., part 2, 145.

SMITH, G. E. 1925. Australopithecus, the man-like ape from Bechuanaland. Illust. London News, Feb. 14, 240.

STRAUS, W. L., JR. 1948. The humerus of Paranthropus robustus. Amer. Jour. Phys. Anthropol., no. 3, Sept., 285.

Fossil Man: General

BOULE, M. and J. PIVETEAU. 1935. Les fossiles: Éléments de paléontologie. Paris.

FIELD, H. 1933. Prehistoric man, hall of the Stone Age of the Old World. Field Mus. Nat. Hist., Chicago.

GIESELER, W. 1936. Abstammungs-und Rassenkunde des Menschen (Anthropologie). Oehringen, Germany.

KEITH, A. 1925. The antiquity of man. Vols. I, II. London.—1930. New discoveries relating to the antiquity of man. New York.—1931. [Fossil evidence of man's evolution.] Brit. Assoc. Adv. Sci., Centenary Meeting . . . Sect. C, 35.

OSBORN, H. F. 1915. Men of the Old Stone Age: their environment, life and art. New York.—1944. The age of man. Amer. Mus. Nat. Hist. Science Guide, No. 52, revised by W. K. Gregory and G. Pinkley.

WEIDENREICH, F. 1940. Some problems dealing with ancient man. Amer. Anthropol., vol. 42, 375.—1946. Apes, giants and man. Univ. Chicago Press.

WEINERT, H. 1930. Menschen der Vorzeit. Stuttgart.

Fossil Man: Asia (including Malaysia), Meganthropus, Pithecanthropus, Sinanthropus, Homo soloensis, the Wadjak Skulls

ADLOFF, P. 1938. Das Gebiss von Sinanthropus pekinensis . . . Zeits. f. Morphol. u. Anthropol., part 3, 490.

BLACK, D. 1926. Tertiary man in Asia: The Chou Kou Tien discovery. Nature, Nov. 20.—1927. The lower molar hominid tooth from the Chou Kou Tien deposit. Paleont. Sinica, vol. 7, fasc. 1, Geol. Surv. China.—1929. Preliminary notice of the discovery of an adult Sinanthropus skull at Chou Kou Tien. Bull. Geol. Soc. China, no. 3, 207.—1931a. Preliminary report on the Sinanthropus lower jaw specimens recovered from the Choukoutien Cave deposit in 1930 and 1931. Ibid., no. 3, 241.—1931b. On an adolescent skull of Sinanthropus pekinensis in comparison with an adult skull of the same species and with other hominid skulls, recent and fossil. Palaeont. Sinica, vol. 7, fasc. 2.—1932. Skeletal remains of Sinanthropus other than skull parts. Bull. Geol. Soc. China, vol. 11, no. 4, 365.—1933. The brain cast of Sinanthropus- a review. Jour. Comp. Neurol., no. 2, 361.—1934. On the discovery, moiphology and environment of Sinanthropus pekinensis. Philos. Trans. Roy. Soc. London, 57.

BLACK, D., P. TEILHARD DE CHARDIN, C. C. YOUNG and W. C. PEI; with a Foreword by W. W. HAO. 1933. Fossil man in China: The Choukoutien Cave deposits with a synopsis of our present knowledge of the late Cenozoic in China. Mem. Geol. Surv. China, no. 11, 1.

CLARK, W. E. LE G. 1945. Pithecanthropus in Peking. Antiquity, 1.

DUBOIS, E. 1894. Pithecanthropus erectus, eine menschenaehnliche Uebergangsform aus Java. Batavia, Landesdruckerei, 1.—1896. On Pithecanthropus erectus: A transitional form between man and the apes. Scient. Trans. Roy. Dublin Soc., 1.—1922. The proto-Australian fossil man of Wadjak, Java. K. Akad. Wetensch. Amsterdam, Sect. Sci., part 2, 1013.—1924. On the principal characters of the cranium and brain of the mandible and the teeth of Pithecanthropus erectus. Ibid., 265.—1926a. On the principal characters of the femur of Pithecanthropus erectus. Ibid., 730.—1926b. Figures of the femur of Pithecanthropus erectus. Ibid., 1275.—1927. The so-called new Pithecanthropus skull. Ibid., no. 1, 134.—1933. The shape and the size of the brain in Sinanthropus and in Pithecanthropus. Ibid., no. 4, 415.—1935. On the gibbon-like appearance of Pithecanthropus erectus. Ibid., no. 6, 578.—1936. Racial identity of Homo soloensis Oppenoorth (including Homo modjokertensis von Koenigswald) and Sinanthropus pekinensis Davidson Black. Ibid., no. 10.—1937a. The osteone arrangement of the thigh-bone compacta of man identical with that, first found, of Pithecanthropus. Ibid., no. 10, 864.—1937b. On the fossil human skulls recently discovered in Java and Pithecanthropus erectus. Man, a Monthly Record of Anthropological Sci., no. 1.—1938a. The mandible recently described and attributed to the Pithecanthropus by G. H. R. von Koenigswald, compared with the mandible of Pithecanthropus erectus described in 1924 by Eug. Dubois. K. Akad. Wetensch. Amsterdam, no. 2, 139.—1938b. On the fossil human skull recently described and attributed to Pithecanthropus erectus by G. H. R. von Koenigswald. Ibid., no. 4, 1.

KAPPERS, C. U. A. 1933. The fissuration on the frontal lobe of Sinanthropus pekinensis Black, compared with the fissuration in Neanderthal man. K. Akad. Wetensch. Amsterdam, no. 9, 1.

KAPPERS, C. U. A. and K. H. BOUMAN. 1939. Comparison of the endocranial casts of the Pithecanthropus erectus skull found by Dubois and Von Koenigswald's Pithecanthropus skull. K. Akad. Wetensch. Amsterdam, no. 1.

KOENIGSWALD, G. H. R. VON. 1935. Ein fossile Säugetierfauna mit Simia aus Südchina. K. Akad. Wetensch. Amsterdam, 872.—1936a. Der gegenwärtige Stand des Pithecanthropus-Problems. Handel. 7de Ned.-Ind. Naturwetensch. Congres, 724.—1936b. Ein fossiler Hominide aus dem Altpleistocän Ostjavas. De Ingenieur in Nederlandsch-Indië, no. 8, 151. [Homo modjokertensis.]—1937. Ein Unterkieferfragment des Pithecanthropus aus den Trinilschichten Mitteljavas. K. Akad. Wetensch. Amsterdam, no. 10, 883.—1938a. Ein neuer Pithecanthropus-Schädel. Ibid., no. 2, 4.—1938b. Neue Pithecanthropus-Funde. Forschungen und Fortschritte, no. 19, 218.—1938c. Nieuwe Pithecanthropusvondsten uit Midden-Java. Natuur. Tijdsch. v. Nederlandsch.-Indië, van Deel 98,

195.—1939. Das Pleistocän Javas. Quartär, vol. 2, 28. [Homo modjokertensis, Pithecanthropus, pl. 10.]—1940. Neue Pithecanthropus-Funde 1936–1938: Ein Beitrag zur Kenntnis der Praehominiden. Wetensch. Mededeel., no. 28, 282. —1942. The South African man-apes and Pithecanthropus. Carnegie Inst. Washington Publ. 530, 205.—1949. The discovery of early man in Java and southern China. In: Studies in Phys. Anthropol., Viking Fund Symposium, no. 1., Ed. W. W. Howells, p. 83.

KOENIGSWALD, G. H. R. VON and F. WEIDENREICH. 1939. The relationship between Pithecanthropus and Sinanthropus. Nature, 926.

LICENT, E., P. TEILHARD DE CHARDIN and D. BLACK. 1927. On a presumably Pleistocene human tooth from the Sjara-Osso-Gol (Southeastern Ordos) deposits. Bull. Geol. Soc. China, nos. 3, 4, 285.

MACCURDY, G. G. 1936. Prehistoric man in Palestine. Proc. Amer. Philos. Soc., no. 4, 523.

MONTANDON, G. 1938. Perspectives de phylogénie humaine suggérées par la morphologie auriculaire et dentaire du Sinanthrope. Revue Scient., nos. 12–15, 561.

MOVIUS, H. L., JR. 1949. Lower Paleolithic Archaeology in Southern Asia and the Far East. In: Studies in Phys. Anthrop., Viking Fund Symposium, no. 1, Ed. W. W. Howells, p. 17. [Large chart, Tentative Correlation of Late Cenozoic events, prehistoric archaeology and early man in Southern and Eastern Asia.]

OPPENOORTH, W. F. F. 1932. Homo (Javanthropus) soloensis, een Plistoceene mensch van Java. Wetensch. Mededeel. Dienst. Mijnbouw Nederl.-Indië, no. 20, 49.

PATTE, E. 1947. Des Pithécanthropes aux hommes modernes. La Revue Scientifique, 85 Année, fasc. 7, 409.

PEI, W. C. 1936. Peking man. Nature, 1056.—1937. The fifth skull of Peking man. Ibid., no. 3507, vol. 139, 109.—1939. The upper cave industry of Choukoutien. Palaeont. Sinica, n.s., no. 9, 1.

PINKLEY, G. 1936. The significance of Wadjak man, a fossil Homo sapiens from Java. Peking Nat. Hist. Bull., part 3, 183.

SCHWALBE, G. 1899. Studien über Pithecanthropus erectus Dubois. Zeitschr. f. Morphol, u. Anthropol., part 1, 16.

SHELLSHEAR, J. L. and G. E. SMITH. 1934. A comparative study of the endocranial cast of Sinanthropus. Philos. Trans. Roy. Soc. London, 469.

SMITH, G. E. 1931. Sinanthropus- Peking man: Its discovery and significance. Scient. Monthly, Sept., 193.

TEILHARD DE CHARDIN, P. 1926. Fossil man in China and Mongolia. Nat. Hist. Mag., no. 3, 238.—1936. Fossil mammals from Locality 9 of Choukoutien. Palaeontologia Sinica, fasc. 4.

TEILHARD DE CHARDIN, P. and W. C. PEI. 1932. The lithic industry of the Sinanthropus deposits in Choukoutien. Bull. Geol. Soc. China, no. 4, 315.—1934. New discoveries in Choukoutien, 1933–1934. Ibid., no. 3, 369.

TEILHARD DE CHARDIN, P. and C. C. YOUNG. 1929. Preliminary report on the Chou Kou Tien fossiliferous deposit. Bull. Geol. Soc. China, no. 3, 173.

TERRA, H. DE. 1943. Pleistocene geology and early man in Java. Trans. Amer. Philos. Soc., n.s., part 3, 437. Part V in: Research on early man in Burma.—1949. Geology and climate as factors in human evolution in Asia. In: Studies on Phys. Anthropol. No. 1. Early man in the Far East, 7. Viking Fund Symposium, W. W. Howells, Ed.

TERRA, H. DE and H. L. MOVIUS, JR. 1943. Research on early man in Burma. Parts I–V. Trans. Amer. Philos. Soc., part 3, 265.

VALLOIS, H. V. 1935. Le Javanthropus. L'Anthropologie, nos. 1, 2. [Homo soloensis.]

WEIDENREICH, F. 1931. Sinanthropus pekinensis und seine Bedeutung für die Abstammungsgeschichte des Menschen. Die Naturw., part 40, 817.—1932. Über pithekoide Merkmale bei Sinanthropus pekinensis und seine stammesgeschichtliche Beurteilung. Zeitschr. f. Anat. u. Entwick., parts 1, 2, 212.—1935. The Sinanthropus population of Choukoutien (Locality 1) with a preliminary report on new discoveries. Bull. Geol. Soc. China, no. 4, 427.—1936a. The mandibles of Sinanthropus pekinensis: A comparative study. Palaeont, Sinica, fasc. 3.—1936b. Observations on the form and proportions of the endocranial casts of Sinanthropus pekinensis, other hominids and the great apes: A comparative study of brain size. Ibid., fasc. 4—1936c. Sinanthropus pekinensis—a distinct primitive hominid. Proc. Joint Meeting Anthropol. Soc. Tokyo and Japanese Soc. Ethnol., 1st Session.—1936–1937. The new discoveries of Sinanthropus pekinensis and their bearing on the Sinanthropus and Pithecanthropus problems. Bull. Geol. Soc. China, 439.—1937a. The new discovery of three skulls of Sinanthropus pekinensis. Nature, 269.—1937b. The dentition of Sinanthropus pekinensis: A comparative odontography of the hominids. Palaeont. Sinica, no. 1.—1937c. The forerunner of Sinanthropus pekinensis. Bull. Geol. Soc. China, no. 2, 137.—1937d. The relation of Sinanthropus pekinensis to Pithecanthropus, Javanthropus and Rhodesian man. Jour. Roy. Anthropol. Inst., 51.—1937e. Reconstruction of the entire skull of an adult female individual of Sinanthropus pekinensis. Nature, 1010.—1938a. Pithecanthropus and Sinanthropus. Ibid., 378.—1938b. Discovery of the femur and the humerus of Sinanthropus pekinensis. Ibid., 614.—1938c. The face of Peking woman. Nat. Hist. Mag., no. 5, 358.—1939a. On the earliest representatives of modern mankind recovered on the soil of East Asia. Peking Nat. Hist. Bull., part 3, 161.—1939b. Six lectures on Sinanthropus pekinensis and related problems. Bull. Geol. Soc. China., no. 1.—1940. Was "Java Man" a man or an ape? Nat. Hist. Mag., Jan., 32.—1941. The extremity bones of Sinanthropus pekinensis. Palaeont. Sinica, no. 5.—1943. The skull of Sinanthropus pekinensis: A comparative study on a primitive hominid skull. Ibid., no. 10.—1944. Giant early man from Java and South China. Science, June 16, 479.—1945a. The puzzle of Pithecanthropus. Science and Scientists in the Netherlands Indies, 380.—1945b. Giant early man from Java and South China. Anthropol. Papers, Amer. Mus. Nat. Hist., part 1.

WEINERT, H. 1928. Pithecanthropus erectus. Zeitschr. f. Anat., u. Ent., 429.—1931. Der "Sinanthropus pekinensis" als Bestätigung des Pithecanthropus erectus. Zeitschr. f. Morphol, u. Anthropol., part 1, 159.—1936. Eine Rekonstruktion des Pithecanthropus-Schädels auf Grund der von Eugen Dubois, 1891, bei Trinil auf Java gefundenen Calotte. Ibid., parts 1–2, 446.—1937a. Eine rekonstruktion des Sinanthropus-Schädels auf Grund der Calvaria Nr. I (Locus E) und des Unterkiefers (Locus A), gefunden 1929–30 bei Chou-Kou-Tien (Peking). Ibid., part 3, 367.—1937b. Die bedeutung der bisherigen Sinanthropus-Funde von Chou-Kou-Tien bei Peking. Der Biologe, part 7, 209.

Fossil Man: Europe (General)

(See also Heidelberg and Neanderthal, Piltdown, Homo sapiens, Cro-Magnon, Swanscombe, Foxhall, below)

BOULE, M. 1923. Les hommes fossiles: Éléments de paléontologie humaine. 2d ed. Paris.

MACCURDY, G. G. 1931. Recent progress in the field of Old World prehistory. Smiths. Rept. for 1930, 495.

MOIR, J. R. 1939. The earliest men. Huxley Memorial Lecture, 1939.

Fossil Man: Europe, Heidelberg and Neanderthal

ANTHONY, R. 1913. L'encéphale de l'homme fossile de la Quina. Bull. et Mém. Soc. d'Anthropol. Paris, 117.

BOULE, M. 1908. L'homme fossile de La Chapelle-aux-Saints. Comptes rendus, Acad. Sci. Paris, 1349.—1909. L'homme fossile de La Chapelle-aux-Saints. L'Anthropologie, 519.—1911–1913. L'homme fossile de La Chapelle-aux-Saints. Ann. d. Paléont., vols. 6, 7.

BUXTON, L. H. D. and G. R. DE BEER. 1932. Neanderthal and modern man. Nature, vol. 129, 940.

FIELD, H. 1927. The early history of man: with reference to the Cap-Blanc skeleton. Field Mus. Nat. Hist., Leaflet 26 (Anthropology), 1.

FREUDENBERG, W. 1923. Plastische Rekonstruktion des Urmenschen von Heidelberg (Homo heidelbergensis). Verhandl. d. Anat. Gesellsch., vols. 23–26, 122.—1938. Vorläufer und Nachfolger des Homo heidelbergensis, neue Funde aus Heidelbergs Umgebung. Beiträge zur Natur- und Urgeschichte Westdeutschlands, part 2.

GORJANOVIC-KRAMBERGER, K. 1924. Neue Beiträge zum Kiefergelenk des diluvialen Menschen von Krapina. I. Izvestia, Bull. d. Travaux d. Classe d. Sci. . . . (1923–1924), 118.—1926. Nachtrag zu "Neue Beiträge zum Kiefergelenk des diluvialen Menschen von Krapina." II. Ibid. (1925), 1.—1927. Das Schulterblatt des diluvialen Menschen von Krapina in seinem Verhaltnis zu dem Schulterblatt des rezenten Menschen und der Affen. Bull. d. l'Inst. Geol. de Zagreb, I, 67.

HOPWOOD, A. T. 1939. Excavations at Brundon, Suffolk (1935–37). Part II. Fossil mammals. Proc. Prehistoric Soc. for 1939 (Jan.–July), 13.—1940. Fossil mammals and Pleistocene correlation. Proc. Geologists' Assoc. London, vol. 51, part 1, 79.

HRDLIČKA, A. 1927. The neanderthal phase of man (Huxley Memorial Lecture for 1927). Jour. Anthropol. Soc. Great Britain and Ireland, 249. London.

JAEKEL, O. 1926. Über einen Neandertaloiden aus Sibirien. Palaeont. Zeitschr., part 1, 143.

KAPPERS, C. U. A. 1936. The endocranial casts of the Ehringsdorf and Homo soloensis skulls. Jour. Anat., part 1, 61.

KEITH, A. 1944. Pre-Neanderthal man in the Crimea. Nature, 515.

LOTH, E. 1938. Beiträge zur Kenntnis der Weichteilanatomie des Neanderthalers. Zeitschr. f. Rassenkunde, part 1, 13.

LULL, R. S. 1910. Restoration of Paleolithic man. Amer. Jour. Sci., art. 11, 171.

MARTIN, H. 1923. L'homme fossile de La Quina. Arch. d. Morphol. Gen. et Exper., 1.

MOIR, J. R. 1939. Excavations at Brundon, Suffolk (1935–37). Part I. Stratigraphy and Archaeology. Proc. Prehistoric Soc. for 1939 (Jan.-July). Paper no. 1 (Palaeolithic implements).

MOLLISON, T. 1931. Eine neue Rekonstruktion des Homo primigenius. Anthropol. Anz., nos. 3, 4, 285.—1937. Eine Rekonstruktion des Menschen von Steinheim von Hermann Friese. Ibid., nos. 3, 4, 309.

MONTANDON, G. 1943. L'homme préhistorique et les préhumains. Bibliothèque Scientifique, 355 pp. Paris.

OBERMAIER, H. 1924. Fossil man in Spain. New Haven.

SCHOETENSACK, O. 1908. Der Unterkiefer des Homo heidelbergensis aus den Sanden von Mauer bei Heidelberg. Ein Beitrag zur Paläontologie des Menschen. Leipzig. 4to.

SCHWALBE, G. 1914. Kritische Besprechung von Boule's Werk: "L'homme fossile de La Chapelle-aux-Saints" mit eigenen Untersuchungen. Zeitschr. f. Morphol. u. Anthropol., part 3, 527.

VIRCHOW, H. 1924. Der Unterkiefer von Ochos. Zeitschr. f. Ethnol., parts 5, 6, 197.

WEIDENREICH, F. 1927. Der Schädel von Weimar-Ehringsdorf. Verhandl. d. Gesellsch. f. Phys. Anthropol., 34.

WEINERT, H. 1925. Der Schädel des eiszeitlichen Menschen von Le Moustier in neuer Zusammensetzung. Berlin.—1936. Der Urmenschenschädel von Steinheim. Zeitschr. f. Morphol. u. Anthropol., part 3, 463.—1937. Dem Unterkiefer von Mauer zur 30 jährigen Wiederkehr seiner Entdeckung. Ibid., part 1, 102.

Fossil Man: Europe, Piltdown

DAWSON, C. and A. S. WOODWARD. 1913. On the discovery of a Palaeolithic human skull and mandible in a flint-bearing gravel overlying the Wealden (Hastings Beds) at Piltdown, Fletching (Sussex). With an Appendix by Prof. G. Elliot Smith. Quart. Jour. Geol. Soc. London, part 1, 117.—1914. Supplementary note on the discovery of a Palaeolithic human skull and mandible at Piltdown (Sussex), Abstr. Proc. Geol. Soc. London, no. 949, 28.

FRIEDERICHS, H. F. 1932. Schädel und Unterkiefer von Piltdown ("Eoanthropus Dawsoni Woodward") in neuer Untersuchung. Zeitschr. f. Anat. u. Entwickl., parts 1, 2, 199.

GIUFFRIDA-RUGGERI, V. 1919. La controversia sul fossile di Piltdown e l'origine del philum umano. Monitore Zoologico Italiano, Anno 30, nos. 1, 2.

GREGORY, W. K. 1914. The Dawn man of Piltdown, England. Amer. Mus. Jour., 189.

HRDLIČKA, A. 1922. The Piltdown jaw. Amer. Jour. Phys. Anthropol., vol. 5, no. 4, 337.—1923. Dimensions of the first and second lower molars with their bearing on the Piltdown jaw and on man's phylogeny. Ibid., vol. 6, no. 2.

HUNTER, J. I. 1918. New light on the controversy of the Piltdown jaw and cranium. Soc. Dental Sci., Sydney, N.S.W., 1.

KEITH, A. 1939. A resurvey of the anatomical features of the Piltdown skull, with some observations on the recently discovered Swanscombe skull. Parts I, II. Jour. Anat., parts 1, 2, 155.

MILLER, G. S., JR. 1915. The jaw of the Piltdown man. Smiths. Miscell. Coll., no. 12.—1918. The Piltdown jaw. Amer. Jour. Phys. Anthropol., no. 1, 25.

OSBORN, H. F. 1928. The antiquity of man in East Anglia. Science, 273.

PYCRAFT, W. O. 1917. The jaw of the Piltdown man. A reply to Mr. Gerrit S. Miller. Science Progress, no. 43, 389.

SMITH, G. E. 1913. The controversies concerning the interpretation and meaning of the remains of the Dawn-man found near Piltdown. Manchester Lit. and Philos. Soc., Nov. 25.—1914. On the exact determination of the median plane of the Piltdown skull. Quart. Jour. Geol. Soc., 93.

WEINERT, H. 1933. Das Problem des "Eoanthropus" von Piltdown. Ein Untersuchung der original Fossilien. Zeitschr. f. Morphol. u. Anthropol., parts 1, 2.

WOODWARD, A. S. 1913. The Piltdown man (Eoanthropus dawsoni). Geol. Mag., 433.—1922. A guide to the fossil remains of man in the department of geology and palaeontology in the British Museum (Natural History). London.—1933. The second Piltdown skull. Nature, 242.

Homo Sapiens, including Cro-Magnon, Swanscombe, Foxhall

BOULE, M. 1906, 1910, 1929. Les Grottes de Grimaldi (Baoussé-Roussé). Tome I, fasc. 2, 4. Géologie et Paléontologie. Monaco. [Two types of human skeletons: (a) "Cro-Magnon" (ref.), (b) negroid.]

CARTAILHAC, E. 1912. Les Grottes de Grimaldi (Baoussé-Roussé). Tome 2, fasc. 2, part 4. Monaco.

COLLYER, R. H. 1924. The fossil human jaw from Suffolk. Amer. Jour. Phys. Anthropol., no. 4.

HRDLIČKA, A. 1924. Critical notes on the Foxhall jaw. Ibid.

MOIR, J. R. 1924a. The human jaw-bone found at Foxhall. Ibid.—1924b. Tertiary man in England. With a note by E. Ray Lancaster. Nat. Hist. Mag., vol. 24, no. 6, 636.

MONTAGU, M. F. A. 1944. An early Swanscombe skull. Nature, 347.

SALLER, K. 1925. Die Cromagnonrasse und ihre Stellung zu anderen jungpalaeolithischen Langschädelrassen. Zeitschr. f. induktive Anstammungs-und Vererbungslehre, part 2, 190.—1927a. Die Menschenrassen im oberen Paläolithikum. Mitteil. d. Anthropol. Gesselsch. in Wien, 81.—1927b. Die Entstehung der "nordischen Rasse." Zeitschr. f. Anat. u. Entwickl., part 4, 411.—1928. Untersuchungen über Konstitutions- und Rassenformen an Turnern der deutschen Nordmark. Zeitschr. f. Konstitutionslehre, part 1.

SMITH, G. E. 1925. The London skull. Brit. Med. Jour., 7, 1.

SWANSCOMBE COMMITTEE. 1938. Report on the Swanscombe skull. Jour. Roy. Anthropol. Inst., vol. 68, 17.

VERNEAU, R. 1906. Les Grottes de Grimaldi (Baoussé-Roussé). Tome 2, fasc. 1, part 3. Anthropologie. Monaco.

WEIDENREICH, F. 1928. Entwicklungs-und Rassetypen des Homo primigenius. Natur und Museum, part 1.

YOUNG, M. 1938. The London skull. Biometrika, parts 3, 4, 277.

Fossil Man: Africa

ARAMBOURG, C. 1947a. Mission scientifique de l'Omo, 1932–1933. Tome I. Géologie-Anthropologie. Fasc. III. Paléontologie. Éditions du Muséum, Paris.—1947b. Les mammifères Pleistocènes d'Afrique. Bull. Soc. Géol. de France, 301.

BROOM, R. 1918. The evidence afforded by the Boskop skull of a new species of primitive man (Homo capensis). Anthropol. Papers, Amer. Mus. Nat. Hist., part 2, 67.

CLARK, W. E. LE G. 1928. Rhodesian man. Man, no. 12, 206.

COOKE, H. B. S., B. D. MALAN and L. H. WELLS. 1945. Fossil man in the Lebombo Mountains, South Africa . . . Man, Jan.-Feb., no. 3, 6.

DART, R. A. 1936. Fossil man and contemporary faunas in southern Africa. Rept. 16th Internat. Geol. Cong., Washington, (1933), 1249.—1940. Recent discoveries bearing on human history in southern Africa. Jour. Roy. Anthropol. Inst., part 1, 13.

DIETRICH, W. O. 1933. Zur Altersfrage der Oldowaylagerstätte. Centralb. f. Min. etc., no. 5, 299.

DREYER, T. F. 1935a. Early man in South Africa. Nature, 620.—1935b. A human skull from Florisbad, Orange Free State, with a note on the endocranial cast, by C. U. A. Kappers. Kon. Akad. van Wetensch. te Amsterdam, no. 1, 119. [Homo "Africanthropus" helmei.]

DREYER, T. F. and A. LYLE. 1931. New fossil mammals and man from South Africa. Grey Univ. Coll., Dept. Zool., 5.

GALLOWAY, A. 1937a. The nature and status of the Florisbad skull as revealed by its nonmetrical features. Amer. Jour. Phys. Anthropol., no. 1, 1.—1937b. The characteristics of the skull of the Boskop physical type. Ibid., no. 1, 31.—1937c. Man in Africa in the light of recent discoveries. South African Jour. Sci., 89.

HOOTON, E. A. 1925. The ancient inhabitants of the Canary Islands. Harvard African Studies, vol. 7. Peabody Museum of Harvard Univ. 4to. Cambridge.

HRDLIČKA, A. 1926. The Rhodesian man. Amer. Jour. Phys. Anthropol., 173.

LEAKEY, L. S. B. 1933. A reported ancestor of the human species. Science Suppl., 8. [Homo kanamensis.]—1935. Human remains from Kanam and Kanjera, Kenya Colony. Nature, 1041.—1948. Fossil and sub-fossil Hominoidea in East Africa. In: The Robert Broom Commemorative Volume, 165.

MEIRING, A. J. D. 1937. The "Wilton" skulls of the Matjes River Shelter. Soölogiese Navorsing van die Nasionale Museum, Bloemfontein, I, 51.

MOLLISON, T. 1929. Der Fossilzustand und der Schädel. In: Untersuchungen über den Oldowayfund. Verhandl. d. Gesellsch. f. physische Anthropol., 60.

PYCRAFT, W. P., G. E. SMITH and others. 1928. Rhodesian man and associated remains . . . Brit. Mus. (Nat. Hist.), London.

SCHEPERS, G. W. H. 1941. The mandible of the Transvaal fossil human skeleton from Springbok Flats. Ann. Transvaal Mus., part 3, 253.

SCHWARZ, R. 1932. Anthropologie. Die Fortschr. der Zahnheilkunde, part 9, 681. [Homo Rhodesiensis.]

SHAW, J. C. M. 1931. The teeth, the bony palate and the mandible in Bantu race of South Africa. London.

WEINERT, H. 1925. Neues vom Homo Rhodensiensis. Kosmos, Handweiser f. Naturfreunde, part 5, 163.—1938a. Der erste afrikanische Affenmensch "Africanthropus njarasensis." Der Biologe, part 4, 125.—1938b. Afrikanthropus, der erste Affenmenschen-Fund aus dem Quartär Deutsch-Ostafrikas. Quartär, 177.—1939. Africanthropus njarasensis. Beschreibung und phyletische Einordnung des ersten Affenmenschen aus Ostafrika. Zeitschr. f. Morphol. u. Anthropol., part 2, 252.

WELLS, L. H. 1941. Africa and the ancestry of man. South African Jour. Sci., 58.

Fossil Man: Australia

BURKITT, A. ST. N. 1928. Neanderthal man and the natives of New Caledonia. Nature, Sept. 29, 4 pp. (reprint).—1929. Further observations upon the "Talgai" skull, more especially with regard to the teeth. Australian Assoc. Adv. Sci. (Hobart) (1928), 366.

BURKITT, A. ST. N. and J. L. HUNTER. 1922. The description of a neanderthal Australian skull, with remarks on the production of the facial characteristics of Australian skulls in general. Jour. Anat., part 1, 31.

HELLMAN, M. 1934. The form of the Talgai palate. Amer. Jour. Phys. Anthropol., no. 1, 1.

JONES, F. W. 1944. The antiquity of man in Australia. Nature, 211.

MAHONY, D. J. 1943a. The problem of antiquity of man in Australia. Mem. Nat. Mus. Melbourne, no. 13, 7.—1943b. The Keilor skull: Geological evidence of antiquity. Ibid., 79.

SCOTT, H. H. 1925. Some notes upon a Tasmanian aboriginal skull. Roy. Soc. Tasmania, Papers and Proc., 5.

SMITH, S. A. 1918. The fossil human skull found at Talgai, Queensland. Philos. Trans. Roy. Soc. London, 351.

TURNER, W. 1909. The craniology, racial affinities and descent of the aborigines of Tasmania. Trans. Roy. Soc. Edinburgh, 365.

WEIDENREICH, F. 1945. The Keilor skull: A Wadjak type from Southeast Asia. Amer. Jour. Phys. Anthropol., no. 1, 21.

WUNDERLY, J. 1943. The Keilor skull: Anatomical description. Mem. Nat. Mus. Melbourne, no. 13, 57.

ZEUNER, F. E. 1944. Homo sapiens in Australia contemporary with Homo neanderthalensis in Europe. Nature, 622.

Fossil Man: North America

ANTEVS, E. 1936. Dating records of early man in the southwest. Amer. Nat., 331.

BARBOUR, E. H. and C. B. SCHULTZ. 1932. The mounted skeleton of Bison occidentalis and associated dart-points. Bull. Nebraska State Mus., 263.—1936. Palaeontologic and geologic consideration of early man in Nebraska. Ibid., no. 45, 431.

BROWN, B. 1932. The buffalo drive. Nat. Hist. Mag., no. 1, 75. [As practised by early men in western North America.]

COLBERT, E. H. 1942. The association of man with extinct mammals in the Western Hemisphere. Proc. Eighth Amer. Scient. Cong. (1940), Anthropol. Sci., 17.

COOK, H. J. 1927. New geological and palaeontological evidence bearing on the antiquity of mankind in America. Nat. Hist. Mag., no. 3, 240.—1928. Further evidence concerning man's antiquity at Frederick, Oklahoma. Science, 371. —1931a. The antiquity of man as indicated at Frederick, Oklahoma: A reply. Jour. Washington Acad. Sci., no. 8, 161.—1931b. More evidence of the "Folsom Culture" race. Scient. Amer., Feb., 102.

COUNT, E. W. 1939. Primitive Amerinds and the Australo-Melanesians. Rev. del Inst. de Anthropol. de la Univ. nac. d. Tucumán, no. 4, 91.

EISELEY, L. C. 1942. The Folsom mystery: Its solution is largely contingent on the solution of another mystery. Scient. Amer., Dec., no. 6, 260.

FIGGINS, J. D. 1927. The antiquity of man in America. Nat. Hist. Mag., no. 3, 229.

HARRINGTON, M. R. 1932. When was America discovered? Scient. Amer., July, 7.

HAY, O. P. 1918. Further consideration of the occurrence of human remains in the Pleistocene deposits at Vero, Florida. Amer. Anthropologist, no. 1, 1.

HAY, O. P. and H. J. COOK. 1930. Fossil vertebrates collected near, or in association with, human artifacts . . . [in Colorado, Texas, Oklahoma, New Mexico]. Proc. Colorado Mus. Nat. Hist., no. 2, 4.

HOOTON, E. A. 1930. The Indians of Pecos Pueblo: A study of their skeletal remains. Yale Univ. Press.

HOWARD, E. B. 1935. Evidence of early man in North America. Univ. Pennsylvania, Mus. Jour., nos. 2–3, 61.—1936. The association of a human culture with an extinct fauna in New Mexico. Amer. Nat., 314.

HOWARD, E. B. and others. 1936. Early man in America with particular reference to the southwestern United States. Amer. Nat., 313.

HRDLIČKA, A. 1918. Recent discoveries attributed to early man in America. Smiths. Inst., Bur. Amer. Ethnol., Bull. 66.—1928. The origin and antiquity of man in America. Bull. N. Y. Acad. Med., no. 7, 802.

HUÉ, E. 1917. L'homme préhistorique dans l'Amerique du Nord. Bull. Soc. Préhist. française, no. 4, 205; no. 5, ibid., 269; no. 6, ibid, 317; no. 7, ibid., 365.

JENKS, A. E. 1936. Pleistocene man in Minnesota: A fossil Homo sapiens. Univ. Minnesota Press. 4to.

LOOMIS, F. B. and J. W. GIDLEY. 1926. Fossil man in Florida. Amer. Jour. Sci., 254.

NELSON, N. C. 1933. The antiquity of man in America in the light of archaeology. In: The American Aborigines, their origin and antiquity, 87. Univ. Toronto Press.

ROMER, A. S. 1933. Pleistocene vertebrates and their bearing on the problem of human antiquity in North America. Ibid., 49.

SCHULTZ, C. B. 1932. Association of artifacts and extinct mammals in Nebraska. Nebraska State Mus., Bull. 33, 271.

SCHULTZ, C. B. and L. EISELEY. 1935. Palaeontological evidence for the antiquity of the Scottsbluff bison quarry and its associated artifacts. Amer. Anthropologist, no. 2, April-June, 306.

SCHULTZ, C. B. and W. D. FRANKFORTER. 1948. Preliminary report on the Lime Creek sites: New evidence of early man in southwestern Nebraska. Bull. Univ. Nebraska State Mus., Nov., no. 4, part 1, 31. [Artifacts of Scottsbluff type and fossil bones found in situ at base of cliff, 47½ feet below surface.]

SIMPSON, G. G. 1933. A Nevada fauna of Pleistocene type and its probable association with man. Amer. Mus. Novitates, no. 667.

SPIER, L. 1918. The Trenton argillite culture. Anthropol. Papers. Amer. Mus. Nat. Hist., part 4, 167.—1928. Concerning man's antiquity at Frederick, Oklahoma. Science, 160.

STOCK, C. 1936. The succession of mammalian forms within the period in which human remains are known to occur in America. Amer. Nat., July-August, 324.

WISSLER, C. 1938. The American Indian. New York.

Fossil Man: South America

AMEGHINO, F. 1907. Notas preliminares sobre el Tetraprothomo Argentinus, un précursor del hombre de Mioceno superior de Monte Hermoso. An. del Mus. nac. de Buenos Aires, vol. 16, 107. [See Hrdlicka, 1912, and Schwalbe, 1911.]—1909. Le Diprothomo Platensis, un precurseur de l'homme du pliocène inférieur de Buenos Aires. Ibid., vol. 19, 107. [See Hrdlicka, 1912, and Schwalbe, 1911.]

CASTELLANOS, A. 1922. La presencia del hombre fósil . . . Bol. de la Acad. Nac. Cienc. de Córdoba, 369.—1923. La limite plio-pleistocène et le problème de l'homme tertiare dans le Republique Argentine. Rev. de la Univ. Nac. de Córdoba, nos. 1, 2, 5.—1924. Contribucion al Estudio de la Paleoantropologia Argentina: Restos en el Arroyo Cululu (Prov. de Santa Fé). Ibid., vol. 11, nos. 7, 8, 9.—1926. La existencia del hombre fósil en la Argentina. Atti del XXII Congr. Intern. d. Americanisti, Roma,

277.—1927. Contribucion al Estudio de la Paleoantropologia Argentina: Apuntes sobre el "Homo chapadmalensis" n.sp. Rev. Méd. del Rosario, no. 8.—1930. Nuevos restos del hombre fósil. Physis (Rev. de la Soc. Argentina de Cienc. Nat.), (1929), 175.—1934. El hombre prehistorico en la Provincia de Córdoba (Argentina). Rev. de la Soc. "Amigos de la Arqueológia," vol. 7, 1.—1938. Nuevos restos del hombre fósil y de "Hornos de Tierra Cocida" en Santiago del Estero (Argentina) . . . Publ. del Inst. de Fisiográfia y Geológia, Rosario, 1. —1943. Antigüedad geológica del Yacimiento de los Restos humanos de la "Gruta de Candonga" (Córdoba). Ibid., no. 14, 1.

EATON, G. F. 1916. The collection of osteological material from Manchu Picchu. Mem. Connecticut Acad. Arts and Sci., 1.

HRDLIČKA, A. and others. 1912. Early man in South America. Smiths. Inst. Bur. Amer. Ethnol., Bull. 52.

KRAGLIEVICH, L. 1928. Singular concordancia del sabio inglés Pilgrim con las ideas antropogénicas del Florentino Ameghino. La Semana Medica, no. 36.

RUSCONI, C. 1932a. Huesos fósiles roidos y huesos trabajados. Publ. de. Mus. Antropol. y Etnográfico . . . ser. A, vol. 2, 149.—1932b. Probable posición estratigráfica de la calota de "Diprothomo platensis" Amegh. Ibid., 177.—1935. Restos humanos fósiles procedentes de la provincia de Santiago del Estero. Rev. Argentina de Paleontol. y Antropol., Ameghinia, vol. 1, nos. 5, 6, 135.—1939. Restos humanos subfósiles de Mendoza. An. de la Soc. Cient. Argentina, vol. 126, 460.—1941. Cronologia de los terrenos neoterciarios de la Argentina en relatión con el hombre. Bol. de la Acad. Nac. de Cienc. en Córdoba, nos. 2, 3, 151.

SCHWALBE, G. 1911. Nachtrag zu meiner Arbeit: Über Ameghino's Diprothomo platensis. Zeitschr. f. Morphol, u. Anthropol., part 3, 533.

SULLIVAN, L. R. and M. HELLMAN. 1925. The Punin calvarium. Anthropol. Papers, Amer. Mus. Nat. Hist., part 7, 313.

VIGNATI, M. A. 1920. Cuestiones de paleoanthropologia Argentina. 2d ed. Buenos Aires.—1921a. À propos du Diprothomo platensis. Une observation a la critique du Professeur Schwalbe. An. del Mus. Nac. de Buenos Aires, vol. 31, 25.—1921b. El hombre fósil de Chapadmalal. Physis (Rev.) de la Soc. Arg. de Cienc. Nat., 80, 98.—1922. Nota preliminar sobre el hombre fósil de Miramar. Ibid., 216.—1934. El hombre fósil de Esperanza. Notas Prelim. del Mus. de la Plata, 7.—1941. Descripción de los molares humanos fósiles de Miramar (Provincia de Buenos Aires). Rec. del Mus. de La Plata, 271.

Man: Geographic Distribution, Dispersal, Migration

BLACK, D. 1925. Asia and the dispersal of primates: A study in ancient geography of Asia and its bearing on the ancestry of man. Bull. Geol. Soc. China, no. 2, 133.

CASTELLANOS, A. 1934. Conexiones Sudamericanas en relación con las migraciones humanas. "Quid Novi?", vol. 11, no. 6.

GIUFFRIDA-RUGGERI, V. 1920. Le prime migrazioni umane. Scientia, Rev. di Scienza, vol. 27, 201.

GRABAU, A. 1935. Tibet and the origin of man. Geografiska Annaler, 317.

GREGORY, W. K. 1927. Did man originate in Central Asia? In: Mongolia, the new world, Part V. Scient. Monthly, May, 385.

HRDLIČKA, A. 1921. The peopling of Asia. Proc. Amer. Philos. Soc., no. 4, 535.—1926. The peopling of the earth. Ibid., no. 3, 150.

LEAKEY, L. S. B. 1946. Was Kenya the centre of human evolution? Illustrated London News, Aug. 24, 197.

SUSCHKIN, P. P. 1933. Die Hochlandgebiete der Erde und die Frage über den Ursprung des Menschen. Palaeobiologica, 275.

TAYLOR, G. 1921. The evolution and distribution of race, culture, and language. Geograph. Rev., no. 1, 52.

TERRA, H. DE. 1937. Cenozoic cycles in Asia and their bearing on human prehistory. Proc. Amer. Philos. Soc., no. 3, 289.—1940. Geologic dating of human evolution in Asia. Scient. Monthly, no. 2, 112.

Man: Races, Speciation, Genetics

COUNT, E. W. (Ed.) 1950. This is race. New York.

DOBZHANSKY, T. 1941. Genetics and the origin of species. 2d ed. revised. Columbia Univ. Biol. Ser., no. 11.—1942. Races and methods of their study. Trans. N. Y. Acad. Sci., 115.—1944. On species and races of living and fossil man. Amer. Jour. Phys. Anthropol., no. 3, 251.

EMERSON, A. E. 1945. Taxonomic categories and population genetics. Entomol. News, no. 1, 14.

FIELD, H. 1946. Contributions to the anthropology of the Soviet Union. Smiths. Misc. Coll., vol. 110, no. 13, 1.

GATES, R. R. 1944. Phylogeny and classification of hominids and anthropoids. Amer. Jour. Phys. Anthropol., no. 3, 279.—1946. Human heredity in relation to animal genetics. Amer. Nat., 68.

GIESELER, W. 1936. Abstammungs-und Rassenkunde des Menschen (Anthropologie). Anhang. Oehringen.

HADDON, A. C. 1925. The races of man and their distribution. New York.

HILL, W. C. O. 1941. The physical anthropology of the existing Veddahs of Ceylon. Part I. Ceylon Jour. Sci., sect. G, vol. 3, part 2, 37.—1942. Part II. Ibid., part 3, 147.

HOOTON, E. A. 1933. Racial types in America and their relation to Old World types. In: The American Aborigines, 133. Toronto Univ. Press.—1936. Plain statements about race. Science, 511.

HOWELLS, W. W. 1942. Fossil man and the origin of races. Amer. Anthropologist, no. 2, 182.

KAPPERS, C. U. A. 1930–1931. Contributions to the anthropology of the Near-East. I. The Armenians. II. The spread of brachycephalic races. Konink. Akad. v. Wetensch. Amsterdam, vol. 33 (1930), no. 8.—III. Phoenician and Palmyrene skulls. IV. The Semitic races. V. Kurds, Circassians and Persians. VI. Turks and Greeks. Ibid., vol. 34 (1931), nos. 1, 4, 8.—1932. The anthropology of the Near East. Publ. Amer. Univ. Beirut, Social Sci. Ser., no. 2.

KEITH, A. 1931. Ethnos, or the problem of race considered from a new point of view. London.

LUSCHAN, F. VON. 1927. Voelker Rassen Sprachen: Anthropol. Betrachtungen. Berlin.

MUNRO, J. 1909. The British race. London.

SALLER, K. 1929. Konstitution und Rasse beim Menschen. Ergeb. d. Anat. u. Entwickl., 250.

SHAPIRO, H. L. 1929. Descendants of the mutineers of the Bounty. Mem. Bernice P. Bishop Mus., Honolulu.—1935. Mystery island of the Pacific. Nat. Hist. Mag., May, no. 5, 365.—1936. The heritage of the Bounty. New York.—1940. The distribution of blood groups in Polynesia (based on data obtained by Dr. George P. Lyman). Amer. Jour. Phys. Anthropol., 409.

TODD, T. W. 1929. Entrenched negro physical features. Human Biol., no. 1, 57.—1930. Racial features in the American negro cranium. Amer. Jour. Phys. Anthrop., vol. 15, 53.

WEIDENREICH, F. 1946. Generic, specific and subspecific characters in human evolution. Amer. Jour. Phys. Anthropol., no. 4, 413.

Man: Anatomy, General, Descriptive and Miscellaneous

CUNNINGHAM, D. J. 1903. Text-book of anatomy. New York.—1927. Cunningham's Text-book of anatomy. 5th ed. revised. (Arthur Robinson, Ed.) New York.

DETWILER, S. R. 1929. Anatomy as a science. Science, 563.

GRAY, H. 1936. Anatomy of the human body. 23rd ed. (Warren H. Lewis, Ed.)

HUNTINGTON, G. S. 1903. The anatomy of the human peritoneum and abdominal cavity. Philadelphia. 4to.

LORIN-EPSTEIN, M. J. 1932. Evolution und Bedeutung des Wurmfortsatzes und der Valvula ileocoecalis im Zusammenhang mit der Aufrichtung des Rumpfes. Zeitschr. f. Anat. u. Entwickl., 68.

MAXIMOW, A. A. and W. BLOOM. 1942. A textbook of histology. 4th ed. Philadelphia and London.

MOLLIER, S. 1938. Plastische Anatomie: die konstruktive Form des menschlichen Körpers. 2nd ed. Munich.

PIERSOL, G. A. 1930. Piersol's human anatomy. 9th ed., revised. (G. Carl Huber, Ed.) Philadelphia.

SINGER, C. 1925. The evolution of anatomy: A short history of anatomical and physiological discovery to Harvey . . . New York.

SOBOTTA, J. and J. P. MCMURRICH. 1930. Atlas of human anatomy. Revised ed. Vols. I, II, III. New York.

SPALTEHOLZ, W. 1907. Hand atlas of human anatomy. 5th ed. in English. Vols. I, II, II. Philadelphia.

Man: Embryology, Growth, Seasonal Rhythms, Ossification

CLARK, W. E. LE G. 1943. Seasonal rhythms in man. Nature, 66.

COUNT, E. W. 1942. A quantative analysis of growth in certain human skull dimensions. Human Biol., May, no. 2, 143.—1943. Growth patterns of the human physique: An approach to kinetic anthropometry. Ibid., no. 4, 1.

DAVENPORT, C. B. 1932. The growth of the human foot . . . [See below under Locomotor Skeleton.]

HELLMAN, M. 1928. Ossification of epiphysial cartilages in the hand. Amer. Jour. Phys. Anthropol., vol. 11, ser. 2, 223.

HUXLEY, J. 1932. Problems of relative growth. New York.

KEITH, A. 1913. Human embryology and morphology. 3rd ed. London.

MINOT, C. S. 1892. Human embryology. New York.

NOBACK, C. V. 1930. Digital epiphyses and carpal bones in the growing infant female gorilla, with sitting height, weight and estimated weight. Zoologica (N.Y.), no. 5, 117.

SCHULTZ, A. H. 1949. The palatine ridges of primates. Carnegie Inst. Washington, Contrib. to Embryology. vol. 32, 45.

SENSENIG, E. C. 1949. The early development of the human vertebral column. Carnegie Inst. Washington Publ. 583, Contrib. to Embryology, vol. 32, 21.

TODD, T. W. 1921. Age changes in the public bone. V. Mammalian pubic metamorphosis. Amer. Jour. Phys. Anthropol., no. 4, 333.—1930. Age changes in the pubic bone. Ibid., no. 2, 225.—1931. The natural history of human growth. Brush Found. Publ., no. 12, 1.—1932. Growth and development. Ibid., no. 19.—1935. Anthropology and growth. Science, Mar. 15, 259; Brush Found. Publ. no. 27.—1938. The anatomy of growth. Northwest Medicine, Jan. and April.

Man and Apes: Locomotor Skeleton and Musculature

ABEL, W. 1930. Beiträge zur Kenntnis der Anpassungserscheinungen der distalen Hinterfussmuskulatur der Säugetiere bei einem Wechsel der Lebensweise. Morphol. Jahrb., parts 3, 4, 558.

APPLETON, A. B. 1922. On the hypotrochanteric fossa and accessory adductor groove of the primate femur. Jour. Anat., parts 3–4, 295.—1929. On the morphology of the cervico-costo-humeralis muscle of Gruber. Ibid., part 4, 434.

BOAS, J. E. V. 1919 Einige Bemerkungen über die Hand des Menschen. Kgl. Danske Vidensk. Selskab Biol. Meddel., no. 1, 3.

DAVENPORT, C. B. 1932. The growth of the human foot. Amer. Jour. Phys. Anthropol., no. 2, 167. —1933. The crural index [of man]. Ibid., no. 3, 333.—1944. Postnatal development of the human extremities. Proc. Amer. Philos. Soc., no. 5, 375.

DETWILER, S. R. 1929. An experimental study of the mechanism of coordinated movements in heterotopic limbs. Jour. Comp. Neurol., no. 4, 437.

EATON, T. H., JR. 1944. Modifications of the shoulder girdle related to reach and stride in mammals. Jour. Morphol., no. 1, 167.

ELFTMAN, H. 1932. The evolution of the pelvic floor of primates. Amer. Jour. Anat., no. 2, 307. —1934. A cinematic study of the distribution of pressure in the human foot. Anat. Rec., no. 4, 481.—1938. The measurement of the external force in walking. Science, 1.—1939a. The function of the arms in walking. Human Biol., no. 4, 529.—1939b. Forces and energy changes in the leg during walking. Amer. Jour. Physiol., no. 2, 339.—1939c. The function of muscles in locomotion. Ibid., no. 2, 357.—1939d. The rotation of the body in walking. Arbeits-physiologie, part 5, 477.—1939e. The force exerted by the ground in walking. Ibid., 485.—1940. The work done by muscles in running. Amer. Jour. Physiol., no. 3, 672.—1941. The action of muscles in the body. Biol. Symposia, 191.—1943. Experimental studies on the dynamics of human walking. Trans. N. Y. Acad. Sci., no. 1, 1.—1944. The bipedal walking of the chimpanzee. Jour. Mammal., no. 1, 67.—1945. Torsion of the lower extremity. Amer. Jour. Phys. Anthropol., no. 3, 255.

ELFTMAN, H. and J. T. MANTER. 1934. The axis of the human foot. Science, 484.—1935a. Chimpanzee and human feet in bipedal walking. Amer. Jour. Phys. Anthropol., no. 1, 69.—1935b. The evolution of the human foot, with especial reference to the joints. Jour. Anat., part 1, 56.

EVANS, F. G. and V. E. KRAHL. 1945. The torsion of the humerus: A phylogenetic survey from fish to man. Amer. Jour. Anat., no. 3, 303.

FRECHKOP, S. 1936. Le pied de l'homme (essai anthropologique). Mém. de Mus. roy d'Hist. nat. de Belgique, part 3, 319.

GREGORY, W. K. 1928a. Were the ancestors of man primtive brachiators? Proc. Amer. Philos. Soc., no. 2, 129.—1928b. The upright posture of man: A review of its origin and evolution. Ibid., no. 4, 339.

HAINES, R. W. 1934. The homologies of the flexor

and adductor muscles of the thigh. Jour. Morphol., no. 1, 21.

HALLISY, J. E. C. 1930. The muscular variations in the human foot. A quantitative study. General results of the study. I. Muscles of the inner border of the foot and the dorsum of the great toe. Amer. Jour. Anat., no. 3, 411.

HRDLIČKA, A. 1934. The hypotrochanteric fossa of the femur. Smiths. Miscell. Coll., no. 1, 1.

JONES, F. W. 1920. Principles of anatomy as seen in the hand. London.—1929. The distinctions of the human hallux. Jour. Anat., part 4, 408.—1944. Structure and function as seen in the foot. London.

KEITH, A. 1920. The engines of the human body. Philadelphia and London.—1923a. The adaptational machinery concerned in the evolution of man's body. Nature, Aug. 18, Suppl.—1923b. Man's posture: Its evolution and disorders. Hunterian Lectures. Parts 1–6. British Med. Jour., March 17–April 21.—1929. The history of the human foot and its bearing on orthopaedic practice. Jour. Bone and Joint Surgery, no. 1, 10.

KRAHL, V. E. and F. G. EVANS. 1945. Humeral torsion in man. Amer. Jour. Phys. Anthropol., no. 3, 229.

KROGMAN, W. M. 1935. Life histories recorded in skeletons. Amer. Anthropologist, no. 1, 92.

MCMURRICH, J. P. 1927. The evolution of the human foot. Amer. Jour. Phys. Anthropol., no. 2, 165.

MILLER, R. A. 1932. Evolution of the pectoral girdle and forelimb in the primates. Amer. Jour. Phys. Anthropol., no. 1.

MONTAGU, M. F. A. 1931. On the primate thumb. Amer. Jour. Phys. Anthropol., no. 2, 291.—1939. Anthropological significance of the musculus pyramidalis and its variability in man. Ibid., no. 3, 435.

MORTON, D. J. 1922. Evolution of the human foot. I. Amer. Jour. Phys. Anthropol., no. 4, 305.—1924a. II. Ibid., no. 1, 1.—1924b. The peroneus tertius muscle in gorillas. Anat. Rec., no. 5, 323.—1926. Evolution of man's erect posture. Jour. Morph. and Physiol., no. 1, 147.—1927. Human origin. Correlation of previous studies of primate feet and posture with other morphologic evidence. Amer. Jour. Phys. Anthropol., no. 2, 173.

OSBORN, H. F. 1928a. The influence of habit in the evolution of man and the great apes. Bull. N. Y. Acad. Med., no. 2, 216.—1928b. The influence of bodily locomotion in separating man from the monkeys and apes. Scient. Monthly, no. 5, 385.

RAVEN, H. C. 1936. Comparative anatomy of the sole of the foot. Amer. Mus. Novitates, no. 871. [Comparative dissections of the foot of bear, chimpanzee, gorilla, man.]

REYNOLDS, E. 1931. The evolution of the human pelvis in relation to the mechanics of the erect posture. Papers of Peabody Mus., Harvard Univ., no. 5, 255.

REYNOLDS, E. and E. A. HOOTON. 1936. Relation of the pelvis to erect posture. An exploratory study. Amer. Jour. Phys. Anthropol., 253.

REYNOLDS, E. L. 1945. The bony pelvic girdle in early infancy: A roentgenometric study. Amer. Jour. Phys. Anthropol., no. 4, 321.

RIED, H. A. 1925. Über die Beziehungen der tibialen Gelenkfläche des Femur zur Schaftkrümmung. Anthropol. Anz., part 2, 113.

SCHULTZ, A. H. 1918. The position of the insertion of the pectoralis major and deltoid muscles on the humerus of man. Amer. Jour. Anat., no. 1, 155.—1930. The skeleton of the trunk and limbs of higher primates. Human Biol., no. 3, 303.—1937. Proportions, variability and asymmetries of the long bones of the limbs and the clavicles in man and apes. Ibid., no. 3, 281.—1938. The relative length of the regions of the spinal column in Old World primates. Amer. Jour. Phys. Anthropol., no. 1, 1.—1941. Chevron bones in adult man. Ibid., no. 1, 91.—1942. Conditions for balancing the head in primates. Ibid., no. 4, 483.

SCHULTZ, A. H. and W. L. STRAUS, JR. 1945. The numbers of vertebrae in primates. Proc. Amer. Philos. Soc., no. 4, 601.

SEIB, G. A. 1934. Incidence of the m. psoas minor in man. Amer. Jour. Phys. Anthropol., no. 2, 229.

STEWART, T. D. 1936. The musculature of the anthropoids. I. Neck and trunk. Amer. Jour. Phys. Anthropol., no. 2, 141.

STRAUS, W. L., JR. 1927a. Growth of the human foot and its evolutionary significance. Carnegie Inst. Washington Publ., no. 380, 93.—1927b. The human ilium: Sex and stock. Amer. Jour. Phys. Anthropol., vol. 11, 1.—1929. Studies on primate ilia. Amer. Jour. Anat., no. 3, 403.—1930. The foot musculature of the Highland gorilla (Gorilla beringei). Quart. Rev. Biol., no.

3, 261.—1940. The posture of the great ape hand in locomotion and its phylogenetic implications. Amer. Jour. Phys. Anthropol., no. 2, 199.—1941a. The phylogeny of the human forearm extensors. Human Biol., no. 1, 23.—1941b. The phylogeny of the human forearm extensors (concluded). Ibid., no. 2, 203.—1941c. Locomotion of gibbons. Amer. Jour. Phys. Anthropol., no. 3, 354.—1942a. The homologies of the forearm flexors: urodeles, lizards, mammals. Amer. Jour. Anat., no. 2, 281.—1942b. Rudimentary digits in primates. Quart. Rev. Biol., no. 3, 228.

SULLIVAN, W. E. and C. W. OSGOOD. 1927. The musculature of the superior extremity of the orang-utan, Simia satyrus. Anat. Rec., no. 3, 193.

VALLOIS, H. V. 1920. L'epiphyse inférieure de femur chez les primates. I, II. Bull. et Mém. Soc. d'Anthropol. Paris, April 3, July 17, 1919. —1921a. La formation progressive du biceps femoral chez les anthropoides et chez l'homme. Ibid., Jan. 6, 1.—1921b. La signification des ligaments du genou humain au point de vue de l'anatomie comparée (1). L'Assoc. des Anatomistes, 16ième Reunion (Paris), 229.—1921c. Étude d'un muscle présternal (M. sternalis). Bull. et Mém. Soc. d'Anthropol. Paris, April 21, 59.—1926a. Valeur et signification du muscle pyramidal de l'abdomen. Arch. d'Anat., d'Histol., et d'Embryol. (1925–1926), 497.—1926b. La sustentation de la tête et le ligament cervical postérieur chez l'homme et les anthropoides. L'Anthropologie, 191.—1928. Les muscles spinaux chez l'homme et les anthropoides. Contribution à l'étude de l'adaptation à l'attitude verticale. Ann. d. Sci. nat., Zool., vol. 11, 1.

WASHBURN, S. L. 1942. Skeletal proportions of adult langurs and macaques. Human Biol., no. 4, 444.

WATERMAN, H. C. 1929. Studies on the evolution of the pelvis in man and other primates. Bull. Amer. Mus. Nat. Hist., 585.

WEIDENREICH, F. 1913. Über das Huftbein und das Becken der Primaten und ihre Umformung durch den aufrechten Gang. Anat. Anz., nos. 20–21, 497.—1921–1922. Der Menschenfuss. Zeitschr. f. Morphol. u. Anthropol., vol. 22, 51. —1923. Evolution of the human foot. Amer. Jour. Phys. Anthropol., no. 1.—1931. Der primäre Greifcharakter der menschlichen Hände und Füsse und seine Bedeutung für das Abstammungsproblem. Verhandl. d. Gesellsch. f. Physische Anthropol., 97.

WELLS, L. H. 1935. A peroneus tertius muscle in a chacma baboon (Papio porcarius). Jour. Anat., part 4, 508.

WILDER, H. H. 1924. The phylogeny of the human foot; the testimony presented by the configuration of the friction ridges. Zeitsch. f. Morphol. u. Anthropol., 111.

WILLIS, T. A. 1923. The lumbo-sacral vertebral column in man; its stability of form and function. Amer. Jour. Anat., vol. 32, no. 1, 95.

Man: Head Form, Cranium, Cranial Musculature, Brain-cast, Relations between Cranium and Brain (See also Brain, below)

ABBIE, A. A. 1947. Headform and human evolution. Jour. Anat., part 3, 233.

ABEL, W. 1934. Die Vererbung von Antlitz und Kopfform des Menschen. Zeitschr. f. Morphol. u. Anthropol., part 2, 261.

BOLK, L. 1912. Über die Obliteration der Nähte am Affenschädel, zugleich ein Beitrag zur Kenntnis der Nahtanomalien. Zeitschr. f. Morphol. u. Anthropol., vol. 15, part 1, 1.—1915. Über Lagerung, Verschiebung und Neigung des Foramen magnum am Schädel der Primaten. Ibid., vol. 17, part 2, 611.—1918. On the topographical relations of the orbits in infantile and adult skulls in man and apes. Proc. Konink. Akad. van Wetensch. Amsterdam, no. 3, 277.

BUTLER, H. 1949. A rare suture in the anterior cranial fossa of the human skull. Man, no. 26. [Post-ethmoid suture of the frontals in several Bombay skulls.]

CAVE, A. J. E. 1931. The craniopharyngeal canal in man and anthropoids. Jour. Anat., part 3, 365.

CLARK, W. E. LE G. 1934. The asymmetry of the occipital region of the brain and skull. Man: A Monthly Rec. of Anthropol. Sci., no. 50, 1.

COBB, W. M. 1940. The cranio-facial union in man. Amer. Jour. Phys. Anthropol., 87.

COLLINS, H. B., JR. 1925. The pterion in the primates. Amer. Jour. Phys. Anthropol., vol. 8, 261.—1926. The temporo-frontal articulation in man. Ibid., vol. 9, 343.—1930. Notes on the pterion. Ibid., vol. 14, 41.

COMAS, J. 1942. Contribution à l'étude du métopisme. Archiv. Suisse d'Anthropol. gen., t. 10, 273. [Includes Wormian bones, Inca bones, etc.]

COUNT, E. W. 1941. The australoid problem and

the peopling of America. Second contribution: A consideration of the three cardinal cranial dimensions. Revista del Inst. d. Anthropol. de la Univ. nac. de Tucuman, vol. 2, no. 7, 121.—1942. A quantitative analysis of growth in certain human skull dimensions. Human Biol., no. 2, 143.

DAVENPORT, C. B. 1943. The development of the head. Amer. Jour. Orthodontics and Oral Surgery, no. 9, 541.

GREGORY, W. K. 1925. The biogenetic law and the skull form of primitive man. Amer. Jour. Phys. Anthrop., no. 4, 373.—1927. The palaeomorphology of the human head: ten structural stages from fish to man. Part I. The skull in norma lateralis. Quart. Rev. Biol., no. 2, 267.—1928. Reply to Professor Wood-Jones's note: "Man and the anthropoids." Amer. Jour. Phys. Anthropol., no. 2, 253. [Discussion of contacts between ethmoid, sphenoid and descending flanges of the frontal.]—1929. The palaeomorphology of the human head: ten structural stages from fish to man. Part II. The skull in norma basalis. Quart. Rev. Biol., no. 2, 233.

HARRIS, H. A. 1927. The skull, the face, and the teeth of primates, with special reference to dolichocephaly and the centers of growth in the face. Part I. Proc. Zool. Soc. London, part 3, 491.—1928. The closure of the cranial sutures in relation to the evolution of the brain. Univ. Coll. Hospital Mag., no. 3, 84.

HERMANN, F. 1908. Gehirn und Schädel: Eine topographisch-anatomische Studie in photographischer Darstellung. Jena.

HRDLIČKA, A. Divisions of the parietal bone in man and other mammals. Bull. Amer. Mus. Nat. Hist., 231.—1906. Anatomical observations on a collection of orang skulls from western Borneo; with a bibliography. Proc. U. S. Nat. Mus. 539.

KEITH, A. 1914. The reconstruction of fossil human skulls. Jour. Roy. Anthropol. Inst., 1.

KROGMAN, W. M. 1930. The problem of growth changes in the face and skull as viewed from a comparative study of anthropoids and man. Dental Cosmos.—1930–1931. Studies in growth changes in the skull and face of anthropoids. I. The eruption of the teeth in anthropoids and Old World apes. Amer. Jour. Anat., vol. 46, no. 2, 303.—II. Ectocranial and endocranial suture closure in anthropoids and Old World apes. Ibid., 315.—III (1931). Growth changes in the skull and face of the gorilla. Ibid., vol. 47, no. 1, 89.—IV. Growth changes in the skull and face of the chimpanzee. Ibid., no. 2, 325.—V. Growth changes in the skull of the orang-utan. Ibid., 343.

MACKLIN, C. C. 1921. The skull of a human fetus of 43 millimeters greatest length. Carnegie Inst. Washington Publ. 273, 58.

MARELLI, C. A. 1914. Otros datos acerca de los huesos fontanelarios y suturales. Bol. de la Soc. Physis, 370.

MONTAGU, M. F. A. 1931. On a post-frontal and orbital elements in a young gorilla skull. Jour. Anat., part 4, 446.—1933. The anthropological significance of the pterion in the primates. Amer. Jour. Phys. Anthropol., no. 2, 159.—1935. The premaxilla in the primates. Part I. Quart. Rev. Biol., vol. 10, no. 1, 32.—Part II. Ibid., no. 2, 181.—1936. The premaxilla in man. Jour. Amer. Dent. Assoc., Nov., 2943.—1937. The medio-frontal suture and the problem of metopism in the primates. Jour. Roy. Anthropol. Inst., 157.

NAEF, A. 1926a. Zur Morphologie und Stammesgeschichte des Affenschädels. Die Naturw., part 6, 89.—1926b. Über die Urformen der Anthropomorphen und die Stammesgeschichte des Menschenschädels. I. Ibid., part 20, 445. II. Ibid., part 21, 472.

REMANE, A. 1925. Das Stirnnahtproblem . . . Zeitschr. f. Morphol. u. Anthropol., part 3, 153.

RUSCONI, C. 1935. Sobre morfogénesis basicraneana de algunos primates actuales y fósiles. Revista Argentina de Paleont. y Anthropol. (Ameghinia), vol. 1, no. 1.

SCHULTZ, A. H. 1915a. Einfluss der Sutura occipitalis transversa auf Grösse und Form des Occipitale und des ganzen Gehirnschädels. Arch. suisses d'Anthropol. gén., no. 3, 184.—1915b. Form, Grösse und Lage der Squama temporalis des Menschen. Zeitschr. f. Morphol. u. Anthropol., vol. 19, part 2, 353.—1916. Der Canalis cranio-pharyngeus persistens beim Mensch und bei den Affen. Morphol. Jahrb., part 2, 1.—1917. Anthropologische Untersuchungen an der Schädelbasis. Archiv. f. Anthropol., vol. 16, 1.—1918. The fontanella metopica and its remnants in an adult skull. Amer. Jour. Anat., no. 2, 259.—1929. The metopic fontanelle, fissure, and suture. Ibid., no. 3, 475.—1941. The relative size of the cranial capacity in primates. Amer. Jour. Phys. Anthropol., no. 3, 273.

SCHWALBE, G. 1903. Fontanella metopica und supranasales Feld. Anat. Anz., no. 1, 1.—1910. Studien zur Morphologie der südamerikanischen Primatenformen. Zeitschr. f. Morphol. u. Anthropol., part 2, 209.

SOLLAS, W. J. 1933. The sagittal section of the human skull. Jour. Roy. Anthropol. Inst., 289.

SUTTON, J. B. 1884. On the relation of the orbitosphenoid to the pterion region in the side wall of the skull. Jour. Anat. and Physiol., 219.

TODD, T. W. and D. W. LYON. 1924. Endocranial suture closure; its progress and age relationship. Part I. Adult males of white stock. Amer. Jour. Phys. Anthropol., vol. 7, no. 3, 325.—1925a. Part II. Ibid., vol. 8, no. 1, 23.—1925b. Part III. Ibid., 47.—1925c. Part IV. Ibid., no. 2, 149.

WEIDENREICH, F. 1925. Domestikation und Kultur in ihrer Wirkung auf Schädelform und Körpergestalt. Zeitschr. f. Konstitutionslehre, vol. 11, part 1.—1940. The torus occipitalis and related structures and their transformations in the course of human evolution. Bull. Geol. Soc. China, no. 4, 479.—1941. The brain and its role in the phylogenetic transformation of the human skull. Trans. Amer. Philos. Soc., part 5, 321.—1945. The brachycephalization of recent mankind. Southwestern Jour. Anthropol., no. 2. —1947. Some particulars of skull and brain of early hominids and their bearing on the problem of the relationship between man and anthropoids. Amer. Jour. Phys. Anthropol., no. 4, 387.

Man: Face, External Features, Upper Face (excluding Jaws), Facial Musculature, Facial Nerves, Expression of Emotions

BRIGGS, H. H. 1928. The evolution of the human face: Its significance in the fields of rhinology and stomatology. Ann. Otology, Rhinology and Laryngology, no. 4, 1110.

BURKITT, A. N. 1926. Observations on the facial characteristics of the Australian aboriginal. Rept. Australian Assoc. Adv. Sci., 521.

BURKITT, A. N. and G. S. LIGHTOLLER. 1926. The facial musculature of the Australian aboriginal. Jour. Anat., part 1, 14.

CAVE, A. J. E. and R. W. HAINES. 1940. The paranasal sinuses of the anthropoid apes. Jour. Anat., part 4, 493.

DAVENPORT, C. B. 1937. Postnatal growth of the external nose. Proc. Amer. Philos. Soc., no. 1, 61.—1939. Postnatal development of the human outer nose. Ibid., no. 2, 177.

FOLEY, J. P. 1935. Judgment of facial expression of emotion in the chimpanzee. Jour. Social Psychol., 31.

GOLDSTEIN, M. S. 1936. Changes in dimensions and form of the face and head with age. Amer. Jour. Phys. Anthropol. no. 1, 37.

GREGORY, W. K. 1917. Evolution of the human face. Chief stages in its development from the lowest forms of life to man. Amer. Mus. Jour., no. 6, 376.—1929. Our face from fish to man. New York.—1935. The origin of the human face: A study in paleomorphology and evolution. Dental Cosmos, no. 4, 1.

HARRIS, H. A. 1927. The skull, the face, and the teeth of primates, with special reference to dolichocephaly and the centres of growth in the face. Part I. Proc. Zool. Soc. London, part 3, 491.

HELLMAN, M. 1935. The face in its developmental career. Dental Comos, nos. 7–8.

HUBER, E. 1930. Evolution of facial musculature and cutaneous field of trigeminus. Part I. Quart. Rev. Biol., no. 2, 133. Part II. Ibid., no. 4, 389.

HUBER, E. and W. HUGHSON. 1926. Experimental studies on the voluntary motor innervation of the facial musculature. Jour. Comp. Neurol., no. 1, 113.

JONES, F. W. 1947. The premaxilla and the ancestry of man. Nature, 439.

KEITH, A. 1906. The results of an anthropological investigation of the external ear. Proc. Anat. and Anthropol. Soc., Univ. Aberdeen (1904–1906), 217.

KEITH, A. and G. G. CAMPION. 1922. A contribution to the mechanism of growth of the human face. Dental Rec., London, Feb. 1.

KROGMAN, W. M. 1930. The problem of growth changes in the face and skull as viewed from a comparative study of anthropoids and man. Dental Cosmos, June, 1.—1934. Racial and individual variation in certain facio-dental relationships. Jour. Dent. Research, no. 4, 277.

LIGHTOLLER, G. H. S. 1925. Facial muscles: the modiolus and muscles surrounding the rima oris, with some remarks about the panniculus adiposus. Jour. Anat., part 1, 1.—1928. The facial muscles of three orangutans and two Cercopithecidae. Ibid., part 1, 19.—1934. The facial musculature of some lesser primates and a

Tupaia. A comparative study. Proc. Zool. Soc. London, 259.

REMANE, A. 1927. Der Verschluss der Intermaxillarnaht bei den Anthropoiden. Anthropol. Anz., vol. 4, part 1, 46.

SCHULTZ, A. H. 1935. The nasal cartilages in higher primates. Amer. Jour. Phys. Anthropol., no. 2, 205.—1940. The size of the orbit and of the eye in primates. Ibid., 389.

SULLIVAN, W. E. and C. W. OSGOOD. 1925. The facialis musculature of the orang, Simia satyrus. Anat. Rec., no. 3, 195.

TODD, T. W. 1929. Recent studies in the development of the face. Int. Jour. Orthodontia, Oral Surgery, and Radiography, no. 12, 1.—1932. Hereditary and environmental factors in facial development. Ibid., vol. 18, no. 8, 799.

WEN, I. C. 1930. Ontogeny and phylogeny of the nasal cartilages in primates. Carnegie Inst. Washington Publ. 414, 109.

Man: Jaws, Jaw Muscles
(See also Mammals: Jaws, Teeth)

BOLK, L. 1922. The problem of orthognathism. Proc. Konink. Akad. Wetensch. Amsterdam, nos. 7, 8, 371.—1924. The chin problem. Ibid., nos. 3, 4, 1.

BLUNTSCHLI, H. 1929a. Anthropologie der Kiefer und Zähne. Handwörterbuch der Gesamten Zahnheilkunde, 148.—1929b. Die Kaumuskulatur des Orang-Utan und ihre Bedeutung für die Formung des Schädels. I. Teil: Das morphologische Verhalten. Morphol. Jahrb., 531.—1936. Schädel und Gebiss in ihren funktionellen Beziehungen. Schweiz. Medizinischen Jahrb., 35.

BLUNTSCHLI, H. and H. SCHREIBER. 1929. Über die Kaumuskulatur (Anatomie). Die Fortschr. d. Zahnheilkunde, part 1.

BRASH, J. C. 1928. Growth of the alveolar bone. Intern. Jour. Orthodontics, 196.

FRECHKOP, S. 1940. Notes sur les mammifères. XXVI. Considérations preliminaires sur l'évolution de la dentition des primates. Bull. Mus. roy. d'Hist. nat. de Belgique, no. 11.—1941. Ethologie et morphologie dentaire des primates. Ann. Soc. roy. Zool. de Belgique, part 1, 30.

GREGORY, W. K. 1931. Certain critical stages in the evolution of the vertebrate jaws. Internat. Jour. Orthodontia, Oral Surgery and Radiography, no. 12, 1138.—1943. The earliest known fossil stages in the evolution of the oral cavity and jaws. Amer. Jour. Orthodontics and Oral Surgery, no. 5, 253.

HRDLIČKA, A. 1930. Mental fossae. Amer. Jour. Phys. Anthropol., no. 2.—1935. Jaws and teeth. Jour. Dent. Research, no. 1.

MARKUS, M. B. 1933. The paleontology, evolution, embryology and postnatal development of the human face, jaws and teeth. A synopsis. Int. Jour. Orthodontia and Dentistry for Children, no. 5, 459.

MONTAGU, M. F. A. 1936. The premaxilla in man. Jour. Amer. Dent. Assoc., 2043.

RABKIN, S. 1937. Cranial structural variations and their association with dental conditions. Jour. Dent. Research, no. 3, 203.

SCHULTZ, A. H. 1915. Form, Grösse und Lage der Squama temporalis des Menschen. Zeitschr. f. Morphol. u. Anthropol., vol. 19, part 2, 353.

SENYUREK, M. S. 1939. Pulp cavities of molars in primates. Amer. Jour. Phys. Anthropol., no. 1, 119.

SIMONTON, F. V. 1923. Mental foramen in the anthropoids and in man. Amer. Jour. Phys. Anthropol., no. 3, 413.

SMITH, G. E. 1931. Evolutionary tendencies in the jaws. Brit. Med. Jour., Aug. 1.

STEIN, M. R. 1934. Some variations of the upper third molar. Jour. Amer. Dent. Assoc., 1815.

SVED, A. 1943. Growth of the jaws and the etiology of malocclusion. Reprinted from Internat. Jour. Orthodontics and Dentistry for Children. New York.

TERRA, M. DE. 1905. Beiträge zur einer Odontographie der Menschenrassen. Berlin. 4to. [Pp. 240, 241, p^1 (p^3) with 3 roots in 5 cases; p^2 (p^4) with 3 roots in 1 case; very rare.]

TERRA, P. DE. 1911. Vergleichende Anatomie des menschlichen Gebisses und der Zähne der Vertebraten. Jena.

TODD, T. W. 1926. Skeletal adjustment in jaw growth. Dental Cosmos. Dec. 1.—1930. Facial growth and mandibular adjustment. Intn. Jour. Orthodontia, Oral Surgery and Radiography, no. 12.—1932a. Prognathism: A study in development of the face. Jour. Amer. Dent. Assoc., 2172.—1932b. Hereditary and environmental factors in facial development. Intn. Jour. Orthodontia, Oral Surgery and Radiography, vol. 18, no. 8, 799.

WALLIS, W. D. 1926. The evolution of the human mandible and correlations with features of the skull. Dental Cosmos. Feb.

WEGNER, R. N. 1915. Zur Kenntnis des Gaumenbeins der Anthropoiden. Zeitschr. f. Morphol. u. Anthropol., part 1.

WEIDENREICH, F. 1926. Über die Beziehungen zwischen Zahn, Alveolarwand und Kiefer. Knochenstudien VI. Teil. Zeitschr. f. Anat. u. Entwickl. parts 3–4, 420.—1934. Das Menschenkinn und seine Entstehung: eine Studie und Kritik. Ergeb. d. Anat. u. Entwickl., vol. 31, 1.

WILLIAMS, J. L. 1914. A new classification of human tooth forms, with a special reference to a new system of artificial teeth. Jour. Allied Dent. Soc., vol. 9, no. 1.

WINKLER, R. 1921. Über den funktionellen Bau des Unterkeifers. Zeitschr. f. Stomatologie, part 7, 1.

Man: Teeth (comparative)
(See also Mammals: Teeth)

ADLOFF, P. 1927a. Studien über die Phylogenie des menschlichen. Eckzähnes. Zeitschr. f. Anat. u. Entwickl., 391.—1927b. Das Gebiss des Menschen und der Anthropoiden und das Abstammungsproblem. Zeitsch. f. Morphol. u. Anthropol., part 3, 431.—1934. Das Tuberculum molare des ersten Milchmolaren und die Tubercula paramolaria der Molaren des Menschen nebst einigen Bemerkungen zur Abstammungsfrage. Anat. Anz., nos. 15–19, 289–368.—1937. Über die primitiven und die sogenannten "pithekoiden" Merkmale im Gebiss des rezenten und fossilen Menschen und ihre Bedeutung. Zeitschr. f. Anat. u. Entwickl., part 1, 68.

BENNEJEANT, C. 1936. Anomalies et variations dentaires chez les primates. Paris.

BLACK, G. V. 1897. Descriptive anatomy of the human teeth. 4th ed. Philadelphia.

BLUNTSCHLI, H. 1911a. Das Platyrrhinengebiss und die Bolksche Hypothese von der Stammesgeschichte des Primatengebiss. Verhandl. d. Anat. Gesellsch. 23–26, April, 120.—1911b. Zur Phylogenie des Gebisses der Primaten mit Ausblicken auf jenes der Säugetiere überhaupt. Naturf. Gesellesch. in Zürich, 351.

BLUNTSCHLI, H. and H. SCHREIBER. 1928. Zur allgemeinen Gebisslehre (Anatomie). Die Fortschr. d. Zahnheilkunde . . . part 1.

BOLK, L. 1914. Les dents surnuméraires dans la region molaire de l'homme. L'Odontologie, Feb. —1916. Problems of human dentition. Amer. Jour. Anat., no. 1, 91.

BROOM, R. 1941. The milk molars of man and the anthropoids. South African Dent. Jour., 314.

BUSCH, V. 1896. Ueber die Verschiedenheit in der Zahl der Würzeln bei den Zähnen des menschlichen Gebisses. Verhandl. d. deutsch. odontolog. Gesellsch., 164. [First upper bicuspid with 2 buccal and one lingual root in 36 cases.]

DAHLBERG, A. A. 1945a. The paramolar tubercle (Bolk). Amer. Jour. Phys. Anthropol., n.s., no. 1, 97.—1945b. The changing dentition of man. Jour. Amer. Dent. Assoc., 676.

DIAMOND, M. and J. P. WEINMANN. 1940. The enamel of human teeth. An inquiry into the formation of normal and hypoplastic enamel matrix and its calcification. School. Dent. and Oral Surgery, Colorado Univ.

EIDMANN, H. 1923. Die Entwicklungsgeschichte der Zähne des Menschen. Berlin.

ERAUSQUIN, J. 1936. Sobre la formacion del substrato organico de la substancia fundamental de la dentina. Revista Odontologica de Buenos Aires, Sept.

FRIEL, S. 1927. Occlusion: Observations on its development from infancy to old age. Intn. Jour. Orthodontia, Oral Surgery and Radiography, no. 4, 322.

GAUDRY, A. 1901a. Sur la similitude des dents de l'homme et quelques animaux. L'Anthropologie, Janv.-Avr.—1901b. . . .(Deuxième note). Ibid., vol. 12, 513.

GLAESSNER, M. F. 1931. Neue Zähne von Menschenaffen aus dem Miozän des Wiener Beckens. Ann. Naturhist. Mus. in Wien, vol. 46, 15.

GOLDSTEIN, M. S. and F. L. STANTON. 1937. Facial growth in relation to dental occlusion. Intn. Jour. Orthodontia, Oral Surgery and Radiography, vol. 23, no. 9, 3.

GREGORY, W. K. 1916. Studies on the evolution of the primates. I. The Cope-Osborn "Theory of Trituberculy" and the ancestral molar patterns of the primates. Bull. Amer. Mus. Nat. Hist., 239. —1918. The evolution of orthodonty. Dental Cosmos, May 3.—1920–1921. The origin and evolution of the human dentition. Jour. Dent. Research, vols. 2, 3.—1922. The origin and evolution of the human dentition: A palaeontological review. Parts I–V. Baltimore.—1926a. Palaeontology of the human dentition. Ten structural stages in the evolution of the cheek teeth. Amer. Jour. Phys. Anthropol., no. 4, 401.

—1926b. Some critical stages in the evolution of the human dental apparatus. Jour. Dent. Research (1924–1926), no. 1, 71.—1934a. Polyisomerism and anisomerism in cranial and dental evolution among vertebrates. Proc. Nat. Acad. Sci., no. 1.—1934b. A half century of trituberculy. The Cope-Osborn theory of dental evolution, with a revised summary of molar evolution from fish to man. Proc. Amer. Philos. Soc., no. 4, 169.—1934c. Some new models illustrating the evolution of the human dentition. Intn. Jour. Orthodontia and Dentistry for Children, no. 11, 1077.—1941. Evolution of dental occlusion from fish to man. In: Development of Occlusion by W. K. Gregory, B. H. Broadbent and M. Hellman. Univ. Pennsylvania Press.

GREGORY, W. K. and M. HELLMAN. 1926a. The dentition of Dryopithecus and the origin of man. Anthropol. Papers, Amer. Mus. Nat. Hist., part 1, 1.—1926b. The crown patterns of fossil and recent human molar teeth and their meaning. Nat. Hist. Mag., no. 3, 300.—1929. Paleontology of the human dentition. Family tree of man. The evolution and relationships of the principal branches of mankind and of anthropoid apes. Intn. Jour. Orthodontia . . . no. 7, 642.—1937. The evidence of the dentition on the origin of man. In: Early Man (Ed. G. G. MacCurdy). Philadelphia.

HELLMAN, M. 1918. Observations on the form of the dental arch of the orang. Intn. Jour. Orthodontia . . . no. 2, 3.—1919a. The relationship of form to position in teeth and its bearing on occlusion. Dental Items of Interest, 1.—1919b. Dimensions versus form in teeth and their bearing on the morphology of the dental arch. Intn. Jour. Orthodontia . . . no. 11.—1920. An interpretation of Angle's classification of malocclusion of the teeth supported by evidence from comparative anatomy and evolution. Dental Cosmos, April.—1926. Some changes in the human face as influenced by the teeth. Nat. Hist. Mag., no. 1.—1927. The face and occlusion of the teeth in man. Intn. Jour. Orthodontia . . . no. 11.—1929. The face and teeth of man: A study of growth and position. Jour. Dent. Research, no. 2, 179.

HRDLIČKA, A. 1911. Human dentition and teeth from the evolutionary and racial standpoint. Dominion Dent. Jour.—1920. Shovel-shaped teeth. Amer. Jour. Phys. Anthropol., no. 4, 429. —1923. Variations in the dimensions of lower molars in man and anthropoid apes. Ibid., vol. 6, no. 4, 423.

KADNER, A. 1929. Das Problem der Vererbung von Zahnstellungsanomalien. Deutsche Zahnärztliche Wochenschr., nos. 15–16.

KEITH, A. 1913. Problems relating to the teeth of the earlier forms of prehistoric man. Proc. Roy. Soc. Medicine, 103.

KROGMAN, W. M. 1927. Anthropological aspects of the human teeth and dentition. Jour. Dent. Research, no. 1.—1930. Studies in growth changes in the skull and face of anthropoids. I. The eruption of the teeth in anthropoids and Old World apes. Amer. Jour. Anat., no. 2, 303.

LONNBERG, E. 1930. The homologies of the incisors of the higher primates in the light of some "anomalies" in the dentition of gibbons. Arkiv. f. Zoologi, vol. 22A, no. 6.

MONTAGU, M. F. A. 1940. The significance of the variability of the upper lateral incisor teeth in man. Human Biol., no. 3, 323.—1943. Variation of the diastemata in the dentition of the anthropoid apes and its significance for the origin of man. Amer. Jour. Phys. Anthropol., no. 4, 325.

MUMMERY, J. H. 1924. The microscopic and general anatomy of the teeth. 2nd ed. Oxford Univ. Press.

OPPENHEIM, A. 1912. Tissue changes, particularly of the bone, incident to tooth movement. Amer. Orthodontist, 113.

OSBORN, H. F. 1892. The history and homologies of the human molar cusps. Anat. Anz., vol. 7, 740.

OSBURN, R. C. 1911. The evolution of the occlusion of the teeth. Amer. Orthodontist, July, 1.—1913. The evolution of occlusion with special reference to that of man. Dental Cosmos.

PEDERSEN, P. O. 1949. The East Greenland Eskimo dentition. Numerical variations and anatomy. Medell. om Grönland, vol. 142. Kobenhavn.

REMANE, A. 1921. Beiträge zur Morphologie des Anthropoidengebisses. Archiv f. Naturges., 87 Jahr., part 11, 1.—1924a. Einige Bemerkungen zur Eckzahnfrage. Anthropol. Anz., vol. 1, part 1.—1924b. Ueber das Eckzahnproblem. Zahnarztliche Rundschau, vol. 33, no. 35, 463.

RÖSE, C. 1909. Über die mittlere Durchbruchszeit der bleidenden Zähne des Menschen. Deutsche Monatsschr. Zahnheilk., 553.

RYDER, J. A. 1878. On the mechanical genesis of tooth-forms. Proc. Acad. Nat. Sci. Philadelphia, 45.

SCHLOSSER, M. 1901. Die menschenähnlichen Zähne aus dem Bohnerz der schwäbischen Alb. Zool. Anz., no. 643, 261.

SCHULTZ, A. H. 1925. Studies on the evolution of human teeth. Dental Cosmos, Oct.—1930. Notes on the growth of anthropoid apes with special reference to deciduous dentition. Rept. Lab. and Mus. Comp. Pathology, Zool. Soc. Philadelphia, 34.—1932. The hereditary tendency to eliminate the upper lateral incisors. Human Biol., no. 1, 34.—1935. Eruption and decay of the permanent teeth in primates. Amer. Jour. Phys. Anthropol., no. 4, 489.

SCHWARZ, R. 1925. Kiefer und Zähne der Melanesier in morphologischer und morphogenetischer Beziehung. Schweiz. Monats. f. Zahnheilkunde, part 2, 43.—1927. Die Zähne des Menschen und der Anthropoiden. Die Fortschr. der Zahnheilkunde, part 8, 718.—1931. Das Abstammungsproblem des Menschen. Ibid., part 9, 753. [Excellent figures of Australopithecus teeth.]

STEIN, M. R. 1935. A critique of Bolk's "Problems of Human Dentition." Jour. Dent. Research, no. 2, 101.

TERRA, M. DE. 1905. Beiträge zur einer Odontographie der Menschenrassen. Berlin. 4to. [Pp. 240, 241: p^1 (p^3) with 3 roots in 5 cases; p^2 (p^4) with 3 roots in 1 case; very rare.]

VALLOIS, H. 193?. Les théories sur l'origine des molaires. L'Orthodontie française, vol. 11, 185.

WEIDENREICH, F. 1937. The dentition of Sinanthropus pekinensis: A comparative odontography of the hominids. Palaeontologia Sinica, no. 1. [Bearing on human phylogeny.]

Man: Nervous System, including Sense Organs, Brain, Brain and Body Weight, Behavior and Response, Psychology, Origin of Speech, Imagination

ANTHONY, R. 1916. Le développement du cerveau chez les singes. Ann. d. Sci. Nat.—Zool., no. 1.

ANTHONY, R. and A. S. DE SANTA-MARIA. 1912a. Le territoire central du néopallium chez les primates. I. Considérations sur la signification morphologique générale et l'operculisation de l'insula antérieure chez les anthropoïdes et chez l'homme. Revie Anthropol., no. 4, 141.—1912b. II. Le circulaire supérieur de Reil et la suprasylvia chez les lémuriens, les singes et l'homme. Ibid., no. 7, 275.—1912c. Le territoire périphérique du néopallium chez les primates. I. Le système operculaire supérieur du complexe sylvien chez les lémuriens, les singes et l'homme. Bull. et Mém. Soc. d'anthropol., Paris, 293.—1913. Recherches sur la morphologie télencéphalique du Lepilemur a l'état adulte et au cours du développement ontogénique. Nouv. Arch. d. Mus., vol. 5, 1.

BEACH, F. A. 1941. Instinct and intelligence. Trans. N. Y. Acad. Sci., no. 1, 32.

BORING, E. G. 1932. The physiology of consciousness. Science, 32.

BOWEN, R. E. 1932. The ampullar organs of the ear. Jour. Comp. Neurol., no. 2, 273.—1933. The cupula of the membranous labyrinth. Ibid., vol. 58, no. 2, 517.

CARPENTER, C. R. 1934. A field study of the behavior and social relations of howling monkeys. Comp. Psychol. Monogr., no. 2.

CLARK, W. E. LE G. 1925. The visual cortex of primates. Jour. Anat. (London), part 4, 350.—1927. Description of the cerebral hemispheres of the brain of a gorilla (John Daniels II). Ibid., part 4, 467.—1932. A morphological study of the lateral geniculate body. Brit. Jour. Opthalmology, 264.

CLARK, W. E. LE G., D. M. COOPER and S. ZUCKERMAN. 1936. The endocranial cast of the chimpanzee. Jour. Roy. Anthropol. Inst., 249.

CONNOLLY, C. J. 1933. The brain of a mountain gorilla, Okero (G. beringei). Amer. Jour. Phys. Anthrop., no. 3, 291.—1936a. The fissural pattern of the primate brain. Ibid., no. 3, 301.—1936b. Origins of speech and the orang-utan. Nature, Dec. 5, 977.

COUNT, E. W. 1947. Brain and body weight in man: their antecedents in growth and evolution. Ann. N. Y. Acad. Sci., 993.

CRAIGIE, E. H. 1937. The comparative anatomy and embryology of the capillary bed of the central nervous system. Proc. Assoc. Research in nervous and Mental Disease, 3.—1945. The architecture of the cerebral capillary bed. Biol. Rev. (London), 133.

CRILE, G. and D. P. QUIRING. 1940. A record of the body weight and certain organ and gland weights of 3,690 animals. Ohio Jour. Sci., no. 5, 219.

DART, R. A. 1934. The dual structure of the neopallium; Its history and significance. Jour. Anat., part 1.

DETWILER, S. R. 1926. The resonance theory of Weiss. Jour. Comp. Neurol., no. 3, 465.

DUBOIS, E. 1897. De verhouding van het gewicht der hersenen tot d grootte van het lichaam bij zoogdieren. Verh. Kon. Akad. Wetensch. Amsterdam, 1.—1898. Über die Abhängigkeit des Hirngewichts von der Körpergrösse beim Menschen. Archiv. f. Anthropol., 123.—1913. On the relation between the quantity of brain and the size of the body in vertebrates. Verh. Kon. Akad. Wetensch. Amsterdam, 647.—1914. Die gesetzmässige Beziehung von Gehirnmasse zur Körpergrösse bei den Wirbeltieren. Zeitschr. Morphol. u. Anthropol., 323.—1918. On the relation between the quantities of the brain, the neurone and its parts, and the size of the body. Verh. Kon. Akad. Wetensch. Amsterdam, vol. 20.—1923a. Phylogenetic and ontogentic increase of the volume of brain in the Vertebrata. Ibid., vol. 25, 230.—1923b. Phylogenetische en ontogenetische toeneming van het volumen der hersenen bij de gewerbelde dieren. Ibid., vol. 31, 307.—1924. On the brain quantity of specialized genera of mammals. Ibid., vol. 27.—1928. The law of the necessary phylogenetic perfection of the psychoencephalon. Ibid., vol. 31, no. 3.—1930. Die phylogenetische Grosshirnzunahme autonome Vervollkommnung der animalen Funktionen. Biologia Generalis, part 2, 247.—1934. Phylogenetic cerebral growth. Congrès Int. des Sci. d'Anthropol. et Ethnol., 71. Roy. Anthropol. Inst. London.

FRAIPONT, C. 1931. L'évolution cérébrale des primates et en particulier des hominiens. Arch. l'Inst. d. Paléontol. humaine, Mém. 8.

FULTON, J. F. and C. F. JACOBSEN. 1935. The functions of the frontal lobes: A comparative study in monkeys, chimpanzees and man. Advances in Modern Biol., nos. 4–5, 113.

GALT, W. E. 1939. The capacity of the rhesus and cebus monkeys and the gibbon to acquire differential response to complex visual stimuli. Gen. Psychol. Monogr., no. 3.

GERARD, R. W. 1946. The biological basis of imagination. Scient. Monthly, June, no. 6, 447.

GRETHER, W. F. 1940. Chimpanzee color vision. Jour. Comp. Psychol., no. 2, 167.

HERRICK, C. J. 1933. Morphogenesis of the brain. Jour. Morphol., no. 2, 233.—1943. The cranial nerves: A review of fifty years. Bull. Scient. Lab., Denison Univ., 41.

HOCHSTETTER, F. 1942. Harte Hirnhaut und ihre Fortsätze bei den Säugetieren. Akad. d. Wiss. in Wien, vol. 106, part 2, 4to. [Sagittal sections of skull, brain, brain-stem, anterior end of vertebral column, in all orders of mammals.]—1943. Beiträge zur Entwicklungsgeschichte der Kraniocerebralen Topographie des Menschen. Ibid., part 3.

JACOBSEN, C. F. and others. 1936. Studies of cerebral function in primates. Comp. Psychol. Monogr., vol. 13, no. 3.

KAPPERS, C. U. A. 1929a. The frontal fissures on the endocranial casts of some Predmost men. Kon. Akad. Wetensch. Amsterdam. no. 5, 552.—1929b. The evolution of the nervous system in invertebrates, vertebrates and man. Haarlem.—1931. The development of the cortex and the functions of its layers. Jour. Mental Sci.—1934. The phylogenetic development of the cerebellum. Psychiatrische en Neurologische Bladen, no. 5.—1935. Development of the different layers of the cerebral cortex with reference to some pathological cases. The forebrain of prehistoric races. Trans. Coll. Physicians, Philadelphia, vol. 3, no. 1, 20.—1936. Brain-body-weight relation in human ontogenesis. Kon. Akad. Wetensch. Amsterdam, vol. 39, no. 7.

KAPPERS, C. U. A., G. C. HUBER and E. C. CROSBY. 1936. The comparative anatomy of the nervous system of vertebrates. Vols. I, II. New York.

KLAAUW, C. J. VAN DER. 1945. Cerebral skull and facial skull. Arch. Néerlandaises de Zool., parts 1–2.

KLÜVER, H. 1933. Behavior mechanisms in monkeys. Chicago Univ. Press.

KOHTS, N. 1928. Adaptive motor habits of the Macacus rhesus under experimental conditions: A contribution to the problem of "labor processes" of monkeys. Moscow. [English summary, pp. 326–352.]—1935. Infant ape and human child (instincts, emotions, play, habits). Scient. Mem. Mus. Darwinianum, Moscow, vol. 3. [English summary, pp. 524–596.]

LASHLEY, K. S. 1931. Mass action in cerebral function. Science, 245.—1949. Persistent problems in the evolution of mind. Quart. Rev. Biol., 28.

LEBOUCQ, G. 1929. Le rapport entre le poids et la surface de l'hémisphère cérébral chez l'homme et les singes. Mém. l'Acad. Roy. de Belgique, part 9.

MAY, R. M. and S. R. DETWILER. 1925. The relation of transplanted eyes to developing nerve centers. Jour. Exp. Zool. no. 1, 83.

McCULLOCH, T. L. 1941. Discrimination of lifted weights by chimpanzees. Jour. Comp. Psychol., no. 3, 507.

MONTAGU, M. F. A. 1944. On the relation between body size, walking activity and the origin of social life in the primates. Amer. Anthropologist, no. 1, 141.

PAGET, R. 1930. Human speech. New York.—1943. A world language. Nature, 80.

PAPEZ, J. W. 1929. Comparative neurology: A manual and text for the study of the nervous system of vertebrates. New York.

PEARL, R. 1939. Patterns for living together. Harper's Magazine, May, 595.

RICHTER, C. P. 1942. Physiological psychology. Ann. Rev. Physiol., vol. 4, 561.

SCHEPERS, G. W. H. 1938. Evolution of the forebrain . . . Cape Town, South Africa.

SCHULTZ, A. H. 1941. The relative size of the cranial capacity in primates. Amer. Jour. Phys. Anthrop., no. 3, 273.

SHELLSHEAR, J. L. 1937. The brain of the aboriginal Australian: A study in cerebral morphology. Philos. Trans. Roy. Soc. London, 293.

SMITH, G. E. 1901. Notes upon the natural subdivision of the cerebral hemisphere. Jour. Anat. Physiol., 431.—1903a. Further observations on the natural mode of subdivision of the mammalian cerebellum. Anat. Anz., vol. 23, nos. 14–15, 368.—1903b. Zuckerkandl on the phylogeny of the corpus callosum. Ibid., 384.—1903c. The so-called "Affenspalte" in the human (Egyptian) brain. Ibid., vol. 24, nos. 2–3, 74.—1903d. The morphology of the human cerebellum. Rev. Neurol. Psychiatry, vol. 1, no. 10, 629.—1904. A note on an exceptional human brain, presenting a pithecoid abnormality of the sylvian region. Jour. Anat. Physiol., vol. 38, 158.—1919. A preliminary note on the morphology of the corpus striatum and the origin of the neopallium. Jour. Anat., part 4, 271.—1930. New light on vision. Nature, May 31, 3.

TILNEY, F. 1927. The brain of prehistoric man. Arch. Neurol. Psychiatry, 723.—1928. The brain from ape to man. 2 vols. New York.—1937. The brain from fish to man: A series of culminating phases in evolution. Scient. Monthly, Oct.-Nov., 289, 415.—1938. The evolution of the human brain. In: "Milestones in Medicine," N. Y. Acad. Medicine, 121. New York.

TILNEY, F. and F. H. PIKE. 1925. Muscular coordination experimentally studied in its relation to the cerebellum. Arch. Neurol. Psychiatry, vol. 13, 289.

TILNEY, F. and H. A. RILEY. 1921. The form and functions of the central nervous system. New York.

TODD, T. W. 1927. A liter and a half of brains. Science, 122.

WARDEN, C. J. and W. E. GALT. 1941. Instrumentation in Cebus and Rhesus monkeys on a multiple-platform task. Jour. Psychol., 3.

WEIDENREICH, F. 1936. Über das phylogenetische Wachstum des Hominidengehirns. Kaibogaku Zasshi (Anat. Gesell. in Japan), vol. 9, part 5.

WHITE, L. A. 1932. The mentality of primates. Scient. Monthly, Jan., 69.—1942. On the use of tools by primates. Jour. Comp. Psychol., no. 3, 369.

WOLTERS, A. W. 1936. The patterns of experience. Nature, 670.

YERKES, R. M. 1916. The mental life of monkeys and apes: A study of ideational behavior. Behavior Monogr., vol. 3, no. 1.—1925. Almost human. New York.—1927–1928. The mind of a gorilla. Part I. Genetic Psychol. Monogr., vol. 2, 1. Part II. Mental development. Ibid., no. 6, 381. Part III. Memory. Ibid., vol. 5 (1928), no. 2, 1.—1936. The significance of chimpanzee-culture for biological research. The Harvey Lectures, 1935–1936.

YERKES, R. M. and B. W. LEARNED. 1925. Chimpanzee intelligence and its vocal expressions. Baltimore.

YERKES, R. M. and A. W. YERKES. 1929. The great apes: A study of anthropoid life. Yale Univ. Press.—1935. Social behavior in infrahuman primates. Handbook of Social Psychology, Chapter 21, 973.

ZUCKERMAN, S. 1932. The social life of monkeys and apes. London.—1938. Observations on the autonomic nervous system and on vertebral and neural segmentation in monkeys. Trans. Zool. Soc. London, part 6, 315.

ZUCKERMAN, S. and R. B. FISHER. 1938. Growth of the brain in the Rhesus monkey. Proc. Zool. Soc. London, part 4, 529.

Man: Origin, Phylogeny
(See also Anthropology)

ABEL, O. 1934. Das Verwandschaftsverhältnis zwischen dem Menschen und den höheren fossilen Primaten. Zeitschr. f. Morphol. u. Anthropol., vol. 34.

ADLOFF, P. 1931. Über den Ursprung des Menschen im Lichte der Gebissforschung. Königsb. Gelehrten Gesellsch., part 6, 299.

ARAMBOURG, C. 1943. La genèse de l'humanité. Paris.—1947. L'état actuel de nos connaissances sur les origines de l'homme. Ann. Biol., t. 23, fasc. 11, 12, 21.

BOLK, L. 1921. The part played by the endocrine glands in the evolution of man. Lancet, Sept. 10, 588.

BROOM, R. 1932. Evolution as the palaeontologist sees it. South African Jour. Sci., vol. 29, 54.—1933. Evolution—is there intelligence behind it: Ibid., vol. 30, 1.—1938. New light on the origin of man. Forum, vol. 1, no. 23, Sept. 5, 22.—1943. South Africa's part in the solution of the problem of the origin of man. South African Jour. Sci., 68.
[See also South African man-apes (Australopithecinae) above.]

BURKITT, A. N. 1946. The zoological position of man. Med. Jour. Australia, Nov. 2, 613.

CLARK, W. E. LE G. 1934. Early forerunners of man: A morphological study of the evolutionary origin of the primates. London.—1940. Palaeontological evidence bearing on human evolution. Biol. Reviews (London), vol. 15, 202.—1941. History of the primates. An introduction to the study of fossil man. Brit. Mus. (Nat. Hist.) London.

COLBERT, E. H. 1949. Some paleontological principles significant in human evolution. Studies in Phys. Anthropol. No. 1. Viking Fund Symposium (W. W. Howells, Ed.), 103. [Adaptive radiation, parallelism, irreversibility of evolution, orthogenesis, extinction, transformation.]

GIUFFRIDA-RUGGERI, V. 1921. Su l'origine dell' Uomo. Nuove teorie e documenti. Bologna.—1922. La phylogénie humaine. Scientia, May, 361.

GRABAU, A. W. 1930. Asia and the evolution of man. The China Journal, no. 3, 152.

GREGORY, W. K. 1916. Studies on the evolution of the primates. Part II. Phylogeny of recent and extinct anthropoids with special reference to the origin of man. Bull. Amer. Mus. Nat. Hist., 258.—1920. Facts and theories of evolution, with special reference to the origin of man. Dental Cosmos, March, 3.—1927a. How near is the relationship of man to the chimpanzee-gorilla stock? Quart. Rev. Biol., no. 4, 549.—1927b. The origin of man from the anthropoid stem; when and where? Proc. Amer. Philos. Soc., Bicentenary No., vol. 66, 439.—1927c. Two views of the origin of man. Science, vol. 65, 601.—1928. Were the ancestors of man primitive brachiators? Proc. Amer. Philos. Soc., no. 2, 129.—1929. Is the pro-dawn man a myth? Human Biol., May, no. 2, 153.—1930a. A critique of Professor Osborn's theory of human origin. Amer. Jour. Phys. Anthropol., no. 2, 133.—1930b. The origin of man from a brachiating anthropoid stock. Science, 645.—1933. The new anthropogeny: Twenty-five stages of vertebrate evolution from Silurian chordate to man. Ibid., 29.—1934a. The origin, rise and decline of Homo sapiens. Scient. Monthly, 481.—1934b. Man's place among the anthropoids. Oxford.—1935. The roles of undeviating evolution and transformation in the origin of man. Amer. Nat., 385.—1944. The hall of the age of man (By Henry Fairfield Osborn; 8th ed. revised by W. K. Gregory and G. Pinkley). Amer. Mus. Nat. Hist. Guide Leaflet Ser. no. 52.

GREGORY, W. K. and M. HELLMAN. 1926. The dentition of Dryopithecus and the origin of man. Anthropol. Papers, Amer. Mus. Nat. Hist., part 1.

GREGORY, W. K. and J. H. MCGREGOR. 1926. A dissenting opinion as to dawn men and ape men. Nat. Hist. Mag., no. 3, 270.
[See also South African man-apes (Australopithecinae) above.]

GREGORY, W. K. and M. ROIGNEAU. 1942. Introduction to human anatomy: Guide to Section I of the Hall of Natural History of Man (3rd ed. revised by W. K. Gregory and H. C. Raven). Amer. Mus. Nat. Hist. Guide Leaflet Ser., no. 86.

HARTMAN, C. G. 1939. The use of the monkey and ape in the studies of human biology, with special reference to primate affinities. Amer. Nat., 139.

HOOTON, E. 1946. Up from the ape. Revised ed. New York.

HUBRECHT, A. A. W. 1897. The descent of the primates. Lecture at Princeton Univ. ["Tarsius theory."]

HUXLEY, J. 1942. The biologist looks at man. Fortune, no. 6, 139.

JONES, F. W. 1916. Arboreal Man. London.—1918. The problem of man's ancestry. London.—1923. The ancestry of man. Douglas Price Mem. Lecture, no. 3, Brisbane.—1928. Man and the anthropoids. Amer. Jour. Phys. Anthropol, 245.—1929. Man's place among the mammals, New York.

KEITH, A. 1922. The evolution of human races in the light of the hormone theory. I. Johns Hopkins Hospital Bull., May, 155. II. Ibid., June, 195.—1925. The antiquity of man. New and enlarged ed. Vols. I, II. London.—1927. Concering man's origin. Presidential Addr. Brit. Assoc. Adv. Sci., Leeds, and recent essays on Darwinian subjects. London.—1929. New discoveries relating to the antiquity of man. New York.—1934. The construction of man's family tree. Forum Ser., no. 18. London.—1944. Evolution of modern man (Homo sapiens). Nature, 742.—1946. Essays on human evolution. London.

KELLER, A. G. 1922. Societal evolution. In: The evolution of man, by R. S. Lull and others, chapt. 5, 120. Yale Univ. Press.—1932. Man's rough road. New York.

LEAKEY, L. S. B. 1934. Adam's ancestors: An up-to-date outline of what is known about the origin of man. New York.

MILLER, G. S., JR. 1920. Conflicting views on the problem of man's ancestry. Amer. Jour. Phys. Anthropol., no. 2, 213.—1929. The controversy over human "missing links." Smiths. Rept. (for 1928), 413.

MOLLISON, T. 1933. Phylogenie des menschen. (Festschrift, Hans Virchow.) Handb. d. Vererbungswissenschaft, vol. 3.

MONTAGU, M. F. A. 1930. The tarsian hypothesis and the descent of man. Jour. Roy. Anthropol. Inst., 335.

MORGAN, T. H. 1925. Evolution and genetics. Princeton Univ. Press.

NAEF, A. 1933. Die Vorstufen der Menschwerdung. Jena.

OSBORN, H. F. 1928a. The influence of habit in the evolution of man and the great apes. Bull. N. Y. Acad. Med., 216.—1928b. Recent discoveries relating to the origin and antiquity of man. Palaebiologica, 189.—1929. Is the ape-man a myth? Human Biol., no. 1, 4.

SCHLAIKJER, E. M. 1938. The road to man. Amer. Mus. Nat. Hist. Guide Leaflet Ser., no. 97.

SCHULTZ, A. H. 1936. Characters common to higher primates and characters specific for man. I. Quart. Rev. Biol., no. 3, 259. II. no. 4, 425.

SMITH, G. E. 1912. [Ancestry of man.] Presidential Address, Anthropol. Sect., Brit. Assoc. Adv. Sci. Dundee.—1929. Human history. New York.—1931. The search for man's ancestors. London.

SONNTAG, C. F. 1924. The morphology and evolution of the apes and man. London.

STRAUS, W. L. 1931. Review of Man's place among the mammals, by Frederic Wood Jones. Jour. Mammal., no. 2, 171.

VALLOIS, H. V. 1927. Y a-t-il plusiers souches humaines? Revue générale des Sciences pures et appliquées, no. 7, 201. [Monophyletic vs. polyphyletic theory of man's origin.]—1929. Les preuves anatomiques de l'origine monophylétic de l'homme. L'Anthropologie, 77.—1938. Les pygmées et l'origine de l'homme. Revue Scient., June 15.

WATSON, D. M. S. 1928. Palaeontology and the evolution of man. Romanes Lecture. Oxford.

TEILHARD DE CHARDIN, P. 1943. Fossil men: Recent discoveries and present problems. Lecture at the Catholic University of Peking. Peking.

WEIDENREICH, F. 1939. The drifts of human phylogenetic evolution. Peking Nat. Hist. Bull. (1938–1939), part 4, 227.—1946. Apes, giants and man. Univ. Chicago Press.—1947a. Some particulars of skull and brain of early hominids and their bearing on the problem of the relationship between man and anthropoids. Amer. Jour. Phys. Anthropol., 387.—1947b. The trend of human evolution. Evolution (Internat. Jour. Organic Evol.), vol. 1, no. 4, 221.
[See also numerous papers and books cited above under Pithecanthropus, Sinanthropus, etc. for this author's views on origin and evolution of man.]

WEINERT, H. 1932a. Ursprung der Menschheit: über den Engeren Anschluss des Menschengeschlechts an die Menschenaffen. Stuttgart.—1932b. Das heutige "missing link." Jena. Zeitsch. f. Naturw. (Festschrift, Ludwig Plate), vol. 67, 245.

ZEUNER, F. E. 1946. Dating the past. London.—

1947. Time and the anthropologist. Discovery, no. 9, 274.

Man: His Present and His Future

CARLSON, A. J. 1940. Science versus life. Sigma Xi Quarterly, no. 4, 147.
CONKLIN, E. G. 1930a. Science and the future of man. Brown Univ. Papers, no. 9. Providence. —1930b. Principles and possibilities of human evolution. Internat. Clinics, vol. 1, ser. 40.—1941. What is man? Three lectures delivered at Rice Inst., Sharp Foundation, May 7, 8, 9. Rice Inst. Pamphlet, vol. 28, no. 4, 153.
COUNT, E. W. 1948. The twilight of science. Age of dinosaurs? Educ. Forum, Jan., 199.
DIETZ, D. Science and the future. Amer. Scholar, 292.
GREGORY, W. K. 1934. The origin, rise and decline of Homo sapiens. Scient. Monthly, 481.
HOOTON, E. A. 1935. Homo sapiens—whence and whither? Science, 19.—1937. Man as director of human evolution. Harvard Dental Rec., no. 2, 60.
HUXLEY, J. 1942. The biologist looks at man. Fortune, no. 6, 139.
KANDEL, I. L. 1941. The end of an era. Teachers' Coll., Columbia Univ.
LIPPMANN, W. 1941. Education versus civilization. Amer. Scholar, 184.
OSBORN, F. 1948. Our plundered planet. Boston.
VOGT, W. 1948. Road to survival. New York.

INDEX OF AUTHORS

INDEX, GENERAL*

* Compiled by the Author and Mildred Kaymore

NATURAL SCIENCES IN AMERICA

An Arno Press Collection

Allen, J[oel] A[saph]. **The American Bisons,** Living and Extinct. 1876

Allen, Joel Asaph. **History of the North American Pinnipeds:** A Monograph of the Walruses, Sea-Lions, Sea-Bears and Seals of North America. 1880

American Natural History Studies: The Bairdian Period. 1974

American Ornithological Bibliography. 1974

Anker, Jean. **Bird Books and Bird Art.** 1938

Audubon, John James and John Bachman. **The Quadrupeds of North America.** Three vols. 1854

Baird, Spencer F[ullerton]. **Mammals of North America.** 1859

Baird, S[pencer] F[ullerton], T[homas] M. Brewer and R[obert] Ridgway. **A History of North American Birds:** Land Birds. Three vols., 1874

Baird, Spencer F[ullerton], John Cassin and George N. Lawrence. **The Birds of North America.** 1860. Two vols. in one.

Baird, S[pencer] F[ullerton], T[homas] M. Brewer, and R[obert] Ridgway. **The Water Birds of North America.** 1884. Two vols. in one.

Barton, Benjamin Smith. **Notes on the Animals of North America.** Edited, with an Introduction by Keir B. Sterling. 1792

Bendire, Charles [Emil]. **Life Histories of North American Birds** With Special Reference to Their Breeding Habits and Eggs. 1892/1895. Two vols. in one.

Bonaparte, Charles Lucian [Jules Laurent]. **American Ornithology:** Or The Natural History of Birds Inhabiting the United States, Not Given by Wilson. 1825/1828/1833. Four vols. in one.

Cameron, Jenks. **The Bureau of Biological Survey:** Its History, Activities, and Organization. 1929

Caton, John Dean. **The Antelope and Deer of America:** A Comprehensive Scientific Treatise Upon the Natural History, Including the Characteristics, Habits, Affinities, and Capacity for Domestication of the Antilocapra and Cervidae of North America. 1877

Contributions to American Systematics. 1974

Contributions to the Bibliographical Literature of American Mammals. 1974

Contributions to the History of American Natural History. 1974

Contributions to the History of American Ornithology. 1974

Cooper, J[ames] G[raham]. **Ornithology.** Volume I, Land Birds. 1870

Cope, E[dward] D[rinker]. **The Origin of the Fittest:** Essays on Evolution and **The Primary Factors of Organic Evolution.** 1887/1896. Two vols. in one.

Coues, Elliott. **Birds of the Colorado Valley.** 1878

Coues, Elliott. **Birds of the Northwest.** 1874

Coues, Elliott. **Key To North American Birds.** Two vols. 1903

Early Nineteenth-Century Studies and Surveys. 1974

Emmons, Ebenezer. **American Geology:** Containing a Statement of the Principles of the Science. 1855. Two vols. in one.

Fauna Americana. 1825-1826

Fisher, A[lbert] K[enrick]. **The Hawks and Owls of the United States in Their Relation to Agriculture.** 1893

Godman, John D. **American Natural History:** Part I — Mastology and **Rambles of a Naturalist.** 1826-28/1833. Three vols. in one.

Gregory, William King. **Evolution Emerging:** A Survey of Changing Patterns from Primeval Life to Man. Two vols. 1951

Hay, Oliver Perry. **Bibliography and Catalogue of the Fossil Vertebrata of North America.** 1902

Heilprin, Angelo. **The Geographical and Geological Distribution of Animals.** 1887

Hitchcock, Edward. **A Report on the Sandstone of the Connecticut Valley,** Especially Its Fossil Footmarks. 1858

Hubbs, Carl L., editor. **Zoogeography.** 1958

[Kessel, Edward L., editor]. **A Century of Progress in the Natural Sciences: 1853-1953.** 1955

Leidy, Joseph. **The Extinct Mammalian Fauna of Dakota and Nebraska,** Including an Account of Some Allied Forms from Other Localities, Together with a Synopsis of the Mammalian Remains of North America. 1869

Lyon, Marcus Ward, Jr. **Mammals of Indiana.** 1936

Matthew, W[illiam] D[iller]. **Climate and Evolution.** 1915

Mayr, Ernst, editor. **The Species Problem.** 1957

Mearns, Edgar Alexander. **Mammals of the Mexican Boundary of the United States.** Part I: Families Didelphiidae to Muridae. 1907

Merriam, Clinton Hart. **The Mammals of the Adirondack Region,** Northeastern New York. 1884

Nuttall, Thomas. **A Manual of the Ornithology of the United States and of Canada.** Two vols. 1832-1834

Nuttall Ornithological Club. **Bulletin of the Nuttall Ornithological Club:** A Quarterly Journal of Ornithology. 1876-1883. Eight vols. in three.

[Pennant, Thomas]. **Arctic Zoology. 1784-1787.** Two vols. in one.

Richardson, John. **Fauna Boreali-Americana;** Or the Zoology of the Northern Parts of British America, Containing Descriptions of the Objects of Natural History Collected on the Late Northern Land Expeditions Under Command of Captain Sir John Franklin, R. N. Part I: Quadrupeds. 1829

Richardson, John and William Swainson. **Fauna Boreali-Americana:** Or the Zoology of the Northern Parts of British America, Containing Descriptions of the Objects of Natural History Collected by the Late Northern Land Expeditions Under Command of Captain Sir John Franklin, R. N. Part II: The Birds. 1831

Ridgway, Robert. **Ornithology.** 1877

Selected Works By Eighteenth-Century Naturalists and Travellers. 1974

Selected Works in Nineteenth-Century North American Paleontology. 1974

Selected Works of Clinton Hart Merriam. 1974

Selected Works of Joel Asaph Allen. 1974

Selections From the Literature of American Biogeography. 1974

Seton, Ernest Thompson. **Life-Histories of Northern Animals: An Account of the Mammals of Manitoba.** Two vols. 1909

Sterling, Keir Brooks. **Last of the Naturalists:** The Career of C. Hart Merriam. 1974

Vieillot, L. P. **Histoire Naturelle Des Oiseaux de L'Amerique Septentrionale,** Contenant Un Grand Nombre D'Especes Decrites ou Figurees Pour La Premiere Fois. 1807. Two vols. in one.

Wilson, Scott B., assisted by A. H. Evans. **Aves Hawaiienses:** The Birds of the Sandwich Islands. 1890-99

Wood, Casey A., editor. **An Introduction to the Literature of Vertebrate Zoology.** 1931

Zimmer, John Todd. **Catalogue of the Edward E. Ayer Ornithological Library.** 1926